CD ROM INCLUDED

APPLIED
COMPLEX ANALYSIS
with
PARTIAL
DIFFERENTIAL EQUATIONS

NAKHLÉ H. ASMAR

University of Missouri

with the assistance of

GREGORY C. JONES

Graduate Physics Program

Harvard University

Prentice
Hall

PRENTICE HALL, Upper Saddle River, New Jersey 07458

Library of Congress Cataloging-in-Publication Data

Asmar, Nakhlé H.
Applied complex analysis and partial differential equations/ Nakhlé H. Asmar with the assistance
of Gregory C. Jones.
p. cm.
ISBN: 0-13-089239-4
1. Functions of complex variables. 2. Differential equations, Partial. I. Jones, Gregory C. II. Title.

QA331.7 .A85 2002
515'9-dc21   2002018846

> With affection, I dedicate this book to my parents Habib and Mounira, my wife Gracia, and my children Julia and Thomas.
>
> *-Nakhlé H. Asmar*

Acquisition Editor: *George Lobell*
Editor-in-Chief: *Sally Yagan*
Vice President/Director of Production and Manufacturing: *David W. Riccardi*
Executive Managing Editor: *Kathleen Schiaparelli*
Senior Managing Editor: *Linda Mihatov Behrens*
Assisting Managing Editor: *Bayani Mendoza de Leon*
Production Editor: *Steven Pawlowski*
Manufacturing Buyer: *Alan Fischer*
Manufacturing Manager: *Trudy Pisciotti*
Marketing Manager: *Angela Battle*
Marketing Assistant: *Rachel Beckman*
Editorial Assistant: *Melanie Van Benthuysen*
Art Director: *Jayne Conte*
Cover Designer: *Bruce Kenselaar*
Cover Photo Credit: "Shanghai Science Museum, Shanghai, China." RTKL Associates Inc.,
"Inside central glazed egg of the Shanghai Science and Technology Museum,"
Shanghai, China, RTKL Associates Inc., architecture, interior architecture ID8 (division of RTKL), graphics,
Shanghai Modern Architectural Design (Group) Co., Ltd., structural and m/e/p engineering.

Learning Resources
Centre

12999865

©2002 by Prentice-Hall, Inc.
Upper Saddle River, New Jersey 07458

Printed in the United States of America

10 9 8 7 6 5 4 3 2

ISBN 0-13-089239-4

Pearson Education LTD., London
Pearson Education Australia PTY, Limited, Sydney
Pearson Education Singapore, Pte. Ltd
Pearson Education North Asia Ltd, Hong Kong
Pearson Education Canada, Ltd., Toronto
Pearson Educacion de Mexico, S.A. de C.V.
Pearson Education - Japan, Tokyo
Pearson Education Malaysia, Pte. Ltd

Contents

APPENDIXES

Preface

This book is intended to serve as a textbook for the following courses.

- Chapters 1–6 are intended for a one-term introductory course on the theory and applications of complex functions.

- Chapters 7–12 are intended for a one-term introductory course on partial differential equations and boundary value problems, that incorporates basic tools from complex variables such as contour integrals, residues, and Laurent series.

- Topics from Chapters 1–12 can be chosen to form a one-term introductory course on complex variables with applications to partial differential equations and boundary value problems.

The Book's Content and Organization

This book differs in many ways from other traditional textbooks on complex variables and partial differential equations. In outlining its contents, I will present some of these differences in four components of the book: The examples and exercises, the applications, the graphics, and the proofs.

The Proofs

In writing this book, my goal was to reach out to all students with varying abilities to do mathematical proofs. For that reason, I have included complete proofs of most results so as to give the instructor the flexibility to choose the proofs that he or she wants to present to the class, while skipping others and referring the students to the book. While recognizing the importance of training students in proof writing, I have tried not to let this learning process hinder their ability to see and appreciate the applications behind the theory.

In this book, all the proofs are written in a style that is very flexible for classroom presentation. They are carefully arranged and illustrated in such a way that they are accessible to students at the undergraduate level. Some proofs are very basic (for example, those found in the early sections of each chapter); others require a deeper understanding of calculus (for example, use of differentiability in Section 2.3); and yet others propel the students to the graduate level of mathematics. The latter advanced proofs are found in optional sections, such as Sections 2.6, 3.5, 3.9, and 7.6. Typically, in my courses, I would cover some of each level of proofs. Even the most advanced proofs found a place in my courses, as assigned group projects and classroom presentation by students.

The Graphics

Drawing from my teaching experience, I learned that even the most abstract notions can benefit from graphics; and so I did use graphics liberally in this book. I inserted as many pictures in the book as I felt is necessary to clarify an argument, or a statement of a problem, or the result of an example. As a result, this book contains over seven hundred figures.

In the exercises, I found that a statement such as "Solve the Dirichlet problem in Figure X," when accompanied by Figure X, is a much more inviting statement than, say, trying to describe by words the Dirichlet problem. Moreover, it requires from the student to think about the problem from the applied perspective and to write down the mathematical equations that describe the problem.

The figures are extremely useful in visualizing the applications that are at the heart of the matter: heat flow, isotherms, vibrations of strings and membranes. Graphics are also extremely useful in visualizing more abstract concepts, such as the maximum modulus principle (Figure 6, Section 3.7), the invariance of Laplace's equation (Figure 9, Section 2.5), the analyticity of composed mappings (Figure 2, Section 2.3), the more complicated topics of conformal mappings (Section 6.2), changes of variables using analytic functions, Green's functions (Figure 2, Section 6.5), the convergence of Fourier series, Gibbs phenomenon, and Bessel and Legendre series.

The Applications

I started writing this book with what is now Section 2.5. My goal was to show students the applications of complex analysis as quickly as possible. More importantly, I wanted students to realize the significance of the abstract notions from complex analysis before taking up their detailed study. As a result, the book is written in the style of Section 2.5, where the applications go hand in hand with the theory. Whenever possible, I tried to describe the methods in a sequence of steps that a student can follow systematically. For example, in Section 2.5 (following Theorem 3), I describe step by step how to solve a Dirichlet problem using conformal mappings. Then immediately after that I solve an interesting Dirichlet problem by following these steps. I have used the same approach to other important applications throughout the book.

The Examples and Exercises

I have included far more examples than can be covered in one course. The examples are presented in full detail. As with the proofs, the objective is to give the instructor the option to choose the examples that are best suitable for classroom discussion, while at the same time giving students a variety of completely worked examples to assist them in doing the exercises.

The exercises vary in difficulty from the straightforward ones that call on the application of a formula to the more involved **Project Problems**. All of the problems have been tested in the classroom, and the harder ones come with detailed hints to make them accessible to all students. Some of the longer exercises can be used by individual students, or as group projects, or as further illustrations by the instructor. Many sections in the book contain far more exercises than one would typically assign in a course. This allows a greater flexibility in the instructor's choice of problems, depending on the needs and backgrounds of the students.

Several exercises present interesting formulas and noteworthy results that are not found in many comparable books and that are more tractable with the availability of computer systems. Hopefully, even the experienced instructor will enjoy them as new material.

Exercises that require the use of the computer are preceded by a computer icon, such as the one in the margin. Typically these exercises ask the student to investigate problems using computer-generated graphics, and to generate numerical data that cannot be computed by hand with a reasonable effort. Occasionally, the computer is used to compute transforms, verify difficult identities, and aid in algebraic manipulations. Based on my teaching experience, I am convinced that such exercises are a great aid in understanding even the most abstract notions that are covered in the course.

Possible Course Outlines

The following are possible outlines of courses based on this book.

Basic undergraduate course in complex analysis
Chapter 1.
Chapter 2, (Section 2.6 is optional).
Chapter 3 (Sections 3.5, 3.8 and 3.9 are optional).
Chapter 4 (Section 4.7 is optional). Omit the proofs in Section 4.6.
Chapter 5 (Sections 5.5, 5.6, and 5.7 are optional).
Chapter 6 (Sections 6.4, 6.5, 6.6 are optional).
Depending on the background and need of the students, these sections can be covered at different speed. In a course with less emphasis on proofs, more applications from the optional Sections 3.8, 4.7, 6.4-6.6 can be presented.

A course in partial differential equations (to follow the basic course on complex analysis, as outlined previously).

Chapter 7 (Section 7.6 is optional).
Chapter 8, (Section 8.8 is optional).
Chapter 9: Sections 9.1–9.5 (appealing to Sections 9.6 and 9.7 as needed).
Chapter 10 Section 10.1–10.4 (appealing to Sections 10.5–10.7 as needed).
Chapters 11 and 12.
Refer to Sections 6.5 and 6.6 as needed to cover the related material on Green's functions and Neumann functions in Chapters 7–11.

A one-term course in complex analysis and partial differential equations

Complex Analysis Part:

Chapter 1

Chapter 2: Section 2.3 (refer to Sections 2.1 and 2.2 as needed), Section 2.4, Section 2.5. Cover Section 2.5 in detail as a substitute for Chapter 6.

Chapter 3: Sections 3.1 and 3.2. Section 3.4 (do only the version of Cauchy's theorem for simple paths, Theorems 5 and 6. Sections 3.6, 3.7, and 3.8.

Chapter 4: Section 4.1, Section 4.2 (Theorem 4 and its corollary). Skip the proofs in Sections 4.3-4.6, and just present the results and their applications from the examples. Do Section 4.7 followed by Section 7.2.

Chapter 5: Sections 5.1–5.3.

Partial Differential Equations Part:

Chapter 8: Refer to Section 7.4 for the half-range expansions.

Chapter 9: Sections 9.1 and 9.2 (refer to Sections 9.6 and 9.7 as needed).

Chapter 11: Sections 11.1–11.3.

Chapter 12: Sections 12.1–12.3.

Depending on the background and need of the students, these sections can be covered at different speed. In a course with less emphasis on proofs, more applications from Chapters 9 and 10 can be presented.

Basic graduate course in complex analysis

Cover the same topics as the basic undergraduate course, with more emphasis on proofs. In particular, I would cover Sections 2.6, 3.5, 4.6, and 5.7.

Acknowledgements

I owe my students a great deal of recognition and thanks for their criticism, comments, and corrections. My former student Gregory C. Jones played a special role in writing this book. To him I owe most of the graphics in Chapters 1, 2, 3, 5, and 6. He also worked with me on the first six chapters of the book, suggested problems and examples, made several corrections, and improved some proofs. Greg graduated *Summa cum laude* with honors in mathematics and physics from the University of Missouri in 2001. He is currently pursuing a Ph.D. in high energy theoretical physics at Harvard University.

I am grateful to my friend Professor Stephen Montgomery-Smith, who is always available to discuss mathematical topics. Many proofs in this book have been improved after talking with him. I am also grateful to Professor Loukas Grafakos for using a preliminary version of the first six chapters of this book in a graduate course intended to prepare our graduate students for the Ph.D. qualifying exams.

It is also a pleasure to thank Richard Winkel for his help in the area of computer technology.

It is a pleasure to thank George Lobell, my Prentice Hall Editor, for his support, advice, encouragement, and choice of the cover for this book.

I want to take this opportunity to express my gratitude to the following mathematicians who have reviewed earlier versions of my book and gave valuable suggestions: Stephen J. Greenfield (Rutgers University), Samih A. Obaid (San Jose State University), Clifford Queen (Lehigh University), Sergei K. Suslov (Arizona State University), Saleem Watson (California State University, Long Beach), Yuesheng Xu (North Dakota State University). Special thanks to Dave McKay (California State University, Long Beach, and Orange Coast College, Costa Mesa) for his many suggestions to improve the text, the proofs, and some of the pictures. In particular, Figure 8, Section 3.5 is his idea.

I am especially grateful for the support that I received from my family and my parents. It is with great pleasure that I dedicate this book to my wonderful and patient wife Gracia, our children, Julia and Thomas, and my parents Habib and Mounira.

–Nakhlé H. Asmar

About the Author

Nakhlé H. Asmar received his Ph.D in mathematics from the University of Washington in 1986. After spending two years on the faculty of California State University, Long Beach, he joined the University of Missouri, Columbia in 1988, where he is currently Professor of Mathematics. He is the author of the book "Partial Differential Equations and Boundary Value Problems," published by Prentice Hall in 1999. He is also the author or co-author of over forty research articles in the areas of harmonic analysis, Fourier series, and functional analysis. His research received support from the National Science Foundation (U.S.A.). He has received several teaching awards from the University of Missouri, including the William T. Kemper Fellowship Award, the Arts and Science Student Government Purple Chalk Award, and the Provost's Outstanding Junior Faculty Teaching Award. He is a member of the Research Board of the University of Missouri and a member of the College of Reviewers for the Canada Research Chairs program.

The author can be contacted by e-mail at the following address:

nakhle@math.missouri.edu

Topics to Review

This chapter requires basic knowledge from calculus. Most functions that you will encounter here are named after the functions that they generalize from calculus. For example, the complex exponential e^z generalizes the real exponential function e^x. The same is true for many other complex functions such as $\cos z$, $\sin z$, $\tan z$, $\log z$. Thus a good knowledge of these elementary functions and their properties is helpful in studying their complex counterparts.

Looking Ahead

The results of this chapter will be used throughout the rest of the book. A good knowledge of the properties of complex numbers, as presented in Sections 1.1–1.3 is essential to all that follows. In particular, you should familiarize yourself with the triangle inequality and understand its meaning and geometric interpretation. Section 1.5 introduces the most important function in complex analysis: the complex exponential function e^z. Its relationship to trigonometric functions, as illustrated by Euler's identity

$$e^{i\theta} = \cos\theta + i\sin\theta,$$

makes complex analysis a powerful tool for studying Fourier series, solving differential equations, and important problems in other areas of applied mathematics. Mapping properties of complex functions are presented in Section 1.4 and discussed with each function as it is introduced. These properties will be needed in solving boundary value problems, starting in Section 2.5.

1

COMPLEX NUMBERS AND FUNCTIONS

Dismissing mental tortures, and multiplying $5 + \sqrt{-15}$ by $5 - \sqrt{-15}$, we obtain $25 - (-15)$. Therefore the product is 40. ... and thus far does arithmetical subtlety go, of which this, the extreme, is, as I have said, so subtle that it is useless.   -Girolamo Cardan
[First explicit use of complex numbers, which appeared around 1545 in Cardan's solution of the problem of finding two numbers whose sum is 10 and whose product is 40.]

This chapter starts with the early discovery of complex numbers and their role in the solution of algebraic equations. From the expression $z = x + iy$, we will represent complex numbers as points or vectors (x, y) in the plane and take advantage of our geometric intuition in working with complex numbers. We will see how complex numbers can be added and subtracted much like we add and subtract vectors in two dimensions. We also have a notion of distance between complex numbers that satisfies the familiar triangle inequality.

Complex numbers also have a polar form, based on polar coordinates (r, θ). This representation enables us to give a very concrete geometric meaning to multiplication and division of complex numbers.

The last four sections of this chapter are devoted to the study of complex-valued functions of a complex variable. The most important of these functions is the complex exponential e^z. We will introduce it in Section 1.5 and use it in Sections 1.6 and 1.7 to define the trigonometric and logarithmic functions. Since we cannot plot the graph of a complex-valued function (this would require four dimensions), we will visualize these functions as mappings from one complex plane, the z-plane, into another plane, the w-plane. Mapping properties will be explored in this chapter and will be used in later sections to solve interesting partial differential equations.

Every once in a while an application is presented to remind us that complex numbers are studied for the purpose of solving real-world problems. The applications in this chapter include finding *real* roots of algebraic equations, simplifying the computation of certain integrals, and finding the solutions of ordinary differential equations.

1.1   Complex Numbers

Complex numbers were discovered in the sixteenth century for the purpose of solving algebraic equations that do not have real solutions. As you know, the equation

$$x^2 + 1 = 0$$

has no real roots, because there is no real number x such that $x^2 = -1$; or equivalently, we cannot take the square root of -1. The Italian mathematician Girolamo Cardano (1501–1576), better known as Cardan, stumbled upon and used the square roots of negative numbers in his work. While Cardan was reluctant to accept these "imaginary" numbers, he did realize their role in solving algebraic equations.

Two centuries later, the Swiss mathematician Leonhard Euler (1707–1783) introduced the symbol i by setting

$$i = \sqrt{-1}, \text{ or equivalently, } i^2 = -1.$$

Though Euler used numbers of the form $a + ib$ routinely in computations, he was skeptical about their meaning and referred to them as imaginary numbers. It took the authority of the great German mathematician Karl Friedrich Gauss (1777–1855) to definitively recognize the importance of these numbers, introducing for the first time the term that we now use-**complex numbers**.

DEFINITION OF COMPLEX NUMBERS

A complex number z is an expression of the form $z = a + i\,b$, where a and b are real numbers and $i^2 = -1$. The number a is called the **real part** of z and denoted $\operatorname{Re} z$. The number b is called the **imaginary part** of z and denoted $\operatorname{Im} z$.

Two complex numbers are **equal** if they have the same real and imaginary parts. That is, $z_1 = z_2$ if and only if $\operatorname{Re} z_1 = \operatorname{Re} z_2$ and $\operatorname{Im} z_1 = \operatorname{Im} z_2$.

For example, if $z = 3 + i$ then

$$\operatorname{Re} z = \operatorname{Re}(3 + i) = 3 \quad \text{and} \quad \operatorname{Im} z = \operatorname{Im}(3 + i) = 1.$$

Note that the imaginary part of a complex number is itself a real number. The imaginary part of $a + ib$ is just b, not ib.

We do not distinguish between the forms $a + ib$ and $a + bi$; for example, $-2 + i4$ and $-2 + 4i$ are the same complex number. When a complex number has a zero imaginary part like $a + i0$ we simply write it as a. These are known as **purely real** numbers and are just new interpretations of real numbers. Sometimes we shall even abuse the language and say that z is "real" when z has a zero imaginary part. When a complex number has a zero real part like $0 + ib$ we simply write it as ib. These numbers are called **purely imaginary**.

For example, i, $2i$, πi, $-\frac{2}{3}i$ are all purely imaginary numbers. The unique complex number with zero real and imaginary parts will be denoted simply as 0, instead of $0 + 0\,i$.

Algebraic Properties

We define **addition** among complex numbers as if i obeyed the same basic algebraic relations that real numbers do. To add complex numbers, add their real and imaginary parts:

$$(a + i\,b) + (c + i\,d) = (a + c) + i\,(b + d).$$

For example, $(3 + 2i) + (-1 - 4i) = (3 - 1) + i(2 - 4) = 2 - 2i$.

Subtraction is done the same way:

$$(a + i\,b) - (c + i\,d) = (a - c) + i\,(b - d).$$

For example, $(3 + 2i) - (-1 + 4i) = (3 - (-1)) + i(2 - 4) = 4 - 2i$.

Multiplication is defined by treating a complex number $a + i\,b$ as a binomial expression and using the identity $i^2 = -1$. So

$$
\begin{aligned}
(a + i\,b)(c + i\,d) &= ac + i\,ad + i\,bc + i^2 bd \\
&= (ac - bd) + i\,(ad + bc).
\end{aligned}
$$

For example, $(-1 + i)(2 + i) = -2 - i + 2i + i^2 = -3 + i$. Also, $-i(4 + 4i) = -4i - 4(i)^2 = -4i + 4 = 4 - 4i$.

For $z = a + i\,b$ we define the **complex conjugate** $\bar{z}$ of z by

$$\bar{z} = \overline{a + ib} = a - i\,b.$$

Conjugation changes the sign of the imaginary part of a complex number but leaves the real part the same. Thus

$$\operatorname{Re}\bar{z} = \operatorname{Re}z \quad \text{and} \quad \operatorname{Im}\bar{z} = -\operatorname{Im}z.$$

For any complex number z, we have

$$z\bar{z} = (a + i\,b)(a - i\,b) = a^2 + b^2 \geq 0.$$

Hence $z\bar{z}$ is *always* a nonnegative real number. We can now introduce division and compute the quotient of two complex numbers $\frac{a+ib}{c+id}$, where $c + i\,d \neq 0$. First observe that

$$c + i\,d \neq 0 \Leftrightarrow c - i\,d \neq 0 \Leftrightarrow c^2 + d^2 \neq 0.$$

Now

$$\frac{a+ib}{c+id} = \frac{(a+ib)(c-id)}{(c+id)(c-id)} \qquad \text{[multiply and divide by } c-id \neq 0\text{]}$$

$$= \frac{(ac+bd)+(bc-ad)\,i}{c^2+d^2}$$

$$(1) \qquad\qquad = \frac{ac+bd}{c^2+d^2} + i\,\frac{bc-ad}{c^2+d^2}.$$

As an illustration, if $z = x + iy \neq 0$, then from (1)

$$(2) \qquad\qquad \frac{1}{z} = \frac{x}{x^2+y^2} - i\,\frac{y}{x^2+y^2}.$$

In particular, if $z = i$, then, since $x = 0$ and $y = 1$,

$$(3) \qquad\qquad \frac{1}{i} = -i.$$

We can now define integer powers of complex numbers: $z^1 = z$, $z^2 = z \cdot z$, and for any positive integer n,

$$z^n = \overbrace{z \cdot z \cdots z}^{n \text{ of these}}.$$

For nonzero z, z^0 is defined to be 1. Also, $z^{-n} = \frac{1}{z^n}$. As a consequence of the definition, we have the familiar results for exponents such as $z^m z^n = z^{m+n}$, $(z^m)^n = z^{mn}$, and $(zw)^m = z^m w^m$.

EXAMPLE 1   Basic operations
Express the given complex number in the form $a + ib$, where a and b are real numbers.

(a) $(2 - 7i) + \overline{(3+i)}$.   (b) $(4 - 2i)(\frac{3}{2} - \frac{7}{i})$.   (c) i^n, $n = 0, 1, 2, \ldots$.

(d) $\dfrac{2 - 7i}{\pi - i}$.   (e) $\dfrac{i}{7 - i}$.   (f) $\dfrac{2 + i}{3i}$.

Solution   (a) We have

$$(2 - 7i) + \overline{(3+i)} = (2 - 7i) + (3 - i) = 5 - 8i.$$

(b) Using (3), we have $\frac{3}{2} - \frac{7}{i} = \frac{3}{2} + 7i$, and so

$$(4 - 2i)(\frac{3}{2} - \frac{7}{i}) = (4 - 2i)(\frac{3}{2} + 7i) = 6 + 28i - 3i - 14i^2 = 20 + 25i.$$

(c) From the identities

$$i^0 = 1, \quad i^1 = i, \quad i^2 = -1, \quad i^3 = -i, \quad i^4 = 1, \quad i^5 = i, \quad \ldots$$

we conclude that for $n = 0, 1, 2, \ldots$

$$
i^n = \begin{cases}
1 & \text{if } n = 4k, \\
i & \text{if } n = 4k + 1, \\
-1 & \text{if } n = 4k + 2, \\
-i & \text{if } n = 4k + 3,
\end{cases}
$$

where $k = 0, 1, 2, \ldots$. We say that the sequence of complex numbers $\{i^n\}$ is periodic with period 4, since it repeats every four terms.

(d)   In computing quotients, instead of appealing to (1), just remember to multiply and divide by the conjugate of the denominator to make it real. So

$$
\frac{2 - 7i}{\pi - i} = \frac{(2 - 7i)}{(\pi - i)} \frac{(\pi + i)}{(\pi + i)} = \frac{(2\pi + 7) + i(2 - 7\pi)}{(\pi^2 + 1)} = \frac{7 + 2\pi}{\pi^2 + 1} + i\frac{2 - 7\pi}{\pi^2 + 1}.
$$

For (e), we have

$$
\frac{i}{7 - i} = \frac{i}{(7 - i)} \frac{(7 + i)}{(7 + i)} = \frac{-1}{50} + i\frac{7}{50}.
$$

For (f), we could multiply and divide by the conjugate of the denominator, $-3i$, but $-i$ will do the trick:

$$
\frac{2 + i}{3i} = \frac{(2 + i)}{(3i)} \frac{(-i)}{(-i)} = \frac{1 - 2i}{3} = \frac{1}{3} - i\frac{2}{3}. \qquad \blacksquare
$$

Several useful properties of complex numbers can be established by manipulating definitions. The proofs are not difficult in some cases. However, it is important to write out the proof formally, recognizing those statements that are definitions or hypotheses and recognizing those statements whose validity is still in need of proof. We illustrate with some examples.

EXAMPLE 2   Algebraic proofs
(a)   Let z_1, z_2, and z_3 be complex numbers. Prove that $\operatorname{Im}(z_1 + z_2 + z_3) = \operatorname{Im} z_1 + \operatorname{Im} z_2 + \operatorname{Im} z_3$.
(b)   Show that for any complex number z, we have $\overline{\overline{z}} = z$.
(c)   For any complex number z, show that

$$
\operatorname{Re} z = \frac{z + \overline{z}}{2} \quad \text{and} \quad \operatorname{Im} z = \frac{z - \overline{z}}{2i}.
$$

Solution   Write $z_1 = x_1 + iy_1$, $z_2 = x_2 + iy_2$, and $z_3 = x_3 + iy_3$. Then

$$
z_1 + z_2 + z_3 = (x_1 + x_2 + x_3) + i(y_1 + y_2 + y_3),
$$

and so $\operatorname{Im}(z_2 + z_2 + z_3) = y_1 + y_2 + y_3$. But $\operatorname{Im} z_1 = y_1$, $\operatorname{Im} z_2 = y_2$, and $\operatorname{Im} z_3 = y_3$, so

$$
\operatorname{Im} z_1 + \operatorname{Im} z_2 + \operatorname{Im} z_3 = y_1 + y_2 + y_3 = \operatorname{Im}(z_2 + z_2 + z_3),
$$

which proves (a). To prove (b) and (c), write $z = a + ib$. Conjugating once, we get $\overline{z} = a - ib$. Conjugating again, we get

$$
\overline{\overline{z}} = \overline{a - ib} = a + ib = z,
$$

which proves (b). Now

$$z + \overline{z} = a + ib + (a - ib) = 2a = 2\operatorname{Re}z,$$

and the first identity in (c) follows upon dividing both sides by 2. Also,

$$z - \overline{z} = a + ib - (a - ib) = 2ib = 2i\operatorname{Im}z,$$

and the second identity in (c) follows upon dividing both sides by $2i$. ■

The complex number system, endowed with the binary operations of addition, subtraction, multiplication, and division, enjoys the same basic algebraic properties as the real numbers. These include the commutative and associative properties of addition and multiplication; the distributive property; the existence of additive inverses; and the existence of multiplicative inverses for nonzero complex numbers (see Exercises 24–26). Consequently, all the algebraic identities that are true for the real numbers remain true for the complex numbers. For example, if z_1 and z_2 are complex numbers, then

$$
\begin{aligned}
(z_1 + z_2)^2 &= z_1^2 + 2z_1z_2 + z_2^2; \\
(z_1 - z_2)^2 &= z_1^2 - 2z_1z_2 + z_2^2; \\
z_1^2 - z_2^2 &= (z_1 - z_2)(z_1 + z_2);
\end{aligned}
$$

and so on.

Square Roots of Negative Numbers

We end this section by revisiting the topic of quadratic equations. As you know, solving such equations involves computing square roots. The problem arises when we attempt to take the square root of a negative number. In the complex number system, a negative number has two square roots. For example, the square roots of -7 are $i\sqrt{7}$ and $-i\sqrt{7}$, because $(i\sqrt{7})^2 = i^2 7 = -7$ and also $(-i\sqrt{7})^2 = i^2 7 = -7$. We need a convention to distinguish between the two roots. Given $r > 0$, when we write $\sqrt{-r}$ we will be referring to the **principal value** of the square root of $-r$, defined as $i\sqrt{r}$, with a positive imaginary part. When we write $-\sqrt{-r}$ we will be referring to the second root, $-i\sqrt{r}$.

As an application, let a, b, and c be real coefficients and consider the quadratic equation

$$a\,x^2 + b\,x + c = 0 \quad (a \neq 0).$$

Dividing by $a \neq 0$ and completing the square, we get

$$
\begin{aligned}
x^2 + \frac{b}{a}x + \left(\frac{b}{2a}\right)^2 &= \left(\frac{b}{2a}\right)^2 - \frac{c}{a}, \\
\left(x + \frac{b}{2a}\right)^2 &= \left(\frac{b}{2a}\right)^2 - \frac{c}{a}, \\
x + \frac{b}{2a} &= \pm\sqrt{\left(\frac{b}{2a}\right)^2 - \frac{c}{a}} = \pm\frac{\sqrt{b^2 - 4ac}}{2a},
\end{aligned}
$$

and we arrive at the **quadratic formulas**

$$x_1 = \frac{-b + \sqrt{b^2 - 4ac}}{2a} \quad \text{and} \quad x_2 = \frac{-b - \sqrt{b^2 - 4ac}}{2a}.$$

If the discriminant $b^2 - 4ac > 0$, then these formulas yield two distinct real solutions. If $b^2 - 4ac = 0$, then we have one double root. If $b^2 - 4ac < 0$, then $4ac - b^2 > 0$ and the solutions are in this case

$$x_1 = \frac{-b + i\sqrt{4ac - b^2}}{2a} \quad \text{and} \quad x_2 = \frac{-b - i\sqrt{4ac - b^2}}{2a}.$$

Note that the solutions are mutually conjugate. That is, $x_1 = \overline{x_2}$.

EXAMPLE 3   The quadratic formulas
What are the roots of $x^2 + x + 1 = 0$?

Solution   Applying the quadratic formula for the first root, we find

$$x_1 = \frac{-1 + \sqrt{-3}}{2} = \frac{-1 + i\sqrt{3}}{2}.$$

Since the roots are conjugate, we immediately have the other root,

$$x_2 = \frac{-1 - i\sqrt{3}}{2}. \qquad \blacksquare$$

Here we have discussed only the square roots of negative numbers for the purpose of solving the quadratic equation with real coefficients. You may wonder about the square roots of arbitrary complex numbers. Indeed, all complex numbers will have two square roots-these will be discussed in Exercises 31–32 and Section 1.3.

Exercises 1.1

In Exercises 1–15, write the given complex expression in the form $a + ib$, where a and b are real numbers. Take x and y to be real when they occur.

1. $-i$.

2. $4(5 + i)$.

3. $(2 - i)(3 + i)$.

4. $(2 - i)^2$.

5. $(\overline{2 - i})^2$.

6. $(3 + 2i) - (ei - \pi)$.

7. $(x + iy)^2$.

8. $i\overline{(2 + i)^2}$.

9. $(\frac{1}{2} + \frac{i}{7})(\frac{3}{2} - i)$.

10. $\left(\frac{1}{2} + \frac{\sqrt{3}}{2}i \right)^3$.

11. $(2i)^5$.

12. $i^{12} + i^{25} - 7i^{111}$.

13. $\dfrac{14 + 13i}{2 - i}$.

14. $\dfrac{x + iy}{x - iy}$.

15. $\dfrac{101 + i}{100 + i}$.

16.   Verify the given property of complex conjugation. Take z, z_1, and z_2 to be complex numbers.

(a) $\overline{z_1 + z_2} = \overline{z_1} + \overline{z_2}$.

(b) $\overline{z_1 - z_2} = \overline{z_1} - \overline{z_2}$.   (c) $\overline{z_1 \, z_2} = \overline{z_1} \, \overline{z_2}$.

(d) $\overline{\left(\frac{z_1}{z_2}\right)} = \frac{\overline{z_1}}{\overline{z_2}}$ $(z_2 \neq 0)$.

(e) $\overline{(z^n)} = \overline{z}^n$, $n = 1, 2, \ldots$.
[Hint: Use induction.]

(f) $z = \overline{z} \Leftrightarrow z$ is purely real.   (g) $z = -\overline{z} \Leftrightarrow z$ is purely imaginary.

In Exercises 17–22, solve the given equation. Express your answers in the form $a + ib$, where a and b are real.

17. $x^2 + 6 = 0$.

18. $x^2 + 2x = 1$.

19. $x^2 + x + 1 = 0$.

20. $3x^2 + x = -2$.

21. $x^4 + 2x^2 + 1 = 0$.

22. $x^4 + 6x^2 + 3 = 0$.

23.   Find two numbers that sum to ten and whose product is 40. (This problem is of some historical value. It is said that Cardan first stumbled upon complex numbers while solving it.)

24.   Let z_1, z_2, z_3 be complex numbers. Prove, using the definition of complex numbers and the commutative, associative, and distributive properties of real numbers that

(a) $z_1 + z_2 = z_2 + z_1$ (commutativity of addition)
(b) $z_1 z_2 = z_2 z_1$ (commutativity of multiplication)
(c) $(z_1 + z_2) + z_3 = z_1 + (z_2 + z_3)$ (associativity of addition)
(d) $(z_1 z_2) z_3 = z_1 (z_2 z_3)$ (associativity of multiplication)
(e) $z_1 (z_2 + z_3) = z_1 z_2 + z_1 z_3$ (distribution)

25.   (a) Let $z = a + ib$ be any complex number. Find the **additive inverse** for z; that is, the complex number w such that $z + w = 0$.
(b)   Let e_s represent a complex number such that $z + e_s = z$ for all complex z. Show that $e_s = 0$; that is, $\text{Re}(e_s) = 0$ and $\text{Im}(e_s) = 0$. Thus $e_s = 0$ is the unique **additive identity** for complex numbers.

26.   (a) Let $z = a + ib \neq 0$ be a complex number. Find z^{-1} in terms of a and b. By z^{-1} we mean the **multiplicative inverse** of z, which satisfies $z \, z^{-1} = z^{-1} z = 1$.

(b)   Let e_m represent a complex number such that $z \, e_m = e_m z = z$ for all complex z. Prove that $e_m = 1$; that is $\text{Re}(e_m) = 1$ and $\text{Im}(e_m) = 0$. Thus $e_m = 1$ is the unique **multiplicative identity** for complex numbers.
(c)   Show that if $z_1 z_2 = 0$, then either $z_1 = 0$ or $z_2 = 0$. [Hint: Suppose $z_1 \neq 0$. Use (a).]

27.   Prove that $c + id \neq 0 \Leftrightarrow c - id \neq 0 \Leftrightarrow c^2 + d^2 \neq 0$. [Hint: It is enough to prove $c + id \neq 0 \Rightarrow c - id \neq 0 \Rightarrow c^2 + d^2 \neq 0 \Rightarrow c + id \neq 0$.]

28.   (a) Show that $i^{-n} = i^n$ for n even, and $i^{-n} = -i^n$ for n odd. (b) Show that $\overline{i^n} = i^{-n}$ for all n.

29.   (a) Let z be a complex number and n a positive integer. A **polynomial** of degree n is an expression of the form $a_n z^n + a_{n-1} z^{n-1} + \cdots + a_1 z + a_0$, where $a_n \neq 0$. The polynomial is said to have real coefficients if all the coefficients a_j are real. For such a polynomial, show that

$$\overline{a_n z^n + a_{n-1} z^{n-1} + \cdots + a_1 z + a_0} = a_n (\overline{z})^n + a_{n-1} (\overline{z})^{n-1} + \cdots + a_1 \overline{z} + a_0.$$

[Hint: Exercise 16(e).]

(b) Recall that z_0 is a **root** of a polynomial $p(z)$ if $p(z_0) = 0$. Show that if z_0 is a root of a polynomial with real coefficients, then $\overline{z_0}$ is also a root. Thus the complex roots of polynomials with real coefficients always appear in conjugate pairs.

30. In each case, use the given roots to find other roots with the help of Exercise 29. Then factor the polynomial and find all its roots.

(a) $z^3 + z^2 + z + 1 = 0, \quad z = i.$

(b) $z^3 + 10 z^2 + 29 z + 30 = 0, \quad z = -2 + i.$

(c) $z^4 + 4 = 0, \quad z_1 = 1 + i, \ z_2 = -1 + i.$

(d) $z^4 - 6 z^3 + 15 z^2 - 18 z + 10 = 0, \quad z_1 = 1 + i, \ z_2 = 2 + i.$

31. Project Problem: Computing square roots.   The problem of finding nth roots of complex numbers will be discussed later in this chapter. The case of square roots is particularly interesting and can be solved by reducing to two equations in two unknowns. In this exercise, you are asked to compute $\sqrt{1+i}$ to illustrate the process.

(a) Finding $\sqrt{1+i}$ is equivalent to solving $z^2 = 1 + i$. Let $z = x + iy$ and obtain

$$\begin{cases} x^2 - y^2 & = 1, \\ 2xy & = 1. \end{cases}$$

(b) Derive the equation in x: $4x^4 - 4x^2 - 1 = 0$. This is essentially a quadratic equation in x^2.

(c) Keep in mind that x^2 is nonnegative and obtain that $x^2 = \frac{1+\sqrt{2}}{2}$. Thus

$$x = \pm \sqrt{\frac{1 + \sqrt{2}}{2}} \quad \text{and} \quad y = \pm \frac{1}{\sqrt{2 + 2\sqrt{2}}}.$$

(d) Conclude that the square roots of $1 + i$ are

$$\sqrt{\frac{1 + \sqrt{2}}{2}} + \frac{i}{\sqrt{2 + 2\sqrt{2}}} \quad \text{and} \quad -\sqrt{\frac{1 + \sqrt{2}}{2}} - \frac{i}{\sqrt{2 + 2\sqrt{2}}}.$$

32. Compute the given square roots.

(a) $\sqrt{i}.$ (b) $\sqrt{-2 + i}.$ (c) $\sqrt{\frac{-1 - i\sqrt{3}}{2}}.$

33. Project Problem: The cubic equation.   In this exercise, we will derive the solution of the cubic equation

(4) $x^3 + ax^2 + bx + c = 0.$

(a)   Use the change of variables $x = y - \frac{a}{3}$ to transform the equation to the following reduced form

(5) $y^3 + py + q = 0,$

which does not contain a quadratic term in y, where $p = b - \frac{a^2}{3}$ and $q = \frac{2a^3}{27} - \frac{ab}{3} + c$. (This trick is due to the Italian mathematician Niccolò Tartaglia (1500–1557).)

(b)   Let $y = u + v$, and show that $u^3 + v^3 + (3uv + p)(u + v) + q = 0$.

(c)   Require that $3uv + p = 0$; then directly we have $u^3 v^3 = -\frac{p^3}{27}$, and from the equation in part (b) we have $u^3 + v^3 = -q$.

(d)   Suppose that U and V are numbers satisfying $U + V = -\beta$ and $UV = \gamma$. Show that U and V are solutions of the quadratic equation $X^2 + \beta X + \gamma = 0$.

(e)   Use (c) and (d) to conclude that u^3 and v^3 are solutions of the quadratic equation $X^2 + qX - \frac{p^3}{27} = 0$. Thus,

$$u = \sqrt[3]{-\frac{q}{2} + \sqrt{\left(\frac{q}{2}\right)^2 + \left(\frac{p}{3}\right)^3}} \quad \text{and} \quad v = \sqrt[3]{-\frac{q}{2} - \sqrt{\left(\frac{q}{2}\right)^2 + \left(\frac{p}{3}\right)^3}}.$$

(f)   Derive a solution of (4),

$$x = \sqrt[3]{-\frac{q}{2} + \sqrt{\left(\frac{q}{2}\right)^2 + \left(\frac{p}{3}\right)^3}} + \sqrt[3]{-\frac{q}{2} - \sqrt{\left(\frac{q}{2}\right)^2 + \left(\frac{p}{3}\right)^3}} - \frac{a}{3}.$$

This is **Cardan's formula**, named after him because he was the first one to publish it. In the case $\left(\frac{q}{2}\right)^2 + \left(\frac{p}{3}\right)^3 \geq 0$, the formula clearly yields one real root of (4). You can use this root to factor (4) down into a quadratic equation, which you can solve to find all the roots of (4). The case $\left(\frac{q}{2}\right)^2 + \left(\frac{p}{3}\right)^3 < 0$ baffled the mathematicians of the sixteenth century. They knew that the cubic equation (4) must have at least one real root, yet the solution in this case involves square roots of negative numbers, which are imaginary numbers. It turns out in this case that u and v are complex conjugate numbers, hence their sum is a real number and the solution x is real! This was discovered by the Italian mathematician Rafael Bombelli (1527–1572) (see Exercise 38).

Not only was Bombelli bold enough to work with complex numbers; by using them to generate real solutions, he demonstrated that complex numbers were not merely the product of our imagination but tools that are essential to derive real solutions. This theme will occur over and over again in this book when we will appeal to complex variable techniques to solve real-life problems calling for real-valued solutions.

For an interesting account of the history of complex numbers, we refer you to the book *The History of Mathematics, An Introduction*, 3rd edition, by David M. Burton (McGraw-Hill, 1997).

In Exercises 34–37, use Cardan's formula to find one real root of the given cubic equation. Then factor the cubic and find the remaining solutions (some roots may be repeated).

34. $x^3 - x + 6 = 0$.

35. $x^3 + 4x^2 + x + 4 = 0$.

36. $x^3 - 2x^2 + x - 12 = 0$.

37. $x^3 - 2x^2 + 3 = 0$.

38. Bombelli's equation.   A problem of historical interest is $x^3 - 15x - 4 = 0$, which was investigated by Bombelli.

(a) Show that, by applying Cardan's formula, we arrive at the solution

$$x = u + v = \sqrt[3]{2 + 11i} + \sqrt[3]{2 - 11i},$$

where u is the first cube root and v is the second.

(b) Bombelli had the incredible insight that u and v have to be conjugate for $u + v$ to be real. Set $u = a + ib$ and $v = a - ib$, where a and b are to be determined. Cube both sides of the equations and note that $a = 2$, $b = 1$ will work for both equations.

(c) What is the real solution, x, of Bombelli's equation? What are the other two solutions?

1.2   The Complex Plane

A useful way of visualizing complex numbers is to plot them as points in a plane. To do this, we associate to each complex number $z = x + iy$ the ordered pair (x, y), then plot the ordered pair (x, y) as a point $P = (x, y)$ in the Cartesian xy-plane. Since x and y uniquely determine z, we thus obtain a one-to-one correspondence between complex numbers $z = x + iy$ and points (x, y) in the Cartesian plane. The horizontal axis will be called the **real axis**, since the abscissa of a complex number is its real part; and complex numbers lying on the horizontal axis are purely real. The vertical axis will be called the **imaginary axis**, since the ordinate of a complex number is its imaginary part; and complex numbers lying on the vertical axis are purely imaginary. The Cartesian plane will be referred to as the **complex plane**, also commonly called the z-**plane**. It is not unusual to denote a point (x, y) in the complex plane by the corresponding complex number $x + iy$ (see Figure 1). We can also think of a complex number $z = x + iy$ as a two-dimensional vector in the complex plane, with its tail at the origin and its head at $P = (x, y)$.

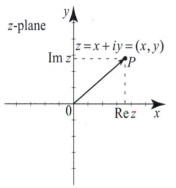

Figure 1 The complex plane.

Historically, the geometric representation of complex numbers is due to Gauss and two lesser known mathematicians: the Frenchman Jean Robert Argand (1768–1822) and a Norwegian surveyor Caspar Wessel (1745–1818). This relatively simple idea dispelled the mystery and skepticism surrounding complex numbers. Much as real numbers are represented as points on a line, complex numbers are represented as points in the plane. As you will soon see, our ability to visualize complex numbers will greatly enhance our understanding of facts concerning complex numbers, and in some cases greatly simplify their proofs.

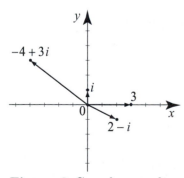

Figure 2 Complex numbers as points and vectors.

EXAMPLE 1   Points and vectors in the plane

Label the following points in the complex plane: 3, 0, i, $2 - i$, $-4 + 3i$. Draw their associated vectors, emanating from the origin.

Solution   The points and the vectors are depicted in Figure 2. The complex number 3, being purely real, lies on the horizontal axis; while i, being purely imaginary, lies on the vertical axis. Note that one cannot really draw a vector to represent 0, whose point is at the origin.   ∎

Geometric Interpretation of Algebraic Rules

The vector representation allows us to add complex numbers by adding their vectors in the plane, with the usual head-to-tail or parallelogram methods (see Figure 3).

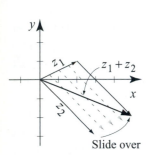

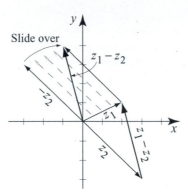

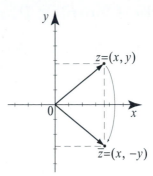

Figure 3 Vector addition of complex numbers: the head-to-tail method. Slide z_2 over so its tail lies atop z_1's head. The resultant vector is then $z_1 + z_2$.

Figure 4 Subtraction: We will find $z_1 - z_2$. We could take $-z_2$, then add it to z_1. Or we could take the vector with its tail at z_2's head, and its head at z_1's head.

Figure 5 Complex conjugation reflects a point about the horizontal axis.

Multiplying a complex number by -1 has the effect of reflecting it about the origin. If $z = x + i y = (x, y)$, then $-z = -x - i y = (-x, -y)$. This will allow us to subtract complex numbers. The complex subtraction $z_1 - z_2$ can be performed by first multiplying the z_2 vector by -1, then adding the resultant to z_1. Alternatively, we can draw both vectors with their tails at the origin, and the difference will be the vector that points from the head of z_2 to the head of z_1 (see Figure 4).

Conjugation has an interesting interpretation in the complex plane. Since the conjugate of $z = x + iy = (x, y)$ is $\overline{z} = x - iy = (x, -y)$, conjugates are reflections of each other across the real axis (see Figure 5).

EXAMPLE 2   Vector addition and subtraction of complex numbers

Let $z_1 = 2 + i$ and $z_2 = 3 - 3 i$. Find graphically $z_1 + z_2$ and $z_1 - z_2$.

Solution   First draw the points in the plane and their associated vectors (see Figure 3). Then take z_2's vector and slide it so its tail lies on z_1's head. This gives $z_1 + z_2 = 5 - 2 i$.

Take z_2 and reflect it about the origin to get $-z_2$. Now add $-z_2$ to z_1 as vectors to get $z_1 - z_2$. The result is $z_1 - z_2 = -1 + 4 i$ (see Figure 4).   ∎

The Absolute Value

We now discuss an important geometric aspect of complex numbers. For a complex number $z = x + iy$, we define the **norm** or **modulus** or **absolute value** of z by

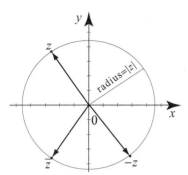

$$(1) \qquad |z| = \sqrt{x^2 + y^2}.$$

If $z = x$ is real, then $|z| = \sqrt{x^2} = |x|$. Thus the absolute value of a complex number z reduces to the familiar absolute value when z is real. Just as the absolute value $|x|$ of a real number x represents the distance from x to the origin on the real line, the absolute value $|z|$ of a complex number z represents the *distance* from the point z to the origin in the complex plane (see Figure 6). It is easy to see from (1) or Figure 6 that

Figure 6 The distance from z to the origin is $|z|$.

$$(2) \qquad |z| = 0 \Leftrightarrow z = 0;$$

and

$$(3) \qquad |-z| = |z|.$$

If $z_1 = x_1 + iy_1$ and $z_2 = x_2 + iy_2$, applying (1) to the complex number $z_1 - z_2 = (x_1 - x_2) + i(y_1 - y_2)$, we obtain

$$(4) \qquad |z_1 - z_2| = \sqrt{(x_1 - x_2)^2 + (y_1 - y_2)^2},$$

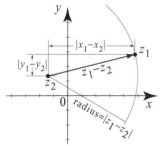

Figure 7 The distance from z_1 to z_2 is $|z_1 - z_2|$.

which is the familiar formula for distance between the points (x_1, y_1) and (x_2, y_2). Thus $|z_1 - z_2|$ has a concrete geometric interpretation as the **distance** between (the points) z_1 and z_2 (see Figure 7).

EXAMPLE 3   The absolute value as a distance

(a)   For $z_1 = 2 + 4i$ and $z_2 = 5 + i$, we have

$$|z_1| = \sqrt{2^2 + 4^2} = 2\sqrt{5} \approx 4.472, \qquad |z_2| = \sqrt{5^2 + 1^2} = \sqrt{26} \approx 5.099.$$

Thus we see that $|z_1| < |z_2|$. Geometrically, this means that z_2 lies farther from the origin in the complex plane (see Figure 8).

(b)   The distance between z_1 and z_2 is given by

$$|z_1 - z_2| = \sqrt{(2 - 5)^2 + (4 - 1)^2} = \sqrt{18} = 3\sqrt{2} \approx 4.24.$$

This distance is also the length of the vector $z_1 - z_2$, as shown in Figure 8.   ■

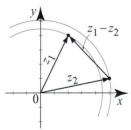

Figure 8

Note that in Example 3(a) we compared the sizes of z_1 and z_2 and *not* the numbers themselves. In general, it does not make sense to write an inequality such as $z_1 \leq z_2$ or $z_2 \leq z_1$, unless z_1 and z_2 are real. This is because the complex numbers do not have a **linear ordering** like the real numbers.

We can use the geometric interpretation of the absolute value to our advantage in describing subsets of the complex numbers as subsets of the plane.

EXAMPLE 4   Circles, disks, and ellipses

(a)   Find and plot the points z satisfying

(5) $$|z + 4 - i| = 2.$$

(b)   Find and plot the points z satisfying

(6) $$|z + 4 - i| \leq 2.$$

(c)   Find and plot all complex numbers z such that

(7) $$|z + 2 + 2i| + |z + 1 + i| = 4.$$

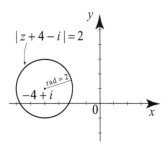

Figure 9
The circle in Example 4(a).

Solution   (a)   In answering the questions, we will write an absolute value in the form $|z - z_0|$ and interpret it as a distance between z and z_0. Thus (5) is equivalent to

$$|z - (-4 + i)| = 2.$$

Reading the absolute value as a distance, the question becomes: What are the points z whose distance to $-4 + i$ is 2? The answer now is obvious:

$|z - (-4 + i)| = 2 \Leftrightarrow$ z lies on the circle centered at $-4 + i$, with radius 2.

(See Figure 9.) As you know, the Cartesian equation of the circle is $(x + 4)^2 + (y - 1)^2 = 4$. To derive this equation, write $z = x + i\,y$ and use (4). Thus

$$|z - (-4 + i)| = 2 \Leftrightarrow |(x + 4) + i(y - 1)| = 2 \Leftrightarrow \sqrt{(x + 4)^2 + (y - 1)^2} = 2,$$

and the Cartesian equation of the circle follows upon squaring both sides.

(b)   Reading the absolute value as a distance, we ask: What are the points z whose distance to $-4 + i$ is less than or equal to 2? The answer is clear:

$|z - (-4 + i)| \leq 2 \Leftrightarrow$ z lies inside or on the circle centered at $-4 + i$, with radius 2.

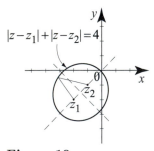

Figure 10
The ellipse in Example 4(c).

In other words, z lies in the disk centered at $-4 + i$, with radius 2 (Figure 9).

(c)   Write (7) in the form

$$|z - (-2 - 2i)| + |z - (-1 - i)| = 4.$$

This time we are looking for all points z the sum of whose distances to the two points $z_1 = -2 - 2i$ and $z_2 = -1 - i$ is constant and equals 4. From geometry, we know that the answer is the ellipse with **foci** (plural of **focus**) located at z_1 and z_2, and semi-major axis 2. (See Figure 10.) ∎

The absolute value of complex numbers enjoys many interesting properties, including all those of the usual absolute value of real numbers.

IDENTITIES INVOLVING THE ABSOLUTE VALUE

Let $z, z_1, z_2, \dots$ be any complex numbers. We have

$$(8) \qquad |z| = \sqrt{z\overline{z}}, \quad \text{and also} \quad |z|^2 = z\overline{z}.$$

Furthermore

$$(9) \qquad |z| = |\overline{z}|.$$

(10) states that *the absolute value of a product is the product of the absolute values.*

The absolute value of a product satisfies

$$(10) \qquad |z_1\, z_2| = |z_1|\,|z_2|;$$
$$(11) \qquad |z_1\, z_2 \cdots z_n| = |z_1|\,|z_2| \cdots |z_n|;$$
$$(12) \qquad |z^n| = |z|^n \quad (n = 1, 2, \dots).$$

(13) states that *the absolute value of a quotient is the quotient of the absolute values.*

The absolute value of a quotient satisfies

$$(13) \qquad \left| \frac{z_1}{z_2} \right| = \frac{|z_1|}{|z_2|} \qquad (z_2 \neq 0).$$

Recall that, for any complex number z, $z\overline{z}$ is always a nonnegative real number. So there is no problem in taking the square root in (8).

Proof   Write $z = x + i\,y$. Starting from (1), we have

$$|z| = \sqrt{x^2 + y^2} = \sqrt{\underbrace{(x + i\,y)}_{z}\,\underbrace{(x - i\,y)}_{\overline{z}}} = \sqrt{z\overline{z}},$$

and (8) follows. Using (1) with $\overline{z} = x - i\,y$, we find

$$|\overline{z}| = \sqrt{x^2 + (-y)^2} = \sqrt{x^2 + y^2} = |z|.$$

Using (8) to compute $|z_1\, z_2|$, we find

$$|z_1\, z_2| = \sqrt{(z_1\, z_2)\,\overline{(z_1\, z_2)}} = \sqrt{(z_1 \overline{z_1})(z_2 \overline{z_2})} = \sqrt{z_1\, \overline{z_1}}\,\sqrt{z_2\, \overline{z_2}} = |z_1|\,|z_2|.$$

(For the second equality, we used $\overline{z_1\, z_2} = \overline{z_1}\,\overline{z_2}$ and the associativity of multiplication.) The proofs of (11) and (12) are straightforward and are left as an exercise. Replacing z_1 by $\frac{z_1}{z_2}$ ($z_2 \neq 0$) in (10), we obtain

$$\left| \frac{z_1}{z_2}\, z_2 \right| = \left| \frac{z_1}{z_2} \right|\,|z_2| \quad \Rightarrow \quad |z_1| = \left| \frac{z_1}{z_2} \right|\,|z_2|,$$

and (13) follows upon dividing by $|z_2| \neq 0$.   ∎

EXAMPLE 5   Moduli of products and quotients

Compute the given absolute value. Take n to be a positive integer.

(a) $|(1+i)\overline{(2-i)}|$.   (b) $|(1-i)^4(1+2i)(1+\sqrt{2}\,i)|$.   (c) $|i^n|$.

(d) $\left|\dfrac{1+2i}{1-i}\right|$.   (e) $\left|\dfrac{(3+4i)^2(3-i)^{10}}{(3+i)^9}\right|$.

Solution   We will use as much as possible the properties of the absolute value to avoid excessive computations.

(a)   Using (10) and (9), we have

$$|(1+i)\overline{(2-i)}| = |1+i|\,|\overline{(2-i)}| = |1+i|\,|2-i| = \sqrt{2}\,\sqrt{5} = \sqrt{10}.$$

(b)   Using (11) and (12), we have

$$|(1-i)^4(1+2i)(1+\sqrt{2}\,i)| = |(1-i)^4|\,\overbrace{|1+2i|}^{=\sqrt{5}}\,\overbrace{|1+\sqrt{2}\,i|}^{=\sqrt{3}} = \overbrace{|1-i|^4}^{=(\sqrt{2})^4}\,\sqrt{15} = 4\sqrt{15}.$$

(c)   From (12) and the fact that $|i| = 1$, we have

$$|i^n| = |i|^n = 1.$$

(d)   Using (13), we have

$$\left|\frac{1+2i}{1-i}\right| = \frac{|1+2i|}{|1-i|} = \sqrt{\frac{5}{2}}.$$

(e)   We have

$$\left|\frac{(3+4i)^2(3-i)^{10}}{(3+i)^9}\right| = \frac{\overbrace{|3+4i|^2}^{=25}\,|3-i|^{10}}{|3+i|^9} = 25|3-i| = 25\sqrt{10},$$

because $|3+i| = |3-i|$ by (9).   ∎

In addition to the identities that we just proved, the absolute value satisfies fundamental inequalities, which are in some cases immediate consequences of elementary facts from geometry. Let us recall two facts and then list the inequalities that will be obtained as a consequence.

- In a right triangle, the hypotenuse is larger than either one of the other two sides.

- In any triangle, the sum of two sides is larger than the third side.

INEQUALITIES INVOLVING THE ABSOLUTE VALUE

Let $z, z_1, z_2, \ldots$ be complex numbers. We have

(14) $$|\operatorname{Re} z| \leq |z|, \qquad |\operatorname{Im} z| \leq |z|;$$

(15) $$|z| \leq |\operatorname{Re} z| + |\operatorname{Im} z|.$$

The absolute value of a sum satisfies the fundamental inequality

(16) $$|z_1 + z_2| \leq |z_1| + |z_2|,$$

known as the **triangle inequality**. More generally,

(17) $$|z_1 + z_2 + \cdots + z_n| \leq |z_1| + |z_2| + \cdots + |z_n|.$$

The absolute value of a difference satisfies

(18) $$|z_1 - z_2| \leq |z_1| + |z_2|.$$

A lower bound for $|z_1 \pm z_2|$ is provided by

(19) $$|z_1 \pm z_2| \geq ||z_1| - |z_2||.$$

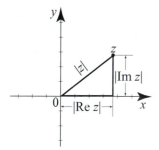

Figure 11
For inequalities (14) and (15).

Proof   Consider a nondegenerate right triangle with vertices at 0, $z = (x, y)$, and $\operatorname{Re} z = x$, as shown in Figure 11. The sides of the triangle are $|\operatorname{Re} z| = |x|$, $|\operatorname{Im} z| = |y|$, and the hypothenuse is $|z|$. Since the hypothenuse is larger than either of the other two sides, we obtain (14). Since the sum of two sides in a triangle is larger than the third side, we obtain (15). Of course, (14) and (15) are also consequences of the inequalities

$$|x| \leq \sqrt{x^2 + y^2}, \quad |y| \leq \sqrt{x^2 + y^2}, \text{ and } \quad \sqrt{x^2 + y^2} \leq |x| + |y|,$$

which are straightforward to prove.

For the moment, we skip (16) and (17) and prove (18). Consider the triangle in Figure 12 with vertices at 0, z_1, and z_2. The lengths of the sides in that triangle are $|z_1|$, $|z_2|$, and $|z_1 - z_2|$. Because the sum of two sides is larger than the third side, we obtain (18). Replacing z_2 by $-z_2$ and noticing that $|-z_2| = |z_2|$, we obtain (16). Inequality (17) follows by repeated applications of (16).

Replacing z_1 by $z_1 - z_2$ in (16), we obtain $|z_1| \leq |z_1 - z_2| + |z_2|$, and so

$$|z_1 - z_2| \geq |z_1| - |z_2|.$$

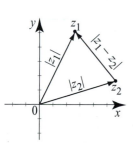

Figure 12
For inequality (18).

Reversing the roles of z_1 and z_2, and realizing that $|z_2 - z_1| = |z_1 - z_2|$, we also have

$$|z_1 - z_2| \geq |z_2| - |z_1|.$$

Putting these two together, we conclude $|z_1 - z_2| \geq ||z_1| - |z_2||$, which proves the "minus-sign" inequality in (19). To get the inequality with a plus sign, namely, $|z_1 + z_2| \geq ||z_1| - |z_2||$, replace z_2 by $-z_2$. ∎

Because the triangle inequality is fundamental in the development of complex analysis, we will also offer an algebraic proof. This will afford us the opportunity to practice some of the properties of complex numbers that we have already derived. Start by observing that

$$\overline{z_1 \, \overline{z_2}} = \overline{z_1} \, \overline{\overline{z_2}} = \overline{z_1} \, z_2.$$

Since for any complex number w, $w + \overline{w} = 2 \operatorname{Re} w$, we obtain

$$z_1 \, \overline{z_2} + \overline{z_1} \, z_2 = z_1 \, \overline{z_2} + \overline{z_1 \, \overline{z_2}} = 2 \operatorname{Re}(z_1 \, \overline{z_2}).$$

Using this interesting identity, along with (8) and basic properties of complex conjugation, we obtain

$$
\begin{aligned}
|z_1 + z_2|^2 &= (z_1 + z_2)\overline{(z_1 + z_2)} = (z_1 + z_2)(\overline{z_1} + \overline{z_2}) \\
&= z_1 \, \overline{z_1} + z_2 \, \overline{z_2} + z_1 \, \overline{z_2} + \overline{z_1} \, z_2 = |z_1|^2 + |z_2|^2 + z_1 \, \overline{z_2} + \overline{z_1} \, z_2 \\
&= |z_1|^2 + |z_2|^2 + 2 \operatorname{Re}(z_1 \, \overline{z_2}) \\
&\leq |z_1|^2 + |z_2|^2 + 2|z_1 \, \overline{z_2}| \qquad \text{[by (15)]} \\
&= |z_1|^2 + |z_2|^2 + 2|z_1| \, |\overline{z_2}| \\
&= |z_1|^2 + |z_2|^2 + 2|z_1| \, |z_2| \quad (\text{[by (9)]}, \ |\overline{z_2}| = |z_2|) \\
&= (|z_1| + |z_2|)^2 ,
\end{aligned}
$$

and (16) follows now upon taking square roots on both sides.

The triangle inequality will be used extensively in proofs to provide estimates on the sizes of complex-valued expressions. We next illustrate this type of applications with examples.

EXAMPLE 6   Estimating the size of an absolute value

What is an upper bound for $|z^5 - 4|$ if $|z| \leq 1$?

Solution   Applying the triangle inequality, we get

$$|z^5 - 4| \leq |z^5| + 4 = |z|^5 + 4 \leq 1 + 4 = 5,$$

because $|z| \leq 1$. Hence if $|z| \leq 1$, an upper bound for $|z^5 - 4|$ is 5.

Can we find a number smaller than 5 that is also an upper bound, or is 5 the *least upper bound*? It is easy to see that for $z = -1$, we have $|z^5 - 4| = |(-1)^5 - 4| = |-1 - 4| = 5$. Thus, the upper bound 5 is best possible. You should be cautioned that, in general, the triangle inequality is considered as a crude inequality, which means that it will not yield least upper bound estimates as it did in this case. See Exercise 33 for an illustration.   ∎

EXAMPLE 7   Inequalities

Show that for any complex numbers z and a

(20)
$$\frac{1}{|a| + |z|} \le \left| \frac{1}{a + z} \right| \le \frac{1}{||a| - |z||}.$$

Solution   The triangle inequality (16) tells us that $|a + z| \le |a| + |z|$, while (19) implies that $||a| - |z|| \le |a + z|$. Hence the inequalities

$$||a| - |z|| \le |a + z| \le |a| + |z|.$$

Taking reciprocals reverses the inequalities and yields

$$\frac{1}{||a| - |z||} \ge \frac{1}{|a + z|} \ge \frac{1}{|a| + |z|},$$

which implies (20), since $\left| \frac{1}{a+z} \right| = \frac{1}{|a+z|}$ by (13).   ■

Our final example illustrates a classical trick in dealing with inequalities. It consists of adding and subtracting a number in order to transform a given expression into a form that contains familiar terms.

EXAMPLE 8   Tricks with the absolute value

(a)   What is an upper bound for $|z - 3|$ if $|z - i| \le 1$?
(b)   What is a lower bound for $|z - 3|$ if $|z - i| \le 1$?

Solution   (a) We want to know some thing about $|z - 3|$ given some information about $|z - i|$. The trick is to add and subtract i, then use the triangle inequality as follows:

$$\begin{aligned} |z - 3| &= |z - i + i + 3| = |(z - i) + (3 + i)| &&\text{(Add and subtract } i.) \\ &\le |z - i| + \underbrace{|3 + i|}_{=\sqrt{10}} \le 1 + \sqrt{10}, &&\text{(triangle inequality)} \end{aligned}$$

since $|z - i| \le 1$. Thus an upper bound is $1 + \sqrt{10}$.

(b) In finding a lower bound, we will proceed as in (a) but use (19) instead of the triangle inequality. We have

$$\begin{aligned} |z - 3| &= |(z - i) + (3 + i)| &&\text{(Add and subtract } i.) \\ &\ge ||z - i| - |3 + i|| = \left| |z - i| - \sqrt{10} \right|, \end{aligned}$$

by (19). Now since $|z - i|$ is at most 1 and $\sqrt{10} > 1$, we see that

$$\left| |z - i| - \sqrt{10} \right| \ge \sqrt{10} - 1.$$

Hence, $|z - 3| \ge \sqrt{10} - 1$ if $|z - i| \le 1$.   ■

Exercises 1.2

In Exercises 1–6, plot the points z, $-z$, $\bar{z}$ and the associated vectors emanating from the origin. In each case, compute the modulus of z.

1. $1-i$.   **2.** $\frac{\sqrt{2}}{2}+i\frac{\sqrt{2}}{2}$.   **3.** $3i+5$.   **4.** i^7.   **5.** $\overline{1-i}$.   **6.** $(1+i)^2$.

7.   Let

$$z_1 = i, \ z_2 = \frac{\sqrt{2}}{2}+i\frac{\sqrt{2}}{2}, \ z_3 = \frac{\sqrt{2}}{2}-i\frac{\sqrt{2}}{2}, \ z_4 = \frac{1}{2}+i\frac{3}{2}.$$

(a)   Plot z_2, z_3, and z_2+z_3 on the same complex plane. Explain in words how you constructed z_2+z_3.

(b)   Plot the points z_1, z_2, z_3, z_4, compute their moduli, and decide which point or points are closest to the origin.

(c)   Find graphically z_1-z_2, z_2-z_3, and z_3-z_4.

(d)   Which one of the points z_2 or z_4 is closer to z_1?

8.   Let $z_1 = i$, $z_2 = 1+2i$.

(a)   On the same complex plane, plot z_1, z_2, z_1+z_2, z_1-z_2, z_2-z_1.

(b)   How does vector z_1-z_2 compare with z_2-z_1?

(c)   On the same complex plane, plot the vectors z_2 and iz_2. How do these vectors compare? Describe in general the vector iz in comparison with vector z.

(d) Describe in general the vector z/i in comparison with z.

In Exercises 9–14, compute the given modulus.

9. $|(1+i)(1-i)(1+3i)|$.   **10.** $\left|(2+3i)^8\right|$.   **11.** $\left|\left(\frac{\sqrt{2}}{2}+i\frac{\sqrt{2}}{2}\right)^{27}\right|$.

12. $\left|\frac{1+i}{(1-i)(1+3i)}\right|$.   **13.** $\left|\frac{i}{2-i}\right|$.   **14.** $\left|\frac{(1+i)^5}{(-2+2i)^5}\right|$.

In Exercises 15–20, describe and plot the set of points satisfying the given equation.

15. $|z-4|=3$.   **16.** $|z+2+i|=1$.   **17.** $|z-i|=-1$.

18. $|z+1|+|z-1|=4$.   **19.** $|z-i|+|z|=2$.   **20.** $|2z|+|2z-1|=4$.

21.   Derive the equation of the ellipse in Exercise 20.

22.   Derive the equation of the circle in Exercise 16.

23.   **Lines in the complex plane.**   (a) Show that the set of points z in the complex plane with $\operatorname{Re}z = a$ (a real) is the vertical line $x = a$.

(b) Show that the set of points z in the complex plane with $\operatorname{Im}z = b$ (b real) is the horizontal line $y = b$.

(c) Let $z_1 \neq z_2$ be two points in the complex plane. Show that the set of points z such that $z = z_1 + t(z_2 - z_1)$, where t is real, is the line going through z_1 and z_2. Illustrate your answer graphically by plotting the vectors z_1, $z_1 + z_2$, and $z_1 + t(z_2 - z_1)$ for several values of t.

24.   Show that three distinct points z_1, z_2, and z_3 lie on the same line if and only if

$$\frac{z_1 - z_2}{z_1 - z_3} = t,$$

where t is real. (Compare with Exercise 23(c).)

25.   Parabolas. Recall from geometry that a **parabola** is the set of points in the plane that are equidistant from a fixed line (called the **directrix**) and a fixed point not on the line (called the **focus**).

(a) Using this description and the geometric interpretation of the absolute value, argue that the set of points z satisfying

$$|z - 1 - i| = \text{Re}\,(z) + 1$$

is a parabola. Find its directrix and its focus and then plot it.

(b) More generally, describe the set of points z such that

$$|z - z_0| = \text{Re}\,(z) - a,$$

where a is a real number.

26.   (a) With the help of Exercise 25, describe the set of points z such that

$$|z - i| = \text{Im}\,(z) + 2.$$

(b) More generally, describe the set of points z such that

$$|z - z_0| = \text{Im}\,(z) - a,$$

where a is a real number.

27.   The multiplicative inverse.   Show that if $z \neq 0$, then

$$z^{-1} = \frac{\overline{z}}{|z|^2}.$$

28.   Use Figure 13 to argue a proof for (16).

29.   Use Figure 14 to argue a proof for (19).

30.   Equality in the triangle inequality.   When do we have

$$|z_1 + z_2| = |z_1| + |z_2|?$$

To answer this question, review the algebraic proof of the triangle inequality and note that the string of equalities in the proof was broken when we replaced $2\,\text{Re}\,(z_1\,\overline{z_2})$ by $2|z_1||z_2|$. Thus we have an equality in the triangle inequality if and only if $\text{Re}\,(z_1\,\overline{z_2}) = |z_1||z_2|$. Show that this is the case if and only either z_1 or z_2 is zero or $z_1 = \alpha z_2$, where α is a positive real number. Geometrically, this states that z_1 and z_2 are on the same side of the ray issued from the origin.

31.   (a) Give an example to show that, in general, $z^2 \neq |z|^2$.

(b) Show that $z^2 = |z|^2 \Leftrightarrow \text{Im}\,z = 0$.

32.   Parallelogram identity.   (a) Consider an arbitrary parallelogram as in Figure 15, with sides a, b, and diagonals c and d. Prove the following identity:

$$c^2 + d^2 = 2(a^2 + b^2).$$

(The identity states the well-known fact that, in a parallelogram, the sum of the squares of the diagonals is equal to the sum of the squares of the sides.) In your

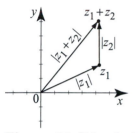

Figure 13 Triangle inequality, Exercise 28.

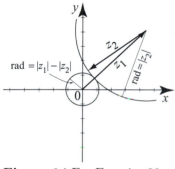

Figure 14 For Exercise 29.

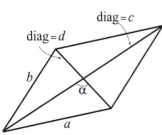

Figure 15 Parallelogram identity, Exercise 32.

proof, use the law of cosines $r^2 + s^2 - 2rs \cos \alpha = a^2$, where α is the angle opposite the side a, and r and s are the other two sides.

(b)   Let u and v be arbitrary complex numbers. Form the parallelogram with sides u and v. Plot $u + v$ and $u - v$ as diagonals of this parallelogram.

(c)   Using parts (a) and (b), show from geometric considerations that for any complex numbers u and v

$$|u + v|^2 + |u - v|^2 = 2\left(|u|^2 + |v|^2\right).$$

This is known as the **parallelogram identity** for complex numbers.

(d) Prove the parallelogram identity algebraically using (8).

33.   The triangle inequality is crude.   (a) Justify each step in the following estimate

$$|\cos\theta + i\sin\theta| \leq |\cos\theta| + |i\sin\theta| = |\cos\theta| + |\sin\theta| \leq 2.$$

Thus, a straightforward application of the triangle inequality yields the estimate $|\cos\theta + i\sin\theta| \leq 2$.

(b)   Use the definition of the absolute value to show $|\cos\theta + i\sin\theta| = 1$. Hence the estimate in (a) is not the best possible estimate.

34.   (a)   Use the triangle inequality to show that $|z - 1| \leq 2$ for $|z| \leq 1$.

(b)   Explain your result in (a) geometrically.

(c)   Is the upper bound in (a) best possible?

35.   Find an upper bound for $|z - 4|$ if $|z - 3i| \leq 1$.

36.   Find a lower bound for $|z - 4|$ if $|z - 3i| \leq 1$.

37.   Find an upper bound for $\left|\frac{1}{z-4}\right|$ if $|z - 1| \leq 1$. [Hint: Find a lower bound for $|z - 4|$.]

38.   Find an upper bound for $\left|\frac{1}{1-z}\right|$ if $|z - i| \leq \frac{1}{2}$.

39.   Find an upper bound for $\left|\frac{1}{z^2+z+1}\right|$ if $|z| \leq \frac{1}{2}$.

40.   Find a lower bound for $\left|\frac{1}{z^2+z+1}\right|$ if $|z| \leq \frac{1}{2}$.

41. Project Problem: Cauchy-Schwarz inequality.   Suppose that v_1, v_2, ... , v_n and w_1, w_2, ... , w_n are arbitrary complex numbers. Our goal in this exercise is to prove the **Cauchy-Schwarz inequality**, which states that

(21)
$$\left|\sum_{j=1}^{n} \overline{v_j} w_j\right| \leq \sqrt{\sum_{j=1}^{n}|v_j|^2}\sqrt{\sum_{j=1}^{n}|w_j|^2}.$$

(a)   Show that in order to prove (21) it is enough to prove

(22)
$$\sum_{j=1}^{n}|v_j|\,|w_j| \leq \sqrt{\sum_{j=1}^{n}|v_j|^2}\sqrt{\sum_{j=1}^{n}|w_j|^2}.$$

(b)   Prove (22) in the case $\sum_{j=1}^{n}|v_j|^2 = 1$ and $\sum_{j=1}^{n}|w_j|^2 = 1$. [Hint: Start with the inequality $0 \leq \sum_{j=1}^{n}\left(|v_j| - |w_j|\right)^2$. Expand and simplify.]

(c)   Prove (22) in the general case. [Hint: Let $v = (v_1, v_2, \ldots, v_n)$ and $w = (w_1, w_2, \ldots, w_n)$, and think of v and w as vectors. Without loss of generality, v and w are not identically 0. Let $\|v\| = \sqrt{\sum_{j=1}^{n} |v_j|^2}$ and define $\|w\|$ similarly. Show that you can apply the result of (b) to the vectors $\frac{1}{\|v\|}v$ and $\frac{1}{\|w\|}w$.]

1.3   Polar Form

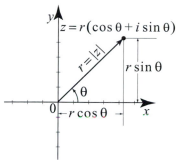

Figure 1 Polar coordinates.

In the previous section, we represented complex numbers as points in the plane and used Cartesian coordinates. In this section, we will use polar coordinates and obtain the so-called *polar representation* or *polar form* of complex numbers. The polar form will introduce trigonometry into the picture, enabling us to take fuller advantage of our geometric insight.

If $P = (x, y) \neq (0, 0)$ is a point given in Cartesian coordinates, we will write it in polar coordinates as $P = (r, \theta)$, where r is the distance from P to the origin O, and θ is the angle between the polar axis (which points toward positive x) and the ray OP. As the usual convention for θ, the counterclockwise direction is positive, while the clockwise direction is negative. We have the familiar change of coordinates formulas:

(1) $$x = r \cos \theta \quad \text{and} \quad y = r \sin \theta, \qquad \text{where } r \geq 0.$$

Given a complex number $z = x + iy$, we think of z as a point (x, y) in the complex plane. Using polar coordinates. we obtain the following polar representation.

POLAR FORM OF COMPLEX NUMBERS

Let $z = x + iy$ be a complex number. We have the **polar representation**

(2) $$z = r(\cos \theta + i \sin \theta),$$

where r is the **modulus** of z and θ is the is the **argument** of z. We have

(3) $$r = \sqrt{x^2 + y^2} \geq 0,$$

and θ is any angle such that

(4) $$\cos \theta = \frac{x}{\sqrt{x^2 + y^2}} = \frac{x}{r}, \quad \sin \theta = \frac{y}{\sqrt{x^2 + y^2}} = \frac{y}{r} \qquad (r \neq 0).$$

The argument of z is not defined when $z = 0$; equivalently, when $r = 0$.

From Figure 1, we have

(5) $$\boxed{\operatorname{Re} z = r \cos \theta, \qquad \operatorname{Im} z = r \sin \theta,}$$

and

(6)
$$\tan\theta = \frac{y}{x} \qquad (x \neq 0).$$

From (3) in this section and (8) in Section 1.2,

(7)
$$r = \sqrt{x^2 + y^2} = |z| = \sqrt{z\,\overline{z}}.$$

It is easy to see that

(8) $\qquad |z| = 1 \Leftrightarrow r = 1 \Leftrightarrow z = \cos\theta + i\,\sin\theta,$ for some real θ.

Any complex number z such that $|z| = 1$ is called **unimodular**. See Exercise 30 for basic properties of unimodular numbers.

We now shift our attention to the more delicate subject of the argument. If θ is any angle that satisfies (4), then $\theta + 2k\pi$ ($k = 0, \pm 1, \pm 2, \dots$) will also satisfy (4), because the cosine and sine are 2π-periodic functions. Thus, relations (4) do not determine a unique value of argument z. If we restrict the choice of θ to the interval $-\pi < \theta \leq \pi$, then there is a unique value of θ that satisfies (4). We call this value the **principal value** of the argument and denote it by $\text{Arg}\,z$. Thus, by definition, if $z = x + i\,y$, then

(9)
$$-\pi < \text{Arg}\,z \leq \pi, \quad \cos(\text{Arg}\,z) = \frac{x}{|z|}, \quad \sin(\text{Arg}\,z) = \frac{y}{|z|}.$$

The set of all values of the argument will be denoted by

(10)
$$\arg z = \{\theta + 2k\pi : \; k = 0, \pm 1, \pm 2, \dots \},$$

where θ is any angle that satisfies (4). In particular, we have

$$\arg z = \{\,\text{Arg}\,z + 2k\pi : \; k = 0, \pm 1, \pm 2, \dots \}.$$

Unlike $\text{Arg}\,z$, which is single-valued, $\arg z$ is multi-valued or set-valued. Sometimes we will abuse notation and write $\arg z = \theta + 2k\pi$ or simply $\arg z = \theta$ to denote a specific value of the argument from the set $\arg z$.

Table 1 contains some special values that will be useful in the examples and exercises.

θ	0	$\frac{\pi}{6}$	$\frac{\pi}{4}$	$\frac{\pi}{3}$	$\frac{\pi}{2}$	$\frac{2\pi}{3}$	$\frac{3\pi}{4}$	$\frac{5\pi}{6}$	π	$\frac{7\pi}{6}$	$\frac{5\pi}{4}$	$\frac{4\pi}{3}$	$\frac{3\pi}{2}$	$\frac{5\pi}{3}$	$\frac{7\pi}{4}$	$\frac{11\pi}{6}$
$\cos\theta$	1	$\frac{\sqrt{3}}{2}$	$\frac{\sqrt{2}}{2}$	$\frac{1}{2}$	0	$-\frac{1}{2}$	$-\frac{\sqrt{2}}{2}$	$-\frac{\sqrt{3}}{2}$	-1	$-\frac{\sqrt{3}}{2}$	$-\frac{\sqrt{2}}{2}$	$-\frac{1}{2}$	0	$\frac{1}{2}$	$\frac{\sqrt{2}}{2}$	$\frac{\sqrt{3}}{2}$
$\sin\theta$	0	$\frac{1}{2}$	$\frac{\sqrt{2}}{2}$	$\frac{\sqrt{3}}{2}$	1	$\frac{\sqrt{3}}{2}$	$\frac{\sqrt{2}}{2}$	$\frac{1}{2}$	0	$-\frac{1}{2}$	$-\frac{\sqrt{2}}{2}$	$-\frac{\sqrt{3}}{2}$	-1	$-\frac{\sqrt{3}}{2}$	$-\frac{\sqrt{2}}{2}$	$-\frac{1}{2}$
$\tan\theta$	0	$\frac{\sqrt{3}}{3}$	1	$\sqrt{3}$	not defined	$-\sqrt{3}$	-1	$-\frac{\sqrt{3}}{3}$	0	$\frac{\sqrt{3}}{3}$	1	$\sqrt{3}$	not defined	$-\sqrt{3}$	-1	$-\frac{\sqrt{3}}{3}$

Table 1. Special trigonometric values.

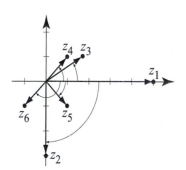

Figure 2 The points in Example 1.

Note that r is always ≥ 0 in the polar representation.

EXAMPLE 1   Polar form

Find the modulus, argument, principal value of the argument, and polar form of the given complex number.

(a) $z_1 = 5.$ (b) $z_2 = -3\,i.$ (c) $z_3 = \sqrt{3} + i.$

(d) $z_4 = 1 + i.$ (e) $z_5 = 1 - i.$ (f) $z_6 = -1 - i.$

Solution   Several parts of our answer are obvious from Figure 2, where we have plotted the given complex numbers.

(a) From (3), $r = |z_1| = 5$. An argument of z_1 is clearly $\theta = 0$. Thus, $\arg z_1 = \{2k\pi : k = 0, \pm 1, \pm 2, \dots \}$. Since 0 is in the interval $(-\pi, \pi]$, we also have $\operatorname{Arg} z_1 = 0$. The polar representation is

$$5 = 5(\cos 0 + i\,\sin 0).$$

(b) Here $r = |z_2| = |-3i| = 3$, and, from Figure 2, $\arg z_2 = \frac{3\pi}{2} + 2k\pi$. To determine $\operatorname{Arg} z_2$, we must pick the unique value of $\arg z_2$ that lies in the interval $(-\pi, \pi]$. From Figure 2, we clearly have $\operatorname{Arg} z_2 = -\frac{\pi}{2}$. Thus, the polar representation

$$-3i = 3\left(\cos\frac{3\pi}{2} + i\,\sin\frac{3\pi}{2}\right) = 3\left(\cos\left(\frac{-\pi}{2}\right) + i\,\sin\left(\frac{-\pi}{2}\right)\right).$$

(c) Here the answers are not obvious from Figure 2, so we will supply all the details. We have $r = |z_3| = |\sqrt{3} + i| = \sqrt{3 + 1} = 2$. So from (4), we have

$$\cos\theta = \frac{x}{r} = \frac{\sqrt{3}}{2} \quad \text{and} \quad \sin\theta = \frac{y}{r} = \frac{1}{2}.$$

From Table 1, we find that $\theta = \frac{\pi}{6}$. Thus, $\arg z_3 = \frac{\pi}{6} + 2k\pi$, $\operatorname{Arg} z_3 = \frac{\pi}{6}$, and the polar representation is

$$\sqrt{3} + i = 2\left(\cos\frac{\pi}{6} + i\,\sin\frac{\pi}{6}\right).$$

(d) We will repeat the steps in (c), even though you may be able to guess the answers from Figure 2. We have $r = |z_4| = \sqrt{1 + 1} = \sqrt{2}$. Factoring the modulus of $1 + i$, we obtain

$$\begin{aligned}
1 + i &= \sqrt{2}\left(\frac{1}{\sqrt{2}} + i\,\frac{1}{\sqrt{2}}\right) = \sqrt{2}\left(\frac{\sqrt{2}}{2} + i\,\frac{\sqrt{2}}{2}\right) \\
&= \sqrt{2}\left(\cos\frac{\pi}{4} + i\,\sin\frac{\pi}{4}\right) \qquad \text{(from Table 1),}
\end{aligned}$$

which is the polar form of z_4, with $\operatorname{Arg} z_4 = \frac{\pi}{4}$.

(e) It is clear from Figure 2 that z_5 is the reflection of $z_4 = 1 + i$ about the x-axis (equivalently, $z_5 = \overline{z_4}$). Hence z_5 and z_4 have the same moduli and opposite arguments. So $r = \sqrt{2}$, $\arg z_5 = -\frac{\pi}{4}$, $\operatorname{Arg} z_5 = -\frac{\pi}{4}$, and the polar representation is

$$1 - i = \sqrt{2}\left(\cos\left(-\frac{\pi}{4}\right) + i\,\sin\left(-\frac{\pi}{4}\right)\right).$$

(f)   Note that z_6 is the reflection of $z_4 = 1 + i$ about the origin (equivalently, $z_6 = -z_4$). Hence z_6 and z_4 have the same moduli and their arguments are related by the identity $\arg z_6 = \arg z_4 + \pi$. So, $r = \sqrt{2}$, $\arg z_6 = \frac{\pi}{4} + \pi = \frac{5\pi}{4}$, $\operatorname{Arg} z_6 = -\frac{3\pi}{4}$, and the polar representation is

$$-1 - i = \sqrt{2}\left(\cos\frac{5\pi}{4} + i\sin\frac{5\pi}{4}\right).$$ ■

The following properties of the argument were illustrated in Example 1(e) and (f) (for proofs, see Exercise 26):

(11) $$\arg \overline{z} = -\arg z,$$

(12) $$\arg(-z) = \arg z + \pi.$$

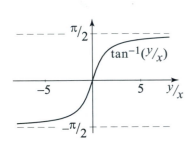

Figure 3 The inverse tangent takes its values in $(\frac{-\pi}{2}, \frac{\pi}{2})$.

It is tempting to compute the argument θ of a complex number $z = x + iy$ by taking the inverse tangent on both sides of (6) and writing $\theta = \tan^{-1}\left(\frac{y}{x}\right)$. This formula is true only if θ is in the interval $(-\frac{\pi}{2}, \frac{\pi}{2})$, because the inverse tangent takes its values in the interval $(-\frac{\pi}{2}, \frac{\pi}{2})$ (see Figure 3); and so it will not yield values of θ that are outside this interval. For example, if $\theta = \frac{4\pi}{3}$, from Table 1 we have $\tan\frac{4\pi}{3} = \sqrt{3}$. However, $\tan^{-1}\sqrt{3} = \frac{\pi}{3}$ and not $\frac{4\pi}{3}$. To overcome this problem, recall that the tangent is π-periodic. That means, for all θ,

$$\tan\theta = \tan(\theta + k\pi) \qquad (k = 0, \pm 1, \pm 2, \dots).$$

Also, since $\tan\theta = \frac{y}{x}$, we conclude that

(13) $$\theta = \tan^{-1}\left(\frac{y}{x}\right) + k\pi \quad (x \neq 0),$$

where the choice of k depends on z. You can check that for $z = x + iy$ with $x \neq 0$,

For $z = x + iy$:
$x > 0 \Leftrightarrow \quad z$ is in the first or fourth quadrants;
$x < 0$ and $y > 0 \Leftrightarrow z$ is in the second quadrant;
$x < 0$ and $y < 0 \Leftrightarrow z$ is in the third quadrant.

(14) $$\operatorname{Arg} z = \begin{cases} \tan^{-1}\left(\frac{y}{x}\right) & \text{if } x > 0; \\[2mm] \tan^{-1}\left(\frac{y}{x}\right) + \pi & \text{if } x < 0 \text{ and } y \geq 0; \\[2mm] \tan^{-1}\left(\frac{y}{x}\right) - \pi & \text{if } x < 0 \text{ and } y < 0. \end{cases}$$

For example, in Figure 4, the point $z_1 = 2 + 3i$ is in the first quadrant. Using a calculator, we find $\operatorname{Arg} z_1 = \tan^{-1}\frac{3}{2} \approx 0.983$. The point $z_2 = -2 - 3i$, is in the third quadrant, $\operatorname{Arg} z_2 = \tan^{-1}\frac{3}{2} - \pi \approx -2.159$. (We remind you that all angles are measured in radians.)

When x is zero, we cannot use (13). In this case,

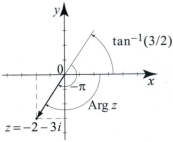

Figure 4 Computing $\operatorname{Arg} z$ using (14).

(15) $$\operatorname{Arg} z = \operatorname{Arg}(iy) = \begin{cases} \frac{\pi}{2} & \text{if } y > 0; \\[2mm] -\frac{\pi}{2} & \text{if } y < 0. \end{cases}$$

Multiplication, Inverses, and Division in Polar Form

The polar form of complex numbers gives us a new insight into multiplication and related operations on complex numbers. Let $z_1 = r_1(\cos\theta_1 + i\sin\theta_1)$ and $z_2 = r_2(\cos\theta_2 + i\sin\theta_2)$ be two nonzero complex numbers. We compute their product directly, using $i^2 = -1$:

$$
\begin{aligned}
z_1 z_2 &= r_1(\cos\theta_1 + i\sin\theta_1)r_2(\cos\theta_2 + i\sin\theta_2) \\
&= r_1 r_2[(\cos\theta_1\cos\theta_2 - \sin\theta_1\sin\theta_2) + i(\sin\theta_1\cos\theta_2 + \cos\theta_1\sin\theta_2)].
\end{aligned}
$$

We recognize the trigonometric expressions as the sum angle formulas for the cosine and sine:

$$\cos(\theta_1 + \theta_2) = \cos\theta_1\cos\theta_2 - \sin\theta_1\sin\theta_2$$

and

$$\sin(\theta_1 + \theta_2) = \sin\theta_1\cos\theta_2 + \cos\theta_1\sin\theta_2.$$

Therefore, the **polar form of the product**

$$(16) \qquad \boxed{z_1 z_2 = r_1 r_2[\cos(\theta_1 + \theta_2) + i\sin(\theta_1 + \theta_2)].}$$

By taking the modulus of both sides of (16), we have

$$(17) \qquad |z_1 z_2| = |z_1||z_2|,$$

which is a fact we already know from Section 1.2. By taking the argument of both sides of (16), we obtain

$$(18) \qquad \arg(z_1 z_2) = \theta_1 + \theta_2 = \arg z_1 + \arg z_2.$$

More precisely,

$$\arg(z_1 z_2) = \{\theta_1 + \theta_2 + 2k\pi : \ k = 0, \pm1, \pm2, \dots\}.$$

Identities (17) and (18) tell us that when we multiply two complex numbers in polar form, we multiply their moduli and add their arguments. See Figure 5.

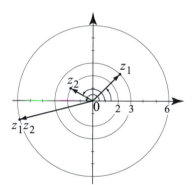

Figure 5 To multiply in polar form, add the arguments and multiply the moduli.

Suppose that $z_1 \neq 0$, and $z_2 = z_1^{-1}$ in (16). Since $z_1 z_2 = z_1 z_1^{-1} = 1$, we obtain

$$1 = r_1 r_2[\cos(\theta_1 + \theta_2) + i\sin(\theta_1 + \theta_2)].$$

Hence from (8), $r_1 r_2 = 1$ and $\theta_1 + \theta_2 = 0$, because the modulus of 1 is 1 and its argument is 0. Hence $r_2 = \frac{1}{r_1}$ and $\theta_2 = -\theta_1$. So the **polar form of the inverse** of a complex number z is given by

$$(19) \qquad \boxed{z^{-1} = \frac{1}{r}\left(\cos(-\theta) + i\sin(-\theta)\right) = \frac{1}{r}(\cos\theta - i\sin\theta).}$$

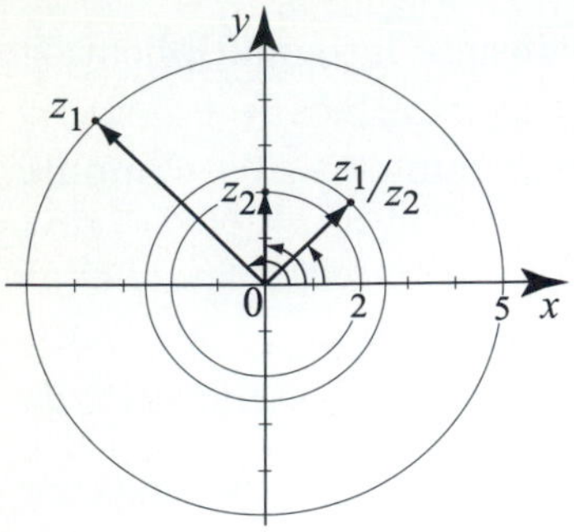

Figure 6 Division in polar form.

Consider now arbitrary z_1 and $z_2 \neq 0$. Since $\frac{z_1}{z_2} = z_1 z_2^{-1}$, then by (16) and (19), we obtain the **polar form of a quotient**

(20)
$$\frac{z_1}{z_2} = \frac{r_1}{r_2}(\cos(\theta_1 - \theta_2) + i \, \sin(\theta_1 - \theta_2)).$$

Thus to divide two complex numbers in polar form, we divide their moduli and subtract their arguments (Figure 6).

EXAMPLE 2   Multiplication and division in polar form

(a)   Let $z_1 = 3(\cos \frac{\pi}{4} + i \, \sin \frac{\pi}{4})$ and $z_2 = 2(\cos \frac{5\pi}{6} + i \sin \frac{5\pi}{6})$. Calculate $z_1 z_2$.

(b)   Let $z_1 = 5(\cos \frac{3\pi}{4} + i \sin \frac{3\pi}{4})$ and $z_2 = 2i$. Calculate $\dfrac{z_1}{z_2}$.

Solution   (a)   Reading off the values $r_1 = 3$, $\theta_1 = \frac{\pi}{4}$, $r_2 = 2$, $\theta_2 = \frac{5\pi}{6}$, we use (16) and get

$$
\begin{aligned}
z_1 z_2 &= 2 \cdot 3[\cos(\frac{\pi}{4} + \frac{5\pi}{6}) + i \, \sin(\frac{\pi}{4} + \frac{5\pi}{6})] \\
&= 6[\cos \frac{13\pi}{12} + i \, \sin \frac{13\pi}{12}].
\end{aligned}
$$

This is the multiplication shown in Figure 5.

(b)   We write z_2 in polar form as $2i = 2(\cos \frac{\pi}{2} + i \, \sin \frac{\pi}{2})$, then perform the division by dividing moduli and subtracting arguments:

$$
\begin{aligned}
\frac{z_1}{z_2} &= \frac{5}{2}[\cos(\frac{3\pi}{4} - \frac{\pi}{2}) + i \, \sin(\frac{3\pi}{4} - \frac{\pi}{2})] \\
&= \frac{5}{2}\left(\cos \frac{\pi}{4} + i \, \sin \frac{\pi}{4}\right) = \frac{5}{2}\left(\frac{\sqrt{2}}{2} + i\frac{\sqrt{2}}{2}\right).
\end{aligned}
$$

This is the division shown in Figure 6.   ∎

De Moivre's Identity

If we take the number $z = 1(\cos \theta + i \, \sin \theta)$ and multiply it by itself, using (16), we get

$$z \, z = (1 \cdot 1)[\cos(\theta + \theta) + i \, \sin(\theta + \theta)] = \cos 2\theta + i \, \sin 2\theta.$$

Computing successive powers of z, a pattern emerges:

$$
\begin{aligned}
z &= \cos \theta + i \, \sin \theta, \\
z^2 &= \cos 2\theta + i \, \sin 2\theta, \\
z^3 &= \cos 3\theta + i \, \sin 3\theta,
\end{aligned}
$$

$$\vdots$$

We have the following useful identity.

DE MOIVRE'S IDENTITY

For any positive integer n and real number θ,

$$(\cos\theta + i\sin\theta)^n = \cos n\theta + i\sin n\theta.$$

Proof   We prove the statement by mathematical induction. For $n=1$, the statement is true, since we have trivially $(\cos\theta + i\sin\theta)^1 = \cos 1\cdot\theta + i\sin 1\cdot\theta$. Now for the inductive step: We assume the statement is true for n, and prove that it is true for $n+1$. Let us compute

$$
\begin{aligned}
(\cos\theta + i\sin\theta)^{n+1} &= (\cos\theta + i\sin\theta)^n(\cos\theta + i\sin\theta)\\
&= (\cos n\theta + i\sin n\theta)(\cos\theta + i\sin\theta)\\
&\qquad\qquad\text{[by induction hypothesis]}\\
&= \cos(n+1)\theta + i\sin(n+1)\theta \qquad \text{[by (16)]}.
\end{aligned}
$$

By induction, the statement holds for all positive integers n.   ∎

We can also use De Moivre's identity to calculate powers of complex numbers of arbitrary modulus $r \neq 0$. For if $z = r(\cos\theta + i\sin\theta)$, then

$$
\begin{aligned}
(21)\qquad z^n &= [r(\cos\theta + i\sin\theta)]^n\\
&= r^n(\cos\theta + i\sin\theta)^n\\
&= r^n(\cos n\theta + i\sin n\theta).
\end{aligned}
$$

EXAMPLE 3   Polar form and powers

Calculate $(2+2i)^{11}$.

Solution   We use De Moivre's identity for a quick calculation. First, write the number $2+2i$ in polar form: $2+2i = 2^{3/2}(\cos\frac{\pi}{4} + i\sin\frac{\pi}{4})$. Then, from (21) we obtain our answer in polar form:

$$(2+2i)^{11} = 2^{33/2}(\cos\frac{11\pi}{4} + i\sin\frac{11\pi}{4}).$$

$$
\begin{aligned}
\cos\frac{11\pi}{4} &= \cos(\frac{11\pi}{4} - 2\pi)\\
&= \cos\frac{3\pi}{4}.
\end{aligned}
$$

By subtracting a multiple of 2π from the angle, we can simplify our answer and put it in Cartesian coordinates as follows:

$$(2+2i)^{11} = 2^{16}\sqrt{2}\left(\cos\frac{3\pi}{4} + i\sin\frac{3\pi}{4}\right) = 2^{16}(-1+i).$$ ∎

EXAMPLE 4   Double-angle identities

Use De Moivre's identity with $n=2$ to recover the double-angle formulas for $\cos 2\theta$ and $\sin 2\theta$.

Solution   From De Moivre's identity,

$$\cos 2\theta + i\sin 2\theta = (\cos\theta + i\sin\theta)^2 = \cos^2\theta - \sin^2\theta + i\,2\sin\theta\cos\theta.$$

Equating real and imaginary parts, we get the double-angle formulas

$$\cos 2\theta = \cos^2 \theta - \sin^2 \theta \quad \text{and} \quad \sin 2\theta = 2 \sin \theta \cos \theta. \qquad \blacksquare$$

Roots of Complex Numbers

We first define an nth root of a complex number.

DEFINITION
OF nth ROOT

Let $w \neq 0$ be a given complex number and n a positive integer. A number z is called an nth **root** of w if $z^n = w$.

Our goal is to find the roots of the given number w. De Moivre's formula for calculating powers can in essence be worked "backward" to find roots of complex numbers.

Write $w = \rho(\cos \phi + i \sin \phi)$ and $z = r(\cos \theta + i \sin \theta)$. The equation $z^n = w$ tells us (using (21)) that

$$(22) \qquad r^n(\cos n\theta + i \sin n\theta) = \rho(\cos \phi + i \sin \phi).$$

Two complex numbers are equal only if their moduli are equal, so $r^n = \rho$. Thus, take $r = \rho^{1/n}$ (meaning the real root of a real number). Also, when two complex numbers are equal, their arguments must differ by $2k\pi$, where k is an integer. So

$$n\theta = \phi + 2k\pi, \quad \text{or} \quad \theta = \frac{\phi}{n} + \frac{2k\pi}{n}.$$

If we take the values $k = 0, 1, \ldots, n-1$, we will get n values of θ that yield n distinct roots of w. Any other value of k will produce a root identical to one of these, since when k increases by n, the argument of z increases by 2π. Thus we have a formula for the n roots of a complex number w.

FORMULA FOR THE
nth ROOTS

Let $w = \rho(\cos \phi + i \sin \phi) \neq 0$. The nth roots of w are the solutions of the equation $z^n = w$. They are given by

$$(23) \qquad z_{k+1} = \rho^{1/n} \left[\cos\left(\frac{\phi}{n} + \frac{2k\pi}{n} \right) + i \sin\left(\frac{\phi}{n} + \frac{2k\pi}{n} \right) \right],$$

$k = 0, 1, \ldots, n-1$.

The unique number z such that $z^n = w$ and $\text{Arg } z = \frac{\text{Arg } w}{n}$ is called the **principal nth root** of w. The principal root is obtained from (23) by taking $\phi = \text{Arg } w$ and $k = 0$.

EXAMPLE 5   Sixth roots of unity

Find and plot all numbers z such that $z^6 = 1$. What is the principal sixth root of 1?

Solution   The modulus of 1 is 1 and the argument of 1 is 0. According to (23), the six roots have $r = 1^{1/6} = 1$ and $\theta = \frac{0}{6} + \frac{k\pi}{3}$, for $k = 0, 1, 2, 3, 4$, and 5. Hence the roots are

$$z_{k+1} = \cos\frac{k\pi}{3} + i\sin\frac{k\pi}{3}, \quad k = 0, 1, \ldots, 5.$$

The principal root is clearly $z_1 = 1$. We can list the roots explicitly, with the help of Table 1:

$$z_1 = 1, \qquad z_2 = \tfrac{1}{2} + i\tfrac{\sqrt{3}}{2}, \qquad z_3 = -\tfrac{1}{2} + i\tfrac{\sqrt{3}}{2},$$

$$z_4 = -1, \quad z_5 = -\tfrac{1}{2} - i\tfrac{\sqrt{3}}{2}, \quad z_6 = \tfrac{1}{2} - i\tfrac{\sqrt{3}}{2}.$$

The six roots are displayed in Figure 7. Since they all have the same modulus, they all lie on the same circle centered at the origin. They have a common angular separation of $\pi/3$. This particular set of roots is symmetric about the horizontal axis because we are finding the roots of a real number; the polynomial equation $z^6 = 1$ has real coefficients, and nonreal solutions come in conjugate pairs (see Section 1, Exercise 29). ∎

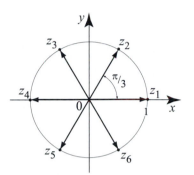

Figure 7 Sixth roots of unity.

EXAMPLE 6   Finding roots of complex numbers

Find and plot all numbers z such that $(z+1)^3 = 2 + 2i$.

Solution   Change variables to $w = z + 1$. We must now solve $w^3 = 2 + 2i$. The polar form of $2 + 2i$ is $2 + 2i = 2^{3/2}(\cos\frac{\pi}{4} + i\sin\frac{\pi}{4})$. Appealing to (23) with $n = 3$, we find

$$w_1 = \sqrt{2}\left(\cos\frac{\pi}{12} + i\sin\frac{\pi}{12}\right), \quad w_2 = \sqrt{2}\left(\cos\frac{9\pi}{12} + i\sin\frac{9\pi}{12}\right),$$

$$w_3 = \sqrt{2}\left(\cos\frac{17\pi}{12} + i\sin\frac{17\pi}{12}\right).$$

Since $z = w - 1$, we conclude that the solutions z_1, z_2, and z_3, of $(z+1)^3 = 2 + 2i$, are

$$z_1 = \sqrt{2}\,\cos\frac{\pi}{12} - 1 + i\sqrt{2}\sin\frac{\pi}{12}, \quad z_2 = \sqrt{2}\cos\frac{9\pi}{12} - 1 + i\sqrt{2}\sin\frac{9\pi}{12},$$

$$z_3 = \sqrt{2}\cos\frac{17\pi}{12} - 1 + i\sqrt{2}\sin\frac{17\pi}{12}.$$

The solutions are illustrated in Figure 8. ∎

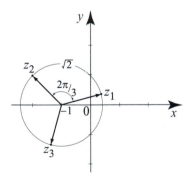

Figure 8 Solutions in Example 6.

As a final application, we revisit the quadratic equation

$$(24) \qquad\qquad a z^2 + b z + c = 0 \qquad (a \neq 0),$$

where a, b, c are now complex numbers. Following the algebraic manipulations in Section 1.1 for solving the equation with real coefficients, we arrive at the solution

(25)
$$z = \frac{-b \pm \sqrt{b^2 - 4ac}}{2a},$$

where now $\pm\sqrt{b^2 - 4ac}$ represents the two complex square roots of $b^2 - 4ac$. These square roots can be computed by appealing to (23) with $n = 2$, thus yielding the solutions of (24).

EXAMPLE 7   Quadratic equation with complex coefficients
Solve $z^2 - 2iz + 3 + i = 0$.

Solution   From (25), we have

$$z = \frac{2i \pm \sqrt{(-2i)^2 - 4(3 + i)}}{2} = i \pm \sqrt{-4 - i}.$$

We have

$$-4 - i = \sqrt{17}\left(-\frac{4}{\sqrt{17}} - \frac{i}{\sqrt{17}}\right) = \sqrt{17}\left(\cos\theta + i\sin\theta\right),$$

where θ is the angle in the third quadrant (we may take $\pi < \theta < \frac{3\pi}{2}$) such that

$$\cos\theta = -\frac{4}{\sqrt{17}}, \quad \text{and} \quad \sin\theta = -\frac{1}{\sqrt{17}}.$$

Appealing to (23) to compute w_1 and w_2, the two square roots of $-4 - i$, we find

$$w_1 = 17^{1/4}\left(\cos\frac{\theta}{2} + i\sin\frac{\theta}{2}\right)$$

and

$$
\begin{aligned}
w_2 &= 17^{1/4}\left(\cos\left(\frac{\theta}{2} + \pi\right) + i\sin\left(\frac{\theta}{2} + \pi\right)\right) \\
&= 17^{1/4}\left(-\cos\frac{\theta}{2} - i\sin\frac{\theta}{2}\right) = -w_1,
\end{aligned}
$$

as you would have expected w_2 to be related to w_1. Thus,

$$z_1 = i + w_1 \quad \text{and} \quad z_2 = i - w_1.$$

To compute z_1 and z_2 explicitly, we must determine the values of $\cos\frac{\theta}{2}$ and $\sin\frac{\theta}{2}$ from the values of $\cos\theta$ and $\sin\theta$. This can be done with the help of the half-angle formulas:

(26)
$$\cos^2\frac{\theta}{2} = \frac{1 + \cos\theta}{2} \quad \text{and} \quad \sin^2\frac{\theta}{2} = \frac{1 - \cos\theta}{2}.$$

Since $\pi < \theta < \frac{3\pi}{2}$, we have $\frac{\pi}{2} < \frac{\theta}{2} < \frac{3\pi}{4}$; so $\cos\frac{\theta}{2} < 0$ and $\sin\frac{\theta}{2} > 0$. Hence from (26)

$$\cos\frac{\theta}{2} = -\sqrt{\frac{1 + \cos\theta}{2}} \quad \text{and} \quad \sin\frac{\theta}{2} = \sqrt{\frac{1 - \cos\theta}{2}}.$$

Using the explicit value of $\cos\theta$, we get

$$\cos\frac{\theta}{2} = -\sqrt{\frac{\sqrt{17}-4}{2\sqrt{17}}} \quad \text{and} \quad \sin\frac{\theta}{2} = \sqrt{\frac{\sqrt{17}+4}{2\sqrt{17}}},$$

and hence

$$z_1 = i + \frac{1}{\sqrt{2}}\left[-\sqrt{\sqrt{17}-4} + i\sqrt{\sqrt{17}+4}\right],$$

$$z_2 = i - \frac{1}{\sqrt{2}}\left[-\sqrt{\sqrt{17}-4} + i\sqrt{\sqrt{17}+4}\right].$$

Observe that the roots z_1 and z_2 are not complex conjugates. This is because the coefficients of the quadratic equation were not all real (see Exercise 29(b), Section 1.1).   ■

Exercises 1.3

In Exercises 1–4, draw the complex number, given in its polar form.

1. $3\left(\cos\frac{7\pi}{12} + i\sin\frac{7\pi}{12}\right).$

2. $\sqrt{2}\left(\cos\frac{-\pi}{2} + i\sin\frac{-\pi}{2}\right).$

3. $\frac{1}{2}\left(\cos\frac{64\pi}{3} + i\sin\frac{64\pi}{3}\right).$

4. $3\left(\cos\frac{-72\pi}{11} + i\sin\frac{-72\pi}{11}\right).$

In Exercises 5–12, represent the given number in polar form.

5. $-3 - 3i.$

6. $-\frac{\sqrt{3}}{2} + \frac{i}{2}.$

7. $-1 - \sqrt{3}i.$

8. $1 + i.$

9. $-\frac{i}{2}.$

10. $\frac{1+i}{1+\sqrt{3}i}.$

11. $\frac{1+i}{1-i}.$

12. $\frac{i}{10+10i}.$

In Exercises 13–16, you are given a complex number z. Find $\operatorname{Arg} z$ with the help of (14) or (15); then find $\arg z$. If needed, use a calculator to compute $\tan^{-1}$.

13. $13 + 2i.$

14. $-3 - 32i.$

15. $-1 + \frac{1}{2}i.$

16. $-\frac{3\pi}{2}i.$

In Exercises 17–20, find the polar form of the given number.

17. $(-\sqrt{3}+i)^3.$

18. $(-2-3i)^{17}.$

19. $\left(\frac{1-i}{1+i}\right)^{10}.$

20. $\frac{i}{(1+2\sqrt{3}i)^5}.$

In Exercises 21–24, find the real and imaginary parts of the given number.

21. $(1+i)^{30}.$

22. $\left(\cos\frac{2\pi}{17} + i\sin\frac{2\pi}{17}\right)^{170}.$

23. $\left(\frac{1-i}{1+i}\right)^4.$

24. $\frac{-i}{(1+i)^5}.$

25. What are the real and imaginary parts of i^n, where n is an integer?

26. Prove (11) and (12).

27. Show that
$$\left(\frac{1+i\tan\theta}{1-i\tan\theta}\right)^n = \frac{1+i\tan n\theta}{1-i\tan n\theta}.$$

28. Show that $\bar{z}$ and z^{-1} are parallel, using their polar forms. Find the positive constant α such that $\bar{z} = \alpha z^{-1}.$

29.   Suppose $z_1, z_2, \ldots, z_n$ are complex numbers with respective moduli r_1, r_2, $\ldots$, r_n and arguments θ_1, θ_2, $\ldots$, θ_n. Show that

$$z_1 z_2 \cdots z_n = r_1 r_2 \cdots r_n [\cos(\theta_1 + \theta_2 + \cdots + \theta_n) + i \sin(\theta_1 + \theta_2 + \cdots + \theta_n)].$$

30.   (a)   Show that if $z_1, z_2, \ldots, z_n$ are all unimodular, then so is $z_1 z_2 \cdots z_n$.
(b)   Let n be a positive integer and z a nonzero complex number. Show that z is unimodular if and only if all its nth roots are unimodular.

31.   (a)   Show that two nonzero complex numbers z_1 and z_2 represent perpendicular vectors only if $\arg\left(\frac{z_1}{z_2}\right) = \frac{\pi}{2} + k\pi$, where k is an integer.
(b)   Show that two nonzero complex numbers z_1 and z_2 represent parallel vectors only if $\arg\left(\frac{z_1}{z_2}\right) = 2k\pi$, where k is an integer.

32.   Suppose $z = r(\cos\theta + i\sin\theta)$. What become of its polar coordinates r and θ if we multiply it by
(a) a positive real number α?
(b) a negative real number $-\alpha$?
(c) a unimodular complex number $\cos\phi + i\sin\phi$?

In Exercises 33–38, solve the given equation and plot the solutions. In each case, determine the principal root.

33. $z^4 = i$. **34.** $z^3 = 1 + i$. **35.** $z^6 = 128$.

36. $z^5 = -30$. **37.** $z^7 = -1$. **38.** $z^8 = -\dfrac{\pi}{2}$.

In Exercises 39–42, solve the given equation.

39. $(z + 2)^3 = 3i$. **40.** $(z - i)^4 = 1$.

41. $(z - 5)^3 = -125$. **42.** $(3z - 2)^4 = 11$.

In Exercises 43–46, solve the given equation.

43. $z^2 - 2(1 + i)z + i = 0$. **44.** $2z^2 - z - 1 - i = 0$.

45. $z^2 + z + \dfrac{i}{4} = 0$. **46.** $z^2 + iz + 1 = 0$.

In Exercises 47–50, use De Moivre's identity to derive the given trigonometric identity. (More generally, see Exercise 56.)

47. $\cos(3\theta) = \cos^3\theta - 3\cos\theta\sin^2\theta$.
48. $\sin(3\theta) = 3\cos^2\theta\sin\theta - \sin^3\theta$.
49. $\cos(4\theta) = \cos^4\theta - 6\cos^2\theta\sin^2\theta + \sin^4\theta$.
50. $\sin(4\theta) = 4\cos^3\theta\sin\theta - 4\cos\theta\sin^3\theta$.

51. Roots of unity.   Let n be a positive integer. Solve $z^n = 1$. These n values of z are called the nth **roots of unity** and are denoted by $\omega_1, \ldots, \omega_n$.

52.   Use the fact that

$$\cos\left(\frac{\phi}{n} + \frac{2k\pi}{n}\right) + i\sin\left(\frac{\phi}{n} + \frac{2k\pi}{n}\right) = \left[\cos\left(\frac{\phi}{n}\right) + i\sin\left(\frac{\phi}{n}\right)\right]\left[\cos\left(\frac{2k\pi}{n}\right) + i\sin\left(\frac{2k\pi}{n}\right)\right]$$

to show that the roots of the equation $z^n = w$ are $w_p^{1/n}\omega_j$ where $w_p^{1/n}$ is the principal root of w and ω_j is an nth root of unity, $j = 1, 2, \ldots, n$.

53. Summing roots of unity.   Let ω_1, ω_2, $\ldots$, ω_n denote the nth roots of unity where $n \geq 2$; that is $\omega_j^n = 1$ for $j = 1, 2, \ldots , n$. Pick and fix any root $\omega_0 \neq 1$ from the set $(\omega_j)_{j=1}^n$.

(a)   Show that $\omega_0\,\omega_j$ is an nth root of unity for $j = 1, 2, \ldots , n$. [Hint: Verify the equation $z^n = 1$.]

(b)   Show that $\omega_0\,\omega_j \neq \omega_0\,\omega_k$ if $j \neq k$. Conclude that the set $(\omega_0\,\omega_j)_{j=1}^n$ is the same as the set of all n roots of unity.

(c)   Show that the sum of the n roots of unity is zero. That is, show that

$$\omega_1 + \omega_2 + \cdots + \omega_n = 0.$$

[Hint: $\omega_1 + \omega_2 + \cdots + \omega_n = \omega_0\,\omega_1 + \omega_0\,\omega_2 + \cdots + \omega_0\,\omega_n$; why? Factor ω_0 and conclude that the sum has to be 0.]

(d)   Show directly that

$$1 + \omega_0 + \omega_0^2 + \cdots + \omega_0^{n-1} = 0.$$

[Hint: Multiply the left side by $1 - \omega_0 \neq 0$.]

54. Project Problem.   Our goal in this exercise is to solve the equation

(27) $$z^n = (z+1)^n.$$

This polynomial equation is actually of order $n - 1$, since upon expanding, the terms in z^n will cancel.

(a)   Divide both sides of (27) by $(z+1)^n$ (evidently, $z+1$ cannot be zero) and conclude that $\frac{z}{z+1}$ must be one of the nth roots of unity (see Exercise 51). Hence $z = (z+1)\omega_k$, where $\omega_k = \cos(\frac{2k\pi}{n}) + i\,\sin(\frac{2k\pi}{n})$, for $k = 1, 2, \ldots , n-1$. We must throw out $k = 0$ because $z = z + 1$ cannot be correct.

(b)   Write $z = x + iy$ and solve for x and y by equating real and imaginary parts in $z = (z+1)\omega_k$. Obtain the answers

$$x = -\frac{1}{2} \quad \text{and} \quad y_k = \frac{\sin(2k\pi/n)}{2(1 - \cos(2k\pi/n))} = \frac{1}{2}\cot(k\pi/n).$$

(c)   Apply the result of (b) to solve $(z+1)^7 = z^7$.

Chebyshev Polynomials

In the remaining problems, we present a family of polynomials, known as the Chebyshev polynomials. These polynomials have useful applications in numerical analysis. Our presentation will use De Moivre's identity and the following well-known formula.

55. The binomial formula.   Use mathematical induction to prove that for any complex numbers a and b and any positive integer n

(28)
$$(a+b)^n = a^n + \binom{n}{1}a^{n-1}b^1 + \binom{n}{2}a^{n-2}b^2 + \cdots + \binom{n}{n-1}a^1 b^{n-1} + b^n,$$

where for $0 \leq m \leq n$, the **binomial coefficient** $\binom{n}{m}$ (read it "n choose m") is defined by

$$\binom{n}{m} = \frac{n!}{(n-m)!\,m!},$$

with $0! = 1$ as a convention. Using the summation notation, the binomial formula becomes

$$(a+b)^n = \sum_{m=0}^{n} \binom{n}{m} a^{n-m}b^m.$$

[Hint: You should come up with two sums. Shift the index on one summation so the summand looks like the other. Pull off the a^{n+1} and b^{n+1} terms. Then use the identity of Pascal's triangle,

$$\binom{n}{m-1} + \binom{n}{m} = \binom{n+1}{m}.]$$

56. (a) Use De Moivre's identity and the binomial formula to show that, for $n = 1, 2, \ldots,$

$$(29) \qquad \cos n\theta = \sum_{k=0}^{\left[\frac{n}{2}\right]} \binom{n}{2k} (\cos\theta)^{n-2k}(-1)^k(\sin\theta)^{2k}$$

and

$$(30) \qquad \sin n\theta = \sum_{k=0}^{\left[\frac{n-1}{2}\right]} \binom{n}{2k+1} (\cos\theta)^{n-2k-1}(-1)^k(\sin\theta)^{2k+1},$$

where, for a real number s, $[s]$ denotes the greatest integer not larger than s.
(b) Show that

$$(31) \qquad \cos n\theta = \sum_{k=0}^{\left[\frac{n}{2}\right]} \binom{n}{2k} (\cos\theta)^{n-2k}(-1)^k(1-\cos^2\theta)^k.$$

(c) Derive the results of Exercises 47 and 48 from (a).

With the Roman alphabet, Chebyshev may be spelled a variety of ways, including Tchebichef. Hence T_n.

57. Chebyshev polynomials.   It is clear from (31) that $\cos n\theta$ can be expressed as a polynomial of degree n in $\cos\theta$. So, for $n = 0, 1, 2, \ldots$, we define the nth **Chebyshev polynomial** T_n by the formula

$$(32) \qquad T_n(\cos\theta) = \cos n\theta, \quad n = 0, 1, 2, \ldots.$$

(a) Obtain the formula

$$(33) \qquad T_n(x) = \sum_{k=0}^{\left[\frac{n}{2}\right]} \binom{n}{2k} (-1)^k x^{n-2k}(1-x^2)^k.$$

(b) Use (33) to derive the following list of Chebyshev polynomials:

$$T_0(x) = 1, \qquad\qquad T_1(x) = x, \qquad\qquad T_2(x) = -1 + 2x^2,$$

$$T_3(x) = -3x + 4x^3, \quad T_4(x) = 1 - 8x^2 + 8x^4, \quad T_5(x) = 5x - 20x^3 + 16x^5.$$

58.   Properties of the Chebyshev polynomials.   Derive the following properties of Chebyshev polynomials. These properties are characteristic of many other families of functions that we will encounter when solving important differential equations such as the Legendre, Laguerre, and Bessel differential equations.

(a) $T_n(1) = 1$ and $T_n(-1) = (-1)^n$.

(b) $T_n(0) = \cos(\frac{n\pi}{2}) = 0$ if n is odd and $(-1)^k$ if $n = 2k$.

(c) $|T_n(x)| \leq 1$ for all x in $[-1, 1]$.

(d) $T_n(x)$ is a polynomial of degree n. [Hint: Induction on n and (33).]

(e)   Derive the **recurrence relation** $T_{n+1}(x) + T_{n-1}(x) = 2x\,T_n(x)$. [Hint: Use the trigonometric identities $\cos(a \pm b) = \cos a \cos b \mp \sin a \sin b$ and (32).]

(f)   Show that $T_n(x)$ has n simple zeros in $[-1, 1]$ at the points $\alpha_k = \cos\left(\frac{2k-1}{2n}\pi\right)$, $k = 1, 2, \dots, n$.

(g)   Show that $T_n(x)$ assumes its absolute extrema in $[-1, 1]$ at the points $\beta_k = \cos\left(\frac{k}{n}\pi\right)$, $k = 0, 1, 2, \dots, n$, with $T_n(\beta_k) = (-1)^k$.

(h)   Illustrate graphically your answers in (f) and (g) by plotting the first ten Chebyshev polynomials.

(i)   Show that for $m \neq n$

$$\int_{-1}^{1} T_m(x)\,T_n(x)\,\frac{dx}{\sqrt{1-x^2}} = 0.$$

[Hint: Change variables: $x = \cos\theta$.]

(j)   Show that for $n = 0, 1, 2, \dots$,

$$\int_{-1}^{1} [T_n(x)]^2 \,\frac{dx}{\sqrt{1-x^2}} = \begin{cases} \pi & \text{if } n = 0; \\ \frac{\pi}{2} & \text{if } n \geq 1. \end{cases}$$

1.4   Complex Functions

So far we have introduced complex numbers, both in Cartesian and polar forms, and we have studied their algebraic and geometric properties. We now move to the topic of functions of a complex variable.

A **complex-valued function** f of a complex variable z is a rule that assigns to each complex number z in a set S a complex number $f(z)$. The set S is a subset of the complex numbers and is called the **domain of definition** of f. The complex number $f(z)$ is called the **value** of f at z and is sometimes written $w = f(z)$. For example the function $f(z) = z^3$ is a rule that assigns to each z the complex number $w = z^3$. When a function is given by a formula and the domain is not specified, the domain is taken to be the largest set on which the formula makes sense. In the case $f(z) = z^3$, the domain is the set of all complex numbers $\mathbb{C}$.

Back in calculus, a real-valued function $y = g(x)$ of a real variable x was represented as a graph in the xy-plane. The graph contains vital information about the function and allows us to easily visualize and deduce properties of $g(x)$. In the complex case, we are not as fortunate. To visualize a complex-valued function $f(z)$ requires four dimensions: two dimensions for the variable z and two for the values $w = f(z)$. Since a four-dimensional picture is not practical, instead we will use two planes, the z-plane and the w-plane, and visualize the function as a **mapping** from a subset of one plane to the other (see Figure 1). As usual, we will write $z = x + i\,y$, and also write $w = u + i\,v$. Thus as a convention, the z-plane axes will be labeled by x and y and those of the w-plane by u and v. The **image** $f[S]$ of a set S under a mapping f is the set of all points w such that $w = f(z)$ for some z in S. We illustrate the mapping process with basic examples including some familiar geometric transformations.

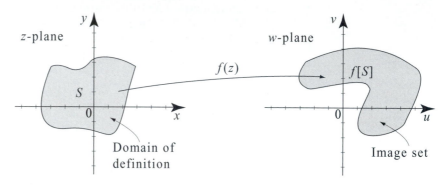

Figure 1 To visualize a mapping by a complex-valued function $w = f(z)$, we use two planes: the z- or xy-plane for the domain of definition and the w- or uv-plane for the image.

EXAMPLE 1   Translation

Let S denote the disk

$$S = \{z : \ |z| \leq 1\}.$$

Find the image of S under the mapping $f(z) = z + 2 + i$.

Solution   For any z in S, the number $f(z)$ is found by adding z to $2 + i$ as vectors. Hence the function translates the point z two units to the right and one unit up. The image of S, then, is the set S translated two units to the right and one unit up (see Figure 2).

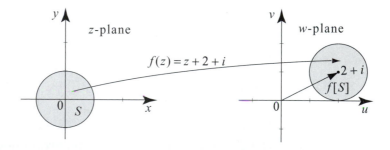

Figure 2 A translation is a mapping of the form $f(z) = z + b$, where b is a complex number. In Example 1, $b = 2 + i$.

Consequently, the image of the disk is a disk of the same radius, centered at $2 + i$ (the image of the original center). Hence,

$$f[S] = \{w : \ |w - 2 - i| \leq 1\}.$$ ∎

Our next example deals with functions of the form $f(z) = az$, where a is a complex constant. To understand these mappings, recall that when we multiply two complex numbers in polar form, we multiply their moduli and add their arguments. Writing a in polar form as $a = r(\cos\theta + i\sin\theta)$, we see that the mapping $f(z) = r(\cos\theta + i\sin\theta)z$ has two effects: It multiplies the modulus of z by $r > 0$ (a dilation) and adds θ to the argument of z (a rotation). Since multiplication is commutative, these operations of dilation and rotation may be applied in either order. When $r = 1$, we obtain the mapping $f(z) = (\cos\theta + i\sin\theta)z$, which is a **rotation** by the angle θ. Mappings of the form $f(z) = rz$, where $r > 0$, are **dilations** by a factor r.

EXAMPLE 2   Dilations and rotations

Let S be the points lying inside and on the square of side length 2 centered at the point 2 on the real axis (Figure 3).
(a) What is the image of S under the mapping $f(z) = 3z$?
(b) What is the image of S under the mapping $g(z) = 2iz$?

Solution   (a)   For any z in S, $f(z) = 3z$ has the effect of tripling the modulus and leaving the argument unchanged. Hence $f(z)$ lies on the ray extending from the origin to z, at three times the distance from the origin to z. In particular, the image of the square is a square whose corners are the images of the four original corners. It is easy to see from Figure 3 that the image of S is a square of side length 6, centered at the point 6 on the real axis.

Figure 3 Images of the corners are computed explicitly using the formula for f:
$$f(1+i) = 3 + 3i,$$
$$f(1-i) = 3 - 3i,$$
$$f(3+i) = 9 + 3i,$$
$$f(3-i) = 9 - 3i.$$
They form the corners of the image.

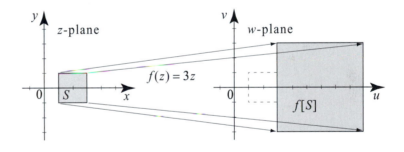

(b)   In polar form, we have $i = \cos\frac{\pi}{2} + i\sin\frac{\pi}{2}$, and so $g(z) = 2(\cos\frac{\pi}{2} + i\sin\frac{\pi}{2})z$. Hence for any z in S, $g(z)$ has the effect of doubling the modulus and adding $\frac{\pi}{2}$ to the argument. To determine $f[S]$, take the set S, dilate it by a factor of 2, then rotate it counterclockwise by $\pi/2$ (Figure 4).   ∎

Figure 4 The mapping $g(z) = 2iz$ is a composition of two mappings: a dilation by a factor of 2, followed by a counterclockwise rotation by an angle of $\pi/2$. The angle of rotation is the argument of $2i$.

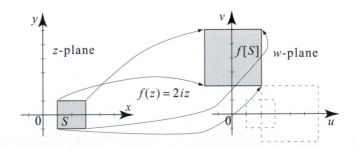

Examples 1 and 2 deal with mappings of the type $f(z) = az+b$, where a and b are complex constants and $a \neq 0$. These are called **linear transformations** and can always be thought of in terms of a dilation, a rotation, and a translation. These transformations map regions to geometrically similar regions. It is important that $a \neq 0$ because otherwise the transformation would be a constant. The transformation $f(z) = z$ is called the **identity transformation** for obvious reasons. The next transformation is not linear.

EXAMPLE 3   Inversion

Find the image of the following sets under the mapping $f(z) = 1/z$.
(a) $S = \{z : \ 0 < |z| < 1, \ 0 \leq \arg z \leq \pi/2\}$.
(b) $S = \{z : \ 2 \leq |z|, \ 0 \leq \arg z \leq \pi\}$.

Solution   (a)   From (19) in Section 1.3, we know that for $z = r(\cos \theta + i \sin \theta) \neq 0$, we have

$$\frac{1}{z} = \frac{1}{r}(\cos(-\theta) + i \sin(-\theta)).$$

According to this formula, the modulus of the number $f(z)$ is the reciprocal of the modulus of z and the argument of $f(z)$ is the negative of the argument of z. Consequently, numbers inside the unit circle ($|z| \leq 1$) get mapped to numbers outside the unit circle ($\frac{1}{|z|} \geq 1$), and numbers in the upper half-plane get mapped to numbers in the lower half-plane. Looking at S, as the modulus of z goes from 1 down to 0, the modulus of $f(z)$ goes from 1 up to infinity. As the argument of z goes from 0 up to $\pi/2$, the argument of $1/z$ goes from 0 down to $-\pi/2$. Hence $f[S]$ is the set of all points in the fourth quadrant, including the border axes, that lie outside the unit circle (see Figure 5):

$$f[S] = \{w : \ 1 < |w|, \ -\pi/2 \leq \arg z \leq 0\}.$$

Figure 5 The inversion
$w = \frac{1}{z}$
has the effect of inverting the modulus and changing the sign of the argument:
$|w| = \frac{1}{|z|}$,
$\text{Arg } w = -\text{Arg } z$.

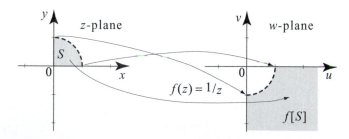

(b)   As the modulus of z increases from 2 up to infinity, the modulus of $1/z$ decreases from $1/2$ down to zero (but never equals zero). As the argument of z goes from 0 up to π, the argument of $1/z$ goes from 0 down to $-\pi$. Hence $f[S]$ is the set of points in the lower half-plane, including the real axis, with $0 < |w| < 1/2$ (see Figure 6): $f[S] = \{w : \ 0 < |w| < 1/2, -\pi \leq \arg z \leq 0\}$.   ■

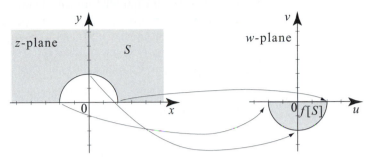

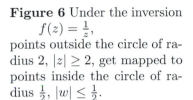

Figure 6 Under the inversion $f(z) = \frac{1}{z}$, points outside the circle of radius 2, $|z| \geq 2$, get mapped to points inside the circle of radius $\frac{1}{2}$, $|w| \leq \frac{1}{2}$.

The function $f(z) = \frac{1}{z}$ in Example 3 is a special case of a general type of mapping of the form

$$(1) \qquad w = \frac{az + b}{cz + d} \qquad (ad \neq bc),$$

known as a **linear fractional transformation**, also known as a **bilinear transformation** or **Möbius transformation**. Here a, b, c, d are complex numbers, and you can check that when $ad = bc$, w is constant. Linear fractional transformations are studied extensively in Chapter 6. They have many important applications that will occur throughout the book.

Real and Imaginary Parts of Functions

Given a complex function $f(z)$, let $u(z) = \operatorname{Re} f(z)$ and $v(z) = \operatorname{Im} f(z)$. The functions u and v are real-valued functions, and we may think of them as functions of two real variables. With a slight abuse of notation, we will sometime write $u(z) = u(x + iy) = u(x, y)$, and $v(z) = v(x + iy) = v(x, y)$. Thus

$$(2) \qquad f(z) = f(x + iy) = u(x, y) + iv(x, y).$$

For example, for $f(z) = z^2 = (x + iy)^2 = x^2 - y^2 + 2ixy$, we have

$$u(x, y) = x^2 - y^2 \qquad \text{and} \qquad v(x, y) = 2xy.$$

As we now illustrate, the functions $u(x, y)$ and $v(x, y)$ may be used to determine algebraically the image of a set when the answer is not geometrically obvious.

EXAMPLE 4   Squaring

Let S be the vertical strip

$$S = \{z = x + iy : \ 1 \leq x \leq 2\}.$$

Find the image of S under the mapping $f(z) = z^2$.

Solution   As before, write $f(z) = z^2 = x^2 - y^2 + 2ixy$. Thus the real part of $f(z)$ is $u(x, y) = x^2 - y^2$ and the imaginary part of $f(z)$ is $v(x, y) = 2xy$. Let us fix $1 \leq x_0 \leq 2$, and find the image of the vertical line $x = x_0$. A point (x_0, y) on the line maps to the point (u, v), where $u = x_0^2 - y^2$, $v = 2x_0 y$. To determine the equation of the curve that is traced by the point (u, v) as y varies from $-\infty$ to ∞, we will eliminate y and get an algebraic relation between u and v. From $v = 2x_0 y$, we obtain $y = \frac{v}{2x_0}$. Plugging into the expression for u, we obtain

$$u = x_0^2 - \frac{v^2}{4x_0^2}.$$

This gives u as a quadratic function of v. Hence the graph is a leftward-facing parabola with a vertex at $(u, v) = (x_0^2, 0)$ and v-intercepts at $(0, \pm 2x_0^2)$. As x_0 ranges from 1 up to 2, the corresponding parabolas in the w-plane sweep out a parabolic region, which is described as follows (see Figure 7). Since all points lie to the right of the parabola, where $x_0 = 1$, and to the left of the parabola, where $x_0 = 2$, we have

$$f[S] = \{w = u + iv : \ 1 - \frac{v^2}{4} \leq u \leq 4 - \frac{v^2}{16}\}. \qquad \blacksquare$$

Figure 7 For $1 \leq x_0 \leq 2$, the image of a line $x = x_0$ under the mapping $f(z) = z^2$ is a parabola
$$u = x_0^2 - \frac{v^2}{4x_0^2}.$$
As we vary x_0 from 1 to 2, these parabolas sweep out a parabolic region, which determines $f[S]$.

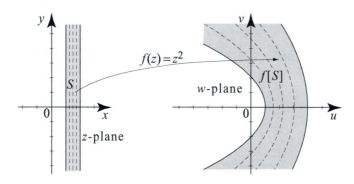

Mappings in Polar Coordinates

Some complex functions and regions are more naturally suited to polar coordinates. Hence it may be to our advantage to write a function as

$$f(z) = f(r\,e^{i\theta}) = u(r, \theta) + i\,v(r, \theta).$$

We may even write $w = f(z)$ in polar coordinates as $w = \rho(\cos\phi + i\,\sin\phi)$. Then we may identify the polar coordinates of w as functions of the polar coordinates of z: $\rho(r, \theta) = |f(r\,e^{i\theta})|$ and $\phi(r, \theta) = \arg f(r\,e^{i\theta})$. The next example uses such polar coordinates to track the mapping of circular sectors.

EXAMPLE 5   Mapping sectors
Let S be the sector

$$S = \{z : \ |z| \leq \frac{3}{2}, \ 0 \leq \arg z \leq \frac{\pi}{4}.\}.$$

Find the image of S under the mapping $f(z) = z^3$.

Solution   If we write $z = r(\cos\theta + i\,\sin\theta)$, then $z^3 = r^3(\cos 3\theta + i\,\sin 3\theta)$. Hence the polar coordinates of $w = f(z) = \rho(\cos\phi + i\,\sin\phi)$ are $\rho = r^3$ and $\phi = 3\theta$. As r increases from 0 up to $\frac{3}{2}$, ρ increases from 0 up to $\frac{27}{8}$; and as θ goes from 0 up to $\frac{\pi}{4}$, ϕ goes from 0 up to $\frac{3\pi}{4}$. Hence the image of S is the set of all w with modulus less than $\frac{27}{8}$ and arguments between 0 and $\frac{3\pi}{4}$ (see Figure 8):

$$f[S] = \{w :\ |w| \le \frac{27}{8},\ 0 \le \arg w \le \frac{3\pi}{4}\}. \qquad \blacksquare$$

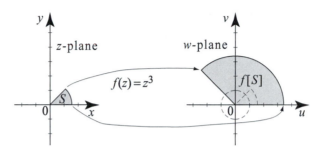

Figure 8   The mapping
$$w = z^3$$
has the effect of cubing the norm and tripling the argument:
$$|w| = |z|^3;$$
$$\arg w = 3\operatorname{Arg} z.$$

All the mappings considered in the examples have been **one-to-one**, in that distinct points z_1 and z_2 always map to distinct points $f(z_1)$ and $f(z_2)$. It is possible that more than one point on the z-plane will map to the same point on the w-plane. Such mappings are not one-to-one, and we are already familiar with some of them. For example, the function $f(z) = z^2$ with domain of definition $\mathbb{C}$ will map z and $-z$ to the same point in the w-plane.

Exercises 1.4

In Exercises 1–6, evaluate $w = f(z)$ for the given value of z.

1. $f(z) = z + i$, $z = 3 + 4i$.

2. $f(z) = 3iz$, $z = -5i$.

3. $f(z) = \operatorname{Re}(z) - 2i \operatorname{Im} z$, $z = -4 + i$.

4. $f(z) = z^{10} + \bar{z}$, $z = -\sqrt{3} + i$.

5. $f(z) = (z^2 + 2)^{\frac{1}{2}}$ (principal root), $z = 1 + i$.

6. $f(z) = \dfrac{z - 1}{z + 1}$, $z = i + 1$.

In Exercises 7–12, express $f(z)$ in the form $u(x, y) + iv(x, y)$ where u and v are the real and imaginary parts of f.

7. $f(z) = iz + 2 - i$.

8. $f(z) = z^2 + 3z + 1 + 3i$.

9. $f(z) = z^3$.

10. $f(z) = 2\bar{z} + \dfrac{1}{2 + i}$.

11. $f(z) = \dfrac{1}{z + 1}$.

12. $f(z) = |z|^3$.

In Exercises 13–18, find the largest subset of $\mathbb{C}$ on which the given formula makes sense and hence defines a function.

13. $f(z) = \dfrac{1}{z}$.

14. $f(z) = 3 + iz^2$.

15. $f(z) = \dfrac{1}{1 + z^2}$.

16. $f(z) = 2 + i \operatorname{Arg} z$.

17. $f(z) = z^{\frac{1}{2}}$ (principal root).

18. $f(z) = \dfrac{z - 2i}{2z + i}$.

19.   Find a linear transformation $f(z)$ such that $f(1) = 3 + i$ and $f(3i) = -2 + 6i$.

20. Find a linear transformation $f(z)$ such that $f(2-i) = -3 - 3i$ and $f(2) = -2 - 2i$.

In Exercises 21–24, find the image $f[S]$ under the given linear transformation. Draw a picture of S and $f[S]$, and depict arrows mapping select points.

21. $S = \{z : |z| < 1\}, \quad f(z) = 4z$.

22. $S = \{z : \operatorname{Re} z > 0\}, \quad f(z) = z + i$.

23. $S = \{z : \operatorname{Re} z > 0, \ \operatorname{Im} z > 0\}, \quad f(z) = -z + 2i$.

24. $S = \{z : |z| \le 2, \ 0 \le \operatorname{Arg} z \le \frac{\pi}{2}\} \quad f(z) = iz + 2$.

25. (a) Suppose that $f(z) = az + b$ and $g(z) = cz + d$ are linear transformations. Show that $f(g(z))$ is also a linear transformation.
(b) Show that every linear transformation $f(z) = az + b$, $a \ne 0$ can be written in the form $g_1(g_2(g_3(z)))$, where g_1 is a translation, g_2 is a rotation, and g_3 is a dilation.

26. Construct a linear transformation that rotates all points in the plane an angle ϕ about a point z_0.

27. Let $f(z) = az$ and $g(z) = z + b$. Show that $f(g(z)) = g(f(z))$ for all z if and only if $a = 1$ or $b = 0$.

28. Show that for each linear transformation $f(z) = az + b$, there exists a linear transformation $g(z)$ such that $f(g(z)) = g(f(z)) = z$ for all complex z.

29. Find the image of the set $S = \{z : z \text{ is real}\}$ under the mapping $f(z) = \operatorname{Arg} z$.

30. Find the image of the set $S = \{z : |z| \le 1\}$ under the mapping $f(z) = z + \overline{z}$.

In Exercises 31–34, find the image $f[S]$ under the inversion $f(z) = \frac{1}{z}$. Draw a picture of S and $f[S]$, and depict arrows mapping select points.

31. $S = \{z : 0 < |z| \le 1\}$.   **32.** $S = \{z : |z| \ge 1, \ \}$.

33. $S = \{z : 0 < |z| \le 3, \ \frac{\pi}{3} \le \operatorname{Arg} z \le \frac{2\pi}{3}\}$.

34. $S = \{z : z \ne 0, \ 0 \le \operatorname{Arg} z \le \frac{\pi}{2}\}$.

35. Let $f(z) = \frac{1}{z}$.
(a) Show that the image of the circle $|z| = a > 0$ is the circle $|w| = \frac{1}{a}$.
(b) Show that the image of the ray $\operatorname{Arg} z = \alpha$ $(z \ne 0)$ is the ray $\operatorname{Arg} w = -\alpha$ $(w \ne 0)$.

36. Show that for $f(z) = 1/z$, $f(z) = g(h(z)) = h(g(z))$, where $g(z) = \overline{z}$ and $h(z) = \frac{z}{|z|^2}$.

37. (a) Show that $f(z) = \frac{1}{z}$ is never zero for any complex z.
(b) Let S be any subset of the complex numbers that does not include zero. Show that $f[S]$ is a subset of the complex numbers that does not include zero.
(c) Show that $f(f(z)) = z$ for all $z \ne 0$.
(d) Show that $f[f[S]] = S$.

38. Find a linear fractional transformation $f(z) = \frac{az+b}{cz+d}$ such that $f(0) = -5 + i$, $f(2) = -\frac{8}{5} - i\frac{11}{5}$, $f(-2i) = 5 - 2i$. [Hint: In solving for a, b, c, d, keep in mind that these are not uniquely determined. Once you determine that a coefficient is not zero, say $a \ne 0$, you may set it equal to 1.]

39. Find a linear fractional transformation $f(z) = \frac{az+b}{cz+d}$ such that $f(0) = -2i$, $f(9i) = -\frac{i}{5}$, $f(4-i) = \frac{1}{2}$.

40. Find the function $f(z) = \frac{1}{cz+d}$ so that $f(1) = \frac{2-i}{5}$ and $f(-\frac{i}{2})$ is not defined.

A point z_0 is called a **fixed point** *of a function $f(z)$ if $f(z_0) = z_0$. In Exercises 41–44, determine the fixed points of the given function.*

41. $f(z) = \dfrac{1}{z}$.

42. $f(z) = az + b$.   [Hint: Take two cases: $a = 1$ and $a \neq 1$]

43. $f(z) = 2\left(z + \dfrac{1}{z}\right)$.

44. $f(z) = \dfrac{-6i + (2+3i)z}{z}$.

45. Define the set of **lattice points** in the plane as

$$L = \{z : \; z = m + in, \quad m \text{ and } n \text{ integers}\}.$$

Consider the mapping $f(z) = z^2$ and the image $f[L]$.
(a) Show that if w is in $f[L]$, $-w$, $\overline{w}$, and $-\operatorname{Re} w + i \operatorname{Im} w$ are in $f[L]$.
(b) Show that if w is in $f[L]$, w is also in L.
(c) Show that if w is in $f[L]$, $f(w)$ is in $f[L]$.

1.5 The Complex Exponential

In this section, we introduce the complex exponential function e^z. Unlike in calculus, where e^x was introduced separately from the trigonometric functions, here we will use the exponential function e^z to define the trigonometric functions of a complex variable z. This makes e^z one of the most important functions in this book.

As a first step toward defining e^z, ask yourself how you might compute e^x for a given real number x. One way is to use the power series

$$e^x = 1 + \frac{x}{1!} + \frac{x^2}{2!} + \frac{x^3}{3!} + \cdots \qquad (-\infty < x < \infty).$$

In fact, we can take this series to be the definition of e^x and use properties of power series to study e^x. Guided by this approach, we define the **complex exponential function** by a power series

$$(1) \qquad \boxed{e^z = \sum_{n=0}^{\infty} \frac{z^n}{n!} = 1 + \frac{z}{1!} + \frac{z^2}{2!} + \frac{z^3}{3!} + \cdots, \quad \text{for all complex } z.}$$

Sometimes the exponential function is written as $\exp(z)$. Of course, we have not yet defined what we mean by an infinite series of complex numbers. However, we are familiar with infinite sums of real numbers, and as you will see in Chapter 4, the theory of series extends to complex numbers. As a consequence, we will show that the series in (1) converges for all z and defines a function that enjoys some of the familiar properties of the exponential function. In particular, we will show that the rule

(2)
$$e^{z_1+z_2} = e^{z_1}e^{z_2}$$

holds for all complex numbers z_1 and z_2. Because we have used the exponential power series to define e^z, it is clear that e^z reduces to the real exponential when $z = x$ is real.

Our next step will be to unlock the mystery of e^z by expressing it in terms of familiar functions. Write $z = x + iy$, where x and y are real. By (2), we have

$$e^z = e^{x+iy} = e^x e^{iy}.$$

The first factor, e^x, is the familiar real exponential function from calculus. To compute the second factor, we use (1) and get

$$
\begin{aligned}
e^{iy} &= 1 + iy + \frac{(iy)^2}{2!} + \frac{(iy)^3}{3!} + \frac{(iy)^4}{4!} + \frac{(iy)^5}{5!} + \cdots \\
&= 1 + iy - \frac{y^2}{2!} - i\frac{y^3}{3!} + \frac{y^4}{4!} + i\frac{y^5}{5!} + \cdots.
\end{aligned}
$$

If we regroup the terms in the series, a step that will be justified in Chapter 4, we obtain

$$
\begin{aligned}
e^{iy} &= \left(1 - \frac{y^2}{2!} + \frac{y^4}{4!} - \cdots\right) + i\left(y - \frac{y^3}{3!} + \frac{y^5}{5!} - \cdots\right) \\
&= \cos y + i \sin y,
\end{aligned}
$$

where we have recognized the power series expansions for the cosine and sine functions (converging for all y),

$$\cos y = \sum_{n=0}^{\infty} \frac{(-1)^n y^{2n}}{(2n)!} \quad \text{and} \quad \sin y = \sum_{n=0}^{\infty} \frac{(-1)^n y^{2n+1}}{(2n+1)!}.$$

Thus the desired expression

$$e^z = e^x e^{iy} = e^x \left(\cos y + i \sin y\right).$$

Consequently,

(3)
$$\operatorname{Re} e^z = e^x \cos y \qquad \text{and} \qquad \operatorname{Im} e^z = e^x \sin y.$$

We summarize these results for ease of reference.

THE COMPLEX
EXPONENTIAL

The complex exponential e^z is defined for all z as in (1). For $z = x + i\,y$, we have

$$(4) \qquad\qquad e^z = e^x \left(\cos y + i\, \sin y \right).$$

In particular, for $z = i\,\theta$, with θ real, we have the identity

$$(5) \qquad\qquad e^{i\theta} = \cos\theta + i\, \sin\theta,$$

known as **Euler's identity**.

EXAMPLE 1   Finding e^z

Compute e^z for the given z.
(a) $2 + i\,\pi$.
(b) $3 - i\frac{\pi}{3}$.
(c) $-1 + i\frac{\pi}{2}$.
(d) $i\frac{5\pi}{4}$.

Solution   (a)   From (4),

$$e^{2+i\pi} = e^2(\cos\pi + i\,\sin\pi) = -e^2.$$

This number is purely real and negative.
(b)   From (4),

$$e^{3-i\frac{\pi}{3}} = e^3(\cos\frac{\pi}{3} + i\,\sin(-\frac{\pi}{3})) = e^3(\frac{1}{2} - i\,\frac{\sqrt{3}}{2}).$$

(c)   From (4),

$$e^{-1+i\frac{\pi}{2}} = e^{-1}(\cos\frac{\pi}{2} + i\,\sin\frac{\pi}{2}) = \frac{i}{e}.$$

This number is purely imaginary.
(d)   Here z is purely imaginary; we may use Euler's identity. From (5),

$$e^{i\frac{5\pi}{4}} = \cos\frac{5\pi}{4} + i\,\sin\frac{5\pi}{4} = -\frac{\sqrt{2}}{2} - i\,\frac{\sqrt{2}}{2}.$$

This number is unimodular.   ∎

Figure 1   For $z = x + i\,y$, e^z has modulus e^x and argument y:
$$|e^z| = e^x;$$
$$\arg(e^z) = y.$$

Example 1 shows that the complex exponential function can take on negative real values and complex values. This is in sharp contrast with the real exponential function, which is always positive.

Looking back at (4), and recalling the well-known fact from calculus that $e^x > 0$ for all x, it follows immediately that (4) is the polar form of e^z (see (2), Section 1.3), where the modulus of e^z is e^x and its argument is $y + 2k\pi$, where k is an integer (see Figure 1). We have the following results.

MODULUS AND ARGUMENT OF THE COMPLEX EXPONENTIAL

For $z = x + iy$, the modulus of e^z is

$$(6) \qquad |e^z| = e^x > 0.$$

Consequently, e^z is never zero. The argument of e^z is

$$(7) \qquad \arg(e^z) = y + 2k\pi \qquad (z = x + iy),$$

where k is an integer.

Using the exponent rule (2) and the fact that $e^0 = 1$, we can write

$$1 = e^{z-z} = e^z e^{-z}.$$

Thus, for any complex number z, the multiplicative inverse of e^z is e^{-z}; equivalently,

$$\boxed{e^{-z} = \frac{1}{e^z}.}$$

Exponential and Polar Representations

Euler's identity (5) provides us with yet another convenient way of representing complex numbers. Indeed, if $z = r(\cos\theta + i\sin\theta)$ is a complex number in polar form, then since $e^{i\theta} = \cos\theta + i\sin\theta$, we obtain the **exponential representation** or polar form $z = re^{i\theta}$.

From this representation it is clear that

$$|z| = 1 \Leftrightarrow r = 1 \Leftrightarrow z = e^{i\theta},$$

where θ is the argument of z. Thus the complex numbers $e^{i\theta}$ are the unimodular complex numbers. Because their distance to the origin always equals 1, all the complex numbers $e^{i\theta}$ lie on the unit circle. All other nonzero complex numbers are positive multiples of some $e^{i\theta}$. This fact is illustrated geometrically in Figure 2, where the ray from the origin to $z = re^{i\theta}$ intersects the unit circle at the point $e^{i\theta}$. To go from the origin to z, we move in the direction of $e^{i\theta}$ by a distance $r = |z|$.

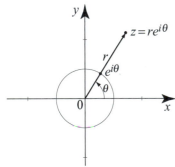

Figure 2 To plot a point in exponential notation $z = re^{i\theta}$, move a distance r along the ray extending from the origin to $e^{i\theta}$.

The exponential notation, $z = r\,e^{i\theta}$, enables us to operate on complex numbers in a very convenient way, as we now show.

EXPONENTIAL REPRESENTATION

The exponential representation of $z = r(\cos\theta + i\sin\theta)$ is

$$(8) \qquad z = r\,e^{i\theta}, \qquad |z| = r \qquad \text{and} \qquad \arg z = \theta + 2k\pi.$$

We have

$$(9) \qquad\qquad \overline{z} \;=\; re^{-i\theta};$$

$$(10) \qquad\qquad z^{-1} \;=\; \frac{1}{r}e^{-i\theta} \qquad (z \neq 0).$$

If $z_1 = r_1 e^{i\theta_1}$ and $z_2 = r_2 e^{i\theta_2}$, then

$$(11) \qquad\qquad z_1 z_2 \;=\; r_1 r_2 e^{i(\theta_1 + \theta_2)};$$

$$(12) \qquad\qquad \frac{z_1}{z_2} \;=\; \frac{r_1}{r_2} e^{i(\theta_1 - \theta_2)} \qquad (z_2 \neq 0).$$

Proof   As we remarked earlier, (8) is a consequence of Euler's identity (5). For (9), we have

$$\overline{z} = \overline{r(\cos\theta + i\sin\theta)} = r(\cos\theta - i\sin\theta) = r(\cos(-\theta) + i\sin(-\theta)) = re^{-i\theta}.$$

To prove (10), write $z = r\,e^{i\theta}$ with $r \neq 0$. Then

$$z\,\frac{1}{r}e^{-i\theta} = re^{i\theta}\frac{1}{r}e^{-i\theta} = e^{i(\theta - \theta)} = e^0 = 1.$$

The proofs of (11) and (12) are immediate from (2) and (10). We leave the details to the exercises.   ∎

Multiplication and division of unimodular numbers are particularly easy to describe using the complex exponential. Indeed, if $z_1 = e^{i\theta_1}$ and $z_2 = e^{i\theta_2}$, then from (11) and (12) we get

$$z_1 z_2 = e^{i\theta_1} e^{i\theta_2} = e^{i(\theta_1 + \theta_2)},$$

and

$$\frac{z_1}{z_2} = \frac{e^{i\theta_1}}{e^{i\theta_2}} = e^{i(\theta_1 - \theta_2)}.$$

Thus, to multiply two unimodular numbers we add their arguments, and to divide them we subtract their arguments. (See Figure 3.)

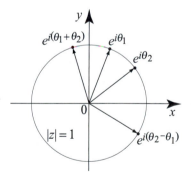

Figure 3 To multiply two unimodular numbers we add their arguments, and to divide them we subtract their arguments.

EXAMPLE 2   Exponential representations

Let $z_1 = -7\sqrt{3} + 7i$ and $z_2 = 1 + i$. Express the following complex expressions in exponential form. Also give your answer in Cartesian form in (d) and (e).

(a) z_1.   (b) z_2.   (c) $z_1 z_2$.   (d) $\dfrac{1}{z_2}$.   (e) $\dfrac{z_1}{z_2}$.

Solution   (a)   We will use (8). We have $|z_1| = \sqrt{(-7\sqrt{3})^2 + 7^2} = 14$. To compute a value of $\arg z_1$, we appeal to (14), Section 1.3. Since z_1 is in the second quadrant, we have $\operatorname{Arg} z_1 = \tan^{-1}(\frac{7}{-7\sqrt{3}}) + \pi = \frac{5\pi}{6}$. Hence

$$z_1 = 14\left(\cos\frac{5\pi}{6} + i\sin\frac{5\pi}{6}\right) = 14\,e^{i\frac{5\pi}{6}}.$$

(b)   Following the steps in (a), we find $|z_2| = \sqrt{1^2 + 1^2} = \sqrt{2}$ and $\operatorname{Arg} z_2 = \tan^{-1}(1) = \frac{\pi}{4}$. Hence

$$z_2 = \sqrt{2}\left(\cos\frac{\pi}{4} + i\sin\frac{\pi}{4}\right) = \sqrt{2}\,e^{i\frac{\pi}{4}}.$$

(c)   We use (a), (b), and (11) and get

$$z_1\,z_2 = 14\,e^{i\frac{5\pi}{6}}\,\sqrt{2}\,e^{i\frac{\pi}{4}} = 14\sqrt{2}\,e^{i\left(\frac{5\pi}{6}+\frac{\pi}{4}\right)} = 14\sqrt{2}\,e^{i\frac{13\pi}{12}}.$$

(d)   We use (b) and (10) and get

$$\frac{1}{z_2} = z_2^{-1} = \frac{1}{\sqrt{2}}e^{-i\frac{\pi}{4}}.$$

This is the polar form of $\frac{1}{z_2}$. To get the Cartesian form, we use Euler's identity:

$$\frac{1}{z_2} = \frac{1}{\sqrt{2}}\left(\cos\frac{-\pi}{4} + i\sin\frac{-\pi}{4}\right) = \frac{1}{\sqrt{2}}\left(\frac{\sqrt{2}}{2} - i\frac{\sqrt{2}}{2}\right) = \frac{1}{2} - i\frac{1}{2}.$$

Checking our answer, we have

$$z_2\,\frac{1}{z_2} = (1+i)(\frac{1}{2} - i\frac{1}{2}) = (\frac{1}{2} + \frac{1}{2}) + i\,(\frac{1}{2} - \frac{1}{2}) = 1,$$

as it should be.

(e)   We use (a) and (b) and (12) and get

$$\frac{z_1}{z_2} = \frac{14}{\sqrt{2}}e^{i\left(\frac{5\pi}{6}-\frac{\pi}{4}\right)} = 7\sqrt{2}\,e^{i\frac{7\pi}{12}}.$$

To get the Cartesian form, we use Euler's identity:

$$\begin{aligned}
\frac{z_1}{z_2} &= 7\sqrt{2}\left(\cos\frac{7\pi}{12} + i\sin\frac{7\pi}{12}\right)\\
&= 7\sqrt{2}\left(\frac{1-\sqrt{3}}{2\sqrt{2}} + i\frac{1+\sqrt{3}}{2\sqrt{2}}\right) = \frac{7}{2}\left((1-\sqrt{3}) + i(1+\sqrt{3})\right),
\end{aligned}$$

where we have used the subtraction formulas for the cosine and sine to compute the exact values of $\cos\frac{7\pi}{12}$ and $\sin\frac{7\pi}{12}$. For example,

$$\begin{aligned}
\cos\frac{7\pi}{12} &= \cos(\frac{5\pi}{6} - \frac{\pi}{4}) = \cos\frac{5\pi}{6}\cos\frac{\pi}{4} + \sin\frac{5\pi}{6}\sin\frac{\pi}{4}\\
&= -\frac{\sqrt{3}}{2}\frac{\sqrt{2}}{2} + \frac{1}{2}\frac{\sqrt{2}}{2} = \frac{1-\sqrt{3}}{2\sqrt{2}}.
\end{aligned}$$

The value of $\sin \frac{7\pi}{12} = \frac{1+\sqrt{3}}{2\sqrt{2}}$ can be derived similarly.   ■

The Exponential as a Mapping

It is important to explore how the exponential function maps complex numbers. The following result shows that the exponential function, unlike its real counterpart, is not one-to-one.

MAPPING PROPERTIES OF THE EXPONENTIAL

We have

$$(13) \qquad e^z = 1 \qquad \text{if and only if} \qquad z = 2k\pi i, \text{ for some integer } k.$$

Also,

$$(14) \qquad e^{z_1} = e^{z_2} \qquad \text{if and only if} \qquad z_1 = z_2 + 2k\pi i, \text{ for some integer } k.$$

Proof   Write $z = x + iy$, and suppose $e^z = 1$. Then $e^x e^{iy} = 1$. Thinking of this as the exponential representation of the complex number 1, we see that $e^x = 1$ and $y = 2k\pi$ for some integer k. But $e^x = 1$ implies that $x = 0$. Hence $z = iy = 2k\pi i$ for some integer k. Conversely, if $z = 2k\pi i$, then $e^{2k\pi i} = \cos 2k\pi + i \sin 2k\pi = 1$. We have proved (13). To prove (14), notice that $e^{z_1} = e^{z_2} \Leftrightarrow e^{z_1 - z_2} = 1$, which by (13) is if and only if $z_1 - z_2 = 2k\pi i$, or $z_1 = z_2 + 2k\pi i$.   ■

A complex-valued function $f(z)$ is **periodic** with period $\tau \neq 0$ if for all z in the domain of definition of f, we have $f(z + \tau) = f(z)$. From (14), it follows that, for all complex numbers z,

$$(15) \qquad\qquad\qquad\qquad e^z = e^{z + 2\pi i}.$$

Hence the exponential function e^z is periodic with period $2\pi i$.

Now we show an example of the exponential function mapping a specific region.

EXAMPLE 3   An exponential mapping

Consider the rectangular area

$$S = \{z = x + iy : \ -1 \leq x \leq 1, \ 0 \leq y \leq \pi\}.$$

Find the image of S under the mapping $f(z) = e^z$.

Solution   Fix x_0 in the interval $[-1, 1]$ and consider EF, the vertical line segment $x = x_0$ inside S. Points on the segment EF are of the form $z = x_0 + iy$, where $0 \leq y \leq \pi$. For such z,

$$f(z) = e^z = e^{x_0} e^{iy} = e^{x_0}(\cos y + i \sin y).$$

The point $w = e^{x_0} e^{iy}$ has modulus e^{x_0} and argument y. In particular, w lies on the circle of radius e^{x_0} with center at 0. As y varies from 0 to π, the point w traces the upper semicircle. Thus the image of EF by f is the upper semicircle with center at 0 and radius e^{x_0}.

Figure 4 As usual, we denote the image of a point P in the xy-plane by the point P' in the uv-plane. The mapping $w = e^z$ takes the vertical line segment EF to a semicircle in the uv-plane with u-intercepts at E' and F'.

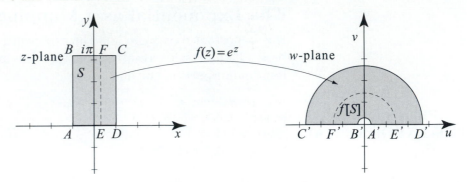

Now, as we vary x_0 from -1 to 1, e^{x_0} varies from e^{-1} to e. As a consequence, the corresponding semicircles increase in radius and fill the semiannular area between the semicircle of radius e^{-1} with center at 0 and the semicircle of radius e with center at 0. (See Figure 4.) ∎

You should check in Example 3 that the rectangular boundary of the area S is mapped to the boundary of the semiannular area $f[S]$. The fact that f maps boundary to boundary is an important property that will be investigated in Section 2.5 and the chapter on conformal mappings.

In Example 3, if we enlarge the rectangular region S to become the infinite horizontal strip, where $-\infty < x < \infty$ and $0 \le y < 2\pi$, the image of S will cover the entire w-plane minus the origin. Since e^z is never 0, e^z will map the entire z-plane onto the w-plane minus the origin. In fact, by periodicity, e^z assumes every nonzero w an infinite number of times; by (14), the solutions z to $e^z = w$ must differ by integer multiples of $2\pi i$.

Trigonometric Functions via Euler's Identity

Let θ be an arbitrary real number. Euler's identity tells us that

$$e^{i\theta} = \cos\theta + i\sin\theta$$

and

$$e^{-i\theta} = \cos\theta - i\sin\theta.$$

Adding these two identities and dividing by 2, we get

$$(16) \qquad \cos\theta = \frac{e^{i\theta} + e^{-i\theta}}{2}.$$

Similarly, subtracting and dividing by $2i$, we get

$$(17) \qquad \sin\theta = \frac{e^{i\theta} - e^{-i\theta}}{2i}.$$

These expressions enable us in certain cases to use to our advantage elementary properties of the exponential function in handling tricky problems involving products of the cosine and sine functions. For example, suppose p is a positive integer. You can express the product $\cos^m \theta \sin^n \theta$, where m and n are nonnegative integers such that $m+n = p$, as a linear combination of terms involving $\cos(j\theta)$ and $\sin(k\theta)$, where $1 \leq j,\ k \leq p$. This process is called **linearization** and has many useful applications in calculus. We illustrate with an example.

EXAMPLE 4   Linearizing powers of the cosine

Linearize $\cos^3 \theta$.

Solution   Appealing to (16), we have

$$\cos^3 \theta = \left(\frac{e^{i\theta} + e^{-i\theta}}{2} \right)^3 = \frac{1}{8} \left(e^{3i\theta} + 3e^{2i\theta}e^{-i\theta} + 3e^{i\theta}e^{-2i\theta} + e^{-3i\theta} \right)$$

$$= \frac{1}{4} \left[\frac{e^{3i\theta} + e^{-3i\theta}}{2} + 3\, \frac{e^{i\theta} + e^{-i\theta}}{2} \right] = \frac{1}{4} \left(\cos 3\theta + 3 \cos \theta \right). \qquad \blacksquare$$

As a typical application of linearization, we evaluate $\int \cos^3 \theta\, d\theta$. By using the result of Example 4, we have

$$\int \cos^3 \theta\, d\theta = \frac{1}{4} \int \left(\cos 3\theta + 3 \cos \theta \right)\, d\theta = \frac{1}{12} \sin 3\theta + \frac{3}{4} \sin \theta + C.$$

In the following section, we will use the expressions (16) and (17) for the cosine and sine in terms of the exponential to generalize trigonometric functions to complex variables. Finally, we mention that (16) and (17) are extremely useful in the theory of Fourier series, where they are used to relate the real form to the complex form of Fourier series. This is developed in Chapter 7 on Fourier series.

Exercises 1.5

In Exercises 1–10, write the given complex number in the form $a + ib$.

1. $e^{i\pi}$.   **2.** $e^{2i\pi}$.   **3.** $e^{200i\pi}$.   **4.** $e^{201\,i\pi}$.   **5.** $e^{i\frac{3\pi}{4}}$.

6. $e^{2-i\frac{\pi}{4}}$.   **7.** $e^{-1-i\frac{\pi}{6}}$.   **8.** $-2\,e^{i+\pi}$.   **9.** $3\,e^{3+i\frac{\pi}{2}}$.   **10.** $e^{701\,i\frac{\pi}{4}}$.

In Exercises 11–18, express the given complex number in the exponential form $re^{i\theta}$. (You may notice that these are the same complex numbers as in Exercises 5–12 in Section 1.3.)

11. $-3 - 3\,i$.   **12.** $-\dfrac{\sqrt{3}}{2} + \dfrac{i}{2}$.   **13.** $-1 - \sqrt{3}\,i$.   **14.** $1 + i$.

15. $-\dfrac{i}{2}$.   **16.** $\dfrac{1+i}{1+\sqrt{3}\,i}$.   **17.** $\dfrac{1+i}{1-i}$.   **18.** $\dfrac{i}{10+10i}$.

In Exercises 19–24, write the given complex number in the form $a + ib$, where a and b are real.

19. e^i.   **20.** $e^{1+\sqrt{2}i\pi}$.   **21.** $\left| e^{-i\frac{\pi}{12}} \right|$.

22. $e^{-1-i}\,\overline{e^{1+2i}}$.

23. $\dfrac{e^{3-i}}{e^{-1+2i}}$.

24. $e^{1-60i\pi}\left|e^{-1+200\,i}\right|$.

25. Find the real and imaginary parts of the following functions.

(a) e^{3z}.
  (b) e^{z^2}.
  (c) $e^{\overline{z}}$.
  (d) e^{iz}.

26. (a) For all complex numbers z, show that $(e^z)^n = e^{n\,z}$, $n = 0, \pm1, \pm2, \ldots$.
(b) Show that $\overline{e^z} = e^{\overline{z}}$.
  (c) Show that $e^{z+i\pi} = -e^z$.

In Exercises 27–32, show that the shaded area S in the z-plane is mapped to the shaded area in the w-plane by the given mapping $f(z)$.

27.

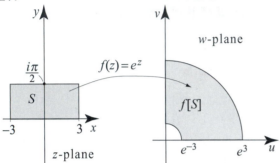

Figure 5

28.

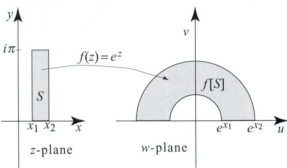

Figure 6

29.

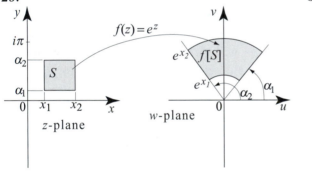

Figure 7

30.

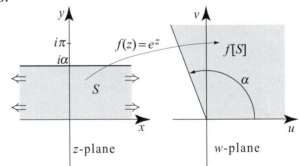

Figure 8

31.

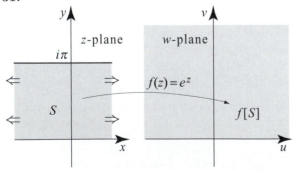

Figure 9

32.

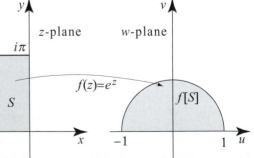
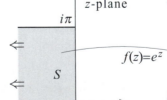

Figure 10

In Exercises 33–36, (a) linearize the integrand; (b) evaluate the given integral.

33. $\displaystyle\int \sin^4 \theta \, d\theta$.
  34. $\displaystyle\int \cos^5 \theta \, d\theta$.
  35. $\displaystyle\int \cos^6 \theta \, d\theta$.
  36. $\displaystyle\int \sin^3 \theta \cos^5 \theta \, d\theta$.

37. Show that for all complex numbers z, $|e^z| \le e^{|z|}$. When does equality hold?

38. Differences and similarities. In the text, we pointed out a few similarities and differences between e^z and e^x. In what follows we consider some additional ones. For each part, either prove your answer or provide an example to show that the statement is false.
(a) The function e^x is one to one. Is e^z one to one?
(b) The function e^x is increasing (if $x_1 < x_2$ then $e^{x_1} < e^{x_2}$). If $|z_1| < |z_2|$ do we have $|e^{z_1}| < |e^{z_2}|$?
(c) The function e^x never vanishes. Can e^z vanish?
(d) The function e^x is always positive. Is e^z always real and positive?
(e) The modulus or absolute value of e^x is e^x. What is the absolute value of e^z?
(f) We have $e^x = 1 \Leftrightarrow x = 0$. Do we have $e^z = 1 \Leftrightarrow z = 0$?

39. Show that $|e^z| \le 1$ if and only if $\operatorname{Re} z \le 0$. When does equality hold?

40. Project Problem: The Dirichlet kernel. (a) For $z \ne 1$ and $n = 1, 2, \ldots$, show that

$$1 + z + z^2 + \cdots + z^n = \frac{1 - z^{n+1}}{1 - z}.$$

(b) Take $z = e^{i\theta}$, where θ is a real number $\ne 2k\pi$ (k an integer), and obtain

$$1 + e^{i\theta} + e^{2i\theta} + \cdots + e^{in\theta} = \frac{i}{2}\,\frac{(1 - e^{i(n+1)\theta})e^{-i\frac{\theta}{2}}}{\sin\frac{\theta}{2}}.$$

[Hint: After substituting $z = e^{i\theta}$, multiply and divide by $e^{-i\frac{\theta}{2}}$; then use (17).]
(c) Take the real and imaginary parts of the identity in (b) to obtain

$$1 + \cos\theta + \cos 2\theta + \cdots + \cos n\theta = \frac{1}{2} + \frac{\sin[(n + \frac{1}{2})\theta]}{2\sin\frac{\theta}{2}},$$

and

$$\sin\theta + \sin 2\theta + \cdots + \sin n\theta = \frac{\cos\frac{\theta}{2} - \sin\left[(n+1)\frac{\theta}{2}\right]}{2\sin\frac{\theta}{2}}.$$

(d) Derive the identity

$$\frac{1}{2} + \cos\theta + \cos 2\theta + \cdots + \cos n\theta = \frac{\sin[(n + \frac{1}{2})\theta]}{2\sin\frac{\theta}{2}}.$$

This finite sum of cosines is a constant multiple of the **Dirichlet kernel** (see Section 7.6). It plays an important role in the theory of Fourier series. The sum of sines in (c) is known as the **conjugate Dirichlet kernel** of order n and is also useful in the theory of Fourier series.

1.6   Trigonometric and Hyperbolic Functions

In the previous section, we defined the complex exponential function e^z as an extension of the real exponential function e^x. The function e^x is one of the so-called elementary functions from calculus. Elementary functions comprise among other functions the trigonometric functions, the hyperbolic functions, the logarithmic functions, and raising numbers to powers. Our goal in this and the following section is to extend some of the elementary functions to complex numbers and study their basic properties. These new functions will provide us with ample examples to test the theory of derivatives and integrals that will be presented in the following chapters.

Trigonometric Functions

Let θ be any real number. Using Euler's identity, we showed in Section 1.5, (16) and (17), that

$$(1) \qquad \cos\theta = \frac{e^{i\theta} + e^{-i\theta}}{2} \quad \text{and} \quad \sin\theta = \frac{e^{i\theta} - e^{-i\theta}}{2i}.$$

Motivated by these identities, we define the complex cosine and sine functions for all complex numbers z by the formulas

$$(2) \qquad \boxed{\cos z = \frac{e^{iz} + e^{-iz}}{2},}$$

and

$$(3) \qquad \boxed{\sin z = \frac{e^{iz} - e^{-iz}}{2i}.}$$

You should keep in mind that these are new functions, even though they are named after familiar functions. As you will soon see, they share similar properties with the usual cosine and sine functions, but they assume complex values and are not bounded in absolute value.

EXAMPLE 1   Finding $\cos z$ and $\sin z$

(a)   Compute $\cos(2 + i\pi)$.

(b)   Compute $\sin(i\frac{5\pi}{4})$.

Solution   We use the definitions. For (a), we have from (2),

$$
\begin{aligned}
\cos(2 + i\pi) &= \frac{1}{2}\left(e^{i(2+i\pi)} + e^{-i(2+i\pi)}\right) \\
&= \frac{1}{2}\left(e^{-\pi+2i} + e^{\pi-2i}\right) = \frac{1}{2}\left(e^{-\pi}e^{2i} + e^{\pi}e^{-2i}\right) \\
&= \frac{1}{2}\left(e^{-\pi}[\cos(2) + i\sin(2)] + e^{\pi}[\cos(2) - i\sin(2)]\right) \\
&= \cos(2)\frac{e^{\pi} + e^{-\pi}}{2} - i\sin(2)\frac{e^{\pi} - e^{-\pi}}{2} \\
&= \cos(2)\cosh\pi - i\sin(2)\sinh\pi,
\end{aligned}
$$

where $\cosh\pi$ and $\sinh\pi$ are the real hyperbolic functions evaluated at π.

For (b), we use (3) and proceed in a similar way:

$$\sin\left(i\,\frac{5\pi}{4}\right) = \frac{1}{2i}\left(e^{i\left(i\frac{5\pi}{4}\right)} - e^{-i\left(i\frac{5\pi}{4}\right)}\right)$$

$$= \frac{-i}{2}\left(e^{-\frac{5\pi}{4}} - e^{\frac{5\pi}{4}}\right) = i\,\sinh\!\left(\frac{5\pi}{4}\right).$$

The appearance of the hyperbolic functions in the expressions of the real and imaginary parts of the complex cosine and sine functions was not a coincidence. In fact, general formulas of this nature will be derived later in this section.   ◼

A function $f(z)$ is said to be **even** if $f(z) = f(-z)$ and **odd** if $f(-z) = -f(z)$, for all z in the domain of definition of f. We can show from their definitions that the cosine is even while the sine is odd; also, both of them are 2π-periodic. In fact, the complex trigonometric functions satisfy many identities that we are familiar with for real trigonometric functions.

PROPERTIES OF TRIGONOMETRIC FUNCTIONS

Let $z = x + iy$ be a complex number. Then

(4) $$\cos(-z) = \cos z, \qquad \sin(-z) = -\sin z;$$

(5) $$\cos(z + 2\pi) = \cos z \qquad \sin(z + 2\pi) = \sin z;$$

(6) $$\sin\!\left(z + \tfrac{\pi}{2}\right) = \cos z;$$

(7) $$e^{iz} = \cos z + i\sin z;$$

(8) $$\cos^2 z + \sin^2 z = 1.$$

Proof   To prove the first identity in (4), we appeal to (2):

$$\cos(-z) = \frac{e^{i(-z)} + e^{-i(-z)}}{2} = \frac{e^{-iz} + e^{iz}}{2} = \frac{e^{iz} + e^{-iz}}{2} = \cos z.$$

The second identity in (4) is proved similarly by appealing to (3) (Exercise 25). In proving (5), we will use the fact that $e^{\pm 2\pi i} = 1$. We have

$$\cos(z + 2\pi) = \frac{e^{i(z+2\pi)} + e^{-i(z+2\pi)}}{2} = \frac{e^{iz}e^{2\pi i} + e^{-iz}e^{-2\pi i}}{2} = \frac{e^{iz} + e^{-iz}}{2} = \cos z.$$

This proves the first identity in (5). The proof of second the identity in (5) is similar (Exercise 26). To prove (6), we calculate

$$\sin\!\left(z + \frac{\pi}{2}\right) = \frac{e^{i(z+\pi/2)} - e^{-i(z+\pi/2)}}{2i} = \frac{ie^{iz} - (-i)e^{-iz}}{2i} = \cos z.$$

You should recognize (7) as Euler's identity (5), Section 1.3, where we have replaced the real argument θ by a complex argument z. To prove (7), we simply multiply (2) by 2 and (3) by $2i$ and add the resulting identities.

Identity (8) is the analog of the famous Pythagorean identity relating the real cosine and sine functions. We prove it with a trick based on the complex exponential function, which emphasizes the relationships between the trigonometric and exponential functions. Using (7), we have

$$1 = e^{iz}e^{-iz} = \overbrace{(\cos z + i \sin z)}^{e^{iz}}\overbrace{(\cos z - i \sin z)}^{e^{-iz}} = \cos^2 z + \sin^2 z. \qquad \blacksquare$$

The familiar angle-addition and half-angle formulas also apply to the complex cosine and sine.

TRIGONOMETRIC IDENTITIES

Let z, z_1, z_2 be a complex numbers. Then

(9) $$\cos(z_1 + z_2) = \cos z_1 \cos z_2 - \sin z_1 \sin z_2;$$
(10) $$\sin(z_1 + z_2) = \sin z_1 \cos z_2 + \cos z_1 \sin z_2;$$
(11) $$\cos^2 z = \frac{1 + \cos(2z)}{2};$$
(12) $$\sin^2 z = \frac{1 - \cos(2z)}{2}.$$

Proof   Expanding the right side of (9), it equals

$$\frac{(e^{iz_1} + e^{-iz_1})(e^{iz_2} + e^{-iz_2})}{2^2} - \frac{(e^{iz_1} - e^{-iz_1})(e^{iz_2} - e^{-iz_2})}{(2i)^2}.$$

Expanding the numerators and adding the fractions, all terms in $e^{i(z_1 - z_2)}$ and $e^{i(z_2 - z_1)}$ cancel and we are left with $\dfrac{2e^{i(z_1 + z_2)} + 2e^{-i(z_1 + z_2)}}{4}$, which is the same as $\cos(z_1 + z_2)$. The proof of (10) is similar. Now, setting $z_1 = z_2 = z$ in (9) yields

$$\cos 2z = \cos^2 z - \sin^2 z.$$

Replacing $\sin^2 z$ by $1 - \cos^2 z$, we conclude (11). Replacing $\cos^2 z$ by $1 - \sin^2 z$, we conclude (12). $\qquad \blacksquare$

Up to this point our statements about the complex trigonometric functions have been no different than those statements for real trigonometric functions. Now we show how they can behave differently. From (2), we have, for any real y,

(13) $$\cos(iy) = \frac{e^{i(iy)} + e^{-i(iy)}}{2} = \frac{e^y + e^{-y}}{2} = \cosh y$$

and

(14) $$\sin(iy) = \frac{e^{i(iy)} - e^{-i(iy)}}{2i} = i\frac{e^y - e^{-y}}{2} = i \sinh y,$$

where $\cosh y$ and $\sinh y$ are the real hyperbolic functions from calculus. We are now in a position to express $\cos z$ and $\sin z$ in terms of their real and imaginary parts, and also to compute their moduli.

REAL AND IMAGINARY PARTS AND MODULI OF TRIGONOMETRIC FUNCTIONS

Let $z = x + iy$ be a complex number. Then

$$(15) \qquad \cos z = \cos x \cosh y - i \sin x \sinh y,$$

$$(16) \qquad \sin z = \sin x \cosh y + i \cos x \sinh y,$$

$$(17) \qquad |\cos z| = \sqrt{\cos^2 x + \sinh^2 y},$$

$$(18) \qquad |\sin z| = \sqrt{\sin^2 x + \sinh^2 y}.$$

Proof To prove (15), we appeal to (9) and (13)–(14) and write

$$
\begin{aligned}
\cos z &= \cos(x + iy) \\
&= \cos x \cos(iy) - \sin x \sin(iy) \\
&= \cos x \cosh y - i \sin x \sinh y.
\end{aligned}
$$

The proof of (16) is similar and is left to Exercise 21. To prove (17), we use (15) and the definition of the modulus of a complex number ((1), Section 1.2). We also use the identity $\cosh^2 y - \sinh^2 y = 1$ for real hyperbolic functions. We get

$$
\begin{aligned}
|\cos z|^2 &= \cos^2 x \cosh^2 y + \sin^2 x \sinh^2 y \\
&= \cos^2 x (1 + \sinh^2 y) + \sin^2 x \sinh^2 y \\
&= \cos^2 x + \sinh^2 y (\cos^2 x + \sin^2 x) \\
&= \cos^2 x + \sinh^2 y.
\end{aligned}
$$

The proof of (18) is similar and is left to Exercise 21. ∎

The following example is intended to show you that, unlike $\cos x$ and $\sin x$, the complex functions $\cos z$ and $\sin z$ are not bounded.

EXAMPLE 2 $\cos z$ **and** $\sin z$ **are not bounded**

Show that $\cos z$ and $\sin z$ are not bounded over the complex plane.

Solution   For $z = x + iy$, using (17), we obtain $|\cos z| = \sqrt{\cos^2 x + \sinh^2 y} \geq \sqrt{\sinh^2 y} = |\sinh y|$. Similarly, using (18), we obtain $|\sin z| = \sqrt{\sin^2 x + \sinh^2 y} \geq \sqrt{\sinh^2 y} = |\sinh y|$. As $y \to \infty$, we have $\sinh y \to \infty$; and as $y \to -\infty$, we have $\sinh y \to -\infty$. Hence $|\cos z|$ and $|\sin z|$ blow up to infinity as $|\operatorname{Im} z|$ tends to infinity. ∎

EXAMPLE 3   **Zeros of the sine and cosine**

(a)   Show that $\sin z = 0 \Leftrightarrow z = k\pi$, for some integer k.
(b)   Show $\cos z = 0 \Leftrightarrow z = \frac{\pi}{2} + k\pi$, for some integer k.

Thus $\cos z$ and $\sin z$ have the same zeros as their real counterparts, $\cos x$ and $\sin x$.

Solution   (a) Suppose that $z = x + iy$ is a point in the plane and that $\sin z = 0$. We know then that $|\sin z| = 0$, so by (18) we have $\sin x = 0$ and $\sinh y = 0$. The real function $\sinh y$ equals zero $\Leftrightarrow y = 0$, and the real function $\sin x$ equals zero $\Leftrightarrow x = k\pi$ for some integer k. Hence (a) holds.

(b) Now that we have the zeros of the sine, we can easily find the zeros of the cosine using (6). We have

$$\cos z = 0 \quad \Leftrightarrow \quad \sin(z + \frac{\pi}{2}) = 0$$

$$\Leftrightarrow \quad z + \frac{\pi}{2} = k\pi, \text{ for some integer } k,$$

$$\Leftrightarrow \quad z = -\frac{\pi}{2} + k\pi, \text{ for some integer } k.$$

Replacing k by $k + 1$, we obtain (b). ∎

In our next example, we consider a mapping by the function $\sin z$.

EXAMPLE 4   The mapping $w = \sin z$

Find the image under the mapping $f(z) = \sin z$ of the semi-infinite strip

$$S = \left\{ z = x + iy : -\frac{\pi}{2} \leq x \leq \frac{\pi}{2}, \, y \geq 0 \right\}.$$

Solution   As in previous examples of mappings, we will first find the image under f of a simple curve in the domain of definition, often a line segment or line. Then we will sweep the domain of definition with this curve and keep track of the area that is swept by the image. Fix $0 \leq y_0 < \infty$ and consider the horizontal line segment EF defined by: $y = y_0$, $-\frac{\pi}{2} \leq x \leq \frac{\pi}{2}$. Let $u + iv$ denote the image of a point $z = x + iy_0$ on EF. Using (16), we get

$$u + iv = \sin(x + iy_0) = \sin x \cosh y_0 + i \cos x \sinh y_0.$$

Hence

$$u = \sin x \cosh y_0 \quad \text{and} \quad v = \cos x \sinh y_0.$$

If $y_0 = 0$, we see that $v = 0$ and $u = \sin x$, which shows that the image of the interval $-\frac{\pi}{2} \leq x \leq \frac{\pi}{2}$ under the mapping $\sin z$ is the interval $-1 \leq u \leq 1$. The case $y_0 > 0$ is more interesting. In this case, we have

(19) $$\frac{u}{\cosh y_0} = \sin x \quad \text{and} \quad \frac{v}{\sinh y_0} = \cos x.$$

Note that $v \geq 0$ because $\cos x \geq 0$ for $-\frac{\pi}{2} \leq x \leq \frac{\pi}{2}$. Squaring both equations in (19) then adding them, we get

$$\left(\frac{u}{\cosh y_0} \right)^2 + \left(\frac{v}{\sinh y_0} \right)^2 = \sin^2 x + \cos^2 x = 1.$$

Hence as x varies in the interval $-\frac{\pi}{2} \leq x \leq \frac{\pi}{2}$, the point (u, v) traces the upper semi-ellipse

$$\left(\frac{u}{\cosh y_0} \right)^2 + \left(\frac{v}{\sinh y_0} \right)^2 = 1, \qquad v \geq 0.$$

The u-intercepts of the ellipse are at $u = \pm \cosh y_0$ and the v-intercept is at $v = \sinh y_0$. As $y_0 \to \infty$, $\cosh y_0$ and $\sinh y_0$ tend to ∞. And as $y_0 \to 0$, $\sinh y_0 \to 0$ and $\cosh y_0 \to 1$. So, as y_0 varies in the interval $0 < y_0 < \infty$, the upper semi-ellipses fill the upper half w-plane $v \geq 0$, including the u-axis (Figure 1). You should verify (Exercise 23) that the boundary of S gets maped to the boundary of $f[S]$, namely, the u-axis. ∎

Figure 1 The mapping $w = \sin z$ takes the horizontal line segment $y = y_0 > 0$, $-\frac{\pi}{2} \leq x \leq \frac{\pi}{2}$ onto the upper semi-ellipse

$$\left(\frac{u}{\cosh y_0} \right)^2 + \left(\frac{v}{\sinh y_0} \right)^2 = 1,$$

$v \geq 0$.

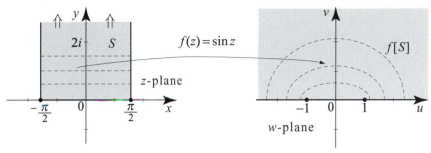

The other trigonometric functions are defined for complex variables in terms of the cosine and sine in accordance with the real definitions.

OTHER TRIGONOMETRIC FUNCTIONS

With $\cos z = \frac{e^{iz}+e^{-iz}}{2}$ and $\sin z = \frac{e^{iz}-e^{-iz}}{2i}$, the other trigonometric functions are defined by

$$(20) \qquad \tan z = \frac{\sin z}{\cos z} \quad (\cos z \neq 0),$$

$$(21) \qquad \cot z = \frac{\cos z}{\sin z} \quad (\sin z \neq 0),$$

$$(22) \qquad \sec z = \frac{1}{\cos z} \quad (\cos z \neq 0),$$

$$(23) \qquad \csc z = \frac{1}{\sin z} \quad (\sin z \neq 0).$$

Like the complex cosine and sine functions, these functions share several properties with their real counterparts. The following is one illustration.

EXAMPLE 5 $\tan z$ is π-periodic

Show that $\tan z_1 = \tan z_2$ if and only if $z_1 = z_2 + k\pi$, where k is an integer.

Solution   Note that $\tan z$ is not defined for $z = \frac{\pi}{2} + k\pi$. For $z_1, z_2 \neq \frac{\pi}{2} + k\pi$, we have

$$\tan z_1 = \tan z_2 \quad \Leftrightarrow \quad \frac{\sin z_1}{\cos z_1} = \frac{\sin z_2}{\cos z_2}$$
$$\Leftrightarrow \quad \sin z_1 \cos z_2 - \cos z_1 \sin z_2 = 0$$
$$\Leftrightarrow \quad \sin(z_1 - z_2) = 0 \qquad \text{(use (10) with } z_2 \text{ replaced by } (-z_2))$$
$$\Leftrightarrow \quad z_1 - z_2 = k\pi \quad \Leftrightarrow \quad z_1 = z_2 + k\pi,$$

where the step before last follows from Example 3(a). ∎

Hyperbolic Functions

We define the hyperbolic functions for complex variables exactly as we do for real variables:

(24)
$$\cosh z = \frac{e^z + e^{-z}}{2} \quad \text{and} \quad \sinh z = \frac{e^z - e^{-z}}{2}.$$

The hyperbolic functions satisfy interesting identities. Some of them are extensions of familiar identities for the real hyperbolic functions, and some are new. We illustrate with a brief list.

PROPERTIES OF HYPERBOLIC FUNCTIONS

For any complex number $z = x + i\,y$, we have

(25) $$\cosh iz = \cos z;$$

(26) $$\sinh iz = i \sin z;$$

(27) $$\cosh^2 z - \sinh^2 z = 1;$$

(28) $$\cosh z = \cosh x \cos y + i \sinh x \sin y;$$

(29) $$\sinh z = \sinh x \cos y + i \cosh x \sin y.$$

These and many more identities (Exercises 35–52) can be proved from the definitions (24). Finally, we define the other hyperbolic functions in terms of $\cosh z$ and $\sinh z$.

OTHER HYPERBOLIC FUNCTIONS

With $\cosh z = \frac{e^z + e^{-z}}{2}$ and $\sinh z = \frac{e^z - e^{-z}}{2}$, the other hyperbolic functions are defined by

(30) $$\tanh z = \frac{\sinh z}{\cosh z} \quad (\cosh z \neq 0),$$

(31) $$\operatorname{sech} z = \frac{1}{\cosh z} \quad (\cosh z \neq 0),$$

(32) $$\operatorname{csch} z = \frac{1}{\sinh z} \quad (\sinh z \neq 0),$$

(33) $$\coth z = \frac{\cosh z}{\sinh z} \quad (\sinh z \neq 0).$$

Exercises 1.6

In Exercises 1–4, (a) evaluate $\cos z$ and $\sin z$ for the given z, using the definitions (2) and (3). (b) Verify that your answers satisfy (15) and (16).

1. i.   **2.** $-2i$.   **3.** $\dfrac{\pi}{2} + 2i$.   **4.** $\pi - i$.

In Exercises 5–8, for the given z, (a) evaluate $\cos z$, $\sin z$, and $\tan z$, using (15) and (16). (b) Compute $|\cos z|$ and $|\sin z|$. (c) Plot the points $\cos z$, $\sin z$, and $\tan z$.

5. $1 + i$.   **6.** $1 - i$.   **7.** $\dfrac{3\pi}{2} + i$.   **8.** $\dfrac{\pi}{6} - i$.

In Exercises 9–14, express the given function $f(z)$ in the form $f(z) = u(x, y) + iv(x, y)$, where u and v are the real and imaginary parts of $f(z)$.

9. $\sin(2z)$. **10.** $\cos(z^2)$. **11.** $\sin(z) + 2z$.

12. $z \cos z$. **13.** $\tan z$. **14.** $\sec z$.

In Exercises 15–20, show that the shaded area S in the z-plane is mapped to the shaded area in the w-plane by the given mapping $f(z)$.

15.

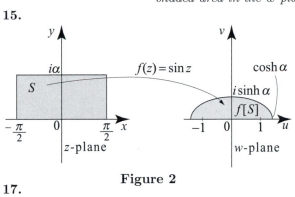

Figure 2

16.

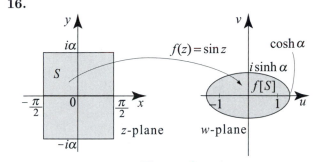

Figure 3

17.

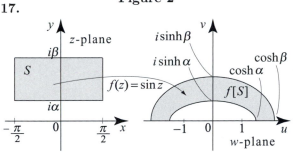

Figure 4

18.

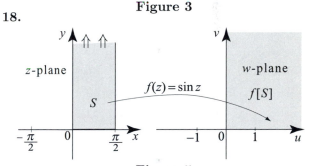

Figure 5

19.

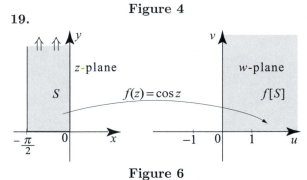

Figure 6

20.

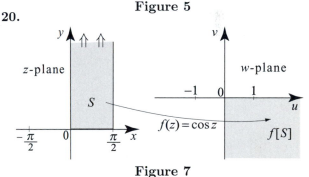

Figure 7

21. Establish (16) and (18).

22. Explain why $\cos z$ and $\sin z$ are not bounded in absolute value.

23. **More on Example 4.** In this exercise, we study further the mapping $f(z) = \sin z$ of Example 4.
(a) Show that the half-line $x = \frac{\pi}{2}$, $y \geq 0$, is mapped to the half-line $u \geq 1$, $v = 0$.
(b) Show that the half-line $x = \frac{-\pi}{2}$ is mapped to the half-line $u \leq -1$, $v = 0$.
(c) Conclude that the boundary of the set S in Example 4 is mapped to the boundary of the set $f[S]$.

(d)   Recall from your calculus course that an ellipse of the form $\frac{x^2}{a^2} + \frac{y^2}{b^2} = 1$ with $0 < b < a$ has its foci at $x = \pm\sqrt{a^2 - b^2}$. Show that all the ellipses in Example 4 have the same foci located on the u-axis at $u = \pm 1$.

24. Zeros of hyperbolic functions.   Show that

$$\sinh z = 0 \quad \Leftrightarrow \quad z = ik\pi, \ k \text{ an integer;}$$

and

$$\cosh z = 0 \quad \Leftrightarrow \quad z = i(k + \frac{1}{2})\pi, \ k \text{ an integer.}$$

[Hint: z is a zero of $\sin z \Leftrightarrow iz$ is a zero of the hyperbolic sine (why?). Reason the same way for the cosine.]

In Exercises 25–52, establish the given identity. In establishing an identity with hyperbolic functions, you may want to use the corresponding one for trigonometric functions and the identities (25) and (26).

25. $\sin(-z) = -\sin z.$ **26.** $\sin(z + 2\pi) = \sin z.$

27. $\cos(z + \pi) = -\cos z.$ **28.** $\sin(z + \pi) = -\sin z.$

29. $\sin(z_1 + z_2) = \sin z_1 \cos z_2 + \cos z_1 \sin z_2.$

30. $\cos 2z = \cos^2 z - \sin^2 z = 2\cos^2 z - 1 = 1 - 2\sin^2 z.$

31. $\sin 2z = 2 \sin z \cos z.$

32. $2 \cos z_1 \cos z_2 = \cos(z_1 - z_2) + \cos(z_1 + z_2).$

33. $2 \sin z_1 \sin z_2 = \cos(z_1 - z_2) - \cos(z_1 + z_2).$

34. $2 \sin z_1 \cos z_2 = \sin(z_1 + z_2) + \sin(z_1 - z_2).$

35. $\cosh(-z) = \cosh z$ and $\sinh(-z) = -\sinh z.$

36. $\cosh(z + 2\pi i) = \cosh z$ and $\sinh(z + 2\pi i) = \sinh z.$

37. $\cosh(z + \pi i) = -\cosh z$ and $\sinh(z + \pi i) = -\sinh z.$

38. $\sinh(z + \frac{i\pi}{2}) = i \cosh z$ and $\cosh(z + \frac{i\pi}{2}) = i \sinh z.$

39. $e^z = \cosh z + \sinh z.$ **40.** $\cosh^2 z - \sinh^2 z = 1.$

41. $\cosh 2z = \cosh^2 z + \sinh^2 z = 2\cosh^2 z - 1 = 1 + 2\sinh^2 z.$

42. $\sinh 2z = 2 \sinh z \cosh z.$

43. $\cosh^2 z = \dfrac{1 + \cosh 2z}{2}.$ **44.** $\sinh^2 z = \dfrac{-1 + \cosh 2z}{2}.$

45. $\cosh(z_1 + z_2) = \cosh z_1 \cosh z_2 + \sinh z_1 \sinh z_2.$

46. $\sinh(z_1 + z_2) = \sinh z_1 \cosh z_2 + \cosh z_1 \sinh z_2.$

47. $2 \cosh z_1 \cosh z_2 = \cosh(z_1 + z_2) + \cosh(z_1 - z_2).$

48. $2 \sinh z_1 \sinh z_2 = \cosh(z_1 + z_2) - \cosh(z_1 - z_2).$

49. $2 \sinh z_1 \cosh z_2 = \sinh(z_1 + z_2) + \sinh(z_1 - z_2).$

50. $\cosh z = \cosh x \cos y + i \sinh x \sin y$, and $\sinh z = \sinh x \cos y + i \cosh x \sin y.$

51. $|\cosh z| = \sqrt{\sinh^2 x + \cos^2 y}.$ **52.** $|\sinh z| = \sqrt{\sinh^2 x + \sin^2 y}.$

53.   Show that either $\tan z = i$ or $\tan z = -i$ has no solution. [Hint: Use (15) and (16).]

1.7   Logarithms and Powers

In this section we define the complex logarithms and define what it means to raise a complex number to a complex power. Thus we will be able to compute expressions like $\operatorname{Log} i$ and i^i.

The logarithm was defined in elementary algebra as the inverse of the exponential function. We will follow this idea in defining the complex logarithm, $\log z$ for $z \neq 0$. Thus,

$$(1) \qquad\qquad w = \log z \quad \Leftrightarrow \quad e^w = z.$$

To determine w in terms of z, write $w = u + i\,v$ and $z = r\,e^{i\theta}$, with $|z| = r > 0$ and $\theta = \arg z$. Then (1) becomes

$$e^u e^{iv} = r e^{i\theta},$$

and hence

$$(2) \qquad\qquad e^u = r \quad \text{and} \quad e^{iv} = e^{i\theta}.$$

The first equation gives $u = \ln r$, where here $\ln r$ denotes the usual natural logarithm of the positive number r. The second equation in (2) tells us that v and θ differ by an integer multiple of 2π, because the complex exponential is $2\pi i$ periodic. So $v = \theta + 2k\pi$, where k is an integer, or simply $v = \arg z$. Putting this together, we obtain the formula for the **complex logarithm**:

$$(3) \qquad\qquad \boxed{\log z = \ln |z| + i \arg z \qquad (z \neq 0).}$$

Unlike the real logarithm, this formula defines a *multiple-valued* function, because $\arg z$ is multiple-valued. The complex logarithm is not a function in our usual sense of the word, since in our understanding a function is a rule that assigns to a given complex number z a single number. We can make the function defined by (3) single-valued by specifying a single-valued $\arg z$. For example, we can use the principal value of the argument by specifying that $-\pi < \arg z \leq \pi$ (see (9), Section 1.3). The corresponding function that we obtain in (3) is called the **principal value** or **principal branch** of the logarithm and is denoted by $\operatorname{Log} z$. Thus

$$(4) \qquad\qquad \boxed{\operatorname{Log} z = \ln |z| + i \operatorname{Arg} z \qquad (z \neq 0).}$$

Recalling that $\arg z = \operatorname{Arg} z + 2k\pi$, where k is an integer, we see from (3) and (4) that all the values of $\log z$ differ from the principal value by $2k\pi i$. Thus

$$(5) \qquad\qquad \boxed{\log z = \operatorname{Log} z + 2k\pi\, i \qquad (z \neq 0).}$$

EXAMPLE 1   Evaluating logarithms

Find the following logarithms.

(a) $\log i$.   (b) $\operatorname{Log} i$.   (c) $\operatorname{Log}(1+i)$.   (d) $\operatorname{Log}(-2)$.

Solution   (a) The polar form of i is $i = e^{i\frac{\pi}{2}}$. So $|i| = 1$ and $\arg i = \frac{\pi}{2} + 2k\pi$, where k is an integer. Hence, by (3),

$$\log i = \overbrace{\ln(1)}^{=0} + i\left(\frac{\pi}{2} + 2k\pi\right) = i\left(\frac{\pi}{2} + 2k\pi\right).$$

As expected, $\log i$ takes on an infinite number of values, all of which happen to be purely imaginary.

(b)   The principal argument of i is $\operatorname{Arg} i = \frac{\pi}{2}$. Thus, according to (4),

$$\operatorname{Log} i = \ln(1) + i\frac{\pi}{2} = i\frac{\pi}{2}.$$

Note that $\operatorname{Log} i$ is single-valued and equal to one of the values of $\log i$.

(c)   We will apply (4) after putting $1 + i$ in polar form. As you can easily check, $1 + i = \sqrt{2}e^{i\frac{\pi}{4}}$, where $\frac{\pi}{4}$ is the principal argument of $1 + i$. Applying (4), we get

$$\operatorname{Log}(1+i) = \ln(\sqrt{2}) + i\frac{\pi}{4} = \frac{1}{2}\ln 2 + i\frac{\pi}{4}.$$

(d)   You should have no problem evaluating $\operatorname{Log}(-2)$ by applying (4). Appreciate, however, that the complex logarithm, unlike its real counterpart, is defined for negative numbers. We have

$$-2 = 2e^{i\pi} \quad \Rightarrow \quad \operatorname{Log}(-2) = \ln(2) + i\pi. \qquad \blacksquare$$

As you can imagine, we could have specified a different range of values of $\arg z$ in defining a logarithmic function out of (3). In fact, given any real number α, we can specify that $\alpha < \arg z \leq \alpha + 2\pi$. This will assign a single value to $\arg z$, denoted by $\arg_\alpha z$. For $z \neq 0$, using $\arg_\alpha z$ in (3), we obtain the corresponding function, called a **branch** of the logarithm,

(6) $$\boxed{\log_\alpha z = \ln|z| + i\arg_\alpha z, \qquad \text{where } \alpha < \arg_\alpha z \leq \alpha + 2\pi.}$$

For example, for $\alpha = \frac{3\pi}{4}$, we have $\frac{3\pi}{4} < \arg_{\frac{3\pi}{4}} z \leq \frac{3\pi}{4} + 2\pi$. If we want to compute $\log_{\frac{3\pi}{4}}(i)$, we first compute

$$\arg_{\frac{3\pi}{4}}(i) = \frac{5\pi}{2}.$$

Then from (6)

$$\log_{\frac{3\pi}{4}}(i) = \ln 1 + i\frac{5\pi}{2} = i\frac{5\pi}{2}.$$

To discuss the mapping properties of the logarithm, you should check that for any real α, the exponential function maps the **period strip** or **fundamental region**

(7) $$S_\alpha = \{z = x + iy : \ \alpha < y \le \alpha + 2\pi\}$$

one-to-one onto the complex plane minus the origin, which we write as $\mathbb{C} \setminus \{0\}$. Thus the branch of the logarithm $\log_\alpha z$ maps the punctured plane $\mathbb{C} \setminus \{0\}$ back onto the fundamental region S_α (see Figure 1).

Figure 1 For α real, the fundamental region S_α of e^z is the infinite horizontal strip

$$S_\alpha = \{z : \ \alpha < \operatorname{Im} z \le \alpha + 2\pi\}.$$

The upper boundary line, $\operatorname{Im} z = \alpha + 2\pi$, is included in S_α, but the lower boundary line, $\operatorname{Im} z = \alpha$, is not. If we were to include the line $\operatorname{Im} z = \alpha$, then e^z would cease to be one-to-one in the region.

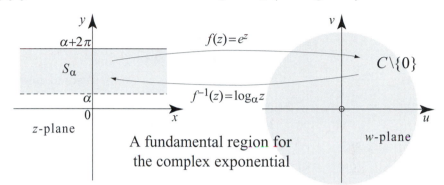

A fundamental region for the complex exponential

Complex Powers

In direct analogy with calculus, for a complex number $z \ne 0$, we define the **complex power**

(8) $$\boxed{z^a = e^{a \log z} \qquad (z \ne 0),}$$

where $\log z$ is the complex logarithm (3). Since $\log z$ is multiple-valued, it follows from (8) that z^a is in general multiple-valued. By specifying a branch of the logarithm, we obtain a single-valued branch of the complex power function from (8). In particular, if we choose the principal logarithm (4), we obtain the **principal value** of z^a:

(9) $$\boxed{z^a = e^{a \operatorname{Log} z} \qquad (z \ne 0),}$$

EXAMPLE 2 Evaluating complex powers

Compute $(-i)^{1+i}$ using (a) the principal branch of the logarithm; (b) the branch of the logarithm with a branch cut at angle $\alpha = 0$.

Solution (a) Using (9), we find

$$(-i)^{1+i} = e^{(1+i) \operatorname{Log}(-i)} = e^{(1+i)(-\frac{i\pi}{2})} = e^{\frac{\pi}{2}} e^{-\frac{i\pi}{2}} = -ie^{\frac{\pi}{2}}.$$

(b) Using the logarithm with a branch cut at angle 0 in (8), we have

$$(-i)^{1+i} = e^{(1+i) \log_0(-i)} = e^{(1+i) \frac{3i\pi}{2}} = e^{-\frac{3\pi}{2}} e^{3\frac{i\pi}{2}} = -ie^{-\frac{3\pi}{2}},$$

which is a different value from the one we found in (a).   ∎

For $z \neq 0$, is the function z^a, defined by (8), always multiple-valued? To answer this question, let us use the formula for $\log z$ from (5) and write

$$z^a = e^{a \log z} = e^{a(\operatorname{Log} z + 2k\pi i)} = e^{a \operatorname{Log} z} e^{2ka\pi i},$$

where k is an integer. To determine the number of distinct values of z^a, we must determine the number of distinct values taken by $e^{2ka\pi i}$ as k varies over the integers. We distinguish three cases.

Case (i): a is a (real) integer. Then $2ka\pi i$ is an integer multiple of $2\pi i$, and hence $e^{2ka\pi i} = 1$ for all integers k. The expression z^a has only one value. This result is in concordance with our notion of z^n, z^{-1}, etc., as being single-valued functions.

Case (ii): a is a (real) rational number. Write $a = \frac{p}{q}$, where p and q are integers and have no common factors. The quantity $e^{2ka\pi i} = e^{2\pi i \frac{pk}{q}}$ will have q distinct values, for $k = 0, 1, \ldots, q - 1$ (see Exercise 38). Thus, for each value of $k = 0, 1, \ldots, q - 1$, we obtain a distinct power function

$$(10) \qquad z^{\frac{p}{q}} = e^{\frac{p}{q} \operatorname{Log} z} e^{2\pi i \frac{pk}{q}}$$

called a **branch** of $z^{\frac{p}{q}}$. The branch for $k = 0$ is called the **principal branch** of $z^{\frac{p}{q}}$. This result is in concordance with our notion of nth roots $z^{1/n}$; there are n of them. Note also that case (i) is just case (ii) with $q = 1$.

Case (iii): a is a complex number not of either of the preceding two types. This is the case when a is an irrational real number or a complex number with a nonzero imaginary part. Then the quantities $e^{2ka\pi i}$ are distinct for all integers k (Exercise 39), and z^a has an infinite number of values. As in case (ii), each value of k determines a **branch** of z^a, except that here we have infinitely many distinct branches.

Note that our definition of a complex power is inconsistent with our definition of the complex exponential e^z. According to (8), we can take e and raise it to the power a, resulting in

$$e^a = e^{a \log e} = e^{a \ln e} e^{a 2k\pi i},$$

which is, in general, multiple-valued. We must distinguish this concept of "raising e to the power a" from our previous, single-valued definition of the "exponential function." As a convention, e^z will always refer to the exponential function, unless otherwise stated.

The last example of this section deals with inverse trigonometric functions. These will be computed using complex powers and logarithms, and so they will be multiple-valued in general. For example, the inverse sine of z is any complex number $w = \sin^{-1} z$ such that $\sin w = z$. As you will see, the solutions of the last equation form an infinite set, and so $\sin^{-1} z$ is infinite-valued.

EXAMPLE 3   Computing the inverse sine function

Show that

(11) $$\sin^{-1} z = -i \log \left(iz + (1 - z^2)^{\frac{1}{2}} \right),$$

where the complex logarithm is given by (3). (For a given z, the square root takes two values in general, and the complex logarithm is infinite-valued. As a result the inverse sine is infinite-valued.)

Solution   We want to solve $\sin w = z$. Recalling the definition of $\sin w$ from (3), Section 1.6, we have

$$z = \frac{e^{iw} - e^{-iw}}{2i},$$

or, after multiplying both sides by e^{iw} and simplifying,

$$\left(e^{iw} \right)^2 - 2ize^{iw} - 1 = 0.$$

This is a quadratic equation in e^{iw}, which we can solve by appealing to the quadratic formula (25), Section 1.3:

$$e^{iw} = iz + (1 - z^2)^{\frac{1}{2}},$$

where the square root has two values in general. The desired identity (11) follows now upon taking logarithms.   ∎

Formulas for the inverses of the other trigonometric functions and hyperbolic functions can be derived with varying degrees of difficulty by using the logarithm. See the exercises for an extensive list of these functions.

Exercises 1.7

In Exercises 1–4, evaluate $\log z$ *for the given* z.
1. $2i$. **2.** $-3 - 3i$. **3.** $5e^{i\frac{\pi}{7}}$. **4.** $e^{2+i\frac{2\pi}{11}}$.

In Exercises 5–8, evaluate $\text{Log } z$ *for the given* z.
5. $3 + i\sqrt{3}$. **6.** $e^{3-i\frac{3\pi}{2}}$. **7.** $e^{i\frac{5\pi}{7}}e^{i\frac{3\pi}{7}}$. **8.** $-\alpha$, where $\alpha > 0$.

In Exercises 9–12, evaluate the given logarithm.
9. $\log_\pi 1$. **10.** $\log_{\sqrt{3}}(1 + i)$. **11.** $\log_5(-i\,e)$. **12.** $\log_{\frac{\pi}{2}} i$.

In Exercises 13–18, solve the given equation.
13. $e^z = 3$. **14.** $e^{-z} = 1 + i$. **15.** $e^{z+3} = i$.
16. $e^{2z} + 3e^z + 2 = 0$. **17.** $e^{2z} + 5 = 0$. **18.** $e^z = \frac{1+i}{1-i}$.

19. (a) Compute $\text{Log}(e^{i\pi})$, $\text{Log}(e^{3i\pi})$, and $\text{Log}(e^{5i\pi})$.
(b) Show that $\text{Log}(e^z) = z$ if and only if $-\pi < \text{Im } z \leq \pi$.

20. Show that $\log z = i \arg z$ if and only if z is unimodular.

21. Compute $\text{Log}(-1)$, $\text{Log } i$, and $\text{Log}(-i)$ and conclude that, unlike the identity for the usual real logarithm, in general, $\text{Log}(z_1 z_2)$ is not equal to $\text{Log } z_1 + \text{Log } z_2$. The situation is not all that hopeless. See Project Problem 37.

22. For which α is $\log_\alpha 1 = 0$?

In Exercises 23–26, evaluate the principal value of the given power.

23. 5^i.   **24.** $(1+i)^{3+i}$.   **25.** $(-5)^{1-i}$.   **26.** $\left(\frac{1+i}{1-i}\right)^i$.

In Exercises 27–30, state how many values the given power takes, and find them.

27. $(3i)^4$.   **28.** $(1+i\sqrt{3})^{\frac{2}{7}}$.   **29.** i^i.   **30.** $(-e)^{\frac{i}{2}}$.

31. Find all solutions of $\cos z = \sin z$.

32. (a) Use the definition (2) from Section 1.6 to find the formula

(12) $$\cos^{-1} z = -i\log\left(z + (z^2 - 1)^{\frac{1}{2}}\right).$$

(b) Derive an expression for $\cos^{-1} z$ more quickly using (6) from Section 1.6 and (11) from this section.

33. (a) Verify from the definition (20) in Section 1.6 that, for suitable w,

$$1 + i\tan w = \frac{2e^{iw}}{e^{iw} + e^{-iw}}.$$

(b) Verify similarly that

$$1 - i\tan w = \frac{2e^{-iw}}{e^{iw} + e^{-iw}}.$$

(c) Divide the quantities in parts (a) and (b) and conclude the formula

(13) $$\tan^{-1} z = \frac{i}{2}\log\frac{1 - iz}{1 + iz} \quad (z \neq \pm i).$$

34. Verify that

(14) $$\sinh^{-1} z = \log\left(z + (z^2 + 1)^{\frac{1}{2}}\right).$$

35. Verify that

(15) $$\cosh^{-1} z = \log\left(z + (z^2 - 1)^{\frac{1}{2}}\right).$$

36. Verify that

(16) $$\tanh^{-1} z = \frac{1}{2}\log\frac{1 + z}{1 - z} \quad (z \neq \pm 1).$$

[Hint: Consider $1 + \tanh w$ and $1 - \tanh w$.]

37. Project Problem: Set equations and log operations. In this problem we define what we mean by an equation like $\log(z_1 z_2) = \log z_1 + \log z_2$ and see when equations like this are true. For $z \neq 0$, recall that $\log z$ is multiple-valued. Let us denote the set of all values of $\log z$ using set notation, and so by (5)

(17) $$\log z = \{w : w = \operatorname{Log} z + 2k\pi i \text{ for some integer } k\}.$$

Our first example is fully worked out, to help you get the idea. We define the **sum** of two sets S_1 and S_2 to be the set

(18) $$S_1 + S_2 = \{z : z = \zeta_1 + \zeta_2 \text{ for some } \zeta_1 \in S_1,\ \zeta_2 \in S_2\}.$$

With this definition, we will show that

(19) $$\log(z_1 z_2) = \log z_1 + \log z_2.$$

Each side of the equation is a set; we must show that they have precisely the same elements. The left side is the set of all $\ln|z_1 z_2| + i \arg(z_1 z_2)$. From calculus we know that $\ln|z_1 z_2| = \ln|z_1| + \ln|z_2|$, and since $\arg(z_1 z_2) = \operatorname{Arg}(z_1 z_2) + 2k\pi$, hence

$$\log(z_1 z_2) = \{z : z = \ln|z_1| + \ln|z_2| + i \operatorname{Arg}(z_1 z_2) + 2k\pi i \text{ for some integer } k\}.$$

Using (18) to figure out the right side, we get

$$\log z_1 + \log z_2 = \{z : z = \ln|z_1| + \ln|z_2|$$
$$+ i \operatorname{Arg} z_1 + i \operatorname{Arg} z_2 + 2k\pi i + 2j\pi i \text{ for some integers } j, \ k\}.$$

Since $\operatorname{Arg}(z_1 z_2)$ differs from $\operatorname{Arg} z_1 + \operatorname{Arg} z_2$ by an integer multiple of 2π, we see that the sets $\log z_1 + \log z_2$ and $\log(z_1 z_2)$ are the same, and so (19) holds. Now you can apply this technique to the following problems.

(a) The **negative** of a set S is the set

(20) $$-S = \{z : -z \in S\}.$$

With this definition, show that

(21) $$\log(z^{-1}) = -\log z.$$

(b) The **difference** of two sets S_1 and S_2 is the set

(22) $$S_1 - S_2 = \{z : \ z = \zeta_1 - \zeta_2 \text{ for some } \zeta_1 \in S_1, \ \zeta_2 \in S_2\}.$$

With this definition, show that

(23) $$\log\left(\frac{z_1}{z_2}\right) = \log z_1 - \log z_2.$$

(c)   The **scalar product** of a set S with the complex number a is the set

(24) $$aS = \{z : z = a\zeta \text{ for some } \zeta \in S\}.$$

We know from (19) that $\log(z^2) = \log z + \log z$. Show that it is *not* true that $\log(z^2) = 2\log z$. If we use $S_1 \supset S_2$ to mean that S_2 is a subset of S_1, show that

(25) $$\log(z^2) \supset 2\log z.$$

38.   Project Problem: Rational powers.   In this problem we prove that for rational $a = \frac{p}{q}$, p and q having no common factors, the expression z^a has exactly q distinct values.

(a)   Show that all values for z^a are of the form $e^{(p/q)\operatorname{Log} z} e^{2kp\pi i/q}$.

(b)   Define $E_n = e^{2np\pi i/q}$, for all integers n. Argue that $E_n = E_{n+q}$ for all n and hence there can be at most q distinct values for z^a. Without loss of generality we need only consider $0 \le n \le q - 1$.

(c) Suppose that $E_j = E_l$ for some $0 \leq j < l \leq q-1$. Use (14) from Section 1.5 to conclude that

$$(26) \qquad \frac{p(l-j)}{q} = k$$

for some integer k.

(d) Argue that (26) is impossible; for $\frac{p(l-j)}{q}$ to be an integer, all the prime factors of q must be canceled by terms in the numerator. However, p has none of them, and $l - j$ cannot have all of them since $l - j < q$. Conclude that all E_n are distinct for $0 \leq n \leq q-1$, and hence $z^{\frac{p}{q}}$ has q distinct values.

39.   Project Problem: Nonreal and irrational powers.   In this problem we prove that if $\operatorname{Im} a \neq 0$ or $\operatorname{Re} a$ is irrational, z^a has an infinite number of values.
(a)   Write $a = \alpha + i\beta$ and show that all values for z^a are of the form

$$e^{a \operatorname{Log} z} e^{-\beta 2k\pi} e^{\alpha 2k\pi i}.$$

(b)   Define $E_n = e^{\beta 2n\pi} e^{\alpha 2n\pi i}$ for all integers n. Argue that if $\beta \neq 0$, then $|E_n|$ are all distinct, and hence all values of E_n are distinct. Thus z^a has an infinite number of values.

(c)   Otherwise, if $\beta = 0$ and α is irrational, we have $E_n = e^{\alpha 2n\pi i}$. Suppose $E_j = E_l$ for some $j < l$. Use (14), Section 1.5, to conclude that $\alpha(l-j) = k$ for some integer k. However, this is impossible because α is irrational. Hence all E_n are distinct and z^a has an infinite number of values.

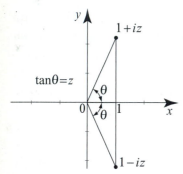

Figure 2 for Exercise 40.

40. Show that the formula for the inverse tangent (13) holds for real z, by using the geometric interpretation of the tangent function and Figure 2.

2

ANALYTIC FUNCTIONS

*I get up at 4 o'clock each morning and I am busy from then on ...
Today I drew the plans for forges that I am to have built in granite. I
am also constructing two lighthouses, one on each of the two piers that
are located at the entrance of the harbor. I do not get tired of working;
on the contrary, it invigorates me and I am in perfect health ...*
                                                    -Augustin Louis Cauchy

In the previous chapter, we introduced complex numbers and complex functions. In this chapter, we begin our analysis of complex functions, which will occupy us through Chapter 6. Most of what we will present is due to Cauchy and was prepared for a course that he taught at the Institut de France in 1814 and later at the Ecole Polytechnique. Cauchy single-handedly defined the derivative and integral of complex functions and developed one of the most fruitful theories of mathematics. As Cauchy developed his theory, he defined for the first time the notion of limit for functions and gave rigorous definitions of continuity and differentiability for real-valued functions, as we now know them in calculus. He also developed on solid foundation the theory of definite integrals and series. With Cauchy began a new age in mathematics, an age of rigor and insistence on mathematical proof.

So who was Cauchy? Augustin Louis Cauchy was born on August 21, 1789 in Paris. He received his early education from his father, Louis-François Cauchy, a master of classical studies. Cauchy entered the Ecole Polytechnique in Paris in 1805 and continued his education as a civil engineer at the Ecole des Ponts et Chaussées. He began his career as a military engineer, working in Napoleon's administration from 1810 to 1813. His mathematical talents were soon discovered by leading mathematicians, among them was Joseph Louis Lagrange, who persuaded Cauchy to leave his career as an engineer and devote himself to mathematics. Cauchy's mathematical output was phenomenal. He is considered to be one of the greatest mathematicians. His contributions cover many areas of pure and applied mathematics, including the theory of heat, the theory of light, the mathematical theory of elasticity, and fluid dynamics. Cauchy's contributions to modern calculus are so fundamental that he "has come to be regarded as the creator of calculus in the modern sense," from *The History of Mathematics, An Introduction*, 3rd edition, by David M. Burton (McGraw-Hill, 1997).

2.1   Regions of the Complex Plane

In the previous chapter, we defined some elementary functions of a complex variable. An important part of a function is its domain of definition. In calculus, functions were usually defined over intervals. In dealing with functions of a complex variable, intervals will be replaced by subsets of the complex plane. The picture is no longer as simple as the one on the real line. For this reason, it is necessary to understand basic properties of subsets of the complex plane, which will assist us in our analytical study of functions of a complex variable.

Open Sets

One very useful definition in this book is that of a neighborhood of a point. It is the analog of an open interval in one dimension.

DEFINITION 1
NEIGHBORHOOD

Let $r > 0$ be a positive real number and z_0 a point in the plane. The r-**neighborhood** of z_0 is the set of all complex numbers z satisfying

$$(1) \qquad\qquad |z - z_0| < r.$$

We denote this set by $B_r(z_0)$.

By interpreting the absolute value as a distance, we see from (1) that $B_r(z_0)$ is the disk centered at z_0 with radius $r > 0$. The fact that the inequality in (1) is strict tells us that we should not include the points on the circle $|z - z_0| = r$ as part of the r-neighborhood of z_0. This is indicated in Figure 1, where we used a dashed line to plot the circle $|z - z_0| = r$.

Sometimes we will not specify the value of r and will simply refer to $B_r(z_0)$ as a neighborhood of z_0.

A neighborhood of z_0 from which we have deleted the center z_0 is called a **deleted neighborhood** or **punctured neighborhood** of z_0 and is sometimes denoted by $B_r'(z_0)$. Thus

$$B_r'(z_0) = \{z : \ 0 < |z - z_0| < r\}.$$

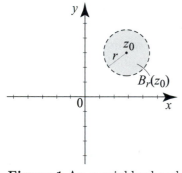

Figure 1 An r-neighborhood $B_r(z_0)$ does not include the points on the circle $|z - z_0| = r$.

DEFINITION 2
INTERIOR AND
BOUNDARY POINTS

Let S be a subset of the complex numbers. A point z_0 in S is called an **interior** point of S if we can find a neighborhood of z_0 that is wholly contained in S. A point z in the complex plane is called a **boundary** point of S if every neighborhood of z contains at least one point in S and at least one point not in S. The set of all boundary points of S is called the **boundary** of S.

From the definitions, every point in S is either an interior point or a boundary point. If a point is an interior point of S, then it cannot be a boundary point of S. Also, while an interior point of S is necessarily a point in S, a boundary point of S need not be in S.

The geometric concepts in Definition 2 are intuitively clear; however, dealing with them often requires delicate handling of the absolute value.

EXAMPLE 1   Interior and boundary points

Consider an r-neighborhood $B_r(z_0)$, where $r > 0$.

(a)   Show that every point z of $B_r(z_0)$ is an interior point.

(b)   Show that the boundary of $B_r(z_0)$ is the circle $|z - z_0| = r$.

Solution   (a)   Pick a point z in $B_r(z_0)$. It is clear from Figure 2 that we can find a disk centered at z, which lies wholly in $B_r(z_0)$. Hence z is an interior point of $B_r(z_0)$.

If we want to give an analytic proof of this simple geometric argument, here is one way to proceed. Let $\delta = |z - z_0|$. By definition of $B_r(z_0)$, we have $0 \leq \delta < r$. Let $\delta' = r - \delta$. For any w in the neighborhood $B_{\delta'}(z)$, we have $|w - z| < \delta'$ and so, by the triangle inequality,

$$|w - z_0| \leq |w - z| + |z - z_0| < \delta' + \delta = r.$$

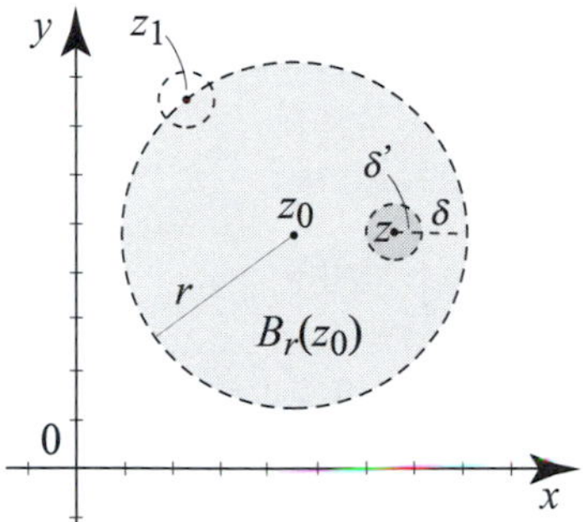

Hence w belongs to $B_r(z_0)$, and since this is true of every w in $B_{\delta'}(z)$, we conclude that $B_{\delta'}(z)$ is contained in $B_r(z_0)$

(b)   Pick a point z_1 on the circle $|z - z_0| = r$. It is clear from Figure 2 that every disk centered at z_1 will contain (infinitely many) points in $B_r(z_0)$ and (infinitely many) points not in $B_r(z_0)$. Hence every point on the circle $|z - z_0|$ is a boundary point of $B_r(z_0)$.

We now show that no other points are boundary points. Since points inside the circle are interior points, they cannot be boundary points. Also, given a point outside the circle we can enclose it in a disk that does not intersect the disk $B_r(z_0)$. Hence such a point is not a boundary point.

Note that in this example, none of the boundary points of $B_r(z_0)$ belong to $B_r(z_0)$. ∎

We come now to an important definition.

Figure 2 The point z is an interior point of $B_r(z_0)$, while z_1 is a boundary point.

DEFINITION 3
OPEN SETS

A subset S of the complex numbers is called **open** if every point in S is an interior point of S.

Thus S is open if around each point z in S you can find a neighborhood $B_r(z)$ that is wholly contained in S. The radius of $B_r(z)$ depends on z. Here are some useful examples to keep in mind.

- The empty set, denoted as usual by $\emptyset$, is open. Because there are no points in $\emptyset$, the definition of open sets is vacuously satisfied.

- The set of all complex numbers $\mathbb{C}$ is open.

The letter B is used for an open disk because its higher-dimensional analog is an "open ball."

- Any r-neighborhood, $B_r(z_0)$, is open. We just verified in Example 1(a) that every point in $B_r(z_0)$ is an interior point.

- The set of all z such that $|z - z_0| > r$ is open. This set is called a **neighborhood of** ∞. To justify this terminology, see Exercise 23.

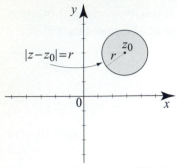

Figure 3 A closed disk includes its boundary, the circle $|z - z_0| = r$.

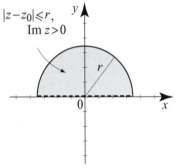

Figure 4 A set that is neither open nor closed.

An r-neighborhood, $B_r(z_0)$, is more commonly called an **open disk** of radius r, centered at z_0.

You can show that a set is open if and only if it contains none of its boundary points (Exercise 18). Sets that contain all of their boundary points are called **closed**. The complex plane $\mathbb{C}$ and the empty set $\emptyset$ are closed since they trivially contain their empty sets of boundary points. The disk

$$\{z : \; |z - z_0| \leq r\}$$

is closed because it contains all its boundary points consisting of the circle $|z - z_0| = r$ (Figure 3). We will refer to such a disk as the **closed disk** of radius r, centered at z_0.

Some sets are neither open nor closed. For example, the set

$$\{z : \; |z - z_0| \leq r; \; \operatorname{Im} z > 0\}$$

contains the boundary points on the upper semicircle, but it does not contain the points on the x-axis. Hence, this set is neither open nor closed (Figure 4).

For our next topic and for use throughout this book, it will be convenient to introduce some set notation. If a point z is in a set S, we say that z is an **element** of S and write $z \in S$. If z does not belong to S, we will write $z \notin S$. Let A and B be two sets of complex numbers. The **union** of A and B, denoted $A \cup B$, is the set

$$A \cup B = \{z : \; z \in A \text{ or } z \in B\}.$$

The **intersection** of A and B, denoted $A \cap B$, is the set

$$A \cap B = \{z : \; z \in A \text{ and } z \in B\}.$$

Two sets A and B are **disjoint** if $A \cap B = \emptyset$. The **set difference** between A and B is the set

$$A \setminus B = \{z : \; z \in A \text{ and } z \notin B\}.$$

We will say that A is a **subset** of B or that B contains A and write $A \subset B$ or $B \supset A$ if every element of A is also an element of B: $z \in A \Rightarrow z \in B$.

Connected Sets

A basic result from calculus states that if $f'(x) = 0$ for all x in (a, b), then f is a constant. This result is not true if the domain of definition of the function is not connected. For example, consider the function

$$f(x) = \begin{cases} 1 & \text{if } 0 < x < 1, \\ -1 & \text{if } 2 < x < 3, \end{cases}$$

whose domain of definition is $(0, 1) \cup (2, 3)$. We have $f'(x) = 0$ for all x in $(0, 1) \cup (2, 3)$, but clearly f is not constant. This example shows you how crucial connectedness is in calculus. For subsets of the plane, one way to define connectedness is as follows.

DEFINITION 4
POLYGONALLY
CONNECTED SETS

A subset S of the complex plane is called **polygonally connected** or **connected** if any two points in S can be joined by a polygonal line consisting of a finite number of line segments joined end to end and wholly contained in S (Figure 6).

The following definition combines two important properties.

DEFINITION 5
REGIONS

A nonempty subset S of the complex plane is called a **region** (or **domain**) if it is open and connected.

Caution: Some authors use the term **domain** instead of region and use the term **region** to denote a subset of the complex plane that may be open or closed. In this book, it is important to keep in mind that regions are always open. The term domain will be used only in connection with its usual meaning as in "domain of definition" of a function.

Connected open sets or regions have the following useful characterization, which is intuitively clear. Its proof is left as an exercise.

PROPOSITION 1
CHARACTERIZATION
OF REGIONS

A nonempty open subset S of the complex plane is a region if and only if it cannot be written as the disjoint union of two nonempty open subsets. Equivalently, if S is a region (nonempty connected open subset of $\mathbb{C}$) and $S = A \cup B$, where A and B are open and disjoint, then either $A = \emptyset$ or $B = \emptyset$.

Here are some useful examples of regions.

- An open disk $B_r(z_0)$ is a region.

- A punctured disk centered at z_0, $B'_r(z_0) = \{z : \ 0 < |z - z_0| < r\}$, is a region.

- An open **annulus** centered at z_0,

$$A_{r_1, r_2}(z_0) = \{z : \ r_1 < |z - z_0| < r_2\},$$

is a region (Figure 5).

- The open upper half-plane $\{z : \ \text{Im}\, z > 0\}$ is a region.

- The complex plane is a region.

Here are sets that are not regions.

- A closed disk is not a region because it is not open.

- The union of two disjoint open disks (for example, $B_1(0) \cup B_{\frac{1}{2}}(2i)$) is not a region because it is not connected.

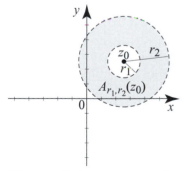

Figure 5 An open annulus $A_{r_1, r_2}(z_0)$ is a region. Note how two arbitrary points A and B in $A_{r_1, r_2}(z_0)$ can be joined by a polygonal line contained in the annulus.

- An interval (a, b) is not a region because it is not an open subset of the complex plane.

We end the section with an application of connectedness to functions of two variables. This is our generalization to two dimensions of the fact that a function with zero derivative is constant.

Suppose that $u(x, y)$ is a real-valued function of (x, y) defined on a nonempty open set Ω. If we fix y, we can think of u as a function of x alone and take its derivative with respect to x. This is called the **partial derivative** of u with respect to x and is denoted by $\frac{\partial u}{\partial x}$ or sometimes u_x. Thus

The limit in (2) involves the values of u at the point $(x + h, y)$. You should convince yourself that this point belongs to Ω if h is sufficiently small, because Ω is open. It is in this sense that we will interpret expressions involving limits.

$$(2) \qquad \frac{\partial u}{\partial x} = \lim_{h \to 0} \frac{u(x + h, y) - u(x, y)}{h}.$$

Similarly, we can fix x, think of $u(x, y)$ as a function of y, and take its partial derivative with respect to y. Thus

$$(3) \qquad \frac{\partial u}{\partial y} = \lim_{h \to 0} \frac{u(x, y + h) - u(x, y)}{h}.$$

THEOREM 1

Suppose that u is a real-valued function defined over a region Ω such that $u_x(x, y) = 0$ and $u_y(x, y) = 0$ for all (x, y) in Ω. Then $u(x, y)$ is constant for all (x, y) in Ω.

Proof   As a motivation, let us first recall the proof of the corresponding one dimensional result: If $f'(x) = 0$ for all x in (a, b), then $f(x) = c$ for all x in (a, b). Fix a point x_0 in (a, b). Let x be in (a, b), and say $a < x < x_0 < b$. The mean value theorem asserts that there is a point x_1 in (x, x_0) such that

$$f(x_0) - f(x) = f'(x_1)(x_0 - x).$$

Since f' is identically zero in (x, x_0), we conclude that $f(x_0) - f(x) = 0$ or $f(x) = f(x_0)$. The case $x < x_0 < b$ is treated similarly, and we obtain that $f(x) = f(x_0)$ for all x in (a, b). In other words, f is constant in (a, b).

This simple proof can be repeated in higher dimensions. What we need is a mean value theorem in higher dimensions. One form of this theorem is proved in Section 2.6, Theorem 5. According to that theorem, if $A = (x_1, y_1)$ and $B = (x_2, y_2)$ are two points in Ω such that the line segment AB is also in Ω, then there exists a point $C = (x_3, y_3)$ on the line segment AB such that

$$(4) \qquad u(x_2, y_2) - u(x_1, y_1) = u_x(x_3, y_3)(x_2 - x_1) + u_y(x_3, y_3)(y_2 - y_1).$$

It is now easy to prove Theorem 1. Fix a point (x_0, y_0) in Ω. Given a point (x, y) in Ω, connect (x_0, y_0) to (x, y) by a finite number of line segments joined end to end and wholly contained in Ω. (Here we have used the fact that Ω is a region.) Let (x_j, y_j), $j = 0, 1, \ldots, n$ denote the endpoints of the consecutive line segments, starting with (x_0, y_0) and ending with $(x_n, y_n) = (x, y)$ (see Figure 6). Apply (4) to each line segment and use the fact that the partial derivatives are zero to conclude that $u(x_{j-1}, y_{j-1}) = u(x_j, y_j)$, and hence that $u(x_0, y_0) = u(x, y)$.   ∎

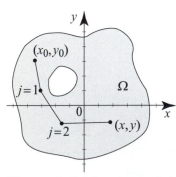

Figure 6 Joining two points in a connected region by a polygonal line.

Exercises 2.1

In Exercises 1–4, identify the interior points and boundary points of the given set.

1. $\{z : |z| \leq 1\}$.
2. $\{z : 0 < |z| \leq 1\}$.
3. $\{z = x + iy : 0 < x < 1, \ y = 0\}$.
4. $\{z : 1 < |z - i| \leq 2\}$.

In Exercises 5–12, draw the given set of points. Is the set open? closed? connected? a region? Justify your answers.

5. $\{z : \operatorname{Re} z > 0\}$.
6. $\{z : \operatorname{Im} z \leq 1\}$.
7. $B_1(i) \cup B_1(0)$.
8. $\{z : z \neq 0, \ |\operatorname{Arg} z| < \frac{\pi}{4}\}$.
9. $\{z : z \neq 0, \ |\operatorname{Arg} z| < \frac{\pi}{4}\} \cup \{0\}$.
10. $\{z : |z + 5 + i| < 1\}$.
11. $\{z : |\operatorname{Re}(z + 3 + i)| > 1\}$.
12. $\{z : |z - 3i| > 1\}$.

In Exercises 13–16, construct an example to illustrate the given statement.

13. The union of two connected sets need not be connected.

14. A set with an infinite number of points need not have interior points.

15. If A is a subset of B, then the boundary of A need not be contained in the boundary of B.

16. The boundary of a region could be empty.

17. Prove that a set S is open if and only if its **complement**, $\mathbb{C} \setminus S$, is closed.

18. Show that a set S is open if and only if it contains none of its boundary points.

19. Suppose that $A_1, A_2, \ldots$ are open sets. Show that their union

$$\bigcup_{n=1}^{\infty} A_n = \{z : z \in A_n \text{ for some } n\}$$

is also open.

20. (a)   Suppose that $A_1, A_2, \ldots,$ are open sets. Show that a finite intersection

$$\bigcap_{n=1}^{N} A_n = \{z : z \in A_n \text{ for all } 1 \leq n \leq N\}$$

is also open. [Hint: Pick a neighborhood that is contained in all the A_n's.]
(b)   Show that the infinite intersection

$$\bigcap_{n=1}^{\infty} A_n = \{z : z \in A_n \text{ for all } n\}$$

need not be open. [Hint: Consider $A_n = \{z : |z| < \frac{1}{n}\}$.]

21. Suppose that A and B are two regions with nonempty intersection. Show that $A \cup B$ is also a region.

22. Project Problem: Is it true that if $u_y(x, y) = 0$ for all (x, y) in a region Ω, then $u(x, y) = \phi(x)$; that is, u depends only on x? The answer is no in general, as the following counterexamples show.
(a)   For (x, y) in the region Ω shown in Figure 7, consider the function

$$u(x, y) = \begin{cases} 0 & \text{if } x > 0, \\ \operatorname{sgn} y & \text{if } x \leq 0, \end{cases}$$

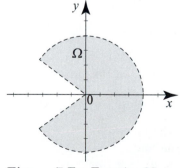

Figure 7 For Exercise 22.

where the **signum function** is defined by $\operatorname{sgn} y = -1, 0, 1$, according as $y < 0$, $y = 0$, or $y > 0$. Show that $u_y(x, y) = 0$ for all (x, y) in Ω but that u is not a function of x alone.

(b)   Note that in the previous example u_x does not exist for $x = 0$. We now construct a function over the same region Ω for which the partials exist, $u_y = 0$, and u is not a function of x alone. Show that these properties hold for

$$u(x, y) = \begin{cases} 0 & \text{if } x > 0, \\ e^{-1/x^2} \operatorname{sgn} y & \text{if } x \leq 0. \end{cases}$$

(c)   Come up with a general condition on Ω that guarantees that whenever $u_y = 0$ on Ω then u depends only on x. [Hint: Use the mean value theorem as applied to vertical line segments in Ω.]

23. Project Problem: Stereographic projection.   Suppose that a sphere of radius one, called a **Riemann sphere**, is positioned on the complex plane with its equator coinciding with the unit circle (Figure 8). Let N be the north pole of the sphere and let z be any point in the complex plane. The line from N to z intersects the sphere at one other point $z^\star$. Conversely, if $z^\star$ is any point of the sphere other than the north pole, then the line from N to $z^\star$ will intersect the plane at a single point z. It is not difficult to convince yourself that the mapping P that takes $z^\star$ to z is one-to-one from the sphere minus the north pole onto the complex plane. This mapping, known as the **stereographic projection**, was introduced by the German mathematician Bernhard Riemann (1826–1866). It enables us to represent points in the complex plane by points on the sphere, and vice versa. This also suggests that we introduce the **point at infinity**, $z = \infty$, as the image of the north pole by the stereographic projection. Thus $P(N) = \infty$. The complex plane together with this point at infinity is called the **extended complex plane** and written $\mathbb{C} \cup \{\infty\}$. It is in one-to-one correspondence with the whole sphere. As you will now discover, several issues concerning ∞ can be clarified by thinking in terms of the sphere and its projections. Answer parts (a)–(e) by using geometric reasoning with the help of Figure 8.

(a)   Consider a circle C on the sphere that is parallel to the complex plane (these are called parallels of latitude). What is its image under P?

(b)   Which points on the sphere are mapped to the set of all z in the plane such that $|z| > R$. Can you now justify the terminology "neighborhood of infinity"?

(c)   What is the image under P of a great circle passing through the poles?

(d)   What is the image under P of a circle passing through the north pole but not the south pole?

(e)   Argue geometrically that $z^\star$ approaches N if and only if $P(z^\star) \to \infty$.

Your answers in (a)–(e) can be justified also with the help of the formulas that you will derive in parts (f)–(h).

(f)   Let $z^\star = (x_1, x_2, x_3)$ and $P(z^\star) = x + i\,y = (x, y)$. Show that the equation of the line through $z^\star$ and z is

$$\frac{x_1 - 0}{x - 0} = \frac{x_2 - 0}{y - 0} = \frac{x_3 - 1}{0 - 1}.$$

(g)   Use (f) and the equation of the Riemann sphere $x_1^2 + x_2^2 + x_3^2 = 1$ to derive

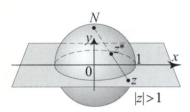

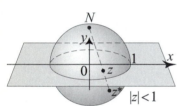

Figure   8   Stereographic projection and the Riemann sphere.   For $|z| > 1$, the point z^* is in the northern hemisphere. For $|z| < 1$, the point z^* is in the southern hemisphere.   For $|z| = 1$, the points z and z^* are coincident.

$$x_1 = \frac{2x}{x^2 + y^2 + 1}, \qquad x_2 = \frac{2y}{x^2 + y^2 + 1}, \qquad x_3 = 1 - \frac{2}{x^2 + y^2 + 1}.$$

(h)   Conversely, solve for x and y in (f) and get

$$x = \frac{x_1}{1 - x_3}, \qquad y = \frac{x_2}{1 - x_3}.$$

2.2   Limits and Continuity

When we define the derivative of a complex-valued function in Section 2.3, we will model our definition after the familiar derivative of a real-valued function from calculus. As you recall, such a derivative was defined by taking limits. Therefore, before we study differentiation, we must define limits of complex functions. We will also define continuous functions by appealing to limits.

DEFINITION 1
LIMITS OF
COMPLEX-VALUED
FUNCTIONS

We say that a complex-valued function $f(z)$ has a **limit** L as z approaches z_0, and we write

$$\lim_{z \to z_0} f(z) = L \quad \text{or} \quad f(z) \to L \text{ as } z \to z_0,$$

if given any $\epsilon > 0$, there exists a $\delta > 0$ such that

(1) $\qquad |f(z) - L| < \epsilon \quad \text{whenever} \quad 0 < |z - z_0| < \delta.$

If the limit of a function exists at a point, then it is unique (Exercise 27). This is referred to as the **uniqueness property** of limits.

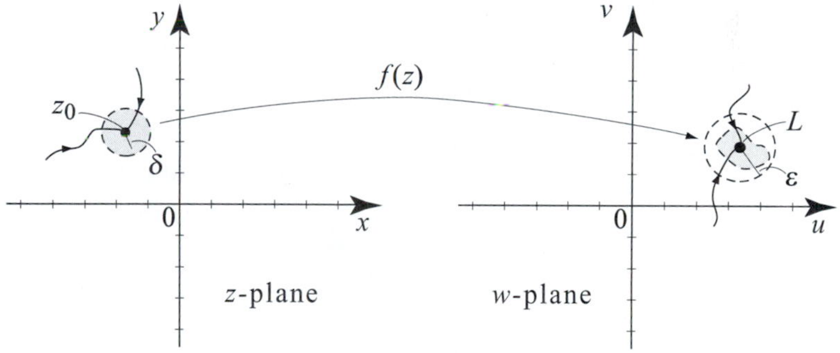

Figure 1 To say that $f(z) \to L$ as $z \to z_0$ is a strong assertion; it states that no matter how z approaches z_0 (and there are many possible ways in the plane), the distance from $f(z)$ to L will tend to zero. By contrast, for limits of functions on the real line, a point x can approach x_0 from only two directions.

Geometrically, interpreting the absolute value $|f(z) - L|$ as the distance between $f(z)$ and L, we see from (1) that the function $f(z)$ has limit L as $z \to z_0$ if and only if the distance from $f(z)$ to L tends to zero as z tends to z_0. Thus, $\lim_{z \to z_0} f(z) = L$ if and only if

(2) $\qquad \lim_{z \to z_0} |f(z) - L| = 0.$

Note that in (1) the value of f at z_0 is immaterial, and in fact f need not even be defined at z_0. Note also that the expression $|f(z) - L|$ that appears in (1) and (2) is real-valued. So even though we will be computing limits of complex-valued functions, we will be working with real quantities.

EXAMPLE 1   Two simple limits

Prove that:

(a) $\lim_{z \to z_0} z = z_0$;   and   (b) $\lim_{z \to z_0} c = c$, where c is a constant.

Solution   (a) Given $\epsilon > 0$, we want to pick a $\delta > 0$ so that

$$0 < |z - z_0| < \delta \quad \Rightarrow \quad |f(z) - L| < \epsilon.$$

Identifying $f(z) = z$ and $L = z_0$, this becomes

$$0 < |z - z_0| < \delta \quad \Rightarrow \quad |z - z_0| < \epsilon.$$

Clearly, the choice $\delta = \epsilon$ will do.

For (b), the inequality $|f(z) - L| = |c - c| < \epsilon$ holds for any choice of δ, which shows that $\lim_{z \to z_0} c = c$. ■

Computing more complicated limits by recourse to the $\epsilon\delta$-definition (1) is not always easy. To simplify this task, we will use properties of limits.

THEOREM 1
OPERATIONS
WITH LIMITS

Suppose $\lim_{z \to z_0} f(z)$ and $\lim_{z \to z_0} g(z)$ both exist and c_1, c_2 are complex constants. Then

(3) $$\lim_{z \to z_0} [c_1 f(z) + c_2 g(z)] = c_1 \lim_{z \to z_0} f(z) + c_2 \lim_{z \to z_0} g(z),$$

(4) $$\lim_{z \to z_0} [f(z)g(z)] = \lim_{z \to z_0} f(z) \lim_{z \to z_0} g(z),$$

(5) $$\lim_{z \to z_0} \left[\frac{f(z)}{g(z)} \right] = \frac{\lim_{z \to z_0} f(z)}{\lim_{z \to z_0} g(z)}, \quad \text{provided } \lim_{z \to z_0} g(z) \neq 0.$$

If $\lim_{z \to z_0} g(z) = w_0$ and $\lim_{w \to w_0} f(w) = A$, then

(6) $$\lim_{z \to z_0} f(g(z)) = A = f\left(\lim_{z \to z_0} g(z) \right).$$

The function $f(g(z))$ is called the **composition** of f and g and is also denoted $(f \circ g)(z)$. The proofs of (3)–(6) are similar to the proofs of the corresponding results from calculus. They are left to the exercises. Next, we illustrate these properties with some applications.

EXAMPLE 2   Operations on limits

Suppose that $\lim_{z \to i} f(z) = 2 + i$ and $\lim_{z \to i} g(z) = 3 - i$. Find

(7)
$$L = \lim_{z \to i} \left[(f(z))^2 + \frac{(3+i)g(z)}{z} \right].$$

Solution   Since the limits of $f(z)$, $g(z)$, and z all exist as $z \to i$, and the denominator in the expression (7) tends to $i \neq 0$, we conclude that

$$L = \left(\lim_{z \to i} f(z) \right)^2 + (3+i) \frac{\lim_{z \to i} g(z)}{\lim_{z \to i} z} = (2+i)^2 + (3+i) \overbrace{\frac{3-i}{i}}^{-i(10)}$$

$$= 3 + 4i - 10i = 3 - 6i.$$ ∎

Our next result is an analog of the squeeze theorem from calculus. To state it we will need the following definition. A function $g(z)$ is **bounded** in a set S if there is a positive real number M such that $|g(z)| \leq M$ for all z in S.

THEOREM 2
THE SQUEEZE
THEOREM

(i) Suppose that $f(z) \to 0$ as $z \to z_0$ and $|g(z)| \leq |f(z)|$ in a deleted neighborhood of z_0. Then $g(z) \to 0$ as $z \to z_0$.
(ii) Suppose that $f(z) \to 0$ as $z \to z_0$ and $g(z)$ is bounded in a deleted neighborhood of z_0. Then $f(z)g(z) \to 0$ as $z \to z_0$.

Proof   (i)   Since $f(z) \to 0$, given $\epsilon > 0$, there is a δ such that $|f(z)| < \epsilon$ whenever $0 < |z - z_0| < \delta$. But since $|g(z)| \leq |f(z)|$, we also have $|g(z) - 0| < \epsilon$ whenever $0 < |z - z_0| < \delta$, which implies that $g(z) \to 0$ as $z \to z_0$.
(ii)   Since g is bounded in a deleted neighborhood of z_0, we can find $M > 0$ and $r > 0$ such that $|g(z)| \leq M$ for $0 < |z - z_0| < r$. For $0 < |z - z_0| < r$, we have

$$0 \leq |f(z)g(z)| \leq M|f(z)|.$$

Since $M|f(z)| \to 0$, it follows from (i) that $f(z)g(z) \to 0$. ∎

EXAMPLE 3   An application of the squeeze theorem

Compute $\lim_{z \to 0} y e^{i/|z|}$.

It is interesting to note that if $h(z)$ is *real-valued*, then $\left| e^{ih(z)} \right| = 1$ no matter how large $h(z)$ is. In Example 3, $h(z) = 1/|z| \to \infty$ as $z \to 0$, and still $\left| e^{i/|z|} \right| = 1$.

Solution   Let $f(z) = y$ and $g(z) = e^{i/|z|}$. As $z \to 0$, $f(z) \to 0$. Also, for $z \neq 0$, since $1/|z|$ is a purely real number, we have $|e^{i/|z|}| = 1$, by (6), Section 1.5. Thus we can apply (ii) from the squeeze theorem and conclude that

$$\lim_{z \to 0} y e^{i/|z|} = 0.$$ ∎

Consider a complex-valued function $f(z) = u(x, y) + i v(x, y)$. It is often advantageous to study the limit of f by using properties of the limits of the real and imaginary parts of f. This is possible because of the following fact.

THEOREM 3
REAL AND
IMAGINARY PARTS
OF LIMITS

Given a complex-valued function $f(z) = u(x, y) + i\,v(x, y)$ and a complex number $L = a + i\,b$, then

$$(8) \qquad \lim_{z \to z_0} f(z) = L \quad \Longleftrightarrow \quad \lim_{z \to z_0} u(x, y) = a \text{ and } \lim_{z \to z_0} v(x, y) = b.$$

Thus the existence of one complex-valued limit is equivalent to the existence of two real-valued limits.

Proof   Suppose that $\lim_{z \to z_0} f(z) = L$ as $z \to z_0$. By (2), we have

Inequalities (14), Section 1.2, state that for any complex number w,
$$|\operatorname{Re} w| \le |w|$$
and
$$|\operatorname{Im} w| \le |w|.$$
Inequality (15), Section 1.2, states that
$$|w| \le |\operatorname{Re} w| + |\operatorname{Im} w|.$$

$$(9) \qquad \lim_{z \to z_0} |f(z) - L| = 0.$$

Now

$$|f(z) - L| = |\overbrace{(u(x, y) - a)}^{\operatorname{Re}\,(f(z)-L)} + i\,\overbrace{(v(x, y) - b)}^{\operatorname{Im}\,(f(z)-L)}|.$$

Appealing to inequalities (14), Section 1.2, we obtain

$$(10) \qquad 0 \le |u(x, y) - a| \le |f(z) - L| \quad \text{and} \quad 0 \le |v(x, y) - b| \le |f(z) - L|.$$

Since $|f(z) - L| \to 0$ as $z \to z_0$, it follows from Theorem 2 and (10) that $|u(x, y) - a| \to 0$ and $|v(x, y) - b| \to 0$ as $z \to z_0$. This shows that $u(x, y) \to a$ and $v(x, y) \to b$ as $z \to z_0$. For the other direction, we use the inequality

$$|f(z) - L| \le |\operatorname{Re}\,(f(z) - L)| + |\operatorname{Im}\,(f(z) - L)| = |u(x, y) - a| + |v(x, y) - b|,$$

which is a consequence of (15), Section 1.2. As $z \to z_0$, the right side goes to 0, implying that $|f(z) - L| \to 0$.   ∎

We have avoided thus far dealing with limits that involve ∞. What do we mean by statements such as $\lim_{z \to z_0} f(z) = \infty$ or $\lim_{z \to \infty} f(z) = L$ or even $\lim_{z \to \infty} f(z) = \infty$? We answer these questions and complete our discussion of limits by introducing the following definitions.

DEFINITION 2
LIMITS INVOLVING
INFINITY

(i)   We write $\displaystyle\lim_{z \to z_0} f(z) = \infty$ to mean that for any $M > 0$ there is a $\delta > 0$ such that $|z - z_0| < \delta \Rightarrow |f(z)| > M$.
(ii)   We write $\displaystyle\lim_{z \to \infty} f(z) = L$ to mean that for any $\epsilon > 0$ there is an $R > 0$ such that $|z| > R \Rightarrow |f(z) - L| < \epsilon$.
(iii)   We write $\displaystyle\lim_{z \to \infty} f(z) = \infty$ to mean that for any $M > 0$ there is an $R > 0$ such that $|z| > R \Rightarrow |f(z)| > M$.

Looking at these definitions, we see that $z \to \infty$ means that the real quantity $|z| \to \infty$, and similarly $f(z) \to \infty$ means that $|f(z)| \to \infty$. Hence

$$(11) \qquad \lim_{z \to z_0} f(z) = \infty \quad \Leftrightarrow \quad \lim_{z \to z_0} |f(z)| = \infty;$$

$$(12) \qquad \lim_{z \to \infty} f(z) = L \quad \Leftrightarrow \quad \lim_{z \to \infty} |f(z) - L| = 0;$$

$$(13) \qquad \lim_{z \to \infty} f(z) = \infty \quad \Leftrightarrow \quad \lim_{z \to \infty} |f(z)| = \infty.$$

Limits at infinity can also be reduced to limits at $z_0 = 0$ by means of the inversion $1/z$. The idea is that taking the limit as $z \to \infty$ of $f(z)$ is the same thing as taking the limit as $z \to 0$ of $f\left(\frac{1}{z}\right)$. So you can check that

$$
(14) \qquad \lim_{z \to \infty} f(z) = L \quad \Leftrightarrow \quad \lim_{z \to 0} f\left(\frac{1}{z}\right) = L;
$$

and

$$
(15) \qquad \lim_{z \to \infty} f(z) = \infty \quad \Leftrightarrow \quad \lim_{z \to 0} f\left(\frac{1}{z}\right) = \infty.
$$

These equivalent statements are sometimes useful. For example, appealing to (15), we have

$$
(16) \qquad \lim_{z \to \infty} \frac{1}{z} = \lim_{z \to 0} \frac{1}{1/z} = \lim_{z \to 0} z = 0.
$$

Similarly, for any constant c and positive integer n, we have

$$
\lim_{z \to \infty} \frac{c}{z^n} = \lim_{z \to 0} \frac{c}{1/z^n} = \lim_{z \to 0} c z^n = 0.
$$

EXAMPLE 4   Limits at ∞

Find:

(a) $\displaystyle \lim_{z \to \infty} \frac{z-1}{z+i}$;   and   (b) $\displaystyle \lim_{z \to \infty} \frac{2z+3i}{z^2+z+1}$.

Solution   (a) Since we are concerned with the behavior of the function for $|z|$ large, it is safe to divide both numerator and denominator of $\frac{z-1}{z+i}$ by z, and we conclude

$$
\begin{aligned}
\lim_{z \to \infty} \frac{z-1}{z+i} &= \lim_{z \to \infty} \frac{1 - \frac{1}{z}}{1 + \frac{i}{z}} \\[2mm]
&= \frac{1 - \lim_{z \to \infty} \frac{1}{z}}{1 + i \lim_{z \to \infty} \frac{1}{z}} \qquad \text{[by (5) and (3)]} \\[2mm]
&= 1 \qquad \text{[by (16)].}
\end{aligned}
$$

(b) Similarly, we divide both numerator and denominator of $\frac{2z+3i}{z^2+z+1}$ by z^2 and conclude

$$
\lim_{z \to \infty} \frac{2z+3i}{z^2+z+1} = \lim_{z \to \infty} \frac{\frac{2}{z} + \frac{3i}{z^2}}{1 + \frac{1}{z} + \frac{1}{z^2}} = \frac{0+0}{1+0+0} = 0. \qquad \blacksquare
$$

While we have successfully used skills from calculus to guide us in taking complex-valued limits, you should be cautioned in using real-variable intuition. For example, the limit $\lim_{z \to \infty} e^{-z}$ is not 0; in fact, the limit does not exist (Exercise 21).

Continuous Functions

Often, the limit of $f(z)$ as z approaches z_0 will equal $f(z_0)$. Functions that satisfy this requirement are said to be continuous.

DEFINITION 3
CONTINUOUS
FUNCTIONS

Suppose $f(z)$ is defined on a neighborhood of z_0. We say that $f(z)$ is **continuous** at the point z_0 if $\lim\limits_{z \to z_0} f(z)$ exists and equals $f(z_0)$. We say f is continuous on a set S if it is continuous at every point in S.

We see from Example 1 that the functions $f(z) = z$ and $f(z) = c$ are continuous at all points in the plane. To check the continuity of more complicated functions, we can appeal to the properties of limits. Since continuity is defined in terms of limits, many properties of limits extend to continuous functions.

THEOREM 4
PROPERTIES OF
CONTINUOUS
FUNCTIONS

(i)   If $f(z)$ and $g(z)$ are continuous at z_0, and c_1, c_2 are complex constants, then the following functions are continuous at z_0:

$$(17) \qquad c_1 f(z) + c_2 g(z), \quad f(z)g(z), \quad \frac{f(z)}{g(z)} \ \text{(provided } g(z_0) \neq 0).$$

(ii)   If g is continuous at z_0 and f is continuous at $g(z_0)$, then the composition $h(z) = f(g(z))$ is continuous at z_0.

(iii)   If $g(z)$ is real-valued and continuous at z_0 and $f(x)$ is continuous at $x_0 = g(z_0)$, then the composition $h(z) = f(g(z))$ is continuous at z_0.

(iv)   The function $f(z) = u(z) + i\,v(z)$ is continuous if and only if u and v are continuous.

The proofs of (i), (ii), and (iv) follow from Theorems 1 and 3. The proof of (iii) follows from the definitions of continuity for complex- and real-valued functions. We apply these results and check the continuity of polynomials and rational functions.

EXAMPLE 5   Polynomial and rational functions

(a) Show that a polynomial $p(z) = a_n z^n + a_{n-1} z^{n-1} + \cdots + a_0$ is continuous at all points in the plane.

(b) A **rational** function is a function of the form

$$r(z) = \frac{p(z)}{q(z)},$$

where p and q are polynomials. Show that a rational function is continuous at all points where $q(z) \neq 0$.

Solution   (a) Since the function $f(z) = z$ is continuous, we repeatedly use the fact that the product of two continuous functions is continuous to conclude that $z^2, z^3, \ldots, z^n$ are continuous. Then, by repeated applications of the fact that a linear combination of continuous functions is continuous, we conclude that the function $a_n z^n + a_{n-1} z^{n-1} + \cdots + a_0$ is continuous.

(b) By part (a), the polynomials $p(z)$ and $q(z)$ are continuous in the entire plane. Hence $r(z) = \frac{p(z)}{q(z)}$ is continuous in the entire plane except those points where $q(z) = 0$. ∎

EXAMPLE 6   Limits and continuity of rational functions
For the given rational function $f(z) = p(z)/q(z)$, find the limit and determine if the function is continuous at the given point.

(a) $\displaystyle\lim_{z \to 2i} \frac{2z^2 - i}{z + 2}$.   (b) $\displaystyle\lim_{z \to i} \frac{z - i}{z^2 + 1}$.

Solution   (a)   Since $q(2i) = 2i + 2 \neq 0$, $f(z)$ is continuous at $z = 2i$, by Example 5(b). To find the limit as $z \to 2i$, we simply evaluate $f(2i)$:

$$\lim_{z \to 2i} \frac{2z^2 - i}{z + 2} = f(2i) = \frac{2(2i)^2 - i}{2i + 2} = \frac{-8 - i}{2 + 2i} = \frac{1}{4}(-9 + 7i).$$

(b)   The denominator $q(z) = z^2 + 1$ vanishes at $z = i$ and so $f(z) = p(z)/q(z)$ is not continuous at $z = i$. Does the limit exist at $z = i$? We have

$$f(z) = \frac{z - i}{z^2 + 1} = \frac{z - i}{(z - i)(z + i)}.$$

For $z \neq i$, we cancel the factor $z - i$ and get

$$\lim_{z \to i} \frac{z - i}{z^2 + 1} = \lim_{z \to i} \frac{1}{z + i} = \frac{1}{2i} = \frac{-i}{2}. \qquad ∎$$

In Example 6(b), because $\lim_{z \to i} f(z)$ exists at the point of discontinuity $z = i$, we can remove the discontinuity of $f(z)$ and make it continuous at $z = i$ by redefining $f(i) = -i/2$. Such a point of discontinuity is called a **removable discontinuity**. If the discontinuity at a point cannot be removed, then it is called a **nonremovable discontinuity**.

Our next example involves a function with infinitely many nonremovable discontinuities. In the example, we will use the uniqueness property of limits to show that a limit fails to exist. The method works as follows. If you can show that $f(z) \to L$ as z approaches z_0 on curve C (Figure 2), but $f(z) \to L'$ as z approaches z_0 on curve C', and $L \neq L'$, then, by the uniqueness of limits, we conclude that $\lim_{z \to z_0} f(z)$ does not exist.

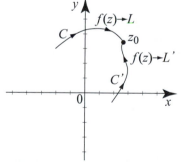

Figure 2 If $L \neq L'$, then the limit at z_0 cannot exist.

EXAMPLE 7   The nonremovable discontinuities of $\operatorname{Arg} z$
The principal branch of the argument $\operatorname{Arg} z$ takes the value of argument z that is in the interval $-\pi < \operatorname{Arg} z \leq \pi$. It is not defined at $z = 0$ and hence $\operatorname{Arg} z$ is not continuous at $z = 0$. We will show that $z = 0$ is not a removable discontinuity by showing that $\lim_{z \to 0} \operatorname{Arg} z$ does not exist. Indeed, if $z = x > 0$, then $\operatorname{Arg} z = 0$ and so $\lim_{z = x \downarrow 0} \operatorname{Arg} z = 0$, where the down-arrow denotes the limit from the right, also denoted by $\lim_{z = x \to 0^+} \operatorname{Arg} z$. However, if $z = x < 0$, then $\operatorname{Arg} z = \pi$ and so $\lim_{z = x \uparrow 0} \operatorname{Arg} z = \pi$, where the up-arrow denotes the limit from the left, also denoted

by $\lim_{z=x\to 0^-} \operatorname{Arg} z$. By the uniqueness of limits, we conclude that $\lim_{z\to 0} \operatorname{Arg} z$ doe not exist.

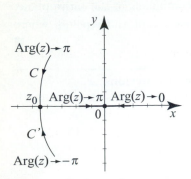

Also, for a point on the negative x-axis, $z_0 = x_0 < 0$, we have $\operatorname{Arg} z_0 = \pi$. If z approaches z_0 from the second quadrant, say along curve C in Figure 3, we have $\lim_{z\to z_0} \operatorname{Arg} z = \pi = \operatorname{Arg} z_0$. But if z approaches z_0 from the third quadrant, say along curve C' in Figure 3, we have $\lim_{z\to z_0} \operatorname{Arg} z = -\pi$. Hence $\operatorname{Arg} z$ is not continuous at z_0 and the discontinuity is not removable, because $\lim_{z\to z_0} \operatorname{Arg} z$ does not exist for such z_0.

It is not hard to show, using geometric considerations, that for $z \neq 0$ and z not on the negative x-axis, $\operatorname{Arg} z$ is continuous. Since the set of points of continuity of $\operatorname{Arg} z$ is the complex plane $\mathbb{C}$ minus the interval $(-\infty, 0]$ on the real line, the principal branch of the argument is continuous on $\mathbb{C} \setminus (\infty, 0]$.   ∎

Figure 3 $\operatorname{Arg} z$ has nonremovable discontinuities at $z = 0$ and at all negative real z. For all other z, $\operatorname{Arg} z$ is continuous.

Many important functions of several variables are made up of products, quotients and linear combinations of functions of a single variable. For example, the function $u(x, y) = e^x \cos y$ is the product of two functions of a single variable each; namely, e^x and $\cos y$. The exponential function $e^z = e^x(\cos y + i \sin y)$ is a linear combination of two products of functions of a single variable. In establishing the continuity of such functions, the following simple observations are very useful.

PROPOSITION 1

> Suppose that $\phi(x)$ is a continuous function of a single variable x over an interval (a, b). Then the function of two variables $f(x, y) = \phi(x)$ is continuous at (x_0, y_0) whenever x_0 is in (a, b). Similarly, $g(x, y) = \phi(y)$ is continuous at (x_0, y_0) whenever y_0 is in (a, b).

Proof   If $(x, y) \to (x_0, y_0)$, then $x \to x_0$ and so $\phi(x) \to \phi(x_0)$ and, consequently, $f(x, y) = \phi(x) \to \phi(x_0) = f(x_0, y_0)$. Thus f is continuous at (x_0, y_0) as claimed. The second part of the proposition follows similarly.   ∎

Combined with Theorem 4, this proposition becomes a very powerful tool. Here are some interesting applications.

EXAMPLE 8   Exponential and trigonometric functions
Show that the following are continuous functions of z.
(a) e^z.   (b) $\cos z$.

Solution   (a)   We know from calculus that the functions e^x, $\cos x$, and $\sin x$ are continuous for all x. By Proposition 1, the functions $f_1(x, y) = e^x$, $f_2(x, y) = \cos y$, and $f_3(x, y) = \sin y$ are continuous for all (x, y). Appealing to Theorem 4, we see that $e^x \cos y + i e^x \sin y = e^z$ is continuous for all (x, y).
(b)   The function e^{iz} is continuous because it is the composition of two continuous functions; namely, $i z$ and e^z. Similarly, e^{-iz} is continuous, and hence the linear combination $\frac{e^{iz}+e^{-iz}}{2} = \cos z$ is also continuous.   ∎

We give one more example of a continuous function.

EXAMPLE 9   The absolute value $|z|$ is continuous

Let $f(z) = |z|$. By the triangle inequality, $0 \leq |z| = |x + i y| \leq |x| + |y|$. As $z \to 0$, we have $x \to 0$ and $y \to 0$; consequently, $|x|$, $|y|$, and $|x|+|y|$ all tend to 0, implying that $|z| = |x+iy|$ tends to 0, by the squeeze theorem. This proves the continuity at 0. We now show that $\lim_{z \to z_0} |f(z) - f(z_0)| = \lim_{z \to z_0} ||z| - |z_0|| = 0$, for arbitrary z_0. By the lower estimate (19), Section 1.2, we have $||z| - |z_0|| \le |z - z_0|$. As $z \to z_0$, $(z - z_0) \to 0$, hence $|z - z_0| \to 0$, by the continuity of the absolute value at 0, and so $||z| - |z_0|| \to 0$, by the squeeze theorem. ∎

Continuity of the Logarithms

Understanding the behavior of the logarithm is crucial to certain applications. We will prove the following important result, which should not surprise you in view of what we already know about Arg z and $|z|$.

THEOREM 5
CONTINUITY OF
THE LOGARITHM

> The principal branch of the logarithm
>
> (18) $\text{Log}\,(z) = \ln|z| + i\,\text{Arg}\,(z), \quad -\pi < \text{Arg}\,z \le \pi \quad (z \ne 0),$
>
> is continuous for all z in $\mathbb{C} \setminus (-\infty, 0]$. For z in $(-\infty, 0]$, the discontinuities of Log z are not removable (Figure 4).

Proof We just showed in Example 9 that $|z|$ is a continuous function of z. Composing the continuous function $\ln x$, for $x > 0$, with the real-valued function $|z|$, it follows from Theorem 4(iii) that $\ln|z|$ is continuous for all $z \ne 0$. We showed in Example 7 that Arg z is continuous except for nonremovable discontinuities at $z = 0$ and z on the negative x-axis. Appealing to Theorem 4(iv), we see that a discontinuity at $z = z_0$ of a function $f(z) = u(z) + iv(z)$ is removable if and only if z_0 is a removable discontinuity of both u and v (Exercise 43). Thus because of the nonremovable discontinuities of Arg z, it follows that $z = 0$ and z on the negative x-axis are also nonremovable discontinuities of Log z. Thus the set of points of continuity of Log z is $\mathbb{C} \setminus (-\infty, 0]$. ∎

A discussion similar to the preceding one shows that a branch of the logarithm, $\log_\alpha z$, is continuous at all z except for nonremovable discontinuities at $z = 0$ and z on the ray at angle α. The set of nonremovable discontinuities of $\log_\alpha z$ is called a **branch cut**. Thus, for example, the branch cut of Log z is $(-\infty, 0]$ (Figure 4), and the branch cut of $\log_\alpha z$ is the ray at angle α (Figure 5).

Exercises 2.2

In Exercises 1–12, evaluate the given limit and justify each step by using properties of limits from this section. [Hint: In evaluating limits involving elementary functions such as exponential or trigonometric functions, you may want to express them in terms of their real and imaginary parts and use results from calculus.]

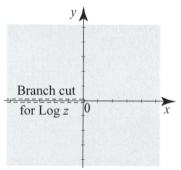

Figure 4 Log z has nonremovable discontinuities at $z = 0$ and at all negative real z. For all other z, Log z is continuous.

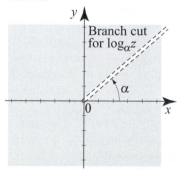

Figure 5 The branch cut of $\log_\alpha z$ is the ray at angle α. The branch cut is the set of nonremovable discontinuities of $\log_\alpha z$.

1. $\lim\limits_{z \to i} 3z^2 + 2z - 1.$

2. $\lim\limits_{z \to 2+i} z + \dfrac{1}{z}.$

3. $\lim\limits_{z \to 0} \dfrac{z}{\cos z}.$

4. $\lim\limits_{z \to 2} \dfrac{z^4 - 16}{z - 2}.$

5. $\lim\limits_{z \to i} \dfrac{1}{z - i} - \dfrac{1}{z^2 + 1}.$

6. $\lim\limits_{z \to 0} z\,\text{Arg}\,z.$

7. $\lim\limits_{z \to 0} ze^{i\,\text{Re}\,z}.$

8. $\lim\limits_{z \to i} \text{Re}\,(z)\sin z.$

9. $\lim\limits_{z \to -3} (\text{Arg}\,z)^2.$

10. $\lim\limits_{z \to 1} (z+1) \operatorname{Im}(iz)$.   **11.** $\lim\limits_{z \to 0} \sin \bar{z}$.   **12.** $\lim\limits_{z \to 0} z\, e^{i/|z|^2}$.

In Exercises 13–18, evaluate the given limit involving ∞. Justify your steps.

13. $\lim\limits_{z \to \infty} \dfrac{z+1}{3iz+2}$.   **14.** $\lim\limits_{z \to \infty} \dfrac{z^2+i}{z^3+3z^2+z+1}$.   **15.** $\lim\limits_{z \to \infty} \left(\dfrac{z^3+i}{z^3-i} \right)^2$.

16. $\lim\limits_{z \to i} \dfrac{1}{z^2+1}$.   **17.** $\lim\limits_{z \to 1} \dfrac{-1}{(z-1)^2}$.   **18.** $\lim\limits_{z \to \infty} \dfrac{\operatorname{Log} z}{z}$.

In Exercises 19–26, show that the given limit at z_0 does not exist by approaching z_0 from different directions. If the limit involves $z \to \infty$, try some of the following directions: the positive x-axis, the negative x-axis, the positive y-axis ($z = iy$, $y > 0$), or the negative y-axis ($z = iy$, $y < 0$).

19. $\lim\limits_{z \to -3} \operatorname{Arg} z$.   **20.** $\lim\limits_{z \to -1} \operatorname{Log} z$.   **21.** $\lim\limits_{z \to \infty} e^{-z}$.

22. $\lim\limits_{z \to 0} \dfrac{\bar{z}}{z}$.   **23.** $\lim\limits_{z \to 0} e^{1/z}$.   **24.** $\lim\limits_{z \to 0} \dfrac{\operatorname{Re} z}{|z|^2}$.

25. $\lim\limits_{z \to 0} \dfrac{z}{|z|}$.   **26.** $\lim\limits_{z \to 0} \dfrac{\operatorname{Im} z}{z}$.

27. Show that if the limit of a function exists then it is unique. [Hint: For the case of a finite limit, suppose L_1 and L_2 are two limits. Then $|L_1 - L_2| \le |L_1 - f(z)| + |L_2 - f(z)|$ (why?). As $z \to z_0$, what happens to the right side?]

28. Use the triangle inequality and (2) to prove (3).

29. Prove (14).   **30.** Prove (15).

31. Show that $\lim_{z \to z_0} f(z) = 0 \iff \lim_{z \to z_0} \frac{1}{f(z)} = \infty$.

32. Use the result of Exercise 31 to evaluate $\lim_{z \to 0} \frac{\cos z}{z}$.

In Exercises 33–40, determine the set of points where the given function is continuous. For a point of discontinuity, determine whether it is removable or not. Whenever possible, use properties from this section; in particular, use Proposition 1 and Theorem 4.

33. $\dfrac{z-i}{z+1+3i}$.   **34.** $\dfrac{2z+1}{z^2+3z+2}$.   **35.** $\bar{z}$.

36. $\operatorname{Log}(z+1)$.   **37.** $\sin z$.   **38.** $(\operatorname{Arg} z)^2$.

39. $z(\operatorname{Arg} z)^2$.   **40.** $\dfrac{z}{|z|}$.

41.   Pre-image of sets.   Let f be a complex-valued function defined on a subset S of $\mathbb{C}$. If A is a set of complex numbers, the **pre-image** or **inverse image** of A by f is the set $f^{-1}[A] = \{z \in S : f(z) \text{ belongs to } A\}$. The following statements are true for arbitrary sets S. To simplify the proofs, take $S = \mathbb{C}$.
(a)   Show that f is continuous if and only if $f^{-1}[A]$ is open whenever A is open.
(b)   Show that f is continuous if and only if $f^{-1}[A]$ is closed whenever A is closed.
(If S is a proper subset of $\mathbb{C}$, (a) and (b) still hold, but we have to define what we mean for a set such as $f^{-1}[A]$ to be open or closed in S. These topics are part of elementary topology and will not be emphasized in this book.)

42.   Continuity and boundedness.   Show that if f is continuous at z_0, then it is bounded in a neighborhood of z_0.

43. Show that a discontinuity at $z = z_0$ of a function $f(z) = u(z) + iv(z)$ is removable if and only if z_0 is a removable discontinuity of both u and v.

2.3   Analytic Functions

For a real-valued function $f(x)$ defined on an open interval containing the point x_0, we defined the derivative at x_0 to be $f'(x_0) = \lim_{x \to x_0} \frac{f(x) - f(x_0)}{x - x_0}$, when the limit exists. Our definition for the derivative of a complex function $f(z)$ is a natural extension of the real case.

DEFINITION 1
COMPLEX
DERIVATIVE

Let $f(z)$ be defined on a neighborhood of z_0. If

$$\lim_{z \to z_0} \frac{f(z) - f(z_0)}{z - z_0}$$

exists, then f is said to be **differentiable** at the point z_0, and the number

(1)
$$f'(z_0) = \lim_{z \to z_0} \frac{f(z) - f(z_0)}{z - z_0}$$

is called the **derivative** of f at z_0.

We can also use the Leibniz notation for the derivative, $\frac{df}{dz}(z_0)$, or $\frac{df}{dz}\Big|_{z=z_0}$. We can recast the difference quotient and define the derivative as

(2)
$$f'(z_0) = \lim_{\Delta z \to 0} \frac{f(z_0 + \Delta z) - f(z_0)}{\Delta z}.$$

While this extension looks similar to the definition of a derivative for functions defined on the real line, we are asking a lot more in the complex case. In the real case, x can only approach x_0 from either the right or the left. In the complex case, z can approach z_0 from any number of directions. For the derivative to exist, we are requiring that the limit exists no matter how we approach z_0 in (1).

We now introduce a fundamental definition in the theory of complex variables.

DEFINITION 2
ANALYTIC
FUNCTIONS

A function $f(z)$ defined on an open set S is said to be **analytic** on S if $f'(z)$ exists and is continuous for all z in S. We say $f(z)$ is analytic at a point if $f(z)$ is analytic on some open set containing that point. We say $f(z)$ is **entire** if $f(z)$ is analytic on $\mathbb{C}$.

It is important to note that while differentiability is defined at a specific point, analyticity is defined on an open set. Even when we say "analytic at a point," we really mean analytic on some open set containing this point.

In Section 3.9, we will prove a theoretical result, called Goursat's theorem. It asserts that the mere existence of $f'(z)$ in an open set implies that $f'(z)$ is continuous. Thus our requirement that $f'(z)$ be continuous in the definition of analytic functions is redundant; we have included it to make some initial proofs easier to understand.

EXAMPLE 1   Some simple entire functions

Show the following functions are entire (analytic at every point in the complex plane), and find their derivatives.

(a) $f(z) = 2 + 4i$.   (b) $f(z) = (3 - i)z$.   (c) $f(z) = z^2$.

Solution   We use the difference quotient as in (1) to calculate the derivatives.

(a)   Fix any z_0 in the plane. Since $f = 2 + 4i$ is constant, $f(z) = f(z_0)$ for any z. Hence $\frac{f(z)-f(z_0)}{z-z_0} = 0$. Taking the limit as $z \to z_0$, we get $f'(z_0) = 0$. Thus $f'(z) = 0$ for all z. Since $f'(z)$ is obviously continuous, it follows that $f(z) = 2+4i$ is analytic on $\mathbb{C}$, or entire.

(b)   Fix any z_0 in the plane. We have

$$f'(z_0) = \lim_{z \to z_0} \frac{f(z) - f(z_0)}{z - z_0} = \lim_{z \to z_0} \frac{(3-i)(z) - (3-i)(z_0)}{z - z_0} = 3 - i.$$

Thus $f'(z) = 3-i$ for all z. Since $f'(z)$ is continuous, it follows that $f(z) = (3-i)z$ is entire.

(c)   Fix any z_0 in the plane. We have

$$f'(z_0) = \lim_{z \to z_0} \frac{z^2 - z_0^2}{z - z_0} = \lim_{z \to z_0} \frac{(z - z_0)(z + z_0)}{z - z_0} = \lim_{z \to z_0} z + z_0 = 2z_0.$$

Thus $f'(z) = 2z$ for all z. Since $f'(z)$ is continuous, it follows that $f(z) = z^2$ is entire.   ∎

The following useful formulas can be derived by the methods of Example 1(a) and (b):

(3) $$f(z) = c \text{ (a constant)} \quad \Rightarrow \quad f'(z) = 0;$$

(4) $$f(z) = z \quad \Rightarrow \quad f'(z) = 1$$

Our first theorem is the analog of the well-known fact from calculus that states that if a function has a derivative then it is continuous. The proof is also similar to the real variable case.

THEOREM 1
DIFFERENTIABLE
IMPLIES
CONTINUOUS

Suppose $f(z)$ is differentiable at z_0. Then $f(z)$ is continuous at z_0.

Proof   We must show that $\lim_{z \to z_0} f(z) = f(z_0)$. Equivalently, we will show that $\lim_{z \to z_0} (f(z) - f(z_0)) = 0$. Using the fact that the limit of a product is the product of the limits ((4), Section 2.2), we have

$$\lim_{z \to z_0} f(z) - f(z_0) = \lim_{z \to z_0} \frac{f(z) - f(z_0)}{z - z_0}(z - z_0)$$

$$= \lim_{z \to z_0} \frac{f(z) - f(z_0)}{z - z_0} \lim_{z \to z_0} (z - z_0) = f'(z_0) \cdot 0 = 0. \quad ∎$$

Equivalently, Theorem 1 states that if a function is not continuous, it cannot be differentiable. The converse of Theorem 1, however, is not true. For example, $\bar{z}$ is continuous but nowhere differentiable (see Example 4 in this section).

Because the definition of the derivative in the complex case is modeled after the definition of the derivative in the real case, it should not surprise you that many of the properties of derivatives that you are familiar with from calculus hold for analytic functions.

**THEOREM 2
PROPERTIES OF
ANALYTIC
FUNCTIONS**

Suppose that f and g are analytic on an open set S and c_1, c_2 are complex constants. Then $c_1 f + c_2 g$ and fg are analytic on S with

$$(5) \qquad (c_1 f + c_2 g)'(z) = c_1 f'(z) + c_2 g'(z) \text{ and}$$
$$(6) \qquad (fg)'(z) = f'(z)g(z) + f(z)g'(z).$$

Also, $\frac{f}{g}$ is analytic on S minus the points where $g = 0$, and

$$(7) \qquad \left(\frac{f}{g}\right)'(z) = \frac{f'(z)g(z) - f(z)g'(z)}{(g(z))^2} \quad (g(z) \neq 0).$$

If g is analytic at z_0 and f is analytic at $g(z_0)$, then the composition $(f \circ g)(z)$ is analytic at z_0 and we have the **chain rule**

$$(8) \qquad (f \circ g)'(z_0) = f'(g(z_0))g'(z_0).$$

Proof   The proof of each part involves two steps. First we must establish the existence of a derivative, then show that it is continuous. Because the right side of each formula in the theorem is constructed using continuous functions according to rules that preserve continuity, the continuity of the derivatives follows immediately once we establish the formulas (5)–(8).

We will prove the product rule (6) to illustrate to you that the methods from calculus apply here. The proofs of (5) and (7) are relegated to the exercises. The chain rule (8) is more delicate–we will prove it in the appendix to this section, where we use a new formalism to express differentiability.

Appealing to the definition of the derivative and using the continuity of g (Theorem 1), we have

$$
\begin{aligned}
(fg)'(z_0) &= \lim_{z \to z_0} \frac{f(z)g(z) - f(z_0)g(z_0)}{z - z_0} \\
&= \lim_{z \to z_0} \frac{f(z)g(z) - f(z_0)g(z) + f(z_0)g(z) - f(z_0)g(z_0)}{z - z_0} \\
&= \lim_{z \to z_0} g(z)\frac{f(z) - f(z_0)}{z - z_0} + \lim_{z \to z_0} f(z_0)\frac{g(z) - g(z_0)}{z - z_0} \\
&= g(z_0)f'(z_0) + f(z_0)g'(z_0). \qquad \blacksquare
\end{aligned}
$$

EXAMPLE 2   Analyticity of z^n for $n = 1, 2, \ldots$

Use (1) to show that for $n = 1, 2, \ldots$

$$(9) \qquad \frac{d}{dz}z^n = nz^{n-1}.$$

Conclude that $f(z) = z^n$ is entire.

Solution   We give a proof by induction. The case $n = 1$ was already stated in (4). Suppose as an induction hypothesis that (9) holds for n; we will prove that it holds for $n + 1$. Given $h(z) = z^{n+1}$, write it as a product, $h(z) = z^n z$. Applying the product rule for differentiation (6) and the induction hypothesis, we get

$$h'(z) = (z^n)'z + z^n z' = nz^{n-1}z + z^n = (n+1)z^n,$$

as desired. Hence (9) holds for all n. Looking at the right side of (9), it is clear that the derivative of $f(z) = z^n$ is continuous for all z. Hence z^n is entire.   ■

EXAMPLE 3   Analyticity of a rational function

Find the derivative of

$$f(z) = \frac{(z+1)(z+i)^2}{z+1-3i},$$

and determine where $f(z)$ is analytic.

Solution   The formal manipulations are exactly as if we were working with a real function and treating the complex numbers as real constants. We use the quotient and product rules for differentiation, (7) and (6), and get

$$
\begin{aligned}
f'(z) &= \frac{((z+i)^2 + (z+1)2(z+i))(z+1-3i) - (z+1)(z+i)^2}{(z+1-3i)^2} \\
&= \frac{2z^3 + (4-7i)z^2 + (14-2i)z + (6+5i)}{(z+1-3i)^2}.
\end{aligned}
$$

The function is analytic at all z except at $z = -1 + 3i$, where the denominator vanishes.   ■

By using linear combinations of powers of z and appealing to the result of Example 2 and the linearity of the derivative (5), we conclude that a polynomial is an entire function. By appealing now to the quotient rule, as we did in Example 3, we see that a rational function is analytic at all z where $g(z) \neq 0$.

THEOREM 3
POLYNOMIALS AND
RATIONAL
FUNCTIONS

Let n be a nonnegative integer. A polynomial of degree n, $p(z) = a_n z^n + a_{n-1}z^{n-1} + \cdots + a_1 z + a_0$, is entire. Its derivative is given by

(10) $$p'(z) = n\, a_n z^{n-1} + (n-1)\, a_{n-1} z^{n-1} + \cdots + a_1 z.$$

A rational function $f(z) = p(z)/q(z)$, where $p(z)$ and $q(z)$ are polynomials, is analytic at all points z where $q(z) \neq 0$. Its derivative is given by

(11) $$f'(z) = \frac{p'(z)q(z) - p(z)q'(z)}{q(z)^2}.$$

Typically, any function that algebraically manipulates z will be differentiable; however, not every complex-valued function is analytic. Functions like $\operatorname{Re} z$, $\operatorname{Im} z$, $\bar{z}$, and $|z|$ will have at best limited differentiability.

EXAMPLE 4   Functions that are nowhere analytic

Show that the functions

(a) $f(z) = \overline{z}$ and   (b) $f(z) = \operatorname{Re} z$

are not analytic at any point.

Solution   In order to show that a function $f(z)$ is not analytic at a point z_0, we will show that the limit (1) that defines the derivative does not exist. For this purpose, we will approach z_0 from two different directions and show that the limits that we obtain are not equal. Since the value of a limit must be unique, we will then conclude that the limit and hence the derivative do not exist.

(a)   Fix a point $z_0 = x_0 + i\, y_0$ in the plane. Our goal is to show that the limit

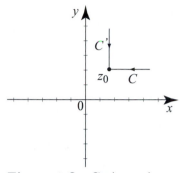

Figure 1 On C, $\Delta z = \Delta x$.
On C', $\Delta z = i \Delta y$.

$$(12) \qquad \lim_{z \to z_0} \frac{f(z) - f(z_0)}{z - z_0}$$

does not exist. We will approach z_0 from the two directions as indicated in Figure 1. For z on C, we have

$$z = z_0 + \Delta x; \quad z - z_0 = \Delta x; \quad f(z) - f(z_0) = \overline{z} - \overline{z_0} = \overline{z - z_0} = \overline{\Delta x} = \Delta x,$$

because Δx is real. Thus,

$$\lim_{\substack{z \to z_0 \\ z \text{ on } C}} \frac{f(z) - f(z_0)}{z - z_0} = \lim_{\substack{z \to z_0 \\ z \text{ on } C}} \frac{\Delta x}{\Delta x} = \lim_{\substack{z \to z_0 \\ z \text{ on } C}} 1 = 1.$$

For z on C', we have

$$z = z_0 + i\, \Delta y; \quad z - z_0 = i\, \Delta y; \quad \overline{z} - \overline{z_0} = \overline{z - z_0} = \overline{i\, \Delta y} = -i\, \Delta y,$$

because $i\, \Delta y$ is purely imaginary. Thus,

$$\lim_{\substack{z \to z_0 \\ z \text{ on } C'}} \frac{f(z) - f(z_0)}{z - z_0} = \lim_{\substack{z \to z_0 \\ z \text{ on } C'}} \frac{-i\, \Delta y}{i\, \Delta y} = \lim_{\substack{z \to z_0 \\ z \text{ on } C'}} -1 = -1.$$

Since the limit along C is not equal to the limit along C', we conclude that the limit in (12) does not exist. Hence the function $f(z) = \overline{z}$ is nowhere analytic.

(b)   We take the same approach as in (a) and use the same directions along C and C'. It is easy to check that for z on C $f(z) - f(z_0) = \operatorname{Re} z - \operatorname{Re} z_0 = x_0 + \Delta x - x_0 = \Delta x$, while for z on C' $f(z) - f(z_0) = \operatorname{Re} z - \operatorname{Re} z_0 = x_0 - x_0 = 0$. Using this information, we obtain

$$\lim_{\substack{z \to z_0 \\ z \text{ on } C}} \frac{f(z) - f(z_0)}{z - z_0} = \lim_{\substack{z \to z_0 \\ z \text{ on } C}} \frac{\Delta x}{\Delta x} = 1$$

and

$$\lim_{\substack{z \to z_0 \\ z \text{ on } C'}} \frac{f(z) - f(z_0)}{z - z_0} = \lim_{\substack{z \to z_0 \\ z \text{ on } C'}} \frac{0}{i\, \Delta y} = 0.$$

This shows that the limit defining the derivative of $\operatorname{Re} z$ does not exist at any point; and hence $\operatorname{Re} z$ is nowhere analytic.

There is another quick proof of (b) based on the result of (a) and the identity $\bar{z} = 2\,\mathrm{Re}\,z - z$. In fact, if $\mathrm{Re}\,z$ has a derivative at z_0, then by the properties of the derivative it would follow that $\bar{z}$ has a derivative at z_0, which contradicts (a).   ∎

Other interesting examples of functions that fail to be analytic at certain points in the plane are presented in the exercises.

So far we have been successful at differentiating polynomials and rational functions. To go beyond these examples we need more tools. In Section 2.4 we will present the Cauchy-Riemann equations, which are the most important tools for checking analyticity. The following result, which is an inside-out chain rule of sorts, is useful when dealing with inverse functions such as logarithms and powers.

THEOREM 4
ANALYTICITY OF
COMPOSED
FUNCTIONS

Suppose that $h(z) = f(g(z))$ is analytic on a region Ω, and f is analytic at $g(z)$ with $f'(g(z)) \neq 0$ for all z in Ω. Suppose further that $g(z)$ is continuous on Ω (Figure 2). Then g is analytic on Ω and

(13)
$$g'(z) = \frac{h'(z)}{f'(g(z))}.$$

As in the case of the chain rule, the proof of this theorem will be greatly simplified by using the formalisms presented in the appendix. We postpone the proof and give instead an application.

Figure 2 Unlike the chain rule, where we suppose that f and g are analytic and conclude that $h = g \circ f$ is analytic, in Theorem 4, we suppose that g is continuous, and f and the composed function h are analytic, and then we conclude that g is analytic.

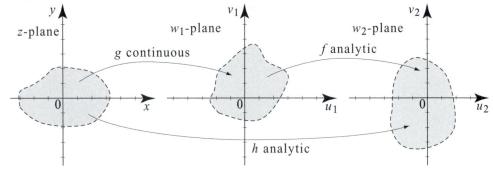

Figure 3 Branch cut of $\mathrm{Log}\,z$.

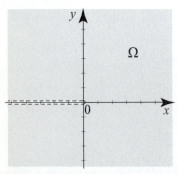

EXAMPLE 5 Analyticity of nth roots

Show that the principal branch of the nth root, $g(z) = z^{1/n}$ ($n = 0, 1, 2, \ldots$), is analytic in the region Ω that consists of all $z \neq 0$ and not on the negative real axis. Also show that for z in Ω, we have

$$g'(z) = \frac{1}{n} z^{(1-n)/n}.$$

Solution   From (10), Section 1.7, we have

$$g(z) = e^{\frac{1}{n}\,\mathrm{Log}\,z},$$

where $\mathrm{Log}\,z$ is the principal branch of the logarithm. We showed in the previous section that e^z is continuous for all z, and since $\mathrm{Log}\,z$ is continuous in Ω (see

Figure 3), it follows that $g(z)$ is continuous in Ω, being the composition of two continuous functions. Taking $f(z) = z^n$, $h(z) = f(g(z)) = (z^{1/n})^n = z$, we see clearly that f and h are analytic, and thus the hypotheses of Theorem 4 are satisfied. Consequently, $g(z)$ is analytic on Ω and

$$g'(z) = \frac{h'(z)}{f'(g(z))} = \frac{1}{n(z^{1/n})^{n-1}} = \frac{1}{n}z^{(1-n)/n}. \qquad \blacksquare$$

Appendix: Proofs Related to Differentiation

Suppose that $f(z)$ has a derivative at z_0 and let

$$(14) \qquad \epsilon(z) = \frac{f(z) - f(z_0)}{z - z_0} - f'(z_0).$$

Then $\epsilon(z) \to 0$ as $z \to z_0$, because the difference quotient in (14) tends to $f'(z_0)$. Solving for $f(z)$ in (14) we obtain

$$(15) \qquad f(z) = \overbrace{f(z_0) + f'(z_0)(z - z_0)}^{\text{linear function of } z} + \epsilon(z)(z - z_0).$$

This expression shows that, near a point where f is differentiable, $f(z)$ is approximately a linear function. The converse is also true. We summarize these results as follows.

THEOREM 5
DIFFERENTIABILITY

> A function $f(z)$ is differentiable at z_0 if and only if it can be written in the form
>
> $$(16) \qquad f(z) = f(z_0) + A(z - z_0) + \epsilon(z)(z - z_0),$$
>
> where $A = f'(z_0)$ and $\epsilon(z) \to 0$ as $z \to z_0$.

Proof   We have already established the theorem in one direction. For the other direction, suppose that $f(z)$ can be written as in (16). Then, for $z \neq z_0$,

$$(17) \qquad \frac{f(z) - f(z_0)}{z - z_0} = A + \epsilon(z).$$

Taking the limit as $z \to z_0$ and using the fact that $\epsilon(z) \to 0$, we conclude that $f'(z_0)$ exists and equals A. $\qquad \blacksquare$

The formalism of Theorem 5 simplifies greatly proofs related to differentiation.

Proof of the chain rule   Suppose g is analytic at z_0 and f is analytic at $g(z_0)$. We want to show that

$$(18) \qquad (f \circ g)'(z_0) = f'(g(z_0))g'(z_0).$$

Since g is differentiable at z_0, appealing to Theorem 5, we can write

$$(19) \qquad \frac{g(z) - g(z_0)}{z - z_0} = g'(z_0) + \epsilon(z), \qquad \epsilon(z) \to 0 \text{ as } z \to z_0.$$

Also, f is differentiable at $g(z_0)$, so by Theorem 5 we can write

$$f(w) - f(g(z_0)) = f'(g(z_0))(w - g(z_0)) + \eta(w)(w - g(z_0)), \qquad \eta(w) \to 0 \text{ as } w \to g(z_0).$$

Replacing w by $g(z)$, dividing by $z - z_0$, and using (19), we obtain

$$(20) \qquad \frac{f(g(z)) - f(g(z_0))}{z - z_0} = f'(g(z_0))\Big(g'(z_0) + \epsilon(z)\Big) + \eta(g(z))\Big(g'(z_0) + \epsilon(z)\Big).$$

As $z \to z_0$, $\epsilon(z) \to 0$, $g(z) \to g(z_0)$ by continuity, and so $\eta(g(z)) \to 0$. Using this in (20), we conclude that

$$\lim_{z \to z_0} \frac{f(g(z)) - f(g(z_0))}{z - z_0} = f'(g(z_0))g'(z_0),$$

as asserted by the chain rule. ∎

Proof of Theorem 4 Let z_0 be in Ω. Given $h(z) = f(g(z))$ analytic at z_0, f analytic at $g(z_0)$ with $f'(g(z_0)) \neq 0$, and g continuous at z_0, once we show that

$$(21) \qquad g'(z_0) = \frac{h'(z_0)}{f'(g(z_0))},$$

then since h', f' and g are continuous and $f'(g(z_0)) \neq 0$, it will follow that g' is continuous at z_0 and hence g is analytic at z_0. Applying Theorem 5 to $h(z) = f(g(z))$, we have

$$(22) \qquad f(g(z)) = f(g(z_0)) + h'(z_0)(z - z_0) + \epsilon(z)(z - z_0), \quad \epsilon(z) \to 0 \text{ as } z \to z_0.$$

Applying Theorem 5 to f at $g(z_0)$, we have

$$(23) \qquad f(g(z)) = f(g(z_0)) + f'(g(z_0))(g(z) - g(z_0)) + \eta(g(z))(g(z) - g(z_0)),$$

where $\eta(g(z)) \to 0$ as $g(z) \to g(z_0)$ or, equivalently, as $z \to z_0$ by continuity of g at z_0. Subtract (23) from (22) and rearrange the terms to get

$$(24) \qquad \frac{g(z) - g(z_0)}{z - z_0} = \frac{h'(z_0) + \epsilon(z)}{f'(g(z_0)) + \eta(g(z))}.$$

As $z \to z_0$, $\epsilon(z) \to 0$ and $\eta(g(z)) \to 0$, implying (21). ∎

Exercises 2.3

In Exercises 1–12, determine the set on which the given function is analytic and compute its derivative. In Exercises 9–12, use the principal branch of the power.

1. $3(z - 1)^2 + 2(z - 1)$.

2. $z^3 + \dfrac{z}{1 + i}$.

3. $\operatorname{Im} z$.

4. $\left(\dfrac{z - 2 + i}{z - 1 + i}\right)^2$.

5. $\dfrac{1}{z^3 + 1}$.

6. $8\bar{z} + i$.

7. $\dfrac{1}{z^2 - (1 - 2i)z - 3 - i}$.

8. $\dfrac{1}{z^2 + (1 + 2i)z + 3 - i}$.

9. $z^{2/3}$.

10. $(z - 1)^{\frac{1}{2}}$.

11. $(z - 3 + i)^{1/10}$.

12. $\dfrac{1}{(z + 1)^{1/2}}$.

In Exercises 13–16, evaluate the given limit by identifying it with a derivative at a point. In Exercises 15–16, use the principal branch of the power.

13. $\displaystyle\lim_{z \to 1} \frac{z^{100} - 1}{z - 1}$.

14. $\displaystyle\lim_{z \to i} \frac{z^{99} + i}{z - i}$.

15. $\lim_{z \to 0} \dfrac{1}{z\sqrt{1+z}} - \dfrac{1}{z}.$

16. $\lim_{z \to 1} \dfrac{z^{1/3} - 1}{z - 1}.$

17. Determine the set on which

$$f(z) = \begin{cases} z & \text{if } |z| \leq 1, \\ z^2 & \text{if } |z| > 1, \end{cases}$$

is analytic and compute its derivative. Justify your answer.

18. (a)   Show that the derivative of $f(z) = |z|^2$ exists at $z = 0$ and does not exist at all other points. [Hint: Proceed as in Example 4.]
(b)   At which points is f analytic?

19. Show that $f(z) = |z|$ is nowhere differentiable. [Hint: Compute the limit in (1) by letting $z = \alpha z_0$ with $\alpha > 0$; then $\alpha < 0$.]

20. For this exercise, refer to (10), Section 1.7.
(a)   Show that the three branches of $z^{1/3}$ are

$$b_1(z) = e^{\frac{1}{3} \operatorname{Log} z}; \qquad b_2(z) = e^{\frac{1}{3} \operatorname{Log} z} e^{\frac{2\pi i}{3}}; \qquad b_3(z) = e^{\frac{1}{3} \operatorname{Log} z} e^{\frac{4\pi i}{3}}.$$

(b)   Use Theorem 4 to show that

$$b_j'(z) = \frac{b_j(z)}{3z} \quad (j = 1,\, 2,\, 3).$$

21. Refer to (10), Section 1.7. Use Theorem 4 to show that for any integer p and positive integer q,

$$\frac{d}{dz} z^{p/q} = \frac{p}{qz} z^{p/q},$$

where we are using the same branch of the power on both sides. [Hint: In applying Theorem 4, take $g(z) = z^{p/q}$ and $f(z) = z^q$.]

22. Linearity of the derivative.   Prove (5) using the definition (1) and appealing to the linearity of limits, (3), Section 2.2.

23. Quotient rule. (a)   Prove the following special case of the quotient rule (7):

(25)
$$\left(\frac{1}{f}\right)'(z_0) = -\frac{f'(z_0)}{(f(z_0))^2} \quad (f(z_0) \neq 0).$$

[Hint: Start by writing the difference quotient for $\frac{1}{f(z)}$ as

$$\frac{\frac{1}{f(z)} - \frac{1}{f(z_0)}}{z - z_0} = -\frac{1}{f(z)f(z_0)} \frac{f(z) - f(z_0)}{z - z_0}.]$$

(b)   Prove the quotient rule (7) by using the product rule (6) and (25).

24. Product rule.   We proved the product rule (6) in the text by mimicking the usual proof from calculus. Provide a shorter proof by using Theorem 5.

25. Derivative of z^n.   In the text we showed that for positive integers n, $\frac{d}{dz} z^n = nz^{n-1}$. Construct a different proof starting with the definition (1) and using the identity

$$z^n - z_0^n = (z - z_0)(z^{n-1} + z^{n-2} z_0 + \cdots + z_0^{n-1}).$$

26. Derivative of z^n, n negative. Show that the formula in Example 2 holds for negative n where $z \neq 0$. Conclude that z^n is analytic for all $z \neq 0$, when n is a negative integer.

27. L'Hospital's rule. Prove the following version of L'Hospital's rule. If $f(z)$ and $g(z)$ are differentiable at z_0 and $f(z_0) = g(z_0) = 0$, but $g'(z_0) \neq 0$, then

$$\lim_{z \to z_0} \frac{f(z)}{g(z)} = \frac{f'(z_0)}{g'(z_0)}.$$

[Hint: $\dfrac{f(z)}{g(z)} = \dfrac{f(z) - f(z_0)}{z - z_0} \dfrac{1}{\frac{g(z) - g(z_0)}{z - z_0}}$. Another way to proceed is to use Theorem 5.]

28. Find the following limits by using L'Hospital's rule.

(a) $\displaystyle \lim_{z \to i} \frac{(z^2 + 1)^7}{z^6 + 1}$.   (b) $\displaystyle \lim_{z \to i} \frac{z^3 + (1 - 3i)z^2 + (i - 3)z + 2 + i}{z - i}$.

2.4   The Cauchy-Riemann Equations

The fact that the derivative of $\bar{z}$ does not exist (Example 4, Section 2.3) should tell you that there is something special about complex-valued functions with derivatives. As you will see, the existence of the derivative implies special relationships between the real and imaginary parts of the functions. These relationships are known as the Cauchy-Riemann equations, which we now derive.

Suppose that $f(z) = f(x + iy) = u(x, y) + iv(x, y)$ is analytic in an open set S and let $z_0 = x_0 + iy_0 = (x_0, y_0)$ be a point in S. Consequently, the derivative at z_0 exists and is given by

(1) $f'(z_0) = \displaystyle \lim_{z \to z_0} \frac{f(z) - f(z_0)}{z - z_0} = \lim_{\Delta z \to 0} \frac{f(z_0 + \Delta z) - f(z_0)}{\Delta z}.$

To derive the Cauchy-Riemann equations, we will simply compute this limit as z approaches z_0 from two different directions and equate the results, both being equal to $f'(z_0)$.

Suppose that z approaches z_0 along the direction of the x-axis, as in Figure 1. Then $z = z_0 + \Delta x = (x_0 + \Delta x, y_0)$, $\Delta z = z - z_0 = \Delta x$, and (1) becomes

(2) $f'(z_0) = \displaystyle \lim_{\Delta x \to 0} \frac{f(x_0 + \Delta x + iy_0) - f(x_0 + iy_0)}{\Delta x}$

$= \displaystyle \lim_{\Delta x \to 0} \left(\frac{u(x_0 + \Delta x, y_0) - u(x_0, y_0)}{\Delta x} + i \frac{v(x_0 + \Delta x, y_0) - v(x_0, y_0)}{\Delta x} \right)$

$= \displaystyle \lim_{\Delta x \to 0} \frac{u(x_0 + \Delta x, y_0) - u(x_0, y_0)}{\Delta x} + i \lim_{\Delta x \to 0} \frac{v(x_0 + \Delta x, y_0) - v(x_0, y_0)}{\Delta x},$

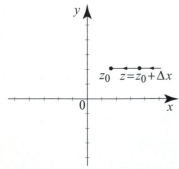

Figure 1 For z approaching z_0 in the direction of the x-axis, $\Delta z = \Delta x$.

where the last step is justified by Theorem 3, Section 2.2, which asserts that the limit of a complex-valued function exists if and only if the limits of its real and imaginary parts exist. Recognizing the last two limits on the right as the partial derivatives with respect to x of u and v (see (2) and (3), Section 2.1), we obtain

$$(3) \qquad \boxed{f'(z_0) = \frac{\partial u}{\partial x}(x_0, y_0) + i\frac{\partial v}{\partial x}(x_0, y_0),}$$

which is an expression of the derivative of f in terms of the partial derivatives with respect to x of u and v. We now repeat the preceding steps, going back to (1) and taking the limit as z approaches z_0 from the direction of the y-axis, as in Figure 2. Then $z = z_0 + i\,\Delta y = (x_0,\, y_0 + \Delta y)$, $\Delta z = z - z_0 = i\,\Delta y$. Proceed as in (2), note that $i\Delta y \to 0$ if and only if $\Delta y \to 0$, and obtain

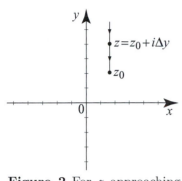

Figure 2 For z approaching z_0 in the direction of the y-axis, $\Delta z = i\,\Delta y$.

$$(4) \quad f'(z_0) = \lim_{\Delta y \to 0} \frac{f(x_0 + i(y_0 + \Delta y)) - f(x_0 + i\,y_0)}{i\,\Delta y}$$

$$= \lim_{\Delta y \to 0} \left(\frac{u(x_0,\, y_0 + \Delta y) - u(x_0,\, y_0)}{i\,\Delta y} + i\,\frac{v(x_0,\, y_0 + \Delta y) - v(x_0,\, y_0)}{i\,\Delta y} \right)$$

$$= \lim_{\Delta y \to 0} \frac{v(x_0,\, y_0 + \Delta y) - v(x_0,\, y_0)}{\Delta y} - i \lim_{\Delta y \to 0} \frac{u(x_0,\, y_0 + \Delta y) - u(x_0,\, y_0)}{\Delta y},$$

where in the last step we have used $1/i = -i$ and rearranged the terms. Recognizing the partial derivatives with respect to y of v and u, we obtain

$$(5) \qquad \boxed{f'(z_0) = \frac{\partial v}{\partial y}(x_0, y_0) - i\frac{\partial u}{\partial y}(x_0, y_0),}$$

which is this time an expression of the derivative of f in terms of the partial derivatives with respect to y of u and v. Now comes the startling result. Equate real and imaginary parts in (3) and (5) and get

THE CAUCHY-RIEMANN EQUATIONS

$$(6) \qquad \frac{\partial u}{\partial x} = \frac{\partial v}{\partial y} \qquad \text{and} \qquad \frac{\partial u}{\partial y} = -\frac{\partial v}{\partial x}.$$

Thus, in order for f to be analytic, u and v must satisfy the Cauchy-Riemann equations. Moreover, since f is analytic, f' is continuous, and it follows from (3) and (5) and Theorem 4(iv), Section 2.2, that $\frac{\partial u}{\partial x}$, $\frac{\partial v}{\partial x}$, $\frac{\partial u}{\partial y}$, and $\frac{\partial v}{\partial y}$ are all continuous. The converse of these statements is also true. In applications, we will need the converse, since to establish the analyticity of a function, it is often easier to check the Cauchy-Riemann equations and the continuity of the partial derivatives. Because the proof of the converse requires advanced topics from calculus of several variables, we postpone it until Section 2.6, where it will be treated in detail. Let us now summarize our discussion and proceed with the applications. To simplify the notation, we will denote partial derivatives with subscripts, $\frac{\partial u}{\partial x} = u_x$, $\frac{\partial u}{\partial y} = u_y$, and so on.

**THEOREM 1
CAUCHY-RIEMANN
EQUATIONS**

The function $f(z) = u(x, y) + i\, v(x, y)$ is analytic in an open set S if and only if the partial derivatives of u and v are continuous and satisfy the Cauchy-Riemann equations

$$(7) \qquad\qquad u_x = v_y \qquad \text{and} \qquad u_y = -v_x.$$

The derivative $f'(z)$ is given as either of

$$(8) \qquad f'(z) = u_x(x, y) + i\, v_x(x, y) \text{ or } f'(z) = v_y(x, y) - i\, u_y(x, y).$$

The Cauchy-Riemann equations appeared in 1821 in the early work of Cauchy on integrals of complex-valued functions. Their connection to the existence of the derivative, as stated in Theorem 1, appeared in 1851 in the doctoral dissertation of the great German mathematician, Bernhard Riemann (1826–1866). As it should be obvious to any one who has studied calculus, Riemann's work shaped the development of modern calculus. His study of complex functions was motivated by his work in hydrodynamics, fluid flow and other applied problems. Some of these applications and their connection to the Cauchy-Riemann equations will be explored in the following section and Chapter 6.

Our first example is perhaps the most important example of an analytic function.

EXAMPLE 1 e^z is entire.
Show that e^z is entire and

$$(9) \qquad\qquad \frac{d}{dz}e^z = e^z.$$

Solution   We will use Theorem 1. From (4), Section 1.5, we have $e^z = e^x \cos y + i\, e^x \sin y$. Thus, $u(x, y) = e^x \cos y$ and $v(x, y) = e^x \sin y$. Differentiating u with respect to x and y, we find

$$\frac{\partial u}{\partial x} = e^x \cos y, \qquad \frac{\partial u}{\partial y} = -e^x \sin y.$$

Differentiating v with respect to x and y, we find

$$\frac{\partial v}{\partial x} = e^x \sin y, \qquad \frac{\partial v}{\partial y} = e^x \cos y.$$

Comparing these derivatives, we see clearly that $\frac{\partial u}{\partial x} = \frac{\partial v}{\partial y}$ and $\frac{\partial u}{\partial y} = -\frac{\partial v}{\partial x}$. Hence the Cauchy-Riemann equations are satisfied at all points. The continuity of the partial derivatives follows from Proposition 1, Section 2.2 (see the discussion preceding the proposition). Appealing to Theorem 1, we conclude that e^z is analytic at all points, or entire. For the derivative, we appeal to either formula from (8), say the first, and use the formulas that we derived for the partial derivatives of u and v. We get

$$\frac{d}{dz}e^z = \frac{\partial u}{\partial x} + i\frac{\partial v}{\partial x} = e^x \cos y + i\, e^x \sin y = e^z. \qquad \blacksquare$$

Combining the chain rule with the result of Example 1, we see that $e^{f(z)}$ is analytic wherever $f(z)$ is analytic and

(10)
$$\boxed{\frac{d}{dz}e^{f(z)} = f'(z)e^{f(z)}.}$$

EXAMPLE 2 $\sin z$ **is entire.**
Show that $\sin z$ is entire and

(11)
$$\frac{d}{dz}\sin z = \cos z.$$

Solution   There are at least two ways to do this problem. Since

$$\sin z = \frac{e^{iz} - e^{-iz}}{2i},$$

(see (3), Section 1.6), we can appeal to (10) and conclude that $\sin z$ is entire and

$$\frac{d}{dz}\sin z = \frac{d}{dz}\left(\frac{e^{iz} - e^{-iz}}{2i}\right) = \frac{ie^{iz} - (-i)e^{-iz}}{2i} = \frac{e^{iz} + e^{-iz}}{2} = \cos z,$$

by (2), Section 1.6. The second method is a little longer but it will afford us the opportunity to practice the Cauchy-Riemann equations. From (16), Section 1.6, we have

$$\sin z = \sin x \cosh y + i \cos x \sinh y.$$

We have

$$u(x, y) = \sin x \cosh y \quad \Rightarrow \quad u_x(x, y) = \cos x \cosh y, \quad u_y(x, y) = \sin x \sinh y;$$
$$v(x, y) = \cos x \sinh y \quad \Rightarrow \quad v_x(x, y) = -\sin x \sinh y, \quad v_y(x, y) = \cos x \cosh y.$$

Comparing partial derivatives, we see that the Cauchy-Riemann equations are satisfied at all points. Moreover, the partial derivatives are continuous. Appealing to Theorem 1, we see that $\sin z$ is entire and

$$\frac{d}{dz}\sin z = u_x(x, y) + i\, v_x(x, y) = \cos x \cosh y - i\, \sin x \sinh y = \cos z,$$

by (15), Section 1.6. $\blacksquare$

Following the methods of Example 2, we can check the analyticity and compute the derivatives of $\cos z$, $\tan z$, and all other trigonometric and hyperbolic functions. Among the elementary functions, we are then left with the logarithm and those functions defined using the logarithm, such as complex powers. To handle them, we will appeal to Theorem 4, Section 2.3.

EXAMPLE 3   Log z is analytic except on the branch cut.

Show that the principal branch of the logarithm, Log z, is analytic on $\mathbb{C} \setminus (-\infty, 0]$, and that

$$
(12) \qquad \boxed{\frac{d}{dz} \operatorname{Log} z = \frac{1}{z}.}
$$

Thus the familiar formula from calculus still holds.

Solution   The result is straightforward from Theorem 4, Section 2.3, and the fact that e^z is entire with $\frac{d}{dz} e^z = e^z$. In the notation of Theorem 4, Section 2.3, set $f(z) = e^z$, $g(z) = \operatorname{Log} z$, and $h(z) = z$. Since $h(z)$ is analytic, $f(z)$ is analytic with $f'(z) = e^z \neq 0$, and $g(z)$ is continuous everywhere except on the branch cut, we conclude that $g(z)$ is analytic there with

$$
\frac{d}{dz} \operatorname{Log} z = \frac{h'(z)}{f'(g(z))} = \frac{1}{e^{\log z}} = \frac{1}{z}.
$$

The logarithm cannot be analytic on the branch cut, because it is not continuous there (see Theorem 5, Section 2.2).   ∎

Using the method of Example 3, we can show that any branch of the logarithm, $\log_\alpha z$, is analytic everywhere except at its branch cut (the ray at angle α), and

$$
(13) \qquad \boxed{\frac{d}{dz} \log_\alpha z = \frac{1}{z}.}
$$

As a consequence, any function constructed from the logarithms according to rules that preserve analyticity will be analytic on an appropriate domain. As an illustration, take the principal branch of a power,

$$
z^a = e^{a \operatorname{Log} z} \qquad \text{(where } a \neq 0 \text{ is a complex number)}.
$$

Since Log z is analytic except at its branch cut, and e^z is entire, it follows that z^a is analytic, except at the branch cut of Log z. To compute its derivative, we use the chain rule and the derivatives of e^z and Log z and get

$$
(14) \qquad \boxed{\frac{d}{dz} z^a = a\, z^{a-1},}
$$

with principal branches of the power on both sides.

We started this section with the example of $\bar{z}$ and how it fails to be analytic at all points. This should be obvious now from the Cauchy-Riemann equations. If you write $\bar{z} = x - i\,y$, then $u_x = 1$, $v_y = -1$, showing that the

Cauchy-Riemann equations do not hold at any point. As our next example illustrates, we can use the Cauchy-Riemann equations to show the failure of analyticity in less obvious situations.

EXAMPLE 4   Failure of analyticity

Use the Cauchy-Riemann equations to show that $f(z) = x^2 + i\,(2y + x)$ fails to be analytic at all points.

Solution   We have $u(x, y) = x^2$ and $v(x, y) = 2y + x$. Since $u_y = 0$ and $v_x = 1$, the Cauchy-Riemann equations are not satisfied at any point and hence the function cannot be analytic at any point.   ∎

The Cauchy-Riemann equations may hold only at one point (see Exercise 13 for an illustration). This does not imply that the function is analytic at that point, since analyticity requires that these equations be satisfied in a neighborhood.

The final applications show how the Cauchy-Riemann equations can be used to derive important results of theoretical nature.

THEOREM 2   Suppose that $f(z) = u(x, y) + i\,v(x, y)$ is analytic in a region (open and connected) Ω and $f'(z) = 0$ for all z in Ω. Then f is constant in Ω.

Proof   By Theorem 1, we have

$$f'(z) = u_x(x, y) + i\,v_x(x, y) \text{ and } f'(z) = v_y(x, y) - i\,u_y(x, y), \quad z \text{ in } \Omega.$$

Since $f'(z) = 0$ it follows that $u_x = u_y = 0$ and $v_x = v_y = 0$ in Ω. Appealing to Theorem 1, Section 2.1, we conclude that u is constant and v is constant, and hence $f = u + i\,v$ is constant.   ∎

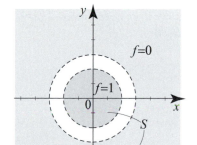

Figure 3 A nonconstant function with zero derivative. Its domain of definition is the nonconnected shaded area.

The connectedness of Ω in the theorem is essential. For example, the function f, defined on the set S in Figure 3, by

$$f(z) = \begin{cases} 1 & \text{if } |z| < 2, \\ 0 & \text{if } |z| > 3, \end{cases}$$

has zero derivative but is not constant.

COROLLARY 1   Suppose that f and g are analytic in a region Ω. If either $\operatorname{Re} f = \operatorname{Re} g$ on Ω or $\operatorname{Im} f = \operatorname{Im} g$ on Ω, then $f(z) = g(z) + c$ on Ω, where c is a constant.

Proof   We do the case $\operatorname{Re} f = \operatorname{Re} g$ on Ω. The case $\operatorname{Im} f = \operatorname{Im} g$ is very similar and is left to Exercise 32. Let $h(z) = f(z) - g(z) = u(z) + iv(z)$. We want to show that $h(z) = c$ on Ω. Since h is analytic, it is enough by Theorem 1 to show that $h'(z) = 0$ on Ω. We have $u = \operatorname{Re} h = \operatorname{Re} f - \operatorname{Re} g = 0$ on Ω, and so $u_x = u_y = 0$ on Ω. By the Cauchy-Riemann equations, $v_x = -u_y = 0$. Consequently, by (8), $h'(z) = u_x + iv_x = 0$ on Ω.   ∎

Exercises 2.4

In Exercises 1–14, use the Cauchy-Riemann equations (Theorem 1) to determine the set on which the given function is analytic and compute its derivative using either one of equations (8).

1. z.

2. z^2.

3. e^{z^2}.

4. $2x + 3i\,y$.

5. $e^{\bar{z}}$.

6. $\dfrac{y - i\,x}{x^2 + y^2}$.

7. $\left(\dfrac{1}{z + 1}\right)^2$.

8. $\dfrac{z}{z - i}$.

9. ze^z.

10. $\cos z$.

11. $\tan z$.

12. $\cosh z$.

13. $|z|^2$.

14. $\dfrac{x^4 + i2xy(x^2 + y^2) - y^4 + x - i\,y}{x^2 + y^2}$.

In Exercises 15–26, use properties of the derivative to compute the derivative of the given function and determine the set on which it is analytic. In Exercises 23 – 26, use the principal branch of the power.

15. ze^{z^2}.

16. $(1 + e^z)^5$.

17. $\sin z \cos z$.

18. $\text{Log}\,(z + 1)$.

19. $\dfrac{\text{Log}\,(3z - 1)}{z^2 + 1}$.

20. $\sinh(3z + i)$.

21. $\cosh(z^2 + 3i)$.

22. $\log_{\frac{\pi}{2}}(z + 1)$.

23. z^i.

24. $(z + 1)^{1/2}$.

25. $\dfrac{1}{(z - i)^{1/2}}$.

26. z^z.

In Exercises 27–30, compute the given limit by identifying it as a derivative; alternatively, you may use L'Hospital's rule (Exercise 27, Section 2.3).

27. $\displaystyle\lim_{z \to 0} \dfrac{\sin z}{z}$.

28. $\displaystyle\lim_{z \to 0} \dfrac{e^z - 1}{z}$.

29. $\displaystyle\lim_{z \to 0} \dfrac{\text{Log}\,(z + 1)}{z}$.

30. $\displaystyle\lim_{z \to i} \dfrac{1 + iz}{z(z - i)}$.

31. Define the **principal branch** of the inverse tangent by taking the principal branch of the logarithm in (13), Section 1.7:

$$\tan^{-1} z = \frac{i}{2} \text{Log}\left(\frac{1 - iz}{1 + iz}\right).$$

Compute the derivative of $\tan^{-1} z$.   **32.**   Complete the proof of Corollary 1 by treating the case $\text{Im}\,f = \text{Im}\,g$ on Ω.

33. Suppose that $f = u + iv$ is analytic in a region Ω. Show that
(a) $f'(z) = u_x - i\,u_y$, also $f'(z) = v_y + i\,v_x$; and
(b) $|f'(z)|^2 = u_x^2 + u_y^2 = v_x^2 + v_y^2$.
(c)   Conclude from (a) or (b) that if either $\text{Re}\,f$ or $\text{Im}\,f$ is constant in Ω, then f is constant in Ω.

34. Suppose that $f(z)$ and $\overline{f(z)}$ are analytic in a region Ω. Show that $f(z)$ must be constant in Ω. [Hint: Consider $f(z) + \overline{f(z)}$ and use Exercise 33.]

35. Let us define the partial derivatives of a complex-valued function $f = u + iv$ as $f_x = u_x + i\,v_x$ and $f_y = u_y + i\,v_y$. Show that the Cauchy-Riemann equations are equivalent to $f_x + i\,f_y = 0$.

36.   Suppose that f is analytic in a region Ω and $f[\Omega]$ is a subset of a line. Show that f must be constant in Ω. [Hint: Rotate the line to make it horizontal or vertical and apply one of Exercise 33.]

37.   Suppose that $f = u + iv$ is analytic in a region Ω and $\operatorname{Re} f = \operatorname{Im} f$. Show that f must be constant in Ω. [Hint: Use Exercise 36 or prove it directly from the Cauchy-Riemann equations.]

38.   Suppose that $f = u + iv$ is analytic in a region Ω and $|f|$ is constant in Ω. Show that f must be constant in Ω as follows.
(a)   Show that $u^2 + v^2 = c$ where c is a nonnegative constant, and that we need only consider the case $c > 0$.
(b)   Obtain the following system of equations in the unknowns u_x, u_y:

$$\begin{cases} u\,u_x - v\,u_y = 0 \\ v\,u_x + u\,u_y = 0. \end{cases}$$

[Hint: Differentiate $u^2 + v^2 = c$ with respect to x then differentiate it with respect to y and use the Cauchy-Riemann equations.]
(c)   Show that the only solutions of the equations in (b) are $u_x = 0$ and $u_y = 0$. [Hint: The determinant of the system is > 0.]
(d)   Show that v_x, v_y and f' are zero on Ω, and conclude that f is constant.

39. Project Problem: Cauchy-Riemann equations in polar form.   In this problem we express the Cauchy-Riemann equations in polar coordinates. Recall the relationships between Cartesian and polar coordinates:

$$x = r\cos\theta \quad \text{and} \quad y = r\sin\theta.$$

For convenience, we write $u(r, \theta) = u(r\cos\theta, r\sin\theta) = u(x, y)$ and $v(r, \theta) = v(r\cos\theta, r\sin\theta) = v(x, y)$.
(a)   The multivariable chain rule from calculus (see also (13), Section 2.6) states that

$$\frac{\partial u}{\partial r} = \frac{\partial u}{\partial x}\frac{\partial x}{\partial r} + \frac{\partial u}{\partial y}\frac{\partial y}{\partial r}, \quad \frac{\partial u}{\partial \theta} = \frac{\partial u}{\partial x}\frac{\partial x}{\partial \theta} + \frac{\partial u}{\partial y}\frac{\partial y}{\partial \theta},$$

$$\frac{\partial v}{\partial r} = \frac{\partial v}{\partial x}\frac{\partial x}{\partial r} + \frac{\partial v}{\partial y}\frac{\partial y}{\partial r}, \quad \frac{\partial v}{\partial \theta} = \frac{\partial v}{\partial x}\frac{\partial x}{\partial \theta} + \frac{\partial v}{\partial y}\frac{\partial y}{\partial \theta}.$$

Show that

$$\frac{\partial u}{\partial r} = \cos\theta\,\frac{\partial u}{\partial x} + \sin\theta\,\frac{\partial u}{\partial y}, \quad \frac{\partial u}{\partial \theta} = -r\sin\theta\,\frac{\partial u}{\partial x} + r\cos\theta\,\frac{\partial u}{\partial y},$$

$$\frac{\partial v}{\partial r} = \cos\theta\,\frac{\partial v}{\partial x} + \sin\theta\,\frac{\partial v}{\partial y}, \quad \frac{\partial v}{\partial \theta} = -r\sin\theta\,\frac{\partial v}{\partial x} + r\cos\theta\,\frac{\partial v}{\partial y}.$$

(b)   Derive the **polar form of the Cauchy-Riemann equations**:

(15) $$u_r = \frac{1}{r}v_\theta \quad \text{and} \quad v_r = -\frac{1}{r}u_\theta.$$

Thus we can state Theorem 1 in polar form as follows. The function $f(z) = u(r, \theta) + i\,v(r, \theta)$ is analytic at $z_0 \neq 0$ if and only if, in a neighborhood of z_0,

u_r, u_θ, v_r, and v_θ are continuous and satisfy the polar form of the Cauchy-Riemann equations.

(c)   Show that the derivative $f'(z)$ is given by

$$(16) \qquad\qquad f'(z) = e^{-i\theta}(u_r + i\,v_r).$$

This represents $f'(z)$ in terms of a radial directional derivative of u and v divided by the unimodular number $e^{i\theta}$ that gives the direction of $f'(z)$.

In Exercises 40–42, use the polar form of the Cauchy-Riemann equations to check the analyticity and find the derivative of the given function.

40. $f(z) = z^n = r^n(\cos(n\theta) + i\,\sin(n\theta))$ $(n = \pm 1, \pm 2, \dots)$.

41. $f(z) = \operatorname{Log} z = \ln|z| + i\operatorname{Arg} z$.   **42.** $f(z) = \sin(\operatorname{Log} z)$.

43. Jacobian of a transformation.   The **Jacobian** of the mapping $(x,\, y) \mapsto (u(x,\, y),\, v(x,\, y))$ is the following determinant:

$$J = \det \begin{vmatrix} \dfrac{\partial u}{\partial x} & \dfrac{\partial u}{\partial y} \\[2mm] \dfrac{\partial v}{\partial x} & \dfrac{\partial v}{\partial y} \end{vmatrix}.$$

Suppose that $f = u + iv$ is analytic. Show that the Jacobian equals $|f'(z)|^2$. You may recall from calculus of several variables that a mapping is locally invertible at every point where the Jacobian is nonzero. This exercise suggests that a similar result holds for f. Indeed, there is an **inverse function theorem** that states that if $f'(z_0) \neq 0$, then in some neighborhood of z_0, f is one-to-one and has an inverse function that is itself analytic in a neighborhood of $f(z_0)$ (see Section 5.7).

2.5   Harmonic Functions and Laplace's Equation

There are many important applications of complex analysis that highlight its pivotal place in the solution of real-world problems. The ones that we present in this section deal with a fundamental equation of applied mathematics, known as Laplace's equation. This equation models important phenomena in engineering and physics, such as steady-state temperature distributions, electrostatic potentials, and fluid flow, to name just a few. For clarity's sake, we will base our presentation around steady-state temperature problems. The solutions that we present use almost all the material that we have covered thus far, and more important, they stress the need for further development of the theory.

Laplace's Equation and Harmonic Functions

Consider a two-dimensional plate of homogeneous material, with insulated lateral surfaces. We represent this plate by a region Ω in the complex plane (see Figure 1). Suppose that the temperature of the points on the boundary of the plate is given by a function of position $b(x,\, y)$ that does not change with time. It is a fact from thermodynamics that the temperature inside the plate will eventually reach and remain at an equilibrium distribution, known as the **steady-state distribution**. For (x, y) in Ω, let $u(x, y)$ denote this steady-state temperature distribution. It is also a fact from thermodynamics that $u(x, y)$ satisfies the (two-dimensional) **Laplace's equation**

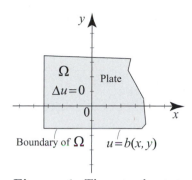

Figure 1 The steady-state temperature distribution satisfies Laplace's equation.

$$(1) \qquad \Delta u = \frac{\partial^2 u}{\partial x^2} + \frac{\partial^2 u}{\partial y^2} = 0.$$

Here Δu is the **Laplacian** of u, which by definition is $\frac{\partial^2 u}{\partial x^2} + \frac{\partial^2 u}{\partial y^2}$. Since Laplace's equation involves partial derivatives, it is called a **partial differential equation**. Any real-valued function that satisfies Laplace's equation on a region Ω and has continuous first and second partial derivatives is called **harmonic** on Ω. Constant functions are clearly harmonic on the entire plane, and so are the functions $u(x, y) = x$, $u(x, y) = y$, $u(x, y) = xy$. Less obvious examples of harmonic functions will be derived from Theorem 1 later in this section.

To determine the steady-state temperature inside the plate, we must solve Laplace's equation inside of Ω subject to the condition $u(x, y) = b(x, y)$ on the boundary of Ω, known as a **boundary condition**. A problem consisting of a partial differential equation along with specified boundary conditions is known as a **boundary value problem**. The special case involving Laplace's equation with specified boundary values is known as a **Dirichlet problem**. Solving Dirichlet problems is of paramount importance in applied mathematics, engineering, and physics. Many methods have been developed. The ones that we will present in this section provide a beautiful application of complex analysis.

The Laplacian is named after the great French mathematician and physicist Pierre-Simon de Laplace (1749–1827). The Laplacian appeared for the first time in a memoir of Laplace in 1784, in which he completely determined the attraction of a spheroid on the points outside it. The Laplacian of a function measures the difference between the value of the function at a point and the average value of the function in a neighborhood of that point. Thus a function that does not vary abruptly has a very small Laplacian. Harmonic functions have a zero Laplacian; they vary in a very regular way. Examples of such functions include the temperature distribution in a plate, the potential of the attractive force due to a sphere, the function that gives the brightness of colors in an image.

Connection with Analytic Functions

Let $f = u + iv$ be an analytic function. We know that u and v satisfy the Cauchy-Riemann equations (Theorem 1, Section 2.4),

$$\frac{\partial u}{\partial x} = \frac{\partial v}{\partial y} \qquad \text{and} \qquad \frac{\partial u}{\partial y} = -\frac{\partial v}{\partial x}.$$

Let us suppose for a moment that we can differentiate u and v twice and interchange the order of partial derivatives. Then

$$\frac{\partial^2 u}{\partial x^2} = \frac{\partial}{\partial x}\left(\frac{\partial u}{\partial x}\right) = \frac{\partial}{\partial x}\left(\frac{\partial v}{\partial y}\right) = \frac{\partial}{\partial y}\left(\frac{\partial v}{\partial x}\right) = \frac{\partial}{\partial y}\left(-\frac{\partial u}{\partial y}\right) = -\frac{\partial^2 u}{\partial y^2},$$

and hence $\frac{\partial^2 u}{\partial x^2} + \frac{\partial^2 u}{\partial y^2} = 0$. In other words, u satisfies Laplace's equation. Reasoning in a similar way, we can show that v also satisfies Laplace's equation.

Recall from calculus that in order to justify the interchange of partial derivatives as we needed to do above, it is enough to know that all first and second partial derivatives are continuous. One of the major results that we will develop in the following chapter guarantees that the real and imaginary parts of an analytic function have continuous partial derivatives of all orders. Thus, the steps that we used above are justified and u and v are harmonic.

**THEOREM 1
ANALYTIC AND
HARMONIC
FUNCTIONS**

Suppose that $f = u + iv$ is analytic on an open set S. Then its real and imaginary parts $u(x, y)$ and $v(x, y)$ are harmonic on S.

What this result effectively does for us is provide us with many examples of harmonic functions; simply take the real or imaginary part of any analytic function.

EXAMPLE 1　Harmonic functions

Show that the following are harmonic functions in the stated region.

(a)　$u(x, y) = x^2 - y^2$ on $\mathbb{C}$.
(b)　$u(x, y) = e^x \sin y$ on $\mathbb{C}$.
(c)　$u(x, y) = \operatorname{Arg} z$ $(z = (x, y))$ on the region $\Omega = \mathbb{C} \setminus (-\infty, 0]$.
(d)　$u(x, y) = \ln |z| = \ln \sqrt{x^2 + y^2}$ on the region $\mathbb{C} \setminus \{0\}$.

Solution　(a)　It is easy to see that u is harmonic here by direct verification of Laplace's equation; but we recognize that $u(x, y) = x^2 - y^2$ is the real part of the entire function $f(z) = z^2 = (x^2 - y^2) + 2ixy$. Thus from Theorem 1 we conclude that u is harmonic on $\mathbb{C}$.

(b)　Recognizing $u(x, y) = e^x \sin y$ as the imaginary part of the entire function $f(z) = e^z$ (see (16), Section 1.6), we conclude from Theorem 1 that u is harmonic on $\mathbb{C}$.

(c)　We have $\operatorname{Arg} z = \operatorname{Im}(\operatorname{Log} z)$, where $\operatorname{Log} z = \ln |z| + i \operatorname{Arg} z$ is the principal branch of the logarithm (see (4), Section 1.7). Since $\operatorname{Log} z$ is analytic on Ω, we conclude that $\operatorname{Arg} z$ is harmonic on Ω, by Theorem 1.

(d)　This part is interesting because we will need two different analytic functions to establish the desired result. Arguing as in (c), it follows that $\ln |z| = \ln \sqrt{x^2 + y^2}$ is harmonic on $\Omega = \mathbb{C} \setminus (-\infty, 0]$, being the real part of $\operatorname{Log} z$, which is analytic on $\mathbb{C} \setminus (-\infty, 0]$. This establishes that $\ln |z|$ is harmonic on the desired region except on the negative x-axis, $(-\infty, 0)$. To show that $\ln |z|$ is harmonic on $(-\infty, 0)$, we will use $\log_\alpha z$, which is a branch of $\log z$ with branch cut at angle α, with $-\pi < \alpha < \pi$. We have from (6), Section 1.7, $\log_\alpha z = \ln |z| + \arg_\alpha z$, and from our discussion in Section 2.4 leading to (13), $\log_\alpha z$ is analytic except at the branch cut and, in particular, is analytic on the negative x-axis. Hence, its real part, $\ln |z|$, is harmonic on the negative x-axis.

You can also answer part (d) directly by verifying that $\ln \sqrt{x^2 + y^2} = \frac{1}{2} \ln(x^2 + y^2)$ satisfies Laplace's equation and the first and second partial derivatives are continuous everywhere except at the origin.　∎

The following property of harmonic functions is very useful. Its proof is left to Exercise 18.

**PROPOSITION 1
LINEARITY**

Suppose that u and v are harmonic on an open set S, and a, b are real constants. Then $a u(x, y) + b v(x, y)$ is harmonic on S.

For example, $\phi(x, y) = 2(x^2 - y^2) + 7$ is harmonic, being of the form $\phi(x, y) = 2\, u(x, y) + 7$ where u is the harmonic function of Example 1(a). Also the function $\phi(x, y) = (a\, x + b)(c\, y + d)$, where a, b, c, d are real constants, is harmonic, being a linear combination of the harmonic functions x, y, xy, and constants.

You should be cautioned that the product of two harmonic functions need not be harmonic. For example, $u(x, y) = x$ is harmonic, but $\left(u(x, y)\right)^2 = x^2$ is not.

Harmonic Conjugates

Reading Theorem 1, it is natural to ask the following question: Given a harmonic function u in a region Ω, can we find another harmonic function v in Ω such that $f = u + i\, v$ is analytic in Ω? Such a function v is called a **harmonic conjugate** of u. Conditions for the existence of a harmonic conjugate will be established in Chapter 3. For now, just keep in mind that the existence of a harmonic conjugate depends on the nature of the region under consideration. For example, the function $\ln |z|$ is harmonic in $\Omega = \mathbb{C} \setminus \{0\}$ (Example 1(d)); but $\ln |z|$ has no harmonic conjugate in that region (Exercise 33). It does, however, have a harmonic conjugate in $\mathbb{C} \setminus (-\infty, 0]$, namely $\operatorname{Arg} z$. As our next example shows, by integrating the Cauchy-Riemann equations, we can always find the harmonic conjugate in a region such as the entire complex plane, a disk, or a rectangle. (More generally, we will show in Chapter 3 that we can find a harmonic conjugate if the region is *simply connected*. This more restrictive concept of connectedness is extremely important and will be at the heart of Chapter 3.)

EXAMPLE 2   Finding harmonic conjugates
Show that $u(x, y) = x^2 - y^2 + x$ is harmonic in the entire plane and find a harmonic conjugate.

Solution   That u is harmonic follows from $u_{xx} = 2$ and $u_{yy} = -2$. To find a harmonic conjugate v, we use the Cauchy-Riemann equations as follows. We want $u + i\, v$ to be analytic. Hence v must satisfy the Cauchy-Riemann equations

$$(2) \qquad \frac{\partial u}{\partial x} = \frac{\partial v}{\partial y}, \quad \text{and} \quad \frac{\partial u}{\partial y} = -\frac{\partial v}{\partial x}.$$

Since $\frac{\partial u}{\partial x} = 2x + 1$, the first equation implies that

$$2x + 1 = \frac{\partial v}{\partial y}.$$

To get v we will integrate both sides of this equation with respect to y. However, because v is a function of x and y, the constant of integration in your answer may be a function of x. Thus integrating with respect to y yields

$$v(x, y) = (2x + 1)y + c(x),$$

where $c(x)$ is a function of x alone. Plugging this into the second equation in (2), we get

$$-2y = -\left(2y + \frac{d}{dx}c(x)\right),$$

and hence $c(x)$ has zero derivative and must be a constant. Let us pick any such constant and write $c(x) = C$. Substituting into the expression for v we get $v(x, y) = (2x + 1)y + c(x) = 2xy + y + C$. You should verify the Cauchy-Riemann equations for the pair of functions u and v and conclude that $(x^2 - y^2 + x) + i(2xy + y + C)$ is entire. ∎

So, from Example 2, a harmonic conjugate of $u(x, y) = x^2 - y^2 + x$ is $v(x, y) = 2xy + y$. What is a harmonic conjugate of $v(x, y) = 2xy + y$? Surely it is related to u. Indeed, you can check that a conjugate of v is $-u(x, y) = -x^2 + y^2 - x$. More generally, we have the following useful result.

PROPOSITION 2

> Suppose that u is harmonic and that v is a harmonic conjugate of u. Then $-u$ is a harmonic conjugate of v.

Proof   We know that $f = u + iv$ is analytic. It follows that the function $(-i)f = v - iu$ is analytic, and hence $-u$ is a harmonic conjugate of v. ∎

In Example 2, we found the harmonic conjugate of u up to an arbitrary additive real constant. In fact, the following properties of the harmonic conjugate are not hard to prove (Exercise 18).

PROPOSITION 3

> Suppose that u is a harmonic function in a region Ω. Then
> (i)   if v_1 and v_2 are harmonic conjugates of u in Ω, then v_1 and v_2 must differ by a real constant.
> (ii)   If v is a harmonic conjugate of u, then v is also a harmonic conjugate of $u + c$ where c is any real constant.

The function u and its conjugate v have a very interesting geometric relationship based on the notion of orthogonal curves.

Suppose that two curves C_1 and C_2 meet at a point A. The curves are said to be **orthogonal** if their respective tangent lines L_1 and L_2 (at A) are orthogonal (Figure 2). We also say that C_1 and C_2 intersect at a right angle at A. Recall that if m_1 and m_2 denote the respective slopes of the tangent lines, and if neither is zero, then L_1 and L_2 are orthogonal if and only if

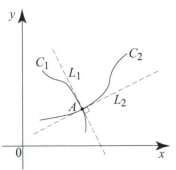

Figure 2 The curves C_1 and C_2 are orthogonal at A if the tangent lines L_1 and L_2 (at A) are orthogonal.

(3) $$m_1 m_2 = -1.$$

Two families of curves are said to be **orthogonal** if each curve from one family intersects the curves from the other family at right angles.

Consider the level curves of a harmonic function $u(x, y)$ in a region Ω. These are the curves determined by the implicit relation

(4) $$u(x, y) = C_1,$$

where C_1 is a constant (in the range of u). Since u is harmonic, it has continuous partial derivatives, and hence we can use the chain rule, Theorem 4, Section 2.6, to differentiate both sides of (4) with respect to x and get

$$\frac{\partial u}{\partial x}\frac{dx}{dx} + \frac{\partial u}{\partial y}\frac{dy}{dx} = 0.$$

But $dx/dx = 1$, so if $\frac{\partial u}{\partial y} \neq 0$, we can solve for dy/dx and get

$$(5) \qquad \frac{dy}{dx} = -\frac{\frac{\partial u}{\partial x}}{\frac{\partial u}{\partial y}}.$$

This gives the slope of the tangent line at a point on a level curve. Now suppose that we can find a harmonic conjugate $v(x, y)$ of $u(x, y)$ in Ω and let us consider the level curves

$$(6) \qquad v(x, y) = C_2,$$

where C_2 is a constant (in the range of v). Since v is harmonic, arguing as we did with u, we find that the slope of the tangent line at a point on a level curve is

$$(7) \qquad \frac{dy}{dx} = -\frac{\frac{\partial v}{\partial x}}{\frac{\partial v}{\partial y}} = \frac{\frac{\partial u}{\partial y}}{\frac{\partial u}{\partial x}},$$

since by the Cauchy-Riemann equations, $\partial v/\partial x = -\partial u/\partial y$ and $\partial v/\partial y = \partial u/\partial x$. Comparing (6) and (7), we see that the slopes of the tangent lines satisfy the orthogonality relation (3), and hence the level curves of u are orthogonal to the level curves of v. This orthogonality relation also holds when the tangents are horizontal and vertical. We thus have the following result.

THEOREM 2
ORTHOGONALITY
OF LEVEL CURVES

Suppose that u is a harmonic function in a region Ω and let v be a harmonic conjugate of u in Ω, so that $f = u + iv$ is analytic in Ω. Then, the two families of level curves, $u(x, y) = C_1$ and $v(x, y) = C_2$, are orthogonal at every point where $f'(z) \neq 0$.

As an illustration, we show in Figure 3(a) and (b) the level curves of the harmonic function $u = x^2 - y^2 + x$ and its conjugate $v(x, y) = 2xy + y$ (see Example 2). The graphs of the two families are superposed in Figure 3(c) to illustrate their orthogonality.

Figure 3   (a) Level curves of
$u(x, y) = x^2 - y^2 + x$.
(b) Level curves of
$v(x, y) = 2xy + y$.
(c) The level curves of u and
v are orthogonal.

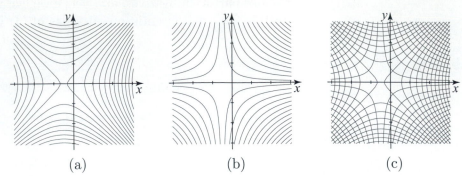

(a)   (b)   (c)

Solving and Interpreting Dirichlet Problems

We now return to our main topic: that of solving Dirichlet problems. Let us first mention an interesting example of a harmonic function, $u(x, y) = a \operatorname{Arg} z + b$, which is harmonic by Example 1(c) and Proposition 1. Because $\operatorname{Arg} z$ is constant on rays (independent of r), it follows that $u = a \operatorname{Arg} z + b$ is also constant on rays. (In fact, this is the only harmonic function with such a property. See Exercise 49.) Thus u is a good candidate for a solution of Dirichlet problems in which the boundary data is constant on rays or independent of r. We illustrate these ideas with an example.

EXAMPLE 3   A Dirichlet problem in a quadrant
The boundary of a large sheet of metal is kept at the constant temperatures $100°$ on the bottom and $50°$ on the left, as illustrated in Figure 4. After a long enough period of time, the temperature inside the plate reaches an equilibrium distribution. Find this steady-state temperature $u(x, y)$.

Solution   The steady-state temperature is a solution of the Dirichlet problem, which consists of Laplace's equation

$$\Delta u = 0, \qquad \text{inside } \Omega;$$

along with the boundary conditions

$$u(x, 0) = 100, \ x > 0, \qquad u(0, y) = 50, \ y > 0.$$

Based on our discussion preceding the example, because the boundary data is independent of r, we will try for a solution the harmonic function

$$u(x, y) = a \operatorname{Arg} z + b,$$

where a and b are real constants to be determined so as to satisfy the boundary conditions. From the first condition

$$u(x, 0) = 100 \Rightarrow a \operatorname{Arg} x + b = 100 \Rightarrow b = 100,$$

because $\operatorname{Arg} x = 0$ for $x > 0$. From the second condition

$$u(0, y) = 50 \Rightarrow a \operatorname{Arg}(iy) + b = 50 \Rightarrow a\frac{\pi}{2} + 100 = 50 \Rightarrow a = -\frac{100}{\pi},$$

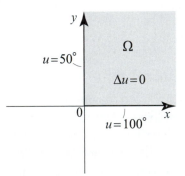

Figure 4 Dirichlet problem in Example 3.

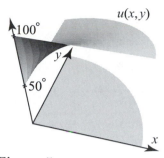

Figure 5
A three-dimensional picture representing the temperature distribution of the plate. Note the boundary values on the graph.

because $\operatorname{Arg}(iy) = \frac{\pi}{2}$ for $y > 0$, and $b = 100$. Thus the steady-state temperature inside the plate is $u(x, y) = -\frac{100}{\pi} \operatorname{Arg}(z) + 100$. Now for $z = x + iy$ with $x > 0$, we have $\operatorname{Arg} z = \tan^{-1}\left(\frac{y}{x}\right)$, and so another way of expressing the solution is

$$u(x, y) = -\frac{100}{\pi} \tan^{-1}\left(\frac{y}{x}\right) + 100.$$

The graph of u is shown in Figure 5. Note the temperature on the boundary; it matches the boundary conditions. ∎

In contrast to $\operatorname{Arg} z$ we can find harmonic functions which are independent of the argument and depend only on $r = |z|$. An example of such a function is $u(z) = a \ln |z| + b$, where a and b are real constants. By Example 1(d), this function is harmonic in $\mathbb{C} \setminus \{0\}$. It is a good candidate for a solution of Dirichlet problems in which the boundary data is constant on circles. See Exercises 29–32 for illustrations.

EXAMPLE 4   A Dirichlet problem in an infinite strip
Solve the Dirichlet problem shown in Figure 6.

Solution   Since the boundary data does not depend on x, it is plausible to guess that the solution will also not depend on x. In this case, $u(x, y)$ is a function of y alone, hence $u_x = 0$ and $u_{xx} = 0$, and Laplace's equation becomes $u_{yy} = 0$, implying that u is a linear function of y. Hence $u(x, y) = ay + b$. Using the boundary conditions, we have

$$u(x, 0) = 0 \Rightarrow b = 0;$$
$$u(x, 1) = 100 \Rightarrow a = 100.$$

Hence the solution of the problem is $u(x, y) = 100 y$, which is clearly harmonic and satisfies the boundary conditions. ∎

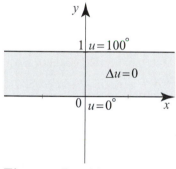

Figure 6 Dirichlet problem in Example 4.

In Example 4 we used a harmonic function that was independent of x. Similarly, we can find harmonic functions that are independent of y (hence $u_y = 0$ and $u_{yy} = 0$). You can show in this case that $u = ax + b$, where a and b are real constants. This function is a good candidate for solving Dirichlet problems in infinite vertical strips with constant boundary data. See Exercises 25 and 26 for illustrations.

Harmonic Conjugates, Isotherms, and Heat Flow

In Example 3, the temperature of the boundary is kept at two constant values, 100° and 50°. Our physical intuition tells us that, because the plate is insulated, the temperature of the points inside the plate will vary between these two values and will equal those values only at the boundary. It is natural to ask for those points inside the plate with the same temperature $u(x, y) = T$, where $50 < T < 100$. These points lie on curves inside the plate, called curves of constant temperature or **isotherms**. Isotherms have many practical applications. Computing them will lead us to interesting properties of harmonic functions.

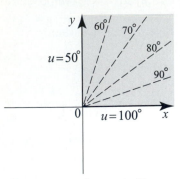

Figure 7 Isotherms in Example 5.

EXAMPLE 5   Isotherms

Find the isotherms in Example 3.

Solution   Since the temperature inside the plate will vary between $50°$ and $100°$, to find the isotherms, we must solve

$$(8) \qquad u(x, y) = T, \quad \text{where} \quad 50 < T < 100,$$

where (x, y) is a point inside the first quadrant, not on the boundary. Appealing to the solution from Example 3, (8) becomes

$$(9) \qquad -\frac{100}{\pi} \tan^{-1}\left(\frac{y}{x}\right) + 100 = T, \quad \text{where} \quad 50 < T < 100, \quad x > 0, \ y > 0.$$

Thus,

$$\tan^{-1}\left(\frac{y}{x}\right) = \pi - \frac{\pi T}{100} \quad \Rightarrow \quad \frac{y}{x} = \tan\left(\pi - \frac{\pi T}{100}\right)$$

$$\Rightarrow \quad \frac{y}{x} = -\tan\frac{\pi T}{100}$$

$$\Rightarrow \quad y = -\overbrace{\left(\tan\frac{\pi T}{100}\right)}^{\text{a positive constant}} x,$$

where we have used the identity $\tan(\pi - \alpha) = -\tan\alpha$ (check it!). Since $50 < T < 100$, $\frac{\pi}{2} < \frac{\pi T}{100} < \pi$ and so $-\tan\frac{\pi T}{100} > 0$. Thus the equation of the isotherms $y = -\left(\tan\frac{\pi T}{100}\right) x$ corresponds to rays in the first quadrant emanating from the origin. As $T \to 100$, the slope of the ray $y = -\left(\tan\frac{\pi T}{100}\right) x$ goes to 0, showing that the ray tends to the positive x-axis. As $T \to 50$, the slope of the ray $-\tan\frac{\pi T}{100} \to \infty$, showing that the ray tends to the positive y-axis. This agrees with our intuition, since points near the x-axis have temperature close to $100°$, and points near the y-axis have temperature close to $50°$. The isotherms corresponding to $T = 90°$, $80°$, $70°$, and $60°$ are shown in Figure 7.   ∎

Related to the topic of isotherms is the topic of **curves of heat flow**. These are the curves along which the heat is flowing inside the plate. To determine these curves, we use **Fourier's law of heat conduction**, which states that heat flows from hot to cold in the direction in which the temperature difference is the greatest. If along the isotherms the temperature difference is zero, then it should make sense that in the direction perpendicular to the isotherms the temperature difference is the greatest. Hence the curves of heat flow are orthogonal to the isotherms.

You should recall from vector calculus that the gradient vector $\nabla u = (u_x, u_y)$ points in the direction of greatest change in a function. The gradient is perpendicular to level curves of $u(x, y)$. Fourier's law states that heat flows along $-\nabla u$, and thus curves of heat flow are orthogonal to level curves of u, and hence coincident with level curves of v.

So to find the curves of heat flow in a plate, it is enough to find a harmonic conjugate $v(x, y)$ of the steady-state temperature distribution $u(x, y)$, since by Theorem 2 the level curves of v are orthogonal to the level curves of u. We illustrate this process with an example.

EXAMPLE 6   Curves of heat flow

Find the curves of heat flow in Example 3.

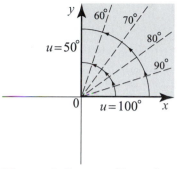

Figure 8 Curves of heat flow in Example 6, along with the isotherms.

Solution   The isotherms were found in Example 5 and are given by $u(x, y) = C_1$, where $u(x, y) = -\frac{100}{\pi} \text{Arg}(z) + 100$. To determine the curves of heat flow, we must find a harmonic conjugate of u. By Proposition 3(ii), it is enough to find a harmonic conjugate of $-\frac{100}{\pi} \text{Arg}(z)$. From our knowledge of analytic functions, we see that $f(z) = -\frac{100}{\pi}(\ln|z| + i\,\text{Arg}(z)) = -\frac{100}{\pi} \text{Log}\, z$ is analytic in $\mathbb{C} \setminus (-\infty, 0]$. Hence by Theorem 1, a harmonic conjugate of $-\frac{100}{\pi} \ln|z|$ is $-\frac{100}{\pi} \text{Arg}\, z$. By Proposition 2, it follows that a harmonic conjugate of $-\frac{100}{\pi} \text{Arg}\, z$ is $\frac{100}{\pi} \ln|z|$. Hence the curves of heat flow are given by

$$\frac{100}{\pi} \ln|z| = \text{constant} \quad \Leftrightarrow \quad |z| = \text{constant} \quad \Leftrightarrow \quad x^2 + y^2 = c^2.$$

Thus the curves of heat flow are arcs of circles centered at the origin. In Figure 8, we show the isotherms along with the curves of heat flow to illustrate their orthogonality.   ∎

Conformal Mappings

We conclude our panoramic overview of Dirichlet problems and their solutions by mentioning the method of conformal mappings, which is of wide use in complex analysis. In the past you have certainly used methods to transform a difficult problem to one with a known solution or whose solution is easier to find. For example, some integrals are simplified by using changes of variables, some differential equations are simplified by applying a Laplace transform, and so on. In solving Dirichlet problems, it is sometimes advantageous to map the region under consideration to a simpler region or one on which the transformed problem is easier to solve. This is the idea behind the method of conformal mappings, which we now explain.

Let a Dirichlet problem be given on a region Ω with boundary Γ. Suppose that we want to solve this problem by somehow transforming it first to the w-plane by means of a mapping $w = f(z)$, where f is analytic in Ω. If $f'(z) \neq 0$ for all z in Ω, we call f a **conformal mapping** of Ω. Conformal mappings have important properties that will be studied in detail in Chapter 6. For example, if f is conformal, then the image of Ω, $\Omega' = f[\Omega]$, is also a region (open, connected set); moreover, if f is one-to-one, then f will map Γ onto Γ', the boundary of Ω'. In the examples of this section, these properties can be checked on a case-by-case basis.

In transforming the Dirichlet problem, we need to know what happens to Laplace's equation under our transformation $w = f(z)$ and what happens to the boundary conditions. Because f maps boundary to boundary, the boundary conditions on Γ will be transformed into boundary conditions on Γ' as we will explain shortly. However, the most important feature of this method is stated in the next theorem and tells us that Laplace's equation is invariant under a change of variables using a conformal mapping.

THEOREM 3
INVARIANCE OF
LAPLACE'S
EQUATION

Suppose that f is an analytic mapping of Ω into Ω' and U is a harmonic function on Ω'. Then $U \circ f$ is harmonic in Ω. Thus, if U satisfies $\Delta U = 0$ on Ω', then $u = U \circ f$ satisfies $\Delta u = 0$ on Ω.

To understand the meaning of $U \circ f(z)$ where f is complex-valued and U is a function of two variables, write $U \circ f(z) = U(\operatorname{Re} f(z), \operatorname{Im} f(z))$. For example, if $f(z) = e^z = e^x \cos y + i e^x \sin y$ and $U(s, t) = st$, then $U \circ f(z) = e^{2x} \cos y \sin y$.

Proof   Let z_0 be a point in Ω and $w_0 = f(z_0)$. We will show in Section 3.8 that U has a harmonic conjugate V in a disk around w_0. Then $U + iV$ is analytic in this disk, and by the composition of analytic functions, $(U + iV) \circ f$ is analytic at z_0. Hence by Theorem 1, $\operatorname{Re}[(U+iV) \circ f] = \operatorname{Re}[U \circ f + i(V \circ f)] = U \circ f$ is harmonic at z_0. Since z_0 was arbitrary, $U \circ f$ is harmonic in Ω. An alternate proof may be given by direct application of the chain rule and Cauchy-Riemann equations.   ∎

Now suppose that you want to use a conformal mapping $w = f(z)$ to solve the Dirichlet problem $\Delta u = 0$ in Ω and $u(z) = b(z)$ on the boundary Γ of Ω. Suppose also that f is one-to-one on Ω and its boundary Γ. Here is how the method works (see Figure 9 as you read through the steps).

Figure 9 The conformal mapping method. If $f(z)$ is analytic and one-to-one on Ω and its boundary Γ, then $\Omega' = f[\Omega]$ is a region with boundary $\Gamma' = f[\Gamma]$. The boundary function $b(z)$ (z on Γ) is used to define a boundary function $b \circ f^{-1}(w)$ for all w on Γ, where f^{-1} is the inverse of f.

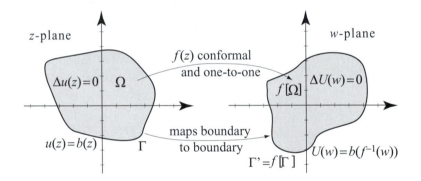

Step 1: Describe clearly the region $\Omega' = f[\Omega]$ and its boundary $\Gamma' = f[\Gamma]$ in the w-plane.

Step 2: Since f is one-to-one, we have an inverse function f^{-1} defined on Ω' and Γ'. For w on Γ', $f^{-1}(w)$ is on Γ and so we can define the function $b \circ f^{-1}(w)$ for all w on Γ'. This determines the boundary values on Γ'.

Step 3: Set up and solve the Dirichlet problem on Ω' consisting of Laplace's equation $\Delta U(w) = 0$ for all w in Ω' and $U(w) = b \circ f^{-1}(w)$ for all w on Γ'. (This is our transformed Dirichlet problem.)

Step 4 Let $u(z) = U \circ f(z)$ for all z in Ω. Then $u(z)$ is a solution of our original Dirichlet problem on Ω. Indeed, by Theorem 3, u is harmonic in Ω. For z on the boundary Γ, we have $f(z)$ on the boundary Γ', and using the fact that $U(w) = b \circ f^{-1}(w)$ on Γ', we obtain for z on Γ, $u(z) = U \circ f(z) = b \circ f^{-1}(f(z)) = b(z)$. Hence u satisfies the desired boundary condition.

In most examples, Steps 2 and 3 can be done without actually computing f^{-1}, as we illustrate in our next example.

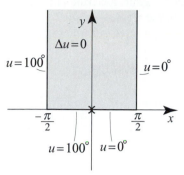

Figure 10 Dirichlet problem in Example 7.

EXAMPLE 7   Conformal mapping solution of a Dirichlet problem
(a) Solve the Dirichlet problem in the semi-infinite strip Ω shown in Figure 10.
(b) Discuss the isotherms and lines of heat flow of the solution.

Solution   (a) We will transform the given problem on the region Ω into a problem in the upper half-plane. As we will see momentarily, the transformed problem is easy to solve. We will follow the basic four steps of the solution.
Step 1: Recall from Example 4, Section 1.6, that the mapping $f(z) = \sin z$ takes Ω in the z-plane onto the upper half of the w-plane (Figure 11). Moreover, the boundary of Ω is mapped to the boundary of the upper half-plane as follows. The line segment on the x-axis, $[0, \frac{\pi}{2}]$, and the vertical half-line $y = \frac{\pi}{2}$ are mapped onto $[0, \infty)$, the nonnegative real axis in the w-plane. The line segment on the x-axis, $[-\frac{\pi}{2}, 0)$, and the vertical half-line $y = -\frac{\pi}{2}$ are mapped onto $(-\infty, 0)$, the negative real axis in the w-plane. (As you can verify in this case, $\frac{d}{dz} \sin z = \cos z$ and since the zeros of $\cos z$ are at the points $\frac{\pi}{2} + k\pi$, k an integer, we conclude that $f'(z) \neq 0$ for all z strictly in Ω. Hence f is a conformal mapping of Ω onto the upper half-plane. Moreover, using the fact that $\sin z$ is 2π-periodic, it follows that $\sin z$ is one-to-one on Ω.)
Step 2: Describe the boundary function in the Dirichlet problem in the w-plane. The boundary function in the w-plane is $b \circ f^{-1}(w)$, where $b(z)$ is the boundary function in the z-plane. From Figure 11, we have $b(z) = 0$ if z is real and positive or $z = \frac{\pi}{2} + i y$, where $y > 0$; and $b(z) = 100$ if z is real and negative or $z = -\frac{\pi}{2} + i y$, where $y > 0$. With this definition of b and the way the boundary is mapped to the boundary, it becomes clear that the boundary function in the w-plane is $b \circ f^{-1}(w) = 0$ if $w > 0$ and $b \circ f^{-1}(w) = 100$ if $w < 0$.
Step 3: The transformed Dirichlet problem in the upper half-plane is described by Figure 11 and is given by

$$\Delta U(w) = 0, \ w \text{ in the upper half-plane,}$$
$$U(w) = 0, \ w > 0, \quad U(w) = 100, \ w < 0.$$

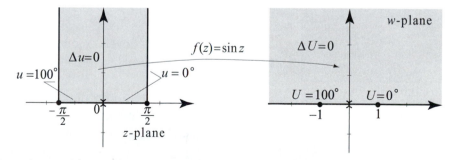

Figure 11 Mapping the semi-infinite vertical strip onto the upper half-plane. Note the boundary correspondence.

Since the transformed boundary data is constant along rays, we can solve the problem in the w-plane as we did in Example 3 by trying for a solution the harmonic function $U(w) = a \operatorname{Arg} w + b$, where a and b are real constants. Using the boundary conditions, we see that $100 = a\pi$ or $a = \frac{100}{\pi}$, and $0 = b$. Hence $U(w) = \frac{100}{\pi} \operatorname{Arg} w$.
Step 4: The solution of the original Dirichlet problem in the z-plane is $u(z) = U(f(z)) = \frac{100}{\pi} \operatorname{Arg} (\sin z)$. If we want to express our answer in terms of x and y, we use the real and imaginary parts of $\sin z$ ((16), Section 1.6): $\sin z = \sin x \cosh y +$

$i \cos x \sinh y$. It is not convenient to use the inverse tangent formula (14), Section 1.3, to express $\operatorname{Arg}(\sin z)$, because the real part of $\sin z$ takes on positive and negative values. It is more convenient in this case to use the inverse cotangent. We have

$$\operatorname{Arg}(\sin z) = \cot^{-1}\left(\frac{\sin x \cosh y}{\cos x \sinh y}\right) = \cot^{-1}(\tan x \coth y),$$

where the inverse cotangent takes its values between 0 and π (see Figure 12). Hence

$$u(z) = u(x, y) = \frac{100}{\pi} \cot^{-1}(\tan x \coth y).$$

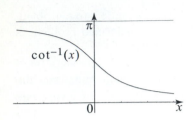

Figure 12 The inverse cotangent takes its values in $(0, \pi)$.

(b) The isotherms are the level curves $u(x, y) = T$, where $0 < T < 100$. The curves of heat flow are the level curves of a harmonic conjugate v of u. To find v, it is enough to find a harmonic conjugate V of U and then set $v(z) = V(\sin z)$ (Exercise 34). Since $\operatorname{Arg} w$ is the harmonic conjugate of $\ln|w|$, it follows from Proposition 2 that $-\ln|w|$ is the harmonic conjugate of $\operatorname{Arg} w$. Hence $V(w) = -\ln|w|$ and $v(x, y) = -\ln|\sin z|$. Some isotherms and curves of heat flow are shown in Figure 13.

A few geometric observations can be made. Note that at the top of Figure 13, the isotherms look like vertical lines. This is to be expected, since the lower boundary, being remote, can be ignored, and we may think of the problem as one on a doubly infinite vertical strip, in which the isotherms are vertical lines. Near the origin, the isotherms look like rays, and the curves of heat flow like circles. These are the isotherms and curves of heat flow in a Dirichlet problem with boundary data constant on rays (see Examples 3, 5, and 6). This too is to be expected since near the origin the vertical boundary data have less effect as compared with the boundary data on the x-axis. So near the origin, we can ignore the vertical boundary data and consider only the horizontal boundary data, which is constant on rays in this case, and so the isotherms and curves of heat flow are like those in Examples 5 and 6. ∎

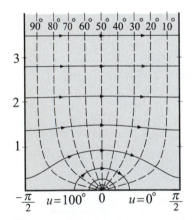

Figure 13 Isotherms and curves of heat flow in Example 7. Note the orthogonality of the curves.

So far we have used our knowledge of analytic functions to solve Dirichlet problems by guessing the solution. Guessing is certainly a legitimate method that you have used in solving differential equations and computing indefinite integrals. Further development of the theory of analytic and harmonic functions will be necessary to tackle more general Dirichlet problems. Topics such as the Poisson integral formula, Fourier series, and conformal mappings are examples of theories and tools that will provide us with systematic ways for solving Dirichlet problems.

Other issues that we need to address concern the uniqueness of the solution of a Dirichlet problem. Consider the problem in Example 3, whose solution is $u(x, y) = -\frac{100}{\pi} \tan^{-1}\left(\frac{y}{x}\right) + 100$. If we add to this solution the harmonic function $\phi(x, y) = xy$, we obtain the harmonic function $\psi(x, y) = u(x, y) + xy$, which solves Laplace's equation. Moreover, because xy vanishes on the boundary of the first quadrant, $\psi(x, y)$ and $u(x, y)$ have the same boundary values. Thus, ψ is another solution of the Dirichlet problem in Example 3. This is quite disturbing since we want to think of the solution as representing a temperature distribution and as such it must be unique. As it turns out, if in addition to the boundary conditions in a Dirichlet problem we add the condition that the solution be bounded (which is effectively a boundary condition at infinity), then the solution will be unique. For example, with this additional boundedness assumption, we can no longer add the function xy to the solution in Example 3 to obtain another solution.

Uniqueness results for the solution of Dirichlet problems will be derived from theoretical properties of analytic and harmonic functions, including results such as the maximum principle. All these topics will be addressed as part of the remainder of this book.

Exercises 2.5

In Exercises 1–12, determine the set on which the given function is harmonic. You may verify Laplace's equation (1) directly or use Theorem 1.

1. $x^2 - y^2 + 2x - y$.
2. $\dfrac{1}{x^2 - y^2}$.
3. $e^x \cos y$.

4. $\dfrac{y}{x^2 + y^2}$.
5. $\dfrac{1}{x + y}$.
6. $\sinh x \cos y$.

7. $\cos x \cosh y$.
8. $e^{2x} \cos(2y)$.
9. $e^{x^2 - y^2} \cos(2xy)$.
10. $\ln(x^2 + y^2)$.
11. $\ln((x-1)^2 + y^2)$.
12. $\arg_{\frac{\pi}{2}} z$.

In Exercises 13–16, find a harmonic conjugate $v(x, y)$ of the given harmonic function $u(x, y)$ by using the method of Example 2. Check your answer by verifying the Cauchy-Riemann equations.

13. $x + 2y$.
14. $x^2 - y^2 - xy$.
15. $e^y \cos x$.
16. $\cos x \sinh y$.

17. Let n be any integer. (a) Show that $u(r, \theta) = r^n \cos(n\theta)$ and $v(r, \theta) = r^n \sin(n\theta)$ are harmonic on $\mathbb{C}$ if $n \geq 0$ and on $\mathbb{C} \setminus \{0\}$ if $n < 0$. (b) Find their respective harmonic conjugates. [Hint: Consider $f(z) = z^n$ in polar coordinates.]

18. (a) Prove Proposition 1. (b) Prove Proposition 3.

19. Show that if u and u^2 are both harmonic in a region Ω, then u must be constant. [Hint: Plug u^2 into Laplace's equation and show that $u_x = u_y = 0$.]

20. Show that if u, v, and $u^2 + v^2$ are harmonic in a region Ω, then u and v must be constant. [Hint: Plug $u^2 + v^2$ into Laplace's equation and show that $u_x = u_y = 0$ and $v_x = v_y = 0$.]

21. Suppose that u is harmonic and v is a harmonic conjugate of u. Show that $u^2 - v^2$ and uv are both harmonic. [Hint: Consider $(u + iv)^2$.]

22. Use Exercises 20 and 21 to show that if f is analytic in a region Ω and $|f|$ is constant in Ω, then f must be constant in Ω. (A different proof was outlined in Exercise 38, Section 2.4.)

23. Translating and dilating a harmonic function. Suppose that u is harmonic. Show that the following functions are also harmonic:
(a) $u(x - \alpha, y - \beta)$, where α and β are real numbers;
(b) $u(\alpha x, \alpha y)$, where $\alpha \neq 0$ is a real number.

24. (a) Suppose that $u(x, y)$ is harmonic. Show that $u(x, -y)$ is also harmonic.

(b)   Suppose that $u(x, y)$ is harmonic. Show that $u(\frac{x}{x^2+y^2}, \frac{y}{x^2+y^2})$ is harmonic. [Hint: Compose u with the analytic function $\frac{1}{z}$; use Theorem 3 and part (a).]

25.   Harmonic functions independent of y.   (a) Suppose that $u(x, y)$ is a harmonic function whose values depend only on x and not on y. Using Laplace's equation, show that $u(x, y) = a\,x + b$, where a and b are real constants.

(b)   Consider the Dirichlet problem in the infinite vertical strip in Figure 14. Because the boundary values do not depend on y, it is plausible to try for a solution a harmonic function whose values do not depend on y. Solve the problem, using (a).

(c)   Determine and plot the isotherms and curves of heat flow.

26.   A Dirichlet problem in an infinite vertical strip.   (a)   Solve the Dirichlet problem in Figure 15.

(b)   Determine and plot the isotherms and curves of heat flow.

27.   Harmonic functions independent of r.   (a)   Solve the Dirichlet problem in Figure 16. Because the boundary values do not depend on r, you should try for a solution a harmonic function whose values do not depend on r.

(b)   Determine and plot the isotherms and curves of heat flow.

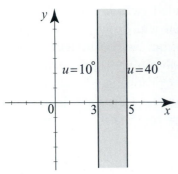

Figure 14 Dirichlet problem in Exercise 25.

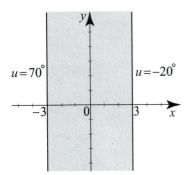

Figure 15 Dirichlet problem in Exercise 26.

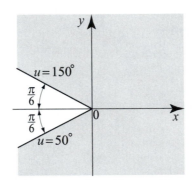

Figure 16 Dirichlet problem in Exercise 27.

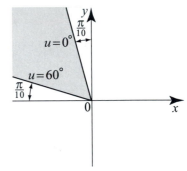

Figure 17 Dirichlet problem in Exercise 28.

28.   A Dirichlet problem in a wedge.   (a)   Solve the Dirichlet problem in Figure 17.

(b)   Determine and plot the isotherms and curves of heat flow.

29.   Harmonic functions independent of θ.   Suppose that $u(r, \theta)$ is a harmonic function, in polar coordinates.

(a)   Suppose that u depends only on r and not θ; hence $u(r, \theta) = u(r)$. Using the polar form of the Laplacian (12) in Exercise 47, show that $u(r)$ satisfies the (ordinary) differential equation in r

$$u_{rr} + \frac{1}{r}\,u_r = 0,$$

known as an **Euler equation**. This is perhaps the simplest second order linear differential equation with nonconstant coefficients. To find its general solution, we need two linearly independent solutions.

(b)   Multiply the Euler equation by the integrating factor r and notice that the left side is now exact. Integrate to conclude $ru_r = c_1$, where c_1 is the constant of integration. Integrate again and find the general solution

$$u(r) = c_1 \ln r + c_2.$$

30. Dirichlet problems in annular regions. The annular region A_{R_1,R_2} in Figure 18 is centered at the origin with inner radius R_1 and outer radius R_2. Consider the Dirichlet problem in A_{R_1,R_2} with constant boundary conditions $u(R_1, \theta) = T_1$ and $u(R_2, \theta) = T_2$ for all θ. Show that the solution of the problem is

$$u(r, \theta) = u(r) = T_1 + (T_2 - T_1)\frac{\ln(r/R_1)}{\ln(R_2/R_1)}$$

[Hint: Since the boundary conditions are independent of θ, you should try a harmonic function independent of θ. According to Exercise 29, try $u(r) = c_1 \ln r + c_2$.]

Exercises 31–32 are Dirichlet problems described by the corresponding figures. In each case, (a) solve the Dirichlet problem. (b) Determine the isotherms. (c) Determine the curves of heat flow. (d) Plot the isotherms and curves of heat flow.

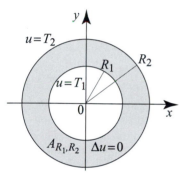

Figure 18 Dirichlet problem in Exercise 30.

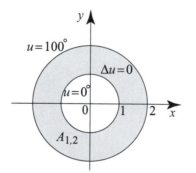

Figure 19 Dirichlet problem in Exercise 31.

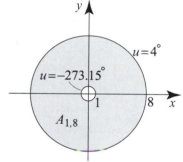

Figure 20 Dirichlet problem in Exercise 32.

33. Nonexistence of a harmonic conjugate. In Example 1(d), we showed that $\ln |z|$ is harmonic in $\mathbb{C} \setminus \{0\}$. We also know that $\ln |z|$ has a harmonic conjugate in $\mathbb{C} \setminus (-\infty, 0]$. In fact, since $\ln |z|$ is the real part of the analytic function $\log_\alpha z$, it has a harmonic conjugate in the complex plane minus the ray emanating from the origin at angle α. In this exercise, we will show that $\ln |z|$ does not have a harmonic conjugate in $\mathbb{C} \setminus \{0\}$.
(a) Suppose that $\phi(z)$ is a harmonic conjugate of $\ln |z|$ in $\mathbb{C} \setminus \{0\}$. Show that $\phi(z) = \text{Arg}\,(z) + c$ for all z in $\mathbb{C} \setminus (-\infty, 0]$. [Hint: The functions $\ln |z| + i\,\phi(z)$ and $\text{Log}\, z$ are analytic in the region $\mathbb{C} \setminus (-\infty, 0]$ and have the same real parts. Use Corollary 1, Section 2.4.]
(b) Argue that, since $\phi(z)$ is harmonic in $\mathbb{C} \setminus \{0\}$, $\phi(z)$ is continuous on $(-\infty, 0)$. Obtain a contradiction using (a) and the fact that the discontinuities of $\text{Arg}\, z$ are not removable on the negative x-axis (Example 7, Section 2.2).

34. (a) Suppose that f is analytic on Ω, U is harmonic on $f[\Omega]$, and $u = U \circ f$. Then u is harmonic in Ω by Theorem 3. Suppose that V is a harmonic conjugate of U. Show that $V \circ f$ is a harmonic conjugate of u.
(b) Derive the isotherms and curves of heat flow in Example 7.

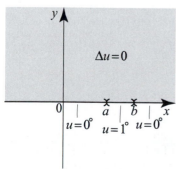

Figure 21 Dirichlet problem for Exercise 35.

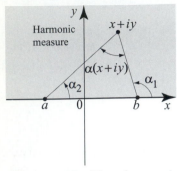

Figure 22 The harmonic measure of an interval is the difference of two translated arguments:
$$\alpha(z) = \text{Arg}\,(z - b)$$
$$\quad\quad - \text{Arg}\,(z - a).$$

35. Harmonic measure of an interval. A very important Dirichlet problem is described in the upper half-plane with boundary data on the x-axis given by

$$u(x, 0) = \begin{cases} 1 & \text{if } a < x < b, \\ 0 & \text{otherwise}, \end{cases}$$

where $a < b$ are fixed real numbers (see Figure 21). We will outline a solution of this problem using basic geometry and properties of harmonic functions.

(a)   For z in the upper half-plane, let $\alpha(z)$ denote the angle at z subtended by the interval $[a, b]$ (Figure 22). Show that $\alpha(z) = \text{Arg}\,(z - b) - \text{Arg}\,(z - a)$. The function $\alpha(z)$ is called the **harmonic measure** of the interval (a, b). [Hint: We have $\alpha_1 = \text{Arg}\,(z - b)$, $\alpha_2 = \text{Arg}\,(z - a)$, and $\alpha(z) = \alpha_1 - \alpha_2$ (Figure 22).]

(b)   Show that $\alpha(z)$ is harmonic. [Hint: Exercise 23(a).]

(c)   Show geometrically that

$$\lim_{y \downarrow 0} \alpha(x + i\,y) = \begin{cases} 0 & \text{if } x > b, \\ \pi & \text{if } a < x < b, \\ 0 & \text{if } x < a. \end{cases}$$

(d)   Conclude that $u(x, y) = \frac{1}{\pi}\alpha(x + i\,y)$ is a solution of the Dirichlet problem in Figure 21.

36.   (a)   Solve the Dirichlet problem in Figure 23.

(b)   Show that the isotherm corresponding to the temperature $0 < T < 100$ consists of the arc in the upper half-plane of the circle with center $(0, \cot \frac{\pi T}{100})$ and radius $|\csc \frac{\pi T}{100}|$. Justify your answer using facts from plane geometry.

(c)   What are the curves of heat flow? [Hint: You know a harmonic conjugate of $\text{Arg}\,z$. Show that if v is a harmonic conjugate of u, then a harmonic conjugate of $u(z - a) \pm u(z - b)$ is $v(z - a) \pm v(z - b)$.]

(d)   Plot the isotherms and curves of heat flow.

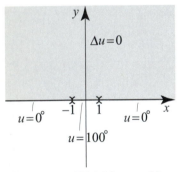

Figure 23 Dirichlet problem for Exercise 36.

37.   Solve the Dirichlet problem in the upper half-plane with boundary data on the x-axis given by

$$u(x, 0) = \begin{cases} T_1 & \text{if } a < x < b, \\ T_2 & \text{otherwise}, \end{cases}$$

where $a < b$ are fixed real numbers. [Hint: Use Exercise 35 and the fact that constant functions are harmonic.]

38. Project Problem: Harmonic measures of two disjoint intervals. In this exercise, we generalize the result of Exercise 35 by solving the Dirichlet problem in the upper half-plane with boundary data

$$u(x, 0) = \begin{cases} T_1 & \text{if } a < x < b, \\ T_2 & \text{if } c < x < d, \\ 0 & \text{otherwise}, \end{cases}$$

where $a < b \le c < d$ (Figure 24).

(a)   Show that if u_1 is a solution of the Dirichlet problem in the upper half-plane with boundary conditions

$$u_1(x, 0) = \begin{cases} T_1 & \text{if } a < x < b, \\ 0 & \text{otherwise}, \end{cases}$$

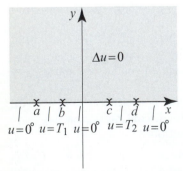

Figure 24 Dirichlet problem for Exercise 38.

and u_2 is a solution of the Dirichlet problem in the upper half-plane with boundary conditions

$$u_2(x, 0) = \begin{cases} T_2 & \text{if } c < x < d, \\ 0 & \text{otherwise,} \end{cases}$$

then the solution of the Dirichlet problem in the upper half-plane with boundary data $u(x, 0)$ is $u(x, y) = u_1(x, y) + u_2(x, y)$.

(b)   Show that $u(x, y) = \frac{T_1}{\pi}\alpha_1(z) + \frac{T_2}{\pi}\alpha_2(z)$ where $\alpha_1(z)$, respectively, $\alpha_2(z)$, is the angle at z subtended by the interval (a, b), respectively, (c, d).

39.   (a)   Solve the Dirichlet problem in Figure 25.

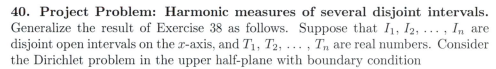

   (b)   Plot the isotherms.

40. Project Problem: Harmonic measures of several disjoint intervals. Generalize the result of Exercise 38 as follows. Suppose that $I_1, I_2, \ldots, I_n$ are disjoint open intervals on the x-axis, and $T_1, T_2, \ldots, T_n$ are real numbers. Consider the Dirichlet problem in the upper half-plane with boundary condition

$$u(x, 0) = \begin{cases} T_j & \text{if } x \in I_j, \\ 0 & \text{otherwise.} \end{cases}$$

Show that the solution is

$$u(x, y) = \frac{1}{\pi} \sum_{j=1}^{n} T_j \alpha_j(z),$$

where $\alpha_j(z)$ is the angle at z subtended by the interval I_j.

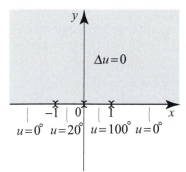

Figure 25 Dirichlet problem for Exercise 39.

41. Project Problem: Approximation of steady-state solutions.   Consider the Dirichlet problem in the upper half-plane with boundary values $u(x, 0) = f(x)$ $(-\infty < x < \infty)$, where $f(x)$ is an arbitrary function that represents the temperature of the points on the x-axis. Such a temperature distribution will be bounded and piecewise continuous. There is an analytical solution of the Dirichlet problem, given by an integral involving $f(x)$, known as the Poisson integral of f (see Exercise 42). In this exercise, we will show how we can use the result of Exercise 40 to approximate the solution for a given $f(x)$. The approach that we take has some merit, since it can be used to obtain approximate numerical values for the steady-state temperature. Moreover, we will use it in Exercise 42 to justify the Poisson integral formula. The rigorous derivation of this formula will be given in a later chapter.

To be able to compare our numerical approximation with the exact solution, let us take $f(x) = \frac{1}{1+x^2}$, $-\infty < x < \infty$. In this case, using properties of the Poisson integral, we will show in a later chapter that the solution is

(10) $$u(x, y) = \frac{1+y}{x^2 + (1+y)^2}.$$

(a)   Verify that u is indeed harmonic in the upper half-plane and that it equals $f(x) = \frac{1}{1+x^2}$ on the x-axis.

(b)   We now pretend that we do not know the exact solution and proceed to find an approximate solution. The idea is to approximate $f(x)$ by a function that takes on constant values on disjoint intervals. Take the interval $(-5, 5)$ and subdivide it into 40 smaller intervals of equal length, (x_j, x_{j+1}), $j = 1, 2, \ldots, 40$. Approximate f on (x_j, x_{j+1}) by $f(x_j)$, and by 0 outside the interval $(-5, 5)$. Thus the boundary values are now replaced by

$$u(x, 0) = \begin{cases} \frac{1}{1+x_j^2} & \text{if } x_j < x < x_{j+1}, \\ 0 & \text{otherwise.} \end{cases}$$

Show that the solution is

$$u(x, y) = \frac{1}{\pi} \sum_{j=1}^{40} \frac{1}{1 + x_j^2} \alpha_j(z),$$

where $\alpha_j(z)$ is the angle at z subtended by the interval (x_j, x_{j+1}).

(c)   With the help of a computer, evaluate your approximate solution at various points, $z_0 = x_0 + i\, y_0$, in the upper half-plane and compare with the exact solution (10).

42. Project Problem: Justifying the Poisson integral formula.   Consider the Dirichlet problem in the upper half-plane with boundary values $u(x, 0) = f(x)$, $-\infty < x < \infty$. One of the major results that we will derive in this book gives the solution as

$$(11) \qquad u(s, y) = \frac{y}{\pi} \int_{-\infty}^{\infty} \frac{f(x)}{(s-x)^2 + y^2}\, dx, \qquad -\infty < s < \infty, \, y > 0.$$

This is known as the **Poisson integral formula** or the **Poisson integral** of f. Our goal in this exercise is to motivate this formula by sketching a proof using the numerical scheme of Exercise 41.

(a)   Based on the approach in Exercise 41, explain how you would derive the approximate solution

$$u_{\text{app}}(s, y) = \frac{1}{\pi} \sum_{j=1}^{n} f(x_j)\alpha_j(s, y),$$

where x_j are equally spaced points on the x-axis and $\alpha_j(s, y)$ is the angle at $z = s + i\,y$ subtended by the interval (x_j, x_{j+1}).

(b)   Use the result of Exercise 35(a) to write

$$u_{\text{app}}(s, y) = \frac{1}{\pi} \sum_{j=1}^{n} f(x_j)\Big(\text{Arg}\,(z - x_{j+1}) - \text{Arg}\,(z - x_j) \Big),$$

where $z = s + i\,y$.

(c)   Use the mean value theorem to show that there exists a ξ_j in the interval (x_j, x_{j+1}) such that $\text{Arg}\,(z - x_{j+1}) - \text{Arg}\,(z - x_j) = \frac{y}{(s-\xi_j)^2+y^2} \Delta x$, where $\Delta x = x_{j+1} - x_j$.

(d)   Thus the approximate solution

$$u_{\text{app}}(s, y) = \frac{y}{\pi} \sum_{j=1}^{n} \frac{f(x_j)}{(s - \xi_j)^2 + y^2} \Delta x.$$

Let $\Delta x \to 0$ and explain why $u_{app}(s, y)$ should approach the Poisson integral (11). [Hint: Interpret $u_{app}(s, y)$ as a Riemann sum that approximates the Poisson integral (11).]

What we have done in Exercise 42 is not exactly a proof of the Poisson integral formula, since Riemann sums are defined for functions on finite intervals and we have used them on the interval $(-\infty, \infty)$. However, if we know that $f(x)$ is continuous and equal to zero outside an interval $[a, b]$ (no matter how large the interval is), then the preceding proof works, giving you a simple proof of an important result in applied mathematics for the class of continuous functions vanishing outside a bounded interval. Often in analysis, knowing that a result is true for this class of continuous functions is a good indication that the result is also true for a larger class of functions. In particular, to extend the Poisson integral formula to a larger class of functions $f(x)$ satisfying some general integrability conditions, we use standard techniques from analysis to approximate $f(x)$ by continuous functions that vanish outside a bounded interval, and then use this approximation to establish the Poisson integral formula for f.

Project Problems: Conformal mapping method. *Exercises 43–46 are Dirichlet problems described by the corresponding figures. In each case, (a) solve the problem by following the four steps, as we did in Example 7.*

(b) Determine and plot the isotherms.

43.

[Hints: In Exercise 43, use the conformal mapping $f(z) = \sin z$.
In Exercise 44, use the conformal mapping $f(z) = z^2$. Solve the transformed problem by using Exercise 36.]

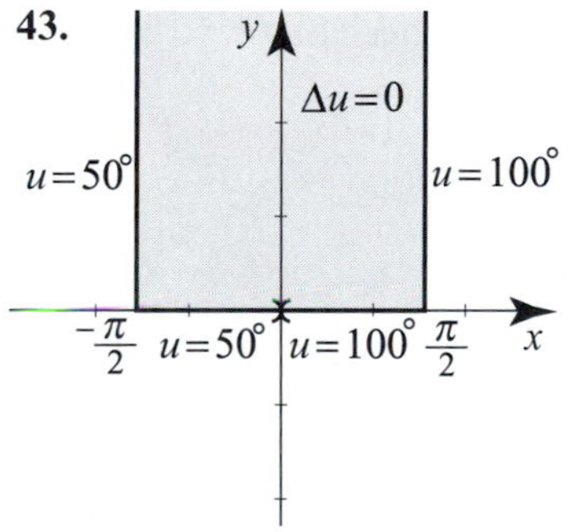

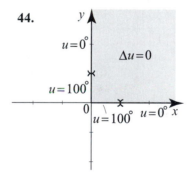

Figure 26 Dirichlet problem in Exercise 43.

44.

Figure 27 Dirichlet problem in Exercise 44.

45.

[Hints: In Exercise 45, use the conformal mapping $f(z) = \sin z$, then the result of Exercise 36.
In Exercise 46, use the conformal mapping $f(z) = z^4$, then the result of Exercise 36.]

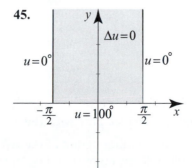

Figure 28 Dirichlet problem in Exercise 45.

46.

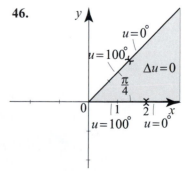

Figure 29 Dirichlet problem in Exercise 46.

47. Project Problem: Laplacian in polar coordinates. In this exercise, you are asked to derive the **polar form of the Laplacian**

(12)
$$\Delta u = \frac{\partial^2 u}{\partial r^2} + \frac{1}{r}\frac{\partial u}{\partial r} + \frac{1}{r^2}\frac{\partial^2 u}{\partial \theta^2}.$$

The derivation is straightforward but a little tedious. To organize your approach, follow the outlined steps.

(a) Recall the relationship between rectangular and polar coordinates

$$x = r\cos\theta, \qquad y = r\sin\theta, \qquad r^2 = x^2 + y^2, \qquad \tan\theta = \frac{y}{x}.$$

Differentiate $r^2 = x^2 + y^2$ with respect to x once, then a second time, and obtain

$$\frac{\partial r}{\partial x} = \frac{x}{r}, \qquad \frac{\partial^2 r}{\partial x^2} = \frac{y^2}{r^3}.$$

(b) Differentiate $\tan\theta = \frac{y}{x}$ with respect to x once, then a second time, and get

$$\frac{\partial \theta}{\partial x} = -\frac{y}{r^2}, \qquad \frac{\partial^2 \theta}{\partial x^2} = \frac{2xy}{r^4}.$$

(c) In a similar way, differentiate with respect to y and obtain

$$\frac{\partial r}{\partial y} = \frac{y}{r}, \qquad \frac{\partial^2 r}{\partial y^2} = \frac{x^2}{r^3}, \qquad \frac{\partial \theta}{\partial y} = \frac{x}{r^2}, \qquad \frac{\partial^2 \theta}{\partial y^2} = -\frac{2xy}{r^4}.$$

(d) From the previous identities, derive

$$\frac{\partial^2 \theta}{\partial x^2} + \frac{\partial^2 \theta}{\partial y^2} = 0 \quad \text{and} \quad \frac{\partial \theta}{\partial x}\frac{\partial r}{\partial x} + \frac{\partial \theta}{\partial y}\frac{\partial r}{\partial y} = 0.$$

(What does the first equation say about the function $\theta(x, y)$?)

(e) Use the chain rule in two dimensions ((13), Section 2.6) to derive

$$\frac{\partial^2 u}{\partial x^2} = \frac{\partial^2 u}{\partial r^2}\left(\frac{\partial r}{\partial x}\right)^2 + 2\frac{\partial^2 u}{\partial \theta \partial r}\frac{\partial r}{\partial x}\frac{\partial \theta}{\partial x} + \frac{\partial u}{\partial r}\frac{\partial^2 r}{\partial x^2} + \frac{\partial^2 u}{\partial \theta^2}\left(\frac{\partial \theta}{\partial x}\right)^2 + \frac{\partial u}{\partial \theta}\frac{\partial^2 \theta}{\partial x^2}.$$

Change x to y and obtain

$$\frac{\partial^2 u}{\partial y^2} = \frac{\partial^2 u}{\partial r^2}\left(\frac{\partial r}{\partial y}\right)^2 + 2\frac{\partial^2 u}{\partial \theta \partial r}\frac{\partial r}{\partial y}\frac{\partial \theta}{\partial y} + \frac{\partial u}{\partial r}\frac{\partial^2 r}{\partial y^2} + \frac{\partial^2 u}{\partial \theta^2}\left(\frac{\partial \theta}{\partial y}\right)^2 + \frac{\partial u}{\partial \theta}\frac{\partial^2 \theta}{\partial y^2}.$$

(f) Add $\frac{\partial^2 u}{\partial x^2}$ and $\frac{\partial^2 u}{\partial y^2}$ and simplify with the help of (d) to derive (12).

48. (a) Use (12) to give a direct proof of the result of Exercise 17(a).
(b) Use (12) to give a direct proof that $\text{Log}\,|z|$ is harmonic for all $z \neq 0$.

49. Show that if u is harmonic and independent of r, then $u_{\theta\theta} = 0$. Conclude that $u = a\theta + b$; equivalently, $u = a\arg_\alpha z + b$, where $\arg_\alpha z$ is a branch of the argument.

2.6   Differentiation of Functions of Several Variables

Our goal in this section is to fulfill our promise of completing the proof of the Cauchy-Riemann theorem, which we stated in Section 2.4. The material that is required for the proof is interesting in its own right. It deals with the concept of differentiation for functions of several variables. We will also apply it to give simple proofs of the chain rule and the mean value theorem in two dimensions.

As a motivation, let us begin by reviewing a geometric interpretation of the derivative of a real-valued function of one variable, $\phi(x)$. When we say that $\phi'(x_0)$ exists, we mean that the limit $\lim_{x \to x_0} \frac{\phi(x) - \phi(x_0)}{x - x_0}$ exists and equals a finite number $\phi'(x_0)$. If we set

$$r(x) = \phi'(x_0) - \frac{\phi(x) - \phi(x_0)}{x - x_0},$$

then $\lim_{x \to x_0} r(x) = 0$. Solving for $\phi(x)$, we obtain

$$\phi(x) = \phi(x_0) + \phi'(x_0)(x - x_0) + r(x)(x - x_0).$$

Let $\epsilon(x) = r(x) \frac{x - x_0}{|x - x_0|}$, then, because $\frac{x - x_0}{|x - x_0|} = \pm 1$ and $r(x) \to 0$ as $x \to x_0$, it follows that $\epsilon(x) \to 0$ as $x \to x_0$, and we have

$$\overbrace{\phi(x) = \phi(x_0) + \phi'(x_0)(x - x_0)}^{\text{tangent line at } x_0} + \epsilon(x)|x - x_0|. \tag{1}$$

This expresses the well-known geometric fact that, near a point $x = x_0$ where the function $\phi(x)$ is differentiable with derivative $\phi'(x_0)$, the tangent line approximates the graph of the function with an error that tends to 0 faster than $|x - x_0|$ (Figure 1).

With (1) in mind, we introduce the notion of differentiability for real-valued functions of two variables. To simplify the notation, we will sometimes denote a point (x, y) by the complex number z.

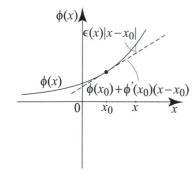

Figure 1 Approximation of a differentiable function by the tangent line.

**DEFINITION 1
DIFFERENTIABILITY
OF REAL-VALUED
FUNCTIONS OF
TWO VARIABLES**

We call a real-valued function $u(x, y)$ **differentiable** at $z_0 = (x_0, y_0)$ if it can be written in the form

$$u(z) = u(z_0) + A(x - x_0) + B(y - y_0) + \epsilon(z)|z - z_0|, \tag{2}$$

where A and B are (real) constants and $\epsilon(z) \to 0$ as $z \to z_0$.

In proofs, we will need to know limits such as

$$(3) \quad \lim_{z \to z_0} \epsilon(z) \frac{|z - z_0|}{z - z_0} = 0, \quad \lim_{z \to z_0} \epsilon(z) \frac{|x - x_0|}{x - x_0} = 0, \quad \lim_{z \to z_0} \epsilon(z) \frac{x - x_0}{|z - z_0|} = 0.$$

These are all of the form $\epsilon(z)$ times a bounded function, and hence by the squeeze theorem they tend to zero as $z \to z_0$, because $\epsilon(z)$ tends to zero. For example,

$$\left| \epsilon(z) \frac{x - x_0}{|z - z_0|} \right| = |\epsilon(z)| \overbrace{\frac{|x - x_0|}{|z - z_0|}}^{\leq 1} \leq |\epsilon(z)|,$$

where the inequality $\frac{|x - x_0|}{|z - z_0|} = \frac{|\operatorname{Re}(z - z_0)|}{|z - z_0|} \leq 1$ follows from (14), Section 1.2.

Our first result states that a differentiable function is continuous and has partial derivatives.

THEOREM 1

> Suppose $u(x, y)$ is differentiable at $z_0 = x_0 + i\, y_0$, so that (2) holds. Then
> (i) $u(x, y)$ is continuous at z_0; and
> (ii) u_x, u_y exist at z_0 and $u_x(z_0) = A$, $u_y(z_0) = B$.

Proof   (i) Taking limits on both sides of (2), we obtain

$$\lim_{z \to z_0} u(z) = \lim_{z \to z_0} \left(u(z_0) + \overbrace{A(x - x_0)}^{\to 0} + \overbrace{B(y - y_0)}^{\to 0} + \overbrace{\epsilon(z)|z - z_0|}^{\to 0} \right) = u(z_0).$$

Hence $u(x, y)$ is continuous at z_0. For (ii), we will only prove that $u_x(x_0, y_0) = A$, the second part being similar. To compute $u_x(x_0, y_0)$, we fix $y = y_0$ and take the derivative of $u(x, y_0)$ with respect to x. From (2) we have

$$
\begin{aligned}
u_x(x_0, y_0) &= \lim_{x \to x_0} \frac{u(x, y_0) - u(x_0, y_0)}{x - x_0} \\
&= A + \lim_{x \to x_0} \epsilon(x, y_0) \frac{|x - x_0|}{x - x_0} = A,
\end{aligned}
$$

since $\displaystyle\lim_{x \to x_0} \epsilon(x, y_0) \frac{|x - x_0|}{x - x_0} = 0$ as a consequence of (3).   ∎

The converse of part (ii) of Theorem 1 is not true. A function of two variables may have partial derivatives and yet fail to be differentiable at a point. In fact, the function may not even be continuous at that point. As an illustration, consider

$$(4) \qquad u(x, y) = \begin{cases} \dfrac{xy}{x^2 + y^2} & (x, y) \neq (0, 0), \\[2mm] 0 & (x, y) = (0, 0). \end{cases}$$

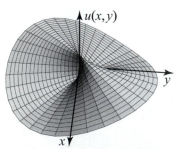

Figure 2 A discontinuous function with partial derivatives at $(0, 0)$. The function is not differentiable at $(0, 0)$.

It is a good exercise to check that $u_x(0,0) = 0$ and $u_y(0,0) = 0$, but u is not continuous at $(0, 0)$. Hence by Theorem 1(i), u is not differentiable at $(0, 0)$. The graph of u is shown in Figure 2.

To obtain differentiability at a point, more is needed than the existence of the partial derivatives. We have the following interesting result.

THEOREM 2
SUFFICIENT
CONDITIONS
FOR DIFFERENTIA-
BILITY

Let u be a real-valued function defined on a neighborhood of z_0. If
(i) $u_x(z_0)$ and $u_y(z_0)$ exist, and
(ii)   either $u_x(z)$ or $u_y(z)$ is continuous at z_0,
then u is differentiable at z_0, and (2) holds with $A = u_x(x_0, y_0)$, $B = u_y(x_0, y_0)$.

Proof   By reversing the roles of x and y, it is enough to prove either one of the cases in (ii). Let us take the case where $u_y(z)$ is continuous at z_0. We have

$$(5) \qquad u(z) - u(z_0) = [u(x, y_0) - u(x_0, y_0)] + [u(x, y) - u(x, y_0)].$$

For fixed y_0, we think of $u(x, y_0)$ as a function of x alone. Since this function (of one variable) has a derivative $u_x(x, y_0)$, it is differentiable and we can write

$$(6) \qquad u(x, y_0) - u(x_0, y_0) = u_x(x_0, y_0)(x - x_0) + \epsilon_1(x)(x - x_0),$$

where $\epsilon_1(x) \to 0$ as $x \to x_0$. Now for fixed x, we think of $u(x, y)$ as a function of y alone, with derivative $u_y(x, y)$. Applying the mean value theorem from single-variable calculus, we obtain

$$(7) \qquad u(x, y) - u(x, y_0) = u_y(x, y_1)(y - y_0),$$

where y_1 lies between y_0 and y (Figure 3). Substituting (6) and (7) into (5), we get

$$(8) \qquad u(z) - u(z_0) = [u_x(x_0, y_0) + \epsilon_1(x)](x - x_0) + u_y(x, y_1)(y - y_0),$$

where y_1 lies between y and y_0. As $z \to z_0$, y tends to y_0 and hence $y_1 \to y_0$, and so $u_y(x, y_1) \to u_y(x_0, y_0)$, by the continuity of u_y at (x_0, y_0). That is, $\lim\limits_{z \to z_0} u_y(x, y_1) = u_y(x_0, y_0)$; equivalently,

$$(9) \qquad u_y(x, y_1) = u_y(x_0, y_0) + \epsilon_2(z), \quad \epsilon_2(z) \to 0 \text{ as } z \to z_0.$$

Putting this in (7) and rearranging,

$$
\begin{aligned}
u(z) - u(z_0) &= [u_x(x_0, y_0) + \epsilon_1(x)](x - x_0) + [u_y(x_0, y_0) + \epsilon_2(z)](y - y_0) \\
&= u_x(z_0)(x - x_0) + u_y(z_0)(y - y_0) + \epsilon_1(x)(x - x_0) + \epsilon_2(z)(y - y_0) \\
(10) \quad &= u_x(z_0)(x - x_0) + u_y(z_0)(y - y_0) + \epsilon(z)|z - z_0|,
\end{aligned}
$$

where

$$\epsilon(z) = \epsilon_1(x)\frac{(x - x_0)}{|z - z_0|} + \epsilon_2(z)\frac{(y - y_0)}{|z - z_0|}.$$

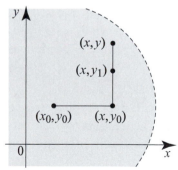

Figure 3 In the expression $u(x, y) - u(x, y_0)$, think of x as fixed, and apply the mean value theorem in the second variable.

Since $\frac{(x - x_0)}{|z - z_0|}$ and $\frac{(y - y_0)}{|z - z_0|}$ are bounded in absolute value (see (3)), $\epsilon(z) \to 0$ as $z \to z_0$, and the differentiability of $u(x, y)$ at z_0 follows from (10). Comparing (10) with (2), we identify $A = u_x(x_0, y_0)$ and $B = u_y(x_0, y_0)$.   ∎

We are now in a position to complete the proof of the Cauchy-Riemann equations theorem.

THEOREM 3

> Let $f(z) = u(x, y) + i\, v(x, y)$ be a complex-valued function, defined in an open set S. If u_x, u_y, v_x, and v_y are continuous and satisfy the Cauchy-Riemann equations in S, then $f(z)$ is analytic on S.

Proof   Let z_0 be any point in S. By Theorem 2, u and v are differentiable at z_0. Using (2), we have

$$
\begin{aligned}
u(z) &= u(z_0) + u_x(z_0)(x - x_0) + u_y(z_0)(y - y_0) + \epsilon_1(z)|z - z_0|, \\
v(z) &= v(z_0) + v_x(z_0)(x - x_0) + v_y(z_0)(y - y_0) + \epsilon_2(z)|z - z_0|,
\end{aligned}
$$

where $\epsilon_1(z)$, $\epsilon_2(z) \to 0$ as $z \to z_0$. Hence by the Cauchy-Riemann equations,

$$
\begin{aligned}
u(z) &= u(z_0) + u_x(z_0)(x - x_0) - v_x(z_0)(y - y_0) + \epsilon_1(z)|z - z_0|, \\
v(z) &= v(z_0) + v_x(z_0)(x - x_0) + u_x(z_0)(y - y_0) + \epsilon_2(z)|z - z_0|.
\end{aligned}
$$

Consequently,

$$
\begin{aligned}
f(z) &= u(z) + i\, v(z) \\
&= u(z_0) + i\, v(z_0) + u_x(z_0)(x - x_0) - v_x(z_0)(y - y_0) + \epsilon_1(z)|z - z_0| \\
&\quad + i\, v_x(z_0)(x - x_0) + i\, u_x(z_0)(y - y_0) + i\, \epsilon_2(z)|z - z_0| \\
&= f(z_0) + (u_x(z_0) + i\, v_x(z_0)) \left[(x - x_0) + i\,(y - y_0)\right] \\
&\quad + (\epsilon_1(z) + i\, \epsilon_2(z))|z - z_0| \\
&= f(z_0) + (u_x(z_0) + i\, v_x(z_0))(z - z_0) + \epsilon(z)(z - z_0),
\end{aligned}
$$

where $\epsilon(z) = (\epsilon_1(z) + i\, \epsilon_2(z)) \frac{|z - z_0|}{z - z_0}$. Since $(\epsilon_1(z) + i\, \epsilon_2(z)) \to 0$ as $z \to z_0$ and $\frac{|z - z_0|}{z - z_0}$ is bounded, it follows that $\epsilon(z) \to 0$ as $z \to z_0$, showing that f is differentiable at z_0 in the sense of Theorem 5, Section 2.3. Moreover, it follows from that theorem that $f'(z_0) = u_x(z_0) + i\, v_x(z_0)$. Since z_0 is arbitrary in S, we have $f'(z) = u_x(z) + i\, v_x(z)$ for all z in S. Also, by Theorem 4(iv), Section 2.2, $f'(z)$ is continuous because u_x and v_x are continuous. Thus f is analytic on S. ∎

Chain Rule and Mean Value Theorems

We can use Definition 1 and basic properties of differentiability to give simple proofs of the chain rule and the mean value theorem in two dimensions. These fundamental results were already used in this chapter and will be used again.

THEOREM 4
THE CHAIN RULE

> Suppose that $u(x, y)$ is a differentiable function of (x, y) in an open set S, where $x = x(t)$ and $y = y(t)$ are both differentiable functions of t. Then the function $U(t) = u(x(t), y(t))$ is a differentiable function of t and
>
> (11)
> $$
> \frac{dU}{dt} = \frac{\partial u}{\partial x} \frac{dx}{dt} + \frac{\partial u}{\partial y} \frac{dy}{dt}.
> $$

In particular, if the partial derivatives of u are continuous, then by Theorem 2, u is differentiable and the chain rule (11) holds.

Proof   Here again, we will use the notation $z = (x, y)$. For $z_0 = (x(t_0), y(t_0))$ in S, by Definition 1 and Theorem 1(ii), we have

$$u(z) - u(z_0) = u_x(z_0)(x - x_0) + u_y(z_0)(y - y_0) + \epsilon(z)|z - z_0|,$$

and so

$$(12) \qquad \frac{u(x(t), y(t)) - u(x(t_0), y(t_0))}{t - t_0}$$

$$= u_x(z_0)\frac{x(t) - x(t_0)}{t - t_0} + u_y(z_0)\frac{y(t) - y(t_0)}{t - t_0} + \epsilon(z)\frac{|z - z_0|}{t - t_0}.$$

As $t \to t_0$, $\frac{x(t)-x(t_0)}{t-t_0} \to \frac{dx}{dt}(t_0)$ and $\frac{y(t)-y(t_0)}{t-t_0} \to \frac{dy}{dt}(t_0)$, and hence (11) will follow from (12) once we prove that $\lim_{t \to t_0} \epsilon(z)\frac{|z - z_0|}{t - t_0} = 0$. As $t \to t_0$, $z \to z_0$ and hence $\epsilon(z) \to 0$. So it suffices to show that $\frac{|z-z_0|}{t-t_0}$ is bounded in a neighborhood of t_0. We have

$$\left|\frac{|z - z_0|}{t - t_0}\right| = \left|\frac{x - x_0}{t - t_0} + i\frac{y - y_0}{t - t_0}\right| \to \left|\frac{dx}{dt}(t_0) + i\frac{dy}{dt}(t_0)\right|,$$

and since this function has a limit, it is bounded in a neighborhood of t_0.   ∎

There is also a version of the chain rule in the situation where x and y are differentiable functions of two variables, s and t. In that case, we set $U(s, t) = u(x(s,t), y(s,t))$, and then

$$(13) \qquad \frac{\partial U}{\partial s} = \frac{\partial u}{\partial x}\frac{\partial x}{\partial s} + \frac{\partial u}{\partial y}\frac{\partial y}{\partial s},$$

$$(14) \qquad \frac{\partial U}{\partial t} = \frac{\partial u}{\partial x}\frac{\partial x}{\partial t} + \frac{\partial u}{\partial y}\frac{\partial y}{\partial t}.$$

The first formula follows by applying (11) to $U(s, t)$ while keeping t fixed, and the second follows by applying (11) while keeping s fixed.

We conclude this section with the mean value theorem in two dimensions.

THEOREM 5
THE MEAN VALUE
THEOREM
IN TWO
DIMENSIONS

Suppose $u(x, y)$ is a differentiable real-valued function of (x, y) in an open set S. Suppose that the line segment $[z_1, z_2]$ joining $z_1 = (x_1, y_1)$ to $z_2 = (x_2, y_2)$ lies entirely in S. Then there exists a point ζ on $[z_1, z_2]$ (see Figure 4) such that

$$(15) \qquad u(z_2) - u(z_1) = u_x(\zeta)(x_2 - x_1) + u_y(\zeta)(y_2 - y_1).$$

Proof   Parametrize the line segment $[z_1, z_2]$ by $x(t) = x_1 + t(x_2 - x_1)$, $y(t) = y_1 + t(y_2 - y_1)$, $0 \le t \le 1$. We have $\frac{dx}{dt} = x_2 - x_1$ and $\frac{dy}{dt} = y_2 - y_1$. Form the function $U(t) = u(x(t), y(t))$ for $0 \le t \le 1$. We have $U(0) = u(z_1)$, $U(1) = u(z_2)$, and by Theorem 4,

$$\frac{dU}{dt} = \frac{\partial u}{\partial x}\frac{dx}{dt} + \frac{\partial u}{\partial y}\frac{dy}{dt}$$

$$= \frac{\partial u}{\partial x}(x_2 - x_1) + \frac{\partial u}{\partial y}(y_2 - y_1).$$

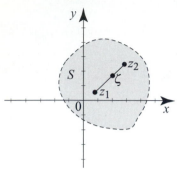

Figure 4 Mean value theorem in two dimensions.

Applying the mean value theorem in one variable to $U(t)$, there is a t_0 in $(0, 1)$ such that $U(1) - U(0) = \frac{dU}{dt}(t_0)(1 - 0)$. Hence

$$u(z_2) - u(z_1) = \frac{\partial u}{\partial x}(x(t_0), y(t_0))(x_2 - x_1) + \frac{\partial u}{\partial y}(x(t_0), y(t_0))(y_2 - y_1),$$

and so (15) follows with $\zeta = (x(t_0), y(t_0))$.   ∎

Exercises 2.6

1. Show that the partial derivatives u_x and u_y of the function given by (4) exist for all (x, y) but that the function is not continuous at $(0, 0)$.

2. The function $\phi(x) = x^2$ is differentiable at $x_0 = 1$. Find the function $\epsilon(x)$ such that

$$\phi(x) = \phi(1) + \phi'(1)(x - 1) + \epsilon(x)|x - 1|,$$

and verify directly that $\epsilon \to 0$ as $x \to 1$.

3. Show directly from Definition 1 that any linear function $u(x, y) = Ax + By + C$ is differentiable.

4. Prove that if $\epsilon(z) \to 0$ as $z \to z_0$, then (a) $\epsilon(z)\frac{x - x_0}{|x - x_0|} \to 0$ as $z \to z_0$, and (b) $\epsilon(z)\frac{z - z_0}{|z - z_0|} \to 0$ as $z \to z_0$.

5. Using (1), (2), and the squeeze theorem, show that if $f(x)$ is differentiable at x_0 and $g(y)$ is differentiable at y_0, then $u(x, y) = f(x)g(y)$ is differentiable at (x_0, y_0).

6. Show that if $u(x, y)$, $v(x, y)$ are differentiable and c_1, c_2 are constants, then $c_1 u(x, y) + c_2 v(x, y)$ and $u(x, y)v(x, y)$ are also differentiable.

7. Recast the function $u(x, y)$ in (4) in polar coordinates by setting $x = r\cos\theta$, $y = r\sin\theta$. Show that $u(x, y) = \frac{1}{2}\sin(2\theta)$, and use this formulation to describe the behavior of the function.

8. By reversing the steps of Theorem 3, show that if $f = u + iv$ is analytic on an open set S, then u and v are differentiable on S.

Topics to Review

The definition of path in Section 3.1 is based on the parametric form of curves from calculus. The definition of path integral in Section 3.2 is based on the definition of definite integrals as a limit of Riemann sums, from calculus. A good understanding of these notions and the fundamental theorem of calculus are required. We will also use basic properties of continuous functions in proofs. (For example, if f is continuous on $[a, b]$, then f attains its maximum and minimum values in $[a, b]$.)

Looking Ahead

Sections 3.1 and 3.2 contain essential definitions and properties pertaining to path integrals. They should be covered in detail. Section 3.3 gives necessary and sufficient conditions for independence of path. These statements are important on a theoretical side, but not very useful from a practical side. So, if proofs are not stressed, this section can be skipped or covered quickly. Section 3.4 is the heart of the subject. A thorough treatment is required. The mathematical definition of deformation of path in Section 3.4 can be skipped. Sections 3.6 and 3.7 are very important. They include striking applications of Cauchy's theorem. They should be covered thoroughly. Section 3.8 is optional; it continues our parallel study of harmonic functions and Dirichlet problems. Section 3.9 contains one theoretical result and can be omitted entirely without affecting the continuity of the course.

3

COMPLEX INTEGRATION

There is no let up! No end to it! ... Innumerable calculations. Endless fighting. Signs. Formulas. Theorems besetting me from dawn to dusk.
-Augustin Louis Cauchy

Having studied the derivative in Chapter 2, our next step is to study the integral of a function of a complex variable. If $f(z)$ is defined on a subset Ω of $\mathbb{C}$ and z_1 and z_2 are points in Ω, how can we define the integral of f from z_1 to z_2? First we must define the notion of path γ in Ω from z_1 to z_2 (Section 3.1); then we will define the integral over this path (Section 3.2), denoted by $\int_\gamma f(z)\,dz$. In Section 3.2, an expression of this integral is given in terms of the usual Riemann integral from calculus, but applied to complex-valued functions on an interval $[a, b]$.

The most important question in this chapter concerns the independence of path: When does $\int_\gamma f(z)\,dz$ depend only on z_1 and z_2 and not on γ? The answer to this question is provided by Cauchy's theorem (Theorem 2, Section 3.4). Cauchy's theorem is to complex analysis what the fundamental theorem of calculus is to calculus. The discussion of Cauchy's theorem in Section 3.4 will lead us to important topics such as simple connectedness of regions in $\mathbb{C}$. We will derive very useful versions of Cauchy's theorem for simply and multiply connected regions.

The proof of Cauchy's theorem is nontrivial. It will be presented in complete detail in Section 3.5.

In Sections 3.6 and 3.7 we start the applications of Cauchy's theorem and derive Cauchy's generalized integral formula, which is the basis for all the applications that will follow. We illustrate the power of this theory by giving a simple proof of the fundamental theorem of algebra (Section 3.7) and by deriving various striking properties of analytic functions, including the mean value property and the maximum modulus principle.

In Section 3.8, we use the theory of analytic functions to study harmonic functions and the solution of Dirichlet problems. In particular, we will give a simple derivation of the famous Poisson integral formula, by using a simple change of variables and the mean value property of analytic functions.

3.1   Contours and Paths in the Complex Plane

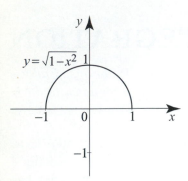

$y = \sqrt{1-x^2}$

Figure 1

You may recall the notion of a curve from calculus as being the graph of a function $y = f(x)$. More generally, you may recall the parametric form of a curve, which represents a curve by expressing x and y as functions of a third variable t. For example, the semi-circle $y = \sqrt{1 - x^2}$, $-1 \leq x \leq 1$, in Figure 1 can be parametrized by

$$x = x(t) = \cos t, \qquad y = y(t) = \sin t, \quad 0 \leq t \leq \pi.$$

A **parametric form** of a curve is thus a representation of the curve by a pair of equations: $x = x(t)$ and $y = y(t)$, where t ranges over a set of real numbers, usually a closed interval $[a, b]$ (see Figure 2). Each value of t determines a point $\gamma(t) = (x(t), y(t))$, which traces the curve as t varies.

Figure 2 A parametric interval mapping to a curve.

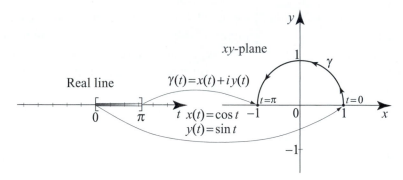

A curve may have more than one parametrization. For example, the interval $[0, 1]$ can be parametrized as $\gamma_1(t) = t$, $0 \leq t \leq 1$ or $\gamma_2(t) = t^2$, $0 \leq t \leq 1$. Both γ_1 and γ_2 represent the same curve. In our analysis, we will always choose and work with a specific parametrization of the curve. For that reason, it will be convenient to refer to a curve by its parametrization $\gamma(t)$ or simply γ, even though the curve may have more than one parametrization.

Let $\gamma(t)$, $a \leq t \leq b$, be a parametrization of a curve. As t varies from a to b the point $\gamma(t)$ traces the curve in a specific direction, starting with $\gamma(a)$, the **initial point** of γ, and ending at $\gamma(b)$, the **terminal point** of γ. For continuous curves, this direction is usually denoted by an arrow on the curve. A curve is **closed** if $\gamma(a) = \gamma(b)$. For circles and circular arcs, if the arrow points in the counterclockwise direction, we will say that the curve has positive orientation. Curves traversed in the clockwise direction have negative orientation (see Figure 3).

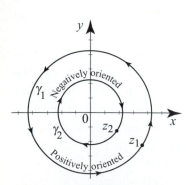

Figure 3 Negative and positive orientation.

Since $z = x + iy$, it makes sense to adopt the notation $z(t) = x(t) + i y(t)$. In particular, we can write the parametric form of a curve γ using complex notation as

$$(1) \qquad \gamma(t) = x(t) + i y(t), \quad a \leq t \leq b,$$

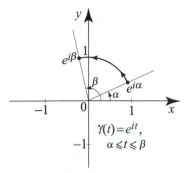

Figure 4(a) A positively oriented arc with initial point $e^{i\alpha}$ and terminal point $e^{i\beta}$.

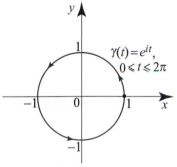

Figure 4(b) A positively oriented circle.

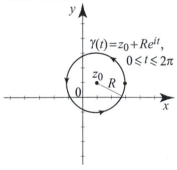

Figure 4(c) Dilating and then translating a circle.

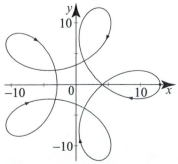

Figure 5 A hypotrochoid.

and think of the curve as the graph of a complex-valued function of a real variable t. The following examples illustrate the use of the complex notation.

EXAMPLE 1   Parametric forms of arcs, circles, and line segments
(a) The arc in Figure 4(a) is conveniently parametrized by $\gamma(t) = e^{it} = \cos t + i \sin t$, $\alpha \le t \le \beta$. Its initial point is $e^{i\alpha}$ and its terminal point is $e^{i\beta}$.
(b) The positively oriented circle in Figure 4(b) is parametrized by

$$\gamma(t) = e^{it} = \cos t + i \sin t, \quad 0 \le t \le 2\pi.$$

The circle is a closed curve with initial point $\gamma(0) = 1$ and terminal point $\gamma(2\pi) = 1$.
(c) The circle in Figure 4(c) is centered at z_0 with radius $R > 0$. We obtain its parametrization by dilating and then translating the equation in (b). This gives

$$\gamma(t) = z_0 + Re^{it} = z_0 + R(\cos t + i \sin t), \quad 0 \le t \le 2\pi.$$

For example, the equation

$$\gamma(t) = 1 + i + 2e^{it}, \quad 0 \le t \le 2\pi,$$

represents a circle centered at the point $(1, 1)$, with radius 2.
(d) Let z_1 and z_2 be arbitrary complex numbers. A **directed line segment** $[z_1, z_2]$ is the path γ over $[0, 1]$ defined by

$$(2) \qquad \gamma(t) = (1 - t)z_1 + tz_2, \qquad 0 \le t \le 1.$$

Its initial point is $\gamma(0) = z_1$ and its terminal point is $\gamma(1) = z_2$.   ∎

As our next example illustrates, there is definitely an advantage in using the complex notation, especially when the parametric representation of a curve involves trigonometric functions.

EXAMPLE 2   A hypotrochoid in complex notation
A **hypotrochoid** is a curve given in parametric form by

$$x(t) = a \cos t + b \cos\left(\frac{at}{2}\right), \qquad y(t) = a \sin t - b \sin\left(\frac{at}{2}\right).$$

Express the equation of the hypotrochoid in complex notation.
Solution In complex form, we have

$$
\begin{aligned}
\gamma(t) &= x(t) + i\, y(t) \\
&= \left(a \cos t + b \cos\left(\frac{at}{2}\right)\right) + i\left(a \sin t - b \sin\left(\frac{at}{2}\right)\right) \\
&= a\left(\cos t + i \sin t\right) + b\left(\cos\left(\frac{at}{2}\right) - i \sin\left(\frac{at}{2}\right)\right) \\
&= ae^{it} + be^{-i\frac{at}{2}}.
\end{aligned}
$$

We have taken $a = 8$ and $b = 5$ and plotted the corresponding hypotrochoid in Figure 5 (where $0 \le t \le 2\pi$).   ∎

If $\gamma(t)$ is a curve parametrized by the interval $[a,\,b]$, the **reverse** of γ is the curve $\gamma_r(t)$ (sometimes written $-\gamma$) parametrized by the same interval $[a,\,b]$ and given by

$$(3) \qquad\qquad \gamma_r(t) = \gamma(b + a - t).$$

The reverse of γ traces the same set of points as γ but in the opposite direction, starting with $\gamma_r(a) = \gamma(b)$ and ending with $\gamma_r(b) = \gamma(a)$.

The reverse of the directed line segment $[z_1,\,z_2]$ in Example 1(d) is clearly the line segment $[z_2,\,z_1]$ with equation

$$\gamma_r(t) = \gamma(1 - t) = t\,z_1 + (1 - t)z_2, \qquad 0 \le t \le 1.$$

The curves in Example 1 are all continuous, as can be seen from their graphs. In fact, these curves have differentiability properties that will be needed when studying integrals along curves. Recall that a curve is given as the graph of a complex-valued function of a real variable. So, in order to study their differentiability properties, we next investigate the derivative of a complex-valued function of a real variable (not to be confused with the derivative of a complex-valued function of a complex variable, $f'(z)$).

Complex-Valued Functions of a Real Variable

Given a complex-valued function of a real variable $f(t) = u(t) + i\,v(t)$, we define the derivative of f in the usual way by

$$(4) \qquad\qquad f'(t) = \frac{d}{dt}f(t) = \lim_{h \to 0} \frac{f(t + h) - f(t)}{h}.$$

As we did in the proof of Theorem 3, Section 2.2, you can show that $f'(t)$ exists if and only if $u'(t)$ and $v'(t)$ both exist, and

$$(5) \qquad\qquad f'(t) = u'(t) + i\,v'(t).$$

The derivative of a complex-valued function of a real variable satisfies many properties similar to those of the usual derivative of a real-valued function. In particular, if f and g are complex-valued differentiable functions, α and β are complex numbers, then

$$(6) \qquad\qquad \Big(\alpha f(t) + \beta g(t)\Big)' = \alpha f'(t) + \beta g'(t),$$

$$(7) \qquad\qquad \big(f(t)\,g(t)\big)' = f'(t)\,g(t) + g'(t)f(t),$$

$$(8) \qquad\qquad \left(\frac{f(t)}{g(t)}\right)' = \frac{f'(t)g(t) - g'(t)f(t)}{g(t)^2} \qquad (g(t) \ne 0),$$

$$(9) \qquad\qquad \big(f\big(g(t)\big)\big)' = f'\big(g(t)\big)g'(t).$$

We leave the verification of these rules as an exercise.

EXAMPLE 3   Derivative of $e^{\alpha t}$

Suppose that $\alpha = a + i b$ is a complex number. Show that $\dfrac{d}{dt} e^{\alpha t} = \alpha\, e^{\alpha t}$. Hence we recover the same formula from calculus. In particular, if $\alpha = i\omega$ is purely imaginary, then

(10)
$$\frac{d}{dt} e^{i\omega t} = i\omega e^{i\omega t}.$$

Solution   We have $e^{\alpha t} = e^{at}\left(\cos bt + i \sin bt\right)$, and so using (5) we obtain

$$
\begin{aligned}
\frac{d}{dt} e^{\alpha t} &= \frac{d}{dt}\left(e^{at}\cos bt\right) + i\frac{d}{dt}\left(e^{at}\sin bt\right) \\
&= \left(ae^{at}\cos bt - be^{at}\sin bt\right) + i\left(ae^{at}\sin bt + be^{at}\cos bt\right) \\
&= \left(a + i b\right)\left(e^{at}\cos bt + i\,e^{at}\sin bt\right) \\
&= \alpha e^{\alpha t}.
\end{aligned}
$$
■

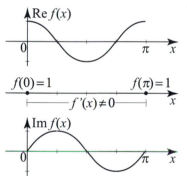

Figure 6 $f(2\pi) - f(0) = 0$, but $f'(x) \neq 0$ for all x in $(0, 2\pi)$ showing that the mean value property fails for complex-valued functions.

Having stated properties of the derivative of a real-valued function that hold for complex-valued functions of a real variable, we should caution you that not all properties of derivatives that hold in the real case hold in the complex case. Most notably, the mean value property does not hold for complex-valued functions. In the real case, the mean value property states that if f is continuous on $[a, b]$ and differentiable on (a, b), then $f(b) - f(a) = f'(c)(b - a)$ for some c in (a, b). This property does not hold for complex-valued functions. To see this, consider $f(x) = e^{ix}$ for x in $[0, 2\pi]$. Then f is continuous on $[0, 2\pi]$ and has a derivative $f'(x) = i\, e^{ix}$ on $(0, 2\pi)$. Also, $f(2\pi) - f(0) = 1 - 1 = 0$, but $f'(x)$ never vanishes since $|f'(x)| = |i\, e^{ix}| = |e^{ix}| = 1$. Hence there is no number c in $(0, 2\pi)$ such that $f(2\pi) - f(0) = (2\pi - 0)f'(c)$, and so the mean value property does not hold (Figure 6).

Before returning to the main topic of curves, we show with an application how complex-valued functions can simplify greatly the solution of problems involving real-valued functions.

EXAMPLE 4   A second order ordinary differential equation

Find a particular solution $y_p(x)$ of

(11)
$$y'' - 2y' + y = \cos 2x.$$

This is a second order linear differential equation with constant coefficients. It is nonhomogeneous because of the term $\cos 2x$ (see Appendix A.2).

Solution   Typically we would solve this problem by using the method of undetermined coefficients, which tells us to try $y_p = a \cos 2x + b \sin 2x$, where a and b are real constants. The solution is completed by plugging y_p into the equation and solving for a and b. Although the steps are straightforward, they require a lot of computations. We can simplify the problem by thinking of $\cos 2x$ as the real part of e^{2ix} and considering the differential equation

> Whereas the solution y_p of (11) has to include a sine and a cosine term to account for the derivatives of $\cos 2x$, in finding a particular solution of (12) we can try $y = Ae^{2ix}$ for the good reason that the derivative of an exponential function is again another exponential function.

$$(12) \qquad\qquad y'' - 2y' + y = e^{2ix}.$$

The trick is to solve (12) and then take real parts to obtain a solution of (11). Dealing with (12) is easier. For a particular solution we try $y = Ae^{2ix}$, where A is a complex number. From Example 3, we have

$$y' = 2iAe^{2ix} \quad \text{and} \quad y'' = -4Ae^{2ix}.$$

Plugging into (12) and simplifying, we obtain

$$-4Ae^{2ix} - 4iAe^{2ix} + Ae^{2ix} = e^{2ix} \Rightarrow e^{2ix}A(-3 - 4i) = e^{2ix}.$$

Dividing by e^{2ix} and simplifying, we find

$$A = \frac{1}{-3 - 4i} = \frac{-3 + 4i}{(-3 - 4i)(-3 + 4i)} = -\frac{3}{25} + i\frac{4}{25}.$$

Hence a particular solution of (12) is

$$\begin{aligned}
y &= \left(-\frac{3}{25} + i\frac{4}{25}\right)e^{2ix} = \left(-\frac{3}{25} + i\frac{4}{25}\right)\left(\cos 2x + i \sin 2x\right) \\
&= \left(-\frac{3}{25} \cos 2x - \frac{4}{25} \sin 2x\right) + i\left(\frac{4}{25} \cos 2x - \frac{3}{25} \sin 2x\right).
\end{aligned}$$

Taking real parts, we obtain a particular solution of (11):

$$y_p(x) = -\frac{3}{25} \cos 2x - \frac{4}{25} \sin 2x.$$

You should verify that this is indeed a solution by plugging it back into (11).   ∎

In the exercises, we illustrate the use of the method of Example 4 in solving other interesting nonhomogeneous differential equations.

Contours and Paths

Returning to our discussion of complex-valued functions of a real variable, we introduce the following useful definitions.

<table>
<tr>
<td>

DEFINITION 1
PIECEWISE
CONTINUOUS AND
PIECEWISE
SMOOTH
FUNCTIONS

</td>
<td>

A complex-valued function of a real variable $f(t)$ is said to be **piecewise continuous** on $[a, b]$ if the following hold:

(i) $f(t)$ exists and is continuous for all but finitely many points in (a, b).

(ii)   At any point c in (a, b) where f fails to be continuous, both the left limit $\lim\limits_{t\uparrow c} f(t)$ and the right limit $\lim\limits_{t\downarrow c} f(t)$ exist and are finite.

(iii)   At the endpoints, the right limit $\lim\limits_{t\downarrow a} f(t)$ and the left limit $\lim\limits_{t\uparrow b} f(t)$ exist and are finite.

The function $f(t)$ is said to be **piecewise smooth** on $[a, b]$ if f and f' are both piecewise continuous on $[a, b]$.

</td>
</tr>
</table>

If f and f' are both continuous on $[a, b]$ we will say that f is **smooth** or **continuously differentiable** on $[a, b]$.

We are now in a position to introduce the notion of path, which is fundamental to our development of the theory of integration. Recall that we have referred to a curve by its parametrization $\gamma(t)$. In addition, we will attribute to the curve the analytical properties of the function $\gamma(t)$, such as continuity and differentiability. Thus, when we talk about a continuous curve or a differentiable curve, we have in mind a specific parametrization with these properties.

DEFINITION 2
PATH OR
CONTOUR

By a **path** or a **contour** we mean a continuous piecewise smooth curve $\gamma(t)$ over a closed interval $[a, b]$.

A path γ is **continuously differentiable** or **smooth** on $[a, b]$ if $\gamma(t)$ and $\gamma'(t)$ are continuous on $[a, b]$. Thus, according to Definition 2, a path or a contour γ is a finite sequence of continuously differentiable curves, $\gamma_1, \gamma_2, \ldots, \gamma_n$, joined together end-to-end. In this case, it is sometimes convenient to write $\gamma = (\gamma_1, \gamma_2, \ldots, \gamma_n)$. The path is **closed** if the initial point of γ_1 is joined to the terminal point of γ_n.

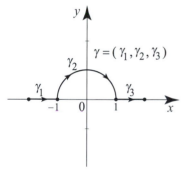

Figure 7 for Example 5.

EXAMPLE 5   Path

The path γ in Figure 7 is made up of three smooth curves: the line segment $\gamma_1 = [-2, -1]$; the semi-circle γ_2; and the line segment $\gamma_3 = [1, 2]$. We can parametrize γ by the interval $[-2, 2]$ as follows:

$$\gamma(t) = \begin{cases} t & \text{if } -2 \leq t \leq -1, \\ e^{i\frac{\pi}{2}(1-t)} & \text{if } -1 \leq t \leq 1, \\ t & \text{if } 1 \leq t \leq 2. \end{cases}$$

The choice of the interval $[-2, 2]$ was just for convenience. Other closed intervals can be used to parametrize γ. ∎

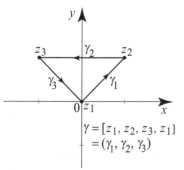

Figure 8 for Example 6.

EXAMPLE 6   Polygonal paths

A **polygonal path**, $\gamma = [z_1, z_2, \ldots, z_n]$, consists of a directed broken line going through the points $z_1, z_2 \ldots, z_n$, with initial point z_1 and terminal point z_n. The path is closed if $z_1 = z_n$. As an illustration, let $z_1 = 0$, $z_2 = 1 + i$, and $z_3 = -1 + i$; then $\gamma = [z_1, z_2, z_3, z_1]$ is a closed polygonal path. To find the equation of γ, we start by finding the equations of the paths γ_1, γ_2, and γ_3, shown in Figure 8. From Example 1(d), we have

$$\begin{aligned} \gamma_1(t) &= (1 + i)\, t, & 0 \leq t \leq 1; \\ \gamma_2(t) &= (1 - t)(1 + i) + t(-1 + i) = (1 + i) - 2t, & 0 \leq t \leq 1; \\ \gamma_3(t) &= (1 - t)(-1 + i), & 0 \leq t \leq 1. \end{aligned}$$

We can now use these equations to parametrize γ over a closed interval, say $[0, 1]$. This can be done by reparametrizing γ_1 over $[0, \frac{1}{3}]$, γ_2 over $[\frac{1}{3}, \frac{2}{3}]$, γ_3 over $[\frac{2}{3}, 1]$, and then pasting the three equations to form the path γ (Figure 9). To parametrize γ_1 over $[0, \frac{1}{3}]$, it suffices to change t to $3t$. This yields

$$\gamma_1(t) = 3t(1+i), \quad 0 \le t \le \frac{1}{3}.$$

To parametrize γ_2 over $[\frac{1}{3}, \frac{2}{3}]$, we first scale down by changing t to $3t$,

$$\gamma_2(t) = (1+i) - 6t, \quad 0 \le t \le \frac{1}{3}.$$

We then shift to the right by $\frac{1}{3}$ units and get

$$\gamma_2(t) = (1+i) - 6(t - \frac{1}{3}) = 3 + i - 6t, \quad \frac{1}{3} \le t \le \frac{2}{3}.$$

For γ_3, we first scale by a factor of $\frac{1}{3}$ and get

$$\gamma_3(t) = (1 - 3t)(-1 + i), \quad 0 \le t \le \frac{1}{3}.$$

We then shift to the right by $\frac{2}{3}$ units and get

$$\gamma_3(t) = (1 - 3(t - \frac{2}{3}))(-1 + i) = (-1 + i)(3 - 3t), \quad \frac{2}{3} \le t \le 1.$$

Pasting the equations together, we obtain

$$\gamma(t) = \begin{cases} 3t(1+i) & \text{if } 0 \le t \le \frac{1}{3}, \\ 3 + i - 6t & \text{if } \frac{1}{3} \le t \le \frac{2}{3}, \\ (-1+i)(3-3t) & \text{if } \frac{2}{3} \le t \le 1. \end{cases}$$

The polygonal path γ is clearly continuous with a piecewise continuous derivative. It is thus a path in the sense of Definition 2.   ∎

Our next example shows two interesting cases of paths.

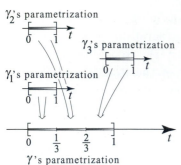

γ_2's parametrization

γ_3's parametrization

γ_1's parametrization

γ's parametrization

Figure 9 for Example 6.

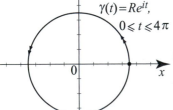

Figure 10 A path that degenerates to a point.

EXAMPLE 7   Degenerate and doubly traced paths

(a) Describe the path given by $\gamma(t) = z_0$, $a \le t \le b$.

(b) Describe the path given by $\gamma_2(t) = Re^{it}$, $0 \le t \le 4\pi$, and describe how it is different from $\gamma_1(t) = Re^{it}$, $0 \le t \le 2\pi$.

Solution   (a) As t ranges through the interval $[a, b]$, the value of $\gamma(t)$ remains fixed at the point z_0. Clearly $\gamma(t)$ is continuous and $\gamma'(t) = 0$ is also continuous, so $\gamma(t)$ is a path, which has degenerated to a single point (Figure 10).

(b) Points on the path $\gamma_2(t)$ are on the circle of radius R, centered at the origin. As t ranges from 0 to 4π, $\gamma(t)$ traces around the circle twice. The path is shown in Figure 11; double arrows show that the path is traced twice in the counterclockwise direction. The path $\gamma_1(t)$ traces around the circle only once. The two paths are not the same but have the same graphs.   ∎

Figure 11 A doubly traced circle.

As a convention, whenever we refer to a closed path, we mean the path that is traversed only once, unless otherwise stated. As an illustration, we will

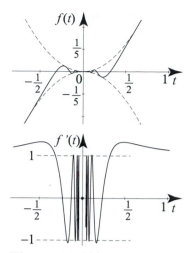

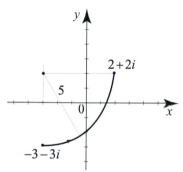

Figure 12 $f(t)$ is continuous and even differentiable, but its graph is not a path.

call the path $\gamma_2(t)$ in Example 7(b) the circle of radius R, centered at 0, traversed twice in the positive direction.

Our final example is a differentiable curve that is not a path.

EXAMPLE 8   A curve that is not a path

Let $f(t) = t^2 \sin \frac{1}{t}$ for $t \neq 0$ and $f(0) = 0$, and define a curve $\gamma(t) = t + i\, f(t)$, where $-\pi \leq t \leq \pi$. The graph of γ is simply the graph of $f(t)$ over the interval $[-\pi, \pi]$. For $t \neq 0$, we have $f'(t) = 2t \sin \frac{1}{t} - \cos \frac{1}{t}$, and $f'(0) = 0$ (use the definition of the derivative and the squeeze theorem). So $f(t)$ is continuous and differentiable on the closed interval $[-\pi, \pi]$. But $f'(t)$ is discontinuous at 0 and neither the left nor the right limits of $f'(t)$ exist at 0 (see Exercise 1, Section 3.9). So $\gamma'(t)$ is not piecewise continuous in $[-\pi, \pi]$ and hence $\gamma(t)$ is not a path. The graphs of f and f' are shown in Figure 12. Note the discontinuity at 0 of $f'(t)$.   ∎

Exercises 3.1

In Exercises 1–14, a curve is given. Parametrize it over a suitable interval $[a, b]$ and plot the curve when the graph is not given.

1. The line segment with initial point $z_1 = 1 + i$ and terminal point $z_2 = -1 - 2i$.
2. The line segment through the origin as initial point and terminal point $z = e^{i\frac{\pi}{3}}$.
3. The counterclockwise circle with center at $3i$ and radius 1.
4. The clockwise circle with center at $-2 - i$ and radius 3.
5. The positively oriented arc on the unit circle such that $-\frac{\pi}{4} \leq \arg z \leq \frac{\pi}{4}$.
6. The negatively oriented arc on the unit circle such that $-\frac{\pi}{4} \leq \arg z \leq \frac{\pi}{4}$.
7. The directed line segment $[z_1, z_2, z_3, z_1]$ where $z_1 = 0$, $z_2 = i$, and $z_3 = -1$.
8. The directed line segment $[z_1, z_2, z_3, z_4]$ where $z_1 = 1$, $z_2 = 2$, $z_3 = i$, and $z_4 = 2i$.
9. The contour in Figure 13.
10. The contour in Figure 14.
11. The contour in Figure 15.
12. The reverse of the contour in Figure 13.
13. The reverse of the contour in Figure 14.
14. The reverse of the contour in Figure 15.

In Exercises 15–18, plot the given path.

15. $\gamma(t) = te^{-it}$, $0 \leq t \leq 2\pi$.
16. $\gamma(t) = 2\cos t + i\,\sin t$, $0 \leq t \leq 2\pi$.
17. $\gamma(t) = t + i\sin(\pi t)$, $0 \leq t \leq 1$.
18. $\gamma(t) = t^2 + it$, $-3 \leq t \leq 3$.

In Exercises 19–24, find the derivative of the given function.

19. $f(t) = te^{-it}$.
20. $f(t) = e^{2it^2}$.
21. $f(t) = (2 + i)\cos(3it)$.
22. $f(t) = \dfrac{2 + i + t}{-i - 2t}$.
23. $f(t) = \left(\dfrac{t + i}{t - i}\right)^2$.
24. $f(t) = \text{Log}\,(it)$.

In Exercises 25–28, find a particular solution of the given differential equation. Use the technique of Example 4. (Hint for Exercise 27: Think of an exponential function whose real part is $e^t \cos t$.)

25. $y'' + y' + y = \cos t$.
26. $y'' + y = \sin 3t$.
27. $y'' - y = e^t \cos t$.
28. $y'' + y' + y = t \cos t$.

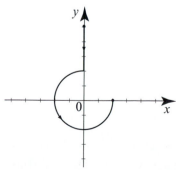

Figure 13 Arc of a circle with center at $-3 + 2i$ and radius 5.

Figure 14 Arc of a circle with center at 0 and radius 2.

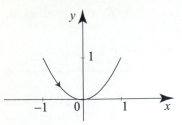

Figure 15 Arc of a parabola for Exercise 11.

Using complex methods. Very often when we use complex methods to solve a real problem, we end up solving two problems: one associated with the real part of the solution and one with its imaginary part. For example, when we solved the differential equation $y'' - 2y' + y = \cos 2x$ in Example 4, we also obtained the solution to $y'' - 2y' + y = \sin 2x$, as can be seen from taking the imaginary part of $y = \left(-\frac{3}{25} + i\frac{4}{25}\right)e^{2ix}$, which is the solution of $y'' - 2y' + y = e^{i2x}$.

In Exercises 29–30, you are given a real differential equation. Find a particular solution by envisioning the equation as the real or imaginary part of a complex differential equation. Also, state the corresponding problem and a particular solution for the other part of the complex differential equation.

29. $y'' - 2y' - 3y = \cos 4t.$ **30.** $y'' + y' + y = 3 + \sin 2t.$

 In Exercises 31–34, a curve is given in parametric form. (a) Find the equation of the curve in complex form. (b) Plot the curve for specific values of a and b of your choice.

31. A **hypocycloid** $(a > b)$

$$x(t) = (a-b)\cos t + b\cos\left(\frac{a-b}{b}t\right),$$
$$y(t) = (a-b)\sin t - b\sin\left(\frac{a-b}{b}t\right).$$

32. An **epicycloid**

$$x(t) = (a+b)\cos t - b\cos\left(\frac{a+b}{b}t\right),$$
$$y(t) = (a+b)\sin t - b\sin\left(\frac{a+b}{b}t\right).$$

33. An **epitrochoid**

$$x(t) = a\cos t - b\cos\left(\frac{at}{2}\right),$$
$$y(t) = a\sin t - b\sin\left(\frac{at}{2}\right).$$

34. A **trochoid**

$$x(t) = at - b\sin t,$$
$$y(t) = a - b\cos t.$$

3.2   Complex Integration

Our goal in this section is to construct the integral of a complex-valued function along a path in the complex plane. Since our construction is modeled after the Riemann integral from calculus, we start the section with a quick review of some notions from integral calculus.

Suppose that f is a real-valued continuous function defined on $[a, b]$. Let P be a partition of $[a, b]$ consisting of closed subintervals of $[a, b]$ with endpoints $a = x_0 < x_1 < x_2 < \cdots < x_m = b$. Let $|P|$ denote the **norm** of the partition, being the largest of the lengths $\Delta x_k = x_k - x_{k-1}$. Now consider the Riemann sum corresponding to P (see Figure 1):

$$(1) \qquad \sum_{k=1}^{m} f(x_k^{\star})(x_k - x_{k-1}),$$

Figure 1 The Riemann integral as a limit of Riemann sums.

where $x_k^{\star}$ is a point in $[x_{j-1}, x_j]$. From calculus, we know that as the partition gets finer, that is, as $|P| \to 0$, the Riemann sum (1) converges to a finite number called the **definite integral** of f and denoted by $\int_a^b f(x)\, dx$. Also, if F is any antiderivative of f, then by the fundamental theorem of calculus we have

$$(2) \qquad \int_a^b f(x)\,dx = F(b) - F(a).$$

Our next step is to extend the Riemann integral to complex-valued functions of a real variable.

Riemann Integral of Complex-Valued Functions

Suppose that $f(x) = u(x) + i\,v(x)$ is a complex-valued continuous function on $[a, b]$. How should we define its integral over $[a, b]$? Motivated by the construction of the Riemann integral of a real-valued function, we consider a partition P of $[a, b]$ given by $a = x_0 < x_1 < x_2 < \cdots < x_m = b$, and form the corresponding Riemann sum analog of (1):

$$\sum_{k=1}^m f(x_k^\star)(x_k - x_{k-1})$$

$$(3) \qquad = \sum_{k=1}^m u(x_k^\star)(x_k - x_{k-1}) + i\sum_{k=1}^m v(x_k^\star)(x_k - x_{k-1}),$$

where $x_k^\star$ is a point in $[x_{k-1}, x_k]$. Since u and v are continuous, the Riemann sums on the right converge to $\int_a^b u(x)\,dx + i\int_a^b v(x)\,dx$. This leads us to the following definition of the definite integral in the complex-valued case:

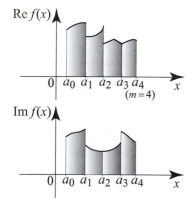

Re $f(x)$

$0 \quad a_0 \ a_1 \ a_2 \ a_3 a_4 \qquad x$

$(m=4)$

Im $f(x)$

$0 \quad a_0 \ a_1 \ a_2 \ a_3 \ a_4 \qquad x$

Figure 2 We integrate a complex-valued function by integrating its real and imaginary parts separately.

$$(4) \qquad \boxed{\int_a^b f(x)\,dx = \int_a^b \Big(u(x) + i\,v(x)\Big)\,dx = \int_a^b u(x)\,dx + i\int_a^b v(x)\,dx.}$$

If $f(x)$ is a piecewise continuous complex-valued function on $[a, b]$, to define its integral over $[a, b]$, we write $[a, b]$ as the finite union of adjacent closed subintervals $[a_0, a_1], [a_1, a_2], \ldots, [a_{m-1}, a_m]$, with $a_0 = a$ and $a_m = b$, such that f is continuous on each subinterval (see Figure 2). Using (4), we define

$$(5) \qquad \boxed{\int_a^b f(x)\,dx = \sum_{j=1}^m \int_{a_{j-1}}^{a_j} u(x)\,dx + i\sum_{j=1}^m \int_{a_{j-1}}^{a_j} v(x)\,dx.}$$

Thus we can treat the integral of a piecewise continuous complex-valued function as a (complex) linear combination of Riemann integrals of real-valued functions. The integral as defined by (4) or (5) inherits many of the properties of the integral for real-valued functions. For example, consider the following properties, whose proofs are left to Exercise 10.

PROPOSITION 1
RIEMANN
INTEGRAL OF
COMPLEX-VALUED
FUNCTIONS

(i) If α and β are complex numbers and f and g are piecewise continuous complex-valued functions on $[a, b]$, then

$$(6) \qquad \int_a^b (\alpha\, f(x) + \beta\, g(x))\, dx = \alpha \int_a^b f(x)\, dx + \beta \int_a^b g(x)\, dx.$$

(ii) If f is a piecewise continuous complex-valued function on $[a, c]$ and $a \le b \le c$, then

$$(7) \qquad \int_a^c f(x)\, dx = \int_a^b f(x)\, dx + \int_b^c f(x)\, dx.$$

EXAMPLE 1   A complex-valued piecewise continuous function
Evaluate $\int_0^2 f(x)\, dx$, where

$$f(x) = \begin{cases} (1+i)\, x & \text{if } 0 \le x \le 1, \\ i\, x^2 & \text{if } 1 < x \le 2. \end{cases}$$

Solution   We use (5) and properties of the integral and get

$$\begin{aligned}
\int_0^2 f(x)\, dx &= \int_0^1 f(x)\, dx + \int_1^2 f(x)\, dx \\
&= (1+i) \int_0^1 x\, dx + i \int_1^2 x^2\, dx \\
&= \frac{1+i}{2} + \frac{i}{3}\,(8-1) = \frac{1}{2} + i\,\frac{17}{6}.
\end{aligned}$$ ∎

Antiderivatives of Complex-Valued Functions

If f is a piecewise continuous complex-valued function on $[a, b]$, we will say that F is an **antiderivative** of f if $F'(x) = f(x)$ at all the points of continuity of f on $[a, b]$, where $F'(x)$ is defined as in (5), Section 3.1. Hence if we write $f(x) = u(x) + i\, v(x)$ and $F(x) = U(x) + i\, V(x)$, then the equalities

$$U'(x) = u(x) \quad \text{and} \quad V'(x) = v(x)$$

hold for all but finitely many x's in $[a, b]$. Using the previous notation, we write $[a, b]$ as the finite union of adjacent closed subintervals $[a_0, a_1]$, $[a_1, a_2], \ldots, [a_{m-1}, a_m]$, with $a_0 = a$ and $a_m = b$, and such that f is continuous on each subinterval. The functions $U(x)$ and $V(x)$ are continuous on each subinterval and this makes F piecewise continuous on $[a, b]$. However, F may not be continuous at the points a_j (see Figure 3). For practical purposes, we want F to be continuous on $[a, b]$. As we now show, a continuous antiderivative can always be found.

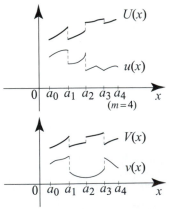

Figure 3 An antiderivative of a piecewise continuous function may not be continuous.

Let f_j denote the restriction of f to $[a_{j-1}, a_j]$, and let F_j denote an antiderivative of f_j over $[a_{j-1}, a_j]$. Each F_j is computed up to an arbitrary complex constant, which can be determined in such a way to make F continuous.

Start by setting the arbitrary constant in F_1 equal 0. Then determine the constant in F_2 so that $\lim_{x\uparrow a_1} F_1(x) = \lim_{x\downarrow a_1} F_2(x)$. (We use the up-arrow to denote a limit from the left and the down-arrow a limit from the right.) This determines F_2 and makes the antiderivative of f continuous on $[a, a_2]$. Continue in this fashion: once you have found F_j, determine the constant in F_{j+1} so that $\lim_{x\uparrow a_j} F_j(x) = \lim_{x\downarrow a_j} F_{j+1}(x)$. By construction, the resulting function F will be continuous on $[a, b]$ (see Figure 4). The following example illustrates the method.

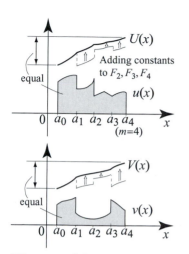

Figure 4 Selecting a continuous antiderivative.

EXAMPLE 2   Finding a continuous antiderivative

Find a continuous antiderivative of the function in Example 1,

$$f(x) = \begin{cases} (1+i)\, x & \text{if } 0 \le x \le 1, \\ i\, x^2 & \text{if } 1 < x \le 2. \end{cases}$$

Solution   By integrating each continuous part of f, we obtain

$$F(x) = \begin{cases} \frac{1+i}{2}\, x^2 & \text{if } 0 \le x \le 1, \\ \frac{i}{3}\, x^3 + C & \text{if } 1 < x \le 2, \end{cases}$$

where C is an arbitrary constant. Note how in the first part of F we already set the arbitrary constant equal 0. To determine C, we evaluate F at 1 using both formulas from the intervals $(0, 1)$ and $(1, 2)$ and get

$$\frac{1+i}{2} = \frac{i}{3} + C \quad \Rightarrow \quad C = \frac{1}{2} + \frac{i}{6}.$$

Hence

$$F(x) = \begin{cases} \frac{1+i}{2}\, x^2 & \text{if } 0 \le x \le 1, \\ \frac{i}{3}\, x^3 + \frac{1}{2} + \frac{i}{6} & \text{if } 1 < x \le 2, \end{cases}$$

is a continuous antiderivative of f on $[0, 2]$.   ∎

The following is an extension of the fundamental theorem of calculus to piecewise continuous complex functions.

THEOREM 1 DEFINITE INTEGRAL OF PIECEWISE CONTINUOUS FUNCTIONS

Suppose that f is a piecewise continuous complex-valued function on the interval $[a, b]$ and let F be a continuous antiderivative of f in $[a, b]$. Then

(8) $$\int_a^b f(x)\, dx = F(b) - F(a).$$

Proof Suppose first that $f = u + iv$ is continuous on $[a, b]$. Let $F(x) = U(x) + iV(x)$ be an antiderivative of f. Using (4) and the fundamental theorem of calculus, we see that

$$
\begin{aligned}
\int_a^b f(x)\, dx &= \int_a^b u(x)\, dx + i \int_a^b v(x)\, dx \\
&= \Big(U(b) - U(a) \Big) + i \Big(V(b) - V(a) \Big) \\
&= \Big(U(b) + iV(b) \Big) - \Big(U(a) + iV(a) \Big) = F(b) - F(a),
\end{aligned}
$$

and hence (8) holds. If f is piecewise continuous on $[a, b]$, we use (5) and the previous case, and get

$$
\begin{aligned}
\int_a^b f(x)\, dx &= \sum_{j=1}^m \int_{a_{j-1}}^{a_j} f(x)\, dx = \sum_{j=1}^m \Big(F(a_j) - F(a_{j-1}) \Big) \\
&= F(a_m) - F(a_0) = F(b) - F(a),
\end{aligned}
$$

which proves (8).   ∎

As an application of Theorem 1, let us evaluate the integral in Example 1 by using the continuous antiderivative that we found in Example 2. In the notation of Example 2, we have

$$
\int_0^2 f(x)\, dx = F(2) - F(0) = \Big(\frac{i}{3} 2^3 + \frac{1}{2} + \frac{i}{6} \Big) - 0 = \frac{1}{2} + \frac{17}{6} i,
$$

which agrees with the result of Example 1.

It is easy to show that two continuous antiderivatives of f differ by a complex constant on $[a, b]$. Motivated by Theorem 1, we will write

$$
(9) \qquad \int f(x)\, dx = F(x) + C,
$$

where C is an arbitrary complex constant, to denote the set of all continuous antiderivatives of f. For example, if $\alpha \neq 0$ is a complex number, then

$$
(10) \qquad \int e^{\alpha t}\, dt = \frac{1}{\alpha} e^{\alpha t} + C \qquad (\alpha \neq 0),
$$

as can be checked by verifying that the derivative of the right side is equal to the integrand on the left side. This simple integral of a complex-valued function has an interesting application to the evaluation of tedious integrals from calculus.

EXAMPLE 3   Integrating $e^{ax} \cos bx$ and $e^{ax} \sin bx$

Let a and b be arbitrary nonzero real numbers. Compute

$$
I_1 = \int e^{ax} \cos bx \, dx \qquad \text{and} \qquad I_2 = \int e^{ax} \sin bx \, dx.
$$

Solution   The idea is to compute $I = I_1 + i\,I_2$ and then obtain I_1 and I_2 by taking real and imaginary parts of I. We have

$$
\begin{aligned}
I &= I_1 + i\,I_2 = \int e^{ax}\cos bx\,dx + i\int e^{ax}\sin bx\,dx \\
&= \int e^{ax}(\cos bx + i\sin bx)\,dx = \int e^{(a+ib)x}\,dx \\
&= \frac{1}{a+ib}e^{(a+ib)x} + C,
\end{aligned}
$$

by (10). Now we need only rewrite $\frac{1}{a+ib}e^{(a+ib)x}$ in terms of its real and imaginary parts. Multiplying and dividing by the denominator's conjugate $a - i\,b$,

$$
\begin{aligned}
\frac{1}{a+ib}e^{(a+ib)x} &= \frac{e^{ax}}{a^2+b^2}(a - ib)(\cos bx + i\sin bx) \\
&= \frac{e^{ax}}{a^2+b^2}(a\cos bx + b\sin bx) + i\frac{e^{ax}}{a^2+b^2}(a\sin bx - b\cos bx).
\end{aligned}
$$

Therefore, we conclude that

(11)
$$
\int e^{ax}\cos bx\,dx = \frac{e^{ax}}{a^2+b^2}(a\cos bx + b\sin bx) + C_1
$$

and

(12)
$$
\int e^{ax}\sin bx\,dx = \frac{e^{ax}}{a^2+b^2}(a\sin bx - b\cos bx) + C_2,
$$

where C_1 and C_2 are arbitrary real constants. As you know, if we were to compute I_1 and I_2 via real calculus, each integral would require two integrations by parts. So there is a clear advantage to using complex-valued functions. ∎

We now have all the tools that we need to construct the integral along a path in the complex plane.

Path or Contour Integrals

Suppose that $\gamma(t)$, $a \leq t \leq b$ is a path; that is, $\gamma(t)$ is a continuous complex-valued function on $[a, b]$ with piecewise continuous derivative $\gamma'(t)$. Suppose that $f(z)$ is a continuous complex-valued function on the graph of $\gamma(t)$; that is, the function $t \mapsto f(\gamma(t))$ is a continuous function from $[a, b]$ into $\mathbb{C}$. Our goal is to construct the integral of f over γ.

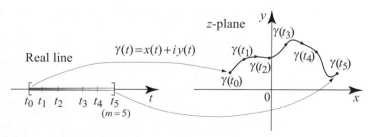

Figure 5 Partitioning a path γ into smaller arcs with endpoints at $\gamma(t_k)$.

To simplify our discussion, we first deal with the case where γ and γ' are both continuous on $[a, b]$. Taking a hint from the Riemann integral, we partition

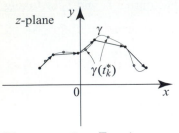

z-plane

Figure 6 Forming a Riemann-like sum of a complex-valued function of a path γ.

the path γ into smaller arcs with endpoints at $\gamma(t_k)$ (Figure 5); evaluate $f(\gamma(t_k^\star))$, where $\gamma(t_k^\star)$ is a point on the arc between $\gamma(t_{k-1})$ and $\gamma(t_k)$; and then sum all the terms of the form $f(\gamma(t_k^\star))(\gamma(t_k)-\gamma(t_{k-1}))$ (Figure 6). This leads us to consider a Riemann-like sum

$$(13) \qquad \sum_{k=1}^{m} f(\gamma(t_k^\star))(\gamma(t_k) - \gamma(t_{k-1})),$$

where $a = t_0 < t_1 < \cdots < t_m = b$ is a partition of $[a, b]$, and $t_k^\star$ is in $[t_{k-1}, t_k]$. Writing $\gamma(t) = x(t) + i\,y(t)$ so that $\gamma'(t) = x'(t) + i\,y'(t)$, and appealing to the mean value theorem from calculus, we get

$$\sum_{k=1}^{m} f(\gamma(t_k^\star))(\gamma(t_k) - \gamma(t_{k-1}))$$

$$= \sum_{k=1}^{m} f(\gamma(t_k^\star))(x(t_k) - x(t_{k-1})) + i \sum_{k=1}^{m} f(\gamma(t_k^\star))(y(t_k) - y(t_{k-1}))$$

$$= \sum_{k=1}^{m} f(\gamma(t_k^\star))x'(\alpha_k)(t_k - t_{k-1}) + i \sum_{k=1}^{m} f(\gamma(t_k^\star))y'(\beta_k)(t_k - t_{k-1}),$$

where $t_{k-1} < \alpha_k, \beta_k < t_k$. Interpreting each sum on the right as a Riemann sum (in the usual sense from calculus) over the interval $[a, b]$, we see that as the partition gets finer, the sum on the left converges to

$$\int_a^b f(\gamma(t))x'(t)\,dt + i \int_a^b f(\gamma(t))y'(t)\,dt = \int_a^b f(\gamma(t))\Big(x'(t) + i\,y'(t)\Big)\,dt$$

$$(14) \qquad\qquad\qquad\qquad\qquad = \int_a^b f(\gamma(t))\gamma'(t)\,dt.$$

For general piecewise smooth γ, the derivative $\gamma'(t)$ may not be continuous but just piecewise continuous. This makes the integrand in (14) piecewise continuous. We evaluate the integral in this case as a finite sum of integrals of continuous functions, as we did in (5). All of this leads us to the following important definition.

DEFINITION 1
PATH OR
CONTOUR
INTEGRALS

Suppose that γ is a path over a closed interval $[a, b]$ and that f is a continuous complex-valued function defined on the graph of γ. The **path** or **contour integral** of f on γ is given by

$$(15) \qquad \int_\gamma f(z)\,dz = \int_a^b f(\gamma(t))\gamma'(t)\,dt.$$

Other notations for the path integral are $\int_\gamma f(z)\,d\gamma$ or simply $\int_\gamma f\,d\gamma$.

Thus, after parametrizing the path by a closed interval $[a, b]$, the path integral becomes a Riemann integral of a piecewise continuous complex-valued function over the interval $[a, b]$.

We now give several examples of path integrals, starting with one that will be needed in the future.

EXAMPLE 4   Path integrals of $(z - z_0)^n$, n any integer

Let $C_R(z_0)$ be the positively oriented circle with center at z_0 and radius $R > 0$ (see Figure 7).

(a)   Show that

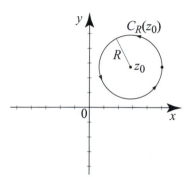

(16)
$$\int_{C_R(z_0)} \frac{1}{z - z_0}\, dz = 2\pi i.$$

(b)   Let $n \neq -1$ be an integer; show that

(17)
$$\int_{C_R(z_0)} (z - z_0)^n\, dz = 0.$$

Figure 7 for Example 4.

Solution   Parametrize $C_R(z_0)$ by $\gamma(t) = z_0 + Re^{it}$, $0 \leq t \leq 2\pi$. Then $\gamma'(t) = iRe^{it}$ and, for $z = z_0 + Re^{it}$ on γ, we have $z - z_0 = Re^{it}$. Using all this in (15), we obtain

$$\int_{C_R(z_0)} \frac{1}{z - z_0}\, dz = \int_0^{2\pi} \frac{1}{Re^{it}} iRe^{it}\, dt = i\int_0^{2\pi} dt = 2\pi i.$$

(b)   We use (15) and some of the information that we found in (a) and get

$$\int_{C_R(z_0)} (z - z_0)^n\, dz = \int_0^{2\pi} \left(Re^{it}\right)^n iRe^{it}\, dt = iR^{n+1}\int_0^{2\pi} e^{i(n+1)t}\, dt$$

$$= \frac{R^{n+1}}{n+1} e^{i(n+1)t}\Big|_0^{2\pi} = \frac{R^{n+1}}{n+1}\left(e^{2\pi(n+1)i} - e^0\right) = 0,$$

because $e^{2\pi(n+1)i} = e^0 = 1$.   ∎

In particular, if $C_1(0)$ denotes the positively oriented unit circle with center at the origin, then (16) and (17) imply that

(18)
$$\int_{C_1(0)} \frac{1}{z}\, dz = 2\pi i \qquad \text{and} \qquad \int_{C_1(0)} z\, dz = 0.$$

Compare (18) with the following integrals involving $\bar{z}$.

EXAMPLE 5   Integrals involving $\bar{z}$

Show that

(19)
$$\int_{C_1(0)} \frac{1}{\bar{z}}\, dz = 0 \qquad \text{and} \qquad \int_{C_1(0)} \bar{z}\, dz = 2\pi i.$$

Solution   To compute the integrals we could go back to the definition of the path integral (15) and proceed as we did in Example 4. Another way to proceed is to use (18) and the following observations. Parametrize $C_1(0)$ by $\gamma(t) = e^{it}$, $0 \leq t \leq 2\pi$. For $z = e^{it}$ on $C_1(0)$, we have

$$\overline{z} = \overline{e^{it}} = e^{-it} = \frac{1}{z}.$$

So, to compute the first integral in (19), we use the second integral in (18) and get

$$\int_{C_1(0)} \frac{1}{\overline{z}}\, dz = \int_{C_1(0)} \frac{1}{1/z}\, dz = \int_{C_1(0)} z\, dz = 0.$$

Likewise, to compute the second integral in (19), we use the first integral in (18) and get

$$\int_{C_1(0)} \overline{z}\, dz = \int_{C_1(0)} \frac{1}{z}\, dz = 2\pi i. \qquad \blacksquare$$

The integrals in Example 5 are generalized in Exercise 11.

Because the path integral is a Riemann integral, several properties of the latter integral carry over to the path integral. We have the following useful properties.

**PROPOSITION 2
PROPERTIES OF
THE PATH
INTEGRAL**

(i)   Suppose that $\gamma(t)$ is a path on $[a, b]$, f and g are continuous functions on γ, and α and β are complex numbers. Then

$$(20) \qquad \int_{\gamma} (\alpha\, f(z) + \beta\, g(z))\, dz = \alpha \int_{\gamma} f(z)\, dz + \beta \int_{\gamma} g(z)\, dz.$$

(ii) Let γ_r denote the reverse of γ. Then

$$(21) \qquad \int_{\gamma_r} f(z)\, dz = - \int_{\gamma} f(z)\, dz.$$

(iii) If $\Gamma = (\gamma_1, \gamma_2, \ldots, \gamma_m)$ is a contour and f is continuous on Γ, then

$$(22) \qquad \int_{\Gamma} f(z)\, dz = \sum_{k=1}^{m} \int_{\gamma_k} f(z)\, dz.$$

Proof   (i) Express the path integrals as Riemann integrals using (15) and then use (6).   (ii) Recall the parametrization of the reverse of γ from (3), Section 3.1: $\gamma_r(t) = \gamma(b + a - t)$, where t runs over the same interval $[a, b]$ that parametrizes γ. Then $\gamma_r'(t) = -\gamma'(b + a - t)$ and so

$$\int_{\gamma_r} f(z)\, dz = - \int_a^b f(\gamma(b + a - t))\gamma'(b + a - t)\, dt.$$

Making the change of variables $T = b + a - t$, we obtain

$$\int_{\gamma_r} f(z)\, dz = \int_b^a f(\gamma(T))\gamma'(T)\, dT = - \int_a^b f(\gamma(T))\gamma'(T)\, dT = - \int_{\gamma} f(z)\, dz,$$

by (15). The proof of (iii) involves parametrizing $\Gamma = (\gamma_1, \gamma_2, \ldots, \gamma_m)$ and using additivity of the Riemann integral over intervals (7). The details are worked in Exercise 42. ∎

In some path integrals, the integrand is written as a function of x and y and not z. We can rewrite the integrand as a function of z by using the relations

$$x = \frac{z + \bar{z}}{2} \quad \text{and} \quad y = \frac{z - \bar{z}}{2i}.$$

Here is an illustration that also uses the linearity of the path integral (20).

EXAMPLE 6   Using linearity

Let $C_1(0)$ denote the positively oriented unit circle, as in Example 5. Compute

$$\int_{C_1(0)} x \, dz = \int_{C_1(0)} \operatorname{Re} z \, dz.$$

Solution   Using (18) and (19), we have

$$\int_{C_1(0)} x \, dz = \int_{C_1(0)} \frac{z + \bar{z}}{2} \, dz$$

$$= \frac{1}{2} \underbrace{\int_{C_1(0)} z \, dz}_{= 0} + \frac{1}{2} \underbrace{\int_{C_1(0)} \bar{z} \, dz}_{= 2\pi i} = \pi i. \quad ∎$$

The integrals in Examples 4, 5, and 6 involved smooth curves. In the following example, we will compute integrals over polygonal paths, which are piecewise smooth.

EXAMPLE 7   Integrals over polygonal paths

Let $z_1 = -1$, $z_2 = 1$, and $z_3 = i$ (Figure 8). Compute

(a) $\displaystyle\int_{[z_1, z_2]} \bar{z} \, dz$,   (b) $\displaystyle\int_{[z_2, z_3]} \bar{z} \, dz$,   (c) $\displaystyle\int_{[z_3, z_1]} \bar{z} \, dz$,

(d) $\displaystyle\int_{[z_1, z_2, z_3]} \bar{z} \, dz$,   (e) $\displaystyle\int_{[z_1, z_2, z_3, z_1]} \bar{z} \, dz$,   (f) $\displaystyle\int_{[z_1, z_3]} \bar{z} \, dz$.

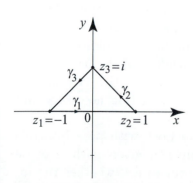

Figure 8 for Example 7.

Solution   Let $\gamma_1(t)$ be a parametrization of $[z_1, z_2]$, $\gamma_2(t)$ be a parametrization of $[z_2, z_3]$, and $\gamma_3(t)$ be a parametrization of $[z_3, z_1]$. There are several possible ways to compute γ_1, γ_2 and γ_3. To be consistent, we will use (2), Section 3.1. Accordingly, we have

$$\gamma_1(t) = -(1 - t) + t = -1 + 2t, \quad 0 \le t \le 1, \quad \Rightarrow \quad \gamma_1'(t) = 2,$$
$$\gamma_2(t) = 1 - t + it, \quad 0 \le t \le 1, \quad \Rightarrow \quad \gamma_2'(t) = -1 + i,$$
$$\gamma_3(t) = (1 - t)i - t = -t + i(1 - t), \quad 0 \le t \le 1, \quad \Rightarrow \quad \gamma_3'(t) = -1 - i.$$

We now appeal to (15) to compute the integrals. For (a), we have

$$\int_{[z_1,\,z_2]} \overline{z}\,dz = \int_0^1 \overline{(-1+2t)}\,(2)\,dt = 2\int_0^1 (-1+2t)\,dt = 2(-t+t^2)\Big|_0^1 = 0.$$

For (b), we have

$$\int_{[z_2,\,z_3]} \overline{z}\,dz = \int_0^1 \big((1-t)-it\big)(-1+i)\,dt = (-1+i)\int_0^1 \Big(1-(1+i)t\Big)\,dt$$

$$= (-1+i)\Big(t-\frac{1+i}{2}t^2\Big)\Big|_0^1 = i.$$

For (c), we have

$$\int_{[z_3,\,z_1]} \overline{z}\,dz = \int_0^1 \Big(-t-(1-t)\Big)(-1-i)\,dt = (-1-i)\int_0^1 \Big(-i+(-1+i)t\Big)\,dt$$

$$= (-1-i)\Big(-it+\frac{-1+i}{2}t^2\Big)\Big|_0^1 = i.$$

The integrals in (d)–(f) follow from (a)–(c) and properties of the path integral. For (d) and (e), we use (22) and get

$$(23) \qquad \int_{[z_1,\,z_2,\,z_3]} \overline{z}\,dz = \int_{[z_1,\,z_2]} \overline{z}\,dz + \int_{[z_2,\,z_3]} \overline{z}\,dz = 0+i = i;$$

$$(24) \qquad \int_{[z_1,\,z_2,\,z_3,\,z_1]} \overline{z}\,dz = \int_{[z_1,\,z_2,\,z_3]} \overline{z}\,dz + \int_{[z_3,\,z_1]} \overline{z}\,dz = i+i = 2\,i.$$

For (f), we use (21) and get

$$(25) \qquad \int_{[z_1,\,z_3]} \overline{z}\,dz = -\int_{[z_3,\,z_1]} \overline{z}\,dz = -i. \qquad \blacksquare$$

Comparing the integrals in (23) and (25), we are led to the following somewhat disturbing conclusion: If f is a continuous function in a region that contains z_1, z_2, and z_3, then the path integral of f on a line directly connecting z_1 to z_3 is not necessarily equal to the path integral of f from z_1 to z_3 along a different path (say, one that goes through z_2). In other words, the path integral of f from z_1 to z_3 is not independent of the path joining z_1 to z_3. Is the path integral ever independent from path? We answer this important question in the next section by formulating necessary and sufficient conditions for independence of path.

Another question that comes to mind as we work with path integrals concerns the parametrization of the path. Since a given path may be parametrized in many different ways, it is natural to ask whether the integral is independent of the choice of the parametrization. Noting that the Riemann-like sum of the path integral (13) can be written in the form

$$\sum_{k=1}^{m} f(z_k^\star)(z_k - z_{k-1}),$$

which makes no reference to t, we would expect that the integral is independent of the parametrization. Indeed, this is true as long as we are describing the same path.

For example, the positively oriented unit circle $C_1(0)$ can be parametrized by $\gamma_1(t) = e^{it}$, $0 \le t \le 2\pi$ or $\gamma_2(t) = e^{2it}$, $0 \le t \le \pi$. If $f(z)$ is a continuous function on $C_1(0)$, we would expect that $\int_{\gamma_1} f(z)\,dz = \int_{\gamma_2} f(z)\,dz$. In fact, using the definition of path integrals and a simple change of variables $s = 2t$, $ds = 2\,dt$, we have

$$\int_{\gamma_2} f(z)\,dz = \int_0^\pi f(e^{2it})2ie^{2it}\,dt = \int_0^{2\pi} f(e^{is})ie^{is}\,ds = \int_{\gamma_1} f(z)\,dz,$$

as expected. The same proof works in general, but we have to explain what we mean by two parametrizations being the same.

We will say that $\gamma_1(t)$, $a \le t \le b$, and $\gamma_2(t)$, $c \le t \le d$, are **equivalent parametrizations** if there is an increasing continuously differentiable function $\phi(t)$ from $[c, d]$ onto $[a, b]$ such that $\phi(c) = a$ and $\phi(d) = b$ and $\gamma_2(t) = \gamma_1 \circ \phi(t)$ for all t in $[c, d]$.

In the case of the unit circle and the two parametrizations $\gamma_1(t) = e^{it}$, $0 \le t \le 2\pi$ and $\gamma_2(t) = e^{2it}$, $0 \le t \le \pi$, we see that γ_1 and γ_2 are equivalent by taking $\phi(t) = 2t$.

The next result confirms a property that we would expect from a path integral.

PROPOSITION 3
INDEPENDENCE OF
PARAMETRIZATION

Suppose that $\gamma_1(t)$, $a \le t \le b$, and $\gamma_2(t)$, $c \le t \le d$, are equivalent parametrizations of the same path γ and let f be a continuous function on γ. Then

$$\int_{\gamma_1} f(z)\,dz = \int_{\gamma_2} f(z)\,dz.$$

Proof Applying the definition of path integrals, and using $\gamma_2(t) = \gamma_1(\phi(t))$ and $\gamma_2'(t) = \gamma_1'(\phi(t))\phi'(t)$, we obtain

$$\int_{\gamma_2} f(z)\,dz = \int_c^d f(\gamma_2(t))\gamma_2'(t)\,dt = \int_c^d f(\gamma_1(\phi(t)))\gamma_1'(\phi(t))\phi'(t)\,dt$$

$$= \int_a^b f(\gamma_1(s))\gamma_1'(s)\,ds = \int_{\gamma_1} f(z)\,dz,$$

where the equality before last follows by setting $s = \phi(t)$, $ds = \phi'(t)\,dt$. ∎

We end the section by revisiting the notion of arc length from calculus and deriving useful estimates on the size of path integrals.

Arc Length and Bounds for Integrals

Given a smooth path $\gamma : [a, b] \to \mathbb{C}$, write $\gamma(t) = x(t) + i\, y(t)$. The **length** of γ, denoted $l(\gamma)$, can be approximated by adding the length of line segments joining consecutive points on the graph of γ as in Figure 9. The sum of the lengths of the line segments is given by

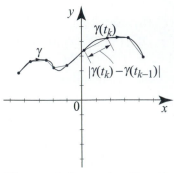

$$(26) \quad \sum_{k=1}^{m} |\gamma(t_k) - \gamma(t_{k-1})| = \sum_{k=1}^{m} \sqrt{(x(t_k) - x(t_{k-1}))^2 + (y(t_k) - y(t_{k-1}))^2},$$

Figure 9 Approximating the length of a path by adding the length of line segments.

where $a = t_0 < t_1 < t_2 < \cdots < t_m = b$ is a partition of $[a, b]$. Thus the length of γ is the limit (when it exists) of the sums on the right side of (26) as the partition of $[a, b]$ gets finer and finer. To find this limit, we use the mean value theorem and write $(x(t_k) - x(t_{k-1}))^2 = (x'(\alpha_k)(t_k - t_{k-1}))^2$ and $(y(t_k) - y(t_{k-1}))^2 = (y'(\beta_k)(t_k - t_{k-1}))^2$, where α_k and β_k are in $[t_{k-1}, t_k]$. Then (26) becomes

$$\sum_{k=1}^{m} \sqrt{x'(\alpha_k)^2 + y'(\beta_k)^2}\,(t_k - t_{k-1}).$$

Recognizing this sum as a Riemann sum and taking limits as the partition gets finer, we recover the formula for arc length from calculus:

$$(27) \qquad \boxed{l(\gamma) = \int_a^b \sqrt{x'(t)^2 + y'(t)^2}\,dt = \int_a^b |\gamma'(t)|\,dt,}$$

where the second equality follows from the complex notation $\gamma'(t) = x'(t) + i\, y'(t)$ and so $\sqrt{x'(t)^2 + y'(t)^2} = |\gamma'(t)|$. For piecewise smooth γ, we can repeat the previous analysis for each smooth part of γ and then add the lengths of all smooth parts. This results in formula (27) for the arc length, where the integrand in this case is piecewise continuous. The element of arc length is usually denoted by ds. Thus,

$$(28) \qquad\qquad ds = \sqrt{x'(t)^2 + y'(t)^2}\,dt.$$

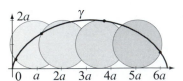

Figure 10 First arch of the cycloid.

EXAMPLE 8   Arc length

Find the length of the arch of the cycloid

$$\gamma(t) = a(t - \sin t) + a\, i\,(1 - \cos t), \qquad 0 \le t \le 2\pi,$$

where a is a positive real number. The curve is illustrated in Figure 10.

Solution   We have

$$x(t) = a(t - \sin t) \quad \Rightarrow \quad x'(t) = a(1 - \cos t);$$
$$y(t) = a\,(1 - \cos t) \quad \Rightarrow \quad y'(t) = a \sin t.$$

Hence

$$ds \;=\; \sqrt{x'(t)^2 + y'(t)^2}\,dt = \sqrt{a^2\Big((1-\cos t)^2 + \sin^2 t\Big)}\,dt$$

$$\;=\; a\,\sqrt{2(1-\cos t)}\,dt = 2a\,\sin\left(\frac{t}{2}\right)\,dt.$$

Applying (27), we obtain the length of the arch

$$l(\gamma) = \int_0^{2\pi} 2a \sin\frac{t}{2}\,dt = \left[-4a\cos\frac{t}{2}\right]_0^{2\pi} = 8a. \qquad \blacksquare$$

Using the notion of length, we can extend to path integrals the familiar inequality for the Riemann integral: If f is a continuous real-valued function of $[a,\,b]$ with $|f(x)| \le M$ on $[a,\,b]$, then

$$\left|\int_a^b f(x)\,dx\right| \le M(b-a).$$

For path integrals, we have the following useful inequality, often called the Ml–inequality.

THEOREM 2
BOUNDS FOR PATH
INTEGRALS

Suppose that γ is a path over $[a,\,b]$ with length $l(\gamma)$, and f is a continuous function on γ such that $|f(z)| \le M$ for all z on γ. Then

(29)
$$\left|\int_\gamma f(z)\,dz\right| \le M\,l(\gamma).$$

Proof   We start with the case of a smooth path γ and consider a Riemann sum defining the path integral (13). Using the triangle inequality and the inequality $|f(z)| \le M$ for z on γ, we get

(30)
$$\left|\sum_{k=1}^m f(\gamma(t_k^\star))(\gamma(t_k)-\gamma(t_{k-1}))\right| \;\le\; \sum_{k=1}^m \left|f(\gamma(t_k^\star))(\gamma(t_k)-\gamma(t_{k-1}))\right|$$

$$\;=\; \sum_{k=1}^m |f(\gamma(t_k^\star))|\,|(\gamma(t_k)-\gamma(t_{k-1}))|$$

$$\;\le\; M\sum_{k=1}^m |(\gamma(t_k)-\gamma(t_{k-1}))|\,.$$

Taking limits as the partition of $[a,\,b]$ gets finer, the sum on the left side converges to $\int_\gamma f(z)\,dz$ while the sum on the right side converges to $l(\gamma)$, implying (29) in the case of a smooth γ. If γ is piecewise smooth, we can write $\gamma = (\gamma_1,\,\gamma_2,\,\ldots,\,\gamma_n)$, where each γ_j is smooth. Then using (22) and applying (29) to each smooth part of γ, we obtain

$$\left| \int_\gamma f(z)\,dz \right| = \left| \sum_{j=1}^n \int_{\gamma_j} f(z)\,dz \right| \le \sum_{j=1}^n \left| \int_{\gamma_j} f(z)\,dz \right|$$

$$\le M \sum_{j=1}^n l(\gamma_j) = Ml(\gamma),$$

where in the last equality we used the fact that the length of γ is equal to the sum of the lengths of its parts. ∎

Remembering that $|(\gamma(t_k) - \gamma(t_{k-1}))|$ is an approximation of the length of the portion of the path γ that joins the points $\gamma(t_{k-1})$ and $\gamma(t_k)$, using an argument based on Riemann sums (recall (26) and (28)), we see that

$$\sum_{k=1}^m |f(\gamma(t_k^\star))|\, |(\gamma(t_k) - \gamma(t_{k-1}))| \longrightarrow \int_a^b |f(\gamma(t))|\, |\gamma'(t)|\, dt,$$

as the partition of $[a, b]$ gets finer. Thus going back to (30) and arguing as we did in the proof of Theorem 2, we obtain the inequalities

$$(31) \qquad \boxed{\left| \int_\gamma f(z)\,dz \right| \le \int_a^b |f(\gamma(t))|\, |\gamma'(t)|\, dt \le M\, l(\gamma),}$$

where f and γ are as in Theorem 2.

EXAMPLE 9   Bounding a path integral
Find an upper bound for

$$\left| \int_{C_1(0)} e^{\frac{1}{z}}\,dz \right|,$$

where $C_1(0)$ is the positively oriented unit circle.

Solution   Since the length of $C_1(0)$ is 2π, it follows from (29) that $\left| \int_{C_1(0)} e^{\frac{1}{z}}\,dz \right| \le 2\pi M$, where M is an upper bound of $|e^{\frac{1}{z}}|$ for z on the unit circle. To find M we proceed as follows. For z on $C_1(0)$, write $z = e^{it}$ where $0 \le t \le 2\pi$. Then

$$\frac{1}{z} = e^{-it} = \cos t - i \sin t,$$

and so, for $z = e^{it}$,

$$\left| e^{\frac{1}{z}} \right| = \left| e^{\cos t - i \sin t} \right| = \left| e^{\cos t} \right| \overbrace{\left| e^{-i \sin t} \right|}^{=1} = e^{\cos t} \le e^1 = e.$$

(In setting $\left| e^{-i \sin t} \right| = 1$ we have used $|e^{iw}| = 1$ for any real w.) Thus $\left| \int_{C_1(0)} e^{\frac{1}{z}}\,dz \right| \le 2\pi e$. Using techniques of integration from Chapter 5, we can evaluate the integral and obtain $\int_{C_1(0)} e^{\frac{1}{z}} \, dz = 2\pi i$. Thus the bound that we obtained by applying (29) is correct but not best possible, since $|2\pi i| = 2\pi$ is the best upper bound.   ■

We close this section with a simple remark about path integrals. The path described by $\gamma(t) = t$ for t in $[a, b]$ is the closed interval $[a, b]$. For such a path, we have $\gamma'(t) \, dt = dt$, and if f is a function defined on γ, the path integral (15) becomes

$$\int_{\gamma} f(z) \, dz = \int_{a}^{b} f(t) \, dt,$$

showing that a Riemann integral of a piecewise continuous complex-valued function over $[a, b]$ is a special case of a path integral, where the path is the line segment $[a, b]$. So results about path integrals apply in particular to Riemann integrals. For example, if f is a piecewise continuous complex-valued function on $[a, b]$ such that $|f(t)| \leq M$ for all t in $[a, b]$, then inequalities (31) become

$$(32) \qquad \boxed{\left| \int_{a}^{b} f(t) \, dt \right| \leq \int_{a}^{b} |f(t)| \, dt \leq M\,(b - a).}$$

Exercises 3.2

In Exercises 1–8, evaluate the given integral.

1. $\displaystyle\int_{0}^{2\pi} e^{3ix} \, dx.$ **2.** $\displaystyle\int_{-1}^{1} (2i + 3 + i\,x)^2 \, dx.$ **3.** $\displaystyle\int_{-1}^{0} \sin(i\,x) \, dx.$

4. $\displaystyle\int_{1}^{2} \mathrm{Log}\,(ix) \, dx.$ **5.** $\displaystyle\int_{-1}^{1} \frac{x + i}{x - i} \, dx.$ **6.** $\displaystyle\int_{1}^{2} x^i \, dx$

(principal branch).

7. $\displaystyle\int_{-1}^{1} f(x) \, dx,$ where   **8.** $\displaystyle\int_{-1}^{1} f(x) \, dx,$ where

$$f(x) = \begin{cases} (3 + 2\,i)x & \text{if } -1 \leq x \leq 0, \\ i\,x^2 & \text{if } 0 < x \leq 1. \end{cases} \qquad f(x) = \begin{cases} e^{i\pi x} & \text{if } -1 \leq x \leq 0, \\ x & \text{if } 0 < x \leq 1. \end{cases}$$

9. Find a continuous antiderivative of the function $f(x)$ in Exercise 7, and then compute $\int_{-1}^{1} f(x) \, dx$ by using Theorem 1.

10. Prove all of Proposition 1. [Hint: Split complex Riemann integrals into real and imaginary parts, then invoke relevant properties of real integrals.]

11. Let n be an integer and let $C_R(z_0)$ denote the positively oriented circle with center at z_0 and radius $R > 0$. Show that

$$\int_{C_R(z_0)} [z - z_0]^n \, dz = \begin{cases} 0 & \text{if } n \neq 1, \\ 2\pi i R^2 & \text{if } n = 1. \end{cases}$$

12. Orthogonality of the 2π-periodic trigonometric and exponential systems.   Let m and n be two arbitrary integers.

(a)   Show that

$$\int_{-\pi}^{\pi} e^{imx} e^{-inx}\, dx = \begin{cases} 0 & \text{if } m \neq n, \\ 2\pi & \text{if } m = n. \end{cases}$$

This identity states that the functions e^{imx} $(m = 0, \pm 1, \pm 2, \dots)$ are **orthogonal** on the interval $[-\pi, \pi]$.

(b)   Now suppose m and n are nonnegative integers. With the help of the identity in (a), show that

$$\int_{-\pi}^{\pi} \cos mx \, \cos nx \, dx = 0 \qquad \text{if } m \neq n;$$

$$\int_{-\pi}^{\pi} \cos mx \, \sin nx \, dx = 0 \qquad \text{for all } m \text{ and } n;$$

$$\int_{-\pi}^{\pi} \sin mx \, \sin nx \, dx = 0 \qquad \text{if } m \neq n;$$

$$\int_{-\pi}^{\pi} \cos^2 mx \, dx = \int_{-\pi}^{\pi} \sin^2 mx \, dx = \pi \quad \text{for all } m \neq 0.$$

These identities state that the functions $1,\ \cos x,\ \cos 2x,\ \cos 3x,\ \dots,\ \sin x,\ \sin 2x,\ \dots,$ are **orthogonal** on the interval $[-\pi, \pi]$.

13. Orthogonality of the $2p$-periodic trigonometric and exponential systems.   Let $p > 0$ be an arbitrary real number, and m and n be arbitrary integers.

(a)   Show that the functions $e^{i\frac{m\pi}{p}x}$ $(m = 0, \pm 1, \pm 2, \dots)$ are $2p$-periodic. The set of these functions is called the $2p$-periodic exponential system. When $p = \pi$, we showed in Exercise 12 that the functions in this system are orthogonal on the interval $[-\pi, \pi]$. The corresponding result for arbitrary $p > 0$ is as follows.

(b)   Show that

$$\int_{-p}^{p} e^{i\frac{m\pi}{p}x} e^{-i\frac{n\pi}{p}x}\, dx = \begin{cases} 0 & \text{if } m \neq n, \\ 2p & \text{if } m = n. \end{cases}$$

Thus the functions in the $2p$-periodic exponential system are **orthogonal** on the interval $[-p, p]$.

(c)   Now suppose m and n are nonnegative integers. With the help of the identity in (a), or by Exercise 12(b), show that

$$\int_{-p}^{p} \cos\left(\frac{m\pi}{p}x\right) \cos\left(\frac{n\pi}{p}x\right) dx = 0 \qquad \text{if } m \neq n;$$

$$\int_{-p}^{p} \cos\left(\frac{m\pi}{p}x\right) \sin\left(\frac{n\pi}{p}x\right) dx = 0 \qquad \text{for all } m \text{ and } n;$$

$$\int_{-p}^{p} \sin\left(\frac{m\pi}{p}x\right) \sin\left(\frac{n\pi}{p}x\right) dx = 0 \qquad \text{if } m \neq n;$$

$$\int_{-p}^{p} \cos^2\left(\frac{m\pi}{p}x\right) dx = \int_{-p}^{p} \sin^2\left(\frac{m\pi}{p}x\right) dx = p \quad \text{for all } m \neq 0.$$

These identities state that the $2p$-periodic functions $1,\ \cos\left(\frac{\pi}{p}x\right),\ \cos\left(\frac{2\pi}{p}x\right), \dots,$ $\sin\left(\frac{\pi}{p}x\right),\ \sin\left(\frac{2\pi}{p}x\right), \dots,$ are **orthogonal** on the interval $[-p, p]$.

14. Let a and b be nonzero real numbers. Derive the formulas

$$\int \cos(ax) \cosh(bx) \, dx = \frac{1}{a^2 + b^2} \left(a \cosh(bx) \sin(ax) + b \cos(ax) \sinh(bx) \right) + c;$$

and

$$\int \sin(ax) \sinh(bx) \, dx = \frac{1}{a^2 + b^2} \left(b \cosh(bx) \sin(ax) - a \cos(ax) \sinh(bx) \right) + c.$$

[Hint: Consider $\int \cos((a + i\,b)x) \, dx$.]

15. Let a and b be nonzero real numbers. Derive the formulas

$$\int \sin(ax) \cosh(bx) \, dx = \frac{1}{a^2 + b^2} \left(- a \cos(ax) \cosh(bx) + b \sin(ax) \sinh(bx) \right) + c;$$

and

$$\int \cos(ax) \sinh(bx) \, dx = \frac{1}{a^2 + b^2} \left(b \cos(ax) \cosh(bx) + a \sin(ax) \sinh(bx) \right) + c.$$

[Hint: Consider $\int \sin(a + i\,b)x \, dx$.]

In Exercises 16–29, (a) plot the given contour. (b) Evaluate the given path integral. The positively oriented circle with center at z_0 and radius $R > 0$ will be denoted $C_R(z_0)$. The directed line segment through the points $z_1, z_2, \ldots, z_n$, will be denoted by $[z_1, z_2, \ldots, z_n]$.

16. $\displaystyle\int_{C_1(0)} (2z + i) \, dz.$

17. $\displaystyle\int_{[z_1, z_2, z_3]} 2\bar{z} \, dz,$ where $z_1 = 1,\ z_2 = i,\ z_3 = 1 + i.$

18. $\displaystyle\int_{[z_1, z_2, z_3, z_1]} \frac{1}{1 + z} \, dz,$ where $z_1 = -i,\ z_2 = 2 - i,\ z_3 = 2 + i.$

19. $\displaystyle\int_{C_1(2+i)} \left((z - 2 - i)^3 + (z - 2 - i)^2 + \frac{i}{z - 2 - i} - \frac{3}{(z - 2 - i)^2} \right) dz.$
[Hint: Example 4.]

20. $\displaystyle\int_{[z_1, z_2, z_3, z_4, z_1]} \left(\operatorname{Re} z - 2(\operatorname{Im} z)^2 \right) dz,$ where $z_1 = 0,\ z_2 = 1,\ z_3 = 1 + i,\ z_4 = i.$

21. $\displaystyle\int_{\gamma} z \, dz,$ where γ is the contour in Exercise 20.

22. $\displaystyle\int_{\gamma} z \, dz,$ where γ is the semi-circle $e^{it},\ 0 \le t \le \pi.$

23. $\displaystyle\int_{\gamma} e^{z} \, dz,$ where γ is the contour in Exercise 20.

24. $\displaystyle\int_{\gamma} \sin z \, dz,$ where γ is the contour in Exercise 20.

25. $\displaystyle\int_{\gamma} z \, dz,$ where γ is the hypotrochoid of Example 2, Section 3.1.

26. $\displaystyle\int_\gamma \overline{z}\,dz$, where γ is the hypotrochoid of Example 2, Section 3.1.

27. $\displaystyle\int_\gamma \sqrt{z}\,dz$, where $\gamma(t) = e^{it}$, $0 \le t \le \frac{\pi}{2}$, and $\sqrt{z}$ denotes the principal value of the square root.

28. $\displaystyle\int_\gamma \mathrm{Log}\,z\,dz$, where $\gamma(t) = e^{it}$, $-\frac{3\pi}{4} \le t \le \frac{3\pi}{4}$.

29. $\displaystyle\int_{[z_1, z_2]} (x^2 + y^2)\,dz$, where $z_1 = 2 + i$, $z_2 = -1 - i$. [Hint: $x^2 + y^2 = z\overline{z}$.]

In Exercises 30–33, find the arc length of the given curve.

30. $\gamma(t) = 2t + \dfrac{2i}{3}t^{3/2}$, $\quad 1 \le t \le 2$. **31.** $\gamma(t) = \dfrac{1}{5}t^5 + \dfrac{i}{4}t^4$, $\quad 0 \le t \le 1$.

32. $\gamma(t) = e^{it}$, $\quad 0 \le t \le \dfrac{\pi}{6}$. **33.** $\gamma(t) = (e^t - t) + 4ie^{\frac{t}{2}}$, $\quad 0 \le t \le 1$.

In Exercises 34–39, derive the given integral estimate.

34. $\left|\displaystyle\int_\gamma e^z\,dz\right| \le \dfrac{\pi}{4}$, where $\gamma(t) = e^{it}$, $\frac{\pi}{2} \le t \le \frac{3\pi}{4}$.

35. $\left|\displaystyle\int_{C_1(0)} \mathrm{Log}\,z\,dz\right| \le 2\pi^2$.

36. $\left|\displaystyle\int_{C_2(0)} \dfrac{1}{z-1}\,dz\right| \le 4\pi$.

37. $\left|\displaystyle\int_{[z_1, z_2, z_3, z_1]} z^{-5}\,dz\right| \le 5^{\frac{5}{2}}(2 + 2\sqrt{5})$, where $z_1 = -1 - i$, $z_2 = 1 - i$, $z_3 = i$.

38. $\left|\displaystyle\int_\gamma \dfrac{1}{z+1}\,dz\right| \le 2\pi$, where $\gamma(t) = 3 + e^{it}$, $0 \le t \le 2\pi$.

39. $\left|\displaystyle\int_{C_1(0)} e^{z^2+1}\,dz\right| \le 2\pi e^2$.

40. Let $f(z)$ be a continuous function on a region Ω, and let $\gamma(t)$ be a path in Ω that has degenerated into a point; that is, $\gamma(t) = z_0$, $a \le t \le b$. Show that $\int_\gamma f(z)\,dz = 0$.

41. Let $f(z)$ be a continuous function on a region Ω, and let γ be a path in Ω. Let $\Gamma = (\gamma, -\gamma)$, that is, γ followed by its reverse. Show that $\int_\Gamma f(z)\,dz = 0$.

42. Project Problem: Additivity over paths. In this exercise, we prove part (iii) of Proposition 2.
(a) We have $\Gamma = (\gamma_1, \gamma_2, \ldots, \gamma_m)$. Let us suppose each $\gamma_k(t)$ maps the interval $[0, 1]$ to γ_k. Show that we can parametrize Γ over the interval $[0, m]$ by defining $\Gamma(t)$ piecewise as

(33) $$\Gamma(t) = \gamma_k(t + 1 - k) \text{ for } k - 1 \le t \le k.$$

(b) Justify the following steps:

$$\int_{\Gamma} f(z)dz = \int_{0}^{m} f(\Gamma(t))\Gamma'(t)\,dt = \sum_{k=1}^{m} \int_{k-1}^{k} f(\Gamma(t))\Gamma'(t)\,dt$$

$$= \sum_{k=1}^{m} \int_{0}^{1} f(\gamma_k(T))\gamma_k'(T)\,dT \text{ [Hint: Let } T = t+1-k \text{ and use (33)]}$$

$$= \sum_{k=1}^{m} \int_{\gamma_k} f(z)\,dz.$$

3.3   Independence of Path

In calculus, the problem of evaluating a definite integral of a function f was often reduced to finding an antiderivative F and evaluating F at the endpoints. In this section, we investigate this procedure for path integrals.

Suppose that $f(z)$ is continuous in a region Ω. We say that $F(z)$ is an **antiderivative** of f in Ω if $F'(z) = f(z)$ for all z in Ω. Since f is continuous, F is analytic in Ω and, in particular, it is continuous. Any two antiderivatives of f differ by a constant. This is a simple consequence of Theorem 2, Section 2.4.

To find an antiderivative of a given $f(z)$, we try if possible to guess our answer, motivated by formulas from calculus. For example, if $f(z) = z\,e^z$, based on results from calculus, we guess an antiderivative $F(z) = z\,e^z - e^z$. Once we guessed an answer, we can verify it by differentiating. However, as our next examples illustrates, as part of checking the answer, we must make sure that F is analytic on the domain under consideration.

EXAMPLE 1   Antiderivatives
(a)   An antiderivative of $f(z) = z^3 + 7z - 2$ is $F(z) = \frac{z^4}{4} + \frac{7}{2}z^2 - 2z$, and the identity $F'(z) = f(z)$ holds for all z in $\mathbb{C}$.
(b)   The function $f(z) = \operatorname{Log} z$ is continuous for z in $\mathbb{C}\backslash(-\infty, 0]$. An antiderivative is $F(z) = z\operatorname{Log} z - z$, and the identity $F'(z) = f(z)$ holds for all z in $\mathbb{C}\backslash(-\infty, 0]$.
(c)   The function $f(z) = \frac{1}{z}$ is continuous in $\mathbb{C}\backslash\{0\}$. We may guess an antiderivative $F(z) = \operatorname{Log} z$. But the equality $F'(z) = f(z)$ holds for all z in $\mathbb{C}\backslash(-\infty, 0]$. So $\operatorname{Log} z$ is an antiderivative of $\frac{1}{z}$ in $\mathbb{C}\backslash(-\infty, 0]$ only, even though f is continuous in $\mathbb{C}\backslash\{0\}$. As the next example shows, other antiderivatives of $\frac{1}{z}$ can be found on different regions.
(d)   Let Ω_α denote the region $\mathbb{C}$ minus the ray at angle α, and let $\log_\alpha z$ denote a branch of the logarithm with a branch cut at angle α. We know from Section 2.4 that $\log_\alpha z$ is analytic in Ω_α and that

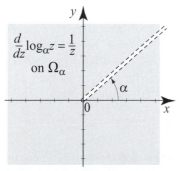

$\frac{d}{dz}\log_\alpha z = \frac{1}{z}$
on Ω_α

Figure 1 $\log_\alpha z$ and   its branch cut at angle α.

$$\frac{d}{dz}\log_\alpha z = \frac{1}{z}, \qquad \text{for all } z \text{ in } \Omega_\alpha.$$

From this we conclude that $F(z) = \log_\alpha z$ is an antiderivative of $\frac{1}{z}$ in Ω_α. In particular, if we choose α in such a way that the ray at angle α is not the negative

x-axis, then $F(z) = \log_\alpha z$ becomes an antiderivative of $\frac{1}{z}$ in a region that contains the negative x-axis (Figure 1).   ∎

As we now show, independence of path for integrals can be used to characterize functions with antiderivatives.

THEOREM 1
INDEPENDENCE OF
PATH

Let f be a continuous complex-valued function on a region Ω, let z_1, z_2 be two points in Ω, and let

$$I = \int_\gamma f(z)\, dz,$$

where γ is a path in Ω joining z_1 to z_2. Then I is independent of the path γ if and only if $f(z) = F'(z)$ for some analytic function F on Ω. In this case, we write

$$I = \int_\gamma f(z)\, dz = \int_{z_1}^{z_2} f(z)\, dz = F(z_2) - F(z_1).$$

We will prove that if f has an antiderivative, then $\int_\gamma f(z)\, dz$ is independent of path, and we will postpone the other direction until the end of the section.

Proof If F is an antiderivative of f in Ω, then the complex-valued function $t \mapsto F(\gamma(t))$ is differentiable at the points in $[a, b]$ where $\gamma'(t)$ exists and we have

(1)
$$\frac{d}{dt} F(\gamma(t)) = F'(\gamma(t))\gamma'(t) = f(\gamma(t))\gamma'(t).$$

This formula is similar to the chain rule for differentiable functions ((8), Section 2.3) and can be established in exactly the same way (see Exercise 35). Now $t \mapsto f(\gamma(t))\gamma'(t)$ is piecewise continuous, because f is continuous and γ' is piecewise continuous. Also, since $F(\gamma(t))$ is continuous, (1) tells us that $F(\gamma(t))$ is a continuous antiderivative of $f(\gamma(t))\gamma'(t)$, in the sense of Theorem 1, Section 3.2. Using this theorem, we obtain

$$\int_\gamma f(z)\, dz = \int_a^b f(\gamma(t))\gamma'(t)\, dt = F(\gamma(b)) - F(\gamma(a)) = F(z_2) - F(z_1),$$

completing the proof of the theorem in one direction.   ∎

We turn now to some applications.

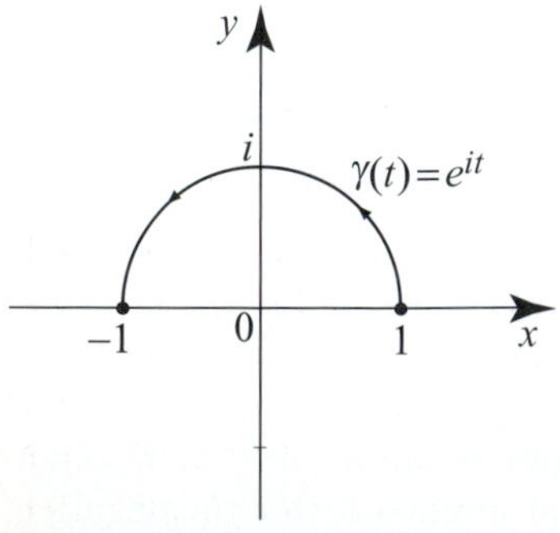

EXAMPLE 2   Integrals involving entire functions
Evaluate

Figure 2(a)

(a) $\displaystyle\int_\gamma e^z\, dz$, where γ is the semi-circle $\gamma(t) = e^{it}$, $0 \le t \le \pi$ (Figure 2(a)).

(b) $\displaystyle\int_{[z_1, z_2, z_3, z_1]} (z^3 + z^2 - 2)\, dz$, where $[z_1, z_2, z_3, z_1]$ is the closed directed line segment with $z_1 = -1$, $z_2 = 1$, $z_3 = i$ (Figure 2(b)).

(c) $\displaystyle\int_\gamma z e^{z^2}\, dz$, where γ is the semi-circle $\gamma(t) = -\frac{i}{2} + \frac{1}{2} e^{it}$, $-\frac{\pi}{2} \le t \le \frac{\pi}{2}$ (Figure 2(c)).

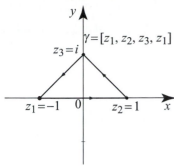

Figure 2(b) A closed triangular path.

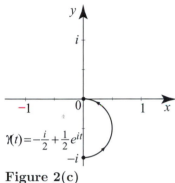

Figure 2(c)

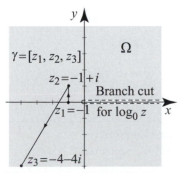

Figure 3 for Example 3.

Solution   We evaluate the integrals by appealing to Theorem 1.

(a)   The function e^z is continuous in the entire plane, with an antiderivative e^z. The initial point of γ is $z_1 = \gamma(0) = 1$ and its terminal point is $z_2 = \gamma(\pi) = -1$. By Theorem 1,

$$\int_\gamma e^z \, dz = e^z \Big|_1^{-1} = e^{-1} - e^1 = -2\sinh(1).$$

(b)   An antiderivative of $z^3 + z^2 - 2$ is $\frac{z^4}{4} + \frac{z^3}{3} - 2z$. The initial and terminal points of the path are the same, $z_1 = -1$. By Theorem 1, we have

$$\int_{[z_1, z_2, z_3, z_1]} (z^3 + z^2 - 2) \, dz = \left[\frac{z^4}{4} + \frac{z^3}{3} - 2z \right]_{-1}^{-1} = 0.$$

(c)   As you can easily verify by direct computation, an antiderivative of ze^{z^2} is $\frac{1}{2} e^{z^2}$. The initial point of γ is $z_1 = -i$ and its terminal point is $z_2 = 0$. By Theorem 1,

$$\int_\gamma ze^{z^2} \, dz = \frac{1}{2} e^{z^2} \Big|_{-i}^0 = \frac{1}{2}\left(1 - e^{-1}\right).$$ ∎

In Example 1, the region that contained the paths was of little concern to us, because the integrands and their antiderivatives were entire. This is not the case in the next two examples, where the region or the antiderivative must be carefully chosen in order to verify all the hypotheses of Theorem 1.

EXAMPLE 3   Choosing an appropriate region

Evaluate $\displaystyle\int_{[z_1, z_2, z_3]} \frac{1}{z} \, dz$, where $[z_1, z_2, z_3]$ is the directed line segment with $z_1 = 1$, $z_2 = 2 + i$, $z_3 = 3$ (Figure 3).

Solution   The function $f(z) = \frac{1}{z}$ is continuous in $\mathbb{C} \setminus \{0\}$. An antiderivative of $\frac{1}{z}$ is $\mathrm{Log}\, z$ in the region $\Omega = \mathbb{C} \setminus (-\infty, 0]$. Since the path $[z_1, z_2, z_3]$ lies entirely in Ω, we may apply Theorem 1 and get

$$\int_{[z_1, z_2, z_3]} \frac{1}{z} \, dz = \mathrm{Log}\, z \Big|_1^3 = \mathrm{Log}\, 3 - \mathrm{Log}\, 1 = \ln 3.$$ ∎

EXAMPLE 4   Choosing an appropriate antiderivative

Evaluate $\displaystyle\int_{[z_1, z_2, z_3]} \frac{1}{z} \, dz$, where $[z_1, z_2, z_3]$ is the directed line segment with $z_1 = -1$, $z_2 = -1 + i$, $z_3 = -4 - 4i$ (Figure 4).

Solution   In order to apply Theorem 1, we must find an antiderivative of $\frac{1}{z}$ that is analytic in a region that contains the path $[z_1, z_2, z_3]$. We cannot use $\mathrm{Log}\, z$ as antiderivative, because it is not analytic in any region that contains the path $[z_1, z_2, z_3]$. Instead, we will use a different branch of the logarithm. We know from Example 1(d) that $\log_\alpha z$ is an antiderivative of $\frac{1}{z}$ in the region Ω_α (the plane minus the ray at angle α). We can apply Theorem 1 if we choose α in such a way that

Figure 4 for Example 4.

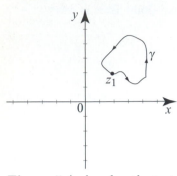

Figure 5 A closed path starting and ending at z_1.

the branch cut of $\log_\alpha z$ does not intersect the path $[z_1, z_2, z_3]$. Take, for example, $\alpha = 0$; then $\log_0 z = \ln |z| + i \arg_0 z$, where $0 < \arg_0 z \le 2\pi$. By Theorem 1,

$$
\begin{aligned}
\int_{[z_1, z_2, z_3]} \frac{1}{z}\, dz &= \log_0(z_3) - \log_0(z_1) \\
&= \frac{1}{2} \ln(32) + i \arg_0(-4 - 4i) - (\ln 1 + i \arg_0(-1)) \\
&= \frac{5}{2} \ln 2 + i \frac{5\pi}{4} - i\pi = \frac{5}{2} \ln 2 + i \frac{\pi}{4}.
\end{aligned}
$$
∎

Integrals over Closed Paths

If the integral of f is independent of path, then integrals around closed paths must be zero. To see this, consider the closed path γ in Figure 5 containing the point z_1. If we start at z_1 and trace γ until we return to z_1, we have from Theorem 1,

$$
\int_\gamma f(z)\, dz = F(z_1) - F(z_1) = 0.
$$

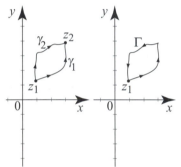

Figure 6 γ_1 followed by γ_2 yields the closed path Γ.

This is illustrated by Example 1(b). Conversely, suppose we know that the integral of f around every closed path is zero. Consider the two paths γ_1 and γ_2 joining z_1 to z_2 in Figure 6, and let Γ be the closed path consisting of γ_1 followed by the reverse of γ_2. Then

$$
0 = \int_\Gamma f(z)\, dz = \int_{\gamma_1} f(z)\, dz - \int_{\gamma_2} f(z)\, dz,
$$

implying that $\int_{\gamma_1} f(z)\, dz = \int_{\gamma_2} f(z)\, dz$. Thus, the integral of f is independent of path.

Combining this discussion with Theorem 1, we have the following useful theorem.

THEOREM 2   Let $f(z)$ be a continuous function defined in a region Ω. Then the following are equivalent.
(i) $f(z)$ has an analytic antiderivative $F(z)$ in Ω.
(ii) The integral of f is independent of path.
(iii) The integral of f around any closed path in Ω is zero.

Theorem 2 has many important applications. We start with an unexpected result.

EXAMPLE 5   A continuous function with no antiderivative
Let $z_1 = -1$, $z_2 = 1$, and $z_3 = i$. We know from Example 7, Section 3.2, that

$$
\int_{[z_1, z_2, z_3]} \overline{z}\, dz = i \qquad \text{and} \qquad \int_{[z_1, z_3]} \overline{z}\, dz = -i.
$$

Since the integral of $\overline{z}$ depends on the path that we choose from z_1 to z_3, we conclude from Theorem 2 that there is no antiderivative of $\overline{z}$ in a region containing the points z_1, z_2, and z_3. Indeed, we will show later that the derivative of an analytic function is itself analytic. In light of this result and the fact that $\overline{z}$ is not analytic at any point (Example 4(a), Section 2.3), it is clear that $\overline{z}$ has no antiderivative. ∎

EXAMPLE 6 Integrals over closed paths

(a) If γ is any closed path, then by Theorem 2,

$$\int_{\gamma} z\, dz = 0,$$

because $f(z) = z$ has an antiderivative $F(z) = \frac{z^2}{2}$ for all z in the plane.

(b) If γ is any closed path, then by Theorem 2,

$$\int_{\gamma} e^{2iz}\, dz = 0,$$

because $f(z) = e^{2iz}$ has an antiderivative $F(z) = \frac{e^{2iz}}{2i}$ for all z in the plane.

(c) Let $C_{\frac{1}{2}}(0)$ denote the positively oriented circle with center at 0 and radius $\frac{1}{2}$. Then by Theorem 2,

$$\int_{C_{\frac{1}{2}}(0)} \frac{1}{1+z}\, dz = 0,$$

because $\operatorname{Log}(1+z)$ is an antiderivative of $\frac{1}{z+1}$ in the region $\mathbb{C} \setminus (-\infty, -1]$, which contains $C_{\frac{1}{2}}(0)$ (see Figure 7).

(d) Let z_0 be a fixed point in the plane, γ be a closed path in a region that does not go through z_0, and $n \geq 2$ be an integer. Then by Theorem 2,

$$\int_{\gamma} \frac{1}{(z-z_0)^n}\, dz = 0,$$

because $F(z) = \frac{1}{(n-1)(z-z_0)^{n-1}}$ is an antiderivative of $\frac{1}{(z-z_0)^n}$ in any region not containing z_0. ∎

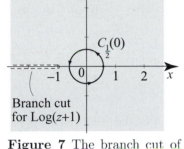

Figure 7 The branch cut of $\operatorname{Log}(z + 1)$ is obtained by translating to the left by one unit the branch cut of $\operatorname{Log} z$.

While Theorems 1 and 2 are very useful, they have their limitations when we do not know an antiderivative of the integrand f. For example, it is not clear from Theorem 1 whether the path integral of a function like e^{z^2} is independent of path, because thus far we do not know whether e^{z^2} has an antiderivative. In the next section we answer these problems and many others by presenting a far-reaching theorem of Cauchy. This result is at the heart of the theory of path integrals and indeed all of complex analysis.

Completion of the Proof of Theorem 1

Recall that the fundamental theorem of calculus states that the derivative of an antiderivative of a continuous function is the function itself. In symbols, if f is continuous and $F(x) = \int_a^x f(t)\, dt$, then

$$f(x) = F'(x) = \lim_{h \to 0} \frac{F(x+h) - F(x)}{h} = \lim_{h \to 0} \frac{1}{h} \int_x^{x+h} f(t)\, dt.$$

For continuous complex-valued functions, we have the following useful lemma.

LEMMA 1   Suppose $f(z)$ is continuous in a region Ω, z and $z + \Delta z$ are in Ω, and the closed line segment $[z, z + \Delta z]$ is also in Ω. Then

$$(2) \qquad \lim_{\Delta z \to 0} \frac{1}{\Delta z} \int_{[z,\, z+\Delta z]} f(\zeta)\, d\zeta = f(z).$$

Proof   Parametrize $[z, z + \Delta z]$ by $\gamma(t) = (1 - t)z + t(z + \Delta z) = z + t\Delta z$, where $0 \le t \le 1$. Then $\gamma'(t) = \Delta z$ and hence $\int_{[z,\, z+\Delta z]} d\zeta = \int_0^1 \Delta z\, dt = \Delta z$. So

$$\frac{1}{\Delta z} \int_{[z,\, z+\Delta z]} f(z)\, d\zeta = f(z), \text{ and}$$

$$\frac{1}{\Delta z} \int_{[z,\, z+\Delta z]} f(\zeta)\, d\zeta - f(z) = \frac{1}{\Delta z} \int_{[z,\, z+\Delta z]} (f(\zeta) - f(z))\, d\zeta.$$

Hence (2) is equivalent to

$$(3) \qquad \lim_{\Delta z \to 0} \frac{1}{\Delta z} \int_{[z,\, z+\Delta z]} (f(\zeta) - f(z))\, d\zeta = 0.$$

To prove (3), given $\epsilon > 0$, since f is continuous at z and Ω is open, we can find $\delta > 0$ such that the disk centered at z with radius δ is contained in Ω and

$$(4) \qquad |\zeta - z| < \delta \quad \Rightarrow \quad |f(\zeta) - f(z)| < \epsilon.$$

For $|\Delta z| < \delta$ and all ζ on $[z, z + \Delta z]$, we have $|\zeta - z| \le |\Delta z|$, and so (4) shows that $|f(\zeta) - f(z)| < \epsilon$ for all ζ on the line segment $[z, z + \Delta z]$. Hence the maximum M of the function $\zeta \mapsto |f(\zeta) - f(z)|$ is less than ϵ for all ζ on $[z, z + \Delta z]$. Applying Theorem 2, Section 3.2, and using the fact that the length of $[z, z + \Delta z]$ is $l([z, z + \Delta z]) = |\Delta z|$, we obtain

$$\left| \frac{1}{\Delta z} \int_{[z,\, z+\Delta z]} (f(\zeta) - f(z))\, d\zeta \right| \le \frac{1}{|\Delta z|} l([z, z + \Delta z])\epsilon = \epsilon,$$

which implies (3) and hence the lemma.   ∎

Proof of Theorem 1   We only need to show that if I is independent of path, then f has an antiderivative F. Fix z_0 in Ω. For z in Ω, define

$$(5) \qquad F(z) = \int_{z_0}^z f(\zeta)\, d\zeta,$$

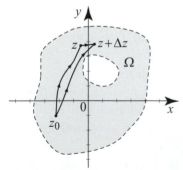

Figure 8

where the integral is taken over any path joining z_0 to z (recall that Ω is connected and the integral is independent of path). Since the integral of f is independent of path, we have $\int_{z_0}^{z+\Delta z} f(\zeta)\, d\zeta = \int_{z_0}^z f(\zeta)\, d\zeta + \int_z^{z+\Delta z} f(\zeta)\, d\zeta$ (see Figure 8), and so

$$F(z + \Delta z) - F(z) = \int_{z_0}^{z+\Delta z} f(\zeta)\, d\zeta - \int_{z_0}^z f(\zeta)\, d\zeta = \int_z^{z+\Delta z} f(\zeta)\, d\zeta.$$

For very small Δz, z and $z + \Delta z$ are in Ω, because Ω is open. So we can choose the path joining z to $z + \Delta z$ to be the line segment $[z, z + \Delta z]$,

$$(6) \qquad \frac{F(z + \Delta z) - F(z)}{\Delta z} = \frac{1}{\Delta z} \int_{[z, z+\Delta z]} f(\zeta) \, d\zeta.$$

Taking the limit as $\Delta z \to 0$ and appealing to Lemma 1, we obtain $F'(z) = f(z)$. ∎

Exercises 3.3

In Exercises 1–14, find an antiderivative of the given function and specify the region Ω where the antiderivative is valid.

1. $z^2 + z - 1$.

2. $ze^z - \sin z$.

3. $\dfrac{\mathrm{Log}\, z}{z}$.

4. $\dfrac{1}{z - 1}$.

5. $\dfrac{1}{(z - 1)(z + 1)}$.

6. $\sec^2 z$.

7. $\cos(3z + 2)$.

8. $ze^{z^2} - \dfrac{1}{z}$.

9. $z \sinh z^2$.

10. $e^z \cos z$.

11. $z \, \mathrm{Log}\, z$.

12. $\log_\alpha z$.

13. $\log_0 z + \log_{\frac{\pi}{2}} z + \dfrac{1}{z}$.

14. $z^{\frac{1}{5}}$ (principal branch).

In Exercises 15–26, evaluate the given path integral. Justify clearly the use of Theorems 1 and 2.

15. $\displaystyle\int_{[z_1, z_2, z_3]} 3(z - 1)^2 \, dz$, where $z_1 = 1$, $z_2 = i$, $z_3 = 1 + i$.

16. $\displaystyle\int_{[z_1, z_2, z_3]} (z^2 - 1)^2 \, z \, dz$, where $z_1 = 0$, $z_2 = 1$, $z_3 = -i$.

17. $\displaystyle\int_\gamma z^2 \, dz$, where $\gamma(t) = e^{it} + 3e^{2it}$, $0 \le t \le \frac{\pi}{4}$.

18. $\displaystyle\int_{C_1(0)} \left((z - 2 - i)^2 + \frac{i}{z - 2 - i} - \frac{3}{(z - 2 - i)^2} \right) dz$.

19. $\displaystyle\int_{[z_1, z_2, z_3]} ze^z \, dz$, where $z_1 = \pi$, $z_2 = -1$, $z_3 = -1 - i\pi$.

20. $\displaystyle\int_{[z_1, z_2, z_3]} e^{i\pi z} \, dz$, where $z_1 = 2$, $z_2 = i$, $z_3 = 4$.

21. $\displaystyle\int_\gamma \sin z \, dz$, where $\gamma(t) = 2e^{it}$, $0 \le t \le \frac{\pi}{2}$.

22. $\displaystyle\int_\gamma \sin^2 z \, dz$, where γ is any closed path.

23. $\displaystyle\int_\gamma \frac{1}{z} \, dz$, where $\gamma(t) = e^{it}$, $0 \le t \le \frac{3\pi}{4}$.

24. $\displaystyle\int_{[z_1, z_2, \ldots, z_n]} dz$, where $z_1, z_2, \ldots, z_n$ are arbitrary.

25. $\displaystyle\int_{[z_1, z_2, z_3, z_1]} z \, \mathrm{Log}\, z \, dz$, where $z_1 = 1$, $z_2 = 1 + i$, $z_3 = -2 + 2i$.

26. $\int_{[z_1,\, z_2,\, z_3]} \dfrac{\text{Log}\, z}{z}\, dz$, where $z_1 = -i$, $z_2 = 1$, $z_3 = i$.

27.   (a) Show that for any complex number α,

$$\frac{d}{dz} z^\alpha = \alpha z^{\alpha-1},$$

where we define both complex powers using a single logarithm branch. Conclude that an antiderivative of z^α is $\frac{1}{\alpha+1} z^{\alpha+1}$, where the same logarithm branch is used.

(b) Evaluate $\int_\gamma \dfrac{1}{\sqrt{z}}\, dz$ (principal branch), where $\gamma(t) = e^{it}$, $-\frac{\pi}{2} \le t \le \frac{\pi}{2}$.

28.   Use the result of Exercise 27(a) to evaluate $\int_\gamma z^{i\pi}\, dz$ (use the branch $\log z = \text{Log}\, z - 2\pi i$), where $\gamma(t) = e^{it}$, $-\frac{\pi}{2} \le t \le 0$.

29.   Let $C_R(z_0)$ denote the positively oriented circle with center at z_0 and radius $R > 0$. Follow the outlined steps to show that

$$\int_{C_R(z_0)} \frac{1}{z}\, dz = \begin{cases} 2\pi i & \text{if } |z_0| < R, \\ 0 & \text{if } |z_0| > R. \end{cases}$$

(a)   If $|z_0| > R$, then the circle $C_R(z_0)$ does not contain 0, and consequently it cannot intersect all four semi-axes at the origin. Pick a semi-axis that $C_R(z_0)$ does not intersect as a branch cut for the logarithm and explain why $\int_{C_R(z_0)} \frac{1}{z}\, dz = 0$, using Theorem 2. Draw a picture to illustrate your proof.

(b)   If $|z_0| < R$, then 0 is in the interior of the circle $C_R(z_0)$. Let z_1 and z_2 be the points of intersection of $C_R(z_0)$ with the positive and negative y-axis, respectively. Let γ_1 be the portion of $C_R(z_0)$ with initial point z_2 and terminal point z_1, and let γ_2 be the portion of $C_R(z_0)$ with initial point z_1 and terminal point z_2. Write

$$\int_{C_R(z_0)} \frac{1}{z}\, dz = \int_{\gamma_1} \frac{1}{z}\, dz + \int_{\gamma_2} \frac{1}{z}\, dz.$$

Show that

$$\int_{\gamma_1} \frac{1}{z}\, dz = \text{Log}\,(z_1) - \text{Log}\,(z_2)$$

and

$$\int_{\gamma_2} \frac{1}{z}\, dz = \log_0(z_2) - \log_0(z_1).$$

Show that $\int_{C_R(z_0)} \frac{1}{z}\, dz = 2\pi i$.

30.   **Replacing the integrand.**   Consider the integral

$$\int_\gamma \text{Log}\, z\, dz, \qquad \gamma(t) = e^{it},\ 0 \le t \le \pi.$$

Since $\text{Log}\, z$ is not continuous at the point -1, it cannot be continuous in any region containing the path, and so we cannot apply Theorem 1 directly. The idea in this problem is to replace $\text{Log}\, z$ by a different branch of the logarithm for which Theorem 1 does apply.

(a)   Show that $\text{Log}\,z = \log_{-\frac{\pi}{2}} z$ for all z on γ.

(b)   Conclude that

$$\int_{\gamma} \text{Log}\,z\,dz = \int_{\gamma} \log_{-\frac{\pi}{2}} z\,dz,$$

and evaluate the integral on the right side by using Theorem 1.

31.   Evaluate $\int_{C_1(0)} \text{Log}\,z\,dz$, where $C_1(0)$ is the positively oriented unit circle. First write the integral as the sum of two integrals over the upper and lower semicircles, then use the ideas of Exercise 30 to evaluate each integral in this sum by appealing to Theorem 2.

32.   Evaluate $\int_{R} \frac{dz}{z}$, where R is the positively oriented square with vertices at $1 \pm i$, $-1 \pm i$. Do not parametrize the integral, but use Theorem 2 as suggested in Exercise 31.

33.   Recall Theorem 2, Section 2.4: If $f(z)$ is analytic in a region Ω and $f'(z) = 0$ for all z in Ω, then f is constant in Ω. Prove this theorem using Theorem 1 of this section. [Hint: Let z_0 and z be points in Ω. What can you say about $\int_{z_0}^{z} f'(z)\,dz$ where the integral is over any path in Ω joining z_0 to z?]

34. L'Hospital's rule.   Prove the following version of L'Hospital's rule. If $f(z)$ and $g(z)$ are analytic in a region Ω, z_0 is in Ω, and $f(z_0) = g(z_0) = 0$, but $g'(z_0) \neq 0$, then

$$\lim_{z \to z_0} \frac{f(z)}{g(z)} = \frac{f'(z_0)}{g'(z_0)}.$$

[Hint: For z in a small neighborhood of z_0 in Ω , write $f(z) = \int_{[z,\, z_0]} f'(\zeta)\,d\zeta$. Do the same for $g(z)$ and then use Lemma 1 to compute

$$\lim_{z \to z_0} \frac{f(z)}{g(z)} = \lim_{\Delta z \to 0} \frac{\frac{1}{\Delta z} f(z)}{\frac{1}{\Delta z} g(z)} .]$$

35. Chain rule for $F(\gamma(t))$.   Suppose $F(z)$ is differentiable at z_0; we have $F(z) = F(z_0) + F'(z_0)(z - z_0) + \epsilon_1(z)(z - z_0)$, where $\epsilon_1(z) \to 0$ as $z \to z_0$. Suppose $\gamma(t)$ is differentiable at t_0 and $\gamma(t_0) = z_0$; we have $\gamma(t) = \gamma(t_0) + \gamma'(t_0)(t - t_0) + \epsilon_2(t)(t - t_0)$ where $\epsilon_2(t) \to 0$ as $t \to t_0$.

(a) Show that $\epsilon_1(\gamma(t)) \to 0$ as $t \to t_0$.

(b) Show that

$$F(\gamma(t)) = F(\gamma(t_0)) + \big(F'(\gamma(t_0)) + \epsilon_1(\gamma(t))\big)\big(\gamma'(t_0) + \epsilon_2(t)\big)\big(t - t_0\big),$$

and then that $\frac{d}{dt} F(\gamma(t))$ exists at t_0 and equals $F'(\gamma(t_0))\gamma'(t_0)$.

3.4   Cauchy's Integral Theorem

Independence of path is a highly desirable property because it simplifies the task of computing integrals. In the previous section we gave necessary and sufficient conditions for independence of path. However, these conditions required finding antiderivatives, which in many cases are not easy to compute. In this section, we introduce the fundamental theorem of Cauchy for path integrals of analytic functions. This result lies at the heart of complex

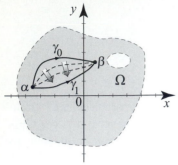

Figure 1 Continuous deformation of γ_0 into γ_1.

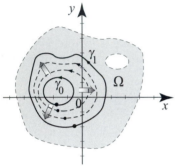

Figure 2 Continuous deformation of γ_0 into γ_1.

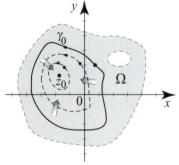

Figure 3 Continuous deformation of γ_0 into a point z_0.

analysis. Among its many important consequences, we will obtain sufficient geometric conditions for independence of path. These conditions are easy to visualize and make of Cauchy's theorem a very useful and practical tool for computing integrals. A complete proof of the theorem will be given in detail in the following section.

Continuous Deformation of Paths

We start by describing three geometric pictures that will serve to illustrate the fundamental concept of deformation of path.

- Suppose that α and β are points in a region Ω, and γ_0 and γ_1 are paths in Ω joining α to β. We say that γ_0 is **continuously deformable** into γ_1 **relative to** Ω if we can continuously move γ_0 over γ_1 while keeping the ends fixed at α and β and without leaving Ω (Figure 1).

- If γ_0 and γ_1 are two closed paths in a region Ω, we say that γ_0 is **continuously deformable** into γ_1 **relative to** Ω if we can continuously move γ_0 without leaving Ω, in such a way that it overlaps with γ_1 in position and direction (Figure 2).

- If γ_0 is a closed path in a region Ω, we say that γ_0 is **continuously deformable into a point relative to** Ω if we can continuously shrink γ_0 into a point z_0 in Ω without leaving Ω (Figure 3). This is the same as the deformation of the closed path γ_0 into the closed path γ_1 where γ_1 is degenerated into a point. That is, $\gamma_1(t)$ is constant for all t.

It is intuitively clear that if we can continuously deform γ_0 into γ_1 in Ω, then we can continuously deform γ_1 into γ_0 in Ω. Hence, we will refer to the paths γ_0 and γ_1 as **mutually deformable** in Ω.

The notion of continuous deformation is central to this section. Even though it is about paths in Ω, we will use it to define properties of the region Ω. For this reason, it is very important to be precise about Ω when talking about continuous deformation. We will give shortly a mathematical definition of continuous deformation, but first let us look at some examples and use our intuition to understand this notion.

EXAMPLE 1   Continuous deformation of paths
(a)   Let Ω be the complex plane. In Figure 4, the upper semi-circle $\gamma_0(t) = e^{it}$, $0 \leq t \leq \pi$, is continuously deformable into the directed line segment joining 1 to -1.
(b)   In the open unit disk $B_1(0)$, the path $[z_1, z_2, z_3]$ is continuously deformable into the directed line segment $[z_1, z_3]$ (Figure 5).
(c)   In the open unit disk $B_1(0)$, the arc γ_0 is continuously deformable into the triangular path $[z_1, z_2, z_3]$ (Figure 6).
(d)   In the annular region Ω in Figure 7, the circle γ_0 is continuously deformable into the circle γ_1 relative to Ω.

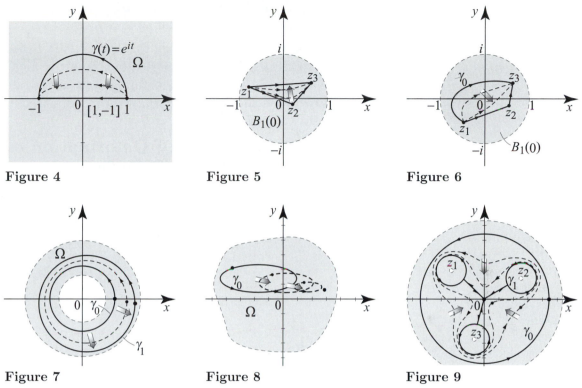

Figure 4   **Figure 5**   **Figure 6**

Figure 7   **Figure 8**   **Figure 9**

(e)   In the region Ω in Figure 8, the ellipse γ_0 is continuously deformable into a point relative to Ω.

(f)   In Figure 9, the region Ω consists of a disk minus the points z_1, z_2, z_3. In Ω the circle γ_0 is continuously deformable into the curve γ_1 consisting of three circles centered at z_1, z_2, z_3, and connected by line segments traversed in both directions, as shown in Figure 9.   ■

In Figure 10, the upper semi-circle of radius 1 is continuously deformed into the lower semi-circle of radius 1 relative to $B_2(0)$. This deformation is impossible in Figure 11, where we have removed the origin from $B_2(0)$. You can imagine a stick protruding at the origin, which will prevent you from continuously deforming the upper curve to lower one.

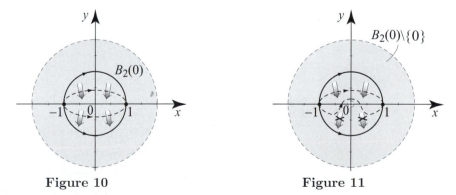

Figure 10   **Figure 11**

To see how important it is to be precise about the region Ω when talking about deformation of paths, consider the situation depicted in Figures 10 and 11. The upper and lower semi-circles joining the points -1 to 1 in Figure 10 are mutually deformable in the disk of radius 2 but are not mutually deformable in the punctured disk of radius 2 (Figure 11). While the disk and punctured disk differ by only one point, from the standpoint of analysis, they are completely different objects.

Note to the reader: We next take up the mathematical theory of deformation of path. This theory will be required in the next section in the proof of the general version of Cauchy's theorem. If you are satisfied with the intuitive treatment of deformation of path, you can skip directly to the next subsection on the Cauchy integral.

Mathematical Definition of Deformation of Paths

In mathematical terms, the path γ_0 is continuously deformable into γ_1 relative to the region Ω if there exist a continuous map H of the unit square $Q = [0,1] \times [0,1]$ into Ω with the following properties:

Figure 12 The path γ_0 is continuously deformed into γ_1 relative to Ω if we can find a continuous mapping H of the square Q into Ω, such that H maps the lower side of Q onto γ_0, its upper side onto γ_1, its left side onto the point α, and its right side onto the point β. As a result, the horizontal line segments in Q will be mapped to curves varying continuously between γ_0 and γ_1, while staying entirely in Ω.

(1) $H(t,\,0) \;=\; \gamma_0(t), \quad 0 \le t \le 1,$

(2) $H(t,\,1) \;=\; \gamma_1(t), \quad 0 \le t \le 1;$

(3) $H(0,\,s) \;=\; \alpha, \quad 0 \le s \le 1,$

(4) $H(1,\,s) \;=\; \beta, \quad 0 \le s \le 1.$

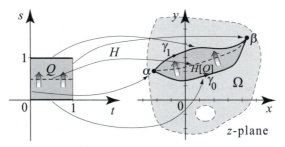

To understand this definition, consider Figure 12. Condition (1) tells us that the image of the lower side of Q is γ_0. Condition (2) tells us that the image of the upper side of Q is γ_1. Conditions (3) and (4) tell us that the vertical sides of Q are mapped to the endpoints of γ_0 and γ_1. The fact that Q is mapped into Ω tells us that the image of any horizontal section in Q is a curve in Ω; and because the vertical sides of Q are mapped to the endpoints of γ_0 and γ_1, this curve has the same endpoints as γ_0 and γ_1. As s varies from 0 to 1, the curve image of a horizontal section of the square Q varies continuously between the two extreme curves, from γ_0 to γ_1. Thus the notion of a continuous deformation of γ_0 into γ_1 relative to Ω.

The definition of a continuous deformation for closed paths γ_0 and γ_1 in Ω is similar to the one we just presented. We require the existence of a continuous mapping H with properties (1) and (2) and, instead of (3) and (4), we require that

(5) $H(0,\,s) = H(1,\,s), \quad 0 \le s \le 1.$

This condition states that the initial point $H(0, s)$ and the terminal point $H(1, s)$ are the same. Thus the images of all horizontal sections of the square Q are closed curves (Figure 13).

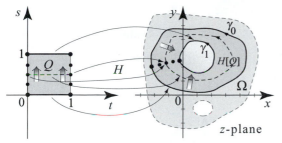

Figure 13 Deformation of closed paths.

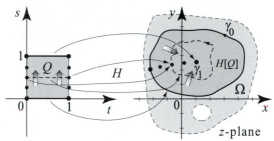

Figure 14 Deformation to one point.

The definition of a continuous deformation of a closed path γ_0 into a point z_1 in Ω requires the existence of a continuous mapping H with properties (1), (2), and (5), where in (2) the path $\gamma_1(t)$ is constant and equals z_1 in Ω (Figure 14).

Although we will not use the following terminology, we should mention that the mapping H that continuously deforms γ_0 into γ_1 is called a **homotopy mapping** of γ_0 into γ_1 and the paths γ_0 and γ_1 are called **homotopically equivalent** relative to Ω. It is pictorially clear that if γ_0 is homotopic to γ_1 relative to Ω, then γ_1 is homotopic to γ_0 relative to Ω. Also, if γ_0 is homotopic to γ_1 relative to Ω and γ_1 is homotopic to γ_2 relative to Ω, then γ_0 is homotopic to γ_2 relative to Ω. The proof of these statements is outlined in Exercise 36.

As you can imagine, in general, it is difficult to construct the deformation mapping H, and so in most problems we will rely on pictures and our intuition to decide whether a continuous deformation is possible. There are a few interesting cases where H can be constructed explicitly. We start with a simple example and generalize the ideas involved in it.

EXAMPLE 2   Deformation of paths by continuous mappings
Show that the positively oriented circle $C_1(0)$ is continuously deformable into the positively oriented circle $C_2(0)$, relative to the open disk of radius 3.

Solution   Intuitively, the result is obvious. To prove it mathematically, we will construct a continuous mapping H of the unit square taking $C_1(0)$ continuously into $C_2(0)$, in the sense that (1), (2), and (5) hold. Parametrize $C_1(0)$ by $\gamma_0(t) = e^{2\pi it}$, $0 \leq t \leq 1$, and $C_2(0)$ by $\gamma_1(t) = 2e^{2\pi it}, 0 \leq t \leq 1$. For a fixed t in $[0, 1]$, consider the corresponding points $\gamma_0(t)$ on $C_1(0)$ and $\gamma_1(t)$ on $C_2(0)$. We can move continuously from $\gamma_0(t)$ to $\gamma_1(t)$ along the line segment

$$(1 - s)\gamma_0(t) + s\gamma_1(t), \qquad 0 \leq s \leq 1,$$

while staying in Ω (Figure 15). Motivated by this idea, we let

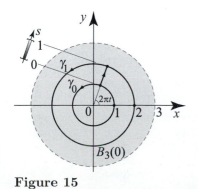

Figure 15

(6) $$H(t, s) = (1 - s)\gamma_0(t) + s\gamma_1(t), \quad 0 \leq t \leq 1, \quad 0 \leq s \leq 1.$$

Then H is a continuous mapping of (t, s) in the unit square, because $\gamma_0(t)$ and $\gamma_1(t)$ are continuous. Moreover, it is easy to check that (1), (2), and (5) hold. Indeed,

$$H(t, 0) = \gamma_0(t) \quad \Rightarrow \quad \text{(1) holds,}$$
$$H(t, 1) = \gamma_1(t) \quad \Rightarrow \quad \text{(2) holds.}$$

Also, since $\gamma_0(0) = 1 = \gamma_0(1)$ and $\gamma_1(0) = 2 = \gamma_1(1)$, we obtain

$$H(0, s) = (1 - s)\gamma_0(0) + s\gamma_1(0) = 1 + s,$$
$$H(1, s) = (1 - s)\gamma_0(1) + s\gamma_1(1) = 1 + s,$$

and so $H(0, s) = H(1, s)$ implying (5).

Figure 16 How the continuous deformation mapping H works: H maps the lower side of Q onto the inner circle $C_1(0)$, the upper side onto $C_2(0)$, and all the in between horizontal segments in Q are mapped to in between circles $C_R(0)$, where $1 < R < 2$.

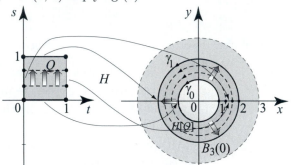

In conclusion, we have explicitly constructed a mapping that continuously deforms $\gamma_0(t)$ into $\gamma_1(t)$ while staying in Ω (Figure 16). ∎

By identifying an important geometric property of the disk, we can generalize the construction in Example 2 as follows. Call a subset S of the complex plane **convex** if whenever z_0 and z_1 are in S, then the line segment joining z_0 to z_1 lies entirely in S. In other words, S is convex if whenever z_0 and z_1 are in S, then the points on the line segment

$$(7) \qquad\qquad (1 - s)z_0 + sz_1, \qquad 0 \le s \le 1,$$

are also in S. The complex plane is a convex set. Any disk is a convex set. Any rectangle is convex (Figure 17). The star-shaped set S in Figure 18 is not convex because we can find points z_1 and z_2 in S such that the line segment $[z_1, z_2]$ is not entirely in S. The punctured disk in Figure 19 is not convex for the same reason.

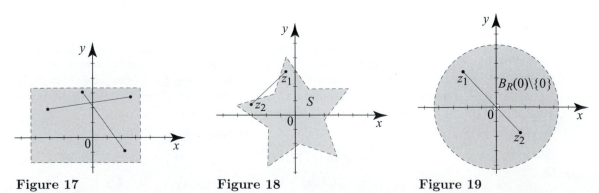

Figure 17   **Figure 18**   **Figure 19**

Even if we know that two paths are mutually deformable relative to a region Ω, finding a deformation mapping can be very difficult. However, as we now show, if Ω is convex, we do have a formula for the mapping H.

Suppose that Ω is a convex region and $\gamma_0(t)$ and $\gamma_1(t)$ are paths in Ω, with same initial and terminal points. Suppose further that γ_0 and γ_1 are parametrized by the interval $[0, 1]$. Motivated by formula (6), we consider the continuous mapping

$$(8) \qquad H(t, s) = (1 - s)\gamma_0(t) + s\gamma_1(t), \quad 0 \le t \le 1, \quad 0 \le s \le 1.$$

Let us momentarily fix t; since $z_0 = \gamma_0(t)$ and $z_1 = \gamma_1(t)$ are two points in Ω and Ω is convex, it follows that $H(t, s) = (1 - s)z_0 + sz_1$ is also in Ω. This reasoning holds for all t, and so all points $H(t, s)$ lie in Ω. Thus H is a continuous mapping of the unit square into Ω. (This is a crucial property of H; it is telling you that the deformation is taking place in Ω.) It is clear from (8) that (1) and (2) hold. Properties (3), (4), and (5) are also easy to verify, and so we have the following very useful result.

THEOREM 1
DEFORMATION OF
PATHS IN CONVEX
SETS

Suppose that Ω is a convex region. (i) Any two paths γ_0 and γ_1 in Ω, joining two points α and β in Ω, are mutually continuously deformable relative to Ω. The continuous mapping H that deforms γ_0 into γ_1 is given by (8).
(ii) Any two closed paths γ_0 and γ_1 in Ω are mutually continuously deformable relative to Ω. The continuous mapping H that deforms γ_0 into γ_1 is given by (8).
(iii) Any closed path γ_0 in Ω is continuously deformable into a point in Ω. The continuous mapping H that deforms a closed path γ_0 into a point z_1 in Ω is given by (8), where $\gamma_1(t) = z_1$ is constant for all t in $[0, 1]$.

EXAMPLE 3   Deformation of paths in convex sets

In the region Ω consisting of the open disk centered at the origin with radius 4, consider the closed paths $\gamma_0(t) = e^{2\pi it} + e^{4\pi it}$, $0 \le t \le 1$, and $\gamma_1(t) = 3e^{2\pi it}$, $0 \le t \le 1$. Because Ω is convex, we can continuously deform γ_0 into γ_1 by using the mapping

$$H(t, s) = (1 - s)\gamma_0(t) + s\gamma_1(t) = (1 - s)(e^{2\pi it} + e^{4\pi it}) + s(3e^{2\pi it}).$$

In Figure 20, we plotted the graphs of $H(t, s)$ for $s = 0$, $\frac{1}{2}$, $\frac{2}{3}$, and 1. Note how the corresponding graphs of $H(t, s)$ vary continuously from γ_0 to γ_1. ∎

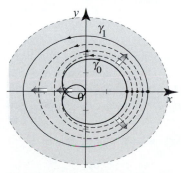

Figure 20

Cauchy's Theorem

Having discussed deformation of path, we are now ready to state a fundamental result in complex analysis. This result holds for *analytic* functions.

THEOREM 2
CAUCHY'S
INTEGRAL
THEOREM FOR
DEFORMABLE
PATHS

Let f be a function analytic in a region Ω. (i) If γ_0 and γ_1 are two paths joining α to β in Ω, and if γ_0 is continuously deformable into γ_1 relative to Ω, then

$$(9) \qquad \int_{\gamma_0} f(z)\,dz = \int_{\gamma_1} f(z)\,dz.$$

(ii) If γ_0 and γ_1 are closed paths in Ω and γ_0 is continuously deformable into γ_1 relative to Ω, then

$$(10) \qquad \int_{\gamma_0} f(z)\,dz = \int_{\gamma_1} f(z)\,dz.$$

In particular, if γ_0 is a closed path in Ω that is continuously deformable into a point in Ω, then

$$(11) \qquad \int_{\gamma_0} f(z)\,dz = 0.$$

Note that (11) follows from (10) because the integral of a function over a point is zero.

The proof of this theorem is quite long and technical. We will present it in full detail in the following section. For now, let us continue with some important applications.

COROLLARY 1
PATH INTEGRALS
OF FUNCTIONS
ANALYTIC ON
DISKS

Suppose that γ is a closed path and f is analytic on an open disk containing γ. Then

$$(12) \qquad \int_{\gamma} f(z)\,dz = 0.$$

In particular, if f is entire then (12) holds for any closed path γ.

Proof Any closed path γ in a disk is continuously deformable to a point in that disk. This is intuitively clear and follows from Theorem 1(iii) because a disk is a convex set. So (12) follows from Theorem 2. The second part of the corollary clearly follows from the first: If f is entire and γ is a closed path, then we can always enclose γ in an open disk on which f is analytic. Alternatively, the second part follows from the fact that $\mathbb{C}$ itself is convex, and so any two closed paths are mutually deformable. ∎

From Corollary 1, we see immediately that if γ is any closed path, then

$$\int_{\gamma} e^{z^2}\,dz = 0,$$

because e^{z^2} is entire. Note that we are able to compute the integral without knowing an antiderivative of e^{z^2}.

Simple Paths and Simply Connected Regions

With Cauchy's theorem in hand, our main task is now to identify paths that are mutually continuously deformable. Instead of dealing with this issue on a case-by-case basis, we will try to identify regions in which any two closed paths are mutually deformable. For such regions, we have the following.

PROPOSITION 1

> Suppose that Ω is a region in $\mathbb{C}$. The following properties are equivalent.
> (i) Any two paths γ_0 and γ_1 in Ω joining two points α and β in Ω are mutually continuously deformable relative to Ω.
> (ii) Any two closed paths γ_0 and γ_1 in Ω are mutually continuously deformable relative to Ω.
> (iii) Any closed path γ_0 in Ω is continuously deformable into a point in Ω.

The proof of the proposition is relegated to Exercise 37. It is based on arguments similar to the ones we used in the proof of Theorem 2, Section 3.3.

A region Ω for which (i), and hence (ii) and (iii) of Proposition 1, hold has a special name.

DEFINITION 1
SIMPLY
CONNECTED
REGION

> A region Ω is called **simply connected** if all paths joining α to β in Ω are mutually continuously deformable relative to Ω. If a region is not simply connected, then it is called **multiply connected**.

Geometrically, a region is simply connected if it has no holes in it. Figure 21 shows several simply connected regions. Theorem 1 proves that every convex set is simply connected; consequently, a disk and a rectangle are simply connected. (As you can imagine, a convex set cannot have holes in it. So this result agrees with our geometric intuition.) But, as clearly illustrated in the second and third regions in Figure 21, a region can be simply connected without being convex.

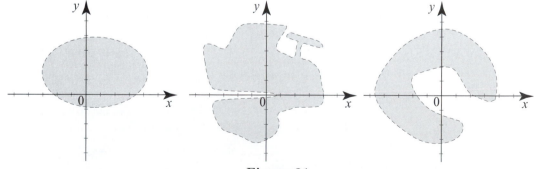

Figure 21

Figure 22 shows several multiply connected regions. In each case, we show two paths that are not mutually deformable relative to the region.

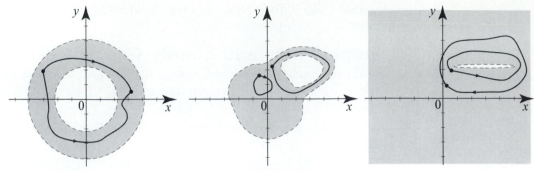

Figure 22

The following fundamental theorem is an immediate consequence of Theorem 2 and the definition of simply connected regions.

THEOREM 3
CAUCHY'S
THEOREM FOR
SIMPLY
CONNECTED
REGIONS

Suppose that Ω is a simply connected region and that f is analytic in Ω.
(i) If γ_0 and γ_1 are two paths joining α to β in Ω, then

$$\int_{\gamma_0} f(z)\,dz = \int_{\gamma_1} f(z)\,dz.$$

(ii) If γ is a closed path in Ω, then

$$\int_{\gamma} f(z)\,dz = 0.$$

Consider the function $\frac{1}{z-z_0}$, which is analytic in any punctured disk centered at z_0. We have $\int_{C_R(z_0)} \frac{dz}{z-z_0} = 2\pi i \neq 0$ (see Example 4(a), Section 3.2); and so by Theorem 3(ii), the punctured disk is not simply connected. This confirms our geometric understanding of simply connected regions as being regions without holes.

The following is another important property of simply connected regions.

THEOREM 4
EXISTENCE OF
ANTIDERIVATIVES

Suppose that Ω is a simply connected region and that f is analytic in Ω. Then there is an analytic function F in Ω such that $F' = f$. In other words, every analytic function in Ω has an antiderivative in Ω. Up to an arbitrary additive constant, we have

$$F(z) = \int_{z_0}^{z} f(\zeta)\,d\zeta,$$

where z_0 is a fixed point in Ω and the integral is taken over any path in Ω joining z_0 to z.

Proof By Theorem 3, the integral of f is independent of path. Hence by Theorem 2, Section 3.3, f has an antiderivative given by (5), Section 3.3. ◼

Our next topic about paths is closely related to the notion of simply connected regions.

DEFINITION 2
SIMPLE PATH

> A closed path γ is called **simple** or a **simple curve** if γ intersects itself only at the endpoints. That is, if $\gamma(t)$, $a \leq t \leq b$, is a parametrization of γ, and $a \leq t_1 < t_2 \leq b$, then $\gamma(t_1) = \gamma(t_2) \Leftrightarrow t_1 = a$ and $t_2 = b$.

The graph of a simple curve can loop around and get very complicated, but it cannot cross itself. A simple path is also called a **Jordan curve**, after the French mathematician Camille Jordan (1838–1922), who discovered that a simple closed path C divides the plane into two regions: a bounded region, called the **interior** of C; and an unbounded region, called the **exterior** of C. This is the famous **Jordan curve theorem** from topology, which is easy to picture but quite difficult to prove. We omit the proof.

Because we can identify the interior and the exterior regions relative to a simple curve, we can use this to define the positive and negative orientations of a simple curve. The **positive orientation** of a simple curve is the one that when followed puts the interior region to our left. The **negative orientation** of a simple curve is the one that when followed puts the interior region to our right. Our definition of positive versus negative orientation thus generalizes our earlier idea of counterclockwise versus clockwise orientation of circles.

The following two very important properties of simple curves are also consequences of the Jordan curve theorem.

- The interior region of a simple curve is simply connected.

- Let Ω be any region containing a simple curve C and the interior of C. Then C can be continuously deformed into a point interior to C relative to Ω.

When we combine these properties with Theorems 2 and 3, we obtain a very useful integral theorem for simple paths. First recall that f is analytic at a point if it is analytic in a neighborhood of that point. So to say that f is analytic on a curve C and its interior means that f is analytic in a region Ω containing C and its interior.

THEOREM 5
CAUCHY'S
THEOREM FOR
SIMPLE PATHS

> Suppose that f is analytic on a simple closed path C and its interior.
> (i) We have
> $$\int_C f(z)\,dz = 0.$$
> (ii) If γ is a closed path (not necessarily simple) contained in the interior of C, then
> $$\int_\gamma f(z)\,dz = 0.$$

Proof Since f is analytic on a region Ω that contains C and its interior, and since C can be continuously deformed into a point relative to Ω, (i) follows from Cauchy's Theorem 2. Part (ii) follows from Theorem 3, because the interior of C is simply connected.   ∎

Multiply Connected Regions

Recall that a region Ω is multiply connected if it is not simply connected. In this section, we are interested in a particular kind of multiply connected regions, for which we have a very useful version of Cauchy's theorem. A typical region is described as follows.

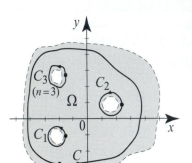

Figure 23

Let C be a simple closed path and let $C_1,\ C_2,\ \ldots,\ C_n$ be simple closed paths, contained in the interior of C and such that the interior regions of any two C_j's have no common points. We also require that C and all $C_j's$ have the same orientation (say, positive). Let Ω be any region containing the paths $C,\ C_1,\ C_2,\ \ldots,\ C_n$, and the region interior to C and exterior to the C_j's. The region Ω is in general multiply connected, as illustrated in Figure 23.

THEOREM 6
CAUCHY'S
THEOREM FOR
MULTIPLY
CONNECTED
REGIONS

Suppose that f is analytic on a region Ω with simple paths C and C_j ($j = 1, 2, \ldots, n$) as described above. Then

(13)
$$\int_C f(z)\,dz = \sum_{j=1}^n \int_{C_j} f(z)\,dz.$$

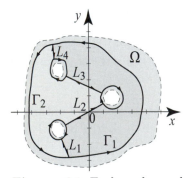

Figure 24 Each polygonal path L_j is traversed in both directions, so the integral over L_j is 0.

Proof Join the outer path C to C_1 using a polygonal path L_1 traversed in two opposite directions. Join C_1 to C_2 by a similar polygonal path L_2. Continue in this fashion, joining the path C_j to the path C_{j+1} by a polygonal path L_{j+1} traversed in two opposite directions, and finally join C_n to C by a polygonal path L_{n+1} traversed in two opposite directions. This construction yields two simple closed paths Γ_1 and Γ_2, as illustrated in Figure 24. (Note that, along either Γ_1 or Γ_2, we traverse portions of the C_j's in a direction opposite to their original orientation.) By Theorem 5, since f is analytic on Γ_1 and Γ_2 and their interiors, it follows that

$$\int_{\Gamma_1} f(z)\,dz = 0 \qquad \text{and} \qquad \int_{\Gamma_2} f(z)\,dz = 0.$$

Adding these two equalities, we obtain

$$\int_{\Gamma_1} f(z)\,dz + \int_{\Gamma_2} f(z)\,dz = 0,$$

or

$$\int_C f(z)\,dz + \sum_{j=1}^n \left(\int_{-C_j} f(z)\,dz + \int_{L_j} f(z)\,dz \right) = 0.$$

Since each L_j consists of a polygonal path traversed in two opposite directions, we have $\int_{L_j} f(z)\, dz = 0$ and hence

$$\int_C f(z)\, dz + \sum_{j=1}^{n} \int_{-C_j} f(z)\, dz = 0,$$

which is equivalent to (13).   ■

EXAMPLE 4   A useful integral

Let C be a positively oriented simple closed path and let z_0 be a point not on C. Show that

$$\int_C \frac{1}{z - z_0}\, dz = \begin{cases} 0 & \text{if } z_0 \text{ is in the exterior of } C; \\ 2\pi i & \text{if } z_0 \text{ is in the interior of } C. \end{cases}$$

Solution   If z_0 is in the exterior of C, then the function $f(z) = \frac{1}{z - z_0}$ is analytic on and inside C and hence its integral along C is 0, by Theorem 5(i). For the second case, where z_0 is in the interior of C, we can no longer claim that the function $f(z) = \frac{1}{z - z_0}$ is analytic on and *inside* C, and so we cannot use Theorem 5. To deal with this case, let us recall that

$$\int_{C_R(z_0)} \frac{1}{z - z_0}\, dz = 2\pi i,$$

where $C_R(z_0)$ is a positively oriented circle with center at z_0 and radius $R > 0$. (See Example 4(a), Section 3.2.) Let $C_\delta(z_0)$ be a positively oriented circle centered at z_0 with radius δ and wholly contained in the interior of C. The function $f(z) = \frac{1}{z - z_0}$ is analytic in $\mathbb{C} \setminus \{z_0\}$ and, in particular, it is analytic in a region that contains $C_\delta(z_0)$ and C and the region interior to C and exterior to $C_\delta(z_0)$. Applying (13) of Theorem 6, we see that

$$\int_C \frac{1}{z - z_0}\, dz = \int_{C_\delta(z_0)} \frac{1}{z - z_0}\, dz = 2\pi i,$$

as desired.   ■

Example 4 can be used to compute interesting path integrals of rational functions. You will also need to review the partial fraction decomposition of a rational function.

EXAMPLE 5   Path integrals of rational functions

Compute

$$\int_C \frac{dz}{(z - 1)(z + i)(z - i)},$$

where C is a simple closed path in the following three cases.

(a) The point 1 is in the interior of C, and the points $\pm i$ are in the exterior of C.

(b) The points 1 and i are in the interior of C, and the point $z = -i$ is in the exterior of C.

(c) All three points 1, $\pm i$ are in the interior of C.

Solution   We start by finding the partial fraction decomposition of the integrand,

$$\frac{1}{(z-1)(z+i)(z-i)} = \frac{a}{z-1} + \frac{b}{z+i} + \frac{c}{z-i},$$

where a, b, and c are complex numbers. Combining, we obtain

$$\frac{1}{(z-1)(z+i)(z-i)} = \frac{a(z+i)(z-i) + b(z-1)(z-i) + c(z-1)(z+i)}{(z-1)(z+i)(z-i)},$$

so

(14) $$1 = a(z+i)(z-i) + b(z-1)(z-i) + c(z-1)(z+i).$$

Alternatively, we can expand the right side of (14) and match coefficients of z^2, z, and 1. We proceed to solve the linear equations for a, b, and c. This method is more time-consuming, but you should keep it in mind when you solve Exercise 33(a).

Taking $z = 1$, we get

$$1 = a(1+i)(1-i) \quad \Rightarrow \quad a = \frac{1}{(1+i)(1-i)} = \frac{1}{2}.$$

Similarly, setting $z = -i$ yields $b = -\frac{1}{4} - \frac{i}{4}$. Setting $z = i$ yields $c = -\frac{1}{4} + \frac{i}{4}$. Thus the partial fraction decomposition

$$\frac{1}{(z-1)(z+i)(z-i)} = \frac{1}{2(z-1)} - \frac{1+i}{4(z+i)} + \frac{-1+i}{4(z-i)}.$$

So

(15)
$$\int_C \frac{dz}{(z-1)(z+i)(z-i)}$$
$$= \frac{1}{2}\int_C \frac{1}{z-1}\,dz - (\frac{1}{4} + \frac{i}{4})\int_C \frac{1}{z+i}\,dz + (\frac{-1}{4} + \frac{i}{4})\int_C \frac{1}{z-i}\,dz.$$

We can now compute the desired integrals using Example 4. For (a), since 1 is the only point in the interior of C, the last two integrals on the right side of (15) are zero, and hence

$$\int_C \frac{dz}{(z-1)(z+i)(z-i)} = \frac{1}{2}\int_C \frac{1}{z-1} = \frac{1}{2}2\pi i = \pi i.$$

For (b), since 1 and i are in the interior of C and $-i$ is in the exterior of C, the middle integral on the right side of (15) is zero, and the desired integral in this case is equal to

$$\frac{1}{2}\int_C \frac{1}{z-1} + (\frac{-1}{4} + \frac{i}{4})\int_C \frac{1}{z-i}\,dz = \left(\frac{1}{2} + (\frac{-1}{4} + \frac{i}{4})\right)2\pi i = (\frac{1}{2} + \frac{i}{2})\pi i.$$

Finally, for (c), since 1 and $\pm i$ are all in the interior of C, all the integrals on the right side of (15) must be accounted for. The answer in this case is

$$\left(\frac{1}{2} + (\frac{-1}{4} + \frac{i}{4}) + (\frac{-1}{4} - \frac{i}{4})\right)2\pi i = 0.$$

That the answer is zero in this case is also a consequence of a more general result. See Exercise 33.   ∎

As our final example shows, in some cases, it is necessary to combine various versions of Cauchy's theorem to evaluate the integral.

EXAMPLE 6   A path that is not simple
Evaluate the integral

(16)
$$\int_C \frac{z}{(z-i)(z-1)} dz,$$

where C is the figure eight path shown in Figure 25.

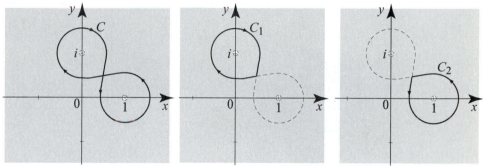

The path C in Figure 25 is not simple because it intersects itself. To use Cauchy's integral theorem for simple paths, we break up C into two simple paths, as shown in Figure 26.

Figure 25   **Figure 26**

Solution We will first break the path into two simple closed paths, C_1 and C_2, shown in Figure 26. Note that the orientation of C_1 is negative, while the orientation of C_2 is positive. The integral (16) becomes

(17)
$$\int_{C_1} \frac{z}{(z-i)(z-1)} dz + \int_{C_2} \frac{z}{(z-i)(z-1)} dz = I_1 + I_2.$$

You can verify that the partial fraction decomposition of the integrand is

$$\frac{z}{(z-i)(z-1)} = \frac{\frac{1}{2} - \frac{i}{2}}{z-i} + \frac{\frac{1}{2} + \frac{i}{2}}{z-1}.$$

Since $\frac{1}{z-1}$ is analytic on and inside C_1, its integral along C_1 is 0, by Theorem 5(i). Also $\int_{C_1} \frac{dz}{z-i} = -2\pi i$, by Example 4. Hence

$$I_1 = \left(\frac{1}{2} - \frac{i}{2}\right) \int_{C_1} \frac{dz}{z-i} + \left(\frac{1}{2} + \frac{i}{2}\right) \int_{C_1} \frac{dz}{z-1} = \left(\frac{1}{2} - \frac{i}{2}\right)(-2\pi i) + 0 = -\pi - i\pi.$$

Arguing similarly to evaluate the integral along C_2, we find

$$I_2 = \left(\frac{1}{2} - \frac{i}{2}\right) \overbrace{\int_{C_2} \frac{dz}{z-i}}^{=0} + \left(\frac{1}{2} + \frac{i}{2}\right) \overbrace{\int_{C_2} \frac{dz}{z-1}}^{2\pi i} = \left(\frac{1}{2} + \frac{i}{2}\right) 2\pi i = -\pi + i\pi.$$

Adding the two integrals, we find the answer $I_1 + I_2 = -2\pi$.   ∎

The applications of this section have clearly illustrated the power of Cauchy's theorem in evaluating integrals and deriving theoretical properties of analytic functions. This was just the beginning. In what follows we will derive formulas and results based on Cauchy's theorem that will have spectacular consequences at both the theoretical and applied levels.

Exercises 3.4

In Exercises 1–6, a figure is given describing two paths γ_0 and γ_1 in a region Ω. Decide whether the two paths are mutually continuously deformable relative to Ω. Justify your answer based on the picture.

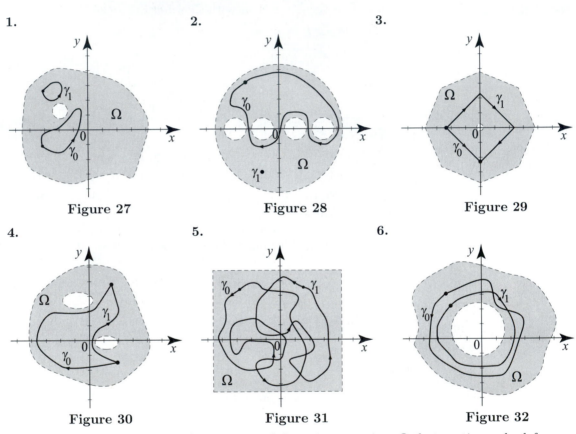

1.   **2.**   **3.**

Figure 27   Figure 28   Figure 29

4.   **5.**   **6.**

Figure 30   Figure 31   Figure 32

7.   Describe the mapping of the unit square into Ω that continuously deforms γ_0 into γ_1 relative to Ω in Figure 33.

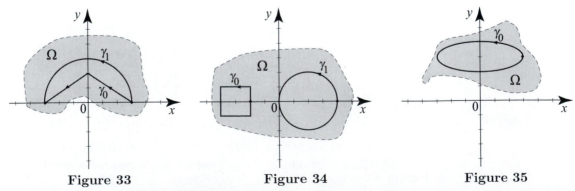

Figure 33   Figure 34   Figure 35

8. (a)   Describe the mapping $H(t, s)$ of the unit square that continuously deforms γ_0 into γ_1 relative to Ω in Figure 34.

(b)   Illustrate the deformation process in (a) by drawing γ_0 and γ_1 and the graph of $H(t, s)$ as a function of t alone, for various fixed values of s.

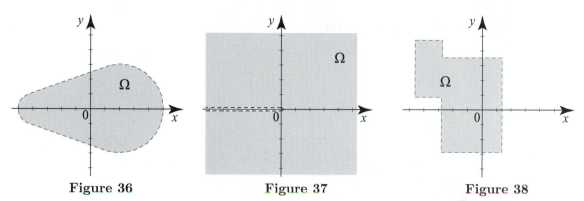

Figure 36   **Figure 37**   **Figure 38**

9.   Describe the mapping of the unit square that continuously deforms the path γ_0 into a point relative to Ω in Figure 35. You may pick any point in Ω.

10.   Of the regions shown in Figures 36-38, determine the ones that are convex. Justify your answer with arguments based on the figure.

11.   Of the regions shown in Figures 39-44, determine the ones that are simply connected. Justify your answer with arguments based on the figure.

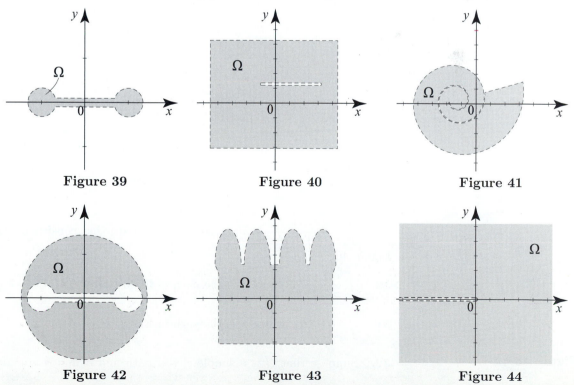

Figure 39   **Figure 40**   **Figure 41**

Figure 42   **Figure 43**   **Figure 44**

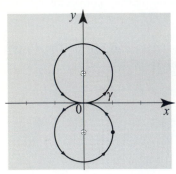

Figure 45

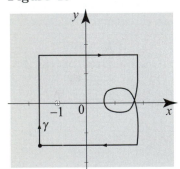

Figure 46

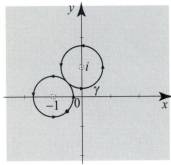

Figure 47

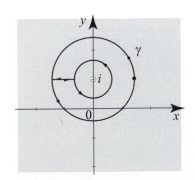

Figure 48

In Exercises 12–31, evaluate the given integral. Indicate clearly the version of Cauchy's theorem or the result from Section 3.3 that you are using.

12. $\int_{[z_1, z_2, z_3, z_1]} \sin(z^2)\, dz$, where $z_1 = 0$, $z_2 = -i$, $z_3 = 1$.

13. $\int_{C_1(0)} \dfrac{e^z}{z+2}\, dz$.

14. $\int_\gamma (z^2 + 2z + 3)\, dz$, where γ is any path joining 0 to 1.

15. $\int_{C_1(i)} \left(\dfrac{z-1}{z+1}\right)^2 z\, dz$.

16. $\int_\gamma \dfrac{3i}{z - 2i}\, dz$, where $\gamma(t) = e^{it} + \frac{e^{2it}}{2}$, $0 \le t \le 2\pi$. [Hint: Show that the path is closed and the integrand is analytic in a region containing the path.]

17. $\int_\gamma \dfrac{e^z}{z+i}\, dz$, where $\gamma(t) = i + e^{it}$, $0 \le t \le 2\pi$.

18. $\int_{C_1(0)} \dfrac{1}{z - \frac{1}{2}}\, dz$.   **19.** $\int_{C_1(0)} \dfrac{1}{(z - \frac{1}{2})^2}\, dz$.

20. $\int_{C_4(0)} \left((z - 2 + i)^2 + \dfrac{i}{z-2+i} - \dfrac{3}{(z-2+i)^2}\right) dz$.

21. $\int_{[z_1, z_2, z_3, z_1]} z^2 \operatorname{Log} z\, dz$, where $z_1 = 1$, $z_2 = 1+i$, $z_3 = -1+i$.

22. $\int_{C_1(i)} \dfrac{1}{(z-i)(z+i)}\, dz$.   **23.** $\int_{C_3(i)} \dfrac{1}{(z-i)(z+i)}\, dz$.

24. $\int_{C_{\frac{3}{2}}(1+i)} \dfrac{1}{(z-1)(z-i)(z+i)}\, dz$.   **25.** $\int_{C_2(0)} \dfrac{z}{z^2-1}\, dz$.

26. $\int_{C_2(0)} \dfrac{1}{z^2+1}\, dz$.   **27.** $\int_{C_{\frac{3}{2}}(0)} \dfrac{z^2+1}{(z-2)(z+1)}\, dz$.

28. $\int_\gamma \dfrac{z}{(z-i)(z+i)}\, dz$, where γ is the path that consists of the two circles in Figure 45.

29. $\int_\gamma \dfrac{1}{(z+1)^2(z^2+1)}\, dz$, where γ is the path that consists of the two circles in Figure 46.

30. $\int_\gamma \dfrac{1}{z+1}\, dz$, where γ is the path in Figure 47.

31. $\int_\gamma \dfrac{z+1}{z-i}\, dz$, where γ is the path in Figure 48.

32. Suppose that C is a simple closed path and α and β are complex numbers not on C. What are the possible values of

$$\int_C \frac{1}{(z-\alpha)(z-\beta)}\, dz\, ?$$

(Distinguish all possible locations of the points α and β relative to the path C.)

33.   (a)   Let $z_1, z_2, \ldots, z_n$ be distinct complex numbers ($n \geq 2$). Show that in the partial fraction decomposition

$$\frac{1}{(z-z_1)(z-z_2)\cdots(z-z_n)} = \frac{A_1}{z-z_1} + \frac{A_2}{z-z_2} + \cdots + \frac{A_n}{z-z_n}$$

we must have $A_1 + A_2 + \cdots + A_n = 0$.
(b)   Suppose that C is a simple closed path that contains the points $z_1, z_2, \ldots, z_n$ in its interior. Use the result in part (a) to show that

$$\int_C \frac{1}{(z-z_1)(z-z_2)\cdots(z-z_n)}\, dz = 0.$$

34.   Let $p(z)$ be a polynomial of degree $n \geq 0$, C be a simple closed path, and z_0 be a point in the interior of C. Show that

$$\frac{1}{2\pi i}\int_C \frac{p(z)}{z-z_0}\, dz = p(z_0).$$

This is a special case of Cauchy's integral formula (Section 3.6). [Hint: Expand $p(z)$ about z_0 (Taylor expansion), $p(z) = p(z_0) + A_1(z-z_0) + \cdots + A_n(z-z_n)^n$.]
(b)   What is the value of the integral in (a) if z_0 is in the exterior of C?

35.   Let $p(z)$ and C be as in Exercise 34, and z_0 be in the interior of C. Show that

$$\frac{n!}{2\pi i}\int_C \frac{p(z)}{(z-z_0)^{n+1}}\, dz = p^{(n)}(z_0),$$

where $p^{(n)}$ denotes the nth derivative of p. This is a special case of Cauchy's generalized integral formula (Section 3.6). [Hint: What are the A_j's in the Taylor expansion in the hint of Exercise 34(a)?]

36. Homotopy is an equivalence relation.   (a) Given a path $\gamma_0(t)$ in a region Ω, show that $H(t, s) = \gamma_0(t)$ will continuously deform γ_0 to itself, relative to Ω. Thus all curves are homotopic to themselves. (In other words, homotopy is a reflexive relation.)
(b)   Show that if $H(t, s)$ is a mapping of the unit square that continuously deforms a path γ_0 into the path γ_1 relative to a region Ω, then $H(t, 1-s)$ is a mapping of the unit square that continuously deforms γ_1 into γ_0 relative to Ω. Hence if γ_0 is homotopic to γ_1 relative to Ω, then γ_1 is homotopic to γ_0 relative to Ω. (In other words, homotopy is a symmetric relation.)
(c)   Suppose that $H_1(t, s)$ continuously deforms γ_0 into γ_1 relative to Ω, and $H_2(t, s)$ continuously deforms γ_1 into γ_2 relative to Ω. Show that

$$H(t, s) = \begin{cases} H_1(t, 2s) & \text{if } 0 \leq s \leq \frac{1}{2}, \\ H_2(t, 2(s - \frac{1}{2})) & \text{if } \frac{1}{2} \leq s \leq 1, \end{cases}$$

continuously deforms γ_0 into γ_2 relative to Ω. Hence if γ_0 and γ_1 are homotopic relative to Ω and γ_1 and γ_2 are homotopic relative to Ω, then γ_0 and γ_2 are homotopic relative to Ω. (In other words, homotopy is a transitive relation.)

(d) A relation that is reflexive, symmetric, and transitive is called an **equivalence relation**. Conclude that homotopy is an equivalence relation.

37. Proof of Proposition 1.   Here we prove that properties (i), (ii), and (iii) in Proposition 1 are equivalent; that is, if one of them holds for a region Ω, then all of them must hold. We show that (i)$\Rightarrow$(iii)$\Rightarrow$(ii)$\Rightarrow$(i).

(a) Suppose any two paths in Ω joining α to β are mutually continuously deformable relative to Ω. Show that any closed path γ_0 is continuously deformable into a point in Ω. [Hint: The closed path γ_0 must start and end at the same point z_0; pick this as the point.]

(b) Suppose any closed path in Ω is continuously deformable into a point in Ω. Show that any two closed paths γ_0 and γ_1 are continuously deformable relative to Ω. [Hint: Deform γ_0 to a point z_0. Deform z_0 to z_1 by ordinary connectedness of Ω. Deform z_1 to γ_1. Use the transitive and symmetric properties of continuous deformation.]

(c) Suppose any two closed paths are mutually continuously deformable relative to Ω. Show that two paths γ_0 and γ_1 connecting α to β are continuously deformable relative to Ω. [Hint: Let Γ be the closed path which traverses γ_0 and then the reverse of γ_1. Parametrize Γ by t in $[0, 1]$ such that $\Gamma(0) = \Gamma(1) = \alpha$ and $\Gamma(\frac{1}{2}) = \beta$. The curve Γ can be continuously deformed into its reverse, where the homotopy mapping H satisfies $H(0, s) = H(1, s) = \alpha$ and $H(\frac{1}{2}, s) = \beta$. What does this imply for γ_0 and γ_1?]

38.   (a)   Let $p(z) = a_n z^n + a_{n-1}z^{n-1} + \cdots + a_1 z + a_0$, $(a_n \neq 0)$ be a polynomial of degree $n \geq 2$. Show that, for any z with $|z| = R > 0$,

$$|p(z)| \geq \left||a_n|R^n - |a_{n-1}|R^{n-1} - \cdots - |a_1|R - |a_0|\right|.$$

Conclude that for z on $C_R(0)$, the circle centered at 0 with radius $R > 0$, we have

$$\left|\frac{1}{p(z)}\right| \leq \frac{1}{\left||a_n|R^n - |a_{n-1}|R^{n-1} - \cdots - |a_1|R - |a_0|\right|}.$$

(b)   Use (a) to show that

$$\left|\int_{C_R(0)} \frac{1}{p(z)}\,dz\right| \to 0, \quad \text{as} \quad R \to \infty.$$

[Hint: Use Theorem 2, Section 3.2.]

(c) Let C be a simple closed path that contains all the roots of $p(z)$ in its interior. Show that

$$\int_C \frac{1}{p(z)}\,dz = 0.$$

[Hint: Take C to be positively oriented and choose R so large that $C_R(0)$ contains C in its interior. Explain why $\int_C \frac{1}{p(z)}\,dz = \int_{C_R(0)} \frac{1}{p(z)}\,dz$, where the right side is in fact independent of R. Then let $R \to \infty$ and use (b).]

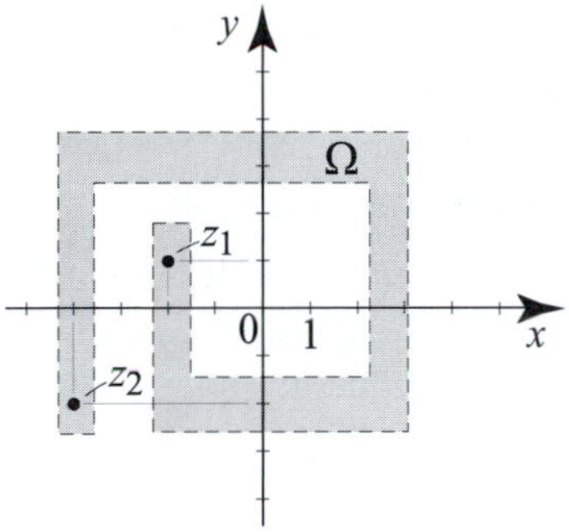

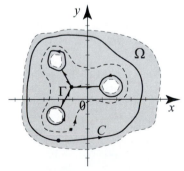

Figure 49

39. The spiraling region Ω in Figure 49 is simply connected. Theorem 4 guarantees us an antiderivative of $f(z) = \frac{1}{z}$ in Ω. Find an antiderivative explicitly in terms of branches of the logarithm.

40. Project Problem: Green's theorem. In this problem we investigate a connection between Cauchy's theorem and Green's theorem from vector calculus. This theorem states the following.

> Suppose that Ω is a simply connected region and C is a simple closed path in Ω with positive orientation. Let D denote the region interior to C. Suppose that $u(x, y)$ and $v(x, y)$ are two continuously differentiable functions on Ω. Then

$$(18) \qquad \int_C \left(u\, dx + v\, dy\right) = \iint_D \left(\frac{\partial v}{\partial x} - \frac{\partial u}{\partial y}\right) dx\, dy.$$

In what follows, we will show how Green's theorem implies the version of Cauchy's integral theorem for simply connected regions (Theorem 5).

(a)   Let $f(z) = u(x, y) + i\,v(x, y)$ be defined on a path γ, where u and v are continuously differentiable. Parametrize γ by $\gamma(t) = x(t) + i\,y(t)$, $a \leq t \leq b$. Going back to the definition of the path integral in Definition 1, Section 3.2, show that

$$\int_\gamma f(z)\, dz = \int_a^b \left(u(\gamma(t))x'(t) - v(\gamma(t))y'(t)\right) dt + i \int_a^b \left(v(\gamma(t))x'(t) + u(\gamma(t))y'(t)\right) dt.$$

(b)   Using the notation $x'(t)\, dt = dx$ and $y'(t)\, dt = dy$, conclude that

$$(19) \qquad \int_\gamma f(z)\, dz = \int_\gamma \left(u\, dx - v\, dy\right) + i \int_\gamma \left(v\, dx + u\, dy\right).$$

(c)   Prove Theorem 5(i) using Green's theorem and (b). [Hint: Since f is analytic, u and v satisfy the Cauchy-Riemann equations and have continuous first partial derivatives.]

41.   Here, we give a less rigorous but simpler proof of Theorem 6 (Cauchy's theorem for multiply connected regions). Take the case $n = 3$ and the particular instance shown in Figure 50. The outer curve C is continuously deformable in Ω to Γ, being the three smaller simple curves C_1, C_2, C_3 and the three line segments L_1, L_2, L_3 traced in both directions. Conclude from Theorem 2 of this section, and Proposition 2 of Section 3.2, that

$$\int_C f(z)\, dz = \sum_{j=1}^3 \int_{C_j} f(z)\, dz + \sum_{j=1}^3 \int_{L_j} f(z)\, dz$$

$$= \sum_{j=1}^3 \int_{C_j} f(z)\, dz.$$

Figure 50

Repeat this same type of reasoning for a larger n (say, $n = 5$) and a more complicated arrangement of C_j's. It will work, but the connecting pieces L_j might be curved.

3.5   Proof of Cauchy's Theorem

The proof is divided in two major parts. In the first part, we prove Cauchy's theorem for star regions; these include disks. In the second part, we prove the general version of Cauchy's theorem, using the version in a disk.

A region Ω is called a **star region** if it contains a point z_0 such that any other point z in Ω can be connected to z_0 by a line segment wholly contained in Ω. Thus $[z_0, z]$ is contained in Ω for any z in Ω. Figure 1 depicts several examples of star regions along with suitable points to serve as z_0.

Figure 1
Star regions.

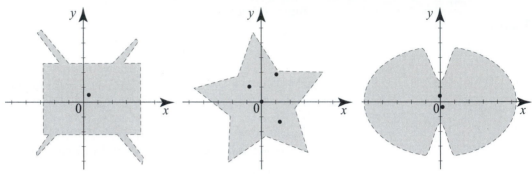

The proof of Cauchy's theorem on star regions is a simple consequence of Theorem 2, Section 3.3, and a general result on differentiation of path integrals, which we state and prove at the end of the section.

THEOREM 1
CAUCHY'S
INTEGRAL
THEOREM
FOR STAR REGIONS

Suppose that f is analytic on a star region Ω. Then

$$\int_\Gamma f(z)\,dz = 0$$

for all closed paths Γ in Ω.

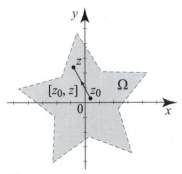

Figure 2

Proof   According to Theorem 2, Section 3.3, it is enough to construct an antiderivative of f in Ω. Let z_0 be a point in Ω that connects to all other points in Ω via line segments (see Figure 2). For z in Ω, define $F(z) = \int_{[z_0,\,z]} f(\zeta)\,d\zeta$. Parametrizing the line segment by $\gamma(t) = (1-t)z_0 + tz$, $0 \le t \le 1$, we have $\gamma'(t) = (z - z_0)dt$, and thus

(1) $$F(z) = (z - z_0)\int_0^1 f\big((1-t)z_0 + tz\big)\,dt.$$

To complete the proof, we will show that $F'(z) = f(z)$. Using the product rule for differentiation and the fact that $(z - z_0)' = 1$, we obtain

(2) $$\frac{d}{dz}F(z) \;=\; \int_0^1 f\big((1-t)z_0 + tz\big)\,dt + (z - z_0)\frac{d}{dz}\int_0^1 f\big((1-t)z_0 + tz\big)\,dt.$$

To compute the last derivative, we appeal to Theorem 4, below. Accordingly,

$$(z - z_0)\frac{d}{dz}\int_0^1 f\big((1-t)z_0 + tz\big)\,dt = (z - z_0)\int_0^1 \frac{d}{dz}f\big((1-t)z_0 + tz\big)\,dt$$

(3)
$$= (z - z_0)\int_0^1 t\,f'\big((z - z_0)t + z_0\big)\,dt.$$

The last integral is simple and can be evaluated by parts. Setting $u = t$, $dv = (z - z_0)f'\big((z - z_0)t + z_0\big)\,dt$, then $du = dt$, and $v = f\big((z - z_0)t + z_0\big)$. So

$$(z - z_0)\int_0^1 t\,f'\big((z - z_0)t + z_0\big)\,dt$$

$$= tf\big((z - z_0)t + z_0\big)\Big|_0^1 - \int_0^1 f\big((z - z_0)t + z_0\big)\,dt$$

(4)
$$= f(z) - \int_0^1 f\big((z - z_0)t + z_0\big)\,dt.$$

Using (4) and (3) in (2), it follows that $\frac{d}{dz}F(z) = f(z)$. ∎

Since a disk is a star region, Theorem 1 implies that Cauchy's integral theorem holds on a disk.

To prove the general version of Cauchy's theorem, stated in Theorem 2, Section 3.4, in addition to the version on a disk, we will need a few facts about continuous functions of one or two complex variables. Indeed, these properties hold for continuous functions defined on any space where we have a notion of a distance, a so-called **metric space**. For example, in the space $\mathbb{C} \times \mathbb{C}$, the product of two copies of $\mathbb{C}$, we have a notion of a distance using the following definition of the absolute value on $\mathbb{C} \times \mathbb{C}$: If $z_1 = a_1 + i\,b_1$, $w_1 = c_1 + i\,d_1$, $z_2 = a_2 + i\,b_2$, $w_2 = c_2 + i\,d_2$, then

(5)
$$\begin{aligned}|(z_1,\, w_1) - (z_2,\, w_2)| \\ = \sqrt{(a_1 - a_2)^2 + (b_1 - b_2)^2 + (c_1 - c_2)^2 + (d_1 - d_2)^2}.\end{aligned}$$

The following properties of continuous functions are generalizations of well-known results from calculus. As you read them, try to formulate the corresponding results from calculus that they generalize.

> Suppose that S is a closed and bounded set and f is a continuous complex-valued function on S. Then $f[S]$ is a closed and bounded subset of $\mathbb{C}$. In other words, the continuous image of a closed and bounded set is closed and bounded.

> Suppose that S is a closed and bounded set and f is a continuous complex-valued function on S. Then $|f|$ is attains its maximum and minimum values in S. That is, we can find z_1 and z_2 in S such that $|f(z_1)| \le |f(z)| \le |f(z_2)|$ for all z in S.

For our next result, we define the notion of distance between two sets. If K and S are two nonempty sets, the **distance** between K and S is defined to be the smallest value of $|z - w|$ where z is in K and w is in S (Figure 3). If a smallest value is not attained, we take the distance to be the largest number δ such that $|z - w|$ is always greater than or equal to δ.

With the help of the results that we just stated about continuous functions, we can prove the following fact, which is geometrically obvious.

> Suppose that K is a closed and bounded subset of $\mathbb{C}$ and S is a closed subset of $\mathbb{C}$. Suppose that K and S are disjoint. Then the distance between K and S is strictly positive.

In the next result, we use the notion of uniform continuity. A function f on a set S is called **uniformly continuous** if given $\epsilon > 0$, there is a $\delta > 0$ such that, for all z_1 and z_2 in S, $|z_1 - z_2| < \delta \Rightarrow |f(z_1) - f(z_2)| < \epsilon$.

Clearly, any uniformly continuous function is continuous. The converse is not true. The key words in the definition of uniform continuity are "for all," which imply that for a given ϵ the choice of δ works for all points in S. Compare this with the definition of continuity at a point, where the choice of δ depends on ϵ and the given point.

The following is one of the most basic results in analysis.

> If f is continuous on a closed and bounded set S, then f is uniformly continuous.

We now have all the necessary ingredients to prove Cauchy's theorem.

Proof of Theorem 2, Section 3.4 Let γ_0 and γ_1 be given in Ω, either joining α to β (part (i) of Cauchy's theorem) or closed (part (ii) of Cauchy's theorem). If Ω is the entire plane, then we are done by Theorem 1, because the complex plane is clearly a star region. If Ω is not the entire plane, its boundary, which we will denote as $\partial\Omega$, is closed (Exercise 6) and nonempty. Let H denote the mapping of the unit square $Q = [0, 1] \times [0, 1]$ into Ω that continuously deforms γ_0 into γ_1, and let $K = H[Q]$. Since Q is closed and bounded, and H is continuous, it follows that K is closed and bounded. Since K is contained in the (open region) Ω, it is disjoint from the closed boundary of Ω. So if δ denotes the distance from $\partial\Omega$ to K, then δ is positive.

Since H is continuous on Q and Q is closed and bounded, it follows that H is uniformly continuous on Q. So we can pick a positive integer n so that

$$(6) \qquad |(t, s) - (t', s')| \leq \frac{\sqrt{2}}{n} \Rightarrow |H(t, s) - H(t', s')| < \delta.$$

Having chosen n, construct a grid over Q by subdividing the sides into n equal subintervals (Figure 4). This is a **simplicial network**, where each smaller square in this network is called a **simplicial square**. Label the points on the grid by

$$p_{jk} = \left(\frac{j}{n}, \frac{k}{n}\right), \quad j, k = 0, 1, \dots, n,$$

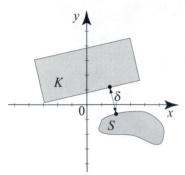

Figure 3 Distance between sets.

and let

$$z_{jk} = H(p_{jk}).$$

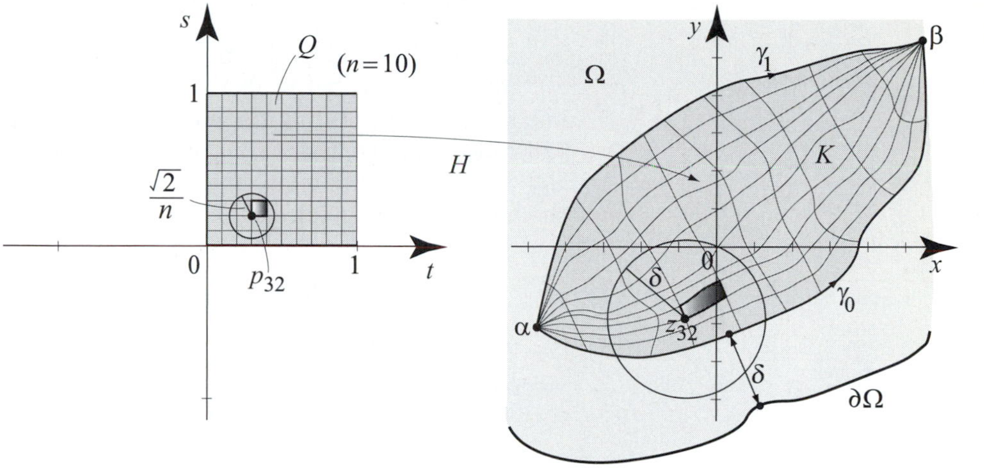

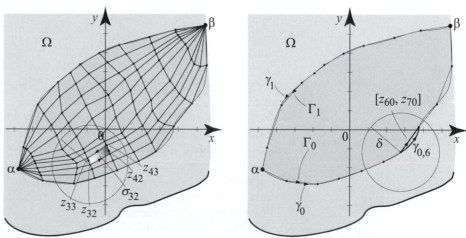

Figure 4 A simplicial network on the square Q with a sample simplicial square with one corner at p_{32}. The simplicial squares are so small that the image of each is contained in a disk in Ω of radius $\delta > 0$, where δ is the distance between the boundary of Ω, $\partial\Omega$, and the closed and bounded set $H[Q]$.

Condition (6) guarantees that the image of each simplicial square lies in the open disk $B_\delta(z_{jk})$, which in turn is contained in Ω. This is a very useful fact, since we are going to apply Cauchy's theorem in the disk $B_\delta(z_{jk})$ as we integrate f over the *closed* polygonal path

$$(7) \qquad\qquad \sigma_{jk} = [z_{jk},\, z_{j+1,k},\, z_{j+1,k+1},\, z_{j,k+1},\, z_{jk}],$$

which is contained in $B_\delta(z_{jk})$ (see Figure 5). Since f is analytic on $B_\delta(z_{jk})$, Cauchy's theorem in a disk yields

$$\int_{\sigma_{jk}} f(z)\,dz = 0.$$

In Figure 5, we show a typical closed polygonal path obtained by joining the images of the four corners of one simplicial square. By construction, these polygonal paths are contained in disks of radius δ in Ω. In Figure 6, after summing all the path integrals and canceling those that are traversed in opposite direction, the only ones that remain are those that correspond to the boundary of Q.

Figure 5

Figure 6

Adding the integrals over all the σ_{jk}'s, we note the following. The integrals over the segments that correspond to internal sides of the simplicial squares cancel, since each internal segment is traversed twice in opposite directions. Thus the only noncanceling integrals are those over line segments that correspond to external sides of the simplicial squares, that is, the boundary of Q.

For part (i) of Cauchy's theorem, the integrals on the line segments connecting z_{jk} going up the right side and down the left side of Q are each zero because the z_{jk} are fixed at α and β. For part (ii) of Cauchy's theorem, the integrals on the line segments connecting z_{jk} going up the right side and down the left side of Q will cancel because they trace the same path in opposite directions (here we have $z_{0j} = z_{nj}$). The only remaining integrals are on the line segments corresponding to the bottom and top sides of Q. So

$$(8) \qquad 0 = \sum_{j,k=0}^{n-1} \int_{\sigma_{jk}} f(z)\,dz = \int_{\Gamma_0} f(z)\,dz - \int_{\Gamma_1} f(z)\,dz,$$

where

$$\Gamma_0 = [z_{00},\, z_{10},\, \dots,\, z_{n0}], \quad \text{and} \quad \Gamma_1 = [z_{0n},\, z_{1n},\, \dots,\, z_{nn}].$$

Applying Cauchy's theorem in the disk $B_\delta(z_{j0})$, we see that

$$(9) \qquad \int_{[z_{j0},\, z_{j+1,0}]} f(z)\,dz = \int_{\gamma_{0,j}} f(z)\,dz,$$

where $\gamma_{0,j}$ is the portion of the path γ_0 that joins the points z_{j0} and $z_{j+1,0}$ (Figure 6). Adding equations (9) as j runs from 0 to $n-1$, we get

$$\int_{\Gamma_0} f(z)\,dz = \int_{\gamma_0} f(z)\,dz.$$

Similarly,

$$\int_{\Gamma_1} f(z)\,dz = \int_{\gamma_1} f(z)\,dz.$$

Comparing with (8), we find that

$$\int_{\gamma_0} f(z)\,dz = \int_{\Gamma_0} f(z)\,dz = \int_{\Gamma_1} f(z)\,dz = \int_{\gamma_1} f(z)\,dz,$$

which completes the proof of the theorem.   ∎

Appendix: Differentiation of Path Integrals

We will present a general principle for differentiation under the integral sign for functions defined by a path integral. This result was needed in the proof of Theorem 1. It will be needed also in the proof of the generalized Cauchy's formula in the following section. To simplify the proof, we have divided it into several steps, each one consisting of a result that is interesting in its own right.

Our first result is a generalization of the well-known Fubini theorem from calculus. This theorem states that if $u(x, y)$ is continuous on $[a, b] \times [c, d]$, then

$$(10) \qquad \int_a^b \int_c^d u(x, y)\,dy\,dx = \int_c^d \int_a^b u(x, y)\,dx\,dy.$$

We can easily extend this result to the case where $u(x, y)$ is not quite continuous over the rectangle $[a, b] \times [c, d]$, but the functions $x \mapsto u(x, y)$ and $y \mapsto u(x, y)$ are merely piecewise continuous in x and y. In this case we can write the iterated integrals in (10) as the sum of finitely many iterated integrals over subintervals of $[a, b]$ and $[c, d]$, such that in each term the integrand is a continuous function of (x, y) over the corresponding domain of integration. Applying (10) to each term separately and then adding the terms together, it follows that (10) holds in this more general situation. With this in mind, we can prove the following version of Fubini's theorem.

THEOREM 2
FUBINI'S THEOREM
FOR PATH
INTEGRALS

Suppose that $f(z, \zeta)$ is continuous for z on a path γ_1 and ζ on a path γ_2. Then

$$(11) \qquad \int_{\gamma_1} \int_{\gamma_2} f(z, \zeta) \, d\zeta \, dz = \int_{\gamma_2} \int_{\gamma_1} f(z, \zeta) \, dz \, d\zeta.$$

Proof   Parametrize γ_1 by $[a, b]$ and γ_2 by $[c, d]$. Then (11) is equivalent to

$$\int_a^b \int_c^d f(\gamma_1(x), \gamma_2(y)) \gamma_1'(x) \gamma_2'(y) \, dy \, dx$$

$$(12) \qquad = \int_c^d \int_a^b f(\gamma_1(x), \gamma_2(y)) \gamma_1'(x) \gamma_2'(y) \, dx \, dy.$$

The function $x \mapsto f(\gamma_1(x), \gamma_2(y)) \gamma_1'(x) \gamma_2'(y)$ is piecewise continuous in x, and the function $y \mapsto f(\gamma_1(x), \gamma_2(y)) \gamma_1'(x) \gamma_2'(y)$ is piecewise continuous in y. Applying Fubini's theorem for piecewise continuous functions, we see that (12) holds.   ■

THEOREM 3
CONTINUITY OF
PATH INTEGRALS

Suppose that $f(z, \zeta)$ is continuous for z in a region Ω and ζ on a path γ. For z in Ω, consider

$$(13) \qquad \phi(z) = \int_\gamma f(z, \zeta) \, d\zeta.$$

Then ϕ is a continuous function of z in Ω.

Proof   Fix z_0 in Ω. For small Δz, $z_0 + \Delta z$ lies in Ω. By Theorem 2, Section 3.2, we have

$$|\phi(z_0 + \Delta z) - \phi(z_0)| = \left| \int_\gamma \left(f(z_0 + \Delta z, \zeta) - f(z_0, \zeta) \right) d\zeta \right| \leq l(\gamma) M,$$

where M is the maximum of $|f(z_0 + \Delta z, \zeta) - f(z_0, \zeta)|$ for ζ on γ (see Figure 7). We will show that $M \to 0$ as $\Delta z \to 0$. Since the set of all (z, w) with $z = z_0$ and w on γ is closed and bounded, it follows that f is uniformly continuous on this set. Since the distance from $(z_0 + \Delta z, \zeta)$ to (z_0, ζ) is $|\Delta z|$, it follows from the uniform continuity of f that $M \to 0$ as $\Delta z \to 0$.   ■

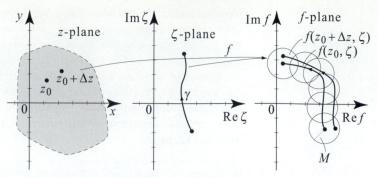

Figure 7 Picturing a continuous function f of two variables z and ζ.

We are now ready to state and prove a general result that justifies differentiation under the integral sign for path integrals.

THEOREM 4
DIFFERENTIATION OF
PATH INTEGRALS

Suppose that $f(z, \zeta)$ is a complex-valued function, defined for z in a region Ω and ζ on a path γ. If both $f(z, \zeta)$ and $\frac{d}{dz}f(z, \zeta)$ are continuous functions of (z, ζ) (hence, for fixed ζ, $f(z, \zeta)$ is analytic in z), then the function

$$(14) \qquad\qquad F(z) = \int_\gamma f(z, \zeta)\, d\zeta$$

is analytic on Ω and

$$(15) \qquad\qquad F'(z) = \int_\gamma \frac{d}{dz} f(z, \zeta)\, d\zeta.$$

Proof   Let z_0 be in Ω and Δz be so small that $z_0 + \Delta z$ is also in Ω. To compute $F'(z_0)$, we must find the limit as $\Delta z \to 0$ of

$$
\begin{aligned}
\frac{1}{\Delta z}\Big(F(z_0 + \Delta z) - F(z_0)\Big) &= \frac{1}{\Delta z}\left(\int_\gamma f(z_0 + \Delta z, \zeta)\, d\zeta - \int_\gamma f(z_0, \zeta)\, d\zeta\right) \\
(16) \qquad &= \frac{1}{\Delta z}\int_\gamma \Big(f(z_0 + \Delta z, \zeta) - f(z_0, \zeta)\Big)\, d\zeta.
\end{aligned}
$$

For fixed ζ, the function $\frac{d}{dz}f(z, \zeta)$ is continuous and has an antiderivative in Ω, namely, $f(z, \zeta)$. We thus infer from Theorem 1, Section 3.3, that

$$(17) \qquad f(z_0 + \Delta z, \zeta) - f(z_0, \zeta) = \int_{z_0}^{z_0 + \Delta z} \frac{d}{dz} f(z, \zeta)\, dz,$$

where the integral is independent of the path from z_0 to $z_0 + \Delta z$. Using (17) in (16), we obtain

$$
\begin{aligned}
\frac{1}{\Delta z}\Big(F(z_0 + \Delta z) - F(z_0)\Big) &= \frac{1}{\Delta z}\int_\gamma \int_{z_0}^{z_0 + \Delta z} \frac{d}{dz} f(z, \zeta)\, dz\, d\zeta \\
&= \frac{1}{\Delta z}\int_{z_0}^{z_0 + \Delta z} \int_\gamma \frac{d}{dz} f(z, \zeta)\, d\zeta\, dz,
\end{aligned}
$$

by Fubini's theorem for path integrals. The function $\phi(z)$ defined by the inner path integral is continuous, by Theorem 3. Hence using Lemma 1, Section 3.3,

$$F'(z_0) = \lim_{\Delta z \to 0} \frac{1}{\Delta z} \int_{z_0}^{z_0 + \Delta z} \phi(z)\, dz = \phi(z_0) = \int_\gamma \frac{d}{dz} f(z_0, \zeta)\, d\zeta,$$

which proves (15). The continuity of $F'(z)$ follows from Theorem 3 with f replaced by $\frac{d}{dz} f(z, \zeta)$. Thus, F is analytic on Ω.   ■

We end the section with a variant of Theorem 4, concerning differentiation under the integral sign for functions of two variables. The proof of this theorem parallels that of Theorem 4 and is somewhat simpler. It is outlined in the exercises.

THEOREM 5
CONTINUITY AND
DIFFERENTIATION OF
INTEGRALS

(i)   Let u be a continuous complex-valued function on the rectangle $R = [a, b] \times [c, d]$. For $c \le y \le d$, define

$$\phi(y) = \int_a^b u(x, y)\, dx.$$

Then $\phi(y)$ is continuous on $[c, d]$ (Figure 8).   (ii)   If u is real-valued and continuous on R and $\frac{\partial u}{\partial y}$ is also continuous on R, then $\phi(y)$ is differentiable on (c, d) with

$$\phi'(y) = \frac{d}{dy} \int_a^b u(x, y)\, dx = \int_a^b \frac{\partial u}{\partial y}(x, y)\, dx \qquad (c < y < d).$$

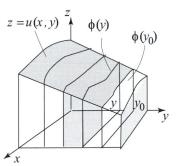

Figure 8 Visualizing the continuity of $\phi(y)$, when u is real and ≥ 0. For a given y, $\phi(y)$ is the area of the cross section above the xy-plane, under the surface $z = u(x, y)$, $a \le x \le b$. To say that ϕ is continuous at y_0 means that the area $\phi(y)$ tends to the area $\phi(y_0)$ as $y \to y_0$.

Exercises 3.5

1.   Give an example of a closed and bounded subset K of the real line and a bounded subset S of the real line such that K and S are disjoint but the distance from K to S is 0. [Hint: The set S is necessarily not closed.]

2.   Give an example of two closed and disjoint subsets K and S of the plane such that the distance from K to S is 0. [Hint: Both sets are necessarily unbounded.]

3.   Show that every convex region is a star region.

4.   Show that the distance formula (5) on $\mathbb{C} \times \mathbb{C}$ is equivalent to

$$|(z_1, w_1) - (z_2, w_2)| = \sqrt{|z_1 - z_2|^2 + |w_1 - w_2|^2}.$$

5.   In this exercise, we show that the distance formula (5) on $\mathbb{C} \times \mathbb{C}$ satisfies the three **norm** axioms. We define the norm of an element (z, w) of $\mathbb{C} \times \mathbb{C}$ by its distance from $(0, 0)$.
(a) Show that $|(z, w)| \ge 0$ with equality only if $(z, w) = (0, 0)$.
(b) Show that the norm is homogeneous with respect to scalar multiplication; that is, for all complex α, with $\alpha(z, w) = (\alpha z, \alpha w)$, we have $|\alpha(z, w)| = |\alpha||(z, w)|$.
(c) Prove the triangle inequality: $|(z_1, w_1) + (z_2, w_2)| \le |(z_1, w_1)| + |(z_2, w_2)|$.

6.   In this exercise, we prove that the boundary $\partial\Omega$ of a set Ω is closed. You are encouraged to review Section 1.2 for relevant definitions. Assume that $\partial\Omega$ is nonempty.
(a) Show that any point in $\mathbb{C}$ is either an interior point of Ω, an interior point of the complement of Ω, or a boundary point of Ω.

(b) Show that $\Omega_1 = \Omega \cup \partial\Omega$ is closed and that $\Omega_2 = (\mathbb{C} \setminus \Omega) \cup \partial\Omega$ is closed. [Hint: Take Ω_1. If z is a boundary point of Ω_1, then every disk around z contains a point w in Ω_1 and an interior point w' of $\mathbb{C} \setminus \Omega$. If w is in Ω, then we are almost done; if w is in $\partial\Omega$, then we may form a smaller disk around w and still conclude that the disk around z contains a point in Ω.]

(c) Show that the intersection of closed sets is a closed set. [Hint: Let F_j be closed sets. If z is a boundary point of $\bigcap F_j$, then every disk around z contains a point that is in each F_j. So z is either an interior or boundary point of each F_j. Since the F_j's are closed, z is in each F_j.]

(d) Show that $\partial\Omega = \Omega_1 \cap \Omega_2$, and hence $\partial\Omega$ is closed.

Project Problem: Differentiation under the integral sign. In Exercises 7–9 you are asked to establish results leading to the proof of Theorem 5.

7. Differentiation of integrals. Prove the following version of Lemma 1, Section 3.3. Suppose that $f(x)$ is a continuous real-valued function on an interval $[a, b]$, where $a < b$. For x in (a, b), we have

$$\lim_{h \to 0} \frac{1}{h} \int_x^{x+h} f(s)\, ds = f(x).$$

[Hint: Use the fundamental theorem of calculus.]

8. Continuity of integrals. Prove part (i) of Theorem 5.

9. Prove Theorem 5(ii) using Exercises 7 and 8. [Hint: Study the proof of Theorem 4.]

10. Project Problem: Cauchy's integral theorem on a disk. The following is an outline of an alternative proof of Theorem 1 in the case where Ω is a disk. Given $f = u + iv$ analytic on $B_R(z_0)$, where $z_0 = x_0 + i y_0$, we will construct an antiderivative of $F(z)$ of $f(z)$. Then the desired result will follow from Theorem 2, Section 3.3.

For $z = x + iy$ in $B_R(z_0)$ (see Figure 9), define

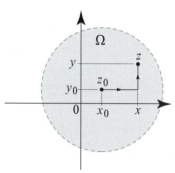

Figure 9

(18)
$$F(z) = \int_{x_0}^x f(s, y_0)\, ds + i \int_{y_0}^y f(x, t)\, dt.$$

(a) Write $F = U + iV$. Using (18), show that

$$U(x, y) = \int_{x_0}^x u(s, y_0)\, ds - \int_{y_0}^y v(x, t)\, dt$$

and

$$V(x, y) = \int_{x_0}^x v(s, y_0)\, ds + \int_{y_0}^y u(x, t)\, dt.$$

(b) With the help of Theorem 5 and the fundamental theorem of calculus, show

$$U_x = u, \qquad U_y = -v, \qquad V_x = v, \qquad V_y = u.$$

Conclude that the Cauchy-Riemann equations hold for U and V, and that U and V have continuous first partial derivatives. Hence F is analytic with derivative $F' = U_x + i V_x = u + iv = f$.

3.6   Cauchy's Integral Formula

In this section we develop one of the most important consequences of Cauchy's integral theorem. It is the Cauchy integral formula, which will enable us to compute many interesting integrals, and more importantly, we will use it to derive fundamental properties of analytic functions.

**THEOREM 1
CAUCHY'S
INTEGRAL
FORMULA**

Suppose that f is analytic inside and on a simple closed path C with positive orientation. If z is any point inside C, then

(1)
$$f(z) = \frac{1}{2\pi i} \int_C \frac{f(\zeta)}{\zeta - z} \, d\zeta.$$

Recall: The closed disk of radius R centered at z_0 is defined as $S_R(z_0) = \{z : |z - z_0| \le R\}$. This set includes its boundary.

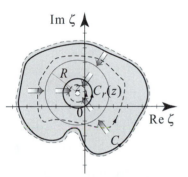

Figure 1 The path C can be continuously deformed into the circle $C_r(z)$, which explains the equality (2).

Proof   Given z inside C, let $R > 0$ be such that the closed disk $S_R(z)$ is contained in the region inside C. The function $\zeta \mapsto \frac{f(\zeta)}{\zeta - z}$ is analytic in the region inside C and outside the circle $C_r(z)$, where $0 < r \le R$ (see Figure 1). Applying Cauchy's theorem for multiply connected regions, Theorem 6, Section 3.4 (with the variable of integration being ζ and the integrand being $\frac{f(\zeta)}{\zeta - z}$), we obtain

(2)
$$\frac{1}{2\pi i} \int_C \frac{f(\zeta)}{\zeta - z} \, d\zeta = \frac{1}{2\pi i} \int_{C_r(z)} \frac{f(\zeta)}{\zeta - z} \, d\zeta,$$

where C and $C_r(z)$ are positively oriented. Since this equality holds for all $0 < r \le R$, taking limits on both sides, we get

(3)
$$\frac{1}{2\pi i} \int_C \frac{f(\zeta)}{\zeta - z} \, d\zeta = \lim_{r \to 0^+} \frac{1}{2\pi i} \int_{C_r(z)} \frac{f(\zeta)}{\zeta - z} \, d\zeta.$$

To finish off the proof, we must show that the limit is $f(z)$. Parametrize $C_r(z)$ by $\gamma(t) = z + re^{it}$, $0 \le t \le 2\pi$, $\gamma'(t) = ire^{it} dt$. Then

$$\frac{1}{2\pi i} \int_{C_r(z)} \frac{f(\zeta)}{\zeta - z} \, d\zeta = \frac{1}{2\pi i} \int_0^{2\pi} \frac{f(z + re^{it})}{re^{it}} ire^{it} \, dt = \frac{1}{2\pi} \int_0^{2\pi} f(z + re^{it}) \, dt.$$

For fixed z, the function $t \mapsto f(z + re^{it})$ is a continuous complex-valued function of t and r in $[0, 2\pi] \times [0, R]$. So by Theorem 5(i), Section 3.5, if we integrate it with respect to t, we get a continuous function of r: $\phi(r) = \frac{1}{2\pi} \int_0^{2\pi} f(z + re^{it}) \, dt$. Thus, $\lim_{r \to 0^+} \phi(r) = \phi(0) = \frac{1}{2\pi} \int_0^{2\pi} f(z) \, dt = f(z)$, as required.   ∎

Before we move to the applications of (1), let us note that if in Cauchy's formula z is any point outside C, then

(4)
$$\frac{1}{2\pi i} \int_C \frac{f(\zeta)}{\zeta - z} \, d\zeta = 0.$$

This is an immediate consequence of Cauchy's theorem for simple paths (Theorem 5, Section 3.4), since in this case the integrand $\zeta \mapsto \frac{f(\zeta)}{\zeta - z}$ is analytic on and inside C, and so its integral along C is zero. We combine (1) and (4) in one convenient formula in which the variable of integration is denoted z:

(5)
$$\frac{1}{2\pi i}\int_C \frac{f(z)}{z - z_0}\,dz = \begin{cases} f(z_0) & \text{if } z_0 \text{ is inside } C, \\ 0 & \text{if } z_0 \text{ is outside } C. \end{cases}$$

(It is traditional to change the variables of integration from ζ to z when applying Theorem 1.)

EXAMPLE 1   Cauchy's integral formula

Let $C_R(z_0)$ denote the positively oriented circle with center at z_0 and radius $R > 0$. Compute the following integrals.

(a) $\displaystyle\int_{C_2(0)} \frac{e^z}{z+1}\,dz,$ (b) $\displaystyle\int_{C_2(1)} \frac{z^2 + 3z - 1}{(z+3)(z-2)}\,dz,$

Solution   (a)   Write the integral as $\int_{C_2(0)} \frac{e^z}{z-(-1)}\,dz$. Since -1 is inside the circle $C_2(0)$, Cauchy's integral formula (5) with $f(z) = e^z$ and $z_0 = -1$ implies

$$\int_{C_2(0)} \frac{e^z}{z - (-1)}\,dz = 2\pi i e^{-1}.$$

(b)   In evaluating $\int_{C_2(1)} \frac{z^2 + 3z - 1}{(z+3)(z-2)}\,dz$, we first note that the integrand is not analytic at the points $z = -3$ and $z = 2$. Only the point $z = 2$ is inside the curve $C_2(0)$. So if we let $f(z) = \frac{z^2 + 3z - 1}{z+3}$ the integral takes the form

$$\int_{C_2(1)} \frac{f(z)}{z - 2}\,dz = 2\pi i f(2) = \frac{18\pi}{5}i,$$

by Cauchy's integral formula, applied at $z_0 = 2$.   ∎

Some integrals require multiple applications of Cauchy's formula along with applications of Cauchy's theorem. We illustrate with one example.

EXAMPLE 2   Multiple applications of Cauchy's formula

Compute

$$\int_{C_2(0)} \frac{e^{\pi z}}{z^2 + 1}\,dz$$

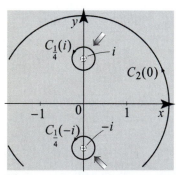

Figure 2 The integral over the outer path C is equal to the sum of the integrals over the inner non-overlapping circles. This reduction allows us to use Cauchy's formula on the individual inner circles.

Solution   Since $z^2 + 1 = (z + i)(z - i)$, the integral cannot be computed directly from Cauchy's formula, since the path contains both $\pm i$ in its interior. To overcome this difficulty, draw small nonintersecting circles inside $C_2(0)$ around $\pm i$, say $C_{1/4}(i)$ and $C_{1/4}(-i)$, as illustrated in Figure 2. Since $\frac{e^{\pi z}}{z^2+1}$ is analytic in a region containing the interior of $C_2(0)$ and the exterior of the smaller circles, by Cauchy's theorem for multiply connected regions (Theorem 6, Section 3.4), we have

$$\int_{C_2(0)} \frac{e^{\pi z}}{z^2 + 1}\,dz = \int_{C_{1/4}(i)} \frac{e^{\pi z}}{z^2 + 1}\,dz + \int_{C_{1/4}(-i)} \frac{e^{\pi z}}{z^2 + 1}\,dz.$$

Now, the two integrals on the right can be evaluated with the help of Cauchy's integral formula (5). For the first one, we apply Cauchy's formula (5) with $f(z) = \frac{e^{\pi z}}{z+i}$ and $z_0 = i$, and obtain

$$\int_{C_{1/4}(i)} \frac{e^{\pi z}}{(z-i)(z+i)} \, dz = \int_{C_{1/4}(i)} \frac{f(z)}{z-i} \, dz = 2\pi i f(i).$$

Since $f(i) = \frac{e^{i\pi}}{2i} = \frac{-1}{2i} = \frac{i}{2}$, we get

$$\int_{C_{1/4}(i)} \frac{f(z)}{z-i} \, dz = -\pi.$$

For the second integral, we have

$$\int_{C_{1/4}(-i)} \frac{e^{\pi z}}{(z-i)(z+i)} \, dz = \int_{C_{1/4}(-i)} \frac{g(z)}{z+i} \, dz = 2\pi i g(-i),$$

where $g(z) = \frac{e^{\pi z}}{z-i}$, and so $g(-i) = \frac{e^{-i\pi}}{-2i} = \frac{1}{2i} = -\frac{i}{2}$. Hence

$$\int_{C_{1/4}(-i)} \frac{g(z)}{z-i} \, dz = \pi.$$

Adding the two integrals together, we find that

$$\int_{C_2(0)} \frac{e^{\pi z}}{(z-i)(z+i)} \, dz = 0. \qquad \blacksquare$$

Two observations concerning Cauchy's formula (1) deserve mentioning. The formula shows that the values of an analytic function $f(z)$ for z inside a simple curve C can be reconstructed from the values of f *on* the curve C. By just knowing f on C, we can determine the values of f inside C.

The second observation is that (1) expresses an analytic function inside a simple path as a function defined by a path integral. Theorem 4 of the previous section tells us that, under appropriate conditions that are met in the present situation, a function defined by a path integral can be differentiated under the integral sign. Thus, under the conditions of Theorem 1, we have

$$\begin{aligned}
f'(z) &= \frac{d}{dz} f(z) = \frac{d}{dz} \frac{1}{2\pi i} \int_C \frac{f(\zeta)}{\zeta - z} \, d\zeta \\
&= \frac{1}{2\pi i} \int_C \frac{d}{dz} \frac{f(\zeta)}{\zeta - z} \, d\zeta = \frac{1}{2\pi i} \int_C \frac{f(\zeta)}{(\zeta - z)^2} \, d\zeta.
\end{aligned}$$

Note that by being able to differentiate under the integral sign, we were able to compute the derivative of $f(z)$ by computing $\frac{d}{dz} \frac{1}{\zeta - z}$, and not $\frac{d}{dz} f(z)$. More importantly, we were able to express $f'(z)$ as a function defined by a path integral, and so it too can be differentiated by differentiating under the integral sign. This yields

$$f''(z) = \frac{2}{2\pi i} \int_C \frac{f(\zeta)}{(\zeta - z)^3} \, d\zeta.$$

Clearly this process can be continued indefinitely, by appealing to Theorem 4, Section 3.5, to differentiate under the integral sign at each step. This yields the following important generalization of Cauchy's integral formula.

THEOREM 2
GENERALIZED
CAUCHY INTEGRAL
FORMULA

Suppose that f is analytic on and inside a simple closed path C with positive orientation, and let $n = 0, 1, 2, \ldots$. Then f has derivatives of all order at all points z in the region inside C given by

(6) $$f^{(n)}(z) = \frac{n!}{2\pi i} \int_C \frac{f(\zeta)}{(\zeta - z)^{n+1}} \, d\zeta.$$

(For $n = 0$, we take by definition $f^{(0)} = f$ and $0! = 1$.)

Here again we note that the integral in (6) is 0 if z is outside C.

EXAMPLE 3   Generalized Cauchy integral formula

Compute the following integrals:

(a) $\dfrac{1}{2\pi i} \displaystyle\int_{C_2(0)} \dfrac{z^{10}}{(z-1)^{11}} \, dz,$ (b) $\displaystyle\int_\gamma \dfrac{e^{iz}}{(z-\pi)^3} \, dz,$

where γ is the ellipse in Figure 3.

Solution   (a)   By (6), we have

$$\frac{10!}{2\pi i} \int_{C_2(0)} \frac{z^{10}}{(z-1)^{11}} \, dz = \frac{d^{10}}{dz^{10}} z^{10} \bigg|_{z=1} = 10!,$$

and so the desired integral is

$$\frac{1}{2\pi i} \int_{C_2(0)} \frac{z^{10}}{(z-1)^{11}} \, dz = 1.$$

(b)   By (6), we have

$$\frac{2!}{2\pi i} \int_\gamma \frac{e^{iz}}{(z-\pi)^3} \, dz = \frac{d^2}{dz^2} e^{iz} \bigg|_{z=\pi} = -e^{i\pi} = 1.$$

Hence the desired integral is πi.   ∎

In the following, you should note the orientation of the paths as we decompose a given figure-eight into two simple paths.

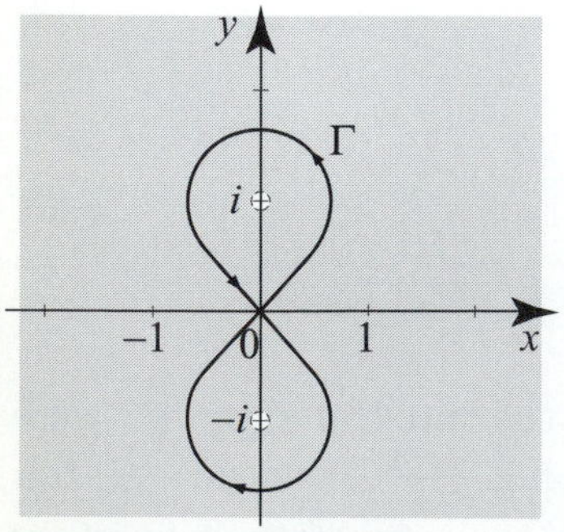

EXAMPLE 4   A path that intersects itself

Compute

$$\int_\Gamma \frac{z}{(z-i)(z^2+1)} \, dz,$$

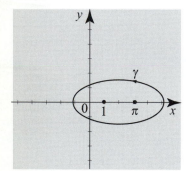

Figure 3 Ellipse in Example 3.

Figure 4 The figure-eight path Γ is not a simple path.

where Γ is the figure-eight in Figure 4.

Solution   Because the path intersects itself, it is not simple. So we cannot appeal to Cauchy's formulas directly. We will first decompose the path Γ into two simple paths Γ_1 and Γ_2, as shown in Figure 5. Noting that $(z-i)(z^2+1) = (z-i)^2(z+i)$, we have

$$\int_\Gamma \frac{z}{(z-i)(z^2+1)}\,dz = \int_{\Gamma_1} \frac{z}{(z-i)^2(z+i)}\,dz + \int_{\Gamma_2} \frac{z}{(z-i)^2(z+i)}\,dz.$$

The integrals on the right can now be evaluated with the help of Cauchy's generalized integral formula (6). We must be careful with the orientation of the paths: The orientation of Γ_1 is positive, while the orientation of Γ_2 is negative. On Γ_1, we apply (6) with $n=1$, $f(z) = \frac{z}{z+i}$ at $z=i$, and get

$$\int_{\Gamma_1} \frac{z}{z+i} \frac{dz}{(z-i)^2} = 2\pi i \left(\frac{z}{z+i}\right)'\Bigg|_{z=i} = 2\pi i \frac{i}{(z+i)^2}\Bigg|_{z=i} = \frac{\pi}{2}.$$

On Γ_2, we apply (6) with $n=0$, $f(z) = \frac{z}{(z-i)^2}$ at $z=-i$, and remembering that the orientation of Γ_2 is negative, we get

$$\int_{\Gamma_2} \frac{z}{(z-i)^2} \frac{dz}{(z+i)} = -2\pi i \frac{z}{(z-i)^2}\Bigg|_{z=-i} = \frac{\pi}{2}.$$

Adding the values of the integrals along Γ_1 and Γ_2 yields the value π for the integral along Γ. ∎

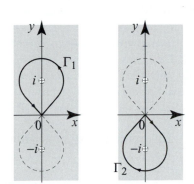

Figure 5 Decomposition of a figure-eight into two simple paths.

Cauchy's formula has many important applications to analytic functions, which in turn will be used to derive properties of harmonic functions and solutions of Dirichlet problems. The first result is already contained in Theorem 2, but it deserves a separate statement.

COROLLARY 1   Suppose that f is analytic in an open set Ω. Then f has derivatives of all order, f', f'', f''', ..., which are analytic in Ω.

Proof   We will apply Theorem 2 locally for all points in Ω. That is, given z_0 in Ω, let $S_R(z_0)$ be a closed disk contained in Ω. Pick C to be the boundary of the closed disk. By Theorem 2, all derivatives of $f(z)$ exist inside C and are given by (6). Each derivative is of course further differentiable and so it must be continuous; hence each derivative is analytic. ∎

Corollary 1 is a striking result that has no analog in the theory of functions of a real variable. Consider the function $f(x) = x^{\frac{5}{3}}$, $-\infty < x < \infty$. Its derivative, $f'(x) = x^{\frac{2}{3}}$, exists and is continuous for all x; however, $f''(x)$ does not exist at $x=0$.

Since the derivatives of an analytic function $f = u + iv$ can be expressed in terms of the partial derivatives of u and v, Corollary 1 has the following immediate consequence.

COROLLARY 2

Suppose that $f = u + iv$ is analytic in an open set Ω. Then all the partial derivatives of u and v exist and are harmonic in Ω.

Proof From Corollary 1, we know that f has analytic derivatives of all orders. Since the derivatives of f exist, we can obtain them by partial differentiation with respect to either x or y (recall the Cauchy-Riemann equations, (8), Section 2.4). This shows that the partial derivatives of u and v exist and are the real or imaginary parts of analytic functions, and hence they are harmonic. For example, differentiating $f = u + iv$ with respect to x yields $f'(z) = u_x(z) + i v_x(z)$. Since f' is analytic, we see that u_x and v_x are harmonic. Differentiating with respect to y, we obtain $f''(z) = v_{xy}(z) - iu_{xy}(z)$, showing that v_{xy} and u_{xy} are harmonic, and so forth. Since all $f^{(n)}(z)$ are continuously differentiable, any partial derivative of u or v will have continuous second-order partials; these partial derivatives commute, and the conditions for harmonicity in Section 2.5 are now clearly established. ∎

The following is a converse of sorts to Cauchy's theorem. It has substantial theoretical implications.

THEOREM 3
MORERA'S
THEOREM

Let f be a continuous complex-valued function on a region Ω. If

$$\int_\gamma f(z)\, dz = 0$$

for all closed paths γ in Ω, then f is analytic on Ω.

It suffices, in fact, to restrict γ to triangular paths lying in arbitrarily small disks in Ω (see Exercise 37).

Proof The fact that the integral of f is zero around closed paths in Ω is equivalent to the fact that f has an analytic antiderivative in Ω (Theorem 2, Section 3.3). Thus there is an analytic function F such that $F'(z) = f(z)$ for all z in Ω. But by Corollary 1, the derivatives of an analytic function are themselves analytic, and so f is analytic on Ω. ∎

The following application will be needed in the study of isolated singularities of analytic functions in the following chapter.

THEOREM 4

Suppose that f is analytic on a region Ω and let z_0 be in Ω. Define

$$(7) \qquad g(z) = \begin{cases} \frac{f(z) - f(z_0)}{z - z_0} & \text{if } z \neq z_0, \\ f'(z_0) & \text{if } z = z_0. \end{cases}$$

Then g is analytic in Ω.

Proof It is enough to establish the analyticity of g on an open disk $B_R(z_0)$ contained in Ω. For $z \neq z_0$ in $B_R(z_0)$, parametrize $[z_0, z]$ the usual way by $\zeta(t) = z_0(1 - t) + tz$, $0 \leq t \leq 1$. Using the fact that f' is analytic in $B_R(z_0)$

(hence its integral is independent of path), we can rewrite $g(z)$ as

$$g(z) = \frac{f(z) - f(z_0)}{z - z_0} = \frac{1}{z - z_0}(f(z) - f(z_0)) = \frac{1}{z - z_0}\int_{[z_0,\, z]} f'(\zeta)\, d\zeta$$

(8)
$$= \int_0^1 f'((z - z_0)t + z_0)\, dt.$$

But the right side also makes sense at $z = z_0$; it is $\int_0^1 f'(z_0)\, dt = f'(z_0)$. So our formula (8) for $g(z)$ holds for all z in $B_R(z_0)$, including z_0. Since f' is analytic in $B_R(z_0)$, its derivative f'' is also analytic in $B_R(z_0)$, and so we can differentiate under the integral sign (Theorem 4, Section 3.5) and infer that $g(z)$ is analytic on $B_R(z_0)$. ∎

As you would expect, Theorem 4 can be used to establish the analyticity of certain functions that we could not establish before.

EXAMPLE 5   Extending $\frac{\sin z}{z}$ to an entire function

Apply Theorem 4 with the function $f(z) = \sin z$ at $z_0 = 0$. Since $f(0) = 0$ and $f'(0) = \cos 0 = 1$, it follows that

$$g(z) = \begin{cases} \frac{\sin z}{z} & \text{if } z \neq 0, \\ 1 & \text{if } z = 0, \end{cases}$$

is analytic at $z = 0$. But $g(z)$ is clearly analytic at all other complex numbers, so $g(z)$ is entire. ∎

The picture with the derivatives of an analytic function will be complete with Goursat's theorem (Section 3.9), which tell us that the mere existence of $f'(z)$ implies that $f'(z)$ is continuous. So we can go back to the definition of analyticity in Section 2.3 and improve it by not requiring that $f'(z)$ be analytic. All the results that we have derived subsequently will still hold.

Further applications to the theory of analytic functions will be presented in Sections 3.7 and 3.9.

Exercises 3.6

In Exercises 1–20, evaluate the given integral. State clearly which version of Cauchy's theorem you are using and justify its application. It would help to plot the path in each case and describe exactly the points of interest in each problem. As usual, $C_R(z_0)$ denotes the positively oriented circle with center at z_0 and radius $R > 0$.

1. $\displaystyle\int_{C_1(0)} \frac{\cos z}{z}\, dz.$

2. $\displaystyle\int_{C_3(0)} \frac{e^{z^2}\cos z}{z - i}\, dz.$

3. $\displaystyle\frac{1}{2\pi i}\int_{C_2(1)} \frac{1}{z^2 - 5z + 4}\, dz.$

4. $\displaystyle\frac{1}{2\pi i}\int_{C_3(1)} \frac{\cos z}{(z - \pi)^4}\, dz.$

5. $\displaystyle\int_{C_{\frac{1}{2}}(i)} \frac{\text{Log } z}{-z + i}\, dz.$

6. $\displaystyle\frac{1}{2\pi i}\int_{C_2(1)} \frac{z^5 - 1}{(z + 3i)(z - 2)}\, dz.$

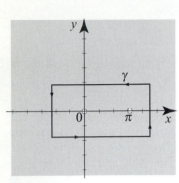

Figure 6

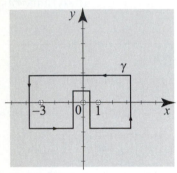

Figure 7

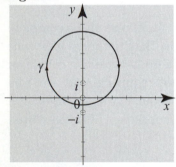

Figure 8 (negative orientation).

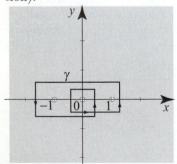

Figure 9

7. $\int_{[z_1, z_2, z_3, z_1]} \dfrac{z^{19}}{(z-1)^{19}}\, dz$, where $z_1 = 0$, $z_2 = -i$, $z_3 = 3 + i$.

8. $\int_{[z_1, z_2, z_3, z_1]} \dfrac{z^{19}}{(z-1)^{20}}\, dz$, where $z_1 = 0$, $z_2 = -i$, $z_3 = 3 + i$.

9. $\int_\gamma \dfrac{\sin z}{(z-\pi)^3}\, dz$, where γ is the positively oriented ellipse $|z-\pi|+|z+\pi| = 2\pi + 1$.

10. $\int_\gamma \dfrac{\sin z}{(z^2 - \pi^2)^2}\, dz$, where γ is the positively oriented ellipse $|z - \pi| + |z + \pi| = 2\pi + 1$.

11. $\int_\gamma \dfrac{e^z \sin z}{z^2(z - \pi)}\, dz$, where γ is as in Figure 6.

12. $\int_\gamma \dfrac{dz}{z^2(z-1)^3(z+3)}$, where γ is as in Figure 7.

13. $\int_\gamma \dfrac{z + \cos(\pi z)}{z(z^2 + 1)}\, dz$, where γ is as in Figure 8.

14. $\int_\gamma \dfrac{1}{z(z - 1)^2(z^2 - 1)}\, dz$, where γ is as in Figure 9.

15. $\int_{C_2(0)} \dfrac{z^2 + z + 1}{z^2 - 1}\, dz$.

16. $\dfrac{1}{2\pi i} \int_{C_2(1)} \dfrac{1}{z^2 - z}\, dz$.

17. $\int_{C_{\frac{3}{2}}(0)} \dfrac{1}{z^3 - 3z + 2}\, dz$.

18. $\dfrac{1}{2\pi i} \int_{C_{\frac{5}{2}}(1)} \dfrac{1}{z^3 + 2z^2 - z - 2}\, dz$.

19. $\int_{C_{\frac{3}{2}}(1)} \dfrac{1}{z^4 - 1}\, dz$.

20. $\int_{C_2(0)} \dfrac{1}{z^4 - 1}\, dz$.

21. For $|z| < 1$, let $f(z) = \dfrac{1}{2\pi} \int_0^{2\pi} \dfrac{e^{it}}{e^{it} - z}\, dt$.
(a) Show that f is analytic in the unit disk. [Hint: Theorem 4, Section 3.5.]
(b) Express the integral as a path integral and conclude that $f(z) = 1$ for all $|z| < 1$. [Hint: Let $e^{it} = \zeta$, $dt = \dfrac{d\zeta}{i\zeta}$.]

22. Compute $\dfrac{1}{2\pi} \int_0^{2\pi} \dfrac{1}{2 + e^{it}}\, dt$. (See the hint in Exercise 21.)

23. Show that $\dfrac{1}{2\pi} \int_0^{2\pi} e^{e^{int}}\, dt = 1$ for $n = \pm 1, \pm 2, \ldots$.

24. Show that $\dfrac{1}{2\pi} \int_0^{2\pi} \cos(e^{it})\, dt = 1$ and $\dfrac{1}{2\pi} \int_0^{2\pi} \sin(e^{it})\, dt = 0$.

25. Define $f(z) = \int_0^1 \cos(zt)\, dt$. Explain why f is entire and then find f.

26. Define $f(z) = \int_0^1 e^{z^2 t}\, dt$. Explain why f is entire and then find f.

27. Define the following functions at $z = 0$ in order that they become entire:
(a) $\dfrac{1 - e^z}{2z}$ and   (b) $\dfrac{\cos z - 1}{z^2}$. Justify your answers using Theorem 4.

28. (a)   Compute $\frac{1}{2\pi i} \int_{C_1(0)} \frac{e^z}{z}\, dz$.

(b)   Use your answer in (a) to show that $\int_0^\pi e^{\cos t} \cos(\sin t)\, dt = \pi$. [Hint: Parametrize $C_1(0)$ by the interval $[-\pi, \pi]$.]

29.   Suppose that f and g are analytic inside and on a simple path C. Suppose that $f = g$ on C. Show that $f = g$ inside C.

30.   Suppose that f is analytic inside and on $C_1(0)$. For $|z| < 1$, show that

$$\int_{C_1(0)} \frac{f(\zeta)}{\zeta - \frac{1}{\bar{z}}}\, d\zeta = 0.$$

[Hint: Where does $\frac{1}{\bar{z}}$ lie if $|z| < 1$?]

31.   Suppose that f is analytic inside and on $C_1(0)$. For $|z| < 1$, show that

$$\frac{1}{2\pi i} \int_{C_1(0)} \frac{f(\zeta)}{(\zeta - z)\zeta}\, d\zeta = \frac{f(z) - f(0)}{z}.$$

32. Project Problem: Approximation of the derivative.   (a)   Suppose that f is analytic on a region Ω. Let $S_R(z_0)$ be a closed disk in Ω. For z in $S_R(z_0)$, show that

$$\frac{f(z) - f(z_0)}{z - z_0} - f'(z_0) = \frac{1}{2\pi i} \int_{C_R(z_0)} \frac{(z - z_0)}{(\zeta - z)(\zeta - z_0)^2} f(\zeta)\, d\zeta.$$

(b)   Suppose that $|f(z)| \le M$ for all z in Ω. Show that, for z in $S_R(z_0)$,

$$\left| \frac{f(z) - f(z_0)}{z - z_0} - f'(z_0) \right| \le M\, \frac{|z - z_0|}{R}\, \frac{1}{R - |z - z_0|}.$$

[Hint: Use (a) and Theorem 2, Section 3.2. Note that $|\zeta - z_0| = R$ and $\frac{1}{|\zeta - z|} \le \frac{1}{R - |z - z_0|}$. To see the last inequality, draw a picture. What is the smallest value of $|\zeta - z|$ if ζ belongs to $C_R(z_0)$?]

(c)   Conclude from (b) that, for $0 < |z - z_0| < \frac{R}{2}$,

$$\left| \frac{f(z) - f(z_0)}{z - z_0} - f'(z_0) \right| \le 2M\, \frac{|z - z_0|}{R^2}.$$

33. Project Problem: Differentiation under the integral sign. Theorem 4, Section 3.5, is a very useful tool. We used it to prove Cauchy's theorem and many other results. In this exercise, we will offer a new proof based on Exercise 32. Even though this proof cannot replace the one that we presented in Section 3.5, since it is based on results that use differentiation under the integral sign, it does offer yet another justification based on Cauchy's formula.

(a)   In the notation of Theorem 4, Section 3.5, fix $S_R(z_0)$ in Ω and take M to be the maximum value of $|f(z, \zeta)|$ for z in $S_R(z_0)$ and ζ on the graph of γ. Using Exercise 32, show that for $0 < |z - z_0| < \frac{R}{2}$,

$$\left| \frac{F(z) - F(z_0)}{z - z_0} - \int_\gamma \frac{d}{dz} f(z_0, \zeta)\, d\zeta \right| \le \frac{2M}{R^2}\, |z - z_0|\, l(\gamma),$$

where $l(\gamma)$ is the length of γ.

(b)   Complete the proof of Theorem 4, Section 3.5, by letting $z \to z_0$ in (a).

34. Project Problem: Wallis's formulas.   (a)   If n is an integer, recall or prove once more the useful identity

$$\frac{1}{2\pi i} \int_{C_1(0)} \frac{1}{z^n} \, dz = \begin{cases} 1 & \text{if } n = 1, \\ 0 & \text{if } n \neq 1. \end{cases}$$

(b)   Parametrize the circle $C_1(0)$ and show that

$$\frac{1}{2\pi i} \int_{C_1(0)} \left(z + \frac{1}{z} \right)^n \frac{dz}{z} = \frac{2^n}{2\pi} \int_0^{2\pi} \cos^n t \, dt.$$

(c)   Expand $\left(z + \frac{1}{z} \right)^n$ using the binomial formula and use (a) to prove that

$$\frac{1}{2\pi} \int_0^{2\pi} \cos^{2k} t \, dt = \frac{(2k)!}{2^{2k}(k!)^2} \quad \text{and} \quad \int_0^{2\pi} \cos^{2k+1} t \, dt = 0,$$

where $k = 0, 1, 2, \ldots$. These are some of **Wallis's formulas**.

35. Show that for $k = 0, 1, 2, \ldots$,

$$\frac{1}{2\pi} \int_0^{2\pi} \sin^{2k} t \, dt = \frac{(2k)!}{2^{2k}(k!)^2} \quad \text{and} \quad \int_0^{2\pi} \sin^{2k+1} t \, dt = 0.$$

[Hint: You can use the approach of Exercise 34 or you can use the result of Exercise 34 and argue geometrically comparing areas under curves.]

36. Project Problem: Logarithms of functions.   Suppose that $f(z)$ is analytic and nonvanishing on a simply connected region Ω. By a **branch of the logarithm** of $f(z)$ we mean any continuous function $g(z)$ on Ω satisfying $e^{g(z)} = f(z)$ for all z in Ω. Such a function will be denoted by $\log f(z)$.

(a)   Prove that if g exists, then g is in fact analytic and $g'(z) = \frac{f'(z)}{f(z)}$. Thus g is an antiderivative of $\frac{f'(z)}{f(z)}$. [Hint: Theorem 4, Section 2.3.]

(b)   Show that if $g(z)$ is an antiderivative of $\frac{f'(z)}{f(z)}$, and if $e^{g(z_0)} = f(z_0)$ at a point in Ω, then in fact $g(z)$ is a branch of the logarithm of $f(z)$. [Hint: To show that $e^{g(z)} = f(z)$, let $h(z) = \frac{e^{g(z)}}{f(z)}$. Show that $h'(z) = 0$ for all z in Ω.]

(c)   Conclude that a branch of $\log f(z)$ exists on Ω and is unique up to an integer multiple of $2\pi i$. [Hint: Theorem 4, Section 3.4.]

37. A stronger Morera's theorem.   Follow the outlined steps to show that in Morera's theorem it is sufficient to restrict the path γ to triangular paths lying in arbitrarily small disks in Ω.

(a)   Argue that it is enough to show that f has an analytic antiderivative in every disk $B_R(z_0)$ contained in Ω.

(b)   For z in $B_R(z_0)$, define $F(z) = \int_{[z_0, z]} f(\zeta) \, d\zeta$. Use the fact that the integral of f over a closed triangular path is 0 to show that $\frac{F(z + \Delta z) - F(z)}{\Delta z} = \frac{1}{\Delta z} \int_{[z, z + \Delta z]} f(\zeta) \, d\zeta$.

(c)   Use Lemma 1, Section 3.3, to compute the limit in (b) as $\Delta z \to 0$, and conclude that $F'(z) = f(z)$, as desired.

38. Cauchy's formula for multiply connected regions. Use Cauchy's integral theorem for multiply connected regions (Theorem 6, Section 3.4) to obtain the following version of Cauchy's formula.

Suppose that f is analytic on a region Ω containing the region interior to the outer simple path C and exterior to the inner simple paths C_j's $(j = 1, 2, \ldots, n)$, as well as the paths themselves. Suppose that the paths all share the same orientation (Figure 10). Let z be any point interior to C and exterior to all C_j. Then

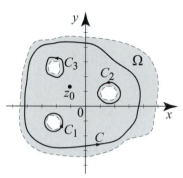

y

C_3 C_2 Ω

z_0

0

C_1 C

x

Figure 10

$$(9) \qquad f(z) = \frac{1}{2\pi i} \int_C \frac{f(\zeta)}{\zeta - z}\, d\zeta - \frac{1}{2\pi i} \sum_{j=1}^{n} \int_{C_j} \frac{f(\zeta)}{\zeta - z}\, d\zeta,$$

and for $n = 1, 2, \ldots$, we have

$$(10) \qquad f^{(n)}(z) = \frac{n!}{2\pi i} \int_C \frac{f(\zeta)}{(\zeta - z)^{n+1}}\, d\zeta - \sum_{j=1}^{n} \frac{n!}{2\pi i} \int_{C_j} \frac{f(\zeta)}{(\zeta - z)^{n+1}}\, d\zeta.$$

39. Suppose $f(z)$ is analytic in a region Ω and C is a simple closed positively oriented curve in Ω with its terminal point (initial and final) at α. Let z_0 be a point inside C.

(a) Use integration by parts to show that

$$\int_C \frac{f(z)}{(z - z_0)^{n+1}}\, dz = \left. \frac{f(z)}{-n(z - z_0)^n} \right|_{\alpha}^{\alpha} - \int_C \frac{f'(z)}{-n(z - z_0)^n}\, dz$$

$$= \frac{1}{n} \int_C \frac{f'(z)}{(z - z_0)^n}\, dz.$$

(b) Use induction and the basic Cauchy integral formula to conclude that

$$\int_C \frac{f(z)}{(z - z_0)^{n+1}}\, dz = \frac{1}{n!} \int_C \frac{f^{(n)}(z)}{z - z_0}\, dz = \frac{2\pi i}{n!} f^{(n)}(z_0).$$

40. Project Problem: Factoring zeros of analytic functions. This exercise contains useful facts about zeros of analytic functions, which will be derived again in Chapter 4 using power series. We include them at this early stage as interesting applications of Theorem 4, and we use them to derive a useful version of l'Hospital's rule. Throughout this problem, f is analytic at a point z_0.

(a)   Suppose that $f(z_0) = 0$, that is, z_0 is a **zero** of f. Use Theorem 4 to show that $f(z) = (z - z_0)g(z)$, where $g(z)$ is analytic at z_0.

(b)   Recall the **Leibniz product rule for differentiation**: If f and g are differentiable functions, then

$$\frac{d^n}{dz^n}(fg) = \sum_{k=0}^{n} \binom{n}{k} \frac{d^k f}{dz^k} \frac{d^{n-k} g}{dz^{n-k}},$$

where $\binom{n}{k} = \frac{n!}{k!(n-k)!}$ and $0! = 1$. Show that if f and g are analytic at z_0 and $f(z) = (z - z_0)^m g(z)$, then $f^{(m)}(z_0) = m! g(z_0)$. Conclude that $f^{(m)}(z_0) = 0$ if and only if $g(z_0) = 0$.

(c) Suppose that $f(z_0) = f'(z_0) = \cdots = f^{(m-1)}(z_0) = 0$. Using (a) and (b), show that $f(z) = (z - z_0)^m g(z)$, where g is analytic at z_0, and $g(z_0) = \frac{f^{(m)}(z_0)}{m!}$.

41. Generalized l'Hospital's rule. Suppose that f and g are analytic at z_0 such that $f(z_0) = f'(z_0) = \cdots = f^{(m-1)}(z_0) = 0$ and $g(z_0) = g'(z_0) = \cdots = g^{(m-1)}(z_0) = 0$, but $g^{(m)}(z_0) \neq 0$. Then $\lim\limits_{z \to z_0} \dfrac{f(z)}{g(z)} = \dfrac{f^{(m)}(z_0)}{g^{(m)}(z_0)}$.

[Hint: Use Exercise 40.]

42. Use the generalized l'Hospital's rule to compute $\lim_{z \to 0} \dfrac{e^z\left((1+e^z)z - 2e^z + 2\right)}{(e^z - 1)^3}$.

3.7   Bounds for Moduli of Analytic Functions

In this section, we will present several landmark results in the theory of analytic functions, which are consequences of Cauchy's formula. In addition to their intrinsic beauty, these results have important applications to harmonic functions (see Section 3.8) and other areas of mathematics. For example, we will present a simple proof of the fundamental theorem of algebra.

The first result states that if an analytic function f is bounded in a neighborhood of a point z_0, then the derivatives of f cannot be arbitrarily large at z_0.

THEOREM 1
CAUCHY'S
ESTIMATE

Suppose that f is analytic on an open disk $B_R(z_0)$ with center at z_0 and radius $R > 0$. Suppose that $|f(z)| \leq M$ for all z in $B_R(z_0)$. Then

(1)
$$\left|f^{(n)}(z_0)\right| \leq M \frac{n!}{R^n}, \qquad (n = 1, 2, \ldots).$$

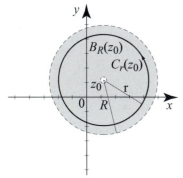

Figure 1

Proof Since we are not assuming that f is analytic on the circle $C_R(z_0)$, we will fix $0 < r < R$ and work on a disk of radius r on which f is analytic (see Figure 1). Applying the generalized Cauchy formula (6), Section 3.6, we have

(2)
$$f^{(n)}(z_0) = \frac{n!}{2\pi i} \int_{C_r(z_0)} \frac{f(\zeta)}{(\zeta - z_0)^{n+1}} \, d\zeta.$$

For ζ on $C_r(z_0)$, we have $|\zeta - z_0| = r$ and so

(3)
$$\left|\frac{f(\zeta)}{(\zeta - z_0)^{n+1}}\right| = \frac{|f(\zeta)|}{|\zeta - z_0|^{n+1}} = \frac{|f(\zeta)|}{r^{n+1}} \leq \frac{M}{r^{n+1}}.$$

Applying the integral inequality (29), Section 3.2, to the integral on the right side of (2) and using (3), we find

$$\left|f^{(n)}(z_0)\right| \leq \frac{n!}{2\pi} \frac{M}{r^{n+1}} \times (\text{length of } C_r(z_0)) = M \frac{n!}{r^n}.$$

Since this holds for all $0 < r < R$, let $r \to R$ and (1) follows. ∎

Here is a surprising application of Cauchy's estimate. Recall that f is entire if it is analytic on all $\mathbb{C}$.

THEOREM 2
LIOUVILLE'S
THEOREM

If f is entire and bounded, then f is constant.

Proof We have that $|f(z)| \leq M$ for all z. Given z_0, apply Cauchy's estimate to f' on a disk of radius $R > 0$ around z_0 and get $|f'(z_0)| \leq \frac{M}{R}$. Letting $R \to \infty$, we obtain that $f'(z_0) = 0$. Since z_0 is arbitrary, it follows that $f'(z) = 0$ for all z, and hence f is constant by Theorem 2, Section 2.4. ∎

There are several variants of Liouville's theorem that will be presented in the exercises. Here is one useful application.

COROLLARY 1

If f is entire and $|f(z)| \to \infty$ as $|z| \to \infty$, then f must have at least one zero.

Proof Suppose to the contrary that f has no zeros in $\mathbb{C}$. Then $g(z) = \frac{1}{f(z)}$ is also entire and $|g(z)| \to 0$ as $|z| \to \infty$. It is easy to see that the latter property implies that g is bounded on $\mathbb{C}$ (Exercise 14). By Liouville's theorem, g is constant and consequently f is constant. ∎

We are now ready to prove the fundamental theorem of algebra. Gauss proved this theorem in 1799. Recognizing its important role in mathematics, Gauss offered two more proofs of this result in 1816.

THEOREM 3
FUNDAMENTAL
THEOREM OF
ALGEBRA

Every polynomial of degree $n \geq 1$ has exactly n zeros counted according to multiplicity.

Proof It is enough to show that every polynomial $p(z)$ of degree $n \geq 1$ has at least one zero. For if we know that z_0 is a zero of $p(z)$, then we can write $p(z) = (z - z_0)q(z)$, where $q(z)$ is a polynomial of degree $n - 1$ (Exercise 13). We continue factoring until we have written $p(z)$ as the product of n linear terms times a constant, which shows that $p(z)$ has exactly n roots. So let us show that $p(z)$ has at least one zero. Write

$$\begin{aligned} p(z) &= a_n z^n + a_{n-1} z^{n-1} + \cdots + a_1 z + a_0 \qquad (a_n \neq 0) \\ &= z^n \left(a_n + \frac{a_{n-1}}{z} + \cdots + \frac{a_1}{z^{n-1}} + \frac{a_0}{z^n} \right). \end{aligned}$$

As $z \to \infty$, the quantity inside parentheses approaches $a_n \neq 0$, while $|z^n| \to \infty$, and hence $|p(z)| \to \infty$. Apply Corollary 1. ∎

Maximum and Minimum Principles

We now turn our attention to a different kind of application, regarding the modulus of analytic functions. Suppose that f is analytic in a region Ω and let $S_R(z)$ be a closed disk in Ω, centered at z, with radius $R > 0$. Parametrize the circle $C_R(z)$ by $\zeta(t) = z + Re^{it}$, $0 \leq t \leq 2\pi$, $d\zeta = Rie^{it}\, dt$. Cauchy's integral formula implies that

$$f(z) = \frac{1}{2\pi i} \int_{C_R(z)} \frac{f(\zeta)}{\zeta - z}\, d\zeta = \frac{1}{2\pi} \int_0^{2\pi} \frac{f(z + Re^{it})}{Re^{it}}\, Re^{it}\, dt,$$

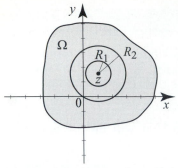

Figure 2 The mean value property of f states that the value of f at z is equal to the average value of f around any circle in Ω centered at z.

and after simplifying we obtain

(4)
$$f(z) = \frac{1}{2\pi} \int_0^{2\pi} f\left(z + Re^{it}\right) dt.$$

The integral on the right is a Riemann integral of a complex-valued function of t. Recalling that the integral is an average, this formula shows that the value of an analytic function at a point z in Ω is equal to the average value of f on any circle centered at z and contained in Ω (Figure 2). This important property is expressed by saying that an analytic function f has the **mean value property**. If we take absolute values on both sides of (4), we obtain

(5) $$|f(z)| = \frac{1}{2\pi} \left| \int_0^{2\pi} f\left(z + Re^{it}\right) dt \right| \le \frac{1}{2\pi} \int_0^{2\pi} \left| f\left(z + Re^{it}\right) \right| dt,$$

which is expressed by saying that the absolute value of an analytic function has the **sub-mean value property** on Ω. This crucial property will be needed to prove the maximum principle for the modulus of an analytic function. We need the following lemma, which states an obvious fact: If the values of a function are less than or equal to some constant M and if the average of the function is equal to M, then the function must be identically equal to M.

LEMMA 1

(i)   Suppose that $h(t)$ is a continuous real-valued function such that $h(t) \ge 0$ for all t in $[a, b]$ $(a < b)$. If $\int_a^b h(t)\, dt = 0$, then $h(t) = 0$ for all t in $[a, b]$.
(ii)   Suppose that $h(t)$ is a continuous real-valued function such that $h(t) \le M$ (alternatively, $h(t) \ge M$) for all t in $[a, b]$. If $\frac{1}{b-a} \int_a^b h(t)\, dt = M$, then $h(t) = M$ for all t in $[a, b]$.

Proof   We prove (ii), since (i) follows from the alternative case in (ii) with $M = 0$. Suppose that $h(t) \le M$. For x in $[a, b]$, define

$$f(x) = \frac{1}{b-a} \int_a^x (M - h(t))\, dt.$$

Then $f(a) = 0$, $f(b) = \frac{1}{b-a} \int_a^b (M - h(t))\, dt = M - \frac{1}{b-a} \int_a^b h(t)\, dt$, and by the fundamental theorem of calculus, $f'(x) = \frac{1}{b-a}(M - h(x)) \ge 0$. Hence, $f(x)$ is increasing, and so for all x in $[a, b]$, $0 = f(a) \le f(x) \le f(b) = M - \frac{1}{b-a} \int_a^b h(t)\, dt$. If $M = \frac{1}{b-a} \int_a^b h(t)\, dt$, then $f(b) = fa) = 0$ and hence $f(x) = 0$ for all x in $[a, b]$, implying that $f'(x) = 0$, and hence $M = h(x)$ for all x in $[a, b]$. This proves (ii) in the case $h(t) \le M$. The case $h(t) \ge M$ is done similarly.   ∎

In the proof, we will also need a result that states that if f is analytic on a region Ω and $|f|$ is constant on Ω, then f must be constant on Ω. This result is stated as Exercise 41 in Section 2.2; also as Exercises 19–22,

Section 2.5. For the sake of completeness, we will sketch a proof, based on the following observation: If u and u^2 are harmonic, then u must be constant (Exercise 19, Section 2.5). Write $f = u + iv$. It suffices to show that u is constant. If $|f| = C$, then $|f|^2 = u^2 + v^2 = C^2$. In particular, $u^2 + v^2$ is harmonic. But f is analytic implies that $f^2 = (u^2 - v^2) + 2iuv$ is analytic; in particular, $u^2 - v^2$ is harmonic. Adding two harmonic functions, we obtain that $2u^2$ is harmonic, and since u is harmonic, it follows that u and u^2 are harmonic and hence u is constant, as desired.

THEOREM 4
THE MAXIMUM
MODULUS
PRINCIPLE

> Suppose that f is analytic on a region Ω. If $|f|$ attains a maximum in Ω, then f is constant in Ω.

Proof   The connectedness property of Ω is crucial in the proof. Suppose that $|f|$ attains a maximum in Ω. We will show that $|f|$ is constant. Let $M = \max_{z \in \Omega} |f(z)|$, $\Omega_0 = \{z \in \Omega : |f(z)| < M\}$, and $\Omega_1 = \{z \in \Omega : |f(z)| = M\}$. Clearly, $\Omega = \Omega_0 \cup \Omega_1$, and Ω_0 and Ω_1 are disjoint, and Ω_1 is nonempty because $|f|$ is assumed to attain its maximum in Ω. The set Ω_0 is open because $|f|$ is continuous (Exercise 41, Section 2.2). Our goal is to show that Ω_1 is also open. Then, because Ω is open and connected, it cannot be written as the union of two open, disjoint, nonempty sets. This will force Ω_0 to be empty. Consequently, $\Omega = \Omega_1$, implying that $|f| = M$ is constant in Ω.

So let us prove that Ω_1 is open. Pick z in Ω_1. Since Ω is open, we can find an open disk $B_\delta(z)$ in Ω, centered at z with radius $\delta > 0$. Pick a smaller disk $B_r(z)$, $0 < r < \delta$ (Figure 3); we will show that $B_r(z)$ is actually contained in Ω_1. This will imply that Ω_1 is open. Using (5) and the fact that $|f(z)| = M$, we obtain

$$M = |f(z)| \le \frac{1}{2\pi} \int_0^{2\pi} \overbrace{\left| f(z + re^{it}) \right|}^{\le M} dt \le \frac{1}{2\pi} \int_0^{2\pi} M\, dt = M.$$

Hence $\frac{1}{2\pi} \int_0^{2\pi} \left| f(z + re^{it}) \right| dt = M$, and Lemma 1(ii) implies that $\left| f(z + re^{it}) \right| = M$ for all t in $[0, 2\pi]$. This shows that the circle of radius r and center at z is contained in Ω_1, completing the proof.   ∎

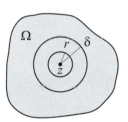

Figure 3

Suppose that f is analytic inside Ω and continuous on the boundary of Ω. By Theorem 4, $|f|$ cannot attain its maximum inside Ω unless f is constant. This leads us to the following two questions.

- Does $|f|$ attain its maximum on the boundary of Ω?

- If $|f(z)| \le M$ on the boundary of Ω, can we infer that $|f(z)| \le M$ for all z in Ω?

The following example shows that in general the answers to both questions are negative.

EXAMPLE 1   Failure of the maximum principle on unbounded regions

Let $\Omega = \{z : \operatorname{Re} z > 0, \text{ and } \operatorname{Im} z > 0\}$ be the first quadrant, bounded by the semi-infinite nonnegative x and y-axes. Let $f(z) = e^{-iz^2} = e^{-i(x^2 - y^2 + 2ixy)} =$

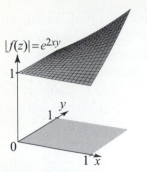

|f(z)| = e^{2xy}

Figure 4 The maximum of $|f|$ does not occur on the boundary of the first quadrant.

$e^{-i(x^2-y^2)}e^{2xy}$. We have

$$|f(z)| = \left| e^{i(x^2-y^2)}e^{2xy} \right| = e^{2xy}.$$

On the boundary, we have $x = 0$ or $y = 0$ and so $|f(z)| = 1$; however, it is clear that $|f(z)| = e^{2xy}$ is not bounded in Ω. To see this, take $x = y$ and let $x, y \to \infty$; then $|f(x + i\,y)| = e^{2x^2} \to \infty$ (see Figure 4). ∎

Example 1 shows that the modulus of an analytic function need not attain its maximum on the boundary, and the maximum value of the modulus on the boundary may not be the maximum value inside the region. As we now show, the situation is different if Ω is bounded. In this case, the answers to both previous questions are affirmative.

COROLLARY 2 (MAXIMUM MODULUS PRINCIPLE)

Suppose that Ω is a bounded region and f is analytic on Ω and continuous on the boundary of Ω. Then
(i) $|f|$ attains its maximum M on the boundary of Ω, and
(ii) either f is constant or $|f| < M$ for all z in Ω.

Proof If f is constant then all statements hold, so suppose f is not constant. The set consisting of Ω and its boundary is closed and bounded. Since $|f|$ is continuous on this set, it attains its maximum M on this set. But Theorem 4 says $|f|$ cannot attain its maximum in Ω, so $|f|$ attains its maximum on the boundary of Ω and $|f| < M$ in Ω. ∎

The modulus of a nonconstant analytic function can attain its minimum on a region Ω. Consider, for example, $f(z) = z$ on the open disk $|z| < 1$. Then the minimum of $|f(z)|$ is 0 and it is attained at $z = 0$. However, if the function is never zero in Ω, then we have the following useful principle.

THEOREM 5 THE MINIMUM MODULUS PRINCIPLE

Suppose that f is nonvanishing and analytic on a region Ω. If $|f|$ attains a minimum in Ω, then f is constant in Ω.

Proof Apply the maximum modulus principle to $g = \frac{1}{f}$. ∎

Combining the previous two results, we obtain the following principle.

COROLLARY 3 (MAXIMUM-MINIMUM MODULUS PRINCIPLE)

Suppose that Ω is a bounded region and f is analytic and nonvanishing on Ω and continuous on the boundary of Ω. Then
(i) $|f|$ attains a maximum M and minimum m on the boundary of Ω, and
(ii) either f is constant or $m < |f| < M$ for all z in Ω.

Proof We follow the proof of Corollary 2; take the case when f is not constant. Since the set consisting of Ω and its boundary is closed and bounded, the continuous function $|f|$ attains its minimum m and maximum M on this set. From Theorems 4 and 5 we know that it cannot attain its minimum or maximum on Ω; hence it attains each on the boundary of Ω and $m < |f| < M$ in Ω. ∎

EXAMPLE 2   Maximum and minimum values

Let $f(z) = \frac{e^z}{z}$, where z ranges over the annulus $\frac{1}{2} \le |z| \le 1$. Find the points where the maximum and minimum values of $|f(z)|$ occur and determine these values.

Solution   By Corollary 3, we must look for these values on the boundary of the annular region: $|z| = R$, where $R = \frac{1}{2}$ and $R = 1$. It will be convenient to use polar coordinates, $z = R(\cos\theta + i\sin\theta)$, where $0 \le \theta < 2\pi$. For z on the boundary, we have

$$|f(z)| = \left| \frac{e^{R(\cos\theta + i\sin\theta)}}{R(\cos\theta + i\sin\theta)} \right| = \frac{\left| e^{R\cos\theta} e^{iR\sin\theta} \right|}{R} = \frac{e^{R\cos\theta}}{R}.$$

It is clear that the maximum will occur when $\cos\theta = 1$ (or $\theta = 0$) and the minimum will occur when $\cos\theta = -1$ (or $\theta = \pi$). When $\theta = 0$, we must check both the inner and outer boundary (see Figure 5). For $\theta = 0$ and $R = \frac{1}{2}$, we have

$$|f(1/2)| = \frac{e^{1/2}}{1/2} = 2\sqrt{e} \approx 3.3.$$

For $\theta = 0$ and $R = 1$, we have

$$|f(1)| = \frac{e}{1} = e \approx 2.7.$$

Thus the maximum value of $|f(z)|$ is $2\sqrt{e} \approx 3.3$ and it occurs at $z = \frac{1}{2}$. We now look for the minimum value on the boundary. Again we have two possibilities. If $\theta = \pi$ and $R = \frac{1}{2}$, then

$$|f(-1/2)| = \frac{e^{-1/2}}{1/2} = \frac{2}{\sqrt{e}} \approx 1.2$$

For $\theta = \pi$ and $R = 1$, we have

$$|f(-1)| = \frac{e^{-1}}{1} = \frac{1}{e} \approx 0.4.$$

Thus the minimum value of $|f(z)|$ is $\frac{1}{e} \approx 0.4$ and it occurs at $z = -1$. For reference's sake, function $|f(z)|$ is plotted in Figure 6.   ∎

In Example 2, the maximum value of the modulus was attained at precisely one point on the boundary, and similarly for the minimum value. This will not always be the case. Consider the function $f(z) = \frac{1}{z}$ for $1 \le |z| \le 2$. The maximum value of $|f(z)|$ is attained at all points on the inner circle $C_1(0)$, and the minimum value is attained at all points of the outer circle $C_2(0)$.

Our next example deals with mappings of the form

$$(6) \qquad \phi_\alpha(z) = \frac{z - \alpha}{1 - \overline{\alpha}z},$$

where α is a complex number such that $|\alpha| < 1$. These are special cases of linear fractional transformations that we introduced in Section 1.4.

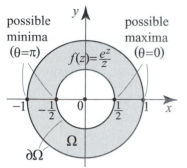

possible minima ($\theta = \pi$)   possible maxima ($\theta = 0$)

$f(z) = \frac{e^z}{z}$

Ω

$\partial\Omega$

Figure 5

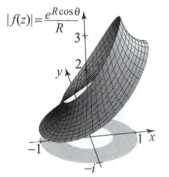

$|f(z)| = \frac{e^{R\cos\theta}}{R}$

Figure 6 Graph of $|f|$ illustrating the minimum value at $z = -1$ and the maximum value at $z = \frac{1}{2}$.

EXAMPLE 3   Linear fractional transformations

(a)   Show that $|\phi_\alpha(e^{i\theta})| = 1$ for all θ. Conclude that $|\phi_\alpha(z)| \leq 1$ for $|z| \leq 1$. That is, $\phi_\alpha(z)$ maps the open unit disk D into itself.
(b)   Check that for any $|\alpha| < 1$, the inverse of ϕ_α is $\phi_{-\alpha}$. That is, check that $\phi_\alpha(\phi_{-\alpha}(z)) = \phi_{-\alpha}(\phi_\alpha(z)) = z$.
(c)   Conclude that ϕ_α is a one-to-one analytic mapping of D onto itself, and that ϕ_α maps the unit circle onto the unit circle. (Up to a unimodular constant multiple, the ϕ_α's are the only analytic one-to-one mapping of D onto itself. See Section 6.2.)

Solution   Note first that ϕ_α is analytic everywhere except where $1 - \overline{\alpha}z = 0$ or $z = \frac{1}{\overline{\alpha}}$. Since $|\alpha| < 1$ it follows that $\left|\frac{1}{\overline{\alpha}}\right| = \frac{1}{|\alpha|} > 1$. Consequently, ϕ_α is analytic inside and on the unit circle $C_1(0)$. To prove (a), let $z = e^{i\theta}$. We have

$$|\phi_\alpha(e^{i\theta})| = \frac{|e^{i\theta} - \alpha|}{|1 - \overline{\alpha}e^{i\theta}|} = \frac{|e^{i\theta} - \alpha|}{|e^{i\theta}(e^{-i\theta} - \overline{\alpha})|} = \frac{|e^{i\theta} - \alpha|}{|e^{-i\theta} - \overline{\alpha}|} = \frac{|e^{i\theta} - \alpha|}{|\overline{e^{i\theta} - \alpha}|} = 1,$$

because the modulus of a complex number $|e^{i\theta} - \alpha|$ is equal to the modulus of its conjugate $|\overline{e^{i\theta} - \alpha}|$. This proves that $|\phi_\alpha(z)| = 1$ for all $|z| = 1$, and from Corollary 2 it follows that $|\phi_\alpha(z)| \leq 1$ for all $|z| \leq 1$. That is ϕ_α maps D into D. Part (b) is straightforward. We have

$$\phi_\alpha(\phi_{-\alpha}(z)) = \frac{\phi_{-\alpha}(z) - \alpha}{1 - \overline{\alpha}\phi_{-\alpha}(z)} = \frac{\frac{z+\alpha}{1+\overline{\alpha}z} - \alpha}{1 - \overline{\alpha}\frac{z+\alpha}{1+\overline{\alpha}z}} = \frac{z(1 - |\alpha|^2)}{1 - |\alpha|^2} = z.$$

Since this is true for all $|\alpha| < 1$, replacing α by $-\alpha$, we get that $\phi_{-\alpha}(\phi_\alpha(z)) = z$, which proves (b). Part (c) is immediate from (a) and (b). To prove the onto part, let w be any point in D, then by (a) (applied to $\phi_{-\alpha}$) we have that $\phi_{-\alpha}(w)$ is in D, and by (b), we find $\phi_\alpha(\phi_{-\alpha}(w)) = w$. Thus ϕ_α maps D onto D. To prove that ϕ_α maps the unit circle onto the unit circle, we argue similarly. To prove the one-to-one part, suppose that $\phi_\alpha(z_1) = \phi_\alpha(z_2)$. Applying $\phi_{-\alpha}$ to both sides, we get $z_1 = z_2$, thus ϕ_α is one-to-one.   ∎

Let us compare the results of Example 3 with some of the maximum-minimum principles of this section. Clearly, ϕ_α is not constant, but we just showed in Example 3(a) that $|\phi_\alpha(z)|$ is constant on $C_1(0)$. By studying Corollary 3, we conclude that $\phi_\alpha(z)$ must vanish somewhere inside the unit disk. Looking back at (6), we see that $\phi_\alpha(z) = 0 \Leftrightarrow z = \alpha$. Thus $z = \alpha$ is the only zero of ϕ_α and since $|\alpha| < 1$, this zero is inside the unit disk, as expected. Linear fractional transformations have many interesting properties that will be investigated in Section 6.2.

We conclude this section with another well-known consequence of the maximum principle.

LEMMA 2	Suppose that f is analytic in the open unit disk $B_1(0)$ with $f(0) = 0$ and $|f(z)| \leq 1$ for all z in $B_1(0)$. Then
SCHWARZ'S LEMMA	

$$(7) \qquad |f'(0)| \leq 1 \qquad \text{and} \qquad |f(z)| \leq |z| \text{ for all } z \text{ in } B_1(0).$$

Moreover, if $f'(0) = 1$ or $|f(z_0)| = |z_0|$ for some z_0 in $B_1(0)$, then $f(z) = Az$ for all z in $B_1(z)$, where A is a constant with $|A| = 1$.

Proof   Consider the function

$$(8) \qquad g(z) = \begin{cases} \frac{f(z)}{z} & \text{if } z \neq 0, \\ f'(0) & \text{if } z = 0. \end{cases}$$

Since $f(z) = 0$, it follows from Theorem 4, Section 3.6, that g is analytic in $B_1(0)$. Fix $0 < R < 1$. Because $|f(z)| \leq 1$, it follows that $|g(z)| \leq \frac{1}{R}$ for all $|z| = R$. Since g is analytic in the open disk $|z| < R$ and continuous on its boundary, Corollary 2 implies that $|g(z)| \leq \frac{1}{R}$ for all $|z| \leq R$. Letting $R \to 1$, we see that $|g(z)| \leq 1$ for all $|z| < 1$, which proves (7). If $|f'(0)| = 1$ or $|f(z_0)| = |z_0|$ for some z_0 in $B_1(0)$, this means that $|g(0)| = 1$ or $|g(z_0)| = 1$. In either case, this implies that $|g(z)|$ attains its maximum inside the disk $B_1(0)$, and hence $g(z) = A$ is constant on $B_1(0)$ by Theorem 4. Clearly, $|A| = 1$ and the lemma follows. ∎

Exercises 3.7

1.   Find the maximum and minimum values of the modulus of $f(z) = z$, where $|z| \leq 1$, and determine the points where these values occur. Explain your answers in view of Corollary 3.

2.   Consider $f(z) = 2z + 3$, where z is in the closed square area with vertices at $1 \pm i$ and $-1 \pm i$. Find the maximum and minimum values of $|f(z)|$ and determine where these values occur.

3.   Consider $f(z) = e^{-z^2}$, where $1 \leq |z| \leq 2$. Find the maximum and minimum values of $|f(z)|$ and determine where these values occur.

4.   Consider the rectangle R with vertices at 0, π, i, and $\pi + i$. For z in R, let $f(z) = \frac{\sin z}{z}$ if $z \neq 0$, and $f(0) = 1$. Find the maximum and minimum values of $|f(z)|$ and determine where these values occur.

5.   Consider $f(z) = \frac{z}{z^2+2}$, where $2 \leq |z| \leq 3$. Find the maximum and minimum values of $|f(z)|$ and determine where these values occur.

6.   Consider $f(z) = \frac{3z}{1-z^2}$, where $|z| \leq \frac{1}{2}$. Find the maximum and minimum values of $|f(z)|$ and determine where these values occur.

7.   Consider $f(z) = \frac{2z-1}{-z+2}$, where $|z| \leq 1$. Find the maximum and minimum values of $|f(z)|$ and determine where these values occur. [Hint: Example 3.]

8.   Consider $f(z) = \frac{2z-i}{iz+2}$, where $|z| \leq 1$. Find the maximum and minimum values of $|f(z)|$ and determine where these values occur. [Hint: Example 3.]

9.   Consider $f(z) = \text{Log } z$, where $1 \leq |z| \leq 2$ and $0 \leq \text{Arg } z \leq \frac{\pi}{4}$. Find the maximum and minimum values of $|f(z)|$ and determine where these values occur.

10.   Consider $f(z) = \text{Log } z$, where $\frac{1}{2} \leq |z| \leq 2$ and $-\frac{\pi}{4} \leq \text{Arg } z \leq \frac{\pi}{4}$. Find the maximum and minimum values of $|f(z)|$ and determine where these values occur.

11. Consider $f(z) = e^{e^z}$, where z belongs to the infinite horizontal strip $-\frac{\pi}{2} \leq$ Im $z \leq \frac{\pi}{2}$. Show that $|f(z)| = 1$ for all z on the boundary, Im $z = \pm\frac{\pi}{2}$. Is $f(z)$ bounded inside the region? Does this contradict Theorem 4? Explain.

12. Suppose that $p(z) = a_n z^n + a_{n-1} z^{n-1} + \cdots + a_1 z + a_0$ $(a_n \neq 0)$ is a polynomial of degree $n \geq 1$. (a) Show that $a_j = \frac{p^{(j)}(0)}{j!}$. (b) Suppose that $|p(z)| \leq M$ for all $|z| \leq R$. Show that $|a_j| \leq \frac{M}{R^j}$ for $j = 0, 1, \ldots, n$. Can we just assume that $|p(z)| \leq M$ for all $|z| < R$ and still get that $|a_j| \leq \frac{M}{R^j}$?

13. Factoring roots. (a) Verify the algebraic identity for complex numbers z and w and positive integers $n \geq 2$,

$$z^n - w^n = (z - w)(z^{n-1} + z^{n-2}w + z^{n-3}w^2 + \cdots + z w^{n-2} + w^{n-1}).$$

(b) Show that if $p = p_n z^n + p_{n-1} z^{n-1} + \cdots + p_1 z + p_0$ is a polynomial of degree $n \geq 2$, and if $p(z_0) = 0$, then

$$
\begin{aligned}
p(z) = p(z) - p(z_0) &= p_n(z^n - z_0^n) + p_{n-1}(z^{n-1} - z_0^{n-1}) + \cdots + p_1(z - z_0) \\
&= (z - z_0)q(z),
\end{aligned}
$$

where $q(z)$ is a polynomial of degree $n - 1$.

14. Suppose that f is continuous on $\mathbb{C}$ and $\lim_{z \to \infty} |f(z)| = c$ exists and is finite. Show that f is bounded. [Hint: Make the following argument rigorous. For large values of $|z|$, say $|z| > M$, $|f(z)|$ is near c and so it is bounded. For $|z| \leq M$, $|f(z)|$ is bounded because it is a continuous function on a closed and bounded set.]

15. Suppose that f is entire and $\lim_{z \to \infty} f(z) = 0$. Show that f is identically 0.

16. (a) Suppose that f is entire and $f'(z)$ is bounded in $\mathbb{C}$. Show that $f(z) = az + b$.
(b) More generally, show that if f is entire and $f^{(n)}$ is bounded, then f is a polynomial of degree n.

17. Suppose that f is entire and that it omits an open nonempty set. Hence there is an open disk $B_R(w_0)$ with $R > 0$ in the w-plane such that $f(z)$ is not in $B_R(w_0)$ for all z. Show that f is constant. [Hint: Consider $g(z) = \frac{1}{f(z)-w_0}$ and show that you can apply Liouville's theorem.] (Indeed, a deep result in complex analysis known as Picard's great theorem asserts that an entire nonconstant function can omit at most one value. Results of this nature will be presented in the next chapter.)

18. Suppose that f is entire. Show that if either Re f or Im f are bounded, then f is constant. [Hint: Use Exercise 17.]

19. Suppose that f is entire and $\lim_{z \to \infty} \frac{f(z)}{z} = 0$. Show that f is constant. [Hint: Use Cauchy's generalized integral formula to show that $f'(z) = 0$. Alternatively, show that $\frac{f(z)-f(0)}{z}$ is entire (use Theorem 4, Section 3.6) and tends to 0 as $z \to \infty$. Then use Exercise 15.]

20. Suppose that f is entire and $\lim_{z \to \infty} \frac{f(z)}{z} = c$, where c is a constant. Show that $f(z) = cz + b$. [Hint: Consider $g(z) = \frac{f(z)-f(0)}{z}$. Show that g is entire and bounded. You will need Theorem 4, Section 3.6.]

21.   A function $f(z) = f(x + iy)$ is called **periodic in x and y** if there are real numbers $T_1 > 0$ and $T_2 > 0$ such that $f((x + T_1) + i(y + T_2)) = f(x + iy)$ for all $z = x + iy$. Show that if f is entire and periodic in x and y, then f is constant. Can an entire function f be periodic in one of x or y without being constant?

22. Knowing that $f(z) = e^z$ has no zeros, what conclusion do you draw from Corollary 1 concerning this function?

23. (a) Suppose that f is analytic in a bounded region Ω and continuous on the boundary of Ω. Suppose that $|f(z)|$ is constant on the boundary of Ω. Show that either f has a zero in Ω or f is constant in Ω.
(b) Find all analytic functions f on the unit disk such that $|z| < |f(z)|$ for all $|z| < 1$ and $|f(z)| = 1$ for all $|z| = 1$. Justify your answer.

24.   Suppose that $f(z)$ and $g(z)$ are analytic on the open unit disk $|z| < 1$ and continuous on the circle $|z| = 1$. Suppose that $|f(z)| = |g(z)|$ for all $|z| = 1$ and $f(z) \neq 0$ for all $|z| = 1$. Show that $f(z) = A g(z)$ for all $|z| \leq 1$, where A is a constant such that $|A| = 1$. [Hint: Consider $h(z) = \frac{g(z)}{f(z)}$.]

25.   Suppose that f is analytic on $|z| < 1$ and continuous on $|z| \leq 1$. Suppose that $f(z)$ is real-valued for all $|z| = 1$. Show that f is constant for all $|z| \leq 1$. [Hint: Consider $g(z) = e^{if(z)}$.]

26.   Suppose that f and g are analytic in a bounded region Ω and continuous on the boundary of Ω. Suppose that g does not vanish in Ω and $|f(z)| \leq |g(z)|$ for all z on the boundary of Ω. Show that $|f(z)| \leq |g(z)|$ for all z in Ω.

27.   Find all analytic functions f such that $|f(z)| \leq 1$ for all $|z| \leq 1$, $f(0) = 0$, and $f\left(\frac{i}{2}\right) = \frac{1}{2}$.

28.   Suppose that f is analytic, $|f(z)| \leq 3$ for all $|z| = 1$, and $f(0) = 0$. Can $|f'(0)| > 3$?

29.   Suppose that f is analytic with $|f(z)| \leq 1$ for all $|z| \leq 1$. Show that for all $|z| < 1$, we have $|f^{(n)}(z)| \leq \frac{n!}{(1-|z|)^n}$. [Hint: Apply Cauchy's estimate to $f^{(n)}(z)$. Consider the radius of the disk that is contained in $B_1(0)$ with center at z.]

3.8   Applications to Harmonic Functions

This section continues our study of harmonic functions that we started in Section 2.5. We will answer some fundamental questions about harmonic functions that were raised in Section 2.5 and derive new properties that will strengthen our understanding of harmonic functions and solutions of Dirichlet problems.

Recall that a real-valued function $u(x, y)$ defined in a region Ω is called harmonic if u has continuous partial derivatives of first and second order and if u satisfies Laplace's equation

(1) $$\Delta u = \frac{\partial^2 u}{\partial x^2} + \frac{\partial^2 u}{\partial y^2} = 0 \quad \text{on } \Omega.$$

In fact, harmonic functions have derivatives of all order. To prove this result, we need a lemma.

LEMMA 1
CONJUGATE
GRADIENT

Suppose that u is harmonic on a region Ω. Let $\phi = u_x - i\,u_y$. Then ϕ is analytic on Ω. The function ϕ is called the **conjugate gradient of** u.

Proof Write $\phi = \mathrm{Re}\,(\phi) + i\,\mathrm{Im}\,(\phi) = u_x - i\,u_y$. Because u has continuous second partial derivatives, it follows that $\mathrm{Re}\,\phi$ and $\mathrm{Im}\,\phi$ have continuous first partial derivatives. To show that ϕ is analytic, it suffices by Theorem 1, Section 2.4, to show that $\mathrm{Re}\,\phi$ and $\mathrm{Im}\,\phi$ satisfy the Cauchy-Riemann equations. We have

$$\frac{\partial}{\partial x}\,\mathrm{Re}\,\phi = \frac{\partial}{\partial x}u_x = u_{xx} \quad \text{and} \quad \frac{\partial}{\partial y}\,\mathrm{Im}\,\phi = \frac{\partial}{\partial y}(-u_y) = -u_{yy}.$$

But since u is harmonic, $u_{xx} = -u_{yy}$, and so $\frac{\partial}{\partial x}\,\mathrm{Re}\,\phi = \frac{\partial}{\partial y}\,\mathrm{Im}\,\phi$. Thus, the first of the Cauchy-Riemann equations is satisfied. Now, since u has continuous second partial derivatives, we have $u_{xy} = u_{yx}$. Thus

$$\frac{\partial}{\partial y}\,\mathrm{Re}\,\phi = u_{xy} \quad \text{and} \quad \frac{\partial}{\partial x}\,\mathrm{Im}\,\phi = -u_{yx} = -u_{xy}.$$

So $\frac{\partial}{\partial y}\,\mathrm{Re}\,\phi = -\frac{\partial}{\partial x}\,\mathrm{Im}\,\phi$ and the second of the Cauchy-Riemann equations holds. Thus ϕ is analytic. ∎

EXAMPLE 1   Conjugate gradient

Consider $u(x,\,y) = \frac{x}{x^2+y^2}$ in the upper half-plane, $y > 0$.
(a) Show that u is harmonic.   (b) Find the conjugate gradient of u.

Solution   (a) The function u has a simpler expression in polar coordinates (see Figure 1):

$$u(r,\,\theta) = \frac{r\,\cos\theta}{r^2} = \frac{\cos\theta}{r} = r^{-1}\cos\theta.$$

From this, it is easy to see that u is the real part of the function

$$f(z) = \frac{1}{z} = z^{-1} = r^{-1}e^{-i\theta} = r^{-1}\big(\cos\theta - i\,\sin\theta\big).$$

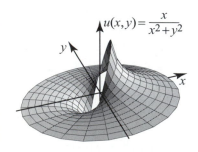

Figure 1

Since f is analytic for all $z \neq 0$, it follows that its real part $u = r^{-1}\cos\theta$ is harmonic for all $z \neq 0$ by Theorem 1, Section 2.5. In particular, u is harmonic in the upper half-plane.
(b) We have

$$u_x = \frac{y^2 - x^2}{(x^2 + y^2)^2} \quad \text{and} \quad u_y = \frac{-2xy}{(x^2 + y^2)^2}.$$

Thus the conjugate gradient in the upper half-plane is

$$\phi = u_x - i\,u_y = \frac{(y^2 - x^2) + 2ixy}{(x^2 + y^2)^2} = -\frac{(x - i\,y)^2}{(x^2 + y^2)^2} = -\frac{(\overline{z})^2}{(z\,\overline{z})^2} = -\frac{1}{z^2}. \quad ∎$$

Suppose that f is analytic and write $f = u + i\,v$. Using the Cauchy-Riemann equations, it follows that $f'(z) = u_x - iu_y$ (see (8), Section 2.4). Thus the derivative of f is the conjugate gradient of u; equivalently, f is an antiderivative of the conjugate gradient of u. The latter fact is very useful, since it allows us to use the conjugate gradient to construct f when only $\operatorname{Re} f = u$ is known. In Example 1, we have $u(x, y) = \frac{x}{x^2+y^2} = \operatorname{Re}(f(z))$ where $f(z) = \frac{1}{z} = \frac{1}{x+i\,y}$ and the conjugate gradient of u is $\phi = -\frac{1}{z^2} = f'(z)$.

Using the conjugate gradient $\phi = u_x - i\,u_y$, we can express any partial derivative of u as the real or imaginary part of an analytic function, namely a derivative of ϕ. For example, to get u_{xxx}, differentiate ϕ with respect to x twice and get $\phi'' = u_{xxx} - i\,u_{yxx}$. (Note that since the derivatives of ϕ exist, we can obtain them by differentiating with respect to any one of the variables x, y, or z. This was done in the proof of the Cauchy-Riemann equations.) Since ϕ is analytic, all its higher-order derivatives are analytic (Corollary 1, Section 3.6), and so u_{xxx} and u_{yxx} are both harmonic by Theorem 1, Section 2.5. We can carry these ideas further and arrive at the following useful result.

COROLLARY 1

Suppose that u is harmonic on a region Ω. Then u has continuous partial derivatives of all order in Ω.

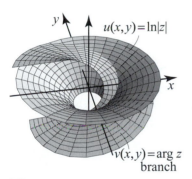

y

$u(x,y) = \ln|z|$

x

$v(x,y) = \arg z$ branch

Figure 2

Recall that a function v is a harmonic conjugate of u on Ω if $f = u + i\,v$ is analytic on Ω. Harmonic conjugates are determined up to an additive constant (Proposition 3, Section 2.5). The existence of harmonic conjugates is an important question since the harmonic conjugate allows us to relate harmonic functions to analytic functions. We know that harmonic conjugates may fail to exist on certain regions (Exercise 33, Section 2.5). For example, $\ln|z|$ is harmonic on the punctured plane, but the only candidate for a harmonic conjugate is a branch of $\arg z$, which has a branch cut and cannot be harmonic on the punctured plane (see Figure 2). So $\ln|z|$ does not have a harmonic conjugate on the punctured plane.

Our next result guarantees the existence of a harmonic conjugate if the region is simply connected.

THEOREM 1
HARMONIC
CONJUGATE

Suppose that u is harmonic on a simply connected region Ω. Then u has a harmonic conjugate in Ω given up to an additive constant by

$$(2) \qquad v(z) = \int_{z_0}^{z} -u_y \, dx + u_x \, dy \qquad (z_0 \text{ fixed in } \Omega),$$

where the integral is independent of path.

Proof Consider the analytic conjugate gradient $\phi = u_x - i\,u_y$. The integral of ϕ is independent of path in Ω (Theorem 3, Section 3.4). Define

$$f(z) = \int_{z_0}^{z} \phi(\zeta) \, d\zeta.$$

Then f is analytic and $f' = \phi$. Write $\zeta = x + iy$, $d\zeta = dx + i\,dy$. Then

$$f(z) = \int_{z_0}^{z} u_x\,dx + u_y\,dy + i \overbrace{\int_{z_0}^{z} -u_y\,dx + u_x\,dy}^{v(z)}.$$

We claim that $\int_{z_0}^{z} u_x\,dx + u_y\,dy = u(z) - u(z_0)$. From this it will follow that $v(z) = \int_{z_0}^{z} -u_y\,dx + u_x\,dy$ is a harmonic conjugate of $u(z)$, since $(u(z) - u(z_0)) + i\,v(z) = f(z)$ is analytic and the additive constant $u(z_0)$ does not affect analyticity. To prove the claim, parametrize the path from z_0 to z by $\zeta(t) = x(t) + i\,y(t)$, $a \leq t \leq b$. Then

$$u_x\,dx + u_y\,dy = \left(u_x \frac{dx}{dt} + u_y \frac{dy}{dt} \right) dt = \frac{d}{dt} u(\zeta(t))dt,$$

by the chain rule in two dimensions. Hence

$$\int_{z_0}^{z} u_x\,dx + u_y\,dy = \int_{a}^{b} \frac{d}{dt} u(\zeta(t))\,dt = u(\zeta(t)) \Big|_{a}^{b} = u(z) - u(z_0),$$

as claimed. ■

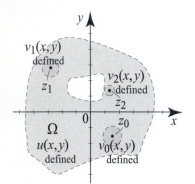

Figure 3 Local existence of the harmonic conjugate.

Keeping in mind the example of the harmonic function $\ln|z|$ with no harmonic conjugate in the punctured plane $\mathbb{C} \setminus \{0\}$, we see that simple connectedness is a crucial property in Theorem 1, and the result is not true on arbitrary regions. What can we say on arbitrary regions? Suppose that u is harmonic on a region Ω and let z_0 be in Ω. Since Ω is open, we can find an open disk $B_R(z_0)$ in Ω. Since $B_R(z_0)$ is simply connected, u has a harmonic conjugate in $B_R(z_0)$. This means that Theorem 1 holds *locally* in Ω (Figure 3). This is a very useful fact that we record separately.

COROLLARY 2

> Suppose that u is harmonic on a region Ω. Then u admits a harmonic conjugate locally in Ω.

Using the local existence of a harmonic conjugate, we obtain the mean value property of harmonic functions from the corresponding property for analytic functions.

COROLLARY 3
GAUSS'S MEAN VALUE PROPERTY

> Suppose that u is harmonic on a region Ω, then u satisfies the **mean value property**, in the following sense. If z is in Ω, and the closed disk $S_r(z)$ $(r > 0)$ is contained in Ω, then
>
> (3) $$u(z) = \frac{1}{2\pi} \int_{0}^{2\pi} u(z + re^{it})\,dt.$$

Proof Let B be an open disk in Ω containing the closed disk $S_r(z_0)$. Since B is simply connected, u admits a harmonic conjugate v in B. So $f = u + iv$ is analytic in B. By the mean value property of analytic functions ((4), Section 3.7), we have

$$f(z) = \frac{1}{2\pi} \int_{0}^{2\pi} f(z + re^{it})\,dt = \frac{1}{2\pi} \int_{0}^{2\pi} u(z + re^{it})\,dt + \frac{i}{2\pi} \int_{0}^{2\pi} v(z + re^{it})\,dt.$$

Now take real parts on both sides to get (3).   ∎

Just as we used the mean value property of analytic functions to prove the maximum modulus principle, we could use the mean value property of harmonic functions to prove a corresponding maximum-minimum principle. Such a proof would be more or less identical to that of Theorem 4, Section 3.7. Instead, we offer a proof that uses what we know about harmonic conjugates and analytic functions. You should pay attention to the role of connectedness.

THEOREM 2
MAXIMUM-
MINIMUM
PRINCIPLE

Suppose that u is a harmonic function on a region Ω. If u attains a maximum or a minimum in Ω, then u is constant in Ω.

Proof By considering $-u$, we need only prove the statement for maxima. We first prove the result under the assumption that Ω is simply connected. Apply Theorem 1 to find an analytic function $f = u + i\,v$ on Ω. Consider the function

$$g(z) = e^{f(z)} = e^{u(z)}e^{iv(z)}.$$

Then g is analytic in Ω and $|g(z)| = e^{u(z)}$. Since the real exponential function is strictly increasing, a maximum of $e^{u(z)}$ corresponds to a maximum of $u(z)$. By Theorem 4 of Section 3.7, if $|g(z)|$ attains a maximum or a minimum in Ω, then $g(z)$ is constant, implying that $u(z)$ is constant in Ω.

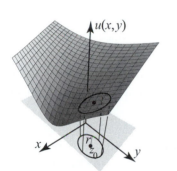

Figure 4

We now deal with the case of an arbitrary region Ω. The proof has the same basic form as the proof of Theorem 4, Section 3.7. Suppose that u attains a maximum M at a point in Ω. Let $\Omega_0 = \{z \in \Omega : u(z) < M\}$ and $\Omega_1 = \{z \in \Omega : u(z) = M\}$. We have $\Omega = \Omega_0 \cup \Omega_1$, Ω_0 is open, and Ω_1 is nonempty by assumption. It is enough to show that Ω_1 is open (see the proof of Theorem 4, Section 3.7). Suppose that z_0 is in Ω_1 and let $B_r(z_0)$ be an open disk in Ω centered at z_0 (Figure 4). Since $B_r(z_0)$ is simply connected and the restriction of u to $B_r(z_0)$ is a harmonic function that attains its maximum at z_0 inside $B_r(z_0)$, it follows from the previous case that u is constant in $B_r(z_0)$. Thus $u(z) = M$ for all z in $B_r(z_0)$, implying that $B_r(z_0)$ is contained in Ω_1. Hence Ω_1 is open.   ∎

Note in Theorem 2 that the minimum and maximum principles are equally strong for harmonic functions; for analytic functions the minimum principle required an additional hypothesis that $f(z)$ was never zero.

The following corollaries of Theorem 2 are similar to results that we proved regarding the modulus of an analytic function. We relegate their proofs to the exercises.

COROLLARY 4

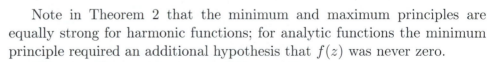

Suppose Ω is a bounded region, and u is harmonic on Ω and continuous on the boundary of Ω. Then
(i) u attains its maximum M and minimum m on the boundary of Ω, and
(ii) either u is constant or $m < u < M$ for all points in Ω.

COROLLARY 5

> Suppose Ω is a bounded region, and u is harmonic on Ω and continuous on the boundary of Ω. If u is constant on the boundary of Ω, then u is constant in Ω.

COROLLARY 6

> Suppose Ω is a bounded region, and u_1 and u_2 are harmonic on Ω and continuous on the boundary of Ω. If $u_1 = u_2$ on the boundary of Ω, then $u_1 = u_2$ in Ω.

Dirichlet Problems on a Disk

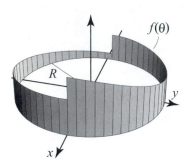

We focus now on some applications of the theory of harmonic functions to the solution of Dirichlet problems on a disk. To simplify the presentation, we will work on a disk centered at the origin with radius $R > 0$. In polar coordinates, the problem is stated as follows. Suppose we are given a piecewise continuous function $f(\theta)$, $0 < \theta \leq 2\pi$ that represents boundary data on the points $Re^{i\theta}$ (Figure 5). Find a function $u(r, \theta)$, $0 \leq r \leq R$, $0 < \theta \leq 2\pi$, such that

Figure 5 $f(\theta)$ gives the boundary data: $f(\theta) = u(R, \theta)$.

$$(4) \qquad \Delta u(r, \theta) = 0, \qquad 0 \leq r < R,\ 0 < \theta \leq 2\pi;$$

$$(5) \qquad \lim_{r \uparrow R} u(r, \theta) = u(R, \theta) = f(\theta), \qquad 0 < \theta \leq 2\pi,$$

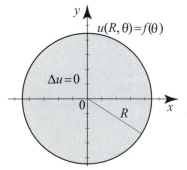

Figure 6 A Dirichlet problem on a disk.

where the limit holds at all points $Re^{i\theta}$ where $f(\theta)$ is continuous (Figure 6). Since θ and $\theta + 2\pi$ represent the same polar angle, we may remove the restriction on θ to lie in the interval $[0, 2\pi]$ and instead require f to be 2π-periodic; that is $f(\theta + 2\pi) = f(\theta)$ for all θ.

In general, the boundary data are given by a piecewise continuous f. If the boundary data is continuous, then an immediate consequence of Corollary 6 is the **uniqueness of the solution** of a Dirichlet problem on a disk. Thus once we have found a solution to a Dirichlet problem with continuous boundary data, this is necessarily the only solution of the problem.

We next consider a Dirichlet problem with special but important type of boundary data:

$$(6) \qquad u(R, \theta) = f(\theta) = a_0 + \sum_{n=1}^{N} \Big(a_n \cos n\theta + b_n \sin n\theta \Big).$$

This is a linear combination of functions from the 2π-periodic trigonometric system: $1,\ \cos x,\ \cos 2x,\ \dots,\ \sin x,\ \sin 2x,\ \dots$.

PROPOSITION 1
A DIRICHLET
PROBLEM ON THE
DISK

> The solution of the Dirichlet problem on the disk $|z| \leq R$ with boundary condition (6) is
>
> $$(7) \qquad u(r, \theta) = a_0 + \sum_{n=1}^{N} \Big(\frac{r}{R} \Big)^n \Big(a_n \cos n\theta + b_n \sin n\theta \Big).$$

Proof For $|z| < R$, write $z = re^{i\theta} = r(\cos\theta + i\sin\theta)$. The function $f(z) = z^n = r^n(\cos n\theta + i\sin n\theta)$ is analytic on the disk $|z| \leq R$. Hence its real and imaginary parts are harmonic (Theorem 1, Section 2.5). This shows that the functions 1, $r\cos\theta$, $r^2\cos 2\theta$, $\ldots$, $r\sin\theta$, $r^2\sin 2\theta$, $\ldots$ are harmonic on the disk $|z| \leq R$. By Proposition 1, Section 2.5, any linear combination of such functions is also harmonic on the unit disk, and so (7) is harmonic. Setting $r = R$ in (7), we see that $u(R, \theta) = f(\theta)$, where f is as in (6). Hence (7) is the solution of the Dirichlet problem with boundary data (6).   ∎

Each term in the finite sum (7) is a constant multiple of the harmonic functions 1, $r\cos\theta$, $r^2\cos 2\theta, \ldots$, $r\sin\theta$, $r^2\sin 2\theta, \ldots$. With the exception of the constant function 1, the graphs of these functions over the disk $|z| < R$ are saddle-shaped. This confirms our expectation that the maxima and minima of these functions must occur on the boundary.

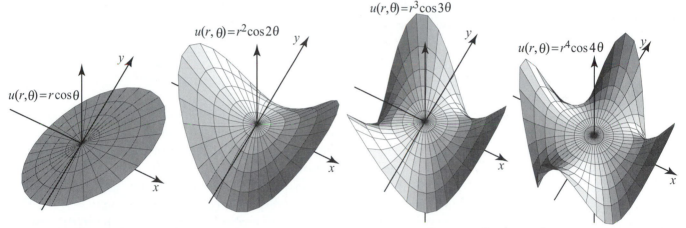

Figure 7 The saddle-shaped graphs of the harmonic functions $r\cos\theta$, $r^2\cos 2\theta$, $r^3\cos 3\theta$, $r^4\cos 4\theta$ seem to confirm the important property that the maximum and minimum of a harmonic function occur on the boundary of the (bounded) region.

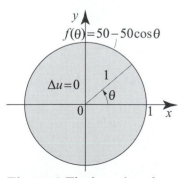

Figure 8 The boundary function, as a function of θ.

EXAMPLE 2   A steady-state problem on a disk

The temperature on the boundary of a circular plate of unit radius, with insulated lateral surface and center placed at the origin, is a function of the radial angle θ and varies between $0°$ and $100°$ according to the formula $f(\theta) = 50 - 50\cos\theta$ (see Figure 8).

(a) Find the steady-state temperature inside the plate.

(b) Describe the isotherms and lines of heat flow.

Solution   (a) To find the steady-state temperature, we must solve the Dirichlet problem with boundary data given by $f(\theta)$. According to (7), the solution is $u(r, \theta) = 50 - 50\,r\cos\theta$, $0 \leq r < 1$. This is a harmonic function, which is equal to $f(\theta)$ when $r = 1$ (Figure 9).

(b) Since the temperature on the boundary varies between 0 and 100, by the maximum-minimum principle for harmonic functions, the temperature inside the plate will vary between these two limits. To find the isotherms, let $0 < T < 100$ and solve the equation $u(r, \theta) = T$ or $50 - 50r\cos\theta = T$. Using $x = r\cos\theta$, the equation becomes $50 - 50x = T$ or $x = \frac{50-T}{50}$. Thus the isotherms are the intersections with the unit disk of the vertical lines $x = \frac{50-T}{50}$. The isotherms vary between $x = 1$ and $x = -1$ as T varies between 0 and 100. We know from Section 2.5 that heat flows along the curves that are orthogonal to the isotherms. In this case, it is easy to see that heat flows inside the unit disk in the direction of the horizontal lines $y = b$ (Figure 10). ∎

In Figure 9, we show the solution of the Dirichlet problem. Note how on the boundary the values of this solution coincide with the values of the boundary function. In Figure 10, we illustrate the isotherms and lines of heat flow. The orthogonality of these two families of curves is obvious in this case.

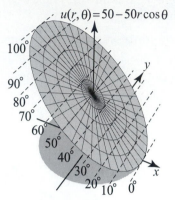

Figure 9

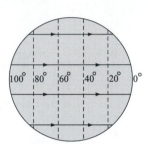

Figure 10

The Dirichlet problem in Proposition 1 is certainly not the most general kind that we can solve on a disk. But here is an amazing fact. Using Fourier series, we will show in Chapter 4 that the solution of the Dirichlet problem with arbitrary piecewise continuous boundary data $f(\theta)$ can be expressed in the form (7) if we allow the series to have infinitely many terms. Moreover, the coefficients a_0, a_n, and b_n are precisely the Fourier coefficients of the boundary function $f(\theta)$, and may be obtained by integrating $f(\theta)$ against trigonometric functions. This connection between harmonic functions and Fourier series has many fruitful consequences in applied mathematics.

Poisson Integral Formula

What is the Poisson integral formula? This is a formula for the solution of the Dirichlet problem on the unit disk. (In later sections, we will discover similar formulas on different regions of the plane.) It is also a formula that allows you to generate the values of a harmonic function $u(z)$ for z inside the unit disk, by using (integrating) the values of u on the boundary of the unit disk $|z| = 1$. We already know an example of such a situation: If u is harmonic in a region Ω containing the closed unit disk $S_1(0)$, then the mean value property of u at 0 implies that

$$u(0) = \frac{1}{2\pi} \int_0^{2\pi} u(e^{it}) \, dt.$$

This is a special application of the Poisson formula at the point $z = 0$. Interestingly, we now show how to derive the Poisson integral formula from the mean value property. Given a point z_0 in the open unit disk, if we can construct a harmonic function U on the unit disk such that $U(0) = u(z_0)$, then by applying the mean value property to U, we will recapture $U(0) = u(z_0)$ from the values on the boundary. The construction goes as follows. Given $|z_0| < 1$, consider the linear fractional transformation

$$\phi_{-z_0}(z) = \frac{z + z_0}{1 + \overline{z_0}z}, \quad |z| < 1.$$

We know from Example 3, Section 3.7, that $\phi_{-z_0}(z)$ is analytic, one-to-one, and maps the unit disk onto itself and $C_1(0)$ onto itself. From Theorem 3, Section 2.5, we know that if we compose a harmonic function with an analytic function, we obtain a harmonic function; so $U(z) = u \circ \phi_{-z_0}(z)$ is a harmonic function on the unit disk. Moreover, $U(0) = u \circ \phi_{-z_0}(0) = u(z_0)$. Applying the mean value property of U at 0, we find that

$$(8) \qquad u(z_0) = U(0) = \frac{1}{2\pi} \int_0^{2\pi} U(e^{it})\, dt = \frac{1}{2\pi} \int_0^{2\pi} u \circ \phi_{-z_0}(e^{it})\, dt.$$

Thus the value of u at the interior point z_0 is expressed as an integral involving the boundary values of $u \circ \phi_{-z_0}$. To get the desired formula that involves the boundary values of u, we will perform the change of variables $e^{is} = \phi_{-z_0}(e^{it})$. Recall that ϕ_{-z_0} maps the unit circle into itself and the inverse of ϕ_{-z_0} is ϕ_{z_0}. So

$$\phi_{z_0}(e^{is}) = e^{it} \quad \Rightarrow \quad \phi'_{z_0}(e^{is})ie^{is}\, ds = ie^{it}\, dt \quad \Rightarrow \quad \frac{\phi'_{z_0}(e^{is})}{e^{it}}e^{is}\, ds = dt$$

$$\Rightarrow \quad dt = \frac{\phi'_{z_0}(e^{is})}{\phi_{z_0}(e^{is})}e^{is}\, ds.$$

Now $\phi_{z_0}(z) = \frac{z - z_0}{1 - \overline{z_0}z}$, hence $\phi'_{z_0}(z) = \frac{1 - |z_0|^2}{(1 - \overline{z_0}z)^2}$, and so

$$\frac{\phi'_{z_0}(e^{is})}{\phi_{z_0}(e^{is})}e^{is} = e^{is}\frac{1 - |z_0|^2}{(1 - \overline{z_0}e^{is})(e^{is} - z_0)} = \frac{1 - |z_0|^2}{(e^{-is} - \overline{z_0})(e^{is} - z_0)} = \frac{1 - |z_0|^2}{|e^{is} - z_0|^2}.$$

Substituting into (8), we obtain

$$(9) \qquad u(z_0) = \frac{1 - |z_0|^2}{2\pi} \int_0^{2\pi} \frac{u(e^{is})}{|e^{is} - z_0|^2}\, ds \qquad (|z_0| < 1).$$

This is the **Poisson integral formula** on the unit disk. If u is harmonic in a disk of radius $R > 0$, centered at the origin, we consider the function $u(Rz)$, which is harmonic in $|z| < 1$, and so according to (9),

$$u(Rz_0) = \frac{1 - |z_0|^2}{2\pi} \int_0^{2\pi} \frac{u(Re^{is})}{|e^{is} - z_0|^2}\, ds \qquad (|z_0| < 1).$$

Let $z = Rz_0$, $z_0 = z/R$, $|z| = r < R$, then

$$(10) \qquad u(z) = \frac{R^2 - r^2}{2\pi} \int_0^{2\pi} \frac{u(Re^{is})}{|Re^{is} - z|^2} \, ds \qquad (|z| < R).$$

This is the **Poisson integral formula** on the disk of radius $R > 0$, centered at the origin. Another common way of expressing the Poisson integral formula is obtained by realizing that for $z = re^{i\theta}$,

$$|Re^{i\phi} - z|^2 = (Re^{i\phi} - re^{i\theta})(Re^{-i\phi} - re^{-i\theta}) = R^2 - 2rR\cos(\theta - \phi) + r^2;$$

and so from (10) (with the variable s replaced by ϕ) we obtain the alternative Poisson integral formula

$$(11) \qquad \boxed{u(re^{i\theta}) = \frac{R^2 - r^2}{2\pi} \int_0^{2\pi} \frac{u(Re^{i\phi})}{R^2 - 2rR\cos(\theta - \phi) + r^2} \, d\phi.}$$

In deriving (11), we were given a harmonic function $u(z)$ in a region containing $S_R(0)$. The importance of (11) is that it can be used to construct $u(z)$ inside the disk $B_R(0)$, when only its (piecewise continuous) values on the circle $C_R(0)$ are known. More precisely, we have the following result.

THEOREM 3
POISSON INTEGRAL
FORMULA

Consider the Dirichlet problem (4) on the disk $|z| \leq R$, with boundary conditions (5), where $f(\theta)$ is piecewise continuous for θ in $[0, 2\pi]$. Then the solution exists and is given by the Poisson integral formula

$$(12) \quad u(r, \theta) = \frac{R^2 - r^2}{2\pi} \int_0^{2\pi} \frac{f(\phi)}{R^2 - 2rR\cos(\theta - \phi) + r^2} \, d\phi \qquad (0 \leq r < R).$$

More precisely, $u(r, \theta)$ is harmonic in the open disk $|z| < R$ and tends to $f(\theta)$ as $r \uparrow R$ at all points of continuity of f.

Theorem 3 guarantees the existence of a solution of the Dirichlet problem on a disk. Its proof uses many of the results that we have derived thus far, along with properties of the function

$$(13) \qquad P(r, \theta) = \frac{R^2 - r^2}{R^2 - 2rR\cos\theta + r^2} \qquad (0 \leq r < R),$$

known as the **Poisson kernel** on a disk. We will present the proof in the exercises. (See Project Problem 26.)

The Poisson formula is difficult to evaluate, even for simple boundary data. For this reason, we will develop later alternative forms of the solution, including some that are based on Fourier series.

We end the section with a useful result obtained by evaluating (12) at $r = 0$. The corollary asserts that we can compute the mean value of a harmonic function at the center of a disk, without necessarily requiring that u be harmonic on the boundary.

COROLLARY 7
MEAN VALUE
PROPERTY

Suppose that $u(re^{i\phi})$ is harmonic for $0 \leq r < R$ and $0 \leq \phi \leq 2\pi$, and $u(Re^{i\phi}) = f(\phi)$, where f is piecewise continuous on the boundary. Then

$$u(0) = \frac{1}{2\pi} \int_0^{2\pi} f(\phi)\,d\phi.$$

Exercises 3.8

In Exercises 1–4, (a) verify that the function u is harmonic on the given region Ω. (b) Find the conjugate gradient of u and check that it is analytic on Ω.

1. $u(x, y) = xy$, $\Omega = \mathbb{C}$.

2. $u(x, y) = e^y \cos x$, $\Omega = \mathbb{C}$.

3. $u(x, y) = \frac{y}{x^2+y^2}$, $\Omega = \{z : \ \text{Im}\, z > 0\}$.

4. $u(x, y) = \ln(x^2 + y^2)$, $\Omega = \mathbb{C} \setminus (-\infty, 0]$.

5. Suppose that $f = u + iv$ is analytic on a region Ω. Show that the conjugate gradient of u is $f'(z)$. [Hint: The Cauchy-Riemann equations.]

6. Suppose that u is harmonic on $\mathbb{C}$ and there is an interval (a, b) with $a < b$ such that $u(x, y)$ is not in (a, b) for all (x, y). Show that u is constant. [Hint: Use Exercise 17, Section 3.7.]

7. (a) Suppose that u is harmonic and bounded on $\mathbb{C}$. Show that u is constant. [Hint: Exercise 6.]
(b) Suppose that u is harmonic on $\mathbb{C}$ and bounded above or below. Show that u is constant. [Hint: Exercise 6.]

8. Find a harmonic conjugate of the function in Exercise 3, using Theorem 1.

9. Suppose that f is analytic on a region Ω. Show that $|f|$ is harmonic on Ω if and only if f is constant. [Hint: You can use Exercise 20, Section 2.5, or you can use the mean value property of harmonic functions and argue as in the proof of Theorem 4, Section 3.7.]

10. Show in detail how to prove Corollaries 4–6 using Theorem 2.

11. Consider $u(z) = u(x, y) = e^x \cos y$, where z is in the square with vertices at $\pm\pi \pm i\pi$. Find the maximum and minimum values of $u(z)$ and determine where these values occur.

12. Show that $u(x, y) = xy$ is harmonic in the upper half-plane. Does u attain a maximum or a minimum on the boundary of the upper half-plane? Does this contradict Corollary 4?

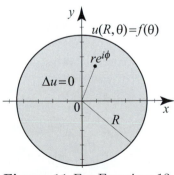

Figure 11 For Exercises 13–16.

In Exercises 13–16, solve the Dirichlet problem (4)–(5) for the given boundary function $f(\theta)$ on the disk with center at the origin and given radius $R > 0$ (Figure 11).

13. $f(\theta) = 1 - \cos\theta + \sin 2\theta$, $R = 1$.

14. $f(\theta) = \cos\theta - \frac{1}{2}\sin 2\theta$, $R = 1$.

15. $f(\theta) = 100\cos^2\theta$, $R = 2$.

16. $f(\theta) = \sum_{n=1}^{10} \frac{\sin n\theta}{n}$, $R = 1$.

17. Find the isotherms in Exercise 13.

18. For $n = 1, 2, \ldots, 0 \le r < 1$, show that

$$\frac{1-r^2}{2\pi} \int_0^{2\pi} \frac{\cos n\phi}{1 - 2r\cos(\theta - \phi) + r^2}\, d\phi = r^n \cos n\theta,$$

and

$$\frac{1-r^2}{2\pi} \int_0^{2\pi} \frac{\sin n\phi}{1 - 2r\cos(\theta - \phi) + r^2}\, d\phi = r^n \sin n\theta.$$

[Hint: Identify the integrals as solutions of Dirichlet problems on the unit disk and use Proposition 1.]

19. For $|z| < 1$, let

$$u(z) = \begin{cases} \mathrm{Arg}\,(z - 1) - \mathrm{Arg}\,(z + 1) & \text{if } \mathrm{Im}\,(z) \ge 0, \\ 2\pi + \mathrm{Arg}\,(z - 1) - \mathrm{Arg}\,(z + 1) & \text{if } \mathrm{Im}\,(z) < 0. \end{cases}$$

(a) Write $z = re^{i\theta}$, $-\pi < \theta \le \pi$. Using basic facts from plane geometry, show that

$$\lim_{r \uparrow 1} u(z) = \begin{cases} \frac{\pi}{2} & \text{if } 0 \le \theta \le \pi, \\[2mm] \frac{3\pi}{2} & \text{if } -\pi < \theta < \pi. \end{cases}$$

(b)   Argue that u is harmonic in the disk $|z| < 1$ and describe the Dirichlet problem that u satisfies.

20.   **Antiderivative of the Poisson kernel.**   For $0 < r < 1$, show that

$$(14) \qquad \int \frac{1}{1 - 2r\cos\theta + r^2}\, d\theta = \frac{2}{1-r^2} \tan^{-1}\left(\frac{1+r}{1-r}\tan\frac{\theta}{2}\right) + C.$$

[Hint: Use the substitution $t = \tan\frac{\theta}{2}$, $\cos\frac{\theta}{2} = \frac{1}{\sqrt{1+t^2}}$, $\sin\frac{\theta}{2} = \frac{t}{\sqrt{1+t^2}}$, and $d\theta = \frac{2}{1+t^2}\, dt$.]

21.   Solve the Dirichlet problem on the unit disk with boundary data

$$f(\theta) = \begin{cases} 100 & \text{if } 0 \le \theta \le \pi, \\ 0 & \text{if } \pi < \theta < 2\pi. \end{cases}$$

[Hint: Apply the Poisson integral formula, then use Exercise 20 to evaluate the integral.]

22.   **Project Problem: The Poisson kernel.** In this exercise, we establish several basic properties of the Poisson kernel (13). Let $z = re^{i\theta}$, where $0 \le r < R$ and $\zeta = Re^{i\phi}$.

(a)   For fixed ϕ, consider the function

$$U(z, \phi) = \frac{Re^{i\phi} + z}{Re^{i\phi} - z}.$$

Show that, for fixed ϕ, $U(z, \phi)$ is analytic in $|z| < R$.

(b)   Show that

$$U(z, \phi) = \frac{R^2 - r^2}{R^2 - 2rR\cos(\theta - \phi) + r^2} + i\frac{2rR\sin(\theta - \phi)}{R^2 - 2rR\cos(\theta - \phi) + r^2}.$$

Conclude that the Poisson kernel satisfies

$$\text{Re}\,(U(z,\phi)) = P(r,\theta-\phi) = \frac{R^2 - r^2}{R^2 - 2rR\cos(\theta-\phi) + r^2} \qquad (0 \le r < R).$$

The function

(15) $$Q(r,\theta-\phi) = \frac{2rR\sin(\theta-\phi)}{R^2 - 2rR\cos(\theta-\phi) + r^2} \qquad (0 \le r < R)$$

is called the **conjugate Poisson kernel** and it has interesting applications in the theory of harmonic functions.

(c)   Show that, for fixed ϕ, the function $(r,\theta) \mapsto P(r,\theta-\phi)$ is harmonic in the disk $|z| < R$. [Hint: It is the real part of an analytic function.]

(d)   For $0 < r < R$, show that $P(r,\theta-\phi)$ is a positive 2π-periodic function of θ. [Hint: $R^2 + r^2 - 2rR\cos(\theta-\phi) \ge R^2 + r^2 - 2rR = (R-r)^2 > 0$.]

(e)   Show that $P(r,\theta) = P(r,-\theta)$. Thus $P(r,\theta)$ is an even function of θ (Figure 12). Conclude from (c) that $(r,\theta) \mapsto P(r,\phi-\theta)$ is harmonic in $|z| < R$.

(f)   Show that $P(r,\theta)$ is decreasing on the interval $0 \le \theta \le \pi$. [Hint: Compute the derivative with respect to θ and show that it is negative on $(0,\pi)$.]

(g) Using the mean value property of harmonic functions, show that, for $0 \le r < R$,

$$\frac{R^2 - r^2}{2\pi} \int_0^{2\pi} \frac{d\phi}{R^2 - 2rR\cos(\theta-\phi) + r^2} = \frac{1}{2\pi} \int_0^{2\pi} P(r,\theta-\phi)\,d\phi = P(0,\theta) = 1,$$

and, in particular,

$$\frac{R^2 - r^2}{2\pi} \int_0^{2\pi} \frac{d\phi}{R^2 - 2rR\cos\phi + r^2} = \frac{R^2 - r^2}{2\pi} \int_{-\pi}^{\pi} \frac{d\phi}{R^2 - 2rR\cos\phi + r^2} = 1.$$

(h)   Establish the inequality

$$\frac{R-r}{R+r} \le P(r,\theta-\phi) \le \frac{R+r}{R-r}.$$

[Hint: See the hint in part (d).]

23.   Project Problem: Integrals involving the Poisson kernel. We continue deriving properties of the Poisson integral, using the notation of the previous exercise. (a)   Let $0 < \delta < \pi$. Show that

$$\lim_{r \to R} \int_\delta^\pi P(r,\theta)\,d\theta = 0.$$

[Hint: $P(r,\theta)$ is decreasing on $(0,\pi)$, so $\int_\delta^\pi P(r,\theta)\,d\theta \le (\pi-\delta)P(r,\delta) \to 0$ as $r \uparrow R$.]

(b)   Use (a), and (g) of the previous exercise to show that

$$\lim_{r \to R} \frac{1}{2\pi} \int_{-\delta}^{\delta} P(r,\theta)\,d\theta = 1.$$

So while part (g) of Exercise 23 tells us that the area under the graph of $\frac{1}{2\pi}P(r,\theta)$ and above the interval $[-\pi,\pi]$ is 1, this part tells us that the area is more and more concentrated around 0 as $r \uparrow R$.

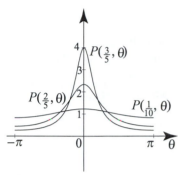

Figure 12

24. Illustrate the mean value property in Corollary 7 with the function $f(\theta)$ as in Exercise 13.

25. Different ways to express the Poisson integral formula. Show that the Poisson integral formula (12) can be expressed in the following equivalent forms:

$$(16) \qquad u(r, \theta) \;=\; \frac{R^2 - r^2}{2\pi} \int_{-\pi}^{\pi} \frac{f(\phi)}{R^2 - 2rR\cos(\theta - \phi) + r^2}\, d\phi;$$

$$(17) \qquad u(r, \theta) \;=\; \frac{R^2 - r^2}{2\pi} \int_{a}^{2\pi+a} \frac{f(\phi)}{R^2 - 2rR\cos(\theta - \phi) + r^2}\, d\phi;$$

$$(18) \qquad u(r, \theta) \;=\; \frac{R^2 - r^2}{2\pi} \int_{a}^{2\pi+a} \frac{f(\theta - \phi)}{R^2 - 2rR\cos\phi + r^2}\, d\phi,$$

where a is any real number. [Hint: The integrand is 2π-periodic so the integral does not change as long as we integrate over an interval of length 2π. See Theorem 1, Section 7.1.]

26.   Project Problem: Proof of Theorem 3. We will prove Theorem 3 under the assumption that $f(\theta)$ is continuous. The proof in the general case of a piecewise continuous f uses similar ideas but is more technical. We will use the notation of Exercises 22 and 23.

(a)   Show that the function defined by the integral (12) is harmonic. [Hint: Write $u(r, \theta) = \frac{1}{2\pi} \int_0^{2\pi} P(r, \theta - \phi)\, d\phi = \frac{1}{2\pi} \mathrm{Re}\left(\int_0^{2\pi} U(z, \phi)\, d\phi\right)$. Show that u is the real part of an analytic function. You will need Theorem 4, Section 3.5.]

(b)   Justify the following steps in the proof of $\lim_{r \to R} u(r, \theta) = f(\theta)$. We have

$$
\begin{aligned}
|u(r, \theta) - f(\theta)| \;&=\; \left| \frac{1}{2\pi} \int_0^{2\pi} P(r, \theta - \phi) f(\phi)\, d\phi - f(\theta) \right| \\[2mm]
&=\; \left| \frac{1}{2\pi} \int_{-\pi}^{\pi} \Big(P(r, \phi) f(\theta - \phi) - f(\theta) \Big)\, d\phi \right| \\[2mm]
&\leq\; \frac{1}{2\pi} \int_{-\pi}^{\pi} P(r, \phi) |f(\theta - \phi) - f(\theta)|\, d\phi.
\end{aligned}
$$

(Use Exercise 22(g) in the second step and (d) in the third step.)   Since f is continuous on $[-\pi, \pi]$, it is bounded. Hence $|f(\phi)| \leq M < \infty$ for all ϕ. Moreover, f is uniformly continuous on $[-\pi, \pi]$. So given $\epsilon > 0$, we can find $\delta > 0$ such that $|\phi| < \delta$ implies that $|f(\theta - \phi) - f(\theta)| < \epsilon$ for all θ. We have

$$
\begin{aligned}
|u(r, \theta) - f(\theta)| \;&\leq\; \frac{1}{2\pi} \int_{-\delta}^{\delta} P(r, \phi) \overbrace{|f(\theta - \phi) - f(\theta)|}^{<\epsilon}\, d\phi + \\[2mm]
&\qquad \frac{1}{2\pi} \int_{\delta \leq |\phi| \leq \pi} P(r, \phi) \overbrace{|f(\theta - \phi) - f(\theta)|}^{\leq 2M}\, d\phi \\[2mm]
&\leq\; \epsilon\frac{1}{2\pi} \int_{-\delta}^{\delta} P(r, \phi)\, d\phi + \frac{M}{\pi} \int_{\delta \leq |\phi| \leq \pi} P(r, \phi)\, d\phi
\end{aligned}
$$

Since $P(r, \phi)$ is positive, its integral increases as we enlarge the interval of integration. Hence $\int_{-\delta}^{\delta} P(r, \phi)\, d\phi \leq \int_{-\pi}^{\pi} P(r, \phi)\, d\phi$, and so

$$
\begin{aligned}
|u(r, \theta) - f(\theta)| \;&\leq\; \epsilon \, \overbrace{\frac{1}{2\pi} \int_{-\pi}^{\pi} P(r, \phi)\, d\phi}^{=1} + \frac{M}{\pi} \int_{\delta \leq |\phi| \leq \pi} P(r, \phi)\, d\phi \\
&=\; \epsilon + \frac{M}{\pi} \int_{\delta \leq |\phi| \leq \pi} P(r, \phi)\, d\phi.
\end{aligned}
$$

As $r \uparrow R$, the last integral tends to 0 (Exercise 23(a). Since ϵ is arbitrary, it follows that $|u(r, \theta) - f(\theta)|$ tends to 0 independently of θ, as $r \uparrow R$.

We have effectively shown that $u(r, \theta)$, the Poisson integral of f, converges *uniformly* to $f(\theta)$ as $r \uparrow R$, if (and only if) f is continuous. The concept of uniform convergence is extremely important in the theory of analytic and harmonic functions. We will study it in great detail in the following chapter.

3.9   Goursat's Theorem

This section contains one theoretical result, Goursat's theorem, named after the French mathematician and member of the French Academy of Science, Edouard Goursat (1858–1936).

Goursat's theorem says that if a function $f(z)$ has a derivative $f'(z)$ for all z in an open set Ω (that is, if f is differentiable in Ω), then $f'(z)$ is necessarily continuous on Ω (that is, f is analytic on Ω). Remember that when we defined analytic functions in Section 2.3, we required the existence and the continuity of $f'(z)$. Goursat's theorem gives us automatically the continuity of $f'(z)$, as soon as we know that $f'(z)$ exists. With this result, we can go back to the definition of analytic functions and restate it without the additional assumption on the continuity of the derivative. This is truly an achievement not only for obvious aesthetical improvements to the theory, but it will broaden the scope of the applications by allowing us to use this theory without checking the continuity of the derivative, which is difficult to realize in some cases.

THEOREM 1
GOURSAT'S
THEOREM

Let f be a complex-valued function on an open set Ω. If f is differentiable on Ω; that is, if the derivative

$$
f'(z_0) = \lim_{z \to z_0} \frac{f(z) - f(z_0)}{z - z_0}
$$

exists for all z_0 in Ω, then f is analytic on Ω (that is, f' is continuous on Ω).

Before we prove the theorem, let us make some additional remarks. Let us record the following important improvement to the definition of analytic functions.

THEOREM 2
ANALYTIC
FUNCTIONS

Let f be a complex-valued function on an open set Ω. Then f is analytic on Ω if and only if

$$f'(z_0) = \lim_{z \to z_0} \frac{f(z) - f(z_0)}{z - z_0}$$

exists for all z_0 in Ω.

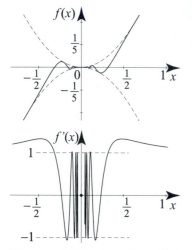

Figure 1 The function $f(x)$ is defined on the real line and has derivative $f'(x)$ that exists everywhere, but $f'(x)$ is not continuous at $x = 0$. Goursat's theorem tells us that nothing like this happens with functions defined on regions Ω in $\mathbb{C}$: If f' exists on Ω, then it must be continuous on Ω.

You should contrast Theorem 1 with the real variable case, where a derivative $f'(x)$ may exist without being continuous. For example, you can check that the function

$$f(x) = \begin{cases} x^2 \sin \frac{1}{x} & \text{if } x \neq 0, \\ 0 & \text{if } x = 0 \end{cases}$$

has a derivative for all x, but $f'(x)$ is not continuous at 0 (see Figure 1).

In the proof of Goursat's theorem, we will use the following notion. If A is a closed and bounded subset of $\mathbb{C}$, we define the **diameter** of A to be the largest value of $|z - z'|$, where z and z' are in A. A basic theorem from topology, known as **Cantor's intersection theorem**, states that if $\{A_n\}$ is an infinite sequence of nested closed and bounded subsets of $\mathbb{C}$ (that is, $A_1 \supset A_2 \supset \cdots \supset A_n \supset \cdots$), such that the diameter of A_n tends to 0 as $n \to \infty$, then the intersection of all the A_n's is nonempty and equals precisely one point. In symbols, there is a point z_0, such that $\bigcap_{n=1}^{\infty} A_n = \{z_0\}$. The statement of Cantor's theorem is intuitively clear. Its proof depends on properties of the complex plane. We will omit it and refer the interested reader to any book on advanced calculus.

Proof of Theorem 1 We will apply the strong version of Morera's theorem (Exercise 37, Section 3.6). Let Δ be an arbitrary triangle lying in some disk in Ω and let γ denote the boundary of Δ. Subdivide Δ into four congruent subtriangles by taking the midpoints of the sides and connecting them with line segments. Call the subtriangles Δ_1^j, $j = 1, 2, 3, 4$ and let γ_1^j denote the boundary of Δ_1^j, with the same orientation as γ (say positive; see Figure 2). Let

$$I = \int_{\gamma} f(z)\, dz, \quad \text{and} \quad I_1^j = \int_{\gamma_1^j} f(z)\, dz, \quad j = 1, 2, 3, 4.$$

Figure 2 This figure illustrates the case when $|I_1^3|$ is larger than $|I_1^j|$ for $j = 1, 2, 3, 4$. In this case, we pick the triangle Δ_1^3 and set $\Delta_2 = \Delta_1^3$. Next, we subdivide Δ_2 into four congruent triangles and proceed as before to determine Δ_3. Continue *ad infinitum*.

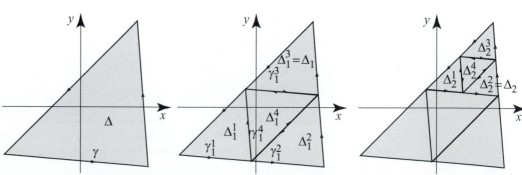

Our goal is to show that $I = 0$. Denote by Δ_1 that Δ_1^j for which $|I_1^j|$ ($j = 1, 2, 3, 4$) is the largest, and let γ_1 and I_1 denote the corresponding boundary and integral, respectively. Thus,

$$|I_1| = \left| \int_{\gamma_1} f(z)\, dz \right| \geq \left| \int_{\gamma_1^j} f(z)\, dz \right|, \quad j = 1, 2, 3, 4.$$

We have

$$|I| = \left| \sum_{j=1}^{4} \int_{\gamma_1^j} f(z)\, dz \right| \leq \sum_{j=1}^{4} \left| \int_{\gamma_1^j} f(z)\, dz \right| \leq 4|I_1|.$$

Now we repeat the process starting with Δ_1 to get Δ_2, and so on, and generate an infinite sequence of triangles. By the way we constructed the triangles, we have

$$l(\gamma) = 2l(\gamma_1), \quad l(\gamma_1) = 2l(\gamma_2), \quad \text{etc.,}$$

and so

(1)
$$\frac{|I|}{l(\gamma)^2} \leq \frac{4|I_1|}{l(\gamma)^2} = \frac{|I_1|}{l(\gamma_1)^2} \leq \frac{|I_2|}{l(\gamma_2)^2} \leq \cdots \leq \frac{|I_n|}{l(\gamma_n)^2} \leq \cdots.$$

The triangles Δ_n form a sequence of nested closed sets with diameters tending to 0. By Cantor's intersection theorem, there is exactly one point, say z_0, that belongs to all Δ_n (Figure 3). We now appeal to the differentiability of f at z_0 and write

$$f(z) = f(z_0) + f'(z_0)(z - z_0) + \epsilon(z)(z - z_0),$$

where $\epsilon(z) \to 0$ as $z \to z_0$. So the maximum value of $|\epsilon(z)|$ can be made arbitrarily small by restricting z to small disks around z_0. In particular, if we let M_n denote the maximum value of $|\epsilon(z)|$ for z in Δ_n, then $M_n \to 0$ as $n \to \infty$, because Δ_n contains z_0 and its diameter tends to zero. We have

$$\begin{aligned}
I_n &= \int_{\gamma_n} f(z)\, dz = \int_{\gamma_n} \Big(f(z_0) + f'(z_0)(z - z_0) + (z - z_0)\epsilon(z) \Big)\, dz \\
&= f(z_0) \underbrace{\int_{\gamma_n} dz}_{=0} + f'(z_0) \underbrace{\int_{\gamma_n} (z - z_0)\, dz}_{=0} + \int_{\gamma_n} (z - z_0)\epsilon(z)\, dz \\
&= \int_{\gamma_n} (z - z_0)\epsilon(z)\, dz,
\end{aligned}$$

where the first two integrals on the left side of the equality before the last are 0 by Cauchy's theorem, since γ_n is a closed curve, and the functions 1 and $z - z_0$ are analytic with continuous derivatives (thus we may apply our version of Cauchy's theorem for analytic functions with continuous derivatives, from Section 3.4). To approximate the last integral, note that for z on γ_n, we have $|z - z_0| \leq l(\gamma_n)$ (see Figure 4), and so

$$|I_n| = \left| \int_{\gamma_n} (z - z_0)\epsilon(z)\, dz \right| \leq l(\gamma_n)^2 M_n.$$

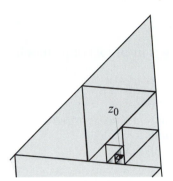

Figure 3 Cantor's intersection theorem applied to the sequence of nested triangles with diameters shrinking to 0 yields the point z_0 that belongs to all the triangles in this sequence.

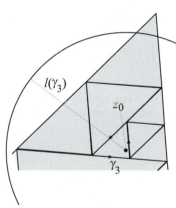

Figure 4 The inequality $|z - z_0| \leq l(\gamma_n)$ for z on γ_n and z_0 inside γ_n, illustrated with $n = 3$.

Consequently,

$$\frac{|I_n|}{l(\gamma_n)^2} \le M_n \to 0, \quad \text{as } n \to \infty,$$

and it follows from (1) that $|I| = 0$, completing the proof.   ∎

Goursat's theorem will be needed at crucial stages in proofs of results in the following chapter.

Exercises 3.9

1.   Consider calculus of a real variable and take the function $f(x) = x^2 \sin(1/x)$.
(a) Show that $f'(x)$ exists for all $x \ne 0$, and find a formula for $f'(x)$.
(b) Show that $f'(0)$ also exists by explicitly computing the limit

$$\lim_{x \to 0} \frac{f(x) - f(0)}{x - 0}.$$

What is $f'(0)$? [Hint: Use the squeeze theorem.]
(c) Show that $f'(x)$ is not continuous at zero, because $\lim_{x \to 0} f'(x)$ does not exist. [Hint: Consider sequences of real numbers $x_n = \frac{1}{2n\pi}$ and $\xi_n = \frac{1}{(2n+1)\pi}$. Each sequence tends to zero. Compute $\lim_{n \to \infty} f'(x_n)$ and $\lim_{n \to \infty} f'(\xi_n)$. If $f'(x)$ had a limit, these sequential limits would have to be the same.]

2.   In complex analysis, a function $f(z)$ that is differentiable on an open set must have a continuous derivative. In real analysis, as we saw in Exercise 1, this is not the case. However, the derivative of a real function does satisfy an intermediate value theorem: If $f(x)$ is differentiable on $[a, b]$ and α is any real number in between $f'(a)$ and $f'(b)$ (say, $f'(a) < \alpha < f'(b)$), then $\alpha = f'(c)$ for some c in (a, b). This theorem is attributed to Duhamel.
(a) Let $g(x) = f(x) - \alpha x$. Show that $g(x)$ is differentiable on $[a, b]$, that $g'(a) < 0$, and that $g'(b) > 0$.
(b) The function $g(x)$ is differentiable, so it must be continuous on $[a, b]$. By the extreme value theorem, it must attain a minimum. Argue from part (a) that this minimum cannot occur at the endpoints $x = a$ or $x = b$. Hence the minimum occurs at $x = c$ with $a < c < b$.
(c) Show that if $g(x) \ge g(c)$ in a neighborhood of c, and if $g'(c)$ exists, then $g'(c) = 0$. [Hint: Take limits of the difference quotient as $x \downarrow c$ and $x \uparrow c$ to show that $g'(c) \ge 0$ and $g'(c) \le 0$, respectively.]
(d) Conclude that $f'(c) = \alpha$ for some c in (a, b).

Topics to Review

Sections 4.1–4.4 parallel the development of series from calculus. A review of the basic definitions of convergence and the tests of convergence and divergence (nth term test, root test, ratio test, comparison test) would help. Unlike calculus, where series were studied almost independently from integrals, here the path integral and, in particular, Cauchy's theorem, will play a major role in proving facts about series.

Looking Ahead

The most important results of this chapter are Theorems 4, Section 4.3 and Theorem 1, Section 4.4. Together they assert that a function is analytic in a disk if and only if it has a Taylor series representation in that disk. For the proof, we introduce the notion of uniform convergence and derive many of its pleasant consequences in integration and differentiation of analytic functions. Section 4.5 extends the main results to analytic functions that have an isolated problem point (singularity) by using Laurent series. The definition, properties, and classification of singular points in Section 4.6 will be needed in later sections and are very important. The proofs in Sections 4.1–4.6 can be covered briefly, in favor of the applications. Section 4.7 is optional. It continues the study of harmonic functions and shows a connection with Fourier series. Fourier series turn out to be at the heart of the solution of Dirichlet problems on the disk. Their treatment in Chapter 7 is independent of Section 4.7.

4

COMPLEX SERIES

If, for every increasing value of n, the sum S_n indefinitely approaches a certain limit S, the series will be called convergent, and the limit in question will be called the sum of the series. If, on the contrary, while n increases indefinitely, the sum S_n does not approach a fixed limit, the series will be divergent and will no longer have a sum.

-Augustin Louis Cauchy
(Cauchy's definition of a convergent series, from his *Cours d'analyse*.)

In calculus, we used Taylor series to represent functions in intervals centered at given points with a radius of convergence that could be 0, positive, or infinite, depending on the remainder associated with the function. For example, $\cos x$, e^x, $\frac{1}{1+x^2}$, and the function defined by e^{-1/x^2} for $x \neq 0$ and 0 if $x = 0$ are all infinitely differentiable for all real x. The radius of convergence of the Taylor series representation around zero is ∞ for the first two, 1 for the third one, and 0 for the last one. Nothing like that will happen in complex analysis! In complex analysis, Taylor series are much nicer, in the sense that the remainder will play no role in determining the convergence of the series. In complex analysis, if $f(z)$ has a derivative in a disk around z_0 of radius R, then f has a Taylor series representation centered at z_0 with radius at least R. So, for example, in the case of $f(z) = \frac{1}{1+z^2}$, this function is not differentiable at $z = \pm i$, so we do not expect the Taylor series around 0 to have a radius of convergence larger than 1.

Taylor series and more generally Laurent series will be used to study important properties of analytic functions, concerning their zeros and isolated problem points (singularities). They are also useful in studying properties of special functions, such as Bessel functions.

The theory of power series as you will see it in this chapter owes a lot to the German mathematician Karl Weierstrass (1815–1897). Weierstrass introduced to analysis the $\epsilon\delta$-notation in proofs, replacing Cauchy's terminology, such as "indefinitely approaches a certain limit" and "increases indefinitely." Weierstrass's contributions to analysis are evidenced by the number of fundamental results that you will encounter in this chapter that bear his name.

4.1   Sequences and Series of Complex Numbers

In previous chapters we studied the derivative and integral of a function of a complex variable. Our next step is to study series of complex functions. If you browse through the results of this section, you will find that most of them are direct analogues of well-known results from calculus. This is true; and in fact, it is good to keep in mind the theory of series from calculus as a guide. In the next section, we will introduce analytic functions to the picture, and then we will see results that have no analogues in the theory of real-valued functions. We will even answer questions that were raised but could not be answered in calculus. For example, why does the power series expansion about 0 of the function $f(x) = \frac{1}{1+x^2}$ converge only in the interval $(-1, 1)$, even though the function has no problem on the real line outside this interval? Why does the function $f(x) = e^{-\frac{1}{x^2}}$, if $x \neq 0$ and $f(0) = 0$, have no power series expansion in a neighborhood of 0 yet have derivatives of all order at 0? These questions and many others have a natural place and obvious answers in complex analysis.

We start by investigating the notions of convergence and divergence of complex sequences. A **sequence** of complex numbers is a function whose domain of definition is the set of positive integers $\{1, 2, \dots, n, \dots\}$ and whose range is a subset of $\mathbb{C}$. Thus a sequence is an ordered list of complex numbers $a(1), a(2), a(3), \dots, a(n), \dots$. It is customary to write a_n instead of $a(n)$ and to denote the sequence by $\{a_n\}_{n=1}^{\infty}$ or simply $\{a_n\}$.

Many analytical expressions involving sequences of complex numbers will look identical to those for real sequences. The difference is that in the complex case, the absolute value refers to distance in the plane, and sequences of complex numbers can be thought of as sequences of points in the plane, which converge by eventually staying inside small disks centered at the limit point (Figure 1).

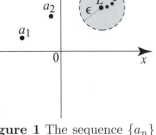

Figure 1 The sequence $\{a_n\}$ converges to the complex number L.

DEFINITION 1
LIMIT OF A
SEQUENCE

We say that a sequence $\{a_n\}$ **converges** to a complex number L or has **limit** L as n tends to infinity and write

$$\lim_{n \to \infty} a_n = L$$

if given any $\epsilon > 0$ there is an integer N such that

$$(1) \qquad\qquad |a_n - L| < \epsilon \qquad \text{for all } n \geq N.$$

If the sequence $\{a_n\}$ does not converge, then we say that it **diverges**.

If a limit of a complex sequence exists, then it is unique (Exercise 7).

If $L = \lim_{n \to \infty} a_n$, then we will adopt the notation $a_n \to L$ as $n \to \infty$ or merely $a_n \to L$. It is immediate from the definition that a sequence $a_n \to L$ if and only if the real sequence $|a_n - L| \to 0$. For sequences that converge to zero, we have

(2) $$\lim_{n\to\infty} a_n = 0 \quad \Leftrightarrow \quad \lim_{n\to\infty} |a_n| = 0.$$

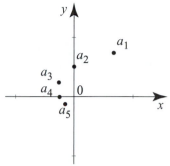

Figure 2 The first five terms of the sequence $a_n = \dfrac{e^{in\frac{\pi}{4}}}{n}$, and the limit $L = 0$. Note how the arguments of two successive terms differ by $\frac{\pi}{4}$. Do you see why?

EXAMPLE 1   Sequences

(a)   The first few terms of the sequence $\left\{e^{in\frac{\pi}{4}}\right\}_{n=1}^{\infty}$ are $\frac{\sqrt{2}}{2} + i\frac{\sqrt{2}}{2}$, i, $-\frac{\sqrt{2}}{2} + i\frac{\sqrt{2}}{2}$, -1, $-\frac{\sqrt{2}}{2} - i\frac{\sqrt{2}}{2}$, $-i$, $\frac{\sqrt{2}}{2} - i\frac{\sqrt{2}}{2}$, 1, $\frac{\sqrt{2}}{2} + i\frac{\sqrt{2}}{2}$, $\ldots$. The sequence is clearly not converging, since its terms will cycle over the first eight terms indefinitely.

(b)   The first few terms of the sequence $\left\{\dfrac{e^{in\frac{\pi}{4}}}{n}\right\}_{n=1}^{\infty}$ are shown in Figure 2. The figure suggests that the sequence converges to $L = 0$. Let us prove this using the definition. Given $\epsilon > 0$, we have

$$|a_n - L| = \left| \frac{e^{in\frac{\pi}{4}}}{n} - 0 \right| = \frac{\left|e^{in\frac{\pi}{4}}\right|}{n} = \frac{1}{n} < \epsilon$$

for all $n > \frac{1}{\epsilon}$, and so the sequence converges to 0.   ∎

Limits of sequences have properties similar to those of functions, because the definitions of limits are similar in both cases. We list some of these properties for ease of reference and omit most of the proofs that can be derived by using the same techniques as in Section 2.2.

THEOREM 1
LIMIT LAWS FOR
SEQUENCE

If $\{a_n\}$ and $\{b_n\}$ are convergent sequences and α and β are complex numbers, then

$$\lim_{n\to\infty}(\alpha a_n + \beta b_n) = \alpha \lim_{n\to\infty} a_n + \beta \lim_{n\to\infty} b_n;$$

$$\lim_{n\to\infty}(a_n b_n) = \lim_{n\to\infty} a_n \lim_{n\to\infty} b_n;$$

$$\lim_{n\to\infty} \frac{a_n}{b_n} = \frac{\lim_{n\to\infty} a_n}{\lim_{n\to\infty} b_n} \quad \text{if } \lim_{n\to\infty} b_n \neq 0.$$

The following useful result is an immediate consequence of continuity.

THEOREM 2
SUBSTITUTION LAW

If $\lim_{n\to\infty} a_n = L$ and f is continuous at L, then

$$\lim_{n\to\infty} f(a_n) = f(L).$$

Applying Theorem 2 with the continuous functions $f(z) = \overline{z}$ and $g(z) = |z|$, we obtain the following.

COROLLARY 1

If $\{a_n\}$ is a convergent sequence with $L = \lim_{n\to\infty} a_n$, then

$$\lim_{n\to\infty} \overline{a_n} = \overline{\lim_{n\to\infty} a_n} = \overline{L}, \quad \text{and} \quad \lim_{n\to\infty} |a_n| = \left| \lim_{n\to\infty} a_n \right| = |L|.$$

In particular, if $\{a_n\}$ converges, then $\{|a_n|\}$ converges. The converse is not true, as illustrated by Example 1(a), where $|a_n| = 1$ and so $\{|a_n|\}$ obviously converges to 1, but $\{a_n\}$ is divergent.

A sequence of complex numbers $\{a_n\}$ is said to be **bounded** if there is a positive number $M > 0$ such that $|a_n| \leq M$ for all n. The following theorem states that all convergent sequences are bounded.

THEOREM 3
BOUNDEDNESS

A convergent sequence is bounded. That is, if $\{a_n\}$ is a convergent sequence, then there is a positive real number $M > 0$ such that $|a_n| \leq M$ for all n.

We also have a squeeze theorem, like the one from calculus.

THEOREM 4
SQUEEZE THEOREM

(i)   Suppose that $\lim_{n \to \infty} a_n = 0$ and $|b_n| \leq |a_n|$ for all $n \geq n_0$. Then $\lim_{n \to \infty} b_n = 0$.
(ii)   Suppose that $\lim_{n \to \infty} a_n = 0$ and $\{b_n\}$ is a bounded sequence, then $\lim_{n \to \infty} a_n b_n = 0$.

The following result is useful in establishing properties of complex-valued sequences by using the corresponding ones for real-valued sequences.

THEOREM 5
REAL AND
IMAGINARY PARTS
OF SEQUENCES

Suppose that $\{a_n\}$ is a sequence of complex numbers and write $a_n = x_n + i\, y_n$, where $x_n = \operatorname{Re} a_n$ and $y_n = \operatorname{Im} a_n$. Then

$$\lim_{n \to \infty} a_n = L = \alpha + i\beta \quad \Leftrightarrow \quad \lim_{n \to \infty} x_n = \alpha \text{ and } \lim_{n \to \infty} y_n = \beta.$$

Our next example shows how we can use the preceding theorems along with our knowledge of real-valued sequences to compute limits of complex-valued sequences.

EXAMPLE 2   A useful limit
Show that

$$\lim_{n \to \infty} z^n = \begin{cases} 0 & \text{if } |z| < 1, \\ 1 & \text{if } z = 1. \end{cases}$$

Show that the limit does not exist for all other values of z; that is, if $|z| > 1$, or $|z| = 1$ and $z \neq 1$, then $\lim_{n \to \infty} z^n$ does not exist.

Solution   Recall that for any real number $r \geq 0$, we have

$$\lim_{n \to \infty} r^n = \begin{cases} 0 & \text{if } 0 \leq r < 1, \\ 1 & \text{if } r = 1, \\ \infty & \text{if } r > 1. \end{cases}$$

Consequently, for a complex number $|z| < 1$, we have $\lim_{n \to \infty} |z|^n = 0$. Since $|z|^n = |z^n|$, we conclude from (2) that $\lim_{n \to \infty} z^n = 0$. For $|z| > 1$, we have $|z|^n \to \infty$ as $n \to \infty$. Hence, by Theorem 3, the sequence $\{z^n\}$ cannot converge because it is not bounded. Now we deal with the case $|z| = 1$. If $z = 1$, the sequence $\{z^n\}$ is the constant sequence 1, which is trivially convergent. The case $|z| = 1$, $z \neq 1$ is not as simple because the sequence, though bounded, does not converge. Write $z = e^{i\theta}$;

then $z^n = e^{in\theta}$. We will prove that if this sequence converges, then $e^{i\theta} = 1$. First note that if $\lim_{n\to\infty} e^{in\theta} = L$, then $|L| = |\lim_{n\to\infty} e^{in\theta}| = \lim_{n\to\infty} |e^{in\theta}| = 1$ and, in particular, $L \neq 0$. Also, if $z^n \to L$, then $z^{n+1} \to L$, as $n \to \infty$. So $e^{i(n+1)\theta} \to L$, as $n \to \infty$. Taking limits on both sides of the equality $e^{i(n+1)\theta} = e^{i\theta}e^{in\theta}$, we obtain $L = e^{i\theta}L$. Dividing by $L \neq 0$, we get $e^{i\theta} = 1$, as we claimed. ∎

You can use Example 2 to show that for θ not an integer multiple of π, $\lim_{n\to\infty} \cos n\theta$ and $\lim_{n\to\infty} \sin n\theta$ do not exist (Exercise 8).

We now introduce a fundamental concept, which is crucial in establishing convergence when the limit is not given.

DEFINITION 2
CAUCHY SEQUENCE

A sequence $\{a_n\}$ is said to be a **Cauchy sequence** if given any $\epsilon > 0$ there is an integer N such that

$$(3) \qquad |a_n - a_m| < \epsilon \qquad \text{for all } m, n \geq N.$$

Thus the terms of a Cauchy sequence become arbitrarily close together.

It is not hard to see that a convergent sequence is a Cauchy sequence; because if the terms all get close to a limit, they must get close to each other. The converse (that a Cauchy sequence converges) is also true but not as obvious. To prove it, we will appeal to the well-known fact that a Cauchy sequences of real numbers must converge. This is a consequence of the **completeness property** of real numbers, which states that if S is a nonempty subset of the real line with an **upper bound** M ($x \leq M$ for all x in S), then S has a **least upper bound** b. That is, b is an upper bound for S and if M is any other upper bound for S, then $b \leq M$. The completeness property is an axiom in the construction of the real number system, and it is equivalent to the statement that all real Cauchy sequences are convergent. Using this property of real numbers, we can prove a corresponding one for complex numbers.

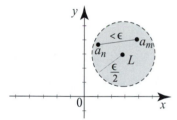

Figure 3 For the proof of Theorem 6, the inequality $|a_m - a_n| < \epsilon$ if $|a_m - L| < \frac{\epsilon}{2}$ and $|a_n - L| < \frac{\epsilon}{2}$.

THEOREM 6
CAUCHY SEQUENCES

A sequence of complex numbers $\{a_n\}$ converges if and only if it is a Cauchy sequence.

Proof Suppose that $\{a_n\}$ converges to a limit L. Given $\epsilon > 0$, let N be such that $n \geq N \;\Rightarrow\; |a_n - L| < \frac{\epsilon}{2}$. For $m, n \geq N$, we have by the triangle inequality (Figure 3):

$$|a_m - a_n| = |(a_m - L) + (L - a_n)| \leq |a_m - L| + |L - a_n| < \frac{\epsilon}{2} + \frac{\epsilon}{2} = \epsilon.$$

Hence $\{a_n\}$ is a Cauchy sequence. Conversely, suppose that $\{a_n\}$ is a Cauchy sequence, so for given $\epsilon > 0$ we have $|a_n - a_m| < \epsilon$ for all $m, n \geq N$. Write $a_n = x_n + i\,y_n$. The inequalities $|\operatorname{Re} z| \leq |z|$ and $|\operatorname{Im} z| \leq |z|$ imply that $|x_n - x_m| \leq |a_n - a_m| < \epsilon$ and $|y_n - y_m| \leq |a_n - a_m| < \epsilon$, which in turn implies that $\{x_n\}$ and $\{y_n\}$ are Cauchy sequences of real numbers. By the completeness property of real numbers, $\{x_n\}$ and $\{y_n\}$ are convergent sequences. Hence $\{a_n\}$ is convergent by Theorem 5. ∎

We are now ready to take up the study of series with complex terms.

Complex Series

An **infinite complex series** is an expression of the form

$$(4) \qquad \sum_{n=1}^{\infty} a_n,$$

where $\{a_n\}$ is an infinite sequence of complex numbers. The indexing set may not always start at $n = 1$; for example, we will often see expressions of the form $\sum_{n=0}^{\infty} a_n$. For simplicity we will sometimes write $\sum a_n$. The number a_n is called the n**th term** of the series. To each series $\sum_{n=1}^{\infty} a_n$ we associate a **sequence of partial sums** $\{s_n\}$, where

$$(5) \qquad s_n = \sum_{j=1}^{n} a_j = a_1 + a_2 + \cdots + a_n.$$

DEFINITION 3
LIMIT OF A SERIES

We say that a series $\sum_{n=1}^{\infty} a_n$ **converges** or is **convergent** to a complex number s and we write

$$s = \sum_{n=1}^{\infty} a_n$$

if the sequence of partial sums $\{s_n\}$ converges to s: $\lim_{n \to \infty} s_n = s$. Otherwise, we say that the series $\sum_{n=1}^{\infty} a_n$ **diverges** or is **divergent**.

So in order to establish the convergence or divergence of a series, we must study the behavior of the sequence of partial sums.

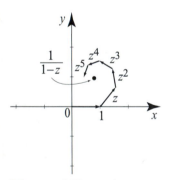

Figure 4 Terms in a convergent geometric series ($|z| < 1$). To get a partial sum s_n, add the vectors $1, z, \ldots, z^n$ using the head-to-tail method.

EXAMPLE 3   Geometric series

Show that the **geometric series** $\sum_{n=0}^{\infty} z^n$ converges and

$$\sum_{n=0}^{\infty} z^n = \frac{1}{1-z} \quad \text{if } |z| < 1.$$

Show that the series diverges for all other values of z.

Solution   Consider the partial sum (a typical case is shown in Figure 4):

$$(6) \qquad s_n = 1 + z + z^2 + \cdots + z^n.$$

If $z = 1$ this partial sum is clearly equal to $n + 1$ and hence the series diverges. For $z \neq 1$, we multiply and divide the right side of (6) by $1 - z \neq 0$, simplify, and get

$$s_n = \frac{(1 + z + z^2 + \cdots + z^n)(1 - z)}{1 - z} = \frac{1 - z^{n+1}}{1 - z}.$$

From Example 2, the sequence $\{z^{n+1}\}$ converges to 0 if $|z| < 1$ and diverges if $|z| > 1$ or $|z| = 1$ and $z \neq 1$. This implies that $\{s_n\}$ converges to $\frac{1}{1-z}$ if $|z| < 1$ and diverges for all other values of z, which is what we wanted to show. ∎

Geometric series may appear in disguise. Basically, whenever you see a series of the form $\sum w^n$ you should be able to use the geometric series to sum it. However, you have to be careful with the region of convergence.

EXAMPLE 4   Geometric series in disguise

Determine the largest region in which the series

$$\sum_{n=0}^{\infty} \frac{1}{(4+2z)^n}$$

is convergent and find its sum.

Solution   The series is a geometric series of the form $\sum w^n$ where $w = \frac{1}{4+2z}$. It converges to $\frac{1}{1-w}$ if and only if $|w| < 1$. Expressing these results in terms of z, we find that the series converges to

$$\frac{1}{1 - \frac{1}{4+2z}} = \frac{4+2z}{3+2z}$$

if and only if

$$\left| \frac{1}{4+2z} \right| < 1 \quad \Leftrightarrow \quad 1 < |4+2z|.$$

To better understand the region of convergence, we write the inequality using expressions of the form $|z - z_0|$ and interpret the latter as a distance in the usual way. We have

$$1 < |4+2z| \quad \Leftrightarrow \quad \frac{1}{2} < |2+z| \quad \Leftrightarrow \quad \frac{1}{2} < |z - (-2)|.$$

This describes the set of all z whose distance to -2 is strictly larger than $\frac{1}{2}$. Thus the series converges outside the closed disk shown in Figure 5, with center at -2 and radius $\frac{1}{2}$.   ∎

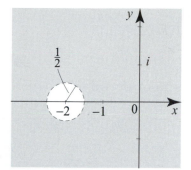

Figure 5 The shaded region describes all z such that $\frac{1}{2} < |z + 2|$. These are the points where the series in Example 4 converges.

Properties of Series and Tests of Convergence

For geometric series we are lucky in that we are able to find a closed expression for the partial sums. In most cases, this will not be possible. We will have to rely on properties of series to establish the convergence or divergence of a given series. Because a series is really a sequence of partial sums, all of the results about sequences can be restated for series. For convenience and for ease of reference, we will state some of these results along with several tests of convergence that are similar to ones for real series. The proofs will be omitted in most cases.

THEOREM 7
LINEARITY,
CONJUGATION,
REAL AND
IMAGINARY PARTS

If $\sum_{n=1}^{\infty} a_n$ and $\sum_{n=1}^{\infty} b_n$ are convergent series and α and β are complex numbers, then
(i) $\sum_{n=1}^{\infty} (\alpha a_n + \beta b_n) = \alpha \sum_{n=1}^{\infty} a_n + \beta \sum_{n=1}^{\infty} b_n$;
(ii) $\overline{\sum_{n=1}^{\infty} a_n} = \sum_{n=1}^{\infty} \overline{a_n}$;
(iii) $\mathrm{Re}\left(\sum_{n=1}^{\infty} a_n \right) = \sum_{n=1}^{\infty} \mathrm{Re}\,(a_n)$ and $\mathrm{Im}\left(\sum_{n=1}^{\infty} a_n \right) = \sum_{n=1}^{\infty} \mathrm{Im}\,(a_n)$.

We can use complex series to sum real series, like we did with integrals in Example 3, Section 3.2.

EXAMPLE 5   Summing real series using complex series
Show that the series $\sum_{n=0}^{\infty} \frac{\cos n\theta}{2^n}$ is convergent for all θ and find its sum.

Solution   We immediately recognize $\cos n\theta$ as the real part of $e^{in\theta} = \left(e^{i\theta}\right)^n$, and so the given series is the real part of the geometric series $\sum_{n=0}^{\infty} z^n$, where $z = \frac{e^{i\theta}}{2}$. From Example 3, since $\left|\frac{e^{i\theta}}{2}\right| = \frac{1}{2} < 1$, we have

$$\sum_{n=0}^{\infty} \left(\frac{e^{i\theta}}{2}\right)^n = \frac{1}{1 - \frac{e^{i\theta}}{2}} = \frac{2}{2 - e^{i\theta}} = \frac{2(2 - e^{-i\theta})}{(2 - e^{i\theta})(2 - e^{-i\theta})} = \frac{4 - 2\cos\theta + 2i\sin\theta}{5 - 4\cos\theta}.$$

Taking real parts and using Theorem 7(iii), we obtain

$$\sum_{n=0}^{\infty} \frac{\cos n\theta}{2^n} = \text{Re}\left(\frac{4 - 2\cos\theta + 2i\sin\theta}{5 - 4\cos\theta}\right) = \frac{4 - 2\cos\theta}{5 - 4\cos\theta}.$$

The series is plotted in Figure 6 as a function of θ.   ■

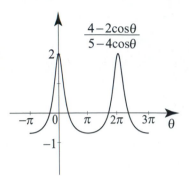

Figure 6 Graph of the series $\sum_{n=0}^{\infty} \frac{\cos n\theta}{2^n}$, which converges for all θ and equals the function $\frac{4 - 2\cos\theta}{5 - 4\cos\theta}$. (This is a 2π-periodic function.)

The following test of divergence is often useful. Its proof for real series works also with complex series.

THEOREM 8
THE nth TERM TEST
FOR DIVERGENCE

> If $\sum_{n=1}^{\infty} a_n$ is convergent, then $\lim_{n \to \infty} a_n = 0$. Equivalently, if either $\lim_{n \to \infty} a_n \neq 0$ or $\lim_{n \to \infty} a_n$ does not exist, then $\sum_{n=1}^{\infty} a_n$ diverges.

Proof Let $s_n = \sum_{m=1}^{n} a_m$. If $s_n \to s$, then also $s_{n-1} \to s$, and so $s_n - s_{n-1} \to s - s = 0$. But $s_n - s_{n-1} = a_n$, and so $a_n \to 0$.   ■

Applying the nth term test, we see right away that the geometric series $\sum_{n=0}^{\infty} z^n$ is divergent if $|z| = 1$ or $|z| > 1$.

For $m \geq 1$, the expression $t_m = \sum_{n=m+1}^{\infty} a_n$ is called a **tail of the series** $\sum_{n=1}^{\infty} a_n$. For fixed m, the tail t_m is itself a series, which differs from the original series by finitely many terms. So it is obvious that a series converges if and only if all its tails converge. As $m \to \infty$, we are dropping more and more terms from the tail series; as a result, we have the following useful fact.

PROPOSITION 1
TAIL OF SERIES

> If $\sum_{n=1}^{\infty} a_n$ is convergent, then $\lim_{m \to \infty} \sum_{n=m+1}^{\infty} a_n = 0$. Hence if a series converges, its tail must go to 0.

Proof   Let $s = \sum_{n=1}^{\infty} a_n$, $t_m = \sum_{n=m+1}^{\infty} a_n$, and $s_m = \sum_{n=1}^{m} a_n$. Since s_m is a partial sum of $\sum_{n=1}^{\infty} a_n$, we have $s_m \to s$ as $m \to \infty$. For each m, we have

$$s_m + t_m = \sum_{n=1}^{\infty} a_n = s \quad \Rightarrow \quad t_m = s_m - s.$$

Let $m \to \infty$ and use $s_m - s \to 0$ to get that $t_m \to 0$, as desired.   ■

A complex series $\sum_{n=1}^{\infty} a_n$ is said to be **absolutely convergent** if the series $\sum_{n=1}^{\infty} |a_n|$ is convergent.

Recall that, for series with real terms, absolute convergence implies convergence. The same is true for complex series.

THEOREM 9
ABSOLUTE
CONVERGENCE
IMPLIES
CONVERGENCE

> If $\sum_{n=1}^{\infty} a_n$ is absolutely convergent, then it is convergent. In symbols,
>
> $$\sum_{n=1}^{\infty} |a_n| < \infty \quad \Rightarrow \quad \sum_{n=1}^{\infty} a_n \text{ converges.}$$

Proof Let s_n denote the nth partial sum of the series $\sum_{j=1}^{\infty} a_j$ and v_n denote the nth partial sum of the convergent series $\sum_{j=1}^{\infty} |a_j|$. By Theorem 6, it is enough to show that the sequence of partial sums, $\{s_n\}$, is a Cauchy sequence. For $n > m \geq 1$, using the triangle inequality, we have

$$|s_n - s_m| = \left| \sum_{j=m+1}^{n} a_j \right| \leq \sum_{j=m+1}^{n} |a_j| = |v_n - v_m|.$$

Since $\sum_{n=1}^{\infty} |a_n|$ converges, the sequence $\{v_n\}$ is convergent and hence it is a Cauchy sequence. Consequently, given $\epsilon > 0$ we can find N so that, $|v_n - v_m| < \epsilon$ for $m, n \geq N$, implying that $|s_n - s_m| < \epsilon$ for $m, n \geq N$. Hence $\{s_n\}$ is a Cauchy sequence. ∎

A complex series $\sum_{n=1}^{\infty} a_n$ that is convergent but not absolutely convergent is called **conditionally convergent**.

Given a complex series $\sum_{n=1}^{\infty} a_n$, consider the series $\sum_{n=1}^{\infty} |a_n|$ whose terms are real and nonnegative. If we can establish the convergence of the series $\sum_{n=1}^{\infty} |a_n|$ using any one of the tests of convergence for series with nonnegative terms, then using Theorem 9, we can infer that the series $\sum_{n=1}^{\infty} a_n$ is convergent. Thus, all the tests of convergence for series with nonnegative terms can be used to test the (absolute) convergence of complex series. A list of such tests follows. We prove the first one just to illustrate the ideas involved.

THEOREM 10
COMPARISON TEST

> Suppose that a_n are complex numbers, b_n are real numbers, $|a_n| \leq b_n$ for all $n \geq n_0$, and $\sum_{n=1}^{\infty} b_n$ is convergent. Then $\sum_{n=1}^{\infty} a_n$ is absolutely convergent.

Proof By the comparison test for real series, we have that $\sum_{n=1}^{\infty} |a_n|$ is convergent. By Theorem 9, it follows that $\sum_{n=1}^{1} a_n$ is convergent. ∎

Here is a simple application of the comparison test, which illustrates the passage from complex to real series in establishing the convergence of a given complex series.

EXAMPLE 6   Comparison test

The series $\displaystyle\sum_{n=0}^{\infty} \frac{2\,e^{in\theta}}{n^2+3}$ is convergent by comparison to the convergent series $\sum_{n=1}^{\infty} \frac{2}{n^2}$, because

$$\left| \frac{2\,e^{in\theta}}{n^2+3} \right| \le \frac{2\left|e^{in\theta}\right|}{n^2} = \frac{2}{n^2}.$$ ∎

THEOREM 11
RATIO TEST

Suppose that

(7) $$\rho = \lim_{n\to\infty} \left| \frac{a_{n+1}}{a_n} \right|$$

exists or is infinite. Then the complex series $\sum_{n=1}^{\infty} a_n$ of nonzero terms converges absolutely if $\rho < 1$ and diverges if $\rho > 1$. If $\rho = 1$ the test is inconclusive.

EXAMPLE 7   Ratio test and the exponential series

We used the series $\displaystyle\sum_{n=0}^{\infty} \frac{z^n}{n!}$ to define the complex exponential function in Section 1.5. We can now show that this series converges absolutely for all z. The series is obviously convergent if $z = 0$. For $z \neq 0$,

$$\rho = \lim_{n\to\infty} \left| \frac{a_{n+1}}{a_n} \right| = \lim_{n\to\infty} \left| \frac{z^{n+1}\,n!}{z^n\,(n+1)!} \right| = \lim_{n\to\infty} \frac{|z|}{n} = 0.$$

Since $\rho < 1$, the series is absolutely convergent by the ratio test. ∎

THEOREM 12
ROOT TEST

Suppose that

(8) $$\rho = \lim_{n\to\infty} |a_n|^{1/n}$$

either exists or is infinite. Then the complex series $\sum_{n=1}^{\infty} a_n$ converges absolutely if $\rho < 1$ and diverges if $\rho > 1$. If $\rho = 1$ the test is inconclusive.

In general, the ratio test is easier to apply than the root test. But there are situations that call naturally for the root test. Here is an example.

EXAMPLE 8   Root test

Test the series $\displaystyle\sum_{n=0}^{\infty} \frac{z^n}{(n+1)^n}$ for convergence.

Solution   The presence of the exponent n in the terms suggests using the root test. We have

$$\rho = \lim_{n\to\infty} \left| \frac{z^n}{(n+1)^n} \right|^{\frac{1}{n}} = \lim_{n\to\infty} \frac{|z|}{n+1} = 0.$$

Since $\rho < 1$, the series is absolutely convergent for all z. ∎

Another well-known consequence of the completeness property of real numbers is that every bounded monotonic sequence (increasing or decreasing) converges. Since the partial sums of a series with nonnegative terms are increasing, we conclude that if these partial sums are bounded, then the series is convergent. This leads to the following result, which is not particularly useful as a test of convergence but will be required in later proofs.

THEOREM 13
BOUNDED PARTIAL SUMS

Suppose that there is a positive number $M > 0$ such that $\sum_{n=1}^{N} |a_n| \leq M$ for all N. Then the series $\sum_{n=1}^{\infty} a_n$ is absolutely convergent.

The two final results of this section are special types of convergence theorems dealing with rearrangements and products of series.

A **rearrangement** of a given complex series $\sum_{n=1}^{\infty} a_n$ is a series of the form $\sum_{n=1}^{\infty} a_{\phi(n)}$, where ϕ is a one-to-one mapping of the indexing set $\{1, 2, 3, \ldots\}$ onto itself. For example, a rearrangement of $a_1 + a_2 + a_3 + \cdots$ could be something like $a_4 + a_1 + a_6 + \cdots$, where each a_n must appear exactly once in the expression.

THEOREM 14
REARRANGEMENTS

If $\sum_{n=1}^{\infty} a_n$ is absolutely convergent, then every rearrangement is absolutely convergent and converges to the same limit.

Proof   For $m \geq 1$, we have $\sum_{n=1}^{m} |a_{\phi(n)}| \leq \sum_{n=1}^{\infty} |a_n| < \infty$. Hence the partial sums of the series $\sum_{n=0}^{m} |a_{\phi(n)}|$ are bounded and so they converge by Theorem 13. Thus any rearrangement converges absolutely. We next show that the rearranged series will still converge to $s = \sum_{n=1}^{\infty} a_n$. Let $\sigma(m) = \max_{n \leq m} \phi(n)$. Consider the expressions $\sum_{n=1}^{m} a_{\phi(n)}$ and $\sum_{n=1}^{\sigma(m)} a_n$. The first is a sum of the first m terms of the rearranged series; the second is a sum of the first $\sigma(m)$ terms in the original series, which will include each term from the first expression. The difference between the expressions is a finite sum of terms that come in the rearranged series after the term $a_{\phi(m)}$. Thus we can derive

$$\left| \sum_{n=1}^{m} a_{\phi(n)} - \sum_{n=1}^{\sigma(m)} a_n \right| \leq \sum_{n=m+1}^{\infty} |a_{\phi(n)}| < \infty.$$

Letting $m \to \infty$, and using $\sum_{n=m+1}^{\infty} |a_{\phi(n)}| \to 0$ (this is the tail of a convergent series) and $\sum_{n=1}^{\sigma(m)} a_n \to s$ (because $\sigma(m) \to \infty$ as $m \to \infty$), we see that $\left| \sum_{n=1}^{\infty} a_{\phi(n)} - s \right| = 0$, which implies that $\sum_{n=1}^{\infty} a_{\phi(n)} = s$. ∎

Our final result concerns products of series. Let us first define how to formally multiply two series and get one product series. It will be convenient to index the terms of a series starting at 0. Given two series $\sum_{n=0}^{\infty} a_n$ and $\sum_{n=0}^{\infty} b_n$, we form a third series $\sum_{n=0}^{\infty} c_n$ according to the formula

(9) $$c_n = a_0 b_n + a_1 b_{n-1} + \cdots + a_{n-1} b_1 + a_n b_0 = \sum_{j=0}^{n} a_j b_{n-j}.$$

The series $\sum_{n=0}^{\infty} c_n$ is called the **Cauchy product** of $\sum_{n=0}^{\infty} a_n$ and $\sum_{n=0}^{\infty} b_n$. To better understand this definition, imagine if you were able to cross multiply all the terms of the series $\sum a_n$ by those of $\sum b_n$. You will get terms of the form $a_j b_k$, where j and k range over $0, 1, 2, \ldots$. We can list the terms $a_j b_k$ in an array as shown in Figure 7.

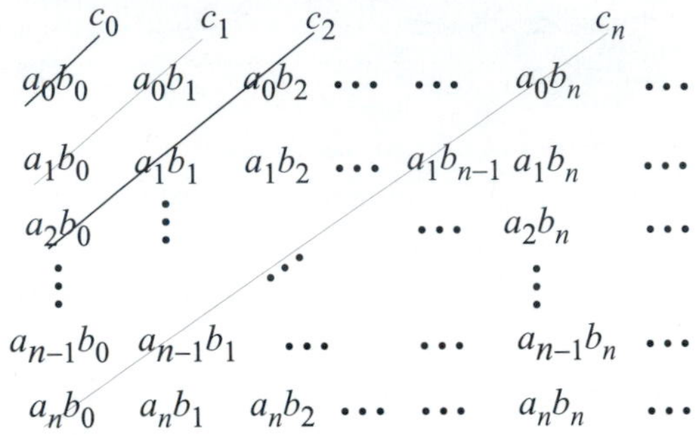

Figure 7 The nth term of a Cauchy product:
$c_n = a_0 b_n + a_1 b_{n-1} + \cdots + a_{n-1} b_1 + a_n b_0 = \sum_{j=0}^{n} a_j b_{n-j}$.
By summing all the c_n's, you will pick up all the terms of form $a_j b_k$, but in a very special order: $a_0 b_0 + a_1 b_0 + a_1 b_1 + \cdots$.

$$
\begin{array}{cccccc}
c_0 & c_1 & c_2 & & c_n & \\
a_0 b_0 & a_0 b_1 & a_0 b_2 & \cdots \ \cdots & a_0 b_n & \cdots \\
a_1 b_0 & a_1 b_1 & a_1 b_2 & \cdots \ a_1 b_{n-1} \ a_1 b_n & & \cdots \\
a_2 b_0 & \vdots & & \cdots & a_2 b_n & \cdots \\
\vdots & & & \ddots & \vdots & \\
a_{n-1} b_0 & a_{n-1} b_1 & \cdots & \cdots & a_{n-1} b_n & \cdots \\
a_n b_0 & a_n b_1 & a_n b_2 & \cdots \ \cdots & a_n b_n & \cdots
\end{array}
$$

The term c_n in the Cauchy product is gotten by summing along the diagonal $j + k = n$ as shown in Figure 7. If you sum all the diagonals, as prescribed by the Cauchy product, you will eventually collect all the terms $a_j b_k$. Does the Cauchy product series converge to the ordinary product of the two series (where we just multiply two complex numbers)? If the two series are absolutely convergent, the answer is yes.

THEOREM 15
CAUCHY
PRODUCTS

Suppose that $\sum_{n=0}^{\infty} a_n$ and $\sum_{n=0}^{\infty} b_n$ are absolutely convergent. Then their Cauchy product $\sum_{n=0}^{\infty} c_n$, where c_n is as in (9), is absolutely convergent and we have

$$(10) \qquad \sum_{n=0}^{\infty} c_n = \left(\sum_{n=0}^{\infty} a_n \right) \left(\sum_{n=0}^{\infty} b_n \right).$$

Proof   First we will show that the Cauchy product is absolutely convergent, then we will show that it converges to the right limit. For the first part, observe that

$$\sum_{k=0}^{n} |c_k| \leq \left(\sum_{k=0}^{n} |a_k| \right) \left(\sum_{k=0}^{n} |b_k| \right) \leq \left(\sum_{k=0}^{\infty} |a_k| \right) \left(\sum_{k=0}^{\infty} |b_k| \right) < \infty.$$

The first inequality follows because all the terms on the left are on or above the diagonal in a $(n+1) \times (n+1)$-array of nonnegative numbers, while the terms on the right are all the terms in the $(n+1) \times (n+1)$-array (see Figure 7). The second inequality follows since for any series with nonnegative terms, a partial sum is smaller than the sum of all terms. Hence the partial sums $\sum_{n=0}^{\infty} |c_n|$ are bounded, and so the series converges by Theorem 13. We next show that $s_n = \sum_{k=0}^{n} c_n$ converges to $\left(\sum_{n=0}^{\infty} a_n \right) \left(\sum_{n=0}^{\infty} b_n \right)$. Because we have already established that s_n converges, it will be enough to prove that a subsequence, $\{s_{2n}\}$, converges to this limit. We have

$$\left| \sum_{k=0}^{2n} c_k - \left(\sum_{k=0}^{n} a_k \right) \left(\sum_{k=0}^{n} b_k \right) \right| \leq \sum_{k=0}^{\infty} |a_k| \sum_{k=n+1}^{\infty} |b_k| + \sum_{k=0}^{\infty} |b_k| \sum_{k=n+1}^{\infty} |a_k|.$$

(To see this, draw a figure like Figure 7 that includes the terms up to $2n$.) Letting $n \to \infty$ and using the fact that the tails of the absolutely convergent series, $\sum_{k=n+1}^{\infty} |b_k|$ and $\sum_{k=n+1}^{\infty} |a_k|$ tend to zero (Proposition 1), we see that the right side tends to 0 as $n \to \infty$, implying (10). ∎

EXAMPLE 9   Exponent rule

Show that $e^{z_1} e^{z_2} = e^{z_1 + z_2}$.

Solution   We have $e^{z_1} = \sum_{n=0}^{\infty} \frac{z_1^n}{n!}$ and $e^{z_2} = \sum_{n=0}^{\infty} \frac{z_2^n}{n!}$, where both series converge absolutely by Example 7. By Theorem 15, we can form the Cauchy product of the two series and get $e^{z_1} e^{z_2} = \sum_{n=0}^{\infty} c_n$, where according to (9)

$$c_n = \sum_{j=0}^{n} \frac{z_1^j}{j!} \frac{z_2^{n-j}}{(n-j)!} = \frac{1}{n!} \overbrace{\sum_{j=0}^{n} \frac{n!}{j!(n-j)!} z_1^j z_2^{n-j}}^{(z_1+z_2)^n} = \frac{(z_1 + z_2)^n}{n!}.$$

We have multiplied and divided c_n by $n!$ to display the binomial expansion of $(z_1 + z_2)^n$ (see the binomial formula, Exercise 55, Section 1.3). Thus $e^{z_1} e^{z_2} = \sum_{n=0}^{\infty} \frac{(z_1+z_2)^n}{n!} = e^{z_1 + z_2}$. ∎

In the exercises you will be asked to derive various properties of the exponential and trigonometric functions by using Theorem 15 as we did in Example 9.

Exercises 4.1

In Exercises 1–6, determine whether or not the sequence $\{a_n\}$ converges, and find its limit if it does converge.

1. $a_n = \dfrac{e^{in\frac{\pi}{2}}}{n}$.   **2.** $a_n = \dfrac{e^{in\frac{\pi}{2}} + in}{n}$.   **3.** $a_n = \dfrac{1}{n+i}$.

4. $a_n = \dfrac{\text{Log}\left(\frac{3}{n} + in\right)}{n+i}$.   **5.** $a_n = \dfrac{\cosh(in)}{n^2}$.   **6.** $a_n = \dfrac{(1+2i)n^2 + 2n - 1}{3in^2 + i}$.

7.   Prove that if $\{a_n\}$ converges to A and to B, then $A = B$.

8.   Using the result of Example 2 show that, for θ not an integer multiple of π, $\lim_{n\to\infty} \cos n\theta$ and $\lim_{n\to\infty} \sin n\theta$ do not exist. [Hint: Use the addition formula for the cosine to show that, for $\theta \neq k\pi$, $\lim_{n\to\infty} \cos n\theta$ exists if and only if $\lim_{n\to\infty} \sin n\theta$ exists. Then use Theorem 5.] What happens when θ is an even multiple of π or an odd multiple of π?

9.   (a)   Show that $\lim_{n\to\infty} a_n = \lim_{n\to\infty} a_{n+1}$ for any convergent sequence $\{a_n\}$.
(b)   Define $a_1 = i$ and $a_{n+1} = \frac{3}{2+a_n}$. Suppose that $\{a_n\}$ is convergent and find its limit.

10.   (a)   Let $a_n = n^{\frac{1}{n}} - 1$. Use the binomial expansion to show that for $n > 1$

$$0 < \frac{n(n-1)}{2} a_n^2 < (1 + a_n)^n = n.$$

(b)   Conclude that $\lim_{n \to \infty} a_n = 0$.

(c)   Derive the useful limit: $\lim_{n \to \infty} n^{\frac{1}{n}} = 1$.

In Exercises 11–20, determine whether the series is convergent or divergent, and find its sum if it is convergent.

11. $\displaystyle\sum_{n=0}^{\infty} \frac{e^{in\frac{\pi}{2}}}{3^n}$.

12. $\displaystyle\sum_{n=0}^{\infty} \left(\frac{1+i}{2}\right)^n$.

13. $\displaystyle\sum_{n=3}^{\infty} \frac{3-i}{(1+i)^n}$.

14. $\displaystyle\sum_{n=0}^{\infty} \frac{\cos n\theta}{3^n}$.

15. $\displaystyle\sum_{n=0}^{\infty} \frac{3 + \sin n\theta}{10^n}$.

16. $\displaystyle\sum_{n=0}^{\infty} \frac{\cos n\theta + (2i)^n}{3^n}$.

17. $\displaystyle\sum_{n=2}^{\infty} \frac{1}{(n+i)((n-1)+i)}$.

18. $\displaystyle\sum_{n=0}^{\infty} \frac{n^2}{(n+i)(n+200+2i)}$.

19. $\displaystyle\sum_{n=2}^{\infty} \frac{\sin(in)}{n^2}$.

20. $\displaystyle\sum_{n=2}^{\infty} \frac{\sin(in)}{e^n}$.

In Exercises 21–32, determine whether the series is convergent or divergent.

21. $\displaystyle\sum_{n=0}^{\infty} \left(\frac{1+3i}{4}\right)^n$.

22. $\displaystyle\sum_{n=1}^{\infty} (-1)^n \frac{2^n + 4^n}{(1+3i)^n}$.

23. $\displaystyle\sum_{n=0}^{\infty} \frac{3i^n}{4+in^2}$.

24. $\displaystyle\sum_{n=0}^{\infty} \left(\frac{3+10i}{4+5in}\right)^n$.

25. $\displaystyle\sum_{n=1}^{\infty} \frac{(1+2in)^n}{n^n}$.

26. $\displaystyle\sum_{n=1}^{\infty} \frac{\tan(in)}{n^2}$.

27. $\displaystyle\sum_{n=1}^{\infty} \frac{\cos(in) + i \sin(in)}{e^{n^3}}$.

28. $\displaystyle\sum_{n=1}^{\infty} \frac{\operatorname{Log}(n+in)}{n^2}$.

29. $\displaystyle\sum_{n=1}^{\infty} \frac{\cos(in)}{e^{n^3}}$.

30. $\displaystyle\sum_{n=1}^{\infty} \frac{(3+10i)n^n}{n!}$.

31. $\displaystyle\sum_{n=0}^{\infty} \frac{(2+3i)^n}{n!}$.

32. $\displaystyle\sum_{n=1}^{\infty} \frac{1}{3+i^n}$.

In Exercises 33–40, use the geometric series to determine the largest region in which the given series converges and find its limit.

33. $\displaystyle\sum_{n=0}^{\infty} \frac{z^n}{2^n}$.

34. $\displaystyle\sum_{n=1}^{\infty} (1+z)^n$.

35. $\displaystyle\sum_{n=0}^{\infty} \left(\frac{(3+i)z}{4-i}\right)^n$.

36. $\displaystyle\sum_{n=0}^{\infty} \frac{(2+i)^n}{z^n}$.

37. $\displaystyle\sum_{n=1}^{\infty} \frac{1}{(2-10z)^n}$.

38. $\displaystyle\sum_{n=0}^{\infty} \frac{2^{n+1}}{(2+i-z)^n}$.

39. $\displaystyle\sum_{n=0}^{\infty} \left\{ \left(\frac{2}{z}\right)^n + \left(\frac{z}{3}\right)^n \right\}$.

40. $\displaystyle\sum_{n=0}^{\infty} \left\{ \frac{1}{(1-z)^n} - z^n \right\}$.

41.   The nth partial sum of a series is $s_n = \frac{i}{n}$. Does the series converge or diverge? If it does converge, what is its limit?

42.   Show that if $\sum a_n$ is absolutely convergent, then $\left| \sum a_n \right| \leq \sum |a_n|$.

In Exercises 43–46, you are asked to prove a result concerning complex series. As in the proof of Theorem 1, find the corresponding result for real series in your calculus text, then use it along with Theorem 9 to prove the desired result.

43.   Prove Theorem 3.

44.   Prove Theorem 11.

45. Prove Theorem 12.   **46.** Prove Theorem 13.

47. The terms of a series are defined recursively by

$$a_1 = 2 + i, \qquad a_{n+1} = \frac{(7 + 3i)n}{1 + 2in^2} a_n.$$

Does the series $\sum a_n$ converge or diverge?

48. The terms of a series are defined recursively by

$$a_1 = i, \qquad a_{n+1} = \frac{e^{\frac{i}{n}}}{\sqrt{n}} a_n.$$

Does the series $\sum a_n$ converge or diverge?

In Exercises 49–52, use Theorem 15 as we did in Example 9 to derive the given identity.

49. $\cos(z_1 + z_2) = \cos z_1 \cos z_2 - \sin z_1 \sin z_2.$

50. $\sin(z_1 + z_2) = \sin z_1 \cos z_2 + \cos z_1 \sin z_2.$

51. $e^z e^{-z} = 1.$ **52.** $\sinh(2z) = 2 \sinh z \cosh z.$

4.2   Sequences and Series of Functions

In the previous section we considered sequences and series of complex numbers. Now we turn our attention to sequences and series of functions.

Suppose that f_n and f are complex-valued functions defined on a subset $E \subset \mathbb{C}$. We say that f_n **converges pointwise** to f on E, if $\lim_{n \to \infty} f_n(z) = f(z)$ for every z in E. Hence f_n converges pointwise to f on E if, given z in E, and given $\epsilon > 0$, we can find $N > 0$ such that for all $n \geq N$, we have $|f(z) - f_n(z)| < \epsilon$. The integer N depends in general on z and ϵ.

Suppose $u_n(z)$ are defined on a set $E \subset \mathbb{C}$. The series of functions $\sum_{n=1}^{\infty} u_n(z)$ is said to **converge pointwise** if the sequence of partial sums $s_n(z) = \sum_{k=1}^{n} u_k(z)$ converges pointwise on E.

Pointwise convergence of sequences of functions is a seemingly natural mode of convergence, since by evaluating the functions we reduce to a sequence of numbers. The problem with pointwise convergence is that it does not preserve some desirable properties of the functions f_n. For example, the pointwise limit of continuous functions may not be continuous; the pointwise limit of integrable functions may not be integrable; and the pointwise limit of analytic functions may not be analytic. So we need a stronger mode of convergence.

DEFINITION 1
UNIFORM
CONVERGENCE

We say that f_n **converges uniformly** to f on E, and we write $\lim_{n \to \infty} f_n = f$ uniformly on E, if given $\epsilon > 0$, we can find $N > 0$ such that for all $n \geq N$ and all z in E, we have $|f(z) - f_n(z)| < \epsilon$.

A series of functions $\sum_{n=1}^{\infty} u_n(z)$ is said to **converge uniformly** on E if the sequence of partial sums $s_n(z) = \sum_{k=1}^{n} u_k(z)$ converges uniformly on

E. The key words in Definition 1 are "for all z in E." These require that $f_n(z)$ be close to $f(z)$ for all z in E simultaneously. Equivalently, let $M_n = \max |f_n(z) - f(z)|$, where the maximum is taken over all z in E. If no maximum is attained, we set M_n to be the least upper bound of $|f_n(z) - f(z)|$ for z in E. Then

$$(1) \qquad f_n \to f \text{ uniformly on } E \quad \Longleftrightarrow \quad M_n \to 0, \text{ as } n \to \infty.$$

Unlike pointwise convergence, uniform convergence preserves continuity and integrability of functions. More importantly, it preserves analyticity and that's the central result of this section (Theorem 5).

As a first example, it would be instructive to consider real-valued functions on intervals of the real line.

EXAMPLE 1   Pointwise versus uniform convergence

For $0 \le x \le 1$ and $n = 1, 2, \ldots$, define

$$f_n(x) = \frac{2nx}{1 + n^2 x^2}.$$

(a)   Does the sequence converge pointwise on $[0, 1]$?
(b)   Does it converge uniformly on $[0, 1]$?
(c)   Does it converge uniformly on $[0.1, 1]$?

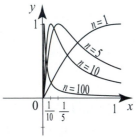

Figure 1 Graphs of $f_n(x)$ for $n = 1, 5, 10, 100$. $f_n(x)$ has a maximum at $x = \frac{1}{n}$ in the interval $[0, 1]$, and $f_n(\frac{1}{n}) = 1$.

Solution   (a)   We have $f_n(0) = 0$ for all n, and so $f_n(x) \to 0$ if $x = 0$. For any $x \ne 0$, we have

$$\lim_{n \to \infty} f_n(x) = \lim_{n \to \infty} \frac{2nx}{1 + n^2 x^2} = \lim_{n \to \infty} \frac{1}{n} \frac{2x}{x^2 + \frac{1}{n^2}} = 0.$$

So for all x in $[0, 1]$, the sequence $\{f_n(x)\}$ converges pointwise to $f(x) = 0$.
(b)   Figure 1 suggests that the sequence does not converge to 0 uniformly on $[0, 1]$. To confirm this, let us see how large $|f_n(x)|$ can get on the interval $[0, 1]$. For this purpose, we compute the derivative

$$f_n'(x) = \frac{2n(1 - n^2 x^2)}{(1 + n^2 x^2)^2}.$$

Thus, for $0 < x \le 1$,

$$f_n'(x) = 0 \quad \Leftrightarrow \quad -n^2 x^2 + 1 = 0 \quad \Leftrightarrow \quad x = \frac{1}{n}.$$

Plugging this value into $f_n(x)$ and simplifying, we find $f_n\left(\frac{1}{n}\right) = 1$. Thus, no matter how large n is, we can always find x in $[0, 1]$, namely $x = \frac{1}{n}$, with $f_n(x) = 1$. This shows that $M_n = \max_{x \in [0, 1]} |f_n(x)| \ge 1$ (in fact, $M_n = 1$), and so f_n does not converge to 0 uniformly over $[0, 1]$, by (1). To see what is going on, note that $f_n(x) = f_1(nx)$. That is, $f_n(x)$ is merely a horizontally shrunken version of the curve $f_1(x)$ and has maximum value of 1. This maximum value moves left as n increases, but it never leaves the interval $[0, 1]$.

(c)   The situation now is much different than in (b). For $n > 10$, we have $\frac{1}{n} < 0.1$, and so the maximum value of $f_n(x)$, which we found to be 1 in (b), is not attained in the interval $[0.1, 1]$; it is attained at $\frac{1}{n}$ outside this interval. So how large is $f_n(x)$ for x in $[0.1, 1]$? Since $f_n'(x) \leq 0$ on $[\frac{1}{n}, 1]$, $f_n(x)$ is decreasing on $[\frac{1}{n}, 1]$ and so the maximum of $f_n(x)$ for x in the interval $[0.1, 1]$ occurs at the left endpoint, $x = \frac{1}{10}$ (Figure 2). We have $M_n = f_n(\frac{1}{10})$ and we know that $f_n(\frac{1}{10}) \to 0$ as $n \to \infty$ from part (a). Thus $M_n \to 0$, implying that $f_n \to 0$ uniformly on $[0.1, 1]$. ∎

There is an important point to be made about part (c) of Example 1. Clearly we could have replaced the left endpoint $x = 0.1$ by any number $0 < a < 1$ and still had uniform convergence on $[a, 1]$. So while the sequence failed to converge uniformly on $[0, 1]$ it does converge uniformly on any proper closed subinterval $[a, 1]$ where $0 < a < 1$. This is a common phenomenon that you will encounter with many sequences or series. They may fail to converge on the whole region of definition, but they will converge uniformly on any closed (and bounded) proper subregion.

We now look at continuity and integrability in the context of uniform limits of functions. This first part of our study does not require analyticity.

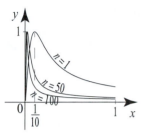

Figure 2 Graphs of $f_n(x)$ for $n = 1, 50, 100$. Note how $f_n(x)$ attains its maximum value in $[0.1, 1]$ at $x = 0.1$.

THEOREM 1
CONTINUITY AND
UNIFORM
CONVERGENCE

(i)   Suppose that $f_n \to f$ uniformly on E and f_n is continuous on E for every n. Then f is continuous on E.

(ii)   Suppose $u(z) = \sum_{n=1}^{\infty} u_n(z)$ converges uniformly on E and u_n is continuous on E for every n. Then u is continuous on E.

Proof (i)   Fix z_0 in E. Given $\epsilon > 0$, by uniform convergence we can find f_N such that $|f_N(z) - f(z)| < \frac{\epsilon}{3}$ for all z in E. Since f_N is continuous at z_0 there is a $\delta > 0$ such that $|f_N(z_0) - f_N(z)| < \frac{\epsilon}{3}$ for all $z \in E$ with $|z - z_0| < \delta$. Putting these two inequalities together and using the triangle inequality, we find that for $|z - z_0| < \delta$ we have

$$|f(z_0) - f(z)| \leq |f(z_0) - f_N(z_0)| + |f_N(z_0) - f_N(z)| + |f_N(z) - f(z)|$$
$$< \frac{\epsilon}{3} + \frac{\epsilon}{3} + \frac{\epsilon}{3} = \epsilon,$$

which establishes the continuity of f at z_0. Part (ii) follows from (i) by taking $f_n(z) = \sum_{k=1}^{n} u_k(z)$ and noting that each f_n is continuous, being a finite sum of continuous functions. ∎

Sometimes we can use Theorem 1 to prove the failure of uniform convergence.

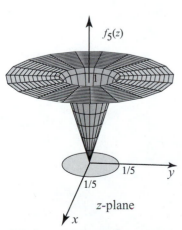

Figure 3 Graph of $f_n(z)$ (with $n = 5$) in Example 2.

EXAMPLE 2   Failure of uniform convergence
A sequence of functions is defined on the closed unit disk by

$$f_n(z) = \begin{cases} n|z| & \text{if } |z| \leq \frac{1}{n}, \\ 1 & \text{if } \frac{1}{n} \leq |z| \leq 1. \end{cases}$$

Does the sequence converge uniformly on $|z| \leq 1$?

Solution   The function $f_n(z)$ with $n = 5$ is depicted in Figure 3. It is clear that each $f_n(z)$ is continuous on $|z| \le 1$ and

$$\lim_{n \to \infty} f_n(z) = \begin{cases} 1 & \text{if } 0 < |z| \le 1, \\ 0 & \text{if } z = 0. \end{cases}$$

Since the limit function is not continuous at $z = 0$, we conclude from Theorem 1 that $\{f_n\}$ cannot converge uniformly on any set containing 0; in particular, it does not converge uniformly in $|z| \le 1$.   ∎

If a sequence of functions $f_n(z)$ are continuous and converge uniformly to a limit $f(z)$, then the limit $f(z)$ is continuous and thus it makes sense to integrate it.

THEOREM 2
SEQUENTIAL
LIMITS AND
INTEGRALS

Let $\{f_n\}$ be a sequence of continuous functions on a region Ω and let γ be a path in Ω. If $f_n \to f$ uniformly on γ, then

$$\lim_{n \to \infty} \int_\gamma f_n(z)\,dz = \int_\gamma f(z)\,dz.$$

Proof   Let $M_n = \max_{z \in \gamma} |f_n(z) - f(z)|$; then $M_n \to 0$. Using the integral inequality, Theorem 2, Section 3.2,

$$\left| \int_\gamma f_n(z)\,dz - \int_\gamma f(z)\,dz \right| = \left| \int_\gamma \left(f_n(z) - f(z) \right) dz \right| \le l(\gamma) M_n \to 0,$$

which proves the theorem.   ∎

Applying Theorem 2 to the partial sums of a uniformly convergent series, we obtain the following important corollary.

COROLLARY 1
TERM-BY-TERM
INTEGRATION OF
SERIES

Suppose that $\{u_n\}$ is a sequence of continuous functions on a region Ω and let γ be a path in Ω. Suppose that $u(z) = \sum_{n=1}^{\infty} u_n(z)$ converges uniformly on γ. Then

$$\int_\gamma u(z)\,dz = \sum_{n=1}^{\infty} \left(\int_\gamma u_n(z)\,dz \right).$$

We are now ready to prove a very useful test for uniform convergence.

THEOREM 3
WEIERSTRASS
M-TEST

Let $\{u_n\}$ be a sequence of functions on $E \subset \mathbb{C}$ and $\{M_n\}$ be a sequence of nonnegative numbers such that for all n

(i) $|u_n(z)| \le M_n$ for all z in E; and

(ii) $\displaystyle\sum_{n=1}^{\infty} M_n < \infty.$

Then $\displaystyle\sum_{n=1}^{\infty} u_n(z)$ converges uniformly and absolutely on E.

Proof   The absolute convergence of $\sum u_n(z)$ follows from (i) and (ii) by comparison to the series $\sum M_n$. Absolute convergence implies convergence, so we set

$\sum u_n(z) = s(z)$ for all z in E. We next prove the uniform convergence. Let $s_m = \sum_{k=1}^{m} u_k(z)$. For $n > m \geq 1$, using the triangle inequality we obtain

$$(2) \quad |s_n(z) - s_m(z)| = \left| \sum_{j=m+1}^{n} u_j(z) \right| \leq \sum_{j=m+1}^{n} |u_j(z)| \leq \sum_{j=m+1}^{n} M_j \leq \sum_{j=m+1}^{\infty} M_j.$$

Letting $n \to \infty$ and using the fact that $s_n(z) \to s(z)$ for all z in E, we obtain from (2), $|s(z) - s_m(z)| \leq \sum_{m+1}^{\infty} M_j$ for all z in E. This means that the maximum value of $|s(z) - s_m(z)|$ (as z ranges over E) is less than or equal to $\sum_{m+1}^{\infty} M_j$, which is the tail of a convergent series and tends to zero as $m \to \infty$. Thus $s_m(z)$ converges to $s(z)$ uniformly. ∎

EXAMPLE 3   Weierstrass M-test

Establish the uniform convergence of the given series on the given set.

(a) $\displaystyle\sum_{n=1}^{\infty} \frac{e^{inx}}{n^2}, \ -\infty < x < \infty.$

(b) $\displaystyle\sum_{n=1}^{\infty} \frac{z^n}{n^2}, \ |z| \leq 1.$

(c) $\displaystyle\sum_{n=1}^{\infty} \left(\frac{z}{2}\right)^n, \ |z| \leq 1.9.$

(d) $\displaystyle\sum_{n=1}^{\infty} \frac{1}{(1-z)^n}, \ 1.01 \leq |1-z|.$

Solution   We will use the notation of the Weierstrass M-test. For (a), $E = (-\infty, \infty)$, $u_n(x) = \frac{e^{inx}}{n^2}$. For all x in E, we have

$$|u_n(x)| = \left| \frac{e^{inx}}{n^2} \right| = \frac{1}{n^2} = M_n.$$

Since $\sum M_n = \sum \frac{1}{n^2}$ is convergent, we conclude from the Weierstrass M-test that $\sum_{n=1}^{\infty} \frac{e^{inx}}{n^2}$ converges uniformly for all x in E.

(b)   Here E is the set $|z| \leq 1$ and $u_n(z) = \frac{z^n}{n^2}$. For all $|z| \leq 1$, we have

$$|u_n(z)| = \left| \frac{z^n}{n^2} \right| \leq \frac{1}{n^2} = M_n.$$

Since $\sum \frac{1}{n^2}$ is convergent, we conclude from the Weierstrass M-test that $\sum_{n=1}^{\infty} \frac{z^n}{n^2}$ converges uniformly for all $|z| \leq 1$.

(c)   Here E is the set $|z| \leq 1.9$ and $u_n(z) = \left(\frac{z}{2}\right)^n$. For all $|z| \leq 1.9$, we have

$$|u_n(z)| = \left| \frac{z}{2} \right|^n \leq \left(\frac{1.9}{2} \right)^n = r^n = M_n,$$

where $r = \frac{1.9}{2} < 1$. Since $\sum M_n = \sum r^n$ is convergent, we conclude that $\sum_{n=1}^{\infty} \left(\frac{z}{2}\right)^n$ converges uniformly for all $|z| \leq 1.9$.

(d)   Here E is the set $1.01 \leq |1-z|$ and $u_n(z) = \frac{1}{(1-z)^n}$. For all z in E, we have

$$|u_n(z)| = \frac{1}{|1-z|^n} \leq r^n = M_n,$$

where $r = \frac{1}{1.01} < 1$. Since $\sum M_n = \sum r^n$ is convergent, we conclude that $\sum_{n=1}^{\infty} \frac{1}{(1-z)^n}$ converges uniformly for all $1.01 \leq |1-z|$. ∎

Our next example is the familiar geometric series. It is such an important example that we treat it separately and in greater detail.

EXAMPLE 4   Uniform convergence and the geometric series

(a)   Show that the geometric series $\sum_{n=0}^{\infty} z^n$ converges uniformly to $\frac{1}{1-z}$ on any closed subdisk $|z| \leq r < 1$ of the open unit disk $|z| < 1$.
(b)   Show that the geometric series $\sum_{n=0}^{\infty} z^n$ does not converge uniformly on the open disk $|z| < 1$.

Solution   (a)   We refer to Example 3, Section 4.1, for results about the geometric series. To establish the uniform convergence of the series for $|z| \leq r < 1$, we will apply the Weierstrass M-test. We have $|u_n(z)| = |z^n| \leq r^n = M_n$ for all $|z| \leq r$. Since $\sum_{n=0}^{\infty} M_n = \sum_{n=0}^{\infty} r^n$ is convergent if $0 \leq r < 1$, we conclude that the series $\sum_{n=0}^{\infty} z^n$ converges uniformly for $|z| \leq r$.
(b)   We now show that uniform convergence is lost when we consider the whole open disk $|z| < 1$. The nth partial sum is $s_n(z) = \frac{1-z^{n+1}}{1-z}$. Let $M_n = \max_{|z|<1} |s_n(z) - \frac{1}{1-z}|$. By (1), it is enough to show that M_n does not converge to 0 as $n \to \infty$. Indeed, for $z = re^{i\theta}$ with $0 \leq r < 1$, we have $|1 - z| \leq 1 + |z| = 1 + r$, $\frac{1}{|1-z|} \geq \frac{1}{1+r}$, and so

$$(3) \qquad M_n \geq \left| s_n(z) - \frac{1}{1-z} \right| = \frac{|z^{n+1}|}{|1-z|} \geq \frac{r^{n+1}}{1+r} \qquad \text{for all } 0 \leq r < 1.$$

As $r \uparrow 1$, $\frac{r^{n+1}}{1+r} \to \frac{1}{2}$, implying that $M_n \geq \frac{1}{2}$. Consequently, M_n does not converge to 0, and uniform convergence fails on $|z| < 1$.   ∎

Sequences and Series of Analytic Functions

The remaining results of this section concern limits of analytic functions. What is a good hypothesis to require when studying sequences and series of analytic functions defined over a region Ω? To answer this question, we take a hint from the behavior of the geometric series. A good hypothesis is to require uniform convergence on closed subdisks of Ω (or on closed and bounded subsets of Ω). This hypothesis is much less restrictive than uniform convergence on Ω and, as we now show, it does preserve analyticity. The following is a central result in the theory of analytic functions. Its proof is a beautiful application of Morera's theorem from Chapter 3.

**THEOREM 4
SEQUENTIAL
LIMITS AND
DERIVATIVES**

Suppose that $\{f_n\}$ is a sequence of analytic functions on a region Ω such that $f_n \to f$ uniformly on every closed disk contained in Ω. Then
(i) f is analytic on Ω, and
(ii)   for any integer $k \geq 1$, $f_n^{(k)}(z) \to f^{(k)}(z)$ for all z in Ω. Thus, the limit of the kth derivative is the kth derivative of the limit.

The proof will be facilitated with the help of the following.

LEMMA 1

> (i)   Suppose that $f_n \to f$ uniformly on a closed and bounded set E, and g is a continuous function on E. Then $f_n g \to fg$ uniformly on E.
>
> (ii)   Suppose that the series $u(z) = \sum u_n(z)$ converges uniformly on a closed and bounded set E, and g is a continuous function on E. Then the series $g(z)u(z) = \sum g(z)u_n(z)$ converges uniformly on E.

Proof   (i) Since g is continuous on E and E is closed and bounded, it follows that g is bounded on E. Let $M = \max_{z \in E} |g(z)|$. For all z in E, we have $|f_n(z)g(z) - f(z)g(z)| = |f_n(z) - f(z)||g(z)| \leq M|f_n(z) - f(z)|$. Thus $\max |f_n(z)g(z) - f(z)g(z)| \leq M \max |f_n(z) - f(z)| \to 0$, because f_n converges uniformly to f on E. Thus $f_n g \to fg$ uniformly on E. To prove (ii), apply (i) to the sequence of partial sums of $\sum u_n$. ∎

Proof of Theorem 4   (i)   The function f is continuous by Theorem 1(i). To prove that f is analytic, we will apply Morera's theorem (Theorem 3 and Exercise 37, Section 3.6). Let γ be an arbitrary closed path lying in a closed disk in Ω. It is enough to show that $\int_\gamma f(z)\,dz = 0$. We have $\int_\gamma f_n(z)\,dz = 0$ for all n, by Cauchy's theorem (Theorem 3, Section 3.4), because f_n is analytic on and inside γ; and by Theorem 2, $\int_\gamma f_n(z)\,dz \to \int_\gamma f(z)\,dz$ as $n \to \infty$. So $\int_\gamma f(z)\,dz = 0$ and (i) follows.

(ii)   Let z_0 be in Ω and let $S_R(z_0)$ be a closed disk contained in Ω, centered at z_0 with radius $R > 0$, with positively oriented boundary $C_R(z_0)$. The generalized Cauchy integral formula (Theorem 2, Section 3.6) tells us that

$$f^{(k)}(z_0) = \frac{k!}{2\pi i} \int_{C_R(z_0)} \frac{f(z)}{(z - z_0)^{k+1}}\,dz \quad \text{and} \quad f_n^{(k)}(z_0) = \frac{k!}{2\pi i} \int_{C_R(z_0)} \frac{f_n(z)}{(z - z_0)^{k+1}}\,dz.$$

Since $f_n(z) \to f(z)$ uniformly for all z on $C_R(z_0)$, and $\frac{1}{(z-z_0)^{k+1}}$ is continuous on $C_R(z_0)$ it follows from Lemma 1(i) that $\frac{f_n(z)}{(z-z_0)^{k+1}} \to \frac{f(z)}{(z-z_0)^{k+1}}$ uniformly for all z on $C_R(z_0)$. Applying Theorem 2, we get

$$f_n^{(k)}(z_0) = \frac{k!}{2\pi i} \int_{C_R(z_0)} \frac{f_n(z)}{(z - z_0)^{k+1}}\,dz \to \frac{k!}{2\pi i} \int_{C_R(z_0)} \frac{f(z)}{(z - z_0)^{k+1}}\,dz = f^{(k)}(z_0),$$

which proves (ii). ∎

Theorem 4 has many applications in the theory of analytic functions and power series, which we will discuss shortly. Theorem 4 may fail if we replace analytic functions by differentiable functions of a real variable. That is, if E is a subset of the real line and $f_n(x) \to f(x)$ uniformly on E, it does not follow in general that f_n' converges to f', as the next example shows.

EXAMPLE 5   Failure of termwise differentiation

For $0 \leq x \leq 2\pi$ and $n = 1, 2, \ldots$, define $f_n(x) = \frac{e^{inx}}{in}$. It is clear that $f_n(x) \to f(x) = 0$ uniformly for all $0 \leq x \leq 2\pi$. But $f_n'(x) = e^{inx}$ and this sequence does not converge except at $x = 0$ or $x = 2\pi$. (See Example 2, Section 4.1.) Consequently, $f_n'(x)$ does not converge to $f'(x) = 0$. Can we understand how this occurs within the larger framework of complex functions? Replace x by z and consider the sequence functions $f_n(z) = \frac{e^{inz}}{in}$. You cannot find a complex neighborhood of the real interval $[0, 2\pi]$ where $f_n(z)$ converges because of those z with negative imaginary part. Thus Theorem 4 does not apply.   ∎

COROLLARY 2
TERM-BY-TERM
DIFFERENTIATION
OF SERIES

Suppose that $\{u_n\}$ is a sequence of analytic functions on a region Ω and that $u(z) = \sum_{n=1}^{\infty} u_n(z)$ converges uniformly on every closed disk in Ω. Then u is analytic on Ω; and for any integer $k \geq 1$, the series may be differentiated term by term k times to yield

$$u^{(k)}(z) = \sum_{n=1}^{\infty} u_n^{(k)}(z) \quad \text{for all } z \text{ in } \Omega.$$

Proof   Apply Theorem 4 to the sequence of partial sums.   ∎

EXAMPLE 6   Term-by-term differentiation of the geometric series
The geometric series $\sum_{n=0}^{\infty} z^n$ converges uniformly to $\frac{1}{1-z}$ on any disk $|z| \leq r < 1$, by Example 4. It will then converge uniformly on any closed disk contained in the open disk $|z| < 1$. Applying Corollary 2, we may differentiate term by term in the open disk $|z| < 1$ and get

$$\frac{1}{(1-z)^2} = \frac{d}{dz} \frac{1}{1-z}$$

$$= \frac{d}{dz} \sum_{n=0}^{\infty} z^n = \sum_{n=0}^{\infty} \frac{d}{dz} z^n = \sum_{n=1}^{\infty} n z^{n-1}, \qquad |z| < 1. \quad ∎$$

There is an important technique of analysis that underlies Theorem 4, Corollary 2, and Example 6. We only have uniform convergence of a series on closed and bounded subsets of a region Ω, but we infer term-by-term differentiability on the whole open set Ω.

EXAMPLE 7   Termwise differentiation and integration

(a)   Let $\rho > 0$ be a positive real number. Show that $u(z) = \sum_{n=0}^{\infty} \frac{1}{n! z^n}$ converges uniformly for all $|z| \geq \rho$.

(b)   Conclude that $u(z)$ is analytic in $\mathbb{C} \setminus \{0\}$.

(c)   Let $C_R(0)$ denote a circle of radius $R > 0$ centered at 0, with positive orientation. Evaluate $\int_{C_R(0)} u(z)\, dz$.

Solution   (a)   For $|z| \geq \rho$, we have $\left|\frac{1}{z^n}\right| \leq \frac{1}{\rho^n}$, and so

$$\left|\frac{1}{n! z^n}\right| \leq \frac{1}{n! \rho^n} \qquad \text{for all } \rho \leq |z|.$$

The series $\sum_{n=0}^{\infty} \frac{1}{n! \rho^n}$ is convergent by the ratio test, since

$$\lim_{n \to \infty} \frac{a_{n+1}}{a_n} = \lim_{n \to \infty} \frac{n! \rho^n}{(n+1)! \rho^{n+1}} = \lim_{n \to \infty} \frac{1}{(n+1)\rho} = 0 < 1.$$

(In fact, $\sum_{n=0}^{\infty} \frac{1}{n! \rho^n} = \sum_{n=0}^{\infty} \frac{\rho^{-n}}{n!} = e^{-\rho}$.) By the Weierstrass M-test, it follows that $u(z) = \sum_{n=0}^{\infty} \frac{1}{n! z^n}$ is uniformly convergent for all $|z| \geq \rho$.

(b) Note from (a) that the series will be uniformly convergent on any closed and bounded subset of the punctured plane $\mathbb{C} \setminus \{0\}$, since any such set can be contained in an annular region $|z| \geq \rho$ (Figure 4). Applying Corollary 2, it follows that $u(z)$ is analytic on $\mathbb{C} \setminus \{0\}$.

(c)   Given the circle $C_R(0)$, pick ρ such that $0 < \rho < R$. Since $\sum_{n=0}^{\infty} \frac{1}{n! z^n}$ converges uniformly on $|z| \geq \rho$, and since $C_R(0)$ is contained in the region $|z| \geq \rho$, it follows that $\sum_{n=0}^{\infty} \frac{1}{n! z^n}$ converges uniformly on $C_R(0)$. Hence, by Corollary 1, the series may be integrated term by term to yield

$$\int_{C_R(0)} u(z)\, dz = \int_{C_R(0)} \left\{ \sum_{n=0}^{\infty} \frac{1}{n! z^n} \right\} dz = \sum_{n=0}^{\infty} \frac{1}{n!} \int_{C_R(0)} \frac{1}{z^n}\, dz.$$

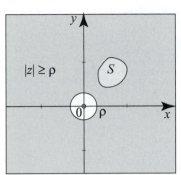

Figure 4 Since S is closed and 0 is not in S, then we can find $\rho > 0$ so that S is contained in the annulus $|z| \geq \rho$.

The integral $\int_{C_R(0)} \frac{1}{z^n}\, dz$ is an all-too-familiar integral (Example 4, Section 3.2). Its value is 0 if $n \neq 1$ and $2\pi i$ if $n = 1$. Thus,

$$\int_{C_R(0)} \left\{ \sum_{n=0}^{\infty} \frac{1}{n! z^n} \right\} dz = 2\pi i.$$

Notice what we have done: Our function is expanded in powers of z, and its integral around the origin is precisely $2\pi i$ times the coefficient of $\frac{1}{z}$. This is but one example of a general technique that forms the basis for Chapter 5. ∎

The convergence theory that we have established in this section will be applied in the next section to study power series and Taylor series expansions of analytic functions.

Exercises 4.2

In Exercises 1–8, (a) find the pointwise limit of the given sequence. (b) Determine if the sequence converges uniformly on the given interval. (c) If the sequence does not converge uniformly on the given interval, describe some subintervals on which uniform convergence takes place.

(d) Use a computer to plot several functions from the sequence to illustrate your answers in (a)–(c).

1. $f_n(x) = \dfrac{\sin nx}{n}$, $0 \leq x \leq \pi$.

2. $f_n(x) = \dfrac{\sin nx}{nx}$, $0 < x \leq \pi$.

3. $f_n(x) = \dfrac{x}{nx + 1}$, $0 \leq x \leq 1$.

4. $f_n(x) = nx^n$, $0 \leq x < .99$.

5. $f_n(x) = \dfrac{nx}{n^2 x^2 - x + 1}$, $0 \leq x \leq 1$.

6. $f_n(x) = \dfrac{nx}{n^2 x^2 - nx + 1}$, $0 \leq x \leq 1$.

7. $f_n(x) = \dfrac{nx}{n^2 x^2 + 2}$, $0 \leq x \leq 1$.

8. $f_n(x) = \displaystyle\int_0^x e^{-n\sqrt{t}}\, dt$, $0 \leq x \leq 1$.

In Exercises 9–12, (a) determine whether or not the sequence of functions converges pointwise and find its limit if it converges. (b) Determine if the sequence converges uniformly on the given subset Ω of $\mathbb{C}$. (c) If uniform convergence fails on Ω, describe some closed and bounded subsets of Ω on which uniform convergence takes place.

9. $f_n(z) = \dfrac{nz + 1}{z + 2n^2}$, $|z| \leq 1$.

10. $f_n(z) = \dfrac{z^n + z}{n + 1}$, $|z| \leq 1$.

11. $f_n(z) = \dfrac{\sin nz}{n^2}$, $|z| \leq 1$.

12. $f_n(z) = \dfrac{z^2 + nz + 1}{n^2 z + 1}$, $2 < |z|$.

In Exercises 13–22, use the Weierstrass M-test to show that the given series converges uniformly on the given region.

13. $\displaystyle\sum_{n=1}^{\infty} \frac{z^n}{n(n+1)}$, $|z| \leq 1$.

14. $\displaystyle\sum_{n=1}^{\infty} \frac{(3z)^n}{n(n+1)}$, $|z| \leq \frac{1}{3}$.

15. $\displaystyle\sum_{n=0}^{\infty} \left(\frac{3z}{4}\right)^n$, $|z| \leq 1.1$.

16. $\displaystyle\sum_{n=0}^{\infty} \left(\frac{z^2 - 1}{4}\right)^n$, $|z| \leq 1$.

17. $\displaystyle\sum_{n=0}^{\infty} \left(\frac{z + 2}{5}\right)^n$, $|z| \leq 2$.

18. $\displaystyle\sum_{n=0}^{\infty} \frac{1}{(5 - z)^n}$, $|z| \leq \frac{7}{2}$.

19. $\displaystyle\sum_{n=0}^{\infty} \frac{(z + 1 - 3i)^n}{4^n}$, $|z - 3i| \leq .5$.

20. $\displaystyle\sum_{n=0}^{\infty} \frac{(z - 1)^n}{4^n}$, $|z| \leq 2$.

21. $\displaystyle\sum_{n=0}^{\infty} \left\{ \frac{(z - 2)^n}{3^n} + \frac{2^n}{(z - 2)^n} \right\}$, $2.01 \leq |z - 2| \leq 2.9$.

22. $\displaystyle\sum_{n=1}^{\infty} \left\{ \left(\frac{z}{5}\right)^n + \frac{1}{z^n} \right\}$, $0.001 \leq |z| \leq 4.9$.

23. (a)   The nth partial sum of a series is given by $s_n(z) = \frac{z^n}{n}$ for $|z| \leq 1$. Does the series converge uniformly on $|z| \leq 1$?
(b)   Construct a series with partial sum $s_n(z)$ as given in (a).

24.   The nth partial sum of a series is given by $s_n(z) = \frac{e^{inz}}{n}$ for $|z| \leq 1$. Does the series converge uniformly on $|z| \leq 1$? Does it converge pointwise?

25. (a)   Does $\sum_{n=0}^{\infty} z^n$ converge uniformly on $|z - \frac{1}{2}| < \frac{1}{6}$?
(b)   Does $\sum_{n=0}^{\infty} z^n$ converge uniformly on $|z - \frac{1}{2}| < \frac{1}{2}$? Justify your answers.

26. (a)   If $\sum u_n(z)$ converges absolutely for all z in a set E, does this imply that $\sum u_n(z)$ converges uniformly on E?
(b)   If $\sum u_n(z)$ converges uniformly on E, does this imply that $\sum u_n(z)$ converge absolutely for all z in E? [Hint: Consider $\sum_{n=0}^{\infty} (-1)^n \frac{x^n}{n}$ for $0 \leq x \leq 1$.]

27. Derivative of the exponential function.   Show that $\frac{d}{dz} e^z = e^z$ by differentiating the series term by term. You must justify this process by showing that you can apply Theorem 4.

28. (a)   Show that $u(z) = \sum_{n=0}^{\infty} \frac{z^n}{1 + z^{2n}}$ is analytic for all $|z| < 1$ and all $|z| > 1$. [Hint: Treat separately the uniform convergence on $|z| \leq r < 1$ and on $|z| \geq R > 1$.]
(b)   What is $u'(z)$ for $|z| < 1$ or $|z| > 1$?

29. Riemann zeta function.   The Riemann zeta function is defined by

$$\zeta(z) = \sum_{n=1}^{\infty} \frac{1}{n^z} \qquad \text{(principal branch of } n^z\text{)}.$$

(a)   Let $\delta > 1$ be a positive real number. Show that the series converges uniformly on every half-plane $H_\delta = \{z : \operatorname{Re} z \geq \delta > 1\}$.
(b)   Conclude that $\zeta(z)$ is analytic on the half-plane $H = \{z : \operatorname{Re} z > 1\}$.
(c)   What is $\zeta'(z)$?
   Riemann used the zeta function to study the distribution of the prime numbers. Although this function was previously known to Euler, Riemann was the first one to consider it over the complex numbers. One of the most important open problems in mathematics today is the **Riemann hypothesis**, which is a conjecture made by Riemann, who stated that the analytic continuation of the zeta function has infinitely many nonreal roots and that all these roots lie on the line $x = \frac{1}{2}$.

30. Uniformly Cauchy sequence.   A sequence of functions $\{f_n\}$ is **uniformly Cauchy** on $E \subset \mathbb{C}$, if given $\epsilon > 0$ we can find a positive integer N so that

$$m, \ n \geq N \quad \Rightarrow \quad |f_n(z) - f_m(z)| < \epsilon \quad \text{for all } z \text{ in } E.$$

Not surprisingly, we have the following.

> **Cauchy criterion:** $\{f_n\}$ converges uniformly on $E \subset \mathbb{C}$ if and only if $\{f_n\}$ is uniformly Cauchy on E.

Prove this criterion using an argument similar to the proof of Theorem 6, Section 4.1.

31. Sums and products of uniformly convergent sequences.   Suppose that $f_n \to f$ uniformly on E and $g_n \to g$ uniformly on E. Show that
(a) $f_n + g_n \to f + g$ uniformly on E; and
(b)   if f and g are bounded on E, then $f_n g_n \to fg$ uniformly on E. [Hint: $f_n g_n - fg = f_n g_n - f_n g + f_n g - fg$.]

32.   Suppose that $\sum u_n(z)$ converges uniformly on $E \subset \mathbb{C}$. Show that $u_n(z) \to 0$ uniformly on E. [Hint: Let $s_n(z)$ denote the nth partial sum. What can you say about $s_{n+1}(z) - s_n(z)$?]

33. Boundary criterion for uniform convergence.   Suppose that C is a simple closed path (for simplicity, you may take C to be a circle) and let Ω denote the region interior to C. Suppose that f_n ($n = 1, 2, \ldots$) is analytic on Ω and continuous on C. Suppose that $\{f_n\}$ converges uniformly to some function f on C. Show that $\{f_n\}$ converges uniformly on Ω. [Hint: Show that $\{f_n\}$ is a uniformly Cauchy sequence on Ω.]

34.   (a) Suppose that f is analytic on the open unit disk, $f(0) = 0$ and $|f(z)| \leq M$ for all $|z| < 1$. Show that $\sum_{n=1}^{\infty} f(z^n)$ converges uniformly on every closed subdisk of $|z| < 1$. Conclude that $g(z) = \sum_{n=1}^{\infty} f(z^n)$ is analytic on $|z| < 1$. [Hint: Use Schwarz's lemma, then the Weierstrass M-test.]
(b) Conclude that if f is entire and $f(0) = 0$, then $g(z) = \sum_{n=1}^{\infty} f(z^n)$ is analytic on $|z| < 1$.
(c) Discuss the results of (a) and (b) in the special cases of $f(z) = z$ and $f(z) = \sin z$.

4.3   Power Series

A **power series** is a series of the form

$$\sum_{n=0}^{\infty} c_n(z - z_0)^n,$$

where z_0 is a complex number called the **center** of the power series, c_n are complex numbers called the **coefficients** of the power series, and z is a complex variable. A power series always converges at the center, because when $z = z_0$ all the terms for $n \geq 1$ are equal to zero. (We take the usual notational convention that $(z - z_0)^0 = 1$ even for $z = z_0$.) For $z \neq z_0$, the power series may converge or diverge. Let us consider a few examples.

The series that we used to define the exponential function, $\sum_{n=0}^{\infty} \frac{z^n}{n!}$, is a power series with center $z_0 = 0$ and coefficients $c_n = \frac{1}{n!}$. This power series converges for all z. The power series $\sum_{n=1}^{\infty} \frac{n^n(z-i)^n}{n}$ converges only at the center $z_0 = i$ (see Example 1 below). The geometric series $\sum_{n=0}^{\infty}(z - 2)^n$ is a power series centered at $z_0 = 2$ with coefficients $c_n = 1$. The geometric series converges for all $|z - 2| < 1$ and diverges for all $|z - 2| > 1$.

These series illustrate the typical behavior of a power series. We have the following trichotomy.

THEOREM 1
CONVERGENCE OF
POWER SERIES

Given a power series $\sum_{n=0}^{\infty} c_n(z - z_0)^n$, there are only three possibilities:
(i)   The series converges absolutely for all z.
(ii)   The series converges only at $z = z_0$.
(iii)   There is a number $R > 0$ such that the series converges absolutely if $|z - z_0| < R$ and diverges if $|z - z_0| > R$.

The number $R > 0$ in case (iii) is called the **radius of convergence** of the series. As a convention, the radius of convergence is ∞ in case (i) and 0 in case (ii). In case (iii), the open disk $|z - z_0| < R$ is called the **disk of convergence** and the circle $|z - z_0| = R$ the **circle of convergence**.

Thus the series converges inside the circle of convergence and diverges outside the circle of convergence. On the circle of convergence, the series may converge at some or all or no points (Figure 1).

The proof of Theorem 1 is a straightforward adaptation of the proof from calculus for real power series. We will present it at the end of the section along with the Cauchy-Hadamard formula for computing the radius of convergence. For now, we can find the radius of convergence of a power series using any one of the tests of convergence from Section 4.1.

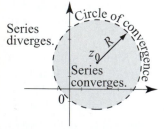

Figure 1 The disk of convergence is open and centered at the center of the power series. The power series converges absolutely inside the disk, diverges outside the disk, and may converge or diverge at points on the circle of convergence.

EXAMPLE 1   Computing the radius of convergence
Find the radii of convergence of the power series

(a) $\sum_{n=0}^{\infty} \frac{(z - 1 + i)^n}{(n!)^2}$,   (b) $\sum_{n=1}^{\infty} \frac{n^n(z - i)^n}{n}$,   (c) $\sum_{n=0}^{\infty}(-2)^n \frac{z^n}{n + 1}$.

Solution   (a)   We apply the ratio test. For $z \neq 1 - i$, we have

$$\rho = \lim_{n \to \infty} \left| \frac{(z - 1 + i)^{n+1}(n!)^2}{(z - 1 + i)^n((n+1)!)^2} \right| = \lim_{n \to \infty} \frac{|z - 1 + i|(n!)^2}{(n+1)^2(n!)^2} = \lim_{n \to \infty} \frac{|z - 1 + i|}{(n+1)^2} = 0.$$

Since $\rho < 1$, we conclude from the ratio test that the series converges for all z. Hence the radius of convergence is $R = \infty$.

(b)   We use the root test (Theorem 12, Section 4.1). For $z \neq i$, we have

$$\rho = \lim_{n \to \infty} \sqrt[n]{|a_n|} = \lim_{n \to \infty} \left(\frac{n^n |z - i|^n}{n} \right)^{\frac{1}{n}} = \lim_{n \to \infty} \frac{n|z - i|}{n^{\frac{1}{n}}} = \infty,$$

because $n|z - i| \to \infty$ and $n^{\frac{1}{n}} \to 1$ as $n \to \infty$ (Exercise 10, Section 4.1). Since $\rho > 1$ for all $z \neq i$, we conclude from the root test that the series diverges for all $z \neq i$. Hence the radius of convergence is $R = 0$.

(c)   We use the ratio test (Theorem 11, Section 4.1). For $z \neq 0$, we have

$$\rho = \lim_{n \to \infty} \left| \frac{a_{n+1}}{a_n} \right| = \lim_{n \to \infty} \left| \frac{2^{n+1} z^{n+1} (n+1)}{2^n z^n (n+2)} \right| = \lim_{n \to \infty} 2|z| \frac{n+1}{n+2} = 2|z|.$$

By the ratio test, the series converges if $2|z| < 1$, and diverges if $2|z| > 1$. Equivalently, the series converges if $|z| < \frac{1}{2}$, and diverges if $|z| > \frac{1}{2}$. Hence the radius of convergence is $R = \frac{1}{2}$.   ∎

Given a power series with radius of convergence $R > 0$, we can think of it as a function $f(z) = \sum_{n=0}^{\infty} c_n (z - z_0)^n$ whose domain of definition is the disk of convergence $|z - z_0| < R$. In the special case when all but finitely many c_n's are 0, this function reduces to a polynomial. So, in a way, we can regard power series as generalizing polynomials. Just as linearity allows us to differentiate polynomials term by term, power series can be differentiated term by term within the disk of convergence $|z - z_0| < R$. The key to these important properties is the uniform convergence of the power series on closed subdisks of the disk of convergence.

THEOREM 2
UNIFORM
CONVERGENCE ON
SUBDISKS

> Suppose that the power series $\sum_{n=0}^{\infty} c_n (z - z_0)^n$ has radius of convergence $R > 0$. Then the power series converges absolutely and uniformly on any closed subdisk $|z - z_0| \leq r$, where $0 < r < R$. Consequently, the power series converges uniformly on any closed and bounded subset of its disk of convergence.

Proof   By definition of the radius of convergence, the power series converges for all $|z - z_0| < R$. In particular, given $0 < r < R$, we can find ζ such that $r < |\zeta - z_0| < R$ and $\sum_{n=0}^{\infty} c_n (\zeta - z_0)^n$ is convergent. By the nth term test, $c_n (\zeta - z_0)^n \to 0$ as $n \to \infty$. Consequently, the sequence $\{c_n (\zeta - z_0)^n\}$ is bounded. Let $M > 0$ be such that $|c_n (\zeta - z_0)^n| \leq M$ for all $n = 0, 1, 2, \ldots$. For $|z - z_0| \leq r$ we have

$$|c_n (z - z_0)^n| = \overbrace{|c_n (\zeta - z_0)^n|}^{\leq M} \left| \frac{z - z_0}{\zeta - z_0} \right|^n \leq M \left(\frac{r}{|\zeta - z_0|} \right)^n = M\rho^n,$$

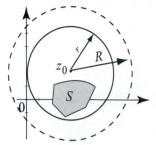

Figure 2 If S is a closed sub-set of the open disk of convergence $B_R(z_0)$, we can find a closed disk $S_r(z_0)$ such that $S \subset S_r(z_0) \subset B_R(z_0)$.

where $\rho = \frac{r}{|\zeta - z_0|} < 1$. Since $\sum M\rho^n$ is convergent, we conclude from the Weierstrass M-test that $\sum_{n=0}^{\infty} c_n(z - z_0)^n$ converges absolutely and uniformly for all $|z - z_0| \leq r$, which proves the first part. Now, given a closed and bounded subset S of the disk of convergence, let $0 < r < R$ be such that the closed disk $S_r(z_0)$ contains S (Figure 2). Since the series converges uniformly on $S_r(z_0)$ and $S \subset S_r(z_0)$ it follows that the series converges uniformly on S. ∎

Theorem 2 opens the door for the applications of the powerful results of the previous section. Each term in a power series is a polynomial of the form $c_n(z - z_0)^n$ and is continuous on $\mathbb{C}$. By Theorem 2, the power series converges uniformly on subdisks $|z - z_0| \leq r < R$, and by Theorem 1, Section 4.2, the series is continuous on these subdisks. Letting r tend up to R, the series is continuous on the whole open disk $|z - z_0| < R$. Applying Theorem 1, Section 4.2, we get the following.

THEOREM 3
TERM-BY-TERM
INTEGRATION

Let $R > 0$ denote the radius of convergence of the power series $\sum_{n=0}^{\infty} c_n(z - z_0)^n$. Let γ be any path in the disk of convergence $|z - z_0| < R$. Then

$$\int_\gamma \left(\sum_{n=0}^{\infty} c_n(z - z_0)^n \right) dz = \sum_{n=0}^{\infty} c_n \int_\gamma (z - z_0)^n \, dz.$$

That is, the power series may be integrated term by term within its radius of convergence.

Proof Since γ is a closed subset of the disk of convergence, by Theorem 2, the power series converges uniformly on γ. Now apply Corollary 1, Section 4.2. ∎

Theorem 3 is particularly interesting when we take advantage of independence of path of integrals of analytic functions on a disk. Let us illustrate with a typical application.

EXAMPLE 2   Term-by-term integration
Let us start with the geometric series

$$\frac{1}{1 + \zeta} = \sum_{n=0}^{\infty} (-1)^n \zeta^n \qquad |\zeta| < 1.$$

Since the integral of an analytic function on a disk is independent of path, we can integrate both sides from 0 to z, where $|z| < 1$, and use Theorem 3 to arrive at

$$\int_0^z \frac{1}{1 + \zeta} \, d\zeta = \sum_{n=0}^{\infty} (-1)^n \int_0^z \zeta^n \, d\zeta.$$

Using independence of path again (Theorem 1, Section 3.3) and the fact that $\frac{d}{dz} \text{Log} \, (1 + z) = \frac{1}{1+z}$ and $\text{Log} \, 1 = 0$, we obtain

$$\text{Log} \, (1 + z) = \sum_{n=0}^{\infty} (-1)^n \frac{z^{n+1}}{n + 1} \qquad |z| < 1.$$

∎

The following straightforward application of Theorem 2 above and Corollary 2, Section 4.2, is perhaps the most important property of power series.

THEOREM 4
ANALYTICITY AND
TERMWISE
DIFFERENTIATION
OF POWER SERIES

Suppose that $f(z) = \sum_{n=0}^{\infty} c_n(z - z_0)^n$ is a power series with radius of convergence $R > 0$. Then f is analytic on the disk of convergence $|z - z_0| < R$. Moreover, for any integer $k \geq 1$, the power series may be differentiated k-times term by term within its radius of convergence to yield, for $|z - z_0| < R$,

$$(1) \qquad f^{(k)}(z) = \sum_{n=k}^{\infty} n(n-1) \cdots (n-k+1)c_n(z-z_0)^{n-k},$$

with radius of convergence R.

Thus a power series is infinitely differentiable (a property of analytic functions) and all its derivatives can be obtained by successive termwise differentiation. Since the power series in (1) converges to $f^{(k)}(z)$ for all $|z - z_0| < R$, its radius of convergence is at least as large as R. A straightforward application of the comparison test shows that the radius of convergence is precisely R (Exercise 23). Another way of writing (1) is

$$(2) \qquad f^{(k)}(z) = \sum_{n=k}^{\infty} \frac{n!}{(n-k)!}c_n(z-z_0)^{n-k}, \qquad \text{for all } |z - z_0| < R.$$

EXAMPLE 3   Term-by-term differentiation

Find the sum $\sum_{n=1}^{\infty} nz^n$, and determine its radius of convergence.

Solution   Let us go straight to the point: The series looks like the derivative of the geometric series. Indeed, term-by-term differentiating the geometric series

$$\frac{1}{1-z} = \sum_{n=0}^{\infty} z^n, \qquad |z| < 1,$$

we obtain the power series

$$\frac{d}{dz}\frac{1}{1-z} = \frac{1}{(1-z)^2} = \sum_{n=1}^{\infty} nz^{n-1}, \qquad |z| < 1.$$

Multiplying both sides of the identity by z, we obtain

$$\frac{z}{(1-z)^2} = \sum_{n=1}^{\infty} nz^n, \qquad |z| < 1.$$

In particular, the radius of convergence is 1.   ∎

EXAMPLE 4   Matching series

Find the sum $\sum_{n=2}^{\infty} \frac{n}{(n-2)!} z^n$, and determine its radius of convergence.

Solution   The factorial in the denominator of the coefficients suggests looking at the exponential series:

$$e^z = \sum_{n=0}^{\infty} \frac{z^n}{n!}, \qquad \text{for all } z.$$

Differentiating the series twice term by term, we obtain

$$e^z = \sum_{n=2}^{\infty} \frac{n(n-1)}{n!} z^{n-2} = \sum_{n=2}^{\infty} \frac{z^{n-2}}{(n-2)!}, \qquad \text{for all } z.$$

Multiplying both sides by z^2, we obtain

$$z^2 e^z = \sum_{n=2}^{\infty} \frac{z^n}{(n-2)!}, \qquad \text{for all } z.$$

Differentiating term by term to get n in the numerator of the coefficients, and then multiplying by z, we obtain

$$\frac{d}{dz}\left(z^2 e^z\right) = 2z e^z + z^2 e^z = \sum_{n=2}^{\infty} \frac{n}{(n-2)!} z^{n-1}, \qquad \text{for all } z;$$

$$(2z^2 + z^3) e^z = \sum_{n=2}^{\infty} \frac{n z^n}{(n-2)!}, \qquad \text{for all } z.$$

In particular, the radius of convergence is ∞. ∎

Many important functions in applied mathematics are defined using power series. The exponential function is one example. In what follows we will introduce the **Bessel functions** using power series and derive some of their interesting properties. Bessel functions are solutions of the Bessel differential equation, which is an equation that arises in solving classical partial differential equations, including Laplace's equation on disks and cylindrical regions (see Chapter 9).

EXAMPLE 5   Bessel functions of integer order

For $n = 0, 1, 2, \ldots$, the **Bessel function of order** n is defined by

$$J_n(z) = \sum_{k=0}^{\infty} \frac{(-1)^k}{k!(k+n)!} \left(\frac{z}{2}\right)^{2k+n} \qquad \text{for all } z.$$

For negative integers n, we set $J_n(z) = (-1)^n J_{-n}(z)$.

(a)   Show that $J_n(z)$ is entire.

(b)   Verify the identity $\dfrac{d}{dz}\left[z^n J_n(z)\right] = z^n J_{n-1}(z)$.

(c)   Prove the recurrence relation: $z J_n'(z) + n J_n(z) = z J_{n-1}(z)$.

Solution   (a)   Since $J_n(z)$ is defined by a power series, to prove that it is entire, it suffices by Theorem 4 to show that the power series converges for all z. Using the ratio test, we have for $z \neq 0$

$$
\begin{aligned}
\rho &= \lim_{k \to \infty} \left| \frac{a_{k+1}}{a_k} \right| = \lim_{k \to \infty} \left| \frac{z^{2(k+1)+n}}{2^{2(k+1)+n}(k+1)!(k+1+n)!} \frac{2^{2k+n}k!(k+n)!}{z^{2k+n}} \right| \\
&= \lim_{k \to \infty} \frac{2^2 |z|^2}{(k+1)(k+1+n)} = 0.
\end{aligned}
$$

Since $\rho < 1$, the series converges for all z.

(b)   We use the series definition of J_n, differentiate term by term, and get

$$
\begin{aligned}
\frac{d}{dz}\left[z^n J_n(z)\right] &= \frac{d}{dz} \sum_{k=0}^{\infty} \frac{(-1)^k 2^n}{k!\,(k+n)!} \left(\frac{z}{2}\right)^{2k+2n} \\
&= \sum_{k=0}^{\infty} \frac{(-1)^k 2^n (k+n)}{k!\,(k+n)!} \left(\frac{z}{2}\right)^{2k+2n-1} \\
&= z^n \sum_{k=0}^{\infty} \frac{(-1)^k}{k!\,(k+n-1)!} \left(\frac{z}{2}\right)^{2k+n-1} \\
&= z^n J_{n-1}(z).
\end{aligned}
$$

(c)   The identity is true if $z = 0$. For $z \neq 0$, expand the left side of the identity in (b) using the product rule and get

$$
z^n J_n'(z) + n z^{n-1} J_n(z) = z^n J_{n-1}(z).
$$

Now (c) follows upon dividing through by $z^{n-1} \neq 0$.   ■

Bessel functions of arbitrary order $p > 0$ will be defined and studied in the exercises. We now record some immediate consequences of Theorem 4.

COROLLARY 1
TAYLOR
COEFFICIENTS

If $f(z) = \sum_{n=0}^{\infty} c_n(z - z_0)^n$ is a power series with radius of convergence $R > 0$, then the coefficients c_n are given by the **Taylor formula**

$$
(3) \qquad c_n = \frac{f^{(n)}(z_0)}{n!} \qquad (n = 0, 1, 2, \ldots),
$$

with the usual convention: $f^{(0)}(z) = f(z)$ and $0! = 1$.

Proof   If we evaluate (2) at $z = z_0$, all the terms in the series vanish except when $n = k$ and we get $f^{(k)}(z_0) = \frac{k!}{(k-k)!} c_k = k! c_k$. Solving for c_k, we get (3) with n replaced by k.   ■

Corollary 1 has an important consequence, which we state as a theorem.

THEOREM 5
UNIQUENESS

If $\sum_{n=0}^{\infty} a_n(z - z_0)^n = f(z) = \sum_{n=0}^{\infty} b_n(z - z_0)^n$ for all $|z - z_0| < R$ where $R > 0$, then $a_n = b_n$ for all n. In particular, if $f(z) = \sum_{n=0}^{\infty} c_n(z - z_0)^n$ for all $|z - z_0| < R$, then $f(z)$ is identically 0 for all $|z - z_0| < R$ if and only if $c_n = 0$ for all n.

Proof   By Corollary 1, $a_n = \frac{f^{(n)}(z_0)}{n!} = b_n$. Now the second part follows from the first part, which guarantees the uniqueness of the coefficients.   ■

Appendix: Proofs related to convergence

The following lemma is needed for the proof of Theorem 1.

LEMMA 1

(i)   If a power series $\sum_{n=0}^{\infty} c_n(z - z_0)$ converges for some $z_1 \neq z_0$ then it converges for all z such that $|z - z_0| < |z_1 - z_0|$.
(ii)   If a power series $\sum_{n=0}^{\infty} c_n(z - z_0)$ diverges for some z_2 then it diverges for all z such that $|z - z_0| > |z_2 - z_0|$.

Proof   We will use an argument similar to the one we used in the proof of Theorem 2.   (i)   Suppose that $\sum_{n=0}^{\infty} c_n(z - z_0)$ converges; then the sequence $\{c_n(z_1 - z_0)^n\}$ is bounded. So there is an $M > 0$ such that $|c_n(z_1 - z_0)^n| \leq M$ for all $n \geq 0$.   For $|z - z_0| < |z_1 - z_0|$, we have $\left| \frac{z - z_0}{z_1 - z_0} \right| = \rho < 1$.   Hence $|c_n(z - z_0)^n| = |c_n(z_1 - z_0)^n| \left| \frac{c_n(z - z_0)^n}{c_n(z_1 - z_0)^n} \right| \leq M\rho^n$, and the series $\sum_{n=0}^{\infty} c_n(z - z_0)$ is absolutely convergent by comparison to the convergent series $\sum M\rho^n$. Part (ii) follows from (i), because if the series converges for some z such that $|z - z_0| > |z_2 - z_0|$, then it will converge for z_2 by (i).   ■

Proof of Theorem 1   Suppose that neither (i) nor (ii) of Theorem 1 is true. Then there are $z_1 \neq z_0$ and z_2 such that $\sum_{n=0}^{\infty} c_n(z_1 - z_0)^n$ converges and $\sum_{n=0}^{\infty} c_n(z_2 - z_0)^n$ diverges. The set of real numbers

$$ S = \left\{ |z - z_0| : \quad \sum_{n=0}^{\infty} c_n(z - z_0)^n \quad \text{is convergent} \right\} $$

is nonempty and bounded above by $|z_2 - z_0|$, by Lemma 1. Hence by the completeness of real numbers, S has a least upper bound R, and $0 < |z_1 - z_0| \leq R \leq |z_2 - z_0| < \infty$. If $|z - z_0| < R$, then because R is the least upper bound of S, there is a z' such that $|z - z_0| < |z' - z_0|$ and $\sum_{n=0}^{\infty} c_n(z' - z_0)^n$ is convergent. By Lemma 1, it follows that $\sum_{n=0}^{\infty} c_n(z - z_0)^n$ is convergent. If $|z - z_0| > R$, then there is a z' such that $|z' - z_0| < |z_1 - z_0|$ such that $\sum_{n=0}^{\infty} c_n(z' - z_0)^n$ is divergent. Again, by Lemma 1, $\sum_{n=0}^{\infty} c_n(z - z_0)^n$ is divergent. As a conclusion, the number $R > 0$ is such that if $|z - z_0| < R$, then the series $\sum_{n=0}^{\infty} c_n(z - z_0)^n$ converges, and if $|z - z_0| > R$, then the series $\sum_{n=0}^{\infty} c_n(z - z_0)^n$ diverges. Thus, (iii) is true.   ■

The final result of this section is a formula, known as the **Cauchy-Hadamard formula**, for computing the radius of convergence of a power series. To present this formula, we need the following definition.

For a sequence of real numbers $\{x_n\}$, define the **limit superior** of $\{x_n\}$, denoted $\limsup_{n \to \infty} x_n$ or simply $\limsup x_n$, to be the smallest number M such that

(4)   for any $\epsilon > 0$, $x_n \leq M + \epsilon$ for all but finitely many n's.

If (4) holds for no real number M, then we define $\limsup x_n = \infty$. If (4) holds for all real numbers M, then we define $\limsup x_n = -\infty$. It is easy to show that if $L = \lim x_n$ exists, then $L = \limsup x_n$, but even if the limit does not exist, the limit superior will still exist (as either a real number or as ∞ or $-\infty$). The existence of a limit superior follows from the completeness of real numbers.

As an illustration, let $x_n = 1 + \frac{1}{n}$ if n is even and $n = \frac{1}{n}$ if n is odd. Then $\{x_n\}$ does not converge, but you can check that $\limsup x_n = 1$.

THEOREM 6
CAUCHY-HADAMARD
FORMULA

The radius of convergence R of the series $\sum_{n=0}^{\infty} c_n(z - z_0)^n$ is given by

(5)
$$\frac{1}{R} = \limsup \sqrt[n]{|c_n|}$$

with the convention $\frac{1}{0} = \infty$ and $\frac{1}{\infty} = 0$.

Proof To simplify the notation, we give the proof for the case $z_0 = 0$. We must show that (i) $\sum_{n=0}^{\infty} c_n z^n$ converges for $|z| < R$; (ii) $\sum_{n=0}^{\infty} c_n z^n$ diverges for $|z| > R$. To prove (i), choose ρ such that $|z| < \rho < R$; then $\frac{1}{R} < \frac{1}{\rho}$ and $\sqrt[n]{|c_n|} < \frac{1}{\rho}$ for all sufficiently large n. Hence $|c_n z^n| < \left(\frac{|z|}{\rho}\right)^n$ for all sufficiently large n, and $\sum_{n=0}^{\infty} c_n z^n$ converges absolutely by comparison to convergent geometric series $\sum \left(\frac{|z|}{\rho}\right)^n$. To prove (ii), note that $\frac{1}{|z|} < \frac{1}{R}$ implies that $\sqrt[n]{|c_n|} > \frac{1}{|z|}$ for infinitely many n's, by definition of R. So, for $|z| > R$, we have $|c_n z^n| > 1$ for infinitely many n's, and the series diverges by the nth term test. ∎

EXAMPLE 6   Cauchy-Hadamard formula
For the series $\sum_{n=0}^{\infty} c_n z^n$, where $c_n = 2^n$ if n is even and $\frac{1}{2^n}$ if n is odd, the Cauchy-Hadamard test gives

$$\frac{1}{R} = \limsup \sqrt[n]{|c_n|} = \limsup\{2, \frac{1}{2}, 2, \frac{1}{2}, \ldots\} = 2.$$

Thus the radius of convergence is $R = \frac{1}{2}$. ∎

Exercises 4.3

In Exercises 1–12, find the radius, disk, and circle of convergence of the given power series. Use the Cauchy-Hadamard formula in Exercises 9–12.

1. $\sum_{n=0}^{\infty} (-1)^n \frac{z^n}{2n+1}.$

2. $\sum_{n=0}^{\infty} \frac{n! z^n}{(2n)!}.$

3. $\sum_{n=0}^{\infty} (2)^n \frac{(z-i)^n}{n!}.$

4. $\sum_{n=0}^{\infty} \frac{(2z+1-i)^{2n}}{n^2+i}.$

5. $\sum_{n=0}^{\infty} \frac{(4iz-2)^n}{2^n}.$

6. $\sum_{n=0}^{\infty} (n-2)! \frac{z^n}{n^2}.$

7. $\sum_{n=0}^{\infty} (n+i)^4 (z+6)^n.$

8. $\sum_{n=0}^{\infty} \left(\frac{z+1}{3-i}\right)^{2n}.$

9. $\sum_{n=0}^{\infty} (1 - e^{in\frac{\pi}{4}})^n z^n.$

10. $\sum_{n=0}^{\infty} z^{n!}.$

11. $\sum_{n=0}^{\infty} (1+i^n)^n z^n.$

12. $\sum_{n=0}^{\infty} (2+2i^n)^n z^n.$

In Exercises 13–18, manipulate the geometric series $\frac{1}{1-z} = \sum_{n=0}^{\infty} z^n$, $|z| < 1$, to find the sum of the given series and its radius of convergence. You may differentiate or integrate term by term, change variables, multiply the series by z^m, take linear combinations of series obtained from the geometric series, and so on. State clearly the operation that you are performing on the series and justify it.

13. $\displaystyle\sum_{n=1}^{\infty} 2n z^{n-1}$.

14. $\displaystyle\sum_{n=2}^{\infty} \frac{z^n}{n(n-1)}$.

15. $\displaystyle\sum_{n=1}^{\infty} (-1)^n \frac{z^{n-1}}{n(n+1)}$.

16. $\displaystyle\sum_{n=1}^{\infty} \frac{(1-(-1)^n)}{n} z^n$.

17. $\displaystyle\sum_{n=0}^{\infty} \frac{(3z-i)^n}{3^n}$.

18. $\displaystyle\sum_{n=1}^{\infty} \frac{(z+1)^n}{n(n+1)}$.

19. Cauchy's estimate. Suppose that $f(z) = \sum_{n=0}^{\infty} c_n (z-z_0)^n$ is a power series with radius of convergence $R > 0$ and $|f(z)| \le M$ for all $|z - z_0| < R$. Show that the coefficients satisfy **Cauchy's estimate**

$$(6) \qquad\qquad |c_n| \le \frac{M}{R^n} \qquad (n = 0, 1, 2, \ldots).$$

[Hint: Use (3) and Cauchy's estimate from Section 3.7.]

20. Suppose that $f(z) = \sum_{n=0}^{\infty} c_n z^n$ converges for all z and that $|f(z)| \le A + B|z|^p$ for some nonnegative real numbers A, B, and p. Show that f is a polynomial of degree $\le p$. (Compare with Liouville's theorem.)

21. Cauchy products of power series. Suppose that $\sum_{n=0}^{\infty} a_n (z - z_0)^n$ has radius of convergence $R_1 > 0$ and $\sum_{n=0}^{\infty} b_n (z - z_0)^n$ has radius of convergence $R_2 > 0$. Show that their Cauchy product is

$$(7) \qquad \left(\sum_{n=0}^{\infty} a_n (z - z_0)^n \right) \left(\sum_{n=0}^{\infty} b_n (z - z_0)^n \right) = \sum_{n=0}^{\infty} c_n (z - z_0)^n, \qquad |z - z_0| < R,$$

where

$$(8) \qquad c_n = \sum_{k=0}^{n} a_k b_{n-k} = a_0 b_n + a_1 b_{n-1} + \cdots + a_{n-1} b_1 + a_n b_0,$$

and R is at least as large as the smallest of R_1 and R_2. [Hint: Justify the application of Theorem 15, Section 4.1.] Does the definition of the Cauchy product seem more natural when considering power series?

22. (a) Compute the Cauchy product of the series $\sum_{n=0}^{\infty} z^n$ and $\sum_{n=0}^{\infty} \frac{z^n}{n!}$. What is the radius of convergence of the Cauchy product series?
(b) Compute the Cauchy product of the series $\sum_{n=0}^{\infty} z^n$ and $1 - z$. What is the radius of convergence of the Cauchy product series? Notice how it is larger than the smaller of the radii of convergence of $\sum z^n$ and $1 + z$.
(c) Compute the Cauchy product of the series $\sum_{n=0}^{\infty} z^n$ with itself. What is the radius of convergence of the product series?

23. Show that the radius of convergence of a power series is equal to the radius of convergence of the k-times term-by-term differentiated power series. [Hint: It is enough to prove the result with $k = 1$ (why?). Use the comparison test or the Cauchy-Hadamard formula.]

24. Project Problem: The gamma function.   For $\operatorname{Re} z > 0$, define

(9)
$$\Gamma(z) = \int_0^\infty e^{-t} t^{z-1}\, dt,$$

where $t^{z-1} = e^{(z-1)\operatorname{Log} t}$ is the principal branch of t^{z-1}. This integral is improper on both ends. It should be interpreted as $\lim_{A\downarrow 0, B\uparrow\infty} \int_A^B e^{-t} t^{z-1}\, dt$. Let Ω denote the half-plane $\Omega = \{z :\ \operatorname{Re} z > 0\}$.
(a) Show that $|e^{-t} t^{z-1}| \le e^{-t} t^{\operatorname{Re} z - 1}$ and conclude that the integral in (9) converges absolutely for all z in Ω.
(b) Write $\Gamma(z) = \sum_{n=0}^\infty I_n(z)$, where $I_n(z) = \int_n^{n+1} e^{-t} t^{z-1}\, dt$. Show that $I_n(z)$ is analytic on Ω. [Hint: Theorem 4, Section 3.5.]
(c) Let S be a bounded subset of Ω, and let $0 < \operatorname{Re} z \le A$ for all z in S. Show that $|I_n(z)| \le M_n = (n+1)^{A-1}\left(e^{-n} - e^{-(n+1)}\right)$ for all z in S. Apply the Weierstrass M-test and use Corollary 2 to conclude that $\Gamma(z)$ is analytic on S. Since S was arbitrary, conclude that $\Gamma(z)$ is analytic on Ω.
(d) Use integration by parts to prove the **basic property** of the gamma function

(10)
$$\Gamma(z+1) = z\,\Gamma(z), \qquad \operatorname{Re} z > 0.$$

(e) Show by direct computation that $\Gamma(1) = 1$. Then use (d) to prove that for any positive integer n, $\Gamma(n) = (n-1)!$. For this reason, the gamma function is sometimes called the **generalized factorial function**.

25. Project Problem: The gamma function (continued).   In this exercise, you are asked to prove the formula

$$\frac{\Gamma(z_1)\Gamma(z_2)}{\Gamma(z_1 + z_2)} = 2\int_0^{\frac{\pi}{2}} \cos^{2z_1-1}\theta \sin^{2z_2-1}\theta\, d\theta, \qquad \operatorname{Re} z_1,\ \operatorname{Re} z_2 > 0.$$

All complex powers in this exercise are principal branches.
(a) Before proceeding with the proof, apply the formula with $z_1 = \frac{1}{2} = z_2$ and show that $\Gamma(\frac{1}{2}) = \sqrt{\pi}$.
(b) Make the change of variables $u^2 = t$ in (9) and obtain

$$\Gamma(z) = 2\int_0^\infty e^{-u^2} u^{2z-1}\, du, \qquad \operatorname{Re} z > 0.$$

(c) Use (b) to show that for $\operatorname{Re} z_1,\ \operatorname{Re} z_2 > 0$,

$$\Gamma(z_1)\Gamma(z_2) = 4\int_0^\infty \int_0^\infty e^{-(u^2+v^2)} u^{2z_1-1} v^{2z_2-1}\, du\, dv.$$

(d) Use polar coordinates, $u = r\cos\theta$, $v = r\sin\theta$, $du\, dv = r\, dr\, d\theta$ and get

$$\Gamma(z_1)\Gamma(z_2) = 2\,\Gamma(z_1 + z_2)\int_0^{\pi/2} \cos^{2z_1-1}\theta \sin^{2z_2-1}\theta\, d\theta.$$

[Hint: Recall (a) as you compute the integral in r.]

4.4   Taylor Series

In the previous section we learned that a power series with a positive radius of convergence is analytic on its disk of convergence. In this section, we will prove a converse of sorts, stating that if $f(z)$ is analytic in some region Ω, then f has a power series representation at every point in Ω. In other words, every analytic function is locally a power series. This is a fundamental result in complex analysis, with important consequences.

THEOREM 1
TAYLOR SERIES

Suppose that f is analytic in a region Ω and z_0 is in Ω. Let $B_R(z_0)$ be the largest open disk centered at z_0 and contained in Ω. Then f has a unique **Taylor series** expansion around z_0 given by

$$(1) \qquad f(z) = \sum_{n=0}^{\infty} \frac{f^{(n)}(z_0)}{n!}(z - z_0)^n, \qquad |z - z_0| < R.$$

The coefficient $\frac{f^{(n)}(z_0)}{n!}$ is called the nth **Taylor coefficient** of f.

Taylor series are named after the English mathematician Brook Taylor (1685–1731), who wrote about them in 1715. The idea of expanding a function by a power series was known to mathematicians before Taylor and appeared in the work of Sir Isaac Newton, the Scottish mathematician James Gregory, and the Swiss mathematician John Bernoulli.

Maclaurin series are named after the Scottish mathematician Colin Maclaurin (1698–1746), who used them and acknowledged that they are special cases of Taylor series.

When $z_0 = 0$ in Theorem 1, the series is commonly referred to as the **Maclaurin series** of f.

Before proceeding with proofs and examples, let us make some simple but useful remarks regarding Theorem 1.

- A Taylor series is a power series. So we may freely use results about power series as we work with Taylor series.

- The Taylor series of $f(z)$ around z_0 is uniquely defined by (1); the coefficients are determined by the derivatives of f. Furthermore, any power series representation of f centered at z_0 must be the Taylor series (Corollary 1, Section 4.3). If $f(z)$ is explicitly written as a power series is some disk, then f is identically its own Taylor series.

- The Taylor series will definitely converge to $f(z)$ for $|z - z_0| < R$, so the radius of convergence is at least R. If $f(z)$ has a serious problem (like a pole or an essential singularity; see Section 4.6) at some point on the circle $|z - z_0| = R$, then the Taylor series cannot converge in a larger disk and its radius of convergence is thus R. If $f(z)$ has a less serious problem like a branch cut or a removable singularity (Section 4.6) or has merely been defined in an artificially small domain, the Taylor series might converge in a circle with radius larger than R.

- Because successive termwise differentiation of a power series is allowed within its radius of convergence, if n is an integer ≥ 0, by changing the dummy index and differentiating (1) we obtain the Taylor series representation of the nth derivative

$$(2) \qquad f^{(n)}(z) = \sum_{j=n}^{\infty} \frac{f^{(j)}(z_0)}{(j-n)!}(z-z_0)^j, \qquad |z-z_0| < R.$$

- Because a power series converges uniformly on subdisks of its disk of convergence, the Taylor series (1) and (2) converge uniformly on any subdisk $|z - z_0| \leq r$ where $0 \leq r < R$.

To understand the statement about the radius of convergence of the Taylor series, let us consider the function $f(z) = \frac{1}{1-z}$ defined on the region $\Omega = \mathbb{C} \setminus \{1\}$. Theorem 1 tells us that the Taylor series expansion of $f(z)$ around $z_0 = 0$ will converge in the largest open disk centered at $z_0 = 0$ and contained in $\mathbb{C} \setminus \{1\}$. Clearly, the largest disk has radius 1. The Taylor series cannot converge in a disk larger than this, because if it did, the Taylor series would be continuous at $z = 1$; but $f(z)$ has unbounded modulus as $z \to 1$, and the Taylor series must equal f on the unit disk.

So the Taylor series of $f(z)$ around 0 has a radius of convergence 1. We can confirm this fact directly by recalling the geometric series

$$(3) \qquad \frac{1}{1-z} = \sum_{n=0}^{\infty} z^n \qquad |z| < 1,$$

which is the Taylor series expansion of $f(z)$ around 0. The advantage of Theorem 1 is that it gives us the radius of convergence without knowing the coefficients in the Taylor series. For example, Theorem 1 tells us that the Taylor series of $f(z) = \frac{1}{1-z}$ around $z_0 = i$ has radius of convergence $R = \sqrt{2}$ (Figure 1).

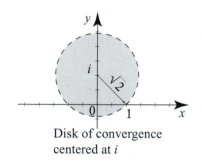

Disk of convergence centered at i

Figure 1 The function $f(z) = \frac{1}{1-z}$ has a serious problem at $z = 1$. If we expand it around $z_0 = i$, the power series will converge in the largest disk around i that does not contain 1. This disk has radius $\sqrt{2}$.

EXAMPLE 1   Maclaurin series of e^z, $\cos z$, and $\sin z$
Find the Maclaurin series expansions of (a) e^z,   (b) $\cos z$, and   (c) $\sin z$.

Solution   Let us first note that all three functions are entire, so the Maclaurin series will converge for all z; that is, $R = \infty$ in all three cases.
(a)   Recall that we defined the exponential function by a power series: $e^z = \sum_{n=0}^{\infty} \frac{z^n}{n!}$ for all z. So this is the Maclaurin series expansion of e^z. Let us reconfirm this using Theorem 1. If $f(z) = e^z$, then $f^{(n)}(z) = e^z$, so $f^{(n)}(0) = 1$ for all n. Therefore, as expected, the Maclaurin series is

$$\sum_{n=0}^{\infty} \frac{f^{(n)}(0)}{n!} z^n = \sum_{n=0}^{\infty} \frac{z^n}{n!}, \qquad \text{for all } z.$$

(b)   We have

$$f(z) = \cos z \quad \Rightarrow \quad f(0) = 1;$$
$$f'(z) = -\sin z \quad \Rightarrow \quad f'(0) = 0;$$
$$f''(z) = -\cos z \quad \Rightarrow \quad f''(0) = -1;$$
$$f'''(z) = \sin z \quad \Rightarrow \quad f'''(0) = 0;$$
$$f^{(4)}(z) = \cos z \quad \Rightarrow \quad f^{(4)}(0) = 1;$$

and so on. The values of the derivatives at 0 will repeat with period 4. Thus

$$\cos z = f(0) + \frac{f'(0)}{1!}z + \frac{f''(0)}{2!}z^2 + \frac{f'''(0)}{3!}z^3 + \frac{f^{(4)}(0)}{4!}z^4 + \cdots$$
$$= 1 - \frac{z^2}{2!} + \frac{z^4}{4!} + \cdots = \sum_{n=0}^{\infty}(-1)^n\frac{z^{2n}}{(2n)!} \qquad \text{for all } z.$$

(c)   We could find the Maclaurin series of $\sin z$ directly, as we did in (b) for $\cos z$. It is much easier to use the result of (b), and the relation $\frac{d}{dz}\cos z = -\sin z$. Then

$$\sin z = -\frac{d}{dz}\cos z = -\frac{d}{dz}\left(1 - \frac{z^2}{2!} + \frac{z^4}{4!} + \cdots\right)$$
$$= \frac{2z}{2!} - \frac{4z^3}{4!} + \cdots = \frac{z}{1!} - \frac{z^3}{3!} + \cdots = \sum_{n=0}^{\infty}(-1)^n\frac{z^{2n+1}}{(2n+1)!}. \qquad ∎$$

Example 1 shows the advantage that complex Taylor series enjoy over their real counterparts. A real Taylor series converges if and only if a certain remainder goes to zero. In the complex case, the remainder is irrelevant; the Taylor series will converge in the largest disk that you can fit inside the domain of definition of the analytic function.

Surely you are not surprised by the Maclaurin series expansions in Example 1. Given that $\cos x = \sum_{n=0}^{\infty}(-1)^n\frac{x^{2n}}{(2n)!}$ and $\sin x = \sum_{n=0}^{\infty}(-1)^n\frac{x^{2n+1}}{(2n+1)!}$ for $-\infty < x < \infty$, isn't it reasonable to expect that the series for $\cos z$ and $\sin z$ are obtained by merely replacing x by z? The answer is affirmative and we have the following useful result, which will allow us to turn well-known Taylor series from calculus into Taylor series of complex-valued function.

PROPOSITION 1
> Suppose that $f(x) = \sum_{n=0}^{\infty}c_nx^n$ is a Taylor series that converges for all $|x| < R$. Suppose that $g(z)$ is analytic on $|z| < R$ and $g(x) = f(x)$ for all $|x| < R$. Then the Taylor series of g is $g(z) = \sum_{n=0}^{\infty}c_nz^n$ for all $|z| < R$.

Proof   Since g is analytic in a neighborhood of 0, it has a Taylor series expansion $g(z) = \sum_{n=0}^{\infty}\frac{g^{(n)}(0)}{n!}z^n$. We can compute the derivatives $g^{(n)}(0)$ by taking limits as $z \to 0$ with $z = x$ real, and since $g = f$ on the real axis, we must have $g^{(n)}(0) = f^{(n)}(0)$ for all n. So the coefficients in the Taylor series of f and g are the same. That the series have the same radius of convergence follows from Lemma 1, Section 4.3, which is also valid for Taylor series of real-valued functions.   ∎

EXAMPLE 2   Manipulating Taylor series

For the given function, find the Maclaurin series expansion, and determine its radius of convergence: (a) $f(z) = ze^{z^2}$,   (b) $f(z) = \frac{1}{1+z^2}$.

Solution   (a) Since $e^z = \sum_{n=0}^{\infty} \frac{z^n}{n!}$ is valid for all z, replacing z by z^2, we obtain

$$e^{z^2} = \sum_{n=0}^{\infty} \frac{(z^2)^n}{n!} = \sum_{n=0}^{\infty} \frac{z^{2n}}{n!}, \qquad \text{for all } z.$$

Multiplying both sides by z, we get the desired Maclaurin series

$$ze^{z^2} = \sum_{n=0}^{\infty} \frac{z^{2n+1}}{n!}, \qquad \text{for all } z.$$

(b) Start with the geometric series $\frac{1}{1-w} = \sum_{n=0}^{\infty} w^n$, which is valid if and only if $|w| < 1$. Replace w by $-z^2$, note that $|w| < 1 \Leftrightarrow |-z^2| < 1 \Leftrightarrow |z| < 1$ and get

$$\frac{1}{1+z^2} = \sum_{n=0}^{\infty} (-z^2)^n = \sum_{n=0}^{\infty} (-1)^n z^{2n}, \qquad \text{for } |z^2| < 1, \text{ equivalently, } |z| < 1. \quad \blacksquare$$

Example 2(b) takes us back to a question that we mentioned in the introduction of Section 4.1. The function $f(x) = \frac{1}{1+x^2}$ has no problem on the real line. It is infinitely differentiable for all x. Yet its Maclaurin series $\sum_{n=0}^{\infty} (-1)^n x^{2n}$ converges only for $|x| < 1$. While it is difficult to explain this from a real analysis point of view, the justification is immediate in complex analysis. Since $\frac{1}{1+x^2}$ is the restriction to the real line of $\frac{1}{1+z^2}$, and since $\frac{1}{1+z^2}$ becomes unbounded at $z = \pm i$, the Maclaurin series of the latter function cannot converge outside the disk $|z| < 1$. So in view of Proposition 1, the Maclaurin series of $\frac{1}{1+x^2}$ cannot converge outside $|x| < 1$.

EXAMPLE 3   Maclaurin series of $\frac{\sin z}{z}$

Find the Maclaurin series expansion of $g(z) = \frac{\sin z}{z}$ $(z \neq 0)$, $g(0) = 1$, and determine its radius of convergence.

Solution   The function g is entire (Example 5, Section 3.6), so its Maclaurin series converges for all z. To find this series, start with $\sin z = \sum_{n=0}^{\infty} (-1)^n \frac{z^{2n+1}}{(2n+1)!}$, which converges for all z. For a given $z \neq 0$, we can multiply the series by $\frac{1}{z}$, and conclude that

$$\frac{\sin z}{z} = \frac{1}{z} \sin z = \frac{1}{z} \sum_{n=0}^{\infty} (-1)^n \frac{z^{2n+1}}{(2n+1)!} = \sum_{n=0}^{\infty} (-1)^n \frac{1}{z} \frac{z^{2n+1}}{(2n+1)!}$$

$$= \sum_{n=0}^{\infty} (-1)^n \frac{z^{2n}}{(2n+1)!} = \frac{1}{1!} - \frac{z^2}{3!} + \frac{z^4}{5!} - \cdots.$$

If $z = 0$, the series equals $1 = g(0)$, and so the Maclaurin series expansion is

$$\frac{\sin z}{z} = \sum_{n=0}^{\infty} (-1)^n \frac{z^{2n}}{(2n+1)!} \qquad \text{for all } z.$$

At $z = 0$, the left side is to be interpreted as its limit, $g(0) = 1$. ■

Recall that a function f is even if $f(-z) = f(z)$, and it is odd if $f(-z) = -f(z)$. The following useful property of Taylor series is obvious. You should verify it with the previous examples of this section.

**PROPOSITION 2
EVEN AND ODD
FUNCTIONS**

Suppose that f is analytic on $|z| < R$ and write $f(z) = \sum_{n=0}^{\infty} c_n z^n$, $|z| < R$, $R > 0$. Then
(i) f is even if and only if $c_{2n+1} = 0$ for all $n = 0, 1, 2, \ldots$.
(ii) f is odd if and only if $c_{2n} = 0$ for all $n = 0, 1, 2, \ldots$.

Proof   Suppose that f is even. Then $f(z) - f(-z) = 0$, so $\sum_{n=0}^{\infty} c_n \left(z^n - (-z)^n \right) = 0$, for all $|z| < R$. But $z^n - (-z)^n = 0$ for all even n and $z^n - (-z)^n = 2z^n$ for all odd n. So $\sum_{n=0}^{\infty} c_n \left(z^n - (-z)^n \right) = \sum_{n=0}^{\infty} 2c_{2n+1} z^{2n+1} = 0$, which implies that $c_{2n+1} = 0$ for all n, by the uniqueness of power series expansion. The case of an odd function is similar, and so we leave it as an exercise.   ■

The following lemma is a consequence of the geometric series. It will facilitate the proof of Theorem 1 and will be needed in the following section.

LEMMA 1

Let z_0, z_1, z_2 be distinct complex numbers such that $|z_1 - z_0| < |z_2 - z_0|$ (Figure 2). Then

(4)
$$\frac{1}{z_2 - z_1} = \sum_{n=0}^{\infty} \frac{(z_1 - z_0)^n}{(z_2 - z_0)^{n+1}}.$$

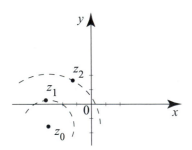

Figure 2

Proof   To expand $\frac{1}{z_2 - z_1}$ around z_0, we will add and subtract z_0 from the denominator and then factor to apply the geometric series $\frac{1}{1-w} = \sum_{n=0}^{\infty} w^n$ ($|w| < 1$). Here are the necessary steps:

$$\frac{1}{z_2 - z_1} = \frac{1}{(z_2 - z_0) - (z_1 - z_0)}$$

$$= \frac{1}{z_2 - z_0} \frac{1}{1 - \frac{z_1 - z_0}{z_2 - z_0}}$$

$$= \frac{1}{z_2 - z_0} \frac{1}{1 - w}$$

$$= \frac{1}{z_2 - z_0} \sum_{n=0}^{\infty} w^n$$

where $w = \frac{z_1 - z_0}{z_2 - z_0}$. The series will converge if and only if $|w| < 1$; equivalently, the series will converge if and only if $\left| \frac{z_1 - z_0}{z_2 - z_0} \right| < 1$ or $|z_1 - z_0| < |z_2 - z_0|$. Replacing w by $\frac{z_1 - z_0}{z_2 - z_0}$, we obtain (4).   ■

Proof of Theorem 1   Let $0 < r < R$. Since f is analytic on and inside $C_r(z_0)$, Cauchy's formula implies that

(5)
$$f(z) = \frac{1}{2\pi i} \int_{C_r(z_0)} \frac{f(\zeta)}{\zeta - z}\, d\zeta \qquad \text{for all } |z - z_0| < r.$$

(See Figure 3.)   The trick now is to expand the integrand in a series and then integrate term by term. Appealing to (4) with $z = z_1$ and $\zeta = z_2$, we see that

(6)
$$\frac{1}{\zeta - z} = \sum_{n=0}^{\infty} \frac{(z - z_0)^n}{(\zeta - z_0)^{n+1}}, \qquad |z - z_0| < |\zeta - z_0| = r.$$

Also, if $|z - z_0| < |\zeta - z_0| = r$, then $\left|\frac{z - z_0}{\zeta - z_0}\right| = \rho < 1$, and so $\left|\frac{(z - z_0)^n}{(\zeta - z_0)^{n+1}}\right| < \frac{\rho^n}{r} = M_n$. Hence, by the Weierstrass M-test, since $\sum M_n < \infty$, the series in (6) converges uniformly in ζ for all $|\zeta - z_0| = r$. Multiplying both sides of (6) by the continuous function $f(\zeta)$, we obtain

(7)
$$\frac{f(\zeta)}{\zeta - z} = \sum_{n=0}^{\infty} (z - z_0)^n \frac{f(\zeta)}{(\zeta - z_0)^{n+1}},$$

where the series converges uniformly on $|\zeta - z_0| = r$ (Lemma 1, Section 4.2). Integrating over $C_r(z_0)$, we get

$$
\begin{aligned}
f(z) &= \frac{1}{2\pi i} \int_{C_r(z_0)} \frac{f(\zeta)}{\zeta - z}\, d\zeta = \frac{1}{2\pi i} \int_{C_r(z_0)} \sum_{n=0}^{\infty} (z - z_0)^n \frac{f(\zeta)}{(\zeta - z_0)^{n+1}}\, d\zeta \\
&= \sum_{n=0}^{\infty} (z - z_0)^n \overbrace{\frac{1}{2\pi i} \int_{C_r(z_0)} \frac{f(\zeta)}{(\zeta - z_0)^{n+1}}\, d\zeta}^{= \frac{f^{(n)}(z_0)}{n!}} \\
&= \sum_{n=0}^{\infty} \frac{f^{(n)}(z_0)}{n!} (z - z_0)^n,
\end{aligned}
$$

where we have used Cauchy's generalized formula (6), Section 3.6. to evaluate the last path integral. This proves (1) for all $|z - z_0| < r$, where $0 < r < R$. Letting $r \to R$, the formula follows for all $|z - z_0| < R$.   ∎

The following example deals with an important family of numbers that arise naturally in many different contexts in mathematics. In the solution we will take full advantage of Theorem 1 and compute the radius of convergence of a Taylor series before knowing the Taylor coefficients.

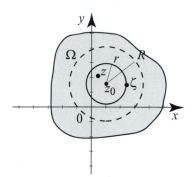

Figure 3 The picture around z_0: We have
$$|z - z_0| < r = |\zeta - z_0|$$
for all ζ on $C_r(z_0)$ and z inside $B_r(z_0)$.

EXAMPLE 4   Bernoulli numbers
Consider the function $f(z) = \frac{z}{e^z - 1}$, $z \neq 0$, $f(0) = 1$.
(a)   Show that f is analytic at 0.
(b)   By Theorem 1, f has a Maclaurin series expansion. Show that its radius of convergence is $R = 2\pi$.
(c)   Write the Maclaurin series in the form

$$f(z) = \sum_{n=0}^{\infty} \frac{B_n}{n!} z^n, \qquad |z| < 2\pi.$$

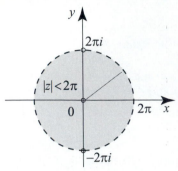

Figure 4 Having shown that $f(z) = \frac{z}{e^z - 1}$ is analytic at $z = 0$, its only problems are at $z = 2k\pi i$ (k integer $\neq 0$), which are the nonzero roots of $e^z - 1$. Thus its Taylor series will converge in the disk $|z| < 2\pi$.

The number B_n is called the nth **Bernoulli number**. Show that $B_0 = 1$, and derive the recursion formula

$$B_n = -\frac{1}{n+1} \sum_{k=0}^{n-1} \binom{n+1}{k} B_k, \quad n \geq 1.$$

(d) Find $B_0, B_1, B_2, \ldots, B_{12}$, with the help of the recursion formula and a calculator.

(e) Show that $B_{2n+1} = 0$ for $n \geq 1$.

Solution   (a) Consider $g(z) = \frac{1}{f(z)} = \frac{e^z - 1}{z}$ for $z \neq 0$, and $g(0) = 1$. By Theorem 4, Section 3.6, g is analytic at 0. Since $g(0) \neq 0$, $\frac{1}{g(z)} = f(z)$ is therefore analytic at $z = 0$.

(b) The Maclaurin series of f converges in the largest disk around $z_0 = 0$ on which f is defined and analytic. For $z \neq 0$, $f(z)$ is analytic except when $e^z - 1 = 0$, where f becomes unbounded. Since $e^z = 1$ precisely when z is an integer multiple of $2\pi i$, we see that the Maclaurin series converges for all $|z| < 2\pi$, and the radius of convergence is 2π.

(c) Multiplying both sides of the Maclaurin series expansion of $\frac{z}{e^z - 1}$ by $e^z - 1$ and using the Maclaurin series $e^z - 1 = z + \frac{z^2}{2!} + \frac{z^3}{3!} + \cdots = \sum_{n=1}^{\infty} \frac{z^n}{n!}$, we obtain

$$(8) \qquad z = (e^z - 1) \sum_{n=0}^{\infty} \frac{B_n}{n!} z^n = \sum_{n=1}^{\infty} \frac{z^n}{n!} \overset{=a_n z^n}{\sum_{n=0}^{\infty} \frac{B_n}{n!} z^n} = \sum_{n=1}^{\infty} c_n z^n, \qquad |z| < 2\pi,$$

where c_n will be computed from the Cauchy product formulas (see Theorem 15, Section 4.1, also, Exercise 21, Section 4.3). Note that because we are multiplying by the power series of $e^z - 1$ whose first term is z, the first term in the Cauchy product will have degree ≥ 1 (thus $c_0 = 0$). We have for each $n \geq 1$

$$c_n = \sum_{k=0}^{n-1} \frac{B_k}{k!} \frac{1}{(n-k)!} \qquad \text{(because } a_0 = 0\text{)}$$

$$= \frac{1}{n!} \sum_{k=0}^{n-1} \frac{n!}{k!(n-k)!} B_k = \frac{1}{n!} \sum_{k=0}^{n-1} \binom{n}{k} B_k.$$

By the uniqueness of the power series expansion, (8) implies that $c_1 = 1$, and $c_n = 0$ for all $n \geq 2$. Thus,

$$c_1 = 1 \quad \Rightarrow \quad \frac{1}{1!} B_0 = 1 \quad \Rightarrow \quad B_0 = 1;$$

$$c_n = 0, \quad n \geq 2 \quad \Rightarrow \quad \frac{1}{n!} \sum_{k=0}^{n-1} \binom{n}{k} B_k = 0, \quad n \geq 2.$$

Changing n to $n+1$ in the last identity, we see that, for $n \geq 1$,

$$\frac{1}{(n+1)!} \sum_{k=0}^{n} \binom{n+1}{k} B_k = 0$$

$$\Rightarrow \quad \frac{1}{(n+1)!} \sum_{k=0}^{n-1} \binom{n+1}{k} B_k + \frac{1}{(n+1)!} \binom{n+1}{n} B_n = 0.$$

Now, realizing that $\dbinom{n+1}{n} = n+1$, we get the desired recursion formula.

(d)   We have used a computer and the recursion formula to generate the Bernoulli numbers shown in Table 1.

n	0	1	2	3	4	5	6	7	8	9	10	11	12
B_n	1	$-\dfrac{1}{2}$	$\dfrac{1}{6}$	0	$-\dfrac{1}{30}$	0	$\dfrac{1}{42}$	0	$-\dfrac{1}{30}$	0	$\dfrac{5}{66}$	0	$-\dfrac{691}{2730}$

Table 1. Bernoulli numbers.

(e)   As the table suggests, $B_{2n+1} = 0$ for $n \geq 1$. This is clearly a fact about an even function. Consider $f(z)$ minus the $B_1 z$ term of its Maclaurin series:

$$(9) \qquad \frac{z}{e^z - 1} + \frac{z}{2} = \frac{z + ze^z}{2(e^z - 1)} = \frac{z(e^{\frac{z}{2}} + e^{-\frac{z}{2}})}{2(e^{\frac{z}{2}} - e^{-\frac{z}{2}})} = \frac{z}{2}\coth\left(\frac{z}{2}\right).$$

This is an even function. Hence all the odd numbered coefficients in its Maclaurin series are 0, which implies that $B_{2n+1} = 0$ for all $n \geq 1$.   ∎

Using (9) and the Maclaurin series of $\frac{z}{e^z - 1}$, we see that for $|z| < 2\pi$

$$\frac{z}{2}\coth\left(\frac{z}{2}\right) = \frac{z}{2} + 1 - \frac{z}{2} + \sum_{n=2}^{\infty} \frac{B_n}{n!} z^n = 1 + \sum_{n=1}^{\infty} \frac{B_{2n}}{(2n)!} z^{2n} = \sum_{n=0}^{\infty} \frac{B_{2n}}{(2n)!} z^{2n};$$

and upon replacing z by $2z$,

$$(10) \qquad z \coth z = \sum_{n=0}^{\infty} \frac{2^{2n} B_{2n}}{(2n)!} z^{2n}, \qquad |z| < \pi.$$

This connection between Bernoulli numbers and the Maclaurin series involving hyperbolic function is truly enchanting. Using it along with various relationships between hyperbolic and trigonometric functions, we will derive important Taylor series such as the ones for $\tan z$ and $\cot z$, in terms of Bernoulli numbers.

Exercises 4.4

In Exercises 1–12, a function $f(z)$ and a point z_0 are given. Without computing the Taylor series of $f(z)$ around z_0, determine the radius of convergence of the Taylor series.

1. e^{z-1}, $z_0 = 0$.

2. $\dfrac{\sin z}{e^z}$, $z_0 = 1 + 7i$.

3. $\dfrac{z}{z - 3i}$, $z_0 = 0$.

4. $\sin\left(\dfrac{z+1}{z-1}\right)$, $z_0 = 0$.

5. $\dfrac{z+1}{z-i}$, $z_0 = 2 + i$.

6. $\dfrac{\sin z}{z^2 + 4}$, $z_0 = 3$.

7. $\dfrac{z \cos z}{2z + 1}$, $z_0 = \dfrac{1}{2}$.

8. $\dfrac{z}{e^z - 1}$, $z_0 = i$.

9. $\tan z$, $z_0 = 0$.

10. $\cot z$, $z_0 = \dfrac{\pi}{2}$.

11. $\dfrac{1}{z^2 + z + 1}$, $z_0 = i$.

12. $\dfrac{\text{Log } z}{z - 1}$, $z_0 = \dfrac{i}{2}$.

In Exercises 13–20, use a known Taylor series to derive the Taylor series of the given function around the indicated point z_0. Determine the radius of convergence in each case.

13. $\dfrac{z}{1-z}$, $\quad z_0 = 0$.

14. $\dfrac{z^2+1}{z-1}$, $\quad z_0 = 0$.

15. $\dfrac{2z}{(z+i)^3}$, $\quad z_0 = 0$.

16. ze^{3z^2}, $\quad z_0 = 0$.

17. ze^z, $\quad z_0 = 1$.

18. $\cos^2 z$, $\quad z_0 = 0$.

19. $z \cos \dfrac{z}{2}$, $\quad z_0 = 0$.

20. $\cos z$, $\quad z_0 = \dfrac{\pi}{2}$.

21. Find the Maclaurin series of $f(z) = \frac{1}{(1-z)(2-z)}$ in two different ways as indicated.

(a) Prove the partial fractions decomposition $\frac{1}{(1-z)(2-z)} = \frac{1}{1-z} - \frac{1}{2-z}$, then use a geometric series expansion of each term in the partial fraction decomposition.

(b) Use a geometric series to expand $\frac{1}{1-z}$ and $\frac{1}{2-z}$ separately, then form their Cauchy product.

(c) Verify that your answers are the same in (a) and (b) and give the radius of convergence of the Maclaurin series of $f(z)$.

22. Let $z_1 \neq z_2$ be two complex numbers and suppose that $0 < |z_1| \leq |z_2|$. Show that

$$\frac{1}{(z_1-z)(z_2-z)} = \frac{1}{z_1-z_2} \sum_{n=0}^{\infty} \frac{(z_1^{n+1} - z_2^{n+1})}{(z_1 z_2)^{n+1}} z^n, \qquad |z| < |z_1|.$$

23. Find the Maclaurin series of $f(z) = \frac{1}{1+z+z^2}$ and determine its radius of convergence. You may use the result of Exercise 22.

24. Let $z_1 \neq 0$. (a) Show that

$$\frac{1}{z_1 - z} = \frac{1}{z_1} \sum_{n=0}^{\infty} \left(\frac{z}{z_1}\right)^n, \qquad |z| < |z_1|.$$

(b) For any positive integer k, show that

$$\frac{1}{(z_1-z)^{k+1}} = \frac{1}{z_1^{k+1}} \sum_{n=k}^{\infty} \binom{n}{k} \left(\frac{z}{z_1}\right)^{n-k} = \frac{1}{z_1^{k+1}} \sum_{n=0}^{\infty} \binom{n+k}{k} \left(\frac{z}{z_1}\right)^n,$$

where $|z| < |z_1|$. [Hint: Differentiate the series in (a) k times.]

In Exercises 25–26, find the Maclaurin series of the given function and determine its radius of convergence. You may use the result of Exercise 24.

25. $f(z) = \dfrac{1}{(z-2i)^3}$.

26. $f(z) = \dfrac{1}{(2z-i+1)^6}$.

In Exercises 27–30, show that the function is analytic at $z_0 = 0$ and find its Maclaurin series. What is the radius of convergence of the series?

27. $f(z) = \begin{cases} \frac{\cos z - 1}{z} & \text{if } z \neq 0, \\ 0 & \text{if } z = 0. \end{cases}$

28. $f(z) = \begin{cases} \frac{e^z - 1}{z} & \text{if } z \neq 0, \\ 1 & \text{if } z = 0. \end{cases}$

29. $f(z) = \begin{cases} \frac{e^{z^2} - 1}{z^2} & \text{if } z \neq 0, \\ 1 & \text{if } z = 0. \end{cases}$

30. $f(z) = \begin{cases} \frac{\sinh z^2}{z} & \text{if } z \neq 0, \\ 0 & \text{if } z = 0. \end{cases}$

31. Maclaurin series of the tangent, cotangent and cosecant.

(a)   Replace z by iz in (10) and simplify to obtain

$$z \cot z = \sum_{n=0}^{\infty} (-1)^n \frac{2^{2n} B_{2n}}{(2n)!} z^{2n}, \qquad |z| < \pi.$$

(b)   Derive the Maclaurin series of the tangent:

$$\tan z = \sum_{n=1}^{\infty} (-1)^{n-1} \frac{2^{2n}(2^{2n}-1)B_{2n}}{(2n)!} z^{2n-1}, \qquad |z| < \frac{\pi}{2}.$$

[Hint: $\tan z = \cot z - 2\cot(2z)$.]

(c)   Prove

$$z \csc z = \sum_{n=0}^{\infty} (-1)^{n-1} \frac{(2^{2n}-2)B_{2n}}{(2n)!} z^{2n}, \qquad |z| < \pi.$$

[Hint: $\cot z + \tan z = 2\csc 2z$.]

32.   Fibonacci numbers.

Discovered in the early thirteenth century by the great mathematician Leonardo of Pisa (1180–1250), who wrote under the name of Fibonacci, this sequence of integers is defined inductively by $c_0 = c_1 = 1$, and $c_n = c_{n-1} + c_{n-2}$ for $n \geq 2$.

(a)   Find $c_0, c_1, \ldots, c_{10}$. In 1843, the French mathematician Jacques-Philippe-Marie Binet (1786–1856) discovered a formula for c_n in terms of n. This formula, known now as **Binet's formula**, states that

$$c_n = \frac{1}{\sqrt{5}} \left[\left(\frac{1+\sqrt{5}}{2} \right)^{n+1} - \left(\frac{1-\sqrt{5}}{2} \right)^{n+1} \right].$$

We will use complex analysis to derive Binet's formula.

(b)   Suppose for a moment that there is a function $f(z)$, analytic at 0 and whose Maclaurin coefficients are the Fibonacci numbers: $f(z) = \sum_{n=0}^{\infty} c_n z^n$, $|z| < R$. Show that f is a solution of the equation $f(z) = 1 + zf(z) + z^2 f(z)$.

(c)   Conclude that $f(z) = \frac{1}{1-z-z^2}$ and find the radius of convergence of its Maclaurin series.

(d)   Compute the Maclaurin series and derive Binet's formula. [Hint: $\frac{1}{1-z-z^2} = \frac{1}{\sqrt{5}} \left(\frac{1}{-z + \frac{-1+\sqrt{5}}{2}} + \frac{1}{z + \frac{1+\sqrt{5}}{2}} \right)$. Now use geometric series expansions.]

33. The Lucas numbers.

This sequence of integers $\{l_n\}$, named after the French mathematician Edouard Lucas (1842–1891), is defined by a recurrence relation similar to the Fibonacci sequence $l_n = l_{n-1} + l_{n-2}$ but with $l_0 = 1$ and $l_1 = 3$. Take an approach similar to Exercise 32 and prove the following.

(a)   Let $f(z) = \sum_{n=0}^{\infty} l_n z^n$, $|z| < R$. Show that f is a solution of the equation $f(z) = 1 + 2z + zf(z) + z^2 f(z)$, and conclude that $f(z) = \frac{1+2z}{1-z-z^2}$.

(b)   Compute the Maclaurin series of f and derive the following formula for the Lucas sequence:

$$l_n = \left(\frac{1+\sqrt{5}}{2} \right)^{n+1} + \left(\frac{1-\sqrt{5}}{2} \right)^{n+1}, \qquad n \geq 0.$$

34. Project Problem: Euler numbers. Like the Bernoulli numbers, the Euler numbers have interesting applications in mathematics and can be generated from the coefficients of special Maclaurin series. In this exercise, you are asked to conduct an analysis similar to the one of Example 4 that will lead you into Euler numbers.

(a)   Show that $f(z) = \sec z$ is analytic at $z_0 = 0$ and its Maclaurin series has radius of convergence $R = \frac{\pi}{2}$.

(b)   Show that the odd numbered coefficients in the Maclaurin series are all 0. Then write

$$\sec z = \sum_{n=0}^{\infty} (-1)^n \frac{E_{2n}}{(2n)!} z^{2n}, \qquad |z| < \frac{\pi}{2}.$$

The numbers E_n are called the **Euler numbers**.

(c)   Derive the recursion formula

$$E_0 = 1, \qquad \sum_{k=0}^{n} \binom{2n}{2k} E_{2k} = 0, \quad n > 0.$$

(d)   Using (c) and induction prove that E_{2n} is a nonzero integer that alternates in sign.

(e)   Use (c) to derive $E_2 = -1$, $E_4 = 5$, $E_6 = -61$, $E_8 = 1385$.

35. Project Problem: Log series and the inverse tangent. In this exercise, we will derive several useful series involving the principal branch of the logarithm and the series of the inverse tangent.

(a)   Show that

$$\frac{1+z}{1-z} = \frac{1-|z|^2}{|1-z|^2} + 2i \frac{\operatorname{Im} z}{|1-z|^2}.$$

(b)   Conclude from (a) that if $|z| < 1$, then $\operatorname{Re}\left(\frac{1+z}{1-z}\right) > 0$ and hence $\operatorname{Log}\left(\frac{1+z}{1-z}\right)$ is analytic on $|z| < 1$.

(c)   Show that for $|z| < 1$, $\operatorname{Log}\left(\frac{1+z}{1-z}\right) = \operatorname{Log}(1+z) - \operatorname{Log}(1-z)$. [Hint: Show that the derivatives are equal.]

In parts (d)-(g), starting with the Log series in Example 2, Section 4.3, derive the Maclaurin series of the given function in the open disk $|z| < 1$:

(d) $\operatorname{Log}(1+z) = \displaystyle\sum_{n=1}^{\infty} \frac{(-1)^{n-1}}{n} z^n$;   (e) $-\operatorname{Log}(1-z) = \displaystyle\sum_{n=1}^{\infty} \frac{z^n}{n}$;

(f) $\operatorname{Log}\left(\frac{1+z}{1-z}\right) = 2 \displaystyle\sum_{n=0}^{\infty} \frac{z^{2n+1}}{2n+1}$;   (g) $\frac{1}{2i} \operatorname{Log}\left(\frac{1+iz}{1-iz}\right) = \displaystyle\sum_{n=0}^{\infty} \frac{(-1)^n}{2n+1} z^{2n+1}$.

(h)   Let $\phi(z) = \frac{1}{2i} \operatorname{Log}\left(\frac{1+iz}{1-iz}\right)$, $|z| < 1$. Using the definition of $\tan z$, verify that $\tan(\phi(z)) = z$. Thus, $\phi(z)$ is the inverse function of $\tan z$. We call $\phi(z)$ the **principal branch of the inverse tangent** and denote it by $\operatorname{Arctan} z$. Thus,

$$\operatorname{Arctan} z = \frac{1}{2i} \operatorname{Log}\left(\frac{1+iz}{1-iz}\right) = \sum_{n=0}^{\infty} \frac{(-1)^n}{2n+1} z^{2n+1}, \qquad |z| < 1.$$

(i)   Using term-by-term differentiation, show that

$$\frac{d}{dz} \operatorname{Arctan} z = \frac{1}{1+z^2}, \qquad |z| < 1.$$

36. Project Problem: Binomial series. In this exercise we study the binomial series with complex exponents. Let α be any complex number. Define the **generalized binomial coefficient** by

(11) $$\binom{\alpha}{0} = 1, \quad \binom{\alpha}{n} = \frac{\alpha(\alpha-1)\cdots(\alpha-n+1)}{n!}, \quad n \geq 1.$$

Use Theorem 1 to show that for $|z| < 1$

(12) $$(1+z)^\alpha = \sum_{n=0}^{\infty} \binom{\alpha}{n} z^n,$$

where $(1+z)^\alpha$ is the principal branch, $(1+z)^\alpha = e^{\alpha \, \text{Log}\,(1+z)}$. Note that if $\alpha = m$ is a positive integer, then from (11), $\binom{m}{n} = 0$ for all $n > m$ (why?) and the series (12) reduces to a finite sum

$$(1+z)^m = \sum_{n=0}^{m} \binom{m}{n} z^n,$$

which is the familiar **binomial theorem**. The finite sum clearly converges for all z, and so the restriction $|z| < 1$ is not necessary if α is a positive integer.

37. Derive the power series expansion

$$(1+z)^{\frac{1}{2}} = \sum_{n=0}^{\infty} \frac{(-1)^{n-1}}{2^{2n}(2n-1)} \binom{2n}{n} z^n, \qquad |z| < 1.$$

38. Project Problem: The Catalan numbers. The Catalan numbers are named after the French mathematician Eugène Charles Catalan (1814–1894). They arise in many combinatorics problems, such as the following:

- The number of ways in which parentheses can be placed in a sequence of numbers to be multiplied, two at a time.

- The number of ways a polygon with $n+1$ sides can be cut into $n-1$ triangles by nonintersecting diagonals.

- The number of paths of length $2(n-1)$ through an n by n grid that connect two diagonally opposite vertices that stay below or on the main diagonal.

Denote the nth Catalan number by c_n. Take the first interpretation of these numbers, and say we have a sequence of three numbers a, b, c. We can multiply them two at a time in exactly two ways: $((ab)c)$ and $(a(bc))$. Thus $c_3 = 2$. With a sequence of four numbers, we can have $(((ab)c)d)$ or $((ab)(cd))$ or $((a(bc))d)$ or $(a((bc)d))$ or $(a(b(cd)))$. Thus $c_4 = 5$. In general, if we have a sequence of n numbers, we can break it into two subsequences of k and $n-k$ numbers. Then each arrangement of parentheses from the k sequence can be combined with an arrangement of parentheses of the $n-k$ sequence to yield an arrangement of parentheses of the n-sequence. Since we have c_k possibilities for the k-sequence and c_{n-k} possibilities for the $(n-k)$-sequence, we get this way $c_k \times c_{n-k}$ possibilities for the n-sequence. But this can be done for $k = 1, 2, \ldots, n-1$, and so we have the recursion relation

$$c_n = \sum_{k=1}^{n-1} c_k c_{n-k}, \quad \text{for } n \geq 2.$$

We have $c_2 = 1$, and we will take $c_1 = 1$ (for reference, $c_0 = 0$). Our goal in this exercise is to derive a formula for c_n in terms of n.
(a) Suppose that there is an analytic function $f(z) = \sum_{n=0}^{\infty} c_n z^n$, where c_n is the nth Catalan number. From the recursion relation, show that f satisfies the equation $f(z) = z + (f(z))^2$. [Hint: Use Cauchy products.]
(b) Solve for f and choose the solution that gives $f(0) = 0$. You should get $f(z) = \frac{1-\sqrt{1-4z}}{2}$.
(c) Expand f in a power series around 0, using Exercise 37. You should get

$$f(z) = \frac{1}{2} + \frac{1}{2} \sum_{n=0}^{\infty} \frac{1}{2n-1} \binom{2n}{n} z^n \quad |z| < \frac{1}{4}.$$

(d) Conclude that for $n \geq 2$,

$$c_n = \frac{1}{2(2n-1)} \binom{2n}{n}.$$

(e) Verify your answer by computing c_3 and c_4 from the formula. What is c_5? Exhibit all c_5 arrangements of parentheses in a sequence of 5 numbers.
(f) Show that $\lim_{n \to \infty} \frac{c_{n+1}}{c_n} = 4$. Thus the Catalan numbers grow at a rate of 4.

39. Weierstrass double series theorem. This theorem is about a series of power series, thus the term *double series*. It states the following. Suppose that the power series $f_k(z) = \sum_{n=0}^{\infty} a_n^{(k)}(z - z_0)^n$ $(k = 0, 1, 2, \ldots)$ are all converging for $|z - z_0| < R$, where $0 < R \leq \infty$. Thus each power series defines a function $f_k(z)$ in the disk $|z - z_0| < R$. Suppose that $F(z) = \sum_{k=0}^{\infty} f_k(z)$ converges uniformly for $|z - z_0| \leq \rho$ for every $\rho < R$. Then F is analytic on $|z - z_0| \leq \rho$ and has a power series expansion, $F(z) = \sum_{n=0}^{\infty} A_n(z - z_0)^n$, that converges for all $|z - z_0| < R$. Moreover, for $n = 0, 1, 2, \ldots$, $A_n = \sum_{k=0}^{\infty} a_n^{(k)}$.
    Prove this theorem. [Hint: F is analytic because of Corollary 2, Section 4.2. F has a power series expansion in $|z - z_0| < \rho$ because of Theorem 1 of this section. We have $A_n = \frac{F^{(n)}(z_0)}{n!}$. Compute the derivatives of F by differentiating the series term by term (Corollary 2, Section 4.2).]

40. (a) Show that the series $f(z) = \sum_{n=1}^{\infty} \frac{z^n}{1+z^{2n}}$ converges uniformly on every closed disk of the form $|z| \leq r < 1$.
(b) Show that $\sum_{n=1}^{\infty} \frac{z^n}{1+z^{2n}} = \sum_{k=0}^{\infty} (-1)^k \frac{z^{2k+1}}{1-z^{2k+1}}$ for $|z| < 1$. [Hint: Expand each function $\frac{1}{1+z^{2n}}$ in a (geometric) power series. Then apply the Weierstrass double series theorem.]

41. Analytic continuation. In this exercise, we will briefly discuss the topic of analytic continuation. By constructing a Taylor series of a function $f(z)$ about z_0, we create a function $g(z)$ analytic on some disk centered at z_0, such that $g(z)$ and $f(z)$ coincide near z_0. If $g(z)$ is defined in a region where $f(z)$ is not, or if

$g(z)$ has different values than $f(z)$ away from z_0, $g(z)$ is said to be an **analytic continuation of** $f(z)$.

As a somewhat contrived example, suppose we are given a function $f(z)$ that is only defined on the unit disk and is given by the formula $f(z) = e^z$ for $|z| < 1$. Can we find a function that coincides with $f(z)$ on the unit disk but is analytic elsewhere? It is clear that we can (just define the exponential function on $\mathbb{C}$), but let us see how we can get it from series expansions. The Maclaurin series of $f(z)$ is $g(z) = \sum_{n=0}^{\infty} \frac{z^n}{n!}$, which converges for all z. Thus $g(z) = e^z$ for all z is an analytic continuation of $f(z)$.

Another example we have already looked at; when we took the Maclaurin expansions for the real trigonometric functions $\cos x$ and $\sin x$, and replaced x by z, we found that $\cos z$ and $\sin z$ are analytic continuations of $\cos x$ and $\sin x$ from the real axis onto the whole complex plane.

We now present a more interesting case of analytic continuation of logarithm branches.

(a) Show that the Taylor expansion of the principal branch of the logarithm $f(z) = \operatorname{Log} z$ about the point $z_0 = -1 + i$ is $g(z) = \operatorname{Log}(-1+i) + \sum_{n=1}^{\infty} \frac{(-1)^{n-1}(z+1-i)^n}{n(-1+i)^n}$.

(b) Show that the radius of convergence of the Taylor series is $\sqrt{2}$. Thus $g(z)$ is analytic in the disk $|z + 1 - i| < \sqrt{2}$ (Figure 5); it is important to note that this crosses the branch cut of f, but not the branch point (the origin). On the upper side of the branch cut, $g(z)$ will coincide with $f(z)$, but on the lower side of the branch cut, $g(z)$ will not coincide with $f(z)$. To see this, consider the branch $f_1(z) = \log_{-\frac{\pi}{4}} z$. Since $f(z) = f_1(z)$ in a neighborhood of z_0, their Taylor expansions about z_0 will be the same; and since $f_1(z)$ is analytic in the disk $|z+1-i| < \sqrt{2}$, Theorem 1 says that $g(z)$ must equal $f_1(z)$ in this disk. Thus, in the part of the disk above the branch cut (quadrants 1 and 2) we have $g(z) = f_1(z) = f(z)$, but in the part of the disk below the branch cut (quadrant 3) we have $g(z) = f_1(z) = f(z) + 2\pi i$. The Taylor expansion of $f(z)$ analytically continues the function into quadrant 3 where $\arg z > \pi$, and the analytic continuation is different from the original function.

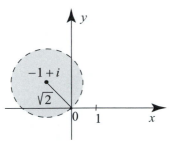

Figure 5 The Taylor series expansion of $\operatorname{Log} z$ around $-1 + i$ has radius of convergence $\sqrt{2}$.

4.5   Laurent Series

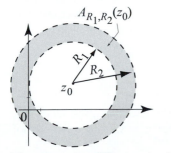

Figure 1 The annulus $A_{R_1,R_2}(z_0)$ centered at z_0 with inner radius R_1 and outer radius R_2.

The Taylor series representation of an analytic function is very useful, especially because of the simple expression of the terms in that series, $c_n(z - z_0)^n$, where $n \geq 0$. There is a similar series representation in terms of both positive and negative powers of $(z - z_0)$ for functions that are analytic in annular regions around a point z_0. These series are known as **Laurent series**, after the French engineer and mathematician Pierre Alphonse Laurent (1813–1854), who discovered them around 1842. Laurent series are of the form $\sum_{n=0}^{\infty} a_n(z - z_0)^n + \sum_{n=1}^{\infty} \frac{b_n}{(z-z_0)^n}$. Note the part with the negative powers of $(z - z_0)$. Without this part, the Laurent series reduces to a power series, which would represent a function analytic at z_0.

Let us recall the notation for an annulus. If $0 \leq R_1 < R_2 \leq \infty$, let

$$A_{R_1,R_2}(z_0) = \{z : R_1 < |z - z_0| < R_2\}.$$

See Figure 1 for an illustration where R_1 and R_2 are nonzero and finite. Note that the annulus $A_{R_1, R_2}(z_0)$ can degenerate into a punctured disk with z_0 removed ($R_1 = 0$ and $R_2 < \infty$), a punctured plane with z_0 removed ($R_1 = 0$, $R_2 = \infty$), or a plane with a disk centered at z_0 cut out of it ($0 < R_1$ and $R_2 = \infty$). These sets still count as annuli by our definition.

THEOREM 1
LAURENT SERIES

Suppose that f is analytic in the annulus $A_{R_1, R_2}(z_0)$ where $0 \leq R_1 < R_2 \leq \infty$. Then f has a unique **Laurent series** representation

$$(1) \qquad f(z) = \sum_{n=0}^{\infty} a_n (z - z_0)^n + \sum_{n=1}^{\infty} \frac{a_{-n}}{(z - z_0)^n}, \qquad R_1 < |z - z_0| < R_2,$$

where the **Laurent coefficients** are given by

$$(2) \qquad a_n = \frac{1}{2\pi i} \int_{C_R(z_0)} \frac{f(\zeta)}{(\zeta - z_0)^{n+1}} \, d\zeta \qquad (n = 0, \pm 1, \pm 2, \dots),$$

with $R_1 < R < R_2$ and $C_R(z_0)$ is any positively oriented circle centered at z_0 and contained in $A_{R_1, R_2}(z_0)$. The Laurent series converges absolutely for all z in $A_{R_1, R_2}(z_0)$ and uniformly on any closed subannulus $R_1 < \rho_1 \leq |z - z_0| \leq \rho_2 < R_2$.

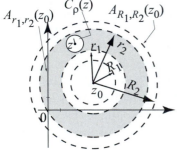

Figure 2 The function
$$\zeta \mapsto \frac{f(\zeta)}{\zeta - z}$$
is analytic inside and on the boundary of the shaded area.

Proof   Given z in the annulus $A_{R_1, R_2}(z_0)$, we can find r_1, r_2 and ρ so that $R_1 < r_1 < r_2 < R_2$ and $C_\rho(z)$ is contained in $A_{r_1, r_2}(z_0)$ (see Figure 2). The function $\zeta \mapsto \frac{f(\zeta)}{\zeta - z}$ is analytic inside and on the boundary of the region outside $C_\rho(z)$ and inside $A_{r_1, r_2}(z_0)$. So by Cauchy's theorem for multiply connected regions (Theorem 6, Section 3.4), we have

$$\frac{1}{2\pi i} \int_{C_{r_2}(z_0)} \frac{f(\zeta)}{\zeta - z} \, d\zeta = \frac{1}{2\pi i} \int_{C_\rho(z)} \frac{f(\zeta)}{\zeta - z} \, d\zeta + \frac{1}{2\pi i} \int_{C_{r_1}(z_0)} \frac{f(\zeta)}{\zeta - z} \, d\zeta,$$

where all circular paths are positively oriented. By Cauchy's integral formula (Theorem 1, Section 3.6), the first integral on the right is equal to $f(z)$, because f is analytic inside and on $C_\rho(z)$. So

$$(3) \qquad f(z) = \frac{1}{2\pi i} \int_{C_{r_2}(z_0)} \frac{f(\zeta)}{\zeta - z} \, d\zeta - \frac{1}{2\pi i} \int_{C_{r_1}(z_0)} \frac{f(\zeta)}{\zeta - z} \, d\zeta.$$

Since z is inside $C_{r_2}(z_0)$, the first integral looks exactly like the integral in (5), Section 4.4. Using a power series expansion of $\frac{1}{\zeta - z}$, exactly as we did in the proof of Theorem 1, Section 4.4, we obtain

$$\frac{1}{2\pi i} \int_{C_{r_2}(z_0)} \frac{f(\zeta)}{\zeta - z} \, d\zeta = \sum_{n=0}^{\infty} a_n (z - z_0)^n, \quad \text{where} \quad a_n = \frac{1}{2\pi i} \int_{C_{r_2}(z_0)} \frac{f(\zeta)}{(\zeta - z_0)^{n+1}} \, d\zeta.$$

To analyze the second integral on the right of (3), we appeal to Lemma 1, Section 4.4

with $z_2 = z$ and $z_1 = \zeta$. Then

(4) $\qquad \dfrac{1}{z - \zeta} = \displaystyle\sum_{n=0}^{\infty} \dfrac{(\zeta - z_0)^n}{(z - z_0)^{n+1}} = \sum_{n=1}^{\infty} \dfrac{(\zeta - z_0)^{n-1}}{(z - z_0)^n}, \qquad |\zeta - z_0| < |z - z_0|.$

Multiplying both sides by $f(\zeta)$ and then integrating term by term, we obtain

$$-\frac{1}{2\pi i} \int_{C_{r_1}(z_0)} \frac{f(\zeta)}{\zeta - z} \, d\zeta = \sum_{n=1}^{\infty} \left\{ \frac{1}{(z - z_0)^n} \frac{1}{2\pi i} \int_{C_{r_1}(z_0)} f(\zeta)(\zeta - z_0)^{n-1} \, d\zeta \right\}$$

$$= \sum_{n=1}^{\infty} \frac{a_{-n}}{(z - z_0)^n}$$

where $a_{-n} = \frac{1}{2\pi i} \int_{C_{r_1}(z_0)} f(\zeta)(\zeta - z_0)^{n-1} \, d\zeta$. The term-by-term integration is justified using uniform convergence, just as in the proof of Theorem 1, Section 4.4 (Exercise 32). This establishes the Laurent series expansion (1). To obtain (2), note that $C_{r_1}(z_0)$ and $C_{r_2}(z_0)$ are continuously deformable into $C_R(z_0)$ relative to the annular region $A_{R_1, R_2}(z_0)$, so we can replace the paths in the integrals defining a_n and b_n by $C_R(z_0)$. These formulas are independent of R so long as $R_1 < R < R_2$, so the Laurent coefficients a_n as given in (2) are uniquely specified. Furthermore, any expansion of the form (1) will have the coefficients given in (2), as is seen by integrating (1) term by term against $\frac{1}{(z-z_0)^{m+1}}$ on a path $C_R(z_0)$. The proofs of the uniform and absolute convergence of the Laurent series are similar to proofs that we did previously with Taylor series (see Exercise 33). ∎

As with power series, often in computing Laurent series, you do not want to use (2) to find the coefficients, but rather resort to manipulations of known series.

EXAMPLE 1   Laurent series
Find the Laurent series expansions of (a) $e^{\frac{1}{z}}$, for $0 < |z|$; and (b) $\frac{1}{1-z}$, for $1 < |z|$.

Solution   (a)   Start with the exponential series $e^z = \sum_{n=0}^{\infty} \frac{z^n}{n!}$, which is valid for all z. In particular, if $z \neq 0$, putting $\frac{1}{z}$ into this series, we obtain

$$e^{\frac{1}{z}} = \sum_{n=0}^{\infty} \frac{1}{n! z^n} = 1 + \sum_{n=1}^{\infty} \frac{1}{n! z^n} = 1 + \frac{1}{1! z} + \frac{1}{2! z^2} + \frac{1}{3! z^3} + \cdots.$$

By the uniqueness of the Laurent series representation in the annulus $0 < |z|$, we have thus found the Laurent series of $e^{\frac{1}{z}}$ in the annulus $0 < |z|$.

(b)   Here we have to use the geometric series $\sum_{n=0}^{\infty} w^n = \frac{1}{1-w}$, for $|w| < 1$. Since $1 < |z|$, we cannot apply this expansion with $w = z$. But we can apply it with $w = \frac{1}{z}$, because $|w| = \frac{1}{|z|} < 1$. This can be done by factoring z from the denominator as follows:

$$\frac{1}{1-z} = \frac{1}{z} \frac{1}{\frac{1}{z} - 1} = \frac{-1}{z} \overbrace{\frac{1}{1 - \frac{1}{z}}}^{= \frac{1}{1-w}} = \frac{-1}{z} \sum_{n=0}^{\infty} \left(\frac{1}{z}\right)^n = \sum_{n=0}^{\infty} \frac{-1}{z^{n+1}}, \qquad 1 < |z|.$$

We could stop here, but if we want to match the Laurent series form (1), we change $n + 1$ to n in the series, adjust the summation limit, and get

$$\frac{1}{1-z} = \sum_{n=1}^{\infty} \frac{-1}{z^n} = -\frac{1}{z} - \frac{1}{z^2} - \frac{1}{z^3} - \cdots, \qquad 1 < |z|. \qquad \blacksquare$$

Combining Example 1(b) with the geometric series, we obtain the following useful identities:

(5)
$$\frac{1}{1-w} = \begin{cases} \displaystyle\sum_{n=0}^{\infty} w^n & \text{if } |w| < 1, \\[3mm] \displaystyle-\sum_{n=1}^{\infty} \frac{1}{w^n} & \text{if } 1 < |w|. \end{cases}$$

Here is an application.

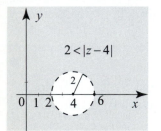

Figure 3 The function $f(z) = \frac{1}{z-6}$ is analytic in the annulus $2 < |z - 4|$ and so has a Laurent series expansion there.

EXAMPLE 2   Manipulating the geometric series

Find the Laurent series expansion of $f(z) = \frac{1}{z-6}$ in the annulus $2 < |z - 4|$ (Figure 3).

Solution   Since we are expanding around 4, we need to see the expression $(z - 4)$ in the denominator of f. For this purpose, and in order to apply (5), let us write

$$\frac{1}{z-6} = \frac{1}{(z-4)-2} = -\frac{1}{2}\frac{1}{1 - \frac{z-4}{2}} = -\frac{1}{2}\frac{1}{1-w}$$

where $w = \frac{z-4}{2}$. If $2 < |z - 4|$ then $|w| = \left|\frac{z-4}{2}\right| > 1$ and so the second identity in (5) implies that

$$\frac{1}{z-6} = -\frac{1}{2}\frac{1}{1-w} = \frac{1}{2}\sum_{n=1}^{\infty}\frac{1}{w^n} = \frac{1}{2}\sum_{n=1}^{\infty}\frac{2^n}{(z-4)^n} = \sum_{n=1}^{\infty}\frac{2^{n-1}}{(z-4)^n},$$

which is the Laurent series of $\frac{1}{z-6}$ in the annulus $2 < |z - 4|$. $\blacksquare$

In working through the next example, you should outline for yourself a general method for finding the Laurent series of a rational function.

EXAMPLE 3   Laurent series of a rational function

Find the Laurent series expansion of $f(z) = \frac{3z^2 - 2z + 4}{z-6}$ in the annulus $2 < |z - 4|$ of the previous example.

Solution   Since the degree of the numerator is larger than the degree of the denominator, the first step is to divide the numerator by the denominator:

$$\frac{3z^2 - 2z + 4}{z-6} = 3z + 16 + \frac{100}{z-6}.$$

The quotient $3z+16$ has a simple expression in terms of powers of $(z-4)$: $3z+16 = 3(z-4)+28$. The next step is to compute the Laurent series of the remainder $\frac{100}{z-6}$. But this part follows easily from Example 2: for $2 < |z-4|$,

$$\frac{100}{z-6} = 100 \sum_{n=1}^{\infty} \frac{2^{n-1}}{(z-4)^n}.$$

So, for $2 < |z-4|$, we have the Laurent series

$$\frac{3z^2 - 2z + 4}{z-6} = 3z + 16 + \frac{100}{z-6} = 3(z-4) + 28 + 100 \sum_{n=1}^{\infty} \frac{2^{n-1}}{(z-4)^n}. \qquad \blacksquare$$

The uniqueness property of the Laurent series in Theorem 1 is very important. However, you should keep in mind that the Laurent series expansion does depend on the region, so different annular regions might give different Laurent series expansions. Clearly, if a given annulus can be expanded within the domain of analyticity of the function, the Laurent series will not change, since the integral formulas (2) do not change. However, when we have disjoint annuli and the function is not analytic in the region between them, the two Laurent series will probably be different.

EXAMPLE 4   The Laurent series expansions of a function
Find all the Laurent series expansions of $f(z) = \frac{3}{(1+z)(2-z)}$ around $z_0 = 0$.

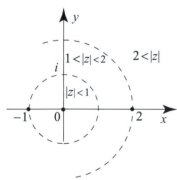

Figure 4 The function
$$f(z) = \frac{3}{(1+z)(2-z)}$$
is not analytic at $z = -1$ and $z = 2$. It has three Laurent series expansions around 0.

Solution   This problem has two parts. First we must determine how many different Laurent series expansions f has around 0. Then we must find these Laurent series. To answer the first question, we ask ourselves, what are the largest disjoint annular regions around $z_0 = 0$ on which f is analytic? The function f is analytic at all points except at $z = -1$ and $z = 2$. So the desired regions around 0 on which f is analytic are $|z| < 1$; $1 < |z| < 2$; and $2 < |z|$ (Figure 4). Note that the first region $|z| < 1$ is really a disk, not an annulus, and so we use a power series expansion there, which is really just a special case of a Laurent series expansion (Theorem 1, Section 4.4). So f has three different Laurent series expansions around 0; one of them being a power series. Let us find them. We will need the partial fraction decomposition

$$\frac{3}{(1+z)(2-z)} = \frac{1}{1+z} + \frac{1}{2-z},$$

which you can easily verify.

For $|z| < 1$, we have from the geometric series expansion

$$\frac{1}{1+z} = \frac{1}{1-(-z)} = \sum_{n=0}^{\infty} (-1)^n z^n, \quad |z| < 1,$$

and

(6) $\qquad \dfrac{1}{2-z} = \dfrac{1}{2} \dfrac{1}{1-\left(\frac{z}{2}\right)} = \dfrac{1}{2} \sum_{n=0}^{\infty} \left(\dfrac{z}{2}\right)^n = \sum_{n=0}^{\infty} \dfrac{z^n}{2^{n+1}}, \quad \left|\dfrac{z}{2}\right| < 1, \text{ or } |z| < 2.$

Adding the two series over their common region of convergence, we obtain

$$\frac{3}{(1+z)(2-z)} = \sum_{n=0}^{\infty}\left((-1)^n + \frac{1}{2^{n+1}}\right)z^n, \quad |z| < 1,$$

which is the Taylor series expansion of f in $|z| < 1$. To find the Laurent series in the annulus $1 < |z| < 2$, we can use (5) or reason as in Example 1(b). We have

(7) $$\frac{1}{1+z} = \frac{1}{z}\,\overbrace{\frac{1}{1-\left(-\frac{1}{z}\right)}}^{=\frac{1}{1-w}} = \frac{1}{z}\sum_{n=0}^{\infty}\left(\frac{-1}{z}\right)^n = \sum_{n=1}^{\infty}\frac{(-1)^{n-1}}{z^n}, \qquad 1 < |z|.$$

For the term $\frac{1}{2-z}$, we can use the previous expansion (6). Adding the two, we obtain for $1 < |z| < 2$,

$$\begin{aligned}
\frac{3}{(1+z)(2-z)} &= \sum_{n=1}^{\infty}\frac{(-1)^{n-1}}{z^n} + \sum_{n=0}^{\infty}\frac{z^n}{2^{n+1}} \\
&= \frac{1}{2} + \frac{z}{2^2} + \frac{z^2}{2^3} + \cdots + \frac{1}{z} - \frac{1}{z^2} + \frac{1}{z^3} - \cdots,
\end{aligned}$$

which is the Laurent series of f in the annulus $1 < |z| < 2$. Finally, let us consider the annulus $2 < |z|$. Since $2 < |z|$, then clearly $1 < |z|$, and for the term $\frac{1}{1+z}$ we can use (7). Also, if $2 < |z|$, then $\left|\frac{2}{z}\right| < 1$, and so

$$\frac{1}{2-z} = \frac{1}{z}\frac{1}{\left(\frac{2}{z}\right) - 1} = \frac{-1}{z}\frac{1}{1 - \left(\frac{2}{z}\right)} = \frac{-1}{z}\sum_{n=0}^{\infty}\left(\frac{2}{z}\right)^n = -\sum_{n=1}^{\infty}\frac{2^{n-1}}{z^n}, \qquad 2 < |z|.$$

Adding the two series, we obtain for $2 < |z|$,

$$\begin{aligned}
\frac{3}{(1+z)(2-z)} &= \sum_{n=1}^{\infty}\frac{(-1)^{n-1}}{z^n} - \sum_{n=1}^{\infty}\frac{2^{n-1}}{z^n} = \sum_{n=1}^{\infty}\frac{(-1)^{n-1} - 2^{n-1}}{z^n} \\
&= -\frac{3}{z^2} - \frac{3}{z^3} - \frac{9}{z^4} - \cdots,
\end{aligned}$$

which is the Laurent series of f in the annulus $2 < |z|$.   ∎

Theorem 1 guarantees the convergence of the Laurent series in the largest annulus where the function is analytic. This is just like the case of a Taylor series which will converge in the largest disk on which the function is analytic.

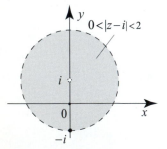

Figure 5 The largest annulus around i on which $f(z) = \frac{1}{1+z^2}$ has a Laurent series expansion.

EXAMPLE 5   Determining the annulus of convergence

Determine the largest annulus around $z_0 = i$ of the form $R_1 < |z - i| < R_2$ on which the function $f(z) = \frac{1}{1+z^2}$ has a Laurent series and then find the Laurent series.

Solution   The function $\frac{1}{1+z^2}$ is analytic at all z except $z = \pm i$. It has a Laurent series in the largest annulus around i on which it is analytic. In order to avoid the singularity at $-i$, we take the annulus $0 < |z - i| < 2$ (Figure 5). We have

$\frac{1}{1+z^2} = \frac{1}{z-i}\frac{1}{z+i}$. Since we are expanding in terms of $(z-i)$, the factor $\frac{1}{z-i}$ is already in a desirable form, and so we keep it as is and work on the factor $\frac{1}{z+i}$. To make $(z-i)$ appear in the latter factor, we add and subtract i from the denominator. Thus

$$\frac{1}{1+z^2} = \frac{1}{z-i}\frac{1}{z+i} = \frac{1}{z-i}\frac{1}{(z-i)+2i} = \frac{1}{2i}\frac{1}{z-i}\frac{1}{1+\left(\frac{z-i}{2i}\right)}.$$

Since $|z-i| < 2$, then $\left|\frac{z-i}{2i}\right| < 1$, and so we can use a geometric series expansion:

$$\frac{1}{1+z^2} = \frac{1}{2i}\frac{1}{z-i}\frac{1}{1-\left(-\frac{z-i}{2i}\right)} = \frac{1}{2i}\frac{1}{z-i}\sum_{n=0}^{\infty}(-1)^n\left(\frac{z-i}{2i}\right)^n$$

$$= \frac{1}{2i}\left[\frac{1}{z-i} - \frac{1}{2i} - \frac{z-i}{2^2} + \frac{(z-i)^2}{2^3 i} + \cdots\right]. \qquad\blacksquare$$

Differentiation and Integration of Laurent Series

Consider a Laurent series (1) in the annulus $A_{R_1,R_2}(z_0)$. The function $a_n(z-z_0)^n$ is analytic on the annulus $A_{R_1,R_2}(z_0)$, since its only possible problem is at z_0, which is excluded from the annulus. So each term of a Laurent series is analytic on the annulus $A_{R_1,R_2}(z_0)$. Since the series converges uniformly on any subannulus of $A_{R_1,R_2}(z_0)$, it follows that it converges uniformly on any closed disk contained in $A_{R_1,R_2}(z_0)$. These observations allow us to apply the useful results of Section 4.2. In particular, from Corollary 2, Section 4.2, we infer that the Laurent series can be differentiated term by term as many times as we want within the annulus $A_{R_1,R_2}(z_0)$. So, for example, differentiating the Laurent series (1) once, we obtain for $R_1 < |z - z_0| < R_2$,

$$(8) \qquad f'(z) = \sum_{n=1}^{\infty} na_n(z-z_0)^{n-1} - \sum_{n=1}^{\infty}\frac{na_{-n}}{(z-z_0)^{n+1}}.$$

We could continue to differentiate term by term to find $f''(z)$, $f'''(z)$, and so forth. Also, by Corollary 1, Section 4.2, if γ is any path contained in $A_{R_1,R_2}(z_0)$, then the Laurent series can be integrated term by term over γ:

$$(9) \qquad \int_\gamma f(z)\,dz = \sum_{n=0}^{\infty} a_n \int_\gamma (z-z_0)^n\,dz + \sum_{n=1}^{\infty} a_{-n}\int_\gamma \frac{dz}{(z-z_0)^n}.$$

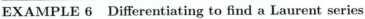

EXAMPLE 6   Differentiating to find a Laurent series
Find the Laurent series for $f(z) = \frac{1}{(1-z)^3}$ in the annulus $1 < |z|$ (Figure 6).

Solution   Starting with the Laurent series

$$\frac{1}{1-z} = -\sum_{n=1}^{\infty}\frac{1}{z^n}, \qquad 1 < |z|,$$

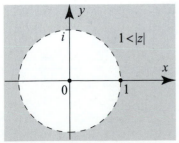

Figure 6 The annulus $1 < |z|$ in Example 6.

if we differentiate both sides we get the Laurent series

$$\frac{1}{(1-z)^2} = \sum_{n=1}^{\infty} \frac{n}{z^{n+1}}, \qquad 1 < |z|.$$

Differentiating a second time, we get

$$\frac{2}{(1-z)^3} = -\sum_{n=1}^{\infty} \frac{n(n+1)}{z^{n+2}}, \qquad 1 < |z|.$$

So the desired Laurent series is

$$\frac{1}{(1-z)^3} = -\frac{1}{2} \sum_{n=1}^{\infty} \frac{n(n+1)}{z^{n+2}}, \qquad 1 < |z|. \qquad \blacksquare$$

EXAMPLE 7   Term-by-term integration of a Laurent series

Let $C_1(0)$ denote the positively oriented circle of radius 1, centered at 0. Evaluate the following integrals:

(a) $\displaystyle\int_{C_1(0)} \frac{e^{\frac{1}{z}}}{z}\, dz,$ 　　　　(b) $\displaystyle\int_{C_1(0)} e^{z+\frac{1}{z}}\, dz.$

Solution   Note that the integrands are continuous on the path of integration. So the integrals do exist; however, they cannot be computed using Cauchy's theorem because of the problem at $z = 0$, which lies inside the path. The idea is to expand the integrand (or part of the integrand) in a Laurent series in an annulus that contains the path, and then integrate term by term.

(a)   In the annulus $0 < |z|$, we have

$$\frac{e^{\frac{1}{z}}}{z} = \frac{1}{z} e^{\frac{1}{z}} = \frac{1}{z} \sum_{n=0}^{\infty} \frac{1}{n! z^n} = \sum_{n=0}^{\infty} \frac{1}{n! z^{n+1}}.$$

So

$$\int_{C_1(0)} \frac{e^{\frac{1}{z}}}{z}\, dz = \int_{C_1(0)} \left(\sum_{n=0}^{\infty} \frac{1}{n! z^{n+1}} \right) dz = \sum_{n=0}^{\infty} \frac{1}{n!} \int_{C_1(0)} \frac{dz}{z^{n+1}} = 2\pi i,$$

where only the $n = 0$ term is nonzero. The technique employed here is the basis of Chapter 5: To compute integrals around circles, we find Laurent expansions and integrate term by term. Only the term involving $\frac{1}{z}$ survives.

(b)   Here again, we work on the annulus $0 < |z|$. With an eye on Cauchy's integral formula, we do not expand e^z. For $0 < |z|$, we have

$$e^{z+\frac{1}{z}} = e^z e^{\frac{1}{z}} = e^z \sum_{n=0}^{\infty} \frac{1}{n! z^n} = e^z + \frac{e^z}{1! z} + \frac{e^z}{2! z^2} + \frac{e^z}{3! z^3} + \cdots = e^z + \sum_{n=0}^{\infty} \frac{e^z}{(n+1)! z^{n+1}}.$$

Integrating term by term yields

$$\int_{C_1(0)} e^{z+\frac{1}{z}}\, dz = \int_{C_1(0)} e^z\, dz + \sum_{n=0}^{\infty} \frac{1}{(n+1)!} \int_{C_1(0)} \frac{e^z}{z^{n+1}}\, dz.$$

Cauchy's theorem implies that $\int_{C_1(0)} e^z\, dz = 0$ because e^z is analytic on and inside $C_1(0)$. Cauchy's generalized integral formula tells us that

$$\int_{C_1(0)} \frac{e^z}{z^{n+1}}\, dz = \frac{2\pi i}{n!} f^{(n)}(0),$$

where $f(z) = e^z$. Hence $f^{(n)}(0) = e^0 = 1$, and so

$$\int_{C_1(0)} e^{z+\frac{1}{z}}\, dz = \sum_{n=0}^{\infty} \frac{1}{(n+1)!} \int_{C_1(0)} \frac{e^z}{z^{n+1}}\, dz = 2\pi i \overbrace{\sum_{n=0}^{\infty} \frac{1}{n!(n+1)!}}^{\approx 1.59} \approx 9.99\, i.$$

We should also mention that $\sum_{n=0}^{\infty} \frac{1}{n!(n+1)!} = -iJ_1(2i)$, where $J_1(z)$ is the Bessel function of order 1 (see Example 5, Section 4.3). This connection with Bessel functions will be explored in the exercises. ∎

Can a function like $\operatorname{Log} z$ have a Laurent series expansion around 0? The answer is no, because a Laurent series can be differentiated term by term within its annulus of convergence, so it is analytic within its annulus of convergence. Since $\operatorname{Log} z$ is not analytic in any annulus around 0, it cannot equal (or be represented by) a Laurent series around 0. In fact, no branch of the logarithm has a Laurent series expansion around 0.

Exercises 4.5

In Exercises 1–12, use a known Taylor series or Laurent series to derive the Laurent series of the given function in the indicated annulus.

1. $\dfrac{1}{1+z^2}$, $1 < |z|$. **2.** $\dfrac{3+z}{2-z}$, $2 < |z|$. **3.** $\dfrac{1+z}{1-z}$, $1 < |z|$.

4. $z + \dfrac{1}{z}$, $1 < |z-1|$. **5.** $\operatorname{Log}\left(1 + \dfrac{1}{z}\right)$, $1 < |z|$. **6.** $\dfrac{\sin z}{z^2}$, $0 < |z|$.

7. $\dfrac{e^{\frac{1}{1-z}}}{1-z}$, $0 < |z-1|$. **8.** $\dfrac{\operatorname{Log}(1+z)}{z^2}$, $0 < |z| < 1$. **9.** $\dfrac{\sin \frac{1}{z} \cos \frac{1}{z}}{z}$, $0 < |z|$.

10. $z^{22} e^{\frac{1}{z^2}}$, $0 < |z|$. **11.** $\coth z$, $0 < |z| < \pi$. **12.** $\cot z$, $0 < |z| < \pi$.

In Exercises 13–20, find the Laurent series of the given function in the indicated annulus.

13. $\dfrac{z}{(z+2)(z+3)}$, $2 < |z| < 3$. **14.** $\dfrac{-2}{(2z-1)(2z+1)}$, $\dfrac{1}{2} < |z|$.

15. $\dfrac{1}{(3z-1)(2z+1)}$, $\dfrac{1}{3} < |z| < \dfrac{1}{2}$. **16.** $\dfrac{1}{2z^2 - 3z + 1}$, $1 < |z|$.

17. $\dfrac{z^2 + (1-i)z + 2}{(z-i)(z+2)}$, $1 < |z| < 2$. **18.** $\dfrac{4z-5}{(z-2)(z-1)}$, $1 < |z-2|$.

19. $\dfrac{2z^3 - 4z^2 - 5z + 11}{z-1}$, $1 < |z-2|$. **20.** $\dfrac{z^2 - (3+2i)z + 2 + 3i}{z - 1 - i}$, $\sqrt{2} < |z|$.

21. Find all the Laurent series of $f(z) = \dfrac{1}{(z-1)(z+i)}$ around $z_0 = -1$.

22. Find all the Laurent series of $f(z) = \dfrac{1}{z^2 + 1}$ around $z_0 = 1$.

23. (a) Derive the Laurent series

$$\frac{1}{1+z} = \sum_{n=1}^{\infty} \frac{(-1)^{n-1}}{z^n} \qquad 1 < |z|.$$

Starting with this Laurent series, find the Laurent series of the following functions in the annulus $1 < |z|$:

(b) $\dfrac{1}{(1+z)^2}$;   (c) $\dfrac{z}{(1+z)^2}$;   (d) $\dfrac{z^2}{(1+z)^3}$.

24. Find the Laurent series of $\csc^2 z$ in the annulus $0 < |z| < \pi$.

In Exercises 25–30, evaluate the given integral by using an appropriate Laurent series. As usual, we denote by $C_R(z_0)$ the positively oriented circle of radius $R > 0$ and center z_0.

25. $\displaystyle \int_{C_1(0)} \sin \frac{1}{z}\, dz.$ **26.** $\displaystyle \int_{C_1(0)} \frac{\cos \frac{1}{z^2}}{z}\, dz.$

27. $\displaystyle \int_{C_1(0)} \cos z \sin \frac{1}{z}\, dz.$ **28.** $\displaystyle \int_{C_1(0)} e^{z^2 + \frac{1}{z}}\, dz.$

29. $\displaystyle \int_{C_4(0)} \mathrm{Log}\left(1 + \frac{1}{z}\right) dz.$ **30.** $\displaystyle \int_{C_1(0)} z^{10} e^{\frac{1}{z}}\, dz.$

31. Suppose that f is analytic in a region Ω, z_0 is in Ω and $S_R(z_0)$ is a closed disk in Ω, centered at z_0. (a) For $n = 0, 1, 2, \ldots$, derive the Laurent series expansion

$$\frac{f(z)}{(z - z_0)^{n+1}} = \sum_{k=0}^{\infty} \frac{f^{(k)}(z_0)}{k!} \frac{1}{(z - z_0)^{n+1-k}}, \qquad 0 < |z - z_0| < R.$$

(b) Prove Cauchy's generalized integral formula using (a) and integration term by term.

32. Refer to (4). Show that the series converges uniformly for all ζ on $C_{r_1}(z_0)$ as follows. Write the series as $\dfrac{1}{z - z_0} \sum_{n=1}^{\infty} \left(\dfrac{\zeta - z_0}{z - z_0}\right)^{n-1}$. Show that $\left|\dfrac{\zeta - z_0}{z - z_0}\right| = \rho < 1$, where z, z_0, and ζ are as described in (4). Obtain the uniform convergence by applying the Weierstrass M-test.

33. Refer to (1). Follow the outlined steps to show that the Laurent series converges absolutely for all z in $A_{R_1, R_2}(z_0)$ and uniformly on any closed subannulus $R_1 < r_1 \le |z - z_0| \le r_2 < R_2$.
(a) Show that the power series in (1) converges absolutely for all $|z - z_0| < R_2$ and uniformly on any subdisk $|z - z_2| \le r_2 < R_2$. [Hint: Use Lemma 2 and Theorem 1, Section 4.3.]
(b) Let R be such that $R_1 < R < r_1$. Since f is analytic in $A_{R_1, R_2}(z_0)$, it is continuous and hence bounded on $C_R(z_0)$. Let M be such that $|f(\zeta)| \le M$ for all ζ on $C_R(z_0)$. Using (2), show that $|a_{-n}| \le MR^n$ for $n = 1, 2, \ldots$. If $R < r_1 \le |z - z_0|$ then $\left|\dfrac{R}{z - z_0}\right| = \rho < 1$. Apply the Weierstrass M-test to show that $\sum_{n=1}^{\infty} \dfrac{a_{-n}}{(z - z_0)^n}$ converges absolutely and uniformly on the subannulus $R_1 < r_1 \le |z - z_0| \le r_2 < R_2$.

34. Uniqueness of Laurent series representation. (a) In the notation of Theorem 1, use (2) to show that f is identically zero in $A_{R_1,R_2}(z_0)$ if and only if $a_n = 0$ for all n.

(b) Suppose that f and g are analytic in an annulus $A_{R_1,R_2}(z_0)$. Show that $f(z) = g(z)$ for all z in $A_{R_1,R_2}(z_0)$ if and only if f and g have the same Laurent series in $A_{R_1,R_2}(z_0)$.

35. Project Problem: Generating function for Bessel functions of integer order. In this project problem, we will derive the **generating function** for Bessel functions of integer order (Example 5, Section 4.3):

$$(10) \qquad e^{\frac{z}{2}\left(\zeta - \frac{1}{\zeta}\right)} = \sum_{n=-\infty}^{\infty} J_n(z)\zeta^n, \qquad 0 < |\zeta|.$$

This formula generates the Bessel functions $J_n(z)$ as the Laurent coefficients of the function $f(\zeta) = e^{z\left(\zeta - \frac{1}{\zeta}\right)}$, which is clearly analytic in the annulus $0 < |\zeta|$.

(a) Proceed as in Example 7(b) and show that for $n = 0, \pm 1, \pm 2, \ldots$,

$$(11) \qquad J_n(z) = \frac{1}{2\pi i} \int_{C_1(0)} e^{\frac{z}{2}\left(\zeta - \frac{1}{\zeta}\right)} \frac{d\zeta}{\zeta^{n+1}}.$$

[Hint: Write $e^{\frac{z}{2}\left(\zeta - \frac{1}{\zeta}\right)} = e^{\frac{z}{2}\zeta} e^{-\frac{z}{2\zeta}}$ and expand $e^{-\frac{z}{2\zeta}}$ in a Laurent series in $0 < |\zeta|$.]

(b) We know that $f(\zeta)$ has a Laurent series expansion in $0 < |\zeta|$. Write this series as $f(\zeta) = \sum_{n=-\infty}^{\infty} c_n(z)\zeta^n$. Express the coefficients $c_n(z)$ by using (2) and integrating over $C_1(0)$. Conclude that $c_n(z)$ is equal to the integral in (11), and hence $c_n(z) = J_n(z)$.

36. Project Problem: Cosine integral representation of Bessel functions. (a) Parametrize the path integral in (11) and obtain the formula

$$(12) \qquad J_n(z) = \frac{1}{2\pi} \int_{-\pi}^{\pi} e^{i(z\sin\theta - n\theta)}\, d\theta = \frac{1}{\pi} \int_0^{\pi} \cos(z\sin\theta - n\theta)\, d\theta.$$

This formula is known as the **cosine integral representation** of $J_n(z)$.

(b) Show that for $z = x$ a real number $|J_n(x)| \le 1$ (Figure 7).

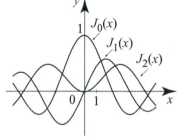

Figure 7 Bessel functions $J_n(x)$, for $n = 0, 1, 2$.

4.6   Zeros and Singularities

In this section, we use Taylor and Laurent series to study the zeros and singular points of an analytic function. We will discover interesting properties that show once more the remarkable characteristics of these functions. We start with a description of the zeros.

THEOREM 1
ORDER OF ZEROS

Suppose that f is analytic on an open set U, z_0 is in U, and $f(z_0) = 0$. Then exactly one of the following holds:

(i) f is identically zero in a neighborhood of z_0.

(ii) There is an integer $m \ge 1$ such that $f(z_0) = f'(z_0) = \cdots = f^{(m-1)}(z_0) = 0$ and $f^{(m)}(z_0) \ne 0$. Moreover, there is a function $\lambda(z)$, analytic and nonvanishing in a neighborhood $B_r(z_0)$ where $r > 0$, such that $f(z) = (z - z_0)^m \lambda(z)$ for all z in $|z - z_0| < r$. In this case, we call z_0 a **zero of order** m of f. If $m = 1$, we call z_0 a **simple zero** of f.

Proof Since f is analytic at z_0, it has a Taylor series, $f(z) = \sum_{n=0}^{\infty} a_n(z - z_0)^n$, around z_0 that converges in the largest open disk $B_R(z_0)$ that is contained in U. By the uniqueness of the Taylor series expansion (Theorem 1, Section 4.4), if f is identically zero in a neighborhood of z_0, then all the Taylor coefficients must be 0. So if (i) holds, (ii) does not hold.

Now suppose that (i) does not hold and let us show that (ii) must hold. Since f is not identically 0 in a neighborhood of z_0, not all the Taylor coefficients are zero, and so there is a first nonvanishing coefficient, say $a_0 = a_1 = \cdots = a_{m-1} = 0$, but $a_m \neq 0$. From Theorem 1, Section 4.4, we see that $f(z_0) = f'(z_0) = \cdots = f^{(m-1)}(z_0) = 0$ and $f^{(m)}(z_0) \neq 0$. For $|z - z_0| < R$, write

$$
\begin{aligned}
f(z) &= a_m(z - z_0)^m + a_{m+1}(z - z_0)^{m+1} + a_{m+2}(z - z_0)^{m+2} + \cdots \\
&= (z - z_0)^m \left(a_m + a_{m+1}(z - z_0) + a_{m+2}(z - z_0)^2 + \cdots \right) \\
&= (z - z_0)^m \lambda(z),
\end{aligned}
$$

where $\lambda(z) = a_m + a_{m+1}(z - z_0) + a_{m+2}(z - z_0)^2 + \cdots$, and $a_m \neq 0$. Since $\lambda(z)$ is a power series, it is analytic for all $|z - z_0| < R$, by Theorem 4, Section 4.3. Also, $\lambda(z_0) = a_m \neq 0$ and λ is continuous at z_0, so we can find a neighborhood $B_r(z_0)$ of z_0 such that $\lambda(z) \neq 0$ for all z in $B_r(z_0)$. ∎

If z_0 is a zero of order m, and we write $f(z) = (z - z_0)^m \lambda(z)$ where $\lambda(z) \neq 0$ for all $|z - z_0| < r$, then $f(z) \neq 0$ for all $0 < |z - z_0| < r$, which means that f has no zeros in the neighborhood $|z - z_0| < r$ other than z_0. This is expressed by saying that z_0 is an **isolated zero** of f. Putting this fact together with Theorem 1, we obtain the following.

THEOREM 2
LOCAL ISOLATION
OF ZEROS

Suppose that f is analytic on an open set U, z_0 is in U, and $f(z_0) = 0$. Then exactly one of the following holds:
(i) f is identically zero in a neighborhood of z_0.
(ii) There is a real number $r > 0$ and a neighborhood $B_r(z_0)$ of z_0 in Ω such that $f(z) \neq 0$ for all $0 < |z - z_0| < r$. That is, z_0 is an isolated zero of f.

EXAMPLE 1   Order of zeros
Find the order m of the zero of $\sin z$ at $z_0 = 0$; then express $\sin z = z^m \lambda(z)$, where $\lambda(z)$ is analytic at 0 with $\lambda(0) \neq 0$. (This proves the obvious fact that the zero of $\sin z$ at $z_0 = 0$ is isolated.)

Solution Clearly, 0 is a zero of $\sin z$. The order of the zero is equal to the order of the first nonvanishing derivative of $f(z) = \sin z$ at 0. Since $f'(z) = \cos z$ and $\cos 0 = 1 \neq 0$, we conclude that the order of the zero at 0 is 1. We have for all z

$$
\sin z = z - \frac{z^3}{3!} + \frac{z^5}{5!} - \cdots = z \left(1 - \frac{z^2}{3!} + \frac{z^4}{5!} - \cdots \right) = z\lambda(z),
$$

where $\lambda(z) = 1 - \frac{z^2}{3!} + \frac{z^4}{5!} - \cdots$. The function $\lambda(z)$ is entire (because it is a convergent power series for all z), and $\lambda(0) = 1$. (Also, for $z \neq 0$, $\lambda(z) = \frac{\sin z}{z}$, which is entire by Example 5, Section 3.6.) ∎

Theorem 2 raises the following question. Suppose that f is analytic and not identically zero on Ω. Can f vanish on an open nonempty subset of Ω? The answer is no; this depends crucially on the fact that Ω is connected.

THEOREM 3
ISOLATION OF
ZEROS

Suppose that $f(z)$ is analytic and not identically 0 on a region Ω (open and connected set). Then all the zeros of f are isolated.

Proof   Let Ω_0 consists of all the zeros of f that are not isolated zeros, and let Ω_1 consists of the remaining points in Ω. Clearly, Ω_0 and Ω_1 are disjoint and their union is Ω. We will show that Ω_0 and Ω_1 are open; then by connectedness either $\Omega = \Omega_0$ or $\Omega = \Omega_1$, which would prove the theorem. Recall that a set is open if it contains a neighborhood of each one of its points. The fact that Ω_0 is open follows from Theorem 2, since if z_0 is not an isolated zero, then f vanishes identically in a neighborhood $B_r(z_0)$ of z_0. Clearly, all the points in $B_r(z_0)$ are not isolated zeros of f, so $B_r(z_0)$ is contained in Ω_0, and hence Ω_0 is open. To prove that Ω_1 is open, let z_0 be given in Ω_1. Either $f(z_0) \neq 0$, in which case we can find a neighborhood of z_0 where f is nonzero; or z_0 is an isolated zero, in which case Theorem 2 guarantees us a neighborhood where f has no zeros except the isolated one at z_0.   ∎

We have two important consequences.

THEOREM 4
IDENTITY
PRINCIPLE

Suppose that $f(z)$ is analytic on a region Ω, and $\{z_n\}$ is an infinite sequence of distinct points in Ω converging to z_0 in Ω. Suppose that $f(z_n) = 0$ for all n. Then f is identically zero on Ω. Consequently, if the set of zeros of an analytic function on Ω contains an infinite sequence of distinct points that converges in Ω, then the function is identically zero in Ω.

Proof   If $z_n \to z_0$ and $f(z_n) = 0$, then $f(z_0) = 0$ and z_0 is not an isolated zero. By Theorem 3, f is identically zero.   ∎

COROLLARY 1

Suppose that $f(z)$ and $g(z)$ are analytic on a region Ω. If the set of points on which $f(z) = g(z)$ contains an infinite sequence of distinct points that converges in Ω, then $f = g$ on Ω.

Proof   Let $h(z) = f(z) - g(z)$ and apply Theorem 4 to h.   ∎

In many interesting applications of Corollary 1, the functions f and g are equal on an interval $[a, b]$ of the real line, or a whole disk $B_R(z_0)$. Such sets clearly contain infinite sequences of points that converge in the set itself.

In Section 1.6, we derived various identities involving complex trigonometric functions, which were the same for the real trigonometric functions. The identity principle can be used to justify these and extend other identities from real functions to complex functions. We illustrate these ideas with an example.

EXAMPLE 2   Applications of the identity principle
(a)   In Section 1.6, we proved that for any complex number z, $\cos^2 z + \sin^2 z = 1$. Let us prove this identity using Corollary 1 and the fact that it holds for real z. Let $f(z) = \cos^2 z + \sin^2 z$ and $g(z) = 1$. Clearly, $f(z)$ and $g(z)$ are entire, and for real $z = x$, we have $f(x) = \sin^2 x + \cos^2 x = 1 = g(x)$. Since $f(z) = g(z)$ for all z

on the real line, which is a set that contains infinite converging sequences, we infer from Corollary 1 that $f(z) = g(z)$ for all z; that is, $\cos^2 z + \sin^2 z = 1$.

(b)   Modifying the method in (a), we can prove identities involving two or more variables. As an illustration, let us prove that for any complex numbers z_1 and z_2,

$$(1) \qquad \cos(z_1 + z_2) = \cos z_1 \cos z_2 - \sin z_1 \sin z_2.$$

In a first step, let $z_2 = x_2$ be an arbitrary real number. Let $f(z) = \cos(z + x_2)$ and $g(z) = \cos z \cos x_2 - \sin z \sin x_2$. Clearly, $f(z)$ and $g(z)$ are entire, and from the addition formula for the cosines, we have $f(x) = g(x)$ for all real x. Hence by Corollary 1, we have $f(z) = g(z)$ for all z; equivalently,

$$(2) \qquad \cos(z + x_2) = \cos z \cos x_2 - \sin z \sin x_2 \qquad \text{for all complex } z.$$

In a second step, fix $z = z_1$ and let $f(z_2) = \cos(z_1 + z_2)$ and $g(z_2) = \cos z_1 \cos z_2 - \sin z_1 \sin z_2$. Here again, $f(z_2)$ and $g(z_2)$ are entire (considered as functions of z_2) and (2) states that $f(z_2)$ and $g(z_2)$ agree on the whole real line. By Corollary 1, $f(z_2) = g(z_2)$ for all z_2, implying that (1) holds.   ∎

Theorem 4 has another useful consequence regarding the number of zeros of an analytic function. In establishing this result, we will need a topological property of complex numbers known as the **Bolzano-Weierstrass theorem**, which states the following:

> Let S denote a closed and bounded subset of $\mathbb{C}$ and let Z denote an infinite subset of S. Then there is an infinite sequence $\{z_n\}$ of distinct elements of Z that converges to a point z_0 in S.

In particular, any infinite sequence of complex numbers in a closed and bounded subset S contains a subsequence that converges in S.

COROLLARY 2

> Suppose that f is analytic on a bounded region Ω and continuous and non-vanishing on the boundary of Ω. Then f can have at most finitely many zeros inside Ω.

Proof   Let S be the set Ω union its boundary. Then S is closed and bounded. Suppose that f has infinitely many zeros in Ω; then by the Bolzano-Weierstrass theorem, there is an infinite sequence of zeros, $\{z_n\}$, that converges to a point z_0 in S. Since f is continuous in S, $f(z_0) = \lim f(z_n) = 0$, and since f is nonvanishing on the boundary, we conclude that z_0 is inside Ω. Hence by Theorem 4, f is identically zero on Ω, and since f is continuous, f must be zero on the boundary, which is a contradiction. Hence f can have at most finitely many zeros in Ω.   ∎

Isolated Singularities

Suppose that f is analytic in a neighborhood of a point z_0, except at z_0; that is, f is analytic in an annulus $0 < |z - z_0| < R$. Then z_0 is called an **isolated singularity** of f. We know from the previous section that f has a Laurent series expansion in the annulus $0 < |z - z_0| < R$. We will see that there are three possibilities for the Laurent series; each of them will give rise to a different type of singularity. To simplify the presentation, let us start with the definitions of the three types of singularities, based on the behavior of f around z_0.

DEFINITION 1
THE THREE TYPES
OF SINGULARITIES

Suppose that z_0 is an isolated singularity of f. Then

(i) z_0 is a **removable singularity** if $f(z)$ can be redefined at z_0 so as to be analytic there.

(ii) z_0 is a **pole** if $\lim_{z \to z_0} |f(z)| = \infty$.

(iii) z_0 is an **essential singularity** if it is neither a pole nor a removable singularity.

When redefining $f(z)$ at a removable singularity z_0, we must set $f(z_0) = \lim_{z \to z_0} f(z)$; otherwise, f would not be continuous and hence could not be analytic at z_0. Note that when z_0 is a removable singularity, f must be bounded near z_0 (Figure 1(a)). Another way to define a removable singularity is as follows: For $f(z)$ analytic in $0 < |z - z_0| < R$, z_0 is a removable singularity if there is a function $g(z)$, analytic in $|z - z_0| < R$, a neighborhood of z_0, such that $f(z) = g(z)$ for all $0 < |z - z_0| < R$. We must have $g(z_0) = \lim_{z \to z_0} f(z)$.

The singularity is a pole if the graph of $|f(z)|$ blows up to infinity as we approach z_0 (Figure 1(b)). An essential singularity is harder to explain at this point. It suffices to say that the graph of $|f(z)|$ near an essential singularity is neither bounded nor tends to infinity (Figure 1(c)). Its behavior is very erratic. We will give several equivalent characterizations of each type of singularity, but first let us look at some examples.

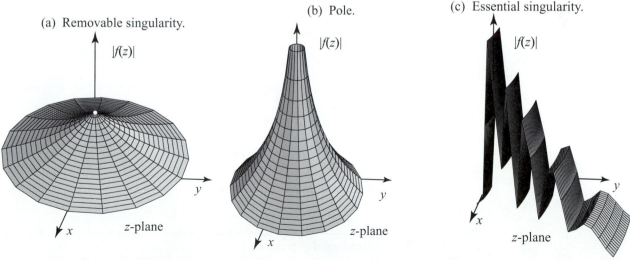

(a) Removable singularity.
$|f(z)|$
x z-plane

(b) Pole.
$|f(z)|$
y
x z-plane

(c) Essential singularity.
$|f(z)|$
x
y
z-plane

Figure 1 The three types of isolated singularities. (a) Near a removable singularity, $|f(z)|$ is bounded. (b) Near a pole, $|f(z)|$ tends to infinity. (c) Near an essential singularity, $|f(z)|$ is neither bounded nor tends to infinity. Its graph behaves erratically.

EXAMPLE 3   Three types of singularities

(a) The function $f(z) = \frac{z^2-1}{z-1}$ ($z \neq 1$) is analytic everywhere, except at $z_0 = 1$. So $z_0 = 1$ is an isolated singularity. For $z \neq 1$, $f(z) = \frac{(z-1)(z+1)}{z-1} = z + 1$. By defining $f(1) = 2$, we make f analytic at $z = 1$. This shows that the singularity at $z_0 = 1$ is removable.

(b) Consider the function

$$f(z) = \frac{z^2}{(z-i)^3}.$$

It has an isolated singularity at $z = i$. Unlike the previous example, the singularity here is not removable; it is a pole. Since

$$\lim_{z \to i} |f(z)| = \lim_{z \to i} \frac{|z|^2}{|z-i|^3} = \infty,$$

the graph of $|f(z)|$ has a pole that blows up to infinity above the singularity $z_0 = i$.

(c)   Consider

$$f(z) = e^{\frac{1}{z}}, \qquad z \neq 0.$$

We have an isolated singularity at $z_0 = 0$. We will show that it is an essential singularity by eliminating the possibility of the other two types. Suppose that $z = x$ is real and tends to 0 from the right. Then

$$\lim_{z = x \downarrow 0} |e^{\frac{1}{z}}| = \lim_{x \downarrow 0} e^{\frac{1}{x}} = \infty.$$

So 0 cannot be a removable singularity. Now suppose that $z = x$ is real and tends to 0 from the left. Then

$$\lim_{z = x \uparrow 0} |e^{\frac{1}{z}}| = \lim_{x \uparrow 0} e^{\frac{1}{x}} = 0;$$

hence 0 cannot be a pole, and so 0 is an essential singularity.   ■

Removable singularities often occur when an analytic function $h(z)$ with a zero at z_0 is divided by a small enough power of $(z - z_0)$.

PROPOSITION 1

Let $h(z)$ be analytic for $|z - z_0| < R$, with a zero of order $m \geq 1$ at z_0. If p is an integer $\leq m$, then $f(z) = \dfrac{h(z)}{(z-z_0)^p}$ has a removable singularity at z_0.

Proof   Write $h(z) = a_m(z - z_0)^m + a_{m+1}(z - z_0)^{m+1} + \cdots$ for $|z - z_0| < R$. Then for $0 < |z - z_0| < R$, $f(z) = a_m(z - z_0)^{m-p} + a_{m+1}(z - z_0)^{m-p+1} + \cdots$. The power series is analytic for $|z - z_0| < R$, and we can redefine $f(z)$ at z_0 to equal the value of this power series at z_0.   ■

Note that Proposition 1 generalizes Theorem 4, Section 3.6. Here are some straightforward examples.

EXAMPLE 4   Dividing by powers

(a) The function $e^{z-2} - 1$ can be expanded in Taylor series about $z_0 = 2$ as $e^{z-2} - 1 = (z - 2) + \frac{1}{2}(z - 2)^2 + \cdots$. Clearly, this function has a zero of order 1 at $z_0 = 2$, so by Proposition 1 (or just by dividing), $\frac{e^{z-2}-1}{z-2}$ has a removable singularity at $z_0 = 2$.

(b) The function $\cos z - 1$ can be expanded in Maclaurin series as $\cos z - 1 = -\frac{1}{2}z^2 + \frac{1}{4!}z^4 - \cdots$. Clearly, this function has a zero of order 2 at 0, so from Proposition 1, each of $\frac{\cos z - 1}{z}$ and $\frac{\cos z - 1}{z^2}$ has a removable singularity at 0.   ∎

There are several ways to characterize each type of singularity. Here is a useful characterization of a removable singularity.

THEOREM 5

Let f be analytic with z_0 an isolated singularity. The singularity at z_0 is removable if and only if

$$(3) \qquad\qquad \lim_{z \to z_0} f(z) = A$$

exists and is finite.

Proof   One direction has already been discussed: If z_0 is removable, then f can be redefined at the point to be analytic; hence the limit must exist. The converse is more interesting. Suppose that f is analytic in $0 < |z - z_0| < R$ and $\lim_{z \to z_0} f(z) = A$. Define $h(z) = (z - z_0)f(z)$ for $z \neq z_0$ and $h(z_0) = 0$. Then for $z \neq z_0$, $h'(z) = (z - z_0)f'(z) + f(z)$. Also,

$$h'(z_0) = \lim_{z \to z_0} \frac{(z - z_0)f(z)}{z - z_0} = \lim_{z \to z_0} f(z) = A.$$

Hence $h(z)$ is differentiable for all $|z - z_0| < R$ and so, by Goursat's theorem, it is analytic in $|z - z_0| < R$. Since h has a zero of order ≥ 1, by Proposition 1, $f(z) = \frac{h(z)}{z - z_0}$ has a removable singularity.   ∎

We can now give several characterizations of a removable singularity.

THEOREM 6
CHARACTERIZATION
OF REMOVABLE
SINGULARITIES

Suppose that f is analytic on $0 < |z - z_0| < R$. The following are equivalent:

(i) f has a removable singularity at z_0.

(ii) $f(z) = \sum_{n=0}^{\infty} a_n(z - z_0)^n$, for $0 < |z - z_0| < R$.

(iii) $\lim_{z \to z_0} f(z)$ exists and is finite.

(iv) $\lim_{z \to z_0} |f(z)|$ exists and is finite.

(v) f is bounded in a neighborhood of z_0.

(vi) $\lim_{z \to z_0} (z - z_0)f(z) = 0$.

Proof   The implication (i) $\Rightarrow$ (ii) follows from the existence of an analytic $g(z) = \sum_{n=0}^{\infty} a_n(z - z_0)^n$ for $|z - z_0| < R$, with $f(z) = g(z)$ for $0 < |z - z_0| < R$. To show (ii) $\Rightarrow$ (iii), use the fact that power series are continuous. To show (iii) $\Rightarrow$ (iv), use continuity of the absolute value function $w \mapsto |w|$. That (iv) $\Rightarrow$ (v) follows from the definition of a limit. To prove (v) $\Rightarrow$ (vi), use the squeeze theorem. Let us now show that (vi) $\Rightarrow$ (i). Let $h(z) = (z - z_0)f(z)$, $z \neq z_0$. By Theorem 5, z_0 is a removable singularity of $h(z)$, and by defining $h(z_0) = \lim_{z \to z_0} (z - z_0)f(z) = 0$, we make $h(z)$ analytic in a neighborhood of z_0. Since h has a zero of order ≥ 1, by Proposition 1, $f(z) = \frac{h(z)}{z - z_0}$ has a removable singularity at z_0.   ∎

EXAMPLE 5   Removable singularities

Show that the given function has a removable singularity at the indicated point:

(a) $f(z) = \dfrac{\sin z}{z}$ at $z_0 = 0$;   (b) $f(z) = \dfrac{e^{z-1} - 1}{z - 1}$ at $z_0 = 1$.

Solution   (a)   This example is not new to us (see Example 5, Section 3.6). We will offer a proof based on Theorem 6(vi). We have

$$\lim_{z \to 0} (z - 0) f(z) = \lim_{z \to 0} z \frac{\sin z}{z} = \lim_{z \to 0} \sin z = 0.$$

Thus $\frac{\sin z}{z}$ has a removable singularity at $z = 0$, by Theorem 6(vi).

(b)   We apply Theorem 6(vi). We have

$$\lim_{z \to 1} (z - 1) f(z) = \lim_{z \to 1} (z - 1) \frac{e^{z-1} - 1}{z - 1} = \lim_{z \to 1} (e^{z-1} - 1) = 0.$$

Hence $z_0 = 1$ is a removable singularity.   ∎

In order to characterize poles, we begin by relating a pole of $f(z)$ to a zero of $\frac{1}{f(z)}$. Suppose that f is analytic in $0 < |z - z_0| < R$, so that z_0 is an isolated singularity. Suppose that z_0 is a pole of f. Then, because $\lim_{z \to z_0} |f(z)| = \infty$, we can find $\rho > 0$ such that $f(z) \neq 0$ for all $0 < |z - z_0| < \rho$. Consider the function

(4)
$$g(z) = \begin{cases} \frac{1}{f(z)} & \text{if } 0 < |z - z_0| < \rho, \\ 0 & \text{if } z = z_0. \end{cases}$$

Clearly, $g(z)$ is analytic and nonzero on $0 < |z - z_0| < \rho$. Also, since $\lim_{z \to z_0} g(z) = 0 = g(z_0)$, it follows from Theorem 6 that g is analytic at $z = z_0$; and since g is not identically 0 in a neighborhood of z_0, it follows that z_0 is a zero of order $m \geq 1$ of g. Appealing to Theorem 1, we write

(5)
$$g(z) = (z - z_0)^m \lambda(z)   \text{for all } |z - z_0| < r,$$

where $m \geq 1$ is the order of the zero of g at z_0, and $\lambda(z)$ is analytic and nonzero on $|z - z_0| < r$. Consequently, if we set $\phi(z) = \frac{1}{\lambda(z)}$, then $\phi(z)$ is analytic and nonzero on $|z - z_0| < r$, and

(6) $f(z) = \dfrac{1}{g(z)} = \dfrac{1}{(z - z_0)^m} \dfrac{1}{\lambda(z)} = \dfrac{1}{(z - z_0)^m} \phi(z), \quad 0 < |z - z_0| < r.$

The positive integer m in (6) is called the **order of the pole** of f at z_0. If $m = 1$, we call z_0 a **simple pole** of f. Thus the order of the pole of f at z_0 is equal to the order of the zero of $\frac{1}{f(z)}$ at $z = z_0$. Let us summarize this discussion and give several characterizations of poles.

THEOREM 7
CHARACTERIZATION
OF POLES

Suppose that f is analytic on $0 < |z - z_0| < R$. The following conditions are equivalent and characterize a pole at z_0.

(i) $\lim_{z \to z_0} |f(z)| = \infty$.

(ii) $f(z) = \frac{1}{(z - z_0)^m} \phi(z)$, where m is an integer ≥ 1 and $\phi(z)$ is analytic and nonzero on $|z - z_0| < r$.

(iii) $\lim_{z \to z_0} (z - z_0)^m f(z) = \alpha \neq 0$ for some integer $m \geq 1$.

(iv) $f(z) = \frac{a_{-m}}{(z - z_0)^m} + \frac{a_{-m+1}}{(z - z_0)^{m-1}} + \cdots + \frac{a_{-1}}{z - z_0} + a_0 + a_1(z - z_0) + \cdots$, for $0 < |z - z_0| < r$ with $a_{-m} \neq 0$, $m \geq 1$.

(v)   The function g in (4) is analytic and has a zero of order m at z_0.

Proof   We have already shown that (i) $\Rightarrow$ (v) $\Rightarrow$ (ii). To prove (ii) $\Rightarrow$ (iv), expand $\phi(z)$ in a power series centered at z_0:

$$\phi(z) = \sum_{n=0}^{\infty} c_n (z - z_0)^n, \quad |z - z_0| < r,$$

where $c_0 = \phi(z_0) \neq 0$. Since $\phi(z_0) \neq 0$, it follows that $a_{-m} = c_0 = \phi(z_0) \neq 0$. That (iii) $\Rightarrow$ (iv) is immediate. To show (iii) $\Rightarrow$ (i), suppose (i) does not hold; then we can find a sequence $z_n \to z_0$ such that $f(z_n)$ is bounded. For this sequence, we have $\lim_{z_n \to z_0} (z - z_0)^m f(z) = 0$, which contradicts (iii). ∎

EXAMPLE 6   Poles

Determine the order of the pole of the given function at the indicated point:

(a) $f(z) = \dfrac{1}{z \sin z}$ at $z_0 = 0$;   (b) $f(z) = \dfrac{e^{z^2} - 1}{z^4}$ at $z_0 = 0$.

Solution   (a)   We use Theorem 7(iii) and our knowledge of the function $\sin z$ around 0. Since

$$\lim_{z \to 0} z^2 \frac{1}{z \sin z} = \lim_{z \to 0} \frac{z}{\sin z} = 1 \neq 0,$$

we conclude that 0 is a pole of order 2 of $f(z)$.

(b)   We use Laurent series. Since

$$e^z = \sum_{n=0}^{\infty} \frac{z^n}{n!}, \quad \text{all } z,$$

then

$$e^{z^2} = \sum_{n=0}^{\infty} \frac{z^{2n}}{n!} = 1 + \frac{z^2}{1!} + \frac{z^4}{2!} + \frac{z^6}{3!} + \cdots, \quad \text{all } z,$$

hence

$$e^{z^2} - 1 = \frac{z^2}{1!} + \frac{z^4}{2!} + \frac{z^6}{3!} + \cdots, \quad \text{all } z,$$

and so for $z \neq 0$

$$\frac{e^{z^2} - 1}{z^4} = \frac{z^2}{1!z^4} + \frac{z^4}{2!z^4} + \frac{z^6}{3!z^4} + \cdots = \frac{1}{1!z^2} + \frac{1}{2!} + \frac{z^2}{3!} + \frac{z^4}{4!} + \cdots.$$

Thus the order of the pole at 0 is 2.   ■

Having characterized removable singularities and poles in terms of Laurent series, this leaves one possibility for essential singularities. They must have infinitely many terms involving negative powers of $(z - z_0)$. For ease of reference, we list all three possibilities together.

**THEOREM 8
LAURENT SERIES
CLASSIFICATION OF
ISOLATED
SINGULARITIES**

Suppose that f is analytic in a region Ω except for an isolated singularity at z_0 in Ω. Let

$$f(z) = \sum_{n=-\infty}^{\infty} a_n (z - z_0)^n$$

denote the Laurent series expansion of f about z_0, which is valid in some annulus $0 < |z - z_0| < R$. Then
(i) z_0 is a removable singularity $\Leftrightarrow a_n = 0$ for all $n = -1, -2, \ldots$.
(ii) z_0 is a pole of order $m \geq 1$ $\Leftrightarrow a_{-m} \neq 0$ for some $m > 0$ and $a_n = 0$ for all $n < -m$.
(iii) z_0 is an essential singularity $\Leftrightarrow a_n \neq 0$ for infinitely many $n < 0$.

EXAMPLE 7   Essential singularities
Classify the isolated singularities of the function $f(z) = e^{\frac{z}{\sin z}}$.

Solution   The function $f(z)$ is analytic at all points except where $\sin z = 0$; that is, except when $z = k\pi$, where k is an integer. It is difficult to find the Laurent series expansion of $f(z)$, so we will use the characterizations of Theorems 6 and 7. When $z = 0$, we have $\lim_{z \to 0} \frac{z}{\sin z} = 1$, hence $\lim_{z \to 0} e^{\frac{z}{\sin z}} = e$, and so the function has a removable singularity at $z = 0$ by Theorem 6(iii). We claim that we have an essential singularity at $z = k\pi$, $k \neq 0$. To prove this, we will eliminate the possibility of a removable singularity or a pole. Suppose that k is even. Then it is easy to see that if $z = x$ is real, then $\lim_{x \downarrow k\pi} \frac{x}{\sin x} = +\infty$ and $\lim_{x \uparrow k\pi} \frac{x}{\sin x} = -\infty$. So if $z = x$ is real, then

$$\lim_{x \downarrow k\pi} e^{\frac{x}{\sin x}} = \infty,$$

implying that $k\pi$ is not a removable singularity of $e^{\frac{z}{\sin z}}$. Also,

$$\lim_{x \uparrow k\pi} e^{\frac{x}{\sin x}} = 0,$$

implying that $k\pi$ is not a pole of $e^{\frac{z}{\sin z}}$. This leaves only one possibility: $k\pi$ is an essential singularity of $e^{\frac{z}{\sin z}}$. A similar argument works for odd k.   ■

Perhaps the best way to determine whether an isolated singularity is an essential singularity is to rule out the possibility of the other two types of singularities. As we just showed in Example 7, this can be achieved by showing that the function is unbounded and has different limits as we approach the isolated singularity in different ways. In fact, the next theorem tells us that given any complex number α, we can approach an essential singularity in such a way that $f(z)$ tends to α. This peculiar phenomenon characterizes an essential singularity.

THEOREM 9
CASORATI-
WEIERSTRASS
THEOREM

Suppose that f is analytic on $0 < |z - z_0| < R$. Then z_0 is an essential singularity of f if and only if the following two conditions hold:

(i) There is a sequence $\{z_n\}$ such that $z_n \to z_0$ and $\lim_{n \to \infty} |f(z_n)| = \infty$.

(ii) For any complex number α, there is a sequence $\{z_n\}$ (that depends on α) such that $z_n \to z_0$ and $\lim_{n \to \infty} f(z_n) = \alpha$.

The theorem was discovered independently by Weierstrass and the Italian mathematician Felice Casorati (1835–1890). Keep in mind that (i) is not saying that $\lim_{z \to z_0} |f(z)| = \infty$. It is just saying that you can approach z_0 in such a way that $|f(z)|$ will tend to infinity. Similarly, part (ii) says that you can approach z_0 in such a way that $f(z)$ will come arbitrarily close to any complex value α. In fact, a deep result in complex analysis, known as Picard's great theorem, states that, in a neighborhood of an essential singularity, a function takes on every complex value, with one possible exception, an infinite number of times.

Proof   If (i) is true, then f is not bounded near z_0, and so z_0 is not a removable singularity. If (ii) is true, then it is not the case that $\lim_{z \to z_0} |f(z)| = \infty$, and so z_0 is not a pole. Thus (i) and (ii) together imply that z_0 is an essential singularity. Conversely, let z_0 be an essential singularity. Since $f(z)$ is not bounded near z_0 (otherwise z_0 would be a removable singularity), it follows that (i) holds. Now suppose that (ii) fails. Then $|f(z) - \alpha| \geq \epsilon > 0$ for some α and all z in some deleted neighborhood $0 < |z - z_0| < r$. Consider the function $g(z) = \frac{1}{f(z) - \alpha}$. Then $|g(z)| \leq \frac{1}{\epsilon}$ for all z in $0 < |z - z_0| < r$, and so z_0 is a removable singularity for g. Moreover, since $\lim_{z \to z_0} |f(z)| \neq \infty$, we conclude that $g(z_0) \neq 0$. Solving for $f(z)$, we find $f(z) = \frac{1}{g(z)} + \alpha$, which is analytic in a neighborhood of z_0, because $g(z_0) \neq 0$. This is a contradiction. Hence (ii) must hold.   ∎

We end the section by extending the definitions of zeros and singularities to the point at infinity. If f is analytic on a neighborhood of infinity–that is, f is analytic for all $|z| > R$–then $f\left(\frac{1}{z}\right)$ is analytic in the annulus $0 < |z| < \frac{1}{R}$ and hence it has an isolated singularity at 0. With this in mind, we make the following definitions.

DEFINITION 2
SINGULARITIES AT
INFINITY

Suppose that f is analytic for all $|z| > R$. Then f has

a **removable singularity at** ∞ if $f\left(\frac{1}{z}\right)$ has a removable singularity at 0;

a **pole of order** m **at** ∞ if $f\left(\frac{1}{z}\right)$ has a pole of order m at 0;

an **essential singularity at** ∞ if $f\left(\frac{1}{z}\right)$ has a essential singularity at 0.

When f has a removable singularity at ∞, $\lim_{z \to \infty} f(z)$ exists. When $\lim_{z \to \infty} f(z) = 0$, we say that f has a **zero** at ∞.

EXAMPLE 8   Singularities at ∞

Characterize all entire functions with a pole of order $m \geq 1$ at ∞.

Solution   Since f is entire, it has a Maclaurin series that converges for all z: $f(z) = \sum_{n=0}^{\infty} c_n z^n$, for all z. For $z \neq 0$, $f\left(\frac{1}{z}\right) = \sum_{n=0}^{\infty} \frac{c_n}{z^n}$. Appealing to Theorem 7(iv), we see that $f\left(\frac{1}{z}\right)$ has a pole of order $m \geq 1$ at 0 if and only if $c_m \neq 0$ and $c_n = 0$ for all $n > m$-that is, if and only if $f(z)$ is a polynomial of degree m. Consequently, $f(z)$ has a pole of order $m \geq 1$ at ∞ if and only if $f(z)$ is a polynomial of degree m.   ∎

Exercises 4.6

In Exercises 1–8, find the isolated zeros of the given function. Also, find the order of each isolated zero.

1. $(1 - z^2) \sin z$.   **2.** $z^3 (e^z - 1)$.   **3.** $\dfrac{z(z-1)^2}{z^2 + 2z - 1}$.   **4.** $\sin \dfrac{1}{z}$.

5. $\dfrac{\sin^7 z}{z^4}$.   **6.** $(z-1)^3 (e^{2z} - 1)^2$.   **7.** $(z^2 + 2z - 1)^3$.   **8.** $\sinh z$.

In Exercises 9–12, find the order of the zero at $z_0 = 0$.

9. $1 - \dfrac{z^2}{2} - \cos z$.   **10.** $z \operatorname{Log}(1 + z)$.   **11.** $z - \sin z$.   **12.** $\tan z$.

In Exercises 13–24, classify the isolated singularities of the given function. Do not include the case at ∞. At a removable singularity, redefine the function in order to make it analytic. If it is a pole, determine its order.

13. $\dfrac{1 - z^2}{\sin z} + \dfrac{z-1}{z+1}$.   **14.** $\dfrac{z-1}{z-i} + \dfrac{z-i}{z-1}$.   **15.** $\dfrac{z(z-1)^2}{\sin(\pi z) \sin z}$.

16. $e^{\frac{1}{1-z}} + \dfrac{1}{1-z}$.   **17.** $z \tan \dfrac{1}{z}$.   **18.** $\dfrac{z}{e^z - 1}$.

19. $\dfrac{z}{z^4 - 1} - \dfrac{\sin(2z)}{z^4}$.   **20.** $z^2 \sin \dfrac{1}{z^2}$.   **21.** $\dfrac{1}{z} - \sin \dfrac{1}{z}$.

22. $\dfrac{1}{(e^z - e^{2z})^2}$.   **23.** $\dfrac{\cot z}{(z - \frac{\pi}{2})^2}$.   **24.** $\dfrac{z \sin z}{\cos z - 1}$.

In Exercises 25–30, determine if the function has an isolated singularity at ∞, and determine its type. Does the function have a zero at ∞?

25. $\dfrac{1}{z+1}$.   **26.** $\dfrac{z^2 - 1}{z^2 + 2z + 3i}$.   **27.** $\dfrac{z}{z^2 + 1} - \dfrac{1}{z}$.

28. $e^z - \cos \dfrac{1}{z}$.   **29.** $\sin \dfrac{1}{z}$.   **30.** $\dfrac{e^z}{e^z - 1}$.

31. Prove the given identity using Corollary 1.

(a) $\sin^2 z = \dfrac{1 - \cos 2z}{2}$.   (b) $\sin(2z) = 2 \sin z \cos z$.

(c) $\tan(2z) = \dfrac{2 \tan z}{1 - \tan^2 z}$, $z \neq \dfrac{\pi}{2} + 2k\pi$ or $\dfrac{\pi}{4} + 2k\pi$.

32. Prove the given identity using Corollary 1.

(a) $e^{z_1 + z_2} = e^{z_1} e^{z_2}$.   (b) $\sin(z_1 + z_2) = \sin z_1 \cos z_2 + \cos z_1 \sin z_2$.

33.   Give a new proof of Liouville's theorem (Theorem 2, Section 3.7) based on Theorem 6 as follows. (a) Suppose $f(z)$ is entire and bounded. Show that $f(\frac{1}{z})$ has a removable singularity at 0.
(b) Express the Laurent series expansion of $f(\frac{1}{z})$ around 0 by replacing z by $\frac{1}{z}$ in the Maclaurin series expansion of $f(z)$. Using (a), conclude that f must be constant.

34.   Show that $f(z)$ has an essential singularity at z_0 if and only if $(z - z_0)^m f(z)$ has an essential singularity at z_0, where m is any integer.

35.   **Basic properties.** Prove the following assertions concerning zeros and isolated singularities.
(a)   If f has a zero of order $m \geq 0$ at z_0 and g has a zero of order $n \geq 0$ at z_0, then fg has a zero of order $m + n$ at z_0. (Here we take a zero of order 0 to mean analytic and nonvanishing in a neighborhood of z_0.)
(b)   If f has a pole of order $m \geq 1$ at z_0 and g has a zero of order $n \geq 1$ at z_0, then at z_0, fg has a pole of order $m - n$ if $m > n$; a zero of order $n - m$ if $n > m$; and a removable singularity if $m = n$.
(c)   If f has a removable singularity at z_0 and g is analytic at z_0, then fg has a removable singularity at z_0.
(d)   If f has an essential singularity at z_0 and g is not identically 0 with a removable singularity or a pole at z_0, then fg has an essential singularity at z_0. [Hint: Multiply or divide by suitable $(z - z_0)^m$ so that $g(z)$ is analytic and nonvanishing in a neighborhood of z_0; use Exercise 34.] Does the assertion remain true if g has an essential singularity at z_0?

36.   Show that f has a removable singularity at z_0 if and only if either $\mathrm{Re}\, f$ or $\mathrm{Im}\, f$ is bounded in a neighborhood of z_0. [Hint: One direction is easy. For the other direction, suppose that $\mathrm{Re}\, f$ is bounded and show that $g(z) = e^{if(z)}$ has a removable singularity at z_0. Compute $g'(z)$ and conclude that f' is analytic at z_0. Conclude that f is analytic at z_0.]

37.   (a) Show that if f has a pole of order $m \geq 1$ at z_0 and n is any nonzero integer then $[f(z)]^n$ has a pole at z_0 of order mn if $n > 0$ or a zero at z_0 of order mn if $n < 0$.
(b) Show that if f has an essential singularity at z_0 and n is any nonzero integer then $[f(z)]^n$ has an essential singularity at z_0.

38.   Suppose that f and g are entire functions such that $f \circ g$ is a polynomial. Show that both f and g must be polynomials. [Hint: If $p(z)$ is a nonconstant polynomial then $\lim_{z \to \infty} |p(z)| = \infty$. Suppose that f is not a polynomial. Then f has an essential singularity at ∞. Use this and the Casorati-Weierstrass theorem to show that $\lim_{z \to \infty} |f \circ g(z)| \neq \infty$. Argue similarly if g is not a polynomial.]

39. Contrasts with the theory of functions of a real variable. (a) Consider $f(x) = \sin \frac{1}{x}$. Show that f is differentiable for all $x \neq 0$ and bounded for all x.
(b)   Show that there is no differentiable function $g(x)$ such that $f(x) = g(x)$ for all $x \neq 0$. Which aspect of the theory of analytic functions did we contrast with this example?
(c)   Define $\phi(x) = x^2 \sin \frac{1}{x}$, $x \neq 0$, $\phi(0) = 0$. Show that ϕ is differentiable for all x. Is the zero of ϕ isolated at $x = 0$? Which aspect of the theory of analytic functions did we contrast?
(d)   Define $\psi(x) = e^{-\frac{1}{x^2}}$, $x \neq 0$, $\psi(0) = 0$. Show that ψ has derivatives of all order

at $x = 0$. Explain why $\psi(x)$ cannot possibly have a Maclaurin series representation. Which aspect of the theory of analytic functions did we contrast?

40.   Verify the great Picard's theorem with the example of the function $f(z) = e^{\frac{1}{z}}$. More precisely, show that in any neighborhood of 0 (an essential singularity) f takes on every complex value with the exception of one an infinite number of times.

41.   Suppose that f and g are analytic in a region Ω and fg is identically zero in Ω. Show that either f or g is identically zero in Ω.

42.   Determine all entire functions with zeros at $\frac{1}{n}$, $n = 1, 2, \ldots$.

43.   Suppose that f is entire such that $f(z)f(\frac{1}{z})$ is bounded. Follow the outlined steps to show that $f(z) = az^n$ for some constant a and nonnegative integer n.
(a)   Let $h(z) = f(z)f(\frac{1}{z})$. Show that h has a removable singularity at 0 and then conclude that $h(z) = c$ is a constant. If $c = 0$, then $f = 0$ and we are done. In what follows, we suppose that $c \neq 0$.
(b)   Show that the only possible zero of f is at $z = 0$.
(c)   If $f(0) \neq 0$, show that f is constant.
(d)   If $f(0) = 0$, show that $f(z) = az^n$, where n is the order of the zero at 0. [Hint: Factor z^n and apply (c) to the entire function $\frac{f(z)}{z^n}$.]

44. World's worst function. (a) For $z \neq 1$ let $I(z) = e^{\frac{z+1}{z-1}}$. Show that $I(z)$ is analytic for all $z \neq 1$ and that $|I(z)| = 1$ for all $|z| = 1$ and $z \neq 1$.
[Hint: $|I(z)| = e^{\operatorname{Re}\left(\frac{z+1}{z-1}\right)} = e^{\frac{|z|^2-1}{|z-1|^2}}$.]
(b) Show that if z is real and $z \to 1^-$, then $I(z) \to 0$.
(c) Show that $|I(z)| < 1$ for all $|z| < 1$.

45.   Suppose that $p(z)$ is a polynomial such that $|p(z)| = 1$ for all $|z| = 1$. Show that $p(z) = Az^m$, where A is a unimodular constant and m is the number of zeros of $p(z)$ inside the unit disk counted according to multiplicity (that means a zero of order $k \geq 1$ is counted k times). [Hint: Recall that if f is analytic on $|z| \leq 1$, $f(z) \neq 0$ for all $|z| < 1$, and $|f(z)| = 1$ for $|z| = 1$, then f is constant. So if p has no zeros in the unit disk, we are done. Otherwise, let $a_1, a_2, \ldots, a_m$ denote the zeros of p inside the open unit disk, repeated according to multiplicity. Multiply $p(z)$ by a product of linear fractional transformations of the form $\phi_{a_j}(z) = \frac{1 - \bar{a}_j z}{a_j - z}$ to reduce to the case of a function without zeros in $|z| < 1$ and with modulus 1 on the circle $|z| = 1$. Recall that $|\phi_{a_j}(z)| = 1$ for all $|z| = 1$ (Example 3, Section 3.7). Then argue that $a_j = 0$ for all j and that $p(z) = Az^m$.]

46.   Denote the product of n functions $f_1(z), f_2(z), \ldots, f_n(z)$ by $\prod_{j=1}^{n} f_j(z)$. Suppose that $f(z)$ is analytic for all $|z| \leq 1$ and $|f(z)| = 1$ for all $|z| = 1$. Suppose further that f has n zeros inside the open unit disk $a_1, a_2, \ldots, a_n$ counted according to multiplicity. Show that $f(z) = A \prod_{j=1}^{n} \phi_{a_j}(z)$, where $\phi_{a_j}(z) = \frac{a_j - z}{1 - \bar{a}_j z}$ is a linear fractional transformation and A is unimodular. (See Exercise 47 for an improved statement.)

47.   Suppose that $f(z)$ is analytic on $|z| < 1$ and continuous on $|z| \leq 1$ such that $|f(z)| = 1$ for all $|z| = 1$. Show that f has finitely many zeros in $|z| \leq 1$ and conclude that $f(z) = A \prod_{j=1}^{n} \phi_{a_j}(z)$, where $\phi_{a_j}(z) = \frac{a_j - z}{1 - \bar{a}_j z}$ is a linear fractional transformation and A is unimodular. [Hint: The first part follows from Corollary 2. The second part follows from Exercise 46.]

48. Failure of the identity principle for harmonic functions.   Give an example of a harmonic function u in a region Ω that vanishes identically on a line segment in Ω, but such that u is not identically zero in Ω. [Hint: Think of the real or imaginary part of a trigonometric function.]

4.7   Harmonic Functions and Fourier Series

In Section 3.8, we solved the Dirichlet problem in a disk and gave the solution in the form of an integral called the Poisson integral formula. In this section, we derive another form of the solution, which will lead us to one of the most powerful tools in applied mathematics, Fourier series.

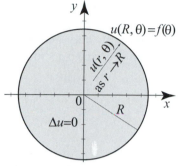

Figure 1 A Dirichlet problem on a disk with radius $R > 0$ and center at the origin.

We will use polar coordinates to state the Dirichlet problem on the disk of radius $R > 0$ and center at the origin. The boundary points are of the form $Re^{i\theta}$, where θ is arbitrary and R is fixed. The boundary values in the Dirichlet problem will be given by a piecewise continuous function $f(Re^{i\theta}) = f(\theta)$. Note how we have written f as a function of θ alone, because R is fixed. This notation is convenient and will allow us to think of f as a function of a real variable θ. Also, since θ and $\theta + 2\pi$ represent the same polar angle, we must have $f(\theta) = f(\theta + 2\pi)$. Thus f is 2π-periodic. The Dirichlet problem on the disk of radius $R > 0$ is the boundary value problem

(1) $$\Delta u(r,\theta) = 0, \qquad 0 \le r < R, \text{ all } \theta;$$

(2) $$\lim_{r\uparrow R} u(r,\theta) = u(R,\theta) = f(\theta),$$

where the limit holds at all points $Re^{i\theta}$ where $f(\theta)$ is continuous (Figure 1). In Section 3.8, Theorem 3, we derived the solution in the form of an integral:

(3) $$u(r,\theta) = \frac{R^2 - r^2}{2\pi} \int_0^{2\pi} \frac{f(\phi)}{R^2 - 2rR\cos(\theta - \phi) + r^2} \, d\phi \qquad (0 \le r < R),$$

known as the Poisson integral formula. The integrand is a function of r, θ, and ϕ. When integrated with respect to ϕ, it yields a function of r and θ. Theorem 3, Section 3.8, tells us that this function $u(r,\theta)$ is harmonic in the open disk $|z| < R$ (thus (1) is satisfied) and tends to $f(\theta)$ as $r \uparrow R$ at all points of continuity of f (thus (2) holds). While the Poisson integral formula offers an elegant solution of the Dirichlet problem on the unit disk, it is difficult to evaluate even for simple boundary values $f(\theta)$. In order to rewrite (3) in a form that is more suitable for numerical computations, we begin by deriving a series form of the Poisson kernel

(4) $$P(r,\theta) = \frac{R^2 - r^2}{R^2 - 2rR\cos\theta + r^2}, \qquad (0 \le r < R),$$

which appears at the heart of the Poisson formula (see (13), Section 3.8).

LEMMA 1
SERIES FORM OF
THE POISSON
KERNEL

For $0 \leq r < R$ and all θ, we have

$$(5) \qquad P(r, \theta) = \text{Re}\left(\frac{R + re^{i\theta}}{R - re^{i\theta}}\right) = 1 + 2\sum_{n=1}^{\infty}\left(\frac{r}{R}\right)^n \cos n\theta.$$

Proof Using $\overline{R - re^{i\theta}} = R - re^{-i\theta}$, we obtain

$$\frac{R + re^{i\theta}}{R - re^{i\theta}} = \frac{(R + re^{i\theta})(R - re^{-i\theta})}{(R - re^{i\theta})(R - re^{-i\theta})} = \frac{R^2 - r^2 + 2irR\sin\theta}{R^2 - 2rR\cos\theta + r^2},$$

and the first equality in (5) follows upon taking real parts on both sides and comparing with (4). To prove the second equality in (5), let $z = re^{i\theta}$. Using a geometric series expansion, we have for $|z| = r < R$,

$$\begin{aligned}
\frac{R + z}{R - z} &= (R + z)\frac{1}{R\left(1 - \frac{z}{R}\right)} = \frac{R + z}{R}\frac{1}{1 - \frac{z}{R}} \\
&= \left(1 + \frac{z}{R}\right)\sum_{n=0}^{\infty}\left(\frac{z}{R}\right)^n = \sum_{n=0}^{\infty}\left(\frac{z}{R}\right)^n + \sum_{n=0}^{\infty}\left(\frac{z}{R}\right)^{n+1} \\
&= 1 + 2\sum_{n=1}^{\infty}\left(\frac{z}{R}\right)^n = 1 + 2\sum_{n=1}^{\infty}\left(\frac{r}{R}\right)^n(\cos n\theta + i\sin n\theta),
\end{aligned}$$

where in the last equality we used $z^n = r^n e^{in\theta} = r^n(\cos n\theta + i\sin n\theta)$, by Euler's identity. Now take real parts on both sides. ∎

The Poisson integral formula will be expressed in terms of the Fourier coefficients of the boundary function. Fourier series and Fourier coefficients will be studied in detail in Chapter 7. Here we will simply use some notation from this chapter to highlight the connection between two important topics: Fourier series and the solution of the Dirichlet problem on the disk.

If f is piecewise continuous on $[0, 2\pi]$, let

$$(6) \qquad a_0 = \frac{1}{2\pi}\int_0^{2\pi} f(\theta)\,d\theta;$$

$$(7) \qquad a_n = \frac{1}{\pi}\int_0^{2\pi} f(\theta)\cos n\theta\,d\theta \qquad (n = 1, 2, \dots);$$

$$(8) \qquad b_n = \frac{1}{\pi}\int_0^{2\pi} f(\theta)\sin n\theta\,d\theta \qquad (n = 1, 2, \dots).$$

The coefficients a_n are known as the **cosine Fourier coefficients** of f and b_n as the **sine Fourier coefficients** of f.

THEOREM 1
FOURIER SERIES
FORM OF THE
POISSON INTEGRAL
FORMULA

Consider the Dirichlet problem (1)–(2) with piecewise continuous boundary data f. Then the solution is given by

$$(9) \qquad u(r,\theta) = a_0 + \sum_{n=1}^{\infty} \left(\frac{r}{R}\right)^n (a_n \cos n\theta + b_n \sin n\theta), \quad 0 \le r < R,$$

where a_0, a_n, and b_n are the Fourier coefficients of f, given by (6)–(8).

Proof   Starting with the solution (3), we expand the Poisson integral in a series by using (5) (replace θ by $\theta - \phi$ in (5)) and get

$$
\begin{aligned}
u(r,\theta) &= \frac{1}{2\pi} \int_0^{2\pi} f(\phi) \left(1 + 2\sum_{n=1}^{\infty} \left(\frac{r}{R}\right)^n \cos n(\theta - \phi)\right) d\phi \\
&= \frac{1}{2\pi} \int_0^{2\pi} f(\phi)\, d\phi + \frac{1}{\pi} \int_0^{2\pi} \sum_{n=1}^{\infty} \left\{ \left(\frac{r}{R}\right)^n f(\phi) \cos n(\theta - \phi) \right\} d\phi.
\end{aligned}
$$

Since f is piecewise continuous, it is bounded on $[0, 2\pi]$. Let $A \ge 0$ be such that $|f(\phi)| \le A$ for all ϕ. For fixed $0 \le r < R$, we have

$$\left| \left(\frac{r}{R}\right)^n f(\phi) \cos n(\theta - \phi) \right| \le A \left(\frac{r}{R}\right)^n,$$

and so the series $\sum_{n=1}^{\infty} \left\{ \left(\frac{r}{R}\right)^n f(\phi) \cos n(\theta - \phi) \right\}$ converges uniformly in ϕ on the interval $[0, 2\pi]$, by the Weierstrass M-test, because $\sum A \left(\frac{r}{R}\right)^n < \infty$. Hence, we can integrate term by term (Corollary 1, Section 4.2). Appealing to (6)–(8), we get

$$
\begin{aligned}
u(r,\theta) &= \frac{1}{2\pi} \int_0^{2\pi} f(\phi)\, d\phi + \sum_{n=1}^{\infty} \left\{ \left(\frac{r}{R}\right)^n \frac{1}{\pi} \int_0^{2\pi} f(\phi) \cos n(\theta - \phi)\, d\phi \right\} \\
&= a_0 + \sum_{n=1}^{\infty} \left\{ \left(\frac{r}{R}\right)^n \frac{1}{\pi} \int_0^{2\pi} f(\phi) \Big(\cos n\theta \cos n\phi + \sin n\theta \sin n\phi \Big)\, d\phi \right\} \\
&= a_0 + \sum_{n=1}^{\infty} \left(\frac{r}{R}\right)^n \Big(a_n \cos n\theta + b_n \sin n\theta \Big),
\end{aligned}
$$

which proves (9).   ■

EXAMPLE 1   A steady-state problem in a disk

The temperature on the boundary of a circular plate with radius $R = 2$, center at the origin, and insulated lateral surface is given by

$$f(\theta) = \begin{cases} 100 & \text{if } 0 \le \theta \le \pi, \\ 0 & \text{if } \pi < \theta < 2\pi. \end{cases}$$

(a)   Find the Fourier series form of the steady-state temperature inside the plate.
(b)   Show that all the points inside the plate on the x-axis have the same temperature. What is this temperature?

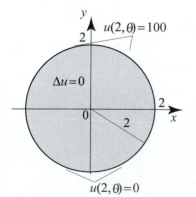

Figure 2 Dirichlet problem in Example 1.

Solution   (a)   According to (9), the solution inside the disk is given by

$$(10) \qquad u(r, \theta) = a_0 + \sum_{n=1}^{\infty} \left(\frac{r}{2}\right)^n (a_n \cos n\theta + b_n \sin n\theta), \qquad 0 \le r < 2,$$

where a_0, a_n, and b_n are the Fourier coefficients of f. Using the formula of f in (6)–(8), we obtain

$$a_0 = \frac{1}{2\pi} \int_0^\pi 100 \, d\theta = 50, \qquad a_n = \frac{1}{\pi} \int_0^\pi 100 \cos n\theta \, d\theta = 0,$$

and

$$b_n = \frac{1}{\pi} \int_0^\pi 100 \sin n\theta \, d\theta = \frac{100}{n\pi}[1 - \cos n\pi].$$

Substituting into (10), we find the solution

$$u(r, \theta) = 50 + \frac{100}{\pi} \sum_{n=1}^{\infty} \frac{1}{n}(1 - \cos n\pi) \left(\frac{r}{2}\right)^n \sin n\theta, \qquad 0 \le r < 2.$$

Notice that $1 - \cos n\pi$ is either 0 or 2 depending on whether n is even or odd. Thus, only odd terms survive, so we put $n = 2k + 1$ for $k = 0, 1, \ldots$, and get

$$(11) \qquad u(r, \theta) = 50 + \frac{200}{\pi} \sum_{k=0}^{\infty} \frac{1}{2k+1} \left(\frac{r}{2}\right)^k \sin(2k+1)\theta, \qquad 0 \le r < 2.$$

(b)   For points on the x-axis, we have $\theta = 0$ or $\theta = \pi$. Either value of θ when inserted into the series solution yields $u(r, \theta) = 50$, because for $\theta = 0$ or $\theta = \pi$ we have $\sin(2k+1)\theta = 0$ for all k. Thus the temperature of the points on the x-axis is constant and equals 50, which is the average temperature of the points on the upper semi-circle and those on the lower semi-circle. This is to be expected since the points on the x-axis are halfway between the points on the upper semi-circle and those on the lower semi-circle. The isotherms corresponding to $T \ne 50$ are found in Exercise 5.   ∎

Fourier Series

Theorem 1 holds a connection to one of the most fruitful areas in applied mathematics: Fourier series. Taking the limit as $r \uparrow R$ in (9) and using the fact that $\lim_{r \uparrow R} u(r, \theta) = f(\theta)$ at the points of continuity of f, we obtain

$$(12) \qquad f(\theta) = a_0 + \lim_{r \uparrow R} \sum_{n=1}^{\infty} \left(\frac{r}{R}\right)^n (a_n \cos n\theta + b_n \sin n\theta).$$

Suppose for a moment that we can take the limit inside the infinite sum. Then because $\lim_{r \uparrow R} \left(\frac{r}{R}\right)^n = 1$ for all n, we get

$$(13) \qquad \boxed{f(\theta) = a_0 + \sum_{n=1}^{\infty} (a_n \cos n\theta + b_n \sin n\theta),}$$

where a_0, a_n, b_n are given by (6)–(8). This representation of f by an infinite sum of cosines and sines is the famous **Fourier series** of f, where the coefficients a_0, a_n, b_n are the Fourier coefficients of f and are given by (6)–(8).

Fourier series were used by many mathematicians before Fourier. In particular, they were known to Euler and Daniel Bernoulli, but both mathematicians were skeptical about the general applicability of these expansions. It took the ingenious work of the French mathematician and engineer Jean Baptiste Joseph Fourier (1768–1830) to dispel the doubts surrounding these series and to recognize their importance.

Fourier series are perhaps the most powerful tools in applied mathematics. They are also the reason for the rise of many modern branches of mathematics. We will use them again in later chapters, when solving partial differential equations. Precise statements about the Fourier series representation will be given in Chapter 7. For now we will limit our discussion to Fourier series arising from the solution of Dirichlet problems, as we now illustrate with the Dirichlet problem of Example 1.

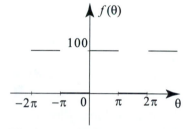

Figure 3 The boundary function in Example 1, as a 2π-periodic function of θ.

EXAMPLE 2   Fourier series of a square wave

The boundary function in Example 1 is plotted in Figure 3. Setting $r = 2$ in the solution (11) and using the fact that $\lim_{r \uparrow 2} u(r, \theta) = f(\theta)$, we would expect to get the Fourier series representation

$$(14) \qquad f(\theta) = 50 + \frac{200}{\pi} \sum_{k=0}^{\infty} \frac{\sin(2k+1)\theta}{2k+1},$$

where

$$f(\theta) = \begin{cases} 100 & \text{if } 0 \le \theta \le \pi, \\ 0 & \text{if } \pi < \theta < 2\pi. \end{cases}$$

To justify this representation, in Figure 4, we plotted $f(\theta)$ and several partial sums of the Fourier series

$$(15) \qquad s_n(\theta) = 50 + \frac{200}{\pi} \sum_{k=0}^{n} \frac{\sin(2k+1)\theta}{2k+1}.$$

The graph of f looks like a square wave that repeats every 2π units.

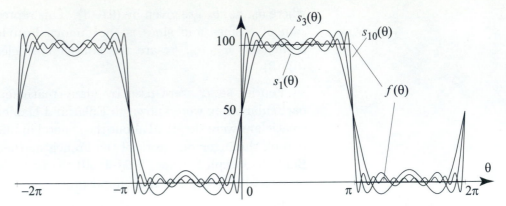

Figure 4. Partial sums of the Fourier series: $s_n(\theta) = 50 + \frac{200}{\pi} \sum_{k=0}^{n} \frac{\sin(2k+1)\theta}{2k+1}$, for $n = 1, 3, 10$. As n increases, the frequencies of the sine terms increase, causing the graphs of the higher partial sums to be more wiggly.

The Fourier series of f converges pointwise to $f(\theta)$ at each point θ where f is continuous. This is illustrated in Figure 4. In particular, we have

$$100 = 50 + \frac{200}{\pi} \sum_{k=0}^{\infty} \frac{\sin(2k+1)\theta}{2k+1} \quad \text{for } 0 < \theta < \pi,$$

and

$$0 = 50 + \frac{200}{\pi} \sum_{k=0}^{\infty} \frac{\sin(2k+1)\theta}{2k+1} \quad \text{for } \pi < \theta < 2\pi.$$

At the points of discontinuity $(\theta = m\pi, m = 0, \pm 1, \pm 2, \dots)$, all terms $\sin(2k+1)\theta$ are zero and so we know that the series converges to 50. The graph of the Fourier series $50 + \frac{200}{\pi} \sum_{k=0}^{\infty} \frac{\sin(2k+1)\theta}{2k+1}$ is shown in Figure 5. It agrees with the graph of the function, except at the points of discontinuity. ∎

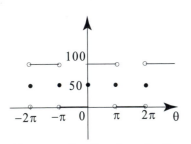

Figure 5 Graph of the Fourier series of a piecewise smooth function in Example 2. It agrees with the function at all the points where $f(\theta)$ is continuous.

So does the Fourier series of a piecewise continuous function always converge to the function as stated in (13) and illustrated by Example 2? Can we justify the step that took us from (12) to (13)? If we do not impose additional properties on f, the answer to these questions is (unfortunately) no. There are examples of continuous 2π-periodic functions with Fourier series diverging at an infinite set of points in $[0, 2\pi]$. In Chapter 7, we will study Fourier series in detail and prove a Fourier series representation that applies to piecewise smooth functions.

Exercises 4.7

1. Project Problem: A steady-state problem with continuous boundary data, Fourier series of a triangular wave. We will apply the results of this section to the Dirichlet problem in the unit disk with boundary data given by

$$f(\theta) = \begin{cases} \pi + \theta & \text{if } -\pi \leq \theta \leq 0, \\ \pi - \theta & \text{if } 0 < \theta < \pi. \end{cases}$$

(a) Think of the boundary function as a 2π-periodic function of θ. Plot its graph over the interval $[-4\pi, 4\pi]$. (Remember that the graph of a 2π-periodic function repeats every 2π units.)

(b) Using (9), show that the solution of the Dirichlet problem is

$$(16) \qquad u(r, \theta) = a_0 + \sum_{n=1}^{\infty} r^n \left(a_n \cos n\theta + b_n \sin n\theta \right) \qquad (0 \le r < 1),$$

where a_0, a_n, and b_n are the Fourier coefficients of f.

(c) Show that $a_0 = \frac{\pi}{2}$, $a_n \frac{2}{\pi} \left\{ \frac{1}{n^2} - \frac{\cos n\pi}{n^2} \right\}$, $b_n = 0$.

(d) Plugging the coefficients into (16), we obtain the solution for $0 < r < 1$,

$$(17) \qquad u(r, \theta) = \frac{\pi}{2} + \sum_{n \text{ odd}} \frac{4}{\pi n^2} r^n \cos n\theta = \frac{\pi}{2} + \frac{4}{\pi} \sum_{k=0}^{\infty} \frac{r^{2k+1}}{(2k+1)^2} \cos(2k+1)\theta.$$

(e) Derive the Fourier series expansion of the triangular wave: for all θ,

$$(18) \qquad f(\theta) = \frac{\pi}{2} + \sum_{n \text{ odd}} \frac{4}{\pi n^2} \cos n\theta = \frac{\pi}{2} + \frac{4}{\pi} \sum_{k=0}^{\infty} \frac{1}{(2k+1)^2} \cos(2k+1)\theta.$$

(f) Illustrate the convergence of the Fourier series to $f(\theta)$ by plotting several partial sums.

Project Problem: Isotherms.   In Exercises 2–5, you are asked to find the isotherms in Example 1. This will require some identities that are interesting in their own right.

2. A useful identity.   Let $z = re^{i\theta} = r(\cos \theta + i \sin \theta)$ be a complex number with $|z| < 1$. (a) Starting with the Maclaurin series (Example 2, Section 4.3)

$$\text{Log}\,(1 + z) = \sum_{n=0}^{\infty} (-1)^n \frac{z^{n+1}}{n+1} \qquad (|z| < 1),$$

derive the series expansion

$$- \text{Log}\,(1 - z) = \text{Log} \left(\frac{1}{1-z} \right) = z + \frac{1}{2} z^2 + \frac{1}{3} z^3 + \dots \qquad (|z| < 1).$$

(b)   Take imaginary parts on both sides of the second identity in (a) to show that

$$\sum_{n=1}^{\infty} r^n \frac{\sin n\theta}{n} = \tan^{-1} \left(\frac{r \sin \theta}{1 - r \cos \theta} \right), \qquad 0 \le r < 1, \text{ all } \theta.$$

3. More useful identities.   Take real and imaginary parts on both sides of the expansion (Example 2, Section 4.3)

$$\text{Log}\,(1 + z) = \sum_{n=1}^{\infty} (-1)^{n+1} \frac{z^n}{n}, \qquad |z| < 1,$$

and obtain that, for $0 \le r < 1$ and all θ,

(a) $\displaystyle \sum_{n=1}^{\infty} (-1)^{n+1} \frac{\sin n\theta}{n} r^n = \tan^{-1} \left(\frac{r \sin \theta}{1 + r \cos \theta} \right);$

and

(b) $\displaystyle\sum_{n=1}^{\infty}(-1)^{n+1}\frac{\cos n\theta}{n}r^n = \frac{1}{2}\ln\left(1 + 2r\cos\theta + r^2\right).$

4.   Add the identities in Exercises 2 and 3(a) and get, for $0 \leq r < 1$ and all θ,

$$\sum_{k=0}^{\infty}\frac{\sin(2k+1)\theta}{2k+1}r^{2k+1} = \frac{1}{2}\left(\tan^{-1}\left(\frac{r\sin\theta}{1-r\cos\theta}\right) + \tan^{-1}\left(\frac{r\sin\theta}{1+r\cos\theta}\right)\right).$$

5.   Follow the outlined steps to compute the isotherms in Example 1.
(a)   Show that in Cartesian coordinates,

$$u(x,\, y) = 50 + \frac{100}{\pi}\left(\tan^{-1}\left(\frac{y}{2-x}\right) + \tan^{-1}\left(\frac{y}{2+x}\right)\right).$$

(b)   Let $0 < T < 100$ be a given temperature. Suppose further that $T \neq 50$. Show that the points $(x,\, y)$ inside the disk $x^2 + y^2 = 4$ such that $u(x,\, y) = T$ lie on the arc of the circle with center at $\left(0,\, -2\cot\left(\frac{\pi}{100}(T-50)\right)\right)$ and radius $2\csc\left(\frac{\pi}{100}(T-50)\right)$. [Hint: Use the formula $\tan(a+b) = \frac{\tan a + \tan b}{1 - \tan a\tan b}$.]
(c)   Plot several isotherms as T varies from 0 to 100.

5

RESIDUE THEORY

After having thought on this subject, and brought together the diverse results mentioned above, I had the hope of establishing on a direct and rigorous analysis the passage from the real to the imaginary; and my research has lead me to this Memoire.

-Augustin Louis Cauchy

[Writing about his Memoire of 1814, which contained the residue theorem and several computations of real integrals by complex methods.]

In the previous four chapters, we introduced complex numbers and complex functions and studied properties of the three essential tools of calculus: the derivative, the integral, and series of complex functions. In this and the next chapter, we use these results to derive some exciting applications of complex analysis. The applications of this chapter are based on one formula, known as Cauchy's residue theorem. We have already used this result many times before when computing integrals. Here we will highlight the main ideas behind it and devise new techniques for computing important integrals that arise from Fourier series and integrals, Laplace transforms, and other applications. For example, integrals like $\int_{-\infty}^{\infty} \frac{\sin x}{x}\, dx$, $\int_0^{\infty} \cos x e^{-x^2}\, dx$ are very difficult to compute using real variable techniques. With the residue theorem and some additional estimates with complex functions, the computations of these integrals are reduced to simple tasks.

In Section 5.7 (an optional section), we will use residues to expand our knowledge of analytic functions. We will use integrals to count the number of zeros of analytic functions and to give a formula for the inverse of an analytic functions. Theoretical results, such as the open mapping property, will be derived and will be used to obtain a fresh and different perspective on concrete results such as the maximum modulus principle.

The residue theorem was discovered around 1814 (stated explicitly in 1831) by Cauchy as he attempted to generalize and put under one umbrella the computations of certain special integrals, some of them involving complex substitutions, that were done by Euler, Laplace, Legendre, and other mathematicians.

5.1   Cauchy's Residue Theorem

In this section, we will derive a useful formula known as Cauchy's residue formula, which combines together powerful techniques from earlier sections. The applications of this formula will be explored throughout the chapter. Let us start by reviewing a few results from Laurent series. Suppose that f has an isolated singularity at z_0. We know from Theorem 1, Section 4.5, that f has a Laurent series in an annulus around z_0: for $0 < |z - z_0| < R$,

$$f(z) = \cdots + \frac{a_{-2}}{(z - z_0)^2} + \frac{a_{-1}}{z - z_0} + a_0 + a_1(z - z_0) + a_2(z - z_0)^2 + \cdots .$$

Moreover,

(1)
$$a_{-1} = \frac{1}{2\pi i} \int_{C_r(z_0)} f(z)\, dz,$$

where $C_r(z_0)$ is any positively oriented circle in the annulus $0 < |z - z_0| < R$. The coefficient a_{-1} is called the **residue of f at** z_0 and is denoted by $\text{Res}\,(f, z_0)$ or simply $\text{Res}\,(z_0)$ when there is no risk of confusing the function f. With the concept of residue in hand, we can state our main result.

THEOREM 1
CAUCHY'S RESIDUE
THEOREM

Let C be a simple closed positively oriented path. Suppose that f is analytic inside and on C, except at finitely many isolated singularities $z_1,\, z_2,\, \ldots,\, z_n$ inside C. Then

(2)
$$\int_C f(z)\, dz = 2\pi i \sum_{j=1}^{n} \text{Res}\,(z_j).$$

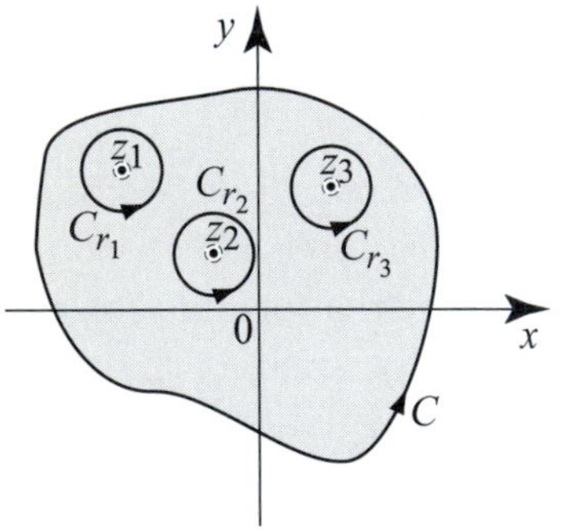

Figure 1 To compute the integral around a path C of a function with multiple isolated singularities in the interior of C, we first carve out the singularities using non overlapping circles inside C, then use Cauchy's theorem for multiply connected regions.

Proof   Take small circles $C_{r_j}(z_j)$ $(j = 1, 2, \ldots, n)$ that do not intersect each other and are contained in the interior of C (Figure 1). Apply Cauchy's integral theorem for multiply connected regions (Theorem 6, Section 3.4) and get

$$\int_C f(z)\, dz = \sum_{j=1}^{n} \int_{C_{r_j}} f(z)\, dz = 2\pi i \sum_{j=1}^{n} \text{Res}\,(z_j),$$

where the last equality follows from (1). So (2) holds.   ■

An analog of Theorem 1 holds if C is negatively oriented; in this case we pick up a negative sign on the right side of (2).

The residue theorem has abounding applications to the evaluation of all sorts of integrals. Our first example illustrates the basic ideas behind this process.

EXAMPLE 1   An application of the residue theorem

Let C be a simple closed positively oriented path such that 1, $-i$, and i are in the interior of C and -1 is in the exterior of C (Figure 2). Find

$$\int_C \frac{dz}{z^4 - 1}.$$

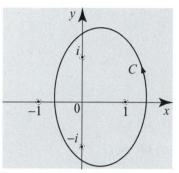

Figure 2 The path C and the poles of $f(z)$ in Example 1.

Solution   The function $f(z) = \frac{1}{z^4-1}$ has isolated singularities at $z = \pm 1$ and $\pm i$. Three of these are inside C, and according to (2) we have

$$(3) \qquad \int_C \frac{dz}{z^4 - 1} = 2\pi i \Big(\operatorname{Res}(1) + \operatorname{Res}(i) + \operatorname{Res}(-i) \Big).$$

To compute the residues, we will find the Laurent series of $f(z)$ around each isolated singularity and obtain the residue as the coefficient a_{-1} in each case. We have $z^4 - 1 = (z-1)(z+1)(z-i)(z+i)$, so ± 1 and $\pm i$ are simple roots of the polynomial $z^4 - 1 = 0$. Hence $f(z) = \frac{1}{z^4-1}$ has simple poles at ± 1 and $\pm i$ (Theorem 7(v), Section 4.6). Let z_0 denote any one of the points 1, $\pm i$. Because z_0 is a simple pole, f has a Laurent series expansion around z_0 of the form

$$f(z) = \frac{a_{-1}}{z - z_0} + a_0 + a_1(z - z_0) + a_2(z - z_0)^2 + \cdots.$$

So

$$(z - z_0)f(z) = a_{-1} + (z - z_0)h(z),$$

where $h(z) = a_0 + a_1(z - z_0) + a_2(z - z_0)^2 + \cdots$ is analytic at z_0. Taking limits as $z \to z_0$ and using $\lim_{z \to z_0}(z - z_0)h(z) = 0$, we obtain the useful formula

$$(4) \qquad \lim_{z \to z_0} (z - z_0)f(z) = a_{-1} = \operatorname{Res}(z_0),$$

which can be used to compute the residue of f at a simple pole. Using the factorization $z^4 - 1 = (z-1)(z+1)(z-i)(z+i)$, we have at $z_0 = 1$

$$\begin{aligned}
\operatorname{Res}(1) &= \lim_{z \to 1}(z - 1)\frac{1}{z^4 - 1} = \lim_{z \to 1} \frac{1}{(z+1)(z-i)(z+i)} \\
&= \frac{1}{(z+1)(z-i)(z+i)}\bigg|_{z=1} = \frac{1}{4}.
\end{aligned}$$

Similarly, at $z_0 = i$, we have

$$\operatorname{Res}(i) = \lim_{z \to i}(z - i)\frac{1}{z^4 - 1} = \frac{1}{(z-1)(z+1)(z+i)}\bigg|_{z=i} = \frac{i}{4},$$

and at $z = -i$,

$$\operatorname{Res}(-i) = \lim_{z \to -i}(z + i)\frac{1}{z^4 - 1} = \frac{1}{(z-1)(z+1)(z-i)}\bigg|_{z=-i} = -\frac{i}{4}.$$

Plugging these values into (3), we obtain

$$\int_C \frac{dz}{z^4 - 1} = \frac{\pi i}{2}.$$

∎

The challenge in applying the residue theorem is in computing residues. Any formula that facilitates this task will be valuable to us. In Example 1, we found one, namely (4). For ease of reference, let us state it separately along with formulas that apply for simple poles. We then derive a formula for the residue at poles of higher order.

PROPOSITION 1
RESIDUE AT A
SIMPLE POLE

(i)   If $f(z)$ has a simple pole at z_0, then

$$(5) \qquad \operatorname{Res}(f, z_0) = \lim_{z \to z_0} (z - z_0)f(z).$$

(ii)   If $f(z) = \frac{p(z)}{q(z)}$, where p and q are analytic at z_0, $p(z_0) \neq 0$, and $q(z)$ has a simple zero at z_0, then

$$(6) \qquad \operatorname{Res}\left(\frac{p(z)}{q(z)}, z_0\right) = \frac{p(z_0)}{q'(z_0)}.$$

(iii)   If $f(z)$ has a simple pole at z_0 and g is analytic at z_0,

$$(7) \qquad \operatorname{Res}(f(z)g(z), z_0) = g(z_0) \operatorname{Res}(f(z), z_0).$$

(iv)   If g is analytic at z_0, then

$$(8) \qquad \operatorname{Res}\left(\frac{g(z)}{z - z_0}, z_0\right) = g(z_0).$$

Proof   (i) follows from the proof of (4). To prove (ii), note that f has a simple pole at z_0. Using (i) and $q(z_0) = 0$, we have

$$\operatorname{Res}\left(\frac{p(z)}{q(z)}, z_0\right) = \lim_{z \to z_0} (z - z_0)\frac{p(z)}{q(z)} = \lim_{z \to z_0} p(z) \lim_{z \to z_0} \frac{z - z_0}{q(z) - q(z_0)} = \frac{p(z_0)}{q'(z_0)}.$$

To prove (iii), we use (i) and the fact that g is analytic at z_0: $\operatorname{Res}(f(z)g(z), z_0) = \lim_{z \to z_0} (z - z_0)f(z)g(z) = \lim_{z \to z_0} g(z) \lim_{z \to z_0} (z - z_0)f(z) = g(z_0) \operatorname{Res}(f, z_0)$. Part (iv) is immediate from (i).   ■

For poles of higher order the situation is more complicated.

THEOREM 2
RESIDUE AT A POLE
OF ORDER m

Suppose that z_0 is a pole of order $m \geq 1$ of f. Then the residue of f at z_0 is

$$(9) \qquad \operatorname{Res}(f, z_0) = \lim_{z \to z_0} \frac{1}{(m-1)!} \frac{d^{m-1}}{dz^{m-1}} \left[(z - z_0)^m f(z)\right],$$

where as usual the derivative of order 0 of a function is the function itself.

You should check that for $m = 1$ formula (9) reduces to (4).

Proof   By the Laurent series characterization of poles (Theorem 8, Section 4.6),

$$f(z) = \frac{a_{-m}}{(z - z_0)^m} + \cdots + \frac{a_{-1}}{z - z_0} + a_0 + a_1(z - z_0) + a_2(z - z_0)^2 + \cdots.$$

To extract the residue a_{-1} out of this expansion, first multiply by $(z - z_0)^m$:

$$(z - z_0)^m f(z) = a_{-m} + \cdots + a_{-1}(z - z_0)^{m-1} + a_0(z - z_0)^m + a_1(z - z_0)^{m+1} + \cdots .$$

Then differentiate $(m - 1)$ times to obtain

$$\frac{d^{m-1}}{dz^{m-1}} [(z - z_0)^m f(z)] = (m - 1)!a_{-1} + m!a_0(z - z_0) + \frac{(m + 1)!}{2} a_1(z - z_0)^2 + \cdots .$$

Finally, take the limit as $z \to z_0$, and get

$$\lim_{z \to z_0} \frac{d^{m-1}}{dz^{m-1}} [(z - z_0)^m f(z)] = (m - 1)!a_{-1} + 0,$$

which is equivalent to (9).   ∎

EXAMPLE 2   Computing residues

Let C be the simple closed path shown in Figure 3. (a) Compute the residues of $f(z) = \frac{z^2}{(z^2+\pi^2)^2 \sin z}$ at all the isolated singularities inside C. (b) Evaluate

$$\int_C \frac{z^2}{(z^2 + \pi^2)^2 \sin z} \, dz.$$

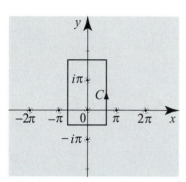

Figure 3 The path C and the poles of $f(z)$ in Example 2.

Solution   (a)   Three steps are involved in answering this question.

Step 1: Determine the singularities of f inside C. The function $f(z) = \frac{z^2}{(z^2+\pi^2)^2 \sin z}$ is analytic except where $z^2 + \pi^2 = 0$ or $\sin z = 0$. Thus f has isolated singularities at $\pm i\pi$ and at $k\pi$ where k is an integer. Only 0 and $i\pi$ are inside C.

Step 2: Determine the type of the singularities of f inside C. Let us start with the singularity at 0. Using $\lim_{z \to 0} \frac{\sin z}{z} = 1$, it follows that $\lim_{z \to 0} \frac{z}{\sin z} = 1$, and so

$$\lim_{z \to 0} f(z) = \lim_{z \to 0} \frac{z}{\sin z} \frac{z}{(z^2 + \pi^2)^2} = 1 \cdot 0 = 0.$$

By Theorem 5, Section 4.6, $f(z)$ has a removable singularity at $z_0 = 0$. To treat the singularities at $i\pi$, we consider $\frac{1}{f(z)} = \frac{(z+i\pi)^2(z-i\pi)^2 \sin z}{z^2}$. Clearly $\frac{1}{f(z)}$ has a zero of order 2 at $i\pi$, and so by Theorem 7(v), Section 4.6, $f(z)$ has a pole of order 2 at $i\pi$.

Step 3: Determine the residues of f inside C. At 0, f has a removable singularity, which means that its Laurent series expansion around zero has no terms with negative powers of z. In particular, $a_{-1} = 0$, and so the residue of f at a removable singularity is 0. At $i\pi$, we can apply Theorem 2, with $m = 2$, $z_0 = i\pi$. Then

$$
\begin{aligned}
\text{Res}\,(i\pi) &= \lim_{z \to i\pi} \frac{d}{dz} \left[(z - i\pi)^2 f(z) \right] \\
&= \lim_{z \to i\pi} \frac{d}{dz} \left[\frac{z^2}{(z + i\pi)^2 \sin z} \right] \\
&= \lim_{z \to i\pi} \frac{2z(z + i\pi) \sin z - z^2((z + i\pi) \cos z + 2 \sin z)}{(z + i\pi)^3 \sin^2 z} \\
&= \frac{2 \sinh \pi + (-\pi \cosh \pi - \sinh \pi)}{-4\pi \sinh^2 \pi} = -\frac{1}{4\pi \sinh \pi} + \frac{\cosh \pi}{4\pi \sinh^2 \pi},
\end{aligned}
$$

where the last line follows by plugging $z = i\pi$ into the previous line and using $\sin(i\pi) = i \sinh \pi$ and $\cos(i\pi) = \cosh \pi$.

(b)   Using Theorem 1 and (a), we obtain

$$\int_C \frac{z^2}{(z^2 + \pi^2)^2 \sin z}\, dz = 2\pi i \Big(\operatorname{Res}(0) + \operatorname{Res}(i\pi) \Big)$$

$$= i \left(-\frac{1}{2 \sinh \pi} + \frac{\cosh \pi}{2 \sinh^2 \pi} \right). \qquad \blacksquare$$

A simple fact worth noting from the previous solution: *If z_0 is a removable singularity, then* $\operatorname{Res}(z_0) = 0$.

As you can see from Example 2, formula (9) becomes increasingly difficult to apply at poles of higher order. Sometimes we can find the residue of a function by algebraically manipulating it into its Laurent series.

EXAMPLE 3   Residue at a second-order pole

Let $f(z) = \frac{(z+1)^3}{(z-1)^2}$. Find $\operatorname{Res}(f(z), 1)$.

Solution We can rewrite the numerator in terms of $z - 1$ and write $f(z)$ as its own Laurent series expansion. We have

$$\frac{(z + 1)^3}{(z - 1)^2} = \frac{(z - 1 + 2)^3}{(z - 1)^2} = \frac{(z - 1)^3 + 6(z - 1)^2 + 12(z - 1) + 8}{(z - 1)^2}$$

$$= \frac{8}{(z - 1)^2} + \frac{12}{z - 1} + 6 + (z - 1).$$

Reading off the coefficient of $\frac{1}{z-1}$, we have $\operatorname{Res}(f(z), 1) = 12$. $\qquad \blacksquare$

Other techniques that apply in place of (9) are discussed in the exercises. Our next example has some interesting applications to series summation by residues (see Section 5.6).

EXAMPLE 4   Residues of the cotangent

(a)   Let k be an integer. Show that $\operatorname{Res}\big(\cot(\pi z), k\big) = \frac{1}{\pi}$.

(b)   Suppose that f is analytic at an integer k. Show that

(10) $$\operatorname{Res}\big(f(z) \cot(\pi z), k\big) = \frac{1}{\pi} f(k).$$

(c)   Evaluate

$$\int_C \frac{\cot(\pi z)}{1 + z^4}\, dz,$$

where C is the positively oriented rectangular path shown in Figure 4.

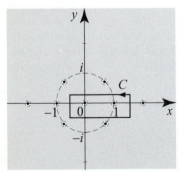

Figure 4 The path C and the poles of $f(z) \cot(\pi z)$ in Example 4(c).

Solution   (a)   We know that the zeros of $\sin(\pi z)$ are precisely the integers. Also, it is immediate from Theorem 1, Section 4.6, that all the zeros of $\sin(\pi z)$ are simple zeros. Hence $\cot(\pi z) = \frac{\cos(\pi z)}{\sin(\pi z)}$ has simple poles at the integers. To find the residue at k, we can apply (9) with $m = 1$ or, better yet, use Proposition 1(ii). We have

$$\operatorname{Res}\big(\cot(\pi z), k\big) = \operatorname{Res}\Big(\frac{\cos(\pi z)}{\sin(\pi z)}, k\Big) = \frac{\cos(k\pi)}{\frac{d}{dz} \sin(\pi z)\big|_{z=k}} = \frac{1}{\pi}.$$

(b)   This is immediate from (a) and Proposition 1(iii). Just be careful with the notation as you appeal to the proposition.

(c)   Since $1 + z^4$ is nonzero inside C and $\cot(\pi z)$ has simple poles at the integers, it follows that $\frac{\cot(\pi z)}{1+z^4}$ has two simple poles inside C at $z = 0$ and $z = 1$. Applying Theorem 1 and using (10) with $f(z) = \frac{1}{1+z^4}$ to compute the residues, we find

$$
\begin{aligned}
\int_C \frac{\cot(\pi z)}{1 + z^4}\, dz &= 2\pi i\left(\operatorname{Res}\left(\frac{\cot(\pi z)}{1 + z^4}, 0\right) + \operatorname{Res}\left(\frac{\cot(\pi z)}{1 + z^4}, 1\right)\right) \\
&= 2\pi i\left(\frac{1}{\pi}\frac{1}{1 + 0^4} + \frac{1}{\pi}\frac{1}{1 + 1^4}\right) = 2i\left(1 + \frac{1}{2}\right) = 3\,i. \qquad \blacksquare
\end{aligned}
$$

So far the examples that we treated involved residues at poles of finite order. There is no formula like (9) for computing the residue at an essential singularity. We have to rely on various tricks to evaluate the coefficient a_{-1} in the Laurent series expansion. We illustrate with several examples, starting with one that contains a useful observation.

EXAMPLE 5   Residue at 0 of an even function

(a)   Suppose that f is an even analytic function with an isolated singularity at 0. Show that

$$
\operatorname{Res}(f, 0) = 0.
$$

(b)   Compute $\operatorname{Res}\left(e^{-\frac{1}{z^2}} \cos \frac{1}{z}, 0\right)$.

Solution   (a)   We proved in Proposition 2, Section 4.4, that if $f(z)$ is even and analytic at 0, then the Taylor coefficients corresponding to the odd powers of z are zero. The same proof applies to Laurent series around 0 and shows that f is even if and only if a_{2n+1} and a_{-2n+1} are zero for all integers n. In particular, if f is even then $a_{-1} = 0$, and so $\operatorname{Res}(f, 0) = 0$.

(b)   The function $e^{-\frac{1}{z^2}} \cos \frac{1}{z}$ is even and has an essential singularity at 0. By (a), its residue at 0 is 0. $\qquad \blacksquare$

Multiplication of series is often useful in computing residues at a singularity, including essential singularities.

EXAMPLE 6   Multiplying series by a polynomial

Compute the residues of $z^2 \sin \frac{1}{z}$ at $z = 0$.

Solution   From the Laurent series

$$
\sin \frac{1}{z} = \frac{1}{z} - \frac{1}{3!}\frac{1}{z^3} + \frac{1}{5!}\frac{1}{z^5} - \cdots,
$$

we get

$$
z^2 \sin \frac{1}{z} = z - \frac{1}{3!}\frac{1}{z} + \frac{1}{5!}\frac{1}{z^3} - \cdots,
$$

and so $\operatorname{Res}\left(z^2 \sin \frac{1}{z}, 0\right) = -\frac{1}{3!}$. $\qquad \blacksquare$

EXAMPLE 7   Using Cauchy products

Find the residue of $\frac{e^{\frac{1}{z}}}{z^2+1}$ at $z = 0$.

Solution   The given function has an essential singularity at $z = 0$. To compute the coefficient a_{-1} in its Laurent series around 0, we use two familiar Taylor and Laurent series as follows. We have, for $0 < |z| < 1$,

$$\frac{e^{\frac{1}{z}}}{z^2+1} = \frac{1}{z^2+1} e^{\frac{1}{z}} = \left(1 - z^2 + z^4 - \cdots\right)\left(1 + \frac{1}{z} + \frac{1}{2!}\frac{1}{z^2} + \frac{1}{3!}\frac{1}{z^3} + \cdots\right).$$

By properties of Taylor and Laurent series, both series are absolutely convergent. So we can multiply them term by term using a Cauchy product. Collecting all the terms in $\frac{1}{z}$, we find that

$$\text{Res}\left(\frac{e^{\frac{1}{z}}}{z^2+1}, 0\right) = a_{-1} = 1 - \frac{1}{3!} + \frac{1}{5!} - \cdots = \sin 1. \qquad \blacksquare$$

Exercises 5.1

In Exercises 1–12, find the residue of the given function at all its isolated singularities.

1. $\dfrac{1+z}{z}$.

2. $\dfrac{1+z}{z^2+2z+2}$.

3. $\dfrac{1+e^z}{z^2} + \dfrac{2}{z}$.

4. $\dfrac{\sin(z^2)}{z^2(z^2+1)}$.

5. $\left(\dfrac{z-1}{z+3i}\right)^3$.

6. $\dfrac{1-\cos z}{z^3} + \sin(z^2+3z)$.

7. $\dfrac{1}{z\sin z}$.

8. $\dfrac{\cot(\pi z)}{z+1}$.

9. $\csc(\pi z)\dfrac{z+1}{z-1}$.

10. $z\sin\left(\dfrac{1}{z}\right)$.

11. $e^{z+\frac{1}{z}}$.

12. $\cos\left(\dfrac{1}{z}\right)\sin\left(\dfrac{1}{z}\right)$.

In Exercises 13–26, evaluate the given path integral. The path R in Exercises 15 and 20 is shown in Figure 5.

13. $\displaystyle\int_{C_1(0)} \dfrac{z^2+3z-1}{z(z^2-3)}\, dz$.

14. $\displaystyle\int_{C_{\frac{1}{10}}(0)} \dfrac{1}{z^5-1}\, dz$.

15. $\displaystyle\int_R \dfrac{z+i}{(z-1-i)^3(z-i)}\, dz$.

16. $\displaystyle\int_{C_3(0)} \dfrac{e^{iz^2}}{z^2+(3-3i)z-2-6i}\, dz$.

17. $\displaystyle\int_{C_{\frac{3}{2}}(0)} \dfrac{dz}{z(z-1)(z-2)\cdots(z-10)}$.

18. $\displaystyle\int_{C_3(0)} \dfrac{z^2+1}{(z-1)^2}\, dz$.

19. $\displaystyle\int_{C_4(0)} z\tan z\, dz$.

20. $\displaystyle\int_R \dfrac{dz}{1+e^{\pi z}}$.

21. $\displaystyle\int_{C_1(0)} \dfrac{e^{z^2}}{z^6}\, dz$.

22. $\displaystyle\int_{C_1(0)} \cos\left(\dfrac{1}{z^2}\right)e^{\frac{1}{z}}\, dz$.

23. $\displaystyle\int_{C_1(0)} z^4\left(e^{\frac{1}{z}}+z^2\right)\, dz$.

24. $\displaystyle\int_{C_{31/2}(0)} z^2\cot(\pi z)\, dz$.

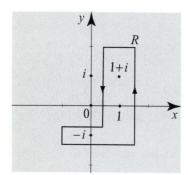

Figure 5 The path R for Exercises 15 and 20.

25. $\displaystyle\int_{C_1(0)} \frac{\sin z}{z^6}\, dz.$

26. $\displaystyle\int_{C_{1/2}(0)} \frac{1}{z^4(e^z - 1)}\, dz.$ [Hint: Show first that the denominator has only one zero inside $C_{1/2}(0)$. Then compute the residue at 0 with the help of a computer.]

27.   (a) Prove (7).
(b) Use (7) to prove (10).

28. Show that $\text{Res}\,(f(z) + g(z),\, z_0) = \text{Res}\,(f(z),\, z_0) + \text{Res}\,(g(z),\, z_0).$

29. Residues of the cosecant.   (a) Show that $\csc(\pi z)$ has simple poles at the integers.
(b)   For an integer k show that

$$\text{Res}\,(\,\csc(\pi z),\, k\,) = \frac{(-1)^k}{\pi}.$$

(c)   Suppose that f is analytic at an integer k. Show that

$$\text{Res}\,(f(z)\csc(\pi z),\, k) = \frac{(-1)^k}{\pi} f(k).$$

30.   Use Exercise 29 to compute
(a) $\displaystyle\int_{C_{25/2}(0)} z\,\csc(\pi z)\, dz,$

(b) $\displaystyle\int_{C_{25/2}(0)} \frac{\csc(\pi z)}{1 + z^2}\, dz.$

[Hint for (b): There are poles due to the zeros of $1 + z^2$.]

31. Explain how the residue theorem implies Cauchy's theorem (Theorem 5, Section 3.4) and Cauchy's integral formula (Theorem 2, Section 3.6).

32. Suppose that f has an isolated singularity at z_0. Show that $\text{Res}\,(f'(z),\, z_0) = 0.$

33.   Given the Laurent series expansions in an annulus around z_0,

$$f(z) = \sum_{n=-\infty}^{\infty} a_n(z - z_0)^n \quad \text{and} \quad g(z) = \sum_{n=-\infty}^{\infty} b_n(z - z_0)^n.$$

Show that $\text{Res}\,(f(z)g(z),\, z_0) = \sum_{n=-\infty}^{\infty} a_n b_{-n-1}.$ [Hint: Write each doubly infinite Laurent series as the sum of two infinite series. Recall that a Laurent series is absolutely convergent within its annulus of convergence. Use Cauchy products to compute the coefficient of $\frac{1}{z}$ (Theorem 15, Section 4.1).]

34.   Use Exercise 33 to compute $\text{Res}\,(e^{\frac{1}{z^2}} \sin\frac{1}{z},\, 0).$

In Exercises 35–36, g and h are analytic at z_0. You are asked to compute a formula for the residue of $f(z) = \frac{g(z)}{h(z)}$ at z_0 under given conditions.

35. If g and h have zeros of the same order at z_0, then $\text{Res}\,\left(\frac{g(z)}{h(z)},\, z_0\right) = 0.$

36. If $g(z_0) \neq 0$ and h has a zero of order 2 at z_0, then

$$\text{Res}\left(\frac{g(z)}{h(z)}, z_0\right) = 2\frac{g'(z_0)}{h''(z_0)} - \frac{2}{3}\frac{g(z_0)h'''(z_0)}{[h''(z_0)]^2}.$$

37. Project Problem: Laplace transform of Bessel functions. In this project you will use residues to derive the formula

(11)
$$\int_0^\infty J_n(t)e^{-st}\,dt = \frac{1}{\sqrt{s^2+1}}\left(\sqrt{s^2+1} - s\right)^n, \quad s > 0,$$

where $J_n(t)$ is the Bessel function of integer order $n \geq 0$. This formula gives the Laplace transform of $J_n(t)$. To illustrate the useful method that is involved in this project and for clarity's sake, we start with the case $n = 0$.
(a)   Using the integral representation of J_0 from Exercise 35(a), Section 4.5, write

$$\int_0^\infty J_0(t)e^{-st}\,dt = \frac{1}{2\pi i}\int_0^\infty \int_{C_1(0)} e^{-t\left(s - \frac{1}{2}(\zeta - \frac{1}{\zeta})\right)}\frac{d\zeta}{\zeta}\,dt.$$

(You should not be bothered by a double integral involving a contour integral. The contour integral can be parametrized and transformed into a definite integral over the interval $[0, 2\pi]$, as shown in Exercise 36, Section 4.5. But this will not be advantageous here.)
(b) Show that for ζ on $C_1(0)$, the complex number $\zeta - \frac{1}{\zeta}$ is purely imaginary and conclude that for real s and t

$$\left|e^{-t\left(s - \frac{1}{2}(\zeta - \frac{1}{\zeta})\right)}\right| = e^{-ts}.$$

(c) Thus the integral in (a) is absolutely convergent. Although we will not give the details here, this fact is enough to justify interchanging the order of integration. Interchange the order of integration, evaluate the integral in t and obtain

$$\int_0^\infty J_0(t)e^{-st}\,dt = \frac{1}{\pi i}\int_{C_1(0)} \frac{d\zeta}{-\zeta^2 + 2s\zeta + 1}, \quad s > 0.$$

(d) Evaluate the integral using the residue theorem and derive

$$\int_0^\infty J_0(t)e^{-st}\,dt = \frac{1}{\sqrt{s^2+1}}, \quad s > 0.$$

(e) Proceed as we did in (a)–(d), using the integral representation for $J_n(t)$, and show that for $s > 0$

$$\int_0^\infty J_n(t)e^{-st}\,dt = \frac{1}{\pi i}\int_{C_1(0)} \frac{d\zeta}{(-\zeta^2 + 2s\zeta + 1)\zeta^n} = \frac{1}{\pi i}\int_{C_1(0)} \frac{\eta^n\,d\eta}{\eta^2 + 2s\eta - 1}.$$

[Hint: To justify the second equality, let $\zeta = \frac{1}{\eta}$, $d\zeta = -\frac{d\eta}{\eta^2}$. As ζ runs through $C_1(0)$, η runs through the reverse of $C_1(0)$.]
(f) Derive (11) by computing the integral in (e) using residues.

5.2   Definite Integrals of Trigonometric Functions

Some of the nicest applications of the residue theorem concern the evaluation of definite integrals of real-valued functions. In this section we consider some straightforward examples that illustrate the underlying method. In later sections, we will develop this method using delicate analytical techniques and evaluate more complicated integrals.

The examples in this section are all of the form

$$(1) \qquad \int_0^{2\pi} F(\cos\theta,\,\sin\theta)\,d\theta,$$

where $F(\cos\theta,\,\sin\theta)$ is a rational function of $\cos\theta$ and $\sin\theta$ with real coefficients and whose denominator does not vanish on the interval $[0,\,2\pi]$. For example, the integrals

$$\int_0^{2\pi} \frac{d\theta}{2+\cos\theta} \quad \text{and} \quad \int_0^{2\pi} \frac{\cos^2(2\theta)}{4+2\sin\theta\cos\theta}\,d\theta$$

are of the form (1). Our goal is to transform the definite integral (1) into a contour integral that can be evaluated using the residue theorem. For this purpose, let us recall Euler's identity:

$$e^{i\theta} = \cos\theta + i\,\sin\theta, \qquad \theta \text{ any real number.}$$

Using $-\theta$ in place of θ, we get

$$e^{-i\theta} = \cos\theta - i\,\sin\theta, \qquad \theta \text{ any real number.}$$

Adding and subtracting the two identities, we obtain the familiar identities

$$(2) \qquad \cos\theta = \frac{1}{2}\left(e^{i\theta} + e^{-i\theta}\right) \quad \text{and} \quad \sin\theta = \frac{1}{2i}\left(e^{i\theta} - e^{-i\theta}\right).$$

As θ varies in the interval $[0,\,2\pi]$, the complex number $z = e^{i\theta}$ traces $C_1(0)$, the unit circle in the positive direction. This suggests transforming the integral (1) into a contour integral, where the variable $z = e^{i\theta}$ traces $C_1(0)$. Observing that $\frac{1}{z} = \frac{1}{e^{i\theta}} = e^{-i\theta}$, we obtain from (2) the key substitutions

$$(3) \qquad \boxed{\cos\theta = \frac{1}{2}\left(z + \frac{1}{z}\right), \quad \text{and} \quad \sin\theta = \frac{1}{2i}\left(z - \frac{1}{z}\right).}$$

This is represented graphically in Figure 1. Also, from $z = e^{i\theta}$, we have $dz = ie^{i\theta}\,d\theta = iz\,d\theta$ or

$$(4) \qquad \boxed{-i\,\frac{dz}{z} = d\theta.}$$

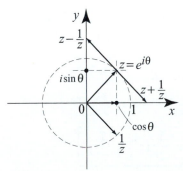

Figure 1 Constructing $\cos\theta$ and $\sin\theta$.

With (3) and (4) in hand, we can now consider some examples.

EXAMPLE 1   Evaluate

$$\int_0^{2\pi} \frac{d\theta}{10 + 8\cos\theta}.$$

Solution   Let $z = e^{i\theta}$. As θ varies from 0 to 2π, z traces $C_1(0)$, the unit circle in the positive direction. Using the substitutions (3) and (4), we transform the given integral into a path integral as follows

$$\int_0^{2\pi} \frac{d\theta}{10 + 8\cos\theta} = \int_{C_1(0)} \frac{-i\frac{dz}{z}}{10 + \frac{8}{2}\left(z + \frac{1}{z}\right)} = -i \int_{C_1(0)} \frac{dz}{4z^2 + 10z + 4}.$$

To compute the last integral using residues, we solve

$$4z^2 + 10z + 4 = 4(z + 2)\left(z + \frac{1}{2}\right) = 0,$$

and get $z = -2$ and $z = -\frac{1}{2}$ (see Figure 3). So the only singularity inside the unit disk is a simple pole at $-\frac{1}{2}$. Applying Proposition 1(ii), Section 5.1, we find

$$\text{Res}\left(\frac{1}{4z^2 + 10z + 4}, -\frac{1}{2}\right) = \frac{1}{\frac{d}{dz}(4z^2 + 10z + 4)\big|_{z=-\frac{1}{2}}} = \frac{1}{8(-1/2) + 10} = \frac{1}{6}.$$

By the residue theorem, we conclude that

$$\int_{C_1(0)} \frac{dz}{4z^2 + 10z + 4} = \frac{2\pi i}{6} = \frac{\pi i}{3},$$

and so

$$\int_0^{2\pi} \frac{d\theta}{10 + 8\cos\theta} = -i \int_{C_1(0)} \frac{dz}{4z^2 + 10z + 4} = \frac{\pi}{3}. \qquad \blacksquare$$

The following observations will facilitate the solution of the next example. Let n be an integer and $z = e^{i\theta}$. De Moivre's identity implies

$$z^n = e^{in\theta} = \cos n\theta + i\sin n\theta$$

and

$$\frac{1}{z^n} = e^{-in\theta} = \cos n\theta - i\sin n\theta.$$

So

(5) $$\boxed{\cos n\theta = \frac{1}{2}\left(z^n + \frac{1}{z^n}\right) \quad \text{and} \quad \sin n\theta = \frac{1}{2i}\left(z^n - \frac{1}{z^n}\right).}$$

As we now illustrate, these formulas are useful when evaluating integrals involving $\cos n\theta$ and $\sin n\theta$.

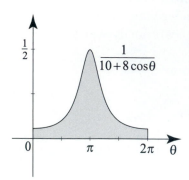

Figure 2 The definite integral in Example 1.

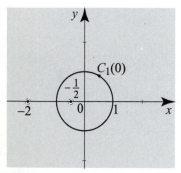

Figure 3 The path and poles for the contour integral in Example 1.

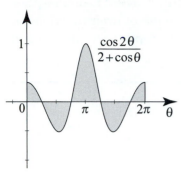

Figure 4 The definite integral in Example 2.

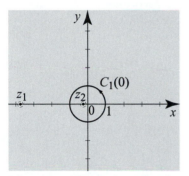

Figure 5 The path and poles for the contour integral in Example 2.

EXAMPLE 2   Compute the definite integral (Figure 4)

$$\int_0^{2\pi} \frac{\cos 2\theta}{2 + \cos \theta} \, d\theta.$$

Solution   Use (3), (4), and (5) with $n = 2$, and get

$$\int_0^{2\pi} \frac{\cos 2\theta}{2 + \cos \theta} \, d\theta = -i \int_{C_1(0)} \frac{\frac{1}{2}\left(z^2 + \frac{1}{z^2}\right)}{2 + \frac{1}{2}\left(z + \frac{1}{z}\right)} \frac{dz}{z} = -i \int_{C_1(0)} \frac{z^4 + 1}{z^2(z^2 + 4z + 1)} \, dz.$$

We now compute the last integral using residues. We clearly have a pole of order 2 at 0. To compute the residue at 0, we use Theorem 2, Section 5.1, with $m = 2$ at $z_0 = 0$, and get

$$\begin{aligned}
\operatorname{Res}\left(\frac{z^4 + 1}{z^2(z^2 + 4z + 1)}, 0\right) &= \lim_{z \to 0} \frac{d}{dz} \frac{z^4 + 1}{z^2 + 4z + 1} \\
&= \lim_{z \to 0} \frac{3z^3(z^2 + 4z + 1) - (z^4 + 1)(2z + 4)}{(z^2 + 4z + 1)^2} = -4.
\end{aligned}$$

The roots of $z^2 + 4z + 1 = 0$ are $z_1 = -2 - \sqrt{3} \approx -3.7$ and $z_2 = -2 + \sqrt{3} \approx -0.27$ (Figure 5). Only z_2 is inside the unit disk. Since z_2 is a simple pole, we can compute the residues at z_2 using Proposition 1(ii), Section 5.1:

$$\begin{aligned}
\operatorname{Res}\left(\frac{z^4 + 1}{z^2} \frac{1}{z^2 + 4z + 1}, -2 + \sqrt{3}\right) &= \frac{(-2 + \sqrt{3})^4 + 1}{(-2 + \sqrt{3})^2} \frac{1}{\frac{d}{dz}(z^2 + 4z + 1)\big|_{-2 + \sqrt{3}}} \\
&= \frac{(-2 + \sqrt{3})^4 + 1}{(-2 + \sqrt{3})^2(2\sqrt{3})} = \frac{7}{\sqrt{3}}.
\end{aligned}$$

By the residue theorem, we conclude that

$$\int_{C_1(0)} \frac{z^4 + 1}{z^2(z^2 + 4z + 1)} \, dz = 2\pi i\left(-4 + \frac{7}{\sqrt{3}}\right),$$

and so

$$\int_0^{2\pi} \frac{\cos 2\theta}{2 + \cos \theta} \, d\theta = -i \int_{C_1(0)} \frac{z^4 + 1}{z^2(z^2 + 4z + 1)} \, dz = 2\pi\left(-4 + \frac{7}{\sqrt{3}}\right) \approx 0.26. \quad \blacksquare$$

In the preceding examples, we needed the interval of integration to be $[0, 2\pi]$, in order for $z = e^{i\theta}$ to trace the whole circle $C_1(0)$. Integrals over the interval $[0, \pi]$ can be handled if the integrand $f(\theta)$ is even, since in this case

$$(6) \qquad \int_0^\pi f(\theta) \, d\theta = \frac{1}{2} \int_{-\pi}^\pi f(\theta) \, d\theta = \frac{1}{2} \int_0^{2\pi} f(\theta) \, d\theta.$$

The first equality follows because the integrand is even and the second one follows because the integrand is 2π-periodic, and so the integral does not change if we integrate over any interval of length 2π (see Theorem 1, Section 7.1). In the next example, we use this technique of even functions, as well as a double-angle identity and the linearity property of the integral. These will simplify the residue calculation.

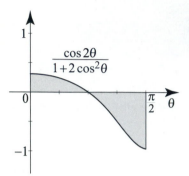

Figure 6 The definite integral in Example 3.

EXAMPLE 3   Techniques of integration
Compute the definite integral (Figure 6)

$$I = \int_0^{\frac{\pi}{2}} \frac{\cos 2\theta}{1 + 2\cos^2\theta}\, d\theta.$$

Solution   First, use the double angle identity $2\cos^2\theta = 1 + \cos 2\theta$, then the change of variables $\theta' = 2\theta$, and get (for convenience, rename θ' by θ)

$$I = \int_0^{\frac{\pi}{2}} \frac{\cos 2\theta}{2 + \cos 2\theta}\, d\theta = \frac{1}{2}\int_0^{\pi} \frac{\cos\theta}{2 + \cos\theta}\, d\theta.$$

This integral is over $[0, \pi]$, but the integrand is even, so according to (6) we have

$$I = \frac{1}{4}\int_0^{2\pi} \frac{\cos\theta}{2 + \cos\theta}\, d\theta.$$

This integral could be evaluated by the use of residues, but we can significantly reduce the amount of residue calculation by using linearity:

$$
\begin{aligned}
I &= \frac{1}{4}\int_0^{2\pi} \frac{2 + \cos\theta - 2}{2 + \cos\theta}\, d\theta \\
&= \frac{1}{4}\int_0^{2\pi} d\theta + \frac{1}{4}\int_0^{2\pi} \frac{-2}{2 + \cos\theta}\, d\theta = \frac{\pi}{2} - \frac{1}{2}\int_0^{2\pi} \frac{d\theta}{2 + \cos\theta}.
\end{aligned}
$$

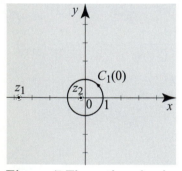

Figure 7 The path and poles for the contour integral in Example 3.

We now evaluate this last integral by the residue method. Letting $z = e^{i\theta}$, $\cos\theta = \frac{1}{2}(z + \frac{1}{z})$, $d\theta = \frac{dz}{iz}$, we obtain

$$I = \frac{\pi}{2} - \frac{1}{i}\int_{C_1(0)} \frac{dz}{z^2 + 4z + 1}.$$

The integrand $\frac{1}{z^2 + 4z + 1} = \frac{1}{(z - z_1)(z - z_2)}$ has simple poles at $z_1 = -2 - \sqrt{3}$ and $z_2 = -2 + \sqrt{3}$, and z_2 is the only pole lying inside $C_1(0)$ (Figure 7). The residue there is

$$
\begin{aligned}
\operatorname{Res}\left(\frac{1}{(z - z_1)(z - z_2)}, z_2\right) &= \lim_{z \to z_2} (z - z_2)\frac{1}{(z - z_1)(z - z_2)} \\
&= \frac{1}{z_2 - z_1} = \frac{1}{2\sqrt{3}},
\end{aligned}
$$

and so $I = \frac{\pi}{2} - \frac{1}{i}2\pi i\frac{1}{2\sqrt{3}} = \pi\left(\frac{1}{2} - \frac{1}{\sqrt{3}}\right).$ ∎

Although the integrals in this section are special, they have important applications, including the computation of certain Fourier series. See Section 7.2 for illustrations.

Exercises 5.2

In Exercises 1–10, use the method of this section to evaluate the given integral.

1. $\displaystyle\int_0^{2\pi} \frac{d\theta}{2 - \cos\theta}$.

2. $\displaystyle\int_0^{2\pi} \frac{d\theta}{5 + 3\cos\theta}$.

3. $\displaystyle\int_0^{2\pi} \frac{d\theta}{10 - 8\sin\theta}$.

4. $\displaystyle\int_0^{2\pi} \frac{1}{\sin^2\theta + 2\cos^2\theta}\, d\theta$.

5. $\displaystyle\int_0^{2\pi} \frac{\cos 2\theta}{5 + 4\cos\theta}\, d\theta$.

6. $\displaystyle\int_0^{2\pi} \frac{\cos\theta - \sin^2\theta}{10 + 8\cos\theta}\, d\theta$.

7. $\displaystyle\int_0^{\pi} \frac{d\theta}{9 + 16\sin^2\theta}$.

8. $\displaystyle\int_0^{\pi} \frac{\cos\theta\,\sin^2\theta}{2 + \cos\theta}\, d\theta$.

9. $\displaystyle\int_0^{2\pi} \frac{d\theta}{7 + 2\cos\theta + 3\sin\theta}$.

10. $\displaystyle\int_0^{2\pi} \frac{d\theta}{7 - 2\cos^2\theta - 3\sin^2\theta}$.

In Exercises 11–16, use the method of this section to derive the given formula, where a, b, c are real numbers.

11. $\displaystyle\int_0^{2\pi} \frac{d\theta}{1 + a\cos\theta} = \frac{2\pi}{\sqrt{1 - a^2}}, \quad 0 < |a| < 1$.

12. $\displaystyle\int_0^{2\pi} \frac{\sin^2\theta}{a + b\cos\theta}\, d\theta = \frac{2\pi}{b^2}\left(a - \sqrt{a^2 - b^2}\right), \quad 0 < |b| < a$.

13. $\displaystyle\int_0^{2\pi} \frac{d\theta}{a + b\cos^2\theta} = \frac{2\pi}{\sqrt{a}\sqrt{a + b}}, \quad 0 < b < a$.

14. $\displaystyle\int_0^{2\pi} \frac{d\theta}{a + b\sin^2\theta} = \frac{2\pi}{\sqrt{a}\sqrt{a + b}}, \quad 0 < b < a$.

15. $\displaystyle\int_0^{2\pi} \frac{d\theta}{a\cos\theta + b\sin\theta + c} = \frac{2\pi}{\sqrt{c^2 - a^2 - b^2}}, \quad a^2 + b^2 < c^2$.

16. $\displaystyle\int_0^{2\pi} \frac{d\theta}{a\cos^2\theta + b\sin^2\theta + c} = \frac{2\pi}{\sqrt{(a + c)(b + c)}}, \quad 0 < c < a,\ c < b$.

5.3   Improper Integrals Involving Rational and Exponential Functions

In the previous section, we transformed certain real trigonometric integrals into contour integrals, which we then computed with the help of the residue theorem. In this section we present another useful technique based on the residue theorem that can be used to evaluate improper integrals involving rational and exponential functions.

Let a and b be arbitrary real numbers. Consider the integrals

(1)
$$\int_{-\infty}^{b} f(x)\, dx, \quad \int_{a}^{\infty} f(x)\, dx,$$

where in each case f is continuous in the interval of integration and at its finite endpoint. These are called **improper integrals**, because the interval of integration is infinite. The integral $\int_a^\infty f(x)\, dx$ is **convergent** if

$\lim_{b \to \infty} \int_a^b f(x)\, dx$ exists as a finite number. Similarly, $\int_{-\infty}^b f(x)\, dx$ is **convergent** if $\lim_{a \to -\infty} \int_a^b f(x)\, dx$ exists as a finite number.

Now let $f(x)$ be continuous on the real line. The integral $\int_{-\infty}^\infty f(x)\, dx$ is also improper since the interval of integration is infinite, but here it is infinite in both the positive and negative direction. Such an integral is said to be **convergent** if both $\int_0^\infty f(x)\, dx$ and $\int_{-\infty}^0 f(x)\, dx$ are convergent. In this case, we set (Figure 1)

Figure 1 Splitting an improper integral over the line.

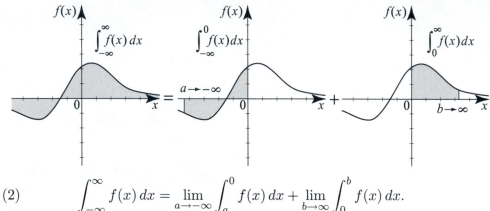

$$(2) \qquad \int_{-\infty}^\infty f(x)\, dx = \lim_{a \to -\infty} \int_a^0 f(x)\, dx + \lim_{b \to \infty} \int_0^b f(x)\, dx.$$

This expression should be contrasted with the following, which integrates the function along expanding symmetric intervals. We define the **Cauchy principal value** of the integral $\int_{-\infty}^\infty f(x)\, dx$ to be (Figure 2)

$$(3) \qquad \text{P.V.} \int_{-\infty}^\infty f(x)\, dx = \lim_{a \to \infty} \int_{-a}^a f(x)\, dx, \text{ if the limit exists.}$$

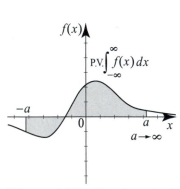

Figure 2 The Cauchy principal value of the integral.

The Cauchy principal value of an integral may exist even though the integral itself is not convergent. For example, $\int_{-a}^a x\, dx = 0$ for all a, which implies that P.V. $\int_{-\infty}^\infty x\, dx = 0$, but the integral itself is clearly not convergent since $\int_0^\infty x\, dx = \infty$.

However, whenever $\int_{-\infty}^\infty f(x)\, dx$ is convergent, then P.V. $\int_{-\infty}^\infty f(x)\, dx$ exists, and the two integrals will be the same. This is because $\lim_{a \to \infty} \int_{-a}^0 f(x)\, dx$ and $\lim_{a \to \infty} \int_0^a f(x)\, dx$ both exist, and so

$$
\begin{aligned}
\text{P.V.} \int_{-\infty}^\infty f(x)\, dx &= \lim_{a \to \infty} \int_{-a}^a f(x)\, dx \\
&= \lim_{a \to \infty} \left(\int_{-a}^0 f(x)\, dx + \int_0^a f(x)\, dx \right) \\
&= \lim_{a \to \infty} \int_{-a}^0 f(x)\, dx + \lim_{a \to \infty} \int_0^a f(x)\, dx \\
&= \int_{-\infty}^\infty f(x)\, dx.
\end{aligned}
$$

Because of this fact, we can compute a convergent integral over the real line by computing its principal value, which can often be obtained by use of complex methods and the residue theorem.

How can we know if an integral is convergent? We can use the following comparison test.

PROPOSITION 1
INEQUALITIES FOR
IMPROPER
INTEGRALS

Let $\int_A^B f(x)\,dx$ represent an improper integral as in (1), where $A = -\infty$ or $B = \infty$.

(i) If $\int_A^B |f(x)|\,dx$ is convergent, then $\int_A^B f(x)\,dx$ is convergent and we have $\left|\int_A^B f(x)\,dx\right| \leq \int_A^B |f(x)|\,dx$.

(ii) If $|f(x)| \leq g(x)$ for all $A < x < B$ and $\int_A^B g(x)\,dx$ is convergent, then $\int_A^B f(x)\,dx$ is convergent and we have $\left|\int_A^B f(x)\,dx\right| \leq \int_A^B g(x)\,dx$.

Proof We prove part (i) in the case A finite and $B = \infty$ and f is real-valued. The proof of the other cases is similar and will be omitted. For any $b > A$, using (32), Section 3.2, we have $\left|\int_A^b f(x)\,dx\right| \leq \int_A^b |f(x)|\,dx$. Given that $I = \int_A^\infty |f(x)|\,dx < \infty$, consider the nonnegative function $|f(x)| - f(x)$. Since it is nonnegative, its integral is nonnegative and increases with the size of the interval of integration. That means that $F(b) = \int_A^b (|f(x)| - f(x))\,dx$ is a nonnegative increasing function of b. Moreover, $F(b) = \int_A^b (|f(x)| - f(x))\,dx \leq 2\int_A^b |f(x)|\,dx \leq 2\int_A^\infty |f(x)|\,dx = 2I$. So, $F(b)$ is nonnegative, increasing and bounded above; therefore it has a limit $L \geq 0$ as $b \to \infty$. Now $L = \lim_{b\to\infty} \int_A^b (|f(x)| - f(x))\,dx = \lim_{b\to\infty} \int_A^b |f(x)|\,dx - \lim_{b\to\infty} \int_A^b f(x)\,dx = I - \lim_{b\to\infty} \int_A^b f(x)\,dx$. So, $\lim_{b\to\infty} \int_A^b f(x)\,dx = I - L$, which shows that the integral $\int_A^\infty f(x)\,dx = I - L$ is finite and $\leq I$, because $L \geq 0$. Repeating this argument with $|f(x)| + f(x)$ in place of $|f(x)| - f(x)$, we obtain that $-\int_A^\infty f(x)\,dx$ exists and is $\leq I$. Hence, $\left|\int_A^\infty f(x)\,dx\right| \leq \int_A^\infty |f(x)|\,dx$.

To prove (ii), let $A \leq b$ and note that $\int_A^b |f(x)|\,dx$ is an increasing function of b, because $|f(x)|$ is nonnegative. It is also bounded above by $\int_A^\infty g(x)\,dx$, because of the inequality $|f(x)| \leq g(x)$, and so $\lim_{b\to\infty} \int_A^b |f(x)|\,dx$ exists and is $\leq \int_A^\infty g(x)\,dx$. Hence $\int_A^\infty |f(x)|dx$ is convergent, and from (i) it follows that $\int_A^\infty f(x)\,dx$ is convergent and is $\leq \int_A^\infty g(x)\,dx$. ◼

You are encouraged to use Proposition 1 or any other test from calculus (such as the limit comparison test) to show that an improper integral is convergent. Once the convergence is determined, we may compute the integral via its principal value, as we will illustrate in the examples.

EXAMPLE 1   Improper integrals and residues: the main ideas
Evaluate
$$I = \int_{-\infty}^{\infty} \frac{x^2}{x^4 + 1}\,dx.$$

Solution   To highlight the main ideas, we present the solution in basic steps.
Step 1: Show that the improper integral is convergent. Because the integrand is continuous on the real line, it is enough to show that the integral outside a

finite interval, say $[-1, 1]$, is convergent. For $|x| \geq 1$, we have $\frac{x^2}{x^4+1} \leq \frac{x^2}{x^4} = \frac{1}{x^2}$, and since $\int_1^\infty \frac{1}{x^2}\, dx$ is convergent, it follows by Proposition 1 that $\int_1^\infty \frac{x^2}{x^4+1}\, dx$ and $\int_{-\infty}^{-1} \frac{x^2}{x^4+1}\, dx$ are convergent. Thus $\int_{-\infty}^\infty \frac{x^2}{x^4+1}\, dx$ is convergent, and so

$$(4) \qquad \int_{-\infty}^\infty \frac{x^2}{x^4+1}\, dx = \lim_{R\to\infty} \int_{-R}^R \frac{x^2}{x^4+1}\, dx.$$

Step 2: Set up the contour integral. We will replace x by z and consider the function $f(z) = \frac{z^2}{z^4+1}$. The general guideline is to integrate this function over a contour that consists partly of the interval $[-R, R]$, so as to recapture the integral $\int_{-R}^R \frac{x^2}{x^4+1}\, dx$ as part of the contour integral on γ_R. Choosing the appropriate contour is not obvious in general. For a rational function, such as $\frac{x^2}{x^4+1}$, where the denominator does not vanish on the x-axis and the degree of the denominator is two more than the degree of the numerator, a closed semi-circle γ_R as in Figure 3 will work. Since γ_R consists of the interval $[-R, R]$ followed by the semi-circle σ_R, using the additive property of path integrals (Proposition 2(iii), Section 3.2), we write

$$(5) \qquad I_{\gamma_R} = \int_{\gamma_R} f(z)\, dz = \int_{[-R,\, R]} f(z)\, dz + \int_{\sigma_R} f(z)\, dz = I_R + I_{\sigma_R}.$$

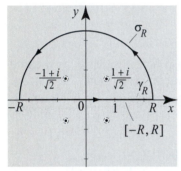

Figure 3 The path and poles for the contour integral in Example 1.

For $z = x$ in $[-R, R]$, we have $f(z) = f(x) = \frac{x^2}{x^4+1}$ and $dz = dx$, and so $I_R = \int_{-R}^R \frac{x^2}{x^4+1}\, dx$, which according to (4) converges to the desired integral as $R \to \infty$. So, in order to compute the desired integral, we must get a handle on the other quantities, I_{γ_R} and I_{σ_R}, in (5). Our strategy is as follows. In Step 3, we compute I_{γ_R} by the residue theorem; and in Step 4, we show that $\lim_{R\to\infty} I_{\sigma_R} = 0$. This will give us the necessary ingredients to complete the solution in Step 5.

Step 3: Compute I_{γ_R} by the residue theorem. For $R > 1$, the function $f(z) = \frac{z^2}{z^4+1}$ has two poles inside γ_R. These are the roots of $z^4 + 1 = 0$ in the upper half-plane. To solve $z^4 = -1$, we write $-1 = e^{i\pi}$; then using the formula for the nth roots from Section 1.3, we find the four roots

$$z_1 = \frac{1+i}{\sqrt{2}}, \quad z_2 = \frac{-1+i}{\sqrt{2}}, \quad z_3 = \frac{-1-i}{\sqrt{2}}, \quad z_4 = \frac{-1-i}{\sqrt{2}}$$

(see Figure 3). Since these are simple roots, we have simple poles at z_1 and z_2 and the residues there are (Proposition 1(ii), Section 5.1)

$$\mathrm{Res}\,(z_1) = \left.\frac{z^2}{\frac{d}{dz}(z^4+1)}\right|_{z=z_1} = \left.\frac{z^2}{4z^3}\right|_{z=z_1} = \left.\frac{1}{4z}\right|_{z=z_1} = \frac{\sqrt{2}}{4(1+i)} = \frac{1-i}{4\sqrt{2}},$$

and similarly

$$\mathrm{Res}\,(z_2) = \left.\frac{1}{4z}\right|_{z=z_2} = \frac{-1-i}{4\sqrt{2}}.$$

So, by the residue theorem, for all $R > 1$

$$(6) \qquad I_{\gamma_R} = \int_{\gamma_R} \frac{z^2}{z^4+1}\, dz = 2\pi i\left(\mathrm{Res}\,(z_1) + \mathrm{Res}\,(z_2)\right) = 2\pi i\frac{-2i}{4\sqrt{2}} = \frac{\pi}{\sqrt{2}}.$$

Step 4: Show that $\lim_{R\to\infty} I_{\sigma_R} = 0$. For z on σ_R, we have $|z| = R$, and so

$$\left| \frac{z^2}{z^4 + 1} \right| \leq \frac{R^2}{R^4 - 1} = M_R.$$

Appealing to the integral inequality (Theorem 2, Section 3.2), we have

$$|I_{\sigma_R}| = \left| \int_{\sigma_R} \frac{z^2}{z^4 + 1}\, dz \right| \leq l(\sigma_R) M_R = \pi R \frac{R^2}{R^4 - 1} = \frac{\pi}{R - 1/R^3} \to 0, \quad \text{as } R \to \infty.$$

Step 5: Compute the improper integral. Using (5) and (6), we obtain

$$\frac{\pi}{\sqrt{2}} = I_R + I_{\sigma_R}.$$

Let $R \to \infty$, then $I_R \to \int_{-\infty}^{\infty} \frac{x^2}{x^4+1}\, dx$ and $I_{\sigma_R} \to 0$, and so

$$\int_{-\infty}^{\infty} \frac{x^2}{x^4 + 1}\, dx = \frac{\pi}{\sqrt{2}}. \qquad \blacksquare$$

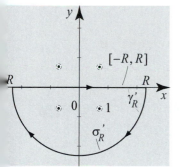

Figure 4 An alternative path for the integral in Example 1.

In Example 1, we could have used the contour γ'_R in the lower half-plane in Figure 4. In this case, it is easiest to take the orientation of γ'_R to be negative in order to coincide with the orientation of the interval $[-R, R]$.

The five steps that we used in the solution of Example 1 are applicable in a general situation described as follows.

PROPOSITION 2
INTEGRALS OF
RATIONAL
FUNCTIONS

Let $f(x) = \frac{p(x)}{q(x)}$ be a rational function with degree $q(x) \geq 2 + \text{degree}\, p(x)$, and let σ_R denote an arc of the circle centered at 0 with radius $R > 0$. Then

(7)
$$\lim_{R\to\infty} \left| \int_{\sigma_R} \frac{p(z)}{q(z)}\, dz \right| = 0.$$

Moreover, if $q(x)$ has no real roots and $z_1, z_2, \ldots, z_N$ denote all the poles of $\frac{p(z)}{q(z)}$ in the upper half-plane, then

(8)
$$\int_{-\infty}^{\infty} \frac{p(x)}{q(x)}\, dx = 2\pi i \sum_{j=1}^{N} \text{Res}\left(\frac{p(z)}{q(z)}, z_j \right).$$

The proof of the proposition is very similar to the solution of Example 1. We will relegate it to the exercises. Here is a straightforward application.

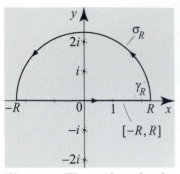

Figure 5 The path and poles for the contour integral in Example 2.

EXAMPLE 2   Improper integral of a rational function

Evaluate

$$\int_{-\infty}^{\infty} \frac{1}{(x^2+1)(x^2+4)}\, dx.$$

Solution   The integrand satisfies the two conditions of Proposition 2: degree $q(x) = 4 \geq 2 + $ degree $p(x) = 2$, and the denominator $q(x) = (x^2+1)(x^2+4)$ has no real roots. So we may apply (8). We have

$$(z^2+1)(z^2+4) = (z+i)(z-i)(z+2i)(z-2i),$$

and hence $\frac{1}{(z^2+1)(z^2+4)}$ has simple poles at i and $2i$ in the upper half-plane (Figure 5). The residues there are

$$\operatorname{Res}(i) = \lim_{z\to i}(z-i)\frac{1}{(z-i)(z+i)(z^2+4)} = \frac{1}{(2i)(-1+4)} = -\frac{i}{6}$$

and

$$\operatorname{Res}(2i) = \lim_{z\to 2i}(z-2i)\frac{1}{(z^2+1)(z-2i)(z+2i)} = \frac{i}{12}.$$

According to (8),

$$\int_{-\infty}^{\infty} \frac{1}{(x^2+1)(x^2+4)}\, dx = 2\pi i\Big(-\frac{i}{6} + \frac{i}{12}\Big) = \frac{\pi}{6}. \qquad \blacksquare$$

The Substitution $x = e^t$

There are interesting integrals of rational functions that are not computable using semi-circular contours as in Example 1. Consider

$$(9) \qquad \int_0^{\infty} \frac{dx}{x^3+1}.$$

If you try to directly apply the method of Example 1 to this integral, you will quickly run into problems. First, we are integrating over just half of the real line, and it is not obvious how to relate this integral to the integral $\int_{-\infty}^{\infty} \frac{dx}{x^3+1}$. Moreover, the latter integral has a problem at $x = -1$. An integral such as (9) can be solved by using the substitution $x = e^t$. We outline this useful technique in Example 3, where we will compute a more general integral involving arbitrary powers of x. Using the exponential function has the advantage of eliminating the need to work with powers of z, which require dealing with branches of the logarithm and can complicate the problem.

Changing variables will affect the interval of integration; here it will change a semi-infinite interval to the real line. If the original integral is convergent, then it will still be convergent after we change variables, so we need only check once.

EXAMPLE 3   The substitution $x = e^t$

Establish the identity

$$(10) \qquad \int_0^{\infty} \frac{1}{x^\alpha+1}\, dx = \frac{\pi}{\alpha \sin \frac{\pi}{\alpha}}, \qquad \alpha \text{ any real number } > 1.$$

Solution   Step 1: Show that the integral converges. The integrand is continuous, so it is enough to show that it has a convergent integral on $[1, \infty)$. We have $\frac{1}{x^\alpha + 1} \leq \frac{1}{x^\alpha}$, where the latter has a convergent integral over $[1, \infty)$.

Step 2: Apply the substitution $x = e^t$. Let $x = e^t$, $dx = e^t\, dt$, and note that as x varies from 0 to ∞, t varies from $-\infty$ to ∞, and so

$$I = \int_0^\infty \frac{1}{x^\alpha + 1}\, dx = \int_{-\infty}^\infty \frac{e^t}{e^{\alpha t} + 1}\, dt = \int_{-\infty}^\infty \frac{e^x}{e^{\alpha x} + 1}\, dx,$$

where, for convenience, in the last integral we have used x as a variable of integration instead of t. As we change x to z, we obtain the function $f(z) = \frac{e^z}{e^{\alpha z}+1}$ with poles at the roots of $e^{\alpha z} + 1 = 0$. Since the exponential function is $2\pi i$-periodic, then

$$e^{\alpha z} = -1 = e^{i\pi} \quad \Leftrightarrow \quad \alpha z = i\pi + 2k\pi i \quad \Leftrightarrow \quad z_k = (2k+1)\frac{\pi}{\alpha}i, \; k = 0, \pm 1, \pm 2, \ldots .$$

Thus $f(z) = \frac{e^z}{e^{\alpha z}+1}$ has infinitely many poles at z_k, all lying on the imaginary axis (Figure 6).

Step 3: Selecting the contour of integration. Our contour should expand in the x-direction in order to cover the entire x-axis. But since $f(z)$ has infinitely many poles on the y-axis, to avoid dealing with an infinite sum of residues, we will not use an expanding semicircle in the upper half-plane. Instead, we will use a rectangular contour γ_R consisting of the paths γ_1, γ_2, γ_3, and γ_4, as in Figure 6, and let I_j denote the path integral of $f(z)$ over γ_j $(j = 1, \ldots, 4)$. As $R \to \infty$, γ_R will expand in the horizontal direction, but the length of the vertical sides remains fixed at $\frac{2\pi}{\alpha}$. To understand the reason for our choice of the vertical length, let us consider I_1 and I_3. On γ_1, we have $z = x$, $dz = dx$,

$$(11) \qquad I_1 = \int_{-R}^R \frac{e^x}{e^{\alpha x} + 1}\, dx,$$

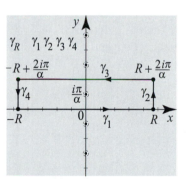

Figure 6 The path and poles for the contour integral in Example 3.

and so I_1 converges to the desired integral I as $R \to \infty$. On γ_3, we have $z = x + i\frac{2\pi}{\alpha}$, $dz = dx$, and using the $2\pi i$-periodicity of the exponential function, we get

$$(12) \qquad I_3 = \int_R^{-R} \frac{e^{x + i\frac{2\pi}{\alpha}}}{e^{\alpha(x + i\frac{2\pi}{\alpha})} + 1}\, dx = -e^{\frac{2\pi}{\alpha}i} \int_{-R}^R \frac{e^x}{e^{\alpha x} + 1}\, dx = -e^{\frac{2\pi}{\alpha}i} I_1.$$

This last equality explains the choice of the vertical sides: They are chosen so that the integral on the returning horizontal side γ_3 is equal to a constant multiple of the integral on γ_1. From here the solution is straightforward.

Step 4: Applying the residue theorem. From Step 2, we have only one pole of $f(z)$ inside γ_R at $z_0 = \frac{\pi}{\alpha}i$. Since $e^{\alpha z} + 1$ has a simple root, this is a simple pole. Using Proposition 1(ii), Section 5.1, we find

$$\operatorname{Res}\left(\frac{e^z}{e^{\alpha z}+1}, \frac{\pi}{\alpha}i\right) = \frac{e^{\frac{\pi}{\alpha}i}}{\alpha e^{\alpha\frac{\pi}{\alpha}i}} = \frac{e^{\frac{\pi}{\alpha}i}}{\alpha e^{i\pi}} = -\frac{e^{\frac{\pi}{\alpha}i}}{\alpha},$$

and so by the residue theorem

$$(13) \qquad I_1 + I_2 + I_3 + I_4 = \int_{\gamma_R} \frac{e^z}{e^{\alpha z}+1}\, dz = 2\pi i \operatorname{Res}\left(\frac{e^z}{e^{\alpha z}+1}, \frac{\pi}{\alpha}i\right) = -2\pi i \frac{e^{\frac{\pi}{\alpha}i}}{\alpha}.$$

Step 5: Show that the integrals on the vertical sides tend to 0 as $R \to \infty$. Let us estimate I_2. For $z = R + iy$ $(0 \le y \le \frac{\pi}{\alpha})$ on γ_2, we have $|e^{\alpha z}| = |e^{\alpha R}e^{i\alpha y}| = e^{\alpha R}$ and so

$$|e^{\alpha z} + 1| \ge |e^{\alpha z}| - 1 = e^{\alpha R} - 1,$$

hence

$$\frac{1}{|e^{\alpha z} + 1|} \le \frac{1}{e^{\alpha R} - 1},$$

and so for z on γ_2

$$|f(z)| = \left| \frac{e^z}{e^{\alpha z} + 1} \right| \le \frac{|e^{R+iy}|}{e^{\alpha R} - 1} = \frac{e^R}{e^{\alpha R} - 1} = \frac{1}{e^{(\alpha-1)R} - e^{-R}}.$$

Consequently, by the inequality for path integrals (Theorem 2, Section 3.2),

$$|I_2| = \left| \int_{\gamma_2} \frac{e^z}{e^{\alpha z} + 1} \, dz \right| \le \frac{l(\gamma_2)}{e^{(\alpha-1)R} - e^{-R}} = \frac{\frac{2\pi}{\alpha}}{e^{(\alpha-1)R} - e^{-R}} \to 0, \text{ as } R \to \infty.$$

The proof that $I_4 \to 0$ as $R \to \infty$ is done similarly; we omit the details.

Step 6: Compute the desired integral (10). Using (11), (12), and (13), we find that

$$I_1\left(1 - e^{\frac{2\pi}{\alpha}i}\right) + I_2 + I_4 = -2\pi i \frac{e^{\frac{\pi}{\alpha}i}}{\alpha}.$$

Letting $R \to \infty$ and using the result of Step 5, we get

$$\left(1 - e^{\frac{2\pi}{\alpha}i}\right) \int_{-\infty}^{\infty} \frac{e^x}{e^{\alpha x} + 1} \, dx = -2\pi i \frac{e^{\frac{\pi}{\alpha}i}}{\alpha},$$

and after simplifying

$$\int_{-\infty}^{\infty} \frac{e^x}{e^{\alpha x} + 1} \, dx = \frac{2\pi i}{\alpha\left(e^{\frac{\pi}{\alpha}i} - e^{-\frac{\pi}{\alpha}i}\right)} = \frac{\pi}{\alpha \sin \frac{\pi}{\alpha}},$$

which implies (10). ∎

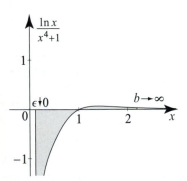

Figure 7 Splitting an improper integral.

The tricky part in Example 3 is choosing the contour. Let us clarify this part with one more example. We will compute the integral $\int_0^{\infty} \frac{\ln x}{x^4+1} \, dx$. The integral is improper because the interval is infinite and because the integrand tends to $-\infty$ as $x \downarrow 0$. To define the convergence of such integrals, we follow the general procedure of taking all one-sided limits one at a time. Thus (Figure 7)

$$\int_0^{\infty} \frac{\ln x}{x^4 + 1} \, dx = \lim_{\epsilon \downarrow 0} \int_{\epsilon}^1 \frac{\ln x}{x^4 + 1} \, dx + \lim_{b \to \infty} \int_1^b \frac{\ln x}{x^4 + 1} \, dx.$$

It is not difficult to show that both limits exist and hence that the integral converges. We will use the substitution $x = e^t$ to solve the problem. If you like, you could instead check the convergence after changing variables.

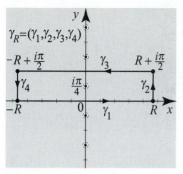

Figure 8 The path and poles for the contour integral in Example 4.

EXAMPLE 4 An integral involving $\ln x$

Derive

$$(14) \qquad \int_0^\infty \frac{\ln x}{x^4 + 1}\, dx = -\frac{\pi^2}{8\sqrt{2}}.$$

Solution   Let $x = e^t$, $\ln x = t$, $dx = e^t dt$. This transforms the integral into

$$\int_{-\infty}^\infty \frac{t}{e^{4t} + 1} e^t\, dt = \int_{-\infty}^\infty \frac{xe^x}{e^{4x} + 1}\, dx,$$

where we have renamed the variable x, just for convenience. In evaluating this integral, we will integrate $f(z) = \frac{ze^z}{e^{4z}+1}$ over the rectangular contour γ_R in Figure 8, and let I_j denote the integral of $f(z)$ on γ_j. Here again, we chose the vertical sides of γ_R so that on the returning path γ_3 the denominator equals to $e^{4x} + 1$. As we will see momentarily, this will enable us to relate I_3 to I_1. Let us now compute $I_{\gamma_R} = \int_{\gamma_R} f(z)\, dz$. As you can check, $f(z)$ has one (simple) pole at $z = i\frac{\pi}{4}$ inside γ_R. By Proposition 1(ii), Section 5.1, the residue there is

$$\mathrm{Res}\left(\frac{ze^z}{e^{4z}+1}, i\frac{\pi}{4}\right) = \frac{i\frac{\pi}{4} e^{i\frac{\pi}{4}}}{4e^{i\pi}} = -\frac{i\pi(1+i)}{16\sqrt{2}}.$$

So by the residue theorem

$$(15) \qquad I_{\gamma_R} = 2\pi i\left(-\frac{i\pi(1+i)}{16\sqrt{2}}\right) = \frac{\pi^2(1+i)}{8\sqrt{2}} = I_1 + I_2 + I_3 + I_4.$$

Moving to each I_j $(j = 1, \ldots, 4)$, we have

$$I_1 = \int_{\gamma_1} \frac{ze^z}{e^{4z}+1}\, dz = \int_{-R}^R \frac{xe^x}{e^{4x}+1}\, dx.$$

On γ_3, $z = x + i\frac{\pi}{2}$, $dz = dx$, so using $e^{i\frac{\pi}{2}} = i$, we get

$$\begin{aligned}
I_3 &= \int_{\gamma_3} \frac{ze^z}{e^{4z}+1}\, dz = \int_R^{-R} \frac{(x + i\frac{\pi}{2})e^x e^{i\frac{\pi}{2}}}{e^{4x}+1}\, dx \\
&= -i\int_{-R}^R \frac{xe^x}{e^{4x}+1}\, dx + \frac{\pi}{2}\int_{-R}^R \frac{e^x}{e^{4x}+1}\, dx \\
&= -iI_1 + \frac{\pi}{2}\int_{-R}^R \frac{e^x}{e^{4x}+1}\, dx.
\end{aligned}$$

To show that I_2 and I_4 tend to 0 as $R \to \infty$, we proceed as in Step 5 of Example 3. For $z = R + iy$ $(0 \le y \le \frac{\pi}{2})$ on γ_2, we have $|z| \le R + y \le R + \frac{\pi}{2}$, and so, as in Example 3,

$$|f(z)| = \left|\frac{ze^z}{e^{4z}+1}\right| = |z|\left|\frac{e^z}{e^{4z}+1}\right| \le (R + \frac{\pi}{2})\frac{e^R}{e^{4R}-1} = \frac{R + \frac{\pi}{2}}{e^{3R} - e^{-R}}.$$

Consequently, by the inequality for path integrals (Theorem 2, Section 3.2),

$$|I_2| = \left|\int_{\gamma_2} \frac{ze^z}{e^{4z}+1}\, dz\right| \le \frac{l(\gamma_2)(R + \frac{\pi}{2})}{e^{3R} - e^{-R}} = \frac{\frac{\pi}{2}(R + \frac{\pi}{2})}{e^{3R} - e^{-R}} \to 0, \text{ as } R \to \infty.$$

The proof that $I_4 \to 0$ as $R \to \infty$ is done similarly; we omit the details. Substituting our finding into (15) and taking the limit as $R \to \infty$, we get

$$
\begin{aligned}
\frac{\pi^2(1+i)}{8\sqrt{2}} &= \lim_{R \to \infty} \left(I_1 + I_2 - iI_1 + \frac{\pi}{2} \int_{-R}^{R} \frac{e^x}{e^{4x}+1} \, dx + I_4 \right) \\
&= (1-i) \int_{-\infty}^{\infty} \frac{xe^x}{e^{4x}+1} \, dx + \frac{\pi}{2} \int_{-\infty}^{\infty} \frac{e^x}{e^{4x}+1} \, dx.
\end{aligned}
$$

Taking imaginary parts of both sides, we obtain our answer $\int_{-\infty}^{\infty} \frac{xe^x}{e^{4x}+1} \, dx = -\frac{\pi^2}{8\sqrt{2}}$. If we take real parts of both sides we get the value of the integral in (10) that corresponds to $\alpha = 4$. ∎

The substitution $x = e^t$ is also useful even when we do not use complex methods to evaluate the integral in t. For example,

$$
\int_0^{\infty} \frac{\ln x}{x^2+1} \, dx = \int_{-\infty}^{\infty} \frac{te^t}{e^{2t}+1} \, dt = 0
$$

since the integral is convergent and the integrand $\frac{te^t}{e^{2t}+1} = \frac{t}{e^t+e^{-t}}$ is an odd function of t.

Exercises 5.3

In Exercises 1–8, evaluate the given improper integral by the method of Example 1.

1. $\displaystyle\int_{-\infty}^{\infty} \frac{dx}{x^4+1} = \frac{\pi}{\sqrt{2}}$.

2. $\displaystyle\int_{-\infty}^{\infty} \frac{dx}{x^4+x^2+1} = \frac{\pi}{\sqrt{3}}$.

3. $\displaystyle\int_{-\infty}^{\infty} \frac{dx}{(x^2+1)(x^4+1)} = \frac{\pi}{2}$.

4. $\displaystyle\int_{-\infty}^{\infty} \frac{dx}{(x-i)(x+3i)} = \frac{\pi}{2}$.

5. $\displaystyle\int_{-\infty}^{\infty} \frac{dx}{(x^2+1)^3} = \frac{3\pi}{8}$.

6. $\displaystyle\int_{-\infty}^{\infty} \frac{dx}{(x^4+1)^2} = \frac{3\pi}{4\sqrt{2}}$.

7. $\displaystyle\int_{-\infty}^{\infty} \frac{dx}{(4x^2+1)(x-i)} = \frac{\pi}{3}i$.

8. $\displaystyle\int_{-\infty}^{\infty} \frac{dx}{(x+i)(x-3i)} = \frac{\pi}{2}$.

In Exercises 9–20, evaluate the given improper integral by the method of Example 3. In some cases, the integral follows from more general formulas that we derived earlier. You may use these formulas to verify your answer, but you must provide a detailed solution.

9. $\displaystyle\int_0^{\infty} \frac{dx}{x^3+1} = \frac{2\pi}{3\sqrt{3}}$.

10. $\displaystyle\int_{-\infty}^{\infty} \frac{dx}{(1+x^2)^{n+1}} = \frac{(2n)!}{2^{2n}(n!)^2}\pi$.

11. $\displaystyle\int_0^{\infty} \frac{x}{x^5+1} \, dx = \frac{\pi}{5\sin\left(\frac{2\pi}{5}\right)}$.

12. $\displaystyle\int_0^{\infty} \frac{x}{x^\alpha+1} \, dx = \frac{\pi}{\alpha\sin\left(\frac{2\pi}{\alpha}\right)}$ $(\alpha > 2)$.

13. $\displaystyle\int_0^{\infty} \frac{\sqrt{x}}{x^3+1} \, dx = \frac{\pi}{3}$.

14. $\displaystyle\int_0^{\infty} \frac{x^\alpha}{(x+1)^2} \, dx = \frac{\pi\alpha}{\sin\pi\alpha}$ $(-1 < \alpha < 1)$.

15. $\displaystyle\int_0^{\infty} \frac{x^2 e^x}{e^{2x}+1} \, dx = \frac{\pi^3}{8}$.

16. $\displaystyle\int_0^{\infty} \frac{x\ln(2x)}{x^3+1} \, dx = \frac{2\pi^2}{27} + \frac{2\pi\ln 2}{3\sqrt{3}}$.

17. $\displaystyle\int_0^\infty \frac{\ln(2x)}{x^2+4}\,dx = \frac{\pi}{2}\ln 2.$

18. $\displaystyle\int_0^\infty \frac{\ln(ax)}{x^2+b^2}\,dx = \frac{\pi}{2b}\ln(ab)\ (a,\,b>0).$

19. $\displaystyle\int_0^\infty \frac{(\ln x)^2}{x^2+1}\,dx = \frac{\pi^3}{8}.$

20. $\displaystyle\int_0^\infty \frac{(\ln x)^2}{x^3+1}\,dx = \frac{10\,\pi^3}{81\sqrt 3}.$

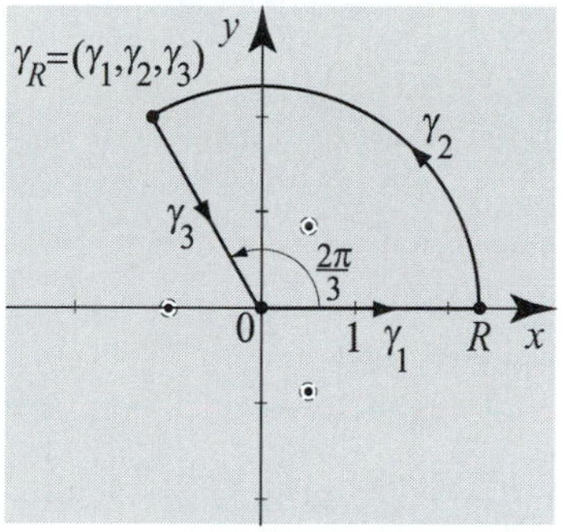

21. Use the contour γ_R in Figure 9 to evaluate

$$\int_0^\infty \frac{1}{x^3+1}\,dx.$$

[Hint: $I_{\gamma_R} = I_1 + I_2 + I_3$; $I_3 = -e^{i\frac{2\pi}{3}}I_1$; and $I_2 \to 0$ as $R \to \infty$.]

22. Follow the steps in Example 1 to prove Proposition 2.

23. A property of the gamma function.   (a) Show that for $0 < \alpha < 1$,

$$\int_0^\infty \frac{x^{\alpha-1}}{1+x}\,dx = \frac{\pi}{\sin \pi\alpha}.$$

(b) Use the definition of the gamma function (Exercise 24, Section 4.3) to derive

$$\Gamma(\alpha)\Gamma(1-\alpha) = \int_0^\infty \int_0^\infty e^{-(s+t)}s^{-\alpha}t^{\alpha-1}\,ds\,dt, \quad 0 < \alpha < 1.$$

(c)   Make the change of variables $x = s+t$, $y = \frac{t}{s}$ in (b), then use (a) to derive

$$\Gamma(\alpha)\Gamma(1-\alpha) = \frac{\pi}{\sin \pi\alpha}, \quad 0 < \alpha < 1.$$

This is known as the **formula of the complements** of the gamma function.
(d)   Use the identity principle (Corollary 1, Section 4.6) to extend the formula of the complements to all z such that $0 < \operatorname{Re}z < 1$.

(e)   Derive the integral $\displaystyle\int_0^{\frac{\pi}{2}} \sqrt{\cot x}\,dx = \frac{\pi}{\sqrt 2}.$ [Hint: Exercise 25(d), Section 4.3.]

$\gamma_R = (\gamma_1, \gamma_2, \gamma_3)$

Figure 9 The path and poles for the contour integral in Exercise 21.

5.4 Improper Integrals of Products of Rational and Trigonometric Functions

In this section, we use residues to evaluate improper integrals of the form

(1) $$\int_{-\infty}^\infty \frac{p(x)}{q(x)}\cos ax\,dx \quad \text{and} \quad \int_{-\infty}^\infty \frac{p(x)}{q(x)}\sin bx\,dx,$$

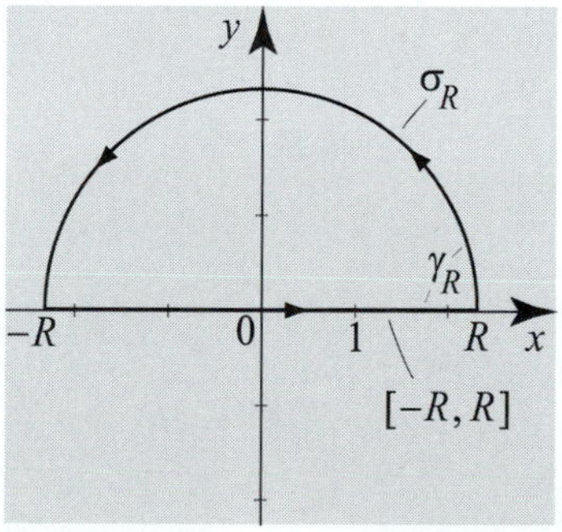

where a and b are real and $\frac{p(x)}{q(x)}$ is a rational function of x. The method is similar to the one we used previously. We will even use the expanding closed semi-circles γ_R in Figure 1, as we did in Section 5.3. But because $|\cos az|$ and $|\sin bz|$ are not bounded as z ranges over the expanding semi-circles σ_R, it will be to our advantage to replace the cosine and sine functions by an exponential function, using the identities:

$$\cos ax = \operatorname{Re}\left(e^{iax}\right) \quad \text{and} \quad \sin bx = \operatorname{Im}\left(e^{ibx}\right),$$

σ_R

γ_R

$[-R, R]$

Figure 1 An expanding closed semi-circle with radius $R \to \infty$.

for any real numbers a, b, x. Then,

$$\int_{-\infty}^{\infty} \frac{p(x)}{q(x)} \cos ax \, dx = \int_{-\infty}^{\infty} \frac{p(x)}{q(x)} \operatorname{Re}\left(e^{iax}\right) dx = \int_{-\infty}^{\infty} \operatorname{Re}\left(\frac{p(x)}{q(x)} e^{iax}\right) dx$$

$$(2) \qquad = \operatorname{Re}\left(\int_{-\infty}^{\infty} \frac{p(x)}{q(x)} e^{iax} \, dx\right),$$

and similarly

$$(3) \qquad \int_{-\infty}^{\infty} \frac{p(x)}{q(x)} \sin bx \, dx = \operatorname{Im}\left(\int_{-\infty}^{\infty} \frac{p(x)}{q(x)} e^{ibx} \, dx\right).$$

This approach will involve functions of the form $\frac{p(z)}{q(z)} e^{iaz}$, which are manageable on the contour γ_R.

EXAMPLE 1 The main ideas

Let $a > 0$ and $s \geq 0$ be real numbers. Derive the identity

$$\int_{-\infty}^{\infty} \frac{\cos sx}{x^2 + a^2} \, dx = \frac{\pi}{a} e^{-sa}.$$

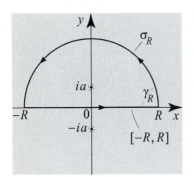

Figure 2 The path and poles for the contour integral in Example 1.

Solution   **Step 1:** Show that the improper integral is convergent. We have $\left|\frac{\cos sx}{x^2 + a^2}\right| \leq \frac{1}{x^2 + a^2}$, and since $\int_{-\infty}^{\infty} \frac{dx}{x^2 + a^2}$ is convergent, our integral converges by Proposition 1, Section 5.3.

Step 2: Set up and evaluate the contour integral. Since

$$(4) \qquad \int_{-\infty}^{\infty} \frac{\cos sx}{x^2 + a^2} \, dx = \int_{-\infty}^{\infty} \operatorname{Re}\left(\frac{e^{isx}}{x^2 + a^2}\right) dx = \operatorname{Re}\left(\int_{-\infty}^{\infty} \frac{e^{isx}}{x^2 + a^2} \, dx\right),$$

we will consider the contour integral

$$(5) \qquad I_{\gamma_R} = \int_{\gamma_R} \frac{e^{isz}}{z^2 + a^2} \, dz = \int_{\sigma_R} \frac{e^{isz}}{z^2 + a^2} \, dz + \int_{-R}^{R} \frac{e^{isx}}{x^2 + a^2} \, dx = I_{\sigma_R} + I_R,$$

where γ_R and σ_R are as in Figure 2. For $R > a$, $\frac{e^{isz}}{z^2+a^2}$ has one simple pole inside γ_R at $z = ia$. The residue there is (Proposition 1, Section 5.1)

$$\operatorname{Res}\left(\frac{e^{isz}}{z^2 + a^2}, ia\right) = \frac{e^{is(ia)}}{2ia} = \frac{e^{-sa}}{2ia}.$$

By the residue theorem, we have for all $R > a$

$$(6) \qquad I_{\gamma_R} = I_{\sigma_R} + I_R = 2\pi i \frac{e^{-sa}}{2ia} = \frac{\pi}{a} e^{-sa}.$$

Step 3: Show that $\lim_{R \to \infty} I_{\sigma_R} = 0$. For $s \geq 0$ and $0 \leq \theta \leq \pi$, we have $\sin \theta \geq 0$, hence $-sR \sin \theta \leq 0$, and so $e^{-sR \sin \theta} \leq 1$. Write z on σ_R, as $z = Re^{i\theta} = R(\cos \theta + i \sin \theta)$, where $0 \leq \theta \leq \pi$. Then

$$(7) \qquad \left|e^{isz}\right| = \left|e^{isR(\cos\theta + i\sin\theta)}\right| = \overbrace{\left|e^{isR\cos\theta}\right|}^{=1} \left|e^{-sR\sin\theta}\right| = e^{-sR\sin\theta} \leq 1.$$

Hence, for $R > a$ and z on the semi-circle σ_R, we have

$$\left| \frac{e^{isz}}{z^2 + a^2} \right| \leq \frac{1}{|z^2 + a^2|} \leq \frac{1}{|z|^2 - a^2} = \frac{1}{R^2 - a^2},$$

and so by the inequality for path integrals (Theorem 2, Section 3.2)

$$\left| \int_{\sigma_R} \frac{e^{isz}}{z^2 + a^2} \, dz \right| \leq l(\sigma_R) \frac{1}{R^2 - a^2} = \frac{\pi R}{R^2 - a^2} \to 0, \text{ as } R \to \infty.$$

This estimate works because the degree of the polynomial in the denominator is two greater than that in the numerator.

Step 4: Compute the desired improper integral. Let $R \to \infty$ in (6), use Step 3, and get

$$\lim_{R \to \infty} I_{\sigma_R} + \lim_{R \to \infty} I_R = 0 + \int_{-\infty}^{\infty} \frac{e^{isx}}{x^2 + a^2} \, dx = \frac{\pi}{a} e^{-sa}.$$

Taking real parts on both sides and using (4), we get the desired integral

$$\int_{-\infty}^{\infty} \frac{\cos sx}{x^2 + a^2} \, dx = \frac{\pi}{a} e^{-sa}. \qquad \blacksquare$$

By simply observing that $\sin \theta \geq 0$ for $0 \leq \theta \leq \pi$, we were able in (7) to obtain the inequality $\left| e^{isz} \right| \leq 1$ for all $s \geq 0$ and $z = Re^{i\theta}$, $0 \leq \theta \leq \pi$. As we will see in the next lemma, a more precise analysis of $\sin \theta$ yields a far better estimate on $\left| e^{isz} \right|$, which in turn will make it possible to compute integrals of the form (1), where degree $q(x) \geq 1 + \text{degree } p(x)$.

**LEMMA 1
JORDAN'S LEMMA**

> Let $f(z)$ be a continuous complex-valued function defined on the circular arcs σ_R consisting of all $z = Re^{i\theta}$, where $R \geq R_0$ (R_0 fixed) and $0 \leq \theta_1 \leq \theta \leq \theta_2 \leq \pi$ (Figure 3), and let $M(R)$ denote the maximum value of $|f(z)|$ for z on σ_R. If $\lim_{R \to \infty} M(R) = 0$, then for all $s > 0$
>
> $$\lim_{R \to \infty} \int_{\sigma_R} e^{isz} f(z) \, dz = 0.$$

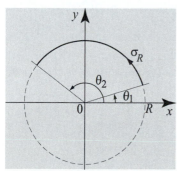

Figure 3

Proof Given $s > 0$, we have from (7), $\left| e^{is\theta} \right| = e^{-sR \sin \theta}$. Note that since $e^{-sR \sin \theta} > 0$, its integral increases if we increase the size of the interval of integration. Thus $\int_{\theta_1}^{\theta_2} e^{-sr \sin \theta} \, d\theta \leq \int_0^{\pi} e^{-sr \sin \theta} \, d\theta$, if $0 \leq \theta_1 \leq \theta_2 \leq \pi$. Parametrize σ_R by $\gamma(\theta) = Re^{i\theta}$, where $\theta_1 \leq \theta \leq \theta_2$. Then $\gamma'(\theta) = Rie^{i\theta}$, $|\gamma'(\theta)| \, d\theta = R \, d\theta$, and hence

$$\left| \int_{\sigma_R} e^{isz} f(z) \, dz \right| \leq \int_{\theta_1}^{\theta_2} \left| e^{isz} f(z) \right| R \, d\theta$$

$$\leq R M(R) \int_{\theta_1}^{\theta_2} e^{-sR \sin \theta} \, d\theta \leq R M(R) \int_0^{\pi} e^{-sR \sin \theta} \, d\theta$$

$$(8) \qquad = R M(R) \left(\int_0^{\frac{\pi}{2}} e^{-sR \sin \theta} \, d\theta + \int_{\frac{\pi}{2}}^{\pi} e^{-sR \sin \theta} \, d\theta \right).$$

In the second integral, make the change of variables $t = \pi - \theta$, $dt = -d\theta$, and notice that $\sin\theta = \sin(\pi - t) = \sin t$. This transforms the integral into $-\int_{\frac{\pi}{2}}^{0} e^{-sR\sin t}\, dt$, which is the same as the first integral in (8); and so from (8) we get

$$(9) \qquad \left| \int_{\sigma_R} e^{isz} f(z)\, dz \right| \leq 2R\, M(R) \int_0^{\frac{\pi}{2}} e^{-sR\sin\theta}\, d\theta.$$

At this point, we need an estimate on $\sin\theta$. On the interval $[0, \frac{\pi}{2}]$, the graph of $\sin\theta$ is concave down, because the second derivative is negative. Hence the graph of $\sin\theta$ for $0 \leq \theta \leq \frac{\pi}{2}$ is above the chordal line that joins any two points on the graph. In particular, it is above the chord that joins the origin to the point $(\frac{\pi}{2}, 1)$, whose equation is $y = \frac{2}{\pi}\theta$. This fact is expressed analytically by the inequality

$$(10) \qquad \sin\theta \geq \frac{2}{\pi}\theta, \qquad 0 \leq \theta \leq \frac{\pi}{2},$$

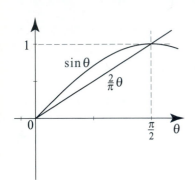

Figure 4
$\sin\theta \geq \frac{2}{\pi}\theta$, $0 \leq \theta \leq \frac{\pi}{2}$.

and is illustrated in Figure 4. From this inequality, it follows immediately that $e^{-sR\sin\theta} \leq e^{-sR\frac{2}{\pi}\theta}$, and so from (9)

$$\left| \int_{\sigma_R} e^{isz} f(z)\, dz \right| \;\leq\; 2R\, M(R) \int_0^{\frac{\pi}{2}} e^{-\frac{2sR}{\pi}\theta}\, d\theta = -2R\, M(R)\frac{\pi}{2sR} e^{-\frac{2sR}{\pi}\theta} \Big|_0^{\frac{\pi}{2}}$$

$$= \frac{\pi}{s} M(R)\left(1 - e^{-sR}\right) \to 0, \quad \text{as} \quad R \to \infty. \qquad \blacksquare$$

Note that an analog of Jordan's lemma will hold for $s < 0$ if the circular arc σ_R is in the lower half-plane.

Applying Jordan's lemma in the special case when $f(z)$ is a rational function, we obtain the following useful result.

COROLLARY 1

Let $f(z) = \frac{p(z)}{q(z)}$, where degree $q(z) \geq 1 + $ degree $p(z)$, Let σ_R denote the semi-circular arc consisting of all $z = Re^{i\theta}$, where $0 \leq \theta \leq \pi$. Then $\lim_{R\to\infty} \int_{\sigma_R} e^{isz} f(z)\, dz = 0$, for all $s > 0$.

Proof   Let $M(R)$ denote the maximum of $|p(z)/q(z)|$ for z on σ_R. Since degree $q(z) \geq 1 + $ degree $p(z)$, $M(R) \to 0$ as $R \to \infty$. Now apply Jordan's lemma. $\blacksquare$

We will next evaluate the improper integral

$$\int_{-\infty}^{\infty} \frac{x\sin x}{x^2 + a^2}\, dx.$$

It is not difficult to show that this integral is convergent using integration by parts (Exercise 19). However, because the degree of $x^2 + a^2$ is only one more than the degree of x, the estimate in Step 3 of Example 1 will not be sufficient to show that the integral on the expanding semi-circle tends to 0. For this purpose, we will appeal to Jordan's lemma.

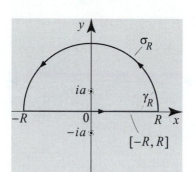

Figure 5 The path and poles for the contour integral in Example 2.

EXAMPLE 2   Applying Jordan's lemma

Derive the identity

$$\int_{-\infty}^{\infty} \frac{x \sin x}{x^2 + a^2}\, dx = \frac{\pi}{e^a}, \quad a \text{ any real number } > 0.$$

Solution   Consider the contour integral over the closed semi-circle γ_R

$$(11) \qquad I_{\gamma_R} = \int_{\sigma_R} \frac{z}{z^2 + a^2} e^{iz}\, dz + \int_{-R}^{R} \frac{x}{x^2 + a^2} e^{ix}\, dx = I_{\sigma_R} + I_R,$$

(Figure 5). By Jordan's lemma (Corollary 1), $\lim_{R \to \infty} I_{\sigma_R} = 0$. For $R > a$, $\frac{z}{z^2 + a^2} e^{iz}$ has a simple pole inside γ_R at ia. By the residue theorem, for all $R > a$,

$$I_{\gamma_R} = 2\pi i \operatorname{Res}\left(\frac{z e^{iz}}{z^2 + a^2},\, ia\right) = 2\pi i\, \frac{(ia) e^{i(ia)}}{2(ia)} = \frac{\pi}{e^a} i.$$

Taking the limit as $R \to \infty$ in (11) and using the fact that $I_{\sigma_R} \to 0$, we get

$$\frac{\pi}{e^a} i = \int_{-\infty}^{\infty} \frac{x}{x^2 + a^2} e^{ix}\, dx = \int_{-\infty}^{\infty} \frac{x \cos x}{x^2 + a^2}\, dx + i \int_{-\infty}^{\infty} \frac{x \sin x}{x^2 + a^2}\, dx,$$

and the desired identity follows upon taking imaginary parts on both sides.   ∎

Indenting Contours

If we let $a \to 0$ in Example 2 and do not worry about justifying taking the limit inside the integral, we get the formula

$$\int_{-\infty}^{\infty} \frac{\sin x}{x}\, dx = \pi.$$

This answer is indeed correct and yields an important integral that arises in many applications. In the remainder of this section, we will develop a method to calculate this and other interesting integrals.

We need to return to the idea of Cauchy principal values of integrals. In Section 5.3 we defined the Cauchy principal value of an improper integral over the real line. However, an integral can also be improper if the integrand becomes unbounded at a point inside the interval of integration. To make our discussion concrete, consider

$$\int_{-1}^{1} f(x)\, dx,$$

where $f(x)$ is continuous on $[-1, 0)$ and $(0, 1]$ but might have infinite limits as x approaches 0 from the left or right. Such an integral is said to be **convergent** if both $\lim_{b \to 0^-} \int_{-1}^{b} f(x)\, dx$ and $\lim_{a \to 0^+} \int_{a}^{1} f(x)\, dx$ are convergent. In this case, we set (Figure 6)

Figure 6 Splitting an improper integral.

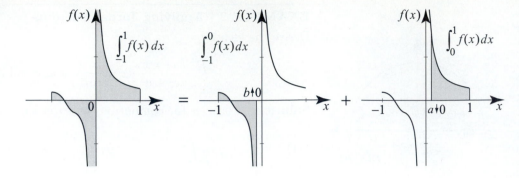

$$(12) \qquad \int_{-1}^{1} f(x)\,dx = \lim_{b \to 0^-} \int_{-1}^{b} f(x)\,dx + \lim_{a \to 0^+} \int_{a}^{1} f(x)\,dx.$$

This expression should be contrasted with the following, which integrates the function on intervals that approach the singular point $x = 0$ in a symmetric fashion. We define the Cauchy principal value of the integral $\int_{-1}^{1} f(x)\,dx$, with a singularity at $x = 0$, to be (Figure 7)

$$(13) \qquad \text{P.V.} \int_{-1}^{1} f(x)\,dx = \lim_{r \to 0^+} \left(\int_{-1}^{-r} f(x)\,dx + \int_{r}^{-1} f(x)\,dx \right).$$

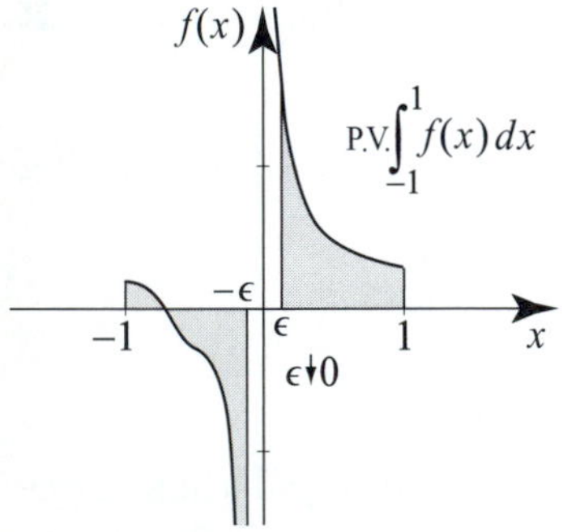

Figure 7 The Cauchy principal value.

The Cauchy principal value of an integral may exist even though the integral itself is not convergent. For example, $\int_{-1}^{-r} \frac{dx}{x} + \int_{r}^{1} \frac{dx}{x} = 0$ for all $r > 0$, so P.V. $\int_{-1}^{1} \frac{dx}{x} = 0$, but the integral itself is clearly not convergent since $\int_{0}^{1} \frac{dx}{x} = \infty$.

However, whenever $\int_{-1}^{1} f(x)\,dx$ is convergent, then P.V. $\int_{-1}^{1} f(x)\,dx$ exists, and the two integrals will be the same, since we can split the limit of a sum in (13) into a sum of limits and recover (12). This fact allows us to compute convergent integrals by computing their principal values, which will be advantageous when using complex integration.

The extension of principal value integrals to more complicated situations is as follows: To compute the principal value of an integral where the integrand has possible singularities at $x_1, x_2, \ldots, x_n$ and the interval of integration may be infinite in both directions, we take limits to approach each singularity (and to the infinite interval) one at a time, yet approach each singularity (or infinity) in a symmetric fashion. If all these limits exist, the principal value exists.

The preceding definitions and formulas for principal value integrals also make sense for complex-valued functions integrated over real intervals.

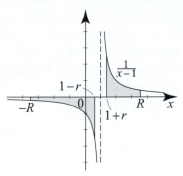

Figure 8 The principal value of the integral in Example 3.

EXAMPLE 3   Cauchy principal values and singular points

Write down P.V. $\int_{-\infty}^{\infty} \frac{1}{x-1} dx$ in terms of limits of integrals on finite intervals.

Solution The integral is improper because it extends over the infinite real line and because the integrand is singular at $x = 1$. Accordingly, the principal value involves two limits; it is

$$(14) \quad \text{P.V.} \int_{-\infty}^{\infty} \frac{1}{x-1} dx = \lim_{R \to \infty} \left(\int_{-R}^{0} \frac{1}{x-1} dx + \int_{2}^{R} \frac{1}{x-1} dx \right)$$

$$+ \lim_{r \to 0^+} \left(\int_{0}^{1-r} \frac{1}{x-1} dx + \int_{1+r}^{2} \frac{1}{x-1} dx \right),$$

where the choices $x = 0$ and $x = 2$ were arbitrary; any pair of numbers where the first is in $(-\infty, 1)$ and the second is in $(1, \infty)$ will work to split the integrals. Using elementary methods, we can see that both limits on the right of (14) exist, so the principal value exists, and we can write (Figure 8)

$$\text{P.V.} \int_{-\infty}^{\infty} \frac{1}{x-1} dx = \lim_{R \to \infty, r \to 0^+} \left(\int_{-R}^{1-r} \frac{1}{x-1} dx + \int_{1+r}^{R} \frac{1}{x-1} dx \right),$$

where the limits may be taken in any manner. Using basic calculus, we find that for large $R > 1$

$$\int_{-R}^{1-r} \frac{1}{x-1} dx = \ln |x-1| \Big|_{-R}^{1-r} = \ln(r) - \ln(R+1);$$

and

$$\int_{1+r}^{R} \frac{1}{x-1} dx = \ln |x-1| \Big|_{1+r}^{R} = \ln(R-1) - \ln(r).$$

So

$$\text{P.V.} \int_{-\infty}^{\infty} \frac{1}{x-1} dx = \lim_{R \to \infty} [\ln(R-1) - \ln(R+1)] = \lim_{R \to \infty} \ln \left(\frac{R-1}{R+1} \right) = 0. \quad \blacksquare$$

Now, we return to the integral $\int_{-\infty}^{\infty} \frac{\sin x}{x} dx$. Since the integrand is well behaved at $x = 0$ and the integral over the infinite interval converges, we can compute it by taking the imaginary part of P.V. $\int_{-\infty}^{\infty} \frac{e^{ix}}{x} dx$. This principal value integral involves the limit as $r \to 0^+$ and $R \to \infty$ of an integral over $[-R, -r]$ and $[r, R]$. We envision the integral as a contour integral of the complex function $\frac{e^{iz}}{z}$ and close the contour with a large positively oriented semicircle of radius R and a small negatively oriented semicircle of radius r, which will bypass the problem at 0. The resulting path $\gamma_{r,R}$ is shown in Figure 9. The limit of the path integral over the small circle will be computed using the following lemma.

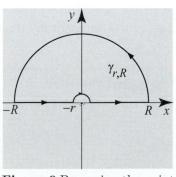

Figure 9 Bypassing the point 0 with the contour $\gamma_{r,R}$.

LEMMA 2
SHRINKING PATH
LEMMA

Suppose that $f(z)$ is a continuous complex-valued function on a closed disk $S_{r_0}(z_0)$ with center at z_0 and radius r_0. For $0 < r \le r_0$, let σ_r denote the positively oriented circular arc at angle α (Figure 10), consisting of all $z = z_0 + re^{i\theta}$, where $\theta_0 \le \theta \le \theta_0 + \alpha$, θ_0 and α are fixed, and $\alpha \ne 0$. Then

$$(15) \qquad \lim_{r \to 0^+} \frac{1}{i\alpha} \int_{\sigma_r} \frac{f(z)}{z - z_0} \, dz = f(z_0).$$

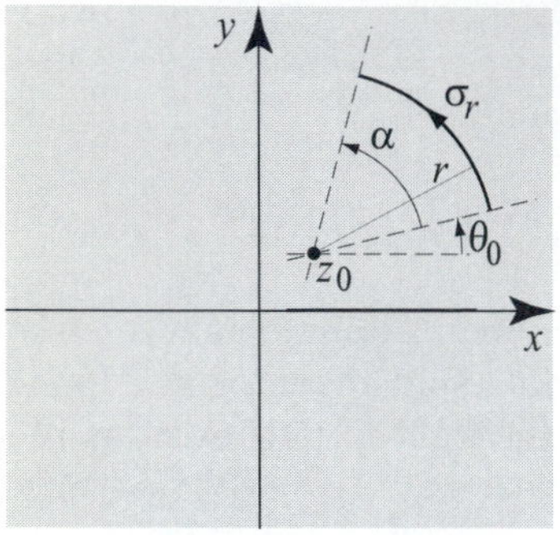

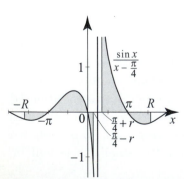

Proof Parametrize the integral in (15) using $z = z_0 + re^{i\theta}$, where $\theta_0 \le \theta \le \theta_0 + \alpha$, $dz = rie^{i\theta} \, d\theta$, $z - z_0 = re^{i\theta}$. Then

$$\frac{1}{i\alpha} \int_{\sigma_r} \frac{f(z)}{z - z_0} \, dz = \frac{1}{i\alpha} \int_{\theta_0}^{\theta_0 + \alpha} \frac{f(z_0 + re^{i\theta})}{re^{i\theta}} \, ire^{i\theta} d\theta = \frac{1}{\alpha} \int_{\theta_0}^{\theta_0 + \alpha} f(z_0 + re^{i\theta}) d\theta.$$

Let $F(r) = \frac{1}{\alpha} \int_{\theta_0}^{\theta_0 + \alpha} f(z_0 + re^{i\theta}) d\theta$. Since $(r, \theta) \mapsto f(z_0 + re^{i\theta})$ is continuous for all $0 \le r \le r_0$ and all θ, it follows from Theorem 5(i), Section 3.5, that $F(r)$ is continuous on $[0, r_0]$. Consequently, $\lim_{r \to 0^+} F(r) = F(0) = \frac{1}{\alpha} \int_{\theta_0}^{\theta_0 + \alpha} f(z_0) d\theta = f(z_0)$, which is equivalent to (15). ∎

Figure 10 Circular arc at angle α.

The following simple consequence of Lemma 2 is useful.

COROLLARY 2

Suppose that $g(z)$ has a simple pole at z_0 and for $0 < r \le r_0$, let σ_r denote the circular arc at angle α (Figure 10). Then

$$(16) \qquad \lim_{r \to 0^+} \int_{\sigma_r} g(z) \, dz = i\alpha \operatorname{Res}(g(z), z_0).$$

Proof Let $f(z) = (z - z_0)g(z)$ for $z \ne z_0$ and define $f(z_0) = \lim_{z \to z_0}(z - z_0)g(z) = \operatorname{Res}(g(z), z_0)$. Since $g(z)$ has a simple pole at z_0, it follows from Theorem 7, Section 4.6, that $f(z)$ is analytic at z_0. Now apply Lemma 2 to $f(z)$ and (16) follows from (15), since for $z \ne z_0$, $f(z)/(z - z_0) = g(z)$. ∎

The improper integrals that we consider next differ from previous ones in that they are not always convergent. However, their Cauchy principal values do exist. Integrals of this sort are important and arise naturally, for example, when computing Hilbert transforms.

EXAMPLE 4   Cauchy principal value: Use of indented contours.

Derive the integral identities

$$\text{P.V.} \int_{-\infty}^{\infty} \frac{\sin x}{x - a} \, dx = \pi \cos a \quad \text{and} \quad \text{P.V.} \int_{-\infty}^{\infty} \frac{\cos x}{x - a} \, dx = -\pi \sin a, \quad -\infty < a < \infty.$$

In particular, when $a = 0$, the first integral yields $\int_{-\infty}^{\infty} \frac{\sin x}{x} \, dx = \pi$.

Solution   The integrals are improper because the interval of integration is infinite

Figure 11 Principal value integral.

and because of the problem at $x = a$. The first integral's principal value is

$$\text{P.V.} \int_{-\infty}^{\infty} \frac{\sin x}{x - a}\, dx = \lim_{\substack{r \to 0^+ \\ R \to \infty}} \left(\int_{-R}^{a-r} \frac{\sin x}{x - a}\, dx + \int_{a+r}^{R} \frac{\sin x}{x - a}\, dx \right)$$

(Figure 11 represents the case $a > 0$). Consider the integral of $f(z) = \frac{e^{iz}}{z-a}$ over the closed contour $\gamma_{r,R}$ in Figure 12, where we have indented the contour around $x = a$. The larger semi-circle σ_R has radius R and positive orientation. The smaller semi-circle has radius r; it is negatively oriented and we shall call it $-\sigma_r$. Since $f(z)$ is analytic on and inside $\gamma_{r,R}$, Cauchy's theorem implies that

$$\begin{aligned}
0 &= \int_{-R}^{a-r} \frac{e^{ix}}{x-a}\, dx + \int_{-\sigma_r} \frac{e^{iz}}{z-a}\, dz + \int_{a+r}^{R} \frac{e^{ix}}{x-a}\, dz + \int_{\sigma_R} \frac{e^{iz}}{z-a}\, dz \\
&= I_1 + I_2 + I_3 + I_4.
\end{aligned}$$

By Jordan's lemma, $I_4 \to 0$ as $R \to \infty$. By the shrinking path lemma, $I_2 \to -i\pi e^{ia}$ as $r \to 0$ (note the negative sign due to the fact that the path of integration is negatively oriented). Thus

$$\lim_{\substack{r \to 0 \\ R \to \infty}} \left(\int_{-R}^{a-r} \frac{e^{ix}}{x-a}\, dx + \int_{a+r}^{R} \frac{e^{ix}}{x-a}\, dx \right) = i\pi e^{ia} = \pi(-\sin a + i\cos a),$$

and the desired integral identities follow upon taking real and imaginary parts. ∎

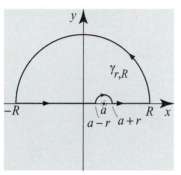

Figure 12 Indenting a contour around a.

Exercises 5.4

In Exercises 1–10, evaluate the given improper integral. Make sure to outline the steps in your solutions. Carry out the details of the solution even if the integral follows from general formulas that we derived previously.

1. $\displaystyle\int_{-\infty}^{\infty} \frac{\cos 4x}{x^2 + 1}\, dx = \pi e^{-4}.$

2. $\displaystyle\int_{-\infty}^{\infty} \frac{\sin \frac{\pi x}{2}}{x^2 + 2x + 4}\, dx = -\frac{\pi}{\sqrt{3}} e^{-\frac{\sqrt{3}}{2}\pi}.$

3. $\displaystyle\int_{-\infty}^{\infty} \frac{x \sin 3x}{x^2 + 2}\, dx = e^{-3\sqrt{2}\pi}.$ **4.** $\displaystyle\int_{-\infty}^{\infty} \frac{\sin \pi x}{x^2 + 2x + 4}\, dx = 0.$

5. $\displaystyle\int_{-\infty}^{\infty} \frac{x^2 \cos 2x}{(x^2 + 1)^2}\, dx = -\frac{\pi}{2} e^{-2}.$ **6.** $\displaystyle\int_{-\infty}^{\infty} \frac{\cos x}{(x^2 + 1)^2}\, dx = \frac{\pi}{e}.$

7. $\displaystyle\int_{-\infty}^{\infty} \frac{\cos(a(x - b))}{x^2 + c^2}\, dx = \frac{\pi}{c} e^{-|a|c} \cos(ab) \quad (a \text{ real, } c > 0).$

8. $\displaystyle\int_{-\infty}^{\infty} \frac{\cos(4\pi x)}{2x^2 + x + 1}\, dx = -2 e^{-\sqrt{2}\pi} \frac{\pi}{\sqrt{7}}.$

9. $\displaystyle\int_{-\infty}^{\infty} \frac{x \cos \pi x}{x^2 + x + 9}\, dx = \pi e^{-\frac{\sqrt{35}}{2}\pi}.$

10. $\displaystyle\int_{-\infty}^{\infty} \frac{\sin \pi x}{x^2 + x + 1}\, dx = -\frac{2\pi}{\sqrt{3}} e^{-\frac{\sqrt{3}}{2}\pi}.$

In Exercises 11–18, use an indented contour to evaluate the Cauchy principal value of the given improper integral.

11. P.V. $\displaystyle\int_{-\infty}^{\infty} \frac{\sin x \cos x}{x}\, dx = \frac{\pi}{2}$.

12. P.V. $\displaystyle\int_{0}^{\infty} \frac{\sin x \cos 2x}{x}\, dx = 0$.

13. P.V. $\displaystyle\int_{-\infty}^{\infty} \frac{1 - \cos x}{x^2}\, dx = \pi$.

14. P.V. $\displaystyle\int_{-\infty}^{\infty} \frac{2x \sin x}{x^2 - a^2}\, dx = 2\pi a$.

15. P.V. $\displaystyle\int_{-\infty}^{\infty} \frac{\sin^2 x}{x^2}\, dx = \pi$.

16. P.V. $\displaystyle\int_{-\infty}^{\infty} \frac{\sin ax}{x - b}\, dx = \pi \cos ab$ and P.V. $\displaystyle\int_{-\infty}^{\infty} \frac{\cos ax}{x - b}\, dx = -\pi \sin ab$.

17. P.V. $\displaystyle\int_{-\infty}^{\infty} \frac{\sin x}{x(x^2 + 1)}\, dx = \pi\left(1 - \frac{1}{e}\right)$.

18. P.V. $\displaystyle\int_{-\infty}^{\infty} \frac{\cos x}{x^2 - a^2}\, dx = -\pi \frac{\sin a}{a}$ $(a \neq 0)$.

19. Show that the improper integral

$$\int_{0}^{\infty} \frac{x \sin x}{x^2 + a^2}\, dx$$

is convergent. [Hint: For $A > 0$, integrate $\int_{0}^{A} \frac{x \sin x}{x^2 + a^2}\, dx$ by parts by letting $u = \frac{x}{x^2 + a^2}$, $dv = \sin x\, dx$.]

20. Show that the improper integral $\int_{0}^{\infty} \frac{\sin x}{x}\, dx$ is convergent. [Hint: For $A > 0$, integrate $\int_{0}^{A} \frac{\sin x}{x}\, dx$ by parts by letting $u = \frac{1}{x}$, $dv = \sin x$, $du = -\frac{dx}{x^2}$ and $v = 1 - \cos x$.]

21. (a) Use Example 3 and a suitable change of variables to establish the formula

$$\frac{2}{\pi} \int_{0}^{\infty} \frac{\sin ax}{x}\, dx = \operatorname{sgn} a,$$

where $\operatorname{sgn} a = -1$ if $a < 0$, 0 if $a = 0$, and 1 if $a > 0$.

(b) Use (a) and a suitable trigonometric identity to prove that

$$\int_{0}^{\infty} \frac{\sin ax \cos bx}{x}\, dx = \begin{cases} 0 & \text{if } 0 < a < b, \\ \frac{\pi}{4} & \text{if } a = b > 0, \\ \frac{\pi}{2} & \text{if } 0 < b < a. \end{cases}$$

22. Use Example 1 to establish the identity

$$\int_{-\infty}^{\infty} \frac{\cos sx}{x^2 + a^2}\, dx = \frac{\pi}{|a|} e^{-|sa|}, \qquad -\infty < s < \infty,\ a \neq 0.$$

23. Use a suitable change of variables in Example 2 to establish the identity

$$\int_{-\infty}^{\infty} \frac{x \sin sx}{x^2 + a^2}\, dx = \pi \operatorname{sgn}(s)e^{-|as|}.$$

24. Project Problem: Fourier transforms of the hyperbolic secant and cosecant. In this exercise, we derive the identities

(17) $$\int_0^\infty \frac{\cos(wx)}{\cosh(\pi x)}\, dx = \frac{1}{2}\operatorname{sech}\frac{w}{2} \quad \text{and} \quad \int_0^\infty \frac{\sin(wx)}{\sinh(\pi x)}\, dx = \frac{1}{2}\tanh\frac{w}{2},$$

where w is a real number. Up to a constant multiple, these integrals give the Fourier cosine transform of the hyperbolic secant $1/\cosh(\pi x)$ and the Fourier sine transform of the hyperbolic cosecant $1/\sinh(\pi x)$.

(a) Let $f(z) = \frac{e^{(\pi+iw)z}}{e^{2\pi z}+1}$ and let $\gamma_{\epsilon,R}$ denote the indented contour in Figure 13. Show that

$$I_{\gamma_{\epsilon,R}} = \int_{\gamma_{\epsilon,R}} f(z)\, dz = 0.$$

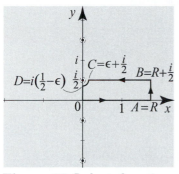

Figure 13 Indented contour for Exercise 24.

(b) Show that $\lim_{R\to\infty} I_{AB} = 0$. [Hint: This part involves estimates similar to those in Step 5, Example 3, Section 5.3.]

(c) Show that

$$I_{OA} = \frac{1}{2}\int_0^R \frac{e^{iwx}}{\cosh(\pi x)}\, dx,$$

$$I_{DO} = -\frac{i}{2}\int_0^{\frac{1}{2}-\epsilon} \frac{e^{-wy}}{\cos(\pi y)}\, dy, \quad \text{and} \quad I_{BC} = i\frac{e^{-\frac{w}{2}}}{2}\int_\epsilon^R \frac{e^{iwx}}{\sinh(\pi x)}\, dx.$$

(d) Use Corollary 2 to show that $\lim_{\epsilon\to 0} I_{CD} = -\frac{1}{4}e^{-\frac{w}{2}}$.

(e) By taking the limit as $R\to\infty$ then $\epsilon\to 0$ and using (a), conclude that

$$\frac{1}{2}\int_0^\infty \frac{e^{iwx}}{\cosh(\pi x)}\, dx - \frac{1}{4}e^{-\frac{w}{2}} + \lim_{\epsilon\to 0}\left[i\frac{e^{-\frac{w}{2}}}{2}\int_\epsilon^\infty \frac{e^{iwx}}{\sinh(\pi x)}\, dx - \frac{i}{2}\int_0^{\frac{1}{2}-\epsilon} \frac{e^{-wy}}{\cos(\pi y)}\, dy\right] = 0.$$

(f) Let $A = \int_0^\infty \frac{\cos(wx)}{\cosh(\pi x)}\, dx$ and $B = \int_0^\infty \frac{\sin(wx)}{\sinh(\pi x)}\, dx$. Take real part in (e) and conclude that $A - e^{-\frac{w}{2}}B = \frac{1}{2}e^{-\frac{w}{2}}$. Use the fact that A is even in w and B is odd in w, replace w by $-w$, and conclude $A + e^{\frac{w}{2}}B = \frac{1}{2}e^{\frac{w}{2}}$. Solve for A and B and derive the desired identities.

25. Project Problem: Bernoulli numbers and residues. In this exercise, we evaluate a sine integral related to the sine Fourier transform of a hyperbolic function and then use our answer to give an integral representation of the Bernoulli numbers, which are defined in Example 4, Section 4.4.

(a) Follow the steps outlined in the previous exercise as you integrate the function $f(z) = \frac{e^{iwz}}{e^{2\pi z}-1}$ on the indented contour in Figure 14, and obtain the identity

$$\int_0^\infty \frac{\sin wx}{e^{2\pi x}-1}\, dx = \frac{1}{4}\frac{1+e^{-w}}{1-e^{-w}} - \frac{1}{2w}, \quad w \neq 0.$$

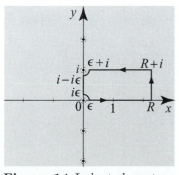

Figure 14 Indented contour for Exercise 25.

(b)   With the help of the Maclaurin series of $\frac{z}{2}\coth\frac{z}{2}$ (see the details following Example 4, Section 4.4), obtain the Laurent series

$$\frac{1}{2}\frac{1+e^{-w}}{1-e^{-w}} - \frac{1}{w} = \sum_{k=1}^{\infty}\frac{B_{2k}}{(2k)!}w^{2k-1}, \qquad |w| < 2\pi.$$

(c)   Replace $\sin wx$ in the integral in (a) by its Taylor series

$$\sin wx = \sum_{k=1}^{\infty}(-1)^{k-1}\frac{w^{2k-1}x^{2k-1}}{(2k-1)!},$$

interchange order of integration, and conclude that

$$\int_0^\infty \frac{x^{2k-1}}{e^{2\pi x}-1}\,dx = \frac{(-1)^{k-1}}{4k}B_{2k}.$$

(The diligent reader can justify the interchange of the sum and the integral.)

5.5   Advanced Integrals by Residues

In this section, we evaluate some classical integrals that arise in applied mathematics. We will use methods from previous sections; however, each integral in this section will require additional care, either in estimating the integrand or in choosing the contour of integration. The examples are independent of each other.

We start with an integral that arises in many applications, including heat problems and computing the Fourier transform of the function e^{-x^2}. We will need the formula

(1)
$$I = \int_{-\infty}^{\infty} e^{-ax^2}\,dx = \sqrt{\frac{\pi}{a}}, \qquad a > 0.$$

To evaluate the integral, the usual trick is to consider

$$I^2 = \int_{-\infty}^{\infty} e^{-ax^2}\,dx \int_{-\infty}^{\infty} e^{-ay^2}\,dy = \int_{-\infty}^{\infty}\int_{-\infty}^{\infty} e^{-a(x^2+y^2)}\,dx\,dy.$$

Now use polar coordinates, $x = r\cos\theta$, $y = r\sin\theta$, $dx\,dy = r\,dr\,d\theta$. Then

$$I^2 = \int_0^{2\pi} d\theta \int_0^\infty e^{-ar^2}\,rdr = 2\pi\left[-\frac{1}{2a}e^{-ar^2}\right]_{r=0}^\infty = \frac{\pi}{a},$$

which is equivalent to (1).

EXAMPLE 1   Derive the **Poisson integral**

(2)
$$\int_{-\infty}^{\infty} e^{-ax^2}\cos wx\,dx = \sqrt{\frac{\pi}{a}}e^{-\frac{w^2}{4a}}, \qquad a > 0,\ -\infty < w < \infty.$$

Solution   The case $w = 0$ follows from (1). Also, it is enough to deal with the case $w > 0$. Since $\sin wx\, e^{-ax^2}$ is an odd function of x, its integral over symmetric intervals is 0. So

$$\int_{-\infty}^{\infty} \cos wx\, e^{-ax^2}\, dx = \int_{-\infty}^{\infty} e^{iwx} e^{-ax^2}\, dx = \int_{-\infty}^{\infty} e^{-ax^2 + iwx}\, dx.$$

Completing the square in the exponent of the integrand, we get

$$e^{-ax^2 + iwx} = e^{-a(x^2 - i\frac{w}{a}x)} = e^{-a(x - i\frac{w}{2a})^2 + a(i\frac{w}{2a})^2} = e^{-\frac{w^2}{4a}} e^{-a(x - i\frac{w}{2a})^2},$$

and since $e^{-\frac{w^2}{4a}}$ is independent of x,

(3)
$$\int_{-\infty}^{\infty} \cos wx\, e^{-ax^2}\, dx = e^{-\frac{w^2}{4a}} \int_{-\infty}^{\infty} e^{-a(x - i\frac{w}{2a})^2}\, dx.$$

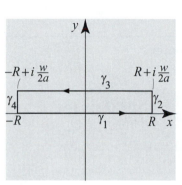

Figure 1 The rectangular contour for Example 1.

To evaluate the integral on the right, we will integrate $f(z) = e^{-a(z - i\frac{w}{2a})^2}$ over the rectangular contour in Figure 1. Let I_j ($j = 1, 2, 3, 4$) denote the integral of $f(z)$ over the path γ_j as indicated in Figure 1. The choice of a rectangular contour should not surprise you in view of our work on path integrals involving exponential functions in Section 5.3. The choice of the y-intercept is related to the shift in the exponent in $f(z)$. It will be justified as we compute I_3.

Since $f(z)$ is entire, Cauchy's theorem implies that

(4)
$$I_1 + I_2 + I_3 + I_4 = 0.$$

We have

$$I_1 = \int_{-R}^{R} e^{-a(x - i\frac{w}{2a})^2}\, dx \rightarrow \int_{-\infty}^{\infty} e^{-a(x - i\frac{w}{2a})^2}\, dx, \text{ as } R \rightarrow \infty.$$

For I_3, we have $z = x + i\frac{w}{2a}$, $dz = dx$, and so

$$I_3 = \int_{R}^{-R} e^{-ax^2}\, dx = -\int_{-R}^{R} e^{-ax^2}\, dx \rightarrow -\sqrt{\frac{\pi}{a}}, \text{ as } R \rightarrow \infty,$$

by (1). For I_2, $z = R + iy$, $0 \le y \le \frac{w}{2a}$, $dz = i\, dy$, and

$$e^{-a(z - i\frac{w}{2a})^2} = e^{-a(R + i(y - \frac{w}{2a}))^2} = e^{-a\left(R^2 - (y - \frac{w}{2a})^2\right)} e^{-2aRi(y - \frac{w}{2a})}.$$

Since $\left| e^{-2aRi(y - \frac{w}{2a})} \right| = 1$, for $0 \le y \le \frac{w}{2a}$,

$$\left| e^{-a(z - \frac{iw}{2a})^2} \right| = e^{-a\left(R^2 - (y - \frac{w}{2a})^2\right)} = e^{-aR^2} e^{a(\frac{w}{2a} - y)^2} \le e^{-aR^2} e^{a(\frac{w}{2a})^2},$$

we obtain using the usual inequality for path integrals,

$$|I_2| = \left| \int_{0}^{\frac{w}{2a}} e^{-a(R + i(y - \frac{w}{2a}))^2} i\, dy \right| \le \int_{0}^{\frac{w}{2a}} \left| e^{-a(R + i(y - \frac{w}{2a}))^2} \right| dy$$

$$\le e^{-aR^2} e^{a(\frac{w}{2a})^2} \frac{w}{2a} \rightarrow 0 \text{ as } R \rightarrow \infty.$$

Similarly, $I_4 \to 0$ as $R \to \infty$. Taking the limit as $R \to \infty$ in (4) and using what we know about the limits of the I_j, we get

$$0 = \int_{-\infty}^{\infty} e^{-a(x - i\frac{w}{2a})^2}\, dx - \sqrt{\frac{\pi}{a}} \quad \text{or} \quad \int_{-\infty}^{\infty} e^{-a(x - i\frac{w}{2a})^2}\, dx = \sqrt{\frac{\pi}{a}},$$

and (2) follows upon using (3).   ■

Integrals Involving Branch Cuts

The remaining examples illustrate the useful technique of integration around branch points and branch cuts.

EXAMPLE 2   Integrating around a branch point

For $0 < \alpha < 1$, derive the integral identities

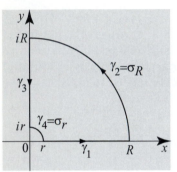

Figure 2 The contour for Example 1.

$$(5) \quad \int_0^{\infty} \frac{\cos x}{x^\alpha}\, dx = \Gamma(1 - \alpha) \sin \frac{\alpha\pi}{2} \quad \text{and} \quad \int_0^{\infty} \frac{\sin x}{x^\alpha}\, dx = \Gamma(1 - \alpha) \cos \frac{\alpha\pi}{2},$$

where $\Gamma(z) = \int_0^{\infty} e^{-t} t^{z-1}\, dt$ is the gamma function (Exercise 24, Section 4.3).

Solution   For the given integrands, it is natural to consider the function $f(z) = \frac{e^{iz}}{z^\alpha}$. The only trouble is that z^α is multiple-valued and so we need to specify a single-valued branch of z^α. The choice of this branch is usually affected by the choice of the contour of integration. In this case, we will use the contour γ in Figure 2, and choose a branch of $z^\alpha = e^{\alpha \log z}$ that equals the real-numbered power x^α on the x-axis and is analytic on the contour. There are several possibilities, but clearly the easiest one would be $z^\alpha = e^{\alpha \,\mathrm{Log}\, z}$ where $\mathrm{Log}\, z$ denotes the principal value branch of the logarithm, with a branch cut on the negative x-axis. Recall that for $z \ne 0$, $\mathrm{Log}\, z = \ln |z| + i \,\mathrm{Arg}\, z$, where $-\pi < \mathrm{Arg}\, z \le \pi$. Since the function

$$f(z) = \frac{e^{iz}}{z^\alpha} = \frac{e^{iz}}{e^{\alpha \,\mathrm{Log}\, z}}$$

is analytic inside and on the contour γ, Cauchy's theorem implies that

$$(6) \qquad 0 = \int_\gamma \frac{e^{iz}}{e^{\alpha \,\mathrm{Log}\, z}}\, dz = I_1 + I_2 + I_3 + I_4,$$

where I_j is the integral of f over γ_j, as indicated in Figure 2. For I_1, $z = x$, $r < x < R$, $dz = dx$, $f(z) = \frac{e^{ix}}{e^{\alpha \,\mathrm{Log}\, x}} = \frac{e^{ix}}{x^\alpha}$, and so

$$\lim_{\substack{r \to 0 \\ R \to \infty}} I_1 = \lim_{\substack{r \to 0 \\ R \to \infty}} \int_r^R \frac{e^{ix}}{x^\alpha}\, dx = \int_0^{\infty} \frac{e^{ix}}{x^\alpha}\, dx = \int_0^{\infty} \frac{\cos x}{x^\alpha}\, dx + i \int_0^{\infty} \frac{\sin x}{x^\alpha}\, dx.$$

For I_3, $z = iy$, $r < y < R$, $dz = i\, dy$,

$$f(z) = \frac{e^{i(iy)}}{e^{\alpha \,\mathrm{Log}\, (iy)}} = \frac{e^{-y}}{e^{\alpha(\ln y + i \,\mathrm{Arg}\, (iy))}} = \frac{e^{-y}}{y^\alpha e^{i\frac{\alpha\pi}{2}}} = e^{-i\frac{\alpha\pi}{2}} e^{-y} y^{-\alpha},$$

and so

$$\lim_{\substack{r \to 0 \\ R \to \infty}} I_3 = \lim_{\substack{r \to 0 \\ R \to \infty}} ie^{-i\frac{\alpha\pi}{2}} \int_R^r e^{-y} y^{-\alpha} \, dy = -ie^{-i\frac{\alpha\pi}{2}} \int_0^\infty e^{-y} y^{-\alpha} \, dy$$

$$= -ie^{-i\frac{\alpha\pi}{2}} \Gamma(1-\alpha) = -i\Big(\cos\frac{\alpha\pi}{2} - i\sin\frac{\alpha\pi}{2}\Big)\Gamma(1-\alpha),$$

where the second-to-last equality follows from the definition of the gamma function.

We now deal with the integrals I_2 and I_4. For z on the circular arc σ_R, write $z = Re^{i\theta}$, where $0 \le \theta \le \frac{\pi}{2}$. Then

$$|z^\alpha| = \left| e^{\alpha(\ln|z| + i\,\mathrm{Arg}\,z)} \right| = e^{\alpha \ln|z|} = |z|^\alpha = R^\alpha;$$

also, $|e^{iz}| = \left| e^{R(-\sin\theta + i\cos\theta)} \right| = e^{-R\sin\theta}$, and so

$$|f(z)| = \left| \frac{e^{iz}}{z^\alpha} \right| = \frac{e^{-R\sin\theta}}{R^\alpha}.$$

Let $M(R)$ denote the maximum of $|f(z)|$ for z on σ_R. Then $M(R) = \frac{1}{R^\alpha}$, which tends to 0 as $R \to \infty$. Thus $I_2 \to 0$ as $R \to \infty$, by Jordan's lemma, Section 5.4. On I_4, we have $\left| \frac{e^{iz}}{z^\alpha} \right| \le \frac{1}{r^\alpha}$, and using the integral inequality in Theorem 2, Section 3.2, we find

$$|I_4| \le l(\sigma_r)\frac{1}{r^\alpha} = \frac{\pi}{4} r^{1-\alpha}.$$

Since $1 - \alpha > 0$, we see that $|I_4| \to 0$ as $r \to 0$. Going back to (6) and taking the limits as $r \to 0$ and $R \to \infty$, we get

$$0 = \int_0^\infty \frac{\cos x}{x^\alpha} \, dx + i \int_0^\infty \frac{\sin x}{x^\alpha} \, dx - i\Big(\cos\frac{\alpha\pi}{2} - i\sin\frac{\alpha\pi}{2}\Big)\Gamma(1-\alpha),$$

which is equivalent to (5).   ∎

Taking $\alpha = \frac{1}{2}$ in (5) and using $\Gamma(\frac{1}{2}) = \sqrt{\pi}$ (see Exercise 25, Section 4.3), we obtain

$$\int_0^\infty \frac{\cos x}{\sqrt{x}} \, dx = \Gamma(\tfrac{1}{2}) \sin\frac{\pi}{4} = \sqrt{\frac{\pi}{2}} \quad \text{and} \quad \int_0^\infty \frac{\sin x}{\sqrt{x}} \, dx = \Gamma(\tfrac{1}{2}) \cos\frac{\pi}{4} = \sqrt{\frac{\pi}{2}}.$$

Letting $x = u^2$, $dx = 2u \, du$, we get the famous **Fresnel integrals**,

$$\int_0^\infty \cos u^2 \, du = \frac{1}{2}\sqrt{\frac{\pi}{2}} \quad \text{and} \quad \int_0^\infty \sin u^2 \, du = \frac{1}{2}\sqrt{\frac{\pi}{2}},$$

named after the French mathematician and physicist Augustin Fresnel (1788-1827), one of the founders of the wave theory of light. The cosine integral and its convergence are illustrated in Figure 3. The integrals can also be obtained by using a different contour integral (see Exercise 5).

In our next example, we illustrate an important method that is based on the fact that different branches of the logarithm differ by an integer

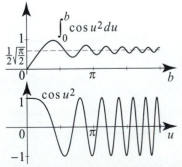

Figure 3 An illustration of the Fresnel cosine integral.

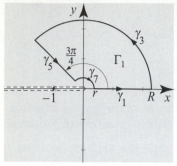

Figure 4 The contour Γ_1 and the branch cut of $f_1(z)$.

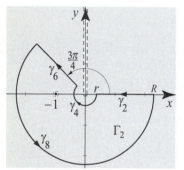

Figure 5 The contour Γ_2 and the branch cut of $f_2(z)$.

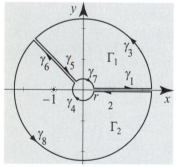

Figure 6 The contours $\Gamma_1 = (\gamma_1, \gamma_3, \gamma_5, \gamma_7)$ and $\Gamma_2 = (\gamma_2, \gamma_4, \gamma_6, \gamma_8)$.

multiple of $2\pi i$. We note that the example can also be evaluated by using the substitution $x = e^t$, as we did with similar integrals in Section 5.3.

EXAMPLE 3   Integrating around a branch cut

For $0 < \alpha < 1$, derive the integral identity

(7)
$$\int_0^\infty \frac{dx}{x^\alpha(x+1)} = \frac{\pi}{\sin \pi\alpha}.$$

Solution   We will integrate branches of $f_1(z) = \frac{1}{z^\alpha(z+1)}$ on the contours $\Gamma_1 = (\gamma_1, \gamma_3, \gamma_5, \gamma_7)$ and $\Gamma_2 = (\gamma_2, \gamma_4, \gamma_6, \gamma_8)$, shown in Figures 4, 5, and 6. On Γ_1, we can take the principal branch of z^α, which will coincide with the real power on the real axis and allow us to recover the integral (7). We cannot use the residue theorem to help integrate this one branch all the way around the origin, because of its branch cut on the negative real axis. Thus we have closed Γ_1, bringing it back at the ray $\theta = \frac{3\pi}{4}$, and will use a different branch of z^α to integrate on Γ_2. For the integral on Γ_2, we choose the branch $z^\alpha = e^{\alpha \log_{\frac{\pi}{2}} z}$. In the second quadrant, this branch coincides with the principal branch (and so integrals over γ_5 and γ_6 will cancel), but the new branch continues to be analytic as we wind around the origin into the third and fourth quadrants. It is important to note that integrals over γ_1 and γ_2 will not cancel because the two branches of z^α are not the same here; the logarithm branches differ by $2\pi i$. With this in mind, let

$$f_1(z) = \frac{1}{e^{\alpha \operatorname{Log} z}(z+1)} \quad \text{and} \quad f_2(z) = \frac{1}{e^{\alpha \log_{\frac{\pi}{2}} z}(z+1)},$$

where $\operatorname{Log} z = \ln|z| + i \operatorname{Arg} z$, $-\pi < \operatorname{Arg} z \leq \pi$, and $\log_{\frac{\pi}{2}} z = \ln|z| + i \arg_{\frac{\pi}{2}} z$, $\frac{\pi}{2} < \arg_{\frac{\pi}{2}} \leq \frac{5\pi}{2}$. We will integrate f_1 on Γ_1 and f_2 on Γ_2. Let I_j denote the integral of the appropriate branch f_1 or f_2 on γ_j. On γ_1, $z = x > 0$, $\operatorname{Log} z = \operatorname{Log} x = \ln x$, $e^{\alpha \ln x} = x^\alpha$, and so

$$I_1 = \int_r^R \frac{dx}{x^\alpha(x+1)};$$

also for $z = x > 0$, $\log_{\frac{\pi}{2}} z = \log_{\frac{\pi}{2}} x = \ln x + 2\pi i$, $e^{\alpha \log_{\frac{\pi}{2}} x} = x^\alpha e^{2\pi i\alpha}$, and so

(8)
$$I_2 = \int_R^r \frac{dx}{e^{2\pi i\alpha}x^\alpha(x+1)} = -e^{-2\pi i\alpha} \int_r^R \frac{dx}{x^\alpha(x+1)} = -e^{-2\pi i\alpha} I_1.$$

For z on γ_5 (and hence γ_6) we have $\operatorname{Log} z = \log_{\frac{\pi}{2}} z$, hence $f_1(z) = f_2(z)$, and consequently $I_5 = -I_6$, and so

(9)
$$I_5 + I_6 = 0.$$

From here on the details of the solution are very much like the previous example. The function $f_1(z)$ is analytic on and inside Γ_1, so by Cauchy's theorem

(10)
$$\int_{\Gamma_1} f_1(z)\, dz = I_1 + I_3 + I_5 + I_7 = 0.$$

The function $f_2(z)$ is analytic on and inside Γ_2 except for a simple pole at $z = -1$, so by the residue theorem

$$(11) \quad \int_{\Gamma_2} f_2(z)\, dz = I_2 + I_4 + I_6 + I_8 = 2\pi i \operatorname{Res}\left(f_2(z),\, -1\right) = \frac{2\pi i}{e^{\alpha \log_{\frac{\pi}{2}}(-1)}} = \frac{2\pi i}{e^{\pi i \alpha}}.$$

Using (8)–(11), we obtain

$$(12) \quad \frac{2\pi i}{e^{\pi i \alpha}} = \int_{\Gamma_1} f_1(z)\, dz + \int_{\Gamma_2} f_2(z)\, dz = (1 - e^{-2\pi i \alpha})I_1 + I_3 + I_4 + I_7 + I_8.$$

We will let $r \to 0$ and $R \to 0$, then $I_1 \to \int_0^\infty \frac{dx}{x^\alpha(x+1)}$. Also, we will show that I_3, I_4, I_7, I_8 tend to 0. Then from (12) it will follow that

$$\frac{2\pi i}{e^{\pi i \alpha}} = (1 - e^{-2\pi i \alpha}) \int_0^\infty \frac{dx}{x^\alpha(x+1)}.$$

Solving for the integral, we find

$$\int_0^\infty \frac{dx}{x^\alpha(x+1)} = \frac{2\pi i}{e^{\pi i \alpha}} \frac{1}{1 - e^{-2\pi i \alpha}} = \pi \frac{2i}{e^{\pi i \alpha} - e^{-\pi i \alpha}} = \frac{\pi}{\sin \pi \alpha},$$

as desired. So let us show that I_3, I_4, I_7, I_8 tend to 0. This part is similar to Example 2; we will just sketch the details. We have $|z^\alpha| = |z|^\alpha$, so $|f_j(z)| \leq \frac{1}{|z|^\alpha |1 - |z||}$ $(j = 1, 2)$. For z on γ_7 or γ_8, $|z| = r$ (we may take $r < 1$), and so

$$|I_7| \leq l(\gamma_7) \frac{1}{r^\alpha(1 - r)} \leq 2\pi r \frac{1}{r^\alpha(1 - r)} = 2\pi \frac{r^{1-\alpha}}{1 - r} \to 0, \text{ as } r \to 0,$$

because $1 - \alpha > 0$. Similarly, $|I_8| \to 0$, as $r \to 0$. For z on γ_3 or γ_4, $|z| = R$ (we may take $R > 1$), and so

$$|I_3| \leq l(\gamma_3) \frac{1}{R^\alpha(R - 1)} \leq 2\pi R \frac{1}{R^\alpha(R - 1)} = 2\pi \frac{R^{1-\alpha}}{R - 1} \to 0, \text{ as } R \to \infty,$$

because the degree of the numerator, $1 - \alpha$, is smaller than the degree of the denominator, which is 1. Similarly, $|I_4| \to 0$, as $R \to \infty$. ∎

In Example 3, the point in introducing different branches of the multiple-valued function $f(z) = \frac{1}{z^\alpha(z+1)}$ was to show explicitly how the function can continuously change as we wind around the origin, and how its value on the real axis is different depending on whether we have gone around the origin or not. Now we will illustrate a useful idea that can be used as an alternative to multiple branches. The idea is to think of a function as taking different values at a point on the branch cut depending on how we approach this point. Consider, for example,

$$(13) \qquad\qquad h(z) = \sqrt{z - a} \qquad (a \text{ real}),$$

where we use the $\log_0$ branch of the square root, so that the branch cut of $h(z)$ is on the part of the positive x-axis $x > a$. Explicitly, $h(z) =$

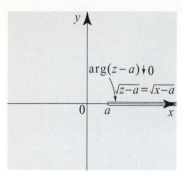

Figure 7 Defining $\sqrt{z-a}$ as z approaches the branch cut from above.

$|z-a|^{\frac{1}{2}}e^{\frac{i}{2}\arg_0(z-a)}$, where $0 < \arg_0(z-a) \le 2\pi$. We will use this function and allow it to take different values on the branch cut, depending on how we approach the branch cut. For example, if z approaches the part of the x-axis $x > a$ from the upper half-plane (Figure 7), then $\arg(z-a)$ approaches 0 and so the values of the function to the right of a and above the x-axis are $\sqrt{z-a} = \sqrt{|x-a|} = \sqrt{x-a}$. If z approaches the part of the x-axis $x > a$ from the lower half-plane (Figure 8), then $\arg(z-a)$ approaches 2π and so the values of the function to the right of a and below the x-axis are taken to be $\sqrt{z-a} = e^{i\frac{1}{2}(2\pi)}\sqrt{|x-a|} = -\sqrt{x-a}$. The mapping $w = h(z)$ is illustrated in Figure 9. Even though we will omit justifying the use of functions with multiple values, this can be done by introducing different branches of the square root, as we did in Example 3.

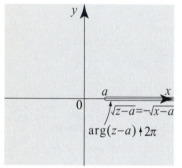

Figure 8 Defining $\sqrt{z-a}$ as z approaches the branch cut from below.

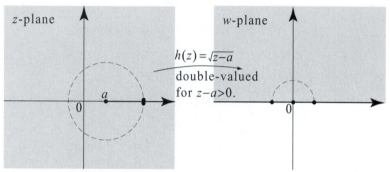

Figure 9 The mapping $w = \sqrt{z-a}$ maps the upper half of the branch cut to the right half of the real axis and the lower half of the branch cut to the left half of the real axis.

In our final example we prove a useful property of **elliptic integrals**, which are integrals of the form $\int \frac{dx}{\sqrt{p(x)}}$, where $p(x)$ is a polynomial of degree ≥ 2. These integrals arise when computing arc length of ellipses, and thus their name. They also arise when computing Schwarz-Christoffel transformations.

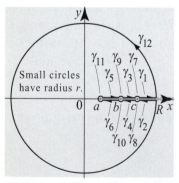

Figure 10 The contour in Example 4 and its twelve components.

EXAMPLE 4   A property of elliptic integrals

Let $a < b < c$ be real numbers. Show that

$$(14) \qquad \int_a^b \frac{dx}{\sqrt{(x-a)(b-x)(c-x)}} = \int_c^\infty \frac{dx}{\sqrt{(x-a)(x-b)(x-c)}}.$$

Solution   Consider the function

$$f(z) = \frac{1}{\sqrt{(z-a)}\sqrt{(z-b)}\sqrt{(z-c)}},$$

where we choose the branches of the square roots with branch cuts along the positive real axis. The function f is analytic for all z except possibly for $z = x \ge a$ (in fact, the singularities are removable along $b < z < c$, see Exercise 23). Since f is

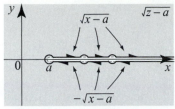

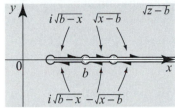

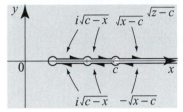

Figure 11 The limiting values of $\sqrt{z-a}$, $\sqrt{z-b}$, and $\sqrt{z-c}$.

analytic inside and on the contour in Figure 10, Cauchy's theorem implies that its integral on this contour is 0. Letting I_j denote the integral of the limiting values of f on γ_j, we obtain

$$(15) \qquad I_1 + I_2 + I_3 + I_4 + I_5 + I_6 + I_7 + I_8 + I_9 + I_{10} + I_{11} + I_{12} = 0.$$

It is now straightforward to show that I_7, I_8, I_9, I_{10}, I_{11}, I_{12} tend to 0 as $r \to 0$ and $R \to \infty$. The details can be found in Examples 2 and 3 and will be omitted. To handle the remaining integrals, we will compute the limiting values of f (denoted still by $f(z)$) in each case with the help of the values in Figure 11:

$$I_1: \quad f(z) = \frac{1}{\sqrt{x-a}\,\sqrt{x-b}\,\sqrt{x-c}} \quad \Rightarrow \quad I_1 = \int_{c+r}^{R} \frac{dx}{\sqrt{x-a}\,\sqrt{x-b}\,\sqrt{x-c}};$$

$$I_2: \quad f(z) = \frac{-1}{\sqrt{x-a}\,\sqrt{x-b}\,\sqrt{x-c}} \quad \Rightarrow \quad I_2 = I_1;$$

$$I_3: \quad f(z) = \frac{-i}{\sqrt{x-a}\,\sqrt{x-b}\,\sqrt{c-x}} \quad \Rightarrow \quad I_3 = \int_{b+r}^{c-r} \frac{(-i)\,dx}{\sqrt{x-a}\,\sqrt{x-b}\,\sqrt{c-x}};$$

$$I_4: \quad f(z) = \frac{-i}{\sqrt{x-a}\,\sqrt{x-b}\,\sqrt{c-x}} \quad \Rightarrow \quad I_4 = -I_3;$$

$$I_5: \quad f(z) = \frac{-1}{\sqrt{x-a}\,\sqrt{b-x}\,\sqrt{c-x}} \quad \Rightarrow \quad I_5 = \int_{a+r}^{b-r} \frac{(-1)\,dx}{\sqrt{x-a}\,\sqrt{b-x}\,\sqrt{c-x}};$$

$$I_6: \quad f(z) = \frac{-1}{\sqrt{x-a}\,\sqrt{b-x}\,\sqrt{c-x}} \quad \Rightarrow \quad I_6 = I_5.$$

Adding these integrals, I_3 and I_4 cancel. Taking the limit as $R \to \infty$ and $r \to 0$ and using (15), we obtain (14). ∎

Exercises 5.5

In Exercises 1–4, evaluate the given integral by using an appropriate contour integral. Carry out the details of the solution even if the integral follows from general formulas that we derived previously.

1. $\dfrac{1}{\sqrt{2\pi}} \displaystyle\int_{-\infty}^{\infty} e^{-\frac{x^2}{2}} \cos wx\, dx = e^{-\frac{w^2}{2}}.$ **2.** $\displaystyle\int_{-\infty}^{\infty} e^{-6x^2 + iwx}\, dx = \sqrt{\dfrac{\pi}{6}}\, e^{-\frac{w^2}{24}}.$

3. $\displaystyle\int_{0}^{\infty} \dfrac{\sin 2x}{\sqrt{x}}\, dx = \dfrac{\sqrt{\pi}}{2}.$ **4.** $\displaystyle\int_{0}^{\infty} \dfrac{dx}{\sqrt{x}(x^2+1)} = \dfrac{\pi}{\sqrt{2}}.$

5. The Fresnel integrals. In this exercise, we present an alternative way to derive the Fresnel integrals. We will need an inequality similar to the one we used in the proof of Jordan's lemma.

(a) Consider the graph of $\cos 2x$ on the interval $[0, \frac{\pi}{4}]$ (Figure 12). Since the graph concaves down, it is above the chord that joins any two points on the graph. Explain how this implies the inequality

$$(16) \qquad \cos 2x \geq 1 - \frac{4}{\pi}x, \quad \text{for } 0 \leq x \leq \frac{\pi}{4}.$$

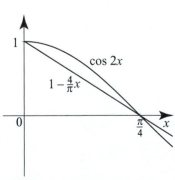

Figure 12 The inequality $\cos 2x \geq 1 - \frac{4}{\pi}x$, $0 \leq x \leq \frac{\pi}{4}$.

Let I_j ($j = 1, 2, 3$) denote the integral of $f(z) = e^{-z^2}$ on the closed contour in Figure 13. Prove the following.

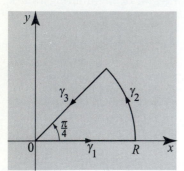

Figure 13

(b) $I_1 + I_2 + I_3 = 0$.

(c) $I_1 \to \int_0^\infty e^{-x^2} dx = \frac{\sqrt{\pi}}{2}$, as $R \to \infty$, and $I_2 \to 0$ as $R \to \infty$. [Hint: Use (16) to show that for $z = Re^{i\theta}$, $\left| e^{-z^2} \right| \le e^{R^2(\frac{4}{\pi}\theta - 1)} = e^{-R^2} e^{\frac{4R^2}{\pi}\theta}$. So $|I_2| \le Re^{-R^2} \int_0^{\frac{\pi}{4}} e^{\frac{4R^2}{\pi}\theta} d\theta = \frac{\pi}{4R} \left(1 - e^{-R^2} \right) \to 0$ as $R \to \infty$.]

(d) $I_3 \to -e^{i\frac{\pi}{4}} \left(\int_0^\infty \cos r^2 dr - i \int_0^\infty \sin r^2 dr \right)$, as $R \to \infty$. Note that this limit incorporates the Fresnel integrals where the variable of integration is r.

(e)   Derive the Fresnel integrals using (b)–(d).

6.   Using the formula of the complements for the gamma function $\Gamma(\alpha)\Gamma(1-\alpha) = \frac{\pi}{\sin \pi\alpha}$ (Exercise 23, Section 5.3), and the result of Example 2, derive the formulas

$$\int_0^\infty \frac{\cos x}{x^{1-\beta}} dx = \Gamma(\beta) \cos \frac{\beta\pi}{2} \quad \text{and} \quad \int_0^\infty \frac{\sin x}{x^{1-\beta}} dx = \Gamma(\beta) \sin \frac{\beta\pi}{2},$$

where $0 < \beta < 1$.

7.   (a) Use the contour in Figure 14 to establish the identity

$$\int_0^\infty \frac{(\ln x)^2}{x^2 + 1} dx = \frac{\pi^3}{8}.$$

(b)   Prove that

$$\int_0^1 \frac{(\ln x)^2}{x^2 + 1} dx = \int_1^\infty \frac{(\ln x)^2}{x^2 + 1} dx = \frac{\pi^3}{16}.$$

[Hint: Change of variables for the first equality, and (a) for the second one.]

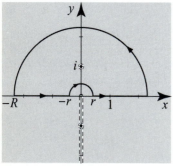

Figure 14 for Exercise 7.

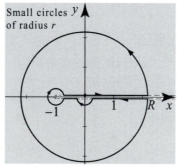

Figure 15 for Exercise 8.

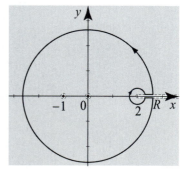

Figure 16 for Exercise 9.

8.   Use the contour in Figure 15 to establish the identity

$$\int_0^\infty \frac{\ln(x + 1)}{x^{1+\alpha}} dx = \frac{\pi}{\alpha \sin \pi\alpha}, \quad (0 < \alpha < 1).$$

9.   Use the contour in Figure 16 to establish the identity

$$\int_2^\infty \frac{dx}{x(x - 1)\sqrt{x - 2}} = 2\pi(1 - \frac{1}{\sqrt{2}}).$$

10.   Establish the identity $\displaystyle\int_3^\infty \frac{dx}{\sqrt{x - 3}\,(x - 2)(x - 1)} = \pi - \frac{\pi}{\sqrt{2}}.$

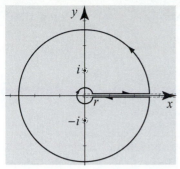

Figure 17 for Exercise 11.

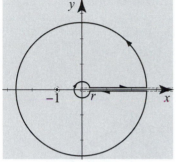

Figure 18 for Exercise 12.

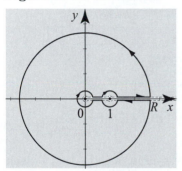

Figure 19 for Exercise 13.

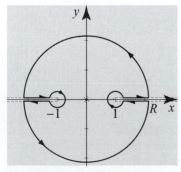

Figure 20 for Exercise 14.

11.   Use the contour in Figure 17 to establish the identity

$$\int_0^\infty \frac{x^{\alpha-1}}{x^2+1}\,dx = \frac{\pi}{2\sin\frac{\alpha\pi}{2}} \qquad (0 < \alpha < 2).$$

12.   Use the contour in Figure 18 to establish the identity

$$\int_0^\infty \frac{x^{\alpha-1}}{(x+1)^2}\,dx = \frac{(1-\alpha)\pi}{\sin\alpha\pi} \qquad (0 < \alpha < 2).$$

13.   (a)   Use the method of Example 4 and the contour in Figure 19 to establish the identity

$$\text{P.V.} \int_0^\infty \frac{x^p}{x(1-x)}\,dx = \pi\cot p\pi \qquad (0 < p < 1).$$

(b)   Use a suitable change of variables to derive the identity

$$\text{P.V.} \int_{-\infty}^\infty \frac{e^{px}}{1-e^x}\,dx = \pi\cot p\pi \qquad (0 < p < 1).$$

(c)   Use (b) to show that for $-1 < w < 1$

$$\text{P.V.} \int_{-\infty}^\infty \frac{e^{wx}}{\sinh x}\,dx = \pi\tan\frac{\pi w}{2}.$$

(c)   Use a suitable change of variables to show that

$$\text{P.V.} \int_{-\infty}^\infty \frac{e^{ax}}{\sinh bx}\,dx = \frac{\pi}{b}\tan\frac{\pi a}{2b} \quad (b > |a|).$$

(c)   Conclude that

$$\int_{-\infty}^\infty \frac{\sinh ax}{\sinh bx}\,dx = \frac{\pi}{b}\tan\frac{\pi a}{2b} \quad (b > |a|).$$

Note that the integral is convergent so there is no need to use the principal value.

14.   Use the contour in Figure 20 to establish the identity

$$\int_1^\infty \frac{dx}{x\sqrt{x^2-1}} = \frac{\pi}{2}.$$

In Exercises 15–20, derive the given identity.

15. $\displaystyle\int_0^\infty \frac{dx}{(x+2)\sqrt{x+1}} = \frac{\pi}{2}.$ **6.** $\displaystyle\int_0^\infty \frac{dx}{(x+2)^2\sqrt{x+1}} = \frac{\pi}{4} - \frac{1}{2}.$

17. $\displaystyle\int_0^\infty \frac{\sqrt{x}}{x^2+x+1}\,dx = \frac{\pi}{\sqrt{3}}.$ **18.** $\displaystyle\int_0^\infty \frac{x^a}{x^2+1}\,dx = \frac{\pi}{2}\sec\frac{a\pi}{2} \quad (-1 < a < 1).$

19. $\displaystyle\int_0^\infty \frac{x}{\sinh x}\,dx = \frac{\pi^2}{4}.$ **20.** $\displaystyle\int_0^\infty \frac{dx}{\cosh x} = \frac{\pi}{2}.$

21. Integral of the Gaussian with complex parameters. We will show that, for α and β complex with $\operatorname{Re}\alpha > 0$,

(17)
$$\int_{-\infty}^{\infty} e^{-\alpha(t-\beta)^2}\,dt = \sqrt{\frac{\pi}{\alpha}},$$

where on the right side we use the principal branch of the square root.

(a) Recognize the left side of (17) as the parametrized form of $\frac{1}{\sqrt{\alpha}}\int_\gamma e^{-z^2}\,dz$, where $\gamma(t) = \sqrt{\alpha}(t-\beta)$, the line at angle $\arg\sqrt{\alpha}$ through $-\beta\sqrt{\alpha}$. Show $|\operatorname{Arg}\sqrt{\alpha}| < \frac{\pi}{4}$.

(b) Consider the contour in Figure 21. Argue that for fixed α and β, there exists $\epsilon > 0$ such that for large enough R, γ_2 lies below the ray $\theta = \frac{\pi}{4} - \epsilon$ in the right half-plane, and γ_4 lies above the ray $\theta = -\frac{3\pi}{4} - \epsilon$ in the left half-plane.

(c) Letting I_j denote the integral of e^{-z^2} on γ_j, show that $I_2 \to 0$ and $I_4 \to 0$ as $R \to \infty$.

(d) Use $\int_{-\infty}^{\infty} e^{-x^2}\,dx = \sqrt{\pi}$ to derive (17).

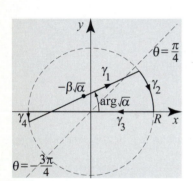

Figure 21 for Exercise 21.

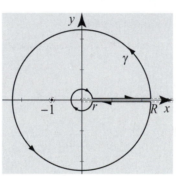

Figure 22 for Exercise 22.

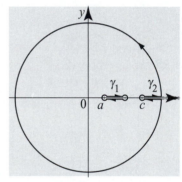

Figure 23 for Exercise 23.

22. Another look at Example 3. Derive (7) by considering the $\log_0$ branch of z^α for $f(z) = \frac{1}{z^\alpha(z+1)}$. Use the contour in Figure 22, and treat $f(z)$ to have different values on the upper and lower sides of the real axis, as in the text following Example 3.

23. Another look at Example 4. It turns out that the branch-cut singularity in $f(z) = \frac{1}{\sqrt{z-a}\sqrt{z-b}\sqrt{z-c}}$ is removable on the real interval (b, c). We will prove this and rederive (14).

(a) Use Figure 11 to help show that the limit of $f(z)$ as z approaches the interval (b, c) from the upper half-plane is the same as from the lower half-plane.

(b) Define $f(z)$ on (b, c) to equal this common value. To show that $f(z)$ is analytic here, use Morera's theorem. [Hint: If a triangle crosses the real axis, subdivide it into a triangle and a quadrilateral, each with one side on the real axis. One lies in the upper and the other in the lower half-plane; integrals will vanish.]

(c) Rederive (14) by using Figure 23 and the fact that the paths γ_1 and γ_2 are continuously deformable to each other.

5.6 Summing Series by Residues

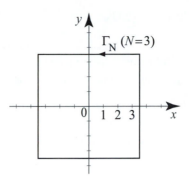

Figure 1 Square contour Γ_N for integers $N \geq 1$.

In this section, we will use residue theory to sum infinite series of the form $\sum_k f(k)$, where $f(z) = \frac{p(z)}{q(z)}$ is a rational function with degree $q(z) \geq 2 +$ degree $p(z)$. For example,

$$\sum_{k=-\infty}^{\infty} \frac{1}{k^2+1}; \qquad \sum_{k=1}^{\infty} \frac{1}{k^6}; \qquad \sum_{k=-\infty}^{\infty} \frac{k^2}{k^4+1}.$$

Our starting point is the result of Example 4, Section 5.1: if k is an integer and $f(z)$ is analytic at k, then

(1)
$$\text{Res}\,(f(z)\cot(\pi z), k) = \frac{1}{\pi} f(k).$$

We will consider integrating $f(z)\cot(\pi z)$ on squares centered at the origin. We define the squares Γ_N for positive integers N to be positively oriented and to have corners at $(N + \frac{1}{2})(1 + i)$, $(N + \frac{1}{2})(-1 + i)$, $(N + \frac{1}{2})(-1 - i)$, and $(N + \frac{1}{2})(1 - i)$, as shown in Figure 1.

LEMMA 1
BOUNDS FOR THE
COTANGENT

Let N be a positive integer and Γ_N be as in Figure 1. For all z on Γ_N, we have

(2)
$$|\cot(\pi z)| \leq 2.$$

Moreover, if $f(z) = \frac{p(z)}{q(z)}$ is a rational function with degree $q(z) \geq 2 +$ degree $p(z)$, then

(3)
$$\lim_{N \to \infty} \left| \int_{\Gamma_N} \frac{p(z)}{q(z)} \cot(\pi z)\, dz \right| = 0.$$

Proof To prove (2), we deal with each side of the square Γ_N separately. On the right vertical side, $z = N + \frac{1}{2} + iy$, where $-N - \frac{1}{2} \leq y \leq N + \frac{1}{2}$. Using the formulas for the absolute values of the cosine and sine, (17) and (18), Section 1.6, we obtain

$$|\cot(\pi z)| = \left| \frac{\cos(\pi z)}{\sin(\pi z)} \right| = \frac{\sqrt{\cos^2\left(\pi(N + \frac{1}{2})\right) + \sinh^2(\pi y)}}{\sqrt{\sin^2\left(\pi(N + \frac{1}{2})\right) + \sinh^2(\pi y)}}$$

$$= \frac{|\sinh(\pi y)|}{\sqrt{1 + \sinh^2(\pi y)}} \leq \frac{|\sinh(\pi y)|}{|\sinh(\pi y)|} = 1 \leq 2.$$

This establishes (2) for z on the right vertical side of Γ_N. We also have the left vertical side is immediate since $\cot(\pi z)$ is an odd function. The horizontal sides are handled similarly (Exercise 19). To prove (3), we use the inequality for path integrals, Theorem 2, Section 3.2, the fact that the perimeter of the square Γ_N is $4(2N + 1)$, and (2) and get

$$\left| \int_{\Gamma_N} \frac{p(z)}{q(z)} \cot(\pi z) \, dz \right| \leq l(\Gamma_N) M_N = 4(2N + 1)(2M_N),$$

where M_N is the maximum value of $\left| \frac{p(z)}{q(z)} \right|$ for z on Γ_N. We have $4(2N+1)(2M_N) \to 0$ as $N \to \infty$, because degree $q(z) \geq 2 + $ degree $p(z)$. Thus (3) holds. ∎

The following result illustrates how (1) and Lemma 1 can be used to sum infinite series. Variations on this result are presented in the exercises. In what follows, all doubly infinite series are to be interpreted as the limit of symmetric partial sums; that is,

$$\sum_{k=-\infty}^{\infty} a_k = \lim_{N \to \infty} \sum_{k=-N}^{N} a_k,$$

whenever the limit exists.

**PROPOSITION 1
SUMMING INFINITE
SERIES BY
RESIDUES**

Suppose that $f(z) = \frac{p(z)}{q(z)}$ is a rational function with degree $q(z) \geq 2 + $ degree $p(z)$. Suppose further that f has no poles at the integers. Then

(4) $$\sum_{k=-\infty}^{\infty} f(k) = -\pi \sum_{j} \mathrm{Res}\left(f(z) \cot(\pi z), \, z_j \right),$$

where the (finite) sum on the right runs over all the poles z_j of $f(z)$.

Proof   The poles of $f(z) \cot(\pi z)$ occur at the integers (where $\cot(\pi z)$ has poles) and at the points z_j (where $f(z)$ has poles). For large enough N, all z_j are on the inside of Γ_N. Applying the residue theorem and (1), we obtain

(5) $$\int_{\Gamma_N} f(z) \cot(\pi z) \, dz = 2\pi i \sum_{k=-N}^{N} \frac{1}{\pi} f(k) + 2\pi i \sum_{j} \mathrm{Res}\left(f(z) \cot(\pi z), \, z_j \right).$$

Letting $N \to \infty$, the left side of (5) goes to zero and we get (4). ∎

EXAMPLE 1   Summing series by residues

Evaluate $\displaystyle\sum_{k=-\infty}^{\infty} \frac{1}{k^2 + 1}$.

Solution   We apply (4) with $f(z) = \frac{1}{z^2+1}$ and find

$$\sum_{k=-\infty}^{\infty} \frac{1}{k^2 + 1} = -\pi \sum_{j} \mathrm{Res}\left(\frac{1}{z^2 + 1} \cot(\pi z), \, z_j \right),$$

where the sum on the right runs over the poles of $\frac{1}{z^2+1}$. The latter function has simple poles at $\pm i$, and the residues of $\frac{1}{z^2+1}\cot(\pi z)$ there are

$$
\begin{aligned}
\operatorname{Res}\left(\frac{1}{z^2+1}\cot(\pi z), i\right) &= \lim_{z\to i}(z-i)\frac{1}{(z-i)(z+i)}\cot(\pi z) = \frac{1}{2i}\cot(\pi i) \\
&= \frac{1}{2i}\frac{\cos(\pi i)}{\sin(\pi i)} = \frac{1}{2i}\frac{\cosh(\pi)}{i\sinh(\pi)} = -\frac{1}{2}\coth\pi,
\end{aligned}
$$

(use (13) and (14), Section 1.6) and similarly

$$
\operatorname{Res}\left(\frac{1}{z^2+1}\cot(\pi z), -i\right) = -\frac{1}{2i}\cot(-\pi i) = \frac{1}{2i}\cot(\pi i) = -\frac{1}{2}\coth\pi.
$$

Thus,

$$
\sum_{k=-\infty}^{\infty}\frac{1}{k^2+1} = -\pi\left(-\frac{1}{2}\coth\pi - \frac{1}{2}\coth\pi\right) = \pi\coth\pi. \qquad \blacksquare
$$

Exercises 5.6

In Exercises 1–12, derive the given identity using Proposition 1.

1. $\displaystyle\sum_{k=-\infty}^{\infty}\frac{1}{k^2+9} = \frac{\pi}{3}\coth(3\pi).$

2. $\displaystyle\sum_{k=-\infty}^{\infty}\frac{1}{(k^2+1)^2} = \frac{\pi}{2}\coth\pi + \frac{\pi^2}{2}\operatorname{csch}^2\pi.$

3. $\displaystyle\sum_{k=-\infty}^{\infty}\frac{1}{k^2+a^2} = \frac{\pi}{a}\coth(a\pi)$ (*ia* not an integer).

4. $\displaystyle\sum_{k=-\infty}^{\infty}\frac{1}{(k^2+a^2)^2} = \frac{\pi}{2a^3}\coth(a\pi) + \frac{\pi^2}{2a^2}\operatorname{csch}^2(a\pi)$ (*ia* not an integer).

5. $\displaystyle\sum_{k=-\infty}^{\infty}\frac{1}{4k^2-1} = 0.$

6. $\displaystyle\sum_{k=-\infty}^{\infty}\frac{k^2}{(k^2-\frac{1}{4})^2} = \frac{\pi^2}{2}.$

7. $\displaystyle\sum_{k=-\infty}^{\infty}\frac{1}{(4k^2-1)^2} = \frac{\pi^2}{8}.$

8. $\displaystyle\sum_{k=-\infty}^{\infty}\frac{1}{(k-\frac{1}{2})^2+1} = \pi\tanh\pi.$

9. $\displaystyle\sum_{k=-\infty}^{\infty}\frac{1}{(k-2)(k-1)+1} = \frac{2\pi\tanh\left(\frac{\sqrt{3}\pi}{2}\right)}{\sqrt{3}}.$

10. $\displaystyle\sum_{k=-\infty}^{\infty}\frac{k^2}{(k^2+1)^2} = \frac{\pi\operatorname{csch}^2\pi}{4}\left(\sinh(2\pi) - 2\pi\right).$

11. $\displaystyle\sum_{k=-\infty}^{\infty} \frac{1}{k^4+4} = \frac{\pi\sinh(2\pi)}{4(\cosh(2\pi)-1)}$.

12. $\displaystyle\sum_{k=-\infty}^{\infty} \frac{1}{k^4+4a^4} = -\pi\frac{\sin(2a\pi)+\sinh(2a\pi)}{4a^3(\cos(2a\pi)-\cosh(2a\pi))}$ $(a>0)$.

13. Project Problem: Summation of series with a pole at 0. If in Proposition 1 the function f has a pole at 0, then (4) has to be modified to account for the residue at 0. In this case, we have the following useful result.

Suppose that $f(z) = \frac{p(z)}{q(z)}$ is a rational function with degree $q(z) \geq 2 + \text{degree } p(z)$. Suppose further that f has no poles at the integers, except possibly at 0. Then

(6) $$\sum_{\substack{k=-\infty \\ k\neq 0}}^{\infty} f(k) = -\pi\sum_{j} \text{Res}\left(f(z)\cot(\pi z), z_j\right),$$

where the (finite) sum on the right runs over all the poles z_j of $f(z)$, including 0.

Prove this result by modifying the proof of Proposition 1.

In Exercises 14–16, use (6) to derive the given identity.

14. $\displaystyle\sum_{k=1}^{\infty} \frac{1}{k^2} = \frac{\pi^2}{6}$.

15. $\displaystyle\sum_{k=1}^{\infty} \frac{1}{k^4} = \frac{\pi^4}{90}$.

16. $\displaystyle\sum_{k=1}^{\infty} \frac{1}{k^2(k^2+4)} = \frac{3+4\pi^2-6\pi\coth(2\pi)}{96}$.

17. Project Problem: Sums of the reciprocals of even powers of integers. In this exercise, we use (6) to derive

(7) $$\sum_{k=1}^{\infty} \frac{1}{k^{2n}} = (-1)^{n-1}\frac{2^{2n-1}B_{2n}\pi^{2n}}{(2n)!},$$

where n is a positive integer, B_{2n} is the Bernoulli number (Example 4, Section 4.4). This remarkable identity sums the reciprocals of the even powers of the integers. We will derive it again in Chapter 7 using Fourier series. There is no known finite expression corresponding to any odd power.

(a)   Show that if $f(z) = \frac{1}{z^{2n}}$ then (6) becomes

$$\sum_{\substack{k=-\infty \\ k\neq 0}}^{\infty} \frac{1}{k^{2n}} = -\pi\,\text{Res}\left(\frac{\cot(\pi z)}{z^{2n}}, 0\right),$$

and so

(8) $$\sum_{k=1}^{\infty} \frac{1}{k^{2n}} = -\frac{\pi}{2}\,\text{Res}\left(\frac{\cot(\pi z)}{z^{2n}}, 0\right).$$

(b) Using the Taylor series expansion of $z \cot z$ from Exercise 31, Section 4.4, obtain

$$\text{Res}\left(\frac{\cot(\pi z)}{z^{2n}}, 0\right) = (-1)^n \frac{2^{2n} B_{2n} \pi^{2n-1}}{(2n)!};$$

then derive (7).

18. Project Problem: Sums with alternating signs. (a) Modify the proof of Proposition 1 to prove the following summation result.

Suppose that $f(z) = \frac{p(z)}{q(z)}$ is a rational function with degree $q(z) \geq 2 + \text{degree } p(z)$. Suppose further that f has no poles at the integers, except possibly at 0. Then,

$$(9) \qquad \sum_{\substack{k=-\infty \\ k \neq 0}}^{\infty} (-1)^k f(k) = -\pi \sum_j \text{Res}\left(f(z) \csc(\pi z), z_j\right),$$

where the (finite) sum on the right runs over all the poles z_j of $f(z)$, including 0.

[Hint: You need a version of Lemma 1 for the cosecant.]
(b)   Show that if $f(z) = \frac{1}{z^{2n}}$ then (9) becomes

$$\sum_{\substack{k=-\infty \\ k \neq 0}}^{\infty} \frac{(-1)^k}{k^{2n}} = -\pi \, \text{Res}\left(\frac{\csc(\pi z)}{z^{2n}}, 0\right),$$

and so

$$(10) \qquad \sum_{k=1}^{\infty} \frac{(-1)^k}{k^{2n}} = -\frac{\pi}{2} \, \text{Res}\left(\frac{\csc(\pi z)}{z^{2n}}, 0\right).$$

(b) Using the Taylor series expansion of $z \csc z$ from Exercise 31, Section 4.4, obtain

$$\text{Res}\left(\frac{\csc(\pi z)}{z^{2n}}, 0\right) = (-1)^{n-1} \frac{(2^{2n} - 2) B_{2n} \pi^{2n-1}}{(2n)!}.$$

(c) Show that for any integer $n \geq 1$

$$\sum_{k=1}^{\infty} \frac{(-1)^k}{k^{2n}} = (-1)^n \frac{(2^{2n-1} - 1) B_{2n} \pi^{2n}}{(2n)!}.$$

19. The horizontal sides of Γ_N. Here we prove inequality (2) for the upper and lower horizontal sides of the square Γ_N.
(a) Argue that we need only consider the upper side, since $\cot(\pi z)$ is an odd function of z.
(b) For $z = x + i(N + \frac{1}{2})$, $-N - \frac{1}{2} \leq x \leq N + \frac{1}{2}$, justify

$$|\cot(\pi z)| = \frac{\sqrt{\cos^2(\pi x) + \sinh^2\left(\pi(N + \frac{1}{2})\right)}}{\sqrt{\cos^2(\pi x) + \sinh^2\left(\pi(N + \frac{1}{2})\right)}} \leq \coth\left(\pi(N + \frac{1}{2})\right).$$

Prove that this is less than or equal to 2, for all $N \geq 1$. This is not a best possible estimate, but it is sufficient to prove (3).

5.7   The Counting Theorem and Rouché's Theorem

In this section, we apply Cauchy's residue theorem to prove several important properties of analytic functions with fruitful applications. We start with a simple but very useful lemma.

LEMMA 1

> Suppose that f is analytic and not identically zero in a region Ω.
> (i) If z_0 is a zero of f of order $m \geq 1$, then $\frac{f'(z)}{f(z)}$ has a simple pole at z_0 and the residue there is m.
> (ii) If z_0 is a pole of f of order $m \geq 1$, then $\frac{f'(z)}{f(z)}$ has a simple pole at z_0 and the residue there is $-m$.

Proof   (i) Write $f(z) = (z - z_j)^m \lambda(z)$, where λ is analytic and nonvanishing in a neighborhood $B_r(z_0)$, (Theorem 1, Section 4.6). For z in $B_r(z_0)$,

$$(1) \qquad \frac{f'(z)}{f(z)} = \frac{m(z - z_0)^{m-1}\lambda(z) + (z - z_0)^m \lambda'(z)}{(z - z_0)^m \lambda(z)} = \frac{m}{z - z_0} + \frac{\lambda'(z)}{\lambda(z)},$$

where $\frac{\lambda'(z)}{\lambda(z)}$ is analytic in $B_r(z_0)$ since $\lambda(z)$ is nonvanishing in $B_r(z_0)$. From this it follows that z_0 is a simple pole of $\frac{f'(z)}{f(z)}$ and the residue there is m.
(ii)   Write $f(z) = (z - z_0)^{-m}\lambda(z)$, where λ is analytic and nonvanishing in a neighborhood $B_r(z_0)$ (Theorem 7, Section 4.6). Then for z in $B_r(z_0)$,

$$(2) \qquad \frac{f'(z)}{f(z)} = \frac{-m(z - z_0)^{m-2}\lambda(z) + (z - z_0)^{-m}\lambda'(z)}{(z - z_0)^{-m}\lambda(z)} = -\frac{m}{z - z_0} + \frac{\lambda'(z)}{\lambda(z)},$$

where $\frac{\lambda'(z)}{\lambda(z)}$ is analytic in $B_r(z_0)$ since $\lambda(z)$ is nonvanishing in $B_r(z_0)$. Hence z_0 is a simple pole of $\frac{f'(z)}{f(z)}$ and the residue there is $-m$. ∎

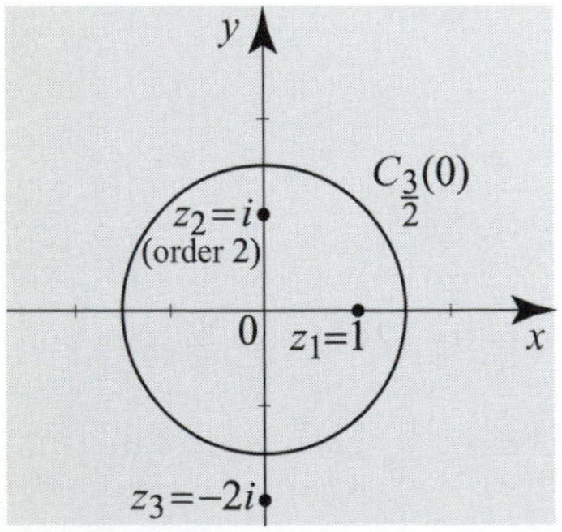

Figure 1 The number of zeros inside the circle according to multiplicity is 3.

Our first theorem is called the **counting theorem** because it counts the number of zeros (according to multiplicity) of an analytic function f inside a simple path C. By "according to multiplicity" we mean that each zero is counted as many times as its order. For example, the function $f(z) = (z - 1)(z - i)^2(z + 2i)$ has zeros at $z_1 = 1$ (order 1), $z_2 = i$ (order 2), and $z_3 = -2i$ (order 1). The number of zeros, according to multiplicity, inside the circle $C_{\frac{3}{2}}(0)$ is three (Figure 1).

**THEOREM 1
COUNTING
THEOREM**

> Suppose that C is a simple closed positively oriented path, Ω is the region inside C, and f is analytic inside and on C and nonvanishing on C. Let $N(f)$ denote the number of zeros of f inside Ω, counted according to multiplicity. Then $N(f)$ is finite and
>
> (3)
> $$N(f) = \frac{1}{2\pi i} \int_C \frac{f'(z)}{f(z)}\, dz.$$

Proof   If f has no zeros in Ω (hence $N(f) = 0$), then $\frac{f'(z)}{f(z)}$ is analytic on Ω and (3) follows from Cauchy's theorem. Note that f cannot have infinitely many zeros in Ω, by Corollary 2, Section 4.6. Let $z_1, z_2, \ldots, z_n$ denote the zeros of f inside Ω. By Lemma 1, $\frac{f'(z)}{f(z)}$ has simple poles in Ω located at z_j and the residue at z_j is the order of the zero at z_j. Thus (3) follows at once from the residue theorem. ■

A function f is called **meromorphic** on a region Ω if f is analytic in Ω except for poles on Ω. For example, $f(z) = \frac{\sin z}{z^2+1}$ is meromorphic in the complex plane. Like zeros, we will count poles according to multiplicity. The following generalizes Theorem 1 to give a counting theorem for meromorphic functions.

**THEOREM 2
MEROMORPHIC
COUNTING
THEOREM**

> Suppose that C is a simple closed positively oriented path, Ω is the region inside C, and f is meromorphic on Ω, analytic and nonvanishing on C. Let $N(f)$ denote the number of zeros of f inside Ω and $P(f)$ denote the number of poles of f inside Ω, counted according to multiplicity. Then $N(f)$ and $P(f)$ are finite and
>
> (4)
> $$N(f) - P(f) = \frac{1}{2\pi i} \int_C \frac{f'(z)}{f(z)}\, dz.$$

Proof   That $N(f)$ and $P(f)$ are finite follows as in the proof of Theorem 1. Apply the residue theorem and use Lemma 1 to see that (4) holds. ■

Either of the preceding theorems is also known as the **argument principle** because the right sides of (3) and (4) can be interpreted as the change in argument as one runs around the image path $f[C]$. We now investigate this.

We know from results in Section 3.6 (Project Problem 36) that if f is analytic and nonvanishing on a region Ω and if the image $f[\Omega]$ is simply connected, then there is a branch of the logarithm of f, $\log f(z)$, whose derivative satisfies the familiar formula $\frac{d}{dz}\log f(z) = \frac{f'(z)}{f(z)}$. The integrand in (4) suggests a connection with the logarithm of $f(z)$. We will explore this connection due to its fruitful consequences.

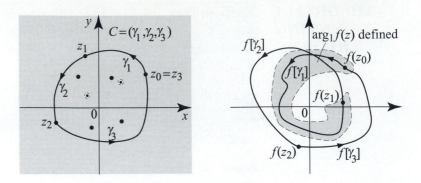

Figure 2 The image $f[\gamma_1]$ is contained in a simply connected region not containing the origin. We can define the branch $\arg_1 z$ on this region.

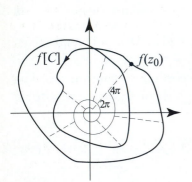

Figure 3 The point $f(z)$ travels twice around the origin and $\Delta_C \arg f = 4\pi$.

Let f and C be as in Theorem 2. We can partition C into subarcs γ_j $(j = 1, \ldots, n)$ such that each image $f[\gamma_j]$ is contained in a simply connected region not including the origin, as shown in Figure 2. (A proof of this statement can be based on the fact that C is closed and bounded.) Denote the initial and terminal points of γ_j by z_{j-1} and z_j, with $z_n = z_0$. For each j, we have a branch of the logarithm of $f(z)$, $\log_j f(z)$, which is an antiderivative of $\frac{f'(z)}{f(z)}$ on γ_j. For each j, we have $\log_j f(z) = \ln|f(z)| + i \arg_j f(z)$, where $\arg_j$ is a branch of the argument, determined up to an additive constant multiple of 2π. These constants are determined in the following way. Starting with $j = 1$, pick and fix $\arg_1 f(z)$. Then choose $\arg_2 f(z)$ in such a way that $\arg_1 f(z_1) = \arg_2 f(z_1)$. Continue in this manner to determine the remaining branches of the argument. To simplify the notation, denote the resulting function by $\arg f(z)$. Note that $\arg f(z)$ is continuous on C, except at $z_0 = z_n$, and we will make the convention that $\arg f(z)$ takes different values at this point depending on whether we approach the point from γ_1 or from γ_n. Thus $\arg z_n = \arg_n z_n$ and $\arg z_0 = \arg_1 z_0$, and the difference is always an integer multiple of 2π that we denote by $\Delta_C \arg f = \arg_n f(z_n) - \arg_1 f(z_0)$. The quantity $\Delta_C \arg f$ measures the net change in the argument of $f(z)$ as z travels once around the curve C (Figure 3). We evaluate the integral on the right side of (3) or (4) with a telescoping sum:

$$\frac{1}{2\pi i} \int_C \frac{f'(z)}{f(z)}\, dz = \sum_{j=1}^{n} \int_{\gamma_j} \frac{f'(z)}{f(z)}\, dz$$

$$= \frac{1}{2\pi i} \sum_{j=1}^{n} \left(\ln|f(z)| + i \arg f(z) \right) \Big|_{z_{j-1}}^{z_j}$$

$$= \frac{1}{2\pi i} \left(\ln|f(z_n)| - \ln|f(z_0)| + i(\arg f(z_n) - \arg f(z_0)) \right)$$

$$= \frac{1}{2\pi} \Delta_C \arg f.$$

Comparing with (4) we find that

(5)
$$\frac{1}{2\pi}\Delta_C \arg f = N(f) - P(f).$$

This unexpected formula links two seemingly unrelated quantities: the net change in the argument of $f(z)$ as we travel once around C, and the number of zeros and poles of f inside C! Beside its esthetic value, this formula has many interesting applications, as we now illustrate.

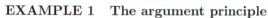

EXAMPLE 1   The argument principle

Find the number of zeros of $f(z) = z^6 + 6z + 10$ in the first quadrant.

Solution   We will apply the argument principle, which in this case is expressed by $\frac{1}{2\pi}\Delta_C \arg f = N(f)$ since f has no poles. The contour C will be a closed quarter of a circle of radius $R > 0$, as shown in Figure 4. Since f has finitely many zeros (at most six), if R is large enough, the contour C will contain all the zeros in the first quadrant. Our goal is to determine the change of the argument of $f(z)$ as z runs through C. We will proceed in steps.

Step 1: Show that there are no roots on the contour C. We can always choose $R > 0$ so that no roots lie on the circular arc γ_R. We need to prove that there are no roots on the nonnegative x- and y-axes. If $z = x \geq 0$, then $f(z) = x^6 + 6x + 10$, and this is clearly positive. If $z = iy$, with $y \geq 0$, then $f(iy) = (iy)^6 + 6iy + 10 = (-y^6 + 10) + 6iy$ and because $\mathrm{Im}\,(f(iy)) = 0 \ \Leftrightarrow \ y = 0 \Rightarrow \mathrm{Re}\,(f(iy)) \neq 0$, we see that $f(iy) \neq 0$.

Step 2: Compute the change of the argument of $f(z)$ as z varies from the initial to the terminal point of $I_1 = [0, R]$. For $z = 0$, $f(z) = 10$, and for $0 \leq z = x \leq R$, $f(z) = x^6 + 6x + 10 > 0$. So the image of the interval $[0, R]$ is the interval $[10, R^6 + 6R + 10]$ and the argument of $f(z)$ does not change on I_1.

Step 3: Compute the change of the argument of $f(z)$ as z varies from the initial to the terminal point of the arc γ_R. Here we are not looking for the exact image of γ_R by f, but only a rough picture that gives us the change in the argument of f. For very large R and z on γ_R, write $z = Re^{i\theta}$, where $0 \leq \theta \leq \frac{\pi}{2}$. Then

$$f(z) = R^6 e^{6i\theta}\left(1 + \frac{6}{R^5}e^{-i5\theta} + \frac{10}{R^6}e^{-6i\theta}\right) \approx R^6 e^{6i\theta},$$

because $\frac{6}{R^5}e^{-i5\theta} + \frac{10}{R^6}e^{-6i\theta} \approx 0$. So as θ varies from 0 to $\frac{\pi}{2}$, the argument of $f(z)$ varies from 0 to $6 \cdot \frac{\pi}{2} = 3\pi$. In fact, the image of the point iR is $f(iR) = -R^6 + 10 + 6iR$, which is a point in the third quadrant with argument very close to 3π. See Figure 5.

Step 4: Compute the change of the argument of $f(z)$ as z varies from iR to 0. As z varies from iR to 0, $f(z)$ varies from $w_3 = -R^6 + 10 + 6iR$ to $w_0 = 10$. Since $\mathrm{Im}\,(f(z)) \geq 0$, this tells us that the point $f(z)$ remains in the upper half-plane as $f(z)$ moves from w_3 to w_0. Hence the change in the argument of $f(z)$ is $-\pi$.

Step 5: Apply the argument principle. The net change of the argument of $f(z)$ as we travel once around C is $3\pi - \pi = 2\pi$. According to (5), the number of zeros of f inside C, and hence in the first quadrant, is $\frac{1}{2\pi}2\pi = 1$. ∎

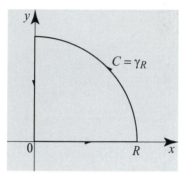

Figure 4 The closed quarter circle C of radius $R > 0$.

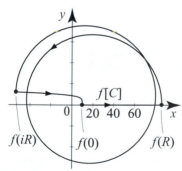

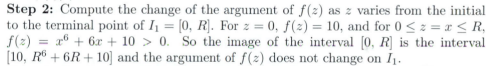

Figure 5 The path $f[C]$ (pictured for $R = 2$) goes around the origin once.

We give one more variant of the counting theorem.

**THEOREM 3
VARIANT OF THE
COUNTING
THEOREM**

Let C, Ω, and f be as in Theorem 2, let g be analytic inside and on C. Let $z_1, z_2, \ldots, z_{n_1}$ denote the zeros of f in Ω and $p_1, p_2, \ldots, p_{n_2}$ denote the poles of f in Ω. Let $m(z_j)$ and $m(p_j)$ denote the orders of the roots and poles. Then

$$(6) \qquad \frac{1}{2\pi i}\int_C g(z)\frac{f'(z)}{f(z)}\,dz = \sum_{j=1}^{n_1} m(z_j)g(z_j) - \sum_{j=1}^{n_2} m(p_j)g(p_j).$$

Proof We modify the proof of the previous theorem as follows. If z_j is a zero of f, then since g is analytic and $\frac{f'(z)}{f(z)}$ has a simple pole at z_j, using Lemma 1, and Proposition 1(iii), Section 5.1,

$$\mathrm{Res}\,(g(z)\frac{f'(z)}{f(z)},\,z_j) = g(z_j)\,\mathrm{Res}\,(\frac{f'(z)}{f(z)},\,z_j) = m(z_j)\,g(z_j).$$

Similarly for the poles p_j,

$$\mathrm{Res}\,(g(z)\frac{f'(z)}{f(z)},\,p_j) = g(p_j)\,\mathrm{Res}\,(\frac{f'(z)}{f(z)},\,p_j) = -m(p_j)\,g(p_j).$$

Now (6) follows from the residue theorem. ∎

EXAMPLE 2   Applying the variant of the counting theorem
Evaluate

$$\int_{C_1(0)} \frac{e^z \cos z}{e^z - 1}\,dz,$$

where $C_1(0)$ is the positively oriented unit circle.

Solution The function $f(z) = e^z - 1$ has a zero at $z = 0$ and, because $f'(0) = 1 \neq 0$, this zero is simple. Also, using the $2\pi i$-periodicity of e^z (see (13), Section 1.5), it is easy to see that $e^z = 1$ inside the unit disk $|z| < 1$ only at $z = 0$. From (6) applied with $f(z) = e^z - 1$ and $g(z) = \cos z$, it follows immediately that

$$\int_{C_1(0)} \frac{e^z \cos z}{e^z - 1}\,dz = 2\pi i \cos 0 = 2\pi i.$$ ∎

As an application of the counting principle we will prove a famous result known as Rouché's theorem, named after the French mathematician and educator Eugène Rouché (1832–1910). We need a simple lemma.

**LEMMA 2
INTEGER-VALUED
FUNCTIONS**

Suppose that ϕ is continuous on a region Ω and $\phi(z)$ is an integer for all z in Ω (in other words, ϕ is integer-valued). Then ϕ is constant on Ω.

Proof For each integer n, let U_n be the set of z in Ω such that $\phi(z) = n$. The U_n's are clearly disjoint and their union is Ω. They are also open, as we now show. Let z_0 be in U_n, so $\phi(z_0) = n$. Since ϕ is continuous at z_0, we can find a neighborhood $B_r(z_0)$ such that $|\phi(z) - n| < \frac{1}{2}$ for all z in $B_r(z_0)$. Since ϕ is integer-valued, this forces $\phi(z) = n$ for all z in $B_r(z_0)$ (how else can two integers differ by at most $1/2$?). This shows the U_n are open. By definition of connectedness, Ω cannot be written as the union of disjoint open sets, so only one of the U_n can be nonempty (say, U_{n_0}), and thus $\phi(z) = n_0$ for all z in Ω. ∎

THEOREM 4
ROUCHÉ'S
THEOREM

Suppose that C is a simple closed path, Ω is the region inside C, f and g are analytic inside and on C. If $|g(z)| < |f(z)|$ for all z on C, then $N(f+g) = N(f)$; that is, $f+g$ and f have the same number of zeros on Ω.

Proof   For z on C and $0 \leq \lambda \leq 1$, if $f(z) + \lambda g(z) = 0$, then $|f(z)| = \lambda|g(z)| \leq |g(z)|$, which contradicts $|f(z)| > |g(z)|$ on C. So $f(z) + \lambda g(z)$ is not equal to zero for all z on C and $0 \leq \lambda \leq 1$. Parametrize C by $\gamma(t)$, $a \leq t \leq b$, and for $0 \leq \lambda \leq 1$ consider

$$\phi(\lambda) = \frac{1}{2\pi i} \int_C \frac{f'(z) + \lambda g'(z)}{f(z) + \lambda g(z)}\, dz = \int_a^b \frac{f'(\gamma(t)) + \lambda g'(\gamma(t))}{f(\gamma(t)) + \lambda g(\gamma(t))} \gamma'(t)\, dt.$$

The integrand is a continuous function of the two variables t and λ over the rectangle $[a, b] \times [0, 1]$. Applying Theorem 5, Section 3.5, it follows that $\phi(\lambda)$ is a continuous function on $[0, 1]$. By Theorem 1, $\phi(\lambda) = N(f + \lambda g)$ and hence it is integer-valued. By Lemma 2, ϕ is constant; and so $N(f) = \phi(0) = \phi(1) = N(f + g)$. ∎

We can give a geometric interpretation of Rouché's theorem, based on the argument principle. According to (5), we are merely claiming that

(7) $$\Delta_C \arg (f + g) = \Delta_C \arg f.$$

As z traces C, $f(z)$ winds around the origin a specific number of times. Since $|g(z)| < |f(z)|$, the point $f(z) + g(z)$ must lie in the disk of radius $|f(z)|$ centered at $f(z)$ (Figure 6), and so $f(z) + g(z)$ must wind around the origin the same number of times as $f(z)$ (see Exercise 24).

The following are typical applications of Rouché's theorem.

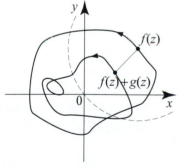

Figure 6 Since $|g(z)| < |f(z)|$, the point $f(z) + g(z)$ goes around the origin the same number of times as $f(z)$.

EXAMPLE 3   Counting zeros with Rouché's theorem
(a)   Show that all zeros of $p(z) = z^4 + 6z + 3$ lie inside the circle given by $|z| = 2$.
(b)   Show that if a is a real number with $a > e$, then the equation $e^z = az^n$ has n roots in $|z| < 1$.

Solution   (a) Since p is a polynomial of degree 4, it is enough to show that $N(p) = 4$ inside the circle $|z| = 2$. Take $f(z) = z^4$ and $g(z) = 6z + 3$ and note that $f(z) + g(z) = p(z)$. For $|z| = 2$, we have $|f(z)| = 2^4 = 32$ and $|g(z)| \leq |6z| + 3 = 15$. Hence $|g(z)| < |f(z)|$ for all z on the circle $|z| = 2$, and so by Rouché's theorem $N(f) = N(f + g) = N(p)$. Clearly f has one zero with multiplicity 4 at $z = 0$. Thus $N(f) = 4$ and so $N(p) = 4$, as desired.
(b) Take $f(z) = az^n$ and $g(z) = -e^z$. Counted according to multiplicity, f has n zeros in $|z| < 1$ and so $N(f) = n$. For $|z| = 1$, $|f(z)| = |az^n| = a$ and

$$|g(z)| = |e^z| = \left|e^{\cos\theta + i\sin\theta}\right| = e^{\cos\theta} \leq e < a.$$

Thus $|g(z)| < |f(z)|$ for all $|z| = 1$ and so by Rouché's theorem $n = N(az^n) = N(az^n - e^z)$, which implies that $az^n - e^z = 0$ has n roots in $|z| < 1$.   ∎

As a further application we give a very simple proof of the fundamental theorem of algebra.

EXAMPLE 4   The fundamental theorem of algebra

Let $p(z) = a_n z^n + a_{n-1} z^{n-1} + \cdots + a_1 z + a_0$ with $a_n \neq 0$ be a polynomial of degree $n \geq 1$. Show that $p(z)$ has exactly n roots, counting multiplicity.

Solution   Take $f(z) = a_n z^n$. Then f has a zero of multiplicity n at $z = 0$. Also, for $|z| = R$, we have $|f(z)| = |a_n| R^n$, which is a polynomial of degree n in R. Now let $g(z) = a_{n-1} z^{n-1} + \cdots + a_1 z + a_0$, then $|g(z)| \leq |a_{n-1}| |z^{n-1}| + \cdots + |a_1| |z| + |a_0|$, and so for $|z| = R$, $|g(z)| \leq |a_{n-1}| R^{n-1} + \cdots + |a_1| R + |a_0|$. Since the modulus of f grows at a faster rate than the modulus of g, in the sense that

$$\lim_{R \to \infty} \frac{1}{|a_n| R^n} (|a_{n-1}| R^{n-1} + \cdots + |a_1| R + |a_0|) = 0,$$

we can find R_0 large enough so that for $R \geq R_0$, we have $|a_{n-1}| R^{n-1} + \cdots + |a_1| R + |a_0| < |a_n| R^n$. This implies that $|g(z)| < |f(z)|$ for all $|z| = R$ with $R \geq R_0$. By Rouché's theorem, $N(f)$, the number of zeros of f in the region $|z| < R$, is the same as $N(f + g)$, the number of zeros of $f + g$. But $N(f) = n$ and $f + g = p$, so $N(p) = n$ showing that p has exactly n zeros.   ∎

The Local Mapping Theorem

In this part, we investigate fundamental properties of the inverse function of an analytic function f. When does it exist? Is it analytic? Do we have a formula for it in terms of f? All these questions can be answered with the help of Rouché's theorem and the counting theorem. Our investigations will lead to interesting new properties of analytic functions and shed new light on some classical results that we have studied earlier. In particular, we will give a simple proof of the maximum modulus principle. Interesting applications of these topics are presented in the exercises, including a formula due to Lagrange for the inversion of analytic functions with some of its applications to the solution to transcendental equations.

Suppose that f is analytic at z_0 and let $w_0 = f(z_0)$. We will say that f has **order** $m \geq 1$ at z_0 if the zero of $f(z) - w_0$ has order m at z_0. The order of $f(z)$ at z_0 is the order of the first nonvanishing term in the Taylor series expansion, past the constant term. Thus f has order m at z_0 if and only if $f^{(j)}(z_0) = 0$ for $j = 1, \ldots, m-1$ and $f^{(m)}(z_0) \neq 0$. In particular, $f(z)$ has order 1 at z_0 if and only if $f'(z_0) \neq 0$.

THEOREM 5
LOCAL MAPPING
THEOREM

Let f be a nonconstant analytic function in a neighborhood of z_0. Let $w_0 = f(z_0)$ and n be the order of f at z_0. There exists $R > 0$ and $\rho > 0$ such that every $w \neq w_0$ in $B_\rho(w_0)$ is attained by f at precisely n distinct points in $B_R(z_0)$,

Proof   Since $f(z) - w_0$ has a zero of order $n \geq 1$, write $f(z) - w_0 = a_n(z - z_0)^n +$

$a_{n+1}(z - z_0)^{n+1} + \cdots$, where $a_n \neq 0$. Then $\lim_{z \to z_0} \frac{f(z) - w_0}{(z - z_0)^n} = a_n \neq 0$. So we can find $R > 0$ such that

$$(8) \qquad \left| \frac{f(z) - w_0}{(z - z_0)^n} \right| > \frac{|a_n|}{2}, \qquad \text{for all } 0 < |z - z_0| \leq R.$$

In particular, for $z \neq z_0$ we have $f(z) \neq w_0$, and for $|z - z_0| = R$ we have

$$(9) \qquad \frac{|f(z) - w_0|}{R^n} > \frac{|a_n|}{2} \qquad \Rightarrow \qquad |f(z) - w_0| > \frac{|a_n| R^n}{2}.$$

By taking a smaller value of R if necessary, we can also assume that $f'(z) \neq 0$ for all $z \neq z_0$ with $|z - z_0| < R$. Now take $\rho = \frac{|a_n| R^n}{2}$, then for $|w_0 - w| < \rho$, set $g(z)$ to be the constant $w_0 - w$. For $|z - z_0| = R$, using (9), we find

$$|g(z)| = |w_0 - w| \leq \frac{|a_n| R^n}{2} < |f(z) - w_0|,$$

and so by Rouché's theorem, $N(f - w) = N(f - w_0 + g) = N(f - w_0)$, and since $N(f - w_0) = n$ it follows that $N(f - w) = n$ for all $|w - w_0| < \rho$. So w has n antecedents in $|z - z_0| < R$. The fact that $f'(z) \neq 0$ for all $0 < |z - z_0| < R$ guarantees that we do not have repeated roots, and so the antecedents are all distinct. ∎

The case $n = 1$ of the theorem deserves a separate statement.

THEOREM 6
INVERSE
FUNCTION
THEOREM

> Suppose that f is analytic on a region Ω and z_0 is in Ω. Then f is one-to-one on some neighborhood $B_r(z_0)$ if and only if $f'(z_0) \neq 0$.

Proof   If $f'(z_0) \neq 0$, f has order 1 and by Theorem 5 we can find R and ρ such that for any $0 < |w - w_0| < \rho$, f takes on the value w exactly once in $B_R(z_0)$. Also, f takes on the value w_0 only at z_0. Let U be the pre-image $f^{-1}[B_\rho(w_0)]$. Since $B_\rho(w_0)$ is open and f is continuous, U is open (Exercise 41, Section 2.2) and it clearly contains z_0. So we can find an open disk $B_r(z_0)$ that is contained in U and $B_R(z_0)$. Then f is one-to-one on $B_r(z_0)$, for if $f(z_1) = f(z_2)$ is a point in $B_\rho(w_0)$, Theorem 5 guarantees $z_1 = z_2$. Conversely, if $f'(z_0) = 0$, given $B_r(z_0)$ we can apply Theorem 5 to find a neighborhood of z_0 where f takes on values more than once; and so f is not one-to-one in this case. ∎

A few comments are in order regarding the inverse function theorem. The theorem can be obtained from the classical inverse function theorem in two variables. The latter states that the mapping $(x, y) \mapsto (u(x, y), v(x, y))$ is one-to-one in a neighborhood of (x_0, y_0) if the Jacobian of the mapping is nonzero at (x_0, y_0), where the Jacobian at (x, y) is given by

$$J(x, y) = \det \begin{vmatrix} u_x(x, y) & u_y(x, y) \\ v_x(x, y) & v_y(x, y) \end{vmatrix} = u_x(x, y) v_y(x, y) - u_y(x, y) v_x(x, y).$$

Hence, using the Cauchy-Riemann equations, we find that

$$J(x, y) = u_x^2(x, y) + u_y^2(x, y) = |f'(x + iy)|^2.$$

So if $f'(x_0 + iy_0) \neq 0$, then $J(x_0, y_0) \neq 0$ and Theorem 6 follows from the inverse function theorem for functions of two variables as claimed.

It is important to keep in mind that the condition $f'(z) \neq 0$ for all z in Ω does not imply that f is one-to-one on Ω. Consider $f(z) = e^z$; then $f'(z) = e^z \neq 0$ for all z, yet f is not one-to-one on the whole complex plane. Theorem 6 only guarantees that f is one-to-one in *some* neighborhood of any given point z_0 where $f'(z_0) \neq 0$. Because this neighborhood depends on z_0 in general, an obvious question is whether we can estimate its size in terms of the sizes of f and f'. An answer to this question was given by the German mathematician Edmund Landau (1877–1938) and is known as the **Landau estimate**. See Exercise 36.

The following two corollaries describe important properties of analytic functions, which are direct applications of the local mapping theorem.

COROLLARY 1
OPEN MAPPING
PROPERTY

A nonconstant analytic function f maps open sets into open sets. A function with this property is said to be **open**.

Proof   Let f map Ω to $f[\Omega]$. Given w_0 in $f[\Omega]$, let z_0 be some point in Ω where $f(z_0) = w_0$. Since Ω is open, we can find a neighborhood $B_R(z_0)$ contained in Ω; by applying Theorem 5 to $B_R(z_0)$, we find that each point in the associated $B_\rho(w_0)$ is assumed by f. Thus $B_\rho(w_0)$ is contained in $f[\Omega]$, and $f[\Omega]$ is open.   ∎

COROLLARY 2
MAPPING OF
REGIONS

If Ω is a nonempty region (open and connected set) and f is a nonconstant analytic function on Ω, then $f[\Omega]$ is a region.

Proof   By Corollary 1, $f[\Omega]$ is open. To show that $f[\Omega]$ is connected, suppose that $f[\Omega] = U \cup V$, where U and V are open and disjoint. Since f is continuous, $f^{-1}[U]$ and $f^{-1}[V]$ are open (note that since f is defined on Ω, when taking a pre-image, we only consider those points in Ω). Clearly, $f^{-1}[U]$ and $f^{-1}[V]$ are disjoint and their union is Ω. Since Ω is connected, either $\Omega = f^{-1}[U]$ or $\Omega = f^{-1}[V]$, and hence $f[\Omega] = U$ or $f[\Omega] = V$, implying that $f[\Omega]$ is connected.   ∎

Another interesting application of the open mapping property of analytic functions is a simple proof of the maximum principle.

COROLLARY 3
MAXIMUM
MODULUS
PRINCIPLE

If f is analytic on a region Ω, such that $|f|$ attains a maximum at some point in Ω, then f is constant in Ω.

Proof   Suppose f is nonconstant. Let z_0 be an arbitrary point in Ω; we will show that $|f(z_0)|$ is not a maximum. By Corollary 1, f is open. Let $B_\rho(w_0)$ be a neighborhood of $w_0 = f(z_0)$, contained in $f[\Omega]$. Clearly, $f(z_0)$ does not have the largest modulus among all the points in $B_\rho(w_0)$ (Figure 7).   ∎

We next use the variant of the counting theorem to give a formula for the inverse function of f in terms of an integral involving f.

**THEOREM 7
INVERSE
FUNCTION
FORMULA**

Suppose that $f(z)$ is analytic at z_0 and $f'(z_0) \neq 0$. Then f has an inverse function $z = g(w)$ in a neighborhood $B_\rho(w_0)$ of w_0 given by

$$(10) \qquad g(w) = \frac{1}{2\pi i} \int_{C_R(z_0)} z \frac{f'(z)}{f(z) - w} \, dz.$$

Furthermore, $g(w)$ is analytic on $B_\rho(w_0)$.

Proof   Let R and ρ be as in Theorem 5. For w in $B_\rho(w_0)$, $f(z) - w$ has exactly one zero in $B_R(z_0)$, which is the pre-image of w by the mapping f. Applying the variant of the counting theorem, we see that the integral on the right of (10) is equal to the pre-image of w. Finally, g is analytic by Theorem 4, Section 3.5   ∎

Theorem 7 can be used to derive a useful formula due to Lagrange for the inversion of power series. See Exercise 33.

Theorem 7 can be viewed as a statement about the local existence of inverse functions. If f is one-to-one on a region Ω, there is no ambiguity in defining the inverse function f^{-1} on the whole region $f[\Omega]$. Indeed, given w in $f[\Omega]$, we define $f^{-1}(w)$ to be the unique z in Ω with $f(z) = w$. For f analytic and one-to-one, we have the following result, which should be compared to Theorem 4, Section 2.3.

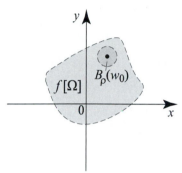

Figure 7

**COROLLARY 4
A GLOBAL INVERSE
FUNCTION**

Suppose that f is analytic and one-to-one on a region Ω. Then its inverse function f^{-1} exists and is analytic on the region $f[\Omega]$. Moreover,

$$(11) \qquad \frac{d}{dw} f^{-1}(w) = \frac{1}{f'(z)}, \qquad \text{where } w = f(z).$$

Proof   By Theorem 7, f^{-1} is analytic in a neighborhood of each point in Ω, and hence it is analytic on Ω. Also, since f is one-to-one, Theorem 6 implies that $f'(z) \neq 0$ for all z in Ω. Differentiating both sides of the identity $z = f^{-1}(f(z))$, we obtain $1 = \frac{d}{dw} f^{-1}(w) f'(z)$, which is equivalent to (11).   ∎

A function f that is analytic and one-to-one is also called a **univalent** function. These functions will play a very important role in the next chapter, on conformal mappings.

Exercises 5.7

In Exercises 1–6, use the method of Example 1 to find the number of zeros in the first quadrant of the given polynomial.

1. $z^2 + 2z + 2$.

2. $z^2 - 2z + 2$.

3. $z^3 - 2z + 4$.

4. $z^3 + 5z^2 + 8z + 6$.

5. $z^4 + 8z^2 + 16z + 20$.

6. $z^5 + z^4 + 13z^3 + 10$.

In Exercises 7–14, use Rouché's theorem to determine the number of zeros of the given function in the given region.

7. $z^3 + 3z + 1$, $|z| < 1$.

8. $z^4 + 4z^3 + 2z^2 + 11$, $|z| < 2$.

9. $7z^3 + 3z^2 + 11$, $|z| < 1$.

10. $7z^3 + z^2 + 11z + 1$, $1 < |z|$.

11. $4z^6 + 41z^4 + 46z^2 + 9$, $2 < |z| < 4$.

12. $z^4 + 50z^2 + 49$, $3 < |z| < 4$.

13. $e^z - 3z$, $|z| < 1$.

14. $e^{z^2} - 4z^2$, $|z| < 1$.

15.   Show that the equation $3 - z + 2e^{-z} = 0$ has exactly one root in the right half-plane $\operatorname{Re} z > 0$. [Hint: Use Rouché's theorem and contours such as the one in Figure 8.]

16.   Suppose that $\operatorname{Re} w > 0$ and let a be any complex number. Show that $w - z + ae^{-z} = 0$ has exactly one root in the right half-plane $\operatorname{Re} z > 0$.

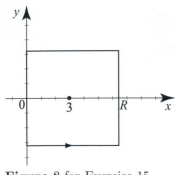

Figure 8 for Exercise 15.

In Exercises 17–20, evaluate the given path integral. As usual, $C_R(z_0)$ stands for the positively oriented circle with radius $R > 0$ and center z_0.

17. $\displaystyle \int_{C_1(0)} \frac{dz}{z^5 + 3z + 5}.$

18. $\displaystyle \int_{C_1(0)} \frac{e^z - 12z^3}{e^z - 3z^4}\, dz.$

19. $\displaystyle \int_{C_2(0)} \frac{ze^{iz}}{z^2 + 1}\, dz.$

20. $\displaystyle \int_{C_1(0)} \frac{z^3 e^{z^2}}{e^{z^2} - 1}\, dz.$

21. Summing roots of unity.   We will give a simple proof based on the variant of the counting theorem of the fact that, for $n \geq 2$, the sum of the n nth roots of unity is 0 (see Exercise 53, Section 1.3). Let S denote this sum.
(a)   Using Theorem 3, explain why

$$S = \frac{1}{2\pi i} \int_{C_R(0)} z \frac{nz^{n-1}}{z^n - 1}\, dz = \frac{n}{2\pi i} \int_{C_{1/R}(0)} \frac{1}{1 - z^n} \frac{dz}{z^2},$$

where $R > 1$. [Hint: To prove the second equality make a suitable change of variables.]
(b)   Evaluate the second integral in (a) using Cauchy's generalized integral formula and conclude that $S = 0$.

22.   Examples concerning Lemma 1. Give an example of a function f with an essential singularity at 0 such that $\frac{f'(z)}{f(z)}$ has
(a)   an essential singularity at 0;
(b)   a pole of order $m \geq 2$ at 0.
Use functions like $e^{1/z}$ in your examples.

23. Meromorphic Rouché's theorem. Suppose that C is a simple closed path, Ω is the region inside C, and f and g are meromorphic inside and on C, having no zeros or poles on C. Show that if $|g(z)| < |f(z)|$ for all z on C, then $N(f+g) - P(f+g) = N(f) - P(f)$. [Hint: Repeat the proof of Rouché's theorem. What can you say about the values of ϕ in the present case?]

24. We complete the geometric argument given in the text concerning Rouché's theorem. (i) Show that for each z, we can find a branch of the argument where $\arg f(z) - \frac{\pi}{2} < \arg\left(f(z) + g(z)\right) < \arg f(z) + \frac{\pi}{2}$.
(ii) Using connectedness, we can show that this inequality holds for the special

argument function $\arg f(z)$, which we used to define $\Delta_C f(z)$. Show that $\Delta_C f(z) - \pi < \Delta_C(f(z) + g(z)) < \Delta_C f(z) + \pi$, and use the fact that $\Delta_C(f(z) + g(z))$ must be an integer multiple of 2π to prove that $\Delta_C f(z) = \Delta_C(f(z) + g(z))$.

25. Project Problem: Hurwitz's theorem. In this exercise, we outline a proof of a useful theorem due to the German mathematician Adolf Hurwitz (1859–1919). The theorem states the following.

> Suppose that $\{f_n\}$ is a sequence of analytic functions on a region Ω converging uniformly on every closed and bounded subset of Ω to a function f. Then either
> (i) f is identically 0 on Ω; or
> (ii) if $B_r(z_0)$ is any open disk in Ω such that f does not vanish on $C_r(z_0)$, then f_n and f have the same number of zeros in $B_r(z_0)$ for all sufficiently large n.

In particular, if f is not identically 0 and f has p distinct zeros in Ω, then so do the functions f_n for all sufficiently large n.

Before we outline the proof, let us observe that f is analytic by Theorem 4, Section 4.2. Also, note that the theorem guarantees that, for large n, f_n and f have the same number of zeros, but these zeros are not necessarily the same for f_n and f. To see this, take $f_n(z) = z - \frac{1}{n}$ and $f(z) = z$. Finally, observe that the possibility that f is identically zero can arise, even if the f_n's are all nonzero. Simply take $f_n(z) = \frac{1}{n}$.

Fill in the details in the following proof. Suppose that f is not identically 0 in Ω, and let z_0 denote a zero of f. Let $S_r(z_0)$ be a closed disk such that f is nonvanishing on $C_r(z_0)$. (Why can we find such a disk?) Let $m = \min|f|$ on $C_r(z_0)$. Then $m > 0$ (why?). Apply uniform convergence to get an index N such that $n > N$ implies that $|f_n - f| < m \le |f|$ on $C_r(z_0)$. Complete the proof by applying Rouché's theorem.

Hurwitz's theorem has many interesting applications. We start with some theoretical properties and then give some applications to counting zeros of analytic functions (Exercises 28–30).

26. Suppose that f_n converges to f uniformly on every closed and bounded subset of Ω with f_n analytic and vanishing nowhere on Ω. Then either f is identically 0 or f vanishes nowhere on Ω.

27. Univalent functions. In this exercise we prove an interesting theorem concerning univalent functions.
(a) Suppose that $\{f_n\}$ is a sequence of univalent functions on Ω that converges to f uniformly on every closed and bounded subset of Ω, and f is not identically constant on Ω. Then f is univalent. [Hint: Let $z \neq z_0$ be in Ω. Apply Exercise 26 to the sequence of functions $\{f_n - f_n(z_0)\}$.]
(b) Give an example of a sequence of univalent functions converging uniformly on the closed unit disk to a constant function.

28. Counting zeros with Hurwitz's theorem. If we want to find the number of zeros inside the unit disk of the polynomial $p(z) = z^5 + z^4 + 6z^2 + 3z + 1$, then we cannot just apply Rouché's theorem since there is not one single coefficient of the polynomial whose absolute value dominates the sum of the absolute values of the other coefficients. Here is how we can handle this problem.

(a)   Consider the polynomials $p_n(z) = p(z) - \frac{1}{n}$. Show that $\{p_n\}$ converges uniformly on the closed unit disk to $p(z)$.

(b)   Apply Rouché's theorem to show that p_n has two zeros in the unit disk.

(c)   Apply Hurwitz's theorem to show that $p(z)$ has two zeros inside the unit disk.

(d)   Verify your answer in (c) by finding a numerical approximation of the roots of $p(z)$.

In Exercises 29–30, modify the steps in Exercise 28 to find the zeros of the given polynomial in the stated region.

29. $z^5 + z^4 + 6z^2 + 3z + 11$, $|z| < 1$.   **30.** $4z^4 + 6z^2 + z + 1$, $|z| < 1$.

31. When there exists a continuous one-to-one function f from a set A onto a set B, such that f^{-1} is also continuous, the two sets are said to be **homeomorphic**. Homeomorphic sets will share topological properties like openness, connectedness, and simple connectedness. Show that if f is one-to-one and analytic on Ω, then Ω and $f[\Omega]$ are homeomorphic.

32. Minimum modulus principle.   Use the fact that a nonconstant analytic function $f(z)$ is open to show that if $f(z)$ is nonzero for z in Ω then $|f|$ does not attain a minimum on Ω.

33.   Project Problem: Lagrange's inversion formula. In this exercise, we outline a proof of the following useful inversion formula for analytic functions due to the French mathematician Joseph-Louis Lagrange (1736–1813).

Suppose that $f(z)$ is analytic at z_0 and $f'(z_0) \neq 0$, and let

$$(12) \qquad\qquad \phi(z) = \frac{z - z_0}{f(z) - w_0}, \quad (z \neq z_0).$$

Then the inverse function $z = g(w)$ has a power series expansion

$$(13) \qquad g(w) = z_0 + \sum_{n=1}^{\infty} b_n (w - w_0)^n, \quad \text{where} \quad b_n = \frac{1}{n!} \frac{d^{n-1}}{dz^{n-1}} [\phi(z)]^n \bigg|_{z=z_0}.$$

Fill in the details in the following proof. Note that $\phi(z)$ is analytic at z_0 (why?). To prove (13), start with the formula for the inverse function (10) and differentiate with respect to w and then integrate by parts to obtain

$$\begin{aligned} g'(w) &= \frac{1}{2\pi i} \int_{C_R(z_0)} z \frac{f'(z)}{(f(z) - w)^2} \, dz \\ &= -\frac{1}{2\pi i} \int_{C_R(z_0)} z \, d((f(z) - w)^{-1}) = \frac{1}{2\pi i} \int_{C_R(z_0)} \frac{dz}{f(z) - w}. \end{aligned}$$

Justify successive differentiation under the integral sign and get

$$g^{(n)}(w) = \frac{(n-1)!}{2\pi i} \int_{C_R(z_0)} \frac{dz}{(f(z) - w)^n} = \frac{(n-1)!}{2\pi i} \int_{C_R(z_0)} [\phi(z)]^n \frac{dz}{(z - z_0)^n}.$$

Hence by Cauchy's generalized integral formula

$$g^{(n)}(w_0) = \frac{d^{n-1}}{dz^{n-1}} [\phi(z)]^n \bigg|_{z=z_0},$$

and (13) follows from the formula for the Taylor coefficients.

34. Let a be an arbitrary complex number. Consider the equation $z = a + we^z$. Show that a solution is given by

$$z = a + \sum_{n=1}^{\infty} \frac{n^{n-1}e^{na}}{n!} w^n,$$

for $|w| < e^{-1-\operatorname{Re} a}$. [Hint: Let $z_0 = a$, $w = (z-a)e^{-z}$, $\phi(z) = \frac{z-a}{(z-a)e^{-z}} = e^z$, and apply Lagrange's formula.]

35. Lambert's w-function. This function has been applied in quantum physics, fluid mechanics, biochemistry, and combinatorics. It is named after the German mathematician Johann Heinrich Lambert (1728–1777). The **Lambert function** or **Lambert w-function** is defined as the inverse function of $f(z) = ze^z$. Using the technique of Exercise 34, based on Lagrange's formula, show that the solution of $w = ze^z$ is

$$z = \sum_{n=1}^{\infty} \frac{(-1)^{n-1}n^{n-2}}{(n-1)!} w^n, \quad |w| < \frac{1}{e}.$$

36. Project Problem: Landau's estimate. In this exercise we present Landau's solution of the following problem: Given f analytic on a closed disk $S_R(z_0)$ with $f'(z_0) \neq 0$, find $r > 0$ such that f is one-to-one on the open disk $B_r(z_0)$. Landau's solution: It suffices to take $r = \dfrac{R^2|f'(z_0)|^2}{4M}$, where M is the maximum value of $|f|$ on $C_R(z_0)$.

  Fill in the details in the following proof. It suffices to choose r so that $f'(z) \neq 0$ for all z in $B_r(z_0)$ (why?). We can without loss of generality take $z_0 = 0$ and $w_0 = f(z_0) = 0$ (why?). Then for $|z| < R$, $f(z) = a_1 z + a_2 z^2 + \cdots$. Write $r = \lambda R$, where $0 < \lambda < 1$ is to be determined so that $f'(z) \neq 0$ for all z in $B_{\lambda R}(0)$. For z_1 and z_2 in $B_{\lambda R}(0)$, we have

$$\left| \frac{f(z_1) - f(z_2)}{z_1 - z_2} \right| = \left| a_1 + \sum_{n=2}^{\infty} a_n \left(z_1^{n-1} + z_1^{n-2}z_2 + \cdots + z_1 z_2^{n-2} + z_2^{n-1} \right) \right|$$

$$\geq |a_1| - \sum_{n=2}^{\infty} n|a_n|\lambda^{n-1}R^{n-1}.$$

If we can choose λ so that

(14)
$$\sum_{n=2}^{\infty} n|a_n|\lambda^{n-1}R^{n-1} < |a_1|,$$

this would make the absolute value of the difference quotient for the derivative > 0, independently of z_1 and z_2 in $B_{\lambda R}(0)$, which in turn would imply that $f'(z) \neq 0$ for all z in $B_{\lambda R}(0)$ and complete the proof. So let us show that we can choose λ so that (14) holds. Let $M = \max|f(z)|$ for z on $C_R(0)$. Cauchy's estimate yields $|a_n| \leq \frac{M}{R^n}$. Then

$$\sum_{n=2}^{\infty} n|a_n|\lambda^{n-1}R^{n-1} \leq \frac{M}{R} \sum_{n=2}^{\infty} n\lambda^{n-1} = \frac{M}{R} \frac{\lambda(2-\lambda)}{(1-\lambda)^2} < \frac{M}{R} \frac{2\lambda}{(1-\lambda)^2}.$$

(Use $\sum_{n=2}^{\infty} n\lambda^{n-1} = \frac{d}{d\lambda}(\lambda^2 + \lambda^3 + \cdots) = \frac{d}{d\lambda}\frac{\lambda^2}{1-\lambda} = \frac{2\lambda - \lambda^2}{(1-\lambda)^2}$.) Consider the choice $\lambda = \frac{R|a_1|}{4M}$. This yields $\lambda \leq \frac{1}{4}$ and $\sum_{n=2}^{\infty} n|a_n|\lambda^{n-1}R^{n-1} < \frac{|a_1|}{4\lambda}\frac{2\lambda}{(1-\lambda)^2} = \frac{|a_1|}{2}\frac{1}{(1-\lambda)^2} \leq \frac{8}{9}|a_1|$. (The maximum of $1/(1-\lambda)^2$ on the interval $[0, 1/4]$ occurs at $\lambda = 1/4$ and is equal to $\frac{16}{9}$.) Hence (14) holds for this choice of λ, and for this choice, we get $r = \lambda R = \frac{R^2|a_1|}{4M}$, which is what Landau's estimate says.

Topics to Review

To motivate the material in Sections 6.1–6.3, you should review Section 2.5. Section 6.2 treats linear fractional transformations. It uses the maximum modulus principle (Theorem 5, Section 3.7) and Example 3, Section 3.7. Section 6.3 expands on the applications from Section 2.5 and uses the Poisson integral formula on the disk in Section 3.8 to derive the corresponding formula in the upper half-plane. Section 6.4 is based on Section 6.3. In addition to the theory of analytic functions that we have developed in previous chapters, in Sections 6.5 and 6.6, we will also derive additional facts from calculus of several variables, in particular, Green's identities. The approach to Green's functions and Neumann functions in Sections 6.5 and 6.6 is based on a change of variables that we present in Section 6.5 and that can be motivated by the change of variables that we performed in Section 3.8 to derive the Poisson integral formula on the disk.

Looking Ahead

This chapter is optional, but is highly recommended as an illustration of the advanced methods of complex analysis in solving applied problems. Sections 6.1 and 6.2 are background or reference material. Sections 6.3, 6.4, and 6.5-6.6 can be covered independently. Sections 6.5 and 6.6 are at a higher level perhaps than the previous sections. They can be covered after acquiring more familiarity with boundary value problems from Chapter 8.

6

CONFORMAL MAPPINGS

First, it is necessary to study the facts, to multiply the number of observations, and then later to search for formulas that connect them so as thus to discern the particular laws governing a certain class of phenomena. In general, it is not until after these particular laws have been established that one can expect to discover and articulate the more general laws that complete theories by bringing a multitude of apparently very diverse phenomena together under a single governing principle
-Augustin Louis Cauchy

This chapter presents a sampling of successful applications of complex analysis in applied mathematics, engineering, and physics. After laying down the theory and methods of conformal mappings in Section 6.1 and expanding our list of conformal mappings in Section 6.2, we revisit in Section 6.3 the Dirichlet problems that we started in Section 2.5 and go far beyond the examples of that section. In particular, we derive Poisson's integral formula in the upper half-plane and many other regions, by performing a change of variables in the formula on the disk. In Section 6.4, we broaden the scope of our applications with the Schwarz-Christoffel transformation, which is a method for finding conformal mappings of regions bounded by polygonal paths. The section contains interesting applications from fluid flow.

Recall that Poisson's formula on the disk (or the upper half-plane) gives the solution of Dirichlet problems in the form of an integral involving the boundary function. A natural question is whether a similar formula, a generalized Poisson integral formula, exists on different regions of the plane. In Section 6.5, we derive this formula on simply connected regions, in terms of the so-called Green's function. The solution is based on a simple change of variables and the mean value property of harmonic functions. Our approach justifies the properties of this important tool, Green's functions, and illustrates in a natural way their applications to the solution of boundary value problems.

In Section 6.6, we tackle other famous boundary value problems, involving Laplace's equation and Poisson's equation. We take the same approach as in Section 6.5, based on changes of variables using conformal mappings, and derive general formulas for the solution in terms of Green's functions and the so-called Neumann function. Both functions are computed explicitly in important special cases.

6.1   Basic Properties

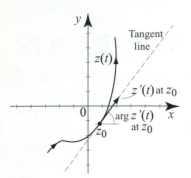

Figure 1 The direction of the tangent line at $z(t)$ is given by $\arg z'(t)$.

Our goal in this section is to present properties of mappings by analytic functions. These basic properties are interesting in their own right and will be very useful to us when solving partial differential equations involving the Laplacian. We start with a review from calculus of the notion of tangent lines to curves in parametric form. We will state these results using the convenient complex notation.

Suppose that γ is a smooth path parametrized by $z(t) = x(t) + i\,y(t)$, $a \le t \le b$. Write $\frac{dz}{dt} = z'(t) = x'(t) + iy'(t)$. Let us also assume that $z'(t) \ne 0$, as this will guarantee the existence of a tangent. The tangent line to the curve at $z(t)$ points in the direction of $z'(t)$; we can characterize this direction by specifying $\arg z'(t)$ (see Figure 1). Thus the direction of the tangent line at a point on a path $z(t)$ is given by the argument of $z'(t)$.

Let z_0 be a point in the z-plane, let γ_1 and γ_2 be two smooth paths that intersect at z_0, and let L_1 and L_2 denote the tangent lines to γ_1 and γ_2 at z_0. We will say that γ_1 and γ_2 intersect at angle α at z_0 if the tangent lines L_1 and L_2 intersect at angle α at z_0 (Figure 2).

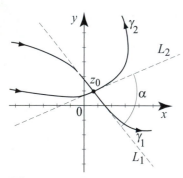

Figure 2 The curves γ_1 and γ_2 intersect at angle α.

To explain the geometric meaning of the mapping properties discussed in this section, let us consider the simple example of a linear mapping $f(z) = az + b$, where $a \ne 0$ and b are complex numbers. As usual, we consider a mapping as taking points in the z-plane to points in the w-plane. Using our geometric interpretation of addition and multiplication of complex numbers, we see that the effect of the linear mapping $f(z) = az + b$ is to rotate by a fixed angle equal to $\arg a$, dilate by a factor equal to $|a|$, and then translate by b. (Note that the rotation and dilation commute, so it does not matter which one you apply first. But you cannot change the order of the translation; it comes last.) In particular, if γ_1 and γ_2 are two smooth paths that intersect at angle α at z_0, then their images by f are two paths in the w-plane that intersect at $w_0 = f(z_0)$; since $f(z)$ has rotated each curve by the same angle $\arg a$, the angle of their intersection in the w-plane is still α. Furthermore, γ_1's orientation as being either clockwise or counterclockwise of γ_2, is preserved (Figure 3).

Figure 3 A linear mapping $f(z) = az + b$ ($a \ne 0$) rotates by an angle $\arg a$, dilates by a factor $|a|$, and translates by b. In particular, it preserves angles between curves.

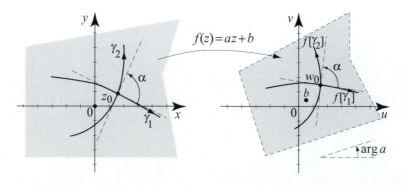

With the example of the linear mapping in mind, we introduce the following basic property. A mapping $w = f(z)$ is said to be **conformal** at z_0 if f preserves angles between curves at z_0 and preserves their angular orientation.

Now suppose that $f(z)$ is analytic at z_0 and $f'(z_0) \neq 0$. From our study of the derivative in Section 2.3 (Theorem 5), we know that f is approximately linear in the neighborhood of z_0. More precisely, we have

$$f(z) = f(z_0) + f'(z_0)(z - z_0) + \epsilon(z)(z - z_0),$$

where $\epsilon(z) \to 0$ as $z \to z_0$. So, near z_0, we can write $f(z) \approx f'(z_0)z + b$, where $b = -f'(z_0)z_0 + f(z_0)$ is a constant. From what we just discovered about linear mappings, this suggests that f is conformal at z_0; it rotates by an angle $\theta = \arg f'(z_0)$ and scales by a factor $|f'(z_0)|$. Indeed, we have the following important result.

**THEOREM 1
CONFORMAL
PROPERTY**

Suppose that f is analytic at z_0 and $f'(z_0) \neq 0$, then f is conformal at z_0.

Proof   Let γ be any smooth path through z_0, parametrized by $z(t) = x(t) + i\,y(t)$, with $z(t_0) = z_0$ and $z'(t_0) \neq 0$. Then the image of γ by f is a path parametrized by $f(z(t))$ that goes through $w_0 = \gamma(z_0)$ in the w-plane. Since $z'(t_0) \neq 0$, the direction of the tangent line to γ at z_0 is $\arg z'(t_0)$. Also, from the chain rule, $\frac{d}{dt}f(z(t))\big|_{t_0} = f'(z(t_0))z'(t_0) \neq 0$ and so the direction of the tangent line to $f(z(t))$ at $w_0 = f(z_0)$ is

$$\arg\left(\frac{d}{dt}f(z(t))\bigg|_{t_0}\right) = \arg\left(f'(z(t_0))z'(t_0)\right) = \arg f'(z(t_0)) + \arg z'(t_0).$$

Hence $f(z)$ rotates the tangent line at z_0 by a fixed angle $\arg f'(z_0)$. Since f rotates any two tangent lines intersecting at z_0 by the same angle $\arg f'(z_0)$, it preserves the angle of intersection and their orientation.   ∎

Boundary Behavior

We will use conformal mappings to transform boundary value problems consisting of Laplace's equation along with boundary conditions. We know from Section 2.5, Theorem 3, that Laplace's equation is not changed by a conformal mapping. To handle the effect of the mapping on the boundary conditions, it would be nice to know that the boundary is mapped to the boundary. But as the following simple example shows, this may fail in general.

EXAMPLE 1   Boundary points mapped to interior points
Consider $f(z) = z^2$ and $\Omega = \{z = re^{i\theta} : \frac{1}{2} < r < 1,\ -\pi < \theta < \pi\}$. Then f is analytic and $f'(z) = 2z \neq 0$ for all z in Ω. Hence $f(z)$ is conformal at each z in Ω. Since $z^2 = r^2 e^{2i\theta}$ it is easy to see that $f[\Omega]$ is the annulus

$$f[\Omega] = \{w = re^{i\theta} : \frac{1}{4} < r < 1,\ -2\pi < \theta < 2\pi\} = \{w = re^{i\theta} : \frac{1}{4} < r < 1,\ 0 \leq \theta \leq 2\pi\}.$$

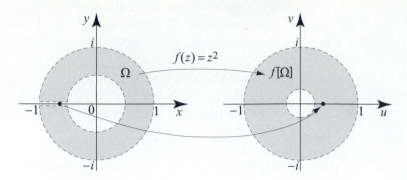

Figure 4 The function $f(z) = z^2$ is conformal in Ω. It takes the interval $(-1, -\frac{1}{2})$ on the boundary of Ω to the interval $(\frac{1}{4}, 1)$ in the interior of $f[\Omega]$. Thus f does not map boundary points to boundary points.

It is also easy to see that the boundary points $z = x$, $-1 \leq x \leq -\frac{1}{2}$ are mapped to the *interior* points $w = u$, $\frac{1}{4} \leq u \leq 1$ (see Figure 4). Thus f does not map the boundary of Ω to the boundary of $f[\Omega]$. ∎

Recall from Section 5.7 that the condition $f'(z) \neq 0$ ensures that f is one-to-one locally at z. The function may fail to be one-to-one on the entire region Ω, which was the case in Example 1. We will show that if $f(z)$ is analytic and one-to-one then it will map boundary to boundary. Before we do so, let us clarify certain issues. We know from Section 5.7 that if f is analytic and nonconstant on a region Ω, then $f[\Omega]$ is a region. So all the points in Ω are mapped to (interior) points in $f[\Omega]$. Now f might not be defined on the boundary of Ω, so we need a special definition to describe how f maps the boundary of Ω. We will say that f **maps the boundary of Ω to the boundary of $f[\Omega]$** if the following condition holds. If $\{z_n\}$ is any sequence in Ω converging to α_0 on the boundary of Ω and β is any point in $f[\Omega]$, then $\{f(z_n)\}$ does not converge to β. So if $\{f(z_n)\}$ converges, it must converge to a boundary point of $f[\Omega]$ or to infinity. (There are examples where $\{f(z_n)\}$ does not converge; see Exercise 21.) If the region Ω is unbounded, we allow putting $\alpha_0 = \infty$.

We will say that f **maps the boundary of Ω onto the boundary of $f[\Omega]$** if for every point w_0 on the boundary of $f[\Omega]$, we can find a sequence z_n in Ω where $z_n \to \alpha_0$, α_0 being on the boundary of Ω, and $f(z_n) \to w_0$.

We have the following important result.

THEOREM 2
BOUNDARY
BEHAVIOR

Suppose that f is analytic and one-to-one in a region Ω. Then f maps the boundary of Ω to the boundary of $f[\Omega]$, and this map is onto.

Proof   We will prove that f maps the boundary of Ω to the boundary of $f[\Omega]$, leaving the "onto" part for Exercise 20. Let $\{z_n\}$ be a sequence in Ω such that $f(z_n) \to \beta$, where β is an interior point of $f[\Omega]$. Since f is analytic and one-to-one, f^{-1} exists and is analytic, and hence continuous. Thus

$$\lim_{n \to \infty} z_n = \lim_{n \to \infty} \left(f^{-1}(f(z_n)) \right) = f^{-1} \left(\lim_{n \to \infty} f(z_n) \right) = f^{-1}(\beta),$$

so z_n converges to an interior point of Ω. ∎

If the conformal mapping can be extended to a continuous function on the boundary, the following version of Theorem 2 will be useful.

COROLLARY 1

> Suppose that f is analytic and one-to-one in a region Ω. Let α be any point of the boundary of Ω such that f is continuous at α. Then $f(\alpha)$ is a point on the boundary of $f[\Omega]$.

Proof   Let $\{z_n\}$ be a sequence in Ω converging to α on the boundary of Ω. Since f is continuous at α, $f(z_n)$ converges to $f(\alpha)$. By Theorem 2, $f(\alpha)$ is a boundary point of $f[\Omega]$.   ∎

The fact that the boundary is mapped to the boundary helps us determine the image of a region from our knowledge of the image of the boundary and one interior point. Let us illustrate this useful technique by revisiting an example from Section 1.6.

EXAMPLE 2   Mapping of regions

Figure 5 The fact that the boundary of $f[\Omega]$ is the u-axis implies that $f[\Omega]$ is either the upper or lower half-plane. We decide which half it is by checking the image of one point. Note how the right angles at $z = \pm\frac{\pi}{2}$ got flattened by f in the w-plane. The function f is still conformal in Ω, even though f is not conformal at two points on the boundary.

Let $f(z) = \sin z$ and $\Omega = \{z = x + iy : -\frac{\pi}{2} < x < \frac{\pi}{2}, \ y > 0\}$. Thus Ω is the semi-infinite vertical strip shown in Figure 5. Since $\sin z_1 = \sin z_2$ if and only if $z_1 = z_2 + 2k\pi$ (k an integer), we see that f is one-to-one on Ω. Also, f is continuous on the boundary of Ω.

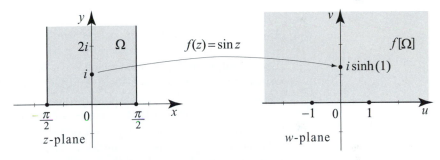

We start by determining the image of the boundary. For $z = x$ real, we have $f(z) = \sin x$, and so f maps the interval $-\frac{\pi}{2} \leq x \leq \frac{\pi}{2}$ onto the interval $[-1, 1]$. For $z = \frac{\pi}{2} + iy$, we have $f(z) = \sin(\frac{\pi}{2} + iy) = \cosh y$, a real number (see (16), Section 1.6). So f maps the vertical semi-infinite line $z = \frac{\pi}{2} + iy$ ($y \geq 0$) onto the semi-infinite interval $[1, \infty)$. For $z = -\frac{\pi}{2} + iy$, we have $f(z) = \sin(-\frac{\pi}{2} + iy) = -\cosh y$ (see (16), Section 1.6). So f maps the vertical semi-infinite line $z = -\frac{\pi}{2} + iy$ ($y \geq 0$) onto the semi-infinite interval $(-\infty, -1]$. Thus, f maps the boundary of Ω to the real axis. According to Theorem 2, the image of the vertical strip has boundary the u-axis, so it is either the upper or the lower half-plane. Checking the value of f at one point in Ω, say $z = i$, we find $f(i) = \sin(i) = i\sinh(1)$, which is a point in the upper half. Thus the image of Ω is the upper half-plane.   ∎

As a further application, we consider the **Joukowski function**

(1)
$$J(z) = \frac{1}{2}\left(z + \frac{1}{z}\right) \quad (z \neq 0).$$

This function has applications in aerospace engineering.

EXAMPLE 3   The Joukowski mapping

(a) Show that $J(z)$ maps the upper unit semicircle $\sigma = \{z : z = e^{i\theta}, 0 \leq \theta \leq \pi\}$ onto the real interval $J[\sigma] = [-1, 1]$, and the semi-infinite intervals $[1, \infty)$ and $(-\infty, -1]$ onto themselves.

(b)   Show that the Joukowski function maps the set $\Omega = \{z : |z| \geq 1, 0 \leq \operatorname{Arg} z \leq \pi\}$ onto the upper half-plane $\{w = u + iv : v \geq 0\}$ (see Figure 6). A more precise description of the Joukowski mapping is outlined in Exercise 17.

Figure 6 The Joukowski function maps the region Ω one-to-one onto the upper half-plane. It also maps the upper semi-circle of radius $R > 1$, $x^2 + y^2 = R^2$, $y \geq 0$, to the upper semi-ellipse
$$\frac{(\operatorname{Re} w)^2}{\left[\frac{1}{2}\left(R + \frac{1}{R}\right)\right]^2} + \frac{(\operatorname{Im} w)^2}{\left[\frac{1}{2}\left(R - \frac{1}{R}\right)\right]^2} = 1,$$
$\operatorname{Im} w \geq 0$. (See Exercise 17.)

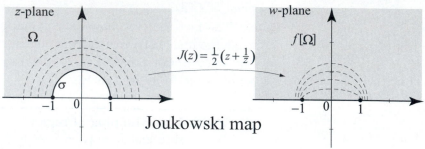

Joukowski map

Solution   (a) For $z = e^{i\theta}$, $0 \leq \theta \leq \pi$, we have

$$w = J(z) = \frac{1}{2}\left(e^{i\theta} + \frac{1}{e^{i\theta}}\right) = \frac{1}{2}(e^{i\theta} + e^{-i\theta}) = \cos\theta.$$

As θ varies from 0 to π, w varies from 1 to -1, showing that the image of the upper semi-circle is the interval $[-1, 1]$. To determine the images of the semi-infinite intervals $[1, \infty)$ and $(-\infty, -1]$, let $z = x$; then $J(z) = \frac{1}{2}(x + \frac{1}{x})$. As x varies through $[1, \infty)$ or $(-\infty, -1]$, $J(x)$ varies through the same interval (in the w-plane).

(b) We showed in (a) that J maps the boundary of Ω onto the real axis. If we can show that J is conformal and one-to-one, it will follow from Corollary 1 that $J[\Omega]$ is either the upper or the lower half of the w-plane. We can then determine which half it is by checking the image of one point in Ω. That J is conformal at all points in Ω is clear from $J'(z) = \frac{1}{2}\left(1 - \frac{1}{z^2}\right) \neq 0$ for all $|z| > 1$. To show that J is one-to-one, let z_1, z_2 be in Ω, and suppose that $J(z_1) = J(z_2)$. Then

$$z_1 + \frac{1}{z_1} = z_2 + \frac{1}{z_2} \quad \Rightarrow \quad \frac{z_1^2 + 1}{z_1} = \frac{z_2^2 + 1}{z_2}$$

$$\Rightarrow \quad z_2 z_1^2 + z_2 - z_1 z_2^2 - z_1 = 0$$

$$\Rightarrow \quad (z_1 - z_2)(z_1 z_2 - 1) = 0.$$

So either $z_1 = z_2$ or $z_1 z_2 = 1$. Since $1 < |z_1|$ and $1 < |z_2|$, we cannot have $z_1 z_2 = 1$. So $z_1 = z_2$, implying that J is one-to-one. We have $J(2i) = \frac{1}{2}(2i + \frac{1}{2i}) = \frac{3i}{4}$. Since $J(2i)$ is in the upper half-plane, we conclude that $J[\Omega]$ is the upper half-plane. ∎

As we observed following Example 1, the condition $f'(z) \neq 0$ for all z in a region Ω is not enough to ensure that f is one-to-one on Ω. However, if f is analytic and one-to-one on the whole region Ω, then $f'(z) \neq 0$ for all z in Ω, so f is a conformal mapping. Moreover, from Section 5.7, we know that the inverse function exists on $\Omega' = f[\Omega]$ and is analytic and one-to-one.

In this situation, we will say that Ω and Ω' are **conformally equivalent** regions.

The following famous theorem of Riemann tells us that any simply connected region of the complex plane other than the plane itself is conformally equivalent to the open unit disk.

THEOREM 3
RIEMANN
MAPPING
THEOREM

> Let Ω be a simply connected region in the complex plane other than the complex plane itself. Then there is a one-to-one analytic function f that maps Ω onto the unit disk $|w| < 1$. The mapping f is unique if we specify that $f(z_0) = 0$ and $f'(z_0) > 0$, for some z_0 in Ω.

The proof of the Riemann mapping theorem is above the level of this book. We refer you to Ahlfors [1] or Conway [10].

Combined with the Poisson integral formula, which solves the Dirichlet problem with arbitrary boundary data on the disk, the Riemann mapping theorem implies that we can at least theoretically solve the Dirichlet problem on any simply connected region Ω. As illustrated in Figure 7, it suffices to conformally map Ω in a one-to-one way onto the unit disk. This gives rise to a new boundary value problem on the disk, which can be solved using the Poisson formula. The solution of the original problem is then obtained by composing the solution on the disk with the conformal mapping. As simple and elegant as it is, this approach has its limitations. Even though we have several powerful tools, such as Fourier series and techniques for evaluating integrals using residues, our experience with the Poisson formula tells us that this formula is not easy to compute in general. Even more difficult is the actual construction of the conformal mapping from Ω onto the unit disk. While the Riemann mapping guarantees its existence, it gives no clue as to how to construct it.

Figure 7 A one-to-one conformal mapping f of Ω onto the unit disk $D = f[\Omega]$ takes boundary to boundary and preserves Laplace's equation. It transforms a Dirichlet problem on Ω into a Dirichlet problem on D, which can be solved using the Poisson formula. The solution in Ω is then $u(z) = U(f(z))$.

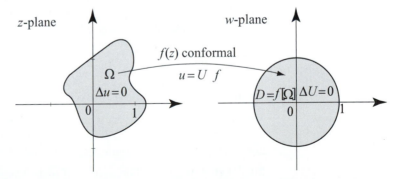

In the following sections, we will take up the construction of important conformal mappings that will lead us to very challenging problems. We will also give several applications to the solution of Laplace's equation.

Exercises 6.1

In Exercise 1 − 6, determine where the given mapping is conformal.

1. $\dfrac{z^2 + 1}{e^z}$.

2. $\dfrac{z + 1}{z^2 + 2z + 1}$.

3. $\dfrac{\sin z}{e^z}$.

4. $\dfrac{e^z + 1}{e^z - 1}$.

5. $z + \dfrac{1}{z}$.

6. $\dfrac{z + 1}{z + i} + \dfrac{2}{z}$.

In Exercise 7 − 12, determine the angle of rotation $\alpha = \arg f'(z)$ and the dilation factor $|f'(z)|$ of the given mapping at the indicated points.

7. $\dfrac{1}{z}$, $z = 1, i, 1 + i$.

8. $\text{Log } z$, $z = 1, i, -i$.

9. $\sin z$, $z = 0, \pi + i a, i\pi$.

10. z^2, $z = 0, 2i, -1 - i$.

11. e^{iz}, $z = \pi, i\pi, \dfrac{\pi}{2}$.

12. $\dfrac{1 + e^z}{1 - e^z}$, $z = 1, i\pi, \ln(3) + 2i$.

Consider the orthogonal lines $x = a$ and $y = b$, where a and b are real numbers. Determine their images by the given mapping $f(z)$. For which values of a and b are the images orthogonal at the point of intersection $f(a + ib)$? Verify your answer by computing the angle between the image curves at the point $f(a + ib)$.

13. e^z.

14. $\sin z$.

15. $(1 + i)z$.

16. $\dfrac{z + 1}{z - 1}$.

17. The Joukowski function.   Refer to Example 3.
(a)   Show that $J(\frac{1}{z}) = J(z)$ for all $z \neq 0$.
(b)   Fix $R > 1$. Show that the upper semicircle of radius R, $S_R = \{z : |z| = R, \ 0 \leq \text{Arg } z\}$, is mapped onto the upper semi-ellipse

$$J[S_R] = \left\{ w = u + iv : \frac{u^2}{\left[\frac{1}{2}\left(R + \frac{1}{R}\right)\right]^2} + \frac{v^2}{\left[\frac{1}{2}\left(R - \frac{1}{R}\right)\right]^2} = 1, \ v \geq 0 \right\}.$$

18.   Let S denote the region in the upper half-plane consisting of all z such that $1 < |z|$. Consider the Dirichlet problem in S with boundary conditions $u(x, y) = 100$ if (x, y) is on the upper unit circle, and $u(x, 0) = 0$ for all $1 < |x|$. Using the result of Exercise 35, Section 2.5, show that a solution is given by

$$u(x, y) = \frac{100}{\pi} \left(\text{Arg}\left(J(z) - 1\right) - \text{Arg}\left(J(z) + 1\right) \right) \quad (z = x + i y),$$

where $J(z)$ is the Joukowski function.

19.   Consider $f(z) = \frac{z}{(1 - z)^2}$. (a) Show that for all $z \neq 1$, $f(z)$ is analytic and $f'(z) \neq 0$ for $z \neq -1$. (b) Show that f is one-to-one in the open unit disk but is not one-to-one in any larger disk centered at 0.

20.   Let f be as in Theorem 2. Fill in the details in the following proof that f maps the boundary of Ω onto the boundary of $f[\Omega]$. Let w_0 be on the boundary of $f[\Omega]$; then we can find points $w_n = f(z_n)$ such that z_n is in Ω and $w_n \to w_0$ (why?). Distinguish two cases. If the sequence $\{z_n\}$ is unbounded, then a subsequence $\{z_{n_j}\}$ tends to ∞, which is a point on the boundary of Ω, and $f(z_{n_j}) \to w_0$. If $\{z_n\}$ is bounded, then by the Bolzano-Weierstrass property, we can find a subsequence $\{z_{n_j}\}$ that converges in Ω union its boundary. Show that $\{z_{n_j}\}$ cannot converge to a point in Ω and so it must converge to a point z_0 on the boundary. Moreover, $f(z_{n_j}) \to w_0$.

21. Let Ω be the slit plane, $\mathbb{C} \setminus \{z : z \le 0\}$, and $f(z) = \text{Log } z$. Show that the sequence $z_n = -1 + i\frac{(-1)^n}{n}$ ($n = 1, 2, \ldots$) converges to -1 on the boundary of Ω and yet $f(z_n)$ does not converge.

6.2   Linear Fractional Transformations

It should be clear by now from our work in Sections 2.5 and 6.1 that our success in solving boundary value problems involving Laplace's equation is closely tied to our ability to construct conformal mappings between regions in the plane. A good place to start our study of special conformal mappings is on the unit disk, since there we have a general formula for the solution of the Dirichlet problem, namely the Poisson integral formula. As we will soon see, the most suitable mappings for regions involving disks and lines are the linear fractional transformations that we introduced in Section 1.4:

$$(1) \qquad\qquad \phi(z) = \frac{az + b}{cz + d} \qquad (ad \ne bc).$$

Since $\phi'(z) = \frac{ad-bc}{(cz+d)^2}$, the condition $ad \ne bc$ ensures that ϕ does not degenerate into a constant. If $c = 0$, the linear fractional transformation reduces to a linear function, which is analytic everywhere or entire. If $c \ne 0$, then ϕ is analytic for all $z \ne -\frac{d}{c}$ and has a simple pole at $z = -\frac{d}{c}$.

PROPOSITION 1
BASIC PROPERTIES

Let $\phi(z)$ be a linear fractional transformation as in (1). Then
(i) ϕ is one-to-one throughout the complex plane and conformal at every point in the complex plane, except at the pole $z = -\frac{d}{c}$ ($c \ne 0$).
(ii) The inverse of $\phi(z)$ is the linear fractional transformation given by

$$(2) \qquad\qquad z = \psi(w) = \frac{d\,w - b}{-c\,w + a}.$$

Proof   The proposition is clear if $c = 0$. Suppose $c \ne 0$. Then

$$\phi'(z) = \frac{ad - bc}{(cz + d)^2} \ne 0 \text{ for all } z \ne -\frac{d}{c}.$$

Hence by Theorem 1, Section 6.1, the mapping is conformal at all $z \ne -\frac{d}{c}$. To get the inverse function, we solve $w = \frac{az+b}{cz+d}$ for z, and get $z = \frac{dw-b}{-cw+a}$. Since ϕ has an inverse, it must be one-to-one, for if $\phi(z_1) = \phi(z_2)$ taking the inverse function of both sides, we get $z_1 = z_2$. ∎

It is not hard to see that every linear fractional transformation (1) with $c \ne 0$ is a composition of a linear mapping $w_1 = cz+d$; an inversion $w_2 = \dfrac{1}{w_1}$; and a linear mapping $w = \dfrac{a}{c} + (b - \dfrac{ad}{c})w_2$ (see Exercise 41). Now linear mappings have a very useful property that is easy to verify: They map a line to a line and a circle to a circle. The inversion $w_2 = \frac{1}{w_1}$ has a somewhat similar property: It maps the collection of lines and circles in the z-plane to the collection of lines and circles in the w-plane. (Unlike linear mappings, the inversion may map a line to a circle or a circle to a line; see Figure 1.)

Figure 1 The inversion $f(z) = \frac{1}{z}$ preserves the collection of lines and circles. To verify the images of the given lines and circles, use the fact that f is conformal (preserves angles) and the values $f(0) = \infty$; $f(\infty) = 0$; $f(\pm 1) = \pm 1$; $f(\pm i) = \mp i$; $f(\frac{3}{4}) = \frac{4}{3}$.

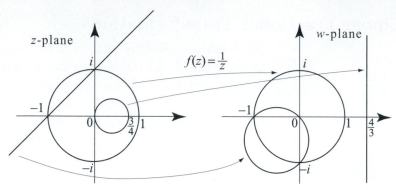

This property of the inversion is not as simple to verify but it is straightforward and is sketched in Exercises 35–40. Since a linear fractional transformation is a composition of linear mappings and an inversion, it inherits this property too. Thus we obtain the following very useful result.

PROPOSITION 2
IMAGES OF LINES
AND CIRCLES

Let $\phi(z)$ be a linear fractional transformation as in (1). Then ϕ maps lines and circles in the z-plane to lines and circles in the w-plane.

It follows immediately from Proposition 1 and Theorem 2 of the previous section that a linear fractional transformation will map boundary to boundary. As we now illustrate, this property is very useful in determining the image of a region.

EXAMPLE 1 Mappings between the unit disk and the upper half-plane

(a) Show that the linear fractional transformation

$$\phi(z) = i\,\frac{1-z}{1+z}$$

maps the unit disk onto the upper half-plane.
(b) Show that the linear fractional transformation

$$\psi(z) = \frac{i-z}{i+z}$$

maps the upper half-plane onto the unit disk.

Solution   (a) By Proposition 2, the image of the circle $C_1(0)$ is either a line or a circle in the w-plane. Since three points will determine either a line or circle, it suffices to check the images of three points on $C_1(0)$. We have

$$\phi(1) = 0; \quad \phi(i) = i\,\frac{1-i}{1+i} = 1; \quad \phi(-i) = i\,\frac{1+i}{1-i} = -1.$$

Thus $\phi(1)$, $\phi(i)$, and $\phi(-i)$ lie on the u-axis (the real axis in the w-plane), and so the image of $C_1(0)$ is the u-axis. Because ϕ is one-to-one, it maps the boundary $C_1(0)$ onto the boundary of the image of the unit disk. Thus the image of the unit disk is either the upper half-plane or the lower half-plane. Checking $\phi(0) = i$ (a point in the upper half-plane), we conclude that ϕ maps the unit disk one-to-one onto the upper half-plane (see Figure 2).

Figure 2 The image of the unit circle is determined by the images of three points. Since these are collinear, the circle is therefore mapped to a line (the real axis, in this case). Note also that because ϕ maps the closed unit disk to an unbounded region (the upper half-plane), ϕ has to be discontinuous somewhere in the closed unit disk. Indeed it is singular at $z = -1$.

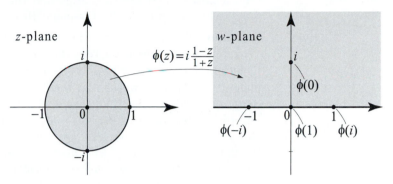

(b) We can do this part in two ways. One way is to use Proposition 1(ii) and notice that ψ is the inverse of ϕ. Another way is to check the image by ψ of the boundary and one interior point. We leave it as an exercise to verify that $\psi(0) = 1$, $\psi(1) = i$, and $\psi(-1) = -i$. Since the images of the three points are not collinear, we conclude that the real axis is mapped onto the circle that goes through the points 1, i, and $-i$, which is clearly the unit circle. (Here again, we are using the fact that three points determine a circle.) Also, $\psi(i) = 0$; hence ψ maps the upper half-plane onto the unit disk. ∎

Another way to realize that the image of the unit circle is a line in Example 1(a) is to consider the point -1 on $C_1(0)$ and note that $\lim_{z \to -1} \phi(z) = \infty$. So the image of $C_1(0)$ is not bounded and since it is either a line or a circle, it has to be a line (which tends to infinity). Sometimes it is convenient to express the fact that the limit at the point $z_0 = -\frac{d}{c}$ is infinity by writing $\phi(z_0) = \infty$. Likewise, it will be convenient to express $\lim_{z \to \infty} \frac{az+b}{cz+d} = \frac{a}{c}$ by simply writing $\phi(\infty) = \frac{a}{c}$.

Before we present our next example, let us note another useful property of linear fractional transformations.

**PROPOSITION 3
COMPOSITION OF
MAPPINGS**

The composition of any two linear fractional transformations is again a linear fractional transformation.

Proof Let $\phi_1(z) = \frac{a_1 z + b_1}{c_1 z + d_1}$ and $\phi_2(z) = \frac{a_2 z + b_2}{c_2 z + d_2}$. Then

$$\phi(z) = \phi_2 \circ \phi_1(z) = \frac{a_2 \frac{a_1 z + b_1}{c_1 z + d_1} + b_2}{c_2 \frac{a_1 z + b_1}{c_1 z + d_1} + d_2}.$$

Multiplying numerator and denominator by $c_1 z + d_1$, we get

$$\phi(z) = \frac{(a_2 a_1 + b_2 c_1)z + a_2 b_1 + b_2 d_1}{(c_2 a_1 + d_2 c_1)z + d_2 c_1 + d_2 d_1},$$

which is a linear fractional transformation. Notice that when we multiplied by $c_1 z + d_1$, we removed the singularity at $z = -\frac{d_1}{c_1}$; the resulting composition $\phi(z)$ has a single pole, and it is not necessarily at the same location as the poles of ϕ_1 or ϕ_2.   ∎

EXAMPLE 2   Composition of linear fractional transformations

Find a linear fractional transformation that maps the disk D with radius 2 and center at -1 onto the right half-plane $\operatorname{Re} w > 0$.

Solution   We describe two methods for doing this problem. Let us start with the quickest one based on the result of Example 1 and a simple application of Proposition 3. We know that $\phi(z) = i\frac{1-z}{1+z}$ maps the unit disk onto the upper half-plane. It is also easy to see that the linear mapping $\tau(z) = \frac{1}{2}(z+1)$ translates the center of D to the origin and then scales the radius by $\frac{1}{2}$. Thus τ maps D onto the unit disk. Consequently, $\phi \circ \tau(z)$ is a linear fractional transformation that maps D onto the upper half-plane. To map onto the right half-plane, it suffices to rotate the upper half-plane by $-\frac{\pi}{2}$. This can be achieved by multiplying by $e^{-i\frac{\pi}{2}} = -i$. So the desired linear fractional transformation (Figure 3) is

$$f(z) = -i\phi \circ \tau(z) = (-i)i\frac{1 - \frac{1}{2}(z+1)}{1 + \frac{1}{2}(z+1)} = \frac{1-z}{3+z}.$$

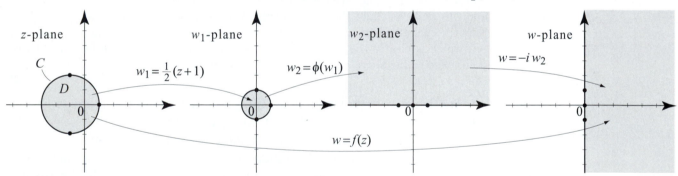

Figure 3 To map a disk to a half-plane, it is always advantageous to map the given disk to the unit disk and then use the transformation ϕ in Example 1.

Another way to do this problem is to start from scratch; we want a linear fractional transformation $g(z) = \frac{az+b}{cz+d}$ to map the boundary of the disk onto the boundary of the right half-plane. We can pick any three points on the circle C and map them to any three points on the imaginary axis. Since our image boundary is a line (which extends to infinity), we may map one of our points to ∞. We pick

$$(3) \qquad\qquad g(1) = 0, \quad g(-3) = \infty, \quad g(i\sqrt{3}) = i.$$

We use these equations to solve for the coefficients a, b, c, and d, and then we will check whether the interior of the disk is mapped to the right half-plane or the left half-plane. Again writing $g(z) = \frac{az+b}{cz+d}$, from $g(1) = 0$ we get $a = -b$. From $g(-3) = \infty$ we get $3c = d$. Thus $g(z) = \frac{az-a}{cz+3c} = \frac{a}{c}\frac{z-1}{z+3}$. From $g(i\sqrt{3}) = i$ we get

$$i = \frac{a}{c}\frac{i\sqrt{3}-1}{i\sqrt{3}+3} \Rightarrow \frac{a}{c} = i\frac{3+i\sqrt{3}}{-1+i\sqrt{3}} = \sqrt{3}.$$

Then $g(z) = \sqrt{3}\,\frac{z-1}{z+3}$ will map the circle C onto the y-axis. Note that any function of the form $\alpha g(z)$, where $\alpha \neq 0$ is real, will also map the circle C onto the y-axis, since multiplying by a nonzero *real* constant leaves a line through the origin unchanged. So for simplicity we divide by $\sqrt{3}$, still calling the function g, and obtain a mapping $g(z) = \frac{z-1}{z+3}$ of the circle C onto the y-axis. Does $g(z)$ take the region inside C onto the right half-plane? We check the image of one point inside C, say -1, and find $g(-1) = \frac{-2}{2} = -1$, which is a point in the left half-plane. So we modify g by multiplying it by -1 and obtain the desired linear fractional transformation $g(z) = \frac{1-z}{3+z}$. Clearly any other *positive* multiple of g will also work, and so the solution to this problem is not unique. ∎

The previous examples illustrate how a linear fractional transformation can be determined from the images of three distinct points. In fact, we have the following useful formula.

PROPOSITION 4
AN IMPLICIT
FORMULA

There is a unique linear fractional transformation $w = \phi(z)$ that maps three distinct points z_1, z_2, and z_3 onto three distinct points w_1, w_2, and w_3. The mapping w is implicitly given by

$$(4) \qquad \frac{z - z_1}{z - z_3}\,\frac{z_2 - z_3}{z_2 - z_1} = \frac{w - w_1}{w - w_3}\,\frac{w_2 - w_3}{w_2 - w_1}.$$

Proof That w is a linear fractional transformation follows by solving for w in (4). To see that w maps z_j to w_j (j=1,2,3) it suffices to note that (4) holds if we replace z by z_j and w by w_j. (For $j = 3$, you must take reciprocals in (4) before replacing z by z_3 and w by w_3.) The uniqueness part is done in Exercises 9–10. ∎

To map a circle onto a line by a linear fractional transformation, as we saw in Example 2, this can be achieved by requiring that $f(z) = \infty$ for some z on the circle. In this case, the formula in Proposition 4 can be simplified as follows. Say you want $f(z_3) = \infty$. As $w_3 \to \infty$, the fraction $\frac{w_2 - w_3}{w - w_3} = \frac{1 - w_2/w_3}{1 - w/w_3} \to 1$. This suggests that we set $\frac{w_2 - w_3}{w - w_3} = 1$ on the right side of (4), obtaining the following formula whose verification is left to Exercise 11.

PROPOSITION 5
MAPPING A POINT
TO INFINITY

Let z_1, z_2, and z_3 be three distinct points. There is a unique linear fractional transformation $w = \phi(z)$ that maps z_1 and z_2 onto two distinct points w_1 and w_2, and maps z_3 to ∞. The mapping w is implicitly given by

$$(5) \qquad \frac{z - z_1}{z - z_3}\,\frac{z_2 - z_3}{z_2 - z_1} = \frac{w - w_1}{w_2 - w_1}.$$

There is also a corresponding identity for a linear fractional transformation mapping ∞ to a point, obtained by reversing the roles of z and w in (5) (see Exercise 11). Our next example uses the conformal property of linear fractional transformations.

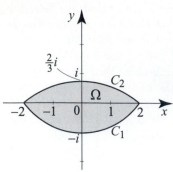

Figure 4 A lens-shaped region.

EXAMPLE 3   Mapping of a lens-shaped region

The lens-shaped region Ω in Figure 4 is bounded by the arcs of two circles.
(a) Use a linear fractional transformation ϕ to map Ω onto a sector in the w-plane in such a way that

$$\phi(-2) = 0, \qquad \phi(-i) = 1, \qquad \phi(2) = \infty.$$

(b) Determine the angle between the circles at the point -2.

Solution   (a) We apply (5) with $z_1 = -2$, $w_1 = 0$, $z_2 = -i$, $w_2 = 1$, $z_3 = 2$, and get

$$\frac{z+2}{z-2}\frac{-i-2}{-i+2} = \frac{w}{1} \quad \Rightarrow \quad w = \phi(z) = \frac{2+i}{2-i}\frac{2+z}{2-z}.$$

(b) Of course we can determine the angle between C_1 and C_2 by finding their equations, then the slopes of the tangent lines at -2, and then the angle between the tangent lines. A better way is to calculate the angle between the images of C_1 and C_2 and use the conformal property of ϕ.

Because ϕ takes lines and circles to lines and circles, it is clear from its action on the points -2, $-i$ and 2 that ϕ maps the circle C_1 onto the u-axis. It also maps the circle C_2 onto a line that goes through the point $\phi(-2) = 0$. To determine this line, it suffices to check the image of another point on C_2. We have $\phi(\frac{2}{3}i) = i$. Thus ϕ maps C_2 onto the v-axis. Because ϕ is conformal at $z = -2$, it preserves the angles at this point. Thus, the angle between the circles at $z = -2$ is equal to the angle between their images, the u- and v-axes, which is clearly $\frac{\pi}{2}$.   ∎

Some applications concerning the electrostatic potential inside a capacitor formed by two cylinders lead to Dirichlet problems in regions bounded by two circles in the plane, which are not necessarily concentric. The problems are easier to solve when the two circles are concentric, giving rise to an annular region. (See for examples the exercises of Section 2.5.) Thus there is a great advantage in using a conformal mapping to center the circles before solving the Dirichlet problem. In what follows, we use a specific example to illustrate this process. More general examples are presented in the exercises.

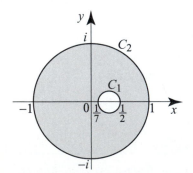

Figure 5 Two nonconcentric circles C_1 and C_2.

Example 4   Centering disks

Find a one-to-one analytic mapping that maps the region between the disks in Figure 5 to an annular region bounded by two concentric circles.

Solution   The idea is to use one of the linear fractional transformations

$$(6) \qquad \phi_\alpha(z) = \frac{z - \alpha}{1 - \overline{\alpha}z},$$

where α is a complex number such that $|\alpha| < 1$. These functions were introduced in Example 3, Section 3.7, where it was shown that $|\phi_\alpha(z)| = 1$ if $|z| = 1$. Thus ϕ_α maps the unit circle onto the unit circle. Because ϕ_α maps boundary to boundary and because $\phi_\alpha(0) = -\alpha$ (a point inside the unit disk), we conclude that ϕ_α maps the unit disk onto itself. Since ϕ_α is a linear fractional transformation, it will map the circle C_1 onto a circle or a line, but because the image has to be inside the image of the unit disk, it follows that $\phi[C_1]$ is bounded and hence it must be a circle. We now ask the following question: Can we find α so that $\phi_\alpha[C_1]$ is a circle centered at the origin? Suppose for a moment that α were real. Then clearly $\phi_\alpha(x)$ is also real, and so ϕ_α maps the real line to the real line. Note that $\phi_\alpha(1/7)$ and $\phi_\alpha(1/2)$ are the points where $\phi_\alpha[C_1]$ meets the u-axis. Also, the circle $\phi_\alpha[C_1]$ must meet the u-axis in a perpendicular fashion (Figure 6), for the following three reasons:

(i) C_1 itself meets the real axis in a perpendicular fashion;

(ii) the x-axis is mapped to the u-axis; and

(iii) the map ϕ_α is conformal.

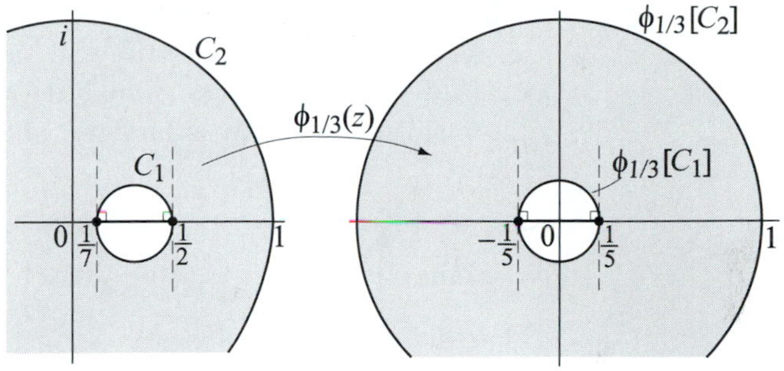

Figure 6 For all $|\alpha| < 1$, $\phi_\alpha(z)$ maps the unit circle C_2 onto itself. But for one special value of α, with $|\alpha| < 1$, ϕ_α will also map the circle C_1 onto a circle centered at the origin, thus centering the images of C_1 and C_2.

So if we want $\phi_\alpha[C_1]$ to be a circle centered at the origin, it is enough to require that $\phi_\alpha(1/7) = -\phi_\alpha(1/2)$. This implies that

$$\frac{\frac{1}{7} - \alpha}{-\frac{\alpha}{7} + 1} = -\frac{\frac{1}{2} - \alpha}{-\frac{\alpha}{2} + 1} \quad \Rightarrow \quad \frac{1 - 7\alpha}{7 - \alpha} = -\frac{1 - 2\alpha}{2 - \alpha} \quad \Rightarrow \quad 9\alpha^2 - 30\alpha + 9 = 0.$$

The last equation in α is equivalent to $3\alpha^2 - 10\alpha + 3 = 0$, with solutions

$$\alpha = \frac{5 \pm \sqrt{16}}{3} = \frac{5 \pm 4}{3} \quad \Rightarrow \quad \alpha = 3 \text{ or } \alpha = \frac{1}{3}.$$

Since we want $|\alpha| < 1$, we take $\alpha = \frac{1}{3}$. Thus

$$\phi(z) = \phi_{\frac{1}{3}}(z) = \frac{z - \frac{1}{3}}{-\frac{1}{3}z + 1} = \frac{3z - 1}{3 - z}$$

will map C_2 onto C_2 and C_1 onto the circle with center at the origin and radius $r = |\phi(1/2)| = \frac{\frac{3}{2} - 1}{3 - \frac{1}{2}} = \frac{1}{5}$. ∎

Composing Elementary Mappings

As the title indicates, we will compose together elementary mappings that we have studied thus far and construct some nontrivial conformal mappings of regions in the plane. We give four examples to illustrate this important process.

EXAMPLE 5   Mapping a lens onto a disk

Construct a sequence of analytic functions that maps the lens-shaped region of Example 3 in a one-to-one way onto the unit disk. Write down the mapping that results from your construction.

Solution   The first step is to use the function of Example 3,

$$w_1 = \phi(z) = \frac{2+i}{2-i}\frac{2+z}{2-z} = \alpha\frac{2+z}{2-z} \qquad \left(\alpha = \frac{2+i}{2-i}\right),$$

which maps the lens to the first quadrant. The first quadrant is then mapped onto the upper half-plane by the function $w_2 = w_1^2$. (Note that our mapping is no longer a linear fractional transformation.) Finally, the linear fractional transformation $w = \psi(w_2)$ in Example 1(ii) will take the upper half-plane onto the unit disk. Each of these mappings is analytic and one-to-one on the region of interest. So the resulting function is analytic and one-to-one. The mapping is illustrated in Figure 7.

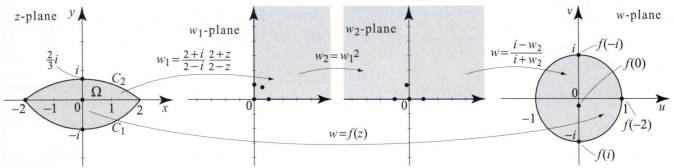

Figure 7 Mapping a lens onto the unit disk. Note the conformal property in the first mapping at the point -2, which preserved the right angle. Note the failure of the conformal property in the second mapping at the point 0, where the angle is doubled.

The explicit formula in terms of z of the conformal mapping of the lens onto the unit disk is

$$w = \frac{i - w_2}{i + w_2} = \frac{i - w_1^2}{i + w_1^2} = \frac{i - [\phi(z)]^2}{i + [\phi(z)]^2},$$

where $\phi(z) = \frac{2+i}{2-i}\frac{2+z}{2-z}$ from Example 3. Replacing $\phi(z)$ by its value in terms of z and simplifying (by hand or with the help of a computer system), we get

$$w = f(z) = -i\,\frac{4 - 28\,i\,z + z^2}{28 + 4\,i\,z + 7\,z^2}.$$

The following values of w confirm the fact that the mapping takes the lens-shaped region onto the unit disk, also taking boundary points to boundary points:

$$f(0) = -\frac{i}{7}, \quad f(-2) = 1, \quad f\left(\tfrac{2}{3}i\right) = -i, \quad f(-i) = i. \qquad \blacksquare$$

EXAMPLE 6   Mapping a half-disk onto a disk

The sequence of one-to-one analytic mappings in Figure 8 takes the upper half of the unit disk onto the unit disk.

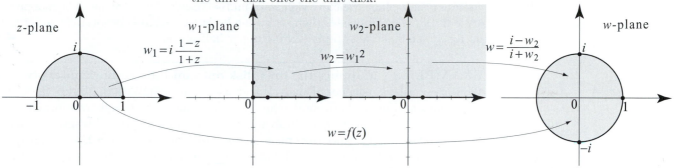

Figure 8 Mapping the upper half of the unit disk onto the unit disk.

The first mapping is the linear fractional transformation $w_1 = \phi(z) = i\frac{1-z}{1+z}$ from Example 1. It takes the unit disk onto the upper half-plane. It also takes the upper half-disk onto the first quadrant, as can be verified by using the conformality at $z = 1$ and checking the image of one interior point, say $\phi(\frac{i}{2}) = \frac{4}{5} + i\frac{3}{5}$, which is in the first quadrant. The action of the second mapping is clear. The third mapping is the mapping ψ from Example 1(ii). The explicit formula of the final mapping $w = f(z)$ is

$$ w = \psi(w_2) = \frac{i - w_2}{i + w_2} = \frac{i - w_1^2}{i + w_1^2} = \frac{i - \left(i\frac{1-z}{1+z}\right)^2}{i + \left(i\frac{1-z}{1+z}\right)^2} = -i\,\frac{1 + 2\,i\,z + z^2}{1 - 2\,i\,z + z^2}. $$

The intermediary mapping $w_2 = w_1^2 = -\left(\frac{1-z}{1+z}\right)^2$ is also of interest. It takes the upper half-disk onto the upper half-plane. ∎

EXAMPLE 7   A crescent-shaped region onto the upper half-plane

The crescent-shaped region in Figure 9 is bounded by two circles that intersect at angle 0 at the origin. We will describe a sequence of one-to-one analytic mappings that takes this region onto the upper half-plane.

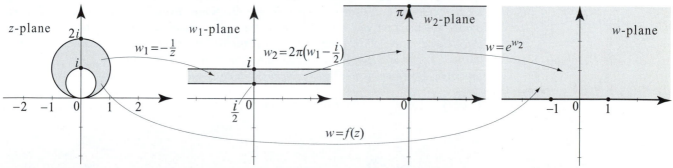

Figure 9 Mapping a crescent onto the upper half-plane.

The first mapping $w_1 = -\frac{1}{z}$, being conformal at $z = i$ and $z = 2i$, will preserve the right angles at these points. Since it maps the imaginary axis onto the imaginary axis, and 0 to ∞, consequently it will map the two circles onto two lines that intersect the imaginary axis at right angle. Thus the images of the circles are parallel horizontal lines as shown in the figure. As we move counterclockwise around the circles in the z-plane, we move rightward on the lines in the w_1-plane. The mapping $w_2 = 2\pi(w_1 - \frac{i}{2})$ translates then scales the horizontal strip appropriately to set us up for an exponential mapping to the upper half-plane.   ∎

EXAMPLE 8   Mapping the unit disk onto an infinite horizontal strip
We will describe a sequence of analytic and one-to-one mappings that takes the unit circle onto an infinite horizontal strip. The first linear fractional transformation, $w_1 = -i\phi(z)$, is obtained by multiplying by $-i$ the linear fractional transformation $\phi(z)$ in Example 1(i). Since $\phi(z)$ maps the unit disk onto the upper half-plane, and multiplication by $-i$ rotates by the angle $-\frac{\pi}{2}$, the effect of $-i\phi(z)$ is to map the unit disk onto the right half-plane.

Figure 10 The principal branch of the logarithm, $\text{Log}\, z$, maps the right half-plane onto an infinite horizontal strip.

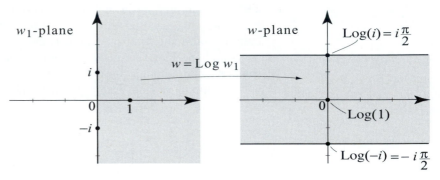

In Figure 10, $\text{Log}\, w_1 = \ln|w_1| + i\,\text{Arg}\, w_1$ is the principal branch of the logarithm. As w_1 varies in the right half-plane, $\text{Arg}\, w_1$ varies between $-\frac{\pi}{2}$ and $\frac{\pi}{2}$, which explains the location of the horizontal boundary of the infinite strip. The desired mapping is

$$w = f(z) = \text{Log}\,(-i\phi(z)) = \text{Log}\,\frac{1-z}{1+z}.$$ ∎

Exercises 6.2

In Exercises 1–4, you are given a linear fractional transformation $\phi(z)$ and three points z_1, z_2, and z_3. Let L_1 denote the line through z_1 and z_2 and L_2 the line through z_2 and z_3. In each case, (a) compute the images w_1, w_2, and w_3 of the points z_1, z_2, and z_3. (b) Describe the images by ϕ of L_1 and L_2. Are they lines or circles? (You will need the images of three points on each line.)

1. $\phi(z) = i\dfrac{1-z}{1+z}$, $z_1 = 1, z_2 = 0, z_3 = i$.

2. $\phi(z) = \dfrac{i+z}{i-z}$, $z_1 = 1+i, z_2 = 0, z_3 = i$.

3. $\phi(z) = \dfrac{1+i-2z}{i-iz}$, $z_1 = i, z_2 = 1, z_3 = -i$.

4. $\phi(z) = \dfrac{1+2z}{i-(1+i)z}$, $z_1 = 1+i, z_2 = 1, z_3 = 1-i$.

5. Find the inverse ψ of the linear fractional transformation in Exercise 1, and verify that ψ maps w_1, w_2, and w_3 to z_1, z_2, and z_3.

6. Repeat Exercise 5 with the linear fractional transformation of Exercise 2.

7. (a) What is the inverse of the function $f(z) = \frac{1}{z}$? Answer this question without using (2), then verify your answer using (2).
(b) Describe the images by $f(z) = \frac{1}{z}$ of the unit circle, the unit disk, and the region outside the unit circle.

8. Let α denote the angle between the two circles in Figure 11 at -2. Show that $\tan \alpha = \frac{4(a+1)(a-4)}{3(a+6)(a-\frac{2}{3})}$. Discuss the cases when $a = -1$, 4, -6, and $\frac{2}{3}$. [Hint: Map the circles to lines using the linear fractional transformation in Example 3.]

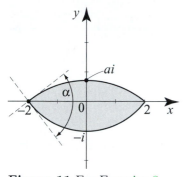

Figure 11 For Exercise 8.

9. Fixed points. Recall that a point z_0 is a fixed point of a function $f(z)$ if $f(z_0) = z_0$. Show that a linear fractional transformation $\phi(z)$ can have at most two fixed points in the complex plane, unless $\phi(z) = z$, in which case all points are fixed points. [Hint: Discuss all possible solutions of $z = \frac{az+b}{cz+d}$.]

10. Uniqueness of a linear fractional transformation. Let z_1, z_2, and z_3 be three distinct points, and let w_1, w_2, and w_3 be three distinct points (we allow ∞). Show that there is a unique linear fractional transformation mapping z_j to w_j. [Hint: The existence is guaranteed by Propositions 4 and 5. To prove uniqueness, suppose that f and g are two linear fractional transformations mapping z_j to w_j. Apply the result of Exercise 9 to $f \circ g^{-1}$. How many fixed points does $f \circ g^{-1}$ have?]

11. (a) **Mapping a point to infinity.** Prove Proposition 5.
(b) **Mapping infinity to a point.** Let z_1 and z_2 be two distinct points, and w_1, w_2, w_3 be three distinct points. Show that there is a unique linear fractional transformation $w = \phi(z)$ that maps z_1 to w_1, z_2 to w_2 and ∞ to w_3. The mapping w is implicitly given by

(7)
$$\frac{z - z_1}{z_2 - z_1} = \frac{w - w_1}{w - w_3} \frac{w_2 - w_3}{w_2 - w_1}.$$

In Exercises 12–24, (a) supply the formulas of the analytic mappings in each sequence of mappings shown in the accompanying figure (Figures 12-24). (b) Verify that the boundary and the interior of the shaded regions are mapped to the boundary and interior of the shaded regions. (c) Derive the given formula for the final composite mapping $w = f(z)$. As usual, we start in the z-plane and end in the w-plane, going through the w_j-planes.

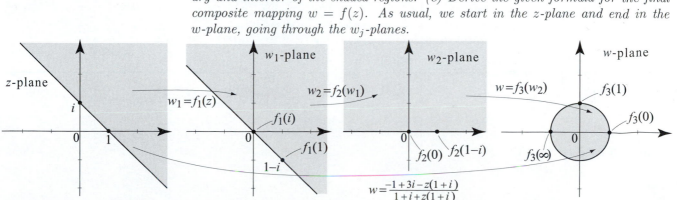

Figure 12 for Exercise 12.

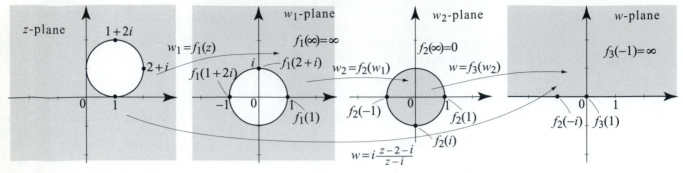

Figure 13 for Exercise 13.

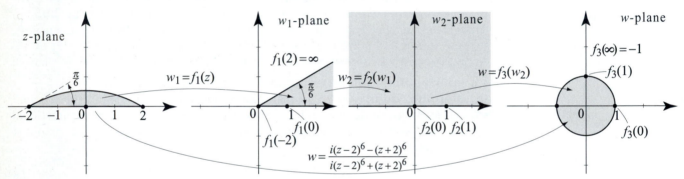

Figure 14 for Exercise 14.

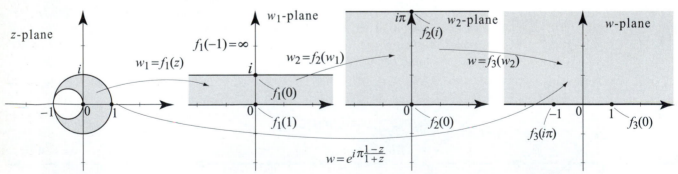

Figure 15 for Exercise 15.

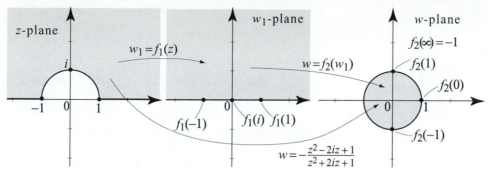

Figure 16 for Exercise 16.

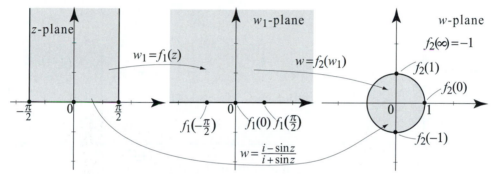

Figure 17 for Exercise 17.

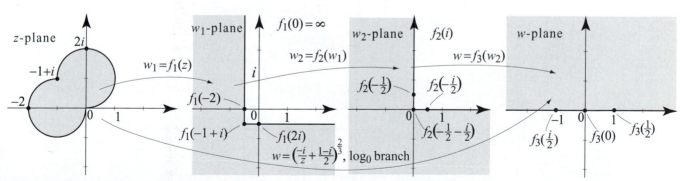

Figure 18 for Exercise 18.

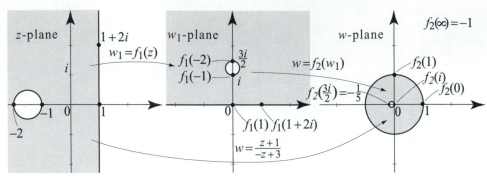

Figure 19 for Exercise 19.

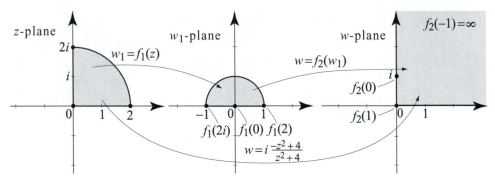

Figure 20 for Exercise 20.

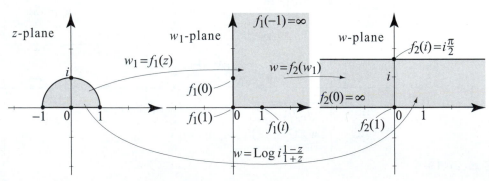

Figure 21 for Exercise 21.

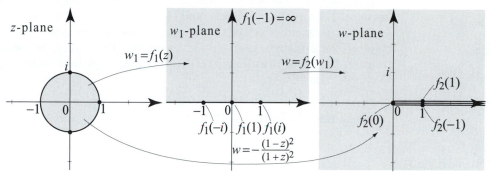

Figure 22 for Exercise 22.

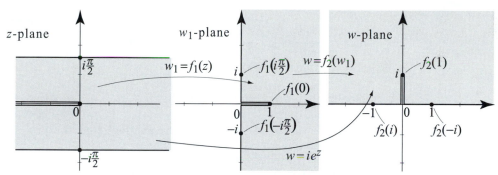

Figure 23 for Exercise 23.

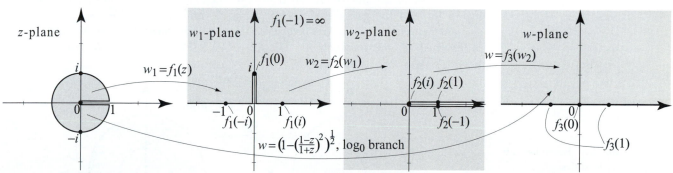

Figure 24 for Exercise 24.

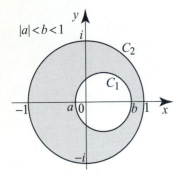

$|a| < b < 1$

Figure 25 for Exercise 25.

25. Project Problem: Centering disks. We generalize the process in Example 4 to any region bounded by two non-intersecting circles, C_2 and C_1, such that C_1 is in the interior of C_2. By translating the center of C_1 to the origin, scaling, and rotating, we can always reduce the picture to the one described in Figure 25, where $|a| < b < 1$, C_2 is the unit circle, and C_1 is centered on the x-axis with x-intercepts a and b. Our goal is to show that we can choose α such that $-1 < \alpha < 1$ and $\phi_\alpha(z)$ maps C_1 onto a circle centered at the origin. Here $\phi_\alpha(z)$ is the linear fractional transformation (6), which maps C_2 onto C_2. As explained in Example 4, it suffices to choose α so that $\phi_\alpha(a) = -\phi_\alpha(b)$.

(a) Show that the latter condition leads to the equation in α:

$$\alpha^2 - 2\,\frac{1+ab}{a+b}\,\alpha + 1 = 0,$$

with roots

$$\alpha_1 = \frac{1+ab}{a+b} + \sqrt{\left(\frac{1+ab}{a+b}\right)^2 - 1} \quad \text{and} \quad \alpha_2 = \frac{1+ab}{a+b} - \sqrt{\left(\frac{1+ab}{a+b}\right)^2 - 1}.$$

(b) Show that if $|a| < 1$ and $|b| < 1$, then $1 + ab \geq a + b$. [Hint: $1 - b \geq a(1 - b)$.]
(c) Show that $\alpha_1 > 1$, while $0 < \alpha_2 < 1$. [Hint: The first inequality follows from (b). For the second inequality, use the fact that the product of the roots of $ax^2 + bx + c = 0$ is $\frac{c}{a}$.]

(d) Conclude that $\phi(z) = \frac{z-\alpha}{1-\alpha z}$ with $\alpha = \frac{1+ab}{a+b} - \sqrt{\left(\frac{1+ab}{a+b}\right)^2 - 1}$ will map C_2 onto C_2, C_1 onto a circle centered at the origin with radius $r = \phi(b)$, and the region between C_2 and C_1 onto the annular region bounded by $\phi[C_1]$ and the unit circle.

In Exercises 26–29, derive the linear fractional transformation that maps the shaded region between the two given circles (or circle and line in Exercise 29) onto an annular region centered at the origin (see the accompanying Figures 26 − 29). Refer to Exercise 25 for instructions. In Exercises 28–29, you need to reduce to the situation described in Exercise 25.

Figure 26 for Exercise 26.
Take $a = -\frac{1}{4}$ and $b = \frac{7}{8}$.

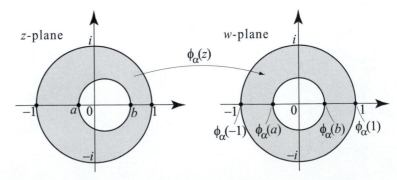

Figure 27 for Exercise 27. Take $a = 0$ and $b = \frac{8}{17}$.

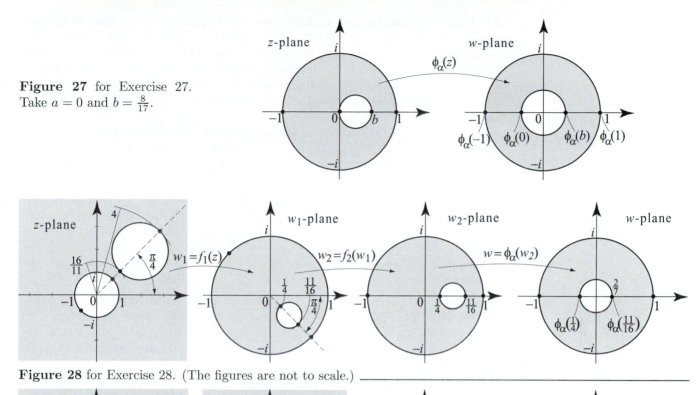

Figure 28 for Exercise 28. (The figures are not to scale.)

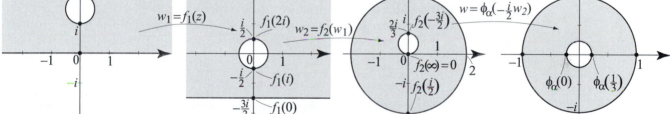

Figure 29 for Exercise 29. [Hint: In the last step, start by rotating the inner circle to center it on the real axis, then scale the outer radius to 1.]

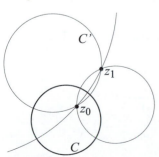

Figure 30 for Exercise 30.

30. A geometric problem. The following is an interesting illustration of the use of linear fractional transformations to prove geometric facts. Consider a circle C and a point z_0 inside C (Figure 30). We will show that all the circles C' through z_0 that intersect C at right angle, also intersect at a common point z_1 as in Figure 30. The point z_1 is called the **reflection** of z_0 in C. We also say that z_0 and z_1 are **symmetric** with respect to C.

(a) There exists a linear fractional transformation ϕ that maps C to the real axis and the interior of C to the upper half-plane. Show that $\phi[C']$ is a circle or a line that intersects the real axis at a right angle.

(b) Conclude that $\phi[C']$ passes through the point $-\phi(z_0)$. Setting $z_1 = \phi^{-1}(-\phi(z_0))$ we see that C passes through z_1.

31. Project problem: one-to-one analytic mappings of the unit disk onto itself. Let $\phi_\alpha(z)$ be as in (6). We know that $\phi_\alpha(z)$ maps the unit disk D onto itself. In this exercise, we show that, up to a unimodular constant multiple, these are the *only* one-to-one analytic mappings of the unit disk onto itself. That is, if f is a one-to-one analytic mapping of D onto D, then $f(z) = c\phi_\alpha(z)$, where $|c| = 1$. This important result will be obtained as a consequence of Schwarz's lemma (Section 3.7).

(a)   Suppose that g is a one-to-one mapping of D onto itself. Show that $f = \phi_{g(0)} \circ g$ is a one-to-one analytic mapping of D onto itself such that $f(0) = 0$. If we can show that f is of the desired form, then because $g = \phi_{g(0)}^{-1} \circ f = \phi_{-g(0)} \circ f$, it will follow that g is of the desired form.

(b) Suppose that g is a one-to-one mapping of D onto itself. By (a), we may without loss of generality assume that $g(0) = 0$. Show that the inverse of g, g^{-1}, is also analytic, one-to-one, and maps D onto D and $g^{-1}(0) = 0$. Apply Schwarz's lemma to g and obtain $|g(z)| \le |z|$. Apply Schwarz's lemma to g^{-1} at the point $g(z)$ and obtain $|g^{-1}(g(z))| \le |g(z)|$, or, equivalently, $|z| \le |g(z)|$. Hence $|g(z)| = |z|$. Apply Schwarz's lemma again and conclude that $g(z) = cz$, where $|c| = 1$.

32. Suppose that Ω is a region and f and g are two analytic one-to-one mappings of Ω onto the unit disk D. Show that there is a linear fractional transformation ϕ_α of the form (6) such that $f(z) = c\phi_\alpha \circ g(z)$, where $|c| = 1$. This shows that all the one-to-one mappings of a region Ω onto the unit disk are the same up to a change of variables effectuated by a linear fractional transformation of the form (6).

33. (a) Characterize all the one-to-one analytic mappings of the upper half-plane onto the unit disk. (b) Characterize all the one-to-one analytic mappings of the upper half-plane onto itself.

34. Matrix correspondence. (a) Prove Proposition 3.

(b) We define a mapping Φ that associates to each linear fractional transformation (1) the 2×2 matrix with complex entries

$$S = \begin{pmatrix} a & b \\ c & d \end{pmatrix}.$$

Thus $\Phi(\phi) = S$. Suppose that ϕ and ψ are two linear fractional transformations with matrices $\Phi(\phi) = S$ and $\Phi(\psi) = T$. Show that the $\Phi(\phi \circ \psi) = ST$, where ST denotes the product of the two matrices S and T.

Project Problem: Lines and circles under inversion, part I.   In Exercises 35–37 we will show that the function $f(z) = \frac{1}{z}$ maps lines and circles to lines and circles.

35.   (a) Show that with $w = u + iv$ and $z = x + iy$, the mapping $w = 1/z$ can be written as $u(x, y) = \frac{x}{x^2+y^2}$, $v(x, y) = -\frac{y}{x^2+y^2}$.

(b)   Deduce that the inverse transformation $z = 1/w$ is given by

(8) $$x(u, v) = \frac{u}{u^2 + v^2}, \qquad y(u, v) = -\frac{v}{u^2 + v^2}.$$

36.   (a)   Show that any circle of the form $(x - x_0)^2 + (y - y_0)^2 = r^2$, $r > 0$ can be written in the form

(9) $$A(x^2 + y^2) + Bx + Cy + D = 0, \quad \text{where } B^2 + C^2 - 4AD > 0.$$

(b)   Show that any line in the plane can be written in the form (9).
(c)   Show that any set of points satisfying $A(x^2 + y^2) + Bx + Cy + D = 0$ with $B^2 + C^2 - 4AD > 0$ is either a circle or a line depending on whether $A = 0$ or $A \neq 0$.
(d)   Show that such circles or lines pass through the origin if and only if $D = 0$.

37.   Suppose a set S is given as those points (x, y) satisfying (9). Use (8) and conclude that points (u, v) in $f[S]$ satisfy an equation of the same form as (9), including the associated constant inequality. Conclude that under the mapping $f(z) = 1/z$ lines and circles are mapped to lines and circles, with the exception of the origin.

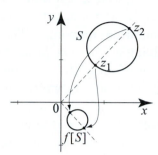

Figure 31 For Exercise 38. Note that S and $f[S]$ are plotted in the same plane.

Project Problem: Lines and circles under inversion, part II.   In Exercises 38–41, we investigate how specific lines and circles are mapped under the function $f(z) = \frac{1}{z}$ and describe a quick method to obtain the images. These exercises depend on Exercises 35–47, and in particular, (8) and (9).

38.   (a) Suppose that S is a circle that does not pass through the origin. Show that $f[S]$ is also a circle that does not pass through the origin.
(b)   Let z_1 and z_2 denote the points in S with the smallest and largest moduli, respectively. Show that $f(z_1)$ and $f(z_2)$ have the largest and smallest moduli, respectively, of those points in $f[S]$. Argue that the circle $f[S]$ is uniquely determined by these two points $f(z_1)$ and $f(z_2)$ (see Figure 31; S and $f[S]$ are plotted on the same plane).

39.   (a)   Suppose S is a line that passes through the origin. Show that with the exception of mapping to and from the origin, $f[S]$ is also a line passing through the origin.
(b)   Argue that the image $f(z_0)$ of any nonzero point z_0 in S uniquely determines the line $f[S]$ (see Figure 32).

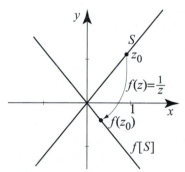

Figure 32 For Exercise 39. Note that S and $f[S]$ are plotted in the same plane.

40.   (a)   Suppose S is a circle that passes through the origin. Show that with the exception of mapping from the origin, $f[S]$ is a line that does not pass through the origin.
(b)   Suppose that S is a line which does not pass through the origin. Show that with the exception of mapping to the origin, $f[S]$ is a circle that passes through the origin.

(c) Let S be a circle that passes through the origin and $f[S]$ be the associated line that does not. Show that the point z_0 of maximum modulus on the circle maps to the point w_0 of minimum modulus on the line, and vice versa. Argue that each of these points uniquely determines the corresponding circle or line. If we let R denote the radius of the circle and a the perpendicular distance from the origin to the line, show that $2R = \frac{1}{a}$ (see Figure 33).

41. Lines and circles under linear fractional transformations.   (a)   Verify that every linear fractional transformation is a composition of a linear transformation, followed by an inversion, followed by a linear transformation, as described in the section.
(b)   Using part (a) and the result of Exercise 37, show that any linear fractional transformation maps lines and circles to lines and circles.

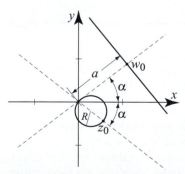

Figure 33 For Exercise 40. Note that S and $f[S]$ are plotted in the same plane.

6.3   Solving Dirichlet Problems with Conformal Mappings

This section continues the applications that we introduced in Section 2.5, using conformal mappings to solve Dirichlet problems. At the heart of the subject is the invariance of Laplace's equation by conformal mappings, Theorem 3, Section 2.5. Very often the difficulty in solving a Dirichlet problem is due to the geometry of the region on which the problem is stated. Conformal mappings can be used to transform a region to one on which the ensuing boundary value problem is easier to solve. Roughly speaking, the conformal mapping method is like a change of variables that leaves Laplace's equation unchanged but transforms the boundary conditions. The success of this method is phenomenal. Not only we will be able to solve specific problems, but we will also take general formulas, such as the Poisson integral formula on a disk, and produce similar formulas for the solution of Dirichlet problems on new regions in the plane.

Recall that for Dirichlet problems where the boundary data is constant along rays, we can find a solution using a branch of the argument. We denote by $\arg_\alpha z$ the branch of the argument with a branch cut at angle α, and by $\operatorname{Arg} z$ the principal branch with a branch cut along the negative real axis. These functions, being the imaginary parts of the corresponding logarithm branches, are harmonic everywhere except on their branch cuts.

Recall also that a linear combination of harmonic functions with real scalars is again a harmonic function (Proposition 1, Section 2.5). So, for example, $u(z) = \frac{100}{\pi}(\pi - \operatorname{Arg} z)$ is harmonic in the upper half-plane with boundary values $u(x) = 100$ if $x > 0$ and $u(x) = 0$ if $x < 0$. (You should review Example 3, Section 2.5, for useful details involving $\operatorname{Arg} z$.) This solution to a very simple Dirichlet problem in the upper half-plane helps us solve a somewhat difficult Dirichlet problem on the unit disk. (We solved a similar problem using Fourier series in Section 4.7.)

EXAMPLE 1   Steady-state temperature distribution in a disk

The boundary of a circular plate of unit radius with insulated lateral surface is kept at a fixed temperature distribution equal to $100°$ on the upper semi-circle and $0°$ on the lower semi-circle (see Figure 1(a)). Find the steady-state temperature inside the plate.

Solution   To answer this question, we must solve $\Delta u = 0$ inside the unit disk with boundary values $u = 100$ on the upper semi-circle and $u = 0$ on the lower semi-circle. While the geometry of the circle makes it difficult to understand the effect of the boundary conditions on the solution inside the unit disk, the corresponding boundary value problem in the upper half-plane has a simple solution. To transform the given problem into a problem in the upper half-plane, we use the linear fractional transformation $\phi(z) = i\frac{1-z}{1+z}$ from Example 1(ii), Section 6.2. It is easy to verify that $\phi(z)$ takes the upper semi-circle onto the positive real axis, and the lower semi-circle onto the negative real axis. Thus ϕ transforms the given Dirichlet problem into a Dirichlet problem in the upper half-plane with boundary values shown in Figure 1(b).

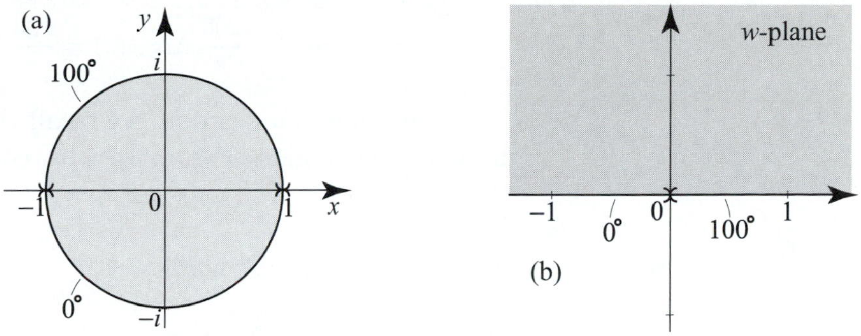

Figures 1(a) and (b)
Transforming the Dirichlet problem in Figure 1(a) into an easier Dirichlet problem in Figure 1(b), by using a linear fractional transformation
$\phi(z) = i\frac{1-z}{1+z}$.

According to our preceding discussion, the solution in the upper half of the w-plane is $U(w) = \frac{100}{\pi}(\pi - \text{Arg}\, w)$. By composing the solution in the w-plane with the conformal map, we get a solution of our original problem, $u(z) = U(\phi(z))$. Hence the solution of the Dirichlet problem in the unit disk is

$$u(z) = \frac{100}{\pi}(\pi - \text{Arg}\, \phi(z)) = \frac{100}{\pi}\left(\pi - \text{Arg}\left(i\frac{1-z}{1+z}\right)\right).$$

For example, the temperature of the center of the circular plate is $u(0) = \frac{100}{\pi}(\pi - \text{Arg}\,(i)) = \frac{100}{\pi}\frac{\pi}{2} = 50$, which is, as you would expect, the average value of the temperature on the circumference. With a little extra work, we can express the solution $u(z)$ in terms of x and y instead of z (see Exercise 18). ■

We next solve a Dirichlet problem in a lens-shaped region.

EXAMPLE 2   Dirichlet problem in a lens-shaped region
Find a harmonic function u in the lens-shaped region Ω in Figure 2(a), with boundary values $u = 100$ on the upper circular arc and $u = 0$ on the lower circular arc.

Solution   The region Ω was discussed in Example 3, Section 6.2, from which we recall the linear fractional transformation $\phi(z) = \frac{2+i}{2-i}\frac{2+z}{2-z}$. It is straightforward to check that ϕ maps the lower boundary of Ω onto the positive real axis, and the upper boundary of Ω onto the positive imaginary axis. By checking the image of one point in Ω, say $z = 0$, we find $\phi(0) = \frac{2+i}{2-i} = \frac{1}{5}(3 + 4i)$, which is a point in the first quadrant. Thus ϕ maps the region Ω onto the first quadrant and transforms the given problem into a Dirichlet problem in the first quadrant of the w-plane, with boundary conditions as shown in Figure 2(b).

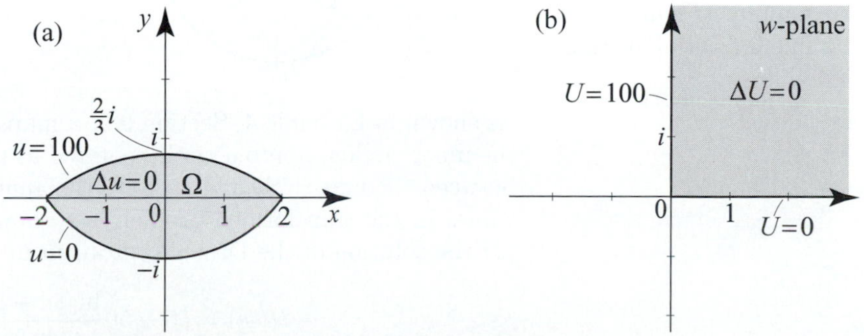

Figures 2(a) and (b) Using a linear fractional transformation to transform a Dirichlet problem in a lens (Figure 2(a)) into a Dirichlet problem in the first quadrant (Figure 2(b)). While the problem in the lens is difficult to solve, the problem in the first quadrant has a very simple solution $U(w) = \frac{200}{\pi}\text{Arg}\, w$.

It is clear that the solution in the w-plane is $U(w) = \frac{200}{\pi} \operatorname{Arg} w$. Thus the solution of the Dirichlet problem on Ω is

$$u(z) = \frac{200}{\pi} \operatorname{Arg} \phi(z) = \frac{200}{\pi} \operatorname{Arg} \left(\frac{2+i}{2-i} \frac{2+z}{2-z} \right). \qquad \blacksquare$$

For our next application, we recall the solution of the Dirichlet problem in an annular region (Figure 3), with constant boundary values:

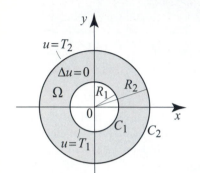

Figure 3 Dirichlet problem in an annulus.

$$(1) \qquad u(z) = T_1 + (T_2 - T_1)\frac{\ln(|z|/R_1)}{\ln(R_2/R_1)}, \quad R_1 < |z| < R_2.$$

This function is harmonic in the complex plane minus the origin and so it is harmonic in the annulus $R_1 < |z| < R_2$. The fact that it takes the values T_1 and T_2 on the boundary can be verified directly. For the derivation of this solution, see Exercise 30, Section 2.5. When the outer circle C_2 is the unit circle ($R_2 = 1$), the solution becomes

$$(2) \qquad u(z) = T_1 + (T_1 - T_2)\frac{\ln|z| - \ln R_1}{\ln R_1}, \quad R_1 < |z| < 1.$$

EXAMPLE 3   A problem on a region between nonconcentric circles
Find a harmonic function u in the nonregular annular region Ω in Figure 4(a), such that $u = 50$ on the inner circle C_1 and $u = 100$ on the outer unit circle C_2.

Solution   We transform the problem into a Dirichlet problem on an annulus using the linear fractional transformation of Example 4, Section 6.2,

$$w = \phi(z) = \frac{3z - 1}{3 - z}.$$

Figures 4(a) and (b)
A Dirichlet problem in a non-regular annulus (Figure 4(a)) is greatly simplified by first transforming the region into a regular annulus (Figure 4(b)), using a linear fractional transformation.

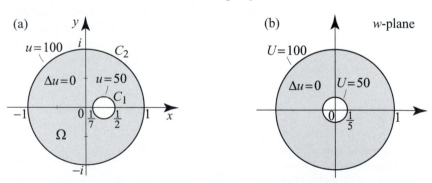

As shown in Example 4, Section 6.2, ϕ maps the unit circle C_2 onto the unit circle, the inner circle C_1 onto a circle centered at the origin with radius $\frac{1}{5}$, and the region between the unit circle and C_2 onto the annular region $\frac{1}{5} < |w| < 1$. The boundary values in the transformed problem are shown in Figure 4(b), and so according to (2) the solution of the Dirichlet problem in the w-plane is

$$U(w) = 50 + 50\frac{\ln|w| + \ln 5}{\ln 5}, \quad \frac{1}{5} < |w| < 1.$$

Substituting $w = \phi(z) = \frac{3z-1}{3-z}$, we obtain the solution in the z-plane

$$u(z) = 50 + 50 \frac{\ln \left| \frac{3z-1}{3-z} \right| + \ln 5}{\ln 5}.$$

With the help of a computer, we have evaluated the solution at various points inside the nonregular annular region in Figure 4(a). The values are shown in Table 1.

(x, y)	$(0, 0)$	$(\frac{1}{7} - 0.001, 0)$	$(\frac{1}{2} + 0.001, 0)$	$(0.99, 0.01)$	$(0.99, -0.01)$	$(\frac{1}{2}, \frac{1}{3})$	$(\frac{1}{2}, -\frac{1}{3})$
$u(x + iy)$	65.87	50.15	50.20	99.38	99.38	74.73	74.73

Table 1. Temperature of various points inside the nonregular annular region in Figure 4(a).

The table seems to confirm the solution that we found. The values of $u(z)$ are between 50 and 100. They are closer to the boundary values as z approaches the boundary (inner and outer). Note also the symmetric property of u, due to the symmetries in the given problem: We have $u(x + iy) = u(x - iy)$. You should expand the table of values and make your own conclusions about the solution. ∎

We next derive the Poisson formula in the upper half-plane by using the corresponding formula in the unit disk. We state the result in a theorem, but its proof is a simple application of the ideas of this section. (The formula was also derived by a different method in the exercises of Section 2.5.)

THEOREM 1
POISSON INTEGRAL
FORMULA IN THE
UPPER
HALF-PLANE

Let $f(x)$ be a bounded piecewise smooth function on the real line. A solution of the Dirichlet problem $\Delta u(x + iy) = 0$ for $x + iy$ in the upper half-plane ($y > 0$) with boundary condition $u(x) = f(x)$ for all $-\infty < x < \infty$ is given by the **Poisson integral formula**

$$(3) \qquad u(x + iy) = \frac{y}{\pi} \int_{-\infty}^{\infty} \frac{f(s)}{(x - s)^2 + y^2} \, ds = \frac{y}{\pi} \int_{-\infty}^{\infty} \frac{f(x - s)}{s^2 + y^2} \, ds.$$

For $y > 0$, the function $P_y(x) = \sqrt{\frac{2}{\pi}} \frac{y}{x^2 + y^2}$ ($-\infty < x < \infty$) is called the **Poisson kernel** on the real line. (The constant $\sqrt{\frac{2}{\pi}}$ is a normalizing constant.) We can write (3) in terms of the Poisson kernel as

$$(4) \quad u(x + iy) = \frac{1}{\sqrt{2\pi}} \int_{-\infty}^{\infty} f(s) P_y(x - s) \, ds = \frac{1}{\sqrt{2\pi}} \int_{-\infty}^{\infty} f(x - s) P_y(s) \, ds.$$

This features the solution of the Dirichlet problem in the upper half-plane as a *convolution* of the boundary function f with the Poisson kernel. The notion of convolution of functions will be discussed in later chapters. The Poisson kernel has a wealth of properties. This fundamental function links complex analysis in the upper half-plane to Fourier analysis of functions on the real line.

Proof We will prove the first equality in (3); the second one follows by making the change of variables $s' = x - s$. Transform the given problem into a Dirichlet problem in the unit disk using the linear fractional transformation $w = \psi(z) = \frac{i-z}{i+z}$ (Example 1(ii), Section 6.2). This mapping takes the real line onto the unit circle, and the upper half-plane onto the unit disk. Denote the image of a point s on the real line by $e^{i\phi}$ on the unit circle, so $e^{i\phi} = \psi(s)$.

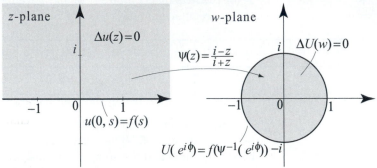

Figure 5 How the boundary values in a Dirichlet problem are transformed, after using a linear fractional transformation that takes boundary to boundary.

As illustrated in Figure 5, the boundary data that we get for the problem on the unit disk is given by $U(e^{i\phi}) = f(\psi^{-1}(e^{i\phi}))$, for all $e^{i\phi}$ on the unit circle. The solution of the Dirichlet problem in the unit disk ($|w| < 1$) with this boundary data is obtained from the Poisson integral formula, which we recall from (10), Section 3.8:

$$(5) \qquad U(w) = \frac{1 - |w|^2}{2\pi} \int_0^{2\pi} \frac{f(\psi^{-1}(e^{i\phi}))}{|e^{i\phi} - w|^2} \, d\phi.$$

Hence the solution in the upper half-plane is (set $w = \psi(z)$ in (5))

$$(6) \qquad u(x + iy) = U(\psi(x + iy)) = U(\psi(z)) = \frac{1 - |\psi(z)|^2}{2\pi} \int_0^{2\pi} \frac{f(\psi^{-1}(e^{i\phi}))}{|e^{i\phi} - \psi(z)|^2} \, d\phi.$$

Our goal is to show that (6) is precisely (3). The details are straightforward but a little tedious. Make the change of variables $s = \psi^{-1}(e^{i\phi})$. Since the integrand in (6) is 2π-periodic, we get the same result if we integrate from $-\pi$ to π instead of 0 to 2π; and as ϕ runs from $-\pi$ to π, s runs from $-\infty$ to ∞. We have

$$s = \psi^{-1}(e^{i\phi}) \;\Rightarrow\; \psi(s) = \frac{i - s}{i + s} = e^{i\phi}.$$

Taking differentials and using $e^{i\phi} = \frac{i-s}{i+s}$, we get

$$\frac{-2i}{(i+s)^2} \, ds = i e^{i\phi} d\phi = i \frac{i-s}{i+s} d\phi \;\Rightarrow\; d\phi = \frac{2}{1 + s^2} \, ds.$$

Substituting what we have so far into (6), we obtain

$$(7) \qquad u(x + iy) = \frac{1 - |\psi(z)|^2}{\pi} \int_{-\infty}^{\infty} \frac{f(s)}{|e^{i\phi} - \psi(z)|^2} \frac{ds}{1 + s^2}.$$

Comparing (7) and (3), it suffices to show that for $z = x + iy$ and $e^{i\phi} = s$,

$$(8) \qquad \frac{1 - |\psi(z)|^2}{|e^{i\phi} - \psi(z)|^2} = (1 + s^2) \frac{y}{(x - s)^2 + y^2}.$$

This part is straightforward and is left to Exercise 21. ∎

EXAMPLE 4 Applying Poisson's integral formula

Solve the Dirichlet problem in the upper half-plane with boundary data on the real line given by

$$f(x) = u(x) = \begin{cases} C & \text{if } a < x < b, \\ 0 & \text{otherwise,} \end{cases}$$

where C is a constant.

Solution We apply the first formula in (3) and get for $-\infty < x < \infty$, $y > 0$,

$$u(x + iy) = \frac{y}{\pi} \int_{-\infty}^{\infty} \frac{f(s)}{(x-s)^2 + y^2}\, ds = \frac{y}{\pi} \int_{a}^{b} \frac{C}{(x-s)^2 + y^2}\, ds,$$

since $f(s)$ is 0 outside the interval $a < s < b$. We evaluate the integral in terms of the inverse tangent, using an obvious change of variables,

$$\begin{aligned}
u(x + iy) &= \frac{y}{\pi} \int_{a}^{b} \frac{C}{(x-s)^2 + y^2}\, ds = \frac{y}{\pi}\frac{C}{y^2} \int_{a}^{b} \frac{1}{\left(\frac{s-x}{y}\right)^2 + 1}\, ds \\
&= \frac{C}{\pi} \int_{\frac{a-x}{y}}^{\frac{b-x}{y}} \frac{ds}{s^2 + 1} \\
&= \frac{C}{\pi}\left[\tan^{-1}\left(\frac{b-x}{y}\right) + \tan^{-1}\left(\frac{x-a}{y}\right) \right].
\end{aligned}$$

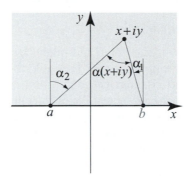

Figure 6 For a given interval (a, b), the angle at $x + iy$ subtended by (a, b) is a harmonic function $\alpha(x + iy)$ called the harmonic measure of (a, b).

We can give a concrete geometric interpretation of this answer. Let $\alpha_1 = \tan^{-1}\left(\frac{b-x}{y}\right)$ and $\alpha_2 = \tan^{-1}\left(\frac{x-a}{y}\right)$ (Figure 6). Then $u(x + iy) = \frac{C}{\pi}(\alpha_1 + \alpha_2) = \frac{C}{\pi}\alpha(x + iy)$, where $\alpha(x+iy)$ is the harmonic measure of the interval (a, b); that is, α is the angle at the point $x + iy$ subtended by the interval (a, b) (Figure 6). You should review Exercise 35, Section 2.5, where this result was derived using guessing methods and properties of the argument function. ∎

Suppose that Ω is a region, f_1 and f_2 are two functions defined on the boundary of Ω. Suppose that u_1 and u_2 are solutions of the Dirichlet problems on Ω with boundary values f_1 and f_2, respectively. Because Laplace's equation $\Delta u = 0$ is linear, it is straightforward to check that $u = u_1 + u_2$ solves the Dirichlet problem on Ω with boundary values $u = f_1 + f_2$ on the boundary of Ω. This useful process of generating a solution by adding solutions of two or more related problems is called **superposition** of solutions. It will appear again in our study of other linear equations.

EXAMPLE 5 Superposing solutions

Let (a_1, b_1) and (a_2, b_2) be two disjoint intervals on the real line, and let T_1 and T_2 be two complex numbers. Solve the Dirichlet problem in the upper half-plane with boundary data on the real line given by $f(x) = T_1$ if x is in (a_1, b_1), T_2 if x is in (a_2, b_2) and 0 otherwise.

Solution For $j = 1, 2$, let $f_j(x) = T_j$ if x is in (a_j, b_j), 0 otherwise. Clearly, $f(x) = f_1(x) + f_2(x)$. From Example 4, the solution of the Dirichlet problem in the upper half-plane with boundary values on the real line given by $f_j(x)$ is $u_j(z) = \frac{T_j}{\pi}\left[\tan^{-1}\left(\frac{b_j-x}{y}\right) - \tan^{-1}\left(\frac{a_j-x}{y}\right)\right]$. Thus the solution to our original problem is $u(z) = u_1(z) + u_2(z)$. ∎

In the next example, we use the Poisson integral formula and the Joukowski mapping from Example 3, Section 6.1.

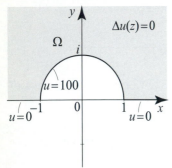

Figure 7 for Example 6.

EXAMPLE 6   Joukowski mapping

Solve the Dirichlet problem in the region Ω shown in Figure 7.

Solution   We transform the problem into a problem in the upper half-plane by using the Joukowski mapping $w = J(z) = \frac{1}{2}(z + \frac{1}{z})$. As shown in Example 3, Section 6.1, $J(z)$ takes the upper-unit circle onto $[-1, 1]$, and the semi-infinite intervals $(-\infty, -1]$ and $[1, \infty)$ onto themselves. Thus the boundary conditions in the transformed Dirichlet problem in the upper half-plane are given by $f(w) = 100$ for real $-1 < w < 1$ and $f(w) = 0$ for real w outside this interval. According to Example 4, the solution in the upper half of the w-plane is given by

$$\frac{100}{\pi}\left[\tan^{-1}\left(\frac{1 - \operatorname{Re}w}{\operatorname{Im}w}\right) + \tan^{-1}\left(\frac{1 + \operatorname{Re}w}{\operatorname{Im}w}\right)\right].$$

Replacing w by $J(z)$ we obtain the solution in the region Ω: for all (x, y) in Ω,

$$u(x + i\,y) = \frac{100}{\pi}\left[\tan^{-1}\left(\frac{1 - \operatorname{Re}J(z)}{\operatorname{Im}J(z)}\right) + \tan^{-1}\left(\frac{1 + \operatorname{Re}J(z)}{\operatorname{Im}J(z)}\right)\right].$$

With a little bit more work, the answer could be expressed in terms of x and y. ∎

Our next example retests our ability to evaluate integrals using residues.

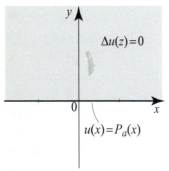

Figure 8 for Example 7.

EXAMPLE 7   A Poisson boundary distribution

Solve the Dirichlet problem (Figure 8) in the upper half-plane with boundary data on the real line given by a Poisson temperature distribution

$$f(x) = u(x) = P_a(x) = \sqrt{\frac{2}{\pi}}\,\frac{a}{x^2 + a^2}, \quad \text{where } a > 0.$$

This problem models the steady-state distribution in a large sheet of metal with insulated lateral surface, whose boundary along the x-axis is kept at a fixed temperature distribution given by a Poisson kernel $P_a(x)$, where $a > 0$ is a positive constant. The temperature at the origin is $f(0) = u(0) = P_a(0) = \sqrt{\frac{2}{\pi}}\,\frac{1}{a}$, which tends to ∞ as a tends to 0. Away from the origin, the temperature decays to 0 like $a/(x^2 + a^2)$. So smaller values of $a > 0$ correspond to temperature distributions concentrated around 0 (Figure 9).

Solution   We apply the first formula in (3) and get for $-\infty < x < \infty$, $y > 0$,

$$u(x + i\,y) = \frac{y}{\pi}\int_{-\infty}^{\infty}\frac{P_a(s)}{(x - s)^2 + y^2}\,ds = \frac{ay}{\pi}\sqrt{\frac{2}{\pi}}\int_{-\infty}^{\infty}\frac{ds}{(s^2 + a^2)((x - s)^2 + y^2)}.$$

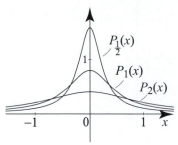

Figure 9 The Poisson kernel $P_y(x)$ for various values of $y > 0$. Properties to note on the graphs of $P_y(x)$ as a function of x:
$P_y(x) > 0$ for all x;
$P_y(x)$ is even;
$P_y(x)$ is a bell-shaped curve;
$\lim_{y \downarrow 0} P_y(0) = \infty$.

We evaluate the last integral using the residue method of Section 5.3, by completing the contour with a semicircle in the upper half-plane. The function $h(z) = \frac{1}{(z^2+a^2)((x-z)^2+y^2)}$ has two simple poles in the upper half-plane, at $z = ia$ and at $z = x + iy$. We compute the residues at these points using Proposition 1, Section 5.1(iii). We have

$$\operatorname{Res}\left(h(z),\, ia\right) = \frac{1}{(x-ia)^2 + y^2} \operatorname{Res}\left(\frac{1}{(z^2+a^2)},\, ia\right) = \frac{1}{(x-ia)^2+y^2}\frac{1}{2ia},$$

and

$$\operatorname{Res}\left(h(z),\, x+iy\right) = \frac{1}{(x+iy)^2+a^2} \operatorname{Res}\left(\frac{1}{(x-z)^2+y^2},\, x+iy\right) = \frac{1}{(x+iy)^2+a^2}\frac{1}{2iy}.$$

Applying Proposition 2, Section 5.3, we obtain

$$
\begin{aligned}
u(x+iy) &= \frac{ay}{\pi}\sqrt{\frac{2}{\pi}}\, 2\pi i \left(\operatorname{Res}\left(h(z),\, ia\right) + \operatorname{Res}\left(h(z),\, x+iy\right)\right) \\
&= 2ayi\sqrt{\frac{2}{\pi}}\left(\frac{1}{(x-ia)^2+y^2}\frac{1}{2ia} + \frac{1}{(x+iy)^2+a^2}\frac{1}{2iy}\right) \\
&= \sqrt{\frac{2}{\pi}}\left(\frac{y}{(x-ia)^2+y^2} + \frac{a}{(x+iy)^2+a^2}\right) \\
&= \sqrt{\frac{2}{\pi}}\frac{a+y}{x^2+(a+y)^2},
\end{aligned}
$$

where the last equality follows by elementary algebraic manipulations. This solves the problem. But note that the last expression is precisely $P_{a+y}(x)$. Thus $u(x+iy) = P_{a+y}(x)$, which shows that the solution of the Dirichlet problem with a Poisson boundary data $P_a(x)$ is another Poisson distribution $P_{a+y}(x)$. This amazing fact about temperature problems with Poisson boundary data has a direct interpretation in terms of convolutions and Fourier transforms, which will be discussed in later chapters. ∎

Exercises 6.3

In Exercises 1–15, (a) solve the Dirichlet problem described by the accompanying figure (Figures $10 - 24$), using the methods of this section. Examples 4 and 5 and superposition of solutions are useful.

(b) Evaluate your answer at various points inside the region and comment on your data, as we did in Example 3.

1.

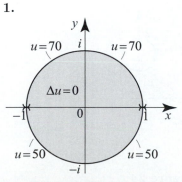

Figure 10

2.

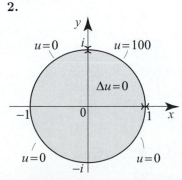

Figure 11

3.

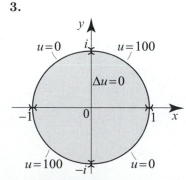

Figure 12

4.

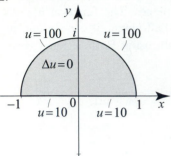

Figure 13

5.

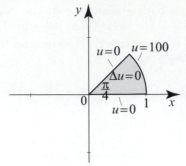

Figure 14

6.

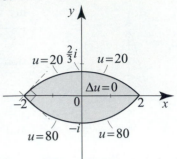

Figure 15

7.

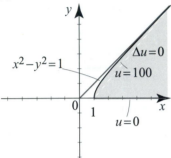

Figure 16

8.

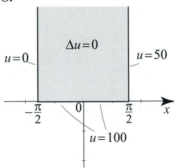

Figure 17

9.

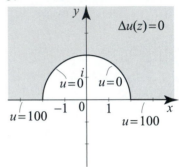

Figure 18

10.

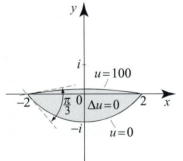

Figure 19

11.

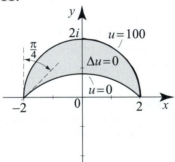

Figure 20

12.

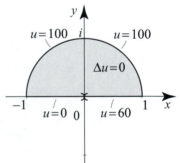

Figure 21

13.

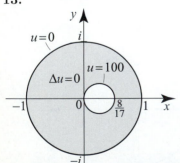

Figure 22

14.

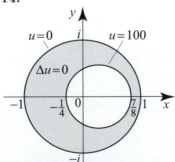

Figure 23

15.

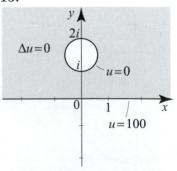

Figure 24

16. Generalize Example 4 as follows. Suppose that $I_1 = (a_1, b_1)$, $I_2 = (a_2, b_2)$, $\ldots$, $I_n = (a_n, b_n)$ are n disjoint intervals on the real line, $T_1, T_2, \ldots, T_n$ are n real or complex values. (a) Show that a solution of the Dirichlet problem in the upper half-plane with boundary data

$$f(x) = \begin{cases} T_j & \text{if } a_j < x < b_j, \\ 0 & \text{otherwise,} \end{cases}$$

is

$$u(x + i\,y) = \frac{1}{\pi} \sum_{j=1}^{n} T_j \left[\tan^{-1}\left(\frac{b_j - x}{y} \right) - \tan^{-1}\left(\frac{a_j - x}{y} \right) \right].$$

If any one of the a_j's is infinite, say $a_1 = -\infty$, then $\tan^{-1}\left(\frac{a_1 - x}{y} \right) = \tan^{-1}(-\infty) = -\frac{\pi}{2}$. Similarly, if one of the b_j's is infinite, say $b_n = \infty$, then $\tan^{-1}\left(\frac{b_n - x}{y} \right) = \tan^{-1}(\infty) = \frac{\pi}{2}$.
(b) Let $z = x + i\,y$. Show that the answer can be written as

$$u(z) = \frac{1}{\pi} \sum_{j=1}^{n} T_j \left(\operatorname{Arg}(z - b_j) - \operatorname{Arg}(z - a_j) \right).$$

17. (a) Solve the Dirichlet Problem in Figure 25(a) by using the Poisson integral formula.

Figures 25(a) and (b) for Exercise 17.

(a)

(b)

(b) Solve the Dirichlet problem in Figure 25(b) by using the conformal map $w = \sin z$ and the result in part (a).

18. Show that the solution in Example 1 is

$$u(x + i\,y) = 50 + \frac{100}{\pi} \left[\tan^{-1}\left(\frac{y}{1 - x} \right) + \tan^{-1}\left(\frac{y}{1 + x} \right) \right], \qquad x^2 + y^2 < 1.$$

19. Isotherms in Example 1. Recall the solution of the corresponding Dirichlet problem in the upper half of the w-plane, $\frac{100}{\pi}(\pi - \operatorname{Arg} w)$.
(a) Show that the isotherms in the w-plane are rays emanating from the origin.
(b) Conclude that the isotherms in the z-plane are arcs of circles in the unit disk. Describe the centers and radii of the circles that define the isotherms in the unit disk. [Hint: Consider the pre-image of the isotherms in the w-plane by the mapping $\phi(z) = i\frac{1-z}{1+z}$.]

20. Isotherms in Example 3. By studying the isotherms of the Dirichlet problem in the w-plane in Example 3, determine the isotherms of the original problem in the z-plane.

21. Complete the proof of Theorem 1 by showing that (8) holds. [Suggestion: Organize your proof as follows. Show $1 - |\psi(z)|^2 = \frac{4y}{x^2+(1+y)^2}$. Then show $|e^{i\pi} - \psi(z)|^2 = |\psi(s) - \psi(z)|^2 = 4\frac{(x-s)^2+y^2}{(1+s^2)(x^2+(1+y)^2)}$.]

22. Show that $P_a(x) = \frac{1}{a}P_1\left(\frac{x}{a}\right)$ for all $a > 0$. Thus the graph of $P_a(x)$ is the graph of $P_1(x)$, scaled vertically by a factor of $\frac{1}{a}$ and scaled horizontally by a factor of a.

23. Show that for any $y > 0$, $\frac{1}{\sqrt{2\pi}}\int_\infty^\infty P_y(x)\,dx = 1$, where $P_y(x)$ is the Poisson kernel.

In Exercises 24–29, solve the Dirichlet problem in the upper half-plane with the boundary values on the x-axis given by $f(x)$ as indicated. Use residues to evaluate the Poisson integral. Take a real and $\neq 0$ in Exercises 28 and 29.

24. $f(x) = \dfrac{x}{x^2+x+1}$.   **25.** $f(x) = \dfrac{1}{x^4+1}$.   **26.** $f(x) = \dfrac{\sin x}{x}$.

27. $f(x) = \cos x$.   **28.** $f(x) = \cos ax$.   **29.** $f(x) = \sin ax$.

Solve the Dirichlet problem depicted in the figure (Figures 26-31) by transforming it into a Dirichlet problem in the upper half-plane. To solve in the upper half-plane, use the Arg function in Exercises 30–32, and the Poisson integral formula in Exercises 33–35. Leave your answer in Exercise 35 in the form of an integral involving the Chebyshev polynomials. Compute the integral when $n = 2$.

30.

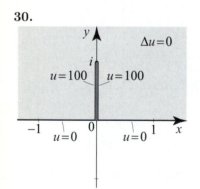

Figure 26

31.

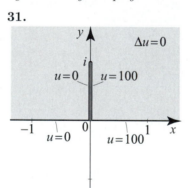

Figure 27

32.

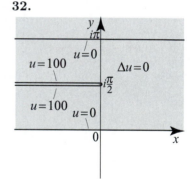

Figure 28

33.

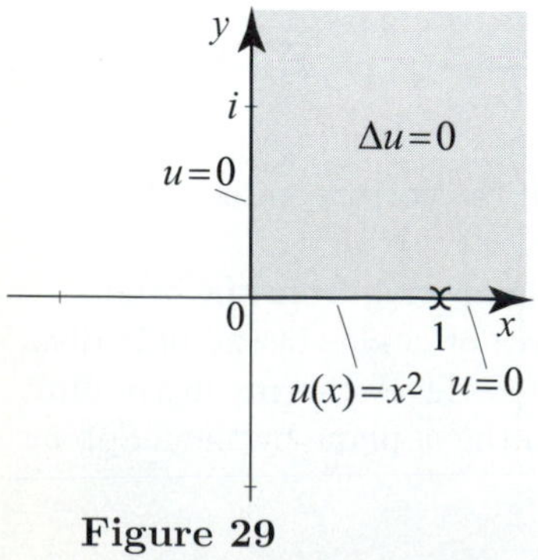

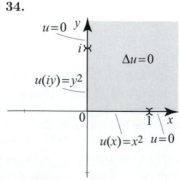

Figure 29

34.

Figure 30

35.

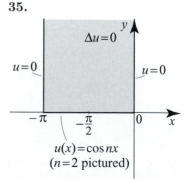

Figure 31

6.4   The Schwarz-Christoffel Transformation

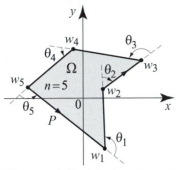

Figure 1 Positively oriented polygonal boundary with corner angles measured from the outside.

In this section we describe a method for constructing one-to-one analytic mappings of the upper half-plane onto polygonal regions. We start by setting the notation.

Suppose that Ω is a region in the w-plane, bounded by a positively oriented polygonal path P with n sides. Let $w_1, w_2, \ldots, w_n$ denote the vertices of P, counted consecutively as we trace the path through its positive orientation (see Figure 1, where $n = 5$). If P is bounded, then the point w_n is taken to be the initial and terminal point of the closed path P. If P is unbounded, we take $w_n = \infty$ and think of P as a polygon with $n-1$ vertices $w_1, w_2, \ldots, w_{n-1}$ (Figure 2). It will be convenient to measure the exterior angle at a vertex, and so we let θ_j denote the angle that we make as we turn the corner of the polygon at w_j. We choose $0 < |\theta_j| < \pi$ ($j = 1, \ldots, n$); a positive value corresponds to a left turn and a negative value corresponds to a right turn. In Figure 1, θ_2 is negative while all other θ_j are positive.

THEOREM 1
SCHWARZ-CHRISTOFFEL TRANSFORMATION

> Let Ω be a region bounded by a polygonal path P with vertices at w_j (counted consecutively) and corresponding exterior angles θ_j. Then there is a one-to-one conformal mapping $f(z)$ of the upper half-plane onto Ω, such that
>
> $$(1) \qquad f'(z) = A\,(z - x_1)^{-\frac{\theta_1}{\pi}} (z - x_2)^{-\frac{\theta_2}{\pi}} \cdots (z - x_{n-1})^{-\frac{\theta_{n-1}}{\pi}},$$
>
> where the x_j's are real, $x_1 < x_2 < \cdots < x_{n-1}$, $f(x_j) = w_j$, $\lim_{z \to \infty} f(z) = w_n$, A is a constant, and principal branches are used to define the powers.

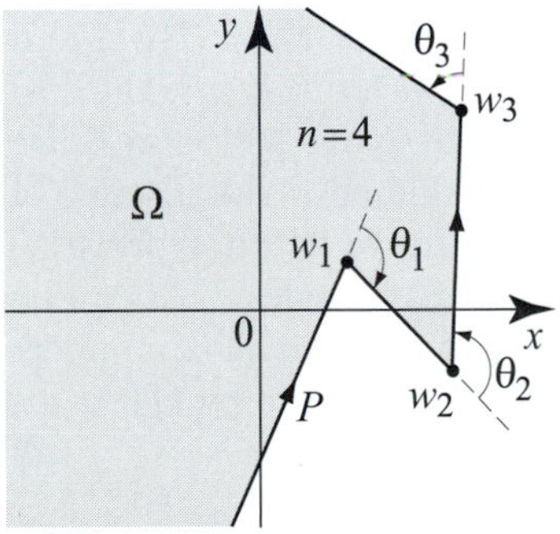

The points x_j ($j = 1, \ldots, n-1$) on the x-axis are the pre-images of the vertices of the polygonal path P in the w-plane. Two of the x_j's may be chosen arbitrarily, so long as they are arranged in ascending order. We can express the fact that $\lim_{z \to \infty} f(z) = w_n$ by writing $f(\infty) = w_n$. In the case of an unbounded polygon P, we have $f(\infty) = \infty$.

The mapping $f(z)$ is called a **Schwarz-Christoffel transformation**, after the German mathematicians Herman Amandus Schwarz (1843–1921) and Elwin Bruno Christoffel (1829–1900). Since $f(z)$ is an antiderivative of (1), we can write

$$(2) \qquad f(z) = A \int (z - x_1)^{-\frac{\theta_1}{\pi}} (z - x_2)^{-\frac{\theta_2}{\pi}} \cdots (z - x_{n-1})^{-\frac{\theta_{n-1}}{\pi}} \, dz + B.$$

Figure 2 Unbounded polygonal region with n sides ($n = 4$) and $n - 1$ vertices.

The constants A and B depend on the size and location of the polygonal path P. The full proof of Theorem 1 is quite complicated. We only sketch a part that illustrates the ideas behind the construction of the transformation.

Sketch of Proof of Theorem 1. Consider a mapping $f(z)$ whose derivative is given by (1) and let w_j denote the image of x_j, where $x_1 < x_2 < \cdots < x_{n-1}$

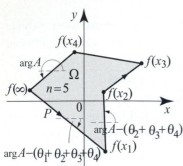

Figure 3 Arguments of the line segments of the polygon P.

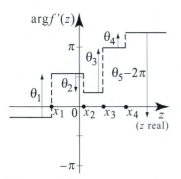

Figure 4 As x crosses x_j from left to right, $\arg f'(x)$ changes abruptly by θ_j then remains constant until it crosses x_{j+1}.

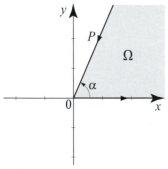

Figure 5 for Example 1.

are real. To understand the effect of the mapping f on the real axis, recall from Section 6.1 that a conformal mapping $f(z)$ at a point z_0 acts like a rotation by an angle $\arg f'(z_0)$. Thus the mapping whose derivative is given by (1) acts like a rotation by an angle

$$(3) \quad \arg f'(z) =$$
$$\arg A - \frac{\theta_1}{\pi} \arg(z - x_1) - \frac{\theta_2}{\pi} \arg(z - x_2) - \cdots - \frac{\theta_{n-1}}{\pi} \arg(z - x_{n-1}).$$

For $z = x$ on the x-axis with $z < x_1$, we have $z - x_j < 0$ for all $j = 1, 2, \ldots, n-1$, hence $\arg(z - x_j) = \pi$ for all $j = 1, 2, \ldots, n-1$, and so from (3) we get

$$(4) \quad \arg f'(z) = \arg A - (\theta_1 + \theta_2 + \cdots + \theta_{n-1}).$$

Thus if $w_n = f(\infty)$, then all the points in the interval $(-\infty, x_1)$ are mapped onto a line segment starting with w_n and ending with $w_1 = f(x_1)$ and at an angle given by (4) (Figure 3). For the points in the interval (x_1, x_2), we have $z - x_1 > 0$ and $z - x_j < 0$ for all $j = 2, \ldots, n-1$, hence $\arg(z - x_1) = 0$ and $\arg(z - x_j) = \pi$ for all $j = 2, \ldots, n-1$, and so from (3)

$$(5) \quad \arg f'(z) = \arg A - (\theta_2 + \cdots + \theta_{n-1}).$$

Thus, at x_1 we have an abrupt change in the argument (the path turns left by angle θ_1) and then all the points in (x_1, x_2) are mapped onto the line segment with initial point w_1 and terminal point $w_2 = f(x_2)$, and at an angle given by (5) (Figure 4). We continue in this fashion, and finally we find that after an abrupt change in the argument, the points in the interval (x_{n-1}, ∞) are mapped onto a line segment with initial point w_{n-1}, and at angle $\arg f'(z) = \arg A$. In the case of a bounded polygon, this line segment will connect back to w_n (Exercise 11). The polygon is then closed; after turning the last corner we have gone through an angle $\theta_1 + \theta_2 + \cdots + \theta_n = 2\pi$. Thus we have shown that the mapping whose derivative is given by (1) takes the real line onto a polygon with vertices $w_j = f(x_j)$ and exterior angles θ_j. Since the upper half-plane is to our left as we traverse the real line rightward, conformality ensures that the image region is to our left as we trace P in the positive sense (that is, f maps the upper half-plane onto the interior of P). The converse of these statements is also true, although we will not prove it. That is, given a polygonal path P with vertices w_j and exterior angles θ_j, it can be shown that we can find ordered real numbers $x_1, \ldots, x_{n-1}$, and complex numbers A and B such that the mapping given by (2) whose derivative is given by (1), takes the real line onto P. Moreover, two of the x_j's can be chosen arbitrarily. ∎

EXAMPLE 1   Schwarz-Christoffel transformation for a sector
Find a Schwarz-Christoffel transformation that maps the upper half-plane onto the sector at angle $0 < \alpha < \pi$ in Figure 5.

Solution   Obviously one answer is $f(z) = z^{\frac{\alpha}{\pi}}$, but let us see how this answer comes out of (2). Since the region has two sides, we have $n = 2$. From Figure 5, the exterior angle at $w_1 = 0$ is $\pi - \alpha$. In the z-plane, choose $x_1 = 0$, then (2) yields

$$f(z) = A \int z^{-\frac{\pi - \alpha}{\pi}} \, dz + B = A \int z^{-1 + \alpha/\pi} \, dz + B = A \frac{\pi}{\alpha} z^{\frac{\alpha}{\pi}} + B,$$

where all branches are principal. In order to have $f(0) = 0$, we take $B = 0$. Obviously any positive value of A will work, so we can take $f(z) = z^{\frac{\alpha}{\pi}}$. ∎

For our next example, we will need the following useful formula, whose proof is sketched in Exercise 12. For z in the upper half-plane ($\operatorname{Im} z > 0$), $0 \leq \alpha \leq \pi$, and all real numbers a, we have

$$(6) \quad \boxed{(z+a)^{\frac{\alpha}{\pi}}(z-a)^{\frac{\alpha}{\pi}} = (-1)^{\frac{\alpha}{\pi}}(a^2 - z^2)^{\frac{\alpha}{\pi}}, \text{ all branches are principal.}}$$

For example, if $\alpha = \frac{\pi}{2}$, $a = 1$, and $\operatorname{Im} z > 0$, then

$$(7) \quad \boxed{(z+1)^{\frac{1}{2}}(z-1)^{\frac{1}{2}} = i(1-z^2)^{\frac{1}{2}}.}$$

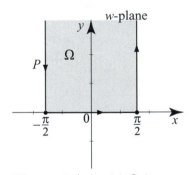

Figure 6 A semi-infinite vertical strip with positively oriented boundary.

EXAMPLE 2   The inverse sine as a Schwarz-Christoffel transformation

Find a Schwarz-Christoffel transformation that maps the upper half-plane onto the the semi-infinite vertical strip in Figure 6.

Solution   We know that $\sin z$ maps the infinite strip in Figure 6 onto the upper half-plane. So the mapping that we are looking for is the inverse of $\sin z$. Let us see how this comes out of (2). In the w-plane, take $w_1 = -\frac{\pi}{2}$, $w_2 = \frac{\pi}{2}$, with exterior angles $\theta_1 = \theta_2 = \frac{\pi}{2}$. In the z-plane, take $x_1 = -1$ and $x_2 = 1$. Then (2) yields

$$f(z) = A \int (z+1)^{-\frac{1}{2}}(z-1)^{-\frac{1}{2}} \, dz + B.$$

Using (7) and a well-known antiderivative, we get

$$f(z) = -Ai \int \frac{1}{\sqrt{1-z^2}} \, dz + B = -Ai \sin^{-1} z + B,$$

where $\sin^{-1} z$ is the principal branch of the inverse sine function; that is,

$$\sin^{-1} z = -i \operatorname{Log}\left(iz + e^{\frac{1}{2}\operatorname{Log}(1-z^2)}\right)$$

(see Example 3, Section 1.7). Setting $f(-1) = -\frac{\pi}{2}$, we find

$$-Ai \sin^{-1}(-1) + B = -\frac{\pi}{2} \quad \Rightarrow \quad Ai\frac{\pi}{2} + B = -\frac{\pi}{2}.$$

Setting $f(1) = \frac{\pi}{2}$, we find

$$-Ai \sin^{-1}(1) + B = \frac{\pi}{2} \quad \Rightarrow \quad -Ai\frac{\pi}{2} + B = \frac{\pi}{2}.$$

Solving for A and B, we find $A = i$ and $B = 0$. Hence $f(z) = \sin^{-1} z$. ∎

Like many constructions involving Schwarz-Christoffel transformations, the next example gives rise to elliptic integrals (see Section 5.5). Although these integrals are very difficult to evaluate, they are extensively tabulated and can be conveniently evaluated numerically using standard functions in most computer systems.

EXAMPLE 3   Schwarz-Christoffel transformation for a triangle

Find a Schwarz-Christoffel transformation that maps the upper half-plane onto the right isosceles triangle in Figure 7.

Solution   It is clear that the triangle is determined by two consecutive vertices and their corresponding angles. In the w-plane, take $w_1 = -1$, $w_2 = 1$, with exterior angles $\theta_1 = \theta_2 = \frac{3\pi}{4}$. In the z-plane, we freely choose the points $x_1 = -1$ and $x_2 = 1$. Then (2) yields

$$f(z) = A \int_0^z (\zeta + 1)^{-\frac{3}{4}} (\zeta - 1)^{-\frac{3}{4}} \, d\zeta + B = A \int_0^z \frac{d\zeta}{(-1)^{\frac{3}{4}}(1 - \zeta^2)^{\frac{3}{4}}} + B,$$

where we have used (6) with $\alpha = \frac{3\pi}{4}$. This integral cannot be expressed in terms of elementary functions for arbitrary z, but we will be able to evaluate it for $z = \pm 1$ in order to determine the constants A and B. Setting $f(-1) = -1$ and using $(-1)^{\frac{3}{4}} = e^{\frac{3\pi i}{4}}$, we get

$$-1 = e^{-\frac{3\pi i}{4}} A \int_0^{-1} \frac{d\zeta}{(1 - \zeta^2)^{\frac{3}{4}}} + B = -e^{-\frac{3\pi i}{4}} A \int_0^1 \frac{d\zeta}{(1 - \zeta^2)^{\frac{3}{4}}} + B,$$

or

$$(8) \qquad -e^{-\frac{3\pi i}{4}} A I + B = -1, \quad \text{where} \quad I = \int_0^1 \frac{d\zeta}{(1 - \zeta^2)^{\frac{3}{4}}}.$$

To evaluate the integral I, we make the change of variables $\zeta = \sin x$, $d\zeta = \cos x \, dx$, then

$$I = \int_0^{\frac{\pi}{2}} \frac{dx}{(\cos x)^{\frac{1}{2}}} = \frac{1}{2} \frac{\Gamma(1/4)\,\Gamma(1/2)}{\Gamma(3/4)} \approx 2.622,$$

where we have appealed to Exercise 25, Section 4.3, to evaluate the integral in terms of the gamma function. The approximate value of the integral was obtained with the help of a computer. Setting $1 = f(1)$, we get

$$(9) \qquad e^{-\frac{3\pi i}{4}} A I + B = 1.$$

Solving for A and B in (9) and (8), we find $B = 0$ and $A = \frac{e^{\frac{3\pi i}{4}}}{I}$, and so

$$f(z) = \frac{e^{\frac{3\pi i}{4}}}{I} e^{-\frac{3\pi i}{4}} \int_0^z \frac{d\zeta}{(1 - \zeta^2)^{\frac{3}{4}}} = \frac{1}{I} \int_0^z \frac{d\zeta}{(1 - \zeta^2)^{\frac{3}{4}}}. \qquad \blacksquare$$

The next two examples illustrate a limiting technique in computing Schwarz-Christoffel transformations.

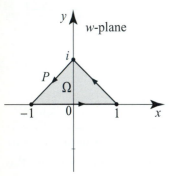

Figure 7 Positively oriented isosceles triangle.

EXAMPLE 4   An L-shaped region
Find a Schwarz-Christoffel transformation that maps the upper half-plane onto the
L-shaped region in Figure 8.

Solution   To determine the orientation of the boundary in such a way that the
region becomes interior to a positively oriented boundary, we think of the region
as a limit of a region with vertices at $w_1 = 0$, $w_2 > 0$, and $w_3 = 1 + i$, and
corresponding exterior angles $\theta_1 = \frac{\pi}{2}$, $\theta_2 = \alpha$, and $\theta_3 = \beta$ (Figure 9).

An L-shaped region with positively oriented boundary. As
we follow the boundary according to this orientation,
the region is to our left.

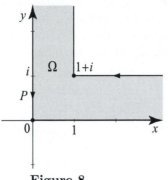

Figure 8   **Figure 9**

As $w_2 \to \infty$, $\theta_2 \to \pi$, and $\theta_3 \to -\frac{\pi}{2}$. Thus, we may think of our region as having a
vertex at infinity with exterior angle $\theta_2 = \pi$ and a vertex at $w_3 = 1 + i$ with exterior
angle $-\frac{\pi}{2}$. In fact, setting $\theta_j = \pi$ will force $f(x_j) = \infty$. Now we may only choose
two of the three x_j's arbitrarily, but in fact a solution can be found in $x_1 = -1$,
$x_2 = 0$, and $x_3 = 1$. Other choices of the three x_j's will typically result in L-shaped
regions where the angle ϕ in Figure 8 is not $\pi/4$. From (2) and (7)

$$f(z) = A \int (z+1)^{-\frac{1}{2}} z^{-1} (z-1)^{\frac{1}{2}} \, dz + B = A \int \frac{z-1}{iz(1-z^2)^{\frac{1}{2}}} \, dz + B.$$

We have

$$\frac{z-1}{iz(1-z^2)^{\frac{1}{2}}} = \frac{-i}{(1-z^2)^{\frac{1}{2}}} + \frac{i}{z(1-z^2)^{\frac{1}{2}}}.$$

We have $\int \frac{-i\,dz}{(1-z^2)^{\frac{1}{2}}} = -i \sin^{-1} z + C$, and letting $z = \frac{1}{\zeta}$, $dz = -\frac{d\zeta}{\zeta^2}$, we have

$$\int \frac{i\,dz}{z(1-z^2)^{\frac{1}{2}}} = -i \int \frac{d\zeta}{\zeta(1-\frac{1}{\zeta^2})^{\frac{1}{2}}} = \int \frac{d\zeta}{(1-\zeta^2)^{\frac{1}{2}}} = \sin^{-1} \zeta + C,$$

where justification involving the square root is left to Exercise 13. Thus

$$f(z) = A\left(-i \sin^{-1} z + \sin^{-1} \frac{1}{z} \right) + B,$$

where inverse sines are principal branches. Setting $f(-1) = 0$, we get $A(-i(-\frac{\pi}{2}) + (-\frac{\pi}{2})) + B = 0$. Setting $f(1) = 1 + i$, we get $A(-i\frac{\pi}{2} + \frac{\pi}{2}) + B = 1 + i$. Solving for
A and B, we get $A = \frac{i}{\pi}$ and $B = \frac{1+i}{2}$, and so

$$f(z) = \frac{1}{\pi}\left(\sin^{-1} z + i \sin^{-1} \frac{1}{z} \right) + \frac{1+i}{2}. \qquad \blacksquare$$

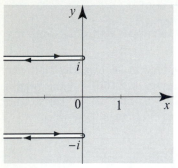

Figure 10 A doubly slit plane with a positively oriented boundary consisting of four sides.

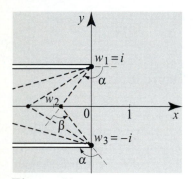

Figure 11

EXAMPLE 5   A doubly slit plane

Find a Schwarz-Christoffel transformation that maps the upper half-plane onto the doubly slit plane in Figure 10, which consists of the w-plane minus the two semi-infinite horizontal lines $\operatorname{Im} w = \pm i$ and $\operatorname{Re} w < 0$.

Solution   To define the region as the interior of a positively oriented boundary requires four sides as shown in Figure 10. We think of the region as a limit of a region with vertices at $w_1 = i$, $w_2 = t$ (where $t < 0$) and $w_3 = -i$, and corresponding exterior angles $\theta_1 = \alpha$, $\theta_2 = \beta$, and $\theta_3 = \gamma$ (Figure 11). As $t \to -\infty$, $\theta_1 \to -\pi$, $\theta_2 \to \pi$, and $\theta_3 \to -\pi$. Thus, we may think of our region as having a vertex at $w_2 = \infty$ with exterior angle $\theta_2 = \pi$ and two vertices at $\pm i$ with each exterior angle being $-\pi$. Taking $x_1 = -1$, $x_2 = 0$, and $x_3 = 1$, we get from (2)

$$f(z) = A \int (z+1)\frac{1}{z}(z-1)\,dz + B = A \int \left(z - \frac{1}{z}\right)\,dz + B = \frac{A}{2}z^2 - A\operatorname{Log} z + B.$$

Setting $f(-1) = i$ and $f(1) = -i$, we get

$$\begin{cases} i & = \frac{A}{2} - A\operatorname{Log}(-1) + B \\ -i & = \frac{A}{2} - A\operatorname{Log}(1) + B, \end{cases} \Rightarrow \begin{cases} i & = \frac{A}{2} - Ai\pi + B \\ -i & = \frac{A}{2} + B. \end{cases}$$

Solving for A and B we get $A = -\frac{2}{\pi}$ and $B = \frac{1}{\pi} - i$, and so

$$f(z) = -\frac{1}{\pi}z^2 + \frac{2}{\pi}\operatorname{Log} z + \frac{1}{\pi} - i. \qquad \blacksquare$$

Image of Level Curves

Suppose that $f(z)$ is a Schwarz-Christoffel transformation taking the upper half-plane onto a region Ω in the w-plane. If we want to solve a Dirichlet problem $\Delta U(w) = 0$, the technique is to map the problem to a corresponding problem in the upper half of the z-plane, where a solution $u(z)$ of Laplace's equation $\Delta u(z) = 0$ with boundary conditions can be obtained more easily. The solution in the w-plane is then $U(w) = u(f^{-1}(w))$. However, the Schwarz-Christoffel transformation gives us f, not f^{-1}, and it is not always possible or easy to invert $f(z)$ in closed form (try Examples 3, 4, and 5). Nevertheless, the conformal map $f(z)$ is still very useful: It allows us to find isotherms of $U(w)$ without actually knowing $U(w)$. As we now show, this is because the image under f of an isotherm $u(z) = C$ (where C is a constant) is an isotherm $U(w) = C$ in the region Ω.

PROPOSITION 1	With the preceding notation, the image under $w = f(z)$ of the level curve
IMAGE OF LEVEL	$u(z) = C$ is a level curve $U(w) = C$. Thus, if $\gamma(t)$ parametrizes an isotherm
CURVES	of u, then $f(\gamma(t))$ parametrizes a corresponding isotherm of U.

Proof Since $\gamma(t)$ is a level curve of u, $u(\gamma(t)) = C$. Since $U(w) = u(f^{-1}(w))$, we conclude that $U(f(\gamma(t))) = u(\gamma(t)) = C$, and hence $f(\gamma(t))$ is a level curve of U. $\blacksquare$

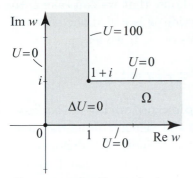

Figure 12 for Example 6.

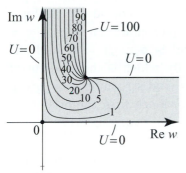

Figure 13 Isotherms in Example 6.

This proposition is of course true for any one-to-one conformal map, not just for Schwarz-Christoffel transformations acting on the upper half-plane. We illustrate with an example.

EXAMPLE 6   Isotherms in the L-shaped region

Find the isotherms of the Dirichlet problem $\Delta U(w) = 0$ in Figure 12.

Solution   From Example 4, we know that the Schwarz-Christoffel transformation $f(z) = \frac{1}{\pi}\left(\sin^{-1} z + i \sin^{-1} \frac{1}{z}\right) + \frac{1+i}{2}$ will map the upper half-plane onto the L-shaped region. Since the boundary data switches from 0 to 100 at $f(1) = 1 + i$, the corresponding Dirichlet problem in the upper half-plane is $\Delta u(z) = 0$, $u(x) = 0$ for $x < 1$, and $u(x) = 100$ for $x > 1$. We immediately write down the solution using the argument function:

$$ u(z) = 100 - \frac{100}{\pi} \operatorname{Arg}(z - 1). $$

The isotherms $u = C$ thus satisfy $\operatorname{Arg}(z - 1) = \pi - \frac{\pi C}{100}$ and are rays emanating from the point $z = 1$. Each ray is parametrized by $\gamma(t) = 1 + te^{i(\pi - \frac{\pi C}{100})}$, where $0 < t < \infty$. By Proposition 1, the image of this ray under $w = f(z)$ is the isotherm $U = C$ in the L- shaped region; and it is parametrized by

$$ f(\gamma(t)) = \frac{1}{\pi}\left(\sin^{-1}\left(1 + te^{i(\pi - \frac{\pi C}{100})}\right) + i\sin^{-1}\left(\frac{1}{1 + te^{i(\pi - \frac{\pi C}{100})}}\right)\right) + \frac{1+i}{2}. $$

Some isotherms are plotted in Figure 13.   ∎

Fluid Flow

We will investigate problems in two-dimensional fluid flow, using our knowledge of harmonic functions and the technique of conformal mapping. In an ideal situation where fluid is flowing over a two-dimensional surface represented by an unbounded region Ω in the z-plane, assuming that the fluid is incompressible (fixed density) and irrotational (circulation around a closed path is zero), it can be shown that there is a harmonic function $\phi(x, y)$ such that the velocity $V(x, y)$ of a point (x, y) in Ω is given by the gradient of ϕ. That is,

$$ (10) \qquad V(x, y) = \nabla\phi(x, y) = \left(\frac{\partial\phi}{\partial x}, \frac{\partial\phi}{\partial y}\right) = (\phi_x, \phi_y). $$

The function ϕ is called the **velocity potential** of the flow. The curves defined by the relation

$$ (11) \qquad\qquad \phi(x, y) = C_1 $$

are called the **equipotential curves** or **equipotential lines**. The **stream-lines** of the flow are the curves that are orthogonal to the equipotential curves. If the streamlines are expressed in the form

$$ (12) \qquad\qquad \psi(x, y) = C_2, $$

for some function ψ, then, from Section 2.5, we know that we can take ψ to be the harmonic conjugate of ϕ in Ω. The function ψ is called the **stream function**. The fluid will flow on the level curves of ψ. If we let

$$(13) \qquad \Phi(z) = \phi(x, y) + i\,\psi(x, y) \quad (z = x + iy \text{ in } \Omega),$$

then Φ is analytic in Ω. It is called the **complex potential** of the flow. Thus the real part of the complex potential is the velocity potential, and the level curves of its imaginary part are the streamlines. If Γ denotes the boundary of Ω, we would expect the fluid to flow along Γ or that Γ is one of the streamlines. Thus the points on Γ should satisfy (12).

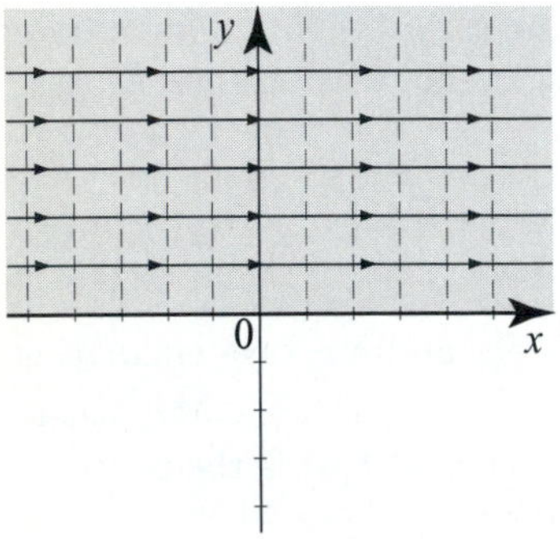

For a simple example of a complex potential, we consider a uniform rightward flow in the upper half-plane with complex potential $\Phi(z) = z = x + iy$ (Figure 14). Here the velocity potential is $\phi(x, y) = x$, and the velocity at each point is given by the vector $(1, 0)$, which is the gradient of $\phi(x, y) = x$. The stream function is $\psi(x, y) = y$, and streamlines are given as $\psi(x, y) = C$ for some constant $C \geq 0$. In accordance with the properties of a flow, the boundary $y = 0$ is one of the streamlines corresponding to the value $C = 0$.

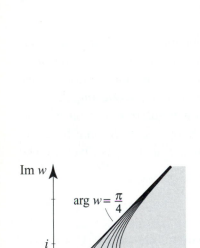

Figure 14 Streamlines in a uniform rightward flow in the upper half-plane.

Given an unbounded region Ω in the w-plane, how can we find the streamlines? We will apply conformal mapping techniques. Let f be a one-to-one conformal mapping of the upper half of the z-plane onto Ω, taking the x-axis onto Γ, the boundary of Ω. Let a stream function $\psi(z)$ be given for the upper half-plane. By the properties of conformal mappings, $\Psi(w) = \psi(f^{-1}(w))$ will be harmonic for all w in Ω. Proposition 1 tells us that the images of streamlines $\psi(z) = C$ under the mapping $f(z)$ are streamlines $\Psi(w) = C$. Since the real axis is a streamline for $\psi(z)$ and Γ is the image of the real axis, we conclude that Γ is a streamline for $\Psi(w)$. Thus $\Psi(w)$ is a stream function for Ω.

In the following examples, we take the simple stream function $\psi(z) = y$ for the upper half-plane. Streamlines for the region Ω are found by using Proposition 1.

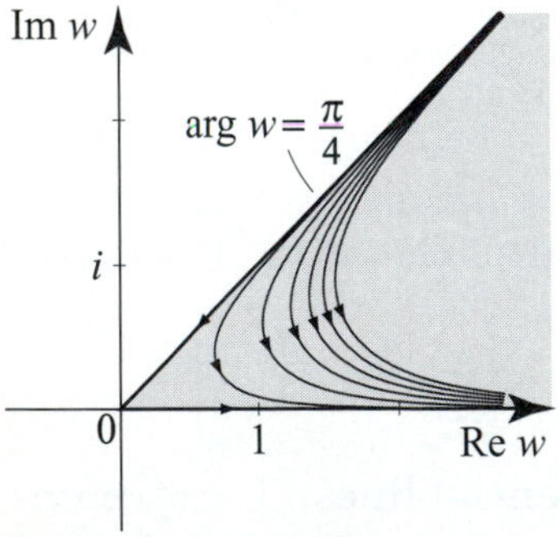

Figure 15 Streamlines in a sector.

EXAMPLE 7   Fluid flow in a sector

Find and plot the streamlines for the sector in Figure 15, where fluid flows in along the line $\operatorname{Arg} w = \frac{\pi}{4}$ and flows out along $\operatorname{Arg} w = 0$.

Solution   From Example 1, the Schwarz-Christoffel transformation $f(z) = z^{\frac{1}{4}}$ maps the upper half-plane to the given sector. We will use the simple stream function in the upper half-plane, $\psi(z) = y$, to generate a solution. Streamlines in the z-plane are parametrized as $\gamma(x) = x + iy_0$ for fixed y_0. Streamlines in the w-plane are images of these under f; we have $f(\gamma(x)) = (x + iy_0)^{\frac{1}{4}}$. As the parameter x increases, the streamlines are traced in the manner shown in Figure 15. Fluid comes in along $\operatorname{Arg} w = \frac{\pi}{4}$ and out along $\operatorname{Arg} w = 0$.   ∎

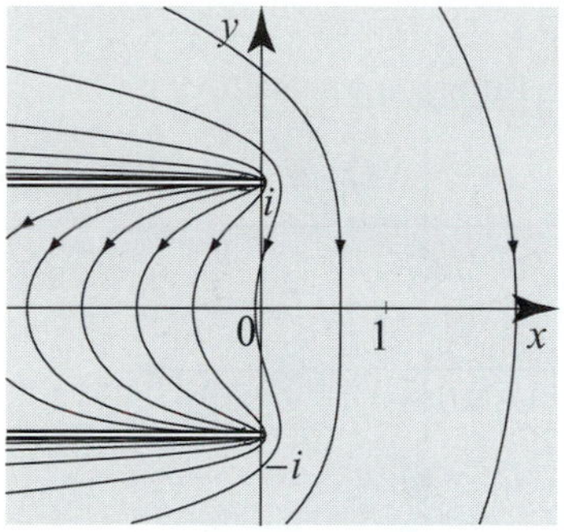

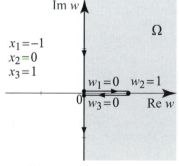

Figure 16 Streamlines in a doubly slit plane.

EXAMPLE 8   Fluid flow in the doubly slit plane

Find and plot the streamlines for the doubly slit plane in Figure 16, where fluid flows in from the upper left, past the double obstacle, and flows out to the lower left.

Solution   In Example 5, we found that $f(z) = -\frac{1}{\pi}z^2 + \frac{2}{\pi}\operatorname{Log} z + \frac{1}{\pi} - i$ is a conformal mapping of the upper half-plane onto the doubly slit plane Ω. Taking $\psi(z) = y$ to be the stream function for the upper half-plane, for each $y_0 \geq 0$ we have a streamline parametrized by $\gamma(x) = x + i y_0$, $-\infty < x < \infty$. By Proposition 1, the images $f(\gamma(x))$ are streamlines in the doubly slit plane. They are

$$f(x + i\,y_0) = -\frac{1}{\pi}(x + i\,y_0)^2 + \frac{2}{\pi}\operatorname{Log}(x + i\,y_0) + \frac{1}{\pi} - i \qquad (-\infty < x < \infty).$$

The streamlines are plotted in Figure 16. Note that the central channel serves neither as a source of fluid nor as a final destination; fluid flows in and then flows out. The fluid far into the central channel is almost stagnant.   ∎

Exercises 6.4

In Exercises 1–6, find the Schwarz-Christoffel transformation of the upper half-plane onto the given region. Use the labeled points to set-up the integral (2).

1. See Figure 17. [Hint: Use (7) and integrate.]

2. See Figure 18. [Hint: $\dfrac{z+1}{(z-1)^{\frac{1}{2}}} = \dfrac{z-1+2}{(z-1)^{\frac{1}{2}}} = (z-1)^{\frac{1}{2}} + \dfrac{2}{(z-1)^{\frac{1}{2}}}.$]

3. See Figure 19. [Hint: $\dfrac{(z+1)^{\frac{1}{2}}}{(z-1)^{\frac{1}{2}}} = \dfrac{z+1}{i(1-z^2)^{\frac{1}{2}}} = \dfrac{z}{i(1-z^2)^{\frac{1}{2}}} + \dfrac{1}{i(1-z^2)^{\frac{1}{2}}}.$]

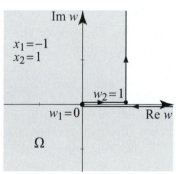
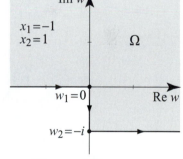

Figure 17

Figure 18

Figure 19

4. See Figure 20. [Hint: $\dfrac{(z-1)^{\frac{1}{2}}}{(z+1)^{\frac{1}{2}}} = \dfrac{z-1}{i(1-z^2)^{\frac{1}{2}}} = \dfrac{z}{i(1-z^2)^{\frac{1}{2}}} - \dfrac{1}{i(1-z^2)^{\frac{1}{2}}}.$]

5. See Figure 21. [Hint: Let $z = \sin\zeta$, where $-\frac{\pi}{2} \leq \operatorname{Re}\zeta \leq \frac{\pi}{2}$ and $\operatorname{Im}\zeta \geq 0$. Then $(1-z^2)^{\frac{1}{2}} = \cos\zeta$. Use $\cos^2\zeta = \frac{1+\cos 2\zeta}{2}$, integrate, then use $\sin 2\zeta = 2\sin\zeta\cos\zeta.$]

6. See Figure 22. [Hints: $\dfrac{(z-1)^{\frac{1}{2}}}{z+1} = \dfrac{1}{(z-1)^{\frac{1}{2}}}\dfrac{z-1}{z+1} = \dfrac{1}{(z-1)^{\frac{1}{2}}} - \dfrac{2}{(z-1)^{\frac{1}{2}}}\dfrac{1}{z+1}.$ In the second term, use $\dfrac{1}{z+1} = \dfrac{i}{2\sqrt{2}}\left(\dfrac{1}{\sqrt{z-1}+i\sqrt{2}} - \dfrac{1}{\sqrt{z-1}-i\sqrt{2}}\right)$, and change variables $u = \sqrt{z-1} - i\sqrt{2}$ and $v = \sqrt{z-1} - i\sqrt{2}$. You cannot find A and B just from $f(1) = 0$. Instead, first argue that $\operatorname{Arg} A = -\frac{\pi}{2}$ (look at the angle of the final line segment;

see the proof of Theorem 1) and thus $A = -i|A|$. Now get $B = i\pi\sqrt{2}|A|$ from $f(1) = 0$. To get $|A|$, either use $\text{Im } f(x) = 1$ for $x < 1$, or use the channel width formula (16) $1 = |s| = \pi|A||x_2 - x_1|^{-\frac{\theta_2}{\pi}}$.]

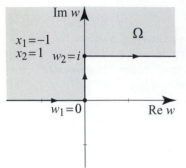

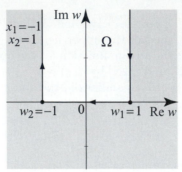

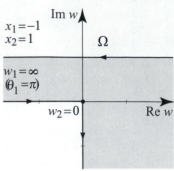

| **Figure 20** | **Figure 21** | **Figure 22** |

7. Consider the Dirichlet problem in the triangular region of Example 3, where the base is kept at temperature $100°$ and the other two sides at temperature $0°$. (a) Determine the isotherms in the triangular region by first studying the isotherms in the upper half-plane of the corresponding Dirichlet problem (see Example 6).

(b) Plot several isotherms to illustrate your answer in (a).

8. Consider the Dirichlet problem in the semi-infinite strip of Example 2, where the base is kept at temperature $100°$ and the other two vertical sides at temperature $0°$. (a) Determine the isotherms in the region by following the method of Example 6.

(b) Plot several isotherms to illustrate your answer in (a).

9. Find and plot the streamlines in the region of Example 1.

10. Find and plot the streamlines in the image of the upper half-plane in Exercise 1.

11. Project Problem: Closure of the polygon. In this exercise, we will show that for a closed polygon where $\theta_1 + \cdots + \theta_n = 2\pi$, the integral formula for $f(z)$, (2), converges to w_n as $z \to \infty$.

(a) Use $\theta_n < \pi$ to show that $\theta_1 + \cdots + \theta_{n-1} > \pi$. Define $\beta_j = \frac{\theta_j}{\pi}$ for $j = 1, \ldots, n-1$. Then $\beta_1 + \cdots + \beta_{n-1} > 1$.

(b) Note that the coefficients A and B in (1) will dilate, rotate, and translate the mapping and do not affect convergence of $f(z)$ or closure of the polygon, so we take $A = 1$ and $B = 0$. For concreteness, pick z_0 real to be the larger of x_{n-1} and 1, and set

$$f(z) = \int_{z_0}^{z} \frac{d\zeta}{(\zeta - x_1)^{\beta_1}(\zeta - x_2)^{\beta_2} \cdots (\zeta - x_{n-1})^{\beta_{n-1}}}.$$

(c) We show that $f(z)$ has a limit on the positive real axis. Restrict $z = x$ real, and use the limit comparison test for the integrand (against $\frac{1}{x^{\beta_1 + \cdots + \beta_{n-1}}}$) to show that $\lim_{x \to \infty} f(x) = \int_{z_0}^{\infty} \frac{d\zeta}{(x - x_1)^{\beta_1}(x - x_2)^{\beta_2} \cdots (x - x_{n-1})^{\beta_{n-1}}}$ is finite. Define w_n to be this number.

(d) To show that $\lim_{z \to \infty} f(z) = w_n$ for any $|z| \to \infty$, write $z = Re^{i\theta}$, $R > 0$. Then

$$|f(Re^{i\theta}) - f(R)| = \left| \int_{R}^{Re^{i\theta}} \frac{dz}{(z - x_1)^{\beta_1}(z - x_2)^{\beta_2} \cdots (z - x_{n-1})^{\beta_{n-1}}} \right| \leq R\pi M(R),$$

where $M(R)$ is the maximum of the absolute value of the integrand on the upper semicircle of radius R. Now $\lim_{R\to\infty}(M(R)\,R^{\beta_1+\cdots+\beta_{n-1}}) = 1$, so $RM(R) = R^{1-\beta_1-\cdots-\beta_{n-1}}(M(R)\,R^{\beta_1+\cdots+\beta_{n-1}}) \to 0$, and so $|f(Re^{i\theta}) - f(R)| \to 0$ uniformly in θ as $R \to \infty$. Since $f(R) \to w_n$, we conclude that $f(z) \to w_n$ as $z \to \infty$.

12. (a) Follow the outlined steps to show that for $\operatorname{Im} z > 0$, a real, and $0 \leq \alpha \leq \pi$,

(14) $$(z-a)^{\frac{\alpha}{\pi}} = (-1)^{\frac{\alpha}{\pi}}(a-z)^{\frac{\alpha}{\pi}}, \text{ all branches principal.}$$

Fix z with $\operatorname{Im} z > 0$. We must prove $e^{\frac{\alpha}{\pi}\operatorname{Log}(z-a)} = e^{i\alpha}e^{\frac{\alpha}{\pi}\operatorname{Log}(a-z)}$, and it will be sufficient to prove $\operatorname{Log}(z-a) = \operatorname{Log}(a-z)+i\pi$, or that $\operatorname{Arg}(z-a) = \operatorname{Arg}(a-z)+\pi$. We know that for each z, $\operatorname{Arg}(z-a) = \operatorname{Arg}(a-z)\pm\pi$. However, $0 < \operatorname{Arg}(z-a) < \pi$ and $-\pi < \operatorname{Arg}(a-z) < 0$, so $0 < \operatorname{Arg}(z-a) - \operatorname{Arg}(a-z) < 2\pi$ and we must use the plus sign. This proves (14).
(b) Follow the outlined steps to prove (6). Use part (a) to show $(z-a)^{\frac{\alpha}{\pi}}(z+a)^{\frac{\alpha}{\pi}} = (-1)^{\frac{\alpha}{\pi}}(a-z)^{\frac{\alpha}{\pi}}(a+z)^{\frac{\alpha}{\pi}}$, and so we must show $(a-z)^{\frac{\alpha}{\pi}}(a+z)^{\frac{\alpha}{\pi}} = (a^2-z^2)^{\frac{\alpha}{\pi}}$. It is sufficient to show that $\operatorname{Arg}(a-z) + \operatorname{Arg}(a+z) = \operatorname{Arg}(a^2-z^2)$. We know that for each z, $\operatorname{Arg}(a-z) + \operatorname{Arg}(a+z) = \operatorname{Arg}(a^2-z^2) + 2k\pi$, where k is 0, 1, or -1. However $-\pi < \operatorname{Arg}(a-z) < 0$ and $0 < \operatorname{Arg}(a+z) < \pi$, so the left side is in $(-\pi, \pi)$, and hence $k = 0$.

13. In this exercise we prove that for $\operatorname{Im}\zeta < 0$,

(15) $$\zeta\left(1-\frac{1}{\zeta^2}\right)^{\frac{1}{2}} = -i(1-\zeta^2)^{\frac{1}{2}}, \text{ all branches principal.}$$

Expanding in terms of the logarithm, it will be sufficient to show that $\operatorname{Arg}\zeta + \frac{1}{2}\operatorname{Arg}\left(1-\frac{1}{\zeta^2}\right) = -\frac{\pi}{2} + \frac{1}{2}\operatorname{Arg}(1-\zeta^2)$, or that $h(\zeta) = 2\operatorname{Arg}\zeta + \operatorname{Arg}\left(1-\frac{1}{\zeta^2}\right) + \pi - \operatorname{Arg}(1-\zeta^2) = 0$. We know that $h(\zeta) = 2k\pi$, where k is an integer that may depend on ζ. However, since the images under $w_1 = \left(1-\frac{1}{\zeta^2}\right)$ and $w_2 = (1-\zeta^2)$ of the lower half-plane $\operatorname{Im}\zeta < 0$ do not include the negative real axis, $h(\zeta)$ is continuous. A continuous function which takes on discrete values must be constant (Lemma 2, Section 5.7), and so k cannot depend on ζ. Pick $\zeta = -i$ and conclude $k = 0$.

14. Project Problem: Channel width formula. We can write the logarithm as a Schwarz-Christoffel transformation $\operatorname{Log} z = \int \frac{dz}{z}$, where there is one vertex, $x_1 = 0$, $\theta_1 = \pi$, $w_1 = \infty$. The selection of $\theta_1 = \pi$ forces a semi-infinite channel in the left half-plane (in this case the channel is also infinite into the right half-plane). For points Δx on the real axis near $x_1 = 0$, we have $f(\Delta x) - f(-\Delta x) = -i\pi$, and so the channel width is π. We now see that this type of behavior is exhibited wherever we set an angle $\theta_{j_0} = \pi$.
(a) Take a Schwarz-Christoffel transformation where a particular $\theta_{j_0} = \pi$, and other $\theta_j \leq \pi$. Then

$$f(z) = A\int \frac{dz}{(z-x_1)^{\frac{\theta_1}{\pi}}\cdots(z-x_{j_0})\cdots(z-x_{n-1})^{\frac{\theta_{n-1}}{\pi}}} + B.$$

Argue briefly that $\lim_{z\to x_{j_0}} f(z) = \infty$ by considering the fact that each factor in the integrand is approximately constant near x_{j_0} except $\frac{1}{z-x_{j_0}}$ which has a divergent integral.
(b) Define the channel separation complex number s by $s = \lim_{r\downarrow 0}(f(x_{j_0} + r) -$

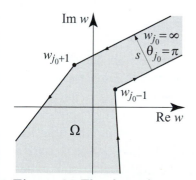

Figure 23 The channel separation s.

$f(x_{j_0} - r))$. A typical case is shown in Figure 23, and its absolute value $|s|$ represents the channel width. Parametrize the upper semicircle from $x_{j_0} - r$ to $x_{j_0} + r$ and get

$$s = A \lim_{r \downarrow 0} \int_0^\pi \frac{-ire^{i(\pi - t)} \, dt}{(x_{j_0} + re^{i(\pi - t)} - x_1)^{\frac{\theta_1}{\pi}} \cdots (re^{i(\pi - t)}) \cdots (x_{j_0} + re^{i(\pi - t)} - x_{n-1})^{\frac{\theta_{n-1}}{\pi}}}$$

$$= \frac{-iA\pi}{(x_{j_0} - x_1)^{\frac{\theta_1}{\pi}} \cdots (x_{j_0} - x_{n-1})^{\frac{\theta_{n-1}}{\pi}}},$$

where in the denominator of the final expression, the term $x_{j_0} - x_{j_0}$ is skipped. Conclude that the channel width is

(16)
$$|s| = \frac{|A|\pi}{|x_{j_0} - x_1|^{\frac{\theta_1}{\pi}} \cdots |x_{j_0} - x_{n-1}|^{\frac{\theta_{n-1}}{\pi}}}$$

where again $|x_{j_0} - x_{j_0}|$ is skipped.

(c) Verify the channel width formula for the L-shaped region of Example 4 and the doubly slit plane of Example 5.

6.5   Green's Functions

What is a Green's function and what can it do for us? To answer these questions, let us review a few facts about the solution of a Dirichlet problem. Suppose that Ω is a simply connected region bounded by a simple path Γ. Let f be a piecewise continuous function on Γ and consider the Dirichlet problem (Figure 1)

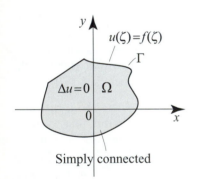

Figure 1 A Dirichlet problem in a simply connected region.

(1)
$$\Delta u(z) = 0 \quad \text{for all } z \text{ in } \Omega;$$

(2)
$$u(\zeta) = f(\zeta) \quad \text{for all } \zeta \text{ on } \Gamma.$$

If Ω is the unit disk D, then the solution of the problem (1)–(2) is given by the Poisson formula (see (12), Section 3.8). The importance of this formula is that it depends only on the region and the boundary function f. For an arbitrary simply connected region Ω, we saw in Section 6.1 that, as a consequence of the Riemann mapping theorem and the Poisson formula on the disk, the Dirichlet problem (1)–(2) on Ω has a solution. It is natural to ask whether we can express this solution as an integral that depends only on the region Ω and works for any piecewise continuous boundary function f. Amazingly, the answer is affirmative! Our goal in this section is to derive a formula that expresses the solution as a path integral over Γ, involving the boundary function f and the so-called **Green's function** of the region Ω, which is a function that depends only on Ω. The formula is named in honor of its discoverer, one of the leading mathematicians and physicists of the nineteenth century, George Green (1793–1841), who was a self-taught mathematician from England. The importance of Green's functions will be appreciated in later sections, where they will be presented as the only tools for solving boundary value problems on certain regions involving Poisson's equation and other important equations in applied mathematics.

If you recall in Section 3.8, we used a change of variables to derive the Poisson integral on the unit disk from the mean value property of harmonic functions. We will take the same approach on Ω, and so we begin by explaining the change of variables that is a key to deriving and understanding Green's functions.

Suppose that Ω and Ω' are two regions bounded by simple paths Γ and Γ'. Let ϕ be a one-to-one analytic map of Ω onto Ω'. We will further suppose that ϕ is analytic and one-to-one on Γ. It follows from Section 6.1 that ϕ maps boundary to boundary. Suppose that F is a real differentiable function of two variables defined on Γ'. We will think of complex numbers as points in the complex plane, and consider $F(z)$ for z on Γ'. We will write $\frac{\partial F}{\partial n_{\Gamma'}}$ or simply $\frac{\partial F}{\partial n}$ to denote the directional derivative of F in the direction of the outward unit normal vector to the path Γ'. By definition, this is the dot product of the gradient of F, $\nabla F = (F_x, F_y)$, with the outward unit normal vector n_{Γ}'. Thus

$$\frac{\partial F}{\partial n_{\Gamma'}} = \nabla F \cdot n_{\Gamma'},$$

where each expression is computed at a given point on Γ' (Figure 2).

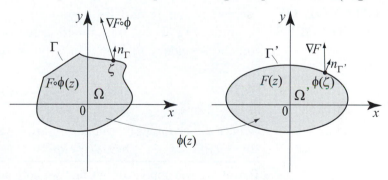

Figure 2 The mapping ϕ is analytic and one-to-one on Ω. It maps boundary to boundary and preserves angles.

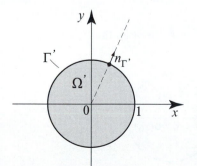

Figure 3 For a circle centered at the origin, the normal derivative is the radial derivative.

Although normal derivatives are tedious to compute in general, they are easy to express in some important special cases. For example, if Γ' is any circle centered at the origin, then $\frac{\partial F}{\partial n_{\Gamma'}}$ is just the radial derivative of F (Figure 3):

$$\frac{\partial F}{\partial n_{\Gamma'}} = \frac{\partial F}{\partial r}.$$

If $F(z) = \ln |z|$ and Γ' is the unit circle, then for all points on the unit circle

$$\left.\frac{\partial F}{\partial n_{\Gamma'}}\right|_{|z|=1} = \left.\frac{\partial}{\partial r} \ln r\right|_{r=1} = \left.\frac{1}{r}\right|_{r=1} = 1.$$

Our goal is to relate the normal derivative of F on Γ' to the normal derivative of $F(\phi(z))$ on Γ. Recall that if ϕ is analytic and $|\phi'(z)| \neq 0$, then ϕ rotates a path through z by a fixed angle and scales by $|\phi'(z)|$. So ϕ will map a normal vector to Γ at ζ to a normal vector to Γ' at $\phi(\zeta)$, and it will scale its modulus by $|\phi'(\zeta)|$. Since the normal derivative measures the rate of change of the function in the direction of the normal vector to the curve, thinking as we do with the chain rule, we expect the normal derivative of $F \circ \phi$ at ζ to equal the normal derivative of F at $\phi(\zeta)$ times $|\phi'(\zeta)|$. This expectation turns out to be correct, and we have the following change of variables formula:

(3)
$$\boxed{\frac{\partial (F \circ \phi)}{\partial n_\Gamma}(\zeta) = |\phi'(\zeta)|\, \frac{\partial F}{\partial n_{\Gamma'}}(\phi(\zeta)).}$$

The proof of (3) is a nice application of conformal properties, the chain rule in two dimensions, and the Cauchy-Riemann equations. We give it at the end of this section in order not to interrupt the presentation. The importance of this formula is that it incorporates the effect of the conformal properties of analytic functions. Let us move a step closer to the desired formula for Green's functions and derive a formula that uses the boundary values of u to reproduce its value at a special point inside Ω. You should note the role of the mean value property of harmonic functions in the proof.

LEMMA 1
CHANGE OF
VARIABLES

Suppose that $w = \phi(z)$ is a one-to-one analytic mapping of a simply connected region Ω and its boundary Γ onto the open unit disk D and its boundary C. Let u be a function harmonic on Ω and piecewise continuous on the boundary Γ. Let z_0 in Ω be the point such that $\phi(z_0) = 0$. Then

(4)
$$u(z_0) = \frac{1}{2\pi} \int_\Gamma u(\zeta) \frac{\partial \ln |\phi(\zeta)|}{\partial n}\, ds,$$

where $ds = |d\zeta|$ is the element of arc length on Γ. Hence if Γ is parametrized by $\gamma(t)$, $a \le t \le b$, then $ds = |\gamma'(t)|\, dt$.

Proof We will apply (3), but first we note one useful result. (Refer to Figure 4 for help with the notation.)

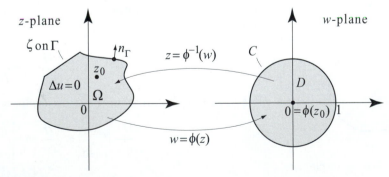

Figure 4 If ϕ is analytic and one-to-one, then ϕ^{-1} is also analytic.

The function ϕ^{-1} is analytic from the closed unit disk (in the w-plane) onto Ω and its boundary (in the z-plane). So the function $u(\phi^{-1}(w))$ is harmonic on the open unit disk, being the composition of a harmonic function u with an analytic function ϕ^{-1} (Theorem 3, Section 2.5). Moreover, $u(\phi^{-1}(w))$ is piecewise continuous on C. Thus, by the mean value property of harmonic functions (Corollary 7, Section 3.8), we have

$$(5) \qquad \frac{1}{2\pi} \int_0^{2\pi} u(\phi^{-1}(e^{it}))\, dt = u(\phi^{-1}(0)) = u(z_0).$$

Our goal now is to show that the integral in (4) is precisely the integral that we just evaluated in (5). Parametrize C by e^{it}, $0 \le t \le 2\pi$. Then Γ will be parametrized by $\phi^{-1}(e^{it})$, $0 \le t \le 2\pi$. The element of arc length on Γ is

$$\left| \frac{d}{dt}\phi^{-1}(e^{it}) \right| dt = \left| \frac{ie^{it}}{\phi'(\phi^{-1}(e^{it}))} \right| dt = \frac{1}{|\phi'(\phi^{-1}(e^{it}))|}\, dt.$$

Using (3) to perform the change of variables $\zeta = \phi^{-1}(e^{it})$, we transform the integral in (4) into

$$\frac{1}{2\pi} \int_0^{2\pi} u(\phi^{-1}(e^{it})) \left. \frac{\partial \ln|w|}{\partial r} \right|_{w=e^{it}} |\phi'(\phi^{-1}(e^{it}))| \frac{dt}{|\phi'(\phi^{-1}(e^{it}))|}$$
$$= \frac{1}{2\pi} \int_0^{2\pi} u(\phi^{-1}(e^{it}))dt = u(z_0),$$

by (5). ∎

Let us note the following interesting property of the logarithm that we derived in the preceding proof: If ϕ is a conformal mapping of Γ and its interior onto the unit circle C and its interior, then for a point ζ on Γ we have

$$(6) \qquad \left. \frac{\partial \ln|\phi(z)|}{\partial n_\Gamma} \right|_{z=\zeta} = |\phi'(\zeta)|.$$

By composing ϕ with an appropriate linear fractional transformation, we will be able to reproduce the values of u at any point inside Ω, not just $z_0 = \phi^{-1}(0)$ as shown in (4). Let z be in Ω and think of $\phi(z)$ as a fixed point inside the unit disk in the w-plane. Consider the linear fractional transformation

$$(7) \qquad \tau_z(w) = \frac{w - \phi(z)}{1 - \overline{\phi(z)}w}.$$

It is one-to-one and maps the unit disk onto the unit disk and the unit circle onto the unit circle (Example 3, Section 3.7). Let us compose τ_z with ϕ and define

$$(8) \qquad \Phi(z, \zeta) = \tau_z(\phi(\zeta)) = \frac{\phi(\zeta) - \phi(z)}{1 - \overline{\phi(z)}\phi(\zeta)}, \qquad z,\ \zeta \text{ in } \Omega.$$

This is a function of two variables z and ζ, but we will often think of it as a function of ζ alone for a fixed value of z. As a function of ζ, it clearly maps z to 0; that is, $\Phi(z, z) = 0$ (Figure 5).

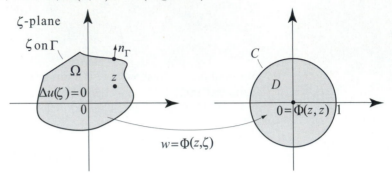

Figure 5 We think of $\Phi(z, \zeta)$ as a function of one variable ζ in Ω, for fixed z in Ω. As a function of ζ, $\Phi(z, \zeta)$ is analytic and one-to-one from Ω onto the unit disk and takes z to 0; that is, $\Phi(z, z) = 0$.

Using $\Phi(z, \zeta)$ in place of $\phi(\zeta)$ in (4), we are able to reproduce the value of u at any point z in Ω.

THEOREM 1
GREEN'S
FUNCTIONS

Suppose that Ω is a simply connected region with boundary Γ, and ϕ is a one-to-one analytic function on Ω and its boundary onto the unit disk and its boundary. Let $u(z)$ be a function harmonic on Ω and piecewise continuous on Γ. For z and ζ in Ω, let $\Phi(z, \zeta)$ be as in (8). Then, for any z in Ω, we have

$$(9) \qquad u(z) = \frac{1}{2\pi} \int_\Gamma u(\zeta) \frac{\partial}{\partial n} \ln |\Phi(z, \zeta)| \, ds,$$

where $ds = |d\zeta|$ is the element of arc length on Γ.

The function

$$(10) \qquad G(z, \zeta) = \ln |\Phi(z, \zeta)| = \ln \left| \frac{\phi(\zeta) - \phi(z)}{1 - \overline{\phi(z)}\,\phi(\zeta)} \right|, \quad z, \zeta \text{ in } \Omega,$$

is called the **Green's function** for the region Ω. It plays a fundamental role in the solution of important partial differential equations (Laplace's equation and Poisson's equation). Formula (9) is a **generalized Poisson integral formula** for the simply connected region Ω.

Like the Poisson formulas on the disk and in the upper half-plane, formula (9) can be used to solve a general Dirichlet problem in a simply connected region Ω, where the boundary data is piecewise continuous. Of course, this solution depends on the explicit formula for the conformal mapping of Ω onto the unit disk. Once this mapping is determined, Green's functions can be used to solve the Dirichlet problem. We illustrate these ideas with several examples and show how we can recapture the Poisson formulas from Green's functions.

We will often write the Green's function $G(z, \zeta)$ in terms of the real and imaginary parts of $z = x + iy$ and $\zeta = s + it$. We will also write the Green's function using polar coordinates of z and ζ, where $z = re^{i\theta}$ and $\zeta = \rho e^{i\eta}$

EXAMPLE 1   Green's function and Poisson formula for the disk

(a)   Show that the Green's function for the unit disk in polar coordinates is

$$(11) \qquad G(z, \zeta) = \ln \left| \frac{\rho e^{i\eta} - re^{i\theta}}{1 - r\rho e^{i(\eta - \theta)}} \right|, \qquad \text{for } z = re^{i\theta} \text{ and } \zeta = \rho e^{i\eta}.$$

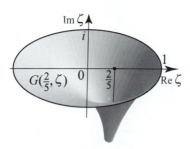

Figure 6 Green's function $G(\frac{2}{5}, \zeta)$ for the unit disk anchored at $z = \frac{2}{5}$. Note that $G(\frac{2}{5}, \zeta) = 0$ for all ζ on the boundary and $G(\frac{2}{5}, \zeta)$ has a singularity at $\zeta = \frac{2}{5}$.

As a specific illustration, we fix $z = \frac{2}{5}$ in the unit disk, and plot in Figure 6 the function $\zeta \mapsto G(\frac{2}{5}, \zeta)$, for ζ in the unit disk. This is Green's function for the unit disk anchored at a specific point $z = \frac{2}{5}$ in the unit disk.

(b)   Derive the Poisson integral formula for the unit disk.

Solution   (a) We will use (10). In this case, the conformal mapping $\phi(z)$ of the unit disk onto itself is simply $\phi(z) = z$, and so

$$G(z, \zeta) = \ln |\Phi(z, \zeta)| = \ln \left| \frac{\phi(\zeta) - \phi(z)}{1 - \overline{\phi(z)}\,\phi(\zeta)} \right| = \ln \left| \frac{\zeta - z}{1 - \overline{z}\zeta} \right|,$$

and (11) follows upon replacing z by $re^{i\theta}$ and ζ by $\rho e^{i\eta}$.

(b) To derive the Poisson integral formula for the unit disk, we must write out (9) when Γ is the unit circle. In this case, $ds = d\eta$, where $0 \le \eta \le 2\pi$. Using (6), we find that

$$\frac{\partial}{\partial n} \ln \left| \frac{\zeta - z}{1 - \overline{z}\zeta} \right|_{|\zeta| = 1} = \left| \frac{d}{d\zeta} \left(\frac{\zeta - z}{1 - \overline{z}\zeta} \right) \right|_{\zeta = e^{i\eta}} = \left| \frac{1 - |z|^2}{(1 - \overline{z}\zeta)^2} \right|_{\zeta = e^{i\eta}}$$

$$= \frac{1 - r^2}{|1 - re^{-i\theta}e^{i\eta}|^2} = \frac{1 - r^2}{1 - 2r\cos(\theta - \eta) + r^2}.$$

Plugging into (9), we find, for $z = re^{i\theta}$ with $0 \le r < 1$,

$$u(z) = \frac{1 - r^2}{2\pi} \int_0^{2\pi} \frac{u(e^{i\eta})}{1 - 2r\cos(\theta - \eta) + r^2} \, d\eta,$$

which is Poisson's formula on the unit disk.   ∎

Before we move to our next example, let us understand the role of $\Phi(z, \zeta)$ in (8). Since Φ is the composition of two conformal mappings, it is itself a conformal mapping of Ω onto the unit disk, and from (8) we have $\Phi(z, z) = 0$. By the Riemann mapping theorem, $\Phi(z, \zeta)$ is uniquely determined by these properties, up to a unimodular multiplicative constant. In particular $|\Phi(z, \zeta)|$ is uniquely determined and so is the Green's function for the region. (The uniqueness part in the Riemann mapping theorem is not difficult to prove, and so we are not appealing to a deep result here.) Consider, for example, the linear fractional transformation

$$(12) \qquad \tau(\zeta) = \frac{z - \zeta}{\overline{z} - \zeta}$$

where z is in the upper half-plane. If ζ is real so that $\overline{\zeta} = \zeta$, then

$$\left|\frac{z-\zeta}{\overline{z}-\zeta}\right| = \left|\frac{z-\zeta}{\overline{z}-\overline{\zeta}}\right| = \frac{|z-\zeta|}{|\overline{z-\zeta}|} = 1.$$

Thus $\tau(\zeta)$ maps the real line onto the unit circle and since it takes z onto the origin, it follows that τ maps the upper half-plane onto the unit disk, and thus $\tau(\zeta) = \Phi(z, \zeta)$ for the upper half-plane.

EXAMPLE 2   Green's function and Poisson's formula in the upper half-plane   (a)   Show that the Green's function for the upper half-plane is

(13) $G(z, \zeta) = \dfrac{1}{2} \ln \dfrac{(x-s)^2 + (y-t)^2}{(x-s)^2 + (y+t)^2}$,   for $z = x + iy$, $\zeta = s + it$ ($y, t > 0$).

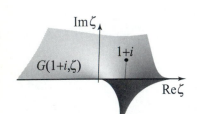

$G(1+i,\zeta)$

Figure 7 Green's function $G(1+i, \zeta)$ for the upper half-plane anchored at $z = 1 + i$. Note that $G(1 + i, \zeta) = 0$ for all ζ on the boundary and $G(1+i, \zeta)$ has a singularity at $\zeta = 1 + i$.

As a specific illustration, we fix $z = 1 + i$ in the upper half-plane, and plot in Figure 7 the function $\zeta \mapsto G(1+i, \zeta)$, for ζ in the upper half-plane. This is Green's function for the upper half-plane anchored at a specific point $z = 1+i$ in the upper half-plane.

(b)   Derive the Poisson integral formula for the upper half-plane.

Solution   According to (10), Green's function for the upper half-plane is $\ln |\Phi(z, \zeta)|$ where $\Phi(z, \zeta)$ is given by (12). Thus,

$$G(z, \zeta) = \ln\left|\frac{z-\zeta}{\overline{z}-\zeta}\right| = \frac{1}{2}\ln\frac{|z-\zeta|^2}{|\overline{z}-\zeta|^2} = \frac{1}{2}\ln\frac{(x-s)^2+(y-t)^2}{(x-s)^2+(-y-t)^2},$$

which is equivalent to (13).

(b)   To derive Poisson's integral formula in the upper half-plane we compute the normal derivative in (9). If Γ is the real s-axis, then the normal derivative is clearly the derivative in the negative direction along the imaginary t-axis. Thus,

$$\frac{\partial}{\partial n}G(z, \zeta) = -\frac{1}{2}\frac{\partial}{\partial t}\ln\frac{(x-s)^2+(y-t)^2}{(x-s)^2+(y+t)^2}.$$

A straightforward calculation of the derivative, then setting $t = 0$, yields

$$\frac{\partial}{\partial n}G(z, \zeta) = \frac{2y}{(x-s)^2+y^2}.$$

Plugging into (9) yields

$$u(z) = \frac{y}{\pi}\int_{-\infty}^{\infty}\frac{u(s)}{(x-s)^2+y^2}\,ds \quad (z = x+iy),$$

which is Poisson's formula for the upper half-plane.   ∎

We give one more example of a Green's function.

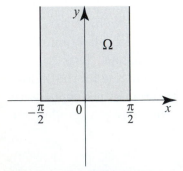

Figure 8 for Example 3.

EXAMPLE 3   Green's function for a semi-infinite vertical strip

We can map the strip Ω in Figure 8 conformally onto the upper half-plane using the mapping $w = \sin z$. Composing the function (12) with this, we obtain a one-to-one analytic mapping of Ω onto the unit disk, taking z in Ω onto the origin. Thus the Green's function for Ω is

$$G(z, \zeta) = \ln \left| \frac{\sin z - \sin \zeta}{\overline{\sin z} - \sin \zeta} \right|.$$ ∎

We prove next some interesting properties of Green's functions.

THEOREM 2
PROPERTIES OF
GREEN'S
FUNCTIONS

> Suppose that Ω is a simply connected region with boundary Γ, and let ϕ, $\Phi(z, \zeta)$, and $G(z, \zeta)$ be as in Theorem 1. Then the Green's function $G(z, \zeta)$ has the following properties:
> (i) $G(z, \zeta) \leq 0$ for all z and ζ in Ω;
> (ii) $G(z, \zeta) = 0$ for all z in Ω and ζ on Γ;
> (iii) $G(z, \zeta) = G(\zeta, z)$ for all z and ζ in Ω (**symmetric property**);
> (iv) for each z in Ω, there is a function $\zeta \mapsto u_1(z, \zeta)$ such that $u_1(z, \zeta)$ is harmonic for all ζ in Ω, $u_1(z, \zeta) = -\ln|z - \zeta|$ for all ζ on the boundary Γ, and $G(z, \zeta) = u_1(z, \zeta) + \ln|z - \zeta|$ for all $\zeta \neq z$ in Ω.

You should verify properties (i) and (ii) on the graphs of the Green's functions in Figures 6 and 7. Before we prove the theorem, we illustrate the properties in Figure 9 for a typical case where Ω is the upper half-plane and Green's function is anchored at $z = 1 + i$.

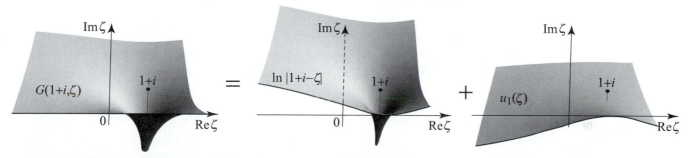

Figure 9   A Green's function $G(z_0, \zeta)$ anchored at $z_0 = 1 + i$ is the sum of a logarithm, $\ln|z_0 - \zeta|$, and a harmonic function, $u_1(\zeta)$, such that $u_1(\zeta) = -\ln|z_0 - \zeta|$ on the boundary. As a result, $G(z_0, \zeta)$ vanishes on the boundary and has a singularity at z_0 like $\ln|z_0 - \zeta|$.

Proof Fix z in Ω. From the definition of ϕ and Φ (see (7) and (8)), we have that $\Phi(z, \zeta)$ is in the open unit disk D (that is, $|\Phi(z, \zeta)| < 1$) for all ζ in Ω and $\Phi(z, \zeta)$ is on the unit circle C (that is, $|\Phi(z, \zeta)| = 1$) for all ζ on Γ. This clearly proves (i) and (ii), because $\ln|x| < 0$ if $|x| < 1$ and $\ln|x| = 0$ if $|x| = 1$. For (iii), we have

$$G(z, \zeta) = \ln \left| \frac{\phi(\zeta) - \phi(z)}{1 - \overline{\phi(z)}\,\phi(\zeta)} \right| = \ln \left| \frac{\phi(z) - \phi(\zeta)}{1 - \overline{\phi(z)}\,\phi(\zeta)} \right| = \ln \left| \frac{\phi(\zeta) - \phi(z)}{1 - \overline{\phi(\zeta)}\,\phi(z)} \right| = G(\zeta, z).$$

To prove (iv), fix z in Ω and consider

$$\psi(z, \zeta) = \frac{\phi(\zeta) - \phi(z)}{\zeta - z} \cdot \frac{1}{1 - \overline{\phi(z)}\phi(\zeta)} \qquad (\zeta \neq z \text{ in } \Omega).$$

Clearly, $\psi(z, \zeta)$ is analytic for all $\zeta \neq z$ in Ω. What happens as ζ approaches z? We have

$$\lim_{\zeta \to z} \psi(z, \zeta) = \lim_{\zeta \to z} \frac{\phi(\zeta) - \phi(z)}{\zeta - z} \frac{1}{1 - \overline{\phi(z)}\phi(\zeta)} = \frac{\phi'(z)}{1 - |\phi(z)|^2},$$

which is finite because $|\phi(z)| < 1$ and nonzero because ϕ is one-to-one and so $\phi'(z) \neq 0$. Hence $\psi(z, \zeta)$ has a removable singularity at z (Theorem 6, Section 4.6). By defining ψ at $\zeta = z$ to be

$$\psi(z, z) = \frac{\phi'(z)}{1 - |\phi(z)|^2},$$

$\psi(z, \zeta)$ becomes analytic and nonzero for all ζ in Ω. Set $u_1(z, \zeta) = \ln|\psi(z, \zeta)|$; then u_1 is harmonic for all ζ in Ω. But for $\zeta \neq z$

$$u_1(z, \zeta) = \ln|\psi(z, \zeta)| = \ln\left|\frac{\phi(\zeta) - \phi(z)}{1 - \overline{\phi(z)}\phi(\zeta)}\right| - \ln|\zeta - z| = G(z, \zeta) - \ln|z - \zeta|.$$

Also, $u_1(z, \zeta) = -\ln|z - \zeta|$ on the boundary because $G(z, \zeta) = 0$ on the boundary, and so (iv) holds. ∎

Because the function $\zeta \mapsto u_1(z, \zeta)$ is the solution of a Dirichlet problem in Ω with boundary values $-\ln|z - \zeta|$, if Ω is bounded, this solution is unique. Thus the representation of Green's function in Theorem 2(iv) is unique when Ω is bounded. Property (iv) in Theorem 2 can be used to define the Green's function of a domain. That is, any function $G(z, \zeta)$ that satisfies (iv) also satisfies (9).

Appendix: Proof of the change of variables formula (3)

Suppose that $\gamma(t) = x(t) + i\,y(t)$ is a parametrization of a smooth path with $\gamma'(t) = x'(t) + i\,y(t) \neq 0$. If we assume our path has a positive orientation, then an outward unit normal may be obtained by rotating the tangent $\gamma'(t)$ clockwise by $\pi/2$ and dividing by its absolute value (Figure 10). Hence $n(t) = \frac{\gamma'(t)}{i|\gamma'(t)|}$ or

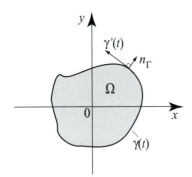

Figure 10

$$(14) \qquad n_\Gamma = \frac{1}{|\gamma'(t)|}\left(y'(t), -x'(t)\right).$$

Let $\phi(z)$ be as in the text preceding (3). To simplify the notation, let us write $\phi(z) = u(x, y) + i\,v(x, y)$, write F as $F(u, v)$, and denote partial derivatives by subscripts. So

$$(15) \qquad (F \circ \phi)_x = \frac{\partial}{\partial x}F(u(x, y), v(x, y)) = F_u u_x + F_v v_x,$$

where the last equality follows from the chain rule in two dimensions. Similarly,

$$(16) \qquad (F \circ \phi)_y = \frac{\partial}{\partial y}F(u(x, y), v(x, y)) = F_u u_y + F_v v_y.$$

Using the definition of the normal derivative, (14), (15), and (16), we get

$$\frac{\partial}{\partial n_\Gamma}F \circ \phi = \nabla(F \circ \phi) \cdot n_\Gamma = \frac{1}{|\gamma'(t)|}\big((F \circ \phi)_x, (F \circ \phi)_y\big) \cdot (y'(t), -x'(t))$$

$$(17) \qquad = \frac{1}{|\gamma'(t)|}\big((F_u u_x + F_v v_x)y'(t) - (F_u u_y + F_v v_y)x'(t)\big).$$

Consider now the path Γ', which is parametrized by

$$\phi(\gamma(t)) = u(x(t),\, y(t)) + i\, v(x(t),\, y(t)).$$

Conformality ensures that the outward normal to $\phi(\gamma(t))$ is still turned clockwise from the tangent; so in analogy with (14) we obtain

$$n_{\Gamma'} = \frac{1}{\left|\frac{d}{dt}\phi(\gamma(t))\right|}\Big(\frac{d}{dt}v(x(t),\, y(t)),\, -\frac{d}{dt}u(x(t),\, y(t))\Big)$$

$$= \frac{1}{|\phi'(\gamma(t))\gamma'(t)|}\big(v_x x'(t) + v_y y'(t),\, -u_x x'(t) - u_y y'(t)\big).$$

Thus

$$\frac{\partial F}{\partial n_{\Gamma'}} = \nabla F \cdot n_{\Gamma'}$$

$$= \frac{1}{|\gamma'(t)|\,|\phi'(\gamma(t))|}(F_u,\, F_v) \cdot \big(v_x x'(t) + v_y y'(t),\, -u_x x'(t) - u_y y'(t)\big)$$

$$(18) \qquad = \frac{1}{|\gamma'(t)|\,|\phi'(\gamma(t))|}\big(F_u(v_x x'(t) + v_y y'(t)) + F_v(-u_x x'(t) - u_y y'(t))\big).$$

Comparing (17) and (18) and using the Cauchy-Riemann equations, $u_x = v_y$, $u_y = -v_x$, we see that (3) holds.

Exercises 6.5

In Exercises 1–8, derive the Green's function for the region depicted in the accompanying figure (Figures $11-18$).

1.

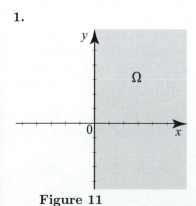

Figure 11

2.

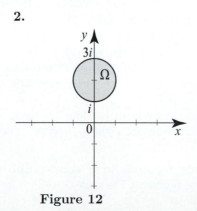

Figure 12

3.

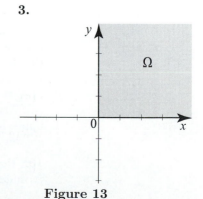

Figure 13

4.

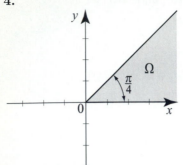

Figure 14

5.

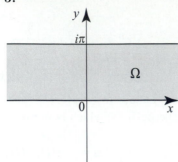

Figure 15

6.

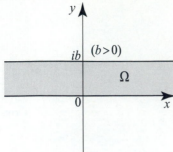

Figure 16

7.

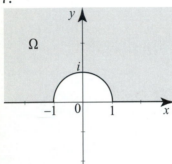

Figure 17

8.

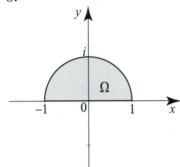

Figure 18

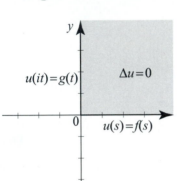

Figure 19 for Exercise 9.

9. Project Problem: Poisson's formula in the first quadrant. (a) Derive the following Poisson formula in the first quadrant for the Dirichlet problem in Figure 19, using Green's function:

$$u(x+iy) = \frac{y}{\pi} \int_0^\infty f(s) \left(\frac{1}{(x-s)^2 + y^2} - \frac{1}{(x+s)^2 + y^2} \right) ds$$
$$+ \frac{x}{\pi} \int_0^\infty g(t) \left(\frac{1}{x^2 + (y-t)^2} - \frac{1}{x^2 + (y+t)^2} \right) dt.$$

(b) Consider the special case in which $g(t) = 0$. Use a symmetry argument to show that the solution in this case is the same as the restriction to the first quadrant of the solution of the Dirichlet problem in the upper half-plane with boundary data on the real axis given by $u(s) = f(s)$ if $s > 0$ and $u(s) = -f(-s)$ if $s < 0$.

(c) Consider the special case in Figure 19 in which $f(s) = 0$. Use a symmetry argument to show that the solution in this case is the same as the restriction to the first quadrant of the solution of the Dirichlet problem in the right half-plane with boundary data on the imaginary axis given by $u(it) = g(t)$ if $t > 0$ and $u(it) = -g(-t)$ if $t < 0$.

(d) Write your answers in (b) and (c) using the Poisson integral formula for the upper half-plane and the right half-plane. Then sum the solutions to rederive your answer in (a).

10. Poisson's formula in a semi-infinite vertical strip. Derive Poisson's formula in the region of Example 3, using Green's function.

11. By composing $\phi_\alpha(z) = \frac{z-\alpha}{1-\bar{\alpha}z}$ with $\psi(z) = \frac{i-z}{i+z}$, you should be able to construct your own map $\Phi(z, \zeta)$ from the upper half-plane to the disk, satisfying $\Phi(z, z) = 0$.
(a) Find the value of α (it will depend on z).
(b) Check that your final map $\Phi(z, \zeta)$ is the same as given in this section, $\frac{z-\zeta}{\bar{z}-\zeta}$, times the unimodular constant $\frac{\bar{z}-i}{z+i}$.

6.6   Poisson's Equation and Neumann Problems

In this section we use Green's functions to solve the Poisson boundary value problem on a region Ω, then apply similar techniques to solve another important type of problem known as a Neumann problem. We start with the Poisson problem, which consists of the **inhomogeneous Laplace's equation** known as **Poisson's equation** along with boundary conditions specifying the values of the function on the boundary Γ of Ω (Figure 1):

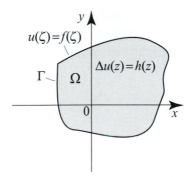

Figure 1 A Poisson problem in a region Ω.

(1) $$\Delta u(z) = h(z) \quad \text{for all } z \text{ in } \Omega;$$

(2) $$u(\zeta) = f(\zeta) \quad \text{for all } \zeta \text{ on } \Gamma.$$

If h is identically 0 in Ω, (1) reduces to Laplace's equation. If h is not identically zero, we obtain the inhomogeneous Laplace's equation. This equation models, for example, the time-independent (or steady-state) temperature distribution in a medium in the presence of heat sources. It also arises in the study of the velocity potential of an incompressible ideal fluid flow in the presence of sources or sinks. We will show that the solution of (1)–(2) has a simple expression in terms of the Green's function of the region Ω. Before we solve the problem, we would like to state the following simple but important **superposition principle**.

THEOREM 1
SUPERPOSITION
PRINCIPLE

Suppose that $u_1(z)$ satisfies (1) in Ω and $u_1(\zeta) = 0$ for all ζ on Γ. Suppose that $u_2(z)$ satisfies Laplace's equation in Ω and $u_2(\zeta) = f(\zeta)$ for all ζ on Γ. Then $u(z) = u_1(z) + u_2(z)$ satisfies Poisson's equation (1) with boundary conditions (2).

The proof is immediate and will be omitted. Theorem 1 allows us to decompose the general Poisson problem (1)–(2) into the sum of a Dirichlet problem with boundary values f and a Poisson problem with zero boundary values as illustrated in Figure 2.

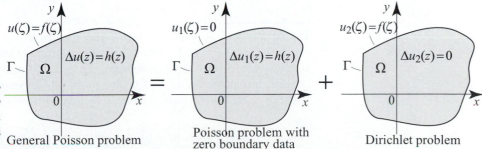

Figure 2 Decomposition of a general Poisson problem into the sum of a Poisson problem with zero boundary data plus a Dirichlet problem.

General Poisson problem

Poisson problem with zero boundary data

Dirichlet problem

We now express the solution of the Poisson problem on Ω in terms of Green's functions. We use the notation of the previous section and suppose that f and h have enough smoothness and integrability properties for the formulas in Theorems 2 and 3 to hold.

THEOREM 2
SOLUTION OF
POISSON PROBLEM

> Suppose that Ω is a region with boundary Γ, and let $G(z, \zeta)$ denote the Green's function for Ω. If $u(z)$ is a solution of Poisson's problem (1)–(2), then
>
> $$(3) \qquad u(z) = \frac{1}{2\pi} \iint_\Omega h(\zeta) G(z, \zeta) \, dA + \frac{1}{2\pi} \int_\Gamma f(\zeta) \frac{\partial}{\partial n} G(z, \zeta) \, ds,$$
>
> where dA is the element of area and ds is the element of arc length.

This form of the solution clearly illustrates the superposition principle. We recognize the second term on the right side of (3) as the solution of the Dirichlet problem on Ω with boundary values f (compare with Theorem 1, Section 6.5). Looking at the other term, we also have

$$(4) \qquad u(z) = \frac{1}{2\pi} \iint_\Omega h(\zeta) G(z, \zeta) \, dA$$

as a solution of Poisson's equation (1) with zero boundary values.

The proof of Theorem 2 is based on the following Green's identities.

THEOREM 3
GREEN'S
IDENTITIES

> Suppose that Ω is a bounded region whose boundary Γ consists of a finite number of simple closed positively oriented paths (as in Theorem 6, Section 3.4). Let $u(x, y)$ and $v(x, y)$ have continuous second partial derivatives on Ω and Γ. Then we have **Green's first identity**
>
> $$(5) \qquad \iint_\Omega (u\Delta v + \nabla u \cdot \nabla v) \, dA = \int_\Gamma u \frac{\partial v}{\partial n} \, ds,$$
>
> and **Green's second identity**
>
> $$(6) \qquad \iint_\Omega (u\Delta v - v\Delta u) \, dA = \int_\Gamma \left(u \frac{\partial v}{\partial n} - v \frac{\partial u}{\partial n} \right) \, ds.$$

Proof We will appeal to Green's theorem from calculus. For simply connected regions, this theorem is stated in Exercise 40, Section 3.4. For multiply connected regions, the version goes as follows. Let $p(x, y)$ and $q(x, y)$ have continuous first partial derivatives in Ω and on its positively oriented boundary Γ. Then, using subscripts to denote partial derivatives, we have

$$(7) \qquad \iint_\Omega (p_x(x, y) + q_y(x, y)) \, dx \, dy = \int_\Gamma (p(x, y) \, dy - q(x, y) \, dx).$$

Apply (7) with

$$p(x, y) = u\, v_x \quad \text{and} \quad q(x, y) = u\, v_y,$$

and get

$$(8) \qquad \iint_\Omega \big(u(v_{xx} + v_{yy}) + (u_x v_x + u_y v_y)\big)\,dx\,dy = \int_\Gamma u\,(v_x\,dy - v_y dx).$$

The integrand on the left is the same as the integrand on the left of (5). To understand the integrand on the right, let us recall that for a positively oriented curve parametrized by $\gamma(t) = x(t) + i\,y(t)$, the outward unit normal may be obtained by rotating the tangent $\gamma'(t) = x'(t) + i\,y'(t)$ clockwise by $\frac{\pi}{2}$ and dividing by its absolute value. Hence

$$n(t) = \frac{\gamma'(t)}{i|\gamma'(t)|} = \frac{1}{|\gamma'(t)|}\big(y'(t),\, x'(t)\big).$$

Thus since the normal derivative $\frac{\partial v}{\partial n}$ is by definition the gradient of v dotted with the outward unit normal vector and $ds = |\gamma'(t)|\,dt$, we have

$$\frac{\partial v}{\partial n}\,ds = (v_x,\, v_y)\cdot (y'(t),\, -x'(t))\,dt = v_x dy - v_y\,dx,$$

which shows that the right side of (8) is the same as the right side of (5), and so (5) follows. To prove (6), we reverse the roles of u and v in (5) and get

$$\iint_\Omega (v\Delta u + \nabla u\cdot\nabla v)\,dx\,dy = \int_\Gamma v\frac{\partial u}{\partial n}\,ds.$$

Subtracting this from (5), we get (6).   ∎

Green's formulas do not hold in general on unbounded regions. Since we will use them to prove (4), we will suppose throughout the proof that Ω is bounded. Nevertheless, formula (3) can be used on unbounded regions and its validity there can be checked on a case by case basis.

Proof of Theorem 2 Fix z in Ω and let $S_\epsilon(z)$ denote the closed disk of radius $\epsilon > 0$ around z in Ω, and $\Omega_\epsilon = \Omega\backslash S_\epsilon$. We are going to apply Green's second identity in Ω_ϵ. By Theorem 2(iv) of Section 6.5, $G(z,\,\zeta)$ is a harmonic function of ζ for all $\zeta \neq z$ in Ω. In particular, for ζ in Ω_ϵ, $\Delta G(z,\,\zeta) = 0$, we also have $\Delta u(\zeta) = h(\zeta)$ because u satisfies Poisson's equation (1). We apply Green's second identity on the region Ω_ϵ whose boundary Γ_ϵ consists of Γ and the circle $C_\epsilon(z)$ (Figure 3), taking $v = G(z,\,\zeta)$ and u equal the solution of the Poisson problem, and get

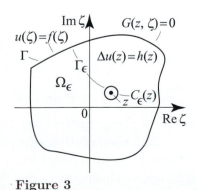

Figure 3

$$-\iint_{\Omega_\epsilon} G(z,\,\zeta)h(\zeta)\,dA \;=\; \int_\Gamma \left(u(\zeta)\frac{\partial G(z,\,\zeta)}{\partial n} - \overbrace{G(z,\,\zeta)}^{=0}\frac{\partial u(\zeta)}{\partial n}\right)\,ds$$

$$+\int_{C_\epsilon(z)}\left(u(\zeta)\frac{\partial G(z,\,\zeta)}{\partial n} - G(z,\,\zeta)\frac{\partial u(\zeta)}{\partial n}\right)\,ds,$$

because $G(z, \zeta) = 0$ for ζ on Γ. We will let $\epsilon \downarrow 0$ and show that

$$(9) \qquad \iint_{\Omega_\epsilon} G(z, \zeta) h(\zeta) \, dA \to \iint_\Omega G(z, \zeta) h(\zeta) \, dA;$$

$$(10) \qquad \int_{C_\epsilon(z)} u(\zeta) \frac{\partial G(z, \zeta)}{\partial n} \, ds \to -2\pi \, u(z);$$

$$(11) \qquad \int_{C_\epsilon(z)} G(z, \zeta) \frac{\partial u(\zeta)}{\partial n} \, ds \to 0.$$

This will imply (3) and complete the proof. Let us start with (11). Write $G(z, \zeta) = u_1(z, \zeta) + \ln|z - \zeta|$, where $u_1(z, \zeta)$ is harmonic, hence bounded by a constant M in some fixed disk centered at z. Also, $\frac{\partial u}{\partial n}$ is bounded in this fixed disk (say, $\left|\frac{\partial u}{\partial n}\right| \le A$), since u has continuous partial derivatives in Ω. For ζ on $C_\epsilon(z)$, we have $|z - \zeta| = \epsilon$, hence $\left|G(z, \zeta)\frac{\partial u}{\partial n}\right| \le (M + |\ln \epsilon|)A$, and so

$$\left| \int_{C_\epsilon(z)} G(z, \zeta) \frac{\partial u(\zeta)}{\partial n} \, ds \right| \le (M + |\ln \epsilon|)A \int_{C_\epsilon(z)} ds = 2\pi\epsilon(M + |\ln \epsilon|)A \to 0, \text{ as } \epsilon \to 0.$$

To prove (10), we note that on $C_\epsilon(z)$,

$$\frac{\partial G(z, \zeta)}{\partial n} = \frac{\partial}{\partial n}\left(u_1(z, \zeta) + \ln|z - \zeta|\right) = \frac{\partial}{\partial n} u_1(z, \zeta) - \frac{1}{\epsilon}.$$

Now

$$\int_{C_\epsilon(z)} u(\zeta) \frac{\partial G(z, \zeta)}{\partial n} \, ds = \int_{C_\epsilon(z)} u(\zeta) \frac{\partial}{\partial n} u_1(z, \zeta) \, ds - \frac{1}{\epsilon} \int_{C_\epsilon(z)} u(\zeta) \, ds.$$

The first integral on the right tends to 0 as $\epsilon \to 0$, because $u(\zeta)\frac{\partial}{\partial n}u_1(z, \zeta)$ is bounded in $S_\epsilon(z)$, as in the proof of (10). To handle the second integral, note that the function $I(\epsilon) = \frac{1}{\epsilon}\int_{C_\epsilon(z)} u(\zeta) \, ds = \int_0^{2\pi} u(z + \epsilon e^{i\theta}) \, d\theta$ is continuous (Theorem 5, Section 3.5). So

$$\lim_{\epsilon \to 0} \int_0^{2\pi} u(z + \epsilon e^{i\theta}) \, d\theta = I(0) = \int_0^{2\pi} u(z) \, d\theta = 2\pi u(z),$$

which completes the proof of (10). The proof of (9) is similar (see Exercise 9). ∎

In practice, (3) is difficult to compute in its present form. In later sections, we will relate it to generalized Fourier series and offer alternative ways for computing the solution of Poisson's equation. We now turn our attention to a different problem, which can be solved using an approach similar to the one we took with Green's functions.

Neumann Condition and Neumann Problems

When modeling heat problems in an insulated plate Ω, where heat exchange on the boundary is prescribed, we are led to a boundary value problem that consists of Laplace's equation along with a condition that describes the values of the normal derivative on the boundary (Figure 4):

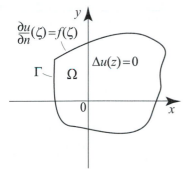

Figure 4 A Neumann problem in a region Ω.

$$(12) \qquad \Delta u(z) = 0 \quad \text{for all } z \text{ in } \Omega;$$

$$(13) \qquad \frac{\partial u}{\partial n}(\zeta) = f(\zeta) \quad \text{for all } \zeta \text{ on } \Gamma.$$

Such a problem is called a **Neumann problem** (after the German mathematician Carl Gottfried Neumann (1832–1925)) and sometime referred to as a **Dirichlet problem of the second kind**. Condition (13) is known as a **Neumann condition**. The normal derivative at the boundary describes the rate of exchange of heat with the surrounding medium or the flux of heat across the boundary. For example, the condition $f(\zeta) = 0$ corresponds to an insulated point where there is no exchange of heat with the surrounding medium. Since $f(\zeta)$ represents the flux of heat across the boundary of u and u represents a steady-state temperature distribution inside Ω, we would expect the total flux across the boundary to be zero; that is, f cannot be arbitrary, it has to satisfy the **compatibility condition**

$$(14) \qquad \int_{\Gamma} f(\zeta)\, ds = 0.$$

Indeed, (14) follows from the following useful property of harmonic functions, by setting $f(\zeta) = \frac{\partial u}{\partial n}$.

**PROPOSITION 1
NORMAL
DERIVATIVE OF
HARMONIC
FUNCTIONS**

Suppose that u is harmonic in a bounded region Ω and its boundary Γ. Then

$$(15) \qquad \int_{\Gamma} \frac{\partial u}{\partial n}\, ds = 0.$$

Proof Reversing the roles of u and v in Green's first identity (5) and then picking $v = 1$, we get

$$(16) \qquad \iint_{\Omega} \Delta u\, dA = \int_{\Gamma} \frac{\partial u}{\partial n}\, ds.$$

Since $\Delta u = 0$, the proposition follows.   ∎

In a Neumann problem, we are asked to find a harmonic function given the values of its normal derivative on the boundary. Such a solution is not unique, since we can add an arbitrary constant to a solution of (12)–(13) and get another solution of (12)–(13). So now we ask: Is the solution unique up to an arbitrary constant? The answer is affirmative. To show this, consider the difference between two solutions, which is harmonic and has zero normal derivative on the boundary. As we now show, a function that is harmonic and has zero normal derivative is a constant.

THEOREM 4
UNIQUENESS OF
NEUMANN
SOLUTION

> Suppose that u is harmonic on a bounded region Ω such that $\frac{\partial u}{\partial n} = 0$ on the boundary Γ of Ω. Then u is identically constant in Ω.

Proof Suppose that u is harmonic in Ω and take $u = v$ in Green's first formula, so that $\Delta v = 0$ in Ω, and get

$$\iint_\Omega (\nabla u \cdot \nabla u)\, dA = \int_\Gamma u \frac{\partial u}{\partial n}\, ds = 0,$$

because $\frac{\partial u}{\partial n} = 0$ on Γ. Now the function $\nabla u \cdot \nabla u = (u_x)^2 + (u_y)^2$ is ≥ 0 and continuous on Ω. The only way for a nonnegative continuous function to integrate to zero is to vanish identically. (This fact is proved in Lemma 1, Section 3.7 in one dimension, but the argument works in higher dimensions.) Hence $(u_x)^2 + (u_y)^2$ is identically zero in Ω, and so $u_x = 0$ and $u_y = 0$ identically in Ω. Applying Theorem 1, Section 2.1, we see that u is constant on Ω. ∎

In order to express the solution of the Neumann problem as an integral, motivated by Green's function and the solution of the Dirichlet problem, we make the following definition.

DEFINITION 1
NEUMANN
FUNCTIONS

> Suppose that Ω is a simply connected region with boundary Γ. A **Neumann function** $N(z, \zeta)$ (z, ζ in Ω) for the region Ω is a function with the following properties:
> (i) for each z in Ω, $N(z, \zeta)$ is harmonic for all $\zeta \neq z$ in Ω;
> (ii) $\frac{\partial N}{\partial n}(z, \zeta) = C$ for all z in Ω and ζ on Γ;
> (iii) for each z in Ω, there is a function $\zeta \mapsto u_1(z, \zeta)$ such that $u_1(z, \zeta)$ is harmonic for all ζ in Ω, and $N(z, \zeta) = \ln|z - \zeta| + u_1(z, \zeta)$ for all ζ in Γ.

Parts (i) and (iii) state that Neumann functions, like Green's functions, are harmonic inside Ω except for a singularity at $z = \zeta$, which is similar to the singularity of $\ln|z - \zeta|$. Part (ii) tells us that the boundary values of the normal derivative of the Neumann function are constant. This is the counterpart of the boundary condition for a Green's function, which states that a Green's function must vanish identically on the boundary. As we now show, the constant C in (ii) depends on the length of the boundary Γ.

PROPOSITION 2

> The constant C in Definition 1(ii), which represents the boundary value of the normal derivative of the Neumann function, is given by
>
> (17) $$C = \frac{2\pi}{L},$$
>
> where $L = \int_\Gamma ds$ is the length of Γ. If L is infinite, we take $C = 0$.

Proof The proof is based on the following interesting property of the logarithm.

For any fixed z inside Ω, we have

(18)
$$\int_\Gamma \frac{\partial}{\partial n} \ln|z - \zeta|\, ds = 2\pi.$$

To see this, let $S_\epsilon(z)$ be a disk of radius $\epsilon > 0$ centered at z and contained in Ω, and let $C_\epsilon(z)$ denote the circle centered at z with radius ϵ. The function $\zeta \mapsto \ln|z - \zeta|$ is harmonic in $\Omega \setminus S_\epsilon$, so according to Proposition 1, we have

$$0 = \int_{C_\epsilon(z)} \frac{\partial}{\partial n} \ln|z - \zeta|\, ds + \int_\Gamma \frac{\partial}{\partial n} \ln|z - \zeta|\, ds.$$

Parametrizing $C_\epsilon(z)$ by $\zeta = z + \epsilon e^{it}$, $0 \le t \le 2\pi$, $\frac{\partial}{\partial n} \ln|z - \zeta| = -\frac{1}{\epsilon}$ and $ds = \epsilon\, dt$, and so

$$0 = -\int_0^{2\pi} dt + \int_\Gamma \frac{\partial}{\partial n} \ln|z - \zeta|\, ds,$$

implying (18). Now, using (ii) and (iii) of Definition 1, write

$$CL = \int_\Gamma C\, ds = \int_\Gamma \frac{\partial N}{\partial n}\, ds = \overbrace{\int_\Gamma \frac{\partial N_1}{\partial n}\, ds}^{=0,\text{ by Proposition 1}} + \int_\Gamma \frac{\partial}{\partial n} \ln|z - \zeta|\, ds = 0 + 2\pi,$$

which implies (17). ∎

Neumann functions are not unique for a region (two Neumann functions can differ by a function of z and not ζ; see Exercise 12). However, any Neumann function satisfying Definition 1 will work to solve a Neumann problem. Before we express the solution of the Neumann problem in terms of the Neumann function, we note that from Theorem 4 if a solution of a Neumann problem exists, then it is unique up to an additive constant.

THEOREM 5
SOLUTION OF
NEUMANN
PROBLEMS

Suppose that Ω is a region bounded by a simple path Γ, and let $N(z, \zeta)$ denote a Neumann function, where z and ζ are in Ω. Then, up to an additive constant, the solution $u(z)$ of the Neumann problem (12)–(14) is given by

(19)
$$u(z) = -\frac{1}{2\pi} \int_\Gamma N(z, \zeta) f(\zeta)\, ds.$$

The proof of Theorem 5 is just like the proof of Theorem 2 (see Exercise 10). Now we derive some classical Neumann functions.

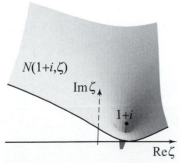

$N(1+i,\zeta)$

Figure 5 A Neumann function for the upper half-plane, anchored at $z = 1 + i$.

EXAMPLE 1   Neumann function for the upper half-plane

Show that the Neumann function for the upper half-plane is

(20) $\begin{aligned} N(z, \zeta) &= \ln|z - \zeta| + \ln|\bar{z} - \zeta| \\ &= \frac{1}{2} \ln\left((x - s)^2 + (y - t)^2\right) + \frac{1}{2} \ln\left((x - s)^2 + (y + t)^2\right), \end{aligned}$

for $z = x + iy$ and $\zeta = s + it$, $y, t > 0$ (Figure 5).

Solution   We know that $N(z, \zeta) = \ln|z - \zeta| + N_1(z, \zeta)$. Computing the normal derivative of $\ln|z - \zeta| = \frac{1}{2}\ln\left((x-s)^2 + (y-t)^2\right)$ along the real axis, we find

$$-\frac{d}{dt}\frac{1}{2}\ln\left((x-s)^2 + (y-t)^2\right)\Big|_{t=0} = \frac{y-t}{(x-s)^2 + (y-t)^2}\Big|_{t=0} = \frac{y}{(x-s)^2 + y^2}.$$

By adding $\frac{1}{2}\ln|\bar{z} - \zeta| = \frac{1}{2}\ln\left((x-s)^2 + (y+t)^2\right)$ to $\ln|z - \zeta|$, this adds $\frac{-y}{(x-s)^2 + y^2}$ to the normal derivative along the real axis, making the normal derivative of $\frac{1}{2}\ln|z - \zeta| + \frac{1}{2}\ln|\bar{z} - \zeta|$ zero along the real axis. This shows that $N(z, \zeta)$ as defined by (20) has 0 normal derivative along the boundary, in accordance with Proposition 2. All other properties of the Neumann function in Definition 1 are easily verified.   ■

The fact that the normal derivative of the Neumann function of an unbounded region is zero on the boundary allows us to construct the Neumann function for this region by using the Neumann function for the upper half-plane and a change of variables via a conformal mapping. More precisely, we have the following construction, which is valid for unbounded regions.

PROPOSITION 3
NEUMANN
FUNCTION FOR
UNBOUNDED
REGIONS

Suppose that Ω is an unbounded region with boundary Γ, and ϕ is a one-to-one analytic mapping of Ω onto the upper half-plane, taking Γ onto the real axis. Then the Neumann function for Ω is given by

$$(21) \qquad N(z, \zeta) = \ln|\phi(z) - \phi(\zeta)| + \ln|\overline{\phi(z)} - \phi(\zeta)| \quad (z, \zeta \text{ in } \Omega).$$

Proof Fix z in Ω and consider the function $F(w) = \ln|\phi(z) - w| + \ln|\overline{\phi(z)} - w|$, where $\phi(z)$ and w are in the upper half-plane. By Example 1, $\frac{\partial F}{\partial n}(w) = 0$, for all w on the real axis. Applying the change of variables formula (3), Section 6.5, it follows that $\frac{\partial F \circ \phi}{\partial n}(\zeta) = 0$ for all ζ on Γ. So $N(z, \zeta)$ has the right normal derivative on the boundary Γ. Does it have the right singularity at $\zeta = z$? Since $\overline{\phi(z)}$ is in the lower half-plane, it follows that $\ln|\overline{\phi(z)} - \phi(\zeta)|$ is harmonic for all ζ in Ω. Now let us compare $\ln|\phi(z) - \phi(\zeta)|$ to $\ln|z - \zeta|$. We have

$$\lim_{\zeta \to z}\left(\ln|\phi(z) - \phi(\zeta)| - \ln|z - \zeta|\right) = \lim_{\zeta \to z}\ln\left|\frac{\phi(z) - \phi(\zeta)}{z - \zeta}\right| = \ln|\phi'(z)|,$$

because ϕ is analytic. Since $\phi'(z) \neq 0$, $\ln|\phi(z) - \phi(\zeta)|$ and $\ln|z - \zeta|$ differ by a finite constant near z. Hence the function $N(z, \zeta) = \ln|z - \zeta|$ plus a harmonic function in Ω.   ■

Let us give an application of Proposition 3.

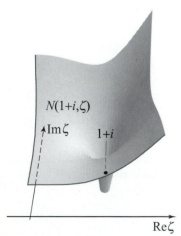

$N(1+i, \zeta)$

$\uparrow$Imζ $1+i$

Reζ

Figure 6 A Neumann function for the first quadrant, anchored at $z = 1 + i$. Note the singularity at $\zeta = 1 + i$.

EXAMPLE 2   Neumann function for the first quadrant

Applying Proposition 3 with $\phi(z) = z^2$, we obtain the Neumann function for the first quadrant (shown in Figure 6 at $z = 1 + i$)

$$\begin{aligned}(22) \qquad N(z, \zeta) &= \ln|z^2 - \zeta^2| + \ln|\bar{z}^2 - \zeta^2| \\ &= \ln|z - \zeta| + \ln|z + \zeta| + \ln|\bar{z} - \zeta| + \ln|\bar{z} + \zeta|.\end{aligned}$$ ■

Finding Neumann functions for bounded regions is more difficult because the condition $\frac{\partial N}{\partial n} = C$ is not preserved by a conformal mapping $\phi(z)$; the normal derivative is scaled by $|\phi'(z)|$. In the exercises, we will compute the Neumann function for the unit disk.

What if we are to solve Poisson's equation (1) with Neumann boundary conditions (13)? The compatibility condition on h and f has already been handled by (16); it is

$$(23) \qquad \iint_\Omega h(\zeta)\, dA = \int_\Gamma f(\zeta)\, ds.$$

This problem is also solvable with the Neumann function, and for reference, we present an analog of Theorem 2.

THEOREM 6
SOLUTION OF
POISSON-NEUMANN
PROBLEM

Suppose that Ω is a region with boundary Γ, let $N(z, \zeta)$ denote a Neumann function for this region, where z and ζ are in Ω. If $u(z)$ is a solution to Poisson's equation (1) subject to a Neumann boundary condition (13) and satisfying (23), then up to an additive constant

$$(24) \qquad u(z) = \frac{1}{2\pi} \iint_\Omega h(\zeta) N(z, \zeta)\, dA - \frac{1}{2\pi} \int_\Gamma N(z, \zeta) f(\zeta)\, ds.$$

Exercises 6.6

In Exercises 1–6, derive the Neumann function for the region depicted in the accompanying figure (Figures 7 − 12).

1.

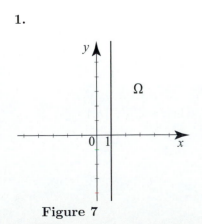

Figure 7

2.

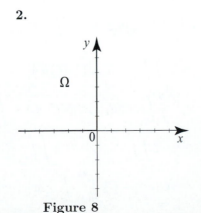

Figure 8

3.

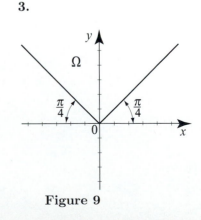

Figure 9

4.

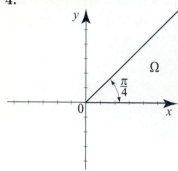

Figure 10

5.

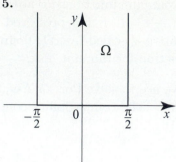

Figure 11

6.

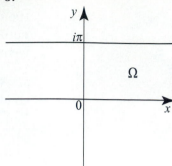

Figure 12

7. Uniqueness of the solution in a Poisson problem. Show that the solution of the Poisson problem in a bounded region Ω is unique. [Hint: The difference between any two solutions is harmonic and has zero boundary values.]

8. Neumann function for the unit disk. Show that this function is defined for z, ζ in the unit disk by

$$N(z, \zeta) = \begin{cases} \ln|z - \zeta| + \ln\left|\frac{1}{\bar{z}} - \zeta\right| + \ln|z| & \text{if } z \neq 0, \\ \ln|\zeta| & \text{if } z = 0. \end{cases}$$

Derive this function by following the outlined steps.
(a)   Write $z = re^{i\theta}$ and $\zeta = \rho e^{i\eta}$. Fix $z \neq 0$, and show that

$$\frac{\partial}{\partial n} \ln|z - \zeta|\Big|_{\rho=1} = \frac{\partial}{\partial \rho}\frac{1}{2}\ln|z - \zeta|^2\Big|_{\rho=1} = \frac{1}{2}\frac{\partial}{\partial \rho}\ln(r^2 + \rho^2 + 2r\rho\cos(\theta - \eta))\Big|_{\rho=1}$$

$$= \frac{1 + r\cos(\theta - \eta)}{1 + r^2 + 2r\cos(\theta - \eta)}.$$

(b) Write $\frac{1}{\bar{z}} = \frac{1}{r}e^{i\theta}$, use (a), and conclude that

$$\frac{\partial}{\partial n}\ln\left|\frac{1}{\bar{z}} - \zeta\right|\Big|_{\rho=1} = \frac{1 + \frac{1}{r}\cos(\theta - \eta)}{(\frac{1}{r})^2 + 1 + 2\cos(\theta - \eta)} = \frac{r^2 + r\cos(\theta - \eta)}{1 + r^2 + 2r\cos(\theta - \eta)}.$$

(c) Use (a) and (b) to show that for $z \neq 0$, $\frac{\partial}{\partial n}N(z, \zeta)\big|_{|\zeta|=1} = 1$.
(d) Verify the remaining properties of the Neumann function for the given $N(z, \zeta)$.

9. Prove (9) by justifying the following steps:

$$\left|\iint_{\Omega_\epsilon} G(z, \zeta)h(\zeta)\,dA - \iint_{\Omega} G(z, \zeta)h(\zeta)\,dA\right|$$

$$= \left|\iint_{B_\epsilon(z)} G(z, \zeta)h(\zeta)\,dA\right|$$

$$\leq \iint_{B_\epsilon(z)} |u_1(z, \zeta)h(\zeta)|\,dA + \iint_{B_\epsilon(z)} |\ln|z - \zeta|h(\zeta)|\,dA$$

$$\leq C_1 \iint_{B_\epsilon(z)} dA + C_2 \iint_{C_\epsilon(z)} \ln|z - \zeta|\,dA$$

$$= C_1\epsilon^2\pi + C_2 \int_0^{2\pi} \int_0^\epsilon r\ln|r|\,dr\,d\theta.$$

Evaluate the last integral and show that the resulting expression on the right side tends to 0 as $\epsilon \to 0$.

10. Proof of Theorem 5. In addition to the hypothesis of the theorem, we further suppose that $\int_\Gamma u(\zeta)\, ds = A$ is finite.
(a) For fixed z in Ω, let $C_\epsilon(z)$, $S_\epsilon(z)$, and Γ_ϵ be as in the proof of Theorem 2. Since $f(\zeta) = \frac{\partial u}{\partial n}$ in (19), we have

$$
\begin{aligned}
\int_\Gamma N(z,\zeta)\frac{\partial u}{\partial n}\, ds &= \int_\Gamma N(z,\zeta)\frac{\partial u}{\partial n}\, ds + \int_{C_\epsilon(z)} N(z,\zeta)\frac{\partial u}{\partial n}\, ds - \int_{C_\epsilon(z)} N(z,\zeta)\frac{\partial u}{\partial n}\, ds \\
&= \int_{\Gamma_\epsilon} N(z,\zeta)\frac{\partial u}{\partial n}\, ds - \int_{C_\epsilon(z)} N(z,\zeta)\frac{\partial u}{\partial n}\, ds \\
&= \int_{\Gamma_\epsilon} u(\zeta)\frac{\partial}{\partial n}N(z,\zeta)\, ds + \int_{C_\epsilon(z)} N(z,\zeta)\frac{\partial u}{\partial n}\, ds.
\end{aligned}
$$

(b) As in the proof of (11)), show that $\int_{C_\epsilon(z)} N(z,\zeta)\frac{\partial u}{\partial n}\, ds \to 0$ as $\epsilon \to 0$.
(c) Justify the following steps:

$$
\int_{\Gamma_\epsilon} u(\zeta)\frac{\partial}{\partial n}N(z,\zeta)\, ds = \int_{C_\epsilon(z)} u(\zeta)\frac{\partial}{\partial n}N(z,\zeta)\, ds + \int_\Gamma u(\zeta)\frac{\partial}{\partial n}N(z,\zeta)\, ds,
$$

$\int_{C_\epsilon(z)} u(\zeta)\frac{\partial}{\partial n}N(z,\zeta)\, ds \to -2\pi u(z)$ (see the proof of (10)), and $\int_\Gamma u(\zeta)\frac{\partial}{\partial n}N(z,\zeta)\, ds = C\int_\Gamma u(\zeta)\, ds = AC = C'$, where C is as in Definition 1(ii).
(d) Complete the proof of Theorem 5.

11. Proof of Theorem 6. The proof mirrors the proof in the text of Theorem 2, using Green's second identity.
(a) Let $\Omega_\epsilon = \Omega \setminus S_\epsilon(z)$, where $S_\epsilon(z)$ is the closed disk of radius $\epsilon > 0$, centered at z. Apply Green's second identity to u and N over the region Ω_ϵ to get

$$
\begin{aligned}
-\iint_{\Omega_\epsilon} N(z,\zeta)h(\zeta)\, dA &= C\int_\Gamma u(\zeta)\, ds - \int_\Gamma N(z,\zeta)\frac{\partial u}{\partial n}\, ds \\
&+ \int_{C_\epsilon} u(\zeta)\frac{\partial N}{\partial n}\, ds - \int_{C_\epsilon} N(z,\zeta)\frac{\partial u}{\partial n}\, ds,
\end{aligned}
$$

where C is the fixed value of the normal derivative of N along Γ.
(b) Argue, as in Exercise 10, that as $\epsilon \to 0$,

$$
\iint_{\Omega_\epsilon} N(z,\zeta)h(\zeta)\, dA \to \iint_\Omega N(z,\zeta)h(\zeta)\, dA.
$$

(c) Note that $C\int_\Gamma u(\zeta)\, ds$ is a constant; in fact, it is 2π times the average value of u on Γ.
(d) Just as we proved (10), show that $\int_{C_\epsilon} u(\zeta)\frac{\partial N}{\partial n}\, ds \to -2\pi u(z)$, as $\epsilon \to 0$.
(e) Just as we proved (11), show that $\int_{C_\epsilon} N(z,\zeta)\frac{\partial u}{\partial n}\, ds \to 0$, as $\epsilon \to 0$.
(f) Complete the proof of Theorem 6.

12. Project Problem: On the uniqueness of the Neumann function. Suppose we have two Neumann functions $N_1(z,\zeta)$ and $N_2(z,\zeta)$ for the same region Ω. They can be written in the form $N_1(z,\zeta) = \ln|z - \zeta| + u_1(z,\zeta)$, $N_2(z,\zeta) =$

$\ln |z - \zeta| + u_2(z, \zeta)$, where u_1 and u_2 are harmonic functions of ζ.

(a) Show that $N_2 - N_1$ is harmonic. What is its normal derivative on Γ, the boundary of Ω?

(b) Apply Theorem 4 and conclude that N_1 and N_2 differ by a constant–that is, an expression independent of ζ. Can this expression depend on z?

(c) Conclude that $N_1(z, \zeta)$ and $N_2(z, \zeta)$ are two Neumann functions for the same region if and only if they differ by any function of z alone.

13. Project Problem: Symmetry of the Neumann function. Unlike Theorem 2 of Section 6.5 for Green's functions, Definition 1 in this section for Neumann functions does not mention that they are symmetric, and in general it is not the case that $N(z, \zeta) = N(\zeta, z)$. In this problem we discover that symmetry can be recaptured by imposing an extra condition on the Neumann function, and that this can always be done without disrupting its role in the solution of the Neumann-Poisson problem.

(a) Refer to Exercise 12. We may add a function of z to a Neumann function and get another Neumann function. Let $N(z, \zeta)$ be a given Neumann function for Ω. Show that we can find a function $F(z)$ and a Neumann function $N_0(z, \zeta) = N(z, \zeta) - F(z)$ such that $\int_\Gamma N_0(z, \zeta) \, ds$ $(ds = |d\zeta|)$ is independent of z. (It is crucial that Γ has finite length.)

(b) Applying Green's second identity to $N(z_1, \zeta)$ and $N(z_2, \zeta)$ over the region Ω_ϵ and taking the limit as $\epsilon \to 0$ (as in the proof of Theorem 2), we get

$$N(z_1, z_2) - N(z_2, z_1) = \int_\Gamma \left(N(z_1, \zeta) \frac{\partial N(z_2, \zeta)}{\partial n} - N(z_2, \zeta) \frac{\partial N(z_1, \zeta)}{\partial n} \right) ds.$$

Use the fact that the normal derivative of $N(z, \zeta)$ is C (a constant) and part (a) to conclude that $N_0(z, \zeta) = N_0(\zeta, z)$.

(c) As a double-check, replace $N(z, \zeta)$ by $N_0(z, \zeta)$ in (24) and show that the equation remains unchanged. [Hint: Remember that $F(z)$ is a constant as far as the integration is concerned, and use the compatibility condition (23).]

14. Neumann problem with odd boundary data. Consider a Neumann problem in the unit disk in which $f(\zeta) = f(e^{i\theta})$ is an odd function of θ, where $-\pi \leq \theta \leq \pi$. Show that $u(z) = 0$ for all z on the real axis inside the unit disk. [Hint: The functions $\ln |z - e^{i\theta}| f(e^{i\theta})$ and $\ln |\frac{1}{z} - e^{i\theta}| f(e^{i\theta})$ are odd functions of θ if z is a real number.]

7

FOURIER SERIES

Mathematics compares the most diverse phenomena and discovers the secret analogies that unite them.

-Joseph Fourier

Like the familiar Taylor series, Fourier series are special types of expansions of functions. With a Taylor series, the expansion is in terms of the special set of functions 1, x, x^2, x^3, $\ldots$. With a Fourier series, we are interested in expanding a function in terms of the special set of functions 1, $\cos x$, $\cos 2x$, $\cos 3x$, $\ldots$, $\sin x$, $\sin 2x$, $\sin 3x$, $\ldots$. Thus, a Fourier series expansion of a function f is an expression of the form

$$f(x) = a_0 + \sum_{n=1}^{\infty} \left(a_n \cos nx + b_n \sin nx \right),$$

where a_0, a_n, and b_n are the Fourier coefficients. Fourier series arose naturally when solving Dirichlet problems on the disk in Section 4.7. As we will see in the remaining chapters, Fourier series are fundamental tools for the implementation of important methods for solving boundary value problems, such as the separation of variables method and the eigenfunction expansions method. Also the theory of Fourier series will serve as a model for theories involving other special functions such as Bessel functions and Legendre polynomials. The latter are the tools of choice for solving boundary value problems involving Laplace's equation on disks, cylinders and spheres.

In this chapter, we will present basic properties of Fourier series that will be used throughout the rest of the book.

7.1   Periodic Functions

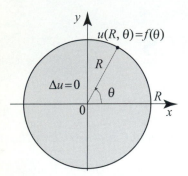

Figure 1 The boundary function $f(\theta)$ in a Dirichlet problem on a disk centered at the origin is 2π-periodic.

One of the problems that we have discussed at length in previous sections was the Dirichlet problem on a disk of radius $R > 0$, centered at the origin. In such a problem, we are supposed to solve Laplace's equation inside the disk, given the values of the function on the boundary of the disk. Using polar coordinates, the boundary data was given by a function $f(\theta)$ of the polar angle θ. Because θ and $\theta + 2\pi$ correspond to the same point on the unit circle, we have $f(\theta) = f(\theta + 2\pi)$. In other words, the function f is 2π-periodic (Figure 1). Periodic functions will arise naturally not only in boundary value problems on a disk but also in wave and heat problems over a finite interval. A function f satisfying the identity

$$(1) \qquad\qquad f(x) = f(x + T) \quad \text{for all } x,$$

where $T > 0$, is called **periodic** or, more specifically, **T-periodic** (Figure 2). The number T is called a **period** of f. If f is nonconstant, we define the **fundamental period**, or simply, the **period** of f to be the smallest positive number T for which (1) holds. For example, the functions 3, $\sin x$, $\sin 2x$ are all 2π-periodic. The period of $\sin x$ is 2π, while the period of $\sin 2x$ is π.

Using (1) repeatedly, we get

$$f(x) = f(x + T) = f(x + 2T) = \cdots = f(x + nT).$$

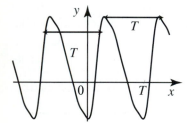

Figure 2 A T-periodic function.

Hence if T is a period, then nT is also a period for any integer $n > 0$. In the case of the sine function, this amounts to saying that 2π, 4π, 6π, ... are all periods of $\sin x$, but only 2π is the fundamental period. Because the values of a T-periodic function repeat every T units, its graph is obtained by repeating the portion over any interval of length T (Figure 2). As a consequence, to define a T-periodic function, it is enough to describe it over an interval of length T. Obviously, the interval can be chosen in many different ways. The following example illustrates these ideas.

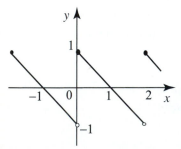

Figure 3 The 2-periodic function in Example 1.

EXAMPLE 1   Describing a periodic function

Describe the 2-periodic function f in Figure 3 in two different ways:
(a) by considering its values on the interval $0 \leq x < 2$;
(b) by considering its values on the interval $-1 \leq x < 1$.

Solution (a) On the interval $0 \leq x < 2$ the graph is a portion of the straight line $y = -x + 1$. Thus

$$f(x) = -x + 1 \quad \text{if } 0 \leq x < 2.$$

Now the relation $f(x + 2) = f(x)$ describes f for all other values of x.

(b) On the interval $-1 \leq x < 1$, the graph consists of two straight lines (Figure 3). We have

$$f(x) = \begin{cases} -x - 1 & \text{if } -1 \leq x < 0, \\ -x + 1 & \text{if } 0 \leq x < 1. \end{cases}$$

As in part (a), the relation $f(x + 2) = f(x)$ describes f for all values of x outside the interval $[-1, 1)$. ∎

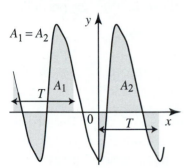

Figure 4 Areas over one period are equal.

Although the formulas in Example 1(a) and (b) are different, they describe the same periodic function. In practice, we use common sense in choosing the most convenient formula. Before we illustrate with an example, we introduce a very useful theorem whose content is intuitively clear. It says that the definite integral of a T-periodic function is the same over any interval of length T (Figure 4).

THEOREM 1
INTEGRAL OVER
ONE PERIOD

Suppose that f is T-periodic. Then, for any real number a, we have

$$\int_0^T f(x)\,dx = \int_a^{a+T} f(x)\,dx.$$

Proof Define

$$F(a) = \int_a^{a+T} f(x)\,dx.$$

By the fundamental theorem of calculus, we have $F'(a) = f(a + T) - f(a) = 0$, because f is periodic with period T. Hence $F(a)$ is constant for all a, and so $F(0) = F(a)$, which implies the theorem. ∎

EXAMPLE 2   Integrating periodic functions
Let f be the 2-periodic function in Example 1. Use Theorem 1 to compute

(a) $\displaystyle\int_{-1}^1 f^2(x)\,dx,$ (b) $\displaystyle\int_{-N}^N f^2(x)\,dx,$ N a positive integer.

Solution (a) Observe that $f^2(x)$ is also 2-periodic. Thus, by Theorem 1, to compute the integral in (a) we may choose any interval of length 2. Using the formula from Example 1 (a), we have

$$\int_{-1}^1 f^2(x)\,dx = \int_0^2 f^2(x)\,dx = \int_0^2 (-x+1)^2\,dx = -\frac{1}{3}(-x+1)^3\Big|_0^2 = \frac{2}{3}.$$

(b) We break up the integral $\int_{-N}^N$ into the sum of N integrals over intervals of length 2, of the form $\int_n^{n+2}$, $n = -N, -N+2, \ldots, N-2$, as follows:

$$\int_{-N}^N f^2(x)\,dx = \int_{-N}^{-N+2} f^2(x)\,dx + \int_{-N+2}^{-N+4} f^2(x)\,dx + \cdots + \int_{N-2}^N f^2(x)\,dx.$$

Since $f^2(x)$ is 2-periodic, by Theorem 1, each integral on the right side is equal to $\int_{-1}^1 f^2(x)\,dx = \frac{2}{3}$, by (a). Hence the desired integral is $N\frac{2}{3} = \frac{2N}{3}$. ∎

The most important periodic functions are those in the (2π-periodic) **trigonometric system**

$$1, \cos x, \cos 2x, \cos 3x, \dots, \cos mx, \dots,$$
$$\sin x, \sin 2x, \sin 3x, \dots, \sin nx, \dots.$$

They are 2π-periodic, and orthogonal on the interval $[0, 2\pi]$. Recall the orthogonality properties of the trigonometric system from Exercise 12, Section 3.2 (in what follows, the indices m and n are nonnegative integers):

$$\int_{-\pi}^{\pi} \cos mx \cos nx \, dx = 0 \quad \text{if } m \neq n,$$

$$\int_{-\pi}^{\pi} \cos mx \sin nx \, dx = 0 \quad \text{for all } m \text{ and } n,$$

$$\int_{-\pi}^{\pi} \sin mx \sin nx \, dx = 0 \quad \text{if } m \neq n.$$

We also have the useful identities:

$$\int_{-\pi}^{\pi} \cos^2 mx \, dx = \int_{-\pi}^{\pi} \sin^2 mx \, dx = \pi \qquad \text{for all } m \neq 0.$$

To prove these, we can use the complex integral, as suggested in Exercise 12, Section 3.2, or just use trigonometric identities. For example, to prove the first one, use

$$\cos mx \cos nx = \frac{1}{2} \left(\cos(m+n)x + \cos(m-n)x \right).$$

Since $m \pm n \neq 0$, we get

$$\int_{-\pi}^{\pi} \cos mx \cos nx \, dx$$

$$= \frac{1}{2} \left[\frac{1}{m+n} \sin(m+n)x + \frac{1}{m-n} \sin(m-n)x \right]_{-\pi}^{\pi} = 0.$$

Exercises 7.1

In Exercises 1–2, find a period of the given function and sketch its graph.

1. (a) $\cos x$,   (b) $\cos \pi x$,   (c) $\cos \frac{2}{3} x$,   (d) $\cos x + \cos 2x$.

2. (a) $\sin 7\pi x$,   (b) $\sin n\pi x$,   (c) $\cos mx$,
(d) $\sin x + \cos x$,   (e) $\sin^2 2x$.

In Exercises 3–6, find a formula that describes the function in the accompanying figure (Figures 5 − 8).

3.

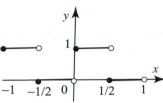

Figure 5

4.

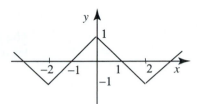

Figure 6

5.

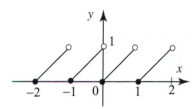

Figure 7

6.

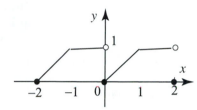

Figure 8

7. Sums of periodic functions. Show that if f_1, f_2, ..., f_n, ... are T-periodic functions, then $a_1 f_1 + a_2 f_2 + \cdots + a_n f_n$ is also T-periodic. More generally, show that if the series $\sum_{n=1}^{\infty} a_n f_n(x)$ converges for all x in $0 < x \leq T$, then its limit is a T-periodic function.

8.   Sums of periodic functions need not be periodic. Let $f(x) = \cos x + \cos \pi x$. (a) Show that the equation $f(x) = 2$ has a unique solution.

(b) Conclude from (a) that f is not periodic. Does this contradict Exercise 7? The function f is called **almost periodic**. These functions are of considerable interest and have many useful applications.

9. Operations on periodic functions. (a) Let f and g be two T-periodic functions. Show that the product $f(x)g(x)$ and the quotient $f(x)/g(x)$ $(g(x) \neq 0)$ are also T-periodic.

(b) Show that if f has period T and $a > 0$, then $f(\frac{x}{a})$ has period aT and $f(ax)$ has period $\frac{T}{a}$.

(c) Show that if f has period T and g is any function (not necessarily periodic), then the composition $g \circ f$ has period T.

10. With the help of Exercise 9, determine the period of the given function.

(a) $\sin 2x$ (b) $\cos \frac{1}{2}x + 3\sin 2x$ (c) $\frac{1}{2+\sin x}$ (d) $e^{\cos x}$

In Exercises 11–14, a π-periodic function is described over an interval of length π. In each case plot the graph over three periods and compute the integral

$$\int_{-\pi/2}^{\pi/2} f(x)\,dx.$$

11. $f(x) = \sin x, \quad 0 \le x < \pi.$

12. $f(x) = \cos x, \quad 0 \le x < \pi.$

13.

14. $f(x) = x^2, \quad -\frac{\pi}{2} \le x < \frac{\pi}{2}.$

$$f(x) = \begin{cases} 1 & \text{if } 0 \le x \le \frac{\pi}{2}, \\ 0 & \text{if } -\frac{\pi}{2} < x < 0. \end{cases}$$

15. Antiderivatives of periodic functions. Suppose that f is 2π-periodic and let a be a fixed real number. Define

$$F(x) = \int_a^x f(t)\, dt, \quad \text{for all } x.$$

Show that F is 2π-periodic if and only if $\int_0^{2\pi} f(t)\, dt = 0$. [Hint: Theorem 1.]

16. Suppose that f is T-periodic and let F be an antiderivative of f, defined as in Exercise 15. Show that F is T-periodic if and only if the integral of f over an interval of length T is 0.

17. (a) Let f be as in Example 1. Describe the function

$$F(x) = \int_0^x f(t)\, dt.$$

[Hint: By Exercise 16, it is enough to consider x in $[0, 2]$.]
(b) Plot F over the interval $[-4, 4]$.

7.2   Fourier Series

Fourier series arose naturally in Section 4.7 from our solution of the Dirichlet problem in a disk centered at 0. They are special expansions of 2π-period functions of the form

(1) $$f(x) = a_0 + \sum_{n=1}^{\infty} (a_n \cos nx + b_n \sin nx),$$

where the coefficients a_0, a_n, and b_n are called the **Fourier coefficients** of f and are given by the following **Euler formulas**.

EULER FORMULAS FOR THE FOURIER COEFFICIENTS

The Fourier coefficients of of a function f are given by

(2) $$a_0 = \frac{1}{2\pi} \int_{-\pi}^{\pi} f(x)\, dx,$$

(3) $$a_n = \frac{1}{\pi} \int_{-\pi}^{\pi} f(x) \cos nx\, dx \quad (n = 1, 2, \ldots),$$

(4) $$b_n = \frac{1}{\pi} \int_{-\pi}^{\pi} f(x) \sin nx\, dx \quad (n = 1, 2, \ldots).$$

For a positive integer N, we denote the Nth partial sum of the Fourier series of f by $s_N(x)$. Thus

$$s_N(x) = a_0 + \sum_{n=1}^{N} (a_n \cos nx + b_n \sin nx).$$

Because all the integrands in (2)–(4) are 2π-periodic, we can use Theorem 1, Section 7.1, to rewrite these formulas using integrals over the interval $[0, 2\pi]$ (or any other interval of length 2π). Such alternative formulas are sometimes useful.

ALTERNATIVE EULER FORMULAS

$$(5) \qquad\qquad\qquad a_0 = \frac{1}{2\pi} \int_0^{2\pi} f(x)\, dx,$$

$$(6) \quad a_n = \frac{1}{\pi} \int_0^{2\pi} f(x) \cos nx\, dx, \quad \text{and} \quad b_n = \frac{1}{\pi} \int_0^{2\pi} f(x) \sin nx\, dx, n \geq 1.$$

The Fourier coefficients were known to Euler before Fourier and for this reason they bear Euler's name. Euler used them to derive particular Fourier series such as the one presented in Example 1 below.

Before we consider some examples of Fourier series, it is instructive to motivate the Euler formulas by deriving them from the Fourier series, using the orthogonality of the trigonometric system. For this purpose, we proceed as Fourier himself did. We integrate both sides of (1) over the interval $[-\pi, \pi]$, assuming term-by-term integration is justified, and get

$$\int_{-\pi}^{\pi} f(x)\, dx = \int_{-\pi}^{\pi} a_0\, dx + \sum_{n=1}^{\infty} \int_{-\pi}^{\pi} (a_n \cos nx + b_n \sin nx)\, dx.$$

But $\int_{-\pi}^{\pi} \cos nx\, dx = \int_{-\pi}^{\pi} \sin nx\, dx = 0$ for $n = 1, 2, \ldots,$ so

$$\int_{-\pi}^{\pi} f(x)\, dx = \int_{-\pi}^{\pi} a_0\, dx = 2\pi a_0 \quad \Rightarrow \quad a_0 = \frac{1}{2\pi} \int_{-\pi}^{\pi} f(x)\, dx.$$

Similarly, starting with (1), we multiply both sides by $\cos mx$ $(m \geq 1)$, integrate term by term, use the orthogonality of the trigonometric system

(Section 7.1), and get

$$\int_{-\pi}^{\pi} f(x)\cos mx\,dx = \overbrace{\int_{-\pi}^{\pi} a_0\cos mx\,dx}^{=0} + \sum_{n=1}^{\infty} \overbrace{\int_{-\pi}^{\pi} a_n\cos nx\cos mx\,dx}^{=0 \text{ for } m\neq n}$$

$$+ \sum_{n=1}^{\infty} \overbrace{\int_{-\pi}^{\pi} b_n\sin nx\cos mx\,dx}^{=0}$$

$$= a_m \overbrace{\int_{-\pi}^{\pi} \cos^2 mx\,dx}^{=\pi} = \pi a_m.$$

Solving for a_m, we obtain (3). By a similar procedure, we derive (4).

Our first example displays many of the peculiar properties of Fourier series.

EXAMPLE 1   Fourier series of the sawtooth function

The sawtooth function, shown in Figure 1, is determined by the formulas

$$f(x) = \begin{cases} \frac{1}{2}(\pi - x) & \text{if } 0 < x \leq 2\pi, \\ f(x + 2\pi) & \text{otherwise.} \end{cases}$$

(a) Find its Fourier series.

(b) With the help of a computer, plot the partial sums $s_1(x)$, $s_7(x)$, and $s_{20}(x)$, and determine the graph of the Fourier series.

Solution (a) Using (5) and (6), we have

$$a_0 = \frac{1}{2\pi}\int_0^{2\pi} f(x)\,dx = \frac{1}{2\pi}\int_0^{2\pi} \frac{1}{2}(\pi - x)\,dx = 0;$$

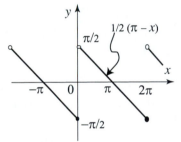

Figure 1 Sawtooth function.

In evaluating a_n, we use the formula $\int x\cos nx\,dx = \frac{1}{n^2}\cos nx + \frac{x}{n}\sin nx$, which is obtained by integrating by parts.

$$a_n = \frac{1}{\pi}\int_0^{2\pi} \frac{1}{2}(\pi - x)\cos nx\,dx$$

$$= \frac{1}{2\pi}\left\{\int_0^{2\pi} \pi\cos nx\,dx - \int_0^{2\pi} x\cos nx\,dx\right\} = 0;$$

$$b_n = \frac{1}{\pi}\int_0^{2\pi} \frac{1}{2}(\pi - x)\sin nx\,dx$$

$$= \frac{1}{2\pi}\left\{\int_0^{2\pi} \pi\sin nx\,dx - \int_0^{2\pi} x\sin nx\,dx\right\}$$

$$= \frac{1}{2\pi}\left\{\frac{-1}{n^2}\sin nx + \frac{x}{n}\cos nx\Big|_0^{2\pi}\right\} \qquad \text{(integration by parts)}$$

$$= \frac{1}{2\pi}\frac{2\pi}{n} = \frac{1}{n}.$$

Figure 2 To distinguish the graphs of the nth partial sums of the Fourier series, $s_n(x) = \sum_{k=1}^{n} \frac{\sin kx}{k}$, note that as n increases, the frequencies of the sine terms increase. This causes the graphs of the higher partial sums to be more wiggly. The limiting graph is the graph of the whole Fourier series, shown in Figure 3. It is *identical* to the graph of the function, except at points of discontinuity.

Substituting these values for a_n and b_n into (1), we obtain $\sum_{n=1}^{\infty} \frac{\sin nx}{n}$ as the Fourier series of f.

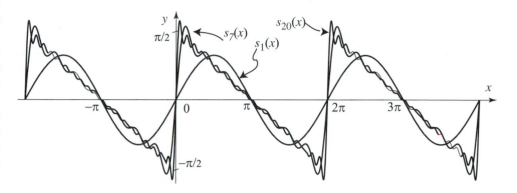

(b)   Figure 2 shows the first, seventh and twentieth partial sums of the Fourier series. We see clearly that the Fourier series of f converges to $f(x)$ at each point x where f is continuous. In particular, for $0 < x < 2\pi$, we have

$$\sum_{n=1}^{\infty} \frac{\sin nx}{n} = \frac{1}{2}(\pi - x).$$

At the points of discontinuity $(x = 2k\pi, k = 0, \pm 1, \pm 2, \dots)$, the series converges to 0. The graph of the Fourier series $\sum_{n=1}^{\infty} \frac{\sin nx}{n}$ is shown in Figure 3. It agrees with the graph of the function, except at the points of discontinuity. ∎

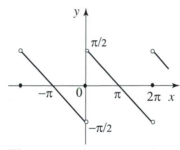

Figure 3 The graph of the Fourier series $\sum_{n=1}^{\infty} \frac{\sin nx}{n}$ coincides with the graph of the function, except at the points of the discontinuity.

Two important facts are worth noting concerning the behavior of Fourier series near points of discontinuity. As we will see shortly, these observations are true in a very general sense.

Note 1: At the points of discontinuity $(x = 2k\pi)$ in Example 1, the Fourier series converges to 0, which is the average value of the function from the left and the right at these points.

Note 2: Near the points of discontinuity, the Fourier series overshoots its limiting values. This is apparent in Figure 2, where humps form on the graphs of the partial sums near the points of discontinuity. This curious phenomenon is known as the **Gibbs (or Wilbraham–Gibbs) phenomenon**. (See the paper [15] for an interesting historical account.)

Fourier Series Representation

To state our main result, we recall from Section 3.1 the definitions of piecewise continuous and piecewise smooth functions. We will write

$$f(c-) = \lim_{x \to c^-} f(x)$$

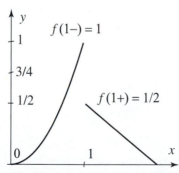

Figure 4 Left and right limits of a function at a point of discontinuity: $f(1+) = \frac{1}{2}$, $f(1-) = 1$.

to denote the fact that f approaches the number $f(c-)$ as x approaches c from below (Figure 4). Similarly, if the limit of f exists as x approaches c

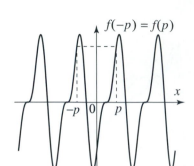

Figure 5 A continuous $2p$-periodic function.

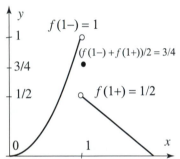

Figure 6 Average of $f(x)$ at $x = 1$.

from above, we denote this limit $f(c+)$ and write

$$f(c+) = \lim_{x \to c^+} f(x).$$

A function f is thus continuous at c if and only if

$$f(c-) = f(c+) = f(c).$$

In this notation, a function f is said to be piecewise continuous on the interval $[a, b]$ if $f(a+)$ and $f(b-)$ exist, and f is defined and continuous on (a, b) except at a finite number of points in (a, b) where the left and right limits exist. A periodic function is said to be **piecewise continuous** if it is piecewise continuous on every interval of the form $[a, b]$. A periodic function is said to be **continuous** if it is continuous on the entire real line. Note that continuity forces a certain behavior of the periodic function at the endpoints of any interval of length one period. For example, if f is $2p$-periodic and continuous, then necessarily $f(-p) = f(p)$ (Figure 5). The function of Example 1 is piecewise continuous, while the function in Example 2 below is continuous. Let us also recall that a function f is said to be piecewise smooth if f and f' are piecewise continuous on $[a, b]$. Similarly, a periodic function is **piecewise smooth** if it is piecewise smooth on every interval $[a, b]$. One more item of terminology is needed. The **average** (or **arithmetic average**) of f at c is

$$\frac{f(c-) + f(c+)}{2}.$$

Clearly if f is continuous at c, then its average at c is $f(c)$. Thus the notion of average will be of interest only at points of discontinuity.

As an illustration, consider the function in Figure 6. It has a discontinuity at $x = 1$ and its average there is $\frac{1+\frac{1}{2}}{2} = \frac{3}{4}$. We can now state a fundamental result in the theory of Fourier series. The proof of this theorem is presented in Section 7.6.

THEOREM 1
FOURIER SERIES
REPRESENTATION

Suppose that f is a 2π-periodic piecewise smooth function. Then for all x we have

$$(7) \qquad \frac{f(x+) + f(x-)}{2} = a_0 + \sum_{n=1}^{\infty}(a_n \cos nx + b_n \sin nx),$$

where the Fourier coefficients a_0, a_n, b_n are given by (2)–(4). In particular, if f is piecewise smooth and continuous at x, then

$$(8) \qquad f(x) = a_0 + \sum_{n=1}^{\infty}(a_n \cos nx + b_n \sin nx).$$

Let us see what (7) is telling us. At a point of continuity of f, the Fourier series converges to $f(x)$. At a point of discontinuity, the Fourier series does its best to converge, and having no reason to favor one side over the other, it converges to the average of the left and right limits (see Figure 7). Note that in (7) the value of the Fourier series of f at a given point x does not depend on $f(x)$ but on the limit of f from the left and right at x. For this reason, we may define (or redefine) f at isolated points without affecting its Fourier series. This is illustrated by the behavior of the Fourier series in Example 1, where, at the points of discontinuity, we could have assigned any values for the function without affecting the behavior of the Fourier series. If we redefine the function at points of discontinuity to be

$$\frac{f(x+) + f(x-)}{2},$$

we then have the equality

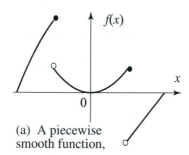

(a) A piecewise smooth function,

$$f(x) = a_0 + \sum_{n=1}^{\infty} (a_n \cos nx + b_n \sin nx)$$

holding at all x. We will often assume such a modification at the points of discontinuity and not worry about the more precise, but cumbersome, equality (7).

It is important to keep in mind that continuity of f alone is not enough to ensure the convergence of its Fourier series. Although we will not encounter such functions, there are continuous functions with Fourier series that diverge at an infinite number of points in $[0, 2\pi]$.

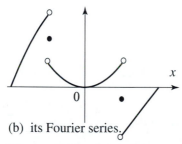

(b) its Fourier series.

Figure 7 At a point of discontinuity of a piecewise smooth function, the Fourier series converges to the average of the function at that point.

The problem of convergence for Fourier series was tackled by Fourier, Cauchy, and many other prominent mathematicians, who tried but could not establish the convergence for arbitrary f. We owe it to Peter Gustav Lejeune Dirichlet (1805–1859), who took a different approach to this problem by first formulating sufficient conditions on f that ensure the convergence of its Fourier series representation. Dirichlet's theorem about Fourier series is basically what we have stated in Theorem 1. Determining conditions on f that ensure the convergence of its Fourier series is an extremely hard problem. The most general results in this direction were obtained in the 1960s by Lennart Carleson (University of Uppsala, Sweden, and University of California, Los Angeles) and Richard Hunt (University of Indiana). These spectacular results are far beyond the level of this book. For a modern account of this theory, see the book by Grafakos [14].

Let us note one more property of Fourier series, which is an immediate consequence of Theorem 1.

COROLLARY 1
UNIQUENESS OF
FOURIER SERIES
REPRESENTATION

Suppose that f and g are 2π-periodic piecewise smooth functions. If f and g have the same Fourier coefficients, then $f(x) = g(x)$ for all x, except possibly at the points where f or g are discontinuous. Consequently, if f and g are continuous and have the same Fourier coefficients, then they are equal everywhere.

Proof Since the Fourier series of a function converges to the value of the function at the points of continuity, it follows that if f and g have the same Fourier series, then they must be equal at all the points of continuity. ■

For all practical purposes in analysis, if f and g have the same Fourier series, then they are considered to be the same function.

EXAMPLE 2   Triangular wave
(a) Find the Fourier series of the 2π-periodic triangular function, which is given on the interval $[-\pi, \pi]$ by

$$f(x) = \begin{cases} \pi + x & \text{if } -\pi \leq x \leq 0, \\ \pi - x & \text{if } 0 \leq x \leq \pi. \end{cases}$$

(b) Plot some partial sums and the Fourier series.

Solution   Figure 8 shows that the function is piecewise smooth and continuous for all x. So, from the second part of Theorem 1, we expect the Fourier series to converge to $f(x)$ for all x. Using (2), we have

$$a_0 = \frac{1}{2\pi} \int_{-\pi}^{\pi} f(x)\,dx = \frac{1}{2\pi}\pi^2 = \frac{1}{2}\pi.$$

(This is the area of the triangular region in Figure 8 with base $[-\pi, \pi]$ divided by 2π.) Using (3), we have

$$
\begin{aligned}
a_n &= \frac{1}{\pi}\int_{-\pi}^{0}(\pi + x)\cos nx\,dx + \frac{1}{\pi}\int_{0}^{\pi}(\pi - x)\cos nx\,dx \\
&= \frac{2}{\pi}\int_{0}^{\pi}(\pi - x)\cos nx\,dx \qquad \text{(change } x \text{ to } -x \text{ in the first integral)} \\
&= \frac{2}{\pi}\left\{\frac{1}{n^2} - \frac{\cos n\pi}{n^2}\right\} \qquad \text{(integration by parts)}.
\end{aligned}
$$

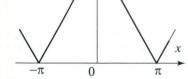

Figure 8 Triangular wave.

Since $\cos n\pi = (-1)^n$, we see that $a_n = 0$ if n is even, and $a_n = \frac{4}{\pi n^2}$ if n is odd. Finally, using (4), we find

$$b_n = \frac{1}{\pi}\int_{-\pi}^{\pi} \overbrace{f(x)\,\sin nx}^{\text{odd function}}\,dx = 0,$$

since we are integrating an odd function over a symmetric interval. Now Theorem 1 implies that

$$(9) \qquad f(x) = \frac{1}{2}\pi + \sum_{n \text{ odd}} \frac{4}{\pi n^2}\cos nx = \frac{1}{2}\pi + \frac{4}{\pi}\sum_{k=0}^{\infty}\frac{1}{(2k+1)^2}\cos(2k+1)x$$

for all x. Since the function and its Fourier series are equal at all points, their graphs coincide (compare Figures 8 and 10).

The partial sums of the Fourier series, illustrated in Figure 9, are converging very fast, much faster than those in Example 1. This is due to the magnitudes of the Fourier coefficients. In Example 1 the coefficients are of the order $1/n$, while in Example 2 the coefficients are of the order $1/n^2$.

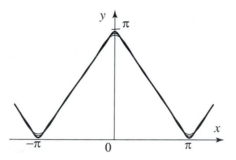

Figure 9 Partial sums of the Fourier series.

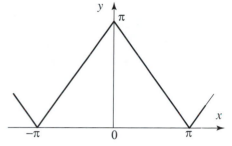

Figure 10 The Fourier series.

In (9), letting k run from 0 to 1, 2, and 5, respectively, we generate the third, fifth, and eleventh partial sums of the Fourier series. These are plotted in Figure 9. Note the fast convergence of the Fourier series. ∎

Figure 9 suggests that the Fourier series in Example 2 converges uniformly to the function. This is indeed true and follows from the following important result.

THEOREM 2
UNIFORM
CONVERGENCE OF
FOURIER SERIES

Suppose that f is a 2π-periodic piecewise smooth function. Then the Fourier series of f converges uniformly for all x to $f(x)$ if and only if f is continuous for all x. Thus, if f is continuous and piecewise smooth, then

$$(10) \qquad f(x) = a_0 + \sum_{n=1}^{\infty}(a_n \cos nx + b_n \sin nx),$$

where the Fourier series converges uniformly for all x.

Proof One direction is immediate from Theorem 1, Section 4.2, which asserts that the uniform limit of continuous functions is continuous. Since a partial sum of a Fourier series is a finite linear combination of sines and cosines, it is clearly continuous. So, if $s_N(x)$ converges uniformly to $f(x)$ for all x, then $f(x)$ is necessarily continuous for all x. For the other direction, we know from Theorem 1 that the Fourier series converges to $f(x)$ for all x. To prove that the convergence is uniform, we will show that there is a constant M such that $|a_n| \le \frac{M}{n^2}$ and $|b_n| \le \frac{M}{n^2}$. Then $|a_n \cos nx + b_n \sin nx| \le \frac{2M}{n^2}$, and the uniform convergence will follow from the Weierstrass M-test (Theorem 3, Section 4.2) since $\sum \frac{2M}{n^2} < \infty$. To simplify the proof, we will further suppose that f' is piecewise smooth. (This condition is satisfied in all the examples in this book, and indeed most applications. A proof that does not depend on it is presented in Section 7.5 and uses Parseval's theorem.)

Integrating by parts, we find that

$$
\begin{aligned}
a_n &= \frac{1}{\pi} \int_{-\pi}^{\pi} f(x) \cos nx \, dx = \frac{1}{n\pi} f(x) \sin nx \Big|_{-\pi}^{\pi} - \frac{1}{n\pi} \int_{-\pi}^{\pi} f'(x) \sin nx \, dx \\
&= -\frac{1}{n\pi} \int_{-\pi}^{\pi} f'(x) \sin nx \, dx,
\end{aligned}
$$

because $\sin n\pi = \sin(-n\pi) = 0$. Similarly, integrating by parts and using the fact that $f(\pi) = f(-\pi)$, we obtain that

$$
b_n = \frac{1}{\pi} \int_{-\pi}^{\pi} f(x) \sin nx \, dx = \frac{1}{n\pi} \int_{-\pi}^{\pi} f'(x) \cos nx \, dx.
$$

The function $f'(x)$ has a piecewise continuous derivative $f''(x)$. So $|f'(x)| \le A$ and $|f''(x)| \le B$ for all x. Let (a, b) be an interval on which f'' is continuous. Then integrating by parts, we find that

$$
\frac{1}{n\pi} \int_a^b f'(x) \sin nx \, dx = \frac{1}{n^2 \pi} \left((f'(a) \cos na - f'(b) \cos nb) + \int_a^b f''(x) \cos nx \, dx \right).
$$

Hence, because $|f'(a)| \le A$, $|f'(b)| \le A$, and $|f''(x) \cos nx| \le B$, we obtain

$$
\frac{1}{n\pi} \left| \int_a^b f'(x) \sin nx \, dx \right| \le \frac{1}{n^2 \pi} (2A + (b - a)B) \le \frac{C}{n^2},
$$

where C is a constant that does not depend on n. Since f'' has a finite number of discontinuities, we can write the integral $\int_{-\pi}^{\pi} f'(x) \sin nx \, dx$ as the finite sum of integrals of the form $\int_a^b f'(x) \sin nx \, dx$ (say, k of them), where $f''(x)$ is continuous on each interval (a, b). It follows from our estimate on the latter integral that $|a_n| \le \frac{kC}{n^2} = \frac{M}{n^2}$, where $M = kC$ is a constant independent of n. The inequality for b_n is obtained in a similar way.   ∎

Theorems 1 or 2 can be used to sum interesting series.

EXAMPLE 3   Using Fourier series to sum series
If we take $x = 0$ on both sides of (9), we get

$$
\pi = f(0) = \frac{1}{2}\pi + \frac{4}{\pi} \sum_{k=0}^{\infty} \frac{1}{(2k + 1)^2}.
$$

Subtracting $\frac{1}{2}\pi$ and then multiplying by $\frac{\pi}{4}$, we get the interesting identity

$$
\frac{\pi^2}{8} = 1 + \frac{1}{3^2} + \frac{1}{5^2} + \frac{1}{7^2} + \cdots,
$$

which can be used to approximate π^2 (and hence also π).   ∎

The Fourier series in Examples 1 and 2 are special in the sense that one of them contains only sine terms, while the other one contains only cosine terms. Fourier series of this type will be studied in the next section. The following Fourier series contains both sine and cosine terms. It is computed using the previous examples and the linearity property of Fourier series.

EXAMPLE 4   A Fourier series with cosine and sine terms

Adding the two functions of Examples 1 and 2, we obtain a 2π-periodic function which is defined on the interval $[0, 2\pi]$ by

$$f(x) = \begin{cases} \frac{3}{2}(\pi - x) & \text{if } 0 < x < \pi \\ \\ \frac{1}{2}(-\pi + x) & \text{if } \pi < x < 2\pi \end{cases}$$

(You should check this formula and plot the function.) To compute the Fourier series, we can use the Euler formulas or, better yet, we can simply add the Fourier series of Examples 1 and 2, thanks to the linearity of the Fourier coefficients (see Exercise 18). We thus obtain the Fourier series representation

$$f(x) = \frac{1}{2}\pi + \sum_{n=1}^{\infty} \left\{ \frac{2}{\pi}\left(\frac{1}{n^2} - \frac{\cos n\pi}{n^2}\right)\cos nx + \frac{\sin nx}{n} \right\}.$$

Figure 11 Note the Gibbs phenomenon at the points of discontinuity $(x = 2k\pi)$. This is due to the fact that the Fourier series consists of a cosine part that is converging very fast (Figure 9) and a sine part that overshoots at the points of discontinuity.

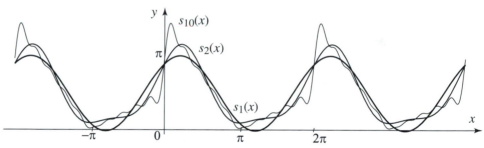

As illustrated in Figure 11, at the points of discontinuity, the Fourier series converges to the average value π. At all other points, the Fourier series converges to $f(x)$. ∎

Complex Methods for Finding Fourier Series

Because of Euler's identity $e^{in\theta} = \cos n\theta + i\sin n\theta$, which relates the complex exponential to the trigonometric functions, we expect the complex exponential function and complex analysis in general to play a role in the theory of Fourier series. This will become clear at many stages in our development of Fourier series and their applications. In what follows, we illustrate the use of major tools from complex analysis, such as residues and Laurent series, in computing Fourier series.

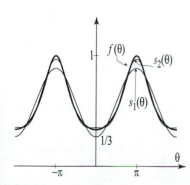

Figure 12 The function in Example 5 and the the partial sums of its Fourier series $s_n(\theta)$ with $n = 1$ and 2.

EXAMPLE 5   Using residues to compute Fourier series

Find the Fourier series of $f(\theta) = \frac{1}{2+\cos\theta}$.

Solution The function is 2π-periodic and even. We have $b_n = \frac{1}{\pi}\int_{-\pi}^{\pi} f(\theta)\sin n\theta\, d\theta = 0$, because $f(\theta)\sin n\theta$ is an odd function so its integral over a symmetric interval is 0. In computing a_n for $n \geq 0$, we will evaluate the integral

$$I_n = \int_0^{2\pi} \frac{\cos n\theta}{2 + \cos\theta}\, d\theta \qquad (n = 0, 1, \ldots),$$

using a slight variation on the methods of Section 5.2. By Theorem 1, Section 7.1,

$$\int_0^{2\pi} \frac{\sin n\theta}{2 + \cos\theta}\, d\theta = \int_{-\pi}^{\pi} \frac{\sin n\theta}{2 + \cos\theta}\, d\theta = 0,$$

because the integrand in the second integral is odd. So

$$I_n = \int_0^{2\pi} \frac{\cos n\theta}{2 + \cos\theta}\, d\theta + i\int_0^{2\pi} \frac{\sin n\theta}{2 + \cos\theta}\, d\theta = \int_0^{2\pi} \frac{e^{in\theta}}{2 + \cos\theta}\, d\theta,$$

where we have used $e^{in\theta} = \cos n\theta + i\sin n\theta$. We now use the method of Section 5.2, as follows. Let $z = e^{i\theta}$, $dz = ie^{i\theta}\,d\theta$ or $d\theta = \frac{dz}{iz}$, $z^n = e^{in\theta}$, and $\cos\theta = \frac{e^{i\theta} + e^{-i\theta}}{2}$. As θ varies over the interval $[0, 2\pi]$, z traverses the unit circle $C_1(0)$, in the positive direction. So

$$\begin{aligned}
I_n &= \int_0^{2\pi} \frac{e^{in\theta}}{2 + \cos\theta}\, d\theta = \int_0^{2\pi} \frac{e^{in\theta}}{2 + \frac{e^{i\theta} + e^{-i\theta}}{2}}\, d\theta \\
&= \int_{C_1(0)} \frac{2\, z^n}{4 + z + \frac{1}{z}}\, \frac{dz}{iz} = -i\int_{C_1(0)} \frac{2\, z^n}{z^2 + 4z + 1}\, dz.
\end{aligned}$$

We compute the last integral using residues. We have $z^2 + 4z + 1 = (z - z_1)(z - z_2)$, where $z_1 = -2 + \sqrt{3}$ and $z_2 = -2 - \sqrt{3}$. We have two simple poles at z_1 and z_2. But since $|z_1| \approx .3$ and $|z_2| \approx 3.7$, only z_1 is inside the unit disk. By Proposition 1(i), Section 5.1, we have

$$I_n = (-2i)2\pi i\, \frac{z^n}{z - z_2}\bigg|_{z = z_1} = 2\pi\frac{\left(-2 + \sqrt{3}\right)^n}{\sqrt{3}}.$$

Using the formula for I_n, we obtain

$$a_0 = \frac{1}{2\pi}I_0 = \frac{1}{\sqrt{3}} \quad\text{and}\quad a_n = \frac{1}{\pi}I_n = 2\frac{\left(-2 + \sqrt{3}\right)^n}{\sqrt{3}}, \quad n = 1, 2, \ldots.$$

Thus the Fourier series representation

$$\frac{1}{2 + \cos\theta} = \frac{1}{\sqrt{3}} + \frac{2}{\sqrt{3}}\sum_{n=1}^{\infty} \left(-2 + \sqrt{3}\right)^n \cos n\theta,$$

which is valid for all θ. Partial sums of the Fourier series are plotted in Figure 12. Note how fast the series is converging, because the coefficients are of the order $(-2 + \sqrt{3})^n = |z_1|^n$, where $0 < |z_1| \approx .3 < 1$. ■

Our next example generalizes the result of Example 5 and illustrates the use of Laurent series in computing Fourier series.

EXAMPLE 6   Using Laurent series to compute Fourier series
Let $a > 1$ be a real number. Derive the Fourier series representation

$$f(\theta) = \frac{1}{a + \cos\theta} = \frac{1}{\sqrt{a^2 - 1}} + \frac{2}{\sqrt{a^2 - 1}}\sum_{n=1}^{\infty} \left(-a + \sqrt{a^2 - 1}\right)^n \cos n\theta,$$

which is valid for all θ.

Solution The function is 2π-periodic and smooth. By Theorem 2, its Fourier series converges uniformly for all θ to $f(\theta)$. To find the Fourier series, we will not compute the Fourier coefficients directly; instead we will match the function with the restriction of a rational function and use Laurent series. As a first step, using the substitution $\cos\theta = \frac{1}{2}\left(z + \frac{1}{z}\right)$, where $z = e^{i\theta}$, we write

$$\frac{1}{a + \cos\theta} = \frac{1}{a + \frac{1}{2}\left(z + \frac{1}{z}\right)} = \frac{2z}{z^2 + 2az + 1}.$$

Next, we expand $\frac{2z}{z^2+2az+1}$ in an annulus that contains the unit circle $|z| = 1$. As we will see in a moment, this will allow us to use the Laurent series with $z = e^{i\theta}$ and obtain the desired Fourier series.

The quadratic equation $z^2 + 2az + 1 = 0$ has two real roots $z_1 = -a - \sqrt{a^2 - 1}$ and $z_2 = -a + \sqrt{a^2 - 1}$. Because $a > 1$, we have $z_1 < -1$ and $-1 < z_2 < 0$ (you should check these assertions). By Theorem 1, Section 4.5, we have a Laurent series expansion in the annulus $|z_2| < |z| < |z_1|$ (Figure 13), and to find it we employ methods from Section 4.5. We have the partial fraction decomposition

$$\frac{2z}{z^2 + 2az + 1} = \frac{A}{z - z_1} + \frac{B}{z - z_2} = \frac{A(z - z_2) + B(z - z_1)}{z^2 + 2az + 1}$$
$$= \frac{z(A + B) - (Az_2 + Bz_1)}{z^2 + 2az + 1}.$$

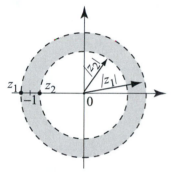

Figure 13 The annulus of convergence of the Laurent series in Example 6.

Solving

$$\begin{cases} A + B & = 2, \\ Az_2 + Bz_1 & = 0, \end{cases}$$

we find $A = \frac{2z_1}{z_1 - z_2} = -\frac{z_1}{\sqrt{a^2 - 1}}$ and $B = -\frac{2z_2}{z_1 - z_2} = \frac{z_2}{\sqrt{a^2 - 1}}$, where we have used $z_2 - z_1 = 2\sqrt{a^2 - 1}$. So we have the partial fraction decomposition

$$(11) \qquad \frac{2z}{z^2 + 2az + 1} = \frac{1}{\sqrt{a^2 - 1}}\left(-\frac{z_1}{z - z_1} + \frac{z_2}{z - z_2}\right).$$

In the annulus $|z_2| < |z| < |z_1|$, the first term inside the parentheses on the right can be expanded in a power series and the second one in a Laurent series, using the geometric series as follows. For $|z| < |z_1|$, we have $|z/z_1| < 1$ and so

$$-\frac{z_1}{z - z_1} = -\frac{z_1}{z_1\left(\frac{z}{z_1} - 1\right)} = \frac{1}{1 - \left(\frac{z}{z_1}\right)} = \sum_{n=0}^{\infty}\left(\frac{z}{z_1}\right)^n, \quad |z| < |z_1|.$$

For $|z_2| < |z|$, we have $|z_2/z| < 1$ and so

$$\frac{z_2}{z - z_2} = \frac{z_2}{z\left(1 - \frac{z_2}{z}\right)} = \frac{z_2}{z}\sum_{n=0}^{\infty}\left(\frac{z_2}{z}\right)^n, \quad |z_2| < |z|.$$

Plugging into (11) and simplifying with the help of the identity $z_1 z_2 = 1$ or $z_2 = \frac{1}{z_1}$,

we get the Laurent series expansion

$$\frac{2z}{z^2 + 2az + 1} = \frac{1}{\sqrt{a^2 - 1}} \sum_{n=0}^{\infty} \left[\left(\frac{z}{z_1} \right)^n + \frac{z_2}{z} \left(\frac{z_2}{z} \right)^n \right]$$

$$= \frac{1}{\sqrt{a^2 - 1}} \left\{ 1 + \sum_{n=1}^{\infty} \left[\left(\frac{z}{z_1} \right)^n + \left(\frac{z_2}{z} \right)^n \right] \right\}$$

$$= \frac{1}{\sqrt{a^2 - 1}} \left\{ 1 + \sum_{n=1}^{\infty} z_2^n \left(z^n + \frac{1}{z^n} \right) \right\},$$

which is valid in the annulus $|z_2| < |z| < |z_1|$. In particular, for $z = e^{i\theta}$ the left side becomes $\frac{1}{a + \cos\theta}$ and the right side reduces to $\frac{1}{\sqrt{a^2 - 1}} \{1 + \sum_{n=1}^{\infty} 2z_2^n \cos n\theta\}$, after using the identity $z^n + \frac{1}{z^n} = 2 \cos n\theta$, which completes the proof. ∎

Exercises 7.2

In Exercises 1–4, a 2π-periodic function is specified on the interval $[-\pi, \pi]$ in the accompanying figure (Figures 14−17). (a) Plot the function on the interval $[-3\pi, 3\pi]$. (b) Plot its Fourier series (without computing it) on the interval $[-3\pi, 3\pi]$.

1.

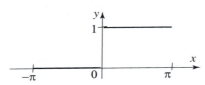

Figure 14

2.

Figure 15

3.

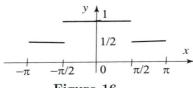

Figure 16

4.

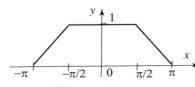

Figure 17

In Exercises 5–16, the equation of a 2π-periodic function is given on an interval of length 2π. You are also given the Fourier series of the function. (a) Sketch the function and its Fourier series on $[-2\pi, 2\pi]$, and decide whether the series converges uniformly for all x. (b) Derive the given Fourier series.

(c) Plot the Nth partial sums of the Fourier series for $N = 1, 2, \ldots, 20$. Discuss the convergence of the partial sums by considering their graphs. Be specific at the points of discontinuity.

5. $f(x) = |x|$ if $-\pi \le x < \pi$.

Fourier series: $\dfrac{\pi}{2} - \dfrac{4}{\pi} \sum_{k=0}^{\infty} \dfrac{1}{(2k+1)^2} \cos(2k+1)x$.

6. $f(x) = \begin{cases} \frac{1}{\pi}x & \text{if} \quad 0 \le x \le \pi, \\ 0 & \text{if} \quad -\pi < x \le 0. \end{cases}$

Fourier series: $\frac{1}{4} - \frac{1}{\pi^2} \sum_{n=1}^{\infty} \left\{ \left(\frac{1}{n^2} - \frac{(-1)^n}{n^2} \right) \cos nx + \frac{\pi(-1)^n}{n} \sin nx \right\}$.

7. $f(x) = |\sin x|$ if $-\pi \le x < \pi$.

Fourier series: $\frac{2}{\pi} - \frac{4}{\pi} \sum_{k=1}^{\infty} \frac{1}{(2k)^2 - 1} \cos 2kx$.

8. $f(x) = |\cos x|$ if $-\pi \le x \le \pi$.

Fourier series: $\frac{2}{\pi} - \frac{4}{\pi} \sum_{k=1}^{\infty} \frac{(-1)^k}{(2k)^2 - 1} \cos 2kx$.

[Hint: You can compute directly, or, if you have done Exercise 7, substitute $x - \pi/2$ for x.]

9. $f(x) = x^2$ if $-\pi \le x \le \pi$.

Fourier series: $\frac{\pi^2}{3} + 4 \sum_{n=1}^{\infty} \frac{(-1)^n}{n^2} \cos nx$.

10. $f(x) = 1 + \sin x + \cos 2x$.
Fourier series: same as $f(x)$.

11. $f(x) = \sin^2 x$; $f(x) = \cos^2 x$.
Fourier series: $\frac{1}{2} - \frac{\cos 2x}{2}$;   Fourier series: $\frac{1}{2} + \frac{\cos 2x}{2}$.

12. $f(x) = \pi^2 x - x^3$ if $-\pi < x < \pi$.

Fourier series: $12 \sum_{n=1}^{\infty} \frac{(-1)^{n+1}}{n^3} \sin nx$.

13. $f(x) = x$ if $-\pi < x < \pi$.

Fourier series: $2 \sum_{n=1}^{\infty} \frac{(-1)^{n+1}}{n} \sin nx$.

[Hint: Let $x = \pi - t$ in Example 1.]

14.

$$f(x) = \begin{cases} 0 & \text{if} \ -\pi \le x \le -d, \\ \frac{c}{d}(x + d) & \text{if} \ -d \le x \le 0, \\ -\frac{c}{d}(x - d) & \text{if} \ 0 \le x \le d, \\ 0 & \text{if} \ d \le x \le \pi, \end{cases}$$

where $c, d > 0$ and $d < \pi$.

Fourier series: $\frac{cd}{2\pi} + \frac{4c}{d\pi} \sum_{n=1}^{\infty} \frac{\sin^2(\frac{dn}{2})}{n^2} \cos nx$.

For part (c), take $c = d = \pi/2$.

15. $f(x) = e^{-|x|}$ if $-\pi \le x \le \pi$.

Fourier series: $\frac{e^\pi - 1}{\pi e^\pi} + \frac{2}{\pi e^\pi} \sum_{n=1}^{\infty} \frac{1}{n^2 + 1} (e^\pi - (-1)^n) \cos nx$.

16.

$$f(x) = \begin{cases} 1/(2c) & \text{if } |x - d| < c, \\ 0 & \text{if } c < |x - d| < \pi, \end{cases}$$

where $c, d > 0$, and $c + d < \pi$.

Fourier series: $\dfrac{1}{2\pi} + \dfrac{1}{c\pi} \displaystyle\sum_{n=1}^{\infty} \left(\dfrac{\sin(nc)\cos(nd)}{n} \cos nx + \dfrac{\sin(nc)\sin(nd)}{n} \sin nx \right)$.

For part (c), take $c = d = \pi/2$.

17. (a) Use the Fourier series of Exercise 9 to obtain

$$\frac{\pi^2}{6} = 1 + \frac{1}{2^2} + \frac{1}{3^2} + \frac{1}{4^2} + \cdots.$$

(b) Use the Fourier series of Exercise 13 to obtain

$$\frac{\pi}{4} = 1 - \frac{1}{3} + \frac{1}{5} - \frac{1}{7} + \cdots.$$

18. Linearity of Fourier coefficients and Fourier series Let α and β be any real numbers. Show that if f and g have Fourier coefficients $a_0, a_1, a_2, \ldots, b_1, b_2,$ $\ldots$, respectively, $a_0^*, a_1^*, a_2^*, \ldots, b_1^*, b_2^*, \ldots$, then the function $\alpha f + \beta g$ has Fourier coefficients $\alpha a_0 + \beta a_0^*, \alpha a_1 + \beta a_1^*, \alpha a_2 + \beta a_2^*, \ldots, \alpha b_1 + \beta b_1^*, \alpha b_2 + \beta b_2^*, \ldots$.

Methods from Complex Analysis

In Exercise 19 − 20, compute the Fourier series of the given function, following the method in Example 5. Verify your answer by using the result of Example 6.

19. $\dfrac{1}{3 + 2\cos\theta}$.   **20.** $\dfrac{3\sin\theta}{10 - 6\cos\theta}$.

In Exercise 21 − 24, compute the Fourier series of the given function, using Laurent series as we did in Example 6.

21. $\dfrac{1}{3 + \cos\theta}$.   **22.** $e^{2\cos\theta}$.

23. $\dfrac{1}{3 + \cos\theta + \sin\theta}$.   **24.** $e^{\cos\theta}\cos(\sin\theta)$.

In Exercise 25 − 26, derive the given Fourier series by using the Fourier series of Example 6 and various manipulations, including appropriate changes of variables.

25. Let $a > 1$ be a real number. Then

$$\frac{1}{a - \cos\theta} = \frac{1}{\sqrt{a^2 - 1}} + \frac{2}{\sqrt{a^2 - 1}} \sum_{n=1}^{\infty} (-1)^n \left(-a + \sqrt{a^2 - 1}\right)^n \cos n\theta,$$

which is valid for all θ.

26. Let $a > 1$ be a real number. Then for all θ,

$$\frac{1}{a + \sin\theta} = \frac{1}{\sqrt{a^2 - 1}} + \frac{2}{\sqrt{a^2 - 1}} \sum_{n=1}^{\infty} \left(-a + \sqrt{a^2 - 1}\right)^n \left(\cos\frac{n\pi}{2}\cos n\theta + \sin\frac{n\pi}{2}\sin n\theta\right);$$

and

$$\frac{1}{a - \sin\theta} = \frac{1}{\sqrt{a^2 - 1}} + \frac{2}{\sqrt{a^2 - 1}} \sum_{n=1}^{\infty} \left(-a + \sqrt{a^2 - 1}\right)^n \left(\cos\frac{n\pi}{2}\cos n\theta - \sin\frac{n\pi}{2}\sin n\theta\right).$$

27. Let $a > 1$ be a real number. Use Example 6 and Exercise 25 to derive the following formulas: for $n \geq 1$,

$$\frac{1}{2\pi} \int_{-\pi}^{\pi} \frac{d\theta}{a + \cos\theta} = \frac{1}{\sqrt{a^2 - 1}};$$

$$\frac{1}{2\pi} \int_{-\pi}^{\pi} \frac{\cos n\theta}{a + \cos\theta} \, d\theta = \frac{1}{\sqrt{a^2 - 1}} \left(-a + \sqrt{a^2 - 1} \right)^n;$$

$$\frac{1}{2\pi} \int_{-\pi}^{\pi} \frac{\cos n\theta}{a - \cos\theta} \, d\theta = \frac{(-1)^n}{\sqrt{a^2 - 1}} \left(-a + \sqrt{a^2 - 1} \right)^n.$$

7.3   Fourier Series of Functions with Arbitrary Periods

In the preceding section we worked with functions of period 2π. The choice of this period was merely for convenience. In this section, we show how to extend our results to functions with arbitrary period by using a simple change of variables. Suppose that f is a function with period $T = 2p > 0$, and let

$$(1) \qquad\qquad g(x) = f\left(\frac{p}{\pi}x\right).$$

Since f is $2p$-periodic, we have

$$g(x + 2\pi) = f\left(\frac{p}{\pi}(x + 2\pi)\right) = f\left(\frac{p}{\pi}x + 2p\right) = f\left(\frac{p}{\pi}x\right) = g(x).$$

Hence g is 2π-periodic. This reduction enables us to extend the main results of Section 7.2 to functions of arbitrary period.

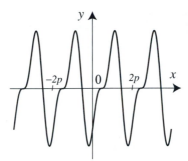

Figure 1 A $2p$-periodic function.

THEOREM 1
FOURIER SERIES
REPRESENTATION:
ARBITRARY
PERIOD

Suppose that f is a $2p$-periodic piecewise smooth function. Then f has a unique Fourier series representation

$$(2) \qquad \frac{f(x-) + f(x+)}{2} = a_0 + \sum_{n=1}^{\infty} \left(a_n \cos\frac{n\pi}{p}x + b_n \sin\frac{n\pi}{p}x \right),$$

where the Fourier coefficients are given by

$$(3) \qquad\qquad a_0 = \frac{1}{2p} \int_{-p}^{p} f(x) \, dx,$$

$$(4) \qquad a_n = \frac{1}{p} \int_{-p}^{p} f(x) \cos\frac{n\pi}{p}x \, dx \quad (n = 1, 2, \dots),$$

$$(5) \qquad b_n = \frac{1}{p} \int_{-p}^{p} f(x) \sin\frac{n\pi}{p}x \, dx \quad (n = 1, 2, \dots).$$

If f is continuous at x, then the Fourier series converges to $f(x)$. The series converges uniformly for all x if and only if f is continuous for all x.

By Theorem 1, Section 7.1, all the integrals $\int_{-p}^{p}$ can be replaced by $\int_{0}^{2p}$ without changing the values of the coefficients.

Proof Since f is piecewise smooth, it follows that the 2π-periodic function g defined by (1) is also piecewise smooth. By Theorem 1 of Section 7.2 we have

$$(6) \qquad \frac{g(x-) + g(x+)}{2} = a_0 + \sum_{n=1}^{\infty}(a_n \cos nx + b_n \sin nx) \quad \text{(for all } x\text{)},$$

where

$$(7) \quad a_0 = \frac{1}{2\pi}\int_{-\pi}^{\pi} g(x)\,dx;\; a_n = \frac{1}{\pi}\int_{-\pi}^{\pi} g(x)\cos nx\,dx;\; b_n = \frac{1}{\pi}\int_{-\pi}^{\pi} g(x)\sin nx\,dx.$$

Replacing x by $\frac{\pi}{p}x$ in (6) and using (1) gives

$$(8) \qquad \frac{f(x-) + f(x+)}{2} = a_0 + \sum_{n=1}^{\infty}(a_n \cos \frac{n\pi}{p}x + b_n \sin \frac{n\pi}{p}x),$$

where the coefficients are given by (7). To express the coefficients in terms of f as in (3)–(5), we use (1) again. For example, to obtain (3), start with the first formula in (7), use (1), then use the change of variables $t = \frac{p}{\pi}x$, and get

$$a_0 = \frac{1}{2\pi}\int_{-\pi}^{\pi} g(x)\,dx = \frac{1}{2\pi}\int_{-\pi}^{\pi} f(\frac{p}{\pi}x)\,dx = \frac{1}{2p}\int_{-p}^{p} f(t)\,dt.$$

Formulas (4) and (5) are derived in a similar way. The details are left to the exercises. The uniqueness and the uniform convergence in the theorem follow from the corresponding results for 2π-periodic functions (Corollary 1 and Theorem 2, Section 7.2).   ∎

EXAMPLE 1   A Fourier series with arbitrary period

Find the Fourier series of the $2p$-periodic function given by $f(x) = |x|$ if $-p \le x \le p$ (Figure 2).

Solution We compute the Fourier coefficients using Theorem 1. The area under the graph of f in Figure 2 gives

$$a_0 = \frac{1}{2p}\int_{-p}^{p} f(x)\,dx = \frac{p}{2}.$$

To compute a_n we take advantage of the fact that $f(x)\cos \frac{n\pi}{p}x$ is an even function and write

$$\begin{aligned} a_n &= \frac{1}{p}\int_{-p}^{p} f(x)\cos \frac{n\pi}{p}x\,dx = \frac{2}{p}\int_{0}^{p} f(x)\cos \frac{n\pi}{p}x\,dx \\ &= \frac{2}{p}\int_{0}^{p} x\cos \frac{n\pi}{p}x\,dx = \frac{-2p}{\pi^2 n^2}(1 - \cos n\pi), \end{aligned}$$

where the last integral is evaluated by parts. Since $\cos n\pi = (-1)^n$, $a_n = 0$ if n is even, and $a_n = \frac{-4p}{\pi^2 n^2}$ if n is odd. A similar computation shows that $b_n = 0$ for all n (since f is even). We thus obtain the Fourier series

$$f(x) = \frac{p}{2} - \frac{4p}{\pi^2}\left(\cos \frac{\pi}{p}x + \frac{1}{3^2}\cos \frac{3\pi}{p}x + \frac{1}{5^2}\cos \frac{5\pi}{p}x + \dots\right).$$

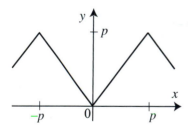

Figure 2 Triangular wave with period $2p$.

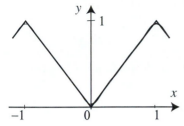

Figure 3 Partial sums of the Fourier series ($p = 1$), in Example 1.

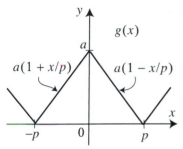

Figure 4 A $2p$-periodic triangular wave.

Because f is continuous and piecewise smooth, Theorem 1 implies that the Fourier series converges uniformly to $f(x)$ for all x, as can be seen in Figure 3.   ∎

Sometimes we can derive a new Fourier series from a known one without performing many additional computations. The following examples illustrate this process.

EXAMPLE 2   Triangular wave with arbitrary period and amplitude

Find the Fourier series of the $2p$-periodic function given by

$$g(x) = \begin{cases} a(1 + \frac{1}{p}x) & \text{if } -p \le x \le 0, \\ a(1 - \frac{1}{p}x) & \text{if } 0 \le x \le p. \end{cases}$$

Solution Comparing Figures 4 and 2 shows that we can obtain the graph of g by reflecting the graph of f in the x-axis, translating it upward by p units, and then scaling it by a factor of $\frac{a}{p}$. This is expressed by writing

$$g(x) = \frac{a}{p}(-f(x) + p) = a - \frac{a}{p}f(x).$$

Now to get the Fourier series of g, all we have to do is perform these operations on the Fourier series of f from Example 1. We get

$$\begin{aligned} g(x) &= a - a\left(\frac{1}{2} - \frac{4}{\pi^2}\left(\cos\frac{\pi}{p}x + \frac{1}{3^2}\cos\frac{3\pi}{p}x + \frac{1}{5^2}\cos\frac{5\pi}{p}x + \dots\right)\right) \\ &= \frac{a}{2} + \frac{4a}{\pi^2}\left(\cos\frac{\pi}{p}x + \frac{1}{3^2}\cos\frac{3\pi}{p}x + \frac{1}{5^2}\cos\frac{5\pi}{p}x + \dots\right). \end{aligned}$$

In compact form we have

$$g(x) = \frac{a}{2} + \frac{4a}{\pi^2}\sum_{k=0}^{\infty}\frac{1}{(2k+1)^2}\cos\frac{(2k+1)\pi}{p}x.$$

(You should check that the special case with $p = a = \pi$ yields the Fourier series of Example 2 of the previous section.)   ∎

Changing variables as we did at the outset of the section can be very useful in deriving new Fourier series from known ones.

EXAMPLE 3   Varying the period in a Fourier series

Find the Fourier series of the function in Figure 5.

Solution Let us start by defining the function in Figure 5. On the interval $0 < x < 2p$, we have $f(x) = c(1 - \frac{x}{p})$. Now, from Example 1, Section 7.2, we have the Fourier series expansion

$$\frac{1}{2}(\pi - x) = \sum_{n=1}^{\infty}\frac{\sin nx}{n}, \quad \text{for } 0 < x < 2\pi.$$

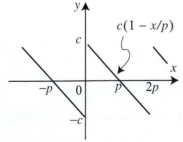

Figure 5 A $2p$-periodic sawtooth function.

Replacing x by $\frac{\pi}{p}x$ in the formula *and* the interval for x, we get

$$\frac{1}{2}(\pi - \frac{\pi}{p}x) = \sum_{n=1}^{\infty} \frac{\sin\frac{n\pi}{p}x}{n}, \quad \text{for } 0 < \frac{\pi}{p}x < 2\pi.$$

Simplifying and multiplying both sides by c to match the formula for f, we get

$$c(1 - \frac{x}{p}) = \frac{2c}{\pi} \sum_{n=1}^{\infty} \frac{\sin\frac{n\pi}{p}x}{n}, \quad \text{for } 0 < x < 2p,$$

which yields the Fourier series of f. ∎

The ideas behind Examples 2 and 3 are quite simple. They are based on the fact that a Fourier series is really a function and can be manipulated as such. As with any infinite series, when you manipulate a formula involving a Fourier series, you must keep in mind the interval on which this formula is valid. In particular, when you perform an operation on a Fourier series, it may affect the interval on which the resulting series is defined. This was the case when we performed a change of variables in Example 3.

Even and Odd Functions

As we noticed already, geometric considerations are helpful in computing Fourier coefficients. This is particularly the case when dealing with even and odd functions. Recall the following definitions:

> A function f is **even** if $f(-x) = f(x)$ for all x.
> A function f is **odd** if $f(-x) = -f(x)$ for all x.

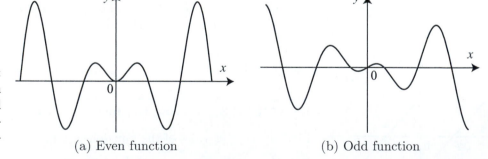

Figure 6 (a) Even function: The graph is symmetric with respect to the y-axis. (b) Odd function: The graph is symmetric with respect to the origin.

(a) Even function   (b) Odd function

It is clear from the graphs in Figure 6 (or by a simple change of variables) that if f is even, then

(9)
$$\int_{-p}^{p} f(x)\, dx = 2 \int_{0}^{p} f(x)\, dx;$$

and if f is odd, then

(10)
$$\int_{-p}^{p} f(x)\,dx = 0.$$

The following useful properties concerning the products of these functions are easily verified.

$$
\begin{array}{rcl}
(Even)(Even) & = & Even \\
(Even)(Odd) & = & Odd \\
(Odd)(Odd) & = & Even
\end{array}
$$

These simple product properties can be used to simplify finding the Fourier coefficients of even and odd functions, as we now show.

THEOREM 2
FOURIER SERIES OF
EVEN AND ODD
FUNCTIONS

Suppose that f is $2p$-periodic and has the Fourier series representation (2). Then (i) f is even if and only if $b_n = 0$ for all n. In this case

$$f(x) = a_0 + \sum_{n=1}^{\infty} a_n \cos \frac{n\pi}{p} x,$$

where

$$a_0 = \frac{1}{p} \int_0^p f(x)\,dx, \quad \text{and} \quad a_n = \frac{2}{p} \int_0^p f(x) \cos \frac{n\pi}{p} x\,dx \quad (n = 1, 2, \ldots).$$

(ii) f is odd if and only if $a_n = 0$ for all n. In this case

$$f(x) = \sum_{n=1}^{\infty} b_n \sin \frac{n\pi}{p} x,$$

where

$$b_n = \frac{2}{p} \int_0^p f(x) \sin \frac{n\pi}{p} x\,dx \quad (n = 1, 2, \ldots).$$

Proof (i) If $f(x) = a_0 + \sum_{n=1}^{\infty} a_n \cos \frac{n\pi}{p} x$, then clearly it is even. Conversely, suppose that f is even. Use (10) and the fact that $f(x) \sin \frac{n\pi}{p} x$ is odd to get that $b_n = 0$ for all n. Use (3), (4), (9) and the fact that $f(x) \cos \frac{n\pi}{p} x$ is even to get the formulas for the coefficients in (i). The proof of (ii) is similar and is left as an exercise. ∎

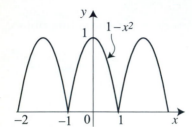

Figure 7 An even function.

EXAMPLE 4   Fourier series of an even function

Find the Fourier series of the 2-periodic function $f(x) = 1 - x^2$ if $-1 < x < 1$.

Solution   The function f is even (see Figure 7); hence $b_n = 0$ for all n.   To compute the a_n's, we use Theorem 2 with $p = 1$ and get

$$a_0 = \int_0^1 (1 - x^2)\, dx = \frac{2}{3};$$

and

$$a_n = 2 \int_0^1 (1 - x^2) \cos n\pi x\, dx = -2 \int_0^1 x^2 \cos n\pi x\, dx = \frac{-4(-1)^n}{\pi^2 n^2}.$$

In computing the last integral we used the formula

$$\int x^2 \cos n\pi x\, dx = \frac{2x \cos n\pi x}{\pi^2 n^2} - \frac{2 \sin n\pi x}{\pi^3 n^3} + \frac{x^2 \sin n\pi x}{\pi n},$$

which can be derived by two integrations by parts.   Since f is continuous and piecewise smooth, we get

$$f(x) = \frac{2}{3} - \frac{4}{\pi^2} \sum_{n=1}^{\infty} \frac{(-1)^n}{n^2} \cos n\pi x,$$

for all x. Since f is continuous and piecewise smooth for all x, its Fourier series converges uniformly to $f(x)$, as illustrated in Figure 8.   ∎

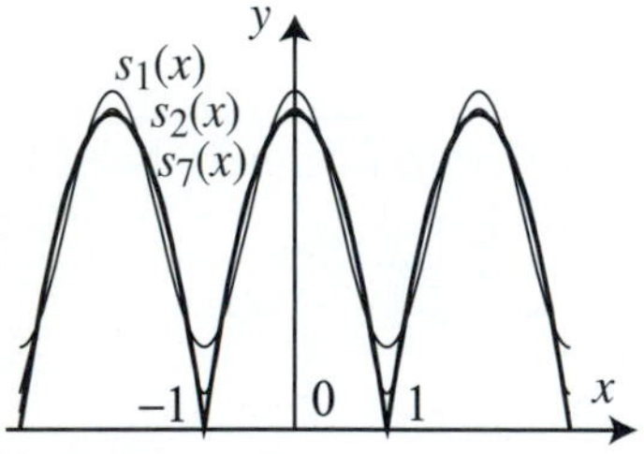

Figure 8 Partial sums of the Fourier series in Example 4.

EXAMPLE 5   Fourier series of an odd function

The function $f(x) = x \cos x$, if $-\frac{\pi}{2} < x < \frac{\pi}{2}$, and $f(x + \pi) = f(x)$ otherwise, is shown in Figure 9.   It is π-periodic and odd.   From Theorem 2, its Fourier series is given by

$$\sum_{n=1}^{\infty} b_n \sin 2nx,$$

where

$$b_n = \frac{4}{\pi} \int_0^{\pi/2} x \cos x \sin 2nx\, dx.$$

In evaluating this integral, we will need the addition formula

$$\cos a \sin b = \frac{1}{2} \left(\sin(a + b) - \sin(a - b) \right)$$

and the integral formula

$$\int u \sin u\, du = \sin u - u \cos u + C.$$

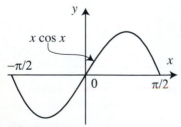

Figure 9 The odd function in Example 5.

Computing with the help of these formulas, we find

$$
\begin{aligned}
b_n &= \frac{2}{\pi} \int_0^{\pi/2} x\left(\sin(2n+1)x + \sin(2n-1)x\right)\, dx \\
&= \frac{2}{\pi(2n+1)^2}\left(\sin(2n+1)x - (2n+1)x\cos(2n+1)x\right)\Big|_0^{\pi/2} \\
&\quad + \frac{2}{\pi(2n-1)^2}\left(\sin(2n-1)x - (2n-1)x\cos(2n-1)x\right)\Big|_0^{\pi/2} \\
&= \frac{2}{\pi(2n+1)^2}\sin(2n+1)\frac{\pi}{2} + \frac{2}{\pi(2n-1)^2}\sin(2n-1)\frac{\pi}{2} \\
&= \frac{2}{\pi}(-1)^n\left[\frac{1}{(2n+1)^2} - \frac{1}{(2n-1)^2}\right] \\
&\quad \left(\text{since } \sin(2n+1)\frac{\pi}{2} = (-1)^n \text{ and } \sin(2n-1)\frac{\pi}{2} = (-1)^{n+1}\right) \\
&= \frac{16}{\pi}(-1)^{n+1}\frac{n}{(2n+1)^2(2n-1)^2}.
\end{aligned}
$$

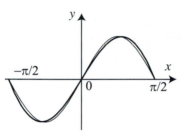

Thus

$$
f(x) = \frac{16}{\pi}\left[\frac{1}{9}\sin 2x - \frac{2}{225}\sin 4x + \ldots\right].
$$

Figure 10 Graphs of $f(x)$, $s_2(x)$ and $s_4(x)$, in Example 5.

Figure 10 illustrates the uniform convergence of the Fourier series to $f(x)$. Along with $f(x)$, we have plotted the partial sums $s_2(x)$ and $s_4(x)$. The graphs of $s_4(x)$ and $f(x)$ can hardly be distinguished from one another, which suggests that the Fourier series converges very fast to $f(x)$.   ■

In the next section we use Fourier series of even and odd functions to periodically extend functions that are defined on finite intervals. As we will see in Chapter 8, this process will be needed in solving partial differential equations by means of Fourier series.

Exercises 7.3

In Exercises 1–10, a 2p-periodic function is given on an interval of length 2p in the accompanying figure (Figures 11 − 20). (a) State whether the function is even, odd, or neither. (b) Derive the given Fourier series, and determine its values at the points of discontinuity. State if the series converges uniformly for all x. (Most of these Fourier series can be derived from earlier examples and exercises, as illustrated by Examples 3 and 4.)

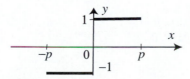

Figure 11

1.

$$
f(x) = \begin{cases} 1 & \text{if } 0 < x < p, \\ -1 & \text{if } -p < x < 0. \end{cases}
$$

Fourier series: $\displaystyle \frac{4}{\pi}\sum_{k=0}^{\infty}\frac{1}{(2k+1)}\sin\frac{(2k+1)\pi}{p}x.$

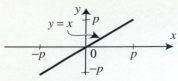

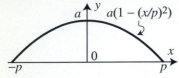

Figure 12

2. $f(x) = x$ if $-p < x < p$. [Hint: Exercise 13, Section 7.2.]

Fourier series: $\dfrac{2p}{\pi} \displaystyle\sum_{n=1}^{\infty} \dfrac{(-1)^{n+1}}{n} \sin(\dfrac{n\pi}{p}x)$.

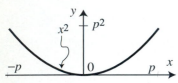

Figure 13

3. $f(x) = a\left(1 - (\frac{x}{p})^2\right)$ if $-p \le x \le p$, $(a \ne 0)$.

Fourier series: $\dfrac{2}{3}a + 4a \displaystyle\sum_{n=1}^{\infty} \dfrac{(-1)^{n+1}}{(n\pi)^2} \cos(\dfrac{n\pi}{p}x)$.

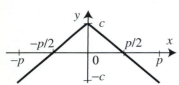

Figure 14

4. $f(x) = x^2$ if $-p \le x \le p$.   [Hint: Use Exercise 3.]

Fourier series: $\dfrac{p^2}{3} - \dfrac{4p^2}{\pi^2}\left[\cos(\dfrac{\pi}{p}x) - \dfrac{1}{2^2}\cos(\dfrac{2\pi}{p}x) + \dfrac{1}{3^2}\cos(\dfrac{3\pi}{p}x) - \cdots\right]$.

5.

$$f(x) = \begin{cases} -\frac{2c}{p}(x - p/2) & \text{if } 0 \le x \le p, \\ \frac{2c}{p}(x + p/2) & \text{if } -p \le x \le 0, \end{cases}$$

where $c \ne 0$ (in the picture $c > 0$.)

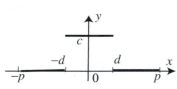

Figure 15

Fourier series: $\dfrac{8c}{\pi^2} \displaystyle\sum_{k=0}^{\infty} \dfrac{1}{(2k+1)^2} \cos((2k+1)\dfrac{\pi}{p}x)$.

6.

$$f(x) = \begin{cases} c & \text{if } |x| < d, \\ 0 & \text{if } d < |x| < p, \end{cases}$$

where $0 < d < p$.

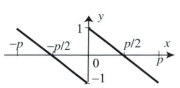

Figure 16

Fourier series: $\dfrac{cd}{p} + \dfrac{2c}{\pi} \displaystyle\sum_{n=0}^{\infty} \dfrac{\sin(\frac{dn\pi}{p})}{n} \cos(\dfrac{n\pi}{p}x)$.

7.

$$f(x) = \begin{cases} -\frac{2}{p}(x - p/2) & \text{if } 0 < x < p, \\ -\frac{2}{p}(x + p/2) & \text{if } -p < x < 0. \end{cases}$$

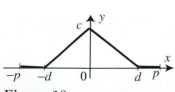

Figure 17

Fourier series: $\dfrac{4}{\pi} \displaystyle\sum_{k=1}^{\infty} \dfrac{1}{2k} \sin(\dfrac{2k\pi}{p}x)$.

8.

$$f(x) = \begin{cases} -\frac{c}{d}(x - d) & \text{if } 0 \le x \le d, \\ 0 & \text{if } d \le |x| \le p, \\ \frac{c}{d}(x + d) & \text{if } -d \le x \le 0, \end{cases}$$

where $0 \le d \le p$.

Figure 18

Fourier series: $\dfrac{cd}{2p} + \dfrac{2cp}{d\pi^2} \displaystyle\sum_{n=1}^{\infty} \dfrac{1}{n^2}(1 - \cos(\dfrac{dn\pi}{p})) \cos(\dfrac{n\pi}{p}x)$.

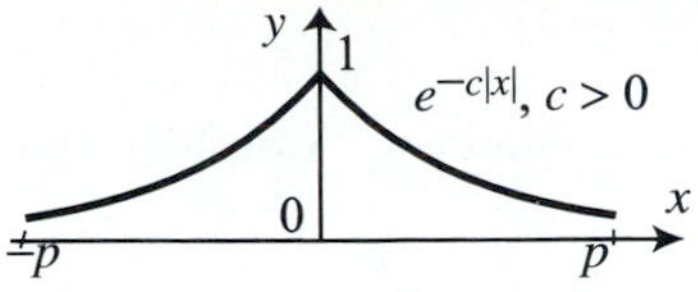

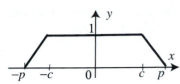

Figure 19

9. $f(x) = e^{-c|x|}$ $(c \neq 0)$ for $|x| \leq p$.

Fourier series: $\dfrac{1}{pc}(1 - e^{-cp}) + 2cp \displaystyle\sum_{n=1}^{\infty} \dfrac{1}{c^2p^2 + (n\pi)^2}(1 - e^{-cp}(-1)^n) \cos(\dfrac{n\pi}{p}x)$.

Figure 20

10.

$$f(x) = \begin{cases} -\frac{1}{p-c}(x - p) & \text{if } c < x < p, \\ 1 & \text{if } |x| < c, \\ \frac{1}{p-c}(x + p) & \text{if } -p < x < -c, \end{cases}$$

where $0 < c < p$.

Fourier series: $\dfrac{p+c}{2p} + \dfrac{2p}{(c-p)\pi^2} \displaystyle\sum_{n=1}^{\infty} \dfrac{1}{n^2}((-1)^n - \cos(\dfrac{cn\pi}{p})) \cos(\dfrac{n\pi}{p}x)$.

11. (a) Find the Fourier series of the 2π-periodic function given on the interval $-\pi < x < \pi$ by $f(x) = x \sin x$.

(b) Plot several partial sums to illustrate the convergence of the Fourier series.

12. (a) Find the Fourier series of the 2π-periodic function given on the interval $-\pi < x < \pi$ by $f(x) = (\pi - x) \sin x$. [Hint: Exercise 11.]

(b) Plot several partial sums to illustrate the convergence of the Fourier series.

In Exercises 13–14 a function is given over one period by a figure (Figures 21 – 22). (a) Find its Fourier series. [Hint: Use Exercise 1.]

(b) *Plot several partial sums to illustrate the convergence of the Fourier series.*

13.   **14.**

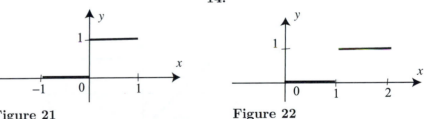

Figure 21   **Figure 22**

15. Obtain the Fourier series of Example 2, Section 7.2, from Example 2 of this section.

16. (a) Illustrate graphically the answer in Exercise 6 by taking $p = 1, c = 1, d = \frac{1}{2}$ and by plotting several partial sums of the Fourier series.

(b) What happens to the Fourier coefficients as d approaches p? Justify your answer.

17. Use the result of Exercise 4 to derive the formulas

(a) $\frac{\pi^2}{12} = 1 - \frac{1}{2^2} + \frac{1}{3^2} - \frac{1}{4^2} + \cdots$

(b) $\frac{\pi^2}{8} = 1 + \frac{1}{3^2} + \frac{1}{5^2} + \frac{1}{7^2} + \cdots$ [Hint: Use (a) also.]

18. Derive (4) and (5) of Theorem 1. [Hint: Study the proof of Theorem 1.]

19. Prove part (ii) of Theorem 2.

Project Problem: Decomposition into even and odd parts. Do Exercise 20 and any one of 21–24. You will discover how an arbitrary function can be decomposed into the sum of an even and odd function.

20. Let f be an arbitrary function defined for all real numbers. Consider the functions

$$f_e(x) = \frac{f(x) + f(-x)}{2} \quad \text{and} \quad f_o(x) = \frac{f(x) - f(-x)}{2}.$$

(a) Show that f_e is even and f_o is odd.

(b) Show that $f = f_e + f_o$. Hence every function is the sum of an even function and an odd function. Moreover, show that this decomposition is unique.

(c) In the remainder of this exercise, we suppose that f is $2p$-periodic. Show that f_e and f_o are both $2p$-periodic.

(d) Let $a_0, a_1, a_2, \ldots, b_1, b_2, \ldots$ denote the Fourier coefficients of f. Show that the Fourier series of f_e is $a_0 + \sum_{n=1}^{\infty} a_n \cos \frac{n\pi}{p} x$, and the Fourier series of f_o is $\sum_{n=1}^{\infty} b_n \sin \frac{n\pi}{p} x$.

In Exercises 21–24, a 2-periodic function is given by its graph over the interval $[-1, 1]$ (Figures 23 − 26). In each case, (a) determine and plot f_e and f_o (see Exercise 20).

(b) Find the Fourier series of f_e and f_o, and then deduce the Fourier series of f.

21.

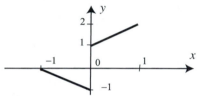

Figure 23

22.

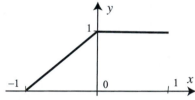

Figure 24

23.

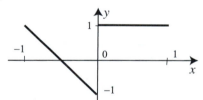

Figure 25

24.

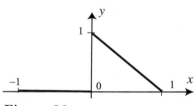

Figure 26

Project Problem: Differentiation of Fourier series. Can a Fourier series be differentiated term by term? The answer is no, in general. Do Exercises 25, 26, and any one of 27–30, and you will learn when you can safely use this process.

25. Fourier series and derivatives. Suppose that f is a $2p$-periodic, piecewise smooth and continuous function such that f' is also piecewise smooth. Let a_n, b_n denote the Fourier coefficients of f and a'_n, b'_n those of f'. Show that

$$a'_0 = 0, \quad a'_n = b_n \frac{n\pi}{p}, \quad \text{and} \quad b'_n = -a_n \frac{n\pi}{p}.$$

[Hint: To compute the Fourier coefficients of f', evaluate the integrals by parts and use $f(p) = f(-p)$.]

26. Term-by-term differentiation of Fourier series. Suppose that f is a $2p$-periodic piecewise smooth and continuous function such that f' is also piecewise smooth. Show that the Fourier series of f' is obtained from the Fourier series of f by differentiating term by term. That is, under the stated conditions, if

$$f(x) = a_0 + \sum_{n=1}^{\infty} (a_n \cos \frac{n\pi}{p} x + b_n \sin \frac{n\pi}{p} x),$$

then

$$f'(x) = \sum_{n=1}^{\infty} (-na_n \frac{\pi}{p} \sin \frac{n\pi}{p} x + nb_n \frac{\pi}{p} \cos \frac{n\pi}{p} x).$$

[Hint: Since f' satisfies the hypothesis of Theorem 1 it has a Fourier series expansion. Use Exercise 25 to compute the Fourier coefficients. Compare with the differentiated Fourier series of f.]

In most cases in this book, f and f' are piecewise smooth. Thus, according to Exercise 26, to differentiate term by term the Fourier series in these cases, it is enough to check that f is continuous. It is important to note that if f fails to satisfy some of the assumptions of Exercise 26, then we cannot in general differentiate the series term by term. See Exercises 31–32.

27. Derive the Fourier series in Exercise 1 by differentiating term by term the Fourier series in Exercise 5. Justify your work.

28. Derive the Fourier series in Exercise 2 by differentiating term by term the Fourier series in Exercise 4. Justify your work.

29. Use the Fourier series of Exercise 8 to find the Fourier series of the $2p$-periodic function in the Figure 27.

30. Use the Fourier series of Exercise 10 to find the Fourier series of the $2p$-periodic function in the Figure 28.

Project Problem: Failure of term-by-term differentiation. Do Exercises 31–32 to show that the Fourier series of the sawtooth function (a piecewise smooth function) cannot be differentiated term by term.

31. Show that the series $\sum_{n=1}^{\infty} \cos nx$ is divergent for all x. [Hint: Apply the nth term test and Exercise 8, Section 4.1.]

32. Failure of term-by-term differentiation. Consider the Fourier series of the sawtooth function $\sum_{n=1}^{\infty} \frac{\sin nx}{n}$.
(a) Show that the function represented by this Fourier series satisfies all the hypotheses of Exercise 26, except that it fails to be continuous.
(b) Show that the series cannot be differentiated term by term. [Hint: Exercise 31.]

33. Project Problem: Term-by-term integration of Fourier series. Let f be as in Theorem 1, and define an antiderivative of f by

$$F(x) = \int_0^x f(t)\, dt\,.$$

From Exercises 15–16, Section 7.1, we know that F is $2p$-periodic if and only if $\int_0^{2p} f(t)\, dt = 0$. Show that, in this case, the Fourier series of F is

$$F(x) = A_0 + \sum_{n=1}^{\infty} (\frac{p}{n\pi} a_n \sin \frac{n\pi}{p} x - \frac{p}{n\pi} b_n \cos \frac{n\pi}{p} x),$$

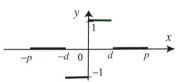

Figure 27 for Exercise 29.

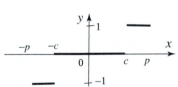

Figure 28 for Exercise 30.

where $A_0 = \frac{p}{\pi} \sum_{n=1}^{\infty} \frac{b_n}{n}$. Hence, as long as F is periodic, with no further assumptions on f other than piecewise smoothness, we can get the Fourier series of F by integrating term by term the Fourier series of f. [Hint: Apply the result of Exercise 25 to $F(x)$ and use $F'(x) = f(x)$. To compute A_0, use $F(0) = 0$ (why?).]

34. Use Exercise 33 to derive the Fourier series of Exercise 4 from that of Exercise 2.

7.4 Half-Range Expansions: The Cosine and Sine Series

In many applications we are interested in representing by a Fourier series a function $f(x)$ that is defined only in a finite interval, say $0 < x < p$. Since f is clearly not periodic, the results of the previous sections are not readily applicable. Our goal in this section is to show how we can represent f by a Fourier series, after extending it to a periodic function.

THEOREM 1
HALF-RANGE
EXPANSIONS

Suppose that $f(x)$ is a piecewise smooth function defined on an interval $0 < x < p$. Then f has a **cosine series expansion**

$$(1) \qquad a_0 + \sum_{n=1}^{\infty} a_n \cos \frac{n\pi}{p} x \quad (0 < x < p),$$

where

$$(2) \qquad a_0 = \frac{1}{p} \int_0^p f(x)\, dx; \qquad a_n = \frac{2}{p} \int_0^p f(x) \cos \frac{n\pi}{p} x\, dx \quad (n \geq 1).$$

Also, f has a **sine series expansion**

$$(3) \qquad \sum_{n=1}^{\infty} b_n \sin \frac{n\pi}{p} x \quad (0 < x < p),$$

where

$$(4) \qquad b_n = \frac{2}{p} \int_0^p f(x) \sin \frac{n\pi}{p} x\, dx \quad (n \geq 1).$$

On the interval $0 < x < p$, the series (1) and (3) converge to $f(x)$ if f is continuous at x and to $\frac{f(x+)+f(x-)}{2}$ otherwise.

The series (1) and (3) are commonly referred to as the **half-range expansions** of f. They are two different series representations of the same function on the interval $0 < x < p$. Theorem 1 will be derived by appealing to the Fourier series representation of even and odd functions (Theorem 2, Section 7.3). For this purpose, we introduce the following notions.

Define the **even periodic extension** of f by $f_1(x) = f(x)$ if $0 < x < p$, $f_1(x) = f(-x)$ if $-p < x < 0$, and $f_1(x) = f_1(x + 2p)$ otherwise. Define the **odd periodic extension** of f by $f_2(x) = f(x)$ if $0 < x < p$, $f_2(x) = -f(-x)$ if $-p < x < 0$, and $f_2(x) = f_2(x + 2p)$ otherwise. (In view of the remark following Theorem 1 of Section 7.2, we will not worry about the definition of the extensions at the points $0, \pm p, \pm 2p, \ldots$)

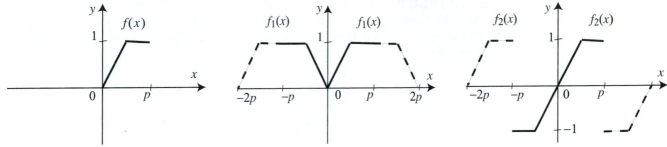

Figure 1 (a) $f(x), 0 < x < p$.   (b) Even $2p$-periodic extension, f_1.   (c) Odd $2p$-periodic extension, f_2.

By the way they are constructed, the function f_1 is even and $2p$-periodic, and the function f_2 is odd and $2p$-periodic. Both functions agree with f on the interval $0 < x < p$ which justifies calling them *extensions* of f (Figure 1). Since f is piecewise smooth, it follows that f_1 and f_2 are both piecewise smooth. Applying Theorem 2 of Section 7.3, we find that f_1 has a cosine series expansion given by (1) with the coefficients (2). Now $f(x) = f_1(x)$ for all $0 < x < p$, and so the cosine series (1) represents f on this interval. Similar reasoning using f_2 yields the sine series expansion of f.

EXAMPLE 1   Half-range expansions

Find the half-range expansions of the function $f(x) = x$ for $0 < x < 1$.

Solution The graphs of the even and odd extensions are shown in Figure 2.

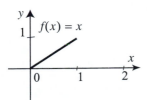 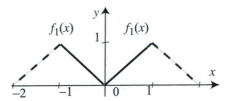 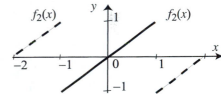

Figure 2 (a) $f(x) = x$, $0 < x < 1$.   (b) Even extension of f, period 2.   (c) Odd extension of f, period 2.

The even extension is a special case of Example 1 of Section 7.3 with $p = 1$. We have

$$x = \frac{1}{2} - \frac{4}{\pi^2} \sum_{k=0}^{\infty} \frac{1}{(2k+1)^2} \cos(2k+1)\pi x, \quad \text{for all } 0 \le x \le 1.$$

The odd extension is a special case of Exercise 2 of Section 7.3, with $p = 1$. However, to illustrate the formulas of Theorem 1, we will derive the sine coefficients using (4). We have

$$b_n = \frac{2}{1} \int_0^1 x \sin n\pi x \, dx = \frac{2(-1)^{n+1}}{n\pi}.$$

Hence

$$x = \frac{2}{\pi} \sum_{n=1}^{\infty} \frac{(-1)^{n+1}}{n} \sin n\pi x, \quad 0 \le x < 1.$$

It is a remarkable fact that the cosine series and the sine series have the same values on the intervals $(0, 1)$, $(2, 3)$, $(-2, -1), \ldots$. ∎

EXAMPLE 2   Half-range expansions

Consider the function $f(x) = \sin x$, $0 \le x \le \pi$. If we take its odd extension, we get the usual sine function, $f_2(x) = \sin x$ for all x. Thus, the sine series expansion is just $\sin x$.

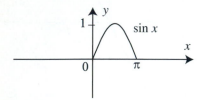
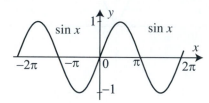
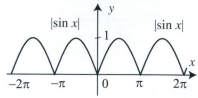

Figure 3 (a) $f(x) = \sin x$, $0 \le x \le \pi$.   (b) Odd extension of f, $\sin x$.   (c) Even extension of f, $|\sin x|$.

If we take the even extension of f, we get the function $|\sin x|$. The Fourier series of this even function can be obtained from Exercise 7, Section 7.2. Thus the cosine series (of $\sin x$) is

$$\sin x = \frac{2}{\pi} - \frac{4}{\pi} \sum_{k=1}^{\infty} \frac{1}{(2k)^2 - 1} \cos 2kx, \quad 0 \le x \le \pi.$$ ∎

Exercises 7.4

In Exercises 1–8, (a) find the half-range expansions of the given function. (Use as much as possible series that you have encountered earlier.)

 (b)   To illustrate the convergence of the cosine and sine series, plot several partial sums of each and comment on the graphs.

1. $f(x) = 1$ if $0 < x < 1$.   **2.** $f(x) = \pi - x$ if $0 \le x \le \pi$.

3. $f(x) = x^2$ if $0 < x < 1$.   **4.**

$$f(x) = \begin{cases} 0 & \text{if } 0 \le x < 1, \\ x - 1 & \text{if } 1 \le x < 2. \end{cases}$$

5.

$$f(x) = \begin{cases} 1 & \text{if } a < x < b, \\ 0 & \text{if } 0 < x < a \\ & \text{or } b < x < p, \end{cases}$$

where $0 < a < b < p < \infty$. For (b), take $p = 1, a = \frac{1}{4}, b = \frac{1}{2}$.

6. $f(x) = \cos x$ if $0 < x < \pi$.   **7.** $f(x) = \cos x$ if $0 \le x \le \frac{\pi}{2}$.

8. $f(x) = x \sin x$ if $0 < x < \pi$.

In Exercises 9–16, find the sine series expansion of the given function on the interval $0 < x < 1$.

9. $x(1 - x)$.   **10.** $1 - x^2$.   **11.** $\sin \pi x$.   **12.** $\sin \frac{\pi}{2} x$.

13. $\sin \pi x \cos \pi x$.   **14.** $(1 + \cos \pi x) \sin \pi x$.   **15.** e^x.   **16.** $1 - e^x$.

Figure 4 for Exercise 17.

17.   Triangular function. Let $f(x)$ denote the shape of a plucked string of length p with endpoints fastened at $x = 0$ and $x = p$, as shown in Figure 4.

(a)   Using the data in the figure, derive the formula

$$f(x) = \begin{cases} \frac{h}{a}x & \text{if } 0 \le x \le a, \\ \frac{h}{a-p}(x-p) & \text{if } a \le x \le p. \end{cases}$$

(b)   Obtain the sine series representation of f:

$$f(x) = \frac{2hp^2}{a(-a+p)\pi^2} \sum_{n=1}^{\infty} \frac{\sin \frac{an\pi}{p}}{n^2} \sin \frac{n\pi}{p} x.$$

(c)   Verify this representation by taking $a = 1/3, p = 1, h = 1/10$ and plotting the resulting function f along with several partial sums of the Fourier series.

7.5   Complex Form of Fourier Series

In previous sections, we have used the formulas

(1) $$\cos u = \frac{e^{iu} + e^{-iu}}{2} \quad \text{and} \quad \sin u = \frac{e^{iu} - e^{-iu}}{2i}$$

several times with tools from complex analysis to compute Fourier series. In this section, we will consider a Fourier series

(2) $$f(x) = a_0 + \sum_{n=1}^{\infty} \left(a_n \cos \frac{n\pi}{p}x + b_n \sin \frac{n\pi}{p}x \right),$$

and then, using (2), we will replace the cosine and sine by their expressions in terms of the complex exponential and derive the complex form of the Fourier series, which is expressed as follows.

**THEOREM 1
COMPLEX FORM OF
FOURIER SERIES**

Let f be a $2p$-periodic piecewise smooth function. The **complex form of the Fourier series** of f is

(3) $$\sum_{n=-\infty}^{\infty} c_n e^{i \frac{n\pi}{p} x},$$

where the **Fourier coefficients** c_n are given by

(4) $$c_n = \frac{1}{2p} \int_{-p}^{p} f(t) e^{-i\frac{n\pi}{p}t} \, dt \quad (n = 0, \pm 1, \pm 2, \ldots).$$

For all x, the Fourier series converges to $f(x)$ if f is continuous at x, and to $\frac{f(x+) + f(x-)}{2}$ otherwise. Moreover, the Fourier series converges to $f(x)$ uniformly for all x if and only if f is continuous for all x (and piecewise smooth).

The Fourier coefficients c_n are also denoted by $\widehat{f}(n)$. The Nth partial sum of (3) is by definition the **symmetric sum**

$$S_N(x) = \sum_{n=-N}^{N} c_n e^{i\frac{n\pi}{p}x}.$$

We will see in a moment that $S_N(x)$ is the same as the usual partial sum of the Fourier series

$$s_N(x) = a_0 + \sum_{n=1}^{N} \left(a_n \cos \frac{n\pi}{p}x + b_n \sin \frac{n\pi}{p}x \right).$$

Proof of Theorem 1   It is enough to show that $S_N = s_N$, then the theorem will follow from Theorem 1, Section 7.3. We clearly have $c_0 = a_0$. For $n > 0$, using (1), we get

$$
\begin{aligned}
a_n \cos \frac{n\pi}{p}x + b_n \sin \frac{n\pi}{p}x &= a_n \frac{e^{i\frac{n\pi}{p}x} + e^{-i\frac{n\pi}{p}x}}{2} + b_n \frac{e^{i\frac{n\pi}{p}x} - e^{-i\frac{n\pi}{p}x}}{2i} \\
&= \frac{1}{2}(a_n + \frac{1}{i}b_n)e^{i\frac{n\pi}{p}x} + \frac{1}{2}(a_n - \frac{1}{i}b_n)e^{-i\frac{n\pi}{p}x} \\
&= \frac{1}{2}(a_n - ib_n)e^{i\frac{n\pi}{p}x} + \frac{1}{2}(a_n + ib_n)e^{-i\frac{n\pi}{p}x}.
\end{aligned}
$$

Using the formulas for a_n and b_n (Theorem 1, Section 7.3) and (4), we have

$$
\begin{aligned}
\frac{1}{2}(a_n - ib_n) &= \frac{1}{2}\frac{1}{p}\int_{-p}^{p} f(t) \cos \frac{n\pi}{p}t \, dt - \frac{i}{2}\frac{1}{p}\int_{-p}^{p} f(t) \sin \frac{n\pi}{p}t \, dt \\
&= \frac{1}{2p}\int_{-p}^{p} f(t)\left(\cos \frac{n\pi}{p}t - i \sin \frac{n\pi}{p}t \right) dt \\
&= \frac{1}{2p}\int_{-p}^{p} f(t)e^{-i\frac{n\pi}{p}t} \, dt = c_n.
\end{aligned}
$$

To simplify the middle integral, we used the identity $e^{-i\theta} = \cos\theta - i\sin\theta$. A similar argument shows that $c_{-n} = \frac{1}{2}(a_n + ib_n)$. Thus, for $n \geq 1$,

$$a_n \cos \frac{n\pi}{p}x + b_n \sin \frac{n\pi}{p}x = c_n e^{i\frac{n\pi}{p}x} + c_{-n}e^{-i\frac{n\pi}{p}x},$$

and so

$$s_N(x) = a_0 + \sum_{n=1}^{N} \left(a_n \cos \frac{n\pi}{p}x + b_n \sin \frac{n\pi}{p}x \right) = c_0 + \sum_{n=1}^{N} c_n e^{i\frac{n\pi}{p}x} + \sum_{n=1}^{N} c_{-n}e^{-i\frac{n\pi}{p}x}.$$

Changing n to $-n$ in the second series on the right and combining, we get that $s_N(x) = S_N(x)$, and the theorem follows.   ∎

We can extract the following useful identities from the previous proof.

(5) $$c_0 = a_0$$
(6) $$c_n = \tfrac{1}{2}(a_n - ib_n), \quad c_{-n} = \tfrac{1}{2}(a_n + ib_n) \quad (n > 0)$$
(7) $$a_n = c_n + c_{-n}, \quad b_n = i(c_n - c_{-n}) \quad (n > 0)$$
(8) $$S_N(x) = s_N(x)$$

If f is real-valued, so that a_n and b_n are both real, then (6) shows that c_{-n} is the complex conjugate of c_n. In symbols,

(9) $$c_{-n} = \overline{c}_n.$$

This identity fails in general if f is not real-valued. Consider $f(x) = e^{ix}$. From the orthogonality relations of the complex exponential functions (Exercise 12, Section 3.2), it follows that $c_1 = 1$ and $c_n = 0$ for all $n \neq 1$ (Exercise 9). Hence, $c_{-1} \neq \overline{c}_1$.

The complex form of the Fourier series is particularly useful when dealing with exponential functions, as we now illustrate.

EXAMPLE 1   A complex Fourier series
Find the complex form of the Fourier series of the 2π-periodic function $f(x) = e^{ax}$ for $-\pi < x < \pi$, where $a \neq 0, \pm i, \pm 2i, \pm 3i, \ldots$. Determine the values of the Fourier series at $x = \pm \pi$.

Solution   From (4), we have

(10) $$c_n = \frac{1}{2\pi} \int_{-\pi}^{\pi} e^{ax} e^{-inx}\, dx = \frac{1}{2\pi}\left[\frac{e^{(a-in)x}}{a - in}\right]_{-\pi}^{\pi} = \frac{(-1)^n}{a - in}\frac{\sinh \pi a}{\pi},$$

where we have used $e^{\pm in\pi} = (-1)^n$ and $\sinh \pi a = \frac{e^{\pi a} - e^{-\pi a}}{2}$. Plugging these coefficients into (3) and simplifying, we obtain the complex form of the Fourier series of f

$$\frac{\sinh \pi a}{\pi} \sum_{n=-\infty}^{\infty} \frac{(-1)^n}{a - in} e^{inx} = \frac{\sinh \pi a}{\pi} \sum_{n=-\infty}^{\infty} \frac{(-1)^n}{a^2 + n^2}(a + in) e^{inx}.$$

(We remind you that here and throughout the section the doubly infinite Fourier series represents the limit of the symmetric partial sums, $\sum_{n=-N}^{N}$.) Applying Theorem 1 to $f(x)$, we obtain the Fourier series representation

$$e^{ax} = \frac{\sinh \pi a}{\pi} \sum_{n=-\infty}^{\infty} \frac{(-1)^n}{a^2 + n^2}(a + in) e^{inx} \quad (-\pi < x < \pi).$$

According to Theorem 1, the values of the Fourier series at the points of discontinuity, and in particular at $x = \pm \pi$, are given by the average of the function at

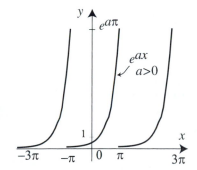

Figure 1 A 2π-periodic function, e^{ax}, $a > 0$, $-\pi < x < \pi$.

these points. With the help of Figure 1, we see that this average is

$$\frac{e^{a\pi} + e^{-a\pi}}{2} = \cosh a\pi.$$

As a specific illustration, if you take $x = \pi$ in the Fourier series, you obtain the interesting identity

$$\cosh a\pi = \frac{\sinh \pi a}{\pi} \sum_{n=-\infty}^{\infty} \frac{a + in}{a^2 + n^2}, \quad a \neq 0, \pm i, \pm 2i, \pm 3i, \ldots .$$

We have used $e^{in\pi} = (-1)^n$ and $(-1)^n(-1)^n = 1$ to simplify the series. (See Exercises 12 and 13 for related results.) Finally, let us note that if $a = \pm in$, then $f(x) = e^{\pm inx}$, and hence f is its own Fourier series.   ∎

EXAMPLE 2   The (usual) Fourier series from the complex form

Obtain the usual Fourier series of the function in Example 1 from its complex form. Take a to be a real number $\neq 0$.

Solution   The point here is not to use the Euler formulas of Section 7.2 to compute the Fourier series. Instead, we will use Example 1 and appropriate formulas relating the Fourier coefficients a_n and b_n to the complex Fourier coefficients c_n. From (5) and (10), we obtain

$$a_0 = c_0 = \frac{1}{a} \frac{\sinh \pi a}{\pi}.$$

From (7) and (10), we have

$$a_n = (-1)^n \frac{\sinh \pi a}{\pi} \left(\frac{1}{a - in} + \frac{1}{a + in} \right) = (-1)^n \frac{\sinh \pi a}{\pi} \frac{2a}{a^2 + n^2},$$

and

$$b_n = i(-1)^n \frac{\sinh \pi a}{\pi} \left(\frac{1}{a - in} - \frac{1}{a + in} \right) = -(-1)^n \frac{\sinh \pi a}{\pi} \frac{2n}{a^2 + n^2}.$$

Thus, the Fourier series of f is

$$\frac{1}{a} \frac{\sinh \pi a}{\pi} + \frac{\sinh \pi a}{\pi} \sum_{n=1}^{\infty} \frac{(-1)^n}{a^2 + n^2} \left(2a \cos nx - 2n \sin nx \right).$$

In particular, for $-\pi < x < \pi$ and $a \neq 0$, we have

$$e^{ax} = \frac{1}{a} \frac{\sinh \pi a}{\pi} + \frac{\sinh \pi a}{\pi} \sum_{n=1}^{\infty} \frac{(-1)^n}{a^2 + n^2} \left(2a \cos nx - 2n \sin nx \right).$$

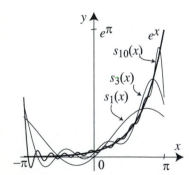

Figure 2   Partial sums of the Fourier series of the 2π-periodic function $f(x) = e^x$, $-\pi < x < \pi$.

We took $a = 1$ and illustrated the convergence of the Fourier series in Figure 2. Note that because the sine coefficients are of the order $1/n$, the series converges relatively slowly like the Fourier series of the sawtooth function.   ∎

Our next example illustrates the use of methods from complex analysis in computing Fourier series. The ideas are similar to those we used in

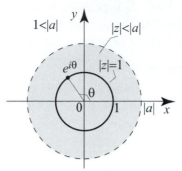

Figure 3 The disk $|z| < |a|$ ($|a| > 1$) contains the circle $|z| = 1$.

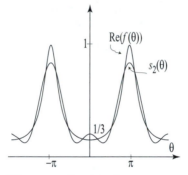

Figure 4 In this figure:
$\mathrm{Re}\left(\frac{1}{2+e^{i\theta}}\right) = \frac{2+\cos\theta}{5+4\cos\theta}$;
$s_2(\theta) = \frac{1}{2}\left(1 - \frac{\cos\theta}{2} + \frac{\cos 2\theta}{4}\right)$.

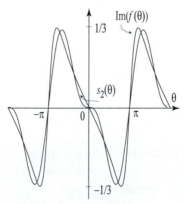

Figure 5 In this figure:
$\mathrm{Im}\left(\frac{1}{2+e^{i\theta}}\right) = \frac{-\sin\theta}{5+4\cos\theta}$;
$s_2(\theta) = -\frac{1}{2}\left(\frac{\sin\theta}{2} - \frac{\sin 2\theta}{4}\right)$.

Section 7.2 and are based on the applications of Laurent series. The example that we present leads to power series, which are special cases of Laurent series.

EXAMPLE 3   Using complex power series to compute Fourier series

Let a be a real number with $|a| > 1$. The complex-valued function

$$f(\theta) = \frac{1}{a + e^{i\theta}} \qquad (\theta \text{ real})$$

is 2π-periodic, and because $|a| > 1$ and $|e^{i\theta}| = 1$ for all θ, the denominator does not vanish, and so f is smooth. The function $f(\theta)$ is the restriction to the unit circle ($z = e^{i\theta}$) of the function $\frac{1}{a+z}$, where z is the variable. Since $|a| > 1$, it follows that $\frac{1}{a+z}$ is analytic in the disk $|z| < |a|$, and so it has a power series expansion in $|z| < |a|$, which can be obtained from the geometric series as follows:

$$\frac{1}{a+z} = \frac{1}{a}\frac{1}{1-(-\frac{z}{a})} = \frac{1}{a}\sum_{n=0}^{\infty}(-1)^n\left(\frac{z}{a}\right)^n \quad (|z| < |a|).$$

Since the unit circle $|z| = 1$ is contained in the disk $|z| < |a|$, the power series expansion is valid for all $z = e^{i\theta}$ (Figure 3). Substituting $z = e^{i\theta}$, then using $z^n = e^{in\theta}$ and simplifying, we get

(11) $$f(\theta) = \frac{1}{a + e^{i\theta}} = \frac{1}{a}\sum_{n=0}^{\infty}\frac{(-1)^n}{a^n}e^{in\theta},$$

which is the complex form of the Fourier series of $f(\theta)$. Out of this expansion, we can derive two interesting real Fourier series, by taking the real and imaginary parts of $f(\theta)$. We have (here we will use the fact that a is real)

$$f(\theta) = \frac{1}{a + e^{i\theta}} = \frac{a + e^{-i\theta}}{(a + e^{i\theta})(a + e^{-i\theta})} = \frac{a + \cos\theta}{1 + a^2 + 2a\cos\theta} - \frac{i\sin\theta}{1 + a^2 + 2a\cos\theta}.$$

Substitute this into (11), use $e^{in\theta} = \cos n\theta + i\sin n\theta$, then equate real and imaginary parts, and get the two Fourier series

$$\frac{a + \cos\theta}{1 + a^2 + 2a\cos\theta} = \frac{1}{a}\left[1 - \frac{\cos\theta}{a} + \frac{\cos 2\theta}{a^2} - \frac{\cos 3\theta}{a^3} + \cdots\right],$$

and

$$\frac{-\sin\theta}{1 + a^2 + 2a\cos\theta} = -\frac{1}{a}\left[\frac{\sin\theta}{a} - \frac{\sin 2\theta}{a^2} + \frac{\sin 3\theta}{a^3} + \cdots\right],$$

which are valid for all θ. In Figures 4 and 5, we plotted the real and imaginary parts of f in the case $a = 2$, along with partial sums of their Fourier series. ■

The Fourier series (11) is very special, in the sense that all the c_n's with $n < 0$ are zero. Such a Fourier series is called **analytic**, because, as we saw in Example 3, it is the restriction of an analytic function to the circle. Note that the functions $f(\theta)$ in Example 3 is complex-valued. Is it possible to have a real-valued analytic Fourier series? It is not hard to show that the only real-valued analytic functions are the constant functions (Exercise 10(b)).

Convolution of Periodic Functions

One of the most important operations in Fourier analysis is the convolution. To define it, consider a pair of $2p$-periodic functions f and g. The **convolution** of f and g, denoted by $f * g(x)$, is defined by

$$(12) \qquad f * g(x) = \frac{1}{2p} \int_{-p}^{p} f(t)g(x-t)\, dt.$$

From its mere definition, it is difficult to explain what the convolution does to a pair of functions. In a moment, we will be able to clearly explain its effect in terms of the complex Fourier coefficients. For now, let us observe that if f and g are piecewise continuous and $2p$-periodic, then, for fixed x, the function $f(t)g(x-t)$ is also piecewise continuous and $2p$-periodic. Its integral can be evaluated over any interval of length $2p$, without affecting its value (Theorem 1, Section 7.1). So

$$(13) \qquad f * g(x) = \frac{1}{2p} \int_{-p}^{p} f(t)g(x-t)\, dt = \frac{1}{2p} \int_{a}^{a+2p} f(t)g(x-t)\, dt,$$

where a is an arbitrary real number. Also, if f and g are continuous with continuous derivatives, we can differentiate under the integral sign (Theorem 5, Section 3.5) and conclude that

$$(14) \qquad \begin{aligned} \frac{d}{dx} f * g(x) &= \frac{1}{2p} \int_{-p}^{p} \frac{d}{dx}(f(t)g(x-t))\, dt \\ &= \frac{1}{2p} \int_{-p}^{p} f(t)g'(x-t)\, dt = f * g'(x). \end{aligned}$$

Similarly, we have $\frac{d}{dx} f * g(x) = f' * g(x)$, and so

$$\frac{d}{dx} f * g(x) = f * g'(x) = f' * g(x).$$

If f and g are merely piecewise smooth, we can write the convolution integral as a sum of integrals over intervals on which f and g are continuous with continuous derivatives, and then differentiate under the integral sign. So, in general, the convolution is piecewise smooth.

Convolutions are important because they correspond to multiplication of the Fourier coefficients. To express this property, it will be convenient to use the notation $\widehat{f}(n)$ for the complex Fourier coefficients instead of c_n.

THEOREM 2
FOURIER
COEFFICIENTS OF
CONVOLUTIONS

Let f and g be $2p$-periodic, piecewise smooth functions. Then

$$\widehat{f * g}(n) = \hat{f}(n)\hat{g}(n) \qquad n = 0, \pm 1, \pm 2, \ldots.$$

Proof We have

$$
\begin{aligned}
\widehat{f * g}(n) &= \frac{1}{2p}\int_{-p}^{p} f * g(x) e^{-i\frac{n\pi}{p}x}\, dx \\
&= \frac{1}{2p}\int_{-p}^{p} \frac{1}{2p}\int_{-p}^{p} f(t)g(x-t)\, dt\, e^{-i\frac{n\pi}{p}x}\, dx \\
&= \frac{1}{2p}\int_{-p}^{p} \frac{1}{2p}\int_{-p}^{p} g(x-t) e^{-i\frac{n\pi}{p}x}\, dx\, f(t)\, dt.
\end{aligned}
$$

Making the change of variables $y = x - t$ in the inner integral and using the fact that the integral does not change over an interval of length $2p$ (Theorem 1, Section 7.1), we obtain

$$
\begin{aligned}
\widehat{f * g}(n) &= \frac{1}{2p}\int_{-p}^{p} \frac{1}{2p}\int_{-p}^{p} g(y) e^{-i\frac{n\pi}{p}y} e^{-i\frac{n\pi}{p}t}\, dy f(t)\, dt \\
&= \frac{1}{2p}\int_{-p}^{p} g(y) e^{-i\frac{n\pi}{p}y}\, dy \frac{1}{2p}\int_{-p}^{p} f(t) e^{-i\frac{n\pi}{p}t}\, dt \\
&= \hat{g}(n)\hat{f}(n) = \hat{f}(n)\hat{g}(n).
\end{aligned}
$$
∎

We will derive several interesting consequences of Theorem 2.

COROLLARY 1
CONTINUITY OF
CONVOLUTIONS

Let f and g be $2p$-periodic, piecewise smooth functions. Then $f * g$ is continuous and piecewise smooth.

Proof We have already showed that $f * g$ is piecewise smooth with derivative given by (14) at all but finitely many points in $[-p, p]$. To show that it is continuous, we note that if $|f|$ and $|f'|$ are bounded, say by M, then $|\hat{f}(n)| \leq \frac{4M}{n}$, which follows by integrating by parts in (4). The details are left as an exercise. So $|\widehat{f * g}(n)| = |\hat{f}(n)||\hat{g}(n)| \leq A\frac{1}{n}\frac{1}{n} = \frac{A}{n^2}$, where A is a constant. We now apply the Weierstrass M-test to the Fourier series of $f * g(x)$, which is $\sum_{-\infty}^{\infty} \hat{f}(n)\hat{g}(n)e^{i\frac{n\pi}{p}}$, and conclude that the series converges uniformly for all x, implying that $f * g(x)$ is continuous, by Theorem 1. ∎

The next result states that convolution is commutative. It can be checked directly by using the definition of convolution. We give a different proof based on Theorem 2.

COROLLARY 2

Let f and g be piecewise smooth $2p$-periodic functions. Then $f * g(x) = g * f(x)$.

Proof The functions $f * g$ and $g * f$ are piecewise smooth and continuous. Moreover, they have the same Fourier coefficients, because

$$\widehat{f * g}(n) = \hat{f}(n)\hat{g}(n) = \hat{g}(n)\hat{f}(n) = \widehat{g * f}(n).$$

Since the Fourier series are the same and converge (uniformly) everywhere to the functions, we conclude that $f * g = g * f$. ∎

**COROLLARY 3
PARSEVAL'S
IDENTITY**

Let f be a real-valued $2p$-periodic piecewise smooth function on $[-p, p]$. Let a_n, b_n, and c_n denote, respectively, the cosine, the sine, and the complex Fourier coefficients of f. Then

$$(15) \qquad \frac{1}{2p} \int_{-p}^{p} f(x)^2 \, dx = \sum_{n=-\infty}^{\infty} |c_n|^2 = a_0^2 + \frac{1}{2} \sum_{n=1}^{\infty} (a_n^2 + b_n^2).$$

Proof Let $g(t) = f(-t)$. Then

$$\hat{g}(n) = \frac{1}{2p} \int_{-p}^{p} f(-t)e^{-i\frac{n\pi}{p}t} \, dt = \frac{1}{2p} \int_{-p}^{p} f(t)e^{i\frac{n\pi}{p}t} \, dt$$

$$= \frac{1}{2p} \int_{-p}^{p} f(t)e^{-i\frac{(-n)\pi}{p}t} \, dt = c_{-n} = \overline{c_n},$$

where the last equality follows from (9). From the Fourier series representation, we have

$$f * g(x) = \sum_{n=-\infty}^{\infty} \hat{f}(n)\hat{g}(n)e^{i\frac{n\pi}{p}x}$$

$$= \sum_{n=-\infty}^{\infty} c_n c_{-n} e^{i\frac{n\pi}{p}x} = \sum_{n=-\infty}^{\infty} |c_n|^2 e^{i\frac{n\pi}{p}x}.$$

Evaluating both sides at $x = 0$, we obtain the first of the Parseval's identities, since $f * g(0) = \frac{1}{2p} \int_{-p}^{p} f(t)g(-t) \, dt = \frac{1}{2p} \int_{-p}^{p} f(t)f(t) \, dt = \frac{1}{2p} \int_{-p}^{p} f^2(t) \, dt$. The second identity follows from the first one by using the relationships between the Fourier coefficients (5)–(7). The details are left as an exercise. ∎

In the following example of a convolution, we will avoid computing directly from the definition. Instead we will compute the Fourier coefficients of the convolution and then identify the convolution function from its Fourier coefficients.

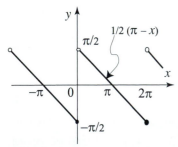

Figure 6 Sawtooth function in Example 4 has the Fourier series $\sum_{n=1}^{\infty} \frac{\sin nx}{n}$.

EXAMPLE 4   Computing convolutions
Find $f * f(x)$, where $f(x)$ denotes the 2π-periodic sawtooth function, defined on $(0, 2\pi)$ by $f(x) = \frac{1}{2}(\pi - x)$ (Figure 6).

Solution Let us first compute $c_n = \hat{f}(n)$. We can use the definition of the Fourier coefficients or, better yet, since we know the Fourier cosine and sine coefficients, $a_n = 0$ and $b_n = \frac{1}{n}$, we can compute c_n by using (6). We have $c_n = \frac{1}{2}(a_n - ib_n) =$

$-\frac{i}{2n}$. From this and Theorem 2, we conclude that $\widehat{f * f}(n) = \left(-\frac{i}{2n}\right)^2 = -\frac{1}{4n^2}$. Computing the cosine and sine Fourier coefficients of $f * f$ from (7), we find that the Fourier sine coefficients are 0, while the Fourier cosine coefficients are $-\frac{1}{2n^2}$. We now ask: Which function has these Fourier coefficients? Consider the function from Exercise 9, Section 7.2. Call this function h. Replace x by $x - \pi$ in the Fourier series and get

$$h(x - \pi) = \frac{\pi^2}{3} + 4 \sum_{n=1}^{\infty} \frac{(-1)^n}{n^2} \cos n(x - \pi) = \frac{\pi^2}{3} + 4 \sum_{n=1}^{\infty} \frac{1}{n^2} \cos nx,$$

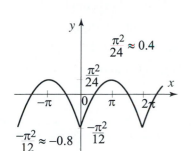

$\frac{\pi^2}{24} \approx 0.4$

$\frac{\pi^2}{24}$

$-\frac{\pi^2}{12} \approx -0.8$ $\frac{-\pi^2}{12}$

Figure 7 The convolution of the sawtooth function with itself is a continuous function, given on $[0, 2\pi]$ by

$$\frac{1}{8}\left(\frac{\pi^2}{3} - (x - \pi)^2\right).$$

where we have used $\cos n(x - \pi) = (-1)^n \cos nx$. Subtract $\frac{\pi^2}{3}$ and divide by -8 to get

$$-\frac{1}{8}\left(h(x - \pi) - \frac{\pi^2}{3}\right) = -\frac{1}{2} \sum_{n=1}^{\infty} \frac{1}{n^2} \cos nx,$$

which is the desired function. Since this function has the same Fourier coefficients as $f * f(x)$, it is thus equal to $f * f(x)$. Using the formula for h, we find that on the interval $(0, 2\pi)$

$$f * f(x) = -\frac{1}{8}\left((x - \pi)^2 - \frac{\pi^2}{3}\right) = \frac{1}{8}\left(\frac{\pi^2}{3} - (x - \pi)^2\right).$$

The graph of $f * f$ is shown in Figure 7. It is continuous and piecewise smooth, even though f is not continuous for all x.   ∎

The final result of this section is an interesting application of Parseval's identity, which implies that the Fourier series of a continuous piecewise smooth function is uniformly and absolutely convergent. We stated this result in Theorem 2, Section 7.2, and proved it under the further assumption that f' is piecewise smooth.

THEOREM 3   Suppose that f is continuous with piecewise continuous derivative f'. Then the Fourier series of f converges uniformly and absolutely for all x.

Proof Denote the Fourier coefficients of f by a_n and b_n and those of f' by a'_n and b'_n. It is enough to show that $\sum_{n=1}^{\infty}(|a_n| + |b_n|) < \infty$. We will prove that $\sum_{n=1}^{\infty} |a_n| < \infty$, since the sum with the b_n is handled similarly. From the proof of Theorem 2, Section 7.2, we have

$$a_n = -\frac{1}{n} b'_n \quad \text{and} \quad b_n = \frac{1}{n} a'_n.$$

So by the Cauchy-Schwarz inequality (Exercise 41, Section 1.2), we have

$$\sum_{n=1}^{N} |a_n| = \sum_{n=1}^{N} |b'_n| \frac{1}{n} \leq \left(\sum_{n=1}^{N} |b'_n|^2\right)^{\frac{1}{2}} \left(\sum_{n=1}^{N} \frac{1}{n^2}\right)^{\frac{1}{2}}.$$

Letting $N \to \infty$ and using the fact that $\sum_{n=1}^{\infty} |b'_n|^2 < \infty$, by Parseval's identity applied to f', it follows that

$$\sum_{n=1}^{\infty} |a_n| \le \left(\sum_{n=1}^{\infty} |b'_n|^2 \right)^{\frac{1}{2}} \left(\sum_{n=1}^{\infty} \frac{1}{n^2} \right)^{\frac{1}{2}} < \infty. \qquad \blacksquare$$

Exercises 7.5

In Exercises 1–6, find the complex form of the Fourier series of the given 2π-periodic function.

1. $f(x) = \cosh ax$ if $-\pi < x < \pi$ ($a \neq 0, \pm i, \pm 2i, \pm 3i, \dots$). [Hint: Example 1.]

2. $f(x) = \sinh ax$ if $-\pi < x < \pi$ ($a \neq 0, \pm i, \pm 2i, \pm 3i, \dots$). [Hint: Example 1.]

3. $f(x) = \cos ax$ if $-\pi < x < \pi$ (a is not an integer). [Hint: (1) and Example 1.]

4. $f(x) = \sin ax$ if $-\pi < x < \pi$ (a is not an integer). [Hint: (1) and Example 1.]

5. $f(x) = \cos 2x + 2\cos 3x$. [Hint: Use (1).]

6. $f(x) = \sin 3x$. [Hint: Use (1).]

In Exercises 7–8, find the Fourier series of the given function by using (5) and (7) or by manipulating the complex form of the Fourier series.

7. $f(x)$ is as in Exercise 3. **8.** $f(x)$ is as in Exercise 4.

9. (a) Use the orthogonality relations of the complex exponential system (Exercise 12, Section 3.2) to show that the 2π-periodic function e^{inx} is its own Fourier series.
(b) Let $m \le n$ be arbitrary integers and c_k be arbitrary complex numbers. What is the Fourier series of the 2π-periodic function $f(x) = \sum_{k=m}^{n} c_k e^{ikx}$?

10. (a) Derive (7) from (6).
(b) Show that if f is real-valued, $2p$-periodic, and piecewise smooth, and all the Fourier coefficients c_n with $n < 0$ are zero, then f is constant. [Hint: Use (9).]

11. For any real number $a \neq 0$, obtain the expansion

$$\frac{\pi}{a \sinh \pi a} = \sum_{n=-\infty}^{\infty} \frac{(-1)^n}{a^2 + n^2}.$$

[Hint: Take $x = 0$ in Example 1.]

12. For any real number $a \neq 0$ and all $-\pi < x < \pi$, obtain the expansion

$$e^{ax} = \frac{\sinh \pi a}{\pi} \sum_{n=-\infty}^{\infty} \frac{(-1)^n}{a^2 + n^2} (a \cos nx - n \sin nx).$$

[Hint: Equate real parts in the Fourier series of Example 1.]

13. (a) Let $a \neq 0$ be a real number. Use Parseval's identity and Exercise 1 to derive the identity

$$\sum_{n=-\infty}^{\infty} \frac{1}{(a^2 + n^2)^2} = \frac{\pi}{2a^2 \sinh^2(\pi a)} \left[\pi + \frac{\sinh(2\pi a)}{2a} \right].$$

(b) With the help of Exercise 2, derive the identity

$$\sum_{n=-\infty}^{\infty} \frac{n^2}{(a^2 + n^2)^2} = \frac{\pi}{2 \sinh^2(\pi a)} \left[\frac{\sinh(2\pi a)}{2a} - \pi \right].$$

14. (a) Use Parseval's identity and the Fourier series expansion $\frac{x}{2} = \sum_{n=1}^{\infty} \frac{(-1)^{n+1}}{n} \sin nx$
for $-\pi < x < \pi$, to obtain $\sum_{n=1}^{\infty} \frac{1}{n^2} = \frac{\pi^2}{6}$.

(b) From (a), obtain that $\sum_{k=1}^{\infty} \frac{1}{(2k)^2} = \frac{\pi^2}{24}$.

(c) Combine (a) and (b) to derive the identity $\sum_{k=0}^{\infty} \frac{1}{(2k+1)^2} = \frac{\pi^2}{8}$.

In Exercises 15–16, find the Fourier series of the given function using Taylor or Laurent series expansions.

15. $f(\theta) = \frac{e^{i\theta}}{2+e^{2i\theta}}$.

16. $f(\theta) = \frac{1}{3+e^{i\theta}+e^{-i\theta}}$.

17. $f(\theta) = e^{e^{i\theta}}$.

18. $f(\theta) = \cos(e^{i\theta})$.

19. Which real Fourier series do you get by taking real and imaginary parts in the Fourier series of Exercise 15?

20. (a) Which real Fourier series do you get by taking real and imaginary parts in the Fourier series of Exercise 17?

(b) Answer (a) with the Fourier series of Exercise 18.

In Exercises 21–24, you are given two 2π-periodic functions f and g on an interval of length 2π. (a) Compute the Fourier coefficients of $f*g$. (b) Find $f*g$ by matching its Fourier coefficients with those of a known function (as we did in Example 4).

21. For $-\pi < x < \pi$, $f(x) = g(x) = x$.

22. For $-\pi < x < \pi$, $f(x) = x$; and $g(x) = -1$ if $-\pi < x < 0$ and $g(x) = 1$ if $0 < x < \pi$.

23. For $-\pi < x < \pi$, $f(x) = e^{ix} + e^{-ix}$, and $g(x)$ is an arbitrary piecewise smooth function.

24. For $-\pi < x < \pi$, $f(x) = \sum_{n=-N}^{N} e^{-inx}$, and $g(x)$ is an arbitrary piecewise smooth function.

Project Problem: Cotangent expansion, Bernoulli numbers, and Fourier series. In Exercises 25–27, we explore a connection between Fourier series and some important complex series expansions, and derive interesting identities.

25.   Consider the 2π-periodic function of Example 2, which is defined on the interval $(-\pi, \pi)$ by $f(x) = e^{ax}$, where $a \neq 0$ is an arbitrary real number.
(a)   Evaluate the Fourier series in Example 2 at $x = \pi$ and obtain for $a \neq 0$

$$\cosh a\pi = \frac{\sinh a\pi}{a\pi} + \frac{\sinh a\pi}{\pi} \sum_{n=1}^{\infty} \frac{2a}{a^2 + n^2}.$$

[Hint: See Example 1.]
(b)   Conclude that for any real number $a \neq 0$,

$$a\pi \coth a\pi = 1 + \sum_{n=1}^{\infty} \frac{2a^2}{a^2 + n^2}.$$

26. Euler's expansion of the cotangent. Let Ω consists of the entire complex plane minus the points $z \neq \pm i, \pm 2i, \ldots$. For z in Ω, consider

$$\phi(z) = 1 + \sum_{n=1}^{\infty} \frac{2z^2}{z^2 + n^2}.$$

(a)   Complete the details of the following argument showing that ϕ is analytic in Ω. It is enough to show that the series converges uniformly on every closed and bounded subset of Ω (Corollary 2, Section 4.2). Let $A \subset \Omega$ be closed and bounded. Let $M > 0$ be an integer such that $|z| < M$ for all z in A. Then for all $n > M + 1$ and all z in A, we have $\left| \frac{2z^2}{z^2+n^2} \right| \le \frac{2M^2}{n^2-M^2}$. So the series converges uniformly on A by the Weierstrass M-test, since $\sum_{n=M+1}^{\infty} \frac{2M^2}{n^2-M^2} < \infty$.

(b)   Show that the function $\psi(z) = \pi z \coth(\pi z)$ is analytic for all $z \ne \pm i, \pm 2i, \dots$.

(c)   By Exercise 25(b), $\phi(z) = \psi(z)$ for all real $z \ne 0$. Using the identity principle (Section 4.6), conclude that $\phi(z) = \psi(z)$ for all z in Ω. Hence

$$\pi z \coth(\pi z) = 1 + \sum_{n=1}^{\infty} \frac{2z^2}{z^2 + n^2} \quad (z \ne \pm i, \pm 2i, \dots).$$

(d)   Replace z by iz in (c) and obtain Euler's expansion of the cotangent

$$\pi z \cot(\pi z) = 1 + \sum_{n=1}^{\infty} \frac{2z^2}{z^2 - n^2} \quad (z \ne \pm, \pm 2, \dots).$$

27. Bernoulli numbers.   (a)   Using Exercise 26(d) and the expansion of the cotangent from Exercise 31, Section 4.4, obtain

$$\sum_{n=1}^{\infty} \frac{2z^2}{z^2 - n^2} = \sum_{n=1}^{\infty} (-1)^n \frac{2^{2n} B_{2n} \pi^{2n}}{(2n)!} z^{2n}, \qquad |z| < 1,$$

where B_{2n} are the Bernoulli numbers (see Example 4, Section 4.4).

(b)   Use the Weierstrass double series theorem (Exercise 39, Section 4.4) to show that for $|z| < 1$,

$$\sum_{n=1}^{\infty} \frac{2z^2}{z^2 - n^2} = \sum_{n=1}^{\infty} \left(\sum_{k=1}^{\infty} \frac{-2}{k^{2n}} \right) z^{2n}.$$

[Hint: For each $n = 1, 2, \dots$, expand $\frac{2z^2}{z^2-n^2} = -2 \sum_{k=1}^{\infty} \left(\frac{z}{n} \right)^{2k}$, $|z| < 1$.]

(c)   Equating the coefficients in the power series expansions in (a) and (b), conclude that for $n = 1, 2, \dots$,

$$\sum_{k=1}^{\infty} \frac{1}{k^{2n}} = (-1)^{n-1} \frac{2^{2n-1} B_{2n} \pi^{2n}}{(2n)!}.$$

(d)   Using the values of the Bernoulli numbers from Table 1, Section 4.4, derive the entries in Table 1, which follows.

n	1	2	3	4	5	6
$\sum_{k=1}^{\infty} \dfrac{1}{k^{2n}}$	$\dfrac{\pi^2}{6}$	$\dfrac{\pi^4}{90}$	$\dfrac{\pi^6}{945}$	$\dfrac{\pi^8}{9450}$	$\dfrac{\pi^{10}}{93555}$	$\dfrac{691\,\pi^{12}}{638512875}$

Table 1.   Sums of reciprocals of even powers of integers.

7.6   Proof of the Fourier Representation Theorem

In this section, we prove Theorem 1, Section 7.2. The proof that we present is based on three preliminary results that are interesting in their own right. We start with the first one, which is an integral representation of the partial sums of Fourier series involving the Dirichlet kernel (see Exercise 40, Section 1.5).

Let N be a positive integer, and consider the Nth partial sum of the Fourier series of a function $f(x)$: $s_N(x) = a_0 + \sum_{n=1}^{N}(a_n \cos nx + b_n \sin nx)$. If we use the Euler formulas to write the Fourier coefficients in terms of f, we see that

$$
\begin{aligned}
s_N(x) &= \frac{1}{2\pi}\int_{-\pi}^{\pi} f(t)\,dt \\
&\quad + \frac{1}{\pi}\sum_{n=1}^{N}\left(\cos nx \int_{-\pi}^{\pi} f(t)\cos nt\,dt + \sin nx \int_{-\pi}^{\pi} f(t)\sin nt\,dt\right) \\
&= \frac{1}{2\pi}\int_{-\pi}^{\pi} f(t)\,dt + \frac{1}{\pi}\int_{-\pi}^{\pi}\left\{f(t)\sum_{n=1}^{N}(\cos nx \cos nt + \sin nx \sin nt)\right\}dt \\
&= \frac{1}{2\pi}\int_{-\pi}^{\pi} f(t)\left(1 + 2\sum_{n=1}^{N}\cos n(x-t)\right)dt,
\end{aligned}
$$

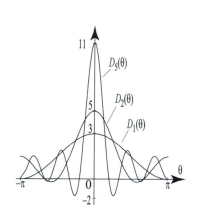

Figure 1 The Nth Dirichlet kernel, $D_N(\theta)$, for $N = 1, 2, 5$. We have $D_N(0) = 2N + 1$.

where we have combined the integrals and used $\cos nx \cos nt + \sin nx \sin nt = \cos(n-t)x$. The sum inside the big parentheses is the Dirichlet kernel evaluated at $x - t$. Thus according to Exercise 40(d), Section 1.5, we have

$$
1 + 2\sum_{n=1}^{N}\cos n(x-t) = \frac{\sin[(N+\frac{1}{2})(x-t)]}{\sin\frac{x-t}{2}} = D_N(x-t),
$$

where we let $D_N(\theta)$ denote the Nth **Dirichlet kernel** (Figure 1):

$$
(1) \qquad D_N(\theta) = 1 + 2\sum_{n=1}^{N}\cos n\theta = \frac{\sin[(N+\frac{1}{2})\theta]}{\sin\frac{\theta}{2}}.
$$

The formula seems to have a problem at $\theta = 2k\pi$, since for those values $\sin\frac{\theta}{2} = 0$. However, for $\theta = 2k\pi$, we have $\cos\theta = 1$, and so $D_N(2k\pi) = 1 + 2\sum_{n=1}^{N}1 = 1 + 2N$, which is also equal to $\lim_{\theta \to 2k\pi}\frac{\sin[(N+\frac{1}{2})\theta]}{\sin\frac{\theta}{2}}$, as you can verify by using l'Hospital's rule. So (1) is true for all θ if we interpret it in the limit at the points $\theta = 2k\pi$. Substituting (1) in the expression for $s_N(x)$, we obtain the first half of the following integral representation of the partial sums of Fourier series.

LEMMA 1
DIRICHLET KERNEL
AND
FOURIER SERIES

If f is a 2π-periodic piecewise continuous function, and $N \geq 1$, then

$$
s_N(x) = \frac{1}{2\pi}\int_{-\pi}^{\pi} f(t)D_N(x-t)\,dt = \frac{1}{2\pi}\int_{-\pi}^{\pi} f(x-t)D_N(t)\,dt,
$$

where D_N is the Nth Dirichlet kernel (1).

Proof The second equality follows from the fact that convolution is a commutative operation. To give a direct proof, start with the first equality and use the change

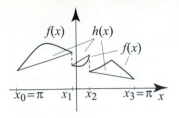

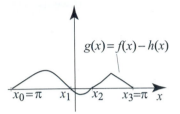

Figure 2 The function $h(x)$ is piecewise linear. Its discontinuities are the same as those of $f(x)$. They are built in order to cancel the discontinuities of $f(x)$ by adding $-h(x)$.

of variables $T = x - t$, $dT = -dt$. Then

$$
\begin{aligned}
s_N(x) &= \frac{1}{2\pi} \int_{-\pi}^{\pi} f(t) D_N(x-t)\,dt = -\frac{1}{2\pi} \int_{x+\pi}^{x-\pi} f(x-T) D_N(T)\,dT \\
&= \frac{1}{2\pi} \int_{x-\pi}^{x+\pi} f(x-T) D_N(T)\,dT = \frac{1}{2\pi} \int_{-\pi}^{\pi} f(x-T) D_N(T)\,dT,
\end{aligned}
$$

where the last equality follows because we are integrating a 2π-periodic function over an interval of length 2π (Theorem 1, Section 7.1). ■

In what follows we will be concerned with piecewise continuous 2π-periodic functions, which may be considered as functions on $[-\pi, \pi]$. The proofs are greatly simplified if we assume further that the functions are continuous on $[-\pi, \pi]$; because functions that are continuous on closed and bounded intervals are in fact *uniformly continuous*. To reduce a proof from a piecewise continuous function to a continuous function, we can add a piecewise linear correction term, which can be handled separately. This useful process will be clarified in the proofs; for now let us describe our linear correction term.

LEMMA 2
LINEAR CORRECTION

> Suppose that f is a 2π-periodic piecewise continuous function. Then there is a piecewise linear function $h(x)$ with finitely many discontinuities in $[-\pi, \pi]$, such that the function $g(x) = f(x) - h(x)$ is 2π-periodic and continuous for all x.

Proof The construction of h is best described by a figure (see Figure 2). It is enough to define h on the interval $[-\pi, \pi]$. Since f is piecewise continuous, it has at most a finite number of discontinuities in $[-\pi, \pi]$, say, $-\pi = x_0 < x_1 < \cdots < x_n = \pi$. Define $h(x)$ on each subinterval (x_j, x_{j+1}) by $h(x_j+) = f(x_j+)$ and $h(x_{j+1}-) = f(x_{j+1}-)$. Then $g(x) = f(x) - h(x)$ is clearly continuous for all $x \neq x_j$. For $x = x_j$, we have $g(x_j+) = f(x_j+) - h(x_j+) = 0$ and $g(x_j-) = f(x_j-) - h(x_j-) = 0$. Hence g is also continuous at x_j and so g is continuous for all x. ■

Our next result states that the Fourier coefficients of a piecewise continuous function tend to 0 as $n \to \infty$.

LEMMA 3
RIEMANN-LEBESGUE
LEMMA

> Suppose that f is a 2π-periodic piecewise continuous function. Then
>
> (2) $\displaystyle \lim_{n\to\infty} \int_{-\pi}^{\pi} f(x) \cos nx\,dx = 0$ and $\displaystyle \lim_{n\to\infty} \int_{-\pi}^{\pi} f(x) \sin nx\,dx = 0.$
>
> More generally, if α is any fixed real number, then
>
> (3)
>
> $\displaystyle \lim_{n\to\infty} \int_{-\pi}^{\pi} f(x) \cos[(n+\alpha)x]\,dx = 0$ and $\displaystyle \lim_{n\to\infty} \int_{-\pi}^{\pi} f(x) \sin[(n+\alpha)x]\,dx = 0.$

Proof We will only establish the first limit in (2); the second one follows similarly. We start by verifying the limit for functions that are piecewise linear. Using integration by parts, we have

$$\int_a^b (cx + d) \cos nx\, dx \;=\; (cx + d)\frac{\sin nx}{n}\Big|_a^b - c\int_a^b \frac{\sin nx}{n}\, dx$$

$$= \frac{(cb + d)\sin nb - (ca + d)\sin na}{n} + c\frac{\cos nb - \cos na}{n^2} \to 0,$$

as $n \to \infty$. If f is piecewise linear, the first integral in (2) is a finite sum of integrals of the form $\int_a^b (cx + d)\cos nx\, dx$, each of which tends to 0 as $n \to \infty$, and so the integral itself tends to 0 as $n \to \infty$. This shows that the first limit in (2) is true if f is piecewise linear. Next we consider the case of a continuous function f. From the identity $\cos a = -\cos(a + \pi)$, we get $\cos nx = -\cos\left(n(x + \frac{\pi}{n})\right)$. Using the substitution $X = x + \frac{\pi}{n}$, we have

$$\int_{-\pi}^{\pi} f(x) \cos nx\, dx \;=\; -\int_{-\pi}^{\pi} f(x) \cos\left(n(x + \frac{\pi}{n})\right)\, dx$$

$$= -\int_{-\pi + \frac{\pi}{n}}^{\pi + \frac{\pi}{n}} f(x - \frac{\pi}{n}) \cos nx\, dx$$

$$= -\int_{-\pi}^{\pi} f(x - \frac{\pi}{n}) \cos nx\, dx,$$

where the last equality follows from Theorem 1, Section 7.1, since all functions are 2π-periodic. So

$$2\int_{-\pi}^{\pi} f(x) \cos nx\, dx = \int_{-\pi}^{\pi} \left(f(x) - f(x - \frac{\pi}{n})\right) \cos nx\, dx,$$

and hence by the inequality for integrals,

$$\left|\int_{-\pi}^{\pi} f(x) \cos nx\, dx\right| \leq \frac{1}{2}\left|\int_{-\pi}^{\pi} \left(f(x) - f(x - \frac{\pi}{n})\right) \cos nx\, dx\right| \leq \frac{1}{2}(2\pi)M_n,$$

where $M_n = \max\left|\left(f(x) - f(x - \frac{\pi}{n})\right)\cos nx\right| = \max\left|f(x) - f(x - \frac{\pi}{n})\right|$ for x in $[-\pi,\, \pi]$. Since f is continuous on the closed interval $[-\pi,\, \pi]$, it is uniformly continuous; hence the difference $|f(x) - f(x - \frac{\pi}{n})|$ tends to 0 uniformly for all x in $[-\pi,\, \pi]$, as $\frac{\pi}{n} \to 0$. So as $n \to \infty$, $M_n \to 0$, implying that $\left|\int_{-\pi}^{\pi} f(x) \cos nx\, dx\right| \to 0$, and thus completing the proof in the case f is continuous. Finally, if f is piecewise continuous, we apply Lemma 2 and write $f(x) = g(x) + h(x)$, where g is continuous and h is piecewise linear. Then $\int_{-\pi}^{\pi} f(x) \cos nx\, dx = \int_{-\pi}^{\pi} g(x) \cos nx\, dx + \int_{-\pi}^{\pi} h(x) \cos nx\, dx \to 0$ as $n \to \infty$, by the previous two cases.

To prove (3), use the addition formula for the sine and cosine and apply (2). For example, using $\cos[(n + \alpha)x] = \cos(nx)\cos(\alpha x) - \sin(nx)\sin(\alpha x)$, we get

$$\int_{-\pi}^{\pi} f(x) \cos[(n + \alpha)x]\, dx$$

$$= \int_{-\pi}^{\pi} [f(x)\cos(\alpha x)] \cos nx\, dx - \int_{-\pi}^{\pi} [f(x)\sin(\alpha x)] \sin nx\, dx.$$

Applying (2) to the functions $f(x)\cos(\alpha x)$ and $f(x)\sin(\alpha x)$, it follows that both terms on the right of the displayed equation tend to 0 as $n \to \infty$. ∎

We are now ready to prove the Fourier representation theorem (Theorem 1, Section 7.2). By assumption, f and f' are piecewise continuous. Thus, f' may have at most a finite number of discontinuities in $[-\pi,\pi]$, otherwise it exists and is continuous. We will prove that $s_N(x)$ converges to $f(x)$ at all points x where $f'(x)$ exists. At the points where f' does not exist, we can add a correction term, as we did in the proof of Lemma 3, and reduce to the case of points where f' does exist. The details are left to the exercises.

Using the fact that $\int_{-\pi}^{\pi}\cos nt\,dt = 0$ for all $n \geq 1$, and $\frac{1}{2\pi}\int_{-\pi}^{\pi}dt = 1$, it follows from (1) that

$$(4) \qquad \frac{1}{2\pi}\int_{-\pi}^{\pi} D_N(t)\,dt = 1, \quad \text{for all } N \geq 1.$$

Using this and Lemma 1, we have

$$|s_N(x) - f(x)| = \left| \frac{1}{2\pi}\int_{-\pi}^{\pi} f(x-t)D_N(t)\,dt - f(x) \right|$$

$$= \left| \frac{1}{2\pi}\int_{-\pi}^{\pi} \big(f(x-t) - f(x)\big)D_N(t)\,dt \right|$$

$$(5) \qquad = \left| \frac{1}{2\pi}\int_{-\pi}^{\pi} \frac{f(x-t) - f(x)}{\sin\frac{t}{2}} \sin[(N+\tfrac{1}{2})t]\,dt \right|.$$

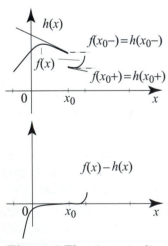

To show that this expression tends to 0 as $N \to \infty$, we use a clever trick. For fixed x, consider the function $g(t) = \frac{f(x-t)-f(x)}{\sin\frac{t}{2}}$, for $t \neq 0$ in $[-\pi,\pi]$, and $g(0) = 2f'(x)$. This function is clearly piecewise continuous for all $t \neq 0$, and

$$\lim_{t\to 0} g(t) = \lim_{t\to 0} \frac{f(x-t)-f(x)}{\sin\frac{t}{2}} = \lim_{t\to 0}\frac{f(x-t)-f(x)}{t}\lim_{t\to 0}\frac{t}{\sin\frac{t}{2}} = 2f'(x) = g(0),$$

where we have used the fact that $f'(x)$ exists and $\lim_{t\to 0}\frac{t}{\sin\frac{t}{2}} = 2$. So the function $g(t)$ is continuous at $t = 0$, and hence it is piecewise continuous on the entire interval $[-\pi,\pi]$. Applying (3) from the Riemann-Lebesgue lemma, we see that $\lim_{N\to\infty}\int_{-\pi}^{\pi} g(t)\sin[(N+\tfrac{1}{2})t]\,dt = 0$, and it follows from (5) that $\lim_{N\to\infty}|s_N(x) - f(x)| = 0$, completing the proof.

Exercises 7.6

Figure 3 The piecewise linear function $h(x)$ has slope equal to $f'(x_0-)$ to the left of x_0 and $f'(x_0+)$ to the right of x_0. The discontinuity of its derivative and its own discontinuity at x_0 are built to cancel those of f and f' in order to make f and f' continuous (and $= 0$) at x_0.

1.   The correction term. Suppose that f and f' are piecewise continuous in $[-\pi,\pi]$ and that f' does not exist at some point x_0 in $[-\pi,\pi]$. Assume without loss of generality that x_0 is in $(-\pi,\pi)$; otherwise work on a different interval of length 2π. Show that there is a piecewise linear function $h(x)$ such that $g(x) = f(x) - h(x)$ is piecewise continuous in $(-\pi,\pi)$ and $g'(x_0)$ exists. (Note: The function $g(x)$ may not be continuous on all of $[-\pi,\pi]$, as was the case in Lemma 2.) [Hint: Using the fact that f and f' are piecewise continuous, define the values of h around x_0 by $h(x_0-) = f(x_0-)$, $h(x_0+) = f(x_0+)$, and the slopes of lines by $h'(x_0-) = f'(x_0-)$, $h'(x_0+) = f'(x_0+)$. See Figure 3.]

(b) Obtain the equation of h:

$$h(x) = \begin{cases} f'(x_0-)(x - x_0) + f(x_0-) & \text{if } -\pi < x < x_0, \\ f'(x_0+)(x - x_0) + f(x_0+) & \text{if } x_0 < x < \pi. \end{cases}$$

2. Fourier series of the correction term. We have already established that the Fourier series of a piecewise smooth function converges to the function at points where the derivative exists. This shows that the Fourier series of $h(x)$ in Exercise 1 converges to $h(x)$, except at $x = x_0$ and $x = \pm\pi$. In this exercise, we will evaluate the Fourier series at x_0 and show that it converges to $\frac{h(x_0+)+h(x_0-)}{2} = \frac{f(x_0+)+f(x_0-)}{2}$.

(a) Replacing x by $x - x_0$, we may assume from here on that $x_0 = 0$. Let $H(x) = h(x) - \frac{h(0+)+h(0-)}{2}$. Note that the Fourier series of H is the same as the Fourier series of h minus the constant $\frac{h(0+)+h(0-)}{2}$. Show that

$$H(x) = \begin{cases} f'(0-)x + \frac{h(0-)-h(0+)}{2} & \text{if } -\pi < x < 0, \\ f'(0+)x + \frac{h(0+)-h(0-)}{2} & \text{if } 0 < x < \pi. \end{cases}$$

(b) Derive the following Fourier coefficients of H:

$$a_0 = \frac{\pi}{4}(f'(0+) - f'(0-)); \qquad a_n = \frac{((-1)^n - 1)(f'(0+) - f'(0-))}{\pi n^2}, \quad n \geq 1.$$

(We will not need the b_n's in the proofs.) (c) Using residues (or other methods of your choice)–more specifically, the results of Exercises 17 and 18, Section 5.6, show that $\sum_{n=1}^{\infty} \frac{(-1)^n - 1}{\pi n^2} = -\frac{\pi}{4}$.

(d) Evaluate the Fourier series of H at 0 and use (c) to show that it converges to 0. Conclude that the Fourier series of h converges to $\frac{h(0+)+h(0-)}{2}$ at $x_0 = 0$.

3. Fourier series at points of discontinuity. Let f be as in Exercise 1 and suppose that f is not continuous at x_0. Add a correction term $h(x)$ as in Exercise 1 so that $g(x) = f(x) - h(x)$ becomes continuous and differentiable at x_0. By construction, we have $g(x_0) = 0$. Let $s_N(g, x)$ denote the partial sums of the Fourier series of g, and define similarly the partial sums of the Fourier series of f and h. By the linearity of Fourier series, we have $s_N(g, x) = s_N(f, x) - s_N(h, x)$. Since $g'(x_0)$ exists, we have $s_N(g, x_0) \to g(x_0) = 0$ (this is the case of the Fourier series representation theorem that we proved in this section). By Exercise 2, we have $s_N(h, x_0) \to \frac{h(x_0+)+h(x_0-)}{2} = \frac{f(x_0+)+f(x_0-)}{2}$, where the second equality follows from the way we defined h. Thus $s_N(f, x) \to \frac{f(x_0+)+f(x_0-)}{2}$, which completes the proof.

8

PARTIAL DIFFERENTIAL EQUATIONS IN RECTANGULAR COORDINATES

... partial differential equations are the basis of all physical theorems. In the theory of sound in gases, liquids and solids, in the investigations of elasticity, in optics, everywhere partial differential equations formulate basic laws of nature which can be checked against experiments.

-Bernhard Riemann

Topics to Review
The main topic of this chapter is the separation of variables method. The application of this method requires solving ordinary differential equations, and expanding functions in Fourier series and other orthogonal expansions. For background material on ordinary differential equations, refer to Appendix A, as needed. Knowledge of Fourier series as presented in Sections 7.1–7.4 is essential to the chapter. Basic topics from complex analysis are used throughout the chapter. These can be reviewed as they occur in the presentation.

Looking Ahead
The topics of this chapter are central to the text. With the applications, the abstract theories that we have presented thus far will come to life. You will see how the notion of expanding a function in a Fourier series (Sections 8.2 and 8.3) or a generalized Fourier series (Section 8.4) becomes a very natural thing to do when solving a partial differential equation with the method of separation of variables. The ideas of this chapter will be further extended and generalized in later chapters. Sections 8.7–8.9 can be omitted without affecting the continuity of a basic course on partial differential equations.

Our study of boundary value problems started in Section 2.5, where we solved Laplace's equation subject to various boundary conditions. The topics of Section 2.5 motivated most of our work on Dirichlet problems in the subsequent chapters, culminating with our advanced applications in Chapter 6.

In this chapter we will revisit Laplace's equation and solve other important partial differential equations that arise in the modeling of physical problems involving functions of more than one variable. The approach on which most of this chapter is based is called the method of separation of variables and is presented in Section 8.2. This powerful method enables us to reduce new problems in several variables to familiar ordinary differential equations. Its success stems from the fact that a function can be expanded in a series in terms of certain special functions. It is at this point that our knowledge of Fourier series will be used. In Section 8.7, we present the method of eigenfunction expansions and use it to solve a variety of problems that cannot be solved directly by the method of separation of variables.

Tools from complex analysis will be used to enhance our ability to solve problems and to derive further properties of the solutions. In particular, qualitative properties of solutions, such as maximum modulus principles and uniqueness, will be derived using tools from complex analysis.

8.1   Partial Differential Equations in Physics and Engineering

In this section we survey the partial differential equations that will be studied in the remainder of the book and set some notation and terminology. Since the wave equation will be used as a basic example in our study, we start by motivating this equation and explaining how it relates to the vibrations of a string. (For a derivation of this equation from basic principles from physics, see [2].)

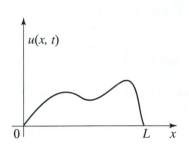

Figure 1 Displacement of the string at time t.

Imagine a string stretched along the x-axis between $x = 0$ and $x = L$, and free to vibrate in a fixed plane. We want to describe the motion of each point of the string as time progresses. For that purpose we use the function $u(x, t)$ to denote the displacement of the string at time t at the point x (Figure 1). Based on this representation, the velocity of the string at position x is $\frac{\partial u}{\partial t}$ and its acceleration there is $\frac{\partial^2 u}{\partial t^2}$. The equation that governs the motion of the stretched string is known as the **one dimensional wave equation** and is given by

$$(1) \qquad \boxed{\frac{\partial^2 u}{\partial t^2} = c^2 \frac{\partial^2 u}{\partial x^2}.}$$

The constant c depends on the physical parameters of the stretched string; $c = \sqrt{\tau/\rho}$, where τ is the tension of the string and ρ its linear mass density. The key to understanding this equation from a physical point of view is to interpret it in light of Newton's second law. The left side represents the acceleration of a small portion of the string centered at a point x, while the right side is telling us that this small portion feels a force whose sign depends on the concavity of the string at that point.

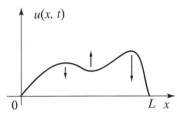

Figure 2 How a string released from rest will start to vibrate.

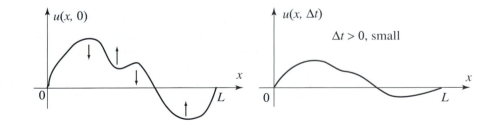

Figure 3 Initial displacement of string released from rest, and snapshot an instant later.

Thus, if we view Figure 2 as giving us the initial displacement of a string that is released from rest, then immediately following release those portions of the string where it is concave up will start to move up, and those where it is concave down will start to move down (see arrows in Figure 2 and Figure 3). This point of view can be applied to any snapshot of the string's position, except that in general each portion of the string will then have a nonzero velocity associated with it and this will also have to be taken into account.

The conditions that the displacement $u(x, t)$ should vanish at the two ends of the string must be supplied in addition to the wave equation itself. These are the **boundary conditions**

$$(2) \qquad u(0,t) = 0 \quad \text{and} \quad u(L,t) = 0, \quad \text{for } t \geq 0.$$

It is also useful to introduce **initial conditions**. For the wave equation, which is second order in time, these constitute the specification of an initial displacement and an initial velocity for some initial time t_0, henceforth taken as $t_0 = 0$:

$$(3) \qquad u(x,0) = f(x), \quad \text{for } 0 \leq x \leq L,$$

and

$$(4) \qquad \frac{\partial u}{\partial t}(x,0) = g(x), \quad \text{for } 0 \leq x \leq L.$$

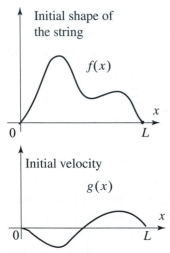

Initial shape of the string

$f(x)$

Initial velocity

$g(x)$

Figure 4 Arbitrary initial displacement and velocity.

The functions $f(x)$ and $g(x)$ represent arbitrary initial displacements and velocities along the string. You should imagine the string starting at time $t = 0$ with initial shape given by $f(x)$ and initial velocity given by $g(x)$ (Figure 4). The problem consisting of the wave equation (1) along with the boundary conditions (2) and the initial conditions (3) and (4) is called a **boundary value problem** associated with the wave equation. Its solution $u(x, t)$ determines the vibrations of the string.

In what follows, we list other classical equations in applied mathematics that arise from modeling heat and vibration phenomena. These include the familiar Laplace and Poisson equations that were studied in previous chapters.

EXAMPLE 1   Classical partial differential equations

Let u denote a function of two or more variables t (time), x and y (spatial coordinates). The following are examples of partial differential equations that you will study in this book.

(a) $\dfrac{\partial^2 u}{\partial t^2} = c^2 \dfrac{\partial^2 u}{\partial x^2}$ (one dimensional wave equation)

(b) $\dfrac{\partial^2 u}{\partial t^2} = c^2 \left(\dfrac{\partial^2 u}{\partial x^2} + \dfrac{\partial^2 u}{\partial y^2} \right)$ (two dimensional wave equation)

(c) $\dfrac{\partial u}{\partial t} = c^2 \dfrac{\partial^2 u}{\partial x^2}$ (one dimensional heat equation)

(d) $\dfrac{\partial u}{\partial t} = c^2 \left(\dfrac{\partial^2 u}{\partial x^2} + \dfrac{\partial^2 u}{\partial y^2} \right)$ (two dimensional heat equation)

(e) $\dfrac{\partial^2 u}{\partial x^2} + \dfrac{\partial^2 u}{\partial y^2} = 0$ (two dimensional Laplace equation)

(f) $\dfrac{\partial^2 u}{\partial x^2} + \dfrac{\partial^2 u}{\partial y^2} = f(x, y)$ $\qquad$ (two dimensional Poisson equation)

(g) $\dfrac{\partial^2 u}{\partial t^2} - c^2 \dfrac{\partial^2 u}{\partial x^2} + 2B \dfrac{\partial u}{\partial t} + Au = 0$ $\qquad$ (telegraph equation)

(h) $i\hbar \dfrac{\partial \psi}{\partial t} = -\dfrac{\hbar^2}{2m} \dfrac{\partial^2 \psi}{\partial x^2} + V(x)\psi$ $\qquad$ (one dimensional Schrödinger equation) ∎

The **order** of the partial differential equation is the highest order of derivative that appears.   The equation is called **linear** if the unknown function and the partial derivatives are of the first degree and at most one of these appears in any given term; otherwise, the equation is called **nonlinear**. The differential equations of Example 1 are all linear of the second order. But the second order equation

$$\frac{\partial^2 u}{\partial t \partial x} + u \frac{\partial u}{\partial x} = e^x$$

is nonlinear because the term $u \frac{\partial u}{\partial x}$ has both u and $\frac{\partial u}{\partial x}$ in it.

To simplify notation, we will use subscripts to denote partial derivatives. With this notation, we can write the **general linear partial differential equation of order two in two variables** as

(5) $\qquad$ $\boxed{\; Au_{xx} + 2Bu_{xy} + Cu_{yy} + Du_x + Eu_y + Fu = G \;}$

where A, B, C, D, E, F, G are functions of x and y.   The equation is called **homogeneous** if $G = 0$.   We shall always assume that u is a nice enough function that its mixed partial derivatives are equal (for example, $u_{xy} = u_{yx}$, $u_{xxy} = u_{xyx} = u_{yxx}$, etc.) for partial derivatives of order less than or equal to the order of the equation in which u appears.   Thus we need not write an explicit term in u_{yx} in (5).   In general, a linear partial differential equation in which each nonzero term contains either the unknown function or one of its partial derivatives is called **homogeneous**; otherwise, the linear equation is called **nonhomogeneous**. Except for the Poisson equation, all the differential equations of Example 1 are homogeneous.

As we saw with the vibrating string, the modeling of a physical problem is often done with a differential equation and a set of other equations that describe the behavior of the solution on the boundary of the region under consideration.   These equations are called **boundary conditions**. Other equations may enter into the description of a physical phenomenon such as the initial values of the solution at time $t = 0$ or **initial conditions**. The partial differential equation along with the boundary and initial conditions is called an **initial boundary value problem**, or simply a **boundary value problem**.

The concepts of linearity and homogeneity extend also to the initial conditions. For example, the initial condition $u(x,0) + u_x(x,0) = 0$ is a linear homogeneous condition, while the condition $u(x,0) + u_x(x,0) = 1$ is linear but not homogeneous.

As in the case of ordinary differential equations, we have the following useful superposition principle whose verification is left as an exercise. (Compare with Theorem 4, Appendix A.1.)

THEOREM 1
SUPERPOSITION
PRINCIPLE

> If u_1 and u_2 are solutions of a linear homogeneous partial differential equation, then any linear combination $u = c_1 u_1 + c_2 u_2$, where c_1 and c_2 are constants, is also a solution. If in addition u_1 and u_2 satisfy a linear homogeneous boundary condition, then so will $u = c_1 u_1 + c_2 u_2$.

We used superposition to breakdown boundary value problems into simpler subproblems in previous sections (see Section 6.6). The fact that a linear combination of harmonic functions is again harmonic is also a fact about superposition of solution of Laplace's equation.

EXAMPLE 2   Superposition of solutions

The wave equation

$$\frac{\partial^2 u}{\partial t^2} = \frac{\partial^2 u}{\partial x^2}$$

is linear of the second order and homogeneous. (It is of the form (5), with the variable t in place of the variable y, and $A = 1$, $C = -1$ and $B = D = E = F = G = 0$.) By plugging into the equation, you can check that the functions

$$u = \sin x \cos t, \; u = \sin 2x \cos 2t, \; u = \sin \frac{x}{2} \cos \frac{t}{2}, \; u = \cos 2x \sin 2t$$

are all solutions. In fact, you should be able to list many more solutions of the wave equation. The superposition principle will be particularly useful for generating further solutions. For example, we can get a solution by taking any linear combination from the preceding list of solutions. A specific illustration would be the function $u(x, t) = -2\sin x \cos t + 4\sin 2x \cos 2t$.

By adding boundary and initial conditions to the wave equation, we narrow down the number of solutions. For example, if we are interested in solutions that satisfy the boundary conditions $u(0, t) = 0$ and $u(\pi, t) = 0$, we see that the only functions from our list that satisfy these two conditions are the first two functions. The effect of the boundary conditions will play a major role in the next section, as we present the solution of the general wave boundary value problem. ∎

Before superposing solutions, you should always check that the equation is linear and homogeneous. As our next example shows, the superposition principle fails badly if the equation is not linear.

EXAMPLE 3   Failure of the superposition principle

The first order equation

$$(6) \qquad\qquad u_t + u u_x = 0$$

is nonlinear because of the term $u u_x$. It can be solved using the method of characteristic curves (see the exercises). You can check by plugging into (6) that

$$u(x,t) = \frac{x}{t+1}$$

is a solution. However, you can see immediately that a scalar multiple of this solution, say

$$\frac{cx}{t+1},$$

is not a solution of (6), unless $c = 1$ or $c = 0$. Thus Theorem 1 fails if the equation is not linear. ∎

In the next section we will introduce the method of separation of variables for solving boundary value problems. The application of this powerful method will lead naturally to Fourier series, and other expansions by special functions.

Exercises 8.1

In Exercises 1–6, decide whether the given partial differential equation and boundary conditions are linear or nonlinear, and, if linear, whether they are homogeneous or nonhomogeneous. Determine the order of the partial differential equation.

1. $u_{xx} + u_{xy} = 2u$, $u_x(0,y) = 0$.   **2.** $u_{xx} + x u_{xy} = 2$, $u(x,0) = 0$, $u(x,1) = 0$.

3. $u_{xx} - u_t = f(x,t)$, $u_t(x,0) = 2$.   **4.** $u_{xx} = u_t$, $u(x,0) = 1$, $u(x,1) = 0$.

5. $u_t u_x + u_{xt} = 2u$, $u(0,t) + u_x(0,t) = 0$.

6. $u_{xx} + e^t u_{tt} = u \cos x$, $u(x,0) + u(x,1) = 0$.

7. Verify that the given functions are solutions of the two dimensional Laplace equation.

(a) $u = x + y$.

(b) $u = x^2 - y^2$.

(c) $u = \frac{x}{x^2+y^2}$.

(d) $u = \frac{y}{x^2+y^2}$.

(e) $u = \ln(x^2 + y^2)$.

(f) $u = e^y \cos x$.

(g) $u = \ln(x^2 + y^2) + e^y \cos x$.

(h) $u = e^y \cos x + x + y$.

8. Verify that the function

$$u = \frac{1}{\sqrt{x^2 + y^2 + z^2}}$$

is a solution of the three dimensional Laplace equation $u_{xx} + u_{yy} + u_{zz} = 0$.

9. Consider the general homogeneous linear second order partial differential equation

$$A u_{xx} + 2B u_{xy} + C u_{yy} + D u_x + E u_y + F u = 0.$$

(a) Show that the function $u(x,y) = e^{ax} e^{by}$ is a solution if and only if a and b satisfy the equation

$$A a^2 + 2 B a b + C b^2 + D a + E b + F = 0.$$

(b) Find at least four solutions of the equation

$$u_{xx} + 2u_{xy} + u_{yy} + 2\,u_x + 2\,u_y + u = 0.$$

10. The telegraph equation

$$u_{tt} + 2Bu_t - c^2 u_{xx} + Au = 0$$

governs the flow of electricity along a cable. Use Exercise 9 to find two solutions assuming $c > 0$ and $A > 0$.

11. Classification of second-order linear equations. A second order linear partial differential equation in two variables, written in the form (1), is called **elliptic** if $AC - B^2 > 0$; **hyperbolic** if $AC - B^2 < 0$; **parabolic** if $AC - B^2 = 0$. Classify equations (a), (c), (e), (f), and (g) of Example 1 according to these definitions (if the equation involves t, treat t as y in equation (1)).

Project Problem: Nonlinear equations of the first order. These equations are of great importance in modeling of traffic flow, shock waves, waves breaking the sound barrier and many other branches of applied mathematics. In Exercises 12–14 you will study the nonlinear equation

(7) $u_t + A(u)u_x = 0,$ with initial condition $u(x,0) = \phi(x),$

where $A(u)$ is a function of u. When $A(u) = u$, (7) reduces to the equation of Example 3.

12. Rewrite the following outlined solution of (7) by providing as many details as possible.
Step 1: Using (7), argue that $u(x(t),\, t)$ is constant on the curves $x = x(t)$, with direction field

(8) $\dfrac{dx}{dt} = A(u(x(t), t)).$

These are the **characteristic curves** of (7). [Hint: Let u be a solution of (7). Use the chain rule in two dimensions to show that $\frac{d}{dt}u(x(t),\, t) = 0$.]
Step 2: Let $x(t)$ denote a characteristic curve (a solution of (8)). Since u is constant along $(x(t), t)$, it follows that $u(x(t), t) = u(x(0), 0) = \phi(x(0))$, by (7). Hence $A(u(x(t), t))$ is constant and equals $A(\phi(x(0)))$. We conclude from (8) that *the characteristic curves are straight lines with slopes $A(\phi(x(0)))$.* We write these lines in the form

$$x = tA(\phi(x(0))) + x(0).$$

Step 3: We complete the solution of (7) by solving for $x(0)$ in the preceding equation to get an implicit relation for the characteristic lines of the form $L(x,t) = x(0)$. The final solution of (7) will be of the form $u(x,t) = f(L(x,t))$, where f is a function chosen so as to satisfy the initial condition in (7). That is, $f(L(x,0)) = \phi(x)$.

As an illustration of this method, let us solve (6) with the initial condition $u(x,0) = x$. Here $A(u) = u$, and $\phi(x) = x$. From Step 2, the characteristic lines are of the form $x = tx(0) + x(0)$. Solving for $x(0)$, we get $x(0) = \frac{x}{t+1}$, which yields the implicit relation $L(x,t) = x(0)$. Thus the solution is

$$u(x,t) = f\left(\frac{x}{t+1}\right).$$

Setting $t = 0$ and using the initial condition, we obtain $f(x) = x$, which yields the solution $u(x,t) = \frac{x}{t+1}$. You can check the validity of this solution by plugging it back into the equation.

In Exercises 13–16, solve the given nonlinear problem, following the method of Exercise 12.

13. $u_t + \ln(u)u_x = 0$, $u(x,0) = e^x$. **14.** $u_t + (u+1)u_x = 0$, $u(x,0) = x^2$.
[Hint: In determining $L(x,t)$ in Step 2, you will have two possible choices for a solution. Take the one that satisfies $\lim_{t\to 0} u(x,t) = u(x,0) = x^2$.]

15. $u_t + (2+u)u_x = 0$, $u(x,0) = x^2$. [Hint: See the hint for Exercise 14.]

16. $u_t + u^2 u_x = 0$, $u(x,0) = x$. **17.** $u_t + u^2 u_x = 0$, $u(x,0) = \sqrt{x}$, $x > 0$.

8.2 Solution of the One Dimensional Wave Equation: The Method of Separation of Variables

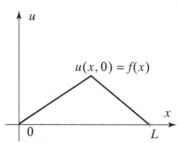

Figure 1 Initial shape of a stretched string, $u(x, 0)$.

In this section we give the full solution of the boundary value problem associated with the one dimensional wave equation with arbitrary initial position (displacement) and velocity using the method of separation of variables. That is, we solve the boundary value problem for the wave equation that describes the vibrations of a string with fixed ends. The string is assumed to be stretched on the x-axis with ends fastened at $x = 0$ and $x = L$ (Figure 1). Let $u(x,t)$ denote the position at time t of the point x on the string. We saw in the previous section that u satisfies the **one dimensional wave equation**

(1)
$$\frac{\partial^2 u}{\partial t^2} = c^2 \frac{\partial^2 u}{\partial x^2}, \qquad 0 < x < L, t > 0.$$

To find u, we will solve this equation subject to the **boundary conditions**

(2)
$$u(0,t) = 0 \qquad \text{and} \qquad u(L,t) = 0 \text{ for all } t > 0,$$

and the **initial conditions**

(3)
$$u(x,0) = f(x) \qquad \text{and} \qquad \frac{\partial u}{\partial t}(x,0) = g(x) \text{ for } 0 < x < L.$$

The boundary conditions state that the ends of the string are held fixed for all time, while the initial conditions give the initial shape of the string $f(x)$, and its initial velocity $g(x)$.

The solution of this problem is based on a general method called the **method of separation of variables**. This powerful method will be used in the solution of many partial differential equations throughout the book. To highlight the principal ideas behind the separation of variables method, we break the solution up into three basic steps.

Step 1: Separating Variables in (1) and (2)

We start by seeking nonzero **product solutions** of (1) of the form

$$(4) \qquad\qquad u(x,t) = X(x)T(t),$$

where $X(x)$ is a function of x alone and $T(t)$ is a function of t alone. The problem is now reduced to finding X and T. Differentiating (4) with respect to t and x we get

$$\frac{\partial^2 u}{\partial t^2} = XT'' \qquad \text{and} \qquad \frac{\partial^2 u}{\partial x^2} = X''T.$$

Plugging these into (1), we obtain

$$XT'' = c^2 X''T,$$

and now, dividing by $c^2 XT$, we get

$$(5) \qquad\qquad \frac{T''}{c^2 T} = \frac{X''}{X}.$$

(We will not worry about XT being 0, and we continue formally with the solution.) In equation (5) the variables are **separated** in the sense that the left side of the equation is a function of t alone, and the right side is a function of x alone. Since the variables t and x are independent of each other, the only way to get equality is to have the functions on both sides of (5) constant and equal. Thus

$$\frac{T''}{c^2 T} = k \quad \text{and} \quad \frac{X''}{X} = k,$$

where k is an arbitrary constant called the **separation constant**. We rewrite the separated equations as two *ordinary* differential equations

At this point we have arrived at two ordinary differential equations in place of our original partial differential equation. This is the gist of the method of separation of variables.

$$(6) \qquad\qquad X'' - kX = 0$$

and

$$(7) \qquad\qquad T'' - kc^2 T = 0.$$

Our next move is to separate the variables in the boundary conditions (2). Using (4) and the boundary conditions, we get

$$X(0)T(t) = 0 \qquad \text{and} \qquad X(L)T(t) = 0, \text{ for all } t > 0.$$

If $X(0) \neq 0$ or $X(L) \neq 0$, then $T(t)$ must be 0 for all t, and so, by (4), u is identically zero. To avoid this trivial solution, we set

$$X(0) = 0 \qquad \text{and} \qquad X(L) = 0.$$

Thus, recalling (6), we arrive at the boundary value problem in X:

$$X'' - kX = 0, \quad X(0) = 0 \text{ and } X(L) = 0.$$

As will be seen in the next step, not all values of the separation constant k yield a nontrivial solution X. Our discussion will revolve around the solutions of simple second order linear ordinary differential equations with constant coefficients. You should refer to Appendix A.2 for a thorough review of these topics.

Step 2: Solving the Separated Equations

We start by solving the equation for X because this equation comes with boundary conditions, whereas the equation for T does not. The boundary conditions allow us to narrow down the possible solutions.

If k is positive, say $k = \mu^2$ with $\mu > 0$, then the equation in X becomes

$$X'' - \mu^2 X = 0,$$

with general solution

$$X(x) = c_1 e^{\mu x} + c_2 e^{-\mu x}.$$

It is straightforward to verify that the only way to satisfy the conditions $X(0) = 0$ and $X(L) = 0$ is to take $c_1 = c_2 = 0$. Indeed, $X(0) = 0$ implies that $c_2 = -c_1$, so that $X(x) = c_1(e^{\mu x} - e^{-\mu x})$. But now $X(L) = 0$ implies that either $c_1 = 0$ or $e^{\mu L} = e^{-\mu L}$, which is impossible since $\mu L > 0$. Hence $c_1 = 0$, implying $X = 0$, and so $u = 0$, by (4). Thus, the case $k > 0$ yields trivial solutions.

Similarly for $k = 0$, the differential equation reduces to $X'' = 0$ with general solution $X(x) = c_1 x + c_2$. The only way to satisfy the boundary conditions on X is to take $c_1 = c_2 = 0$, which again leads to the trivial solution $u = 0$. The only choice left to check is

$$k = -\mu^2 < 0.$$

The corresponding boundary value problem in X is

$$X'' + \mu^2 X = 0, \quad X(0) = 0 \text{ and } X(L) = 0.$$

The general solution of the differential equation is

$$X = c_1 \cos \mu x + c_2 \sin \mu x.$$

The condition $X(0) = 0$ implies that $c_1 = 0$, and hence $X = \sin \mu x$. The condition $X(L) = 0$ implies that

$$c_2 \sin \mu L = 0.$$

We take $c_2 = 1$ for convenience. Any other nonzero value will do.

To avoid the trivial solution $X = 0$, we take $c_2 = 1$, which forces

$$\sin \mu L = 0.$$

Since the sine function vanishes at the integer multiples of π, we conclude that

$$\mu = \mu_n = \frac{n\pi}{L}, \quad n = \pm 1, \pm 2, \ldots,$$

and so

$$X = X_n = \sin \frac{n\pi}{L} x, \quad n = 1, 2, \ldots .$$

Note that for negative values of n we obtain the same solutions except for a change of sign; hence, solutions corresponding to negative n's may be discarded without loss.

We now go back to (7) and substitute $k = -\mu^2 = -\left(\frac{n\pi}{L}\right)^2$ and get

$$\boxed{T'' + \left(c\frac{n\pi}{L}\right)^2 T = 0.}$$

The general solution of this equation is

$$T_n = b_n \cos \lambda_n t + b_n^* \sin \lambda_n t,$$

where we have set

$$\boxed{\lambda_n = c\frac{n\pi}{L}, \quad n = 1, 2, \ldots .}$$

Combining the solutions for X and T as described by (4), we obtain an infinite set of product solutions of (1), all satisfying the boundary conditions (2):

$$\boxed{u_n(x, t) = \sin \frac{n\pi}{L} x \left(b_n \cos \lambda_n t + b_n^* \sin \lambda_n t\right), \quad n = 1, 2, \ldots .}$$

These are the **normal modes** of the wave equation. Their physical significance will be discussed at the end of this section.

Since all the normal modes satisfy the linear and homogeneous equation (1) and boundary conditions (2), by the superposition principle (Theorem 1, Section 8.1), any linear combination will also solve (1) and (2). However, it is not hard to see that in general, such a linear combination may not satisfy the initial conditions (3). So, motivated by the superposition principle, it is natural to try an "infinite" linear combination

$$u(x, t) = \sum_{n=1}^{\infty} \sin \frac{n\pi}{L} x \left(b_n \cos \lambda_n t + b_n^* \sin \lambda_n t\right)$$

as a solution of the boundary value problem (1)–(3).

Step 3: Fourier Series Solution of the Entire Problem

To completely solve our problem, we must determine the unknown coefficients b_n and b_n^* so that the function $u(x,t)$ satisfies the initial conditions (3). Starting with the first condition in (3), and plugging $t = 0$ into the infinite series for u, we get

$$u(x,0) = f(x) = \sum_{n=1}^{\infty} b_n \sin \frac{n\pi}{L} x, \quad 0 < x < L.$$

The series on the right is the half-range sine series expansion of f. We thus conclude that the coefficients b_n are the sine Fourier coefficients of $f(x)$ given by (4), Section 7.4:

$$b_n = \frac{2}{L} \int_0^L f(x) \sin \frac{n\pi}{L} x \, dx, \quad n = 1, 2, \dots .$$

Similarly, we determine b_n^* by using the second initial condition in (3). Differentiating the series for u term by term with respect to t, and then setting $t = 0$, we get

$$g(x) = \sum_{n=1}^{\infty} b_n^* \lambda_n \sin \frac{n\pi}{L} x.$$

Since this should be the half-range sine series expansion of g, it follows that $b_n^* \lambda_n$ are the sine Fourier coefficients of $g(x)$ given by (4), Section 7.4:

$$b_n^* \lambda_n = \frac{2}{L} \int_0^L g(x) \sin \frac{n\pi}{L} x \, dx, \quad n = 1, 2, \dots .$$

Solving for b_n^* and recalling the value of λ_n, we get

$$b_n^* = \frac{2}{cn\pi} \int_0^L g(x) \sin \frac{n\pi}{L} x \, dx, \quad n = 1, 2, \dots .$$

We have thus determined all the unknown coefficients in the series representation of the solution u. We summarize our findings in the following box.

SOLUTION OF THE ONE DIMENSIONAL WAVE EQUATION

The solution of the one dimensional wave equation

$$\frac{\partial^2 u}{\partial t^2} = c^2 \frac{\partial^2 u}{\partial x^2}, \qquad 0 < x < L, \ t > 0,$$

with boundary conditions

$$u(0, t) = 0 \quad \text{and} \quad u(L, t) = 0 \quad \text{for all} \quad t > 0,$$

and initial conditions

$$u(x, 0) = f(x) \quad \text{and} \quad \frac{\partial u}{\partial t}(x, 0) = g(x) \quad \text{for} \quad 0 < x < L,$$

is

$$(8) \qquad u(x, t) = \sum_{n=1}^{\infty} \sin \frac{n\pi}{L} x \left(b_n \cos \lambda_n t + b_n^* \sin \lambda_n t \right),$$

where

$$(9) \quad b_n = \frac{2}{L} \int_0^L f(x) \sin \frac{n\pi}{L} x \, dx, \qquad b_n^* = \frac{2}{cn\pi} \int_0^L g(x) \sin \frac{n\pi}{L} x \, dx,$$

and

$$(10) \qquad \lambda_n = c \frac{n\pi}{L}, \qquad n = 1, 2, \ldots.$$

EXAMPLE 1   Vibration of a stretched string with fixed edges

The ends of a stretched string of length $L = 1$ are fixed at $x = 0$ and $x = 1$. The string is set to vibrate from rest by releasing it from an initial triangular shape modeled by the function

$$f(x) = \begin{cases} \frac{3}{10} x & \text{if } 0 \le x \le \frac{1}{3}, \\ \\ \frac{3(1-x)}{20} & \text{if } \frac{1}{3} \le x \le 1, \end{cases}$$

whose graph is shown in Figure 2. Determine the subsequent motion of the string, given that $c = 1/\pi$.

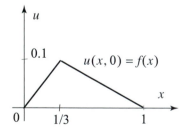

Figure 2 Initial shape the string in Example 1.

Solution Since $g(x) = 0$, we have $b_n^* = 0$. Using (9) and integrating by parts, we get

$$b_n = 2 \int_0^1 f(x) \sin n\pi x \, dx$$

$$= \frac{3}{5} \int_0^{1/3} x \sin n\pi x \, dx + \frac{3}{10} \int_{1/3}^1 (1-x) \sin n\pi x \, dx$$

$$= -\frac{\cos \frac{n\pi}{3}}{5n\pi} + \frac{3}{5} \frac{\sin \frac{n\pi}{3}}{n^2\pi^2} + \frac{\cos \frac{n\pi}{3}}{5n\pi} + \frac{3}{10} \frac{\sin \frac{n\pi}{3}}{n^2\pi^2}$$

$$= \frac{9}{10\pi^2} \frac{\sin \frac{n\pi}{3}}{n^2}.$$

From (10) we have $\lambda_n = n$. Putting all this in (8), we find the solution

$$u(x,t) = \frac{9}{10\pi^2} \sum_{n=1}^{\infty} \frac{\sin \frac{n\pi}{3}}{n^2} \sin n\pi x \cos nt.$$

In some of our examples we may see unrealistically large displacements. This is only a matter of convenience. Since our equations are linear, to obtain realistic displacements, we need only multiply our solution by a scaling factor.

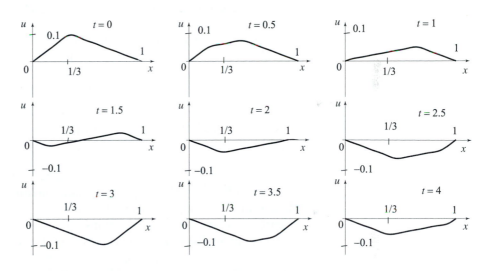

Figure 3 The instantaneous shape of the string at various times, plotted using the 10th partial sum of the series solution.

By plotting $u(x,t)$ for a fixed value of t, we get the shape of the string at that time. An approximation of the graphs is obtained by plotting partial sums of the series. As expected, when $t = 0$ we have the initial shape of the string given by the function $f(x)$ (Figure 3). ∎

As given by (8), the solution of the vibrating string problem is an infinite sum of the normal modes

(11) $\qquad u_n(x,t) = \sin \frac{n\pi}{L} x \, (b_n \cos \lambda_n t + b_n^* \sin \lambda_n t) \qquad n = 1, 2, \ldots .$

When the string vibrates according to one of the u_n's, we say that it is in its nth normal mode of vibration. The first normal mode is called the **fundamental mode**; all other modes are known as **overtones** (Figure 4). In music, the **intensity** of the sound produced by a given normal mode depends on $\sqrt{b_n^2 + (b_n^*)^2}$, the amplitude of the nth normal mode. The **circular** or **natural frequency** of the normal mode, which gives the number of oscillation in 2π units of time, is $\lambda_n = n\pi c/L$. The larger the natural frequency, the higher the pitch of the sound produced. Since $c = \sqrt{\tau/\rho}$, where τ is the tension of the string and ρ is the mass density, the pitch of the sound can be changed by varying the tension or the length of the string. For example, by clamping down the string, the length is shortened and the pitch is increased.

Figure 4 Normal modes Not counting the ends of the string, there are $(n - 1)$ equidistant points on the string that do not vibrate in the nth normal mode.

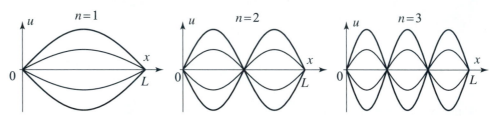

When the string vibrates in a normal mode, some points on the string are fixed at all times (Figure 4). These are the solutions of the equation $\sin \frac{n\pi}{L}x = 0$. Not counting the ends of the string among these points, there are $n - 1$ equidistant points that do not vibrate in the nth normal mode.

We talked about a string vibrating in a normal mode. Which initial conditions cause the string to vibrate this way? We illustrate the answer with the following example.

EXAMPLE 2   Normal modes of vibration
Show that if a string with initial shape $f(x) = \sin \frac{m\pi}{L}x$ for $0 < x < L$ is set to vibrate from rest, then its vibrations are given by the mth mode. (Note that the initial shape of the string is obtained from the mth normal mode (11) by setting $t = 0$.)

Solution We use (8) to find the function $u(x, t)$. Since $g(x) = 0$, we get $b_n^* = 0$. To find b_n we compute the integral

$$b_n = \frac{2}{L}\int_0^L f(x)\sin\frac{n\pi}{L}x\,dx = \frac{2}{L}\int_0^L \sin\frac{m\pi}{L}x\sin\frac{n\pi}{L}x\,dx.$$

By the orthogonality of the trigonometric system, the last expression is zero unless $m = n$, in which case its value is 1. Thus $b_n = 0$ if $n \neq m$ and $b_m = 1$. Putting these values in (8), we find the solution

$$u(x, t) = u_m(x, t) = \sin\frac{m\pi}{L}x\cos c\frac{m\pi}{L}t,$$

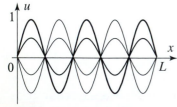

Figure 5 The 5th normal modes $u_5(x, t)$, shown for various values of t.

which is the mth normal mode of the string (see Figure 5 for an illustration when $m = 5$). Thus a string that starts to vibrate from rest with an initial shape given by a normal mode will continue to vibrate in that mode. Figure 5 illustrates this phenomenon for the 5th normal mode.   ■

We close with another example illustrating how we can use normal modes to understand more complicated motions.

EXAMPLE 3   A nonzero initial velocity
Solve for the motion of a string of length $L = \frac{\pi}{2}$ if $c = 1$ and the initial displacement and velocity are given by $f(x) = 0$ and $g(x) = x \cos x$.

Solution   Since $f(x) = 0$, all the b_n's in (8) are 0. In computing b_n^* we first note from (9) that b_n^* is equal to $\frac{1}{\lambda_n}$ times the sine Fourier coefficient of the function $g(x) = x \cos x$, $0 < x < \frac{\pi}{2}$, where $\lambda_n = 2n$. Since $x \cos x$ is odd, its Fourier sine series on $0 < x < \frac{\pi}{2}$ is identical to its Fourier series on $-\frac{\pi}{2} < x < \frac{\pi}{2}$. Appealing to the result of Example 5, Section 7.3, we obtain

$$b_n^* = \frac{1}{\lambda_n} \frac{16(-1)^{n+1}n}{\pi(4n^2 - 1)^2} = \frac{8(-1)^{n+1}}{\pi(4n^2 - 1)^2}, \quad n = 1, 2, \ldots .$$

Therefore, by (8), the solution is

$$(12) \qquad u(x, t) = \frac{8}{\pi} \sum_{n=1}^{\infty} \frac{(-1)^{n+1}}{(4n^2 - 1)^2} \sin 2nx \sin 2nt.$$

Figure 6 Snapshots of the string in Example 3 for $t = \frac{\pi}{12}, \frac{\pi}{6}, \frac{3\pi}{4}$ using the first and fourth partial sums from (12). Note that the first normal mode (shown on the left) gives a very good picture of the string as it moves.

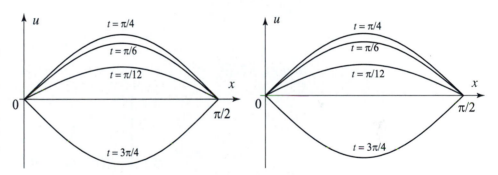

In Figure 6, we show several snapshots of the string as it begins to move under the influence of the initial velocity. Note that the first partial sum

$$\frac{8}{9\pi} \sin 2x \sin 2t$$

already gives a very good picture of how the string moves. This can be justified by observing that the coefficients in the series in (12) decrease rapidly to zero, and so the contributions of additional terms become small.   ■

For problems with nonzero initial displacement and velocity, we have only to work them as in Examples 1 and 3 and put the results together, as specified by equations (8) and (9). That is, from the initial displacement $f(x)$ we find the b_n's as in Example 1, from the initial velocity $g(x)$ we find the b_n^*'s as in Example 3, and then we put these results into (8). We note too that if we have the Fourier sine series of $f(x)$ and/or $g(x)$ at hand, our task is considerably reduced, as illustrated by Example 3.

There is somewhat a disadvantage to the Fourier series solution of the wave equation, because in general this series converges slowly. Also, it is difficult, if not impossible, to justify differentiation of the series solution term by term in order to verify the wave equation. However, using this series solution and properties of Fourier series, we will derive in the exercises a simpler form of the solution, called d'Alembert's solution.

Exercises 8.2

In Exercises 1–10, (a) solve the boundary value problem (1)–(3) for a string of unit length, subject to the given conditions.

(b)   Illustrate the motion of the string by plotting a partial sum of your series solution at various values of t.   To decide how many terms to include in your partial sum, compare the graph at $t = 0$ and the graph of $f(x)$. The graphs should match when you have enough terms in your partial sum.

1. $f(x) = .05 \sin \pi x$, $g(x) = 0$, $c = \frac{1}{\pi}$.

2. $f(x) = \sin \pi x \cos \pi x$, $g(x) = 0$, $c = \frac{1}{\pi}$.

3. $f(x) = \sin \pi x + 3 \sin 2\pi x - \sin 5\pi x$, $g(x) = 0$, $c = 1$.

4. $f(x) = \sin \pi x + \frac{1}{2} \sin 3\pi x + 3 \sin 7\pi x$, $g(x) = \sin 2\pi x$, $c = 1$.

5. $g(x) = 0$, $c = 4$,

$$f(x) = \begin{cases} 2x & \text{if } 0 \le x \le \frac{1}{2}, \\ 2(1 - x) & \text{if } \frac{1}{2} < x \le 1. \end{cases}$$

6. $g(x) = 2$, $c = \frac{1}{\pi}$,

$$f(x) = \begin{cases} 0 & \text{if } 0 \le x \le \frac{1}{3}, \\ \frac{1}{30}(x - 1/3) & \text{if } \frac{1}{3} \le x \le \frac{2}{3}, \\ \frac{1}{30}(1 - x) & \text{if } \frac{2}{3} < x \le 1. \end{cases}$$

7. $g(x) = 1$, $c = 4$,

$$f(x) = \begin{cases} 4x & \text{if } 0 \le x \le \frac{1}{4}, \\ 1 & \text{if } \frac{1}{4} < x \le \frac{3}{4}, \\ 4(1 - x) & \text{if } \frac{3}{4} < x \le 1. \end{cases}$$

8. $f(x) = x \sin \pi x$, $g(x) = 0$, $c = \frac{1}{\pi}$.

9. $f(x) = x(1 - x)$, $g(x) = \sin \pi x$, $c = 1$.

10. $g(x) = 0$, $c = 1$,

$$f(x) = \begin{cases} 4x & \text{if } 0 \le x \le \frac{1}{4}, \\ -4(x - 1/2) & \text{if } \frac{1}{4} < x \le \frac{3}{4}, \\ 4(x - 1) & \text{if } \frac{3}{4} < x \le 1. \end{cases}$$

11. Time period of motion. (a) Show that the nth normal mode (11) is periodic in time with period $2L/nc$. Conclude that for any n, a period of u_n is $2L/c$.

(b) Show that any superposition of normal modes is periodic in time with period $2L/c$. Conclude that the string vibrates with a time period $2L/c$.

(c) **Shape of the string at half a time period.** Using (8), show that for all x and t, $u(x, t + L/c) = -u(L - x, t)$. What does this imply about the shape of the string at half a time period?

Project Problem: Solve a case of the wave equation with damping in Exercise 12 and then apply your solution to a specific problem by doing any one of Exercises 13–15.

12. Damped vibrations of a string. In the presence of resistance proportional to velocity, the one dimensional wave equation becomes

$$\frac{\partial^2 u}{\partial t^2} + 2k \frac{\partial u}{\partial t} = c^2 \frac{\partial^2 u}{\partial x^2}.$$

We will solve this equation subject to conditions (2) and (3) by following the method of this section.

(a)   Assume a product solution of the form $u(x, t) = X(x)T(t)$, and derive the following equations for X and T:

$$X'' + \mu^2 X = 0, \quad X(0) = 0, \ X(L) = 0,$$

$$T'' + 2kT' + (\mu c)^2 T = 0,$$

where μ is the separation constant.

(b) Show that

$$\mu = \mu_n = \frac{n\pi}{L} \quad \text{and} \quad X = X_n = \sin \frac{n\pi}{L} x, \ n = 1, 2, \ldots .$$

(c) To determine the solutions in T we have to solve $T'' + 2k\, T' + (\frac{n\pi}{L} c)^2 T = 0$. Review the general solution of the second order linear differential equation with constant coefficients (Appendix A.2), and explain why three possible cases are to be treated separately: $n < \frac{kL}{\pi c}$, $n = \frac{kL}{\pi c}$, and $n > \frac{kL}{\pi c}$.   The respective solutions for T are

$$T_n = e^{-kt}(a_n \cosh \lambda_n t + b_n \sinh \lambda_n t),$$

$$T_{\frac{kL}{\pi c}} = a_{\frac{kL}{\pi c}} e^{-kt} + b_{\frac{kL}{\pi c}} t e^{-kt},$$

$$T_n = e^{-kt}(a_n \cos \lambda_n t + b_n \sin \lambda_n t),$$

where

$$\lambda_n = \sqrt{\left| k^2 - (\frac{n\pi}{L} c)^2 \right|}.$$

(d) Conclude that when $\frac{kL}{\pi c}$ is not a positive integer, the solution is

$$u(x, t) \ = \ e^{-kt} \sum_{1 \le n < \frac{kL}{\pi c}} \sin \frac{n\pi}{L} x (a_n \cosh \lambda_n t + b_n \sinh \lambda_n t)$$

$$+ e^{-kt} \sum_{\frac{kL}{\pi c} < n < \infty} \sin \frac{n\pi}{L} x (a_n \cos \lambda_n t + b_n \sin \lambda_n t),$$

where these sums run over integers only, and where

$$a_n = \frac{2}{L} \int_0^L f(x) \sin \frac{n\pi}{L} x\, dx, \quad n = 1, 2, \ldots,$$

and the b_n are determined from the equation

$$-ka_n + \lambda_n b_n = \frac{2}{L} \int_0^L g(x) \sin \frac{n\pi}{L} x\, dx, \quad n = 1, 2, \ldots.$$

(e) Conclude that when $\frac{kL}{\pi c}$ is a positive integer, the solution is as in (d) with the one additional term

$$\sin(\frac{k}{c} x)(a_{\frac{kL}{\pi c}} e^{-kt} + b_{\frac{kL}{\pi c}} te^{-kt})$$

with a_n and b_n as in (d), except that $b_{\frac{kL}{\pi c}}$ is determined from the equation

$$-ka_{\frac{kL}{\pi c}} + b_{\frac{kL}{\pi c}} = \frac{2}{L} \int_0^L g(x) \sin \frac{k}{c} x\, dx.$$

13. Solve

$$\frac{\partial^2 u}{\partial t^2} + \frac{\partial u}{\partial t} = \frac{\partial^2 u}{\partial x^2},$$

$$u(0, t) = u(\pi, t) = 0,$$

$$u(x, 0) = \sin x, \quad \frac{\partial u}{\partial t}(x, 0) = 0.$$

[Hint: Since $k = .5$ and $L = \pi$, we have $n > \frac{kL}{\pi}$ for all n. So only one case from the solution of Exercise 12 needs to be considered.]

14. Solve

$$\frac{\partial^2 u}{\partial t^2} + \frac{\partial u}{\partial t} = \frac{\partial^2 u}{\partial x^2},$$

$$u(0, t) = u(\pi, t) = 0,$$

$$u(x, 0) = x \sin x, \quad \frac{\partial u}{\partial t}(x, 0) = 0.$$

15. (a) Solve

$$\frac{\partial^2 u}{\partial t^2} + 3 \frac{\partial u}{\partial t} = \frac{\partial^2 u}{\partial x^2},$$

$$u(0, t) = u(\pi, t) = 0, \quad u(x, 0) = 0, \quad \frac{\partial u}{\partial t}(x, 0) = 10.$$

(b) Illustrate graphically the fact that the solution tends to zero as t tends to infinity.

D'Alembert's Method

In what follows, we derive the solution of the wave boundary value problem (1)–(3) in the form

(13)
$$u(x, t) = \frac{1}{2} [f^*(x - ct) + f^*(x + ct)] + \frac{1}{2c} \int_{x-ct}^{x+ct} g^*(s)\, ds,$$

where f^* and g^* are the odd, $2L$-periodic extensions of f and g. This form of the solution is known as **D'Alembert's solution**, named after the French mathematician Jean le Rond D'Alembert (1717–1783). When the initial velocity is zero, d'Alembert's solution takes on the simpler form

$$(14) \qquad u(x,t) = \frac{1}{2}[f^*(x-ct) + f^*(x+ct)].$$

This has an interesting geometric interpretation. For fixed t, the graph of $f^*(x-ct)$ (as a function of x) is obtained by translating the graph of $f^*(x)$ by ct units to the right. As t increases, the graph represents a wave traveling to the right with velocity c. Similarly, the graph of $f^*(x+ct)$ is a wave traveling to the left with velocity c. It follows from (14) that this solution of the wave equation is an average of two waves traveling in opposite directions with shapes determined from the initial shape of the string.

16. D'Alembert's solution for zero initial velocity. (a) Starting from (8), show that if the initial velocity $g(x) = 0$, then

$$u(x,t) = \frac{1}{2}\sum_{n=1}^{\infty} b_n\left[\sin\frac{n\pi}{L}(x-ct) + \sin\frac{n\pi}{L}(x+ct)\right].$$

(b) Derive d'Alembert's solution (13) using (a). [Hint: $f^*(s) = \sum_{n=1}^{\infty} b_n \sin\frac{n\pi}{L}s.$]

17. Project Problem: D'Alembert's solution (the general case). Let g^* denote the odd $2L$-periodic extension of g, and let

$$G(x) = \int_0^x g^*(s)\, ds.$$

(a) Show that G is even and $2L$-periodic, and

$$G(x) = \sum_{n=1}^{\infty} B_n\left(\cos\frac{n\pi}{L}x - 1\right),$$

where

$$B_n = \frac{-2}{n\pi}\int_0^L g(x)\sin\frac{n\pi}{L}x\, dx = -cb_n^* \quad (n = 1, 2, \ldots).$$

[Hint: Exercise 33, Section 7.3.]
(b) Use (a) to show that

$$G(x+ct) - G(x-ct) = \sum_{n=1}^{\infty} B_n\left[\cos\frac{n\pi}{L}(x+ct) - \cos\frac{n\pi}{L}(x-ct)\right].$$

(c) Using (b) show that $\frac{1}{2c}\int_{x-ct}^{x+ct} g^*(s)\, ds = \sum_{n=1}^{\infty} b_n^* \sin\frac{n\pi}{L}x \sin\lambda_n t.$
(d) Use (c), (8), and Exercise 16 to derive d'Alembert's solution (13).

8.3   The One Dimensional Heat Equation

In this and the following section we study the temperature distribution in a uniform bar of length L with insulated lateral surface and no internal sources of heat, subject to certain boundary and initial conditions.   To describe the problem, let $u(x,t)$ $(0 < x < L,\ t > 0)$ represent the temperature of the point x of the bar at time t (Figure 1).   Given that the initial temperature distribution of the bar is $u(x,0) = f(x)$, and given that the ends of the bar are held at constant temperature 0, we ask: What is $u(x,t)$ for $0 < x < L,\ t > 0$?   As you would expect, to answer this question, we must solve a boundary value problem. It can be shown using laws from physics that u satisfies the **one dimensional heat equation**

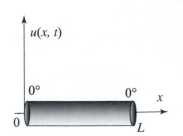

$u(x, t)$

$0°$ $0°$ x

0 L

Figure 1 Insulated bar with ends kept at 0°.

$$\frac{\partial u}{\partial t} = c^2 \frac{\partial^2 u}{\partial x^2}, \quad 0 < x < L, \quad t > 0.$$

In addition, u satisfies the **boundary conditions**

$$u(0,t) = 0 \quad \text{and} \quad u(L,t) = 0 \quad \text{for all } t > 0,$$

Since the problem is first order in t, we only need one initial condition, unlike the wave problem where two conditions were needed.

and the **initial condition**

$$u(x,0) = f(x) \quad \text{for } 0 < x < L.$$

We solve this problem using the method of separation of variables.   Interesting and important variations on these problems are presented in the following section.

Separation of Variables

We start by looking for product solutions of the form

$$u(x,t) = X(x)T(t),$$

where $X(x)$ is a function of x alone and $T(t)$ is a function of t alone. Plugging into the heat equation and separating variables, we obtain

$$\frac{T'}{c^2 T} = \frac{X''}{X}.$$

For the equality to hold, we must have

$$\frac{T'}{c^2 T} = k \quad \text{and} \quad \frac{X''}{X} = k,$$

where k is the **separation constant**. From these equations, we get two *ordinary differential equations*

$$X'' - kX = 0 \quad \text{and} \quad T' - kc^2 T = 0.$$

Separating variables in the boundary conditions, we get

$$X(0)T(t) = 0 \quad \text{and} \quad X(L)T(t) = 0 \text{ for all } t > 0.$$

To avoid trivial solutions, we require

$$X(0) = 0 \quad \text{and} \quad X(L) = 0.$$

We thus obtain the boundary value problem in X:

$$X'' - kX = 0, \quad X(0) = 0 \quad \text{and} \quad X(L) = 0.$$

This problem is exactly the one that we solved in Section 8.2 for the vibrating string. We found that

$$k = -\mu^2, \quad \text{where } \mu = \mu_n = \frac{n\pi}{L}, \quad n = 1, 2, \ldots,$$

and

$$\boxed{X = X_n = \sin \frac{n\pi}{L} x, \quad n = 1, 2, \ldots \, .}$$

Substituting the values of k in the differential equation for T, we get the first order ordinary differential equation

$$T' + \left(c \frac{n\pi}{L} \right)^2 T = 0$$

whose general solution is

See Theorem 1, Appendix A.1.

$$\boxed{T_n(t) = b_n e^{-\lambda_n^2 t}, \quad n = 1, 2, \ldots,}$$

where we set

$$\boxed{\lambda_n = c \frac{n\pi}{L}, \quad n = 1, 2, \ldots \, .}$$

We thus arrive at the product solution, or **normal mode**,

$$u_n(x, t) = b_n e^{-\lambda_n^2 t} \sin \frac{n\pi}{L} x, \quad n = 1, 2, \ldots \, .$$

By construction, each u_n is a solution of the heat equation and the given (homogeneous) boundary conditions. Motivated by the superposition principle (Theorem 1, Section 8.1), we let

$$\boxed{u(x, t) = \sum_{n=1}^{\infty} b_n e^{-\lambda_n^2 t} \sin \frac{n\pi}{L} x.}$$

Our next step is to determine the coefficients b_n so as to satisfy the initial condition $u(x, 0) = f(x)$.

Fourier Series Solution of the Entire Problem

We set $t = 0$, use the initial condition, and get

$$f(x) = u(x, 0) = \sum_{n=1}^{\infty} b_n \sin \frac{n\pi}{L} x.$$

Recognizing this sum as the half-range sine series expansion of f, we get from (4), Section 7.4,

$$b_n = \frac{2}{L} \int_0^L f(x) \sin \frac{n\pi}{L} x \, dx \quad n = 1, 2, \ldots,$$

which completely determines the solution. We summarize our findings as follows.

SOLUTION OF THE ONE DIMENSIONAL HEAT EQUATION

The solution of the one dimensional heat boundary value problem

$$(1) \qquad \frac{\partial u}{\partial t} = c^2 \frac{\partial^2 u}{\partial x^2} \qquad 0 < x < L, \quad t > 0,$$

$$(2) \qquad u(0, t) = 0 \quad \text{and} \quad u(L, t) = 0 \qquad \text{for all } t > 0,$$

$$(3) \qquad u(x, 0) = f(x) \qquad \text{for } 0 < x < L,$$

is

$$(4) \qquad u(x, t) = \sum_{n=1}^{\infty} b_n e^{-\lambda_n^2 t} \sin \frac{n\pi}{L} x,$$

where

$$(5) \qquad b_n = \frac{2}{L} \int_0^L f(x) \sin \frac{n\pi}{L} x \, dx \quad \text{and} \quad \lambda_n = c \frac{n\pi}{L}, \qquad n = 1, 2, \ldots.$$

EXAMPLE 1   Temperature in a bar with ends held at 0°

A thin bar of length π units is placed in boiling water (temperature 100°). After reaching 100° throughout, the bar is removed from the boiling water. With the lateral sides kept insulated, suddenly, at time $t = 0$, the ends are immersed in a medium with constant freezing temperature 0°. Taking $c = 1$, find the temperature $u(x, t)$ for $t > 0$.

Solution The boundary value problem that we need to solve is

$$\frac{\partial u}{\partial t} = \frac{\partial^2 u}{\partial x^2}, \quad 0 < x < \pi, \quad t > 0,$$
$$u(0, t) = 0 \quad \text{and} \quad u(\pi, t) = 0, \quad t > 0,$$
$$u(x, 0) = 100, \quad 0 < x < \pi.$$

From (4), we have

$$u(x,t) = \sum_{n=1}^{\infty} b_n e^{-n^2 t} \sin nx,$$

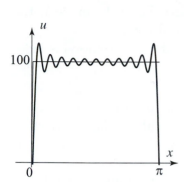

Figure 2 Partial sum of the sine Fourier series expansion of the initial temperature distribution (with k up to 10) $100 = \frac{400}{\pi} \sum_{k=0}^{\infty} \frac{\sin(2k+1)x}{2k+1}$, $0 < x < \pi$. (See Exercise 1, Section 7.3.)

where

$$b_n = \frac{2}{\pi} \int_0^{\pi} 100 \sin nx \, dx = \frac{200}{n\pi}(1 - \cos n\pi).$$

Substituting the values of b_n and using the fact that $(1 - \cos n\pi) = 0$ if n is even and 2 if n is odd, we get

$$u(x,t) = \frac{400}{\pi} \sum_{k=0}^{\infty} \frac{e^{-(2k+1)^2 t}}{2k+1} \sin(2k+1)x.$$

If we plug a given value of t into the series solution, we obtain a function of x alone. This function gives the temperature distribution of the bar at the given time t. In particular, when $t = 0$, $u(x,0)$ yields the half-range sine series expansion of the initial temperature distribution $f(x)$, illustrated in Figure 2. In Figures 3 and 4, we have approximated the series solution by summing it through the terms with $k = 0$ and $k = 10$, respectively, and have shown the temperature distribution at various values of t. It is always a good idea to check that the graph at time $t = 0$ approximates $f(x)$. Does the first picture in Figure 4 meet with your expectation? Compare with Figure 2. ∎

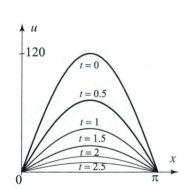

Figure 3 Approximation of the temperature by the first normal mode

$$u_1(x,t) = \frac{400}{\pi} e^{-t} \sin x.$$

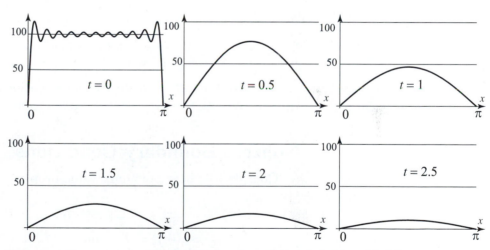

Figure 4 Temperature distribution in a bar with ends held at 0°. The temperature decays to 0 as t increases. Note that for large t, the shape of the graph is dominated by the first normal mode. Indeed, comparison with Figure 3 shows that the two curves are virtually indistinguishable for $t \geq 0.5$.

Steady-State Temperature Distribution

The graphs in Figure 4 show that the temperature in the bar tends to zero as t increases. This is intuitively clear, since the ends of the bar are kept at 0° and there is no internal source of heat. In general, the temperature distribution that we get as $t \to \infty$ is a function of x alone called the **steady-state solution** (or **time-independent solution**). So, in Example 1, the steady-state solution is the function that is identically 0.

For general boundary conditions, since the steady-state solution is independent of t, we must have $\partial u/\partial t = 0$. Substituting this in (1), we see that the steady-state distribution satisfies the differential equation $\frac{\partial^2 u}{\partial x^2} = 0$, or simply $\frac{d^2 u}{dx^2} = 0$, because u, the steady-state solution, is a function of x only. (This should not surprise us in view of what we know about the steady-state solution in two-dimensional plates; it satisfies Laplace's equation.) The general solution of this simple differential equation is $u(x) = Ax + B$, where A and B are constants that are determined using the boundary conditions. We illustrate with an example.

EXAMPLE 2   Steady-state solution

Describe the steady-state solution in a bar of length L with one end kept at temperature T_1 and the other at temperature T_2. Assume that the lateral surface is insulated and that there are no internal sources of heat.

Solution We have $u(0) = T_1$ and $u(L) = T_2$. Hence, from the fact that $u(x) = Ax + B$, it follows that $B = T_1$ and $AL + T_1 = T_2$. Solving for A, we get $A = \frac{T_2 - T_1}{L}$ and so

$$u(x) = \frac{T_2 - T_1}{L} x + T_1.$$

Thus, the graph of the steady-state solution is a straight line that goes through the given boundary values T_1 at 0 and T_2 at L (see Figure 5).   ■

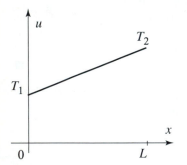

Figure 5 Steady-state or time-independent solution.

We next illustrate how steady-state solutions and the superposition principle can be used to solve certain nonhomogeneous boundary value problems.

Nonzero Boundary Conditions

Consider the heat boundary value problem

$$\text{(6)} \qquad \frac{\partial u}{\partial t} = c^2 \frac{\partial^2 u}{\partial x^2}, \qquad 0 < x < L, \quad t > 0,$$

$$\text{(7)} \qquad u(0,t) = T_1 \quad \text{and} \quad u(L,t) = T_2, \quad t > 0,$$

$$\text{(8)} \qquad u(x,0) = f(x), \quad 0 < x < L.$$

The problem is nonhomogeneous when T_1 and T_2 are not both zero. If you try to solve it in this case using the method of separation of variables, you will encounter difficulties because of the boundary conditions. As we now show, the problem can be reduced to the zero-ends case by subtracting and then adding the steady-state solution.

We begin by finding the steady-state solution, $u_1(x)$, corresponding to the boundary conditions (7). From Example 2, we have

(9)
$$u_1(x) = \frac{T_2 - T_1}{L} x + T_1 .$$

Subtract $u_1(x)$ from the initial temperature distribution in (8) and consider the resulting zero-ends (*homogeneous*) heat boundary value problem

(10) $$\frac{\partial u}{\partial t} = c^2 \frac{\partial^2 u}{\partial x^2}, \qquad 0 < x < L, \quad t > 0,$$

(11) $$u(0, t) = 0 \quad \text{and} \quad u(L, t) = 0, \quad t > 0,$$

(12) $$u(x, 0) = f(x) - u_1(x), \quad 0 < x < L.$$

Let $u_2(x, t)$ be the solution of (10)–(12). According to (4) and (5), we have

(13)
$$u_2(x, t) = \sum_{n=1}^{\infty} b_n e^{-\lambda_n^2 t} \sin \frac{n\pi}{L} x,$$

where $\lambda_n = \frac{cn\pi}{L}$, and

(14)
$$b_n = \frac{2}{L} \int_0^L (f(x) - (\overbrace{\frac{T_2 - T_1}{L} x + T_1}^{u_1(x)})) \sin \frac{n\pi}{L} x \, dx.$$

Now the solution of (6)–(8) is obtained by adding to $u_2(x, t)$ the steady-state solution $u_1(x)$ as follows:

(15)
$$u(x, t) = u_1(x) + u_2(x, t).$$

This can be verified directly by plugging into the equations (6)–(8) and using the properties of u_1 and u_2 (Exercise 10).

EXAMPLE 3   A nonhomogeneous boundary value problem

Consider the experiment in Example 1. Find the solution if after bringing the temperature of the bar to $100°$, the end at $x = 0$ is frozen at $0°$, while the end at $x = \pi$ is kept at $100°$.

Solution We have $T_1 = 0$ and $T_2 = 100$. Thus $u_1(x) = \frac{100}{\pi} x$, and the initial temperature distribution in (12) becomes $100 - \frac{100}{\pi} x$. We now determine the coefficients in the series solution u_2 in (13). Using (14) and integration by parts, we get

$$b_n = \frac{2}{\pi} \int_0^\pi (100 - \frac{100}{\pi} x) \sin nx \, dx = \frac{200}{n\pi}.$$

Finally, appealing to (15), we obtain the solution

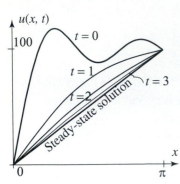

Figure 6 As t increases the graph of $u(x,t)$ approaches that of the steady-state solution $u_1(x)$. When $t = 0$, the graph approximates the initial temperature distribution. Note the Gibbs phenomenon at the endpoints, causing the graph to overshoot the value 100.

(16)
$$u(x,t) = \frac{100}{\pi}x + \frac{200}{\pi}\sum_{n=1}^{\infty}\frac{\sin nx}{n}e^{-n^2t}.$$

Two questions come to mind when we consider this solution. Does it yield the steady-state solution $u_1(x)$ when $t \to \infty$? Does it yield the initial temperature distribution 100 when $t = 0$? We have $\lim_{t\to\infty}\frac{\sin nx}{n}e^{-n^2t} = 0$. Thus each term in the series in (16) tends to 0 as $t \to \infty$, which leads us to conclude that

$$\lim_{t\to\infty} u(x,t) = \frac{100}{\pi}x = u_1(x).$$

This argument lacks rigor (in general, the limit of an *infinite* sum of functions is not equal to the sum of the limits of the functions) but can be justified using properties of the special series in question. Toward answering our second question, we set $t = 0$ in (16) and get

$$u(x,0) = \frac{100}{\pi}x + \frac{200}{\pi}\sum_{n=1}^{\infty}\frac{\sin nx}{n}.$$

Is this equal to 100 for $0 < x < \pi$? Recognizing the infinite series as the Fourier series of the sawtooth function, we obtain

$$u(x,0) = \frac{100}{\pi}x + \frac{200}{\pi}\frac{1}{2}(\pi - x) = 100, \quad \text{for } 0 < x < \pi,$$

which yields an affirmative answer to our second question. The graphs in Figure 6 illustrate both answers. To plot $u(x,t)$, we used (16) and truncated the series after three terms. ∎

From the last example, you could extract a method for solving problems with nonhomogeneous boundary conditions. By subtracting the steady-state solution, we were able to reduce to a problem with zero boundary conditions and then solve using the method of separation of variables. The success of this method, which is a sort of a superposition principle, depends in a crucial way on the fact that the heat equation is linear. These ideas are very important and will be used in later sections to solve more complicated nonhomogeneous higher dimensional problems.

Exercises 8.3

In Exercises 1–6, solve the boundary value heat problem (1)–(3) with the given data. In each case, give a brief physical explanation of the problem.

1. $L = \pi$, $c = 1$, $f(x) = 78$.

2. $L = \pi$, $c = 1$, $f(x) = 30\sin x$.

3. $L = \pi$, $c = 1$,

$$f(x) = \begin{cases} 33x & \text{if } 0 < x \le \frac{\pi}{2}, \\ 33(\pi - x) & \text{if } \frac{\pi}{2} < x < \pi. \end{cases}$$

4. $L = \pi$, $c = 1$,

$$f(x) = \begin{cases} 100 & \text{if } 0 < x \le \frac{\pi}{2}, \\ 0 & \text{if } \frac{\pi}{2} < x < \pi. \end{cases}$$

5. $L = 1$, $c = 1$, $f(x) = x$.

6. $L = 1$, $c = 1$, $f(x) = e^{-x}$.

7. (a) Plot the temperature distributions for various values of $t > 0$ in Example 1. Approximate how long it will take for the maximum temperature to drop to $50°$.

(b) Plot the surface $z = u(x,t)$ $(t > 0)$ and explain what it represents. Be specific in your description of the surface as $t \to 0$ and $t \to \infty$.

8. (a) Generalize Example 1 by solving the problem for arbitrary c.

(b) Plot the solution at $t = 1$, for $c = 1, 2, 3$.   Describe how a change in the value of c affects the rate of heat transfer. Justify your answer by considering the solution given by (4) and (5).

9. Determine the steady-state solution in a bar of length 1 with the given boundary conditions.
(a) $u(0,t) = 0$, $u(1,t) = 100$.
(b) $u(0,t) = 100$, $u(1,t) = 100$.

10. (a) Verify that the solution of (6)–(8) is given by (15).
(b) Show that, in terms of the Fourier coefficients of the initial temperature distribution $f(x)$ in (8), the coefficient in (14) is

$$b_n = \frac{2}{L} \int_0^L f(x) \sin \frac{n\pi}{L} x \, dx - 2 \frac{T_1 + (-1)^{n+1} T_2}{n\pi}.$$

In Exercises 11–14, solve the nonhomogeneous boundary value problem (6)–(8) *for the given data.*

11. $u(0,t) = 100$, $u(1,t) = 0$, $f(x) = 30 \sin(\pi x)$, $L = 1$, $c = 1$.

12. $u(0,t) = 100$, $u(1,t) = 100$, $f(x) = 50 \, x(1 - x)$, $L = 1$, $c = 1$.

13. $u(0,t) = 100$, $u(\pi,t) = 50$, $f(x)$ as in Exercise 3, $L = \pi$, $c = 1$.

14. $u(0,t) = 0$, $u(\pi,t) = 100$, $f(x)$ as in Exercise 4, $L = \pi$, $c = 1$.

15. What is the solution of (6)–(8) if the initial temperature distribution $f(x)$ is equal to the steady-state solution $u_1(x)$? Justify your answer on physical grounds.

16. Project Problem: A nonhomogeneous wave problem. Using the ideas behind the solution of the boundary value problem (6)–(8), solve the nonhomogeneous wave problem

$$u_{tt} = c^2 u_{xx}, \quad 0 < x < L, \quad t > 0,$$
$$u(0,t) = A, \quad u(L,t) = B, \quad t > 0,$$
$$u(x,0) = f(x), \quad u_t(x,0) = g(x), \quad 0 < x < L,$$

where A and B are not both zero. (As you may verify, the time independent solution is still $u_1(x) = \frac{B-A}{L} x + A$. Unlike the case of the heat equation, the solution of the wave equation does not have a limit in general as $t \to \infty$. Hence $u_1(x)$ cannot be interpreted as an equilibrium solution. It is simply a solution of the time independent problem $u_{xx} = 0$.)

17. Show that for fixed $t > 0$, the solution of the heat equation (4) is uniformly convergent for $0 \le x \le L$. (Assume that $f(x)$ is bounded for $0 \le x \le L$.)

8.4   Heat Conduction in Bars: Varying the Boundary Conditions

We continue our study of the one dimensional heat equation, with different boundary conditions. These model temperature distributions in bars with both ends insulated, with one end at zero temperature and the other end exchanging heat with a surrounding medium, and other interesting variations. We will see that the boundary conditions determine the type of functions that are used to build up the solution. In the previous section, the solution was expressed as a sine series; here we will see cosine series and other related expansions known as generalized Fourier series.

EXAMPLE 1   Bar with insulated ends

The problem of heat transfer in a bar of length L with initial heat distribution $f(x)$ and no heat loss at either end (Figure 1) is modeled by the heat equation

$$(1) \qquad \frac{\partial u}{\partial t} = c^2 \frac{\partial^2 u}{\partial x^2}, \quad 0 < x < L, \quad t > 0,$$

along with the boundary conditions (no heat flux in the x-direction at either end)

$$(2) \qquad \frac{\partial u}{\partial x}(0,t) = \frac{\partial u}{\partial x}(L,t) = 0, \quad t > 0,$$

and the initial condition

$$(3) \qquad u(x,0) = f(x), \quad 0 < x < L.$$

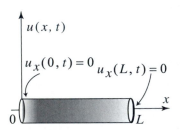

Figure 1 Bar with insulated ends.

Solve this problem using the method of separation of variables.

Solution We assume that $u(x,t) = X(x)T(t)$, follow the method of Section 8.3, and arrive in exactly the same way at the following equations:

$$T' - kc^2 T = 0,$$
$$X'' - kX = 0, \quad X'(0) = 0 \quad \text{and} \quad X'(L) = 0,$$

where k is the separation constant. Note that the differential equations are the same as in Section 8.3, but the boundary conditions are different due to the conditions (2). This difference in the boundary conditions will give rise to different solutions in X. It is easy to check that positive values of k lead only to the trivial solution $X = 0$. For $k = 0$, we have $X'' = 0$, leading to $X = c_1 x + c_2$. Differentiating and imposing the boundary conditions, we find $c_1 = 0$ and c_2 is arbitrary. Thus $k = 0$ leads to a nontrivial solution

$$X_0 = 1.$$

To consider the case of negative separation constants, we set $k = -\mu^2$, solve the resulting equation, and get $X = c_1 \cos \mu x + c_2 \sin \mu x$. Computing

$$X' = -c_1 \mu \sin \mu x + c_2 \mu \cos \mu x$$

and using the first boundary condition $X'(0) = 0$, we obtain $c_2 = 0$. Using the second boundary condition $X'(L) = 0$, we obtain the equation $\sin \mu L = 0$, from which we get

$$\mu = \mu_n = \frac{n\pi}{L}, \quad n = 1, 2, \ldots,$$

and hence

$$X_n = \cos \frac{n\pi}{L} x, \quad n = 1, 2, \dots.$$

The corresponding solutions to the equation for T are

$$T_0 = a_0 \quad \text{and} \quad T_n = a_n e^{-\lambda_n^2 t}, \qquad n = 1, 2, \dots,$$

where $\lambda_n = c \frac{n\pi}{L}$. Superposing the product solutions, we obtain

(4)
$$u(x,t) = a_0 + \sum_{n=1}^{\infty} a_n e^{-\lambda_n^2 t} \cos \frac{n\pi}{L} x.$$

Compare this solution with the solution (4) of the heat problem with zero-ends conditions (Section 8.3). Note that the change in boundary conditions causes sines to be replaced by cosines, and the solution to involve the Fourier cosine expansion of the initial temperature distribution rather than the Fourier sine expansion.

To finish the solution, we must determine the coefficients a_n so as to match the initial condition (3). Setting $t = 0$ yields

$$f(x) = a_0 + \sum_{n=1}^{\infty} a_n \cos \frac{n\pi}{L} x.$$

Recognizing this as a cosine expansion of f, we conclude that

(5)
$$a_0 = \frac{1}{L} \int_0^L f(x)\, dx \quad \text{and} \quad a_n = \frac{2}{L} \int_0^L f(x) \cos \frac{n\pi}{L} x\, dx.$$

(Use Section 7.4, (2).) Thus the solution of the boundary value problem (1)–(3) is given by (4) with the coefficients determined by (5).   ■

The fact that cosine expansions arose in Example 1 while sine expansions arose in the previous section is caused by the change in the boundary conditions: vanishing endpoint conditions force sine expansions, vanishing derivatives force cosine expansions. As the next example shows, other expansions will arise naturally from the imposition of yet other boundary conditions.

EXAMPLE 2   Bar with one radiating end: generalized Fourier series
Determine the temperature $u(x,t)$ in a bar of length L, given that one end is kept at zero temperature and the other end loses heat to the surrounding medium at a rate proportional to its temperature. Thus the boundary conditions are

(6)
$$u(0,t) = 0, \quad \frac{\partial u}{\partial x}(L,t) = -\kappa u(L,t) \quad (t > 0),$$

where κ is a positive constant called the **heat transfer constant** or **convection coefficient**. Denoting the initial temperature distribution by $f(x)$, the boundary value problem to be solved consists of the heat equation (1), the boundary conditions (6), and the initial condition (3).

Solution   Again we assume that $u(x,t) = X(x)T(t)$, follow the method of Section 8.3, and arrive at the equations

$$X'' + \mu^2 X = 0, \quad X(0) = 0 \quad \text{and} \quad X'(L) = -\kappa X(L),$$
$$T' + \mu^2 c^2 T = 0,$$

where we have already excluded the cases of a nonnegative separation constant, since, as you can check, they lead only to trivial solutions. Solving the equation in X and imposing the boundary conditions, we get

$$X = \sin \mu x,$$

where, because of the condition $X'(L) = -\kappa X(L)$, μ must satisfy

$$\mu \cos \mu L = -\kappa \sin \mu L,$$

or equivalently,

(7)
$$\tan \mu L = -\frac{\mu}{\kappa}.$$

Note that the boundary conditions are leading to the complicated equation (7), compared to the much simpler equation $\sin \mu L = 0$ that we had to solve in the heat problem with zero end conditions (Section 8.3).

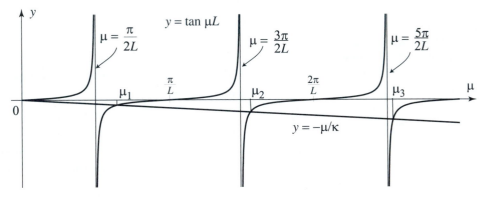

Figure 2 Graphs of $y = \tan \mu L$ and $y = -\frac{\mu}{\kappa}$ as functions of $\mu > 0$. The transcendental equation $\tan \mu L = -\frac{\mu}{\kappa}$ has infinitely many positive roots $\mu_1 < \mu_2 < \mu_3 \ldots$.

In Figure 2 we have plotted the graphs of $y = \tan \mu L$ and $y = -\frac{\mu}{\kappa}$ as functions of μ. The intersections of these graphs determine the solutions of (7). Thus, as suggested by Figure 2, equation (7) has infinitely many positive solutions μ_1, μ_2, $\ldots$. While we do not have explicit forms for the μ_n as we had in the previous example, we can still proceed with the solution much as before. Approximate numerical values of the μ_n's can be obtained with the help of a computer system, as illustrated in Example 3. Corresponding to μ_n, we have

Compare the solution (8) to (4), Section 8.3. Note that the solution here is no longer periodic in x, since there is no common period for the solutions $\sin \mu_n x$.

$$X_n = \sin \mu_n x,$$

and

$$T_n = c_n e^{-c^2 \mu_n^2 t}, \qquad n = 1, 2, \ldots,$$

which are obtained by solving the equation for T. Superposing the product solutions, we get

Note that expansion (9) has some similarity with the sine series expansions of Section 7.4 but differs in some ways (for example, the functions $\sin \mu_n x$, $n = 1, 2, \ldots$ have no common period). This new representation generalizes Fourier sine series expansions to which it reduces when $\kappa \to \infty$ (in which case, we can see graphically from Figure 2 that μ_n tends to $n\pi/L$).

(8)
$$u(x,t) = \sum_{n=1}^{\infty} c_n e^{-c^2 \mu_n^2 t} \sin \mu_n x$$

as a solution of (1) and (2). To finish the solution, we must determine the coefficients c_n so as to match the initial condition. Setting $t = 0$, we see that

(9)
$$f(x) = \sum_{n=1}^{\infty} c_n \sin \mu_n x.$$

This is a so-called **generalized Fourier series** with **generalized Fourier coefficients** c_n. Taking our lead from the case of Fourier series, we observe that if the functions

$$\sin \mu_1 x, \ \sin \mu_2 x, \ \ldots$$

satisfy the following "orthogonality relations"

(10)
$$\int_0^L \sin \mu_m x \sin \mu_n x \, dx = 0, \quad \text{for } m \neq n,$$

then we can solve for the c_n as follows. We multiply (9) through by $\sin \mu_m x$ and integrate from 0 to L. Assuming that we can interchange the sum and the integral and using (10), we get

$$\int_0^L f(x) \sin \mu_m x \, dx = \sum_{n=1}^{\infty} c_n \overbrace{\int_0^L \sin \mu_m x \sin \mu_n x \, dx}^{=0 \text{ unless } n=m} = c_m \int_0^L \sin^2 \mu_m x \, dx.$$

This yields the formula

(11)
$$\boxed{c_n = \frac{1}{\int_0^L \sin^2 \mu_n x \, dx} \int_0^L f(x) \sin \mu_n x \, dx.}$$

This formal discussion is completed by the proof of the orthogonality relations (10) outlined in Exercise 10. Therefore, the complete solution of the boundary value problem (1), (6), (3) is the series (8) with the coefficients given by (11). ∎

To make the previous discussion rigorous, we would need a representation theorem for series of the form (9) analogous to the Fourier representation theorem (Theorem 1, Section 7.3). Such representation theorems belong to a general theory, known as Sturm-Liouville theory. The gist of this theory is that a series like (9) with coefficients determined by (11) will converge for piecewise smooth functions in much the same way Fourier series will. These ideas are illustrated in the following numerical application of Example 2 (see [2] for more details on Sturm-Liouville theory).

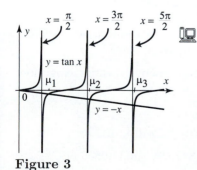

Figure 3

EXAMPLE 3   A numerical application for a bar with one radiating end

Consider the boundary value problem in Example 2 with $L = 1$, $c = 1$, $\kappa = 1$, and with the initial temperature distribution $f(x) = x(1 - x)$.
(a) Compute explicitly the first five μ_n's and the corresponding X_n's.
(b) Compute the first five terms of the series expansion of the function f given by (9). Plot f and several partial sums to check the validity of the expansion.
(c) Write down the first five terms of the solution (8). Plot this partial sum for various values of t, and use these to estimate the time it takes for the maximum temperature in the bar to drop to 0.1.

Solution (a) In Figure 3 we have plotted the graphs of $y = \tan x$ and $y = -x$. According to the solution of Example 2, to find the μ_n's, we must solve the equation

$\tan x = -x$. With the help of a computer system, we find the first five solutions to be approximately

$$\mu_1 = 2.0288, \ \mu_2 = 4.9132, \ \mu_3 = 7.9787, \ \mu_4 = 11.0855, \ \mu_5 = 14.2074.$$

Thus

$$X_1(x) = \sin(2.0288\, x), \ X_2(x) = \sin(4.9132\, x), \ X_3(x) = \sin(7.9787\, x),$$
$$X_4(x) = \sin(11.0855\, x), \ X_5(x) = \sin(14.2074\, x).$$

(b) From (9) and (11), we have

$$x(1-x) = \sum_{n=1}^{\infty} c_n \sin \mu_n x,$$

where

$$c_n = \int_0^1 x(1-x)\sin \mu_n x \, dx \ \Big/ \ \int_0^1 \sin^2 \mu_n x \, dx$$

with the numerical values of the μ_j's given in (a). We evaluate these coefficients with the help of a computer system and find

$$c_1 = .2133, \ c_2 = .1040, \ c_3 = -.0220, \ c_4 = .0187, \ c_5 = -.0083.$$

Thus the expansion of f is

$$f(x) = .2133 \, \sin(2.0288x) + .1040 \, \sin(4.9132x) - .0220 \, \sin(7.9787x)$$
$$+ .0187 \, \sin(11.0855x) - .0083 \, \sin(14.2074x) + \cdots .$$

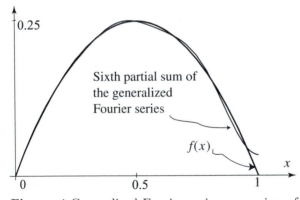

Figure 4 Generalized Fourier series expansion of $f(x) = x(1-x)$, $0 < x < 1$.

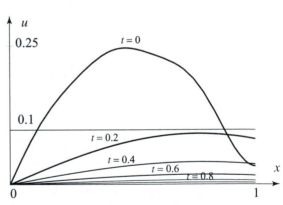

Figure 5 Temperature distribution at various times (using the sixth partial sum).

In Figure 4, we plotted some partial sums of the generalized Fourier series expansion of f to illustrate the convergence of this series.

(c) Combining (8) with part (b), we find that

$$u(x,t) = .2133 \, e^{-\mu_1^2 t} \sin(2.0288\, x) + .1040 \, e^{-\mu_2^2 t} \sin(4.9132\, x)$$
$$- .0220 \, e^{-\mu_3^2 t} \sin(7.9787\, x) + .0187 \, e^{-\mu_4^2 t} \sin(11.0855\, x)$$
$$- .0083 \, e^{-\mu_5^2 t} \sin(14.2074\, x) + \cdots ,$$

where the μ_n's are given in (a). Plotting the sixth partial sum at $t = 0, .2, .4, \ldots,$ we see from Figure 5 that after approximately $t = .2$ the maximum temperature in the bar drops below .1. ∎

Exercises 8.4

In Exercises 1–6, solve the heat problem (1) − (3) *for the given data.*

1. $L = \pi$, $c = 1$, $f(x) = 100$. [Hint: In this case, you can guess the answer based on your physical intuition.]

2. $L = \pi$, $c = 1$, $f(x) = x$.

3. $L = \pi$, $c = 1$,

$$f(x) = \begin{cases} 100x & \text{if } 0 < x \leq \frac{\pi}{2}, \\ 100(\pi - x) & \text{if } \frac{\pi}{2} < x < \pi . \end{cases}$$

4. $L = 1$, $c = 1$,

$$f(x) = \begin{cases} 100 & \text{if } 0 < x \leq \frac{1}{2}, \\ 0 & \text{if } \frac{1}{2} < x < 1 . \end{cases}$$

5. $L = 1$, $c = 1$, $f(x) = \cos \pi x$. **6** $L = 1$, $c = 1$, $f(x) = \sin \pi x$.

7. The **average temperature** in a bar of length L at a given time t is given by

$$\frac{1}{L} \int_0^L u(x, t)\, dx .$$

Suppose that, as in Example 1, the bar is insulated in such a way that there is no exchange of heat with the surrounding medium. Show that the average temperature is constant for all time. What is the average temperature in terms of the initial heat distribution $f(x)$? Does your answer agree with your intuition? Explain. [Hint: Integrate (4) term by term.]

8. In Example 1, show that the steady-state temperature is constant and equals the average temperature.

9. Compute the average temperature in Exercises 1 and 5.

10. Orthogonality. Establish the orthogonality relations (10) following the outlined steps.
(a) For $m \neq n$, show that

$$\int_0^L \sin \mu_m x \sin \mu_n x\, dx = [\mu_m \sin \mu_n L \cos \mu_m L - \mu_n \cos \mu_n L \sin \mu_m L] / (\mu_n^2 - \mu_m^2) .$$

(b) Use (7) to prove that the right side in (a) is 0.

In Exercises 11–14, repeat Example 3 for the given initial temperature distribution.

11. $f(x) = x$. **12.** $f(x) = \sin \pi x$.

13.

$$f(x) = \begin{cases} 100 & \text{if } 0 < x \le \frac{1}{2}, \\ 0 & \text{if } \frac{1}{2} < x < 1. \end{cases}$$

14.

$$f(x) = \begin{cases} 0 & \text{if } 0 < x \le \frac{1}{2}, \\ 100(x - \frac{1}{2}) & \text{if } \frac{1}{2} < x < 1. \end{cases}$$

15. Project problem: Bar with one insulated and one radiating end. Work through Example 2 with the left boundary condition replaced by $u_x(0, t) = 0$ but with all other conditions unchanged. Your answer will involve cosine functions.

16. Project Problem: Heat conduction in a thin circular ring. In this problem we study heat conduction in a thin circular ring of unit radius that is insulated along its lateral sides. The temperature in the ring is governed by the heat equation (1), where x represents arclength along the ring. Thus despite the two dimensional shape, we have a one dimensional problem. The shape does come into play, however, in that the boundary conditions are now periodic; that is,

$$u(-\pi, t) = u(\pi, t) \quad \text{and} \quad u_x(-\pi, t) = u_x(\pi, t).$$

Here we think of the ring to be parameterized by the interval $-\pi < x \le \pi$. The initial condition is given by (3).
(a) Using separation of variables, derive the differential equations and boundary conditions

$$T' - kc^2 T = 0,$$
$$X'' - kX = 0,$$
$$X(-\pi) = X(\pi) \quad \text{and} \quad X'(-\pi) = X'(\pi).$$

(b) Argue that positive choices of k lead only to trivial solutions for X.
(c) Show that for $k = 0$, we obtain the solution $X_0 = a_0$.
(d) Show that for $k = -\mu^2$, nontrivial solutions arise only if $\mu = n$ for $n = 1, 2, 3, \ldots$, and the corresponding solutions are $X_n = a_n \cos nx + b_n \sin nx$.
(e) Conclude that

$$u(x, t) = a_0 + \sum_{n=1}^{\infty} (a_n \cos nx + b_n \sin nx)e^{-n^2 c^2 t},$$

where a_0, a_n, and b_n are the Fourier coefficients of f, given by the Euler formulas in Section 7.2.

The point of the following set of problems is to show you that the boundary conditions can have all sorts of effects on the solution. For example, in the next problem, you will see that positive and negative values of the separation constant have to be included. You will also encounter new equations for μ, and hence new generalized Fourier series expansions.

17. Project Problem: A problem with positive and negative values of the separation constant. Consider the heat boundary value problem

$$u_t = u_{xx}, \quad t > 0, \quad 0 < x < 1,$$
$$u_x(0, t) = -u(0, t), \quad u_x(1, t) = -u(1, t),$$
$$u(x, 0) = f(x).$$

This models a heat problem in a bar which is losing heat at the right end while gaining heat at the left end.

(a) Using separation of variables, obtain

$$T' - kT = 0,$$
$$X'' - kX = 0, \quad X'(0) = -X(0), \quad X'(1) = -X(1),$$

where k is a separation constant.

(b) Argue convincingly that k cannot be 0.

(c) Show that if $k = \mu^2$, then $\mu = 1$ and the corresponding solutions are

$$T_0(t) = e^t \quad \text{and} \quad X_0(x) = e^{-x}.$$

(d) Show that if $k = -\mu^2$, then $\mu = n\pi$, $n = 1, 2, \ldots$, and the corresponding solutions are

$$T_n(t) = e^{-(n\pi)^2 t}$$

and

$$X_n(x) = n\pi \cos n\pi x - \sin n\pi x, \; n = 1, 2, \ldots.$$

(e) Establish the orthogonality of the functions X_0, X_1, X_2, $\ldots$ on the interval $(0, 1)$ by showing that

$$\int_0^1 X_j(x)X_k(x)dx = 0 \quad \text{if } j \neq k.$$

(f) Conclude that

$$u(x, t) = c_0 e^t e^{-x} + \sum_{n=1}^{\infty} c_n T_n(t)X_n(x),$$

where

$$c_0 = \frac{2e^2}{e^2 - 1} \int_0^1 f(x)e^{-x}\, dx,$$

and

$$c_n = \frac{2}{1 + n^2\pi^2} \int_0^1 f(x)X_n(x)\, dx, \quad n = 1, 2, \ldots.$$

18. Illustrate the solution in Exercise 17 when $f(x) = x$.   Plot the temperature distribution at various values of t and discuss the behavior at $x = 0$ and $x = 1$.

19. Project Problem: Bar with two radiating ends, the equation $\tan \mu = \frac{2\mu}{\mu^2 - 1}$. Consider the heat boundary value problem

$$u_t = u_{xx}, \quad t > 0, \quad 0 < x < 1,$$
$$u_x(0, t) = u(0, t), \quad u_x(1, t) = -u(1, t),$$
$$u(x, 0) = f(x).$$

(a) Using separation of variables, obtain

$$T' - kT = 0,$$
$$X'' - kX = 0, \quad X'(0) = X(0), \quad X'(1) = -X(1),$$

where k is a separation constant.

(b) Show that $k = -\mu^2 < 0$.

(c) Show that the corresponding solutions are

$$T_n(t) = e^{-\mu_n^2 t}$$

and

$$X_n(x) = \mu_n \cos \mu_n x + \sin \mu_n x, \quad n = 1, 2, \ldots,$$

where μ_n is the nth positive root of the equation $\tan \mu = \frac{2\mu}{\mu^2 - 1}$.

(d) Show graphically that the last equation has infinitely many positive solutions. Approximate the first five positive roots: $\mu_1, \mu_2, \ldots, \mu_5$.

(e) Check the orthogonality relations on the interval $(0, 1)$ for X_n, $n = 1, 2, \ldots, 5$. (You can actually prove the orthogonality for all n by taking a hint from Exercise 10.)

(f) Conclude that

$$u(x, t) = \sum_{n=1}^{\infty} c_n T_n(t) X_n(x),$$

where

$$c_n = \frac{1}{\kappa_n} \int_0^1 f(x) X_n(x) \, dx$$

and

$$\kappa_n = \int_0^1 X_n^2(x) \, dx.$$

20. Illustrate the solution in Exercise 19 when $f(x) = x$. Plot the temperature distribution at various values of t and discuss the behavior at $x = 0$ and $x = 1$.

8.5   Two Dimensional Wave and Heat Equations

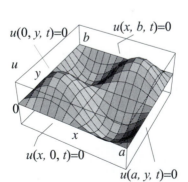

Figure 1 Initial shape of a membrane with edges held fixed.

Suppose that a thin membrane is stretched over a rectangular frame with dimensions a and b, and that the edges are held fixed (Figure 1). The membrane is then set to vibrate by displacing it vertically and then releasing it. The vibrations of the membrane are governed by the **two dimensional wave equation**

$$(1) \qquad \frac{\partial^2 u}{\partial t^2} = c^2 \left(\frac{\partial^2 u}{\partial x^2} + \frac{\partial^2 u}{\partial y^2} \right), \quad 0 < x < a, \quad 0 < y < b, \quad t > 0,$$

where $u = u(x, y, t)$ denotes the deflection at the point (x, y) at time t. The fact that the edges are held fixed is expressed by the condition $u(x, y, t) = 0$ on the boundary for all $t \geq 0$. More explicitly, we have the **boundary conditions**

$$u(0, y, t) = 0 \text{ and } u(a, y, t) = 0 \quad \text{for } 0 \leq y \leq b \text{ and } t \geq 0,$$

$$(2)$$

$$u(x, 0, t) = 0 \text{ and } u(x, b, t) = 0 \quad \text{for } 0 \leq x \leq a \text{ and } t \geq 0.$$

The initial conditions

(3) $$u(x, y, 0) = f(x, y) \quad \text{and} \quad \frac{\partial u}{\partial t}(x, y, 0) = g(x, y)$$

represent, respectively, the shape and the velocity of the membrane at time $t = 0$. To determine the vibrations of the membrane, we must find the function u that satisfies (1)–(3). We solve the boundary value problem using the separation of variables method, following the steps outlined in Section 8.2.

Separating Variables in (1) and (2)

In applying the method of separation of variables in higher dimensions, it helps to keep in mind the one dimensional cases treated earlier.

We first look for product solutions of the form

$$u(x, y, t) = X(x)Y(y)T(t).$$

Differentiating and plugging into (1), we obtain

$$XYT'' = c^2(X''YT + XY''T).$$

Dividing both sides by $c^2 XYT$, we get

$$\frac{T''}{c^2 T} = \frac{X''}{X} + \frac{Y''}{Y}.$$

Since the left side is a function of t alone, and the right side is a function of x and y only, the expressions on both sides must be equal to a constant. Expecting a periodic solution in t, we consider negative separation constants only. (We could also rule out the nonnegative cases by arguing, as has been done in previous sections, that they only lead to trivial solutions.) Thus

$$\frac{T''}{c^2 T} = -k^2 \quad \text{and} \quad \frac{X''}{X} + \frac{Y''}{Y} = -k^2 \qquad (k > 0).$$

The first equation yields

$$T'' + k^2 c^2 T = 0$$

(with periodic solutions), and the second one yields

$$\frac{X''}{X} = -\frac{Y''}{Y} - k^2.$$

Because in this last equation the right side depends only on y and the left side only on x, we infer that

$$\frac{X''}{X} = -\mu^2 \quad \text{and} \quad -\frac{Y''}{Y} - k^2 = -\mu^2, \qquad \mu > 0,$$

or

$$X'' + \mu^2 X = 0 \quad \text{and} \quad Y'' + \nu^2 Y = 0,$$

where $\nu^2 = k^2 - \mu^2$. (Here again we have ruled out all nonnegative values of the separation constant on the basis that they lead to trivial solutions.) Separating variables in the boundary conditions (2), we arrive at the equations

$$X'' + \mu^2 X = 0, \quad X(0) = 0, \quad X(a) = 0,$$
$$Y'' + \nu^2 Y = 0, \quad Y(0) = 0, \quad Y(b) = 0,$$
$$T'' + c^2 k^2 T = 0, \quad k^2 = \mu^2 + \nu^2.$$

Solution of the Separated Equations

The general solutions of the last three differential equations are, respectively,

$$X(x) = c_1 \cos \mu x + c_2 \sin \mu x,$$
$$Y(y) = d_1 \cos \nu y + d_2 \sin \nu y,$$
$$T(t) = e_1 \cos ckt + e_2 \sin ckt \quad (k^2 = \mu^2 + \nu^2).$$

From the boundary conditions for X and Y we get $c_1 = 0$ and $c_2 \sin \mu a = 0$, $d_1 = 0$ and $d_2 \sin \nu a = 0$. Thus

$$\mu = \mu_m = \frac{m\pi}{a} \quad \text{and} \quad \nu = \nu_n = \frac{n\pi}{b} \quad m, n = 1, 2, \ldots,$$

and so

$$X_m(x) = \sin \frac{m\pi}{a} x \quad \text{and} \quad Y_n(y) = \sin \frac{n\pi}{b} y.$$

(Note that if $m = 0$ or $n = 0$, the solutions are identically zero, which are of no interest. Also, negative choices of m and n would only change the signs of the solutions, and hence would not contribute new solutions.) For $m, n = 1, 2, \ldots$, we have

$$k = k_{mn} = \sqrt{\mu_m^2 + \nu_n^2} = \sqrt{\frac{m^2\pi^2}{a^2} + \frac{n^2\pi^2}{b^2}},$$

and so

$$T(t) = T_{mn}(t) = B_{mn} \cos \lambda_{mn} t + B_{mn}^* \sin \lambda_{mn} t,$$

where we put

$$\lambda_{mn} = c\pi \sqrt{\frac{m^2}{a^2} + \frac{n^2}{b^2}}.$$

The λ_{mn}'s are called the **characteristic frequencies** of the membrane. We have thus derived the product solutions satisfying (1) and (2):

$$u_{mn}(x, y, t) = \sin \frac{m\pi}{a} x \sin \frac{n\pi}{b} y (B_{mn} \cos \lambda_{mn} t + B_{mn}^* \sin \lambda_{mn} t).$$

The functions u_{mn} are called the **normal modes** of the two dimensional wave equation.

Double Fourier Series Solution of the Problem

In order to find a solution that also satisfies the initial conditions (3), motivated by the superposition principle, we sum all the product solutions and try

$$(4) \quad u(x, y, t) = \sum_{n=1}^{\infty} \sum_{m=1}^{\infty} (B_{mn} \cos \lambda_{mn} t + B_{mn}^* \sin \lambda_{mn} t) \sin \frac{m\pi}{a} x \sin \frac{n\pi}{b} y.$$

From the initial condition $u(x, y, 0) = f(x, y)$, we get

$$(5) \qquad\qquad f(x, y) = \sum_{n=1}^{\infty} \sum_{m=1}^{\infty} B_{mn} \sin \frac{m\pi}{a} x \sin \frac{n\pi}{b} y.$$

The key to computing the coefficients B_{mn} is to observe that the functions $\sin \frac{m\pi}{a} x \sin \frac{n\pi}{b} y$ are "orthogonal" over the rectangle $0 \leq x \leq a$, $0 \leq y \leq b$. That is,

$$(6) \qquad\qquad \int_0^b \int_0^a \sin \frac{m\pi}{a} x \sin \frac{n\pi}{b} y \sin \frac{m'\pi}{a} x \sin \frac{n'\pi}{b} y \, dx \, dy = 0$$

if $(m, n) \neq (m', n')$. Also, if $(m, n) = (m', n')$, then we get

$$(7) \qquad\qquad \int_0^b \int_0^a \sin^2 \frac{m\pi}{a} x \sin^2 \frac{n\pi}{b} y \, dx \, dy = \frac{ab}{4}.$$

The proofs of (6) and (7) are straightforward and are left to the exercises. Multiplying (5) by $\sin \frac{m'\pi}{a} x \sin \frac{n'\pi}{b} y$, integrating over the $a \times b$ rectangle, and using the orthogonality properties, we find

$$(8) \qquad \boxed{B_{mn} = \frac{4}{ab} \int_0^b \int_0^a f(x, y) \sin \frac{m\pi}{a} x \sin \frac{n\pi}{b} y \, dx \, dy.}$$

The series in (5) with coefficients given by (8) is called the **double Fourier sine series** of f. Similarly, using the second initial condition, we get

$$g(x, y) = \sum_{n=1}^{\infty} \sum_{m=1}^{\infty} B_{mn}^* \lambda_{mn} \sin \frac{m\pi}{a} x \sin \frac{n\pi}{b} y.$$

Arguing as before with the help of orthogonality, we obtain

$$(9) \qquad \boxed{B_{mn}^* = \frac{4}{ab\lambda_{mn}} \int_0^b \int_0^a g(x, y) \sin \frac{m\pi}{a} x \sin \frac{n\pi}{b} y \, dx \, dy.}$$

We have now completely determined the solution of the vibrating rectangular membrane and we summarize our results as follows.

SOLUTION OF THE TWO DIMENSIONAL WAVE EQUATION

The solution of the two dimensional wave equation (1) with boundary conditions (2) and initial conditions (3) is

$$u(x, y, t) = \sum_{n=1}^{\infty} \sum_{m=1}^{\infty} (B_{mn} \cos \lambda_{mn} t + B_{mn}^* \sin \lambda_{mn} t) \sin \frac{m\pi}{a} x \sin \frac{n\pi}{b} y,$$

where

$$\lambda_{mn} = c\pi \sqrt{\frac{m^2}{a^2} + \frac{n^2}{b^2}}$$

and B_{mn} and B_{mn}^* are as in (8) and (9).

To justify the convergence of the series in (5), and for ease of reference, we state the double Fourier sine series representation theorem which holds for continuous functions with continuous first and second partial derivatives in x and y. This important result will also be needed for the solution of Poisson's equation (Section 8.7). Its proof can be obtained by reducing to the one dimensional Fourier series representation theorem. These interesting ideas are presented in the exercises.

THEOREM 1 DOUBLE SINE SERIES REPRESENTATION

Suppose that $f(x, y)$ is defined for all $0 < x < a$, $0 < y < b$. Then we have the double Fourier sine series expansion

$$(10) \qquad f(x, y) = \sum_{n=1}^{\infty} \sum_{m=1}^{\infty} B_{mn} \sin \frac{m\pi}{a} x \sin \frac{n\pi}{b} y,$$

where the **double Fourier sine series coefficient** B_{mn} is given by (8).

EXAMPLE 1   Vibration of a stretched membrane with fixed edges

A square membrane with $a = 1$, $b = 1$, and $c = 1/\pi$, is placed in the xy-plane as shown in the first picture in Figure 2. The edges of the membrane are held fixed, and the membrane is stretched into a shape modeled by the function $f(x, y) = x(x - 1)y(y - 1)$, $0 < x < 1$, $0 < y < 1$. Suppose that the membrane starts to vibrate from rest. Determine the position of each point on the membrane for $t > 0$.

Solution We have $g(x, y) = 0$, and so $B_{mn}^* = 0$. For $m, n = 1, 2, \ldots$, we have

$$\begin{aligned}
B_{mn} &= 4 \int_0^1 \int_0^1 x(x - 1)y(y - 1) \sin m\pi x \sin n\pi y \, dx \, dy \\
&= 4 \int_0^1 y(y - 1) \sin n\pi y \, dy \int_0^1 x(x - 1) \sin m\pi x \, dx.
\end{aligned}$$

Integrating by parts, we get

$$\int_0^1 x(x - 1) \sin m\pi x \, dx = \frac{2((-1)^m - 1)}{\pi^3 m^3}.$$

A similar formula holds for the integral in the y variable. So

$$B_{mn} = 4 \frac{2((-1)^n - 1)}{\pi^3 n^3} \frac{2((-1)^m - 1)}{\pi^3 m^3} \quad \text{for all } m,\ n = 1, 2, \ldots.$$

If either m or n is even, B_{mn} is zero. If both m and n are odd, then $B_{mn} = \frac{64}{\pi^6 m^3 n^3}$.
Hence, the solution is

$$
\begin{aligned}
u(x, y, t) &= \sum_{n \text{ odd}} \sum_{m \text{ odd}} \frac{64}{\pi^6 m^3 n^3} \sin m\pi x \sin n\pi y \cos \sqrt{m^2 + n^2}\, t \\
&= \sum_{l=0}^{\infty} \sum_{k=0}^{\infty} \Bigg\{ \frac{64}{\pi^6 (2k+1)^3 (2l+1)^3} \sin((2k+1)\pi x) \sin((2l+1)\pi y) \\
&\qquad \times \cos \sqrt{(2k+1)^2 + (2l+1)^2}\, t \Bigg\}.
\end{aligned}
$$

$\blacksquare$

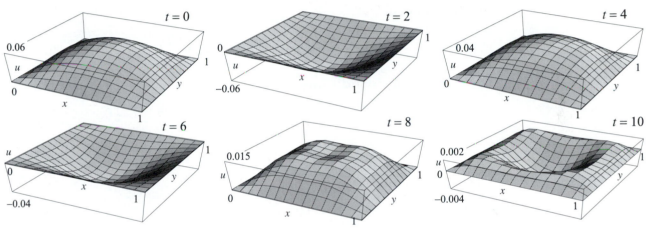

Figure 2 Shape of the membrane in Example 1 at various values of t.

Nodal Lines

As an experiment, we sprinkle sand on the surface of a vibrating membrane. It is observed that for certain frequencies the sand gathers on fixed curves on the surface. These curves, known as **nodal lines**, consist of the points that remain fixed as the membrane vibrates. We illustrate this phenomenon by analyzing the solution when it is given by

$$u_{mn}(x, y, t) = \sin \frac{m\pi}{a} x \sin \frac{n\pi}{b} y (B_{mn} \cos \lambda_{mn} t + B^*_{mn} \sin \lambda_{mn} t).$$

The points (x, y) that remain fixed for all t are solutions of the equation $u_{mn}(x, y, t) = 0$ for all $t > 0$. Thus to determine the nodal lines it is enough to solve the equation

$$\sin \frac{m\pi}{a} x \sin \frac{n\pi}{b} y = 0, \quad \text{for } 0 < x < a, \quad 0 < y < b.$$

For example, when $a = b = 1$, the nodal line for u_{21} is the line $x = \frac{1}{2}$. The nodal lines for u_{22} are $x = \frac{1}{2}$ and $y = \frac{1}{2}$. The nodal lines for u_{32} are $x = \frac{1}{3}$, $x = \frac{2}{3}$, $y = \frac{1}{2}$ (Figure 3).

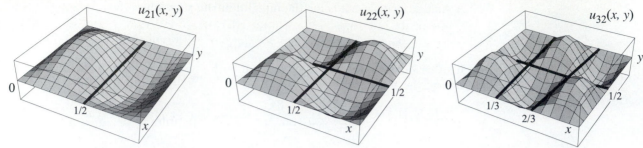

Figure 3 Nodal lines for u_{21}, u_{22}, u_{32}.

Two Dimensional Heat Equation

We end this section by stating the solution of the **two dimensional heat equation**

$$(11) \qquad \frac{\partial u}{\partial t} = c^2 \left(\frac{\partial^2 u}{\partial x^2} + \frac{\partial^2 u}{\partial y^2} \right), \quad 0 < x < a, \quad 0 < y < b, \quad t > 0,$$

with **boundary conditions**

$$u(0, y, t) = u(a, y, t) = 0, \quad 0 < y < b, \quad t > 0,$$

(12)

$$u(x, 0, t) = u(x, b, t) = 0, \quad 0 < x < a, \quad t > 0,$$

and **initial condition**

$$(13) \qquad u(x, y, 0) = f(x, y), \quad 0 < x < a, \quad 0 < y < b.$$

Applying the separation of variables method, as we did with the wave equation, we arrive at the following.

SOLUTION OF THE TWO DIMENSIONAL HEAT EQUATION FOR A RECTANGLE

The solution of the two dimensional heat boundary value problem (11)–(13) is

$$(14) \qquad u(x, y, t) = \sum_{n=1}^{\infty} \sum_{m=1}^{\infty} A_{mn} \sin \frac{m\pi}{a} x \sin \frac{n\pi}{b} y \, e^{-\lambda_{mn}^2 t},$$

where

$$(15) \qquad \lambda_{mn} = c\pi \sqrt{\frac{m^2}{a^2} + \frac{n^2}{b^2}},$$

and

$$(16) \qquad A_{mn} = \frac{4}{ab} \int_0^b \int_0^a f(x, y) \sin \frac{m\pi}{a} x \sin \frac{n\pi}{b} y \, dx \, dy$$
$$m, n = 1, 2, \ldots .$$

The two dimensional heat problem models the distribution of temperature in a thin rectangular plate with insulated sides, edges kept at zero temperature, and with an initial temperature distribution $f(x, y)$. The solution of the problem follows step by step the solution of the two dimensional wave equation. The details are outlined in the exercises.

EXAMPLE 2   A two dimensional heat problem
Solve the heat problem in a square plate with $a = b = 1$, and $c = \frac{1}{\pi}$. Assume that the edges are kept at zero temperature and the initial temperature distribution is $u(x, y, 0) = 100°$.

Solution We have

$$u(x, y, t) = \sum_{n=1}^{\infty} \sum_{m=1}^{\infty} A_{mn} \sin m\pi x \sin n\pi y \, e^{-\lambda_{mn}^2 t}$$

where $\lambda_{mn} = \sqrt{m^2 + n^2}$ and

$$
\begin{aligned}
A_{mn} &= 400 \int_0^1 \int_0^1 \sin m\pi x \sin n\pi \, y \, dx \, dy \\
&= \frac{400}{\pi^2} \frac{[1 - (-1)^m][1 - (-1)^n]}{mn}.
\end{aligned}
$$

Since A_{mn} is zero if either m or n is even, we get

$$u(x, y, t) = \frac{1600}{\pi^2} \sum_{k=0}^{\infty} \sum_{l=0}^{\infty} \frac{\sin(2l+1)\pi x \sin(2k+1)\pi y}{(2l+1)(2k+1)} e^{-\lambda_{(2l+1)(2k+1)}^2 \, t}.$$

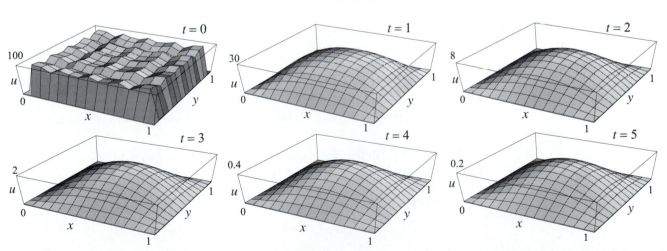

Figure 4 Temperature distribution in Example 2, at various times (using a partial sum with k and l running from 0 to 4). The first picture approximates the initial temperature distribution. Note the zero temperature on the boundary for $t > 0$.

Figure 4 shows the temperature distribution in the plate at various values of t. Note how the temperature smooths out across the membrane and tends to zero.

This is precisely what we would expect in view of the boundary conditions (zero temperature). ∎

Exercises 8.5

In Exercises 1–8, (a) solve the boundary value problem (1)–(3) with $a = b = 1$, $c = 1/\pi$, and the given functions f and g.

(b) *Plot a partial sum of the series solution at various values of t to illustrate the vibrations of the membrane. Include the graph at $t = 0$ and compare it to the graph of $f(x, y)$.*

1. $f(x, y) = \sin 3\pi x \sin \pi y$, $g(x, y) = 0$.   **2.** $f(x, y) = \sin \pi x \sin \pi y$, $g(x, y) = \sin \pi x$.

3. $f(x, y) = x(1 - x)y(1 - y)$, $g(x, y) = 2 \sin \pi x \sin 2\pi y$.

4. $f(x, y) = x \cos \frac{\pi x}{2} y(1 - y)$, $g(x, y) = 1$.

5. $f(x, y) = 0$, $g(x, y) = 1$.   **6.** $f(x, y) = 0$, $g(x, y) = x(1 - y)$.

7. $f(x, y) = x(1 - e^{x-1})y(1 - y^2)$, $g(x, y) = 0$.

8. $f(x, y) = \sin(xy) \cos \frac{\pi x}{2} \sin \pi y$, $g(x, y) = 0$.

In Exercises 9–10, find and plot the nodal lines for the given function.

9. $u(x, y, t) = \sin 4\pi x \sin \pi y \cos t$, $0 < x < 1$, $0 < y < 1$.

10. $u(x, y, t) = \sin \frac{3\pi}{2} x \sin \pi y \sin \sqrt{2}t$, $0 < x < 2$, $0 < y < 1$.

In Exercises 11–14, solve the heat equation (11) in a unit square ($a = b = 1$) with the given initial temperature distribution f. Assume that the edges are kept at zero temperature and that $c = 1$.

11. $f(x, y) = 100$ **12.** $f(x, y) = x(1 - x)y(1 - y)$ **13.** $f(x, y) = \sin \pi x \sin \pi y$

14.

$$f(x, y) = \begin{cases} 1 & \text{if } y \leq x, \\ 0 & \text{otherwise.} \end{cases}$$

15. (a) Take $c = 1$ in Example 2 and plot the solutions for various values of t. Approximate how long it will take for the maximum temperature in the plate to drop to 50°.

(b) What is the answer if $c = 2$? What is your conclusion about the speed of heat transfer as a function of c?

16. Establish relations (6) and (7) (orthogonality and normalization of the functions $\sin \frac{m\pi}{a} x \sin \frac{n\pi}{b} y$).

17. Project Problem: Solution of the two dimensional heat equation. In this exercise we derive the solution (14).

(a) Show that if we assume $u = X(x)Y(y)T(t)$, then the separation of variables method yields

$$X'' + \mu^2 X = 0, \quad X(0) = 0, \quad X(a) = 0,$$
$$Y'' + \nu^2 Y = 0, \quad Y(0) = 0, \quad Y(b) = 0,$$
$$T' + c^2(\mu^2 + \nu^2)T = 0.$$

(b) Obtain the product solutions

$$u_{mn}(x, y, t) = A_{mn}e^{-c^2\left[(\frac{m\pi}{a})^2 + (\frac{n\pi}{b})^2\right]t} \sin \frac{m\pi}{a} x \sin \frac{n\pi}{b} y.$$

(c) Given the initial temperature distribution $f(x, y)$, derive (14)–(17) using Theorem 1.

18. Project Problem: Double Fourier series representation. In this project, you are asked to justify Theorem 1, by showing how it can be obtained from the one dimensional Fourier series representation. The proof is sketched for you; what you need to do is justify the steps by quoting an appropriate result about Fourier series from the previous chapter.

(a) Let $f(x, y)$ be as in Theorem 1, and suppose that f is continuous and has continuous first partial derivatives. For fixed y, the function $x \mapsto f(x, y)$ has a half-range sine series expansion

$$f(x, y) = \sum_{m=1}^{\infty} b_m(y) \sin \frac{m\pi}{a} x,$$

where the Fourier sine coefficients depend on y and are given by

$$b_m(y) = \frac{2}{a} \int_0^a f(s, y) \sin \frac{m\pi}{a} s \, ds.$$

(b) For each m, the function $y \mapsto b_m(y)$ defined by the integral in (a) has a Fourier sine series

$$b_m(y) = \sum_{n=1}^{\infty} B_{mn} \sin \frac{n\pi}{a} y,$$

where the coefficients B_{mn} depend on m and are given by

$$B_{mn} = \frac{2}{b} \int_0^b b_m(t) \sin \frac{n\pi}{a} t \, dt = \frac{4}{ab} \int_0^b \int_0^a f(s, t) \sin \frac{n\pi}{a} s \sin \frac{n\pi}{a} t \, ds dt.$$

Substituting into the Fourier series representation in (a), we obtain the double Fourier series representation of Theorem 1.

8.6   Laplace's Equation in Rectangular Coordinates

Laplace's equation in two dimensions is one of the most studied equations in this book. Starting with Chapter 2 all the way to Chapter 6, we have developed methods from complex analysis, including conformal mappings, and applied them to solve Dirichlet problems involving Laplace's equation on various domains in the plane; our greatest successes were on the disk and the upper half-plane, where in each case we discovered nice closed forms for the solution in terms of the boundary function–the Poisson integral formulas. Yet with all the complex analysis methods, we were not able to express the solution of Laplace's equation over a rectangle in a practical form suitable for numerical computations. The time has come to settle this problem with the help of the separation of variables method and Fourier series. Let us first state the full Dirichlet problem consisting of **Laplace's equation** over an $a \times b$ rectangle,

$$(1) \qquad \frac{\partial^2 u}{\partial x^2} + \frac{\partial^2 u}{\partial y^2} = 0, \quad 0 < x < a, \quad 0 < y < b,$$

along with the **boundary conditions**

$$u(x, 0) = f_1(x), \quad u(x, b) = f_2(x), \quad 0 < x < a,$$
$$u(0, y) = g_1(y), \quad u(a, y) = g_2(y), \quad 0 < y < b,$$

which are illustrated in Figure 1. The functions f_1, f_2, f_3 and f_4 are assumed to be piecewise smooth.

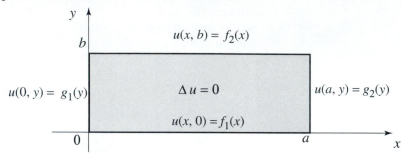

Figure 1 A general Dirichlet problem on a rectangle.

Rather than attacking this problem in its full generality, we will start by solving the special case where f_1, g_1, and g_2 are all zero. We will return to the general problem at the end of the section.

EXAMPLE 1   A Dirichlet problem on a rectangle

Solve the boundary value problem described in Figure 2 using the method of separation of variables.

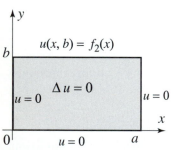

Figure 2 Dirichlet problem in Example 1.

Solution We begin by looking for product solutions $u(x, y) = X(x)Y(y)$. Substituting into (1) and using the separation method, we arrive at the equations

$$X'' + kX = 0, \quad Y'' - kY = 0,$$

where k is the separation constant, with the boundary conditions

$$X(0) = 0, \; X(a) = 0, \; \text{and } Y(0) = 0.$$

For the boundary value problem in X, you can check that the values $k \leq 0$ lead to trivial solutions only. For $k = \mu^2 > 0$, we obtain the solutions $X = c_1 \cos \mu x + c_2 \sin \mu x$. Imposing the boundary conditions on X forces $c_1 = 0$,

$$\mu = \mu_n = \frac{n\pi}{a}, \quad n = 1, 2, \ldots$$

and hence

$$X_n(x) = \sin \frac{n\pi}{a} x, \quad n = 1, 2, \ldots .$$

Turning now to Y with $k = \mu_n^2$, we find

$$Y = A_n \cosh \mu_n y + B_n \sinh \mu_n y .$$

Imposing $Y(0) = 0$, we find that $A_n = 0$, and hence

$$Y_n = B_n \sinh \mu_n y .$$

We have thus found the product solutions

$$B_n \sin \frac{n\pi}{a} x \sinh \frac{n\pi}{a} y \, .$$

Superposing these solutions, we get the general form of the solution

(2)
$$u(x,y) = \sum_{n=1}^{\infty} B_n \sin \frac{n\pi}{a} x \sinh \frac{n\pi}{a} y \, .$$

Finally, the boundary condition $u(x,b) = f_2(x)$ implies that

$$f_2(x) = \sum_{n=1}^{\infty} B_n \sinh \frac{n\pi b}{a} \sin \frac{n\pi}{a} x \, .$$

To meet this last requirement, we choose the coefficients $B_n \sinh \frac{n\pi b}{a}$ to be the Fourier sine coefficients of f_2 on the interval $0 < x < a$. Thus from Theorem 1, Section 7.4, it follows that

(3)
$$B_n = \frac{2}{a \sinh \frac{n\pi b}{a}} \int_0^a f_2(x) \sin \frac{n\pi}{a} x \, dx, \quad n = 1, 2, \dots \, .$$

The solution of the Dirichlet problem described in Figure 2 is therefore given by (2) with coefficients determined by (3).   ■

The following is a specific application of Example 1.

EXAMPLE 2   Steady-state temperature in a square plate
(a)   Determine the steady-state temperature distribution in a 1×1 square plate where one side is held at $100°$ and the other three sides are held at $0°$.

(b)   In particular, find the steady-state temperature at the center of the plate.

Solution (a)   By choosing coordinates appropriately, we can do this problem as a special case of Example 1 where $f_2(x) = 100°$ and $a = b = 1$. From (2) and (3), we have

$$u(x,y) = \sum_{n=1}^{\infty} B_n \sin n\pi x \sinh n\pi y,$$

where

$$B_n = \frac{200}{\sinh n\pi} \int_0^1 \sin n\pi x \, dx = \frac{200}{n\pi \sinh n\pi} (1 - \cos n\pi) \, .$$

Simplifying, we find the solution

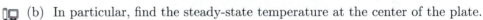

$$u(x,y) = \frac{400}{\pi} \sum_{k=0}^{\infty} \frac{\sin(2k+1)\pi x}{(2k+1)} \frac{\sinh(2k+1)\pi y}{\sinh(2k+1)\pi} \, .$$

Note that when $y = 1$, this reduces to the Fourier sine series expansion of 100, matching the boundary condition. Also, if $0 < y < 1$, the ratio of the hyperbolic sines decays exponentially with k, and hence leads to rapid convergence of the series.

(b) Plugging in $x = y = \frac{1}{2}$ to find the temperature at the center, we get

(4)
$$u\left(\frac{1}{2}, \frac{1}{2}\right) = \frac{200}{\pi} \sum_{k=0}^{\infty} \frac{(-1)^k}{(2k+1)} \frac{1}{\cosh(2k+1)\frac{\pi}{2}}$$

where we have simplified using

$$\sinh u = 2 \sinh \frac{u}{2} \cosh \frac{u}{2}$$

and

$$\sin(2k+1)\frac{\pi}{2} = (-1)^k.$$

By a judicious use of symmetry, we will show later that this series converges to 25. At this point, we can approximate the infinite sum to within 10^{-4} by adding the first four terms, obtaining 25.   ■

We now return to the general problem described in Figure 1. It turns out that this problem can be solved by using the solution to Example 1. The trick is to decompose the original problem into four subproblems, as described by Figure 3.

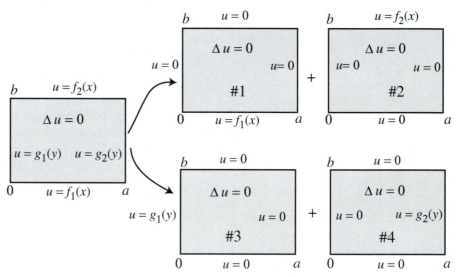

Figure 3 Linearity is used here to decompose the Dirichlet problem into the "sum" of four simpler Dirichlet subproblems.

Let u_1, u_2, u_3, u_4 be the solutions of subproblems 1, 2, 3, 4, respectively. By direct computation, we see that the function

$$u = u_1 + u_2 + u_3 + u_4$$

is the solution to the original problem given in Figure 1. Thus we need only determine u_1, u_2, u_3, u_4. The function u_2 is already computed in Example 1. We have

$$u_2(x, y) = \sum_{n=1}^{\infty} B_n \sin \frac{n\pi}{a} x \sinh \frac{n\pi}{a} y$$

where

$$(5) \qquad B_n = \frac{2}{a \sinh \frac{n\pi b}{a}} \int_0^a f_2(x) \sin \frac{n\pi}{a} x \, dx.$$

The other solutions can be found analogously. In particular, u_4 is the same as u_2 except that a and b are interchanged, as are x and y. Thus

$$u_4(x, y) = \sum_{n=1}^{\infty} D_n \sinh \frac{n\pi}{b} x \sin \frac{n\pi}{b} y,$$

where

$$(6) \qquad D_n = \frac{2}{b \sinh \frac{n\pi a}{b}} \int_0^b g_2(y) \sin \frac{n\pi}{b} y \, dy.$$

The solutions u_1 and u_3 are found similarly. We have

$$u_1(x, y) = \sum_{n=1}^{\infty} A_n \sin \frac{n\pi}{a} x \sinh \frac{n\pi}{a} (b - y),$$

where

$$(7) \qquad A_n = \frac{2}{a \sinh \frac{n\pi b}{a}} \int_0^a f_1(x) \sin \frac{n\pi}{a} x \, dx,$$

and

$$u_3(x, y) = \sum_{n=1}^{\infty} C_n \sinh \frac{n\pi}{b} (a - x) \sin \frac{n\pi}{b} y$$

where

$$(8) \qquad C_n = \frac{2}{b \sinh \frac{n\pi a}{b}} \int_0^b g_1(y) \sin \frac{n\pi}{b} y \, dy.$$

We have thus completely solved the Dirichlet problem in Figure 1. We summarize the solution as follows.

SOLUTION OF THE DIRICHLET PROBLEM IN A RECTANGLE

The solution of the two dimensional Dirichlet problem in Figure 1 is

$$u(x, y) = \sum_{n=1}^{\infty} A_n \sin \frac{n\pi}{a} x \sinh \frac{n\pi}{a} (b - y) + \sum_{n=1}^{\infty} B_n \sin \frac{n\pi}{a} x \sinh \frac{n\pi}{a} y$$

$$(9)$$

$$+ \sum_{n=1}^{\infty} C_n \sinh \frac{n\pi}{b} (a - x) \sin \frac{n\pi}{b} y + \sum_{n=1}^{\infty} D_n \sinh \frac{n\pi}{b} x \sin \frac{n\pi}{b} y,$$

where the coefficients A_n, B_n, C_n, and D_n are given by (5)–(8).

Let us now revisit Example 2(b). We shall derive the value of $u(\frac{1}{2}, \frac{1}{2})$ by using a symmetry argument. Consider the problem where the boundary data 100 is specified on all four sides of the plate. Clearly, this problem has solution $u(x, y) = 100$ throughout the square. On the other hand, we can certainly decompose this problem into the four subproblems described in Figure 3. Because of the symmetry, the four functions u_1, u_2, u_3, and u_4 assume the same value at the center. Equating the two versions of the solution at the center, we see that

$$100 = u_1(\frac{1}{2}, \frac{1}{2}) + u_2(\frac{1}{2}, \frac{1}{2}) + u_3(\frac{1}{2}, \frac{1}{2}) + u_4(\frac{1}{2}, \frac{1}{2}) = 4u_2(\frac{1}{2}, \frac{1}{2}).$$

Hence $u_2(\frac{1}{2}, \frac{1}{2}) = 25$. Note that by this means we have found the exact value of the series in (4). As a challenging exercise, try to sum this series by other means.

You may recall how in Section 8.3 we used steady-state solutions to solve heat problems with nonhomogeneous boundary conditions. As you can imagine, these methods are also useful in solving nonhomogeneous heat problems in higher dimensions. See Exercise 10 for an illustration.

Methods from Complex Analysis

Let $u(x, y)$ denote the solution of the Dirichlet problem in Figure 1. Since u is harmonic inside the rectangle, we can use properties of harmonic functions to derive properties of u, with interesting concrete physical applications.

**THEOREM 1
MAXIMUM-
MINIMUM
MODULUS
PRINCIPLE**

Consider the Dirichlet problem in Figure 1, in which the boundary functions f_1, f_2, f_3 and f_4 are piecewise smooth. Let M and m denote the maximum, respectively, minimum values of the boundary functions. Then for all (x, y) inside the rectangle, we have

$$m \leq u(x, y) \leq M.$$

If equality holds at any point inside the rectangle, then u is constant inside the rectangle.

Proof Apply Theorem 2 and Corollary 4, Section 3.8. Even though the corollary is stated for continuous boundary functions, it does hold in the setting of the present theorem. ∎

As a consequence of Theorem 1, we have the following.

**COROLLARY 1
UNIQUENESS OF
SOLUTION**

The solution of the Dirichlet problem in Figure 1 is unique.

Proof If u_1 and u_2 are solution of the same Dirichlet problem in Figure 1, then $u_1 - u_2$ is a solution of Laplace's equation inside the rectangle with zero boundary values, because $u_1 = u_2$ on the boundary. Thus $u_1 - u_2$ is harmonic inside the rectangle and equal 0 on the boundary and so it is identically 0 inside the rectangle, by Theorem 1. Hence $u_1 = u_2$ and so the solution is unique. ∎

In our next application, we think of the solution of the Dirichlet problem inside a rectangle in Figure 1 as representing a steady-state solution to a heat problem with given boundary temperature distribution. As we know from our study in Section 2.5, the curves of heat flow inside the rectangle are given by the level curves of the harmonic conjugate of u. We also know from Corollary 2, Section 3.8, that, because the rectangle is a simply connected region, u always has a harmonic conjugate inside the rectangle. Indeed, we have the following interesting result.

THEOREM 2
HARMONIC
CONJUGATE OF
THE SOLUTION

A harmonic conjugate of the solution (9) of the two dimensional Dirichlet problem in Figure 1 is

$$v(x, y) = \sum_{n=1}^{\infty} A_n \cos \frac{n\pi}{a} x \cosh \frac{n\pi}{a}(b - y) + \sum_{n=1}^{\infty} B_n \cos \frac{n\pi}{a} x \cosh \frac{n\pi}{a} y$$

(10)

$$+ \sum_{n=1}^{\infty} C_n \cosh \frac{n\pi}{b}(a - x) \cos \frac{n\pi}{b} y + \sum_{n=1}^{\infty} D_n \cosh \frac{n\pi}{b} x \cos \frac{n\pi}{b} y,$$

where the coefficients A_n, B_n, C_n, and D_n are given by (5)–(8).

Thus to get a harmonic conjugate of u in (9), we change every sine into a cosine and every hyperbolic sine into a hyperbolic cosine.

Proof Every term in the series solution (9) is of the form $\phi(x, y) = A \sin[\alpha(\beta - x)] \sinh[\alpha(\gamma - y)]$, where A, α, β, and γ are constants. It is instructive for you at this point to compute the Laplacian of ϕ and see that it is indeed 0, and hence ϕ is harmonic (Exercise 13). To compute a harmonic conjugate of ϕ, we use the Cauchy-Riemann equations as we did in Example 2, Section 2.5. We leave it to Exercise 13 to show that a harmonic conjugate of ϕ is $A \cos[\alpha(\beta - x)] \cosh[\alpha(\gamma - y)]$. Since the harmonic conjugate of a linear combination of harmonic functions is the linear combination of the harmonic conjugates, we see that the theorem holds if the sum in (9) is finite. For infinite sums, we must justify differentiation term by term. This process can be done on a case-by-case basis, if necessary, because the series converge very fast due to the presence of the exponentials in the denominator of the terms. ∎

As we now demonstrate, the theoretical results that we just derived are strikingly realistic, when illustrated by graphics with the help of a computer.

EXAMPLE 3   Isotherms and curves of heat flow

(a) Determine the isotherms and curves of heat flow in the steady-state heat problem in Example 2.

(b) Plot some isotherms and curves of heat flow and comment on the graphs.

Solution (a) The solution was found to be

$$u(x,\, y) = \frac{400}{\pi} \sum_{k=0}^{\infty} \frac{\sin(2k+1)\pi x}{(2k+1)} \frac{\sinh(2k+1)\pi y}{\sinh(2k+1)\pi}.$$

By the maximum-minimum principle, the temperature inside the plate will vary between the maximum and minimum values on the boundary, which are 100 and 0. Thus the isotherms are given by the level curves $u(x,\, y) = T$ where $0 < T < 100$. To determine the curves of heat flow, which are the orthogonal curves to the isotherms (see Section 2.5), we use Theorem 2 to find the harmonic conjugate v of u. Then the curves of heat flow are the level curves of v. As prescribed by Theorem 2, we have

$$v(x,\, y) = \frac{400}{\pi} \sum_{k=0}^{\infty} \frac{\cos(2k+1)\pi x}{(2k+1)} \frac{\cosh(2k+1)\pi y}{\sinh(2k+1)\pi}.$$

(b) To plot the isotherms and the curves of heat flow, we have approximated the infinite series by taking the first five terms from each. In Figure 4 we illustrate the isotherms corresponding to $T = 10, 25, 50,$ and 80. Note how the isotherm $u(x,\, y) = 25$ goes through the point $(\frac{1}{2},\, \frac{1}{2})$, which is confirmed by our computation of the temperature of that point in Example 2. In Figure 5, we plotted various level curves of v. Note how the heat flows from hottest to coldest, as our intuition would tell us. Finally in Figure 6, we superposed the previous two figures to illustrate the fact that the curves of heat flow are orthogonal to the isotherms. ∎

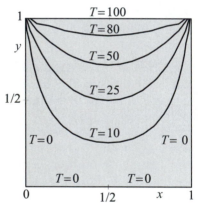

Figure 4 Isotherms in Example 3.

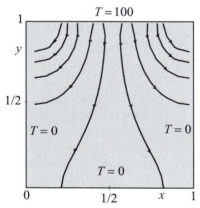

Figure 5 Curves of heat flow in Example 3.

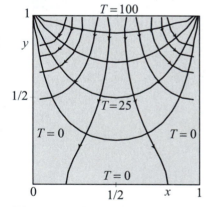

Figure 6 The isotherms and curves of heat flow are orthogonal.

Exercises 8.6

In Exercises 1–6, solve the Dirichlet problem in Figure 1 for the given data.

1. $f_2(x) = x,\ \ f_1 = g_1 = g_2 = 0,\ a = 1, b = 2.$

2. $f_1(x) = 0,\ f_2 = 100,\ g_1 = 0,\ g_2 = 100,\ a = 1, b = 1.$

3. $f_1(x) = 100,\ f_2 = g_1 = 0,\ g_2(y) = 100(1 - y),\ a = 2, b = 1.$

4. $f_1(x) = 1 - x,\ f_2(x) = x,\ g_1 = g_2 = 0,\ a = b = 1.$

5. $f_1(x) = \sin 7\pi x,\ f_2(x) = \sin \pi x,\ g_1(y) = \sin 3\pi y,\ g_2(y) = \sin 6\pi y,\ a = b = 1.$

6. $f_1(x) = \cos x$, $f_2(x) = x \sin x$, $g_1(y) = \pi - y$,

$$g_2(y) = \begin{cases} 3 & \text{if } 0 < y < \pi/2 \\ 0 & \text{if } \pi/2 < y < \pi \end{cases}$$

$a = b = \pi$.

7. (a) Show that the solution to Example 1, when expressed in terms of the Fourier sine coefficients b_n of $f_2(x)$, is

$$u(x, y) = \sum_{n=1}^{\infty} b_n \sin \frac{n\pi}{a} x \frac{\sinh \frac{n\pi}{a} y}{\sinh \frac{n\pi b}{a}}.$$

(b) State the corresponding result for the solution of the general problem (9).

8. Approximate the temperature at the center of the plate in Exercise 1.

9. Derive u_1 and the coefficient A_n in (7), using the method of separation of variables.

10. Project Problem: Two dimensional heat problem with nonzero boundary data. Consider the nonhomogeneous heat problem in Figure 7. Let $v(x, y)$ denote the solution of the Dirichlet problem $\Delta v = 0$ with boundary values as in Figure 7, and let $w(x, y, t)$ denote the solution of the heat problem $w_t = c^2 \Delta w$ with 0 boundary data and initial temperature distribution $w(x, y, 0) = f(x, y) - v(x, y)$.

(a) Show that the solution of the heat problem in Figure 7 is $u(x, y, t) = v(x, y) + w(x, y, t)$, where v is given by (9), and w is given by (14)–(17), Section 8.5, where, in (17), $f(x, y)$ is to be replaced by $f(x, y) - v(x, y)$.

(b) Solve the problem in the figure when $a = b = 1$, $c = 1$, $f_1 = 0$, $f_2 = 100$, $g_1 = g_2 = 0$, and $f(x, y) = 0$.

(c) Plot the temperature of the center in part (b) for $t > 0$ and discuss its behavior as $t \to \infty$.

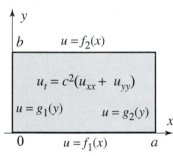

Figure 7 for Exercise 10.

11. Project Problem: Three dimensional Laplace's equation and Dirichlet problems. Steady-state temperature problems inside a three dimensional solid lead to Dirichlet problems associated with Laplace's equation in three dimensions:

$$\Delta u = \frac{\partial^2 u}{\partial x^2} + \frac{\partial^2 u}{\partial y^2} + \frac{\partial^2 u}{\partial z^2} = 0.$$

Consider such a problem as described in Figure 8. (See Exercise 12 for a generalization.) Follow the outlined steps to derive the solution

$$u(x, y, z) = \sum_{n=1}^{\infty} \sum_{m=1}^{\infty} A_{mn} \sin \tfrac{m\pi}{a} x \sin \tfrac{n\pi}{b} y \sinh \lambda_{mn} z,$$

$$\lambda_{mn} = \pi \sqrt{\left(\tfrac{m}{a}\right)^2 + \left(\tfrac{n}{b}\right)^2},$$

$$A_{mn} = \tfrac{4}{ab \sinh(c\lambda_{mn})} \int_0^b \int_0^a f(x, y) \sin \tfrac{m\pi}{a} x \sin \tfrac{n\pi}{b} y \, dx \, dy.$$

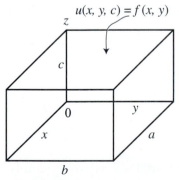

Figure 8 for Exercise 11.

(a) **Step 1:** Look for product solutions of the form $X(x)Y(y)Z(z)$. Use separation of variables and derive the equations

$$X'' + \mu^2 X = 0, \quad Y'' + \nu^2 Y = 0, \quad Z'' - (\mu^2 + \nu^2)Z = 0,$$
$$X(0) = X(a) = 0, \quad Y(0) = Y(b) = 0, \quad Z(0) = 0.$$

(You should justify the signs of the separation constants.)

(b) **Step 2**: Show that $\mu = \frac{m\pi}{a}$, $\nu = \frac{n\pi}{b}$ (m, $n = 1, 2, \ldots$), and derive the product solutions $u_{mn}(x, y, z) = A_{mn} \sin \frac{m\pi}{a} x \sin \frac{n\pi}{b} y \sinh \lambda_{mn} z$.

(c) **Step 3**: Complete the solution using Theorem 1, Section 8.5.

12.   How would you solve the problem in Exercise 11 if the boundary data were nonzero on some of the six faces?

Methods from Complex Analysis

13. Let $\phi(x, y) = A \sin[\alpha(\beta - x)] \, \sinh[\alpha(\gamma - y)]$, where A, α, β, and γ are constants. Show that ϕ is harmonic for all (x, y). Show that a harmonic conjugate of ϕ is $\psi(x, y) = A \cos[\alpha(\beta - x)] \, \cosh[\alpha(\gamma - y)]$.

14. (a) Determine the isotherms and curves of heat flow in the Dirichlet problem in Exercise 1.

(b) Plot some isotherms and curves of heat flow and comment on the graphs, as we did in Example 3.

15. Repeat Exercise 14 with the Dirichlet problem in Exercise 3.

8.7   The Method of Eigenfunction Expansions

In this section, we present a new approach to boundary value problems involving the Laplacian. The method that we describe is called the eigenfunction expansion method. We will use it to derive alternative solutions of previous problems that were solved using the separation of variables method. More importantly, we will solve nonhomogeneous problems that we could not solve by previous methods.

To highlight the power of the eigenfunction expansion method, we will tackle a Poisson boundary value problem on an $a \times b$ rectangle, as described by Figure 1, involving **Poisson's equation**

Note that if $f(x, y) \neq 0$, then the equation is nonhomogeneous. If you try the separation of variables method, you will quickly realize that this method does not apply.

(1)
$$\Delta u = \frac{\partial^2 u}{\partial x^2} + \frac{\partial^2 u}{\partial y^2} = f(x, y).$$

Poisson problems were introduced in Section 6.6 and solved over simply connected regions in terms of the Green's function of the region. So far we do not have an explicit formula for Green's function for a rectangle. The eigenfunction expansion method will be used to derive an alternative solution of Poisson's equation, which in turn can be used to derive Green's function for a rectangle (see Exercise 22).

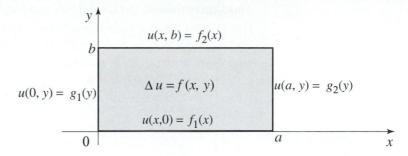

Figure 1 A general Poisson problem on a rectangle.

As we did in Section 6.6, the first step in solving the general Poisson problem in Figure 1 consists of decomposing the problem into two simpler subproblems, using the superposition principle, Theorem 1, Section 6.6. The decomposition in the present case is described by Figure 2.

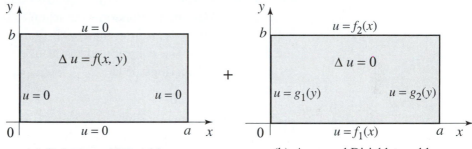

Figure 2 Decomposition of a general Poisson problem.

(a) Poisson problem with zero boundary data.

(b) A general Dirichlet problem.

Figure 2(b) describes a Dirichlet problem which can be solved by the methods of the previous section. Thus, to complete the solution of the problem in Figure 1, we only need to treat Poisson's equation with zero boundary data (Figure 2(a)).

We will take a hint from the solution of the Dirichlet problem, which is a special case of our problem when $f(x, y) = 0$, and consider the functions $\phi_{mn}(x, y) = \sin \frac{m\pi}{a} x \sin \frac{n\pi}{b} y$, which clearly satisfy the 0 boundary conditions in Figure 2(a). What else do they satisfy? Computing the Laplacian of $\phi_{mn}(x, y)$ we find

$$
\begin{aligned}
\Delta(\phi_{mn}) &= \sin \frac{n\pi}{b} y \frac{\partial^2}{\partial x^2} \left(\sin \frac{m\pi}{a} x \right) + \sin \frac{m\pi}{a} x \frac{\partial^2}{\partial y^2} \left(\sin \frac{n\pi}{b} y \right) \\
&= -\left[\left(\frac{m\pi}{a} \right)^2 + \left(\frac{n\pi}{b} \right)^2 \right] \sin \frac{m\pi}{a} x \sin \frac{n\pi}{b} y.
\end{aligned}
$$

So the Laplacian of ϕ_{mn} is a constant multiple of ϕ_{mn}. Using a terminology that is common in linear algebra, we call the constant

$$
\lambda_{mn} = \left(\frac{m\pi}{a} \right)^2 + \left(\frac{n\pi}{b} \right)^2 \quad (m, n = 1, 2, \dots)
$$

an **eigenvalue** of the Laplacian and the function $\phi_{mn}(x, y) = \sin \frac{m\pi}{a}x \sin \frac{n\pi}{b}y$ the corresponding **eigenfunction**. Because the effect of the Laplacian on an eigenfunction is very simple to describe (it just multiplies the eigenfunction by a constant), it makes sense to try for a solution to a problem that involves the Laplacian a series of eigenfunctions. This idea will be clarified as we carry out the solution of the problem in Figure 2(a). Let us try for a solution

Note that summing over all the functions $\sin \frac{m\pi}{a}x \sin \frac{n\pi}{b}y$ leads to a double Fourier sine series form of the solution (Section 8.5).

$$(2) \qquad \boxed{u(x,y) = \sum_{n=1}^{\infty} \sum_{m=1}^{\infty} E_{mn} \sin \frac{m\pi}{a}x \sin \frac{n\pi}{b}y,}$$

where E_{mn} are constants to be determined. It is straightforward to check that u satisfies the zero boundary conditions in Figure 2(a). We now solve for E_{mn} so as to satisfy Poisson's equation (1). As usual, in this process, we will assume that we can interchange the sums and the derivatives. Plugging the series (2) into the equation (1), and using the fact that the terms of the series are eigenfunctions of the Laplacian, we obtain

$$(3) \qquad \sum_{n=1}^{\infty} \sum_{m=1}^{\infty} -E_{mn} \underbrace{\left[\left(\frac{m\pi}{a} \right)^2 + \left(\frac{n\pi}{b} \right)^2 \right]}_{\lambda_{mn}} \sin \frac{m\pi}{a}x \sin \frac{n\pi}{b}y = f(x,y).$$

Thinking of (3) as a double Fourier sine series expansion of $f(x,y)$ (Theorem 1, Section 8.5), we conclude that $-E_{mn}\lambda_{mn}$ is the double sine series coefficient of f; equivalently,

$$(4) \qquad \boxed{E_{mn} = \frac{-4}{ab\lambda_{mn}} \int_0^b \int_0^a f(x,y) \sin \frac{m\pi}{a}x \sin \frac{n\pi}{b}y \, dx \, dy.}$$

This determines E_{mn} and completely solves Poisson's problem in Figure 1. We summarize our findings as follows.

SOLUTION OF POISSON'S EQUATION IN A RECTANGLE

The solution of the Poisson problem in Figure 1 is

$$u(x,y) = u_1(x,y) + u_2(x,y),$$

where u_1 is the solution of the Poisson problem with zero boundary data in Figure 2(a), and u_2 is the solution of the Dirichlet problem in Figure 2(b). The function u_1 is given by (2) and (4), and the function u_2 is given by (5)–(9), Section 8.6.

EXAMPLE 1   A Poisson problem with zero boundary data

Solve $\Delta u = 1$ in a 1×1 square, subject to the boundary condition $u = 0$ on all four sides of the square.

Solution   Note that in this case $u_2 = 0$. The function u_1 is given by (2) and (4) with $a = b = 1$ and $f(x, y) = 1$. Thus

$$E_{mn} = \frac{-4}{\lambda_{mn}} \int_0^1 \int_0^1 \sin m\pi x \sin n\pi y \, dx \, dy \, ,$$

where

$$\lambda_{mn} = (m\pi)^2 + (n\pi)^2 \, .$$

Evaluating the integrals (as we did in Example 2, Section 8.5), and plugging into (2), we obtain the solution

$$u(x, y) = \frac{-16}{\pi^4} \sum_{l=0}^{\infty} \sum_{k=0}^{\infty} \frac{\sin(2k + 1)\pi x \sin(2l + 1)y}{(2k + 1)(2l + 1)\left((2k + 1)^2 + (2l + 1)^2\right)} \, . \qquad \blacksquare$$

The Method of Eigenfunction Expansions

Let us summarize the steps that we used to solve Poisson's problem and describe in the process a general method for solving boundary value problems involving the Laplacian. We considered the functions

$$(5) \qquad \phi_{mn}(x, y) = \sin \frac{m\pi}{a} x \sin \frac{n\pi}{b} y,$$

which are solutions (or **eigenfunctions**) of the following **eigenvalue problem**:

$$(6) \qquad \Delta\phi(x, y) = -\lambda \, \phi(x, y),$$

These are the same boundary conditions as in Figure 2(a).

$$(7) \qquad \phi(0, y) = 0, \ \ \phi(a, y) = 0, \ \ \ \phi(x, 0) = 0, \ \ \phi(x, b) = 0 \, .$$

Equation (6) is known as the **Helmholtz equation**. It arises when separating variables in the heat and wave equations (see Exercise 16).   Each ϕ_{mn} corresponds to the **eigenvalue**

$$(8) \qquad \lambda = \lambda_{mn} = \left(\frac{m\pi}{a}\right)^2 + \left(\frac{n\pi}{b}\right)^2, \quad m, \ n = 1, \ 2, \ 3, \ \dots \, .$$

The method that consists of building up the solution of a boundary value problem as a sum of eigenfunctions of a related Helmholtz problem is known as the **method of eigenfunction expansions**. The success of this method on a given region depends on whether the eigenfunctions of the Helmholtz problem on that region form a **complete set**, in the sense that a function defined on the region can be expanded in a series in terms of the eigenfunctions, called an **eigenseries expansion**. For the rectangle, the eigenseries expansion is the double sine series expansion. Other regions will give rise to different types of series expansions, as we will see in the following chapters.

In our next example, we use the eigenfunction expansion method to solve a nonhomogeneous problems that cannot be tackled directly by the method of separation of variables.

EXAMPLE 2   The eigenfunction expansion method

Solve $\Delta u = u + 3$ in a 1×1 square ($0 < x < 1$, $0 < y < 1$), subject to the boundary condition $u = 0$ on all four sides of the square.

Solution The associated Helmholtz problem is the one given by (6)–(7), with $a = b = 1$. The eigenvalues are $\lambda_{mn} = (m\pi)^2 + (n\pi)^2$, with corresponding eigenfunctions $\phi_{mn}(x, y) = \sin m\pi x \sin n\pi y$. Thus $\Delta(\phi_{mn}(x, y)) = -((m\pi)^2 + (n\pi)^2)\phi_{mn}(x, y)$, as you can easily check. Our candidate for a solution is

$$u(x, y) = \sum_{n=1}^{\infty} \sum_{m=1}^{\infty} E_{mn} \sin m\pi x \sin n\pi y,$$

where E_{mn} are to be determined. Plugging this expression into the equation and assuming that the sums and the derivatives can be interchanged, we get

$$\Delta \left(\sum_{n=1}^{\infty} \sum_{m=1}^{\infty} E_{mn} \sin m\pi x \sin n\pi y \right) = \sum_{n=1}^{\infty} \sum_{m=1}^{\infty} E_{mn} \sin m\pi x \sin n\pi y + 3$$

$$\Leftrightarrow \sum_{n=1}^{\infty} \sum_{m=1}^{\infty} E_{mn} \Delta \left(\sin m\pi x \sin n\pi y \right) = \sum_{n=1}^{\infty} \sum_{m=1}^{\infty} E_{mn} \sin m\pi x \sin n\pi y + 3$$

$$\Leftrightarrow \sum_{n=1}^{\infty} \sum_{m=1}^{\infty} -E_{mn} \lambda_{mn} \sin m\pi x \sin n\pi y = \sum_{n=1}^{\infty} \sum_{m=1}^{\infty} E_{mn} \sin m\pi x \sin n\pi y + 3$$

$$\Leftrightarrow \sum_{n=1}^{\infty} \sum_{m=1}^{\infty} E_{mn}(-1 - \lambda_{mn}) \sin m\pi x \sin n\pi y = 3.$$

Thus $E_{mn}(-1 - \lambda_{mn})$ is the double sine Fourier coefficient of the function $f(x, y) = 3$, $0 < x < 1$, $0 < y < 1$. From Theorem 1, Section 8.5, we find

$$E_{mn}(-1 - \lambda_{mn}) = 4 \int_0^1 \int_0^1 3 \sin m\pi x \sin n\pi y \, dx \, dy.$$

Evaluating the integral and then solving for E_{mn}, we get

$$E_{mn} = -\frac{12}{1 + \lambda_{mn}} \frac{1}{mn\pi^2} ((-1)^m - 1)((-1)^n - 1)$$

$$= \begin{cases} 0 & \text{if either } m \text{ or } n \text{ is even,} \\ \dfrac{-48}{\pi^2 mn \, (1 + \lambda_{mn})} & \text{if both } m \text{ and } n \text{ are odd.} \end{cases}$$

This determines completely the solution.   ■

A One Dimensional Eigenvalue Problem

If you compare the solution (2) to the solution of the Dirichlet problem in Example 1, Section 8.6, you will notice that here we used double series, while in the previous section the solution was expressed as a single series. We will show that it is possible to obtain a single series solution to the Poisson problem, if we start with a related one dimensional homogeneous problem. This alternative approach has merit, since single series usually converge faster and are easier to work with than double series.

To solve the problem in Figure 2(a), consider the related *one dimensional* eigenvalue problem

We continue to use the notation $\Delta\phi(x)$ to show the relation to equation (1), but, of course, the notation $\frac{d^2\phi}{dx^2}$ is more appropriate here.

$$(9) \qquad\qquad \Delta\phi(x) = -\lambda\,\phi(x),$$
$$(10) \qquad\qquad \phi(0) = 0, \quad \phi(a) = 0.$$

It is straightforward to check that the eigenfunctions of this problem are

$$(11) \qquad\qquad \phi_m(x) = \sin\tfrac{m\pi}{a}x,$$

corresponding to the eigenvalues

$$(12) \qquad\qquad \lambda = \lambda_m = \left(\tfrac{m\pi}{a}\right)^2, \quad m = 1,\,2,\,3,\,\dots.$$

Now for a solution of the problem in Figure 2(a), we try

$$(13) \qquad\qquad \boxed{u(x,y) = \sum_{m=1}^{\infty} E_m(y)\sin\frac{m\pi}{a}x.}$$

Here we have to allow the coefficients E_{mn} to be functions of the second variable y. To complete the solution, we must solve for $E_m(y)$. Plugging (13) into (1), we obtain

$$(14) \qquad \boxed{\sum_{m=1}^{\infty} \underbrace{\left(E_m''(y) - \left(\frac{m\pi}{a}\right)^2 E_m(y)\right)}_{b_m(y)} \sin\frac{m\pi}{a}x = f(x,y),}$$

where $0 < x < a,\, 0 < y < b$. For each y in the interval $(0,b)$, think of (14) as a Fourier sine series expansion of the function $x \mapsto f(x,y)$. The Fourier sine coefficients in (14) are functions of y that we denote by $b_m(y)$. From (4), Section 7.4, we have

$$(15) \qquad\qquad b_m(y) = \frac{2}{a}\int_0^a f(x,y)\sin\frac{m\pi}{a}x\,dx.$$

Hence $E_m(y)$ is the unique solution of the second order nonhomogeneous initial value problem

(16) $$E_m''(y) - \left(\tfrac{m\pi}{a}\right)^2 E_m(y) = b_m(y),$$
(17) $$E_m(0) = 0 \quad \text{and} \quad E_m(b) = 0,$$

where the initial conditions (17) follow from (13) and the fact that $u(x,0) = 0$ and $u(x,b) = 0$. We now proceed to solve (16) using the variation of parameters formula (Appendix A.3). Two solutions of the associated homogeneous equation

$$E_m''(y) - \left(\frac{m\pi}{a}\right)^2 E_m(y) = 0$$

are readily found to be

(18) $$\boxed{h_1(y) = \sinh\left(\frac{m\pi}{a}(b-y)\right) \quad \text{and} \quad h_2(y) = \sinh\left(\frac{m\pi}{a}y\right).}$$

A straightforward computation shows that the Wronskian of h_1 and h_2 is

Recall from Appendix A.1:

$$W(h_1, h_2) = h_1 h_2' - h_1' h_2 .$$ (19)

$$\boxed{W(h_1, h_2) = \frac{m\pi}{a}\sinh\left(\frac{m\pi}{a}b\right).}$$

Since $W(h_1, h_2) \neq 0$, the two solutions are linearly independent. Applying the variation of parameters formula ((4), Appendix A.3), we obtain a particular solution of (16) in the form

(20) $$h_1(y)\int \frac{-h_2(y)}{W(h_1, h_2)}b_m(y)dy + h_2(y)\int \frac{h_1(y)}{W(h_1, h_2)}b_m(y)\,dy.$$

This formula determines a solution of (16) up to an arbitrary constant of integration. Different constants yield particular solutions of (16) which differ by linear combinations of h_1 and h_2. The (unique) solution of (16)–(17) is determined by a specific choice of the constants of integration. From (19), we see that $W(h_1, h_2)$ is independent of y, and hence it can be pulled outside the integrals in (20). You can check that

(21) $$\boxed{E_m(y) = \frac{-1}{\frac{m\pi}{a}\sinh\left(\frac{m\pi}{a}b\right)}\left[h_1(y)\int_0^y h_2(s)b_m(s)\,ds + h_2(y)\int_y^b h_1(s)b_m(s)\,ds\right]}$$

satisfies the initial conditions, and hence is the unique solution of the initial value problem (16)–(17). (Use the fact that $h_1(b) = 0$ and $h_2(0) = 0$.) This completely solves the problem in Figure 2(a) and yields a single series solution (13) with coefficients given by (21).

EXAMPLE 3   A single series solution of a Poisson problem

To find a single series expression of the solution in Example 1, we use (13). From (15), we have

$$(22) \qquad b_m(y) = 2 \int_0^1 \sin m\pi x \, dx = \frac{-2(\cos m\pi - 1)}{m\pi} = \begin{cases} \frac{4}{m\pi} & \text{if } m \text{ is odd,} \\ 0 & \text{if } m \text{ is even.} \end{cases}$$

From (18), we have

$$h_1(y) = \sinh(m\pi(1 - y)) \quad \text{and} \quad h_2(y) = \sinh(m\pi y).$$

From (21) and (22), we see that $E_m(y) = 0$ if m is even, and if m is odd

$$\begin{aligned} E_m(y) \;=\; & \frac{-4}{m^2 \pi^2 \sinh(m\pi)} \Big[\sinh(m\pi(1 - y)) \int_0^y \sinh(m\pi s)\, ds \\ & + \sinh(m\pi y) \int_y^1 \sinh(m\pi(1 - s))\, ds \Big]. \end{aligned}$$

Evaluating the integrals, we find that, for $m = 1, 3, 5, \ldots$,

$$\begin{aligned} E_m(y) \;=\; & \frac{-4}{m^3 \pi^3 \sinh(m\pi)} \big[\sinh(m\pi(1 - y))(\cosh(m\pi y) - 1) \\ & + \sinh(m\pi y)(\cosh(m\pi(1 - y)) - 1) \big]. \end{aligned}$$

Putting all this in (13), we get

$$u(x, y) = \sum_{k=0}^{\infty} E_{2k+1}(y) \sin((2k + 1)\pi x).$$

In Exercise 12 you are asked to use a computer to verify that indeed this solution is equal to the double series solution of Example 1.   ∎

Exercises 8.7

In Exercises 1–4, use double Fourier series to solve the Poisson problem in Figure 1 for the given data. Take $a = b = 1$.

1. $f(x, y) = x$, $f_1 = f_2 = 0$, $g_1 = g_2 = 0$.

2. $f(x, y) = \sin 2\pi x$, $f_1 = f_2 = 0$, $g_1 = g_2 = 0$.

3. $f(x, y) = \sin \pi x$, $f_1(x) = 0$, $f_2(x) = x$, $g_1 = g_2 = 0$.

4. $f(x, y) = xy$, $f_1(x) = 0$, $f_2(x) = x$, $g_1 = g_2 = 0$.

5. Use the eigenfunction expansions method to solve $\Delta u = 3u - 1$ inside the unit square $(0 < x < 1, 0 < y < 1)$, given that u is zero on the boundary.

6. Use the eigenfunction expansions method to solve $\Delta u = u_{xx} + u$ inside the unit square $(0 < x < 1, 0 < y < 1)$, given that u is zero on the boundary.

7. Solve the problem in Exercise 2 using one dimensional eigenfunction expansions.

8. Solve the problem in Exercise 1 using one dimensional eigenfunction expansions.

9. Use the eigenfunction expansions method to solve $\Delta u = u + u_{xx}$ inside the unit square $(0 < x < 1, 0 < y < 1)$, given that u is zero on the boundary.

10. Derive (18) and (19).

11. Show that (21) is a solution of (16)–(17).

12. Using a computer, verify that the solutions in Examples 1 and 3 are the same by evaluating them at various points inside the 1×1 square.

13. Project Problem: Dirichlet problem in a rectangle. Derive the result of Example 1, Section 8.6, using the method of eigenfunction expansions. Start by considering the following related one dimensional eigenvalue problem:

$$\frac{d^2\phi}{dx^2} = -\lambda\,\phi(x), \quad \phi(0) = 0, \quad \phi(a) = 0.$$

14. Project Problem: A Poisson problem in a box. Use (triple) Fourier series expansions to solve Poisson's equation

$$\Delta u(x, y, z) = f(x, y, z)$$

inside a rectangular box $(0 < x < a, \ 0 < y < b, \ 0 < z < c)$ subject to the boundary condition $u = 0$ on all six sides. In solving this problem, you should answer the following questions.
(a) What is the related homogeneous problem?
(b) What are the eigenvalues and their corresponding eigenfunctions?
(c) What is the analog of Theorem 1, Section 8.5, for functions of three variables?

15. Describe the solution of the problem in Exercise 14 if u is not zero on all sides of the box. [Hint: Exercises 11 and 12, Section 8.6.]

16. The Helmholtz equation. (a) Separate variables in the heat equation

$$u_t = c^2(u_{xx} + u_{yy} + u_{zz})$$

by letting $u(x, y, z, t) = \phi(x, y, z)T(t)$, where (x, y, z) is the spatial variable and t is the time variable. Show that the spatial part ϕ satisfies the Helmholtz equation.
(b) Repeat part (a) for the one and two dimensional heat equations.
(c) Repeat part (a) for the three dimensional wave equation.

17. Project Problem: The heat equation via the eigenfunction expansions method. Solve the two dimensional heat boundary value problem of Section 8.5 by using the eigenfunction expansions method. Note: The related eigenfunction problem is given by (6) and (7). The coefficients in the series solution should be functions of t.

Project Problem: Nonhomogeneous heat boundary value problem. Do Exercises 18 and 20 to solve a nonhomogeneous heat boundary value problem. To illustrate the solution, do Exercises 19 and 21.

18. Nonhomogeneous heat boundary value problem: a special case. The boundary value problem

$$u_t = c^2 u_{xx} + q(x,t), \quad t > 0, \ 0 < x < L,$$
$$u(0,t) = 0, \ u(L,t) = 0, \quad t > 0,$$
$$u(x,0) = f(x),$$

models heat transfer in a uniform bar of length L with insulated lateral surface and initial temperature distribution $f(x)$, given that the ends of the bar are kept at $0°$ temperature. The addition of the term $q(x,t)$ accounts for a time dependent heat source. We will solve this problem using the eigenfunction expansions method.
(a) Solve the related eigenvalue problem

$$\frac{d^2\phi}{dx^2} = -\lambda\,\phi(x), \quad \phi(0) = 0, \quad \phi(L) = 0.$$

(b) Write the Fourier sine series of f and q as

$$f(x) = \sum_{n=1}^{\infty} b_n \sin \frac{n\pi}{L} x$$

where

$$b_n = \frac{2}{L} \int_0^L f(x) \sin \frac{n\pi}{L} x\, dx;$$

and

$$q(x,t) = \sum_{n=1}^{\infty} q_n(t) \sin \frac{n\pi}{L} x$$

where

$$q_n(t) = \frac{2}{L} \int_0^L q(x,t) \sin \frac{n\pi}{L} x\, dx.$$

To apply the eigenfunction expansions method, let

$$u(x,t) = \sum_{n=1}^{\infty} B_n(t) \sin \frac{n\pi}{L} x.$$

Show that $B_n(t)$ satisfies the first order initial value problem

$$\frac{dB_n}{dt} + c^2 \lambda_n B_n = q_n(t), \quad B_n(0) = b_n,$$

where $\lambda_n = \left(\frac{n\pi}{L}\right)^2$.
(c) Show that

$$B_n(t) = e^{-c^2 \lambda_n t} \left(b_n + \int_0^t q_n(s) e^{c^2 \lambda_n s}\, ds \right).$$

19. (a) Work out all the details in Exercise 18 for the specific case when $c = 1$, $L = \pi$, $q(x,t) = xe^{-t}$, and $f(x) = 0$.
(b) Plot the solution for various values of t and comment on the graphs as $t \to \infty$. Does the behavior of the graphs meet with your expectations?

20. Nonhomogeneous heat boundary value problem: general case. In this exercise, we will solve the heat equation of Exercise 18 with the same initial condition but with the following time dependent boundary conditions

$$u(0, t) = A(t), \quad u(L, t) = B(t), \quad t > 0.$$

To motivate the solution, refer to Section 8.3 (problem (6)–(8)), where we did the case when $A(t)$ and $B(t)$ are constant.

(a) Show that the solution is of the form

$$u(x, t) = v(x, t) + \frac{B(t) - A(t)}{L} x + A(t),$$

where $v(x, t)$ is the solution of the following heat problem:

$$v_t = c^2 v_{xx} + q(x, t) - \frac{B'(t) - A'(t)}{L} x - A'(t), \quad t > 0, \quad 0 < x < L,$$
$$v(0, t) = 0, \quad v(L, t) = 0, \quad t > 0,$$
$$v(x, 0) = f(x) - \frac{B(0) - A(0)}{L} x - A(0).$$

(b) Solve the problem using (a) and Exercise 18.

(c) Describe what you get when $A(t)$ and $B(t)$ are constant. Compare your result to that of Section 8.3.

21. (a) Work out all the details in Exercise 20 for the specific case: $c = 1$, $L = \pi$, $q(x, t) = 0$, $A(t) = te^{-t}$, $B(t) = 0$, and $f(x) = 0$.

(b) Plot the solution for various values of t and comment on the graphs as $t \to \infty$.

22. Project problem: Green's function for a rectangle. Follow the outlined steps to compute this function on the $a \times b$ rectangle in Figure 1. You should review the definition of Green's function from Section 6.5.

(a) Replace the formula for the coefficients (4) into (2), proceed formally to interchange the sums and the integrals, and write the solution of Poisson's problem in Figure 1 in the form

$$u(x, y) = \frac{1}{2\pi} \int_0^b \int_0^a h(s, t) G(x, y, s, t) \, ds dt,$$

where

$$G(x, y, s, t) = -\frac{8}{ab} \sum_{n=1}^{\infty} \sum_{m=1}^{\infty} \frac{\sin \frac{m\pi}{a} s \sin \frac{m\pi}{a} x \sin \frac{n\pi}{b} t \sin \frac{n\pi}{b} y}{\lambda_{mn}},$$

and

$$\lambda_{mn} = \left(\frac{m\pi}{a}\right)^2 + \left(\frac{n\pi}{b}\right)^2.$$

(b) Modify the notation in Theorem 2, Section 6.6, and conclude that the solution of Poisson's problem in Figure 1 is

$$u(x, y) = \frac{1}{2\pi} \int_0^b \int_0^a h(s, t) G(x, y, s, t) \, ds dt,$$

where G is Green's function for the rectangle. Conclude from (a) that Green's function for a rectangle is given by the formula in (a). Some properties of Green's function that we stated in Theorem 2, Section 6.5, are clear from the formula for G in (a). For example, you should verify the symmetric property of Green's function, and the fact that $G(x, y, s, t) = 0$ for all (s, t) on the boundary of R. The other two properties in Theorem 2, Section 6.5, that $G(x, y, s, t) \leq 0$ for all $z = (x, y)$ and $\zeta = (s, t)$ in R, and that G blows up like $\ln|z - \zeta|$ at $z = (x, y)$, are not immediate from the formula. You can verify these facts graphically by plotting G as a function of $\zeta = (s, t)$ for a fixed point $z = (x, y)$ in R. (These properties were established in general in Section 6.5.)

8.8 Neumann Problems and Robin Conditions

So far we have have successfully solved boundary value problems using tools from complex analysis, the separation of variables method, and the eigenfunction expansion methods. We took advantage of linearity and used superposition principles to simplify problems and break them into simpler subproblems. In this section, we devise methods based on a combination of the tools that we just listed, and greatly expand the scope of our applications.

The problems that we present introduce further classical topics, such as Robin and Neumann conditions. More important, they illustrate how simple ideas can be combined together to devise powerful techniques. To simplify the presentation, all the examples in this section will involve Laplace's equation on the $a \times b$-rectangle R, with zero boundary conditions on the two vertical sides,

$$(1) \qquad u(0, y) = 0 \quad \text{and} \quad u(a, y) = 0 \quad \text{for all } 0 < y < b,$$

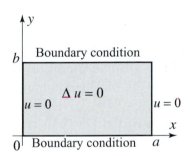

Figure 1 Laplace's equation in a rectangle with zero boundary conditions on the vertical sides.

as shown in Figure 1. It will become clear from the examples that more general problems can be reduced to this situation using superposition.

We start with a simple result that will be used often.

PROPOSITION 1
PRODUCT
SOLUTIONS

If $\phi(x, y) = X(x)Y(y)$ is a product solution of Laplace's equation in the $a \times b$-rectangle of Figure 1, with the zero boundary conditions on the vertical sides, then

$$(2) \qquad \phi(x, y) = \phi_m(x, y) = \sin\frac{m\pi}{a}x \left(A_m \cosh\frac{m\pi}{a}y + B_m \sinh\frac{m\pi}{a}y \right),$$

where $m = 1, 2, \ldots$, and A_m and B_m are arbitrary constants.

In some applications, it will be convenient to write the solution (2) in the form

$$(3) \qquad \phi_m(x, y) = \sin\frac{m\pi}{a}x \left(\alpha_m \cosh\frac{m\pi}{a}(\beta - y) + \beta_m \sinh\frac{m\pi}{a}(\beta - y) \right),$$

where β is a fixed number. You can check directly that (3) is a solution of the problem in Figure 1, or you can verify that (2) and (3) are equivalent by expressing the hyperbolic cosine and sine in terms of the exponential function.

Proof As we have done several times before (see, for example, Section 8.6), using the separation of variables method, we find that if $\phi(x, y) = XY$ is a solution, then X must satisfy $X'' + kX = 0$ and $X(0) = X(a) = 0$. The only non-trivial solutions are $X = X_m = \sin \frac{m\pi}{a} x$. Moving to the Y component, since $\phi(x, y) = \sin\left(\frac{m\pi}{a} x\right) Y(y)$ must satisfy $\phi_{xx} + \phi_{yy} = 0$, we get $-\left(\frac{m\pi}{a}\right)^2 \sin\left(\frac{m\pi}{a} x\right) Y(y) + \sin\left(\frac{m\pi}{a} x\right) Y''(y) = 0$; equivalently, $Y'' - \left(\frac{m\pi}{a}\right)^2 Y = 0$, which implies that $Y = A_m \cosh \frac{m\pi}{a} y + B_m \sinh \frac{m\pi}{a} y$, as desired. ∎

In a boundary value problem, we will call a **Dirichlet condition** any condition that specifies the values of the solution u on the boundary, and a **Neumann condition** any condition that specifies the normal derivative $\frac{\partial u}{\partial n}$ on the boundary. A **Robin condition** is any condition that specifies $\frac{\partial u}{\partial n} + au$ on the boundary, where a is a function of x and y. Thus Dirichlet and Neumann conditions are special cases of Robin conditions. In a rectangular region, the normal derivative on the boundary is a derivative in the positive or negative direction of the x- or y-axes. Thus a Robin condition is a condition that specifies the values of $u_x + au$ or $u_y + au$ on the boundary of the rectangle.

With this terminology, let us solve a problem with mixed Dirichlet and Neumann conditions.

EXAMPLE 1   Mixed Dirichlet and Neumann conditions

Solve the boundary value problem in Figure 2.

Solution Appealing to Proposition 1, the product solutions are of the form

$$
(4) \qquad \phi_m(x, y) = \sin \frac{m\pi}{a} x \left(A_m \cosh \frac{m\pi}{a} y + B_m \sinh \frac{m\pi}{a} y \right).
$$

As we know, these solutions already satisfy the boundary conditions on the vertical sides. We can specify the constants A_m and B_m in order to satisfy the zero Dirichlet condition on the horizontal side $y = 0$. Indeed, $u(x, 0) = 0$ implies that $A_m \sin \frac{m\pi}{a} x = 0$, which in turn implies that $A_m = 0$. So $\phi_m(x, y) = B_m \sin \frac{m\pi}{a} x \sinh \frac{m\pi}{a} y$. To satisfy the nonhomogeneous Neumann condition on the upper horizontal side, we will superpose the product solutions and try

$$
(5) \qquad u(x, y) = \sum_{m=1}^{\infty} B_m \sin \frac{m\pi}{a} x \sinh \frac{m\pi}{a} y.
$$

Proceeding formally to compute u_y by differentiating the series term by term, and then setting $u_y = f(x)$ when $y = b$, we obtain

$$
f(x) = u_y(x, b) = \sum_{m=1}^{\infty} B_m \frac{m\pi}{a} \cosh(\frac{m\pi}{a} b) \sin \frac{m\pi}{a} x, \quad 0 < x < a.
$$

Recognizing this as the Fourier sine series expansion of $f(x)$ on the interval $0 < x < a$, it follows that $B_m \frac{m\pi}{a} \cosh(\frac{m\pi}{a} b)$ is the Fourier sine coefficient of $f(x)$;

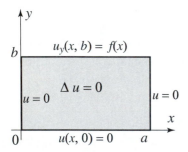

Figure 2 Mixed Dirichlet and Neumann boundary conditions in Example 1. The Neumann condition is nonhomogeneous.

equivalently,

(6)
$$B_m = \frac{2}{\pi m \cosh(\frac{m\pi}{a}b)} \int_0^a f(x) \sin \frac{m\pi}{a} x \, dx.$$

This determines the coefficients B_m and solves the problem completely.   ∎

We next consider a problem with a Robin condition.

EXAMPLE 2   Mixed Dirichlet, Neumann, and Robin conditions

Solve the boundary value problem in Figure 3.

Solution From (2), we have the product solutions

$$\phi(x, y) = \sin \frac{m\pi}{a} x \left(A_m \cosh \frac{m\pi}{a} y + B_m \sinh \frac{m\pi}{a} y \right).$$

In order to satisfy the homogeneous Robin condition, we compute

$$\phi_y = \sin \frac{m\pi}{a} x \left(\frac{m\pi}{a} A_m \sinh \frac{m\pi}{a} y + \frac{m\pi}{a} B_m \cosh \frac{m\pi}{a} y \right).$$

Thus $\phi_y(x, 0) + 2\phi(x, 0) = \sin \frac{m\pi}{a} x \left(\frac{m\pi}{a} B_m + 2A_m \right) = 0$, which implies that $B_m = -\frac{2a}{m\pi} A_m$, and so the product solutions are of the form

$$A_m \sin \frac{m\pi}{a} x \left(\cosh \frac{m\pi}{a} y - \frac{2a}{m\pi} \sinh \frac{m\pi}{a} y \right).$$

Moving to the last nonhomogeneous Neumann boundary condition on the upper horizontal side, we superpose the product solutions and take

(7)
$$u(x, y) = \sum_{m=1}^{\infty} A_m \sin \frac{m\pi}{a} x \left(\cosh \frac{m\pi}{a} y - \frac{2a}{m\pi} \sinh \frac{m\pi}{a} y \right).$$

(We always deal with the nonhomogeneous condition last.) From $u_y(x, b) = f(x)$, we get

$$f(x) = u_y(x, b) = \sum_{m=1}^{\infty} A_m \sin \frac{m\pi}{a} x \left(\frac{m\pi}{a} \sinh(\frac{m\pi}{a}b) - 2 \cosh(\frac{m\pi}{a}b) \right),$$

which is the sine Fourier series of $f(x)$ on the interval $0 < x < a$; and so

(8)
$$A_m = \frac{2}{a \left(\frac{m\pi}{a} \sinh(\frac{m\pi}{a}b) - 2 \cosh(\frac{m\pi}{a}b) \right)} \int_0^a f(x) \sin \frac{m\pi}{a} x \, dx,$$

which determines A_m and solves the problem.   ∎

In the previous examples, we always dealt with the nonhomogeneous condition last, by using the infinite series of product solutions. This approach relied heavily on the fact that, in each case, the partial differential equation and three of the boundary conditions were homogeneous; so the infinite sum

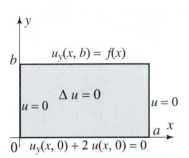

Figure 3 Mixed homogeneous Robin condition and nonhomogeneous Neumann boundary condition in Example 2.

of product solutions still satisfied the equation and the homogeneous conditions. In boundary value problems where more than one boundary condition is nonhomogeneous, we can use linearity to break up the problem into subproblems in which at most one boundary condition is nonhomogeneous. We illustrate with examples.

EXAMPLE 3   Decomposition of boundary conditions

The boundary value problem in Figure 4(a) has two nonhomogeneous boundary conditions on the horizontal sides. We can write it as the sum of two subproblems where three of the boundary conditions are homogeneous, in each case (Figures 4(b) and (c)). You should check that if u_1 is a solution of problem #1 and u_2 is a solution of problem #2, then $u = u_1 + u_2$ is a solution of the original problem in Figure 4(a). We now proceed to solve the subproblems.

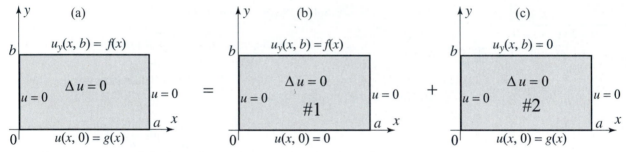

Figure 4 Decomposition of the boundary value problem in Example 3.

Problem # 1 in Figure 4(b) is solved in Example 1. We have from (5) and (6)

$$u_1(x, y) = \sum_{m=1}^{\infty} B_m \sin \frac{m\pi}{a} x \sinh \frac{m\pi}{a} y,$$

where

$$B_m = \frac{2}{\pi m \cosh(\frac{m\pi}{a} b)} \int_0^a f(x) \sin \frac{m\pi}{a} x \, dx.$$

For problem #2, it will be more convenient to use the product solutions in the form (3): $\phi(x, y) = \sin \frac{m\pi}{a} x \left(A_m \cosh \frac{m\pi}{a} (b - y) + B_m \sinh \frac{m\pi}{a} (b - y) \right)$. Applying the homogeneous boundary condition $u_y(x, b) = 0$, we obtain $0 = -B_m \sin \frac{m\pi}{a} x$; which implies that $B_m = 0$. Finally, in order to satisfy the last nonhomogeneous Dirichlet condition on the lower horizontal side, we superpose the product solutions and take

$$u_2(x, y) = \sum_{m=1}^{\infty} A_m \sin \frac{m\pi}{a} x \cosh \frac{m\pi}{a} (b - y).$$

Evaluating at $y = 0$, we get

$$g(x) = u(x, 0) = \sum_{m=1}^{\infty} A_m \cosh(\frac{m\pi}{a} b) \sin \frac{m\pi}{a} x, \quad 0 < x < a.$$

Recognizing this as the Fourier sine series expansion of $g(x)$ on the interval $0 < x < a$, it follows that

$$A_m = \frac{2}{a \cosh(\frac{m\pi}{a}b)} \int_0^a g(x) \sin \frac{m\pi}{a}x\, dx.$$

This determines the coefficients A_m, solves problem #2, and thus completes the solution.   ∎

In the next example, we have to break up a nonhomogeneous boundary condition into two parts.

EXAMPLE 4   Decomposition of boundary conditions

The boundary value problem in Figure 5(a) has two nonhomogeneous boundary conditions on the horizontal sides. To solve the problem, we break it up into two subproblems with three homogeneous boundary conditions each. It is straightforward to check that if u_1 is a solution of problem #1 and u_2 is a solution of problem #2, then $u = u_1 + u_2$ is a solution of the original problem.

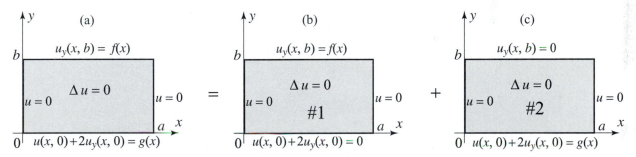

Figure 5 Decomposition of the boundary value problem in Example 4.

The problem in Figure 5(b) was solved in Example 2. The problem in Figure 5(c) has only one nonhomogeneous boundary condition and can be solved by the methods of this section. We leave the details to Exercise 3.   ∎

Exercises 8.8

In Exercises 1–6, solve the boundary value problem described by the figure (Figures 6 − 11).

1.

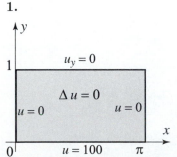

Figure 6 for Exercise 1.

2.

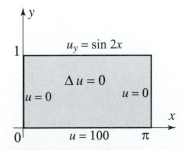

Figure 7 for Exercise 2.

3.

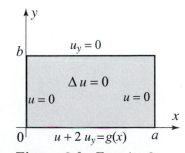

Figure 8 for Exercise 3.

4.

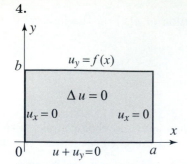

Figure 9 for Exercise 4.

5.

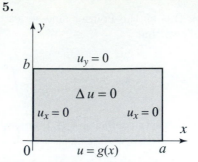

Figure 10 for Exercise 5.

6.

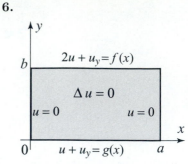

Figure 11 for Exercise 6.

In Exercises 7–10, (a) solve the given boundary value problem. (b) Check the validity of your answer by verifying the boundary conditions.

7. The boundary value problem in Figure 2, with $a = b = \pi$ and $f(x) = \sin 2x$.

8. The boundary value problem in Figure 3, with $a = b = \pi$ and $f(x) = \sin x$.

9. The boundary value problem in Figure 4, with $a = b = \pi$ and $f(x) = g(x) = 1$.

10. The boundary value problem in Figure 8 with $a = b = \pi$ and $g(x) = 1$.

Topics to Review

Fourier series (Sections 7.1–7.4) and the separation of variables method (Section 8.2) are crucial in this chapter. Section 9.1 treats boundary value problems on a disk, similar to those of Sections 8.6 and 8.7 on a rectangular region. The new ordinary differential equation that arises from the separation of variables method in Section 9.1 is Euler's equation, and in Sections 9.2–9.5 it is Bessel's equation. Euler's equation is treated in detail in Appendix A.3. The required knowledge of Bessel's equation and its solutions, the Bessel functions, is presented in Sections 9.6 and 9.7. This material can be recalled as needed or it can be covered before starting Section 9.2. In Section 9.5 we develop the eigenfunction expansions method on the disk. For this section, it is recommended to review the material of Section 8.7.

Looking Ahead

You recall from Section 8.4 how by varying the boundary conditions we were led to new types of series expansions. In this chapter we will solve boundary value problems over circular and cylindrical domains. It should not surprise you that the solutions will entail new series expansions (for example, Bessel series). These series look quite complicated at first, but with the help of a computer system, you will be able to plot them and see that they behave very much like Fourier series.

9

PARTIAL DIFFERENTIAL EQUATIONS IN POLAR AND CYLINDRICAL COORDINATES

One cannot understand . . . the universality of laws of nature, the relationship of things, without an understanding of mathematics. There is no other way to do it.
                                                                    -Richard P. Feynman

In the previous chapter we used our knowledge of Fourier series to solve several interesting boundary value problems by the method of separation of variables. The success of our method depended to a large extent on the fact that the domains under consideration were easily described in Cartesian coordinates. In this chapter we address problems where the domains are easily described in polar and cylindrical coordinates. Specifically, we consider boundary value problems for the wave, heat, Laplace, and Poisson equations over disks or cylinders. Upon restating these problems in suitable coordinate systems and separating variables, we will encounter new ordinary differential equations, such as Euler's equation and Bessel's equation. The full implementation of the separation of variables method will lead us to study expansions of functions in terms of Bessel functions (the solutions of Bessel's equation) in ways analogous to Fourier series expansions.

You do not need to know about Bessel series to start the chapter. As needed, we will refer to Sections 9.6 and 9.7, where you will find a comprehensive treatment of these special series expansions.

Bessel functions of integer order were introduced in Section 4.3 as examples of entire functions, that are defined by a power series. In this chapter, you will see this power series arising as the solution of Bessel's differential equation, when we apply the Frobenius method. Several other interesting facts about Bessel functions, including their integral representation, can be derived using complex methods. This material will be needed in some exercises and proofs to establish many of the peculiar properties of Bessel functions.

9.1   Laplace's Equation in Polar Coordinates

In the previous chapter, we applied the separation of variables method and the eigenfunction expansions method to solve boundary value problems involving the Laplacian in Cartesian coordinates. In this chapter, we present a similar approach to problems involving the Laplacian in polar coordinates (Figure 1). Why polar coordinates? It should be clear by now that the geometry of the region under consideration and the boundary conditions affect the solution in a fundamental way. For problems over regions with circular symmetry, such as disks and cylinders, it is to our advantage to use polar coordinates (r, θ), since the boundaries in these cases have simple expressions. For example, in polar coordinates, the boundary of a disk centered at the origin is the circle $r = a$. Another advantage will become apparent as we proceed with the solutions: In polar coordinates, we can separate the variables in Laplace's equation.

In the first part of this section, we will apply the separation of variables method to derive the general solution of Laplace's equation in a disk, without specifying the boundary conditions.

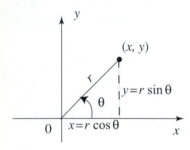

Figure 1 Polar coordinates.

Separation of Variables in Laplace's Equation

Recall from Exercise 47, Section 2.5, the **polar form** of the Laplacian

(1)
$$\Delta u = \nabla^2 u = \frac{\partial^2 u}{\partial r^2} + \frac{1}{r}\frac{\partial u}{\partial r} + \frac{1}{r^2}\frac{\partial^2 u}{\partial \theta^2}.$$

Thus, in polar coordinates, Laplace's equation on a disk centered at the origin with radius a becomes

(2)
$$\Delta u = \frac{\partial^2 u}{\partial r^2} + \frac{1}{r}\frac{\partial u}{\partial r} + \frac{1}{r^2}\frac{\partial^2 u}{\partial \theta^2} = 0, \quad 0 < r < a,\ 0 < \theta < 2\pi.$$

To solve this equation using the method of separation of variables, we start by looking for product solutions of (2) of the form $u(r, \theta) = R(r)\Theta(\theta)$. Before we plug into (2), we observe that Θ is necessarily 2π-periodic, being a function of the polar angle θ. This is essentially a boundary condition, which will be needed to determine the solutions. Now, plugging $u(r, \theta) = R(r)\Theta(\theta)$ into (2), we obtain

$$R''\Theta + \frac{1}{r}R'\Theta + \frac{1}{r^2}R\Theta'' = 0 \quad \Rightarrow \quad R''\Theta + \frac{1}{r}R'\Theta = -\frac{1}{r^2}R\Theta''.$$

To separate the variables, we multiply by r^2 and divide by $R\Theta$, and get

$$r^2\frac{R''}{R} + r\frac{R'}{R} = -\frac{\Theta''}{\Theta}.$$

Since the variables are separated, the only way to get equality is to have both sides constant and equal. Thus

$$r^2 \frac{R''}{R} + r \frac{R'}{R} = k \quad \text{and} \quad \frac{\Theta''}{\Theta} = -k;$$

equivalently,

$$r^2 R'' + rR' - kR = 0 \quad \text{and} \quad \Theta'' + k\Theta = 0.$$

At this point we have to discuss the possibilities for the separation constant k. We start with the equation in θ, since we have a crucial information about the function Θ; namely, that it is 2π-periodic. The solutions of the second order linear differential equation $\Theta'' + k\Theta = 0$ are either exponential or cosine and sine functions. You can check that the only way to get 2π-periodic solutions is to take $k = n^2$, $n = 0, 1, 2, \ldots$, with corresponding solutions $\Theta_n(\theta) = a_n \cos n\theta + b_n \sin n\theta$. With the choice of $k = n^2$, the equation in R becomes

$$r^2 R'' + rR' - n^2 R = 0, \ 0 < r < a,$$

which we recognize as an **Euler equation**. Appealing to results from Appendix A.3, we find that the indicial roots are $\pm n$, and hence the solutions

Recall from Appendix A.3, Euler's equation
$x^2 y'' + \alpha xy' + \beta y = 0,$
with indicial equation
$\rho^2 + (\alpha - 1)\rho + \beta = 0$
and indicial roots ρ_1 and ρ_2.
If $\rho_1 \neq \rho_2$, the general solution is
$y = c_1 x^{\rho_1} + c_2 x^{\rho_2}.$
If $\rho_1 = \rho_2$, the general solution is
$y = c_1 x^{\rho_1} + c_2 x^{\rho_1} \ln x.$

(3)   and

$$R(r) = c_1 \left(\frac{r}{a}\right)^n + c_2 \left(\frac{r}{a}\right)^{-n}, \quad n = 1, 2, \ldots,$$

$$R(r) = c_1 + c_2 \ln\left(\frac{r}{a}\right), \quad n = 0.$$

In each case, the solution has two terms; one bounded for r near 0, and one unbounded for r near 0. For the Dirichlet problem in the disk we take $c_2 = 0$, since the solution should remain bounded near $r = 0$. (Other choices of the constant will be needed for other problems, for example, the Dirichlet problem outside a disk or on an annular region. See Exercises 13 and 16.) We thus arrive at the product solutions

$$u_n(r, \theta) = \left(\frac{r}{a}\right)^n (a_n \cos n\theta + b_n \sin n\theta), \ n = 0, 1, 2, \ldots.$$

Superposing these solutions, we obtain the **general solution** of Laplace's equation in the disk

(4)
$$u(r, \theta) = a_0 + \sum_{n=1}^{\infty} \left(\frac{r}{a}\right)^n [a_n \cos n\theta + b_n \sin n\theta].$$

To go any further and determine the unknown coefficients a_0, a_n, b_n, we need to have boundary conditions. In our first application we consider a Dirichlet problem and derive the Fourier series solution, which we obtained earlier in Section 4.7.

Dirichlet Problems on a Disk

Consider the Dirichlet problem in Figure 2, where the boundary condition is given by

$$(5) \qquad u(a, \theta) = f(\theta), \qquad 0 < \theta < 2\pi.$$

Since f is a function of the polar angle θ, it is necessarily 2π-periodic. For a solution of this problem, we use the infinite series (4). Setting $r = a$ and using the boundary condition (5), we get

$$f(\theta) = u(a, \theta) = a_0 + \sum_{n=1}^{\infty} (a_n \cos n\theta + b_n \sin n\theta).$$

Recognizing this as the Fourier series of f, we conclude that a_0, a_n, and b_n are the Fourier coefficients of f. Thus

$$a_0 = \frac{1}{2\pi} \int_0^{2\pi} f(\theta) \, d\theta,$$

$$(6)$$

$$a_n = \frac{1}{\pi} \int_0^{2\pi} f(\theta) \cos n\theta \, d\theta, \qquad b_n = \frac{1}{\pi} \int_0^{2\pi} f(\theta) \sin n\theta \, d\theta,$$

$n = 1, 2, \ldots$. We have thus determined the coefficients in (4) as being the Fourier coefficients of f and solved completely the Dirichlet problem. (This result was obtained earlier in Theorem 1, Section 4.7.)

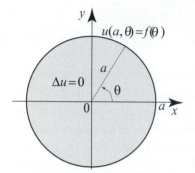

Figure 2 Dirichlet problem in a disk. The function $f(\theta)$ is 2π-periodic.

EXAMPLE 1   A Dirichlet problem on the disk
Solve the Dirichlet problem in Figure 3.
Solution The boundary values are described by the 2π-periodic function

$$u(1, \theta) = f(\theta) = \begin{cases} 100 & \text{if } 0 < \theta < \frac{\pi}{2} \text{ or } \pi < \theta < \frac{3\pi}{2}, \\ 0 & \text{if } \frac{\pi}{2} < \theta < \pi \text{ or } \frac{3\pi}{2} < \theta < 2\pi, \end{cases}$$

shown in Figure 4. Before we compute its Fourier coefficients, let us note that by subtracting 50 from f, we obtain the function $g(\theta) = f(\theta) - 50$, which is clearly an odd function (see Figure 5). Because g is odd, its Fourier coefficients a_0 and a_n are all 0, and

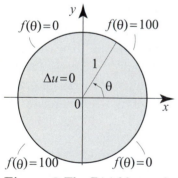

Figure 3 The Dirichlet problem in Example 1.

$$\begin{aligned} b_n &= \frac{2}{\pi} \int_0^{\pi} g(\theta) \sin n\theta \, d\theta = \frac{100}{\pi} \int_0^{\frac{\pi}{2}} \sin n\theta \, d\theta - \frac{100}{\pi} \int_{\frac{\pi}{2}}^{\pi} \sin n\theta \, d\theta \\ &= -\frac{100}{n\pi} \cos n\theta \Big|_0^{\frac{\pi}{2}} + \frac{100}{n\pi} \cos n\theta \Big|_{\frac{\pi}{2}}^{\pi} \\ &= \frac{100}{n\pi} \left(1 + (-1)^n - 2 \cos \frac{n\pi}{2} \right). \end{aligned}$$

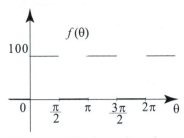

Figure 4 The boundary function $f(\theta)$ is 2π-periodic.

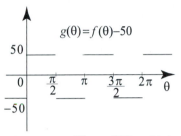

Figure 5 $g(\theta) = f(\theta) - 50$ is an odd function.

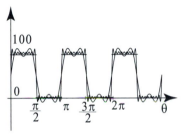

Figure 6 Convergence of the partial sums of the Fourier series of f.

PROPOSITION 1
HARMONIC
CONJUGATE

If n is odd, then $(1 + (-1)^n) = 0$ and $\cos\frac{n\pi}{2} = 0$. If $n = 2k$ is even, then $1 + (-1)^n - 2\cos\frac{n\pi}{2} = 2 - 2\cos k\pi = 2(1 - (-1)^k)$, which is 0 if k is even and 4 if $k = 2m + 1$ is odd. Thus, $b_{4m+2} = \frac{400}{\pi(4m+2)}$, $m = 0, 1, \ldots$, and $b_n = 0$ otherwise; and so the Fourier series of g is

$$g(\theta) = \frac{400}{\pi}\sum_{m=0}^{\infty}\frac{1}{4m+2}\sin(4m+2)\theta.$$

From $f(\theta) = g(\theta) + 50$, we obtain the Fourier series of f

$$f(\theta) = 50 + \frac{400}{\pi}\sum_{m=0}^{\infty}\frac{1}{4m+2}\sin(4m+2)\theta.$$

Some partial sums of this Fourier series are plotted in Figure 6. They illustrate the convergence of the Fourier series to f. To solve the Dirichlet problem in Figure 3, all we have to do now is substitute the Fourier coefficients of f into (4) and obtain

$$u(r, \theta) = 50 + \frac{400}{\pi}\sum_{m=0}^{\infty}\frac{r^{4m+2}}{4m+2}\sin(4m+2)\theta, \quad 0 \le r < 1, \text{ all } \theta. \qquad \blacksquare$$

Before we move to different boundary conditions on the disk, it is instructive to tie up our discussion of the Dirichlet problem with some basic concepts concerning harmonic functions. The solution that we derived in (4) satisfies Laplace's equation in the disk. It is thus harmonic in the disk. This assertion can be justified by differentiating the series term by term and checking that it satisfies Laplace's equation. A more rigorous proof can be based on results from analytic function theory (see Exercise 18). As we know from Section 2.5, the level curves of the solution represent the isotherms in a steady-state problem. It is of interest to describe the curves of heat flow in the disk. Again from Section 2.5, these are given by the level curves of a harmonic conjugate of u. Recall that v is a harmonic conjugate of u if $u + iv$ is analytic. We have the following useful result.

Up to an additive constant, a harmonic conjugate of (4) is

$$(7) \quad v(r, \theta) = \sum_{n=1}^{\infty}\left(\frac{r}{a}\right)^n\left(-b_n\cos n\theta + a_n\sin n\theta\right), \quad 0 \le r < a, \text{ all } \theta.$$

Proof For $n = 1, 2, \ldots$, a harmonic conjugate of $r^n\cos n\theta$ is $r^n\sin n\theta$, because $r^n\cos n\theta + ir^n\sin n\theta = r^n(\cos n\theta + i\sin n\theta) = z^n$, and z^n is analytic. From Proposition 2, Section 2.5, it follows that a harmonic conjugate of $r^n\sin n\theta$ is $-r^n\cos n\theta$. Taking a linear combination of harmonic conjugates, we obtain a harmonic conjugate of the linear combination. Thus, a harmonic conjugate of $\left(\frac{r}{a}\right)^n(a_n\cos n\theta + b_n\sin n\theta)$ is $\left(\frac{r}{a}\right)^n(-b_n\cos n\theta + a_n\sin n\theta)$. To complete the proof of the proposition, we must justify taking an infinite sum of harmonic conjugates. This can be done with the help of tools from analytic function theory (see Exercise 18). $\blacksquare$

EXAMPLE 2   Curves of heat flow

Describe the isotherms and curves of heat flow in Example 1.

Solution Because the temperature of the boundary varies between 0 and 100, the temperature inside the disk varies strictly between these two values (maximum-minimum principle for harmonic functions). The isotherms are the level curves of the solution of the Dirichlet problem: for $0 < T < 100$,

$$T = 50 + \frac{400}{\pi} \sum_{m=0}^{\infty} \frac{r^{4m+2}}{4m+2} \sin(4m+2)\theta.$$

Heat flows along the curves orthogonal to the isotherms. These are the level curves of v, a harmonic conjugate of u, which is obtained by applying Proposition 1:

$$C = -\frac{400}{\pi} \sum_{m=0}^{\infty} \frac{r^{4m+2}}{4m+2} \cos(4m+2)\theta. \qquad \blacksquare$$

Neumann and Robin Conditions

In polar coordinates, the normal derivative at the boundary of a disk centered at the origin is simply the partial derivative with respect to r. Thus for problems on a disk centered at the origin, a Neumann condition specifies u_r on the boundary; and a Robin condition specifies $u_r(r, \theta) + a(\theta)u(r, \theta)$ on the boundary, where a is a function of θ. We can use the series solution of Laplace's equation in a disk (4) to solve such problems, as we now illustrate.

EXAMPLE 3   A Neumann problem on a disk

Solve the Neumann problem $\Delta u = 0$ for $0 \leq r < a$ with boundary condition $u_r(a, \theta) = f(\theta)$, where $f(\theta)$ determines the normal derivative on the circle $r = a$. For this problem to have a solution, we know from Section 6.6 that the normal derivative must satisfy on the boundary the compatibility condition

$$\int_0^{2\pi} f(\theta)\, d\theta = 0.$$

This requirement will become more apparent as we derive the solution.

Solution The solution is given by (4), where the coefficients will be determined so as to satisfy the boundary condition. Differentiate the series (4) term by term with respect to r,

$$u_r(r, \theta) = \sum_{n=1}^{\infty} \frac{n}{a^n} r^{n-1}[a_n \cos n\theta + b_n \sin n\theta];$$

then evaluate at $r = a$, use the Neumann condition, and get

(8) $$f(\theta) = \sum_{n=1}^{\infty} \frac{n}{a}[a_n \cos n\theta + b_n \sin n\theta].$$

Recognizing this as the Fourier series of f, it follows that for $n = 1, 2, \ldots$,

$$\frac{n}{a} a_n = \frac{1}{\pi} \int_0^{2\pi} f(\theta) \cos n\theta\, d\theta \quad \text{and} \quad \frac{n}{a} b_n = \frac{1}{\pi} \int_0^{2\pi} f(\theta) \sin n\theta\, d\theta;$$

equivalently,

$$(9) \qquad a_n = \frac{a}{n\pi} \int_0^{2\pi} f(\theta) \cos n\theta \, d\theta \quad \text{and} \quad b_n = \frac{a}{n\pi} \int_0^{2\pi} f(\theta) \sin n\theta \, d\theta.$$

This determines the coefficients and solves the problem. Note that in our solution, we looked at (8) as the Fourier series of f. But in this series $a_0 = 0$, and since a_0 is a multiple of $\int_0^{2\pi} f(\theta)d\theta$, it follows that the latter integral must be zero, which explains the compatibility condition.   ■

In the exercises, we will use the solution of the Neumann problem to derive the Neumann function for the disk.

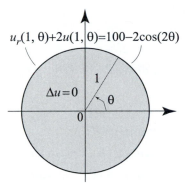

$u_r(1, \theta)+2u(1,\theta)=100-2\cos(2\theta)$

$\Delta u=0$

Figure 7 A Robin boundary condition in Example 4.

EXAMPLE 4   A Robin condition

Solve Laplace's equation on the unit disk, subject to the Robin condition $u_r(1, \theta)+2u(1, \theta) = 100 - 2\cos 2\theta$ (Figure 7).

Solution   Here again, since we are dealing with Laplace's equation, the solution is given by (4) (with $a = 1$). We will determine the coefficients so as to satisfy the given Robin condition. Let us compute $u_r(r, \theta)$, by differentiating term by term,

$$u_r(r, \theta) = \sum_{n=1}^{\infty} nr^{n-1}[a_n \cos n\theta + b_n \sin n\theta].$$

Thus

$$u_r(r, \theta) + 2u(r, \theta) = 2a_0 + \sum_{n=1}^{\infty}(nr^{n-1} + 2r^n)[a_n \cos n\theta + b_n \sin n\theta].$$

Evaluating at $r = 1$, and using the boundary condition, we get

$$100 - 2\cos 2\theta = 2a_0 + \sum_{n=1}^{\infty}(n + 2)[a_n \cos n\theta + b_n \sin n\theta].$$

Since the left side is already in a Fourier series form, by equating the coefficients, we find that $2a_0 = 100$, $4a_2 = -2$, $(n+2)a_n = 0$ if $n \neq 2$, and $(n+2)b_n = 0$ for all n. Thus $a_0 = 50$, $a_2 = -\frac{1}{2}$, and all other coefficients are 0. So the solution is

$$u(r, \theta) = 50 - \frac{1}{2}r^2 \cos 2\theta,$$

as can be verified directly.   ■

Exercises 9.1

In Exercises 1–6, (a) solve the Dirichlet problem on the unit disk for the given boundary values. (b) Describe the isotherms and curves of heat flow in each case.

1. $f(\theta) = 100 + 100\cos\theta$.

2. $f(\theta) = \sin 2\theta$.

3. $f(\theta) = \frac{1}{2}(\pi - \theta)$, $0 < \theta < 2\pi$.

4. $f(\theta) = 50\sin^2\theta$.

5. $f(\theta) = \begin{cases} 100 & \text{if } 0 \leq \theta \leq \pi/4, \\ 0 & \text{if } \pi/4 < \theta < 2\pi. \end{cases}$

6. $f(\theta) = \begin{cases} \pi - \theta & \text{if } 0 \leq \theta \leq \pi, \\ 0 & \text{if } \pi \leq \theta < 2\pi. \end{cases}$

In Exercises 7–12, solve Laplace's equation inside the unit disk subject to the given Neumann or Robin condition on the boundary.

7. $u_r(1,\,\theta) + u(1,\,\theta) = \theta$, $0 < \theta < 2\pi$.

8. $u_r(1,\,\theta) = u(1,\,\theta) - 1$.

9. $u_r(1,\,\theta) = 50 - 50\cos\theta$.

10. $u_r(1,\,\theta) + u(1,\,\theta) = f(\theta)$.

11. $u_r(1,\,\theta) + 2\cos\theta\, u(1,\,\theta) = 1 + \sin\theta$. **12.** $u_r(1,\,\theta) + \cos\theta\, u(1,\,\theta) = 32\cos 3\theta$.

13. Dirichlet problem outside the disk. Show that the bounded solution of the Dirichlet problem outside the disk $r = a$ (Figure 8) is given by the series

$$u(r,\theta) = a_0 + \sum_{n=1}^{\infty}\left(\frac{a}{r}\right)^n(a_n\cos n\theta + b_n\sin n\theta), \quad r > a,$$

where the coefficients are given by (6). [Hint: Proceed as in the solution of the Dirichlet problem inside the disk. Choose the bounded solutions from (4).]

14. Solve the Dirichlet problem outside the unit disk with boundary values as in Example 1.

15. (a) Solve the Dirichlet problem outside the unit disk with boundary values as in Exercise 5.
(b) Show that the isotherms are circles passing through the points $(1,\,0)$ and $\left(\frac{1}{\sqrt{2}},\,\frac{1}{\sqrt{2}}\right)$.

16. Project Problem: Dirichlet problems on annular regions. (a) Show that the steady-state solution in the annular region shown in the Figure 9 is

$$u(r,\theta) = a_0\frac{\ln r - \ln R_2}{\ln R_1 - \ln R_2} + \sum_{n=1}^{\infty}[a_n\cos n\theta + b_n\sin n\theta]\left(\frac{R_1}{r}\right)^n\left(\frac{R_2^{2n} - r^{2n}}{R_2^{2n} - R_1^{2n}}\right),$$

$(R_1 < r < R_2)$ where a_0, a_n, and b_n are the Fourier coefficients of $f_1(\theta)$. [Hint: Proceed as in the solution derived in this section, and use the condition $R(R_2) = 0$.]
(b) What is the steady-state solution if the boundary conditions are $u(R_1,\,\theta) = 0$, and $u(R_2,\,\theta) = f_2(\theta)$, for all θ?
(c) Combine (a) and (b) to find the steady-state solution in the annular region given that $u(R_1,\,\theta) = f_1(\theta)$, and $u(R_2,\,\theta) = f_2(\theta)$ for all θ.

17. Project Problem: Neumann function for the disk. In this exercise, we derive the Neumann function for the disk. A similar formula was computed in Exercise 8, Section 6.6.

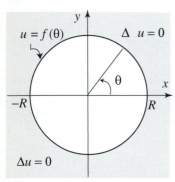

Figure 8 for Exercise 13.

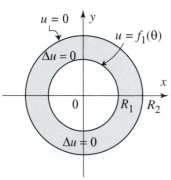

Figure 9 for Exercise 16.

(a) Consider the Neumann problem in Example 3, on a disk of radius $a > 0$: $\Delta u = 0$ and $u_r(a, \theta) = f(\theta)$. Show that the solution can be put in the form

$$u(r, \theta) = \frac{a}{\pi} \int_0^{2\pi} f(\phi) \left[\sum_{n=1}^{\infty} \left(\frac{r}{a}\right)^n \frac{\cos[n(\theta - \phi)]}{n} \right] d\phi.$$

[Hint: Put the coefficients (9) back into the solution (4) and interchange the integral and summation signs.]

(b) Compare your answer in (a) with Theorem 5, Section 6.6, and conclude that the Neumann function for the disk of radius a is

$$N(z, \zeta) = -2 \sum_{n=1}^{\infty} \left(\frac{r}{a}\right)^n \frac{\cos[n(\theta - \phi)]}{n} \quad (z = re^{i\theta}, \ \zeta = ae^{i\phi}).$$

[Hint: Write $f(\zeta) = f(ae^{i\phi}) = f(\phi)$. On the disk of radius a, $ds = ad\phi$.]

(c) Using the series identities in Exercise 3, Section 4.7, show that

$$\sum_{n=1}^{\infty} \frac{\cos n\theta}{n} r^n = -\frac{1}{2} \ln(1 - 2r \cos \theta + r^2).$$

(d) Using the identity in (c), show that the Neumann function in (b) satisfies the property in Proposition 2, Section 6.6: on the boundary of the disk,

$$\frac{\partial N}{\partial r}(z, ae^{i\phi}) = \frac{2\pi}{2\pi a} = \frac{1}{a}.$$

18. Let u be as in (4) and v be as in Proposition 1. Use the Weierstrass M-test to show that the series defining $u + iv$ converges uniformly on every closed disk of the form $|z| \leq r$, where $r < a$. [Hint: Show that $|a_n|, |b_n| \leq M = \frac{1}{\pi} \int_0^{2\pi} |f(t)| dt$.]

(b) Conclude that $u + iv$ is analytic in the disk $|z| < a$, and hence that u and v are harmonic and that v is the harmonic conjugate of u. [Hint: Combine the series for $u + iv$. Each term is analytic. Apply Theorem 4(i), Section 4.2, to conclude that $u + iv$ is analytic in $|z| < a$. The real and imaginary parts of an analytic function are harmonic, so u and v are harmonic; and v is the harmonic conjugate of u.]

9.2   Vibrations of a Circular Membrane: Symmetric Case

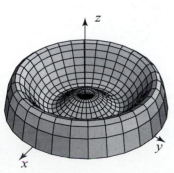

Figure 1 A radially symmetric shape.

In this and the next section we study the vibrations of a thin circular membrane with uniform mass density, clamped along its circumference. We place the center of the membrane at the origin, and we denote the radius by a. The vibrations of the membrane are governed by the two-dimensional wave equation, which will be expressed in polar coordinates, because these are the coordinates best suited to this problem. Using the polar form of the Laplacian ((1), Section 9.1), the **two dimensional wave equation** becomes

(1)
$$\frac{\partial^2 u}{\partial t^2} = c^2 \left(\frac{\partial^2 u}{\partial r^2} + \frac{1}{r} \frac{\partial u}{\partial r} + \frac{1}{r^2} \frac{\partial^2 u}{\partial \theta^2} \right).$$

The initial shape of the membrane will be modeled by the function $f(r, \theta)$, and its initial velocity by $g(r, \theta)$.

In this section we confine our study to the case where f and g are **radially symmetric**, that is, they depend only on the radius r and not on θ. It is reasonable on physical grounds that in this case the solution also does not depend on θ (see Figure 1). Consequently, $\partial u / \partial \theta = 0$, and (1) becomes

$$
(2) \qquad \boxed{\frac{\partial^2 u}{\partial t^2} = c^2 \left(\frac{\partial^2 u}{\partial r^2} + \frac{1}{r} \frac{\partial u}{\partial r} \right),}
$$

where $u = u(r, t)$, $0 < r < a$, and $t > 0$. Since the membrane is clamped at the circumference, we have the **boundary condition**

$$
(3) \qquad\qquad u(a, t) = 0, \qquad t \geq 0.
$$

The radially symmetric **initial conditions** are

The initial conditions are radially symmetric, so they depend only on r.

$$
(4) \qquad u(r, 0) = f(r), \quad \frac{\partial u}{\partial t}(r, 0) = g(r), \quad 0 < r < a.
$$

We solve the boundary value problem (2)–(4) using the separation of variables method, as we did throughout Chapter 3. The goal is to separate the variables in the partial differential equation (2) and reduce the problem to two ordinary differential equations in r and t. As you will see, the equation in t is the same as the one that we obtained after separating variables in the wave equation in rectangular coordinates. Hence the solution in t will consist of sines and cosines. The equation in the spatial variable r is new, and its solution will involve the so-called Bessel functions.

Separating Variables

We start by looking for product solutions of the form $u(r, t) = R(r)T(t)$. After differentiating, plugging into (2), and separating variables, we get

$$
\frac{T''}{c^2 T} = \frac{1}{R} \left(R'' + \frac{1}{r} R' \right) = -\lambda^2.
$$

Because we expect periodic solutions in T, we have set the sign of the separation constant negative. (For a more rigorous argument based on the fact that the solution in R should be bounded in the interval $[0, a]$, see the solution of the Dirichlet problem in Section 9.4.) Plugging $u(r, t) = R(r)T(t)$ into the boundary condition, we obtain $R(a)T(t) = 0$, which implies that $R(a) = 0$. Hence

$$
(5) \qquad\qquad rR'' + R' + \lambda^2 r R = 0, \quad R(a) = 0,
$$
$$
(6) \qquad\qquad T'' + c^2 \lambda^2 T = 0.
$$

Solving the Separated Equations

Here again, we begin by solving the equation with the boundary conditions to narrow down the possible solutions. Equation (5) is known as the **parametric form of Bessel's equation of order zero** (here λ is the parameter). It is second order and homogeneous, with nonconstant coefficients. It can be solved by the method of Frobenius from Appendix A.5. The details are straightforward, but lengthy; we present them in Section 9.6, where we will show that the general solution is

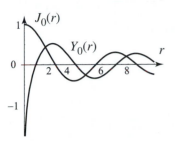

Figure 2 The Bessel functions of order 0.

$$(7) \qquad\qquad R(r) = c_1 J_0(\lambda r) + c_2 Y_0(\lambda r),$$

where J_0 is the Bessel function of the **first kind** of order 0, and Y_0 is the Bessel function of the **second kind** of order 0. In addition to the properties of the Bessel functions that we derived in Chapter 4, we present in Sections 9.6 and 9.7 a detailed treatment of the Bessel functions of the first and second kind. In this section, we recall facts only as needed.

Figure 2 shows the graphs of J_0 and Y_0; in particular, it shows that Y_0 is not bounded near 0. Since on physical grounds the solutions to the wave equation are expected to be bounded, it follows that the spatial part of the solution, $R(r)$, has to be bounded near $r = 0$. This is effectively a second boundary condition on R. Now the fact that Y_0 is unbounded near 0 forces us to choose $c_2 = 0$ in (7). To avoid trivial solutions, we will take $c_1 = 1$ and get

$$(8) \qquad\qquad R(r) = J_0(\lambda r).$$

The condition $R(a) = 0$ (see (5)) implies that

$$J_0(\lambda a) = 0,$$

and so λa must be a root of the Bessel function J_0. As Figure 2 suggests, J_0 has infinitely many positive zeros, which we denote by

$$\alpha_1 < \alpha_2 < \alpha_3 < \cdots < \alpha_n < \cdots .$$

(This important property is proved in Exercise 35, Section 9.7.) Thus λ is given by

$$\lambda_n = \frac{\alpha_n}{a}, \quad n = 1, 2, \ldots ,$$

and the corresponding solutions of (5) are

$$R_n(r) = J_0(\frac{\alpha_n}{a} r), \quad n = 1, 2, \ldots ,$$

where α_n is the nth positive zero of J_0.   These solutions are analogous to the solutions $\sin\frac{n\pi}{L}x$ that we have encountered several times previously, in particular, while solving the one dimensional wave equation.   The only difference is that the function sine and its zeros $n\pi$ are now replaced by the function J_0 and its zeros α_n.   Returning to (6) with $\lambda = \lambda_n$, we find

$$\boxed{T(t) = T_n(t) = A_n \cos c\lambda_n t + B_n \sin c\lambda_n t\,.}$$

We thus obtain the product solutions of (2) and (3):

$$u_n(r,\,t) = (A_n \cos c\lambda_n t + B_n \sin c\lambda_n t)J_0(\lambda_n r), \quad n = 1,\,2,\,\dots\,.$$

Bessel Series Solution of the Entire Problem

To satisfy the initial conditions, motivated by the superposition principle, we let

$$u(r,\,t) = \sum_{n=1}^{\infty}(A_n \cos c\lambda_n t + B_n \sin c\lambda_n t)J_0(\lambda_n r)\,.$$

We determine the unknown coefficients by evaluating the series at $t = 0$ and using the initial conditions.   We get from the first condition in (4)

$$u(r,\,0) = f(r) = \sum_{n=1}^{\infty}A_n J_0(\lambda_n r), \quad 0 < r < a\,.$$

This series representation of $f(r)$ is akin to a Fourier sine series, except that the sine functions are now replaced by Bessel functions.   There are analogous expansion theorems that apply in such cases; the series expansions that arise are known as **Bessel**, or **Fourier-Bessel, expansions** (see Theorem 2, Section 9.7).   For the case at hand, we make use of Theorem 2, Section 9.7, with $p = 0$.   The **Bessel coefficients** A_n are given by

$$A_n = \frac{2}{a^2 J_1^2(\alpha_n)}\int_0^a f(r)J_0(\lambda_n r)r\,dr,$$

where J_1 is the Bessel function of order 1.   Now, differentiating the series for u term by term with respect to t, and then setting $t = 0$, we get from the second initial condition

$$u_t(r,\,0) = g(r) = \sum_{n=1}^{\infty}c\lambda_n B_n J_0(\lambda_n r)\,.$$

Thus $c\lambda_n B_n = c\frac{\alpha_n}{a} B_n$ is the nth Bessel coefficient of g, and so

$$B_n = \frac{2}{c\,\alpha_n\,a J_1^2(\alpha_n)} \int_0^a g(r) J_0(\lambda_n r)\, r\, dr\,.$$

This completely determines the solution.

THEOREM 1
WAVE EQUATION IN POLAR COORDINATES

The solution of the radially symmetric two-dimensional wave equation (2) with boundary and initial conditions (3) and (4) is

$$(9) \qquad u(r,\, t) = \sum_{n=1}^{\infty} (A_n \cos c\lambda_n t + B_n \sin c\lambda_n t) J_0(\lambda_n r),$$

where

$$A_n = \frac{2}{a^2 J_1^2(\alpha_n)} \int_0^a f(r) J_0(\lambda_n r) r\, dr,$$

(10)

$$B_n = \frac{2}{c\,\alpha_n a\, J_1^2(\alpha_n)} \int_0^a g(r) J_0(\lambda_n r)\, r\, dr;$$

$$\lambda_n = \frac{\alpha_n}{a}, \quad \text{and} \quad \alpha_n = n\text{th positive zero of } J_0\,.$$

There is a clear analogy between the solution (9) and the solution of the one-dimensional wave equation (8), Section 8.2. The only difference is that spatial variations are now determined by Bessel functions rather than the simpler sine functions.

When applying (10) in concrete situations, we are required to evaluate integrals involving Bessel functions that are quite complicated. In many interesting cases these integrals can be evaluated with the help of integral formulas developed in the exercises and in Section 9.7. As an illustration, consider the integral

$$\int_0^a x^{p+1} J_p\left(\frac{\alpha}{a} x\right) dx, \quad p \geq 0,\ \alpha > 0.$$

Let $u = \frac{\alpha}{a} x,\ du = \frac{\alpha}{a} dx$, then

From (7), Section 9.7,

$$\int x^{p+1} J_p(x)\, dx =$$

$$x^{p+1} J_{p+1}(x) + C\,.$$

$$\int x^{p+1} J_p\left(\frac{\alpha}{a} x\right) dx \ =\ \frac{a^{p+2}}{\alpha^{p+2}} \int u^{p+1} J_p(u)\, du$$

$$=\ \frac{a^{p+2}}{\alpha^{p+2}} u^{p+1} J_{p+1}(u) + C,$$

where the last equality follows from (7), Section 9.7. Substituting back $u = \frac{\alpha x}{a}$, simplifying, and then evaluating at $x = 0$ and $x = a$, we obtain the very useful identity

$$(11) \qquad \boxed{\int_0^a x^{p+1} J_p\left(\frac{\alpha}{a} x\right) dx = \frac{a^{p+2}}{\alpha} J_{p+1}(\alpha)\,.}$$

EXAMPLE 1 Circular membrane with constant initial velocity

An explosion near the surface of a flexible circular membrane with clamped edges imparts a uniform initial velocity equal to -100 m/sec. Assume the initial shape of the membrane to be flat, take $a = 1$ and $c = 100$, and determine the subsequent vibrations of the membrane.

Solution The solution is given by (9), where $A_n = 0$ for all n, since $f(r) = 0$. From (10) we have

$$
\begin{aligned}
B_n &= \frac{-2}{\alpha_n J_1^2(\alpha_n)} \int_0^1 J_0(\alpha_n r)\, r\, dr \\
&= \frac{-2}{\alpha_n^2 J_1(\alpha_n)} \qquad \text{(by (11) with } p = 0\text{)}.
\end{aligned}
$$

Thus, from (9), we obtain the solution

$$
u(r,t) = \sum_{n=1}^{\infty} \frac{-2}{\alpha_n^2 J_1(\alpha_n)} \sin(100\,\alpha_n t) J_0(\alpha_n r). \qquad \blacksquare
$$

To get numerical values from our answer in Example 1, it is clearly necessary to know the values of the zeros of the Bessel function J_0. Since these values are useful in solving many problems, they have been computed and tabulated to a high degree of accuracy. With the help of a computer system, we approximated the first five positive roots of the equation $J_0(x) = 0$. These and other relevant numerical data are given in Table 1.

j	1	2	3	4	5
α_j	2.4048	5.5201	8.6537	11.7915	14.9309
$J_1(\alpha_j)$	.5191	$-.3403$	.2714	$-.2325$	.2065
$\dfrac{-2}{\alpha_j^2 J_1(\alpha_j)}$	-0.6662	0.1929	$-.0984$	0.0619	-0.0434

Table 1 Numerical data for Example 1

With the help of this table, we find the first three terms of the solution in Example 1:

$$
\begin{aligned}
u(r,t) \approx\ & -0.6662\, J_0(2.40\,r)\sin(240\,t) \\
& +0.1929\, J_0(5.52\,r)\sin(552\,t) - .0984\, J_0(8.65\,r)\sin(865\,t).
\end{aligned}
$$

We used these terms to plot the shape of the membrane at various values of $t > 0$ in Figure 3.

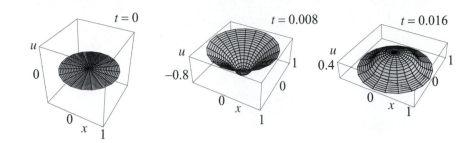

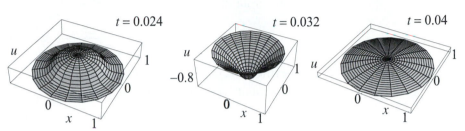

Figure 3 Vibrating circular membrane with radial symmetry in Example 1. As expected, soon after the explosion, the elastic membrane starts to vibrate downward.

The next example treats the case of a vibrating membrane with nonzero initial displacement and zero initial velocity.

EXAMPLE 2 Circular membrane with radially symmetric shape

Solve the boundary value problem (2)–(4), given that

$$f(r) = 1 - r^2, \quad g(r) = 0, \quad a = c = 1.$$

Solution Note that the problem is radially symmetric because of the boundary and initial conditions. The solution is given by (9), where $B_n = 0$ for all n since $g(r) = 0$, and A_n is the Bessel coefficient of the function $1 - r^2$, given by (10). We have

$$
\begin{aligned}
A_n &= \frac{2}{J_1^2(\alpha_n)} \int_0^1 (1 - r^2) J_0(\alpha_n r)\, r\, dr \\
&= \frac{2}{\alpha_n^4 J_1^2(\alpha_n)} \int_0^{\alpha_n} (\alpha_n^2 - s^2) J_0(s) s\, ds \qquad (s = \alpha_n r).
\end{aligned}
$$

Integrating by parts, with $u = \alpha_n^2 - s^2$, $dv = J_0(s)s\, ds$, and hence $du = -2s\, ds$, $v = J_1(s)s$ (by (7), Section 9.7, with $p = 0$), we find

$$
\begin{aligned}
A_n &= \frac{2}{\alpha_n^4 J_1^2(\alpha_n)} \left[(\alpha_n^2 - s^2) J_1(s) s \Big|_0^{\alpha_n} + 2 \int_0^{\alpha_n} J_1(s) s^2\, ds \right] \\
&= \frac{4}{\alpha_n^4 J_1^2(\alpha_n)} \int_0^{\alpha_n} J_1(s) s^2\, ds \,.
\end{aligned}
$$

To evaluate the integral, we appeal to (11) and arrive at

$$A_n = \frac{4}{\alpha_n^2 J_1^2(\alpha_n)} J_2(\alpha_n) \,.$$

At this point, we could appeal to (9) to write the explicit form of the solution. However, before doing so, we mention one more worthy simplification based on yet another property of Bessel functions. Appealing to (6), Section 9.7, with $p = 1$, and recalling that α_n is a zero of J_0, we find

From (6), Section 9.7:
$$J_{p+1}(x) = \frac{2p}{x} J_p(x) - J_{p-1}(x).$$

$$J_2(\alpha_n) = \frac{2}{\alpha_n} J_1(\alpha_n) - J_0(\alpha_n) = \frac{2}{\alpha_n} J_1(\alpha_n),$$

and hence

$$A_n = \frac{8}{\alpha_n^3 J_1(\alpha_n)}.$$

Thus, from (9), we obtain the solution

$$u(r, t) = \sum_{n=1}^{\infty} \frac{8}{\alpha_n^3 J_1(\alpha_n)} \cos(\alpha_n t) J_0(\alpha_n r). \qquad \blacksquare$$

Setting $t = 0$ in the solution of Example 2, we should get the initial displacement, that is, we should get

$$\boxed{1 - r^2 = \sum_{n=1}^{\infty} \frac{8}{\alpha_n^3 J_1(\alpha_n)} J_0(\alpha_n r), \quad 0 < r < 1.}$$

This is the Bessel series of the function $1 - r^2$ that we have computed in passing as we worked out the solution to Example 2. Figure 4 shows some partial sums of this series converging to $1 - r^2$, $0 < r < 1$.

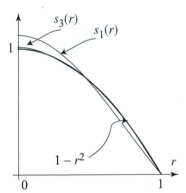

Figure 4 Partial sums of Bessel series.

We end this section with a remark concerning the physical interpretation of the solution of Example 2. It is a fact that the wave equation models the small vibrations of a drum, but you may not be willing to call a unit displacement at the center of a drum of unit radius small. To give the problem a meaningful interpretation, we could rescale the initial data. Because of the linearity of the boundary value problem, this leads only to the same rescaling of the solution.

Exercises 9.2

In Exercises 1–8, solve the vibrating membrane problem (2)–(4) for the given data. If possible, with the help of a computer, find numerical values for the first five nonzero coefficients of the series solution and plot the shape of the membrane at various values of t. (Formula (11) is useful in all these exercises.)

1. $a = 2$, $c = 1$, $f(r) = 0$, $g(r) = 1$.

2. $a = 1$, $c = 10$, $f(r) = 1 - r^2$, $g(r) = 1$.

3. $a = 1$, $c = 1$, $f(r) = 0$,

$$g(r) = \begin{cases} 1 & \text{if } 0 < r < \frac{1}{2}, \\ 0 & \text{if } \frac{1}{2} < r < 1. \end{cases}$$

[Hint: Follow Example 1.]

4. $a = 1$, $c = 1$, $f(r) = 0$, $g(r) = J_0(\alpha_3 r)$.
[Hint: Orthogonality relations in Section 9.7.]

5. $a = 1$, $c = 1$, $f(r) = J_0(\alpha_1 r)$, $g(r) = 0$.
[Hint: Orthogonality relations in Section 9.7.]

6. $a = 2$, $c = 1$, $f(r) = 1 - r$, $g(r) = 0$.

7. $a = 1$, $c = 1$, $f(r) = J_0(\alpha_3 r)$, $g(r) = 1 - r^2$.

8. $a = 1$, $c = 1$, $f(r) = \frac{1}{128}(3 - 4r^2 + r^4)$, $g(r) = 0$.
[Hint: Integration by parts, Example 2.]

9. (a) Find the solution in Exercise 3 for an arbitrary value of $c > 0$.
(b) Describe what happens to the solution as c increases.

10. Project Problem: Radially symmetric heat equation on a disk. Use the methods of this section to show that the solution of the heat boundary value problem

$$\frac{\partial u}{\partial t} = c^2 \left(\frac{\partial^2 u}{\partial r^2} + \frac{1}{r}\frac{\partial u}{\partial r} \right), \quad 0 < r < a, \ t > 0,$$
$$u(a, t) = 0, \quad t > 0,$$
$$u(r, 0) = f(r), \quad 0 < r < a,$$

is

$$u(r, t) = \sum_{n=1}^{\infty} A_n e^{-c^2 \lambda_n^2 t} J_0(\lambda_n r),$$

with

$$A_n = \frac{2}{a^2 J_1^2(\alpha_n)} \int_0^a f(r) J_0(\lambda_n r) r \, dr,$$

where $\lambda_n = \frac{\alpha_n}{a}$, and $\alpha_n = n$th positive zero of J_0.

11. (a) Solve the heat problem of Exercise 10 when $f(r) = 100$, $0 < r < a$. What does your solution represent?

(b) Approximate the temperature of the hottest point on the plate at time $t = 3$, given that $a = 1$ and $c = 1$. Where is this point on the plate? Justify your answer intuitively.

12. Project Problem: Integral identities with Bessel functions.
(a) Use (7) and (8), Section 9.7, to establish the identities

$$\int J_1(x) \, dx = -J_0(x) + C \quad \text{and} \quad \int x J_0(x) \, dx = x J_1(x) + C.$$

In the rest of this problem we generalize these identities.
(b) By integrating (5), Section 9.7, show that

$$\int J_{p+1}(x) \, dx = \int J_{p-1}(x) \, dx - 2 J_p(x).$$

(c) Use the first integral in (a), (b), and induction to establish that

$$\int J_{2n+1}(x) \, dx = -J_0(x) - 2 \sum_{k=1}^{n} J_{2k}(x) + C, \quad n = 0, 1, 2, \ldots.$$

As an illustration, derive the following identities:

$$\int J_3(x)\,dx = -J_0(x) - 2J_2(x) + C,$$

$$\int J_5(x)\,dx = -J_0(x) - 2J_2(x) - 2J_4(x) + C.$$

(d) By integrating (3), Section 9.7, show that

$$xJ_{p+1}(x) + p\int J_{p+1}(x)\,dx = \int xJ_p(x)\,dx.$$

[Hint: Evaluate the integral of $xJ_p'(x)$ by parts.]
(e) Take $p = 2n$ in (d) and use (c) to prove that for $n = 0, 1, 2, \ldots$,

$$\int xJ_{2n}(x)\,dx = xJ_{2n+1}(x) - 2nJ_0(x) - 4n\sum_{k=1}^{n} J_{2k}(x) + C.$$

Derive the following identities:

$$\int xJ_2(x)\,dx = xJ_3(x) - 2J_0(x) - 4J_2(x) + C,$$

$$\int xJ_4(x)\,dx = xJ_5(x) - 4J_0(x) - 8J_2(x) - 8J_4(x) + C.$$

9.3   Vibrations of a Circular Membrane: General Case

We continue our study of the vibrating circular membrane, now without any symmetry assumptions. We will solve the **two dimensional wave equation in polar coordinates**

(1)
$$\boxed{\frac{\partial^2 u}{\partial t^2} = c^2\left(\frac{\partial^2 u}{\partial r^2} + \frac{1}{r}\frac{\partial u}{\partial r} + \frac{1}{r^2}\frac{\partial^2 u}{\partial \theta^2}\right),}$$

where $0 < r < a$, $0 < \theta < 2\pi$, $t > 0$. Here $u = u(r, \theta, t)$ denotes the deflection of the membrane at the point (r, θ) at time t. The **initial conditions** (displacement and velocity) are

(2)
$$u(r, \theta, 0) = f(r, \theta), \qquad \frac{\partial u}{\partial t}(r, \theta, 0) = g(r, \theta),$$

$0 < r < a$, $0 < \theta < 2\pi$. The requirement that the edges of the membrane be held fixed translates into the **boundary condition**

(3)
$$u(a, \theta, t) = 0, \quad 0 < \theta < 2\pi, t > 0.$$

Since θ is a polar angle, (r, θ) and $(r, \theta + 2\pi)$ represent the same point, and hence $u(r, \theta, t) = u(r, \theta + 2\pi, t)$. In other words, u is 2π-periodic in θ. Consequently,

$$(4) \qquad u(r, 0, t) = u(r, 2\pi, t) \quad \text{and} \quad \frac{\partial u}{\partial \theta}(r, 0, t) = \frac{\partial u}{\partial \theta}(r, 2\pi, t).$$

Separation of Variables

We start by deriving the general solution of (1) subject to the boundary condition (3). We use the method of separation of variables and set $u(r, \theta, t) = R(r)\Theta(\theta)T(t)$. Differentiating u, substituting into (1), and separating variables gives

$$\frac{T''}{c^2 T} = \frac{R''}{R} + \frac{R'}{rR} + \frac{\Theta''}{r^2\Theta}.$$

The left side depends only on t and the right side only on r and θ. Therefore each side must equal a constant k. Expecting periodic solutions in T, we take $k = -\lambda^2$. Thus

$$\frac{T''}{c^2 T} = -\lambda^2, \quad \text{and} \quad \frac{R''}{R} + \frac{R'}{rR} + \frac{\Theta''}{r^2\Theta} = -\lambda^2.$$

Separating variables in the second equation, we get

$$\lambda^2 r^2 + \frac{r^2 R''}{R} + \frac{rR'}{R} = \mu^2 \quad \text{and} \quad -\frac{\Theta''}{\Theta} = \mu^2.$$

We have chosen a nonnegative sign for the separating constant μ^2 because the solutions of the equation in Θ have to be 2π-periodic. The boundary condition (3) becomes $R(a)\Theta(\theta)T(t) = 0$ for $0 < \theta < 2\pi$ and $t > 0$. To avoid the trivial solution, we impose the condition $R(a) = 0$. Similarly, using (4), we find that $\Theta(0) = \Theta(2\pi)$ and $\Theta'(0) = \Theta'(2\pi)$. Thus we have arrived at the following separated equations:

$$\Theta'' + \mu^2\Theta = 0, \quad \Theta(0) = \Theta(2\pi), \quad \Theta'(0) = \Theta'(2\pi),$$
$$r^2 R'' + rR' + (\lambda^2 r^2 - \mu^2)R = 0, \quad R(a) = 0,$$
$$T'' + c^2\lambda^2 T = 0.$$

Solving the Separated Equations

Note that we start with the Θ equation, since we have a full complement of boundary conditions for it, and it contains only one separation constant. After determining that $\mu = m$, $m = 0, 1, 2, 3, \ldots$, we can turn to the equation in R and determine which values of the separation constant λ allow for nontrivial solutions. The T equation is dealt with last.

We begin by solving for Θ. For $\mu = 0$ the solution is a constant A_0. If $\mu \neq 0$, the general solution is of the form $\Theta(\theta) = c_1 \cos \mu\theta + c_2 \sin \mu\theta$. To satisfy the boundary conditions, we must take μ to be an integer. Thus

$$\Theta_m(\theta) = A_m \cos m\theta + B_m \sin m\theta, \quad m = 0, 1, 2, \ldots .$$

(Note that negative values of m do not contribute any new solutions.) Setting $\mu = m$ in the equation for R, we get

$$r^2 R'' + rR' + (\lambda^2 r^2 - m^2)R = 0, \quad R(a) = 0 .$$

This is the **parametric form of Bessel's equation of order** m which is treated in Theorem 3, Section 9.7. Quoting from this theorem, we have

We get J_m's here and not Y_m's or a combination of J_m's and Y_m's because, on physical grounds, we insist that our solutions remain bounded at $r = 0$.

$$R(r) = R_{mn}(r) = J_m(\lambda_{mn}r), \quad m = 0, 1, 2, \ldots , \ n = 1, 2, \ldots ,$$

where $\lambda_{mn} = \alpha_{mn}/a$ and α_{mn} is the nth positive zero of the Bessel function J_m. For $\lambda = \lambda_{mn}$ the equation in T becomes $T'' + c^2\lambda_{mn}^2 T = 0$ with solutions

$$A_{mn} \cos c\lambda_{mn}t \quad \text{and} \quad B_{mn} \sin c\lambda_{mn}t .$$

Using the expressions for R, Θ, and, T, we arrive at the product solutions of (1) and (3):

$$(5) \quad u_{mn}(r, \theta, t) = J_m(\lambda_{mn}r)(a_{mn} \cos m\theta + b_{mn} \sin m\theta) \cos c\lambda_{mn}t$$

and

$$(6) \quad u_{mn}^*(r, \theta, t) = J_m(\lambda_{mn}r)(a_{mn}^* \cos m\theta + b_{mn}^* \sin m\theta) \sin c\lambda_{mn}t,$$

where $m = 0, 1, 2, \ldots , \ n = 1, 2, \ldots .$ Note that we have replaced the coefficient $A_m A_{mn}$ by a_{mn}, and similarly for b_{mn}, a_{mn}^*, and b_{mn}^*. While this may appear to be just relabeling of the unknown coefficients, in fact, it allows us greater choice of solutions which will be needed as we proceed. Note too that b_{0n} and b_{0n}^* will never be needed, since $\sin m\theta = 0$ when $m = 0$, and so for the sake of definiteness we take them to be 0.

Superposition Principle and the General Solution

The superposition principle suggests adding all the functions in (5) and (6). The resulting sum is displayed in (16). Because of the complexity of this solution, we consider two cases separately: one in which the initial velocity g is zero, and a second in which the initial displacement f is zero. The general solution is then obtained by combining these two cases.

EXAMPLE 1   Vibrations of a membrane with zero initial velocity

Solve the boundary value problem consisting of (1)–(3) given that $g = 0$.

Solution The initial conditions in this case are

$$u(r,\,\theta,\,0) = f(r,\,\theta), \quad \frac{\partial u}{\partial t}(r,\,\theta,\,0) = 0, \quad 0 < r < a,\; 0 < \theta < 2\pi\,.$$

It is easily seen that the only product solutions that meet the second condition are those given by (5). Thus the superposition principle suggests a solution of the form

$$(7) \qquad u(r,\,\theta,\,t) = \sum_{m=0}^{\infty} \sum_{n=1}^{\infty} J_m(\lambda_{mn} r)\,(a_{mn}\cos m\theta + b_{mn}\sin m\theta)\cos c\lambda_{mn} t\,.$$

Setting $t = 0$, we get

$$(8) \qquad \boxed{\; f(r,\theta) = \sum_{m=0}^{\infty} \sum_{n=1}^{\infty} J_m(\lambda_{mn} r)(a_{mn}\cos m\theta + b_{mn}\sin m\theta)\,. \;}$$

This surely is a sort of a generalized Fourier series of $f(r,\,\theta)$ in terms of the functions $J_m(\lambda_{mn} r)\cos m\theta$ and $J_m(\lambda_{mn} r)\sin m\theta$, and hence a_{mn} and b_{mn} are the corresponding generalized Fourier coefficients of the function f. This fact and many important related applications are explored in Section 9.5 (see in particular Theorems 1 and 2 of that section). We now proceed to determine a_{mn} and b_{mn}, using properties of the usual Fourier series, and Bessel series.

Fix r and think of $f(r,\,\theta)$ as a (2π-periodic) function of θ. To facilitate the use of Fourier series, we write (8) as

$$f(r,\,\theta) = \overbrace{\sum_{n=1}^{\infty} a_{0n} J_0(\lambda_{0n} r)}^{=a_0(r)} + \sum_{m=1}^{\infty}\left\{ \overbrace{\left(\sum_{n=1}^{\infty} a_{mn} J_m(\lambda_{mn} r)\right)}^{=a_m(r)}\cos m\theta \right.$$

$$\left. + \underbrace{\left(\sum_{n=1}^{\infty} b_{mn} J_m(\lambda_{mn} r)\right)}_{=b_m(r)}\sin m\theta \right\}$$

$$= a_0(r) + \sum_{m=1}^{\infty}\left(a_m(r)\cos m\theta + b_m(r)\sin m\theta \right).$$

Now we see clearly that (for fixed r) $a_0(r)$, $a_m(r)$, and $b_m(r)$ are the Fourier coefficients in the Fourier series expansion of the function $\theta \mapsto f(r, \theta)$. Using the Euler formulas for the Fourier coefficients (Section 7.2), we conclude that

$$(9) \qquad a_0(r) = \frac{1}{2\pi} \int_0^{2\pi} f(r, \theta)\, d\theta = \sum_{n=1}^{\infty} a_{0n} J_0(\lambda_{0n} r),$$

$$(10) \qquad a_m(r) = \frac{1}{\pi} \int_0^{2\pi} f(r, \theta) \cos m\theta\, d\theta = \sum_{n=1}^{\infty} a_{mn} J_m(\lambda_{mn} r),$$

$$(11) \qquad b_m(r) = \frac{1}{\pi} \int_0^{2\pi} f(r, \theta) \sin m\theta\, d\theta = \sum_{n=1}^{\infty} b_{mn} J_m(\lambda_{mn} r),$$

for $m = 1, 2, \ldots$ Now let r vary and think of the last three series as the Bessel series expansions of order $m = 0, 1, 2, \ldots$ of the functions $a_0(r)$, $a_m(r)$, and $b_m(r)$, respectively. The coefficients in these series are Bessel coefficients and so from (17), Section 9.7, we obtain

$$a_{0n} = \frac{2}{a^2 J_1^2(\alpha_{0n})} \int_0^a a_0(r) J_0(\lambda_{0n} r) r\, dr,$$

$$a_{mn} = \frac{2}{a^2 J_{m+1}^2(\alpha_{mn})} \int_0^a a_m(r) J_m(\lambda_{mn} r) r\, dr,$$

$$b_{mn} = \frac{2}{a^2 J_{m+1}^2(\alpha_{mn})} \int_0^a b_m(r) J_m(\lambda_{mn} r) r\, dr.$$

Now using (9)–(11), we get

$$(12) \qquad a_{0n} = \frac{1}{\pi a^2 J_1^2(\alpha_{0n})} \int_0^a \int_0^{2\pi} f(r, \theta) J_0(\lambda_{0n} r) r\, d\theta\, dr,$$

$$(13) \qquad a_{mn} = \frac{2}{\pi a^2 J_{m+1}^2(\alpha_{mn})} \int_0^a \int_0^{2\pi} f(r, \theta) \cos m\theta\, J_m(\lambda_{mn} r) r\, d\theta\, dr,$$

$$(14) \qquad b_{mn} = \frac{2}{\pi a^2 J_{m+1}^2(\alpha_{mn})} \int_0^a \int_0^{2\pi} f(r, \theta) \sin m\theta\, J_m(\lambda_{mn} r) r\, d\theta\, dr,$$

for $m, n = 1, 2, \ldots$. Substituting these coefficients in (7) completes the solution of the problem.   ∎

Before giving a numerical application, we present a useful identity involving Bessel functions.

A USEFUL IDENTITY

For any $k \geq 0$, $a > 0$, and $\alpha > 0$, we have

$$(15) \qquad \int_0^a (a^2 - r^2) r^{k+1} J_k\left(\frac{\alpha}{a} r\right) dr = 2 \frac{a^{k+4}}{\alpha^2} J_{k+2}(\alpha).$$

Proof We first make a change of variables, $\frac{\alpha}{a} r = x$, $dr = \frac{a}{\alpha} dx$, and transform the integral into

$$\frac{a^{k+2}}{\alpha^{k+2}} \int_0^\alpha \left(a^2 - \frac{a^2 x^2}{\alpha^2}\right) x^{k+1} J_k(x) dx$$

$$= \frac{a^{k+4}}{\alpha^{k+2}} \int_0^\alpha x^{k+1} J_k(x) dx - \frac{a^{k+4}}{\alpha^{k+4}} \int_0^\alpha x^{k+3} J_k(x) dx.$$

From (7), Section 9.7, with $p = k$, the first term is

$$\frac{a^{k+4}}{\alpha^{k+2}} \left[x^{k+1} J_{k+1}(x)\right]_0^\alpha = \frac{a^{k+4}}{\alpha} J_{k+1}(\alpha).$$

The second integral can be evaluated with the help of (7), Section 9.7, and integration by parts. Let $u = x^2$, $dv = x^{k+1} J_k(x) dx$; then $du = 2x \, dx$, $v = x^{k+1} J_{k+1}(x)$. Hence the second term becomes

$$-\frac{a^{k+4}}{\alpha^{k+4}} \left[x^{k+3} J_{k+1}(x)\right]_0^\alpha + 2\frac{a^{k+4}}{\alpha^{k+4}} \int_0^\alpha x^{k+2} J_{k+1}(x) dx.$$

Using (7), Section 9.7, one more time and simplifying, we get

$$-\frac{a^{k+4}}{\alpha} J_{k+1}(\alpha) + 2\frac{a^{k+4}}{\alpha^2} J_{k+2}(\alpha),$$

and (15) follows. ∎

EXAMPLE 2 A vibrating membrane

Refer to Example 1 and determine the solution $u(r, \theta, t)$ when $a = c = 1$, $f(r, \theta) = (1 - r^2) r \sin \theta$, $g(r, \theta) = 0$.

Solution From (12), we have

$$a_{0n} = \frac{1}{\pi J_1^2(\alpha_{0n})} \int_0^1 \int_0^{2\pi} (1 - r^2) r \sin \theta J_0(\alpha_{0n} r) r \, d\theta \, dr = 0,$$

because $\int_0^{2\pi} \sin \theta \, d\theta = 0$. A similar argument using (13), (14), and the orthogonality of the trigonometric functions shows that $a_{mn} = 0$ for all m and n, and $b_{mn} = 0$, except when $m = 1$, in which case we have

$$b_{1n} = \frac{2}{\pi J_2^2(\alpha_{1n})} \int_0^1 \int_0^{2\pi} (1 - r^2) r \sin^2 \theta \, J_1(\alpha_{1n} r) r \, d\theta \, dr$$

$$= \frac{2}{J_2^2(\alpha_{1n})} \int_0^1 (1 - r^2) r^2 J_1(\alpha_{1n} r) \, dr$$

because $\frac{1}{\pi}\int_0^{2\pi}\sin^2\theta\,d\theta = 1$. We now appeal to (15) with $a = 1$, $k = 1$ and get

$$b_{1n} = \frac{4J_3(\alpha_{1n})}{\alpha_{1n}^2 J_2^2(\alpha_{1n})} = \frac{16}{\alpha_{1n}^3 J_2(\alpha_{1n})},$$

where in the last step we have used (6) from Section 9.7 with $p = 2$, and the fact that $J_1(\alpha_{1n}) = 0$. Recall that α_{1n} denotes the nth positive zero of J_1. See Figure 1 for an illustration and Table 1, Section 9.7, for a list of numerical values of the first five α_{1n}. Substituting b_{1n} into (7), we arrive at the solution

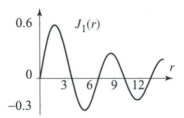

Figure 1 $J_1(r)$ and its first positive zeros.

$$u(r,\theta,t) = \sin\theta \sum_{n=1}^{\infty} \frac{16}{\alpha_{1n}^3 J_2(\alpha_{1n})} J_1(\alpha_{1n}r) \cos\alpha_{1n}t.$$

With the help of a computer system we found approximate numerical values of the first three coefficients in the series and plotted in Figure 2 the partial sum of the series solution (with n up to 3) at various values of t.   ∎

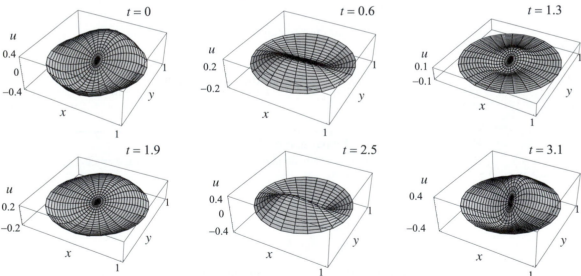

Figure 2 Vibrating circular membrane: a nonradially symmetric case.

To complete the solution of the vibrating membrane, we need to treat the case of a nonzero initial velocity. Save for some minor differences, this case is similar to the one we just treated. The proof is outlined in Exercises 7 and 8. For ease of reference, we state the entire solution in the following box.

THE WAVE EQUATION IN POLAR COORDINATES: GENERAL CASE

The solution of the boundary value problem (1)–(3) is given by

$$u(r,\theta,t) = \sum_{m=0}^{\infty}\sum_{n=1}^{\infty} J_m(\lambda_{mn}r)(a_{mn}\cos m\theta + b_{mn}\sin m\theta)\cos c\lambda_{mn}t$$

$$(16)\qquad + \sum_{m=0}^{\infty}\sum_{n=1}^{\infty} J_m(\lambda_{mn}r)(a_{mn}^*\cos m\theta + b_{mn}^*\sin m\theta)\sin c\lambda_{mn}t,$$

where $\lambda_{mn} = \frac{\alpha_{mn}}{a}$; α_{mn} is the nth positive zero of J_m; a_{mn}, b_{mn} are given by (12)–(14); and

$$(17)\qquad a_{0n}^* = \frac{1}{\pi c\alpha_{0n}aJ_1^2(\alpha_{0n})}\int_0^a\int_0^{2\pi} g(r,\theta)J_0(\lambda_{0n}r)r\,d\theta\,dr,$$

$$(18)\qquad a_{mn}^* = \frac{2}{\pi c\alpha_{mn}aJ_{m+1}^2(\alpha_{mn})}\int_0^a\int_0^{2\pi} g(r,\theta)\cos m\theta J_m(\lambda_{mn}r)r\,d\theta\,dr,$$

$$(19)\qquad b_{mn}^* = \frac{2}{\pi c\alpha_{mn}aJ_{m+1}^2(\alpha_{mn})}\int_0^a\int_0^{2\pi} g(r,\theta)\sin m\theta J_m(\lambda_{mn}r)r\,d\theta\,dr,$$

for $m,\,n = 1, 2, \ldots$.

EXAMPLE 3   **Nonzero initial displacement and velocity**
Determine the solution $u(r,\theta,t)$ of (1)–(3) when

$$a = c = 1,\ f(r,\theta) = (1-r^2)r\sin\theta,\ g(r,\theta) = (1-r^2)r^2\sin 2\theta.$$

Solution   The solution is given by (16). We only need to compute the second double series since the first one is computed in Example 2. Using the orthogonality of the trigonometric functions and arguing as we did in Example 2, we find that $a_{mn}^* = 0$ for all m and n, and $b_{mn}^* = 0$, except when $m = 2$. To compute b_{2n}^* we use (19) with $g(r,\theta) = (1-r^2)r^2\sin 2\theta$, $a = c = 1$, and $\lambda_{mn} = \alpha_{mn}$, and get

$$b_{2n}^* = \frac{2}{\pi\alpha_{2n}J_3^2(\alpha_{2n})}\int_0^1\int_0^{2\pi}(1-r^2)r^2\sin^2 2\theta J_2(\alpha_{2n}r)r\,d\theta\,dr$$

$$= \frac{2}{\alpha_{2n}J_3^2(\alpha_{2n})}\int_0^1(1-r^2)r^3J_2(\alpha_{2n}r)dr$$

because $\frac{1}{\pi}\int_0^{2\pi}\sin^2 2\theta\,d\theta = 1$. To compute the last integral, we apply (15) with $a = 1$, $k = 2$ and obtain

$$b_{2n}^* = \frac{4J_4(\alpha_{2n})}{\alpha_{2n}^3 J_3^2(\alpha_{2n})} = \frac{24}{\alpha_{2n}^4 J_3(\alpha_{2n})},$$

where in the last step we have used (6) from Section 9.7 with $p = 3$, and the fact that $J_2(\alpha_{2n}) = 0$. Substituting in the second double series in (16) and using the solution of Example 2, we get the solution

$$u(r, \theta, t) = \sin \theta \sum_{n=1}^{\infty} \frac{16}{\alpha_{1n}^3 J_2(\alpha_{1n})} J_1(\alpha_{1n} r) \cos \alpha_{1n} t$$

$$+ \sin 2\theta \sum_{n=1}^{\infty} \frac{24}{\alpha_{2n}^4 J_3(\alpha_{2n})} J_2(\alpha_{2n} r) \sin \alpha_{2n} t .$$

The coefficients in the series can be approximated with the help of a computer, as we did in Example 2.   ∎

In the exercises, we will use the methods of this section to solve the general heat problem on the disk.

Exercises 9.3

In Exercises 1–8, solve the vibrating membrane problem (1)–(3) for the given data. If possible, with the help of a computer, find numerical values for the first five nonzero coefficients of the series solution and plot the shape of the membrane at various values of t. (Formula (15) is helpful in doing these problems.)

1. $f(r, \theta) = (1 - r^2)r^2 \sin 2\theta$, $g(r, \theta) = 0$, $a = c = 1$.

2. $f(r, \theta) = (9 - r^2) \cos 2\theta$, $g(r, \theta) = 0$, $a = 3$, $c = 1$.

3. $f(r, \theta) = (4 - r^2)r \sin \theta$, $g(r, \theta) = 1$, $a = 2$, $c = 1$.

4. $f(r, \theta) = J_3(\alpha_{32} r) \sin 3\theta$, $g(r, \theta) = 0$, $a = c = 1$.

5. $f(r, \theta) = 0$, $g(r, \theta) = (1 - r^2)r^2 \sin 2\theta$, $a = c = 1$.

6. $f(r, \theta) = 1 - r^2$, $g(r, \theta) = J_0(r)$, $a = c = 1$.

7. Project Problem: Circular membrane with zero initial displacement.
Follow the steps outlined in this exercise to determine the vibrations of a circular membrane with radius a and fixed boundary, given that the initial displacement of the membrane is 0 and its initial velocity is $g(r, \theta)$.   Review the solution of Example 1 for hints.
(a) Write down explicitly the differential equation, the boundary conditions, and the initial conditions.
(b) Assume a product solution of the form $R(r)\Theta(\theta)T(t)$ and show that $T(0) = 0$. Conclude that

$$u(r, \theta, t) = \sum_{m=0}^{\infty} \sum_{n=1}^{\infty} J_m(\lambda_{mn} r)(a_{mn}^* \cos m\theta + b_{mn}^* \sin m\theta) \sin c\lambda_{mn} t .$$

(c) Use the given initial velocity and (b) to obtain

$$g(r, \theta) = \sum_{n=1}^{\infty} c\lambda_{0n} a_{0n}^* J_0(\lambda_{0n} r) + \sum_{m=1}^{\infty} \left\{ \left(\sum_{n=1}^{\infty} c\lambda_{mn} a_{mn}^* J_m(\lambda_{mn} r) \right) \cos m\theta \right.$$

$$\left. + \left(\sum_{n=1}^{\infty} c\lambda_{mn} b_{mn}^* J_m(\lambda_{mn} r) \right) \sin m\theta \right\}.$$

(d) Derive (17)–(19) by proceeding from here as we did in the derivation of (12)–(14).

8.   General solution of the vibrating circular membrane problem.
(a) Show that the solution of the boundary value problem (1)–(3) can be written as $u(r, \theta, t) = u_1(r, \theta, t) + u_2(r, \theta, t)$, where u_1 and u_2 satisfy (1) and (3) and the following initial conditions:

$$u_1(r, \theta, 0) = f(r, \theta), \quad \frac{\partial u_1}{\partial t}(r, \theta, 0) = 0;$$

$$u_2(r, \theta, 0) = 0, \quad \frac{\partial u_2}{\partial t}(r, \theta, 0) = g(r, \theta).$$

(b) Combine the results of Example 1 and Exercise 7 to derive the general solution (16).

9.   Project Problem: An integral formula for Bessel functions. Follow the outlined steps to prove that for any $k \geq 0$, and any integer $l \geq 0$, we have

$$\int r^{k+1+2l} J_k(r)\, dr = \sum_{n=0}^{l} (-1)^n\, 2^n\, \frac{l!}{(l-n)!} r^{k+1+2l-n} J_{k+n+1}(r) + C\,.$$

(a) Show that the formula holds for $l = 0$ and all $k \geq 0$.   [Hint: Use (7), Section 9.7.]
(b) Complete the proof by induction on l.   [Hint: Assume the formula is true for $l - 1$ and all k.   To establish the formula for l, integrate by parts and use the formula with $l - 1$ and $k + 1$.]
10. Solve the boundary value problem (1)–(3) when $a = c = 1$, $f(r, \theta) = (1 - r^2)(\frac{1}{4} - r^2)r^3 \sin 3\theta$, and $g(r, \theta) = 0$.   [Hint: Exercise 9 is useful.]

11.   Project Problem: Two-dimensional heat equation (general case).
Use the methods of this section to show that the solution of heat boundary value problem

$$\frac{\partial u}{\partial t} = c^2 \left(\frac{\partial^2 u}{\partial r^2} + \frac{1}{r}\frac{\partial u}{\partial r} + \frac{1}{r^2}\frac{\partial^2 u}{\partial \theta^2} \right), \quad 0 < r < a,\ 0 < \theta < 2\pi,\ t > 0,$$

$$u(a, \theta, t) = 0, \quad 0 < \theta < 2\pi,\ t > 0,$$

$$u(r, \theta, 0) = f(r, \theta), \quad 0 < r < a,\ 0 < \theta < 2\pi,$$

is

$$u(r, \theta, t) = \sum_{m=0}^{\infty} \sum_{n=1}^{\infty} J_m(\lambda_{mn} r)(a_{mn} \cos m\theta + b_{mn} \sin m\theta)e^{-c^2 \lambda_{mn}^2 t},$$

where $\lambda_{mn} = \frac{\alpha_{mn}}{a}$; α_{mn} is the nth positive zero of J_m; and a_{mn}, b_{mn} are given by (12)–(14).

12.   (a) Solve the heat problem in Exercise 11 for the following data: $a = c = 1$, $f(r, \theta) = (1 - r^2)r \sin \theta$.   [Hint: Example 2.]
(b) What are the hottest points when $t = 0$?
(c) Locate the hottest point in the plate for $t = 1,\ 2$.   Approximate the temperature of these points at the given times.

13. Solve the heat equation

$$\frac{\partial u}{\partial t} = \frac{\partial^2 u}{\partial r^2} + \frac{1}{r}\frac{\partial u}{\partial r} + \frac{1}{r^2}\frac{\partial^2 u}{\partial \theta^2}, \quad 0 < r < 1,\ 0 < \theta < 2\pi,\ t > 0,$$

given the nonzero boundary condition

$$u(1, \theta, t) = \sin 3\theta$$

and the initial condition $u(r, \theta, 0) = 0$.

14. Project Problem: Two dimensional heat equation with a nonhomogeneous boundary condition. Show that the solution of the two dimensional heat equation

$$u_t = c^2\Big(\frac{\partial^2 u}{\partial r^2} + \frac{1}{r}\frac{\partial u}{\partial r} + \frac{1}{r^2}\frac{\partial^2 u}{\partial \theta^2}\Big), \quad 0 < r < a, \ 0 < \theta < 2\pi, \ t > 0,$$

subject to the **nonhomogeneous boundary condition**

$$u(a, \theta, t) = G(\theta), \quad 0 < \theta < 2\pi, \ t > 0,$$

and the **initial condition**

$$u(r, \theta, 0) = F(r, \theta), \quad 0 < \theta < 2\pi, \ 0 \le r < a,$$

is

$$
\begin{aligned}
u(r, \theta, t) &= a_0 + \sum_{m=1}^{\infty} [a_m \cos m\theta + b_m \sin m\theta]\Big(\frac{r}{a}\Big)^m \\
&+ \sum_{m=0}^{\infty}\sum_{n=1}^{\infty} J_m(\lambda_{mn} r)(a_{mn}\cos m\theta + b_{mn}\sin m\theta)e^{-c^2\lambda_{mn}^2 t},
\end{aligned}
$$

where the coefficients a_0, a_m, b_m are the Fourier coefficients of the 2π-periodic function $G(\theta)$; $\lambda_{mn} = \frac{\alpha_{mn}}{a}$ where α_{mn} is the nth positive zero of the Bessel function of order m, J_m; and a_{mn}, b_{mn} are given by (12)–(14) with $f(r, \theta) = F(r, \theta) - u_1(r, \theta)$, where u_1 represents the steady-state solution (Section 9.1). [Hint: Consider two separate problems. First problem (steady-state solution): $\Delta u_1 = 0$, $0 < r < a$, θ arbitrary, subject to the boundary condition $u_1(a, \theta) = G(\theta)$ for all θ. Second problem: $v_t = c^2\Delta v$, $0 < r < a$, θ arbitrary, and $t > 0$, subject to the boundary condition $v(a, \theta, t) = 0$ for all θ and $t > 0$, and the initial condition $v(r, \theta, 0) = F(r, \theta) - u_1(r, \theta)$ for all θ and $0 \le r < a$. Combine the solution of the Dirichlet problem with that of Exercise 11.]

15. A circular plate of radius 1 was placed in a freezer and its temperature was brought to $0°$. The plate was then insulated and removed from the freezer. The edge of the plate was kept at a temperature varying from $0°$ to $100°$ according to the formula $f(\theta) = 50(1 - \cos\theta)$, $0 \le \theta \le 2\pi$.
(a) Plot the temperature of the boundary as a function of θ, and find its Fourier series.
(b) Find the steady-state temperature,
(c) Plot the steady-state temperature and from your graph determine the points of the plate with temperature $0°$, respectively, $100°$.
(d) Determine and plot the isotherms. Confirm your answer in (c).
(e) Solve the heat problem in the plate given that the thermal conductivity $c = 1$

and given that the initial temperature distribution is $0°$.

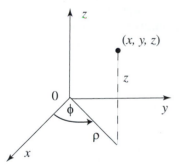 (f) Plot the temperature distribution at various values of $t > 0$ and estimate the time it takes to raise the temperature of the center to $25°$. What is the limiting value of the temperature of the center? Verify your answer using (b).

16. (a) Solve the heat problem in Exercise 14 for the following data: $a = c = 1$, $F(r, \theta) = (1 - r^2) r \sin \theta$, $G(\theta) = \sin 2\theta$ for $0 < \theta < 2\pi$.
(b) What is the steady-state solution in this case?

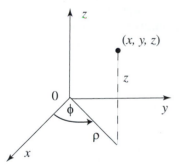 (c) Plot the solution for several values of t and compare with the graph of your answer in (b).

9.4   Steady-State Temperature in a Cylinder

In this section we treat certain radially symmetric Dirichlet problems in cylindrical regions. We will use (ρ, ϕ, θ) to denote cylindrical coordinates (Figure 1). The Laplacian in cylindrical coordinates is obtained by adding u_{zz} to the Laplacian in polar coordinates. Thus Laplace's equation with no ϕ dependence becomes

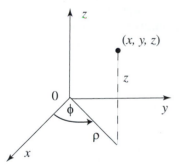

Figure 1 Cylindrical coordinates.

(1)
$$\Delta u = \frac{\partial^2 u}{\partial \rho^2} + \frac{1}{\rho}\frac{\partial u}{\partial \rho} + \frac{\partial^2 u}{\partial z^2} = 0.$$

The first problem that we will consider models the steady-state temperature distribution inside a cylinder with lateral surface and bottom kept at zero temperature and with radially symmetric temperature distribution at the top, as shown in Figure 2.

DIRICHLET PROBLEM IN A CYLINDER WITH ZERO LATERAL TEMPERATURE

The solution of Laplace's equation (1) with boundary conditions

$$u(\rho, 0) = 0, \quad 0 < \rho < a,$$
$$u(a, z) = 0, \quad 0 < z < h,$$
$$u(\rho, h) = f(\rho), \quad 0 < \rho < a,$$

is

(2)
$$u(\rho, z) = \sum_{n=1}^{\infty} A_n J_0(\lambda_n \rho) \sinh \lambda_n z,$$

where

(3) $$A_n = \frac{2}{\sinh(\lambda_n h) a^2 J_1^2(\alpha_n)} \int_0^a f(\rho) J_0(\lambda_n \rho) \rho \, d\rho, \quad \lambda_n = \frac{\alpha_n}{a},$$

and α_n is the nth positive zero of J_0, the Bessel function of order 0.

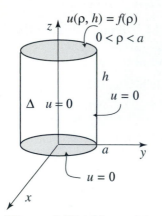

Figure 2 Dirichlet problem in a cylindrical region.

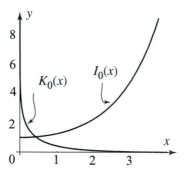

Figure 3 Modified Bessel functions.

Proof Using the method of separation of variables and setting $u(\rho, z) = R(\rho)Z(z)$, we get the equations

$$\rho^2 R'' + \rho R' - k\rho^2 R = 0, \quad R(a) = 0,$$
$$Z'' + kZ = 0, \quad Z(0) = 0,$$

where k is the separation constant. We also require that R be bounded at $\rho = 0$, since we are solving for the temperature inside the cylinder. If $k = 0$, it is straightforward to check that we only get the solution $R = 0$. If $k > 0$, say $k = \lambda^2$, then we get the **parametric form of the modified Bessel equation of order 0** defined in Exercise 7 (see also Exercises 29 and 30, Section 9.6). The general solution in this case is a linear combination of the **modified Bessel functions of the first** and **second kind**, I_0 and K_0, shown in Figure 3 (see Exercise 7). Since the first one is positive and strictly increasing for $\rho > 0$, and the second one is unbounded near zero, we conclude that no nontrivial bounded linear combination of these functions can satisfy the boundary conditions on R. So this leaves the only possibility $k = -\lambda^2 < 0$. In this case we have

$$\rho^2 R'' + \rho R' + \lambda^2 \rho^2 R = 0, \quad R(a) = 0,$$
$$Z'' - \lambda^2 Z = 0, \quad Z(0) = 0.$$

Applying Theorem 3, Section 9.7, we find that

$$R = R_n(\rho) = J_0(\lambda_n \rho), \text{ where } \lambda_n = \alpha_n/a, \; n = 1, 2, \ldots.$$

Solving the equation for Z with $\lambda = \lambda_n$ we find

$$Z_n(z) = \sinh \lambda_n z \quad n = 1, 2, \ldots.$$

Superposing the product solutions we get (2) as a solution. To determine the unknown coefficients A_n, we set $z = h$ and get the Bessel series expansion

$$f(\rho) = \sum_{n=1}^{\infty} A_n J_0(\lambda_n \rho) \sinh \lambda_n h.$$

Thus $A_n \sinh \lambda_n h$ must be the nth Bessel coefficient of $f(\rho)$, and so (3) follows from Theorem 2, Section 9.7. ∎

As a second illustration, we consider a boundary value problem with a nonzero boundary condition on the lateral surface of the cylinder (see Figure 4).

**DIRICHLET
PROBLEM IN A
CYLINDER WITH
NONZERO LATERAL
TEMPERATURE**

The solution of Laplace's equation (1) with boundary conditions

$$u(\rho, 0) = u(\rho, h) = 0, \quad 0 < \rho < a,$$
$$u(a, z) = f(z), \quad 0 < z < h,$$

(see Figure 4) is

(4) $$u(\rho, z) = \sum_{n=1}^{\infty} B_n I_0\left(\frac{n\pi}{h}\rho\right) \sin\frac{n\pi}{h}z,$$

where I_0 is the modified Bessel function of the first kind of order 0, and

(5) $$B_n = \frac{2}{I_0(\frac{n\pi a}{h})h} \int_0^h f(z) \sin\frac{n\pi}{h}z \, dz.$$

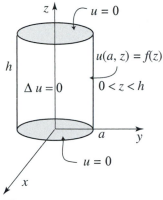

z ↑ $u = 0$

h

$u(a, z) = f(z)$

$\Delta u = 0$ $0 < z < h$

a y

$u = 0$

x

Figure 4

The derivation of the solution is very much like the one we did previously, except that now the interesting case of the separation constant is $k = \nu^2 > 0$. The details are left to Exercise 8.

Exercises 9.4

In Exercises 1–4, find the steady-state temperature in the cylinder of Figure 2 for the given temperature distribution of its top. Take $a = 1$, and $h = 2$.

1. $f(\rho) = 100$.

2. $f(\rho) = 100 - \rho^2$.

3. $f(\rho) = \begin{cases} 100 & \text{if } 0 < \rho < \frac{1}{2}, \\ 0 & \text{if } \frac{1}{2} < \rho < 1. \end{cases}$

4. $f(\rho) = 70\, J_0(\rho)$.

5. (a) Find the steady-state temperature in the cylinder with boundary values as shown in Figure 5.
(b) Solve (1) for the boundary conditions

$$u(\rho, 0) = f_1(\rho), \quad 0 < \rho < a,$$
$$u(a, z) = 0, \quad 0 < z < h,$$
$$u(\rho, h) = f_2(\rho), \quad 0 < \rho < a.$$

[Hint: Combine (a) with the solution in this section.]

6. Solve (1) for the boundary conditions

$$u(\rho, 0) = 100, \quad 0 < \rho < 1,$$
$$u(1, z) = 0, \quad 0 < z < 2,$$
$$u(\rho, 2) = 100, \quad 0 < \rho < 1.$$

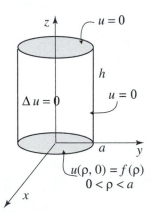

z ↑ $u = 0$

h

$\Delta u = 0$ $u = 0$

a y

$u(\rho, 0) = f(\rho)$
$0 < \rho < a$

x

Figure 5 for Exercise 5.

7. Make the substitution $x = \lambda\rho$ ($\lambda > 0$) in the **parametric form of the modified Bessel equation** $\rho^2 R'' + \rho R' - \lambda^2\rho^2 R = 0$ and obtain that its general solution is $y = c_1 I_0(\lambda\rho) + c_2 K_0(\lambda\rho)$, where I_0 and K_0 are the modified Bessel functions of the first and second kind. [Hint: Use Exercises 29 and 30, Section 9.6.]

Project Problem: Lateral surface with nonzero temperature. Do Exercises 8 and 9.

8. In this exercise we derive (4) and (5).

(a) Refer to the Dirichlet problem in the cylinder with boundary conditions as given just before (4). Use the separation of variables method and obtain

$$Z'' + \nu^2 Z = 0, \quad Z(0) = 0 \text{ and } Z(h) = 0,$$
$$\rho^2 R'' + \rho R' - \nu^2 \rho^2 R = 0.$$

(b) Show that the only possible solutions of the first equation correspond to $\nu_n = \frac{n\pi}{h}$ and hence are

$$Z_n(z) = \sin \frac{n\pi}{h} z, \quad n = 1, 2, \dots.$$

(c) Derive (4) and (5). [Hint: Use Exercise 7.]

9. Solve (1) for the boundary conditions

$$u(\rho, 0) = u(\rho, 2) = 0, \quad 0 < \rho < 1,$$
$$u(1, z) = 10z, \quad 0 < z < 2.$$

10. Solve (1) for the boundary conditions

$$u(\rho, 0) = 100, \quad 0 < \rho < 1,$$
$$u(1, z) = 10z, \quad 0 < z < 2,$$
$$u(\rho, 2) = 100, \quad 0 < \rho < 1.$$

[Hint: Combine the solutions of Exercises 6 and 9.]

11. Find the steady-state temperature in a cylindrical barrel floating in water, as shown in Figure 6.

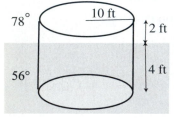

Figure 6 for Exercise 11.

9.5 The Helmholtz and Poisson Equations

In Section 8.7 we used the method of eigenfunction expansions to solve many interesting problems over rectangles. As similar program can be carried out with success over a disk. For this purpose, we start by solving the **eigenvalue problem** consisting of the **Helmholtz equation** on a disk of radius a,

(1)
$$\Delta\phi(r, \theta) = -k\phi(r, \theta), \ 0 < r < a, \ 0 < \theta < 2\pi,$$

with the **boundary condition**

(2)
$$\phi(a, \theta) = 0, \ 0 < \theta < 2\pi.$$

To solve this problem means to determine the values of k (or **eigenvalues**) for which we have nontrivial solutions and find these nontrivial solutions (or **eigenfunctions**).

Substituting $\phi(r, \theta) = R(r)\Theta(\theta)$ into (1), separating variables, and using the fact that Θ is 2π-periodic, we arrive at the equations

(3) $$\Theta'' + m^2\Theta = 0, \quad m = 0, 1, 2, \ldots,$$

(4) $$r^2 R'' + rR' + (kr^2 - m^2)R = 0, \quad R(a) = 0.$$

The solutions of (3) are

$$\cos m\theta \text{ and } \sin m\theta, \quad m = 0, 1, 2, \ldots.$$

Equation (4) is the parametric form of Bessel's equation of order m. We know from Theorem 3, Section 9.7, that the nontrivial solutions of (4) are constant multiples of $J_m(\lambda_{mn} r)$, which is the solution corresponding to the eigenvalue $k = \lambda_{mn}^2$. Piecing together the product solutions, we obtain the following important result.

THEOREM 1

THE HELMHOLTZ EQUATION IN A DISK

The eigenvalues of the Helmholtz problem (1)-(2) are

(5) $$k = \lambda_{mn}^2 = (\alpha_{mn}/a)^2, \quad m = 0, 1, 2, \ldots, \quad n = 1, 2, \ldots,$$

where α_{mn} is the nth positive zero of the Bessel function J_m. To each eigenvalue λ_{mn}^2 correspond the eigenfunctions

(6) $$\cos m\theta J_m(\lambda_{mn} r) \text{ and } \sin m\theta J_m(\lambda_{mn} r).$$

(Note that for $m = 1, 2, \ldots$, we have two distinct eigenfunctions for a given eigenvalue.)

In other words, if $\phi_{mn}(r, \theta) = \cos m\theta J_m(\lambda_{mn} r)$ or $\phi_{mn}(r, \theta) = \sin m\theta J_m(\lambda_{mn} r)$, then $\Delta\phi_{mn} = -\lambda_{mn}^2 \phi_{mn}$ and $\phi(a, \theta) = 0$.

As you will discover, the eigenfunctions satisfy orthogonality relations that can be used to expand functions on the disk, much as we used $\cos nx$ and $\sin nx$ to expand functions in terms of Fourier series. The orthogonality here follows as a consequence of the orthogonality of the Bessel functions and the trigonometric system. Because the orthogonality relations for the Bessel functions are expressed with respect to the weight function r (Theorem 1, Section 9.7), the orthogonality of the eigenfunctions in (6) will be expressed by integrals over the disk with respect to $r\,dr\,d\theta$. For example, we have

(7) $$\int_0^{2\pi} \int_0^a \sin m\theta J_m(\lambda_{mn} r) \cos m\theta J_m(\lambda_{mn} r) r\,dr\,d\theta = 0.$$

Putting these facts together, we obtain the following expansion theorem, which was already used in (8), Section 9.3.

**THEOREM 2
EXPANSIONS IN
TERMS OF THE
EIGENFUNCTIONS
OF THE
HELMHOLTZ
EQUATION**

Suppose that $f(r, \theta)$ is defined for all $0 < r < a$ and $0 < \theta < 2\pi$. Then

$$(8) \qquad f(r, \theta) = \sum_{m=0}^{\infty} \sum_{n=1}^{\infty} J_m(\lambda_{mn} r)(a_{mn} \cos m\theta + b_{mn} \sin m\theta),$$

where (the generalized Fourier coefficients of f) a_{mn} and b_{mn} are given by (12)–(14), Section 9.3.

In Section 9.3, we derived (8) as a consequence of Fourier series and Bessel-Fourier series. A simpler and more direct derivation can be obtained using the orthogonality of the eigenfunctions (6) (see Exercise 4). Now that we know the eigenfunctions of the Laplacian over a disk and that we can use them to expand functions in generalized Fourier series, as expressed in Theorem 2, we are ready to apply the methods that we used in Section 8.7 to solve a variety of interesting problems on the disk, including some non-homogeneous equations for which the separation of variables method is not suitable.

EXAMPLE 1   The method of eigenfunction expansions

Solve $\Delta u(r, \theta) = u(r, \theta) + 3r^2 \cos 2\theta$ in the unit disk, given that $u = 0$ on the boundary.

Solution We look for a solution in the form of an eigenfunction expansion

$$u(r, \theta) = \sum_{m=0}^{\infty} \sum_{n=1}^{\infty} J_m(\alpha_{mn} r)(A_{mn} \cos m\theta + B_{mn} \sin m\theta).$$

Since each eigenfunction satisfies the boundary condition, our candidate for a solution, $u(r, \theta)$, also satisfies the boundary condition. Plugging u into the equation and using the fact that, for an eigenfunction ϕ_{mn}, $\Delta \phi_{mn} = -\alpha_{mn}^2 \phi_{mn}$, we get

$$\sum_{m=0}^{\infty} \sum_{n=1}^{\infty} -\alpha_{mn}^2 J_m(\alpha_{mn} r)(A_{mn} \cos m\theta + B_{mn} \sin m\theta)$$

$$= \sum_{m=0}^{\infty} \sum_{n=1}^{\infty} J_m(\alpha_{mn} r)(A_{mn} \cos m\theta + B_{mn} \sin m\theta) + 3r^2 \cos 2\theta;$$

hence

$$\sum_{m=0}^{\infty} \sum_{n=1}^{\infty} (-1 - \alpha_{mn}^2) J_m(\alpha_{mn} r)(A_{mn} \cos m\theta + B_{mn} \sin m\theta) = 3r^2 \cos 2\theta.$$

Thus $(-1 - \alpha_{mn}^2) A_{mn}$ and $(-1 - \alpha_{mn}^2) B_{mn}$ are the generalized Fourier coefficients of the function $f(r, \theta) = 3r^2 \cos 2\theta$. Note that the orthogonality of the trigonometric system will imply that only A_{2n} is nonzero. All other coefficients will be zero.

Appealing to (13), Section 9.3, with $m = 2$, we find

$$
\begin{aligned}
(-1 - \alpha_{2n}^2)A_{2n} &= \frac{2}{\pi J_3^2(\alpha_{2n})} \int_0^1 \int_0^{2\pi} 3 \cos^2 2\theta \, d\theta r^2 J_2(\alpha_{2n}r) \, r \, dr \\
&= \frac{6}{J_3^2(\alpha_{2n})} \int_0^1 r^3 J_2(\alpha_{2n}r) \, dr \qquad \left(\int_0^{2\pi} \cos^2 2\theta \, d\theta = \pi\right) \\
&= \frac{6}{J_3^2(\alpha_{2n})} \frac{J_3(\alpha_{2n})}{\alpha_{2n}} \qquad \text{(by (11), Section 9.2)} \\
&= \frac{6}{\alpha_{2n} J_3(\alpha_{2n})}.
\end{aligned}
$$

Solving for A_{2n} and plugging into the eigenfunction expansion of u, we find

$$
u(r, \theta) = \cos 2\theta \sum_{n=1}^{\infty} \frac{-6}{(1 + \alpha_{2n}^2)\alpha_{2n} J_3(\alpha_{2n})} J_2(\alpha_{2n}r).
$$

The collapsing of the double sum in (8) to a single sum is due to the fact that only the terms in $\cos 2\theta$ are needed here. In general, you may need the entire double sum. ∎

Poisson's Equation in a Disk

Consider the Poisson problem

(9) $$\Delta u = f(r, \theta), \quad 0 < r < a, \quad 0 < \theta < 2\pi,$$

(10) $$u(a, \theta) = g(\theta), \quad 0 < \theta < 2\pi.$$

Our first step is to decompose the problem into the two simpler subproblems in Figure 1.

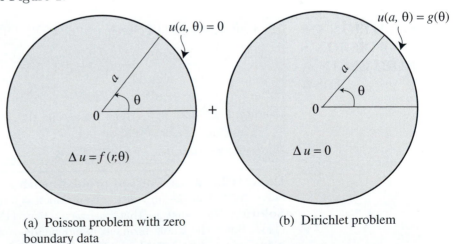

(a) Poisson problem with zero boundary data

(b) Dirichlet problem

Figure 1 Decomposition of a Poisson problem.

The Dirichlet problem in Figure 1(b) can be solved by the methods of Section 9.1. Thus, to complete the solution, we need only solve Poisson's equation with zero boundary data (Figure 1(a)). We will use the method of eigenfunction expansions, and thus look for a solution of (9) (with zero boundary data) in the form

$$(11) \qquad u(r,\theta) = \sum_{m=0}^{\infty} \sum_{n=1}^{\infty} J_m(\lambda_{mn} r)(A_{mn} \cos m\theta + B_{mn} \sin m\theta).$$

Plugging into (9) and using the fact that each term in the series is an eigenfunction, we obtain

$$\sum_{m=0}^{\infty} \sum_{n=1}^{\infty} -\lambda_{mn}^2 J_m(\lambda_{mn} r)(A_{mn} \cos m\theta + B_{mn} \sin m\theta) = f(r,\theta).$$

Thinking of this as being the eigenfunction expansion of $f(r,\theta)$, we apply Theorem 2, solve for A_{mn} and B_{mn}, and obtain, for m, $n = 1, 2, \ldots,$

$$(12) \qquad A_{0n} = \frac{-1}{\pi \alpha_{0n}^2 J_1^2(\alpha_{0n})} \int_0^a \int_0^{2\pi} f(r,\theta) J_0(\lambda_{0n} r) r \, d\theta \, dr,$$

$$(13) \qquad A_{mn} = \frac{-2}{\pi \alpha_{mn}^2 J_{m+1}^2(\alpha_{mn})} \int_0^a \int_0^{2\pi} f(r,\theta) \cos m\theta \, J_m(\lambda_{mn} r) r \, d\theta \, dr,$$

$$(14) \qquad B_{mn} = \frac{-2}{\pi \alpha_{mn}^2 J_{m+1}^2(\alpha_{mn})} \int_0^a \int_0^{2\pi} f(r,\theta) \sin m\theta \, J_m(\lambda_{mn} r) r \, d\theta \, dr.$$

This completely determines the solutions of the problem in Figure 1(a). We summarize our findings as follows.

SOLUTION OF POISSON'S PROBLEM IN A DISK

The solution of the Poisson problem (9)–(10) is given by

$$(15) \qquad u(r,\theta) = u_1(r,\theta) + u_2(r,\theta),$$

where u_1 is the solution of the Poisson problem with zero boundary data in Figure 1(a), and u_2 is the solution of the Dirichlet problem in Figure 1(b). The function u_1 is given by (11)–(14), and the function u_2 is given by (4) and (6), Section 9.1.

EXAMPLE 2   A Poisson problem with zero boundary data
Solve $\Delta u = 1$ in the unit disk, given that $u = 0$ on the boundary.
Solution Note that in this case $u_2 = 0$ in (15). The function u_1 is given by (11). Since the whole problem is independent of θ, we expect the solution to be independent of θ. Indeed, plugging $f(r,\theta) = 1$ into (13) and (14), we get 0 because of the integral in θ. Now (12) yields

$$A_{0n} = \frac{-2}{\alpha_{0n}^2 J_1^2(\alpha_{0n})} \int_0^1 J_0(\alpha_{0n} r) r \, dr.$$

Using (11), Section 9.2, to evaluate the integral and simplifying, we get $A_{0n} = \frac{-2}{\alpha_{0n}^3 J_1(\alpha_{0n})}$. Substituting into (11), we obtain

$$u(r, \theta) = \sum_{n=1}^{\infty} \frac{-2}{\alpha_{0n}^3 J_1(\alpha_{0n})} J_0(\alpha_{0n} r). \qquad \blacksquare$$

Interesting applications of the eigenfunction expansions method are presented in the exercises.

Exercises 9.5

1. Derive (3) and (4) from (1) and (2).

2. State and prove all the orthogonality relations for the eigenfunctions of the Helmholtz problem (1) and (2) ((7) is one of them).

3. Let $\phi_{mn}(r, \theta)$ denote either one of the eigenfunctions in (6). Evaluate

$$\int_0^{2\pi} \int_0^a \phi_{mn}^2(r, \theta)\, r\, dr\, d\theta.$$

Treat the case $m = 0$ separately. [Hint: Use (12), Section 9.7.]

4. Derive the coefficients in Theorem 2 by using the orthogonality of the eigenfunctions (6). [Hint: Multiply both sides of (8) by an eigenfunction, interchange integrals and summation signs, then integrate over the disk with respect to $r\, d\theta\, dr$.]

In Exercises 5–12, use the method of eigenfunction expansions to solve the given problem in the unit disk.

5. $\Delta u = -u + 1$, $u(1, \theta) = 0$.   **6.** $\Delta u = 3\, u + r \sin\theta$, $u(1, \theta) = 0$.

7. $\Delta u = 2 + r^3 \cos 3\theta$, $u(1, \theta) = 0$.   **8.** $\Delta u = r^2$, $u(1, \theta) = 0$.

9.   **10.** $\Delta u = r^m \sin m\theta$, $u(1, \theta) = 0$.

$$\Delta u = \begin{cases} r \sin\theta & \text{if } 0 < r < \frac{1}{2}, \\ 0 & \text{if } \frac{1}{2} < r < 1, \end{cases}$$

$$u(1, \theta) = 0.$$

11. $\Delta u = 1$, $u(1, \theta) = \sin 2\theta$.   **12.** $\Delta u = 1 + r \cos\theta$, $u(1, \theta) = 1$.

13. A heat problem. Do Exercise 11, Section 9.3, using the method of eigenfunction expansions.

14. Project Problem: A nonhomogeneous heat problem. For this project, you are asked to use the eigenfunction expansions method to solve the nonhomogeneous heat boundary value problem, with time-dependent heat source,

$$\frac{\partial u}{\partial t} = c^2 \left(\frac{\partial^2 u}{\partial r^2} + \frac{1}{r} \frac{\partial u}{\partial r} + \frac{1}{r^2} \frac{\partial^2 u}{\partial \theta^2} \right) + q(r, \theta, t),$$
$$0 < r < a, \quad 0 < \theta < 2\pi, \quad t > 0,$$
$$u(a, \theta, t) = 0, \ t > 0,$$
$$u(r, \theta, 0) = f(r, \theta), \quad 0 < r < a, \quad 0 < \theta < 2\pi.$$

Justify the following steps.

(a) Let

$$u(r, \theta, t) = \sum_{m=0}^{\infty} \sum_{n=1}^{\infty} J_m(\lambda_{mn} r)(A_{mn}(t) \cos m\theta + B_{mn}(t) \sin m\theta),$$

$$f(r, \theta) = \sum_{m=0}^{\infty} \sum_{n=1}^{\infty} J_m(\lambda_{mn} r)(a_{mn} \cos m\theta + b_{mn} \sin m\theta),$$

$$q(r, \theta, t) = \sum_{m=0}^{\infty} \sum_{n=1}^{\infty} J_m(\lambda_{mn} r)(c_{mn}(t) \cos m\theta + d_{mn}(t) \sin m\theta).$$

(Why should this be your starting point?) What are a_{mn}, b_{mn}, $c_{mn}(t)$, and $d_{mn}(t)$, in terms of f and q?

(b) Show that A_{mn} and B_{mn} are solutions of the following initial value problems:

$$A'_{mn}(t) + \lambda_{mn}^2 A_{mn}(t) = c_{mn}(t), \quad A_{mn}(0) = a_{mn};$$
$$B'_{mn}(t) + \lambda_{mn}^2 B_{mn}(t) = d_{mn}(t), \quad B_{mn}(0) = b_{mn}.$$

(c) Complete the solution by showing that

$$A_{mn}(t) = e^{-\lambda_{mn}^2 t}\left(a_{mn} + \int_0^t e^{\lambda_{mn}^2 s} c_{mn}(s) ds\right)$$

and

$$B_{mn}(t) = e^{-\lambda_{mn}^2 t}\left(b_{mn} + \int_0^t e^{\lambda_{mn}^2 s} d_{mn}(s) ds\right).$$

15. (a) Work out the details in Exercise 14 for the specific case when $c = 1$, $f = 1$, and $q(r, \theta, t) = e^{-t}$.

(b) Plot the temperature of the center and describe what happens as $t \to \infty$.

16. (a) Work out the details in Exercise 14 for the specific case when $c = 1$, $f = r \sin \theta$, and $q = 1$.

(b) Plot the temperature of the center and describe what happens as $t \to \infty$.

9.6   Bessel's Equation and Bessel Functions

We saw in this chapter that **Bessel's equation of order** $p \geq 0$,

(1)
$$\boxed{x^2 y'' + xy' + (x^2 - p^2)y = 0, \quad x > 0,}$$

arises when solving partial differential equations involving the Laplacian in polar and cylindrical coordinates. Note that Bessel's equation is a whole family of differential equations, one for each value of p.

Bessel's equation also appears in solving various other classical problems. Historically, the equation with $p = 0$ was first encountered and solved by

Daniel Bernoulli in 1732 in his study of the hanging chain problem. Similar equations appeared later in 1770 in the work of Lagrange on astronomical problems. In 1824, while investigating the problem of elliptic planetary motion, the great German astronomer F. W. Bessel encountered a special form of (1). Influenced by the monumental work of Fourier that had just appeared in 1822, Bessel conducted a systematic study of (1).

Solution of Bessel's Equation

We will apply the method of Frobenius from Appendix A.5. It is easy to show that $x = 0$ is a regular singular point of Bessel's equation. So, as suggested by the method of Frobenius, we try for a solution

$$(2) \qquad y = \sum_{m=0}^{\infty} a_m x^{r+m},$$

where $a_0 \neq 0$. Substituting this into (1) yields

We have shifted the index of summation by 2 in the third series so that each series is expressed in terms of x^{r+m}.

$$\sum_{m=0}^{\infty} a_m (r+m)(r+m-1)x^{r+m} + \sum_{m=0}^{\infty} a_m (r+m)x^{r+m}$$

$$+ \sum_{m=2}^{\infty} a_{m-2} x^{r+m} - p^2 \sum_{m=0}^{\infty} a_m x^{r+m} = 0.$$

Writing the terms corresponding to $m = 0$ and $m = 1$ separately gives

$$a_0(r^2 - p^2)x^r + a_1 \left[(r+1)^2 - p^2\right] x^{r+1}$$

$$+ \sum_{m=2}^{\infty} \left(a_m \left[(r+m)^2 - p^2\right] + a_{m-2} \right)x^{r+m} = 0.$$

Equating coefficients of the series to zero gives

$$(3) \qquad a_0(r^2 - p^2) = 0 \quad (m = 0);$$
$$(4) \qquad a_1 \left[(r+1)^2 - p^2\right] = 0 \quad (m = 1);$$
$$(5) \qquad a_m \left[(r+m)^2 - p^2\right] + a_{m-2} = 0 \quad (m \geq 2).$$

From (3), since $a_0 \neq 0$, we get the **indicial equation**

$$(r + p)(r - p) = 0$$

with **indicial roots** $r = p$ and $r = -p$.

First Solution of Bessel's Equation

Setting $r = p$ in (5) gives the recurrence relation

$$a_m = \frac{-1}{m(m+2p)} a_{m-2}, \quad m \geq 2.$$

This is a two-step recurrence relation, so the even- and odd-indexed terms are determined separately. We deal with the odd-indexed terms first. With $r = p$, (4) becomes $a_1 \left[(p+1)^2 - p^2\right] = 0$ which implies that $a_1 = 0$ (recall that $p \geq 0$ in (1)), and so $a_3 = a_5 = \cdots = 0$. To make it easier to find a pattern for the even-indexed terms we rewrite the recurrence relation with $m = 2k$ and get

$$a_{2k} = \frac{-1}{2^2 k(k+p)} a_{2(k-1)}, \quad k \geq 1.$$

This gives

$$a_2 = \frac{-1}{2^2(1+p)} a_0;$$

$$a_4 = \frac{-1}{2^2 2(2+p)} a_2 = \frac{1}{2^4 2!(1+p)(2+p)} a_0;$$

$$a_6 = \frac{-1}{2^2 3(3+p)} a_4 = \frac{-1}{2^6 3!(1+p)(2+p)(3+p)} a_0;$$

and so on. Substituting these coefficients into (2) gives one solution to Bessel's equation:

(6)
$$y = a_0 \sum_{k=0}^{\infty} \frac{(-1)^k}{2^{2k} k!(1+p)(2+p)\cdots(k+p)} x^{2k+p},$$

where $a_0 \neq 0$ is arbitrary. This solution may be written in a nicer way with the aid of the gamma function:

(7)
$$\Gamma(x) = \int_0^{\infty} t^{x-1} e^{-t}\, dt, \quad x > 0.$$

(This is the definition from Exercise 24, Section 4.3, restricted to real values of z.) The integral is improper and converges for all $x > 0$. We choose

$$a_0 = \frac{1}{2^p \Gamma(p+1)}$$

and simplify the terms in the series using the basic property of the gamma function, $\Gamma(x+1) = x\Gamma(x)$ (Exercise 24(d), Section 4.3), as follows:

$$
\begin{aligned}
\Gamma(1+p)\left[(1+p)(2+p)\cdots(k+p)\right] &= \Gamma(2+p)\left[(2+p)\cdots(k+p)\right] \\
&= \Gamma(3+p)\left[\cdots(k+p)\right] \\
&= \cdots = \Gamma(k+p+1).
\end{aligned}
$$

After this simplification, (6) yields the first solution, denoted by J_p and called the **Bessel function of order** p,

$$(8) \qquad \boxed{ J_p(x) = \sum_{k=0}^{\infty} \frac{(-1)^k}{k!\,\Gamma(k+p+1)} \left(\frac{x}{2}\right)^{2k+p}. }$$

When $p = n$, we have $\Gamma(k+p+1) = (k+n)!$ (Exercise 24(e), Section 4.3), and so the Bessel function of order n is

$$\boxed{ J_n(x) = \sum_{k=0}^{\infty} \frac{(-1)^k}{k!\,(k+n)!} \left(\frac{x}{2}\right)^{2k+n}. }$$

To get an idea of the behavior of the Bessel functions, we sketch the graphs of J_0, $J_{1/2}$, J_1, J_2 and J_7 in Figure 1.

Figure 1 Graphs of $J_p(x)$ for $p = 0, \frac{1}{2}, 1, 2, 7$.

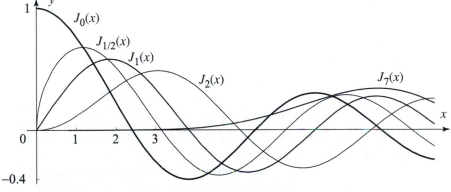

$J_0(0) = 1$,
$J_p(0) = 0$, if $p > 0$;
$|J_p(x)| \le 1$ for all x.

Note that J_p is bounded at 0. In fact, $|J_p(x)| \le 1$ for all x, as a consequence of the integral representation of Bessel functions (Exercise 36, Section 4.5). As we will see shortly, this property is not shared by the second linearly independent solution.

Second Solution of Bessel's Equation

If in (2) we replace r by the second indicial root $-p$, we arrive as before at the solution

$$(9) \qquad J_{-p}(x) = \sum_{k=0}^{\infty} \frac{(-1)^k}{k!\,\Gamma(k-p+1)} \left(\frac{x}{2}\right)^{2k-p}.$$

It turns out that if p is not an integer, then (9) is linearly independent of J_p. Thus when p is not an integer, (8) and (9) determine a fundamental set of solutions of Bessel's equation of order p. Before turning to the case when p is an integer, we compute the Bessel functions J_p and J_{-p} for the value $p = \frac{1}{2}$.

EXAMPLE 1   Bessel functions of order $p = \pm\frac{1}{2}$

Show that for $x \geq 0$

$$J_{1/2}(x) = \sqrt{\frac{2}{\pi x}}\, \sin x \quad \text{and} \quad J_{-1/2}(x) = \sqrt{\frac{2}{\pi x}}\, \cos x \,.$$

Solution   Substituting $p = \frac{1}{2}$ in (8), we get

$$J_{1/2}(x) = \sum_{k=0}^{\infty} \frac{(-1)^k}{k!\,\Gamma(k + \frac{1}{2} + 1)}\left(\frac{x}{2}\right)^{2k+\frac{1}{2}}.$$

To simplify this expression we use the result of Exercise 24(b), which implies that

$$\Gamma\left(k + \frac{1}{2} + 1\right) = \frac{(2k+1)!}{2^{2k+1}\,k!}\,\sqrt{\pi}\,.$$

Thus

$$
\begin{aligned}
J_{1/2}(x) &= \frac{1}{\sqrt{\pi}} \sum_{k=0}^{\infty} \frac{(-1)^k 2^{2k+1} k!}{k!(2k+1)!}\left(\frac{x}{2}\right)^{2k+\frac{1}{2}} \\
&= \sqrt{\frac{2}{\pi x}} \sum_{k=0}^{\infty} \frac{(-1)^k}{(2k+1)!} x^{2k+1} = \sqrt{\frac{2}{\pi x}}\, \sin x.
\end{aligned}
$$

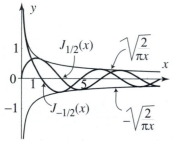

Figure 2 Graphs of $J_{1/2}$, $J_{-1/2}$, and their envelopes $y = \pm\sqrt{\frac{2}{\pi x}}$.

The other part is proved similarly by substituting $p = -\frac{1}{2}$ into (9) and simplifying with the help Exercise 24(b). In Figure 2 we plotted $J_{\frac{1}{2}}$ and $J_{-\frac{1}{2}}$. Clearly these two functions are linearly independent, since the first one is bounded while the second one is not.   ■

It is important to keep in mind that J_p and J_{-p} are linearly independent only when p is not an integer. In fact, when p is a positive integer, we observe that $k - p + 1 \leq 0$ for $k = 0, 1, \ldots, p-1$, and so the coefficients in (9) are not even defined for $k = 0, 1, \ldots, p-1$, because the gamma function is not defined at 0 and negative integers. It is useful, however, to have a definition for J_{-n} for $n = 1, 2, \ldots$. We recall from Example 5, Section 4.3,

$$(10) \qquad J_{-n}(x) = (-1)^n J_n(x) \quad (n \text{ integer } \geq 0)\,.$$

We could use the Frobenius method to derive a second linearly independent solution. However, we will describe an alternative method that is commonly used in applied mathematics. We start again with the case when p is not an integer and define

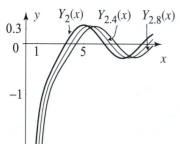

Figure 3 Approximation of Y_2.

$$(11) \qquad Y_p(x) = \frac{J_p(x)\cos p\pi - J_{-p}(x)}{\sin p\pi} \qquad (p \text{ not an integer})\,.$$

Since J_p and J_{-p} are in this case linearly independent solutions of Bessel's equation, it easily follows from (11) that Y_p is also a solution of Bessel's equation that is linearly independent of J_p. The function Y_p is called a **Bessel function of the second kind of order** p. For integer p, this function is constructed by a limiting process from the noninteger values as follows:

$$(12) \qquad\qquad Y_p = \lim_{\nu \to p} Y_\nu \,.$$

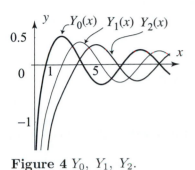

Figure 4 Y_0, Y_1, Y_2.

It can be shown that this limit exists (see Figure 3 for an illustration) and defines a solution of Bessel's equation of order p that is also linearly independent of J_p. As illustrated in Figure 4, we have

$$(13) \qquad\qquad \boxed{\lim_{x \to 0^+} Y_p(x) = -\infty \,.}$$

In particular, the Bessel functions of the second kind are not bounded near 0. We summarize our analysis of (1) as follows.

GENERAL SOLUTION OF BESSEL'S EQUATION OF ORDER p

The general solution of Bessel's equation (1) of order p is

$$y(x) = c_1 J_p(x) + c_2 Y_p(x),$$

where J_p is given by (8) and Y_p is given by (11) or (12). When p is not an integer, a general solution is also given by

$$y(x) = c_1 J_p(x) + c_2 J_{-p}(x),$$

where J_p is given by (8) and J_{-p} is given by (9).

Explicit formulas and computations of the Bessel functions are presented in the exercises.

Exercises 9.6

In Exercises 1–4, determine the order p of the given Bessel equation. Use (8) to write down three terms of the first series solution.

1. $x^2 y'' + xy' + (x^2 - 9)y = 0$.

2. $x^2 y'' + xy' + x^2 y = 0$.

3. $x^2 y'' + xy' + (x^2 - \frac{1}{4})y = 0$.

4. $x^2 y'' + xy' + (x^2 - \frac{1}{9})y = 0$.

In Exercises 5–8, find the general solution of the given differential equation on $(0, \infty)$. Write down two terms of the series expansions of each part of the solution. [Hint: Use (9).]

5. $x^2 y'' + xy' + (x^2 - \frac{9}{4})y = 0$.

6. $x^2 y'' + xy' + (x^2 - \frac{25}{4})y = 0$.

7. $x^2 y'' + x y' + (x^2 - \frac{1}{16}) y = 0$.   **8.** $x^2 y'' + x y' + (x^2 - \frac{1}{25}) y = 0$.

9. Find at least three terms of a second linearly independent solution of the equation of Exercise 1 using the Frobenius method. (A first solution is given by (8).)

10. Verify that $y_1 = x^p J_p(x)$ and $y_2 = x^p Y_p(x)$ are linearly independent solutions of

$$ x y'' + (1 - 2p) y' + x y = 0, \quad x > 0. $$

In Exercises 11–14, use the result of Exercise 10 to solve the given equation for $x > 0$.

11. $x y'' - y' + x y = 0$.   **12.** $y'' + y = 0$.

13. $x y'' - 2 y' + x y = 0$.   **14.** $x y'' - 3 y' + x y = 0$.

15. Establish the following properties:
(a) $J_0(0) = 1$, $J_p(0) = 0$ if $p > 0$;
(b) $J_n(x)$ is an even function if n is even, and odd if n is odd;
(c) $\lim_{x \to 0^+} \frac{J_p(x)}{x^p} = \frac{1}{2^p \Gamma(p+1)}$.

16. Suppose in (8) we replace p by a negative integer $-n$.
(a) Based on the properties of the gamma function, explain why in (8) it makes sense to set $1/\Gamma(k - p + 1) = 0$ for $k = 0, 1, 2, \ldots, p - 1$. [Hint: Exercise 32 below.]
(b) By reindexing the series that you obtain, show that $J_{-n}(x) = (-1)^n J_n(x)$.

In Exercises 17–20, solve the given equation by reducing it first to a Bessel's equation. Use the suggested change of variables and take $x > 0$.

17. $x y'' + (1 + 2p) y' + x y = 0$, $y = x^{-p} u$. **18.** $x y'' + y' + \frac{1}{4} y = 0$, $z = \sqrt{x}$.

19. $y'' + e^{-x} y = 0$, $z = 2 e^{-\frac{x}{2}}$.   **20.** $x^2 y'' + x y' + (4x^4 - \frac{1}{4}) y = 0$, $z = x^2$.

21. Show that $J_{-1/2}(x) = \sqrt{\frac{2}{\pi x}} \cos x$.

22. Establish the identities
(a) $J_{3/2}(x) = \sqrt{\frac{2}{\pi x}} \left[\frac{\sin x}{x} - \cos x \right]$.   (b) $J_{-3/2}(x) = \sqrt{\frac{2}{\pi x}} \left[-\frac{\cos x}{x} - \sin x \right]$.

23. General solution of Bessel's equation of order 0.
(a) Use (8) to derive the first six terms of the series solution J_0 of the Bessel's equation $x^2 y'' + x y' + x^2 y = 0$.
(b) Use (a) and the reduction of order formula to find six terms of y_2, the second series solution. [Hint: See Example 5, Appendix A.5.]
(b) Plot your answers and compare their graphs to those of J_0 and Y_0 for x near zero, say $0 < x < 4$. Describe what you find.
(c) Explain why we must have $Y_0 = a J_0 + b y_2$, where a and b are some constants. Evaluate the functions at two points in the interval $0 < x < 4$, say at $x = .2$ and $x = .3$, and obtain two equations in the unknown coefficients a and b. Solve the equations to determine a and b, and then plot and compare the graphs of Y_0 and $a J_0 + b y_2$.

24. (a) Use the value $\Gamma(\frac{1}{2}) = \sqrt{\pi}$ (Exercise 25 (a), Section 4.3) and the basic property of the gamma function $\Gamma(x + 1) = x \Gamma(x)$ (Exercise 24, Section 4.3) to show that

$$ \Gamma(\frac{3}{2}) = \frac{\sqrt{\pi}}{2} \quad \text{and} \quad \Gamma(\frac{5}{2}) = \frac{3}{2} \frac{\sqrt{\pi}}{2} = \frac{3}{4} \sqrt{\pi}. $$

More generally, show that for $n = 1, 2, \ldots$,

$$\Gamma\left(n + \frac{1}{2}\right) = \frac{(2n)!}{2^{2n}\, n!} \sqrt{\pi} \quad \text{and} \quad \Gamma\left(n + \frac{1}{2} + 1\right) = \frac{(2n+1)!}{2^{2n+1}\, n!} \sqrt{\pi}.$$

(c) Write the basic property as $\Gamma(x) = \frac{1}{x}\Gamma(x+1)$ and then define the value of the gamma function at x from its value at $x+1$. For example, we have $\Gamma(-\frac{1}{2}) = -2\Gamma(\frac{1}{2}) = -2\sqrt{\pi}$. This extends the definition of the gamma function to negative numbers other than $-1, -2, -3, \ldots$. The graph of the gamma function is sketched in Figure 5. Notice the vertical asymptotes at $x = 0, -1, -2, \ldots$. Also notice the alternating sign of the gamma function over negative intervals. Show that $\Gamma(-\frac{3}{2}) = \frac{4}{3}\sqrt{\pi}$.

Figure 5 $\Gamma(x)$, the generalized factorial function.
For $n = 0, 1, 2, \ldots$,
$\Gamma(n+1) = n!$,
$\Gamma(1) = 0! = 1$,
$\Gamma(-n)$ is not defined;
$\Gamma(x) > 0$ for $x > 0$,
$\Gamma(x)$ alternates signs on the negative axis.

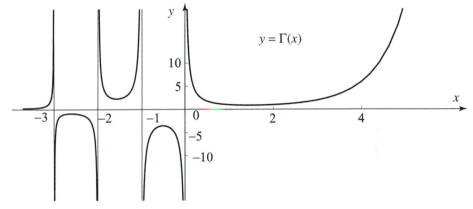

25. Project Problem: The aging spring problem. The equation

$$y''(t) + e^{-at+b} y(t) = 0 \qquad (a > 0), \quad t > 0,$$

models the vibrations of a spring whose spring constant is tending to zero with time.

(a) Show that the change of variables $u = \frac{2}{a} e^{-\frac{1}{2}(at-b)}$ transforms the differential equation into Bessel's equation of order zero (with the new variable u). [Hint: Let $Y(u) = y(t)$, $e^{-at+b} = \frac{a^2}{4} u^2$; $\frac{dy}{dt} = -\frac{a}{2} u \frac{dY}{du}$; $\frac{d^2 y}{dt^2} = \frac{a^2}{4} u \left[\frac{dY}{du} + u \frac{d^2 Y}{du^2} \right]$.]

(b) Obtain the general solution of the differential equation in the form

$$y(t) = c_1 J_0\left(\frac{2}{a} e^{-\frac{1}{2}(at-b)}\right) + c_2 Y_0\left(\frac{2}{a} e^{-\frac{1}{2}(at-b)}\right),$$

where c_1 and c_2 are arbitrary constants, J_0 is the Bessel function of order 0, and Y_0 is the Bessel function of order 0 of the second kind.

(c) Discuss the behavior of the solution as $t \to \infty$ in the following three cases: $c_1 = 0$, $c_2 \neq 0$; $c_1 \neq 0$, $c_2 = 0$; $c_1 \neq 0$, $c_2 \neq 0$. Does it make sense to have unbounded solutions of the differential equation? [Hint: What happens to the differential equation as $t \to \infty$?]

In Exercises 26–27, solve the given aging spring problem. In each case determine whether the solution is bounded or unbounded as $t \to \infty$.

26. $y''(t) + e^{-2t} y(t) = 0$, $y(0) = J_0(1) \approx .765$, $y'(0) = -J_0'(1) \approx .440$.

27. $y''(t) + e^{-2t}y(t) = 0$, $y(0) = .1$, $y'(0) = 0$. [Use $J_0(1) \approx .765$; $J_0'(1) \approx -.440$; $Y_0(1) \approx .088$; $Y_0'(1) \approx .781$.]

28. Bessel's function of the second kind of order zero. The Bessel function of the second kind of order 0 is given explicitly by the formula

$$Y_0(x) = \frac{2}{\pi}\left[J_0(x)\left(\ln x + \gamma\right) + \sum_{m=1}^{\infty} \frac{(-1)^{m-1}h_m}{2^{2m}(m!)^2}x^{2m}\right],$$

where $h_m = 1 + \frac{1}{2} + \frac{1}{3} + \ldots + \frac{1}{m}$ and γ is **Euler's constant**: $\gamma = \lim_{m\to\infty}(h_m - \ln m) \approx 0.577216$.

(a) Approximate the numerical value of Euler's constant.

(b) Justify the property $\lim_{x\to 0+} Y_0(x) = -\infty$.

29.   Modified Bessel function. In some applications the Bessel function J_p appears as a function of the pure imaginary number ix.

(a) Show that $J_p(ix) = i^p \sum_{k=0}^{\infty} \frac{(x/2)^{2k+p}}{k!\,\Gamma(k+p+1)}$.   Thus except for the factor i^p the function that we get is real-valued.   This function defines the so-called **modified Bessel function of order p**,

$$I_p(x) = \sum_{k=0}^{\infty} \frac{(x/2)^{2k+p}}{k!\,\Gamma(k+p+1)}\,.$$

(b)   Verify that the modified Bessel function of order p satisfies the **modified Bessel's differential equation**

$$x^2 y'' + xy' - (x^2 + p^2)y = 0\,.$$

(c)   Plot the modified Bessel function of order 0 and note that it is positive and increasing for $x > 0$.

30.   Modified Bessel functions of the second kind.

(a) Show that $K_p(x) = \frac{\pi}{2\sin p\pi}\left[I_{-p}(x) - I_p(x)\right]$ is also a solution of the modified Bessel's equation of Exercise 29.   This function is called the **modified Bessel function of the second kind** (sometimes called of the third kind).

(b) Show that when p is not an integer, $I_p(x)$ and $K_p(x)$ are linearly independent.

(c) How would you construct $K_p(x)$ when p is an integer?

In Exercises 31–34, use the result of Exercise 25, Section 4.3, to compute the given integral.

31. $\displaystyle\int_0^{\pi/2} \cos\theta \sin\theta \, d\theta$.

32. $\displaystyle\int_0^{\pi/2} \cos^2\theta \sin^3\theta \, d\theta$.

33. $\displaystyle\int_0^{\pi/2} \cos^5\theta \sin^6\theta \, d\theta$.

34. $\displaystyle\int_0^{\pi/2} \cos^8\theta \, d\theta$.

Use the result of Exercise 25, Section 4.3 to establish the following Wallis's formulas.

35. $\displaystyle\int_0^{\pi/2} \sin^{2k}\theta \, d\theta = \frac{\pi}{2}\frac{(2k)!}{2^{2k}(k!)^2}$, $k = 0, 1, 2, \ldots$.

36. $\displaystyle\int_0^{\pi/2} \sin^{2k+1}\theta \, d\theta = \frac{2^{2k}(k!)^2}{(2k+1)!}$, $k = 0, 1, 2, \ldots$.

37.   (a)   Explain with the help of a graph why

$$\int_0^{\pi/2} \cos^{2k} \theta \, d\theta = \int_0^{\pi/2} \sin^{2k} \theta \, d\theta$$

and

$$\int_0^{\pi/2} \cos^{2k+1} \theta \, d\theta = \int_0^{\pi/2} \sin^{2k+1} \theta \, d\theta.$$

(b)   Use (a) and Exercises 35 and 36 to show that for $k = 0, 1, 2, \ldots$,

$$\int_0^{\pi/2} \cos^{2k} \theta \, d\theta = \frac{\pi}{2} \frac{(2k)!}{2^{2k} (k!)^2} \quad \text{and} \quad \int_0^{\pi/2} \cos^{2k+1} \theta \, d\theta = \frac{2^{2k} (k!)^2}{(2k+1)!}.$$

38.   Derive the following formulas using Exercise 37: for $k = 0, 1, 2, \ldots$,

$$\frac{1}{\pi} \int_0^{\pi} \cos^{2k} \theta \, d\theta = \frac{(2k)!}{2^{2k} (k!)^2} \quad \text{and} \quad \int_0^{\pi} \cos^{2k+1} \theta \, d\theta = 0.$$

39.   (a)   Use the result of Exercise 25, Section 4.3, to obtain that the arc length of the lemniscate $r^2 = 2 \cos 2\theta$ is $2\sqrt{2\pi} \Gamma\left(\frac{1}{4}\right) / \Gamma\left(\frac{3}{4}\right)$.   [Hint: Arc length in polar coordinates is $L = \int_a^b \sqrt{r^2 + \left(\frac{dr}{d\theta}\right)^2} \, d\theta$.]

(b)   Approximate the arc length in (a).

40.   The **beta function** is defined for $r, s > 0$ by

$$\beta(r, s) = \int_0^1 t^{r-1} (1-t)^{s-1} \, dt .$$

(a)   Use the change of variables $t = \sin^2 \theta$ to obtain

$$\beta(r, s) = 2 \int_0^{\pi/2} \cos^{2s-1} \theta \sin^{2r-1} \theta \, d\theta .$$

(b)   From Exercise 25, Section 4.3, conclude that

$$\beta(r, s) = \frac{\Gamma(r)\Gamma(s)}{\Gamma(r+s)} .$$

9.7   Bessel Series Expansions

In this section we explore some recurrence relations, orthogonality properties of Bessel functions, and expansions of functions in Bessel series.   Many of these properties are used in solving the boundary value problems occurring in this chapter and throughout this book.

Identities Involving Bessel Functions

We start with two basic identities. For any $p \geq 0$,

(1) $$\frac{d}{dx}\left[x^p J_p(x)\right] = x^p J_{p-1}(x),$$

(2) $$\frac{d}{dx}\left[x^{-p} J_p(x)\right] = -x^{-p} J_{p+1}(x).$$

Note that for $p = 0$, the second identity yields

$$\frac{d}{dx}\left[J_0(x)\right] = -J_1(x).$$

These identities were established in Example 5, Section 4.3, when p is an a integer. The same proof works for arbitrary real number p.

Many other useful identities follow from (1) and (2). We list some of the most commonly used ones:

(3) $$x J_p'(x) + p J_p(x) = x J_{p-1}(x),$$
(4) $$x J_p'(x) - p J_p(x) = -x J_{p+1}(x),$$
(5) $$J_{p-1}(x) - J_{p+1}(x) = 2 J_p'(x),$$
(6) $$J_{p-1}(x) + J_{p+1}(x) = \frac{2p}{x} J_p(x).$$

(We note that the corresponding formulas for Bessel functions of the second kind also hold.)   For a proof of (3), see Example 5, Section 4.3. Identity (4) is proved similarly by starting with (2) and expanding using the product rule. Adding (3) and (4) and simplifying yields (5). Subtracting (4) from (3) and simplifying yields (6).

There are similar identities involving integrals of Bessel functions.   For example, the identities

(7) $$\int x^{p+1} J_p(x)\, dx = x^{p+1} J_{p+1}(x) + C,$$

and

(8) $$\int x^{-p+1} J_p(x)\, dx = -x^{-p+1} J_{p-1}(x) + C$$

follow easily by integrating both sides of (1) and (2) (Exercise 1).

Orthogonality of Bessel Functions

To understand the orthogonality relations of Bessel functions, let us recall the familiar example of the functions $\sin n\pi x$, $n = 1, 2, 3, \ldots$. We know that these functions are orthogonal on the interval $[0, 1]$, in the sense that

$$\int_0^1 \sin(n\pi x) \sin(m\pi x) \, dx = 0 \quad \text{if} \quad n \neq m.$$

The key here is to note that the functions $\sin n\pi x$ are constructed from a single function, namely $\sin x$, and its positive zeros, namely $n\pi$, $n = 1, 2, 3, \ldots$. In constructing systems of orthogonal Bessel functions, we will proceed in a similar way by using a single Bessel function and its zeros. Fix an order $p \geq 0$, and consider the graph of $J_p(x)$ for $x > 0$. Figure 1 shows a graph of a typical Bessel function $J_p(x)$ with $p > 0$.

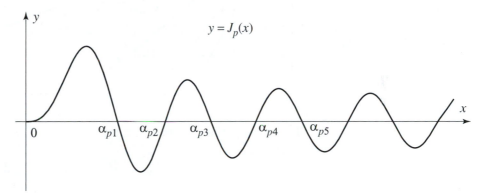

Figure 1 A Bessel function $J_p(x)$ has infinitely many positive zeros.

We see from Figure 1 that the Bessel function J_p has infinitely many zeros on the positive axis $x > 0$ (just like $\sin x$). This important fact is proved in Exercises 14 and 35, when $p = 0$ and $\frac{1}{2}$. We denote these zeros in ascending order:

$$0 < \alpha_{p1} < \alpha_{p2} < \cdots < \alpha_{pj} < \cdots.$$

Hence α_{pj} denotes the jth positive zero of J_p. (Sometimes the notation $\alpha_{p,j}$ will be used.) Unlike the case of the sine function, where the zeros are easily determined by $n\pi$, there is no formula for the positive zeros of the Bessel functions. Since the numerical values of these zeros are very important in applications; they are found in most mathematical tables and computer systems. For later use, we list in Table 1 the first five positive zeros of J_0, J_1, and J_2.

j	1	2	3	4	5
α_{0j}	2.40483	5.52008	8.65373	11.7915	14.9309
α_{1j}	3.83171	7.01559	10.1735	13.3237	16.4706
α_{2j}	5.13562	8.41724	11.6198	14.796	18.9801

Table 1 Positive zeros of J_0, J_1, and J_2.

Let a be a positive number. To generate orthogonal functions on the interval $[0, a]$ from J_p, we proceed as in the case of the sine function, using α_{pj}, the zeros of the Bessel function. We obtain the functions

$$(9) \qquad J_p\left(\frac{\alpha_{pj}}{a}x\right), \qquad j = 1,\, 2,\, 3,\, \dots\, .$$

The first four functions, corresponding to $p = 0$, are shown in Figure 2.

Figure 2 Orthogonal functions generated with $J_0(x)$: $J_0(\alpha_{01}x)$, $J_0(\alpha_{02}x)$, ... It is interesting to note that all these functions are bounded by 1 and their number of zeros increase in the interval $(0, 1)$. These properties are shared by other systems of orthogonal functions encountered earlier, in particular, the trigonometric functions.

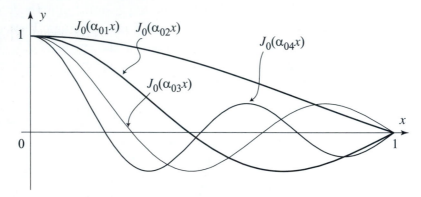

To simplify the notation, we let

$$(10) \qquad \lambda_{pj} = \frac{\alpha_{pj}}{a} \qquad j = 1,\, 2,\, 3,\, \dots\, .$$

So λ_{pj} is the value of the jth positive zero of J_p scaled by a fixed factor $1/a$. We are now in a position to state some fundamental identities.

THEOREM 1
ORTHOGONALITY
OF BESSEL
FUNCTIONS WITH
RESPECT TO A
WEIGHT

Fix $p \geq 0$ and $a > 0$. Let $J_p(\lambda_{pj}\, x)$ $(j = 1,\, 2,\, \dots)$ be as in (9) and (10). Then

$$(11) \qquad \int_0^a J_p(\lambda_{pj}x)J_p(\lambda_{pk}x)x\,dx = 0 \quad \text{for } j \neq k,$$

and

$$(12) \qquad \int_0^a J_p^2(\lambda_{pj}x)x\,dx = \frac{a^2}{2}J_{p+1}^2(\alpha_{pj}) \quad \text{for } j = 1,\, 2,\, \dots\, .$$

Note that (12) involves λ_{pj} and α_{pj}. Property (11) is described by saying that the functions $J_p(\lambda_{pj}x)$, $j = 1,\, 2,\, \dots$, are **orthogonal on the interval** $[0, a]$ **with respect to the weight** x. The phrase "with respect to the weight x" refers to the presence of the function x in the integrand in (11). On the interval $[0, 1]$—that is, when $a = 1$—formulas (11) and (12) take on a simpler form:

$$(13) \qquad \int_0^1 J_p(\alpha_{pj}x)J_p(\alpha_{pk}x)x\,dx = 0 \quad \text{for } j \neq k,$$

$$(14) \qquad \int_0^1 J_p^2(\alpha_{pj}x)x\,dx = \frac{1}{2}J_{p+1}^2(\alpha_{pj}) \quad \text{for } j = 1,\, 2,\, \dots\,.$$

The proof of (11) is found in the proof of Theorem 3. The proof of (12) is outlined in Exercise 36.

Bessel Series and Bessel-Fourier Coefficients

Just as we used the functions $\sin n\pi x$ to expand functions in sine Fourier series, now we will see how we can expand functions using Bessel series. More precisely, a given function f on the interval $[0, a]$ can be expressed as a series

$$(15) \qquad f(x) = \sum_{j=1}^{\infty} A_j J_p(\lambda_{pj}x)$$

called the **Bessel series of order** p of f. Putting aside questions of convergence, let us assume (15) is valid and proceed to find the coefficients in the series. Multiplying both sides of (15) by $J_p(\lambda_{pk}x)\,x$ and integrating term by term on the interval $[0, a]$ gives

$$(16) \qquad \int_0^a f(x)J_p(\lambda_{pk}x)x\,dx = \sum_{j=1}^{\infty} A_j \overbrace{\int_0^a J_p(\lambda_{pj}x)J_p(\lambda_{pk}x)x\,dx}^{=\,0\ \text{except when } j\,=\,k}\,.$$

The orthogonality property (11) shows that all the terms on the right side of (16) are 0 except when $j = k$. Canceling the zero terms and using (12), we get

$$(17) \qquad \boxed{A_j = \frac{\int_0^a f(x)J_p(\lambda_{pj}x)x\,dx}{\int_0^a J_p^2(\lambda_{pj}x)x\,dx} = \frac{2}{a^2\, J_{p+1}^2(\alpha_{pj})} \int_0^a f(x)J_p(\lambda_{pj}x)x\,dx\,.}$$

The number A_j is called the **jth Bessel-Fourier coefficient** or simply the **Bessel coefficient of the function** f.

The next theorem gives conditions under which the Bessel series expansion of a function is valid.

THEOREM 2
BESSEL SERIES OF ORDER p

If f is piecewise smooth on $[0, a]$, then f has a Bessel series expansion of order p on the interval $(0, a)$ given by

$$f(x) = \sum_{j=1}^{\infty} A_j J_p(\lambda_{pj} x),$$

where $\lambda_{p1}, \lambda_{p2}, \ldots$ are the scaled positive zeros of the Bessel function J_p given by (10), and A_j is given by (17). In the interval $(0, a)$, the series converges to $f(x)$ where f is continuous and converges to the average $\frac{f(x+) + f(x-)}{2}$ at the points of discontinuity.

Note the similarity with Fourier series. At the points of discontinuity the Bessel series converges to the average of the function.

Before giving an example of a Bessel series, we make a useful remark about the notation. While the notation α_{pj} and λ_{pj} is appropriate to denote the zeros and scaled zeros of the Bessel function of order p, it is a little cumbersome to work with. For this reason, when it is understood which order we are dealing with, and so there is no risk of confusion, we will drop the index p and write α_j and λ_j instead of α_{pj} and λ_{pj}.

EXAMPLE 1   A Bessel series on the interval $[0, 1]$
Find the Bessel series expansion of order 0 of the function $f(x) = 1$, $0 < x < 1$.
Solution Applying Theorem 2, we get

$$f(x) = \sum_{j=1}^{\infty} A_j J_0(\alpha_j x),$$

where α_j is the jth positive zero of J_0, and

$$
\begin{aligned}
A_j &= \frac{2}{J_1^2(\alpha_j)} \int_0^1 f(x) J_0(\alpha_j x) x \, dx \\
&= \frac{2}{J_1^2(\alpha_j)} \int_0^1 J_0(\alpha_j x) x \, dx \\
&= \frac{2}{\alpha_j^2 J_1^2(\alpha_j)} \int_0^{\alpha_j} J_0(t) t \, dt \qquad (t = \alpha_j x) \\
&= \frac{2}{\alpha_j^2 J_1^2(\alpha_j)} J_1(t) t \big|_0^{\alpha_j} \qquad \text{(by (7) with } p = 0) \\
&= \frac{2}{\alpha_j J_1(\alpha_j)}.
\end{aligned}
$$

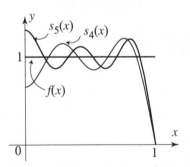

Figure 3 Partial sums of the Bessel series.

Theorem 2 asserts that the Bessel series converges to $f(x)$ at all points in the interval. Thus

$$1 = \sum_{j=1}^{\infty} \frac{2}{\alpha_j J_1(\alpha_j)} J_0(\alpha_j x) \quad 0 < x < 1.$$

j	1	2	3	4	5
α_j	2.4048	5.5201	8.6537	11.7915	14.9309
$J_1(\alpha_j)$	.5191	$-.3403$	.2714	$-.2325$	.2065
$\dfrac{2}{\alpha_j J_1(\alpha_j)}$	1.6020	-1.0648	.8514	$-.7295$	.6487

Table 2 Numerical data for Example 1.

Using the numerical data provided by Tables 1 and 2, we can write explicitly the first few terms of the series:

$$1 = 1.6020\, J_0(2.4048\, x) - 1.0648\, J_0(5.5201\, x) + .8514\, J_0(8.6537\, x)$$
$$- .7295\, J_0(11.7915\, x) + .6487\, J_0(14.9309\, x) + \cdots .$$

It is worth noticing that the Bessel coefficients tend to 0 as $j \to \infty$. This is a property that holds in general.

Note that Theorem 2 tells us nothing about the behavior of the Bessel series at the endpoints of the interval. In this example, if we take $x = 1$ in the series, all the terms become zero, since we are evaluating J_0 at its zeros. This is also clear from the graphs of the partial sums in Figure 3. So, in this example, the Bessel series does not converge to the function at one of the endpoints. ∎

Parametric Form of Bessel's Equation

In the remainder of this section, we explore two important differential equations that are closely related to Bessel's equation.

**THEOREM 3
PARAMETRIC
FORM OF BESSEL'S
EQUATION**

Let $p \geq 0$, $a > 0$, and let α_{pj} denote the jth positive zero of $J_p(x)$. For $j = 1, 2, \ldots$, the functions $J_p(\frac{\alpha_{pj}}{a}\, x)$ are solutions of the **parametric form of Bessel's equation of order** p,

$$(18) \qquad x^2 y''(x) + x y'(x) + (\lambda^2 x^2 - p^2) y(x) = 0,$$

together with the boundary conditions

$$(19) \qquad y(0) \text{ finite}, \ y(a) = 0,$$

when $\lambda = \lambda_{pj} = \frac{\alpha_{pj}}{a}$, and these are the only solutions of (18), aside from scalar multiples, with these properties.   Moreover, these solutions satisfy (11) and (12) and so they are orthogonal on the interval $[0, a]$ with respect to the weight function x.

The condition that $y(0)$ be finite is effectively a second boundary condition on y.

Proof We will make a change of variables in (18) and reduce it to Bessel's equation as follows.   Let $u = \lambda x$, $du = \lambda dx$, and let $y(x) = y(\frac{u}{\lambda}) = Y(u)$.   From the chain rule, $y'(x) = Y'(u)\frac{du}{dx} = \lambda Y'(u)$, and, likewise, $y''(x) = \lambda^2 Y''(u)$.   Substituting in (18), we get Bessel's equation of order p in $Y(u)$:

$$u^2 Y''(u) + u Y'(u) + (u^2 - p^2) Y(u) = 0\,.$$

Thus the general solution is

$$Y(u) = c_1 J_p(u) + c_2 Y_p(u) = c_1 J_p(\lambda x) + c_2 Y_p(\lambda x) = y(x)\,.$$

For $y(0)$ to be finite, we must set $c_2 = 0$, because Y_p blows up at 0. So $y(x) = c_1 J_p(\lambda x)$, and the boundary condition $y(a) = 0$ holds (for $c_1 \neq 0$) if and only if $\lambda a = \alpha_{pj}$. Hence,

$$\lambda = \lambda_{pj} = \frac{\alpha_{pj}}{a}$$

are the only positive values of λ for which there are nontrivial solutions. Therefore, the solutions of (18) and (19) are as claimed.

We come now to the proof of (11). To simplify notation, let us write λ_j for λ_{pj}, and $\phi_j(x)$ for $J_{pj}(\lambda_{pj}x)$. The goal is to show that for $j \neq k$

$$\int_0^a \phi_j(x)\phi_k(x)x\,dx = 0.$$

Let us write (18) in the form $(x\,y')' = -\frac{(\lambda^2 x^2 - p^2)}{x}\,y$. Since the ϕ's satisfy this equation with the corresponding λ's, we have

$$(x\,\phi_j')' = -\frac{(\lambda_j^2\,x^2 - p^2)}{x}\phi_j$$

and

$$(x\,\phi_k')' = -\frac{(\lambda_k^2\,x^2 - p^2)}{x}\phi_k.$$

Multiplying the first equation by ϕ_k and the second one by ϕ_j and then subtracting the resulting equations, we get after simplifying

$$(\lambda_k^2 - \lambda_j^2)\phi_j\phi_k\,x = \phi_k(x\phi_j')' - \phi_j(x\,\phi_k')'.$$

Note that

$$\phi_k(x\phi_j')' - \phi_j(x\phi_k')' = \frac{d}{dx}[\phi_k x\,\phi_j' - \phi_j x\phi_k'].$$

Hence

$$(\lambda_k^2 - \lambda_j^2)\int_0^a \phi_j(x)\phi_k(x)x\,dx = \phi_k x\phi_j' - \phi_j x\phi_k'\Big|_0^a = 0,$$

because $\phi_j(a) = \phi_k(a) = 0$, and the desired result follows, since $\lambda_k^2 - \lambda_j^2 \neq 0$. For the proof of (12), see Exercise 36. ∎

Our last example is a differential equation that gives rise to yet another important family of functions.

EXAMPLE 2   Spherical Bessel functions
The equation

$$(20) \qquad x^2 y'' + 2xy' + (kx^2 - n(n+1))y = 0, \qquad 0 < x < a, \quad y(a) = 0,$$

arises in many important applications in Chapter 10. In this equation, k is a nonnegative real number and $n = 0, 1, 2, \ldots$. We are seeking bounded solutions in the interval $0 \leq x \leq a$. You can check that the substitution

$$(21) \qquad\qquad\qquad y = x^{-1/2}w$$

transforms (20) into the following parametric form of Bessel's equation:

$$x^2 w'' + x w' + (kx^2 - (n + \tfrac{1}{2})^2)w = 0, \quad w(a) = 0$$

(Exercise 37). We know from Theorem 3 (with $p = n + 1/2$) that bounded solutions arise only when

$$k = \lambda^2,$$

and

$$\lambda = \lambda_{n,j} = \frac{\alpha_{n+1/2,j}}{a},$$

where $\alpha_{n+1/2,j}$ denotes the jth positive zero of $J_{n+\frac{1}{2}}$. The corresponding solutions are scalar multiples of

$$w_{n,j}(x) = J_{n+\frac{1}{2}}(\lambda_{n,j}x), \qquad n = 0, 1, 2, \dots, \quad j = 1, 2, \dots.$$

Hence, using (21), it follows that the solutions y_{nj} of (20) are scalar multiples of

$$(22) \qquad x^{-1/2} J_{n+\frac{1}{2}}(\lambda_{n,j}x), \qquad n = 0, 1, 2, \dots, \quad j = 1, 2, \dots.$$

It is customary to express the solutions in terms of the **spherical Bessel functions of the first kind** (plotted in Figure 4)

$$(23) \qquad j_n(x) = (\frac{\pi}{2x})^{1/2} J_{n+\frac{1}{2}}(x), \qquad n = 0, 1, 2, \dots.$$

Choosing a specific multiple of the functions in (22), we obtain the solutions of (20) as the spherical Bessel functions:

$$(24) \qquad y_{n,j}(x) = j_n(\lambda_{n,j}x), \qquad n = 0, 1, 2, \dots, \quad j = 1, 2, \dots. \qquad \blacksquare$$

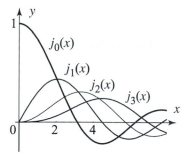

Figure 4 Spherical Bessel functions, j_0, j_1, j_2, j_3.

In the exercises, you are asked to study the spherical Bessel functions, including their orthogonality. In most cases, the results are simple consequences of properties of Bessel functions and (23).

Exercises 9.7

1. (a) Supply the details of the proof of (2), (7), and (8). [Hint: To prove (2), show that the derivative is $\sum_{k=1}^{\infty} \frac{(-1)^k 2k}{k!\Gamma(k+p+1)2^{2k+p}} x^{2k-1}$. Then change k to $k+1$ to start the sum at 0.]
(b) Supply the details of the proof of (4), (5), and (6).

2. (a) On the graphs of J_0 and J_1 in Figure 1, Section 9.6, note that the maxima and minima of J_0 occur at the zeros of J_1. Prove this fact.
(b) Show that the maximum and minimum values of J_p occur when

$$x = \frac{pJ_p(x)}{J_{p+1}(x)} \quad \text{or} \quad x = \frac{pJ_p(x)}{J_{p-1}(x)} \quad \text{or} \quad J_{p-1}(x) = J_{p+1}(x).$$

(c) Illustrate (b) graphically when $p = 2$.
(d) Show that at the zeros of J_p, J_{p-1} and J_{p+1} have equal absolute values but opposite signs.

In Exercises 3–10, evaluate the given integral.

3. $\displaystyle\int x J_0(x)\,dx.$ **4.** $\displaystyle\int x^4 J_3(x)\,dx.$ **5.** $\displaystyle\int J_1(x)\,dx.$

6. $\displaystyle\int x^{-2} J_3(x)\,dx.$ **7.** $\displaystyle\int x^3 J_2(x)\,dx.$ **8.** $\displaystyle\int x^3 J_0(x)\,dx.$

9. $\displaystyle\int J_3(x)\,dx.$ **10.** $\displaystyle\int_0^\infty J_1(x)[J_0(x)]^n\,dx.$ [Hint: Let $u = J_0(x)$.]

11. Use (2) and Example 1 of Section 9.6 to derive the formula

$$J_{3/2}(x) = \sqrt{\frac{2}{\pi x}}\left[\frac{\sin x}{x} - \cos x\right] \quad\text{and}\quad J_{-\frac{3}{2}}(x) = -\left(\frac{2}{\pi x}\right)^{\frac{1}{2}}\left(\frac{\cos x}{x} + \sin x\right).$$

12. Bessel functions of half-integer order.   Derive the identity

$$J_{n+\frac{1}{2}}(x) = (-1)^n \left(\frac{2}{\pi x}\right)^{\frac{1}{2}} x^{n+\frac{1}{2}}\left(\frac{d}{xdx}\right)^n \frac{\sin x}{x}, \qquad n = 0,\,1,\,2,\,\dots.$$

13. Express J_5 in terms of J_0 and J_1.

14. Use the explicit formula for $J_{1/2}$ given in Example 1 of Section 9.6 to show that $J_{1/2}$ has infinitely many positive zeros in the interval $0 < x < \infty$. What are these zeros?

15. (a) Plot the functions $\frac{\sin x}{x}$ and $\cos x$ on the same graph to illustrate the fact that they intersect at infinitely many points. Conclude that the function $J_{3/2}(x)$ has infinitely many zeros in $(0, \infty)$.
(b) Approximate the first three roots of the equation $\frac{\sin x}{x} = \cos x$.

16. (a) Let $0 < \alpha_{p1} < \alpha_{p2} < \cdots < \alpha_{pj} < \cdots$ denote the positive zeros of J_p. How many roots does the function $J_p(\frac{\alpha_{pj}}{a}x)$ have in the interval $0 < x < a$? Justify your answer.
(b) Illustrate your answer graphically with $p = 2$, $a = 1$, and $j = 5$.

17. Let $0 < c < 1$, and let

$$f(x) = \begin{cases} 1 & \text{if } 0 < x < c, \\ 0 & \text{if } c < x < 1. \end{cases}$$

(a) Derive the Bessel series of order 0 of f

$$\sum_{j=1}^{\infty} \frac{2c J_1(c\,\alpha_j)}{\alpha_j\, J_1(\alpha_j)^2} J_0(\alpha_j x), \quad 0 < x < 1.$$

(b) Discuss the convergence of the series in the interval $0 < x < 1$.
(c) Take $c = \frac{1}{2}$. Investigate the convergence of the series at $x = 0$ and $x = 1$ numerically. Does the series seem to converge at one or both points? To what values? (Note that convergence at these points is not covered by Theorem 2.)

18. Take $a = \frac{1}{3}$ in Exercise 17 and write explicitly the first five terms of the series in this case. Plot several partial sums and describe the behavior at and around the point $x = \frac{1}{3}$. Observe how the partial sums overshoot their limiting values regardless of the increase in the number of terms in the approximating partial sums.   This illustrates the **Gibbs phenomenon** at a point of discontinuity for Bessel series expansions.

19. (a) Obtain the Bessel series expansion

$$x^4 = 2 \sum_{j=1}^{\infty} \frac{1}{\alpha_j J_5(\alpha_j)} J_4(\alpha_j x), \quad 0 < x < 1,$$

where the α_j's are the positive zeros of J_4.

(b) Plot some partial sums and discuss their behavior on the interval $0 < x < 1$.

20. Bessel series of order m Fix a number $m \geq 0$ and let $0 < \alpha_1 < \alpha_2 < \ldots < \alpha_j < \ldots$ denote the sequence of positive zeros of J_m. Obtain the Bessel series expansion

$$x^m = 2 \sum_{j=1}^{\infty} \frac{1}{\alpha_j J_{m+1}(\alpha_j)} J_m(\alpha_j x), \quad 0 < x < 1.$$

21. Refer to Exercise 20.

(a) Take $m = \frac{1}{2}$, and determine the exact values of the α_j's in this case? [Hint: Example 1 of Section 9.6.]

(b) Compute explicitly the coefficients in the Bessel series expansion of $f(x) = \sqrt{x}$, $0 < x < 1$.

(c) Recall the formula for $J_{\frac{1}{2}}$ in terms of $\sin x$ and show that the expansion in (b) is really a sine Fourier series expansion of x.

(d) Plot some partial sums of the Bessel series and discuss the convergence on the interval $0 < x < 1$.

22. Study the behavior of the partial sums of the Bessel series in Exercise 20 when $m = \frac{3}{2}$. What do they converge to in the interval $0 < x < 1$?

In Exercises 23–30, approximate the given function by a Bessel series of the given order p. Plot several partial sums of the series and discuss their convergence on the given interval.

23. $f(x) = x^2; \ 0 < x < 1; \ p = 2.$ **24.** $f(x) = x^3; \ 0 < x < 1; \ p = 3.$

25. $f(x) = \begin{cases} 0 & \text{if } 0 < x < \frac{1}{2}, \\ \frac{1}{x} & \text{if } \frac{1}{2} < x < 1, \end{cases}$ **26.** $f(x) = x^2, \ 0 < x < 1, \ p = 0.$

$p = 1$

27. $f(x) = \begin{cases} 0 & \text{if } 0 < x < \frac{1}{2}, \\ \frac{1}{x^2} & \text{if } \frac{1}{2} < x < 1, \end{cases}$ **28.** $f(x) = \begin{cases} 0 & \text{if } 0 < x < \frac{1}{30}, \\ \frac{1}{x^3} & \text{if } \frac{1}{30} < x < 1, \end{cases}$

$p = 2$ $p = 3$

29. $f(x) = 1; \ 0 < x < 2; \ p = 1.$ **30.** $f(x) = J_0(x); \ 0 < x < \alpha_{01}; \ p = 2.$

In Exercises 31–34, find all $\lambda > 0$ for which the given parametric form of Bessel's equation has a solution that is finite at $x = 0$ and satisfies the given boundary condition.

31. $x^2 y''(x) + x y'(x) + (\lambda^2 x^2 - 1) y(x) = 0, \quad y(1) = 0.$

32. $x^2 y''(x) + x y'(x) + \lambda^2 x^2 y(x) = 0, \quad y(2) = 0.$

33. $x^2 y''(x) + x y'(x) + (\lambda^2 x^2 - \frac{1}{4}) y(x) = 0, \quad y(\pi) = 0.$

34. $x^2 y''(x) + x y'(x) + (\lambda^2 x^2 - 1) y(x) = 0, \quad y(\frac{\pi}{2}) = 0.$

35.   Project Problem: Zeros of J_0.   It is a fact that the positive zeros of $J_p(x)$, $p \geq 0$, form an increasing sequence

$$0 < \alpha_1 < \alpha_2 < \cdots < \alpha_k < \cdots .$$

In this exercise, we will prove this fact when $p = 0$.

(a) Plot the graph of $J_0(x)$ and note that there are infinitely many positive zeros that we will denote in ascending order by $0 < \alpha_1 < \alpha_2 < \alpha_3 < \cdots$.

(b) Show that the substitution $y(x) = \frac{1}{\sqrt{x}}u(x)$ transforms Bessel's equation of order 0 into

$$u'' + (1 + \frac{1}{4x^2})u = 0 .$$

Conclude that $u(x) = \sqrt{x}J_0(x)$ is a solution of this differential equation.

(c) Let $v(x) = \sin x$.   Check that

$$-(u'' + u)v(x) = \frac{d}{dx}(uv' - u'v) .$$

(d) Using (b), show that $-(u'' + u) = \frac{u}{4x^2}$.

(e) Show that $\int \frac{u(x)v(x)}{4x^2}\, dx = uv' - u'v + C$ and hence

$$\int_{2k\pi}^{(2k+1)\pi} \frac{u(x)\sin x}{4x^2}\, dx = -[u(2k\pi) + u(2k\pi + \pi)] .$$

(f) Conclude from (e) that $u(x)$ has at least one zero in $[2k\pi, (2k+1)\pi]$.   [Hint: Assume the contrary, say $u(x) > 0$ for all x in the interval, and get a contradiction by studying the signs of the terms on both sides of the last equality in (e).]

(g) Conclude that $J_0(x)$ has infinitely many positive zeros.

36. Project Problem: A proof of (12).   Let $0 < \alpha_1 < \alpha_2 < \cdots < \alpha_k < \cdots$ denote the sequence of zeros of J_p in $(0, \infty)$, and let $\lambda_k = \frac{\alpha_k}{a}$.

(a) Rewrite (18) as

$$(xy')' + (\lambda^2 x - \frac{p^2}{x})y = 0, \quad y(a) = 0 .$$

Multiply both sides of the equation by xy' and put it in the form

$$[(xy')^2]' + (\lambda^2 x^2 - p^2)(y^2)' = 0 .$$

(b) Take $y = J_p(\lambda_k x)$ and $\lambda = \lambda_k$, integrate the equation in (a) over $0 < x < a$, and obtain

$$[ay'(a)]^2 + (\lambda^2 a^2 - p^2)y(a)^2 - 2\lambda^2 \int_0^a y(x)^2 x\, dx = 0 .$$

[Hint: Use integration by parts on the second term in (a).   Justify the equality $p\,y(0) = 0$.]

(c) For $y = J_p(\lambda_k x)$ and $\lambda = \lambda_k$, justify the equality $y(a) = 0$, and get

$$[y'(a)]^2 = \frac{2\lambda^2}{a^2} \int_0^a xy(x)^2\, dx .$$

(d) For $y = J_p(\lambda_k x)$, use (4) to show that $y'(a) = -\lambda_k J_{p+1}(\alpha_k)$. Derive (12) from (c).

Spherical Bessel Functions

37. Use (21) to transform (20) into one of the equations of the form (18).

38. Use (23) and explicit formulas for the Bessel functions to show that

$$j_0(x) = \frac{\sin x}{x},$$

$$j_1(x) = \frac{\sin x - x \cos x}{x^2},$$

$$j_2(x) = \frac{(3 - x^2)\sin x - 3x \cos x}{x^3}.$$

[Hint: See Exercises 11 and 12.]

In Exercises 39–42, derive the given recurrence relation for the spherical Bessel functions. [Hint: Use (23) and the corresponding formulas for the Bessel functions.]

39. $\dfrac{d}{dx}[x^{n+1}j_n(x)] = x^{n+1}j_{n-1}(x).$

40. $\dfrac{d}{dx}[x^{-n}j_n(x)] = -x^{-n}j_{n+1}(x).$

41. $\displaystyle\int x^{n+2}j_n(x)\,dx = x^{n+2}j_{n+1}(x) + C.$

42. $\displaystyle\int x^{-n+1}j_n(x)\,dx = -x^{-n+1}j_{n-1}(x) + C.$

43. Recurrence relations. Use the results of Exercises 39 and 40 to obtain that

$$j_{n-1}(x) + j_{n+1}(x) = \frac{2n + 1}{x}\,j_n(x)$$

and

$$n j_{n-1}(x) - (n + 1)j_{n+1}(x) = (2n + 1)j_n'(x).$$

44. Use the results of Exercises 43 and 38 to obtain that

$$j_3(x) = \frac{(15 - 6x^2)\sin x - (15x - x^3)\cos x}{x^4}.$$

45.   Orthogonality of the spherical Bessel functions. Use (11) and (12) to show that

$$\int_0^a j_n(\lambda_{n,j}x)j_n(\lambda_{n,j'}x)x^2\,dx = 0 \quad \text{for } j \neq j'$$

and

$$\int_0^a j_n^2(\lambda_{n,j}x)x^2\,dx = \frac{a^3}{2}j_{n+1}^2(\alpha_{n+\frac{1}{2},j}),$$

where $\alpha_{n+\frac{1}{2},j}$ is the jth positive zero of j_n (equivalently, of $J_{n+\frac{1}{2}}$).   Hence the spherical Bessel functions are orthogonal on the interval $[0, a]$ with respect to the weight function x^2.

10

PARTIAL DIFFERENTIAL EQUATIONS IN SPHERICAL COORDINATES

Don't just read it; fight it! Ask your own questions, look for your own examples, discover your own proofs...

-Paul Halmos

In this chapter we turn our attention to problems in spheres and other regions, such as the region between two spheres or the region outside a sphere, for which it is natural to use spherical coordinates. For example, to find the temperature in a metallic ball whose surface is kept at a given temperature distribution, we need to solve Laplace's equation inside the ball that takes the given boundary values on its surface. These and related problems can be treated by the same techniques as in the previous two chapters; in particular, the method of separation of variables and the method of eigenfunction expansions can be applied. You may recall from the previous chapters how Fourier series and Bessel series arose from applying the separation of variables method to equations involving the Laplacian in polar coordinates. Here, with equations involving the Laplacian in spherical coordinates, we will encounter other special functions when carrying the method of separation of variables to completion. A comprehensive treatment of this requisite material is presented in the last three sections of the chapter. It covers in detail Legendre polynomials and associated Legendre functions, and their corresponding expansion theories. You do not need to cover this material before starting the chapter; we will refer to it as needed.

Even though the setting in this chapter is three dimensional, the theory of analytic functions in the plane will be very useful in studying the special functions that arise from the solution of boundary value problems. More specifically, we will use Cauchy's formula to prove important properties of Legendre polynomials and sum certain Legendre series. This enriching material is optional and is presented in Sections 10.5 and 10.6.

10.1   Preview of Problems and Methods

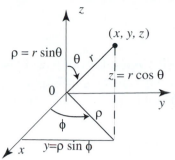

Figure 1 Spherical coordinates.

Here the polar angle in the xy-plane is denoted by ϕ, while in the two-dimensional case (Chapter 9), we used θ.

In this chapter we will solve boundary value problems that involve the Laplacian in spherical coordinates. In this first section we will survey the methods and additional tools that are required for the solutions. As you would expect by now, the method will involve solving certain ordinary differential equations, forming generalized Fourier series, and expressing the boundary or initial data in terms of these series. For example, the solutions of problems in Cartesian coordinates (Chapter 8) involved, among other things, Fourier sine series. In Chapter 9, where we considered problems in polar coordinates, we were led to Bessel series expansions. In this section you will encounter new types of expansions (Legendre series, spherical harmonics expansions, and others) that arise naturally when solving problems in spherical coordinates.

Laplace's equation in spherical coordinates is

$$(1) \qquad \Delta u = \frac{\partial^2 u}{\partial r^2} + \frac{2}{r}\frac{\partial u}{\partial r} + \frac{1}{r^2}\left(\frac{\partial^2 u}{\partial \theta^2} + \cot\theta\frac{\partial u}{\partial \theta} + \csc^2\theta\frac{\partial^2 u}{\partial \phi^2}\right) = 0,$$

where $0 < r < a$, $0 < \phi < 2\pi$, and $0 < \theta < \pi$. (The derivation of the Laplacian in spherical coordinates is straightforward but technical. We present it at the end of this section.) The solution of this equation is quite involved. To clarify the presentation, we will treat simultaneously the simpler case when u is symmetric with respect to the z-axis. In this case, u is independent of the azimuthal angle ϕ, the derivatives with respect to ϕ are all 0, and (1) reduces to

**RADIALLY
SYMMETRIC
LAPLACE'S
EQUATION**

$$(2) \qquad \Delta u = \frac{\partial^2 u}{\partial r^2} + \frac{2}{r}\frac{\partial u}{\partial r} + \frac{1}{r^2}\left(\frac{\partial^2 u}{\partial \theta^2} + \cot\theta\frac{\partial u}{\partial \theta}\right) = 0,$$

where $0 < r < a$ and $0 < \theta < \pi$.

Separating Variables in Laplace's Equation

Let

$$u(r, \theta, \phi) = R(r)\Theta(\theta)\Phi(\phi)\,.$$

Differentiate, plug into (1), divide by $R\Theta\Phi$, and separate variables to arrive at the **Euler equation**:

$$(3) \qquad r^2 R'' + 2rR' - \mu R = 0, \quad 0 < r < a,$$

and

(4)
$$\frac{\Theta''}{\Theta} + \cot\theta\,\frac{\Theta'}{\Theta} + \csc^2\theta\,\frac{\Phi''}{\Phi} = -\mu,$$

where μ is a separation constant. The details of the separation of variables are left to Exercise 1. Recall that when we separated variables in Laplace's equation in polar coordinates (Section 9.1), we also obtained an Euler equation in R. Separating variables in (4), we arrive at the equations

(5)
$$\Phi'' + m^2\Phi = 0, \quad m = 0, 1, 2, \ldots,$$

and

(6)
$$\Theta'' + \cot\theta\,\Theta' + (\mu - m^2\csc^2\theta)\Theta = 0.$$

Expecting 2π-periodic solutions in Φ, since ϕ is a polar angle, we have already determined that the separation constant should be m^2 in (5). We have now separated the variables in (1) and arrived at the three equations (3), (5), and (6). Of these three equations only (6) is new. As we will see shortly, it is related to a family of differential equations known as the associated Legendre differential equations.

In the symmetric case, with no dependence on ϕ, starting with (2), we arrive in a similar way at (3) and the following equation in Θ:

(7)
$$\Theta'' + \cot\theta\,\Theta' + \mu\,\Theta = 0.$$

Note that (7) is a special case of (6) with $m = 0$. We will see shortly that it is related to the so-called Legendre's differential equation.

Product Solutions of Laplace's Equation

We now describe the solutions of (3), (5), and (6) and derive the product solutions of (1). Equation (5) is readily solved and yields

We are using the complex form of the solution to keep the notation compact. You could use $\cos m\phi$ and $\sin m\phi$ instead. Recall that $e^{im\phi} = \cos m\phi + i\sin m\phi$.

(8)
$$\Phi(\phi) = e^{im\phi}, \quad m = 0, \pm 1, \pm 2, \ldots.$$

To solve (6), we make the change of variables

$$(9) \qquad s = \cos\theta; \quad \frac{ds}{d\theta} = -\sin\theta.$$

Hence, by the chain rule,

$$(10) \qquad \Theta' = \frac{d\Theta}{d\theta} = \frac{d\Theta}{ds}\frac{ds}{d\theta} = -\frac{d\Theta}{ds}\sin\theta;$$

$$(11) \qquad \begin{aligned} \frac{d^2\Theta}{d\theta^2} &= -\frac{d}{d\theta}\left(\frac{d\Theta}{ds}\sin\theta\right) = -\frac{d^2\Theta}{ds^2}\frac{ds}{d\theta}\sin\theta - \cos\theta\frac{d\Theta}{ds} \\ &= \sin^2\theta\frac{d^2\Theta}{ds^2} - \cos\theta\frac{d\Theta}{ds} = (1-s^2)\frac{d^2\Theta}{ds^2} - s\frac{d\Theta}{ds}. \end{aligned}$$

Plugging into (6) and simplifying, we arrive at

$$(12) \qquad \boxed{(1-s^2)\frac{d^2\Theta}{ds^2} - 2s\frac{d\Theta}{ds} + (\mu - \frac{m^2}{1-s^2})\Theta = 0, \quad -1 < s < 1.}$$

This second order, linear, ordinary differential equation is known as the **associated Legendre differential equation** (Section 10.7). The difficulty in solving this equation is due to the fact that the coefficients are nontrivial functions of s. In the symmetric case, back to equation (7), if we make the substitution $s = \cos\theta$ and simplify, we arrive at the equation

$$(13) \qquad \boxed{(1-s^2)\frac{d^2\Theta}{ds^2} - 2s\frac{d\Theta}{ds} + \mu\Theta = 0, \quad -1 < s < 1.}$$

This is **Legendre's differential equation** (Section 10.5). It is a special case of (12) with $m = 0$. Legendre's differential equation is treated in detail in Sections 10.5 and 10.6. It is a fact that (12) and (13) have bounded solutions in the interval $[-1, 1]$ if and only if the separation constant has the special form

$$(14) \qquad \boxed{\mu = n(n+1), \quad n = 0, 1, 2, \ldots.}$$

Since for practical reasons we are only interested in bounded solutions, henceforth we take μ as in (14). The corresponding bounded solutions of (12), when properly normalized, are denoted by $P_n^m(s)$ and are called the

associated Legendre functions. Substituting back $s = \cos\theta$, we see that the bounded solutions of (6) are

$$P_n^m(\cos\theta).$$

In the symmetric case, when $m = 0$, P_n^0 reduces to the **Legendre polynomial of degree n**, which is denoted by P_n. This yields

$$P_n(\cos\theta)$$

as solutions of (7). Now that we know the solutions in Θ and Φ, let us solve for R. Substituting $\mu = n(n+1)$ in (3), we obtain the Euler equation

$$(15) \qquad r^2 R'' + 2rR' - n(n+1)R = 0, \quad 0 < r < a.$$

To solve this equation, we appeal to results from Appendix A.3. The indicial equation is

$$\nu^2 + \nu - n(n+1) = 0,$$

with indicial roots $\nu = n$ and $\nu = -(n+1)$. Since the indicial roots are distinct, we are in Case I of Euler's equation, and hence the solutions are

$$(16) \qquad \boxed{R_n(r) = r^n \quad \text{and} \quad R_n^*(r) = r^{-(n+1)}, \quad n = 0,\,1,\,2\,\ldots.}$$

For problems inside the ball with $0 < r < a$, we choose the bounded solutions in (16), $R_n(r) = r^n$, and discard the others. For problems outside the ball, with $r > a$, we take $R_n^*(r) = r^{-(n+1)}$ in (16) and discard $R_n(r) = r^n$.

Summing up, we have found the following product solutions of (1):

The functions $e^{im\phi}P_n^m(\cos\theta)$ are very important in applications. When properly normalized, they are denoted by $Y_{n,m}(\theta, \phi)$ and called the **spherical harmonics** (see (4), Section 10.3).

$$(17) \qquad \boxed{u(r, \theta, \phi) = r^n e^{im\phi} P_n^m(\cos\theta),}$$

where P_n^m are the associated Legendre functions. We have also found the following product solutions of (2):

$$(18) \qquad \boxed{u(r, \theta) = r^n P_n(\cos\theta),}$$

where P_n is the nth Legendre polynomial, $n = 0,\,1,\,\ldots.$

Solutions of boundary value problems involving Laplace's equation in a ball will be expressed as infinite series in terms of the product solutions (17) (superposition principle).   In determining the coefficients in these series, we will use the boundary conditions and appeal to various properties of the associated Legendre functions (and Legendre polynomials).   More specifically, we will require expansion theorems for associated Legendre functions (and Legendre polynomials) that are similar to Bessel series representations and Fourier series.   This requisite material is developed in detail in Sections 10.5–10.7. We will refer to it as needed. We end the section with a derivation of the Laplacian in spherical coordinates.

The Laplacian in spherical coordinates

We will use (r, θ, ϕ) to denote the spherical coordinates of the point (x, y, z) (see Figure 1). We have

$$x = r \cos \phi \sin \theta, \quad y = r \sin \phi \sin \theta, \quad z = r \cos \theta,$$

$$r^2 = x^2 + y^2 + z^2 .$$

From the geometry in Figure 1, we have

$$\rho = r \sin \theta, \quad x = \rho \cos \phi, \quad y = \rho \sin \phi, \quad \rho^2 = x^2 + y^2 .$$

Our goal is to express Δu in terms of r, θ, and ϕ.   From the polar form of the Laplacian, we have

$$(19) \qquad \frac{\partial^2 u}{\partial x^2} + \frac{\partial^2 u}{\partial y^2} = \frac{\partial^2 u}{\partial \rho^2} + \frac{1}{\rho} \frac{\partial u}{\partial \rho} + \frac{1}{\rho^2} \frac{\partial^2 u}{\partial \phi^2}.$$

Observe that the relations

$$z = r \cos \theta, \qquad \rho = r \sin \theta$$

are analogous to those between polar and rectangular coordinates.   So, by using again the polar form of the Laplacian with z and ρ (in place of x and y), we get

$$(20) \qquad \frac{\partial^2 u}{\partial z^2} + \frac{\partial^2 u}{\partial \rho^2} = \frac{\partial^2 u}{\partial r^2} + \frac{1}{r} \frac{\partial u}{\partial r} + \frac{1}{r^2} \frac{\partial^2 u}{\partial \theta^2}.$$

Adding $\frac{\partial^2 u}{\partial z^2}$ to (19) and using (20) gives

$$(21) \qquad \frac{\partial^2 u}{\partial x^2} + \frac{\partial^2 u}{\partial y^2} + \frac{\partial^2 u}{\partial z^2} = \frac{\partial^2 u}{\partial r^2} + \frac{1}{r} \frac{\partial u}{\partial r} + \frac{1}{r^2} \frac{\partial^2 u}{\partial \theta^2} + \frac{1}{\rho} \frac{\partial u}{\partial \rho} + \frac{1}{\rho^2} \frac{\partial^2 u}{\partial \phi^2}.$$

It remains to express $\partial u / \partial \rho$ in spherical coordinates.   From the relation $\theta = \tan^{-1}(\rho/z)$, we get

$$\frac{\partial \theta}{\partial \rho} = \frac{1}{1 + (\rho/z)^2} \frac{1}{z} = \frac{z}{z^2 + \rho^2} = \frac{z}{r^2} = \frac{\cos \theta}{r}.$$

Differentiating $\rho = r \sin\theta$ with respect to ρ, we get

$$1 = \frac{\partial r}{\partial \rho} \sin\theta + r\cos\theta \frac{\partial\theta}{\partial\rho} = \frac{\partial r}{\partial\rho}\sin\theta + \cos^2\theta.$$

Hence

$$\frac{\partial r}{\partial\rho} = \frac{1 - \cos^2\theta}{\sin\theta} = \sin\theta.$$

Now note that ϕ and ρ are polar coordinates in the xy-plane, hence $\partial\phi/\partial\rho = 0$. Using the chain rule, we get

$$\frac{\partial u}{\partial\rho} = \frac{\partial u}{\partial r}\frac{\partial r}{\partial\rho} + \frac{\partial u}{\partial\theta}\frac{\partial\theta}{\partial\rho} + \frac{\partial u}{\partial\phi}\frac{\partial\phi}{\partial\rho} = \frac{\partial u}{\partial r}\frac{\rho}{r} + \frac{\partial u}{\partial\theta}\frac{\cos\theta}{r}.$$

Substituting this in (21) and simplifying, we get the **spherical form of the Laplacian**:

$$(22)\qquad \boxed{\Delta u = \frac{\partial^2 u}{\partial r^2} + \frac{2}{r}\frac{\partial u}{\partial r} + \frac{1}{r^2}\left(\frac{\partial^2 u}{\partial\theta^2} + \cot\theta\frac{\partial u}{\partial\theta} + \csc^2\theta\frac{\partial^2 u}{\partial\phi^2}\right).}$$

Exercises 10.1

1. Carry out the details of the separation of variables method to derive (3), (5), and (6) from (1).

2. Carry out the details of the separation of variables method to derive (3) and (7) from (2).

3. Refer to Section 10.5, where you will find a list of Legendre polynomials. Use this list to compute explicitly $P_n(\cos\theta)$ for $n = 0, 1, 2$. Verify that these functions are solutions of (7) for the corresponding value of μ.

4. Refer to Section 10.7, where you will find a list of associated Legendre functions. Use this list to compute explicitly $P_1^m(\cos\theta)$ for $m = -1, 0, 1$. Verify that these functions are solutions of (6) for the corresponding values of μ and m.

5. Write down (18) explicitly when $n = 1$, and verify that it is a solution of (2).

6. Write down (17) explicitly when $m = n = 1$ and verify that it solves (1).

10.2   Dirichlet Problems with Symmetry

In this section we will solve Laplace's equation inside a ball of radius $a > 0$, with prescribed boundary values on the sphere of radius a (see Figure 1). We will assume throughout this section that the problem is independent of the angle ϕ, and so we will be dealing with the radially symmetric Laplace equation

$$(1)\qquad \Delta u = \frac{\partial^2 u}{\partial r^2} + \frac{2}{r}\frac{\partial u}{\partial r} + \frac{1}{r^2}\left(\frac{\partial^2 u}{\partial\theta^2} + \cot\theta\frac{\partial u}{\partial\theta}\right) = 0,$$

where $0 < r < a$, and $0 < \theta < \pi$ (see (2), Section 10.1). The boundary condition is also radially symmetric

$$(2) \qquad\qquad u(a, \theta) = f(\theta).$$

Let us recall some facts from the previous section. Applying the separation of variables method, we let $u(r, \theta) = R(r)\Theta(\theta)$ plug into (1), and we arrive at the separated equations

$$r^2 R'' + 2r R' - n(n+1)R = 0, \quad 0 < r < a,$$
$$\Theta'' + \cot\theta\,\Theta' + n(n+1)\Theta = 0, \quad 0 < \theta < \pi,$$

where the separation constant is $n(n+1)$ with $n = 0, 1, 2, \ldots$. (All other choices of the separation constant lead to unbounded solutions and hence are discarded on physical grounds.) The equation in R is an Euler equation with bounded solution r^n. Making the change of variables $s = \cos\theta$ in the second equation, we get Legendre's differential equation of order n:

$$(1 - s^2)\Theta'' - 2s\Theta' + n(n+1)\Theta = 0, \quad -1 < s < 1$$

(see Section 10.1, (9)–(11) and (13)). For each n, this equation has a polynomial solution of degree n, denoted by $P_n(s)$ and called the nth Legendre polynomial. Substituting back $s = \cos\theta$, we obtain $r^n P_n(\cos\theta)$ as product solutions of (1). Since any multiple of these functions is also a solution of (1), it will be convenient to denote the product solutions by

$$(3) \qquad\qquad A_n \left(\frac{r}{a}\right)^n P_n(\cos\theta),$$

where A_n is an arbitrary constant and a is the radius of the ball. To complete the solution of the Dirichlet problem (1)–(2), we will superpose the product solutions (3) in an infinite series to form our candidate for a solution, then determine the A_n's in the series so as to satisfy the boundary condition (2). In this last step, we will appeal to the orthogonality of the Legendre polynomials from Section 10.6. We have the following important result.

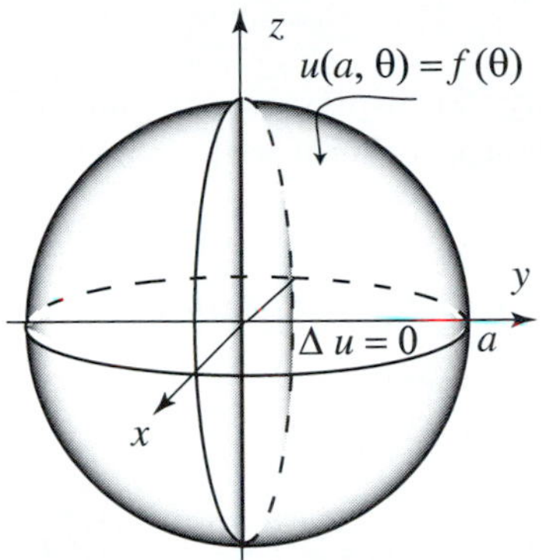

Figure 1 A Dirichlet problem with symmetry.

**THEOREM 1
DIRICHLET
PROBLEM IN A
BALL**

The solution of the Dirichlet problem (1)–(2) is given by

$$(4) \qquad\qquad u(r, \theta) = \sum_{n=0}^{\infty} A_n \left(\frac{r}{a}\right)^n P_n(\cos\theta),$$

where P_n is the nth Legendre polynomial (Section 10.5), and

$$(5) \qquad A_n = \frac{2n+1}{2} \int_0^\pi f(\theta) P_n(\cos\theta)\sin\theta\,d\theta, \quad n = 0, 1, 2, \ldots .$$

Proof Superposing the product solutions (3), we arrive at (4). Setting $r = a$ in (4) and using the boundary condition (2), we get

$$f(\theta) = \sum_{n=0}^{\infty} A_n P_n(\cos\theta), \quad 0 < \theta < \pi.$$

To determine the coefficients A_n, we proceed formally. Multiplying both sides of the equation by $P_m(\cos\theta)\sin\theta$, and then integrating term by term with respect to θ, we get

$$\int_0^\pi f(\theta) P_m(\cos\theta)\sin\theta\, d\theta = \sum_{n=0}^{\infty} A_n \int_0^\pi P_n(\cos\theta) P_m(\cos\theta)\sin\theta\, d\theta$$

$$= \sum_{n=0}^{\infty} A_n \int_{-1}^1 P_n(x) P_m(x)\, dx,$$

where the last equality follows by making the substitution $x = \cos\theta$, $dx = -\sin\theta\, d\theta$. Appealing to the orthogonality of the Legendre polynomials, Theorem 1, Section 10.6, we see that all the terms in the series are zero, except when $m = n$, in which case the term is equal to $A_m \int_{-1}^1 [P_m(x)]^2\, dx = A_m \frac{2}{2m+1}$. Thus

$$\int_0^\pi f(\theta) P_m(\cos\theta)\sin\theta\, d\theta = A_m \frac{2}{2m+1},$$

which implies (5).   ∎

It is clear from Theorem 1 that the functions $P_n(\cos\theta)$ play an important role in the solution of the Dirichlet problem in the sphere. To get an idea of the magnitude of these functions, we show in Figure 2 the graphs of $r = |P_n(\cos\theta)|$, for $0 \le \theta \le \pi$, $n = 0, 1, 2, 3$. The graphs are symmetric with respect to rotation about the z-axis and acquire more lobes as n increases. The latter property is a consequence of the increasing number of zeros of the Legendre polynomials in the interval $[-1, 1]$.

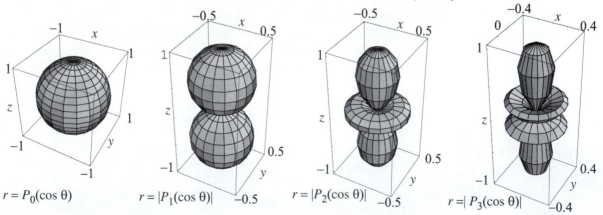

Figure 2 Graphs of $r = |P_n(\cos\theta)|$, $n = 0, 1, 2, 3$. In spherical coordinates, these represent the points $(|P_n(\cos\theta)|, \theta, \phi)$ for $0 < \theta < \pi$, $0 < \phi < 2\pi$. Because r is independent of ϕ, the graphs are symmetric with respect to the z-axis. They aquire more lobes, as n increases, due to the increasing number of zeroes of $P_n(x)$, for $0 < x < 1$.

EXAMPLE 1   A Dirichlet problem inside a sphere

Find the steady-state temperature in a sphere of unit radius, given that the upper hemisphere is kept at $100°$ and the lower one is kept at $0°$ (Figure 3).

Solution The boundary function is given by

$$f(\theta) = \begin{cases} 100 & \text{if } 0 < \theta < \frac{\pi}{2}, \\ 0 & \text{if } \frac{\pi}{2} < \theta < \pi. \end{cases}$$

Since f is independent of ϕ, the problem is covered by Theorem 1.   Accordingly, we have

$$u(r, \theta) = \sum_{n=0}^{\infty} A_n r^n P_n(\cos \theta),$$

where

$$(6) \qquad A_n = 50(2n + 1) \int_0^{\pi/2} P_n(\cos \theta) \sin \theta \, d\theta = 50(2n + 1) \int_0^1 P_n(x) \, dx.$$

At this point, you can use the explicit formulas for the P_n's to compute as many coefficients as you wish.   For example, since $P_0(x) = 1$, we get $A_0 = 50 \int_0^1 dx = 50$. Also, using $P_1(x) = x$, we get $A_1 = 150 \int_0^1 x \, dx = \frac{150}{2} = 75$. Continuing in this manner, by appealing to the explicit formulas of the P_n's, we arrive at

$$A_0 = 50, \quad A_1 = 75, \quad A_2 = 0, \quad A_3 = -\frac{175}{4}, \quad A_4 = 0.$$

Hence the temperature inside the sphere is

$$u(r, \theta) = 50 + 75 \, r P_1(\cos \theta) - \frac{175}{4} \, r^3 P_3(\cos \theta) + \cdots.$$

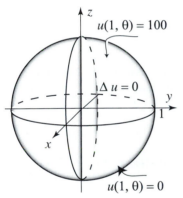

$u(1, \theta) = 100$

$\Delta u = 0$

$u(1, \theta) = 0$

Figure 3 Dirichlet problem.

In Table 1, we have used the first three nonzero terms of the series solution to approximate the temperature inside the sphere at the indicated points.

θ	0	$\pi/4$	$\pi/2$	$3\pi/4$	π
$u(1/4, \theta)$	68.1	63.4	50	36.6	32
$u(1/2, \theta)$	82	77.5	50	22.5	18
$u(3/4, \theta)$	87.8	93	50	7	12.2

Table 1 Approximation of the temperature inside the ball

For fixed r, as θ varies from 0 to π, the points move from the north pole to the south pole.   Note how the temperature decreases as θ increases.   This is to be expected, given the boundary conditions.   Also, note that as r approaches 1, the points in the upper hemisphere have temperature near $100°$.   Better approximation of the steady-state temperature can be obtained by taking more terms from the series solution.

As you will soon discover, you always learn more about a solution of a problem by appealing to properties of the Legendre polynomials from Sections 10.5 and 10.6. Here, for example, we will show how to compute the A_n's in closed form, thus yielding a more satisfactory form of the solution. From Exercise 10, Section 10.6, we have

$$\int_0^1 P_0(x)\, dx = 1, \quad \int_0^1 P_{2n}(x)\, dx = 0, \ n = 1, 2, \ldots,$$

$$\int_0^1 P_{2n+1}(x)\, dx = \frac{(-1)^n (2n)!}{2^{2n+1}(n!)^2(n+1)}, \ n = 0, 1, 2, \ldots .$$

The proofs of these identities are nontrivial, but you can do them if you wish by following the outlined steps in Exercise 10, Section 10.6. Using these identities in (6), we obtain

$$A_0 = 50, \quad A_{2n} = 0, \ n = 1, 2, 3, \ldots,$$

(7) $$A_{2n+1} = 50(4n+3)\frac{(-1)^n(2n)!}{2^{2n+1}(n!)^2(n+1)}, \ n = 1, 2, \ldots .$$

Thus

(8) $$u(r, \theta) = 50 + 25\sum_{n=0}^{\infty}(4n+3)\frac{(-1)^n(2n)!}{2^{2n}(n!)^2(n+1)}r^{2n+1}P_{2n+1}(\cos\theta).$$

See Exercise 7 for further properties of the solution.   ∎

EXAMPLE 2   A polynomial temperature distribution on the surface
The temperature on the surface of a ball of unit radius varies from 0° to 100° as one moves from the north pole to the south pole, according to the formula

$$f(\theta) = 50(1 - \cos\theta), \quad 0 < \theta < \pi .$$

(Thus f is a first degree polynomial in $\cos\theta$.) Find the steady-state temperature inside the ball.
Solution According to (4), the temperature is given by

$$u(r, \theta) = \sum_{n=0}^{\infty} A_n P_n(\cos\theta) r^n,$$

where

(9) $$A_n = \frac{2n+1}{2}\int_0^{\pi} 50(1 - \cos\theta)P_n(\cos\theta)\sin\theta\, d\theta$$

$$= 25(2n+1)\int_{-1}^{1}(1-x)P_n(x)\, dx \quad (x = \cos\theta, \ dx = -\sin\theta\, d\theta).$$

Using the explicit formulas for the P_n's, it is easy to check that

$$A_0 = 50, \quad A_1 = -50, \quad A_2 = 0, \quad A_3 = 0, \quad A_4 = 0 .$$

Indeed, with the help of a computer you can check that $A_n = 0$ for $n \geq 2$. So

$$u(r, \theta) = 50P_0(\cos\theta) - 50\, r P_1(\cos\theta) = 50 - 50\, r\, \cos\theta .$$

Surely there must be a reason for the vanishing of the A_n's when $n \geq 2$. A full justification of this fact is again found by understanding basic properties of Legendre polynomials and Legendre series. According to (9), A_n is the nth Legendre coefficient of the polynomial $50 - 50\,x$ (see (3), Section 10.6 for the definition of the Legendre coefficient). Thus, we are seeking A_n so that

$$50 - 50\,x = A_0 P_0(x) + A_1 P_1(x) + A_2 P_2(x) + A_3 P_3(x) + \cdots .$$

Since $P_0(x) = 1$, and $P_1(x) = x$, it is now clear why we must have $A_0 = 50$, $A_1 = -50$ and $A_n = 0$ for all $n \geq 2$. This argument can be generalized as follows. Suppose that $p(x)$ is a polynomial of degree k. Then the nth coefficient in the Legendre series expansion of $p(x)$ is zero for all $n > k$. For a proof, see Exercise 36, Section 10.6.   ∎

To treat the general case of Laplace's equation without symmetry, we need more than Legendre polynomials: We need the so-called associated Legendre functions. A full account of this theory is presented in the next section.

Exercises 10.2

In Exercises 1–6, solve Laplace's equation (1) inside a sphere of unit radius subject to the given boundary condition.

1. $f(\theta) = 20(\cos\theta + 1)$.

2. $f(\theta) = \cos^2\theta + 2$.

3. $f(\theta) = \begin{cases} 100 & \text{if } 0 < \theta < \pi/2, \\ 20 & \text{if } \pi/2 < \theta < \pi. \end{cases}$

4. $f(\theta) = \begin{cases} 100 & \text{if } 0 < \theta < \pi/3, \\ 0 & \text{if } \pi/3 < \theta < \pi. \end{cases}$

5. $f(\theta) = \begin{cases} \cos\theta & \text{if } 0 < \theta < \pi/2, \\ 0 & \text{if } \pi/2 < \theta < \pi. \end{cases}$
[Hint: Exercise 11, Section 10.6.]

6. $f(\theta) = |\cos\theta|$.
[Hint: Exercise 27, Section 10.6.]

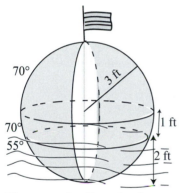

Figure 4 for Exercise 9.

7.   Experimenting with Example 1.   In this exercise, use various properties of Legendre polynomials from Sections 10.5 and 10.6 to derive properties of the solution in Example 1. Throughout this exercise, $u(r, \theta)$ refers to (8).
(a) Using (8), show that $u(r, \pi/2) = 50$. Does this meet with your expectation?
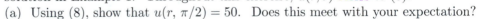
(b) If you take $r = 1$ in (8), you should get the boundary function f. Confirm this fact by plotting several partial sums of the solution. Plot your partial sums as a function of θ over the interval $[0, \pi]$.
(c) If you take $\theta = 0$ in (8), you should get the steady-state temperature of the points on the z-axis in the upper hemisphere. Plot (8) for $\theta = 0$ and $0 < r < 1$, and comment on the graph.
(d) Show that, for $0 < r < 1$, $\frac{u(r,0) + u(r,\pi)}{2} = 50$. Does this make sense on physical grounds?

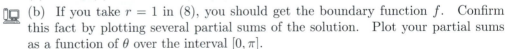
8. Get a better approximation of the temperature inside the sphere in Example 1 and the points indicated in Table 1. Use at least ten terms from (8).

9. Find the steady-state temperature inside a buoy floating on the ocean, as shown in the Figure 4.
Project Problem: Dirichlet problem outside a sphere. Exercises 10 and 11 deal with Laplace's equation in the region outside the sphere $r > a$, subject to the boundary conditions $u(a, \theta) = f(\theta)$. Such equations arise in potential theory,

when studying, for example, the potential outside a spherical capacitor. We will impose the additional condition $\lim_{r\to\infty} u(r,\,\theta) = 0$, which expresses the fact that the potential tends to zero as we move far away from the sphere.

10.   Show that the solution of (1) subject to the conditions $u(a,\,\theta) = f(\theta)$ and $\lim_{r\to\infty} u(r,\,\theta) = 0$, in the region outside the sphere $r = a$, is given by

$$u(r,\,\theta) = \sum_{n=0}^{\infty} A_n \left(\frac{r}{a}\right)^{-(n+1)} P_n(\cos\theta), \quad (r > a),$$

where A_n is as in (5). [Hint: Repeat the proof of Theorem 1, using the solutions R_n^* instead of R_n from (16), Section 10.1. Why should you make this change?]

11.   Solve the Dirichlet problem outside the unit sphere with boundary values on the unit sphere as in Exercise 3.

Project Problem: A Dirichlet problem in the region between two concentric spheres.   In Exercise 12 you are asked to solve this problem with nonzero data on the inner sphere and zero data on the outer sphere. In Exercise 13 you are asked to treat the opposite case, and in Exercise 14 you are asked to combine the results to give a solution of the Dirichlet problem with nonzero data on both spherical surfaces.

12.   Consider the Dirichlet problem consisting of (1) with the boundary conditions described in the Figure 5.
(a)   Use both functions in (16), Section 10.1, and impose the condition $R(r_2) = 0$ to arrive at the solution

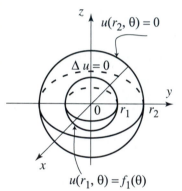

$$u_1(r,\,\theta) = \sum_{n=0}^{\infty} A_n^* \left[\left(\frac{r}{r_2}\right)^n - \left(\frac{r_2}{r}\right)^{n+1}\right] P_n(\cos\theta),$$

where $r_1 < r < r_2$, $0 < \theta < \pi$, and A_n^* are to be determined. Verify that the boundary condition $u_1(r_2,\,\theta) = 0$ is satisfied.
(b)   The other boundary condition implies that

$$u_1(r_1,\,\theta) = f_1(\theta) = \sum_{n=0}^{\infty} A_n^* \left[\left(\frac{r_1}{r_2}\right)^n - \left(\frac{r_2}{r_1}\right)^{n+1}\right] P_n(\cos\theta).$$

Figure 5 for Exercise 12.

Conclude that

$$A_n^* \left[\left(\frac{r_1}{r_2}\right)^n - \left(\frac{r_2}{r_1}\right)^{n+1}\right] = \frac{2n+1}{2} \int_0^\pi f_1(\theta) P_n(\cos\theta) \sin\theta \, d\theta.$$

Determine A_n^* to complete the solution.

13.   Show that the steady-state temperature in the region between the concentric spheres, as shown in the Figure 6, is

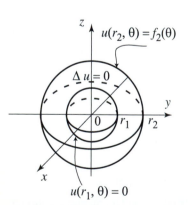

$$u_2(r,\,\theta) = \sum_{n=0}^{\infty} B_n^* \left[\left(\frac{r}{r_1}\right)^n - \left(\frac{r_1}{r}\right)^{n+1}\right] P_n(\cos\theta),$$

where $r_1 < r < r_2$, $0 < \theta < \pi$, and

Figure 6 for Exercise 13.

$$B_n^* \left[\left(\frac{r_2}{r_1}\right)^n - \left(\frac{r_1}{r_2}\right)^{n+1}\right] = \frac{2n+1}{2} \int_0^\pi f_2(\theta) P_n(\cos\theta) \sin\theta \, d\theta.$$

14.   Show that the solution of the Dirichlet problem consisting of (1) and the boundary conditions $u(r_1, \theta) = f_1(\theta)$, and $u(r_2, \theta) = f_2(\theta)$, in the region between the concentric sphere $r_1 < r < r_2$ is $u(r, \theta) = u_1(r, \theta) + u_2(r, \theta)$, where u_1 and u_2 are given in Exercises 12 and 13, respectively.

10.3 Spherical Harmonics and the General Dirichlet Problem

Recall from Section 9.1 that the solution of the Dirichlet problem in a disk was expressed in terms of the Fourier series of the boundary function, which was defined on the circle. The solution of the Dirichlet problem inside a ball shares a similar property: It will be expressed in terms of a "Fourier series" of the boundary function, which is defined on the sphere. Thus solving the Dirichlet problem on the sphere will require developing an analog of Fourier series expansions for functions defined on the sphere, the so-called spherical harmonics expansions. Like Fourier series, these are very useful tools, especially when dealing with boundary value problems in spherical regions, such as heat, wave, Dirichlet, and Poisson problems. We will develop the theory of spherical harmonics as it arises from our solution of the Dirichlet problem.

Let us start by recalling a few facts from Section 10.1. We considered Laplace's equation in spherical coordinates

(1)
$$\frac{\partial^2 u}{\partial r^2} + \frac{2}{r}\frac{\partial u}{\partial r} + \frac{1}{r^2}\left(\frac{\partial^2 u}{\partial \theta^2} + \cot\theta\,\frac{\partial u}{\partial \theta} + \csc^2\theta\,\frac{\partial^2 u}{\partial \phi^2}\right) = 0,$$

where $0 < r < a$, $0 < \theta < \pi$, and $0 < \phi < 2\pi$, with boundary condition

(2)
$$u(a, \theta, \phi) = f(\theta, \phi), \quad 0 < \theta < \pi,\ 0 < \phi < 2\pi.$$

When we applied the method of separation of variables in Section 10.1, we set $u(r, \theta, \phi) = R(r)\Theta(\theta)\Phi(\phi)$, plugged into (1), proceeded in the usual way, and arrived at the three equations

$$r^2 R'' + 2r R' - n(n+1)R = 0, \quad 0 < r < a, \quad n = 0, 1, 2, \dots,$$
$$\Phi'' + m^2\Phi = 0, \quad m = 0, 1, 2, \dots,$$

and
$$\Theta'' + \cot\theta\,\Theta' + (n(n+1) - m^2\csc^2\theta)\Theta = 0.$$

After making the change of variables $s = \cos\theta$; $\frac{ds}{d\theta} = -\sin\theta$ in the last equation and simplifying, we arrived at the **associated Legendre differential equation**

$$(1 - s^2)\frac{d^2\Theta}{ds^2} - 2s\frac{d\Theta}{ds} + \left(n(n+1) - \frac{m^2}{1 - s^2}\right)\Theta = 0, \quad -1 < s < 1.$$

The equation in R is an Euler equation with bounded solutions $R(r) = r^n$, $0 < r < a$. The solutions of the equation in Φ are the 2π-periodic $\cos m\phi$ and $\sin m\phi$, which we combined in complex form as $\Phi(\phi) = e^{im\phi}$. The associated Legendre differential equation is treated in detail in Section 10.7. Its solutions are the **associated Legendre functions** $P_n^m(s)$. We thus have the following product solutions of (1):

$$u(r, \theta, \phi) = r^n e^{im\phi} P_n^m(\cos\theta).$$

The functions $e^{im\phi} P_n^m(\cos\theta)$ appear in the solutions of many other important applications. As we will see shortly, when properly normalized these products are called **spherical harmonics**. Because of their importance, we will formulate their definitions and study their basic properties. After doing so, we will return to the Dirichlet problem and express its solution in terms of the spherical harmonics.

Spherical Harmonics

We start by recalling facts from Section 10.7. For $n = 0, 1, 2, \ldots$ and $m = 0, 1, 2, \ldots$, the associated Legendre function $P_n^m(s)$ is defined in terms of the mth derivative of the Legendre polynomial of degree n by

(3)
$$P_n^m(x) = (-1)^m (1 - x^2)^{m/2} \frac{d^m P_n(x)}{dx^m},$$

(see (1), Section 10.7). Since P_n is a polynomial of degree n, for P_n^m to be nonzero, we must take $0 \le m \le n$. For negative m's, we defined in (2), Section 10.7,

$$P_n^m(x) = (-1)^m \frac{(n + m)!}{(n - m)!} P_n^{-m}(x).$$

This extends the definition of the associated Legendre functions for $n = 0, 1, 2, \ldots$, and $m = -n, -(n - 1), \ldots, n - 1, n$. We now define the **spherical harmonics** $Y_{n,m}(\theta, \phi)$ by

(4)
$$Y_{n,m}(\theta, \phi) = \sqrt{\frac{2n + 1}{4\pi} \frac{(n - m)!}{(n + m)!}}\, P_n^m(\cos\theta) e^{im\phi},$$

where $n = 0, 1, 2, \ldots$, and $m = -n, -n+1, \ldots, n-1, n$. The coefficient in (4) is chosen so that the spherical harmonics become an orthonormal set of functions on the surface of the sphere when using the element of surface area $\sin\theta\, d\theta\, d\phi$. More precisely, we have the following orthogonality relations.

THEOREM 1
ORTHOGONALITY
RELATIONS FOR
SPHERICAL
HARMONICS

Let $n = 0, 1, 2, \ldots$, and $m = -n, -n+1, \ldots, n-1, n$, and let $Y_{n,m}$ be as in (4). Then

$$(5) \quad \int_0^{2\pi} \int_0^{\pi} Y_{n,m}(\theta, \phi)\overline{Y}_{n',m'}(\theta, \phi) \sin \theta \, d\theta \, d\phi = 0 \quad \text{if } n \neq n' \text{ or } m \neq m',$$

and

$$(6) \quad \int_0^{2\pi} \int_0^{\pi} |Y_{n,m}(\theta, \phi)|^2 \sin \theta \, d\theta \, d\phi = 1.$$

The proofs are based on the orthogonality of the associated Legendre functions (Theorem 1, Section 10.7) and the orthogonality of the complex exponentials (Exercise 12, Section 3.2). Simply use (4) to rewrite the integrals in terms of $P_n^m(\cos \theta) \, e^{im\phi}$, and then use the change of variables $s = \cos \theta$, $ds = -\sin \theta \, d\theta$. The details are straightforward and are left to Exercise 3.

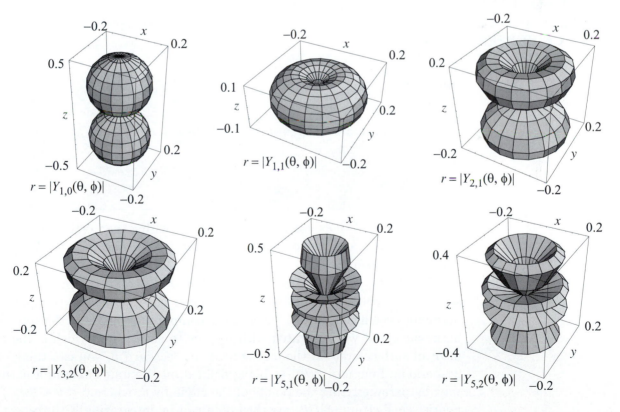

Figure 1 Graphs of $r = |Y_{n,m}(\theta, \phi)|$ in spherical coordinates. Because r is independent of ϕ, the graphs are symmetric with respect to the z-axis.

In Figure 1 we used spherical coordinates to plot the surfaces

$$r = |Y_{n,m}(\theta, \phi)| \quad \text{where } 0 < \theta < \pi, 0 < \phi < 2\pi.$$

These surfaces represent the points $(|Y_{n,m}(\theta, \phi)|, \theta, \phi)$ and hence they show the magnitude of the spherical harmonics over a sphere centered at the origin, in the following sense. Think of a sphere centered at the origin, with radius $a > 0$. Pick a point on the sphere and denote its spherical coordinates by (a, θ_0, ϕ_0). The ray through $(0, 0, 0)$ and (a, θ_0, ϕ_0) intersects the surface $r = |Y_{n,m}(\theta, \phi)|$ at the point $(|Y_{n,m}(\theta_0, \phi_0)|, \theta_0, \phi_0)$. The distance from the origin to that point of intersection is clearly $|Y_{n,m}(\theta_0, \phi_0)|$, which is the magnitude of the spherical harmonics at the point (θ_0, ϕ_0). From (4) and the fact that $|e^{im\phi}| = 1$, it follows that $|Y_{n,m}(\theta, \phi)|$ is independent of ϕ. This explains why the surface $r = |Y_{n,m}(\theta, \phi)|$ is symmetric with respect to the z-axis.

Our next theorem states that the spherical harmonics can be used to expand functions defined on the sphere, much as we used Fourier series and other orthogonal series expansions.

THEOREM 2
SPHERICAL
HARMONICS SERIES
EXPANSIONS

Let $f(\theta, \phi)$ be a function defined for all $0 < \phi < 2\pi$, $0 < \theta < \pi$, and suppose that f is 2π-periodic in ϕ. Then we have the **spherical harmonics series expansion**

$$(7) \qquad f(\theta, \phi) = \sum_{n=0}^{\infty} \sum_{m=-n}^{n} A_{n\,m} Y_{n,m}(\theta, \phi),$$

where the **spherical harmonics coefficients** are given by

$$(8) \qquad A_{nm} = \int_0^{2\pi} \int_0^{\pi} f(\theta, \phi) \overline{Y}_{n,m}(\theta, \phi) \sin\theta \, d\theta \, d\phi.$$

We can justify (8) as we have done before with other orthogonal series expansions, by using the orthogonality relations (5) and (6). We leave the details as an exercise.

You should compare Theorems 1 and 2 to the complex form of Fourier series (Theorem 1, Section 7.5). In particular, when we defined the spherical harmonics coefficient in (8), we used the complex conjugate of the spherical harmonics, and we integrated with respect to $\sin\theta \, d\theta \, d\phi$, which is the element of surface area on the unit sphere. In Section 7.5, you can think of a 2π-periodic function, $f(\theta)$, as being defined on the unit circle in a natural way, by parametrizing the points of the circle by θ, where $0 \le \theta < 2\pi$. The Fourier coefficients of $f(\theta)$ are then obtained by integrating with respect to $d\theta$, which is an element of arc length.

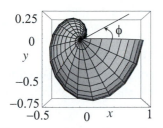

0.25
0
y
−0.5
−0.75
−0.5   0   x   1

Figure 2 View from above of the surface $r = f(\theta, \phi)$. As ϕ increases from 0 to 2π, r increases from 0 to 1.

EXAMPLE 1   A spherical harmonics series expansion

Consider the function $f(\theta, \phi) = \frac{1}{2\pi}\phi$ for $0 < \theta < \pi$, $0 < \phi < 2\pi$, where f is 2π-periodic in ϕ. The surface $r = f(\theta, \phi)$ is shown in Figure 2. As ϕ increases from 0 to 2π, f increases from 0 to 1. Note the discontinuity of the graph at $\phi = 0$ or $\phi = 2\pi$. To expand f in a spherical harmonics series, we appeal to Theorem 2. Using (8) and (4), we find

$$
\begin{aligned}
A_{nm} &= \int_0^{2\pi}\int_0^\pi f(\theta,\phi)\overline{Y}_{n,m}(\theta,\phi)\sin\theta\,d\theta\,d\phi \\
&= \frac{1}{2\pi}\sqrt{\frac{2n+1}{4\pi}\frac{(n-m)!}{(n+m)!}}\int_0^{2\pi}\phi\,e^{-im\phi}\,d\phi\int_0^\pi P_n^m(\cos\theta)\sin\theta\,d\theta.
\end{aligned}
$$

(9)

The inner integral is straightforward to evaluate using integration by parts (see Exercise 5). The change of variables $\cos\theta = s$ transforms the second integral into

$$\int_{-1}^1 P_n^m(s)\,ds.$$

This integral can be evaluated by appealing to properties of the associated Legendre functions (see Exercise 6). For example, from Section 10.7, we know that $P_n^m(s)$ is odd if $n+m$ is odd, and hence its integral as s varies from -1 to 1 is zero. Hence when $n+m$ is odd, we have $A_{nm} = 0$. Table 1 shows the coefficients corresponding to $n = 0$, 1, 2 and $m = -n, \ldots, n$.

$n\backslash m$	-2	-1	0	1	2
0			$\sqrt{\pi}$		
1		$\frac{-i}{4}\sqrt{\frac{3\pi}{2}}$	0	$\frac{-i}{4}\sqrt{\frac{3\pi}{2}}$	
2	$\frac{-i}{2}\sqrt{\frac{5}{6\pi}}$	0	0	0	$\frac{i}{2}\sqrt{\frac{5}{6\pi}}$

Table 1 Spherical harmonics coefficients A_{nm}.

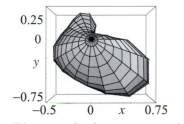

0.25
0
y
−0.75
−0.5   0   x   0.75

Figure 3 Surface corresponding to a partial sum of the Dirichlet series in Example 1.

Partial sums of the spherical harmonics series (7) can now be formed using the coefficients from Table 1. After using the explicit form of the $Y_{n,m}$ from Exercise 1 and simplifying, we obtain

$$f(\theta,\phi) \approx \sum_{n=0}^2\sum_{m=-n}^n A_{nm}Y_{n,m}(\theta,\phi) = \frac{1}{2} - \frac{3}{8}\sin\phi\sin\theta - \frac{5}{8\pi}\sin 2\phi\sin^2\theta.$$

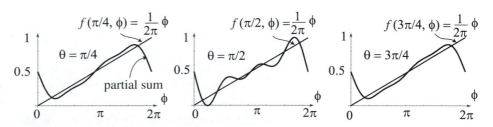

Figure 4 Partial sums of the spherical harmonics series for fixed values of θ.

In Figure 3 we plotted the surface corresponding to the partial sum,

$$r = \frac{1}{2} - \frac{3}{8}\sin\phi\sin\theta - \frac{5}{8\pi}\sin 2\phi\sin^2\theta.$$

Note the resemblance to the surface in Figure 2. You can get much better results by adding more terms to the partial sum. In Figure 4, we plotted the partial sum for fixed values of θ. Both Figures 3 and 4 show that the partial sum of the spherical harmonics series expansion approximates f, as θ varies in $(0, \pi)$ and ϕ varies in $(0, 2\pi)$.   ■

Solution of the Dirichlet Problem in a Ball

We are now in a position to derive the entire solution of the Dirichlet problem (1)–(2), illustrated in Figure 5. Let us recall the product solutions of (1),

$$r^n e^{im\phi} P_n^m(\cos\theta), \quad n = 0, 1, 2, \ldots, \quad |m| \leq n.$$

Since Laplace's equation is homogeneous, we may choose any scalar multiple of these functions. In particular, using (4), we can replace $e^{im\phi} P_n^m(\cos\theta)$ by the spherical harmonics $Y_{n,m}(\theta, \phi)$ and take the following product solutions

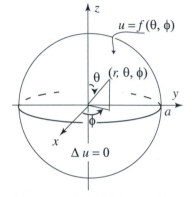

Figure 5 A Dirichlet problem in a ball.

(10)
$$\left(\frac{r}{a}\right)^n Y_{n,m}(\theta, \phi).$$

Superposing scalar multiples of these solutions, we get

(11)
$$u(r, \theta, \phi) = \sum_{n=0}^{\infty} \sum_{m=-n}^{n} A_{nm} \left(\frac{r}{a}\right)^n Y_{n,m}(\theta, \phi),$$

where the coefficient A_{nm} will be determined from the boundary condition (2). Setting $r = a$ in (11) and using (2), we get

$$f(\theta, \phi) = \sum_{n=0}^{\infty} \sum_{m=-n}^{n} A_{nm} Y_{n,m}(\theta, \phi),$$

which is the spherical harmonics expansion of f. Hence A_{nm} is given by (8). We summarize our findings as follows.

THEOREM 3
DIRICHLET
PROBLEM IN A
BALL

The solution of (1) subject to the boundary condition (2) is given by (11), where A_{nm} is the spherical harmonics coefficient of the boundary function f and is given by (8).

You should compare (11), the solution of the Dirichlet problem in a ball, to the solution of the Dirichlet problem in a disk ((4) and (5), Section 9.1). In both cases, the solution has a nice expression in terms of the "Fourier coefficients" of the boundary function.

EXAMPLE 2   A Dirichlet problem inside the unit ball

If we want to solve the Dirichlet problem inside the unit ball with boundary values given by the function f in Example 1, appealing to Theorem 3, we find the solution

$$u(r, \theta, \phi) = \sum_{n=0}^{\infty} \sum_{m=-n}^{n} A_{nm} r^n Y_{n,m}(\theta, \phi),$$

where A_{nm} is given by (9). Using the data from Example 1, we obtain the following partial sum approximation of the solution:

$$
\begin{aligned}
u(r, \theta, \phi) &\approx \sum_{n=0}^{2} \sum_{m=-n}^{n} A_{nm} r^n Y_{n,m}(\theta, \phi) \\
&= \frac{1}{2} - \frac{3}{8} r \sin\phi \sin\theta - \frac{5}{8\pi} r^2 \sin 2\phi \sin^2\theta \,.
\end{aligned}
$$

If $u(r, \theta, \phi)$ represents the steady-state temperature distribution inside the ball, then we can use this partial sum to approximate the temperature of points inside the ball. ■

Differential Equation for the Spherical Harmonics

For future applications, we need to know the differential equation for the spherical harmonics. In deriving this equation, we will work backward from our solution of Laplace's equation (1). We now know that the product solutions of Laplace's equation are of the form $u(r, \theta, \phi) = r^n Y_{n,m}(\theta, \phi)$, where $n = 0, 1, 2, \ldots$ and $|m| \leq n$. We also know that the radial part, r^n, satisfies Euler's equation. We are interested in determining the differential equation for the spherical harmonics. For this purpose, we separate variables in (1) by setting

$$u(r, \theta, \phi) = R(r) Y(\theta, \phi),$$

thus keeping the θ and ϕ variables together. Plugging into (1) and carrying out the usual details of the separation of variables, we arrive at the Euler equation in R, as expected, and the following equation in Y

$$(12) \qquad \frac{\partial^2 Y}{\partial \theta^2} + \cot\theta \, \frac{\partial Y}{\partial \theta} + \csc^2\theta \, \frac{\partial^2 Y}{\partial \phi^2} + \mu Y = 0,$$

where μ is a separation constant. Again, from our knowledge of the solution of Laplace's equation (1), we conclude that (12) has nontrivial solutions when $\mu = \mu_n = n(n+1), \quad n = 0, 1, 2, 3, \ldots$. To each $\mu = n(n+1)$ correspond $2n+1$ spherical harmonics solutions $Y(\theta, \phi) = Y_{n,m}(\theta, \phi), |m| \leq n$. We summarize these results as follows.

THEOREM 4
DIFFERENTIAL
EQUATION FOR THE
SPHERICAL
HARMONICS

The equation

$$\frac{\partial^2 Y}{\partial \theta^2} + \cot \theta \frac{\partial Y}{\partial \theta} + \csc^2 \theta \frac{\partial^2 Y}{\partial \phi^2} + \mu Y = 0,$$

where $0 < \theta < \pi$, $0 < \phi < 2\pi$, admits nontrivial bounded solutions when

$$\mu = n(n+1), \quad n = 0, 1, 2, 3, \ldots .$$

To each acceptable value of μ (or eigenvalue) we have $2n + 1$ nontrivial solutions (or eigenfunctions) given by the spherical harmonics

$$Y(\theta, \phi) = Y_{n,m}(\theta, \phi), \quad |m| \le n.$$

The eigenfunctions are given explicitly by (4) and satisfy the orthogonality relations of Theorem 1.

Exercises 10.3

1. Spherical harmonics. Use (4) and the list of associated Legendre functions from Example 1, Section 10.7, to derive the following list of spherical harmonics for $n = 0, 1, 2, 3$, and $m = -n, \ldots, n$.

(a) $(n = 0)$
$$Y_{0,0}(\theta, \phi) = \frac{1}{2\sqrt{\pi}}.$$

(b) $(n = 1)$
$$Y_{1,-1}(\theta, \phi) = \frac{1}{2}\sqrt{\frac{3}{2\pi}} \sin \theta \, e^{-i\phi}; \quad Y_{1,0}(\theta, \phi) = \frac{1}{2}\sqrt{\frac{3}{\pi}} \cos \theta;$$

$$Y_{1,1}(\theta, \phi) = -\frac{1}{2}\sqrt{\frac{3}{2\pi}} \sin \theta \, e^{i\phi}.$$

(c) $(n = 2)$
$$Y_{2,-2}(\theta, \phi) = \frac{3}{4}\sqrt{\frac{5}{6\pi}} \sin^2 \theta \, e^{-2i\phi}; \quad Y_{2,-1}(\theta, \phi) = \frac{3}{2}\sqrt{\frac{5}{6\pi}} \cos \theta \sin \theta \, e^{-i\phi};$$

$$Y_{2,0}(\theta, \phi) = \frac{1}{4}\sqrt{\frac{5}{\pi}}(-1 + 3\cos^2 \theta);$$

$$Y_{2,1}(\theta, \phi) = -\frac{3}{2}\sqrt{\frac{5}{6\pi}} \cos \theta \sin \theta \, e^{i\phi}; \quad Y_{2,2}(\theta, \phi) = \frac{3}{4}\sqrt{\frac{5}{6\pi}} \sin^2 \theta \, e^{2i\phi}.$$

(d) $(n = 3)$

$$Y_{3,-3}(\theta, \phi) = \frac{1}{8}\sqrt{\frac{35}{\pi}} \sin^3 \theta \, e^{-3i\phi}; \quad Y_{3,-2}(\theta, \phi) = \frac{15}{4}\sqrt{\frac{7}{30\pi}} \cos \theta \sin^2 \theta \, e^{-2i\phi};$$

$$Y_{3,-1}(\theta, \phi) = \frac{1}{8}\sqrt{\frac{21}{\pi}}(-1 + 5\cos^2 \theta) \sin \theta \, e^{-i\phi}; \quad Y_{3,0}(\theta, \phi) = \frac{1}{4}\sqrt{\frac{7}{\pi}}(-3\cos \theta + 5\cos^3 \theta);$$

$$Y_{3,1}(\theta, \phi) = \frac{1}{8}\sqrt{\frac{21}{\pi}}(1 - 5\cos^2 \theta) \sin \theta \, e^{i\phi}; \quad Y_{3,2}(\theta, \phi) = \frac{15}{4}\sqrt{\frac{7}{30\pi}} \cos \theta \sin^2 \theta \, e^{2i\phi};$$

$$Y_{3,3}(\theta, \phi) = -\frac{1}{8}\sqrt{\frac{35}{\pi}} \sin^3 \theta \, e^{3i\phi}.$$

2. Check the orthogonality relations (5) and (6) for $n = 0$, 1, and $m = -n, \ldots, n$, using the explicit formulas from Exercise 1.

3. Prove the orthogonality relations (5) and (6), using the orthogonality of the associated Legendre functions from Section 10.7.

4. (a) Prove that $\overline{Y}_{n,m}(\theta, \phi) = (-1)^m Y_{n,-m}(\theta, \phi)$.
(b) Show that $Y_{n,m}$ is 2π-periodic in ϕ.

5. **Experimenting with Example 1.**
(a) Compute the integral $\int_0^{2\pi} \phi e^{-im\phi}\, d\phi$ that appears in (9).
(b) Compute by hand the entries in the first two rows of Table 1.
(c) Using properties of the Legendre polynomials, explain why $A_{n0} = 0$ for all $n = 1, 2, \ldots$.

(d) Compute A_{nm} for $n = 3$, 4, and m between $-n$ and n. Form a partial sum of the spherical harmonics series of f with $n = 0, 1, \ldots, 4$ and m between $-n$ and n. Use graphics to illustrate the convergence of the series to f.

6.   **Project Problem.**   In evaluating the coefficients (9), we encountered the integral

$$I_{nm} = \int_{-1}^{1} P_n^m(s)\, ds .$$

As it turns out, this integral is quite difficult to evaluate in closed form. Prove the following properties.
(a) I_{nm} is 0 if $n + m$ is odd.
(b) $I_{00} = 2$ and $I_{n0} = 0$ for $n = 1, 2, 3, \ldots$.

(c) Professor Stephen Montgomery-Smith offered the following formula for I_{nm}

$$I_{nm} = c_{nm} \frac{4m \left[\Gamma\left(1 + \frac{n}{2}\right) \right]^2 (-1 + m + n)!}{n(1+n)! \Gamma\left(1 + \frac{-m+n}{2}\right) \Gamma\left(\frac{m+n}{2}\right)},$$

where

$$c_{nm} = \left\{ \frac{-1 + (-1)^n}{2} + \frac{1 + (-1)^m}{2} \operatorname{sgn} m \right\} \frac{1 + (-1)^{n+m}}{2},$$

and $\operatorname{sgn} m = -1$, 0, or 1 according as m is negative, 0, or positive.   Test this formula for $n = 1, 2, \ldots, 10$ and m varying from $-n$ to n in steps of 2.   When $m = -n$, there is a problem with the formula as it is written.   In this case, you should compute $\lim_{s \to -n} I_{ns}$.   You can check that

$$\lim_{s \to -n} \frac{(-1 + s + n)!}{\Gamma\left(\frac{s+n}{2}\right)} = \frac{1}{2},$$

and $c_{n,-n} = -1$, and so you can set

$$I_{n,-n} = \frac{2}{(1+n)! n!} \left[\Gamma\left(1 + \frac{n}{2}\right) \right]^2 .$$

This is a much faster way to compute the integral I_{nm} when $m = -n$.
(d) As a challenging part of this project, you can try to prove the result of (c).

7. A steady-state problem inside a sphere. Follow the outlined steps to find the steady-state temperature in the sphere with radius one given that the surface of the wedge between the lines of longitude $\phi = -\frac{\pi}{4}$ and $\phi = \frac{\pi}{4}$ is kept at $100°$ and the rest of the surface of the sphere is kept at $0°$.

(a) Let $u(r,\,\theta,\,\phi)$ denote the steady-state temperature inside the sphere. Conclude that u is given by (11), where

$$A_{nm} = 100 \int_{-\pi/4}^{\pi/4} \int_0^\pi \overline{Y}_{n,m}(\theta,\,\phi) \sin\theta \, d\theta \, d\phi \,.$$

(b) Using the explicit formulas for the spherical harmonics from Exercise 1, obtain the following table of coefficients corresponding to $n = 0,\, 1,\, 2$, and $m = -n,\, \dots,\, n$.

$n\backslash m$	-2	-1	0	1	2
0			$50\sqrt{\pi}$		
1		$25\sqrt{3\pi}$	0	$-25\sqrt{3\pi}$	
2	$100\sqrt{\frac{5}{6\pi}}$	0	0	0	$100\sqrt{\frac{5}{6\pi}}$

Spherical harmonics coefficients A_{nm} .

(c) Using the coefficients in (b) and the list of spherical harmonics from Exercise 1, obtain the partial sum approximation of the solution

$$u(r,\,\theta,\,\phi) \approx 25 + 75 \frac{\sqrt{2}}{2} r\sin\theta\cos\phi + 125\, r^2 \frac{1}{\pi}\sin^2\theta\cos 2\phi \,.$$

(d) Evaluate this partial sum at various points inside the sphere of radius 1 to get an idea about the temperature distribution. Do these values agree with your intuition? Use graphics as we did in Example 1 to illustrate the convergence of the series solution to the boundary function as $r \to 1$.

8. Repeat Example 1 with $f(\theta,\phi) = \frac{1}{\pi^2}\,\phi(2\pi - \phi)$. Use a computer to evaluate the spherical harmonics coefficients. You should get nicer pictures when you are illustrating the convergence of the spherical harmonics series. Can you justify this fact?

In Exercise 9–12, solve Laplace's equation inside the unit sphere for the given boundary function. Follow the steps that are outlined in Exercise 7.

9. $f(\theta,\,\phi) = Y_{0,0}(\theta,\,\phi)$.

10. $f(\theta,\,\phi) = Y_{1,0}(\theta,\,\phi) + 3\,Y_{1,1}(\theta,\,\phi)$.

11. $f(\theta,\,\phi) = \begin{cases} 50 & \text{if } \frac{-\pi}{3} < \phi < \frac{\pi}{3}, \\ 0 & \text{otherwise.} \end{cases}$

12. $f(\theta,\,\phi) = \begin{cases} 100 & \text{if } 0 < \phi < \frac{\pi}{2}, \\ 0 & \text{otherwise.} \end{cases}$

13. A symmetric case with no dependence on ϕ.
(a) What does Theorem 2 reduce to if f is independent of ϕ?
(b) Obtain Theorem 1 of Section 10.2 from Theorem 3 of this section.

14. Temperature of the center as an average. Since the solution of the Dirichlet problem, (11), represents a steady-state temperature distribution, the temperature of the center of the ball ($r = 0$) should be equal to the average of the temperature of the boundary. Find the temperature of the center by plugging $r = 0$ into (11), and then explain how your answer can be interpreted as an average of the temperature of the boundary.

15.   Project Problem: A Dirichlet problem between two concentric spheres. (a) Show that the Dirichlet problem given by (1) and (2) in the region $a < r$, $0 < \theta < \pi$, $0 < \phi < 2\pi$ is given by

$$u(r,\, \theta,\, \phi) = \sum_{n=0}^{\infty} \sum_{m=-n}^{n} A_{nm} \left(\frac{r}{a}\right)^{-(n+1)} Y_{n,m}(\theta,\, \phi),$$

where A_{nm} are determined by (8). [Hint: Going back to (16), Section 10.1, choose the solutions that are bounded for $r > a$.]
(b) Solve Laplace's equation (1) for $r_1 < r < r_2$, $0 < \theta < \pi$, $0 < \phi < 2\pi$, given the boundary conditions $u(r_1,\, \theta,\, \phi) = f_1(\theta,\, \phi)$, $u(r_2,\, \theta,\, \phi) = f_2(\theta,\, \phi)$. [Hint: See Exercise 14, Section 10.2.]

10.4   The Helmholtz Equation with Applications to the Poisson, Heat, and Wave Equations

In this section we complete our treatment of problems in spherical coordinates by considering the Helmholtz equation and boundary value problems involving the heat, wave, and Poisson equations. We will start by solving an eigenvalue problem involving the Helmholtz equation. The remaining boundary value problems will be solved using the eigenfunction expansions method. This was the approach that we took in Sections 8.7 and 9.5, where we treated similar problems in Cartesian and polar coordinates.

The Helmholtz Equation in a Ball

Consider the eigenvalue problem

(1) $$\Delta \Psi(r,\, \theta,\, \phi) = -k\Psi(r,\, \theta,\, \phi),$$

(2) $$\Psi(a,\, \theta,\, \phi) = 0,$$

where $0 < r < a$, $0 < \theta < \pi$, $0 < \phi < 2\pi$, k is a nonnegative constant and $\Delta \Psi$ denotes the Laplacian of Ψ in spherical coordinates. Equation (1) is the **Helmholtz equation in spherical coordinates**. The eigenvalue problem (1)–(2) is homogeneous and can be solved using the separation of variables method. To this end, let

Our notation already suggests that the solution in Y will involve the spherical harmonics.

(3) $$\Psi(r,\, \theta,\, \phi) = R(r)Y(\theta,\, \phi).$$

(Note that, unlike previous instances, where we used the separation of variables method, here we do not separate the variables θ and ϕ. The reason will become apparent when we derive the equation for Y.) To separate variables in (1), we use (3) and the explicit form of the Laplacian in spherical coordinates ((1), Section 10.1) and arrive at

(4) $$r^2 R'' + 2rR' + (kr^2 - \mu)R = 0, \quad R(a) = 0,$$

and

$$(5) \qquad \frac{\partial^2 Y}{\partial \theta^2} + \cot \theta \, \frac{\partial Y}{\partial \theta} + \csc^2 \theta \, \frac{\partial^2 Y}{\partial \phi^2} + \mu Y = 0,$$

where μ is the separation constant and Y is 2π-periodic in ϕ. (The simple details of the separation of the variables are left to Exercise 1.) Appealing to Theorem 4, Section 10.3, we see that (5) is the **differential equation for the spherical harmonics**. It has nontrivial solutions when

$$(6) \qquad \boxed{\mu = n(n+1), \quad n = 0,\, 1,\, 2,\, \dots \, .}$$

For each $\mu = n(n+1)$, we have $2n+1$ solutions, the **spherical harmonics**

$$(7) \qquad \boxed{Y_{n,m}(\theta,\, \phi) \quad m = -n,\, -n+1,\, \dots,\, 0,\, \dots,\, n-1,\, n\, .}$$

Putting the values of μ from (6) into (4), and letting

$$k = \lambda^2,$$

we arrive at the **spherical Bessel equation** in R:

$$(8) \qquad \boxed{r^2 R'' + 2r R' + (\lambda^2 r^2 - n(n+1))R = 0, \quad R(a) = 0}$$

(see Example 2, Section 9.7). For each n, we have infinitely many nontrivial solutions:

$$(9) \qquad \boxed{R_{n,j}(r) = j_n(\lambda_{n,j}\, r), \quad n = 0,\, 1,\, 2,\, \dots,\, j = 1,\, 2,\, \dots,}$$

where

$$(10) \qquad \boxed{\lambda = \lambda_{n,j} = \frac{\alpha_{n+1/2,j}}{a},}$$

$\alpha_{n+1/2,j}$ denotes the jth positive zero of the Bessel function $J_{n+\frac{1}{2}}$, and j_n is the **spherical Bessel function of the first kind**, which is defined in terms of the Bessel function by

(11)
$$j_n(r) = (\frac{\pi}{2r})^{1/2} J_{n+\frac{1}{2}}(r).$$

The complete solution of (1) and (2) can now be stated in terms of products of spherical harmonics and spherical Bessel functions as follows.

THEOREM 1
HELMHOLTZ
EQUATION IN A
BALL

The eigenvalue problem (1)–(2) has eigenvalues $k = \lambda^2$, where $\lambda = \lambda_{n,j}$ is as in (10). To each eigenvalue corresponds $2n + 1$ eigenfunctions

(12)
$$\Psi_{jnm}(r, \theta, \phi) = j_n(\lambda_{n,j}\, r)\, Y_{n,m}(\theta, \phi),$$

where $|m| \leq n$.

The eigenfunctions (12) enjoy orthogonality relations that will be used to expand functions defined inside the ball. The following results are straightforward to check, using the orthogonality of the spherical harmonics and spherical Bessel functions. (See Exercise 3.)

THEOREM 2
ORTHOGONALITY
OF SOLUTIONS OF
THE HELMHOLTZ
EQUATION

For $j = 1, 2, \dots$, $n = 0, 1, 2, \dots$, and $|m| \leq n$, we have

(13) $\displaystyle\int_0^a \int_0^{2\pi} \int_0^\pi j_n(\lambda_{n,j}\, r) Y_{n,m}(\theta, \phi) j_{n'}(\lambda_{n',j'}\, r) \overline{Y}_{n',m'}(\theta, \phi) r^2 \sin\theta\, d\theta\, d\phi\, dr = 0$

for $(j, n, m) \neq (j', n', m')$, and

(14) $\displaystyle\int_0^a \int_0^{2\pi} \int_0^\pi j_n^2(\lambda_{n,j}\, r)\, |Y_{n,m}(\theta, \phi)|^2\, r^2 \sin\theta\, d\theta\, d\phi\, dr = \frac{a^3}{2} j_{n+1}^2(\alpha_{n+\frac{1}{2},j}),$

where $\alpha_{n+\frac{1}{2},j}$ is the jth positive zero of $J_{n+\frac{1}{2}}$.

Note that the integrals in Theorem 2 are with respect to the element of volume in spherical coordinates, $r^2 \sin\theta\, d\theta\, d\phi\, dr$. Our next result, Theorem 3, states that functions defined in a ball can be expanded in series in terms of the eigenfunctions in Theorem 1. The coefficients in these series are defined by integrals using the element of volume in spherical coordinates. As you would expect, these series have to include all the eigenfunctions; thus three indices of summation are required.

Theorems 2 and 3 of this section are the higher dimensional analogues of Theorems 1 and 2 of Section 9.5. In both situations, the eigenfunctions of the Helmholtz equation are used to expand functions that are defined on the disk (in Section 9.5) and in the ball (in Section 10.4).

**THEOREM 3
SERIES
EXPANSIONS OF
FUNCTIONS
DEFINED IN A BALL**

Let $f(r, \theta, \phi)$ be a function defined for $0 < r < a$, $0 < \theta < \pi$, $0 < \phi < 2\pi$, and 2π-periodic in ϕ. Then we have

(15)
$$f(r, \theta, \phi) = \sum_{j=1}^{\infty} \sum_{n=0}^{\infty} \sum_{m=-n}^{n} A_{jnm} j_n(\lambda_{n,j} r) Y_{n,m}(\theta, \phi),$$

where

$$A_{jnm} = \frac{2}{a^3 j_{n+1}^2(\alpha_{n+\frac{1}{2},j})}$$

(16)
$$\times \int_0^a \int_0^{2\pi} \int_0^{\pi} f(r, \theta, \phi) j_n(\lambda_{n,j} r) \overline{Y}_{n,m}(\theta, \phi) r^2 \sin\theta \, d\theta \, d\phi \, dr.$$

We are now ready to tackle boundary value problems using the eigenfunction expansions method. For motivation, you should review the lower dimensional problems that are treated in Sections 8.7 and 9.5.

Poisson's Equation in a Ball

The steady-state temperature distribution inside a spherically shaped body with time independent heat source, given that the surface is kept at a certain temperature distribution, is modeled by **Poisson's equation**

(17)
$$\Delta u(r, \theta, \phi) = f(r, \theta, \phi),$$

where $0 < r < a$, $0 < \theta < \pi$, $0 < \phi < 2\pi$, along with the **boundary condition**

(18)
$$u(a, \theta, \phi) = g(\theta, \phi).$$

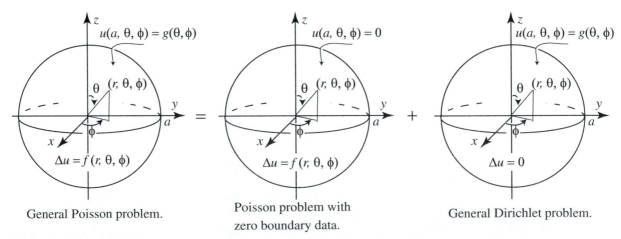

General Poisson problem.   Poisson problem with zero boundary data.   General Dirichlet problem.

Figure 1 Decomposition of a general Poisson problem.

As we did in the lower dimensional cases, our first step is to reduce the given problem to two subproblems: a Dirichlet problem with boundary data given by (18), and a Poisson problem with zero boundary data (see Figure 1). For the Dirichlet problem, you are referred to the previous section. Thus, it remains to solve (17) with the homogeneous boundary condition

$$(19) \qquad\qquad u(a,\,\theta,\,\phi) = 0\,.$$

In solving (17) and (19), we will use the method of eigenfunction expansions. Accordingly, we start by assuming that the solution has a series expansion in terms of the eigenfunctions of the Helmholtz problem (1) and (2). Hence,

$$(20) \qquad \boxed{u(r,\,\theta,\,\phi) = \sum_{j=1}^{\infty}\sum_{n=0}^{\infty}\sum_{m=-n}^{n} B_{jnm}j_n(\lambda_{n,j}r)Y_{n,m}(\theta,\,\phi)\,.}$$

To determine B_{jnm}, we plug the triple series into (17), use the fact that each term satisfies (1) with $k = \lambda_{n,j}^2$, and get

$$\sum_{j=1}^{\infty}\sum_{n=0}^{\infty}\sum_{m=-n}^{n} -\lambda_{n,j}^2 B_{jnm}j_n(\lambda_{n,j}\,r)Y_{n,m}(\theta,\phi) = f(r,\,\theta,\,\phi)\,.$$

Thinking of this last equation as an eigenfunction expansion of f, we get from Theorem 3 that

$$(21) \qquad \boxed{B_{jnm} = \frac{-1}{\lambda_{n,j}^2}A_{jnm},}$$

where A_{jnm} is given by (16). This completely determines the solution of the Poisson problem. We summarize our findings as follows.

THEOREM 4
POISSON PROBLEM
IN A BALL

The solution of the Poisson problem (17)–(18) is

$$u(r,\,\theta,\,\phi) = u_1(r,\,\theta,\,\phi) + u_2(r,\,\theta,\,\phi),$$

where u_1 is the solution of the Poisson problem with zero boundary data, and u_2 is the solution of the Dirichlet problem inside the ball of radius a with boundary condition (18). The function u_1 is given by (20) and (21), and the function u_2 is given by Theorem 3, Section 10.3. (The coefficients in the series solution (11), Section 10.3, are the spherical harmonics coefficients of the boundary function g.)

A Nonhomogeneous Heat Problem

Our next example is a heat problem inside a ball with time independent internal heat source (nonhomogeneous heat equation), given an initial temperature distribution and given that the surface of the sphere is held at $0°$ temperature (homogeneous boundary condition). More general problems involving nonhomogeneous boundary conditions and time dependent heat sources can be solved by the same methods and will be developed in the exercises.

Consider the **nonhomogeneous heat equation in spherical coordinates**:

$$(22) \qquad \frac{\partial u}{\partial t} = c^2 \left\{ \frac{\partial^2 u}{\partial r^2} + \frac{2}{r}\frac{\partial u}{\partial r} + \frac{1}{r^2}\left(\frac{\partial^2 u}{\partial \theta^2} + \cot\theta \frac{\partial u}{\partial \theta} + \csc^2\theta \frac{\partial^2 u}{\partial \phi^2} \right) \right\} + q(r, \theta, \phi),$$

where $0 < r < a$, $0 < \theta < \pi$, $0 < \phi < 2\pi$, $t > 0$, with **boundary condition**

$$(23) \qquad u(a, \theta, \phi, t) = 0,$$

and **initial condition**

$$(24) \qquad u(r, \theta, \phi, 0) = f(r, \theta, \phi).$$

We solve this problem using the method of eigenfunction expansions, and hence start by assuming that u has an expansion in terms of the eigenfunctions of the Helmholtz problem (1)-(2). Thus

$$(25) \qquad u(r, \theta, \phi, t) = \sum_{j=1}^{\infty}\sum_{n=0}^{\infty}\sum_{m=-n}^{n} B_{jnm}(t) j_n(\lambda_{n,j} r) Y_{n,m}(\theta, \phi),$$

where the coefficients $B_{jnm}(t)$ are functions of t, and $\lambda_{n,j} = \frac{\alpha_{n+1/2,j}}{a}$. We now expand f and q using Theorem 3 and obtain

$$(26) \qquad f(r, \theta, \phi) = \sum_{j=1}^{\infty}\sum_{n=0}^{\infty}\sum_{m=-n}^{n} f_{jnm} j_n(\lambda_{n,j} r) Y_{n,m}(\theta, \phi),$$

and

$$(27) \qquad q(r, \theta, \phi) = \sum_{j=1}^{\infty}\sum_{n=0}^{\infty}\sum_{m=-n}^{n} q_{jnm} j_n(\lambda_{n,j} r) Y_{n,m}(\theta, \phi),$$

where f_{jnm} and q_{jnm} are computed with the help of (16) by using f and q, respectively. To complete the solution, we must determine the coefficients

$B_{jnm}(t)$ in (25). As you will see, these turn out to be solutions of simple first order ordinary differential equations. Plug the expression of u in (25) into the heat equation (22). Use the fact that each term in the series is a solution of (1) with $k = \lambda_{n,j}^2$. Also use (27) and get

$$\sum_{j=1}^{\infty}\sum_{n=0}^{\infty}\sum_{m=-n}^{n} B'_{jnm}(t)j_n(\lambda_{n,j}r)Y_{n,m}(\theta,\phi)$$

$$= \sum_{j=1}^{\infty}\sum_{n=0}^{\infty}\sum_{m=-n}^{n} \left(-\lambda_{n,j}^2 c^2 B_{jnm}(t) + q_{jnm}\right)j_n(\lambda_{n,j}r)Y_{n,m}(\theta,\phi).$$

This yields the differential equation for $B_{jnm}(t)$,

$$(28) \qquad\qquad B'_{jnm}(t) + \lambda_{n,j}^2 c^2 B_{jnm}(t) = q_{jnm}.$$

The initial condition for this equation is obtained by using (24), (25), and (26), thus implying

$$(29) \qquad\qquad B_{jnm}(0) = f_{jnm}.$$

As you can check directly, the solution of the initial value problem (28)–(29) is

$$(30) \qquad \boxed{B_{jnm}(t) = e^{-\lambda_{n,j}^2 c^2 t}\left(f_{jnm} - \frac{q_{jnm}}{c^2 \lambda_{n,j}^2}\right) + \frac{q_{jnm}}{c^2 \lambda_{n,j}^2}.}$$

This determines the unknown coefficients in (25) and completes the solution of the heat problem. Note that the final answer involves the coefficients of the eigenfunction expansions of f and q, and the familiar "heat kernel" $e^{-\lambda_{n,j}^2 c^2 t}$.

Wave Equation in a Ball

In our final application, we consider the **wave equation in spherical coordinates**:

$$(31) \qquad \frac{\partial^2 u}{\partial t^2} = c^2\left\{\frac{\partial^2 u}{\partial r^2} + \frac{2}{r}\frac{\partial u}{\partial r} + \frac{1}{r^2}\left(\frac{\partial^2 u}{\partial \theta^2} + \cot\theta\frac{\partial u}{\partial \theta} + \csc^2\theta\frac{\partial^2 u}{\partial \phi^2}\right)\right\},$$

where $0 < r < a$, $0 < \theta < \pi$, $0 < \phi < 2\pi$, $t > 0$, with **boundary condition**

$$(32) \qquad\qquad u(a,\theta,\phi,t) = 0,$$

and **initial conditions**

$$(33) \qquad u(r,\theta,\phi,0) = f(r,\theta,\phi), \quad u_t(r,\theta,\phi,0) = g(r,\theta,\phi).$$

To solve this problem, we use the eigenfunction expansions method and assume that the solution u is a series in terms of the eigenfunctions of the Helmholtz problem (1)–(2). We obtain the following solution.

THEOREM 5
WAVE EQUATION
IN A BALL

The solution of wave boundary value problem (31)–(33) is

(34)
$$u(r, \theta, \phi, t) = \sum_{j=1}^{\infty} \sum_{n=0}^{\infty} \sum_{m=-n}^{n} j_n(\lambda_{n,j} r) Y_{n,m}(\theta, \phi)$$
$$\times \left(C_{jnm} \cos c\lambda_{n,j} t + D_{jnm} \sin c\lambda_{n,j} t \right),$$

where

$$C_{jnm} = \frac{2}{a^3 j_{n+1}^2(\alpha_{n+\frac{1}{2},j})} \int_0^a \int_0^{2\pi} \int_0^{\pi} f(r, \theta, \phi) j_n(\lambda_{n,j} r)$$

(35)
$$\times \overline{Y}_{n,m}(\theta, \phi) r^2 \sin \theta \, d\theta \, d\phi \, dr$$

and

$$D_{jnm} = \frac{2}{c\lambda_{nj} a^3 j_{n+1}^2(\alpha_{n+\frac{1}{2},j})} \int_0^a \int_0^{2\pi} \int_0^{\pi} g(r, \theta, \phi) j_n(\lambda_{n,j} r)$$

(36)
$$\times \overline{Y}_{n,m}(\theta, \phi) r^2 \sin \theta \, d\theta \, d\phi \, dr.$$

The interesting details of the derivation are very similar to those of the heat equation and are left to the exercises.

Exercises 10.4

1. Derive (4) and (5) from (1) and (2).

2. Use the explicit formulas for the spherical Bessel functions and the spherical harmonics to verify (13) and (14) when $j = 1$, $n = 0, 1$, and $m = -1, 0, 1$.

3. (a) Use the orthogonality properties of the spherical Bessel functions (Exercise 45, Section 9.7) and the spherical harmonics (Section 10.3) to prove (13) and (14).
(b) Use Theorem 2 to justify (16).

4. What does Theorem 3 reduce to if f depends only on r?

5. Define a function inside the unit ball by $f(r) = 1$ for $0 < r < 1$. Compute the series expansion (15) for f.

6. Define a function inside the unit ball by $f(r) = 1$ if $0 < r < 1/2$ and $f(r) = 0$ if $1/2 < r < 1$. Compute the series expansion (15) for f.

7. Define a function inside the unit ball by $f(r) = r^2$. Compute the series expansion (15) for f.

8. Define a function inside the unit ball by $f(r, \phi) = r^2\phi$. Compute the series expansion (15) for f.

9. Solve the Poisson problem (17), (19) inside the unit ball with $f = 1$, and $g = 0$.

10.   Solve the Poisson problem (17), (18) inside the unit ball with $f = 1$ and $g = \frac{1}{2\pi}\phi$.

11.   Derive (29); then solve (28)–(29) to get (30).

12.   What is the solution of the heat problem (22)–(24) if $q = 0$?

13.   **Project Problem: A problem with time dependent heat source.** Solve the heat problem (22)–(24) with a time dependent heat source $q(r, \theta, \phi, t)$. [Hint: Modify the solution of (22)–(24) as follows. In (27), you should allow the coefficients q_{jmn} to depend on t. In (28), the right side depends on t. Solve (28)–(29) as in Exercise 18(c), Section 8.7.]

14.   **A nonhomogeneous heat problem.**   Solve the heat equation (22) subject to the initial condition (24), and the nonhomogeneous boundary condition $u(a, \theta, \phi, t) = g(\theta, \phi)$. [Hint: Decompose the problem into two subproblems. See, for example, Exercise 14, Section 9.3.]

15.   **A heat problem with symmetry.** A solid ball at 30°C with radius $a = 1$ is placed in a refrigerator that maintains a constant temperature of 0°C. (a) Take $c = 1$ and determine the temperature $u(r, \theta, \phi, t)$ inside the ball.

(b) The function

$$\theta_4(u, q) = 1 + 2\sum_{n=1}^{\infty}(-1)^n q^{n^2} \cos 2nu$$

is called an **elliptic theta function**. This type of function is built into many computer systems. Express the temperature of the center in part (a) in terms of θ_4, then plot the temperature for $t > 0$.

(c) Use your graph in (b) to determine how long it will take for the temperature at the center to drop to 10°C.

16. (a) Solve the heat problem in the ball of radius 1 with $c = 1$, given that the surface temperature is kept at 0° and that the initial temperature distribution is $f(r, \theta, \phi) = r^2$. Assume that there is no internal source of heat.

(b) Plot the temperature of the center for $t > 0$.

17. (a) Solve the heat problem (22)–(24) inside the unit ball. Take $c = 1$, $q = 20\,r^2$, and $f(r, \theta, \phi) = 100$.

(b) Plot the temperature of the center for $t > 0$.

18.   **Project Problem: The wave equation in a ball.** Derive (34)–(36).

10.5   Legendre's Differential Equation and Legendre Polynomials

Legendre's differential equation,

(1) $$\qquad (1 - x^2)y'' - 2x\,y' + \mu\,y = 0, \quad -1 < x < 1,$$

arose when solving Laplace's equation in spherical coordinates (see (13), Section 10.1). It is thus a very important differential equation. Legendre's differential equation is really a family of equations, one for each value of the parameter μ. Upon dividing by $(1 - x^2)$, the equation becomes

$$y'' - \frac{2x}{1 - x^2}\, y' + \frac{\mu}{1 - x^2}\, y = 0\,.$$

Using the geometric series expansion, we can verify that the functions $p(x) = -\frac{2x}{1-x^2}$ and $q(x) = \frac{\mu}{1-x^2}$ have power series expansions about 0 with both radii of convergence equal to 1. It follows from Theorem 1, Appendix A.4, that (1) has two linearly independent solutions given by power series that converge for $|x| < 1$. To find these solutions we substitute $y = \sum_{m=0}^{\infty} a_m x^m$ and its derivatives into (1) and get

$$(1 - x^2) \sum_{m=2}^{\infty} m(m - 1)a_m x^{m-2} - 2x \sum_{m=1}^{\infty} m a_m x^{m-1} + \mu \sum_{m=0}^{\infty} a_m x^m = 0\,.$$

Expanding and reindexing gives

$$\sum_{m=0}^{\infty} (m + 2)(m + 1)a_{m+2} x^m - \sum_{m=0}^{\infty} m(m - 1)a_m x^m$$

$$- \sum_{m=0}^{\infty} 2m a_m x^m + \sum_{m=0}^{\infty} \mu a_m x^m = 0\,.$$

Note that we have started the second and third series at $m = 0$ instead of $m = 2$ and $m = 1$. This makes no difference, since the added terms have value 0. Combining the series and setting the coefficients equal to 0, we get the two-step recurrence relation

$$(m + 2)(m + 1)a_{m+2} - [m(m + 1) - \mu]a_m = 0,$$

or

$$a_{m+2} = \frac{m(m + 1) - \mu}{(m + 2)(m + 1)} a_m\,.$$

Thus a_0 determines a_2, which determines a_4, a_6, $\ldots$, and a_1 determines a_3, which determines a_5, a_7, $\ldots$. Using these coefficients, we get the two series solutions, known as **Legendre functions**, one consisting of even powers of x only, and the other consisting of odd powers of x only.

Legendre Polynomials

If μ is a nonnegative integer of the form $n(n + 1)$, then the recurrence relation shows that $a_{n+2} = 0$, implying that $a_{n+4} = a_{n+6} = \cdots = 0$. Thus one of the Legendre functions in this case is a polynomial of degree n, while the other one is an infinite series. For all other values of the parameter μ, the Legendre functions are both infinite series with infinitely many nonzero terms. It turns out that, except for the polynomial solutions of (1), any Legendre function diverges at one or both of the endpoints $x = \pm 1$, and so is not bounded in this case. Figure 1 shows the graphs of two Legendre functions when $\mu = \frac{3}{4}$. Note how each function blows up at one endpoint of $[-1, 1]$. Since for practical reasons we are interested in the bounded solutions of (1), from here on we will restrict our study to the case

$$\mu = n(n + 1), \quad n = 0, 1, 2, \ldots .$$

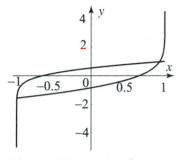

Figure 1 Legendre functions with $\mu = 3/4$.

The recurrence relation becomes

$$(2) \qquad a_{m+2} = -\frac{(n - m)(n + m + 1)}{(m + 2)(m + 1)} a_m,$$

from which we get

$$a_2 = -\frac{n(n + 1)}{2!} a_0, \qquad a_3 = -\frac{(n - 1)(n + 2)}{3!} a_1,$$

$$a_4 = \frac{(n - 2)n(n + 1)(n + 3)}{4!} a_0, \quad a_5 = \frac{(n - 3)(n - 1)(n + 2)(n + 4)}{5!} a_1,$$

$$a_6 = -\frac{(n - 4)(n - 2)n(n + 1)(n + 3)(n + 5)}{6!} a_0,$$

$$a_7 = -\frac{(n - 5)(n - 3)(n - 1)(n + 2)(n + 4)(n + 6)}{7!} a_1.$$

Hence, the general solution of (1) is

$$y(x) = a_0 y_1(x) + a_1 y_2(x),$$

where

$$(3) \qquad y_1(x) = 1 - \frac{n(n + 1)}{2!} x^2 + \frac{(n - 2)n(n + 1)(n + 3)}{4!} x^4$$
$$- \frac{(n - 4)(n - 2)n(n + 1)(n + 3)(n + 5)}{6!} x^6 + \cdots$$

and

$$(4) \qquad y_2(x) = x - \frac{(n - 1)(n + 2)}{3!} x^3 + \frac{(n - 3)(n - 1)(n + 2)(n + 4)}{5!} x^5$$
$$- \frac{(n - 5)(n - 3)(n - 1)(n + 2)(n + 4)(n + 6)}{7!} x^7 + \cdots .$$

If n is an even integer, then, from (2), we have that $a_{n+2} = 0$, and hence $a_{n+4} = a_{n+6} = \cdots = 0$. Thus (3) reduces to a polynomial of even degree n containing only even powers of x:

$$(5) \qquad\qquad y_1 = a_0 + a_2 x^2 + \cdots + a_n x^n \quad (n \text{ even}).$$

Similarly, if n is an odd integer, then (4) reduces to a polynomial of odd degree n containing only odd powers of x:

$$(6) \qquad\qquad y_2 = a_1 x + a_3 x^3 + \cdots + a_n x^n \quad (n \text{ odd}).$$

Thus far the solutions are determined up to an arbitrary constant. It is customary to normalize the polynomial solution by taking

$$(7) \qquad\qquad \boxed{a_n = \frac{(2n)!}{2^n (n!)^2}.}$$

To compute the remaining coefficients, we need a backward recurrence relation that takes us from a_m to a_{m-2}. For that purpose, express a_m in terms of a_{m+2} in (2), and then replace m by $m - 2$ to get

$$(8) \qquad\qquad a_{m-2} = -\frac{m(m-1)}{(n-m+2)(n+m-1)} a_m.$$

We compute a_{n-2} from a_n using (7) and (8)

$$a_{n-2} = -\frac{n(n-1)}{2(2n-1)} a_n = -\frac{n(n-1)}{2(2n-1)} \frac{(2n)!}{2^n (n!)^2} = -\frac{(2n-2)!}{2^n (n-1)!(n-2)!}.$$

Similarly,

$$a_{n-4} = -\frac{(n-2)(n-3)}{4(2n-3)} a_{n-2} = \frac{(2n-4)!}{2^n\, 2!\, (n-2)!(n-4)!},$$

and in general for $2m \leq n$, we have

$$a_{n-2m} = (-1)^m \frac{(2n-2m)!}{2^n m!\, (n-m)!(n-2m)!}.$$

With these coefficients the polynomial solution is called the **Legendre polynomial of degree** n and is denoted by $P_n(x)$. Setting $M = n/2$ if n is even or $(n-1)/2$ if n is odd, the nth Legendre polynomial becomes

$$(9) \qquad\qquad \boxed{P_n(x) = \frac{1}{2^n} \sum_{m=0}^{M} (-1)^m \frac{(2n-2m)!}{m!(n-m)!(n-2m)!} x^{n-2m}.}$$

The first few Legendre polynomials are

$$P_0(x) = 1 \qquad\qquad\qquad\qquad P_1(x) = x$$

$$P_2(x) = \tfrac{1}{2}(3x^2 - 1) \qquad\qquad P_3(x) = \tfrac{1}{2}(5x^3 - 3x)$$

$$P_4(x) = \tfrac{1}{8}(35x^4 - 30x^2 + 3) \qquad P_5(x) = \tfrac{1}{8}(63x^5 - 70x^3 + 15x)$$

$$P_6(x) = \tfrac{1}{16}(231x^6 - 315x^4 + 105x^2 - 5) \quad P_7(x) = \tfrac{1}{16}(429x^7 - 693x^5 + 315x^3 - 35x).$$

Figure 2 Properties of Legendre polynomials: $P_n(x)$ is even if n is even; $P_n(x)$ is odd if n is odd; $P_n(0) = 0$ if n is odd; $P_n(1) = 1$; $P_n(-1) = (-1)^n$; $|P_n(x)| \le 1$, for $-1 \le x \le 1$; $P_n(x)$ has n distinct zeros in $(-1, 1)$.

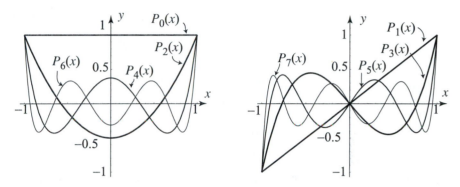

In summary, for $n = 0, 1, 2, \ldots$, the **Legendre equation of order n**

$$(10) \qquad (1 - x^2)y'' - 2xy' + n(n+1)y = 0, \quad -1 < x < 1,$$

has two linearly independent solutions y_1 and y_2 given by (3) and (4). One of them is a polynomial of degree n and the other one is an infinite series solution.   If we normalize the polynomial solution by (7), we obtain the Legendre polynomial of degree n, $P_n(x)$, given by (9).   The other solution of (10), when suitably normalized, is denoted by $Q_n(x)$ and is called a **Legendre function of the second kind**. It converges on the interval $-1 < x < 1$ and diverges at the endpoints (see Exercises 23–26). Since P_n is bounded in $[-1, 1]$ and Q_n is not, it follows that P_n and Q_n are linearly independent. Hence the general solution of Legendre's equation of order n is

$$(11) \qquad\qquad y(x) = c_1 P_n(x) + c_2 Q_n(x),$$

where c_1 and c_2 are arbitrary constants.

EXAMPLE 1   Solution of a Legendre's differential equation
Find the general solution of the differential equation

$$(1 - x^2)y'' - 2xy' + 12y = 0.$$

Solution This is Legendre's equation with $n = 3$. From (11) the general solution
is $y(x) = c_1 P_3(x) + c_2 Q_3(x)$. Since n is odd, the infinite series solution $Q_3(x)$ is
thus given by (3), up to normalization. We have

$$y = c_1 \left(\frac{1}{2} [5x^3 - 3x] \right) + c_2 \left(1 - 6x^2 + 3x^4 + \frac{4}{5} x^6 - \cdots \right).$$ ∎

EXAMPLE 2   An initial value problem
Without deriving the solution, determine whether the initial value problem

$$(1 - x^2)y'' - 2xy' + 6y = 0; \quad y(0) = 1, \ y'(0) = 1,$$

has a bounded solution.

Solution We recognize the differential equation as a Legendre's equation with
$n = 2$. The only way to get a bounded solution from the general solution (11) is
to take $c_2 = 0$. In this case, the solution is a constant multiple of $P_2(x)$, and so it
is of the form $y_1(x) = c_1 P_2(x) = a_0 + a_2 x^2$. It is easy to see that the second initial
condition cannot be satisfied by any polynomial of this form, since $y_1' = 2a_2 x$, and
so $y_1'(0) = 0$. Hence the initial value problem does not have a bounded solution.
The solution must be an infinite series. To find the coefficients of this series, we
start with $a_0 = 1$ and $a_1 = 1$, and generate the rest using (2) with $n = 2$.   ∎

Properties of Legendre Polynomials

Like Bessel functions and other special functions that arise from solving
special ordinary differential equations, Legendre polynomials have a wealth
of properties. In what follows, we establish those properties that are needed
in our applications. We start with a concise formula that could be used as
the definition of the Legendre polynomials. It is apparent from this formula
that the Legendre polynomials are polynomials of degree n.   Formulas of
this type will be used in defining other useful functions (see Section 10.7).

**THEOREM 1
RODRIGUES'
FORMULA**

For $n = 0, 1, 2, \ldots$, we have

$$(12) \qquad\qquad P_n(x) = \frac{1}{2^n n!} \frac{d^n}{dx^n} (x^2 - 1)^n.$$

Proof We have

$$(13) \qquad P_n(x) = \frac{1}{2^n} \sum_{m=0}^{M} (-1)^m \frac{(2n - 2m)!}{m!(n - m)!(n - 2m)!} x^{n-2m},$$

where $M = n/2$ if n is even or $(n-1)/2$ if n is odd. For each $0 \le m \le M$, we have

$$\frac{d^n}{dx^n} x^{2n-2m} = \frac{(2n - 2m)!}{(n - 2m)!} x^{n-2m},$$

and for $M < m \le n$, we have

$$\frac{d^n}{dx^n} x^{2n-2m} = 0.$$

So we can write (13) as

$$P_n(x) = \frac{1}{2^n n!} \sum_{m=0}^{n} (-1)^m \frac{n!}{m!(n-m)!} \frac{d^n}{dx^n} x^{2n-2m},$$

and because differentiation is a linear operation, we get

Binomial formula:
$$(a+b)^n$$
$$= \sum_{m=0}^{n} \frac{n!}{m!(n-m)!} a^{n-m} b^m.$$

$$P_n(x) = \frac{1}{2^n n!} \frac{d^n}{dx^n} \sum_{m=0}^{n} (-1)^m \frac{n!}{m!(n-m)!} (x^2)^{n-m} = \frac{1}{2^n n!} \frac{d^n}{dx^n} (x^2 - 1)^n,$$

where the last equality follows from the binomial formula. ∎

We now use Rodrigues' formula to introduce complex analysis to the picture. The proof of Rodrigues' formula works equally well if x is replaced with a complex variable z and yields

(14)
$$P_n(z) = \frac{1}{2^n n!} \frac{d^n}{dz^n} (z^2 - 1)^n.$$

Since $(z^2-1)^n$ is a polynomial of degree $2n$, it is entire and so we can compute its derivatives using the generalized Cauchy integral formula (Theorem 2, Section 3.6). Consequently, from (14), we obtain an integral representation of the Legendre polynomials,

(15)
$$P_n(z) = \frac{1}{2^{n+1} \pi i} \int_C \frac{(\zeta^2 - 1)^n}{(\zeta - z)^{n+1}} d\zeta,$$

where C is any simple closed path containing z in its interior. This formula has several advantages. By using it and making a judicious choice of the path C, we will derive the important Laplace's integral representation of the Legendre polynomials.

**THEOREM 2
LAPLACE'S
FORMULA**

For $-1 \le x \le 1$, we have

(16)
$$P_n(x) = \frac{1}{\pi} \int_0^{\pi} \left[x + i\sqrt{1-x^2} \cos \phi \right]^n d\phi \quad (n = 0, 1, 2, \ldots).$$

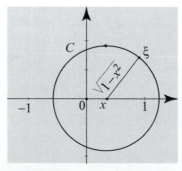

Figure 3

Proof It is enough to prove (16) for $-1 < x < 1$. Then because the functions on both sides of the equality are continuous functions of x, the equality extends to the endpoints by taking limits as $x \to \pm 1$. Fix x in the interval $(-1, 1)$. We will evaluate the integral (15) at $z = x$, along the path C described in Figure 3: C is the positively oriented circle with center at x and radius $r = \sqrt{1-x^2} > 0$. It will be convenient to parametrize this circle by $\zeta = x + ire^{i\phi} = x + re^{i(\phi + \frac{\pi}{2})}$, $-\pi \le \phi \le \pi$. We have

$$\zeta = x + ire^{i\phi}, \ d\zeta = -re^{i\phi} d\phi; \ \zeta - x = ire^{i\phi};$$

and

$$
\begin{aligned}
\zeta^2 - 1 &= (\zeta - 1)(\zeta + 1) = (x - 1 + ire^{i\phi})(x + 1 + ire^{i\phi}) \\
&= \overbrace{(x^2 - 1)}^{=-r^2} + 2ixre^{i\phi} - r^2 e^{2i\phi} = -r^2 + 2ixre^{i\phi} - r^2 e^{2i\phi}.
\end{aligned}
$$

Substituting into (15) and simplifying, we get

$$
\begin{aligned}
P_n(x) &= \frac{1}{2\pi i 2^n} \int_C \left(\frac{\zeta^2 - 1}{\zeta - x} \right)^n \frac{d\zeta}{\zeta - x} \\
&= \frac{1}{2\pi i 2^n} \int_{-\pi}^{\pi} \left(\frac{-r^2 + 2ixre^{i\phi} - r^2 e^{2i\phi}}{ire^{i\phi}} \right)^n \frac{(-r)e^{i\phi} \, d\phi}{ire^{i\phi}} \\
&= \frac{1}{2\pi 2^n} \int_{-\pi}^{\pi} \left(2x + ir(e^{i\phi} + e^{-i\phi}) \right)^n d\phi = \frac{1}{2\pi 2^n} \int_{-\pi}^{\pi} 2^n \left(x + ir \cos \phi \right)^n d\phi,
\end{aligned}
$$

which is equivalent to (16), since $\cos \phi$ is an even function.   ∎

Applying Laplace's formula, we will derive some of the properties that were illustrated in Figure 2.

COROLLARY 1

(i) For $n = 0, 1, 2, \ldots$, we have

(17) $P_n(1) = 1 \quad \text{and} \quad P_n(-1) = (-1)^n.$

(ii) For all x in $[-1, 1]$, we have $|P_n(x)| \leq 1$.

Proof To prove the first identity in (i), substitute $x = 1$ in (16) and get

$$
P_n(1) = \frac{1}{\pi} \int_0^{\pi} (1)^n \, d\phi = 1.
$$

The second identity follows similarly (Exercise 4). To prove (ii), for x in the interval $[-1, 1]$, write $x = \cos \theta$; then

$$
\begin{aligned}
P_n(x) = P_n(\cos \theta) &= \frac{1}{\pi} \int_0^{\pi} \left[\cos \theta + i \sqrt{1 - \cos^2 \theta} \cos \phi \right]^n d\phi \\
&= \frac{1}{\pi} \int_0^{\pi} \left[\cos \theta + i \sin \theta \cos \phi \right]^n d\phi.
\end{aligned}
$$

Note that $|\cos \theta + i \sin \theta \cos \phi| = \sqrt{\cos^2 \theta + \sin^2 \theta \cos^2 \phi} \leq \sqrt{\cos^2 \theta + \sin^2 \theta} = 1$. So

$$
|P_n(x)| \leq \frac{1}{\pi} \int_0^{\pi} |\cos \theta + i \sin \theta \cos \phi|^n \, d\phi \leq \frac{1}{\pi} \int_0^{\pi} d\phi = 1.   \quad ∎
$$

The next result shows a connection between the Legendre polynomials and the Taylor coefficients of the binomial series. The formula was discovered by Legendre in his work on gravitational problems involving three masses. In particular, the function $\frac{1}{\sqrt{1 - 2xu + u^2}}$ has a concrete interpretation in terms of potential function for an attractive force (see Exercise 31).

**THEOREM 3
GENERATING
FUNCTION FOR
LEGENDRE
POLYNOMIALS**

For $|x| \leq 1$ and $|u| < 1$, we have

(18)
$$\frac{1}{\sqrt{1 - 2xu + u^2}} = \sum_{n=0}^{\infty} P_n(x)u^n.$$

This is called the **generating function for the Legendre polynomials**.

Proof Recall the binomial series expansion (Exercise 36, Section 4.4),

$$(1 + v)^\alpha = \sum_{k=0}^{\infty} \binom{\alpha}{k} v^k, \quad |v| < 1,$$

where the kth binomial coefficient is

$$\binom{\alpha}{0} = 1 \quad \text{and} \quad \binom{\alpha}{k} = \frac{\alpha(\alpha - 1)(\alpha - 2)\ldots(\alpha - k + 1)}{k!} \quad \text{for } k \geq 1.$$

Taking $\alpha = -\frac{1}{2}$ and simplifying the binomial coefficients in this case (Exercise 2), we obtain

(19)
$$\frac{1}{\sqrt{1 + v}} = \sum_{k=0}^{\infty}(-1)^k \frac{(2k)!}{2^{2k}(k!)^2} v^k, \quad |v| < 1.$$

Let $v = -2xu + u^2$. A straightforward argument that we omit shows that $|v| < 1$ if $|x| \leq 1$ and $|u| < 1$. Plugging $v = -2xu + u^2$ into (19), and expanding with the help of the binomial theorem, we obtain

$$\frac{1}{\sqrt{1 - 2xu + u^2}} = \sum_{k=0}^{\infty}(-1)^k \frac{(2k)!}{2^{2k}(k!)^2}(u^2 - 2xu)^k$$

$$= \sum_{k=0}^{\infty}(-1)^k \frac{(2k)!}{2^{2k}(k!)^2} u^k \sum_{m=0}^{k} \binom{k}{m} u^m (-2)^{k-m} x^{k-m}$$

$$= \sum_{k=0}^{\infty} \sum_{m=0}^{k}(-1)^m \frac{(2k)!}{2^{k+m} k! m! (k - m)!} u^{k+m} x^{k-m}.$$

We will collect the terms in powers of u. This step can be justified using the Weierstrass double series theorem (Exercise 39, Section 4.4). Let $k + m = n$ so $k - m = n - 2m$. As k runs from 0 to ∞ and m runs from 0 to k, n runs from 0 to ∞. Also, we have $0 \leq m \leq k$ and $k + m = n$, so $0 \leq 2m \leq n$. To collect the powers of u^n in the last double sum, we set $n = k + m$ and obtain

$$\sum_{n=0}^{\infty} u^n \sum_{0 \leq m \leq 2n}(-1)^m \frac{(2n - 2m)!}{2^n (n - m)! m! (n - 2m)!} x^{n-2m} = \sum_{n=0}^{\infty} P_n(x)u^n. \quad \blacksquare$$

Like Laplace's formula, the generating function for Legendre polynomials can be used to simplify the proofs of (nontrivial) properties of Legendre polynomials. As an illustration, we present a recurrence relation, which relates three consecutive Legendre polynomials.

**THEOREM 4
BONNET'S
RECURRENCE
RELATION**

For $n = 1, 2, \ldots$, we have

(20) $$(n+1)P_{n+1}(x) + nP_{n-1}(x) = (2n+1)xP_n(x).$$

Proof The generating function (18) is analytic for $|u| < 1$. Taking the derivative with respect to u on both sides of (18) and differentiating the series term by term, we obtain

$$\frac{x-u}{(1-2xu+u^2)^{\frac{3}{2}}} = \sum_{n=1}^{\infty} nP_n(x)u^{n-1}.$$

Multiplying both sides by $1 - 2xu + u^2$, we get

$$
\begin{aligned}
\frac{x-u}{(1-2xu+u^2)^{\frac{1}{2}}} &= (1-2xu+u^2)\sum_{n=1}^{\infty} nP_n(x)u^{n-1} \\
&= \sum_{n=1}^{\infty} nP_n(x)u^{n-1} - \sum_{n=1}^{\infty} 2nxP_n(x)u^n + \sum_{n=1}^{\infty} nP_n(x)u^{n+1} \\
&= \sum_{n=0}^{\infty} (n+1)P_{n+1}(x)u^n - \sum_{n=1}^{\infty} 2nxP_n(x)u^n \\
&\quad + \sum_{n=2}^{\infty} (n-1)P_{n-1}(x)u^n.
\end{aligned}
$$

But, by (18), the left side is also equal to

$$\frac{x-u}{(1-2xu+u^2)^{\frac{1}{2}}} = (x-u)\sum_{n=0}^{\infty} P_n(x)u^n = \sum_{n=0}^{\infty} xP_n(x)u^n - \sum_{n=1}^{\infty} P_{n-1}(x)u^n.$$

Equating the coefficients of u^n in the two representations, we obtain

$$(n+1)P_{n+1}(x) - 2nxP_n(x) + (n-1)P_{n-1}(x) = xP_n(x) - P_{n-1}(x),$$

which is equivalent to (20).   ∎

Bonnet's relation has several applications. For example, we can use it to generate as many Legendre polynomials as we wish by just knowing P_0 and P_1 (Exercise 1.) There are many other recurrence formulas for the Legendre polynomials that can be established by similar methods or by using Bonnet's recurrence formula. We illustrate with two more examples.

EXAMPLE 3   Recurrence relations

Establish the following recurrence relations for Legendre polynomials:

(21) $$nP_n(x) = xP'_n(x) - P'_{n-1}(x);$$

(22) $$P'_{n+1}(x) = P'_{n-1}(x) + (2n+1)P_n(x).$$

Solution Differentiate both sides of (18) with respect to x,

$$\frac{u}{(1-2xu+u^2)^{\frac{3}{2}}} = \sum_{n=1}^{\infty} P'_n(x)u^n.$$

Divide by u both sides, then multiply by $x - u$, and get

$$\frac{x - u}{(1 - 2xu + u^2)^{\frac{3}{2}}} = \sum_{n=1}^{\infty} xP'_n(x)u^{n-1} - \sum_{n=1}^{\infty} P'_n(x)u^n$$

$$= \sum_{n=1}^{\infty} xP'_n(x)u^{n-1} - \sum_{n=2}^{\infty} P'_{n-1}(x)u^{n-1}.$$

Differentiate both sides of (18) with respect to u,

$$\frac{x - u}{(1 - 2xu + u^2)^{\frac{3}{2}}} = \sum_{n=1}^{\infty} nP_n(x)u^{n-1}.$$

Equate the coefficients of u^{n-1} in both representations, and (21) follows. To prove (22), differentiate both sides of (20) with respect to x,

$$(n + 1)P'_{n+1}(x) + nP'_{n-1}(x) = (2n + 1)P_n(x) + (2n + 1)xP'_n(x).$$

Multiply both sides of (21) by $2n+1$ and subtract from the last displayed equation,

$$(n + 1)P'_{n+1}(x) - (n + 1)P'_{n-1}(x) = (2n^2 + 3n + 1)P_n(x) = (2n + 1)(n + 1)P_n(x).$$

Divide both sides by $n + 1$, and (22) follows. ∎

Exercises 10.5

1. (a) Use (9) to derive the first five Legendre polynomials.
(b) Derive $P_2(x)$, $P_3(x)$,and $P_4(x)$, using (20), $P_0(x) = 1$, and $P_1(x) = x$.
(c) Use (9) to prove that if n is even, then $P_n(x)$ is even and contains only even powers of x; and if n is odd, then $P_n(x)$ is odd and contains only odd powers of x.

2. (a) From the definition of the binomial coefficient, show that

$$\begin{pmatrix} -\frac{1}{2} \\ k \end{pmatrix} = (-1)^k \frac{(2k)!}{2^{2k}(k!)^2}.$$

(b) Show that $P_{2n}(0) = (-1)^n \frac{(2n)!}{2^{2n}(n!)^2}$, and $P_{2n+1}(0) = 0$;
(c) $P'_{2n}(0) = 0$, and $P'_{2n+1}(0) = (2n + 1)P_{2n}(0)$. [Hint: For (b), you can use (9), or you can use (18), set $x = 0$, and then use (a).]

3. Use Exercise 2(b) to obtain the following useful identity for $n = 0, 1, 2, \ldots$:

$$P_{2n}(0) - P_{2n+2}(0) = (-1)^n \frac{(2n)!(4n + 3)}{2^{2n+1}(n!)^2(n + 1)}.$$

4. Show that $P_n(1) = 1$, $P_n(-1) = (-1)^n$ in three different ways: by using Theorems 2, 3, and 4. [Hint: In applying Theorem 3, take $x = \pm 1$, expand a geometric series in u, and compare the coefficients. In applying Theorem 4, use induction on n.]

Evaluate the following integrals.

5. $\displaystyle\int_{-1}^{1} P_3(x)\,dx$.   **6.** $\displaystyle\int_{-1}^{1} P_2(x)P_3(x)\,dx$.   **7.** $\displaystyle\int_{0}^{1} P_2(x)\,dx$.   **8.** $\displaystyle\int_{-1}^{1} P_2^2(x)\,dx$.

In Exercises 9–12, find the general solution of the given differential equation. Write down at least two terms from the series expansions of each part of your answer.

9. $(1 - x^2)y'' - 2xy' + 30y = 0$.

10. $(1 - x^2)y'' - 2xy' + 42y = 0$.

11. $(1 - x^2)y'' - 2xy' = 0$.

12. $y'' - \frac{2x}{(1-x^2)}y' + \frac{6}{(1-x^2)}y = 0$.

In Exercises 13–16, without solving, decide whether the initial value problem has a bounded solution. Find the solution if it is bounded.

13. $(1 - x^2)y'' - 2xy' + 6y = 0$; $y(0) = 0$, $y'(0) = 1$.

14. $(1 - x^2)y'' - 2xy' + 56y = 0$; $y(0) = 0$, $y'(0) = 1$.

15. $(1 - x^2)y'' - 2xy' + 56y = 0$; $y(0) = 1$, $y'(0) = 0$.

16. $(1 - x^2)y'' - 2xy' + 110y = 0$; $y(0) = 1$, $y'(0) = 1$.

In Exercises 17–20, use (3) and (4) to determine the first four terms of the series solution of the given initial value problem.

17. $(1 - x^2)y'' - 2xy' + \frac{3}{4}y = 0$; $y(0) = 1$, $y'(0) = 1$.

18. $(1 - x^2)y'' - 2xy' + 2y = 0$; $y(0) = 1$, $y'(0) = 0$.

19. $(1 - x^2)y'' - 2xy' + 3y = 0$; $y(0) = 0$, $y'(0) = 1$.

20. $(1 - x^2)y'' - 2xy' + 2y = 0$; $y(0) = 1$, $y'(0) = 1$.

21. (a) Evaluate $P_n'(1)$ for several values of n and show that it tends to infinity as $n \to \infty$.
(b) What does (a) imply about the graphs of the P_n's? Illustrate your answer graphically.

Project Problem: Legendre functions of the second kind. Do Exercise 22 and any one of 23 to 26.

22. (a) Use the reduction of order formula (Appendix A.3) to derive the following formula for the Legendre functions of the second kind:

$$Q_n(x) = P_n(x) \int \frac{1}{[P_n(x)]^2(1 - x^2)}\, dx \quad (n = 0, 1, 2, \ldots).$$

(The integral is defined up to an arbitrary constant.)
(b) Compute the Wronskian $W(P_n(x), Q_n(x))$ and show that it vanishes nowhere on $-1 < x < 1$. Conclude that the two solutions are linearly independent.

Use the formula in Exercise 22 to derive the given expression for the Legendre function of the second kind.

23. $Q_0(x) = \frac{1}{2}\ln\left(\frac{1+x}{1-x}\right)$.

24. $Q_1(x) = -1 + \frac{x}{2}\ln\left(\frac{1+x}{1-x}\right)$.

25. $Q_2(x) = -\frac{3}{2}x + \frac{1}{4}(-1 + 3x^2)\ln\left(\frac{1+x}{1-x}\right)$.

26. $Q_3(x) = \frac{1}{6}(4 - 15x^2) + \frac{1}{4}(-3x + 5x^3)\ln\left(\frac{1+x}{1-x}\right)$.

Prove the following recurrence relations for the Legendre polynomials.

27. $P_{n+1}'(x) - xP_n'(x) = (n + 1)P_n(x)$.

28. $(x^2 - 1)P_n'(x) = nxP_n(x) - nP_{n-1}(x)$.

29. Write (22) in the form $P_{n+1}'(x) - P_{n-1}'(x) = (2n+1)P_n(x)$. Apply this formula with $n = 1, 2, \ldots, n - 1$, and add the resulting formulas to prove that

$$P'_{n+1}(x) + P'_n(x) = P_0(x) + 3P_1(x) + \cdots + (2n + 1)P_n(x).$$

30. Project Problem. The **Christoffel formula** of summation states that

$$\sum_{k=0}^{n}(2k + 1)P_k(x)P_k(y) = (n + 1)\frac{P_n(x)P_{n+1}(y) - P_{n+1}(x)P_n(y)}{y - x}.$$

Prove this formula following the outlined steps.
(a) Apply Bonnet's formula with k in place of n, multiply by $P_k(y)$, and get

$$(k + 1)P_{k+1}(x)P_k(y) - (2k + 1)xP_k(x)P_k(y) + kP_{k-1}(x)P_k(y) = 0.$$

(b) Interchange x and y in (a), subtract the formulas, simplify and get

$$\begin{aligned}(2k + 1)(y - x)P_k(x)P_k(y) &= (k + 1)\left[P_{k+1}(y)P_k(x) - P_{k+1}(x)P_k(y)\right] \\ &\quad + k\left[P_{k-1}(y)P_k(x) - P_{k-1}(x)P_k(y)\right].\end{aligned}$$

(c) Apply this formula at $k = 0, 1, 2, \ldots, n$, simplify, and add to obtain the desired formula.

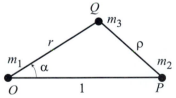

Figure 4 Three masses.

31. A three-mass problem. Consider three masses in a plane m_1, m_2, and m_3, located at the points O, P, and Q, respectively (Figure 4). According to Newton's Law of gravitation, the magnitude of the attractive force between two masses is inversely proportional to their distance. Thus the magnitude of the force between m_1 and m_3 is $F_1 = \frac{Gm_1m_3}{r^2}$, where G is the gravitational constant and $r = |OQ|$. The force between m_2 and m_3 has a similar expression in terms of ρ. To express this force as a function of r, use the law of cosines in a triangle to show that $\rho^2 = 1 + r^2 - 2r\cos\alpha$, and obtain

$$F_2 = \frac{Gm_2m_3}{1 + r^2 - 2r\cos\alpha}.$$

In terms of potential function for the mass m_2, we find that the potential of m_2 is

$$\frac{Gm_2}{\rho} = \frac{Gm_2}{\sqrt{1 + r^2 - 2r\cos\alpha}}.$$

Replacing r by u and letting $\cos\alpha = x$, we see that, up to a constant multiple, this potential is the generating function for the Legendre polynomials.

10.6   Legendre Series

Our goal in this section is to establish the orthogonality of the Legendre polynomials and show how they can be used to expand functions in series in terms of Legendre polynomials.

THEOREM 1
ORTHOGONALITY
OF LEGENDRE
POLYNOMIALS

(i) If $m \neq n$, then

$$\int_{-1}^{1} P_m(x)P_n(x)\,dx = 0.$$

(ii) For each n, we have

$$\int_{-1}^{1} P_n^2(x)\,dx = \frac{2}{2n+1}.$$

Proof (i) Consider Legendre's differential equation

$$(1 - x^2)y'' - 2xy' + \lambda(\lambda + 1)y = 0$$

and rewrite it as

$$\left[(1 - x^2)y'\right]' = -\lambda(\lambda + 1)y.$$

You should compare this proof with that of the orthogonality of Bessel functions, and note the use of the special form of the differential equations in both cases.

Since P_m and P_n are solutions of this differential equation with $\lambda = m$ and $\lambda = n$, respectively, we obtain

$$-\lambda_m P_m = \left[(1 - x^2)P_m'\right]' \qquad \text{where } \lambda_m = m(m+1),$$
$$-\lambda_n P_n = \left[(1 - x^2)P_n'\right]' \qquad \text{where } \lambda_n = n(n+1).$$

Multiplying the first equation by $-P_n$ and the second by P_m and adding gives

$$(\lambda_m - \lambda_n)P_m P_n = -\left[(1 - x^2)P_m'\right]' P_n + \left[(1 - x^2)P_n'\right]' P_m$$
$$= \frac{d}{dx}\left[(1 - x^2)(P_m P_n' - P_m' P_n)\right].$$

Hence,

$$(\lambda_m - \lambda_n)\int_{-1}^{1} P_m(x)P_n(x)\,dx$$
$$= \left[(1 - x^2)(P_m(x)P_n'(x) - P_m'(x)P_n(x))\right]\Big|_{x=-1}^{x=1} = 0,$$

and since $\lambda_m \neq \lambda_n$, (i) follows.
(ii) Bonnet's recurrence relations for n and $n-1$ give

$$(n + 1)P_{n+1} + nP_{n-1} = (2n + 1)xP_n$$

and

$$nP_n + (n - 1)P_{n-2} = (2n - 1)xP_{n-1}.$$

We multiply the first equation by P_{n-1} and the second equation by P_n and then integrate the resulting equations from $x = -1$ to $x = 1$. Using (i), we get

$$n \int_{-1}^{1} P_{n-1}^2(x)\, dx = (2n+1) \int_{-1}^{1} x P_n(x) P_{n-1}(x)\, dx$$

and

$$n \int_{-1}^{1} P_n^2(x)\, dx = (2n-1) \int_{-1}^{1} x P_{n-1}(x) P_n(x)\, dx\,.$$

Comparing the right sides of these equations, we get

$$(2n+1) \int_{-1}^{1} P_n^2(x)\, dx = (2n-1) \int_{-1}^{1} P_{n-1}^2(x)\, dx;$$

equivalently,

$$(1) \qquad \int_{-1}^{1} P_n^2(x)\, dx = \frac{2n-1}{2n+1} \int_{-1}^{1} P_{n-1}^2(x)\, dx\,.$$

Now recall that $P_0(x) = 1$, and so $\int_{-1}^{1} P_0^2(x)\, dx = 2$. Repeated applications of (1) give

$$\int_{-1}^{1} P_1^2(x)\, dx = \frac{(2-1)}{(2+1)} 2 = \frac{2}{3};$$

$$\int_{-1}^{1} P_2^2(x)\, dx = \frac{(4-1)}{(4+1)} \frac{2}{3} = \frac{2}{5};$$

and so on. A simple induction argument completes the proof of (ii). ∎

EXAMPLE 1   Integrals involving Legendre polynomials
Compute the following integral

$$\int_{-1}^{1} x P_2(x) P_3(x)\, dx.$$

Solution Using Bonnet's relation with $n = 2$, we obtain $5x P_2(x) = 3 P_3(x) + 2 P_1(x)$ or $x P_2(x) = \frac{3}{5} P_3(x) + \frac{2}{5} P_1(x)$. Multiplying through by $P_3(x)$ and using Theorem 1, we get

$$\int_{-1}^{1} x P_2(x) P_3(x)\, dx = \frac{3}{5} \overbrace{\int_{-1}^{1} P_3^2(x)\, dx}^{=2/7} + \frac{2}{5} \overbrace{\int_{-1}^{1} P_1(x) P_3(x)\, dx}^{=0} = \frac{3}{5} \frac{2}{7} = \frac{6}{35}. \quad ∎$$

Legendre Series

The orthogonality of the Legendre polynomials can be used to expand a given function f in series of the form

$$(2) \qquad f(x) = \sum_{j=0}^{\infty} A_j P_j(x),$$

called the **Legendre series expansion** of f. The coefficient A_j is called the jth **Legendre coefficient of** f. The following theorem, which you are asked to apply in the exercises, gives sufficient conditions for a function to have a Legendre series and provides a formula for the jth Legendre coefficient.

THEOREM 2
LEGENDRE SERIES

Suppose that f is a piecewise smooth function on the interval $-1 \leq x \leq 1$. Then f has the Legendre series expansion

$$f(x) = \sum_{j=0}^{\infty} A_j P_j(x),$$

where

(3)
$$A_j = \frac{2j+1}{2} \int_{-1}^{1} f(x) P_j(x) \, dx.$$

For x in the (open) interval $(-1, 1)$, the Legendre series converges to $f(x)$ if f is continuous at x, and to $(f(x+) + f(x-))/2$ otherwise.

EXAMPLE 2   A Legendre series
(a) Find the first eight Legendre coefficients A_0, A_1, ..., A_7 of the function

$$f(x) = \begin{cases} 0 & \text{if } -1 \leq x < 0, \\ 1 & \text{if } 0 \leq x \leq 1. \end{cases}$$

(b) On the same coordinate axes, sketch the graphs of f and the partial sums $s_n(x) = \sum_{j=0}^{n} A_j P_j(x)$ for $n = 0, 1, 3, 7$. Describe the behavior of the partial sums over the interval $[-1, 1]$.

Solution   (a)   We use (7) to compute the Legendre coefficients and note that $\int_{-1}^{1} f(x) P_j(x) \, dx = \int_{0}^{1} P_j(x) \, dx$. Using the list of Legendre polynomials found in Section 10.5, we get

$$A_0 = \frac{1}{2} \int_{0}^{1} P_0(x) \, dx = \frac{1}{2} \int_{0}^{1} 1 \, dx = \frac{1}{2},$$

$$A_1 = \frac{3}{2} \int_{0}^{1} P_1(x) \, dx = \frac{3}{2} \int_{0}^{1} x \, dx = \frac{3}{4},$$

$$A_2 = \frac{5}{2} \int_{0}^{1} P_2(x) \, dx = \frac{5}{2} \int_{0}^{1} \frac{3x^2 - 1}{2} \, dx = 0,$$

$$A_3 = \frac{7}{2} \int_{0}^{1} P_3(x) \, dx = \frac{7}{2} \int_{0}^{1} \frac{5x^3 - 3x}{2} \, dx = -\frac{7}{16}.$$

In a similar way, we get

$$A_4 = 0, \quad A_5 = \frac{11}{32}, \quad A_6 = 0, \quad \text{and} \quad A_7 = -\frac{75}{256}.$$

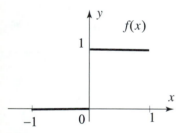

Figure 1 Function in Example 2.

(b) Using the coefficients we obtained in part (a) and simplifying (preferably with a computer system), we get

$$
\begin{aligned}
s_0(x) &= A_0 = \tfrac{1}{2}, \\
s_1(x) &= A_0 + A_1 P_1(x) = \tfrac{1}{2} + \tfrac{3}{4}x, & s_2(x) &= s_1(x), \\
s_3(x) &= \sum_{j=0}^{3} A_j P_j(x) = \frac{-35x^3 + 45x + 16}{32}, & s_4(x) &= s_3(x), \\
s_5(x) &= \sum_{j=0}^{5} A_j P_j(x) = \frac{693x^5 - 1050x^3 + 525x + 128}{256}, & s_6(x) &= s_5(x),
\end{aligned}
$$

$$
s_7(x) = \sum_{j=0}^{7} A_j P_j(x) = \frac{-32,175\,x^7 + 63,063\,x^5 - 40,425\,x^3 + 11,025\,x + 2048}{4096}.
$$

The graphs of f together with these partial sums are shown in Figure 2.   As the number of terms in a partial sum increases, the graph of the partial sum gets closer to the graph of f.   Note that all the graphs of the partial sums have y-intercept $\tfrac{1}{2}$. This confirms the assertion of Theorem 2 concerning the limit of the Legendre series at points of discontinuity, since 0 is a point of discontinuity and $(f(0+) + f(0-))/2 = \tfrac{1}{2}$.   ∎

Figure 2 The partial sums of the Legendre series converge to $f(x)$ for all $0 < x < 1$, except at the point of discontinuity $x = 0$. There, the series converges to

$$
\frac{f(0+) + f(0-)}{2} = \frac{1}{2}.
$$

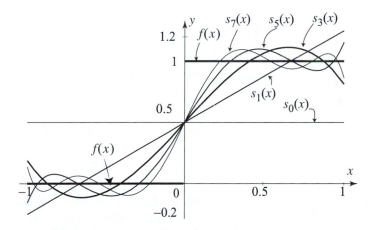

In Exercise 26 you are asked to find a closed form for the Legendre coefficients in Example 2 and to further study the behavior of the partial sums.   You will see that Legendre series exhibit a Gibbs phenomenon near a point of discontinuity.

Interesting examples of Legendre series expansions are provided by the generating function of Legendre polynomials.   We illustrate with two such examples.

EXAMPLE 3   Special Legendre series

(a) Taking $u = \frac{1}{2}$ in the generating function (Theorem 3, Section 10.5) and simplifying, we obtain the Legendre series

$$\frac{2}{\sqrt{5-4x}} = \sum_{n=0}^{\infty} \frac{P_n(x)}{2^n}, \quad -1 \le x \le 1.$$

The Legendre coefficients in this series are $A_n = \frac{1}{2^n}$. The function and some of the partial sums of its Legendre series are illustrated in Figure 3.

(b) Multiplying both sides of the series in (a) by x, we obtain

$$\frac{2x}{\sqrt{5-4x}} = \sum_{n=0}^{\infty} \frac{xP_n(x)}{2^n} = x + \sum_{n=1}^{\infty} \frac{xP_n(x)}{2^n}, \quad -1 \le x \le 1.$$

From this, we can get the Legendre series of $\frac{2x}{\sqrt{5-4x}}$. Indeed, applying Bonnet's formula to express $xP_n(x)$ in terms of $P_{n-1}(x)$ and $P_{n+1}(x)$, and simplifying, we obtain for $-1 \le x \le 1$,

$$\frac{2x}{\sqrt{5-4x}} = x + \sum_{n=1}^{\infty} \frac{1}{2^n} \left[\frac{(n+1)}{(2n+1)} P_{n+1}(x) + \frac{n}{2n+1} P_{n-1}(x) \right]$$

$$= \sum_{n=0}^{\infty} \frac{1}{2^{n-1}} \frac{n}{(2n-1)} P_n(x) + \sum_{n=0}^{\infty} \frac{1}{2^{n+1}} \frac{n+1}{2n+3} P_n(x)$$

$$= \sum_{n=0}^{\infty} \frac{1}{2^{n+1}} \frac{10n^2 + 13n - 1}{4n^2 + 4n - 3} P_n(x).$$

The function and some of the partial sums of its Legendre series are illustrated in Figure 4.   ■

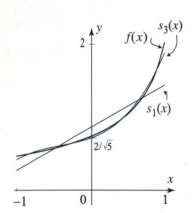

Figure 3 The function $f(x) = \frac{2}{\sqrt{5-4x}}$ and partial sums of its Legendre series over the interval $-1 \le x \le 1$. Note that $s_1(x)$ is linear (why?). The graph of $s_3(x)$ is already very close to $f(x)$.

Exercises 10.6

Evaluate the following integrals using Bonnet's relation and orthogonality.

1. $\displaystyle\int_{-1}^{1} x\, P_7(x)\, dx$.

2. $\displaystyle\int_{-1}^{1} P_2(x)\, P_7(x)\, dx$.

3. $\displaystyle\int_{-1}^{1} xP_2(x)P_3(x)\, dx$.

4. $\displaystyle\int_{-1}^{1} x^2 P_4(x)\, dx$.

5. $\displaystyle\int_{-1}^{1} x^2\, P_7(x)\, dx$.

6. $\displaystyle\int_{-1}^{1} x^2 P_6(x)\, P_7(x)\, dx$.

7. Show that $\int_{-1}^{1} P_n(x)\, dx = 0$ for $n = 1, 2, \ldots$, in two different ways.

(a) Use orthogonality. [Hint: $P_n(x) = P_0(x)P_n(x)$.]

(b) Integrate with respect to x both sides of the generating function (Theorem 3, Section 10.5). Show that the result on the left side is 0. Compare the coefficients of the power series in u.

8. Show that for any real number u with $|u| < 1$ and $m = 0, 1, 2, \ldots$, we have

$$\int_{-1}^{1} \frac{P_m(x)}{\sqrt{1 - 2xu + u^2}}\, dx = \frac{2}{2m+1} u^m.$$

[Hint: Multiply both sides of the generating function by $P_m(x)$ and use orthogonality.]

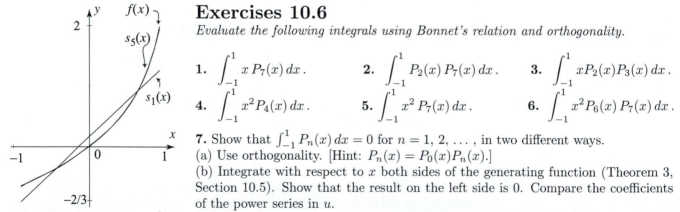

Figure 4 The function $\frac{2x}{\sqrt{5-4x}}$ and its Legendre series over the interval $-1 \le x \le 1$: $s_1(x)$ is linear; $s_5(x)$ is indistinguishable from the function itself.

9. (a) Use (22), Section 10.5, to prove that for $n = 1, 2, \ldots$

$$\int_x^1 P_n(t)\, dt = \frac{1}{2n+1}[P_{n-1}(x) - P_{n+1}(x)].$$

(b) Deduce that $\int_{-1}^1 P_n(t)\, dt = 0$ for $n = 1, 2, \ldots$.

(c) Use (a) and (b) to prove that for $n = 1, 2, \ldots$,

$$\int_{-1}^x P_n(t)\, dt = \frac{1}{2n+1}[P_{n+1}(x) - P_{n-1}(x)].$$

10. Use Exercise 9 to derive the following identities:

(a)
$$\int_0^1 P_{2n}(t)\, dt = 0, \quad n = 1, 2, \ldots,$$

(b)
$$\int_0^1 P_{2n+1}(t)\, dt = \frac{(-1)^n (2n)!}{2^{2n+1}(n!)^2(n+1)}, \quad n = 0, 1, 2, \ldots.$$

[Hint: Exercise 3, Section 10.5.]

11. Derive the following identities. For (b) and (c), use Bonnet's relation and Exercise 10.

(a)
$$\int_0^1 x P_0(x)\, dx = \frac{1}{2}; \qquad \int_0^1 x P_1(x)\, dx = \frac{1}{3}.$$

(b)
$$\int_0^1 x\, P_{2n}(x)\, dx = \frac{(-1)^{n+1}(2n-2)!}{2^{2n}((n-1)!)^2 n(n+1)}; \quad n = 1, 2, \ldots;$$

(c)
$$\int_0^1 x\, P_{2n+1}(x)\, dx = 0; \quad n = 1, 2, \ldots.$$

12. Use Rodrigues' formula and integration by parts to show that

$$\int_{-1}^1 f(x) P_n(x)\, dx = \frac{(-1)^n}{2^n n!} \int_{-1}^1 f^{(n)}(x)(x^2 - 1)^n\, dx, \quad n = 0, 1, 2, \ldots.$$

(As a convention $f^{(0)}(x) = f(x)$.)

In Exercises 13–24, evaluate the integral using Exercise 12. Take $n = 0, 1, 2, \ldots$.

13. $\displaystyle\int_{-1}^1 (1 - x^2)\, P_{13}(x)\, dx$.

14. $\displaystyle\int_{-1}^1 x^4 P_3(x)\, dx$.

15. $\displaystyle\int_{-1}^1 x^n\, P_n(x)\, dx$.

16. $\displaystyle\int_{-1}^1 x^{n+1} P_n(x)\, dx$.

[Hint: In Exercises 15 and 16, use Wallis's formulas.]

17. $\displaystyle\int_{-1}^1 \ln(1 - x) P_2(x)\, dx$.

18. $\displaystyle\int_{-1}^1 \ln(1 + x) P_4(x)\, dx$.

19. $\displaystyle\int_{-1}^1 \ln(1 - x) x P_2(x)\, dx$.

20. $\displaystyle\int_{-1}^1 \ln(1 + x) x P_2(x)\, dx$.

[Hint: In Exercises 19 and 20, use Bonnet's relation.]

21. $\displaystyle\int_{-1}^1 \ln(1 - x) P_n(x)\, dx$.

22. $\displaystyle\int_{-1}^1 \ln(1 + x) P_n(x)\, dx$.

23. $\displaystyle\int_{-1}^1 \ln(1 - x) x P_n(x)\, dx$.

24. $\displaystyle\int_{-1}^1 \ln(1 + x) x P_n(x)\, dx$.

25. Find the Legendre series expansion of $f(x) = \frac{1}{\sqrt{10-6x}}$, $-1 \le x \le 1$.

26. (a) Consider the function in Example 2. Derive the Legendre series expansion

$$f(x) = \frac{1}{2} + \sum_{n=0}^{\infty} (-1)^n \left(\frac{4n+3}{4n+4}\right) \frac{(2n)!}{2^{2n}(n!)^2} P_{2n+1}(x).$$

[Hint: Exercise 10.]

(b) Plot several partial sums of the series and note that near $x = 0$, a point of discontinuity of the function, the partial sums overshoot the graph of the function. Do the overshoots disappear with larger partial sums? State a property or a principle concerning the convergence of partial sums of Legendre series near points of discontinuity.

27. Let $f(x) = |x|$, $-1 \le x \le 1$. (a) Compute the first three nonzero Legendre coefficients of f.
(b) Derive the Legendre coefficients $A_0 = \frac{1}{2}$, $A_{2n+1} = 0$ for all n, and

$$A_{2n} = (-1)^{n+1} \frac{n(2n-2)!}{2^{2n}(n!)^2} \left(\frac{4n+1}{n+1}\right), \qquad n = 1, 2, \ldots.$$

[Hint: Exercise 11.]

(c) Plot the function and the partial sums of the Legendre series $\sum_{j=0}^{n} A_j P_j(x)$ for $n = 1, 2, 5$, and 20.
(d) If $f(x)$ is to be approximated by its Legendre series to within 0.05 on the entire interval $-1 \le x \le 1$, how large should n be?

28. (a) Find the first eight Legendre coefficients $A_0, A_1, \ldots, A_7$ of the function

$$f(x) = \begin{cases} 0 & \text{if } -1 \le x < 0, \\ x & \text{if } 0 \le x \le 1. \end{cases}$$

(b) Illustrate Theorem 2 by plotting on the same coordinate axes the graphs of f and the partial sums

$$\sum_{j=0}^{3} A_j P_j(x), \sum_{j=0}^{5} A_j P_j(x), \sum_{j=0}^{7} A_j P_j(x).$$

29. Derive the Legendre series of the function in Exercise 28. [Hint: Consider the function $\frac{1}{2}(|x| + P_1(x))$ and use Exercise 27.]

30. (a) Obtain the Legendre series expansion

$$\ln(1-x) = \ln 2 - 1 - \sum_{n=1}^{\infty} \frac{2n+1}{n(n+1)} P_n(x), \qquad -1 < x < 1.$$

[Hint: Exercise 21.]

(b) Plot several partial sums to illustrate the convergence of the Legendre series. (Note that the function does not satisfy the hypothesis of Theorem 2, because of its behavior at the endpoint $x = 1$, yet the Legendre series does converge.)

31. (a) Make a suitable change of variables in the series expansion of Exercise 30 to derive the Legendre series of $f(x) = \ln(1 + x)$, $\quad -1 < x < 1$.

(b) Plot several partial sums to illustrate the convergence of the Legendre series.

32. Expand the Legendre function of the second kind $Q_0(x)$ in a Legendre series. The explicit formula for Q_0 is given in Exercise 23, Section 10.5. [Hint: Exercises 30 and 31.]

(b) Plot several partial sums to illustrate the convergence of the Legendre series.

33. (a) Find the Legendre series of the function $f(x) = x \ln(1 - x)$, $-1 < x < 1$. [Hint: Exercise 23.]
(b) Find the Legendre series of the function $f(x) = x \ln(1+x)$, $-1 < x < 1$. [Hint: Change variables in (a).]
34. Find the Legendre series of the Legendre function of the second kind $Q_1(x)$. For the explicit formula of $Q_1(x)$ see Exercise 24, Section 10.5. [Hint: Exercise 32.]

35. The jth Legendre coefficient. In this exercise we justify (3). Multiply both sides of (2) by $P_j(x)$ and then integrate from -1 to 1. Assume term-by-term integration is allowed and derive (3) using Theorem 1.

36. Legendre series of polynomials. **(a)** Use Exercise 12 to show that if $p(x)$ is a polynomial of degree n, then $\int_{-1}^{1} p(x)P_m(x)\,dx = 0$ for all $m > n$.
(b) Conclude that if $p(x)$ is a polynomial of degree n, then the Legendre series of $p(x)$ has only finitely many terms.

In Exercises 37–40, find the Legendre series of the given polynomial. [Hint: Write the polynomial as a linear combination of the Legendre polynomials, and then determine the coefficients.]

37. $p(x) = x^2 + 2x + 1$.

38. $p(x) = 63x^5 - 7x^3 + 15x$.

39. $p(x) = x^4 + 2x^3 + x^2 + x$.

40. $p(x) = x^6$.

10.7 Associated Legendre Functions and Series Expansions

Like Legendre polynomials, the associated Legendre functions arise from the solutions of important problems involving Laplace's equation in spherical coordinates. You should keep in mind the major concepts that you encountered with the Legendre polynomials to guide you through the topics of this section.

Rodrigues' Formula

For each $m = 0, 1, 2, \ldots$, we define a family of functions, called the **associated Legendre functions of order** m, by the formula

(1)
$$P_n^m(x) = (-1)^m(1 - x^2)^{m/2}\frac{d^m P_n(x)}{dx^m},$$

where $P_n(x)$ is the Legendre polynomial of degree n (see Section 10.5). Following the usual convention that the derivative of order 0 of a function is the function itself, we see that

$$P_n^0(x) = P_n(x).$$

Thus the associated Legendre functions are generalizations of the Legendre polynomials. For this reason, whatever property we derive concerning the associated Legendre functions, it should reduce to a property of the Legendre polynomials when you take $m = 0$.

Since $P_n(x)$ is a polynomial of degree n, we see from (1) that, in order to get nonzero functions, we must take $0 \leq m \leq n$. For the applications, we extend the definition of the associated Legendre functions to negative m's by setting

(2)
$$P_n^m(x) = (-1)^m \frac{(n+m)!}{(n-m)!} P_n^{-m}(x).$$

Note that for negative m's, $P_n^m(x)$ is simply a scalar multiple of $P_n^{-m}(x)$. So, for each $n = 0, 1, 2, \ldots$, we have $2n + 1$ associated Legendre functions $P_n^m(x)$, where m runs from $-n$ to n.

EXAMPLE 1   Associated Legendre functions

Using (1) and (2) with $m = -2, -1, 0, 1, 2$, respectively, we get

(a) $P_0^0(x) = 1 = P_0(x)$, $P_1^0(x) = x = P_1(x)$,
$P_2^0(x) = \frac{3x^2-1}{2} = P_2(x)$, $P_3^0(x) = \frac{5x^3-3x}{2} = P_3(x)$.

(b) $P_1^1(x) = -\sqrt{1-x^2}$, $P_2^1(x) = -3x\sqrt{1-x^2}$,
$P_3^1(x) = -\frac{3(5x^2-1)}{2}\sqrt{1-x^2}$, $P_4^1(x) = -\frac{5(7x^3-3x)}{2}\sqrt{1-x^2}$.

(c) $P_2^2(x) = 3(1-x^2)$, $P_3^2(x) = 15x(1-x^2)$,
$P_4^2(x) = \frac{15(7x^2-1)}{2}(1-x^2)$, $P_5^2(x) = \frac{105(3x^3-x)}{2}(1-x^2)$.

(d) $P_1^{-1}(x) = \frac{1}{2}\sqrt{1-x^2}$, $P_2^{-1}(x) = \frac{1}{2}x\sqrt{1-x^2}$,
$P_3^{-1}(x) = \frac{(5x^2-1)}{8}\sqrt{1-x^2}$, $P_4^{-1}(x) = \frac{(7x^3-3x)}{8}\sqrt{1-x^2}$.

(e) $P_2^{-2}(x) = \frac{1}{8}(1-x^2)$, $P_3^{-2}(x) = \frac{1}{8}x(1-x^2)$,
$P_4^{-2}(x) = \frac{(7x^2-1)}{48}(1-x^2)$, $P_5^{-2}(x) = \frac{(3x^3-x)}{16}(1-x^2)$.   ∎

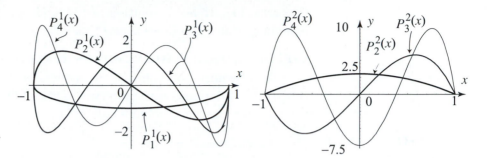

Figure 1 Associated Legendre functions.

It is clear from (1) that if m is odd and nonzero, P_n^m is not a polynomial, because of the factor $(1 - x^2)^{m/2}$. Also, as illustrated by Figure 1, P_n^m is even if $n + m$ is even and it is odd if $n + m$ is odd.

The Associated Legendre Differential Equation

We know from Section 10.5 that the Legendre polynomials satisfy the (Legendre) differential equation $(1 - x^2)y'' - 2xy' + n(n + 1)y = 0$. We now establish a similar result for the associated Legendre functions.

For $n = 0, 1, 2, \ldots$ and $m = 0, \pm 1, \pm 2, \ldots, \pm n$, the **associated Legendre differential equation** is given by

$$(3) \qquad (1 - x^2)y'' - 2xy' + (n(n + 1) - \frac{m^2}{1 - x^2})y = 0, \quad -1 < x < 1.$$

When $m = 0$, the equation reduces to

$$(1 - x^2)y'' - 2xy' + n(n + 1)y = 0,$$

which is Legendre's differential of order n ((1), Section 10.5), and so it is satisfied by the Legendre polynomials $P_n = P_n^0$. Thus in showing that the associated Legendre functions are solutions of (3), it suffices to consider the case $m \neq 0$. Moreover, since for negative m, P_n^m is proportional to P_n^{-m} (see (2)), it suffices to consider $m > 0$. Let us start with Legendre's equation which is satisfied by the nth Legendre polynomial P_n. Using the Leibnitz rule to differentiate this equation m times with respect to x and then plugging P_n for y, we arrive at

$$(4) \qquad (1 - x^2)P_n^{(m+2)} - 2(m + 1)xP_n^{(m+1)} + (n - m)(n + m + 1)P_n^{(m)} = 0$$

(Exercise 14). Thus the mth derivative of P_n, $P_n^{(m)}$ (not to be confused with P_n^m), satisfies the differential equation

$$(5) \qquad (1 - x^2)y'' - 2(m + 1)xy' + (n - m)(n + m + 1)y = 0.$$

Now, you can verify that this equation reduces to (3) if we use the substitution

$$y = (1 - x^2)^{-m/2}v$$

(Exercise 14). Hence a solution of (3) is $(1-x^2)^{m/2}\frac{d^m P_n(x)}{dx^m}$, and since (3) is homogeneous, it follows that P_n^m is also a solution.

Orthogonality and Series Expansions

Like Legendre polynomials, the associated Legendre functions enjoy orthogonality relations on the interval $[-1, 1]$. The proofs are very much like the ones for Legendre polynomials. They will be outlined in the exercises.

THEOREM 1
ORTHOGONALITY
OF ASSOCIATED
LEGENDRE
FUNCTIONS

Let $k \leq n$ be nonnegative integers and let m be an integer such $|m| \leq k$. Then,

$$(6) \qquad \int_{-1}^{1} P_k^m(x) P_n^m(x)\, dx = 0, \quad k \neq n,$$

and

$$(7) \qquad \int_{-1}^{1} [P_n^m(x)]^2\, dx = \frac{2}{2n+1}\frac{(n+m)!}{(n-m)!}, \quad |m| \leq n.$$

We next state a very useful series expansion theorem for associated Legendre functions. You should check that it reduces to Theorem 2, Section 10.6, when $m = 0$.

THEOREM 2
ASSOCIATED
LEGENDRE SERIES
EXPANSIONS

Fix a nonnegative integer m. Suppose that $f(x)$ is piecewise smooth on $[-1, 1]$. Then we have the **associated Legendre series expansion of order** m

$$(8) \qquad f(x) = \sum_{n=m}^{\infty} A_n P_n^m(x),$$

where the **associated Legendre coefficient** A_n is given by

$$(9) \quad A_n = \frac{2n+1}{2}\frac{(n-m)!}{(n+m)!}\int_{-1}^{1} f(x) P_n^m(x)\, dx, \quad n = m,\, m+1,\, m+2,\, \dots\,.$$

For x in the open interval $(-1, 1)$, the series converges to $f(x)$ if f is continuous at x, and to $(f(x+) + f(x-))/2$ otherwise.

The proof of Theorem 2 is beyond the level of this text and will be omitted. You can test the validity of the result by considering concrete applications. In the following example, we have computed the associated Legendre series expansion of a simple function, when $m = 2$. We have used a computer system with built-in associated Legendre functions to carry out the computation of the associated Legendre coefficients and to plot some partial sums of the associated Legendre series. As you can imagine, the explicit expressions of these series are very complicated. It is truly remarkable that they actually converge, as indicated by Theorem 2.

EXAMPLE 2   An associated Legendre series expansion of order $m = 2$

Consider the function

$$f(x) = \begin{cases} 0 & \text{if } -1 < x < 0, \\ 1 & \text{if } 0 < x < 1. \end{cases}$$

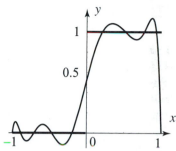

Figure 2 Partial sum of the associated Legendre series $\sum_{n=2}^{\infty} A_n P_n^2(x)$.

We will illustrate its associated Legendre series expansion of order $m = 2$.   For this purpose, we will compute the coefficients A_n for $n = 2, 3, 4, \ldots, 10$ (note that $n \geq m$).   Then using these coefficients, we will form and plot a partial sum of the associated Legendre series with n up to 10.   With the help of a computer, the coefficients are found to be as shown in Table 1.

n	2	3	4	5	6	7	8	9	10
A_n	$\frac{5}{24}$	$\frac{7}{64}$	$\frac{1}{40}$	0	$\frac{13}{1680}$	$\frac{65}{6144}$	$\frac{17}{5040}$	$-\frac{19}{30720}$	$\frac{7}{3960}$

Table 1 Associated Legendre coefficients.

The graphs of f and a partial sum of the associated Legendre series (with n up to 10) are shown in Figure 2. Note that the associated Legendre coefficients tend to zero as n increases.   Also note the overshoot of the partial sum of the associated Legendre series near the points of discontinuity.   As you would expect, both of these facts are true for general associated Legendre series.   ■

Exercises 10.7

In Exercises 1–4, use (1) and (2) to derive P_n^m for the given m and n.

1. $m = \pm 1, \ n = 2$ **2.** $m = 1, \ n = 1, 2, 3$ **3.** $m = 3, \ n = 4$ **4.** $m = -3, \ n = 4$

In Exercises 5–8, (a) determine m and n for the given associated Legendre differential equation.   (b) Use the list of Legendre functions in Example 1 to find one solution.   (c) Verify your answer in (b) by plugging back into the differential equation.

5. $(1 - x^2)y'' - 2xy' + (2 - \frac{1}{1-x^2})y = 0$. **6.** $(1 - x^2)y'' - 2xy' + (6 - \frac{4}{1-x^2})y = 0$.

7. $(1 - x^2)y'' - 2xy' + (6 - \frac{1}{1-x^2})y = 0$. **8.** $(1 - x^2)y'' - 2xy' + (12 - \frac{4}{1-x^2})y = 0$.

9. Verify (6) and (7) with $m = 1$, and $n = 1, 2$.

In Exercises 10–15, you are given an order m and a function $f(x)$ defined on the interval $[-1, 1]$.   (a) Use a computer system with built-in associated Legendre functions to compute the associated Legendre coefficients of $f(x)$ of order m.   Take $n = m, m + 1, \ldots, m + 10$.   (b) Use the coefficients in (a) to construct several partial sums of the associated Legendre series expansion of order m.   Plot these partial sums along with the given function to illustrate Theorem 2.

10. $m = 1, \ f(x) = \sin \pi x$. **11.** $m = 1, \ f(x) = (x - 1)(x + 1)$.

12. $m = 2$, $f(x) = x$.   **13.** $m = 2$, $f(x) = |x|$.

14. (a) Prove (4) using the Leibnitz rule for differentiation.

(b) With the help of a computer system, verify that the substitution $y = (1 - x^2)^{-m/2} v$ transforms (5) into (3).

15. Prove (6) by following the steps of the proof of Theorem 1, (i), Section 10.6. [Hint: Show that (3) can be put in the form

$$\left((1 - x^2)y'\right)' + (n(n+1) - \frac{m^2}{1 - x^2})y = 0.]$$

16. Project Problem: Proof of (7).

(a) Use (1) to show that

$$P_n^{m+1} = -(1 - x^2)^{1/2} \frac{dP_n^m}{dx} - m(1 - x^2)^{-1/2} x P_n^m .$$

(b) Square both sides of (a) and integrate to get

$$\int_{-1}^{1} \left[P_n^{m+1}(x)\right]^2 dx = \int_{-1}^{1} (1 - x^2) \left[\frac{dP_n^m}{dx}\right]^2 dx + 2m \int_{-1}^{1} x P_n^m(x) \frac{dP_n^m}{dx} dx$$
$$+ \int_{-1}^{1} \frac{m^2 x^2}{1 - x^2} [P_n^m(x)]^2 dx .$$

(c) Show that $\left[mx \left[P_n^m(x)\right]^2\right]_{-1}^{1} = 0.$

(d) Use integration by parts to show that the right side of the equation in (b) is equal to

$$-\int_{-1}^{1} P_n^m(x) \frac{d}{dx} \left[(1 - x^2) \frac{dP_n^m}{dx}\right] dx - m \int_{-1}^{1} [P_n^m(x)]^2 dx$$
$$+ \int_{-1}^{1} \frac{m^2 x^2}{1 - x^2} [P_n^m(x)]^2 dx .$$

(e) Explain why

$$\frac{d}{dx} \left[(1 - x^2) \frac{dP_n^m}{dx}\right] = - \left[n(n+1) - \frac{m^2}{1 - x^2}\right] P_n^m(x) .$$

[Hint: See the hint for Exercise 15.]

(f) Combine (d)–(e) to get

$$\int_{-1}^{1} \left[P_n^{m+1}(x)\right]^2 dx = (n - m)(n + m + 1) \int_{-1}^{1} [P_n^m(x)]^2 dx .$$

(g) Use (f) to get

$$\int_{-1}^{1} [P_n^m(x)]^2 dx$$
$$= (n - m + 1)(n - m + 2) \ldots n(n + m)(n + m - 1) \ldots (n + 1) \int_{-1}^{1} [P_n(x)]^2 dx .$$

(h) Use (g) and Theorem 1(ii), Section 10.6, to get (7).

11

THE FOURIER TRANSFORM AND ITS APPLICATIONS

A generalization made not for the vain pleasure of generalizing but in order to solve previously existing problems is always a fruitful generalization.

-Henri Lebesgue

In Chapters 8–10 we used Fourier series and orthogonal expansions as tools to solve boundary value problems over bounded regions such as intervals, rectangles, disks and spheres. As you can imagine, the modeling of certain physical phenomena will give rise naturally to boundary value problems over unbounded regions. For example, to describe the temperature distribution in a very long insulated wire, you can suppose that the length of the wire is infinite, which gives rise to a boundary value problem over an infinite line. To solve this new type of problem we will generalize the notion of Fourier series by developing the Fourier transform. The applications that we present are as diverse and rich as the ones we studied with Fourier series. The methods of this chapter, although mainly tailored for the Fourier transform, are also suitable with other transforms–for example, the Fourier cosine and sine transforms (Sections 11.6-11.7) and the Laplace and Hankel transforms (Chapter 12).

In addition to the basic study of the Fourier transform, we introduce in Section 11.2 the class of generalized functions and consider some of their applications in solving partial differential equations and computing Fourier transforms. This material, along with the applications of complex analysis, is presented at the end of sections and is designed as enrichment in the subject. As a sample of applications, take a look at the exercises in Section 11.2. You will find transforms that involve Legendre polynomials, Chebyshev polynomials, Bessel functions, and other special functions. You will also find interesting improper integrals, whose evaluations illustrate the power of the Fourier analysis techniques.

11.1   The Fourier Transform

In Chapter 7, we used Fourier series to analyze functions that are periodic on the real line or that are defined on a finite interval and so can be considered as the restriction of periodic functions. Our goal in this section is to extend the basic tools of Fourier analysis to functions that are defined on the entire real line but are not periodic. We will replace Fourier series by the Fourier transform and prove results that are direct analogs of results about Fourier series.

We will say that $f(x)$ is **integrable** on the real line if $\int_{-\infty}^{\infty} |f(x)|\, dx < \infty$. Thus by "integrable" we mean that the absolute value of f has a finite improper integral over the real line. We will say that f is **piecewise smooth** on the real line if it is piecewise smooth on every finite interval of the real line.

Like Fourier series, the Fourier transform can be expressed in terms of integrals involving the cosine and sine functions or the complex exponential. We will take the complex form of Fourier series as a model and make the following definition.

DEFINITION 1
THE FOURIER
TRANSFORM

Suppose that f is integrable on the real line, the **Fourier transform** of f, denoted by $\mathcal{F}(f)$ or $\widehat{f}$, is defined by

$$(1) \qquad \widehat{f}(\omega) = \frac{1}{\sqrt{2\pi}} \int_{-\infty}^{\infty} f(x) e^{-i\omega x} dx \qquad (-\infty < \omega < \infty).$$

The notation $\mathcal{F}(f(x))(\omega)$ is also used to be specific about the variables. Like the Fourier coefficients, the Fourier transform of f is defined by integrating f against the exponential function over the domain of definition of f, which is the entire real line in the present case. However, unlike the Fourier coefficients of a periodic function, which are defined over a discrete set of values, the Fourier transform is defined over a continuous set of ω's.

A Fourier series reconstructs a periodic function from its Fourier coefficients. To reconstruct a function on the real line from its Fourier transform, we must sum over a continuous set of Fourier coefficients; thus we must integrate the Fourier transform. The resulting integral is called the inverse Fourier transform.

DEFINITION 2
THE INVERSE
FOURIER
TRANSFORM

The **inverse Fourier transform** of $g(\omega)$ is

$$(2) \qquad \mathcal{F}^{-1}(g)(x) = \frac{1}{\sqrt{2\pi}} \int_{-\infty}^{\infty} g(\omega) e^{i\omega x}\, d\omega \qquad (-\infty < x < \infty).$$

The definition of the inverse Fourier transform is very similar to that of the Fourier transform; the only difference is the negative sign in the exponent of the exponential function inside the integral. The improper integral in (2) is to be interpreted as a principal value:

$$\mathcal{F}^{-1}(g)(x) = \lim_{a \to \infty} \frac{1}{\sqrt{2\pi}} \int_{-a}^{a} g(\omega)e^{i\omega x}\, d\omega.$$

**THEOREM 1
INVERSION OF THE
FOURIER
TRANSFORM**

Suppose that f is piecewise smooth and integrable on the real line. Then, for all x, we have

$$\frac{f(x+) + f(x-)}{2} = \mathcal{F}^{-1}(\widehat{f})(x) = \frac{1}{\sqrt{2\pi}} \int_{-\infty}^{\infty} \widehat{f}(\omega)e^{i\omega x}\, d\omega.$$

In particular, if f is piecewise smooth and continuous at x, then

$$f(x) = \frac{1}{\sqrt{2\pi}} \int_{-\infty}^{\infty} \widehat{f}(\omega)e^{i\omega x}\, d\omega.$$

The proof is similar to the proof of the Fourier series representation theorem, Section 7.6. We will present it at the end of this section.

Theorem 1 asserts that f can be recaptured from its Fourier transform, by performing an inverse Fourier transform. This makes the Fourier transform a very powerful tool, as it will become apparent from the methods and applications in Sections 11.2 and 11.3.

Let us now consider some examples and basic properties of the Fourier transform. First note that putting $\omega = 0$ in (1), we find that

$$\boxed{\widehat{f}(0) = \frac{1}{\sqrt{2\pi}} \int_{-\infty}^{\infty} f(x)\, dx\,.}$$

Thus the value of the Fourier transform at $\omega = 0$ is equal to the signed area between the graph of $f(x)$ and the x-axis, multiplied by a factor of $1/\sqrt{2\pi}$.

EXAMPLE 1   A Fourier transform
(a) Let $a > 0$. Find the Fourier transform of the function

$$f(x) = \begin{cases} 1 & \text{if } |x| < a, \\ 0 & \text{if } |x| > a, \end{cases}$$

shown in Figure 1. What is $\widehat{f}(0)$?
(b) Express f as an inverse Fourier transform.

Solution   For $\omega \neq 0$ we have

$$\begin{aligned}
\widehat{f}(\omega) &= \frac{1}{\sqrt{2\pi}} \int_{-\infty}^{\infty} f(x)e^{-i\omega x}\, dx = \frac{1}{\sqrt{2\pi}} \int_{-a}^{a} e^{-i\omega x}\, dx \\
&= \frac{-1}{\sqrt{2\pi}i\omega} e^{-i\omega x}\Big|_{-a}^{a} = \sqrt{\frac{2}{\pi}}\, \frac{\sin a\omega}{\omega}.
\end{aligned}$$

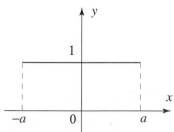

Figure 1 Graph of f in Example 1.

For $\omega = 0$, we have

$$\widehat{f}(0) = \frac{1}{\sqrt{2\pi}} \int_{-a}^{a} dx = a\sqrt{2/\pi}\,.$$

Since

$$\lim_{\omega \to 0} \widehat{f}(\omega) = \lim_{\omega \to 0} \sqrt{\frac{2}{\pi}} \frac{\sin a\omega}{\omega} = a\sqrt{\frac{2}{\pi}} = \widehat{f}(0),$$

it follows that $\widehat{f}(\omega)$ is continuous at 0 (Figure 2), and we may write

$$\widehat{f}(\omega) = \sqrt{\frac{2}{\pi}} \frac{\sin a\omega}{\omega}, \qquad \text{for all } \omega.$$

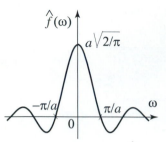

$\widehat{f}(\omega)$

$a\sqrt{2/\pi}$

$-\pi/a$ π/a ω

0

Figure 2 Graph of $\widehat{f}$ in Example 1.

(b) To express f as an inverse Fourier transform, we use (2) and get

$$
\begin{aligned}
f(x) &= \frac{1}{\sqrt{2\pi}} \int_{-\infty}^{\infty} e^{i\omega x} \sqrt{\frac{2}{\pi}} \frac{\sin a\omega}{\omega}\, d\omega = \frac{1}{\pi} \int_{-\infty}^{\infty} e^{i\omega x} \frac{\sin a\omega}{\omega}\, d\omega \\
&= \frac{1}{\pi} \int_{-\infty}^{\infty} (\cos \omega x + i \sin \omega x) \frac{\sin a\omega}{\omega}\, d\omega = \frac{1}{\pi} \int_{-\infty}^{\infty} \frac{\cos \omega x \sin a\omega}{\omega}\, d\omega,
\end{aligned}
$$

because $\sin \omega x \frac{\sin a\omega}{\omega}$ is an odd function of ω and so its integral is zero. This representation is valid at the points of continuity of f; that is, for $x \neq \pm a$. For $x = a$, the inverse Fourier transform converges to the average value $\frac{f(a+)+f(a-)}{2} = \frac{1}{2}$. Similarly, for $x = -a$, the inverse Fourier transform converges to $\frac{1}{2}$. Putting these facts together and writing f explicitly, we obtain the inverse Fourier transform representation of f:

$$(3) \qquad \frac{1}{\pi} \int_{-\infty}^{\infty} \frac{\cos \omega x \sin a\omega}{\omega}\, d\omega = \begin{cases} 0 & \text{if } x < -a \text{ or } x > a, \\ 1 & \text{if } -a < x < a, \\ \frac{1}{2} & \text{if } x = \pm a. \end{cases}$$

A particular case of this integral deserves special attention. If we take $a = 1$ and $x = 0$, we get

$$\frac{1}{\pi} \int_{-\infty}^{\infty} \frac{\sin \omega}{\omega}\, d\omega = 1.$$

This is the sine integral that we computed using residues in Example 3, Section 5.4. In recognition of the residue methods, we should mention that this integral is needed in the proof of Theorem 1, which we quoted when computing (3). So we still need the residue theory in deriving the sine integral. ∎

The Fourier transform in Example 1 is continuous on the entire real line even though the function has jump discontinuities at $x = \pm a$. In fact, it can be shown that the Fourier transform of an integrable function is *always* continuous.

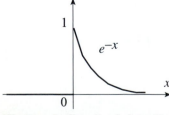

1

e^{-x}

x

0

Figure 3 Graph of f in Example 2, for the case $a = 1$.

EXAMPLE 2   A Fourier transform involving e^{-ax}

Find the Fourier transform of

$$f(x) = \begin{cases} e^{-ax} & \text{if } x > 0, \\ 0 & \text{if } x \leq 0, \end{cases}$$

where $a > 0$. The graph of f is shown in Figure 3 for the case $a = 1$.

Solution We have

$$\widehat{f}(\omega) \;=\; \frac{1}{\sqrt{2\pi}}\int_0^\infty e^{-ax}e^{-i\omega x}\,dx = \frac{1}{\sqrt{2\pi}}\int_0^\infty e^{-x(a+i\omega)}\,dx$$

$$=\; \frac{-1}{\sqrt{2\pi}(a+i\omega)}e^{-i\omega x}e^{-ax}\bigg|_0^\infty .$$

Since $\big|e^{-ix\omega}\big| = 1$, it follows that $\lim_{x\to\infty}\big|e^{-x(a+i\omega)}\big| = \lim_{x\to\infty}e^{-ax} = 0$, and so

$$\widehat{f}(\omega) = \frac{1}{\sqrt{2\pi}(a+i\omega)} = \frac{a - i\omega}{\sqrt{2\pi}(a^2 + \omega^2)}\, .$$

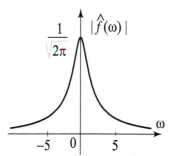

Figure 4 Graph of $|\widehat{f}|$ in Example 2 ($a = 1$).

Figure 4 shows the graph of $|\widehat{f}(\omega)|$ with $a = 1$. Here again, it is worth noting that $\widehat{f}$ and $|\widehat{f}|$ are both continuous even though f is not.   ∎

Example 2 illustrates a noteworthy fact that the Fourier transform may be complex-valued even though the function is real-valued. When is the Fourier transform real-valued? To answer this question we investigate the Fourier transform of $f(-x)$. We have the following useful result.

THEOREM 2
CONJUGATING THE
TRANSFORM

Suppose that f is real-valued and integrable and let $g(x) = f(-x)$. Then

$$\widehat{f}(-\omega) = \overline{\widehat{f}(\omega)} = \widehat{g}(\omega), \qquad \text{for all } \omega.$$

Proof We leave the first equality as an exercise and prove that $\widehat{f}(-\omega) = \widehat{g}(\omega)$. Using (1) and a change of variables, we get

$$\widehat{g}(\omega) \;=\; \frac{1}{\sqrt{2\pi}}\int_{-\infty}^\infty g(x)e^{-i\omega x}\,dx = \frac{1}{\sqrt{2\pi}}\int_{-\infty}^\infty g(-x)e^{i\omega x}\,dx$$

$$=\; \frac{1}{\sqrt{2\pi}}\int_{-\infty}^\infty f(x)e^{-i(-\omega)x}\,dx = \widehat{f}(-\omega).$$ ∎

The following property of the Fourier transform is straightforward to verify and is a consequence of the linearity of the integral.

THEOREM 3
LINEARITY

The Fourier transform is a linear operation; that is, for any integrable functions f and g and any real numbers a and b,

$$\mathcal{F}(af + bg) = a\mathcal{F}(f) + b\mathcal{F}(g)\,.$$

Proof See Exercise 14.   ∎

We can now answer our question about the values of the Fourier transform.

THEOREM 4
VALUES OF THE
FOURIER
TRANSFORM

Suppose that f is real-valued and integrable. Then
(i) $\widehat{f}$ is real-valued if and only if f is even;
(ii) $\widehat{f}$ is purely imaginary if and only if f is odd.

Proof We prove (i) only and leave (ii) to Exercise 14. We have

$$f \text{ is even} \quad \Leftrightarrow \quad f(x) = \frac{f(x) + f(-x)}{2}$$

$$\Leftrightarrow \quad \widehat{f}(\omega) = \frac{\widehat{f}(\omega) + \widehat{f(-x)}(\omega)}{2} \quad \text{(linearity)}$$

$$\Leftrightarrow \quad \widehat{f}(\omega) = \frac{\widehat{f}(\omega) + \overline{\widehat{f}(\omega)}}{2} \quad \text{(Theorem 2)}$$

$$\Leftrightarrow \quad \widehat{f}(\omega) = \mathrm{Re}\left(\widehat{f}(\omega)\right) \quad \Leftrightarrow \quad \widehat{f}(\omega) \text{ is real-valued.} \quad \blacksquare$$

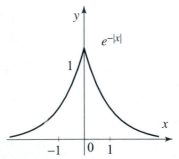

$e^{-|x|}$

Figure 5 Graph of $e^{-|x|}$, $a = 1$ in Example 3.

EXAMPLE 3   Fourier transform of $e^{-a|x|}$ $(a > 0)$

Find the Fourier transform of $h(x) = e^{-a|x|}$ where $a > 0$ (Figure 5).

Solution We can compute directly from the definition of the Fourier transform, or better yet, we can use the transform in Example 2 and properties of the Fourier transform. Since $h(x) = f(x) + f(-x)$, where $f(x)$ is as in Example 2, then by Theorem 2 and linearity,

$$\widehat{h}(\omega) \;=\; \widehat{f}(\omega) + \overline{\widehat{f}(\omega)}$$

$$=\; \frac{a - i\omega}{\sqrt{2\pi}(a^2 + \omega^2)} + \frac{a + i\omega}{\sqrt{2\pi}(a^2 + \omega^2)} = \sqrt{\frac{2}{\pi}}\frac{a}{a^2 + \omega^2}.$$

In accordance with Theorem 4, the Fourier transform of $e^{-a|x|}$ is real-valued, since $e^{-a|x|}$ is even. $\blacksquare$

Example 3 illustrates how properties of the Fourier transform can be used to our advantage to find new transforms from known ones. The more we know about the transform, the easier it is to compute with it. Here are two simple observations which will yield interesting results.

THEOREM 5
RECIPROCITY
RELATIONS

If f and g are integrable, then

(4) $$\mathcal{F}g(\omega) = \mathcal{F}^{-1}g(-\omega);$$

(5) $$\mathcal{F}(\mathcal{F}(f))(\omega) = f(-\omega).$$

Proof In the definitions of the Fourier transform and its inverse, (1) and (2), the exponential functions in these integrals differ only by a negative sign in the exponent, and so (4) is immediate. Applying (4) with $g = \mathcal{F}(f)$, we obtain $\mathcal{F}\mathcal{F}(f)(\omega) = \mathcal{F}^{-1}\mathcal{F}(f)(-\omega)$. But $\mathcal{F}^{-1}\mathcal{F}(f) = f$, and (5) follows. $\blacksquare$

EXAMPLE 4   A transform using reciprocity

From Example 3, the function $\sqrt{\frac{2}{\pi}}\frac{a}{a^2+x^2}$ $(a > 0)$ is the Fourier transform of $f(x) = e^{-a|x|}$. Using the reciprocity relation (5), we obtain

$$\mathcal{F}\left(\sqrt{\frac{2}{\pi}}\frac{a}{a^2+x^2}\right)(\omega) = \mathcal{F}\left(\mathcal{F}\left(e^{-a|x|}\right)\right)(\omega) = f(-\omega) = e^{-a|-\omega|} = e^{-a|\omega|}. \quad \blacksquare$$

If in Example 4 we were to compute the Fourier transform of $\frac{a}{a^2+x^2}$ from the definition (1), we would have to evaluate the integral $\int_{-\infty}^{\infty}\frac{\cos\omega x}{a^2+x^2}\,dx$. This nontrivial integral is one of many that we computed using the residue theorem in Chapter 5 (see Example 1, Section 5.4). The next example is another important Fourier transform, which we will obtain by simply recalling an integral that we computed using residues. A different derivation will be given in the next section using operational properties of the transform.

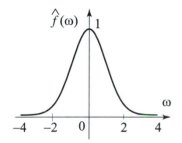

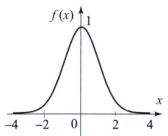

Figure 6 Graphs of $e^{-\frac{x^2}{2}}$ and its Fourier transform $e^{-\frac{\omega^2}{2}}$. They are the same functions (of different variables).

EXAMPLE 5   Fourier transform of the Gaussian

Let $a > 0$. We will derive the Fourier transform

(6)
$$\boxed{\mathcal{F}\left(e^{-ax^2}\right)(\omega) = \frac{1}{\sqrt{2a}}e^{-\frac{\omega^2}{4a}} \quad (-\infty < \omega < \infty).}$$

(See Figure 6 for the case $a = \frac{1}{2}$.) Using (1), we have

$$\mathcal{F}\left(e^{-ax^2}\right)(\omega) = \frac{1}{2\pi}\int_{-\infty}^{\infty}e^{-ax^2}e^{-i\omega x}\,dx = \frac{1}{2\pi}\int_{-\infty}^{\infty}e^{-ax^2}(\cos\omega x - i\sin\omega x)\,dx.$$

The integral of the sine part is 0, because the integrand is odd. So

$$\mathcal{F}\left(e^{-ax^2}\right)(\omega) = \frac{1}{2\pi}\int_{-\infty}^{\infty}e^{-ax^2}\cos\omega x\,dx,$$

and (6) follows from Example 1, Section 5.5. $\quad \blacksquare$

Taking $a = \frac{1}{2}$ in (6), we obtain the interesting formula

(7)
$$\boxed{\mathcal{F}\left(e^{-\frac{x^2}{2}}\right)(\omega) = e^{-\frac{\omega^2}{2}} \quad (-\infty < \omega < \infty).}$$

Thus $e^{-\frac{x^2}{2}}$ is its own Fourier transform; equivalently, formula (7) states that $e^{-\frac{x^2}{2}}$ is an eigenfunction of the Fourier transform corresponding to the eigenvalue 1. (See the next section for transforms with similar properties.)

In our next example, the function is not integrable; however, its Fourier transform does exist. This is one of many Fourier transforms of functions that are not necessarily integrable, but for which the improper integral (1)

does converge. In fact, the theory of Fourier transforms can be extended to a much wider class of functions than the integrable functions, including the so-called generalized functions. We will touch on this subject in the next section and derive some interesting applications.

EXAMPLE 6   Fourier transform of $\sqrt{\frac{2}{\pi}}\frac{\sin ax}{x}$

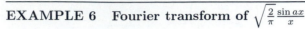

The function $f(x) = \sqrt{\frac{2}{\pi}}\frac{\sin ax}{x}$, where $a > 0$ is not integrable over the real line because the improper integral of its absolute value is not finite (Exercise 14(c)). We can still use (1) to compute its Fourier transform. For any $-\infty < \omega < \infty$, we have

$$
\begin{aligned}
\widehat{f}(\omega) &= \frac{1}{\pi}\int_{-\infty}^{\infty}\frac{\sin ax}{x}e^{-i\omega x}\,dx = \frac{1}{\pi}\int_{-\infty}^{\infty}\frac{\sin ax}{x}(\cos\omega x - i\sin\omega x)\,dx \\
&= \frac{1}{\pi}\int_{-\infty}^{\infty}\frac{\sin ax}{x}\cos\omega x\,dx,
\end{aligned}
$$

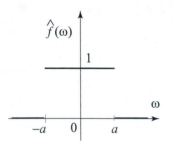

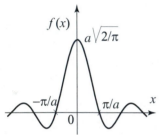

Figure 7 for Example 6. Graphs of $f(x) = \sqrt{\frac{2}{\pi}}\frac{\sin ax}{x}$ and its Fourier transform $\widehat{f}(\omega) = 1$ if $|\omega| < a$ and 0 if $|\omega| > a$. The function is not integrable and its transform is not continuous.

because the imaginary part of the integrand is odd so its integral is zero. To evaluate the last integral, we appeal to (3) and find

$$
\widehat{f}(\omega) = \begin{cases} 0 & \text{if } x < -a \text{ or } x > a, \\ 1 & \text{if } -a < x < a, \\ \frac{1}{2} & \text{if } x = \pm a. \end{cases}
$$

(See Figure 7.) These values can be confirmed by using the reciprocity relations. From Example 1, the function $\sqrt{\frac{2}{\pi}}\frac{\sin ax}{x}$ is the Fourier transform of $g(x) = 1$ if $|x| < a$, and $g(x) = 0$ if $|x| > a$. By (5), we get

$$
\mathcal{F}\left(\sqrt{\frac{2}{\pi}}\frac{\sin ax}{x}\right)(\omega) = \mathcal{F}(\mathcal{F}(g))(\omega) = g(-\omega) = \begin{cases} 1 & \text{if } |\omega| < a, \\ 0 & \text{if } |\omega| > a. \end{cases}
$$

We observed earlier, before Example 2, that the Fourier transform of an integrable function is always continuous. In the present example, the Fourier transform is not continuous. This is not a contradiction, because the function in this example is not integrable.   ■

In our final example we compute some residues as we evaluate a Fourier transform.

EXAMPLE 7   Using residues to compute a Fourier transform

Compute the Fourier transform of $f(x) = \dfrac{2}{\sqrt{\pi}(1+x^4)}$.

Solution Using (1), we compute

$$
\begin{aligned}
\widehat{f}(\omega) &= \frac{\sqrt{2}}{\pi}\int_{-\infty}^{\infty}\frac{1}{1+x^4}(\cos\omega x - i\sin\omega x)\,dx \\
&= \frac{\sqrt{2}}{\pi}\int_{-\infty}^{\infty}\frac{\cos\omega x}{1+x^4}\,dx.
\end{aligned}
$$

Note that the Fourier transform is an even function of ω, so it is enough to compute it for $\omega \geq 0$. So suppose $\omega \geq 0$ and follow the steps in Example 1, Section 5.4, to compute the integral. Accordingly, we have

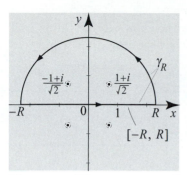

(8)
$$\widehat{f}(\omega) = \frac{\sqrt{2}}{\pi} \int_{-\infty}^{\infty} \frac{\cos \omega x}{1 + x^4} \, dx = \frac{\sqrt{2}}{\pi} \operatorname{Re} \left(\int_{\gamma_R} \frac{e^{i\omega z}}{1 + z^4} \, dz \right),$$

Figure 8 The contour and poles in the upper half-plane in Example 7.

where γ_R is a closed semi-circular path with R large enough to enclose the poles of $\frac{e^{i\omega z}}{1+z^4}$ (Figure 8). (As we will see shortly, R must be > 1.) The function $F(z) = \frac{e^{i\omega z}}{1+z^4}$ has four simple poles at the roots of $z^4 + 1 = 0$. These are $z_1 = e^{i\frac{\pi}{4}}$, $z_2 = e^{i\frac{3\pi}{4}}$, $z_3 = e^{i\frac{5\pi}{4}}$, $z_4 = e^{i\frac{7\pi}{4}}$ (Figure 9). Only z_1 and z_2 are in the upper half-plane and so lie inside of γ_R for $R > 1$. To compute the residues at these poles, we will use Proposition 1(ii), Section 5.1. Accordingly,

$$\operatorname{Res}\left(\frac{e^{i\omega z}}{1 + z^4}, z_j\right) = \left.\frac{e^{i\omega z}}{4z^3}\right|_{z=z_j} = \frac{e^{i\omega z_j}}{4z_j^3}.$$

We now have every thing we need to compute the integral with the help of the residue theorem. We have

$$\int_{\gamma_R} \frac{e^{i\omega z}}{1 + z^4} \, dz = 2\pi i \left(\operatorname{Res}(F(z), z_1) + \operatorname{Res}(F(z), z_2) \right) = 2\pi i \left(\frac{e^{i\omega z_1}}{4z_1^3} + \frac{e^{i\omega z_2}}{4z_2^3} \right).$$

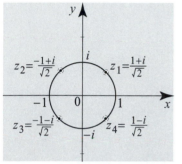

Figure 9 The four unimodular roots of $z^4 + 1 = 0$.

We now plug back into (8), use $z_1 = \frac{\sqrt{2}}{2} + i\frac{\sqrt{2}}{2}$, $z_2 = -\frac{\sqrt{2}}{2} + i\frac{\sqrt{2}}{2}$, $\operatorname{Re}(i\zeta) = -\operatorname{Im}(\zeta)$ for any complex number ζ, simplify, and get, for $\omega \geq 0$,

$$
\begin{aligned}
\widehat{f}(\omega) &= -\frac{\sqrt{2}}{2} \operatorname{Im}\left(\frac{e^{i\omega z_1}}{z_1^3} + \frac{e^{i\omega z_2}}{z_2^3} \right) \\
&= -\frac{\sqrt{2}}{2} \operatorname{Im}\left(e^{-i\frac{3\pi}{4}} \left(e^{i\omega(\frac{\sqrt{2}}{2}+i\frac{\sqrt{2}}{2})} + i e^{i\omega(-\frac{\sqrt{2}}{2}+i\frac{\sqrt{2}}{2})} \right) \right) \\
&= e^{-\frac{\sqrt{2}}{2}\omega} \left(\sin\left(\frac{\sqrt{2}}{2}\omega\right) + \cos\left(\frac{\sqrt{2}}{2}\omega\right) \right).
\end{aligned}
$$

Thus for all ω

$$\widehat{f}(\omega) = e^{-\frac{\sqrt{2}}{2}|\omega|} \left(\sin\left(\frac{\sqrt{2}}{2}|\omega|\right) + \cos\left(\frac{\sqrt{2}}{2}\omega\right) \right).$$

Setting $\omega = 0$ and recalling the definition of the Fourier transform, we get

$$\widehat{f}(0) = 1 = \frac{\sqrt{2}}{\pi} \int_{-\infty}^{\infty} \frac{dx}{1 + x^4},$$

which yields the value of an interesting nontrivial improper integral. ∎

Proof of Theorem 1

We need two lemmas. The first one is similar to the expression of the partial sums of Fourier series as a convolution with the Dirichlet kernel.

**LEMMA 1
PARTIAL INVERSE
FOURIER
TRANSFORM**

Suppose that f is integrable on the real line and $a > 0$. Then

$$(9) \qquad \frac{1}{\sqrt{2\pi}} \int_{-a}^{a} \widehat{f}(\omega) e^{ix\omega} \, d\omega = \frac{1}{\pi} \int_{-\infty}^{\infty} f(x-y) \frac{\sin ay}{y} \, dy.$$

Proof Using $\frac{\sin ay}{y} = \frac{1}{2} \int_{-a}^{a} e^{iy\omega} \, d\omega$, we have

$$
\begin{aligned}
\sqrt{\frac{2}{\pi}} \int_{-\infty}^{\infty} f(x-y) \frac{\sin ay}{y} \, dy &= \sqrt{\frac{2}{\pi}} \int_{-\infty}^{\infty} \frac{1}{2} f(x-y) \int_{-a}^{a} e^{iy\omega} \, d\omega \, dy \\
&= \frac{1}{\sqrt{2\pi}} \int_{-a}^{a} \int_{-\infty}^{\infty} f(x-y) e^{iy\omega} \, dy \, d\omega \\
&= \int_{-a}^{a} \overbrace{\frac{1}{\sqrt{2\pi}} \int_{-\infty}^{\infty} f(y) e^{-iy\omega} \, dy}^{=\widehat{f}(\omega)} e^{ix\omega} \, d\omega \\
&= \int_{-a}^{a} \widehat{f}(\omega) e^{ix\omega} \, d\omega,
\end{aligned}
$$

To go from the first to the second equality, interchange the order of integration. To go from the second to the third equality, change variables: $x - y = Y$, $dy = -dY$.

which is equivalent to (9). ∎

The next result is an analog of the Riemann-Lebesgue lemma for Fourier series.

**LEMMA 2
RIEMANN-LEBESGUE
LEMMA**

Suppose that $\int_{A}^{B} |g(x)| \, dx < \infty$, where $-\infty \leq A < B \leq \infty$. Then

$$(10) \qquad \lim_{\omega \to \infty} \int_{A}^{B} g(y) \sin \omega y \, dy = 0 \quad \text{and} \quad \lim_{\omega \to \infty} \int_{A}^{B} g(y) \cos \omega y \, dy = 0.$$

Proof To simplify the proof, we will suppose that g is bounded and g' is integrable. Integrating by parts, we obtain

$$\int_{A}^{B} g(y) \sin \omega y \, dy = -\frac{1}{\omega} g(y) \cos \omega y \Big|_{A}^{B} + \frac{1}{\omega} \int_{A}^{B} g'(y) \cos \omega y \, dy.$$

Since g is bounded, then so is $g(y) \cos \omega y$, and the first term on the right tends to 0 as $\omega \to \infty$. Also, since g' is integrable, we have $\int_{A}^{B} |g'(y)| \, dy = M < \infty$, and so

$$\frac{1}{\omega} \left| \int_{A}^{B} g'(y) \cos \omega y \, dy \right| \leq \frac{M}{\omega} \to 0 \text{ as } \omega \to \infty.$$

This proves the first limit in (10). The second one is done similarly. ∎

We now sketch a proof of Theorem 1 for the case when f is smooth. For fixed x, since $f'(x)$ exists, the function $g(y) = \frac{f(x-y)-f(x)}{y}$ tends to $f'(x)$ as $y \to 0$; hence it is bounded near 0. For all other values of y, the function g is smooth. Recall from Example 3, Section 5.4, the integral $\frac{1}{\pi}\int_{-\infty}^{\infty}\frac{\sin x}{x}\,dx = 1$. From this integral, it follows that for any $a > 0$, $\frac{1}{\pi}\int_{-\infty}^{\infty}\frac{\sin ay}{y}\,dy = 1$. (Just make a change of variables $ay = x$.) Using Lemma 1, we have

$$
\begin{aligned}
\frac{1}{\sqrt{2\pi}}\int_{-a}^{a}\widehat{f}(\omega)e^{ix\omega}\,d\omega - f(x) &= \frac{1}{\pi}\int_{-\infty}^{\infty}f(x-y)\frac{\sin ay}{y}\,dy - f(x) \\
&= \frac{1}{\pi}\int_{-\infty}^{\infty}\frac{f(x-y) - f(x)}{y}\sin ay\,dy \\
&= \frac{1}{\pi}\int_{-\infty}^{\infty}g(y)\sin ay\,dy = I_1 + I_2 + I_3,
\end{aligned}
$$

where I_1 is the integral over $(-\infty, 1)$, I_2 is the integral over $(-1, 1)$, and I_3 is the integral over $(1, \infty)$. To show that

$$
\lim_{a\to\infty}\frac{1}{\sqrt{2\pi}}\int_{-a}^{a}\widehat{f}(\omega)e^{ix\omega}\,d\omega - f(x) = 0,
$$

we will show that $\lim_{a\to\infty} I_j = 0$ for $j = 1, 2, 3$. Since g is bounded in the interval $(-1, 1)$, the integral of its absolute value is finite on $(-1, 1)$, and so, by Lemma 2, $\lim_{a\to\infty} I_2 = 0$. Write $I_3 = \int_1^\infty \frac{f(x-y)}{y}\sin ay\,dy - f(x)\int_1^\infty\frac{\sin ay}{y}\,dy$. The first integral tends to 0 as $a \to \infty$, by Lemma 2, because $\int_1^\infty\left|\frac{f(x-y)}{y}\right|\,dy \le \int_1^\infty|f(x-y)|\,dy \le \int_{-\infty}^{\infty}|f(y)|\,dy < \infty$. To handle the second integral, we integrate by parts and get

$$
f(x)\int_1^\infty\frac{\sin ay}{y}\,dy = f(x)\frac{-\cos ay}{ay}\Big|_{y=1}^{\infty} - \frac{1}{a}\int_1^\infty\frac{\cos ay}{y^2}\,dy,
$$

which tends to 0 as $a \to \infty$, because $\left|\int_1^\infty\frac{\cos ay}{y^2}\,dy\right| \le \int_1^\infty\frac{1}{y^2}\,dy = 1$. Thus $\lim_{a\to\infty} I_3 = 0$ and similarly for I_1, which completes the proof. ∎

Exercises 11.1

In Exercises 1–10, find the Fourier transform of the given function.

1.

$$f(x) = \begin{cases} x & \text{if } -1 < x < 1, \\ 0 & \text{otherwise.} \end{cases}$$

2.

$$f(x) = \begin{cases} -1 & \text{if } -1 < x < 0, \\ 1 & \text{if } 0 < x < 1, \\ 0 & \text{otherwise.} \end{cases}$$

3.

$$f(x) = \begin{cases} 1 & \text{if } 0 < x < 1, \\ 0 & \text{otherwise.} \end{cases}$$

4.

$$f(x) = \begin{cases} 0 & \text{if } -1 < x < 1, \\ 1 & \text{if } 1 < |x| < 2, \\ 0 & \text{otherwise.} \end{cases}$$

5.

$$f(x) = \begin{cases} 1 - |x| & \text{if } -1 < x < 1, \\ 0 & \text{otherwise.} \end{cases}$$

6.

$$f(x) = \begin{cases} 1 - x^2 & \text{if } -1 < x < 1, \\ 0 & \text{otherwise.} \end{cases}$$

7.

$$f(x) = e^{-|x|+2}.$$

8.

$$f(x) = e^{-2(x^2+1)}.$$

9.

$$f(x) = \begin{cases} \cos x & \text{if } -\pi/2 < x < \pi/2, \\ 0 & \text{otherwise.} \end{cases}$$

10.

$$f(x) = \begin{cases} \sin x & \text{if } -\pi < x < \pi, \\ 0 & \text{otherwise.} \end{cases}$$

In Exercises 11–13, derive the given integral formula by first taking the Fourier transform of the function on the right of the equality and then expressing the function as an inverse Fourier transform.

11.

$$\int_0^\infty \frac{\cos x\omega + \omega \sin x\omega}{1 + \omega^2}\, d\omega = \begin{cases} 0 & \text{if } x < 0, \\ \pi/2 & \text{if } x = 0, \\ \pi e^{-x} & \text{if } x > 0. \end{cases}$$

12.

$$\frac{2}{\pi} \int_0^\infty \frac{\sin \pi\omega}{1 - \omega^2} \sin \omega x\, d\omega = \begin{cases} \sin x & \text{if } |x| \le \pi, \\ 0 & \text{if } |x| > \pi. \end{cases}$$

13.

$$\frac{2}{\pi} \int_0^\infty \frac{\cos \frac{\pi\omega}{2}}{1 - \omega^2} \cos \omega x\, d\omega = \begin{cases} \cos x & \text{if } |x| < \pi/2, \\ 0 & \text{if } |x| > \pi/2. \end{cases}$$

14. (a) Prove Theorem 3.   (b) Prove Theorem 4(ii).
(c) Show that $\frac{\sin x}{x}$ is not integrable on the real line. That is, show that $\int_{-\infty}^\infty \left|\frac{\sin x}{x}\right| dx = \infty$. [Hint: Show that for $n \ge 2$, the area under the arch of the curve, above the x-axis, between $(n-1)\pi$ and $n\pi$ is greater than $1/(n-1)$ (Figure 10).]

15. (a) Use Example 1 to show that

$$\int_0^\infty \frac{\sin \omega \cos \omega}{\omega}\, d\omega = \frac{\pi}{4}.$$

(b) Use integration by parts and (a) to obtain

$$\int_0^\infty \frac{\sin^2 \omega}{\omega^2}\, d\omega = \frac{\pi}{2}.$$

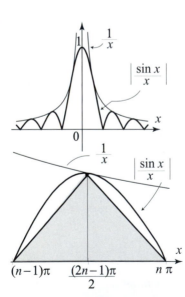

Figure 10 for Exercise 14(c). The graph of $\left|\frac{\sin x}{x}\right|$ and the area under one arch between $x = (n-1)\pi$ and $x = n\pi$. Show that the shaded area is $\ge \frac{1}{n-1}$.

16. Derive the given identities: (a) $\int_0^\infty \dfrac{\sin \omega \cos 2\omega}{\omega}\, d\omega = 0$. [Hint: Example 1.]

(b) $\int_0^\infty \dfrac{1 - \cos t}{t^2}\, dt = \dfrac{\pi}{2}$. [Hint: Integrate by parts and use Example 1.]

Methods from Complex Analysis

17. (a) Let $0 < \alpha < 1$ and define $f(x) = \frac{1}{|x|^\alpha}$. Using Example 2, Section 5.5, derive the Fourier transform

$$\mathcal{F}\left(\frac{1}{|x|^\alpha} \right)(\omega) = \sqrt{\frac{2}{\pi}} |\omega|^{\alpha - 1} \Gamma(1 - \alpha) \sin \frac{\alpha \pi}{2} \quad (-\infty < \omega < \infty).$$

(b) From (a), derive the Fourier transform

$$\mathcal{F}\left(\frac{1}{\sqrt{|x|}} \right)(\omega) = \frac{1}{\sqrt{|\omega|}} \quad (-\infty < \omega < \infty).$$

18. Project Problem: Fourier transform of the Bessel function $J_0(x)$. In this exercise, we will derive the formula

$$(11) \qquad \mathcal{F}(J_0(ax))(\omega) = \begin{cases} \sqrt{\dfrac{2}{\pi}} \dfrac{1}{\sqrt{a^2 - \omega^2}} & \text{if } |\omega| < a, \\[2ex] 0 & \text{if } |\omega| > a, \end{cases}$$

where $a > 0$ (Figure 11). Instead of computing directly, we will compute the Fourier transform of the function on the right side of the equality and then use the reciprocity relation (5). So let $f(x) = \sqrt{\frac{2}{\pi}} \frac{1}{\sqrt{a^2 - x^2}}$ if $|x| < a$ and $f(x) = 0$ if $|x| > a$.
(a) Show that $\widehat{f}(\omega) = \frac{1}{\pi} \int_0^\pi \cos(a\omega \cos \theta)\, d\theta$. [Hint: Apply the definition (1), use the change of variables $x = \cos \theta$, then use one more change of variables $\theta = \theta + \pi$.]
(b) Show that $\widehat{f}(\omega) = J_0(a\omega)$ and then use Theorem 5 to derive (11). [Hint: The first part is nontrivial, but it follows immediately from the cosine integral representation of the Bessel function J_0 (Exercise 36(a), Section 4.5).]

In Exercises 19–23, use residues, as we did in Example 7, to derive the given Fourier transform.

19. $\mathcal{F}\left(\dfrac{\sqrt{2}}{\sqrt{\pi}(1 + x^6)} \right)(\omega) = \dfrac{e^{-\frac{|\omega|}{2}}}{3} \left(e^{-\frac{|\omega|}{2}} + \cos\left(\dfrac{\sqrt{3}}{2}\omega \right) + \sqrt{3} \sin\left(\dfrac{\sqrt{3}}{2}|\omega| \right) \right).$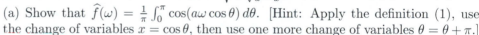

20. $\mathcal{F}\left(\dfrac{\sqrt{2}}{\sqrt{\pi}(1 + x^2)^2} \right)(\omega) = \dfrac{e^{-|\omega|}}{2}(1 + |\omega|).$

21. $\mathcal{F}\left(\dfrac{\sqrt{2}\, x}{\sqrt{\pi}(1 + x^2)^2} \right)(\omega) = \dfrac{i}{2}|\omega| e^{-|\omega|}.$

22. $\mathcal{F}\left(\dfrac{\sqrt{3}}{\sqrt{2\pi}(1 + x + x^2)} \right)(\omega) = e^{-\frac{\sqrt{3}}{2}|\omega|} \left(\cos\left(\dfrac{\omega}{2} \right) - i \sin\left(\dfrac{|\omega|}{2} \right) \right).$

23. $\mathcal{F}\left(\dfrac{\sqrt{2}}{\sqrt{\pi} \cosh x} \right)(\omega) = \operatorname{sech} \dfrac{\pi \omega}{2}.$

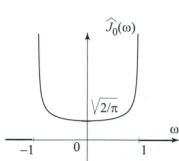

$\widehat{J_0}(\omega)$

$\sqrt{2/\pi}$

ω

$-1 \quad 0 \quad 1$

Figure 11 Graph of $\mathcal{F}(J_0(x))(\omega)$, for Exercise 18. We have

$$\mathcal{F}(J_0(x))(0) = \sqrt{\frac{2}{\pi}},$$

which implies that

$$\int_{-\infty}^\infty J_0(x)\, dx = 2.$$

But J_0 is not integrable on the real line, because the graph of its Fourier transform, $\mathcal{F}(J_0(x))(\omega)$, is not continuous.

24. Project Problem: Transforms related to the Fresnel integrals.
In this exercise, we will derive the Fourier transforms

$$(12) \qquad \mathcal{F}\left(\cos(x^2)\right)(\omega) \;=\; \frac{1}{2}\left(\cos\frac{\omega^2}{4} + \sin\frac{\omega^2}{4}\right);$$

$$(13) \qquad \mathcal{F}\left(\sin(x^2)\right)(\omega) \;=\; \frac{1}{2}\left(\cos\frac{\omega^2}{4} - \sin\frac{\omega^2}{4}\right).$$

(a) Let $f_1(\omega) = \mathcal{F}\left(\cos(x^2)\right)(\omega)$ and $f_2(\omega) = \mathcal{F}\left(\sin(x^2)\right)(\omega)$. Show that both f_1 and f_2 are real-valued and even and that

$$f_1(\omega) + i f_2(\omega) = \frac{1}{\sqrt{2\pi}}\int_{-\infty}^{\infty} e^{i(x^2 - x\omega)}\,dx = \frac{e^{-i\frac{\omega^2}{4}}}{\sqrt{2\pi}}\int_{-\infty}^{\infty} e^{i(x-\frac{\omega}{2})^2}\,dx.$$

(b) Make a change of variables, $u = x - \frac{\omega}{2}$, and then evaluate the resulting integral using the Fresnel integrals (see the discussion following Example 2, Section 5.5). Take real and imaginary parts of your answer and derive (12) and (13).

11.2   Operational Properties

As we saw in the previous section, the Fourier transform takes a function f and produces a new function $\widehat{f}$, and the inverse transform recovers the original function f from $\widehat{f}$. This process makes of transform pairs a powerful tool in solving partial differential equations. The idea, which will be explored in the following sections, is to "Fourier transform" a given equation into one that may be easier to solve. After solving the transformed equation involving $\widehat{f}$, we recover the solution of the original problem with the inverse transform. To assist us in handling the transformed equations, we will develop the operational properties of the Fourier transform. We start with a result that describes the effect of the Fourier transform on derivatives.

THEOREM 1
FOURIER
TRANSFORMS OF
DERIVATIVES

(i) Suppose $f(x)$ and $f'(x)$ are integrable and $f(x) \to 0$ as $|x| \to \infty$; then

$$\mathcal{F}(f') = i\omega\,\mathcal{F}(f).$$

(ii) If, in addition, $f''(x)$ is integrable and $f'(x) \to 0$ as $|x| \to \infty$, then

$$\mathcal{F}(f'') = i\omega\mathcal{F}(f') = -\omega^2\mathcal{F}(f).$$

(iii) In general, if $f^{(k)}(x) \to 0$ $(k = 1,\, 2,\, \ldots,\, n-1)$ as $|x| \to \infty$, and f and its derivatives of order up to n are integrable, then

$$\mathcal{F}(f^{(n)}) = (i\omega)^n \mathcal{F}(f).$$

Proof Parts (ii) and (iii) are obtained by repeated applications of (i). To prove (i), we use the definition of $\mathcal{F}(f')$ and integrate by parts:

$$
\begin{aligned}
\mathcal{F}(f')(\omega) &= \frac{1}{\sqrt{2\pi}} \int_{-\infty}^{\infty} f'(x)e^{-i\omega x}\,dx \\
&= \frac{1}{\sqrt{2\pi}} \left[f(x)\,e^{-i\omega x}\Big|_{-\infty}^{\infty} - (-i\omega) \int_{-\infty}^{\infty} f(x)e^{-i\omega x}\,dx \right] \\
&= 0 + i\omega \mathcal{F}(f) \quad (\text{since } f(x) \to 0 \text{ as } |x| \to \infty, \text{ and } |e^{\pm i\omega x}| = 1). \ \blacksquare
\end{aligned}
$$

EXAMPLE 1   Fourier transform of a derivative
Since $xe^{-x^2} = -\frac{1}{2}\frac{d}{dx}e^{-x^2}$, it follows from Theorem 1(i) that

$$
\mathcal{F}(xe^{-x^2})(\omega) = -\frac{i\omega}{2}\mathcal{F}(e^{-x^2})(\omega).
$$

Applying (6), Section 11.1, we obtain

$$
\mathcal{F}(xe^{-x^2})(\omega) = -\frac{i\omega}{2\sqrt{2}}e^{-\frac{\omega^2}{4}}. \qquad\qquad \blacksquare
$$

THEOREM 2 DERIVATIVES OF FOURIER TRANSFORMS

(i) Suppose $f(x)$ and $xf(x)$ are integrable, then

$$
\mathcal{F}(xf(x))(\omega) = i\left[\widehat{f}\,\right]'(\omega) = i\frac{d}{d\omega}\mathcal{F}(f)(\omega).
$$

(ii) In general, if $f(x)$ and $x^n f(x)$ are integrable, then

$$
\mathcal{F}(x^n f(x)) = i^n \left[\widehat{f}\,\right]^{(n)}(\omega).
$$

Proof Part (ii) follows from (i). To motivate (i) we will assume that we can differentiate under the integral sign as follows

$$
\begin{aligned}
\left[\widehat{f}\,\right]'(\omega) &= \frac{d}{d\omega}\frac{1}{\sqrt{2\pi}} \int_{-\infty}^{\infty} f(x)e^{-i\omega x}\,dx = \frac{1}{\sqrt{2\pi}} \int_{-\infty}^{\infty} f(x)\frac{d}{d\omega}e^{-i\omega x}\,dx \\
&= -\frac{i}{\sqrt{2\pi}} \int_{-\infty}^{\infty} xf(x)e^{-i\omega x}\,dx = -i\mathcal{F}(xf(x))(\omega),
\end{aligned}
$$

and (i) follows upon multiplying both sides by i. $\blacksquare$

EXAMPLE 2   Derivatives of Fourier transforms
(a) We can derive the Fourier transform in Example 1 by appealing to Theorem 2(i), as follows:

$$
\mathcal{F}(xe^{-x^2})(\omega) = i\frac{d}{d\omega}\mathcal{F}(e^{-x^2})(\omega) = i\frac{d}{d\omega}\left(\frac{1}{\sqrt{2}}e^{-\frac{\omega^2}{4}}\right) = -\frac{i\omega}{2\sqrt{2}}e^{-\frac{\omega^2}{4}}.
$$

(b) Let $a > 0$. To compute the Fourier transform of the function

$$
g(x) = \begin{cases} x & \text{if } |x| < a, \\ 0 & \text{if } |x| > a, \end{cases}
$$

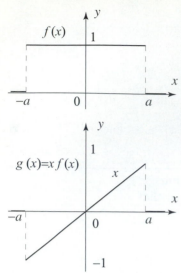

Figure 1 Graphs of $f(x)$ and $g(x) = xf(x)$ in Example 2. The effect of multiplying by $f(x)$ is to truncate the function x for $|x| > a$.

we will use the fact that the Fourier transform of the function

$$f(x) = \begin{cases} 1 & \text{if } |x| < a, \\ 0 & \text{if } |x| > a, \end{cases}$$

is $\mathcal{F}(f)(\omega) = \sqrt{\frac{2}{\pi}} \frac{\sin a\omega}{\omega}$ (Example 1, Section 11.1) and that $g(x) = xf(x)$ (Figure 1). Applying Theorem 2(i), it follows that

$$\mathcal{F}(g)(\omega) = i\frac{d}{d\omega}\sqrt{\frac{2}{\pi}} \frac{\sin a\omega}{\omega} = i\sqrt{\frac{2}{\pi}} \frac{a\omega\cos(a\omega) - \sin a\omega}{\omega^2}. \qquad \blacksquare$$

In Example 5 of the previous section, we used an improper integral that we computed with the help of residue theory to derive the Fourier transform of the Gaussian function. We can now give another interesting indirect derivation based on the operational properties.

EXAMPLE 3   Fourier transform of the Gaussian

Let $f(x) = e^{-ax^2}$, where $a > 0$. A simple verification shows that f satisfies the first order linear differential equation

$$f'(x) + 2axf(x) = 0.$$

Taking Fourier transforms and using Theorems 1 and 2, we get

$$\omega\widehat{f}(\omega) + 2a\frac{d}{d\omega}[\widehat{f}](\omega) = 0.$$

Thus $\widehat{f}$ satisfies a similar first order linear ordinary differential equation. Solving this equation in $\widehat{f}$, we find

$$\widehat{f}(\omega) = A\,e^{-\frac{\omega^2}{4a}},$$

where A is an arbitrary constant. But

$$A = \widehat{f}(0) = \frac{1}{\sqrt{2\pi}}\int_{-\infty}^{\infty} e^{-ax^2}\,dx = \frac{1}{\sqrt{2\pi}}\sqrt{\frac{\pi}{a}} = \frac{1}{\sqrt{2a}},$$

by (1), Section 5.5, and so $\mathcal{F}(e^{-ax^2})(\omega) = \frac{1}{\sqrt{2a}}e^{-\frac{\omega^2}{4a}}$. $\qquad \blacksquare$

One very common operation in analysis is translation or shifting. Given a function $f(x)$ and a real number α, the **translate** of f by α, denoted by $f_\alpha(x)$, is defined by

$$(1) \qquad\qquad f_\alpha(x) = f(x - \alpha) \quad \text{for all } x.$$

Translating a function by α corresponds to multiplying its Fourier transform by $e^{-i\alpha\omega}$. We have the following theorem, whose proof is left as an exercise.

**THEOREM 3
SHIFTING AND
FOURIER
TRANSFORMS**

(i) **Shifting on the x-axis**: Let α be arbitrary, then

$$\mathcal{F}(f(x - \alpha))(\omega) = e^{-i\alpha\omega}\,\widehat{f}(\omega).$$

(ii) **Shifting on the ω-axis**:

$$\mathcal{F}(e^{i\alpha x}f(x))(\omega) = \widehat{f}(\omega - \alpha).$$

Convolution of Functions

We expand our list of operational properties by introducing the convolution of two functions f and g by

CONVOLUTION

(2)
$$f * g(x) = \frac{1}{\sqrt{2\pi}} \int_{-\infty}^{\infty} f(x - t)g(t)\,dt\,.$$

(The factor $\frac{1}{\sqrt{2\pi}}$ is merely for convenience.   If we drop it from the definition of the convolution, it will reappear in its Fourier transform.) It can be shown that $f * g$ is also integrable whenever f and g are (Exercise 23). Computing a convolution is often tedious. This task can be facilitated by using the following important property of the Fourier transform, as illustrated by Example 4 below.

**THEOREM 4
FOURIER
TRANSFORMS OF
CONVOLUTIONS**

Suppose that f and g are integrable, then

$$\mathcal{F}(f * g) = \mathcal{F}(f)\mathcal{F}(g)\,.$$

Theorem 4 is expressed by saying that the Fourier transform takes convolutions into products.

Proof Using the definitions of the Fourier transform and convolutions, and then interchanging the order of integration we get

$$
\begin{aligned}
\mathcal{F}(f * g)(\omega) &= \frac{1}{\sqrt{2\pi}} \int_{-\infty}^{\infty} \frac{1}{\sqrt{2\pi}} \int_{-\infty}^{\infty} f(x - t)e^{-i\omega x}\,dx\,g(t)\,dt \\
&= \frac{1}{\sqrt{2\pi}} \int_{-\infty}^{\infty} \frac{1}{\sqrt{2\pi}} \int_{-\infty}^{\infty} f(u)e^{-i\omega u}\,du\,e^{-i\omega t}g(t)\,dt \\
&\qquad\qquad\qquad\qquad (u = x - t,\ du = dx) \\
&= \mathcal{F}(f)(\omega)\mathcal{F}(g)(\omega)\,. \qquad\blacksquare
\end{aligned}
$$

EXAMPLE 4   Convolution of Gaussian functions
Let $f(x) = e^{-\alpha x^2}$ and $g(x) = e^{-\beta x^2}$ where α and β are positive constants.
(a) Compute $\mathcal{F}(f * g)(\omega)$.   (b) Compute $f * g(x)$.

Solution (a) Using Theorem 4 and Example 3, we get

$$\mathcal{F}(f * g)(\omega) = \mathcal{F}(f)(\omega)\mathcal{F}(g)(\omega) = \frac{1}{\sqrt{2\alpha}}e^{-\frac{\omega^2}{4\alpha}}\frac{1}{\sqrt{2\beta}}e^{-\frac{\omega^2}{4\beta}}$$

(3)
$$= \frac{1}{2\sqrt{\alpha\beta}}e^{-\frac{\alpha+\beta}{4\alpha\beta}\omega^2}.$$

(b) To compute $f * g$, it suffices to find the inverse Fourier transform of (3). It is clear that we should be looking at another Gaussian function, because of the presence of the function $e^{-\omega^2}$. Indeed, if we let

$$\frac{1}{4a} = \frac{\alpha+\beta}{4\alpha\beta} \quad \Rightarrow \quad a = \frac{\alpha\beta}{\alpha+\beta} \quad \text{and} \quad \frac{1}{\sqrt{2a}} = \frac{\sqrt{\alpha+\beta}}{\sqrt{2\alpha\beta}},$$

then from Example 3,

$$\mathcal{F}(e^{-ax^2}) = \frac{1}{\sqrt{2a}}e^{-\frac{\omega^2}{4a}} = \frac{\sqrt{\alpha+\beta}}{\sqrt{2\alpha\beta}}e^{-\frac{\alpha+\beta}{4\alpha\beta}\omega^2}.$$

To match the Fourier transform in (3), we multiply e^{-ax^2} by $\frac{1}{2\sqrt{\alpha\beta}}$ and divide by $\frac{\sqrt{\alpha+\beta}}{\sqrt{2\alpha\beta}}$, thereby obtaining the function

$$\frac{1}{2\sqrt{\alpha\beta}}\frac{\sqrt{2\alpha\beta}}{\sqrt{\alpha+\beta}}e^{-ax^2} = \frac{1}{\sqrt{2(\alpha+\beta)}}e^{-\frac{\alpha\beta}{\alpha+\beta}x^2}$$

as the inverse Fourier transform of (3). Consequently, we obtain the useful convolution formula: for any $\alpha,\ \beta > 0$,

(4)
$$\boxed{e^{-\alpha x^2} * e^{-\beta x^2} = \frac{1}{\sqrt{2(\alpha+\beta)}}e^{-\frac{\alpha\beta}{\alpha+\beta}x^2}.}$$

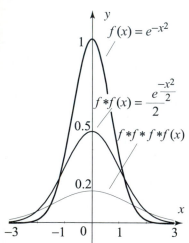

Figure 2 Graphs of $f(x) = e^{-x^2}$, $f * f$ and $f * f * f * f$. The effect of the convolution is to smear out the values of the function.

Thus the convolution of two Gaussian functions is another scaled Gaussian function. This fact is at the heart of the solution of the heat problem on the real line (see Section 11.4). In Figure 2, we show the graphs of $f(x) = e^{-x^2}$, $f * f(x) = \frac{1}{2}e^{-\frac{x^2}{2}}$, and $f * f * f * f(x) = \frac{1}{4\sqrt{2}}e^{-\frac{x^2}{4}}$. Note how the convolution smears out the values of the function. ∎

It is clear from our answer in Example 4(a) that $f * g = g * f$. This equality holds for all f and g as can be seen by taking Fourier transforms:

$$\mathcal{F}(f * g) = \mathcal{F}(f)\mathcal{F}(g) = \mathcal{F}(g)\mathcal{F}(f) = \mathcal{F}(g * f).$$

Now taking inverse Fourier transform, we obtain

(5)
$$f * g = g * f,$$

which expresses the fact that convolution is a commutative operation. This simple technique of the Fourier transform to establish results about convolutions has far reaching applications. We illustrate with another property of convolutions. To simplify the statements of results, in what follows, we will always suppose that the functions in questions have enough nice properties to be able to compute Fourier transforms and inverse Fourier transforms.

THEOREM 5
CONVOLUTIONS
AND DERIVATIVES

Let n, α and β denote nonnegative integers, such that $n = \alpha + \beta$. Then

$$(6) \qquad \frac{d^n}{dx^n}(f * g) = \frac{d^\alpha f}{dx^\alpha} * \frac{d^\beta g}{dx^\beta}.$$

In particular, taking $\alpha = n$ and $\beta = 0$, then $\alpha = 0$ and $\beta = n$, we obtain

$$(7) \qquad \frac{d^n}{dx^n}(f * g) = \frac{d^n f}{dx^n} * g = f * \frac{d^n g}{dx^n}.$$

Proof To prove (6), we compute the Fourier transforms of the functions on both sides of the equality, using Theorems 1(iii) and 4 and the fact that $n = \alpha + \beta$:

$$\begin{aligned}
\mathcal{F}\left(\frac{d^n}{dx^n}(f * g)\right) &= (i\omega)^n \mathcal{F}(f * g) = (i\omega)^\alpha \mathcal{F}(f)(i\omega)^\beta \mathcal{F}(g) \\
&= \mathcal{F}\left(\frac{d^\alpha f}{dx^\alpha}\right)\mathcal{F}\left(\frac{d^\beta g}{dx^\beta}\right) = \mathcal{F}\left(\frac{d^\alpha f}{dx^\alpha} * \frac{d^\beta g}{dx^\beta}\right),
\end{aligned}$$

and the desired result follows by taking inverse Fourier transforms. ∎

Plancherel's and Parseval's Theorems

Recall that a function f defined on the real line is square integrable if $\int_{-\infty}^{\infty} |f(x)|^2 dx < \infty$. The following two important results hold for square integrable functions.

THEOREM 6
PARSEVAL'S
THEOREM

Suppose that f and g are square integrable functions on the real line. Then

$$(8) \qquad \int_{-\infty}^{\infty} f(x)\overline{g(x)}\,dx = \int_{-\infty}^{\infty} \widehat{f}(\omega)\overline{\widehat{g}(\omega)}\,d\omega.$$

By taking $f = g$ in Parseval's theorem, we obtain Plancherel's theorem.

THEOREM 7
PLANCHEREL'S
THEOREM

Suppose that f is a square integrable function on the real line. Then

$$(9) \qquad \int_{-\infty}^{\infty} |f(x)|^2 dx = \int_{-\infty}^{\infty} |\widehat{f}(\omega)|^2 d\omega.$$

Proof of Parseval's theorem Use the inverse Fourier transform to write $g(x) = \frac{1}{\sqrt{2\pi}} \int_{-\infty}^{\infty} e^{i\omega x}\widehat{g}(\omega)\,d\omega$, and recall that $\overline{e^{i\omega x}} = e^{-i\omega x}$. Now, assuming that we can interchange the order of integration, we have

$$\int_{-\infty}^{\infty} f(x)\overline{g(x)}\, dx = \int_{-\infty}^{\infty} f(x) \frac{1}{\sqrt{2\pi}} \int_{-\infty}^{\infty} e^{-i\omega x}\overline{\widehat{g}(\omega)}\, d\omega\, dx$$

$$= \int_{-\infty}^{\infty} \overline{\widehat{g}(\omega)} \overbrace{\frac{1}{\sqrt{2\pi}} \int_{-\infty}^{\infty} e^{-i\omega x} f(x)\, dx}^{=\widehat{f}(\omega)}\, d\omega$$

$$= \int_{-\infty}^{\infty} \widehat{f}(\omega)\overline{\widehat{g}(\omega)}\, d\omega,$$

which proves the theorem.   ∎

Just like Parseval's identity for Fourier series was useful in summing series, Plancherel' theorem is useful in evaluating certain nontrivial integrals.

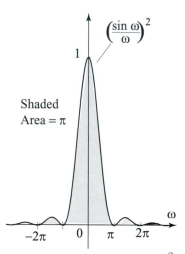

Figure 3 Graph of $\left(\frac{\sin \omega}{\omega}\right)^2$. The function is positive and the total area under the curve, above the ω-axis is π.

Shaded Area $= \pi$

EXAMPLE 5   An application of Plancherel's theorem
Consider the function $f(x) = 1$ if $|x| < a$ and 0 otherwise, and its Fourier transform $\widehat{f}(\omega) = \sqrt{\frac{2}{\pi}} \frac{\sin a\omega}{\omega}$ (see Example 1. Section 11.1). Applying Plancherel's theorem, we obtain

$$\int_{-a}^{a} |f(x)|^2\, dx = \int_{-\infty}^{\infty} \left(\sqrt{\frac{2}{\pi}} \frac{\sin a\omega}{\omega} \right)^2\, d\omega.$$

But $f(x) = 1$ for $|x| < a$, so the integral on the left is $2a$, and hence

$$a\pi = \int_{-\infty}^{\infty} \left(\frac{\sin a\omega}{\omega} \right)^2\, d\omega.$$

The case $a = 1$ is illustrated in Figure 3.   ∎

Generalized Functions

The need to compute Fourier transforms or inverse Fourier transforms of functions that are not integrable on the real line is clear from our first computation of a Fourier transform in Example 1, Section 11.1, all the way to the last result that concerns square integrable functions. In fact, as soon as the function is not continuous, its Fourier transform is not going to be integrable (this is a consequence of the fact that the Fourier transform of an integrable function is continuous). So we need to go beyond integrable functions. The Fourier transform can be defined in a very general setting, which is suitable for solving partial differential equations. This is the setting of generalized functions or distributions. A rigorous treatment of these objects is beyond the level of this book. However, by touching a little bit on this subject, we get a lot out of it, as far as expanding our ability to compute transforms and convolutions, even for functions that are integrable.

To motivate the topic of generalized functions, let's consider the following question: Is there an identity element for the binary operation of convolution? That is, is there a function ϕ such that $f * \phi = \phi * f = f$ for all f? If such a function exists, then taking Fourier transforms, we must have $\widehat{f\phi} = \hat{f}$, which suggests that $\hat{\phi}$ must be identically 1. This cannot happen if ϕ were integrable, by the Riemann-Lebesgue Lemma (Lemma 2, Section 11.1). So if ϕ exists, it is not an integrable function. If we evaluate the identity $f * \phi(x) = \phi * f(x) = f(x)$ at $x = 0$, using the definition of the convolution, we get

$$\frac{1}{\sqrt{2\pi}} \int_{-\infty}^{\infty} f(-y)\phi(y)\,dy = f(0).$$

Renaming the function $f(-y) = g(y)$ and changing the variable of integration to x, we see that ϕ must satisfy

$$\frac{1}{\sqrt{2\pi}} \int_{-\infty}^{\infty} g(x)\phi(x)\,dx = g(0).$$

So the values of $g(x)$ for $x \neq 0$ do not affect the value of the integral of the product $\phi(x)g(x)$, which suggests that $\phi(x) = 0$ for all $x \neq 0$. If you think about it, you will realize that there is no function ϕ that can possibly have this effect. This is one of many *generalized functions* that cannot be defined pointwise, as we usually define functions; instead, they are defined by the values of their integrals against functions. That is, while we do not have $\phi(x)$, for all x, we do have $\phi[f] = \int_{-\infty}^{\infty} f(x)\phi(x)\,dx$ for all f in a class of functions known as the test functions. We have used the notation $\phi[f]$ to suggest that ϕ is a function of the test functions and not a function of real numbers. As far as we are concerned, we will not be specific about the class of test functions and we will assume that $\phi[f]$ is known at least for all continuous integrable functions.

Our first example of a generalized function is the **Dirac delta function**, denoted by $\delta_0(x)$ and defined by the values of its integrals:

$$(10) \qquad \int_{-\infty}^{\infty} f(x)\delta_0(x)\,dx = f(0),$$

for all continuous functions f. By taking f to be continuous and zero outside an interval $[a, b]$, we obtain from (10)

$$(11) \qquad \int_{a}^{b} f(x)\delta_0(x)\,dx = \begin{cases} f(0) & \text{if } a \leq 0 \leq b, \\ 0 & \text{if } 0 < a \text{ or } 0 > b. \end{cases}$$

It is sometimes convenient to think of δ_0 as a function of x, with values $\delta_0(x) = 0$ for all $x \neq 0$, and $\delta_0(0) = \infty$, and depict the function by a graph as in Figure 4.

The delta function can be used to define a formal identity for convolution.

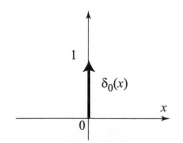

Figure 4 Graphical representation of the Dirac delta function, with support at $x = 0$.

EXAMPLE 6   A formal identity for convolution

Show that for any continuous function f

$$(12) \qquad \delta_0 * f(x) = \frac{1}{\sqrt{2\pi}} f(x) \quad \text{equivalently} \quad \left(\sqrt{2\pi}\delta_0\right) * f(x) = f(x).$$

Solution Using (10), we have

$$\left(\sqrt{2\pi}\delta_0\right) * f(x) = \int_{-\infty}^{\infty} f(x - y)\delta_0(y)\, dy = f(x),$$

by evaluating the function $y \mapsto f(x - y)$ at $y = 0$. ■

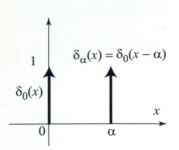

Let α be a real number. The **translate** by α of δ_0 is another generalized function denoted by δ_α and defined by the values of its integral against continuous functions by

$$(13) \qquad \delta_\alpha[f] = \int_{-\infty}^{\infty} f(x)\delta_\alpha(x)\, dx = f(\alpha).$$

Thus integrating a function f against δ_α picks up the value of f at α. Since we also have $f(\alpha) = \int_{-\infty}^{\infty} f(x + \alpha)\delta_0(x)\, dx$, thinking of δ_0 as a function and making the change of variables $x + \alpha = X$, we find that

$$(14) \qquad \int_{-\infty}^{\infty} f(x)\delta_\alpha(x)\, dx = \int_{-\infty}^{\infty} f(X)\delta_0(X - \alpha)\, dX.$$

Thus, $\delta_\alpha(x) = \delta_0(x - \alpha)$. The graph of $\delta_\alpha(x)$ is shown in Figure 5; it is obtained by translating the graph of $\delta_0(x)$ by α. So if we think of δ_0 as being supported at the point $x = 0$, then δ_α is supported at $x = \alpha$.

An antiderivative of δ_0 is defined by

$$(15) \qquad \mathcal{U}_0(x) = \int_{-\infty}^{x} \delta_0(t)\, dt = \begin{cases} 0 & \text{if } x < 0, \\ 1 & \text{if } x \geq 0. \end{cases}$$

This is the **Heaviside unit step function** (Figure 6). Similarly, we introduce the translates of the Heaviside function by

$$(16) \qquad \mathcal{U}_\alpha(x) = \int_{-\infty}^{x} \delta_\alpha(t)\, dt = \begin{cases} 0 & \text{if } x < \alpha, \\ 1 & \text{if } x \geq \alpha. \end{cases}$$

We have the formal derivative relations

$$(17) \qquad \mathcal{U}_0'(x) = \delta_0(x) \quad \text{also} \quad \mathcal{U}_\alpha'(x) = \delta_\alpha(x).$$

We can give some meaning to these identities: The Heaviside is constant so its slope (derivative) is 0, except at $x = \alpha$, where the values of $\mathcal{U}_\alpha$ jump from 0 to 1 causing an infinite slope. This reasoning with slopes allows us to carry

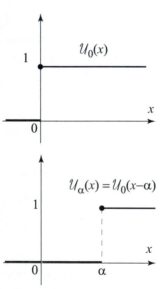

Figure 5 Graphical representation of the Dirac delta function and its translate, with supports at $x = 0$ and $x = \alpha$.

Figure 6 The Heaviside step function $\mathcal{U}_0(x)$ and its translate $\mathcal{U}_\alpha(x) = \mathcal{U}_0(x - \alpha)$.

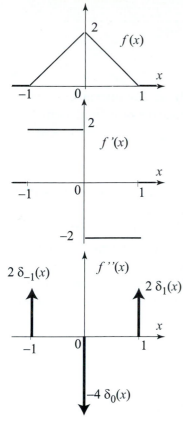

Figure 7 for Example 7.

the notion of derivatives to piecewise continuous functions as illustrated by the following example.

EXAMPLE 7   Derivatives of piecewise continuous functions

The function $f(x)$ is shown in Figure 7. Compute $f'(x)$ and $f''(x)$.

Solution Since the function is zero for $x < -1$ or $x > 1$, we have $f'(x) = f''(x) = 0$ if $x < -1$ or $x > 1$. For $-1 < x < 0$ we have $f'(x) = 2$ and for $0 < x < 1$ we have $f'(x) = -2$, as shown in Figure 7. In terms of Heaviside functions, we have

$$(18) \qquad f'(x) = 2\mathcal{U}_{-1}(x) - 4\mathcal{U}_0(x) + 2\mathcal{U}_1(x).$$

The second derivative is 0 everywhere except at the points -1, 0, and 1. In order to cause $f'(x)$ to jump the way shown in Figure 7, we place weighted δ functions at these points corresponding to the jump on the graph of $f'(x)$. If $f'(x)$ increases at a jump, the weight of the δ function there is positive. If $f'(x)$ decreases at a jump, the weight of the δ function there is negative. The result is a second derivative given by

$$(19) \qquad f''(x) = 2\delta_{-1}(x) - 4\delta_0(x) + 2\delta_1(x)$$

(Figure 7). Note how this can be obtained by differentiating (18) and using (17). You may be wondering about the values of $f'(x)$ at the points -1, 0, and 1. You can insert δ functions at these points, but the weights have to be 0, because $f(x)$ is continuous at these points. ∎

We now introduce the Fourier transform of generalized functions, again by formal manipulations of identities that hold for integrable functions.

FOURIER TRANSFORMS OF GENERALIZED FUNCTIONS

We have:

$$(20) \qquad \mathcal{F}\left(\delta_0(x)\right)(\omega) = \frac{1}{\sqrt{2\pi}};$$

$$(21) \qquad \mathcal{F}\left(\delta_\alpha(x)\right)(\omega) = \frac{1}{\sqrt{2\pi}}e^{-i\alpha\omega};$$

$$(22) \qquad \mathcal{F}\left(\mathcal{U}_0(x)\right)(\omega) = -\frac{i}{\sqrt{2\pi}\,\omega};$$

$$(23) \qquad \mathcal{F}\left(\mathcal{U}_\alpha(x)\right)(\omega) = -\frac{i}{\sqrt{2\pi}\,\omega}e^{-i\alpha\omega};$$

Proof To prove (20), we use the definition of the Fourier transform and (10) and get

$$\mathcal{F}\left(\delta_0\right)(\omega) = \frac{1}{\sqrt{2\pi}}\int_{-\infty}^{\infty} e^{-i\omega x}\delta_0(x)\,dx = \frac{1}{\sqrt{2\pi}}e^{-i\omega \cdot 0} = \frac{1}{\sqrt{2\pi}}.$$

The remaining identities follow from (20) and the properties of the Fourier transform. For example, to prove (21) use Theorem 3. To prove (22) use Theorem 1: we have

$$\mathcal{F}(\mathcal{U}_0') = i\omega\mathcal{F}(\mathcal{U}_0);$$

but $\mathcal{U}_0' = \delta_0$, so

$$i\omega \mathcal{F}(\mathcal{U}_0) = \mathcal{F}(\delta_0) = \frac{1}{\sqrt{2\pi}};$$

which implies (22).   ■

We can add more to these identities by applying the operational properties. For example, we can consider the nth derivative of δ_0 and write

$$(24) \qquad \mathcal{F}\left(\delta_\alpha^{(n)}\right)(\omega) = \frac{(i\omega)^n}{\sqrt{2\pi}} e^{-i\alpha\omega},$$

where we have combined Theorem 1(iii) and (21). As extravagant as they may seem, these identities are very useful when used as operational properties in intermediary steps in computations. We illustrate with several examples, starting with a piecewise linear function that vanishes outside a finite interval. The function is integrable, but as is generally the case with piecewise continuous functions, computing its Fourier transform from the definition is tedious. We can use generalized functions to facilitate the task.

EXAMPLE 8   Fourier transform of a piecewise linear function
Find the Fourier transform of the function $f(x)$ in Example 7.

Solution From (18), we have

$$f'(x) = 2\mathcal{U}_{-1}(x) - 4\mathcal{U}_0(x) + 2\mathcal{U}_1(x).$$

So with the help of (23), we get

$$\mathcal{F}(f') = 2\mathcal{F}(\mathcal{U}_{-1}) - 4\mathcal{F}(\mathcal{U}_0) + 2\mathcal{F}(\mathcal{U}_1) = -\frac{i}{\sqrt{2\pi}\,\omega}\left(2e^{i\omega} - 4 + 2e^{-i\omega}\right).$$

Appealing to Theorem 1(i), we have

$$\mathcal{F}(f') = i\omega\mathcal{F}(f) \quad \Rightarrow \quad \mathcal{F}(f) = \frac{-i}{\omega}\mathcal{F}(f'),$$

and so

$$\mathcal{F}(f) = -\frac{2e^{i\omega} - 4 + 2e^{-i\omega}}{\sqrt{2\pi}\omega^2}.$$

Simplifying with the help of the identity $e^{i\omega} + e^{-i\omega} = 2\cos\omega$, we obtain

$$\mathcal{F}(f) = \frac{4(1 - \cos\omega)}{\sqrt{2\pi}\omega^2},$$

which is continuous at all ω, including $\omega = 0$, as one would expect from the Fourier transform of an integrable function. The value of the transform at $\omega = 0$ is

$$\mathcal{F}(f)(0) = \lim_{\omega \to 0} \mathcal{F}(f)(\omega) = \lim_{\omega \to 0} \frac{4(1 - \cos\omega)}{\sqrt{2\pi}\omega^2} = \frac{2}{\sqrt{2\pi}},$$

as you can check by applying l'Hospital's rule twice.   ■

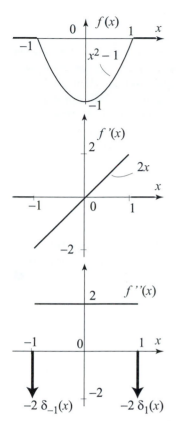

Figure 8 The function $f(x)$ and its derivatives in Example 9.

EXAMPLE 9   Fourier transform of a piecewise smooth function

Find the Fourier transform of the function $f(x)$ in Figure 8.

Solution First we reduce the problem to finding Fourier transforms of generalized functions. Taking derivatives of f (see Figure 8), we find that

$$f'' = 2(\mathcal{U}_{-1} - \mathcal{U}_1) - 2\delta_{-1} - 2\delta_1,$$

where the delta functions are inserted to account for the jumps in f' at ± 1. Computing the Fourier transform with the help of (21) and (23), we get

$$
\begin{aligned}
\mathcal{F}(f'') &= 2\mathcal{F}(\mathcal{U}_{-1}) - \mathcal{F}(\mathcal{U}_1) - 2\left(\mathcal{F}(\delta_{-1}) + \mathcal{F}(\delta_1)\right) \\
&= \frac{-2i}{\sqrt{2\pi}} \frac{e^{i\omega} - e^{-i\omega}}{\omega} - \frac{2}{\sqrt{2\pi}} \left(e^{i\omega} + e^{-i\omega}\right) \\
&= \frac{4}{\sqrt{2\pi}} \frac{\sin\omega}{\omega} - \frac{4}{\sqrt{2\pi}} \cos\omega.
\end{aligned}
$$

Appealing to Theorem 1(ii), we have

$$\mathcal{F}(f'') = -\omega^2 \mathcal{F}(f) \quad \Rightarrow \quad \mathcal{F}(f) = \frac{-1}{\omega^2} \mathcal{F}(f''),$$

and so

$$\mathcal{F}(f) = \frac{4}{\sqrt{2\pi}\,\omega^2} \left(\cos\omega - \frac{\sin\omega}{\omega}\right).$$

Again, you should verify that the transform is continuous. (Consider its Taylor series about 0.) ∎

We now illustrate the use of generalized functions in computing convolutions. We start with the following simple but useful identity:

(25)
$$\delta_a * \delta_b = \frac{1}{\sqrt{2\pi}} \delta_{a+b}.$$

This interesting identity says that the convolution of two delta functions is another scaled delta function supported at the sum of the supports of the delta functions (Figure 9). To prove (25), take Fourier transforms on both sides and use (21) (Exercise 41).

Here is an interesting application of (25).

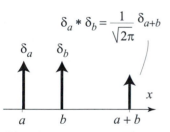

Figure 9 The convolution of two Dirac deltas is another scaled Dirac delta, supported at the sum of the supports.

EXAMPLE 10   Convolution of a step function

Let $f(x)$ be the step function in Figure 10. Compute $f * f(x)$. (For the convolution of f with itself n-times, see Exercise 46.)

Solution We have $f'(x) = \delta_{-1} - \delta_1$. Also, from Theorem 5, and (25),

$$
\begin{aligned}
\frac{d^2}{dx^2}(f * f) &= f' * f' = (\delta_{-1} - \delta_1) * (\delta_{-1} - \delta_1) \\
&= \frac{1}{\sqrt{2\pi}} \left(\delta_{-2} - 2\delta_0 + \delta_2\right).
\end{aligned}
$$

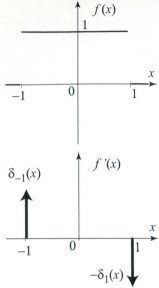

Figure 10 $f(x)$ and $f'(x)$ in Example 10.

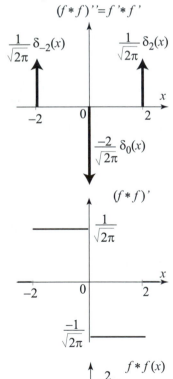

Figure 11 Antidifferentiation in Example 10.

Taking the antiderivative once and using (17), we find

$$(f * f)' = \frac{1}{\sqrt{2\pi}} \left(\mathcal{U}_{-2} - 2\mathcal{U}_0 + \mathcal{U}_2 \right),$$

which is the step function depicted in Figure 11. To get $f * f$, we find the continuous antiderivative of $(f * f)'$, which is 0 outside a large interval. (In fact, it can be shown that if f vanishes outside a set A and g vanishes outside a set B, then $f * g$ will vanish outside the set $A + B$, which consists of all real numbers of the form $x + y$, where x is in A and y is in B. In our example, f vanishes outside $[-1, 1]$, so $f * f$ will vanish outside $[-1, 1] + [-1, 1] = [-2, 2]$.) This is not difficult to do with the help of Figure 11, yielding the function $f * f$ in Figure 11. More explicitly, we have

$$f * f(x) = \begin{cases} 0 & \text{if } x < -2, \\ \frac{1}{\sqrt{2\pi}}(x + 2) & \text{if } -2 \le x \le 0, \\ -\frac{1}{\sqrt{2\pi}}(x - 2) & \text{if } 0 \le x \le 2, \\ 0 & \text{if } x > 2. \end{cases}$$

The method of this example can be applied to evaluate the nth convolution of f by itself, which in turn can be used to evaluate important improper integrals (see Exercise 47). ∎

As you can see, the tricks with generalized functions are endless; some more will be presented in the exercises.

In the next section, we develop the Fourier transform method for solving partial differential equations on the real line. We will use this method to solve boundary value problems associated with the heat and wave equations, and a variety of other important problems on the real line and regions in the two dimensional space.

Exercises 11.2

In Exercises 1–16 use the operational properties and a known Fourier transform to compute the Fourier transform of the given function.

1. $f(x) = x^2 e^{-x^2}$.

2. $f(x) = xe^{-|x|}$.

3. $f(x) = \dfrac{x}{(1 + x^2)^2}$.

4. $f(x) = \dfrac{x^2}{(1 + x^2)^2}$.

5. $f(x) = (1 + 6x)e^{-|x|}$.

6. $f(x) = e^{-x^2 - 2x}$.

7. $f(x) = e^{-\frac{x^2}{2} + x - 3}$.

8. $f(x) = e^{-2x^2 + 3ix}$.

9.
$$f(x) = \begin{cases} x & \text{if } 0 < x < 1, \\ 0 & \text{otherwise.} \end{cases}$$

10.
$$f(x) = \begin{cases} 0 & \text{if } x < 0 \text{ or } x > 1, \\ x^2 & \text{if } 0 < x < 1. \end{cases}$$

11. $f(x) = \dfrac{2x}{(a^2 + x^2)^2}, \quad a \ne 0$.

12. $f(x) = \dfrac{a - ix}{a^2 + x^2}, \quad a \ne 0$.

13. $f(x) = (1 - x^2)e^{-x^2}$.

14. $f(x) = (1 - x)^2 e^{-|x|}$.

15. $f(x) = xe^{-\frac{1}{2}(x-1)^2}$.

16. $f(x) = (1 - x)e^{-|x-1|}$.

In Exercises 17–22, (a) find the Fourier transforms of f and g. (b) Find the Fourier transform of $f * g$. (c) What is $f * g$? [Hint: Compute $f * g(x)$ using the method of Example 4. If needed, use the table of Fourier transforms (Appendix B.1) to assist you in inverting the Fourier transform of $f * g$.]

17. $f(x) = e^{-x^2}$, $g(x) = xe^{-x^2}$.   **18.** $f(x) = xe^{-x^2}$, $g(x) = xe^{-x^2}$.

19. $f(x) = e^{-(x-1)^2}$, $g(x) = e^{-(x+1)^2}$.

20. $f(x) = \mathcal{U}_{-1}(x) - \mathcal{U}_1(x)$, $g(x) = (\mathcal{U}_{-1}(x) - \mathcal{U}_1(x))\,x$.

21. $f(x) = \mathcal{U}_{-a}(x) - \mathcal{U}_a(x)$, $g(x) = \mathcal{U}_{-b}(x) - \mathcal{U}_b(x)$, where $0 < a \leq b$.

22. $f(x) = \mathcal{U}_{-1}(x) - \mathcal{U}_1(x) + \mathcal{U}_2(x) - \mathcal{U}_3(x)$, $g(x) = \mathcal{U}_{-1}(x) - \mathcal{U}_1(x)$.

23.   Basic properties of convolutions. Establish the following properties of convolutions. For (a) and (b), use operational properties of the Fourier transform.
(a) $f * (g * h) = (f * g) * h$ (associativity).
(b)   Let a be a real number and let f_a denote the translate of f by a, that is, $f_a(x) = f(x-a)$. Show that $(f_a) * g = f * (g_a) = (f * g)_a$. This important property says that convolutions commute with translations.
(c) **Integrability of the convolution.**   Show that if f and g are integrable, then so is $f * g$. (It is also continuous.) [Hint: Use the definition of convolution, interchange orders of integration and use $|\int (f(x)\, dx| \leq \int |f(x)|\, dx$.]

24. (a) Show that $x(f * g) = (xf) * g + f * (xg)$. [Hint: Fourier transform.]
(b) Verify the identity in (a) with $f(x) = g(x) = e^{-x^2}$.

25. Convolution with a complex exponential.
(a) Show that $e^{i\alpha x} * f(x) = \widehat{f}(\alpha)e^{i\alpha x}$.
(b) Compute $\cos x * e^{-|x|}$. [Hint: $\cos x = \frac{e^{ix}+e^{-ix}}{2}$.]

26.   Shifting on the ω-axis. Using Theorem 3, prove the following identities:

$$\mathcal{F}(\cos(ax)f(x))(\omega) = \frac{\widehat{f}(\omega - a) + \widehat{f}(\omega + a)}{2};$$

$$\mathcal{F}(\sin(ax)f(x))(\omega) = \frac{\widehat{f}(\omega - a) - \widehat{f}(\omega + a)}{2i};$$

$$\mathcal{F}^{-1}(\cos(a\omega)\widehat{f}(\omega))(x) = \frac{f(x - a) + f(x + a)}{2};$$

$$\mathcal{F}^{-1}(\sin(a\omega)\widehat{f}(\omega))(x) = \frac{f(x + a) - f(x - a)}{2i}.$$

In Exercises 27–32, use Exercise 26 and known transforms to compute the Fourier transform of the given function.

27. $f(x) = \dfrac{\cos x}{e^{x^2}}$.   **28** $f(x) = \dfrac{\sin 2x}{e^{|x|}}$.

29. $f(x) = \dfrac{\cos x + \cos 2x}{1 + x^2}$.   **30.** $f(x) = \dfrac{\sin x + \cos 2x}{4 + x^2}$.

31   **32.**

$$f(x) = \begin{cases} \cos x & \text{if } |x| < 1, \\ 0 & \text{otherwise.} \end{cases}$$ $$f(x) = \begin{cases} \sin x & \text{if } |x| < 1, \\ 0 & \text{otherwise.} \end{cases}$$

In Exercises 33–38, follow the method of Examples 8 and 9 to compute the Fourier transform of the given function. To illustrate your answer, graph the function and its derivatives, including the parts with generalized functions.

33. $\mathcal{U}_2(x) - \mathcal{U}_4(x)$.   **34.** $(\mathcal{U}_{-1} - \mathcal{U}_1(x))|x|$.

35. $\displaystyle\sum_{j=0}^{5}(-1)^j\,(\mathcal{U}_j(x) - \mathcal{U}_{j+1}(x))$.   **36.** $\displaystyle\sum_{j=0}^{5}(-1)^j\,(\mathcal{U}_j(x) - \mathcal{U}_{j+1}(x))\,(x-j)$.

37. $(\mathcal{U}_0(x) - \mathcal{U}_1(x))\,x + (\mathcal{U}_1(x) - \mathcal{U}_2(x))\,(2-x)$.

38. $(\mathcal{U}_{-1}(x) - \mathcal{U}_1(x))\,x^2$.

In Exercises 39–40, follow the method of Example 10 to compute the given convolution.

39. $f * f * f(x)$ where $f(x) = \mathcal{U}_{-1}(x) - \mathcal{U}_1(x)$.

40. $f * g(x)$ where $f(x) = \mathcal{U}_{-1}(x) - \mathcal{U}_1(x)$ and $g(x) = (\mathcal{U}_{-1} - \mathcal{U}_1(x))x$.

41. Prove (25), using the Fourier transform.

42. Fourier transforms and dilations. (a) Show that

$$\mathcal{F}(f(ax))(\omega) = \frac{1}{|a|}\,\mathcal{F}(f)\left(\frac{\omega}{a}\right), \qquad a \neq 0.$$

(b)   Use (a) to derive the Fourier transform of $g(x) = e^{-2|x|}$ from the Fourier transform of $f(x) = e^{-|x|}$.

43. Use Plancherel's theorem to derive the value of $\int_{-\infty}^{\infty}\left(\frac{\sin x}{x}\right)^4 dx$ in Table 1 (see Exercise 47). [Hint: Find a function whose Fourier transform is $\left(\frac{\sin\omega}{\omega}\right)^2$.]

44. Project Problem: Fourier transform of Bessel functions of integer order. This exercise is based on Exercise 18 of the previous section. We will derive the formula

$$(26) \qquad \mathcal{F}\left(J_n(x)\right)(\omega) = \begin{cases} \sqrt{\dfrac{2}{\pi}}\dfrac{(-i)^n}{\sqrt{1-\omega^2}}T_n(\omega) & \text{if } |\omega| < 1, \\[2mm] 0 & \text{if } |\omega| > 1, \end{cases}$$

where T_n is the nth Chebyshev polynomial (see Exercise 57, Section 1.3, for definitions and properties of these polynomials).

(a) Using the operational properties of the Fourier transform and the identity $\frac{d}{dx}J_0 = -J_1$, obtain the Fourier transform of $J_1(x)$:

$$\mathcal{F}\left(J_1(x)\right)(\omega) = \begin{cases} \sqrt{\dfrac{2}{\pi}}\dfrac{-i\omega}{\sqrt{1-\omega^2}} & \text{if } |\omega| < 1, \\[2mm] 0 & \text{if } |\omega| > 1. \end{cases}$$

Now that you have the Fourier transforms of J_0 and J_1, explain how these follow from (26) with $n = 0$ and 1.

(b) Use the identity $J_{p-1} - J_{p+1} = 2J_p'$ (see Section 9.7) to derive that $\mathcal{F}(J_{p-1}) - \mathcal{F}(J_{p+1}) = 2i\omega\mathcal{F}(J_p)$. Conclude from what you know about the Fourier transforms of J_0 and J_1 that $J_n(\omega) = 0$ if $|\omega| > 1$ and that for $|\omega| < 1$ $J_n(\omega) = \sqrt{\dfrac{2}{\pi}}\dfrac{p_n(\omega)}{\sqrt{1-\omega^2}}$, where $p_n(x)$ is a polynomial of degree n in x satisfying the recurrence relation $p_{n-1}(x) - p_{n+1}(x) = 2ixp_n(x)$.

(c) Verify that $p_0(x) = 1 = T_0(x)$, $p_1(x) = -ix = -iT_1(x)$. More generally, show that the $p_n(x) = (-i)^n T_n(x)$. [Hint: Write $p_n(x) = (-i)^n q_n(x)$. We have $q_0 = T_0$ and $q_1 = T_1$. Use the recurrence relation for the p_n's to show that $q_{n-1} + q_{n+1} = 2xq_n$. Since this is the recurrence relation for the Chebyshev polynomials (Exercise 58(e), Section 1.3), it follows that $q_n = T_n$.]

45. Project Problem: Fourier transforms involving Bessel functions of half-integer order and Legendre polynomials. We will derive the following Fourier transform for $n = 0, 1, 2, \ldots$,

$$(27) \qquad \mathcal{F}\left(\frac{J_{n+\frac{1}{2}}(x)}{\sqrt{x}}\right)(\omega) = \begin{cases} (-i)^n P_n(\omega) & \text{if } |\omega| < 1, \\ 0 & \text{if } |\omega| > 1, \end{cases}$$

where P_n is the nth Legendre polynomial (see Section 10.5 for definitions and properties of these polynomials). To establish the identity, proceed as follows.

(a) Show that the functions $f_n(x) = \dfrac{J_{n+\frac{1}{2}}(x)}{\sqrt{x}}$ satisfy the recurrence relation

$$xf_{n-1} + xf_{n+1} = (2n+1)f_n.$$

[Hint: Use (6), Section 9.7.]

(b) Conclude that $i\frac{d}{d\omega}\widehat{f_{n-1}} + i\frac{d}{d\omega}\widehat{f_{n+1}} = (2n+1)\widehat{f_n}$.

(c) Compute $\widehat{f_n}$ for $n = 0$ and 1 directly and show that they fit the formula (27). Conclude from (b) that, for $n \geq 2$, $\widehat{f_n}(\omega)$ is 0 if $|\omega| > 1$ and is a polynomial of degree n for $|\omega| < 1$.

(d) Complete the proof by showing that the functions $(-i)^n P_n(\omega)$ satisfy the recurrence relation in (b).

46. Project Problem: Convolutions of the indicator function. In this exercise we derive the following interesting formula. Let $f(x) = \mathcal{U}_0(x) - \mathcal{U}_1(x)$, and let $n \geq 2$ be an integer. Define the nth convolution of f by itself by

$$(28) \qquad f^{*n}(x) = \overbrace{f * f * \cdots * f}^{n \text{ of these}}.$$

Then $f^{*n}(x) = 0$ if $x < 0$ or $x > n$, and for x in $[0, n]$, we have

$$(29) \qquad f^{*n}(x) = \frac{1}{(2\pi)^{\frac{n-1}{2}}(n-1)!} \sum_{k=0}^{[x]} (-1)^k \binom{n}{k} (x-k)^{n-1},$$

where $[x]$ denotes the greatest integer in x and $\binom{n}{k}$ is the binomial coefficient.

Follow the suggested proof and supply all the missing details by referring to appropriate operational properties of the Fourier transform and generalized functions.

Suggested Proof We have

$$\left(\frac{d}{dx}\right)^n f^{*n}(x) = (f')^{*n}(x) \quad \text{and} \quad f'(x) = \delta_0(x) - \delta_1(x);$$

so

$$(f')^{*n}(x) = \frac{1}{(2\pi)^{n-1}} \sum_{k=0}^{n} (-1)^k \begin{pmatrix} n \\ k \end{pmatrix} \delta_k(x).$$

To obtain (29), integrate from $-\infty$ to x, n-times. (You may want to do the cases $n = 2$ and 3 first.)

47. Project Problem. We will derive the following formula, for $n = 2, 3, \ldots,$

$$(30) \qquad I_n = \int_{-\infty}^{\infty} \left(\frac{\sin x}{x} \right)^n dx = \frac{\pi}{(n-1)!} \sum_{k=0}^{\left[\frac{n}{2}\right]} (-1)^k \begin{pmatrix} n \\ k \end{pmatrix} \left(\frac{n}{2} - k \right)^{n-1}.$$

We have already computed $I_1 = \pi$ (with the help of residue theory) and $I_2 = \pi$ (with the help of Plancherel's theorem). We now use the result of the previous exercise to derive the values of the integral for all other values of $n = 2, 3, \ldots$. It will be convenient to work with the indicator function of a symmetric interval. Let $g(x) = \mathcal{U}_{-\frac{1}{2}}(x) - \mathcal{U}_{\frac{1}{2}}(x)$. Note that $g(x) = f(x + \frac{1}{2})$, where f is as in Exercise 46.
(a) Using the result of Exercise 46, show that for x in $[-\frac{n}{2}, \frac{n}{2}]$,

$$g^{*n} = \frac{1}{(2\pi)^{\frac{n-1}{2}}(n-1)!} \sum_{k=0}^{\left[\frac{n}{2}+x\right]} (-1)^k \begin{pmatrix} n \\ k \end{pmatrix} \left(x + \frac{n}{2} - k \right)^{n-1}.$$

[Hint: The formula is obtained by translating the formula for f^{*n} by $\frac{n}{2}$. To justify this, verify that if ϕ_a and ψ_b are translates of ϕ and ψ, then $\phi_a * \psi_b = (\phi * \psi)_{a+b}$.]
(b) Show that

$$G_n(\omega) = \mathcal{F}(g^{*n}) = \left(\sqrt{\frac{2}{\pi}} \frac{\sin \frac{\omega}{2}}{\omega} \right)^n.$$

(c) Use the reciprocity relations and the fact that $\mathcal{F}(\phi)(0) = \frac{1}{\sqrt{2\pi}} \int_{-\infty}^{\infty} \phi(x) \, dx$ to prove that $\frac{1}{\sqrt{2\pi}} \int_{-\infty}^{\infty} G_n(x) \, dx = g^{*n}(0)$. Conclude that

$$\int_{-\infty}^{\infty} \left(\frac{\sin \frac{x}{2}}{x} \right)^n dx = \frac{\pi}{2^{n-1}(n-1)!} \sum_{k=0}^{\left[\frac{n}{2}\right]} (-1)^k \begin{pmatrix} n \\ k \end{pmatrix} \left(\frac{n}{2} - k \right)^{n-1}.$$

(d) Make a change of variables in the integral in (c) to obtain (30). In Table 1 we used (30) to compute I_n for $n = 2, 3, \ldots, 8$.

n	1	2	3	4	5	6	7	8
$\int_{-\infty}^{\infty} \left(\dfrac{\sin x}{x} \right)^n dx$	π	π	$\dfrac{3\pi}{4}$	$\dfrac{2\pi}{3}$	$\dfrac{115\pi}{192}$	$\dfrac{11\pi}{20}$	$\dfrac{5887\pi}{11520}$	$\dfrac{151\pi}{315}$

Table 1

48. Give an example of two functions f and g such that neither one vanishes identically on any interval but $f * g$ is identically 0. [Hint: Think on the Fourier transform side. Let $\hat{f}$ be a tent function supported over the interval $(-2, 2)$, as in Figure 6. Define $\hat{g}$ by translating $\hat{f}$ to the left (or right) by 4 units.]

11.3   The Fourier Transform Method

In this section we describe a method for solving boundary value problems, where one of the variables, x, y, z or t belongs to the real line. These include, for example, the wave and heat equations on the real line, and Laplace's equation in upper half-plane and others.

We will be computing Fourier transforms of functions of the form $u(x, t)$, where x and t are the variables, and at least one of them varies in the interval $(-\infty, \infty)$, say $-\infty < x < \infty$. Because of the presence of two variables, care is needed in identifying the variable with respect to which the Fourier transform is computed. For example, for fixed t, the function $u(x, t)$ becomes a function of the spatial variable x, and as such, we can take its Fourier transform with respect to the x variable. We denote this transform by $\hat{u}(\omega, t)$. Thus

FOURIER TRANSFORM IN THE x VARIABLE

$$(1) \qquad \mathcal{F}(u(x,t))(\omega) = \hat{u}(\omega, t) = \frac{1}{\sqrt{2\pi}} \int_{-\infty}^{\infty} u(x,t) e^{-i\omega x}\, dx\,.$$

To illustrate the use of this notation we compute some very useful transforms. We will assume that the function $u(x, t)$, as a function of x, has sufficient properties that enable us to use freely the operational properties of the Fourier transform from Section 11.2.

FOURIER TRANSFORM AND PARTIAL DERIVATIVES

Note that on the right sides of (2) and (3) we have used ordinary derivatives in t instead of partial derivatives. The reason is to emphasize the crucial fact that, in applying the Fourier transform method, we will transform a partial differential equation in $u(x,t)$ into an ordinary differential equation in $\hat{u}(\omega, t)$, where t is the variable. This will become more apparent in the examples.

Given $u(x, t)$ with $-\infty < x < \infty$ and $t > 0$, we have

$$(2) \qquad \mathcal{F}(\tfrac{\partial}{\partial t} u(x,t))(\omega) = \tfrac{d}{dt} \hat{u}(\omega, t);$$

$$(3) \qquad \mathcal{F}(\tfrac{\partial^n}{\partial t^n} u(x,t))(\omega) = \tfrac{d^n}{dt^n} \hat{u}(\omega, t), \quad n = 1, 2, \ldots;$$

$$(4) \qquad \mathcal{F}(\tfrac{\partial}{\partial x} u(x,t))(\omega) = i\omega \hat{u}(\omega, t);$$

$$(5) \qquad \mathcal{F}(\tfrac{\partial^n}{\partial x^n} u(x,t))(\omega) = (i\omega)^n \hat{u}(\omega, t), \quad n = 1, 2, \ldots.$$

The last two identities are consequences of (1) and Theorem 1 of Section 11.2. To prove (2) we start with the right side and differentiate under the integral sign with respect to t:

$$\frac{d}{dt} \hat{u}(\omega, t) = \frac{1}{\sqrt{2\pi}} \frac{d}{dt} \int_{-\infty}^{\infty} u(x,t) e^{-i\omega x}\, dx = \frac{1}{\sqrt{2\pi}} \int_{-\infty}^{\infty} \frac{\partial}{\partial t} u(x,t) e^{-i\omega x}\, dx\,.$$

The last expression is the Fourier transform of $\frac{\partial}{\partial t} u(x,t)$ as a function of x, and (2) follows. Repeated differentiation under the integral sign with respect to t yields (3).

The Fourier Transform Method

The use of the Fourier transform to solve partial differential equations is best described by examples. We start with the wave equation.

EXAMPLE 1   The wave equation for an infinite string

Solve the boundary value problem

$$\frac{\partial^2 u}{\partial t^2} = c^2 \frac{\partial^2 u}{\partial x^2} \qquad (-\infty < x < \infty, t > 0),$$

$$u(x,0) = f(x) \qquad \text{(initial displacement)},$$

$$\frac{\partial}{\partial t} u(x,0) = g(x) \qquad \text{(initial velocity)}.$$

Give the answer as an inverse Fourier transform.

Solution We fix t and take the Fourier transform of both sides of the differential equation and the initial conditions. Using (3) and (5) with $n = 2$, we get

$$(6) \qquad\qquad \frac{d^2}{dt^2}\widehat{u}(\omega,t) = -c^2\omega^2\widehat{u}(\omega,t),$$

$$(7) \qquad\qquad \widehat{u}(\omega,0) = \widehat{f}(\omega),$$

$$(8) \qquad\qquad \frac{d}{dt}\widehat{u}(\omega,0) = \widehat{g}(\omega).$$

It is clear that (6) is an ordinary differential equation in $\widehat{u}(\omega,t)$, where t is the variable. Let us write (6) in the standard form

$$\frac{d^2}{dt^2}\widehat{u}(\omega,t) + c^2\omega^2\widehat{u}(\omega,t) = 0.$$

The general solution of this equation is

See Appendix A.2 for the solution of second order linear differential equations (Case III).

$$\widehat{u}(\omega,t) = A(\omega)\cos c\omega t + B(\omega)\sin c\omega t,$$

where $A(\omega)$ and $B(\omega)$ are constant in t. (You should note that while A and B are constant in t, they can depend on ω, which explains writing $A(\omega)$ and $B(\omega)$.) We determine $A(\omega)$ and $B(\omega)$ from the initial conditions (7) and (8) as follows:

$$\widehat{u}(\omega,0) = A(\omega) = \widehat{f}(\omega),$$

$$\frac{d}{dt}\widehat{u}(\omega,0) = c\omega B(\omega) = \widehat{g}(\omega).$$

So

$$\widehat{u}(\omega,t) = \widehat{f}(\omega)\cos c\omega t + \frac{1}{c\omega}\widehat{g}(\omega)\sin c\omega t.$$

To get the solution we use the inverse Fourier transform and get

$$(9) \qquad u(x,t) = \frac{1}{\sqrt{2\pi}}\int_{-\infty}^{\infty}[\widehat{f}(\omega)\cos c\omega t + \frac{1}{c\omega}\widehat{g}(\omega)\sin c\omega t]e^{i\omega x}d\omega. \qquad\blacksquare$$

Formula (9) gives you the solution of the wave boundary value problem in the form of an integral involving the Fourier transform of the initial displacement and velocity. In specific cases, these transforms and the integral can be computed explicitly. Even in its general form, (9) can be computed explicitly in terms of f and g to yield d'Alembert's form of the solution. (See Exercise 21.)

We summarize the Fourier transform method as follows.

Step 1: Fourier transform the given boundary value problem in $u(x, t)$ and get an ordinary differential equation in $\widehat{u}(\omega, t)$ in the variable t.
Step 2: Solve the ordinary differential equation and find $\widehat{u}(\omega, t)$.
Step 3: Inverse Fourier transform $\widehat{u}(\omega, t)$ to get $u(x, t)$.

This simple method is successful in treating a variety of partial differential equations. In our next example we use it to solve the heat equation on the real line. The example models the transfer of heat in a very long rod extending in both directions on the x-axis.

EXAMPLE 2   The heat equation for an infinite rod
Solve the boundary value problem

$$\frac{\partial u}{\partial t} = c^2 \frac{\partial^2 u}{\partial x^2} \qquad (-\infty < x < \infty,\ t > 0),$$
$$u(x, 0) = f(x) \qquad \text{(initial temperature distribution)}.$$

Give your answer in the form of an inverse Fourier transform.

Solution Fourier transforming the boundary value problem, we get

$$\frac{d}{dt}\widehat{u}(\omega, t) = -c^2 \omega^2 \widehat{u}(\omega, t),$$
$$\widehat{u}(\omega, 0) = \widehat{f}(\omega).$$

The general solution of the first order differential equation in t is

$$\widehat{u}(\omega, t) = A(\omega)e^{-c^2\omega^2 t},$$

Once again, since we are dealing with a family of differential equations (one for each ω), we must use a different constant for each equation, which explains the use of the notation $A(\omega)$.

where $A(\omega)$ is a constant that depends on ω. Setting $t = 0$ and using the transformed initial condition, we get

$$\widehat{u}(\omega, 0) = A(\omega) = \widehat{f}(\omega).$$

Hence

$$\widehat{u}(\omega, t) = \widehat{f}(\omega)e^{-c^2\omega^2 t}.$$

Taking inverse Fourier transforms we get the solution

$$(10) \qquad u(x, t) = \frac{1}{\sqrt{2\pi}} \int_{-\infty}^{\infty} \widehat{f}(\omega)e^{-c^2\omega^2 t} e^{i\omega x}\, d\omega. \qquad \blacksquare$$

Formula (10) gives you the solution of the heat boundary value problem in the form of an integral involving the Fourier transform of the initial heat distribution. In the next section, we will compute this integral and give the answer as a convolution of f and a fixed function known as the heat or Gauss's kernel.

Our next two examples illustrate the use of the Fourier transform method in solving problems with mixed and higher order derivatives.

EXAMPLE 3   The Fourier transform method with mixed derivatives
Solve the boundary value problem

$$\frac{\partial^2 u}{\partial t \partial x} = \frac{\partial^2 u}{\partial x^2} \quad (-\infty < x < \infty,\ t > 0),$$

$$u(x,0) = \sqrt{\frac{\pi}{2}} e^{-|x|}.$$

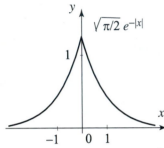

$\sqrt{\pi/2}\, e^{-|x|}$

Figure 1 Graph of $\sqrt{\frac{\pi}{2}} e^{-|x|}$.

The function $f(x) = \sqrt{\frac{\pi}{2}} e^{-|x|}$ is shown in Figure 1.

Solution Using (2) and (4), we obtain

$$\mathcal{F}\left(\frac{\partial^2 u}{\partial t \partial x}\right) = \mathcal{F}\left(\frac{\partial}{\partial t}\frac{\partial u}{\partial x}\right) = \frac{d}{dt}\mathcal{F}\left(\frac{\partial u}{\partial x}\right) = i\omega\frac{d\widehat{u}}{dt}(\omega,t).$$

Fourier transforming the differential equation and using (5), we get

$$i\omega\frac{d\widehat{u}}{dt}(\omega,t) = -\omega^2\widehat{u}(\omega,t).$$

Solving this first order ordinary differential equation, we find

$$\widehat{u}(\omega,t) = A(\omega)e^{i\omega t}.$$

The initial condition implies that

$$\widehat{u}(\omega,0) = \mathcal{F}\left(\sqrt{\frac{\pi}{2}} e^{-|x|}\right) = \frac{1}{1+\omega^2},$$

and so $A(\omega) = \frac{1}{1+\omega^2}$. Hence

$$\widehat{u}(\omega,t) = \frac{e^{i\omega t}}{1+\omega^2},$$

and the solution u is obtained by taking inverse Fourier transforms. In this case, we can determine u explicitly by using the shifting property of the Fourier transform (Theorem 3(i), Section 11.2). We first note that the inverse Fourier transform of $\frac{1}{1+\omega^2}$ is $\sqrt{\frac{\pi}{2}} e^{-|x|}$. Hence by the shifting property,

$$u(x,t) = \mathcal{F}^{-1}\left(\frac{e^{i\omega t}}{1+\omega^2}\right) = \sqrt{\frac{\pi}{2}} e^{-|x+t|}.$$

The solution is illustrated in Figure 2. ∎

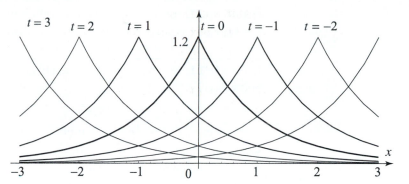

Figure 2 Graphs of $u(x,t)$ in Example 3 at various values of t.

Our next example features an important use of implicit conditions that are not usually given as part of the statement of the problem.

EXAMPLE 4   Use of implicit boundedness assumptions
Solve the boundary value problem

$$\frac{\partial^2 u}{\partial t^2} = \frac{\partial^4 u}{\partial x^4} \qquad (-\infty < x < \infty,\ t > 0),$$
$$u(x,0) = f(x).$$

Give your answer in the form of an inverse Fourier transform.

Solution   Fourier transforming the problem gives

$$\frac{d^2 \widehat{u}}{dt^2}(\omega, t) - \omega^4 \widehat{u}(\omega, t) = 0 \qquad (\text{because } (i\omega)^4 = \omega^4),$$
$$\widehat{u}(\omega, 0) = \widehat{f}(\omega).$$

Solving the second order differential equation in t yields

$$\widehat{u}(\omega, t) = A(\omega)e^{-\omega^2 t} + B(\omega)e^{\omega^2 t}.$$

At this point, we impose certain boundedness conditions on $\widehat{u}(\omega, t)$. These conditions follow from the fact that the solution (in most reasonable cases) should have a bounded Fourier transform $\widehat{u}$ for all $t > 0$ and ω. Consequently, this assumption forces $B(\omega) = 0$ for all ω; otherwise, by letting $t \to \infty$, we get an unbounded Fourier transform. Hence,

$$\widehat{u}(\omega, t) = A(\omega)e^{-\omega^2 t}.$$

Now, using the transformed initial condition, we get

$$\widehat{u}(\omega, t) = \widehat{f}(\omega)e^{-\omega^2 t}.$$

Finally, taking the inverse Fourier transform yields

$$u(x,t) = \frac{1}{\sqrt{2\pi}} \int_{-\infty}^{\infty} \widehat{f}(\omega)e^{-\omega^2 t}e^{i\omega x}\, d\omega.$$

■

So far we have used the Fourier transform method to solve partial differential equations with constant coefficients. As our next example illustrates, the method is also useful in solving problems with nonconstant coefficients.

EXAMPLE 5   An equation with nonconstant coefficients
Solve

$$t\frac{\partial u}{\partial x} + \frac{\partial u}{\partial t} = 0 \quad u(x,0) = f(x),$$

where $-\infty < x < \infty$, $t > 0$. Simplify your answer as much as possible.

Solution   The new feature in this example is the presence of the term $t\frac{\partial u}{\partial x}$ in the equation. Since we will use the Fourier transform with respect to x, the variable t will be treated as a constant. Thus

$$\mathcal{F}\left(t\frac{\partial u}{\partial x}\right) = t\mathcal{F}\left(\frac{\partial u}{\partial x}\right) = i\omega t\widehat{u}(\omega,t).$$

Going back to our problem, we use the Fourier transform and get

$$i\omega t\,\widehat{u}(\omega,t) + \tfrac{d}{dt}\widehat{u}(\omega,t) = 0,$$
$$\widehat{u}(\omega,0) = \widehat{f}(\omega).$$

Solving the first order differential equation in t yields

$$\widehat{u}(\omega,t) = A(\omega)e^{-i\frac{t^2}{2}\omega},$$

where the arbitrary constant, $A(\omega)$, is allowed to be a function of ω. Putting $t = 0$ implies

$$\widehat{u}(\omega,t) = \widehat{f}(\omega)e^{-i\frac{t^2}{2}\omega}.$$

To determine u, we appeal to Theorem 3, Section 11.2, and obtain

$$u(x,t) = f\left(x - \frac{t^2}{2}\right).$$

It is instructive to check our answer at this point. We have $u_x(x,t) = f'(x - \frac{t^2}{2})$; and, by the chain rule, $u_t(x,t) = -tf'(x - \frac{t^2}{2})$. Thus $tu_x + u_t = 0$, verifying the equation. At $t = 0$, we get $u(x,0) = f(x)$, as desired.   ■

Exercises 11.3

In Exercises 1–6, determine the solution of the given wave or heat problem. Give your answer in the form of an inverse Fourier transform. Take the variables in the ranges $-\infty < x < \infty$, $t > 0$.

1.

$$\frac{\partial^2 u}{\partial t^2} = \frac{\partial^2 u}{\partial x^2},$$
$$u(x,0) = \frac{1}{1+x^2}, \quad \frac{\partial u}{\partial t}(x,0) = 0.$$

2.

$$\frac{\partial^2 u}{\partial t^2} = \frac{\partial^2 u}{\partial x^2},$$
$$u(x,0) = \begin{cases} \cos x & \text{if } -\frac{\pi}{2} \leq x \leq \frac{\pi}{2}, \\ 0 & \text{otherwise}, \end{cases}$$
$$\frac{\partial u}{\partial t}(x,0) = 0.$$

3.   **4.**

$$\frac{\partial u}{\partial t} = \frac{1}{4}\frac{\partial^2 u}{\partial x^2},$$

$$u(x,0) = e^{-x^2}.$$

$$\frac{\partial u}{\partial t} = \frac{1}{100}\frac{\partial^2 u}{\partial x^2},$$

$$u(x,0) = \begin{cases} 100 & \text{if } -1 < x < 1, \\ 0 & \text{otherwise.} \end{cases}$$

5.   **6.**

$$\frac{\partial^2 u}{\partial t^2} = c^2\frac{\partial^2 u}{\partial x^2},$$

$$u(x,0) = \sqrt{\frac{2}{\pi}}\frac{\sin x}{x}, \quad \frac{\partial u}{\partial t}(x,0) = 0.$$

$$\frac{\partial u}{\partial t} = \frac{\partial^2 u}{\partial x^2},$$

$$u(x,0) = \begin{cases} 1 - \frac{|x|}{2} & \text{if } -2 < x < 2, \\ 0 & \text{otherwise.} \end{cases}$$

In Exercises 7–20, solve the given problem. Take $-\infty < x < \infty$, and $t > 0$.

7.   **8.**

$$\frac{\partial u}{\partial x} + 3\frac{\partial u}{\partial t} = 0,$$

$$u(x,0) = f(x).$$

$$a\frac{\partial u}{\partial x} + b\frac{\partial u}{\partial t} = 0,$$

$$u(x,0) = f(x).$$

9.   **10.**

$$t^2\frac{\partial u}{\partial x} - \frac{\partial u}{\partial t} = 0,$$

$$u(x,0) = 3\cos x.$$

$$a(t)\frac{\partial u}{\partial x} + \frac{\partial u}{\partial t} = 0,$$

$$u(x,0) = f(x).$$

11.   **12.**

$$\frac{\partial u}{\partial x} = \frac{\partial u}{\partial t},$$

$$u(x,0) = f(x).$$

$$\frac{\partial u}{\partial t} + \sin t\frac{\partial u}{\partial x} = 0,$$

$$u(x,0) = \sin x.$$

13.   **14.**

$$\frac{\partial u}{\partial t} = t\frac{\partial^2 u}{\partial x^2},$$

$$u(x,0) = f(x).$$

$$\frac{\partial u}{\partial t} = a(t)\frac{\partial^2 u}{\partial x^2},$$

$$u(x,0) = f(x),$$

where $a(t) > 0$.

15.   **16.**

$$\frac{\partial^2 u}{\partial t^2} + 2\frac{\partial u}{\partial t} = -u,$$

$$u(x,0) = f(x), \quad u_t(x,0) = g(x).$$

$$\frac{\partial u}{\partial t} = e^{-t}\frac{\partial^2 u}{\partial x^2},$$

$$u(x,0) = 100.$$

17.

$$\frac{\partial^2 u}{\partial t^2} = \frac{\partial^4 u}{\partial x^4}$$

$$u(x,0) = \begin{cases} 100 & \text{if } |x| < 2, \\ 0 & \text{otherwise.} \end{cases}$$

18.

$$\frac{\partial u}{\partial t} = t \frac{\partial^4 u}{\partial x^4},$$

$$u(x,0) = f(x).$$

19.

$$\frac{\partial^2 u}{\partial t^2} = \frac{\partial^3 u}{\partial t \partial x^2},$$

$$u(x,0) = f(x), \quad u_t(x,0) = g(x).$$

20.

$$\frac{\partial^2 u}{\partial t^2} - 4 \frac{\partial^3 u}{\partial t \partial x^2} + 3 \frac{\partial^4 u}{\partial x^4} = 0,$$

$$u(x,0) = f(x), \quad u_t(x,0) = g(x).$$

21. D'Alembert's solution of the wave equation
(a) Verify that

$$u(x,t) = \frac{1}{2}[f(x - ct) + f(x + ct)] + \frac{1}{2c} \int_{x-ct}^{x+ct} g(s)\, ds$$

is a solution of the boundary value problem of Example 1.
(b) Derive d'Alembert's solution from (9) of this section and Theorem 3 and Exercise 26 of Section 11.2.

22. (a) Use D'Alembert's solution to describe the vibration of a very long string with $c = 1$, $f(x) = \cos x$ for $|x| \leq \frac{\pi}{2}$ and 0 otherwise, and $g(x) = 0$.
(b) Draw the shape of the string at $t = 0, \pi/4, \pi/2, \pi$.

Project Problem. Do Exercises 23 and 24 to solve a **heat problem with convection**.

23. Solve the boundary value problem

$$\frac{\partial u}{\partial t} = c^2 \frac{\partial^2 u}{\partial x^2} + k \frac{\partial u}{\partial x}, \quad -\infty < x < \infty, \quad t > 0,$$

$$u(x,0) = f(x).$$

This problem models heat transfer in a long heated bar that is exchanging heat with the surrounding medium. This phenomenon is called **convection**, and k is a positive constant called the **coefficient of convection**. (See Exercise 10, Section 11.4, for the convolution form of the solution.)

24. Specialize Exercise 23 to the case $c = 1$, $k = .5$, $f(x) = e^{-x^2}$.

Project Problem. In Exercises 25 and 26, you are asked to study the vibrations of a very long elastic beam.
25. The vibrations of a very long beam extending in both directions on the x-axis are modeled by the boundary value problem

$$\frac{\partial^2 u}{\partial t^2} = c^2 \frac{\partial^4 u}{\partial x^4}, \quad -\infty < x < \infty, \ t > 0,$$

$$u(x,0) = f(x), \quad \frac{\partial u}{\partial t}(x,0) = g(x).$$

Solve this problem by the Fourier transform method.

26. Specialize Exercise 25 to the case $c = 1$, $f(x) = \delta_0(x)$, $g(x) = 0$, where $\delta_0(x)$ is the Dirac delta function.

Project Problem. Do Exercises 27 and 28.

27. Solve the boundary value problem

$$
\begin{aligned}
\frac{\partial u}{\partial t} &= c^2 \frac{\partial^3 u}{\partial x^3}, & -\infty < x < \infty, \ t > 0, \\
u(x,0) &= f(x).
\end{aligned}
$$

This equation is known as the **linearized Korteweg-de Vries equation**.

28. Specialize Exercise 27 to the case $c = 1$, $f(x) = e^{-x^2/2}$.

11.4   The Heat Equation and Gauss's Kernel

In the previous section we used the Fourier transform to solve familiar boundary value problems, such as heat and wave problems. The solution was expressed as an inverse Fourier transform involving the Fourier transforms of the initial data. For practical, numerical applications, it is desirable to evaluate the inverse Fourier transform and express the solution in terms of the initial data itself. In this and the next section we illustrate these ideas with two important applications associated with the heat equation and Dirichlet problems.

Let us start with the heat problem (Example 2, Section 11.3):

$$
\begin{aligned}
(1) \qquad \frac{\partial u}{\partial t} &= c^2 \frac{\partial^2 u}{\partial x^2}, & (-\infty < x < \infty, \ t > 0) \\
u(x,0) &= f(x).
\end{aligned}
$$

From Example 2 of the previous section, we know that

$$
(2) \qquad \widehat{u}(\omega, t) = \widehat{f}(\omega)e^{-c^2\omega^2 t},
$$

and hence, by applying the inverse Fourier transform,

$$
(3) \qquad u(x, t) = \frac{1}{\sqrt{2\pi}} \int_{-\infty}^{\infty} \widehat{f}(\omega)e^{-c^2\omega^2 t}e^{i\omega x} \, d\omega.
$$

Our goal in this section is to evaluate this inverse Fourier transform in terms of f. We note from (2) that $\widehat{u}$ is the product of two Fourier transforms, one of them being $\widehat{f}$ and the other $e^{-c^2\omega^2 t}$. Remembering that products of Fourier transforms correspond to convolutions, we see that u is the convolution of f with the function whose Fourier transform is $e^{-c^2\omega^2 t}$. As we will see in a moment, this function is the so-called **heat kernel** or **Gauss's kernel**

$$
g_t(x) = \frac{1}{c\sqrt{2t}} e^{-x^2/4c^2 t}.
$$

Figure 1 Gauss's kernel, $g_t(x)$, is one of the most important functions in applied mathematics. For each $t > 0$, the graph of $g_t(x)$ is a bell-shaped curve, symmetric with respect to the y-axis. As t tends to zero, the curves become more and more localized near zero. The area under each curve is constant, even as t varies (see Exercise 29).

(We think of the heat kernel $g_t(x)$ as a family of functions of x; one function for each $t > 0$, as illustrated in Figure 1.) This leads us to the following important result.

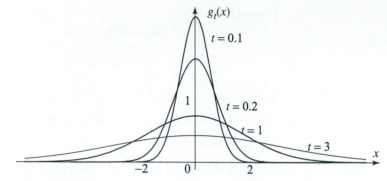

THEOREM 1
SOLUTION OF THE
HEAT EQUATION AS
A CONVOLUTION

The solution of the heat equation (1) with initial temperature distribution f is the convolution of f with the heat kernel. More explicitly,

$$(4) \qquad u(x,t) = \frac{1}{c\sqrt{2t}}\, e^{-x^2/4c^2t} * f = \frac{1}{2c\sqrt{\pi t}} \int_{-\infty}^{\infty} f(s) e^{-(x-s)^2/4c^2t}\, ds.$$

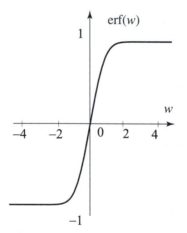

Figure 2 The error function is the integral of a Gaussian function.

Proof We have to show that the function $g_t(x) = \frac{1}{c\sqrt{2t}}\, e^{-x^2/4c^2t}$ has Fourier transform $e^{-c^2\omega^2 t}$. But this is immediate from Example 5 of Section 11.1, with $a = \frac{1}{4c^2t}$. ∎

With the help of (4) we can express the solution of some problems in terms of special functions that can be used to study the solution. We illustrate with the so-called **error function** (Figure 2), which is defined by

$$(5) \qquad \operatorname{erf}(w) = \frac{2}{\sqrt{\pi}} \int_{0}^{w} e^{-z^2}\, dz, \qquad \text{for all } w.$$

The error function is used in different contexts of applied mathematics and probability. Its numerical values are tabulated and it is available as a standard function in most computer systems. For its basic properties, see Exercise 15. In the next example, we express the solution of the given heat problem in terms of the error function.

EXAMPLE 1   An application of the error function

Solve the heat problem on the infinite line with $c = 1$ and initial temperature distribution $f(x) = 100$ if $|x| < 1$ and 0 otherwise. Express your answer in terms of the error function.

Solution Applying (4), we obtain

$$u(x,t) = \frac{50}{\sqrt{\pi t}} \int_{-1}^{1} e^{-(x-s)^2/4t}\, ds.$$

Let $z = \frac{x-s}{2\sqrt{t}}$, $dz = \frac{-1}{2\sqrt{t}}\,ds$, then

$$
\begin{aligned}
u(x,t) &= \frac{100}{\sqrt{\pi}} \int_{\frac{x-1}{2\sqrt{t}}}^{\frac{x+1}{2\sqrt{t}}} e^{-z^2}\,dz = \frac{100}{\sqrt{\pi}} \left(\int_0^{\frac{x+1}{2\sqrt{t}}} e^{-z^2}\,dz - \int_0^{\frac{x-1}{2\sqrt{t}}} e^{-z^2}\,dz \right) \\
&= 50 \left[\operatorname{erf}\left(\frac{x+1}{2\sqrt{t}} \right) - \operatorname{erf}\left(\frac{x-1}{2\sqrt{t}} \right) \right] \qquad \text{(by (5))}.
\end{aligned}
$$

Using this formula, we have plotted in Figure 3 the graphs of $u(x,t)$ at various values of t. The graphs show that initially the temperature drops fast, and then slowly it reaches the equilibrium temperature of 0. ∎

Figure 3 The temperature distribution in Example 1 at various values of $t > 0$. For small values of t, the temperature in the bar is close to the initial temperature distribution. As t increases, the temperature spreads through the bar and eventually tends to 0.

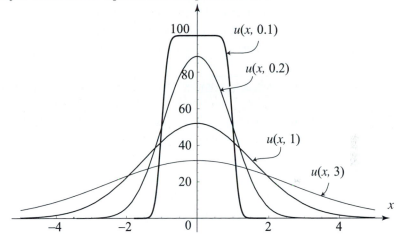

Our next example deals with a heat problem with nonconstant coefficients.

EXAMPLE 2   A heat problem with nonconstant coefficients
Solve

$$
\begin{aligned}
u_t &= t u_{xx}, \quad (-\infty < x < \infty,\ t > 0), \\
u(x,0) &= f(x).
\end{aligned}
$$

Express your answer as a convolution.
Solution We cannot appeal to (4), since the problem here involves a new equation. We will first apply the Fourier transform method and then complete the solution by using the same ideas leading to (4). Fourier transforming the problem, we get

$$
\frac{d}{dt}\,\widehat{u}(\omega,t) = -t\,\omega^2 \widehat{u}(\omega,t), \qquad \widehat{u}(\omega,0) = \widehat{f}(\omega).
$$

Solving this first order differential equation, we find

$$
\widehat{u}(\omega,t) = \widehat{f}(\omega) e^{-\frac{t^2}{2}\omega^2}.
$$

Hence $u(x,t)$ is the convolution of f with the function whose Fourier transform is

$e^{-\frac{t^2}{2}\omega^2}$. Using the result of Example 5, Section 11.1, we find that

$$\mathcal{F}\left(\frac{e^{-\frac{x^2}{2t^2}}}{t}\right)(\omega) = e^{-\frac{t^2}{2}\omega^2}.$$

Thus

$$u(x,t) = f * \frac{e^{-\frac{x^2}{2t^2}}}{t} = \frac{1}{t\sqrt{2\pi}}\int_{-\infty}^{\infty} f(s)e^{-\frac{(x-s)^2}{2t^2}}\,ds\,. \qquad \blacksquare$$

The differential equation in Example 2 can be interpreted as modeling heat transfer in a bar with time-varying thermal diffusivity equal to t. Thus, we expect the rate of heat transfer to vary with t. We next experiment with these ideas by comparing the solutions in Examples 1 and 2.

EXAMPLE 3   Varying the thermal diffusivity

(a) In Example 2, take f as in Example 1, and express the solution in terms of erf(x).

(b) Call $u_1(x,t)$ the solution in Example 1 and $u_2(x,t)$ the solution in part (a). Plot and compare u_1 and u_2 at $t = .2, 1, 1.8, 3.4$ and 4.2.

Solution (a) Using the given f in the result of Example 2, we find

$$
\begin{aligned}
u(x,t) &= \frac{100}{t\sqrt{2\pi}}\int_{-1}^{1} e^{-\frac{(x-s)^2}{2t^2}}\,ds \\[2mm]
&= \frac{100}{t\sqrt{2\pi}}\int_{\frac{x-1}{\sqrt{2}\,t}}^{\frac{x+1}{\sqrt{2}\,t}} e^{-z^2}(-\sqrt{2}\,t)dz \quad \left(\text{let } z = \frac{x-s}{\sqrt{2}\,t},\ dz = -\frac{ds}{\sqrt{2}\,t}\right) \\[2mm]
&= \frac{100}{\sqrt{\pi}}\left(\int_{0}^{\frac{x+1}{\sqrt{2}\,t}} e^{-z^2}\,dz - \int_{0}^{\frac{x-1}{\sqrt{2}\,t}} e^{-z^2}\,dz\right) \\[2mm]
&= 50\left[\mathrm{erf}\left(\frac{x+1}{\sqrt{2}\,t}\right) - \mathrm{erf}\left(\frac{x-1}{\sqrt{2}\,t}\right)\right] \qquad \text{(by (5))}.
\end{aligned}
$$

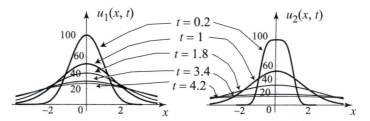

Figure 4 Varying the thermal diffusivity and its effect on the solution.

(b) Using this formula, we have plotted in Figure 4 the solution at the designated values of t. The graphs show that for small values of t, $u_1(x,t)$ (the temperature in Example 1) drops faster than $u_2(x,t)$ (the temperature in Example 2). As t increases, the thermal diffusivity increases and this results in a faster rate of change of the temperature, as illustrated in the graphs at times $t = 3.4$ and 4.2. $\qquad \blacksquare$

The theory of heat that we have developed so far requires some kind of integrability conditions on the initial temperature distribution. In some cases, however, the solution of the problem is obvious, even though the initial temperature distribution is not integrable. Consider the case of the initial temperature distribution $f(x) = 1$ for all x. Clearly $f(x)$ is not integrable on the entire line, but intuitively the problem has a solution $u(x, t) = 1$. This problem and many others can be solved by appealing to (4) directly, which proves another advantage to the convolution form of the solution. We explore these applications in the exercises.

Exercises 11.4

In Exercises 1–14, use convolutions, the error function, and other operational properties of the Fourier transform to solve the boundary value problem. Take $-\infty < x < \infty$, $t > 0$.

1.

$$\frac{\partial u}{\partial t} = \frac{1}{4}\frac{\partial^2 u}{\partial x^2},$$
$$u(x, 0) = \begin{cases} 20 & \text{if } -1 < x < 1, \\ 0 & \text{otherwise.} \end{cases}$$

2.

$$\frac{\partial u}{\partial t} = \frac{1}{100}\frac{\partial^2 u}{\partial x^2},$$
$$u(x, 0) = \begin{cases} 100 & \text{if } -2 < x < 0, \\ 50 & \text{if } 0 < x < 1, \\ 0 & \text{otherwise.} \end{cases}$$

3.

$$\frac{\partial u}{\partial t} = \frac{\partial^2 u}{\partial x^2},$$
$$u(x, 0) = 70\, e^{-x^2/2}.$$

4.

$$\frac{\partial u}{\partial t} = \frac{\partial^2 u}{\partial x^2},$$
$$u(x, 0) = \begin{cases} 100(1 - \frac{|x|}{2}) & \text{if } -2 \le x \le 2, \\ 0 & \text{otherwise.} \end{cases}$$

5.

$$\frac{\partial u}{\partial t} = \frac{\partial^2 u}{\partial x^2},$$
$$u(x, 0) = \frac{100}{1 + x^2}.$$

6.

$$\frac{\partial u}{\partial t} = \frac{\partial^2 u}{\partial x^2},$$
$$u(x, 0) = e^{-|x|}.$$

7.

$$\frac{\partial u}{\partial t} = t^2 \frac{\partial^2 u}{\partial x^2},$$
$$u(x, 0) = f(x).$$

8.

$$\frac{\partial^2 u}{\partial t^2} = \frac{\partial^4 u}{\partial x^4},$$
$$u(x, 0) = f(x).$$

9.

$$\frac{\partial u}{\partial t} = e^{-t} \frac{\partial^2 u}{\partial x^2},$$
$$u(x, 0) = f(x).$$

10.

$$\frac{\partial u}{\partial t} = c^2 \frac{\partial^2 u}{\partial x^2} + k\frac{\partial u}{\partial x} \quad (k > 0),$$
$$u(x, 0) = f(x).$$

11.   **12.**

$$\frac{\partial u}{\partial t} = a(t)\frac{\partial^2 u}{\partial x^2},$$
$$u(x,0) = f(x),$$

$$-\frac{\partial^2 u}{\partial x^2} = \frac{\partial^2 u}{\partial t^2} + 2\frac{\partial^2 u}{\partial t \partial x},$$
$$u(x,0) = f(x).$$

where $a(t) > 0$.

13. Solve Exercise 9 with $f(x)$ as in Example 1. Compare your solution to that of Example 1. Model your answer after Example 3.

14. Solve Exercise 11 with $a(t) = e^t$ and $f(x)$ as in Example 1. Compare your solution to that of Example 1. Model your answer after Example 3.

15.   Project Problem: Basic properties of the error function. Establish the following.
(a) $\text{erf}(-x) = -\text{erf}(x)$ (erf is an odd function).
(b) $\text{erf}(0) = 0$, $\text{erf}(\infty) = 1$.
(c) $\frac{d}{dx}\text{erf}(x) = \frac{2}{\sqrt{\pi}}e^{-x^2}$. Conclude that erf is strictly increasing.
(d) $\int \text{erf}(x)\,dx = x\,\text{erf}(x) + \frac{1}{\sqrt{\pi}}e^{-x^2} + C$. [Hint: Integration by parts.]
(e) $\text{erf}(x) = \frac{2}{\sqrt{\pi}}\sum_{n=0}^{\infty}\frac{(-1)^n}{n!(2n+1)}x^{2n+1}$.

16. The **complementary error function** is defined by

$$\text{erfc}(w) = \frac{2}{\sqrt{\pi}}\int_{w}^{\infty}e^{-z^2}\,dz, \quad \text{for all } w.$$

(a) Show that $\text{erf}(w) + \text{erfc}(w) = \frac{2}{\sqrt{\pi}}\int_{0}^{\infty}e^{-z^2}\,dz = 1$.
(b) Conclude that $\text{erfc}(w) = 1 - \text{erf}(w)$.
(c) Use the graph of the error function to plot the complementary error function.

17. (a) Use convolution to solve the heat problem with given initial temperature distribution $f(x) = T_0$ for $a < x < b$ and 0 otherwise.
(b) Express your answer in terms of the error function.

18. Consider the heat problem of Example 1. Vary c by taking $c = 1$, $c = 2$, $c = 1/2$. Plot the corresponding solution $u(x,t)$ for $-10 < x < 10$ and $0 < t < 20$. What can you say about the propagation of heat as a function of c? Justify your answer using the graphs.

19. Constant function as an initial temperature distribution. It is obvious that if we take the initial temperature distribution $f(x)$ to be identically equal to 1, then the solution of the heat equation (1) is $u(x,t) = 1$ for all x and $t > 0$. Use (4) to confirm this fact.

20.   A step function as an initial temperature distribution.
(a) Use (4) to show that if the initial temperature distribution is $f(x) = T_0$ if $x > 0$ and 0 otherwise, then

$$u(x,t) = \frac{T_0}{2}\left[1 + \text{erf}\left(\frac{x}{2c\sqrt{t}}\right)\right].$$

(b) Plot the solution for several values of $t > 0$. What do you observe? Do the graphs meet with your expectation?

21. Dirac delta function as an initial temperature distribution.
To analyze the temperature response to an application of a welding torch to a rod at a point x_0, we can take the initial temperature distribution to be $f(x) = \delta_0(x - x_0)$, where δ_0 is the Dirac delta function.
(a) Use (4) to show that

$$u(x, t) = \frac{1}{2c\sqrt{\pi t}} e^{-(x-x_0)^2/(4c^2t)}.$$

(b) Take $x_0 = 0$, $c = 1$, and plot the graphs of $u(x, t)$ for $-10 < x < 10$ and $t = .01, .05, 0.1, 0.5, 1, 5, 10, 15$. Describe what you observe on the graphs.

22. Let $f(x) = \frac{\sin x}{x}$ be the initial temperature distribution in an infinite rod with $c = 1$.
(a) Use (3) to obtain that $u(x, t) = \int_0^1 e^{-\omega^2 t} \cos \omega x \, d\omega$.
(b) Verify that the answer in (a) is indeed a solution by plugging into the equation and checking the initial values.
(c) Show that $u(0, t) = \frac{\sqrt{\pi}}{2\sqrt{t}} \operatorname{erf}(\sqrt{t})$. What is the physical interpretation of this function?
(d) How long does it take for the temperature at $x = 0$ to drop by 80 percent?

23. Shifting the initial data. If you shift the initial temperature distribution by a units, that is, if you replace $f(x)$ by $f(x - a)$, you would expect the solution to be shifted by the same amount on the x-axis. Confirm this fact by using (4) to show that if $f(x)$ is replaced by $f(x - a)$, then the solution becomes $u(x - a, t)$. [Hint: Change variables.]

24. Initial Gaussian temperature distribution. Use the result of Example 4, Section 11.2, to show that the solution of the heat equation (1) with initial temperature distribution $f(x) = e^{-kx^2}$, $(k > 0)$, is

$$u(x, t) = \frac{1}{\sqrt{4kc^2t + 1}} e^{\frac{-kx^2}{4kc^2t+1}}.$$

In Exercises 25–28, (a) use the results of Exercise 23 and 24 to solve the heat equation (1) subject to the given initial temperature distribution.
(b) Illustrate your answer by plotting the solution for various values of t.

25. $f(x) = e^{-(x-1)^2}$.   **26.** $f(x) = 100 \, e^{-(x+1)^2}$.

27. $f(x) = e^{-(x-2)^2/2}$.   **28.** $f(x) = e^{-(x-2)^2}$.

29. Project Problem: Properties of Gauss's kernel.
Establish the following properties of Gauss's kernel:

$$g_t(x) = \frac{1}{c\sqrt{2t}} e^{-x^2/4c^2t} \quad (c > 0, \ t > 0, \ -\infty < x < \infty).$$

(a) $g_t(x)$ is an even function of x, and $g_t(x) \geq 0$ for all x.
(b) The graph of $g_t(x)$ is a bell-shaped curve centered at the origin. Verify this assertion by plotting $g_t(x)$ for $c = 1$ and $t = 1, 2, 3$.
(c) $\lim_{t \to 0} g_t(0) = \infty$. (As t tends to 0, the graph of $g_t(x)$ becomes more and more localized near 0, in the sense that most of the total area under the graph and above the x-axis becomes concentrated near 0. Again, you can verify this assertion

graphically.)

(d) The total area under the graph of $g_t(x)$ and above the x-axis is $\int_{-\infty}^{\infty} g_t(x)\,dx = \sqrt{2\pi}$.

(e) The Fourier transform of $g_t(x)$ is $\widehat{g_t}(\omega) = e^{-c^2 t\omega^2}$.

(f) If f is an integrable and piecewise smooth function, then at its points of continuity, we have

$$\lim_{t \to 0} g_t * f(x) = f(x).$$

Justify this assertion on the basis of what it means in terms of the solution of the heat problem. Alternatively, you can use the Fourier transform and the fact that $\lim_{t \to 0} \mathcal{F}(g_t)(\omega) = 1$, which is a consequence of (e).

11.5   The Poisson Integral and the Hilbert Transform

Consider the **Dirichlet problem in the upper half-plane** (Figure 1)

(1) $$\Delta u = \frac{\partial^2 u}{\partial x^2} + \frac{\partial^2 u}{\partial y^2} = 0, \qquad (-\infty < x < \infty,\ y > 0)$$

(2) $$u(x, 0) = f(x), \qquad \text{(boundary values)}.$$

The problem is asking us to find a harmonic function $u(x, y)$ in the upper half-plane, which tends to a given function $f(x)$ as we approach the boundary of the upper-half plane, namely the x-axis. We have solved this problem in Theorem 1, Section 6.3, using conformal mappings and the Poisson integral formula on the disk. For ease of reference, let us recall the solution

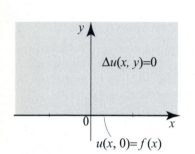

(3) $$u(x, y) = \frac{y}{\pi} \int_{-\infty}^{\infty} \frac{f(s)}{(x - s)^2 + y^2}\,ds = P_y * f(x).$$

Figure 1 Dirichlet problem in the upper half-plane.

This is the Poisson integral formula, or the Poisson integral of f, which expresses $u(x, y)$ as the convolution of the boundary function $f(x)$ with the **Poisson kernel** (Figure 2)

(4) $$P_y(x) = \sqrt{\frac{2}{\pi}} \frac{y}{x^2 + y^2} \qquad (-\infty < x < \infty,\ y > 0).$$

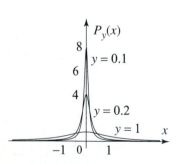

Figure 2 The Poisson kernel.

The fact that the solution is a convolution suggests that the Fourier transform can be used to our advantage. Indeed, our next step is to show how to derive (3), using the Fourier transform method. We will then show how the conjugate function of the solution can be constructed by using the boundary function $f(x)$ and a certain convolution, known as the Hilbert transform. This will establish a connection between analytic functions in the upper half-plane and Fourier analysis of functions on the real line.

EXAMPLE 1   Fourier transform derivation of th Poisson formula

To derive (3), Fourier transform (1) and (2) with respect to the x variable and get

$$-\omega^2 \widehat{u}(\omega, y) + \frac{d^2}{dy^2}\, \widehat{u}(\omega, y) = 0; \quad \widehat{u}(\omega, 0) = \widehat{f}(\omega).$$

Solving the differential equation in $\widehat{u}(\omega, y)$ we get

$$\widehat{u}(\omega, y) = A(\omega)e^{-\omega y} + B(\omega)e^{\omega y}.$$

As in the solution of Example 4 of Section 11.3, we impose a boundedness condition on $\widehat{u}(\omega, y)$. This forces $A(\omega) = 0$ if $\omega < 0$, and $B(\omega) = 0$ if $\omega > 0$. Thus we may write $\widehat{u}(\omega, y) = C(\omega)e^{-y|\omega|}$. Setting $y = 0$, and using the transformed boundary condition we get

$$\widehat{u}(\omega, y) = \widehat{f}(\omega)e^{-y|\omega|}.$$

Recall from Example 4, Section 11.1, the Fourier transform of the Poisson kernel

(5)
$$\widehat{P}_y(\omega) = \mathcal{F}\left(\sqrt{\frac{2}{\pi}}\frac{y}{x^2 + y^2}\right)(\omega) = e^{-y|\omega|}.$$

Hence $\widehat{u}(\omega, y) = \widehat{f}(\omega)\widehat{P}_y(\omega)$, and so by Theorem 4, Section 11.2, it follows that u is the convolution of f with $P_y(x)$, as claimed. ∎

As stated, the Dirichlet problem (1)–(2) does not have a unique solution, due to the failure of the maximum modulus principle on unbounded regions (see Sections 3.7 and 3.8). To see this, consider the function $v(x, y) = y$. You can check that $v(x, y)$ satisfies (1) and equals 0 on the x-axis. So, if $u(x, y)$ is any solution of (1) and (2), then $u(x, y) + y$ is also a solution. Hence the solution of (1) and (2) is not unique. This is troublesome, since we tend to think of a Dirichlet problem as modeling a steady-state problem, and as such, we want the solution to be unique. What is happening here is that by stating (1) and (2), we did not impose enough conditions to ensure the uniqueness of the solution. It can be shown that by adding boundedness conditions to (1) and (2), for example,

$$|u(x, y)| \le M, \quad \text{for all } x \text{ and all } y > 0,$$

then the resulting problem will have a unique bounded solution given by the Poisson integral formula (3), whenever the boundary function $f(x)$ is bounded. In our treatment, we will always assume such boundedness assumptions and thus speak of a unique solution of the Dirichlet problem.

Several applications of the Poisson integral formula were derived in Section 6.3. We add one more application that involves a Dirichlet problem with a Dirac delta applied on the boundary.

EXAMPLE 2   Effect of a Dirac delta on the boundary

Consider the Dirichlet problem in the upper half-plane with boundary data given by

$$u(x, 0) = \sqrt{2\pi}\delta_0(x).$$

Such a problems arises when modeling the steady-state temperature distribution resulting from the application of a welding torch (very high temperature) at a point of the boundary of a long sheet of metal with insulated surface. The constant $\sqrt{2\pi}$ is added for convenience, as you will see from the solution. Solve this problem and describe the isotherms.

Solution According to (3), the solution is $u(x, y) = P_y * (\sqrt{2\pi}\delta_0)(x)$. But $\sqrt{2\pi}\delta_0$ is a unit for the convolution operation (see Example 6, Section 11.2), so

$$u(x, y) = P_y * (\sqrt{2\pi}\delta_0)(x) = P_y(x) = \sqrt{\frac{2}{\pi}}\frac{y}{x^2 + y^2}.$$

Thus the solution is the Poisson kernel.

To find the isotherms in this problem, we must determine the level curves of the Poisson kernel:

$$\sqrt{\frac{2}{\pi}}\frac{y}{x^2 + y^2} = T \quad \Rightarrow \quad x^2 + \left(y - \frac{1}{T\sqrt{2\pi}}\right)^2 = \frac{1}{2T^2\pi}.$$

Thus the isotherm corresponding to $T > 0$ is the portion in the upper half-plane of the the circle centered on the y-axis at $(0, \frac{1}{T\sqrt{2\pi}})$ with radius $\frac{1}{\sqrt{2\pi}T}$. Note that there is no restriction on $T > 0$, which makes sense on physical grounds, since this problem is supposed to model the application of a very high temperature at the point 0. So no matter how large is the value of T, we can always find points in the upper half-plane with temperature T. As $T \to \infty$, the radius of the isotherm tends to 0, which means that the points are closer to the origin. As $T \to 0$, the radius tends to ∞, which corresponds to points that are far away from the origin. ∎

Example 2 raises two questions. Being the solution of a Dirichlet problem with boundary values $\sqrt{2\pi}\delta_0(x)$, $P_y(x)$ must be harmonic in the upper half-plane and must tend to $\sqrt{2\pi}\delta_0(x)$ as $y \to 0$. Is $P_y(x)$ harmonic in the upper half-plane and in what sense does it tend to $\sqrt{2\pi}\delta_0(x)$ as $y \to 0$? The answer to the first question is easy to verify directly. We have the following useful result.

THEOREM 1
HARMONICITY OF
THE POISSON
KERNEL

The Poisson kernel $P_y(x) = \sqrt{\frac{2}{\pi}}\frac{y}{x^2+y^2}$ is harmonic for all $(x, y) \neq (0, 0)$. Moreover, it has a harmonic conjugate (called the **conjugate Poisson kernel**) given by

$$(6) \qquad Q_y(x) = \sqrt{\frac{2}{\pi}}\frac{x}{x^2 + y^2}, \qquad (x, y) \neq (0, 0).$$

Proof To prove both assertions, consider the function $F(z) = \sqrt{\frac{2}{\pi}}\frac{i}{z}$ for $z \neq 0$. This function is analytic for all $z \neq 0$. Writing $z = x + iy$ and expressing F in terms of its real and imaginary parts, we find that

$$F(z) = \sqrt{\frac{2}{\pi}}\frac{i}{x + iy} = \sqrt{\frac{2}{\pi}}\frac{i(x - iy)}{x^2 + y^2} = P_y(x) + i\,Q_y(x).$$

Thus $P_y(x)$ and $Q_y(x)$ are the real and imaginary parts of an analytic function for $(x, y) \neq (0, 0)$ and as such they are harmonic and $Q_y(x)$ is a harmonic conjugate of $P_y(x)$. ∎

The answer to the second question that we raised following Example 2 is more involved, since we are talking about the convergence of a function $P_y(x)$ to a Dirac delta, which is not a function. The concept of convergence that we need is called **weak convergence**. Even though we will not study it in this book, we can motivate it as follows. Look at the graphs of $P_y(x)$, as $y \to 0$ (Figure 2). These graphs become more and more concentrated around 0; while the total area under each graph is

$$\int_{-\infty}^{\infty} P_y(x)\, dx = \sqrt{2\pi} = \int_{-\infty}^{\infty} \sqrt{2\pi}\,\delta_0(x)\, dx.$$

So, as $y \to 0$, the graph of $P_y(x)$ approximates the graph of $\sqrt{2\pi}\,\delta_0(x)$: It is almost 0 for $x \neq 0$ and infinite for $x = 0$, and the total area under the graph is $\sqrt{2\pi}$, which is the integral of $\sqrt{2\pi}\,\delta_0(x)$. It can also be shown that $\lim_{y \to 0} \frac{1}{\sqrt{2\pi}} \int_{-\infty}^{\infty} P_y(x) f(x)\, dx = \int_{-\infty}^{\infty} f(x)\delta_0(x)\, dx = f(0)$, for all integrable and continuous f. This limit can be used to define the notion of weak convergence of $\frac{1}{\sqrt{2\pi}} P_y(x)$ to δ_0 (equivalently, the weak convergence of $P_y(x)$ to $\sqrt{2\pi}\,\delta_0$).

Harmonic Conjugate and Hilbert Transform

In many interesting applications, including finding the curves of heat flow in steady-state problems, we are required to compute the harmonic conjugate of the solution of th Dirichlet problem in the upper half-plane. If $u(x, y)$ is a solution of (1) and (2), recall that by a harmonic conjugate of u, we mean a harmonic function $v(x, y)$ in the upper half-plane, such that $u(x, y) + i\, v(x, y)$ is analytic. Since the upper half-plane is simply connected, the existence of a harmonic conjugate is guaranteed by Theorem 1, Section 3.8, which also provides a formula for computing the harmonic conjugate. Our goal is to describe a formula for constructing a harmonic conjugate of $u(x, y)$ in terms of the boundary function $f(x)$. We start by giving a formula for the harmonic conjugate of $u(x, y)$ in the upper half-plane. As we would expect, this formula involves the conjugate Poisson kernel, Q_y.

THEOREM 2
HARMONIC
CONJUGATE

Let $u(x, y) = P_y * f(x)$ denote the solution of the Dirichlet problem (1)–(2) in the upper half-plane. Then a harmonic conjugate of $u(x, y)$ is given by

$$(7) \qquad v(x, y) = Q_y * f(x) = \frac{1}{\pi} \int_{-\infty}^{\infty} f(x - s) \frac{s}{s^2 + y^2}\, ds \qquad (y > 0),$$

where Q_y is the conjugate Poisson kernel (6).

Formula (7) is known as the **conjugate Poisson integral of** f. We will not prove this theorem, but only motivate (7) by proceeding formally from the Poisson integral formula. Let us write $P_y(x)$ as $P(x, y)$; then the Poisson formula becomes

$$u(x, y) = \frac{1}{\sqrt{2\pi}} \int_{-\infty}^{\infty} P(x - s, y) f(s) \, ds.$$

Similarly, from (7), we have

$$v(x, y) = \frac{1}{\sqrt{2\pi}} \int_{-\infty}^{\infty} Q(x - s, y) f(s) \, ds.$$

From Theorem 1, $Q(x, y)$ is harmonic and so its translate, $Q(x - s, y)$, is also harmonic for any s (see Exercise 23, Section 2.5). Suppose that we can differentiate under the integral signs. Then to compute the Laplacian of Δv, we write

$$\Delta v(x, y) = \frac{1}{\sqrt{2\pi}} \int_{-\infty}^{\infty} \Delta[Q(x - s, y)] f(s) \, ds = 0,$$

because $Q(x - s, y)$ is harmonic and so $\Delta[Q(x - s, y)] = 0$. This shows that $v(x, y)$ is harmonic in the upper half-plane. To show that v is a harmonic conjugate of u, again we can show that u and v satisfy the Cauchy-Riemann equations, by differentiating under the integral signs and using the fact that $P(s, y)$ and $Q(s, y)$ satisfy these equations, because $Q(s, y)$ is a harmonic conjugate of $P(s, y)$ (Theorem 1).

Continuing our formal manipulation of the Poisson and conjugate Poisson integral of f, we ask: What is the limit of $Q_y * f(x)$ as $y \to 0$? That is, we are asking for the boundary values of the harmonic conjugate of the solution of the Dirichlet problem (1)–(2). To answer this question, we can take the limit as $y \to 0$ in (7). This gives

(8) $$\lim_{y \to 0} Q_y * f(x) = \frac{1}{\pi} \int_{-\infty}^{\infty} \frac{f(x - s)}{s} \, ds.$$

This integral is improper at $s = 0$ and $\pm\infty$. In computing it, we will take its principal value. Formula (8) is a convolution of f with the kernel $\frac{1}{\sqrt{2\pi}\, s}$. This important convolution arises in many different contexts of applied mathematics and engineering. It is called the **Hilbert transform of** f, and denoted $H(f)$. Thus

(9) $$H(f)(x) = \frac{1}{\pi} \int_{-\infty}^{\infty} \frac{f(x - s)}{s} \, ds = \lim_{y \to 0} Q_y * f(x),$$

where the integral is to be computed as a principal value integral. There are several difficulties in studying the Hilbert transform, due in part to the fact that the kernel of this transform is not integrable. The results that we present are true in a very general setting, but their rigorous treatment requires a level beyond the level of this book. We will continue our formal presentation and treat this transform as we did with generalized functions.

One of the most important features of the Hilbert transform is its effect on the Fourier transform of f.

THEOREM 3
FOURIER
TRANSFORM OF
THE HILBERT
TRANSFORM

Let $f(x)$ be defined and integrable on the real line, and let $H(f)(x)$ denote its Hilbert transform. Then

(10)
$$\widehat{Hf}(\omega) = -i\operatorname{sgn}(\omega)\widehat{f}(\omega) \quad (-\infty < \omega < \infty).$$

Thus the effect of the Hilbert transform is to multiply the Fourier transform of f by $-i\operatorname{sgn}\omega$.

Proof We have

$$
\begin{aligned}
\widehat{Hf}(\omega) &= \frac{1}{\sqrt{2\pi}}\int_{-\infty}^{\infty} H(f)(x)e^{-i\omega x}\,dx = \frac{1}{\sqrt{2\pi}}\int_{-\infty}^{\infty}\frac{1}{\pi}\int_{-\infty}^{\infty}\frac{f(x-s)}{s}\,ds\,e^{-i\omega x}\,dx \\
&= \frac{1}{\sqrt{2\pi}}\int_{-\infty}^{\infty} f(x)e^{-i\omega x}\,dx\frac{1}{\pi}\int_{-\infty}^{\infty}\frac{e^{-i\omega x}}{s}\,ds = \widehat{f}(\omega)\frac{1}{\pi}\int_{-\infty}^{\infty}\frac{e^{-i\omega s}}{s}\,ds.
\end{aligned}
$$

The last integral is evaluated by using a change of variables and the integral $\frac{1}{\pi}\int_{-\infty}^{\infty}\frac{\sin s}{s}\,ds = 1$ (see Exercise 21, Section 5.4). We have

$$\frac{1}{\pi}\int_{-\infty}^{\infty}\frac{e^{-i\omega s}}{s}\,ds = \frac{-i}{\pi}\int_{-\infty}^{\infty}\frac{\sin(\omega s)}{s}\,ds = -i\operatorname{sgn}(\omega),$$

and (10) follows.   ∎

Let us recap what we have so far. Starting with a function $f(x)$ defined on the real line, we can form its Poisson integral $u(x, y) = P_y * f(x)$. This function is harmonic in the upper half-plane and tends to $f(x)$ as $y \to 0$. If we take the harmonic conjugate of u, we obtain the conjugate Poisson integral of f, $v(x, y) = Q_y * f$. This function is harmonic in the upper half-plane and tends to $H(f)(x)$, the Hilbert transform of f, as $y \to 0$. So the conjugate Poisson integral of f, $Q_y * f(x)$, has boundary values the Hilbert transform of f, $H(f)(x)$. This suggests that the conjugate Poisson integral can be constructed from the Hilbert transform in much the same way we construct $u(x, y)$ from its boundary values $f(x)$. Indeed, we have the following interesting result.

**THEOREM 4
HILBERT
TRANSFORM AND
CONJUGATE
POISSON INTEGRAL**

Let $f(x)$ be defined and integrable on the real line, and let $H(f)(x)$ denote its Hilbert transform. Then

$$(11) \qquad v(x, y) = Q_y * f(x) = P_y * (Hf)(x).$$

In other words, the conjugate Poisson integral of f is the Poisson integral of its Hilbert transform.

Proof We will prove (11) by using the Fourier transform. Fix y and consider the functions in (11) as functions of x. Taking the Fourier transform and using Theorem 4, Section 11.2, we have

$$(12) \qquad \widehat{Q_y * f}(\omega) = \widehat{Q_y}(\omega)\widehat{f}(\omega) \quad \text{and} \quad \widehat{P_y * (Hf)}(\omega) = \widehat{P_y}(\omega)\widehat{(Hf)}(\omega).$$

But $Q_y(x) = \frac{x}{y}P_y(x)$, and so from (5) and Theorem 2, Section 11.2,

$$(13) \qquad \widehat{Q_y}(\omega) = \frac{i}{y}\frac{d}{d\omega}\widehat{P_y}(\omega) = -i\,\mathrm{sgn}\,(\omega)e^{-y|\omega|} = -i\,\mathrm{sgn}\,(\omega)\widehat{P_y}(\omega).$$

From Theorem 3, we have $\widehat{(Hf)}(\omega) = -i\,\mathrm{sgn}\,(\omega)\widehat{f}(\omega)$. Plugging back into (12), it follows that $\widehat{Q_y * f}(\omega) = \widehat{P_y * (Hf)}(\omega)$, and hence (11) holds. ∎

It is time to give an example with the Hilbert transform.

EXAMPLE 3   Hilbert transforms
(a) Find $H(\sin x)$.   (b) Show that $H(P_y)(x) = Q_y(x)$.

Solution (a) We have

$$
\begin{aligned}
H(\sin x) &= \frac{1}{\pi}\mathrm{P.V.}\int_{-\infty}^{\infty}\frac{\sin(x-s)}{s}\,ds \\
&= \frac{1}{\pi}\mathrm{P.V.}\int_{-\infty}^{\infty}\frac{\sin x\cos s - \cos x\sin s}{s}\,ds \\
&= \frac{\sin x}{\pi}\mathrm{P.V.}\int_{-\infty}^{\infty}\frac{\cos s}{s}\,ds - \frac{\cos x}{\pi}\mathrm{P.V.}\int_{-\infty}^{\infty}\frac{\sin s}{s}\,ds = -\cos x,
\end{aligned}
$$

because $\mathrm{P.V.}\int_{-\infty}^{\infty}\frac{\cos s}{s}\,ds = 0$ and $\frac{1}{\pi}\mathrm{P.V.}\int_{-\infty}^{\infty}\frac{\sin s}{s}\,ds = 1$.
(b) The best way to do this part is to show that $H(P_y)(x)$ and $Q_y(x)$ have the same Fourier transforms. Indeed, using Theorem 3, we have

$$\widehat{H(P_y)}(\omega) = -i\,\mathrm{sgn}\,(\omega)\widehat{P_y}(\omega) = \widehat{Q_y}(\omega),$$

where the second equality follows from (13). ∎

Exercises 11.5

In Exercises 1–4, solve the Dirichlet problem (1)–(2) for the given boundary data.

1.

$$f(x) = \begin{cases} 50 & \text{if } -1 < x < 1, \\ 0 & \text{otherwise.} \end{cases}$$

2.

$$f(x) = \begin{cases} 100\,(1+x) & \text{if } -1 < x < 0, \\ 100\,(1-x) & \text{if } 0 < x < 1, \\ 0 & \text{otherwise.} \end{cases}$$

3. $f(x) = \cos x$.   **4.** $f(x) = 1 + \sin x$.

5. Semigroup property of the Poisson kernel. Using the Fourier transform, show that $P_{y_1} * P_{y_2}(x) = P_{y_1+y_2}(x)$. This is known as the semigroup property of the Poisson kernel.

6. (a) Solve the Dirichlet problem for the boundary data $f(x) = \frac{1}{1+x^2}$. [Hint: Express f in terms of the Poisson kernel P_1 and use Exercise 5.]
(b) What are the isotherms in this case?
(c) Plot the isotherms to verify your answer in (b).

7. Repeat Exercise 6 with $f(x) = \frac{1}{4+x^2}$.

8. Properties of the Poisson kernel. As you work through this exercise, compare the results with the properties of the heat kernel from the previous section.
(a) $P_y(x)$ is an even function of x, and $P_y(x) \geq 0$ for all x.
(b) The graph of $P_y(x)$ is a bell-shaped curve centered at the origin. (You may simply illustrate this part graphically.)
(c) We have $P_y(0) = \sqrt{\frac{2}{\pi}} \frac{1}{y}$; and hence $\lim_{y\to 0} P_y(0) = \infty$. Thus, as y tends to 0, the graph of $P_y(x)$ becomes more and more localized near 0, in the sense that most of the total area under the graph and above the x-axis becomes concentrated near 0.
(d) The total area under the graph of $P_y(x)$ and above the x-axis is $\int_{-\infty}^{\infty} P_y(x)\, dx = \sqrt{2\pi}$, for all $y > 0$.
(e) The Fourier transform of $P_y(x)$ is $e^{-y|\omega|}$. (Use the table of Fourier transforms.)
(f) If f is an integrable function and piecewise smooth, then at the points where f is continuous we have $\lim_{y\to 0} P_y * f(x) = f(x)$. (A rigorous proof of this result is difficult, but you should be able to justify it by taking Fourier transforms.)

In Exercises 9–12, find the Hilbert transform of the given function.

9. $f(x) = \cos x + \sin 2x$.   **10.** $\mathcal{U}_0(x+1) - \mathcal{U}_0(x-1)$.

11. $(\mathcal{U}_0(x) - \mathcal{U}_0(x-1))\, x$.   **12.** $(\mathcal{U}_0(x) - \mathcal{U}_0(x-\pi)) \sin x$.

13. Show that

$$\mathcal{F}\left(\frac{f + iH(f)}{2}\right)(\omega) = \begin{cases} \mathcal{F}(f)(\omega) & \text{if } \omega > 0, \\ 0 & \text{if } \omega \leq 0. \end{cases}$$

Thus, by adding $iH(f)$ to f we truncated its Fourier transform at $\omega \leq 0$.

14. Using the Fourier transform, show that $H(H(f)) = -f$.

15. Using the Fourier transform, show that $Q_a * Q_b = -P_{a+b}$.

11.6   The Fourier Cosine and Sine Transforms

So far in this chapter we have considered boundary value problems with initial or boundary data defined on the entire real line. It is easy to imagine similar problems with boundary or initial data defined on half lines. As an example, consider a heat problem in a bar with one end extending to infinity. In this case, the initial temperature distribution will be given by a function defined on half the real line. As a second illustration, consider a Dirichlet problem in the first quadrant. In this case, the boundary consists of two half lines. A variety of such problems will be discussed in the following section. To treat these problems, we introduce in this section two transforms closely related to the Fourier integral: the cosine and sine Fourier transforms. As you will see, these transforms are the analogs of the half-range expansions in Fourier series.

The basic question is: How can we use the Fourier transform when the function $f(x)$ is defined for $x > 0$ only? Having worked with Fourier series and half-range expansions, we know that the answer will involve even and odd (nonperiodic) extensions of f.

If we extend f as an even function on the entire real line (Figure 1) and then take the Fourier transform of the extension, we obtain

$$(1) \qquad \mathcal{F}(f)(\omega) = \frac{1}{\sqrt{2\pi}} \int_{-\infty}^{\infty} f(x)e^{-i\omega x}\, dx = \sqrt{\frac{2}{\pi}} \int_{0}^{\infty} f(x)\cos\omega x\, dx,$$

where we have kept the same notation for f and its extension and used the fact that the extension is an even function.

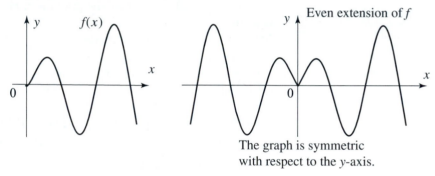

Figure 1

The graph is symmetric with respect to the y-axis.

Applying the inverse Fourier transform, and using the fact that the Fourier transform is even in this case, we obtain

$$(2) \qquad f(x) = \sqrt{\frac{2}{\pi}} \int_{0}^{\infty} \mathcal{F}(f)(\omega)\cos\omega x\, d\omega.$$

The integral in (1) depends only on the values of $f(x)$ for $x \geq 0$, so without mention of the even extension, we define the **cosine Fourier transform** of f by (1), setting

$$(3) \qquad \boxed{\widehat{f}_c(\omega) = \sqrt{\frac{2}{\pi}} \int_{0}^{\infty} f(t)\cos\omega t\, dt \qquad (\omega \geq 0).}$$

Then the inverse Fourier transform in (2) becomes

$$(4) \qquad \boxed{f(x) = \sqrt{\frac{2}{\pi}} \int_{0}^{\infty} \widehat{f}_c(\omega)\cos\omega x\, d\omega \qquad (x > 0).}$$

This formula features the **inverse Fourier cosine transform** and expresses f as an inverse Fourier cosine transform of $\widehat{f_c}(\omega)$. Note that the Fourier cosine transform is the same as the inverse Fourier cosine transform.

Similarly, if we use an odd extension (Figure 2), apply the Fourier transform and then its inverse Fourier transform, we obtain the

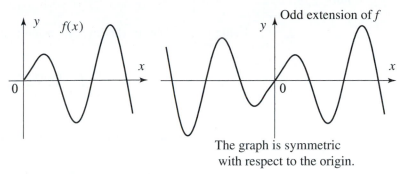

The graph is symmetric
with respect to the origin.

Figure 2

Fourier sine transform of f:

$$(5) \qquad \widehat{f_s}(\omega) = \sqrt{\frac{2}{\pi}} \int_0^\infty f(t) \sin \omega t\, dt \qquad (\omega \ge 0),$$

and its **inverse Fourier sine transform**,

$$(6) \qquad f(x) = \sqrt{\frac{2}{\pi}} \int_0^\infty \widehat{f_s}(\omega) \sin \omega x\, d\omega \qquad (x > 0).$$

Other commonly used notation for these transforms is as follows:

$$\mathcal{F}_c(f) = \widehat{f_c} \quad \text{and} \quad \mathcal{F}_s(f) = \widehat{f_s}\,.$$

Both (4) and (6) hold under the same conditions on f as in the Fourier inversion theorem (Section 11.1). We require that f be integrable on $[0, \infty)$ (by this we mean that $\int_0^\infty |f(x)|\, dx < \infty$) and piecewise smooth. Also, on the left sides of (4) and (6), $f(x)$ is to be replaced by $(f(x+) + f(x-))/2$ at points of discontinuity of f.

EXAMPLE 1   Fourier cosine and sine transforms
Consider the function

$$f(x) = \begin{cases} \sin x & \text{if } 0 \le x \le \pi, \\ 0 & \text{if } x > \pi. \end{cases}$$

Express f using an inverse Fourier cosine and then an inverse sine transform.

Solution We first start by computing the Fourier cosine transform. From (3),

In evaluating the integral, use
the identity
$$\sin a \cos b = \tfrac{1}{2}[\sin(a-b) + \sin(a+b)].$$

$$\widehat{f}_c(\omega) = \sqrt{\frac{2}{\pi}} \int_0^\pi \sin x \cos \omega x \, dx = \sqrt{\frac{2}{\pi}} \, \frac{\cos(\pi \omega) + 1}{1 - \omega^2}.$$

Using (4), we obtain the inverse cosine transform representation

$$f(x) = \frac{2}{\pi} \int_0^\infty \frac{\cos(\pi \omega) + 1}{1 - \omega^2} \cos \omega x \, d\omega \quad (x \geq 0).$$

We compute the sine transform similarly by using (5),

$$\widehat{f}_s(\omega) = \sqrt{\frac{2}{\pi}} \frac{\sin(\pi \omega)}{1 - \omega^2},$$

and thus the inverse sine transform representation

$$f(x) = \frac{2}{\pi} \int_0^\infty \frac{\sin(\pi \omega)}{1 - \omega^2} \sin \omega x \, d\omega \quad (x \geq 0). \qquad \blacksquare$$

EXAMPLE 2   A Fourier cosine transform

Consider the function $f(x) = e^{-ax}$ $(a > 0)$, $x > 0$.
(a) Find the Fourier cosine transform of $f(x)$.
(b) Express $f(x)$ as an inverse Fourier cosine transform.

Solution (a) From (3)

From a table of integrals:
$\int e^{-at} \cos \omega t \, dt =$
$\frac{a}{a^2 + \omega^2} e^{-at}(\frac{\omega}{a} \sin \omega t - \cos \omega t).$

$$
\begin{aligned}
\mathcal{F}_c(f)(\omega) &= \sqrt{\frac{2}{\pi}} \int_0^\infty e^{-at} \cos \omega t \, dt \\
&= \sqrt{\frac{2}{\pi}} \frac{a}{a^2 + \omega^2} \left[e^{-at} \left(\frac{\omega}{a} \sin \omega t - \cos \omega t \right) \right]_0^\infty = \sqrt{\frac{2}{\pi}} \frac{a}{a^2 + \omega^2}.
\end{aligned}
$$

(b) From (4) we obtain for $x > 0$

$$e^{-ax} = \sqrt{\frac{2}{\pi}} \int_0^\infty \widehat{f}_c(\omega) \cos \omega x \, d\omega = \frac{2}{\pi} \int_0^\infty \frac{a \cos \omega x}{a^2 + \omega^2} \, d\omega. \qquad \blacksquare$$

Fourier cosine and sine transforms can sometimes be computed from Fourier transforms.

EXAMPLE 3   A Fourier cosine transform from a Fourier transform

Find the Fourier cosine transform of $f(x) = e^{-ax^2/2} \ (a > 0), x > 0$.

Solution We have

$$
\begin{aligned}
\mathcal{F}_c(f)(\omega) &= \sqrt{\frac{2}{\pi}} \int_0^\infty e^{-ax^2/2} \cos \omega x \, dx \\
&= \frac{1}{2}\sqrt{\frac{2}{\pi}} \int_{-\infty}^\infty e^{-ax^2/2} \cos \omega x \, dx \quad \text{(the integrand is even)} \\
&= \frac{1}{\sqrt{2\pi}} \int_{-\infty}^\infty e^{-ax^2/2} e^{-i\omega x} \, dx \\
&\qquad \text{(the imaginary part integrates to 0 because it is odd)} \\
&= \mathcal{F}(e^{-ax^2/2})(\omega) \quad \text{(by definition of the Fourier transform)} \\
&= \frac{1}{\sqrt{a}} e^{-\omega^2/2a} \quad \text{(by Example 5, Section 11.1)}. \qquad \blacksquare
\end{aligned}
$$

Example 3 illustrates the following useful general rule, whose proof is left as an exercise.

RELATIONSHIPS BETWEEN TRANSFORMS

If $f(x)$ $(x \geq 0)$ is the restriction of an *even* function f_e, then

$$(7) \qquad \qquad \mathcal{F}_c(f)(\omega) = \mathcal{F}(f_e)(\omega) \quad \text{for all } \omega \geq 0.$$

If $f(x)$ $(x \geq 0)$ is the restriction of an *odd* function f_o, then

$$(8) \qquad \qquad \mathcal{F}_s(f)(\omega) = i\,\mathcal{F}(f_o)(\omega) \quad \text{for all } \omega \geq 0.$$

EXAMPLE 4   Using Fourier transforms

To illustrate the applications of (7) and (8), we use known Fourier transforms (see Example 4, Section 11.1) and compute, for $\omega \geq 0$,

$$\mathcal{F}_c\left(\frac{a}{a^2 + x^2}\right)(\omega) = \mathcal{F}\left(\frac{a}{a^2 + x^2}\right)(\omega) = \sqrt{\frac{\pi}{2}} e^{-a\omega} \quad (a > 0),$$

and

$$\mathcal{F}_s\left(\frac{x}{a^2 + x^2}\right)(\omega) = i\mathcal{F}\left(\frac{x}{a^2 + x^2}\right)(\omega) = \sqrt{\frac{\pi}{2}} \omega e^{-a\omega} \quad (a > 0). \qquad \blacksquare$$

Operational Properties

The Fourier cosine and sine transforms have operational properties very similar to those of the Fourier transform. We list those that will be needed in the next section. The proofs will be omitted, being very similar to the case of the Fourier transform (Section 11.2). In what follows, we assume that the functions are integrable, so that all the transforms exist.

THEOREM 1
LINEARITY

If f and g are functions and a and b are numbers, then

$$\mathcal{F}_c(af + bg) = a\mathcal{F}_c(f) + b\mathcal{F}_c(g),$$

and

$$\mathcal{F}_s(af + bg) = a\mathcal{F}_s(f) + b\mathcal{F}_s(g).$$

Compare the following theorem with Theorem 2, Section 11.2.

THEOREM 2
TRANSFORMS OF
DERIVATIVES

Suppose that $f(x) \to 0$ as $x \to \infty$; then

$$(9) \qquad \mathcal{F}_c(f') = \omega \mathcal{F}_s(f) - \sqrt{\frac{2}{\pi}} f(0),$$

$$(10) \qquad \mathcal{F}_s(f') = -\omega \mathcal{F}_c(f).$$

If in addition $f'(x) \to 0$ as $x \to \infty$, then

$$(11) \qquad \mathcal{F}_c(f'') = -\omega^2 \mathcal{F}_c(f) - \sqrt{\frac{2}{\pi}} f'(0),$$

$$(12) \qquad \mathcal{F}_s(f'') = -\omega^2 \mathcal{F}_s(f) + \sqrt{\frac{2}{\pi}} \omega f(0).$$

Note that each formula for the first derivative involves both transforms. The formulas for the second derivatives, however, involve only one transform at a time.

EXAMPLE 5   Transform of a derivative

In Example 3 we found that

$$\mathcal{F}_c(e^{-x^2})(\omega) = \frac{1}{\sqrt{2}} e^{-\omega^2/4}.$$

Applying Theorem 2(ii) with $f(x) = e^{-x^2}$, we obtain

$$\mathcal{F}_s(-2xe^{-x^2}) = -\omega \mathcal{F}_c(e^{-x^2}) = -\omega \frac{1}{\sqrt{2}} e^{-\omega^2/4}.$$

Hence

$$\mathcal{F}_s(xe^{-x^2}) = \frac{\omega}{2\sqrt{2}} e^{-\omega^2/4}. \qquad \blacksquare$$

THEOREM 3
DERIVATIVES OF
TRANSFORMS

We have

$$(13) \qquad \mathcal{F}_c(xf(x)) = \frac{d}{d\omega} \mathcal{F}_s(f(x)),$$

$$(14) \qquad \mathcal{F}_s(xf(x)) = -\frac{d}{d\omega} \mathcal{F}_c(f(x)).$$

EXAMPLE 6   Using operational properties

Find the Fourier sine transform of $f(x) = xe^{-ax}$ $(a > 0)$, $x > 0$.

Solution In Example 2 we found that

$$\mathcal{F}_c(e^{-ax})(\omega) = \sqrt{\frac{2}{\pi}} \frac{a}{a^2 + \omega^2}.$$

From Theorem 3(ii) we have

$$\mathcal{F}_s(xe^{-ax}) = -\frac{d}{d\omega}\mathcal{F}_c(e^{-ax}) = -\sqrt{\frac{2}{\pi}}\frac{d}{d\omega}\frac{a}{a^2 + \omega^2} = \sqrt{\frac{2}{\pi}}\frac{2a\omega}{(a^2 + \omega^2)^2}. \qquad \blacksquare$$

Exercises 11.6

In Exercises 1–6, find the Fourier cosine transform of $f(x)$ $(x > 0)$ and write f as an inverse cosine transform. Use a known Fourier transform and (7) when possible.

1.

$$f(x) = \begin{cases} 1 & \text{if } 0 < x < 1, \\ 0 & \text{otherwise.} \end{cases}$$

2.

$$f(x) = \begin{cases} 1 & \text{if } 0 < a < x < b < \infty, \\ 0 & \text{otherwise.} \end{cases}$$

3. $f(x) = 3\,e^{-2x}$.

4. $f(x) = x^2 e^{-x^2}$.

5.

$$f(x) = \begin{cases} \cos x & \text{if } 0 < x < 2\pi, \\ 0 & \text{otherwise.} \end{cases}$$

6.

$$f(x) = \begin{cases} 1 - x & \text{if } 0 < x < 1, \\ 0 & \text{otherwise.} \end{cases}$$

In Exercises 7–12, find the Fourier sine transform of $f(x)$ $(x > 0)$ and write $f(x)$ as an inverse sine transform. Use a known Fourier transform and (8) when possible.

7.

$$f(x) = \begin{cases} 1 & \text{if } 0 < x < 1, \\ 0 & \text{otherwise.} \end{cases}$$

8. $f(x) = xe^{-x^2}$.

9. $f(x) = e^{-2x}$.

10. $f(x) = xe^{-x}$.

11.

$$f(x) = \begin{cases} \sin 2x & \text{if } 0 < x < \pi, \\ 0 & \text{otherwise.} \end{cases}$$

12. $f(x) = \frac{x}{1 + x^2}$.

In Exercises 13–18, compute the given transform.

13. $\mathcal{F}_c\left(\frac{1}{1 + x^2}\right)$.

14. $\mathcal{F}_c\left(\frac{x^2}{e^{x^2}}\right)$.

15. $\mathcal{F}_s\left(\frac{x}{1 + x^2}\right)$.

16. $\mathcal{F}_s\left(xe^{-x}\right)$.

17. $\mathcal{F}_c\left(\frac{\cos x}{1 + x^2}\right)$.

18. $\mathcal{F}_c\left(\frac{\sin x}{x}\right)$.

19. Prove (7) and (8).

20. Prove (9)–(12).

21.   Reciprocity relations Show that $\mathcal{F}_c\mathcal{F}_c f = f$ and $\mathcal{F}_s\mathcal{F}_s f = f$.

11.7   Problems Involving Semi-Infinite Intervals

We will solve boundary value problems on semi-infinite intervals, and on unbounded regions in the plane such as half-planes, quarter-planes, and strips, by using the Fourier sine and cosine transforms instead of the Fourier transform. The reason is that the functions with which we will be dealing are defined on a subset of the real line, and to use the Fourier transform requires that a function be defined on the entire real line. The method of this section is completely analogous to the Fourier transform method. We will illustrate it with examples.

EXAMPLE 1   Heat equation for a semi-infinite rod

Solve the boundary value problem

$$(1) \qquad \frac{\partial u}{\partial t} = c^2 \frac{\partial^2 u}{\partial x^2}, \qquad 0 < x < \infty, \ t > 0,$$

$$(2) \qquad u(x,0) = f(x), \qquad x > 0,$$

$$(3) \qquad u(0,t) = 0, \qquad t > 0.$$

This problem models heat diffusion in a semi-infinite rod, with initial temperature distribution $f(x)$, and with one end kept at $0°$ temperature (see Figure 1).

Solution Note that since the initial conditions (2) involve the x variable, the problem calls for a transform in the x variable and not t. As you will see shortly, the Fourier sine transform is the right choice in this problem (see the remarks following the solution). We have

$$\mathcal{F}_s\left(\frac{\partial u}{\partial t}\right) = \frac{d}{dt}\,\widehat{u}_s(\omega, t),$$

and, from Theorem 2, Section 11.6,

$$\mathcal{F}_s\left(\frac{\partial^2 u}{\partial x^2}\right) = -\omega^2 \widehat{u}_s(\omega, t) + \sqrt{\frac{2}{\pi}}\,\omega u(0, t) = -\omega^2 \widehat{u}_s(\omega, t)\,.$$

Thus, transforming (1) and (2) we get

$$\frac{d}{dt}\widehat{u}_s(\omega, t) = -c^2\,\omega^2 \widehat{u}_s(\omega, t),$$

$$\widehat{u}_s(\omega, 0) = \widehat{f}_s(\omega)\,.$$

The general solution of the first order ordinary differential equation in t is

$$\widehat{u}_s(\omega, t) = A(\omega)e^{-c^2\omega^2 t},$$

where $A(\omega)$ is a constant that depends on ω. Setting $t = 0$ yields

$$\widehat{u}_s(\omega, 0) = A(\omega) = \widehat{f}_s(\omega),$$

and so

$$\widehat{u}_s(\omega, t) = \widehat{f}_s(\omega)e^{-c^2\omega^2 t}\,.$$

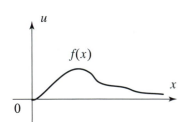

Figure 1 Initial temperature distribution.

Recall from Section 11.4 that $e^{-c^2\omega^2 t}$ is the Fourier transform of the heat kernel. Thus, the solution in this example also involves the heat kernel. See Exercise 7 for an alternative form of the solution.

Taking inverse Fourier sine transforms ((6), Section 11.6), we obtain the solution in the form

(4) $\quad u(x,t) = \sqrt{\dfrac{2}{\pi}} \displaystyle\int_0^\infty \widehat{u}_s(\omega, t) \sin \omega x \, d\omega = \sqrt{\dfrac{2}{\pi}} \displaystyle\int_0^\infty \widehat{f}_s(\omega) e^{-c^2\omega^2 t} \sin \omega x \, d\omega \, .$ ■

You may ask why the Fourier sine transform was used in Example 1 and not the Fourier cosine transform. The choice was suggested by the boundary condition (3). To compute the cosine transform of the second derivative $\frac{\partial^2 u}{\partial x^2}$, the operational property (11), Section 11.6, requires the value of $\frac{\partial u}{\partial x}(0,t)$, a quantity not given in the problem. In general, to successfully apply a transform we must be able to use the initial conditions to supply the values needed in the operational formulas.

The next example involves Dirichlet and Neumann type conditions.

EXAMPLE 2   A Dirichlet-Neumann problem in a semi-infinite strip
Solve the boundary value problem

(5) $\qquad \Delta u = \dfrac{\partial^2 u}{\partial x^2} + \dfrac{\partial^2 u}{\partial y^2} = 0, \quad 0 < x < a, \ y > 0,$

(6) $\qquad \dfrac{\partial u}{\partial y}(x,0) = 0, \quad 0 < x < a,$

(7) $\qquad u(0,y) = 0, \qquad u(a,y) = f(y), \quad y > 0.$

The problem models, for example, the steady–state temperature in a very large sheet of metal, where the boundaries are kept at a given temperature distribution and there is no heat flow across the boundary on the x-axis, as illustrated in Figure 2.

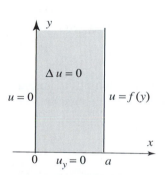

Figure 2 A Dirichlet-Neumann problem.

Solution Since the domain of the variable y is semi-infinite (the domain of the variable x is finite); we choose to transform the equations with respect to the variable y. Also since the boundary condition (6) involves the derivative at $y = 0$ the cosine transform is the right choice in this case. Using this transform with the help of the operational properties of Section 11.6, we get

$$\mathcal{F}_c\!\left(\dfrac{\partial^2 u}{\partial y^2}\right) = -\omega^2 \widehat{u}_c(x,\omega) - \sqrt{\dfrac{2}{\pi}} \dfrac{\partial u}{\partial y}(x,0) = -\omega^2 \widehat{u}_c(x,\omega),$$

(8) $\qquad \dfrac{d^2}{dx^2} \widehat{u}_c(x,\omega) - \omega^2 \widehat{u}_c(x,\omega) = 0 \qquad \text{(transforming (5))},$

(9) $\qquad \widehat{u}_c(0,\omega) = 0, \widehat{u}_c(a,\omega) = \widehat{f}_c(\omega) \qquad \text{(transforming (7))}.$

The general solution of the second order ordinary differential equation (8) is

(10) $\qquad \widehat{u}_c(x,\omega) = A(\omega) \cosh \omega x + B(\omega) \sinh \omega x,$

where $A(\omega)$ and $B(\omega)$ are constants that depend on ω. Setting $x = 0$ and then $x = a$ and using (9), we get

$$A(\omega) = 0, \quad B(\omega) = \dfrac{\widehat{f}_c(\omega)}{\sinh \omega a}.$$

Putting this into (10) and taking inverse Fourier cosine transform ((4), Section 11.6), we get the solution in the form

$$u(x, y) = \sqrt{\frac{2}{\pi}} \int_0^\infty \hat{u}_c(x, \omega) \cos \omega y \, d\omega = \sqrt{\frac{2}{\pi}} \int_0^\infty \frac{\hat{f}_c(\omega)}{\sinh \omega a} \sinh \omega x \cos \omega y \, d\omega \,. \quad \blacksquare$$

The Fourier transform method is a powerful device for solving partial differential equations. Its importance stems from its ability to handle a large variety of problems. Several additional types of problems (on different types of regions) are investigated in the exercises. Choosing the appropriate transform is a crucial step in implementing the method. As we saw in the examples, the choice is suggested by the type of region and the boundary conditions.

Exercises 11.7

In Exercises 1–4 solve the heat equation (1) with $c = 1$, $u(0, t) = 0$, and the given initial temperature distribution. Take $0 < x < \infty$ and $t > 0$.

1.

$$f(x) = \begin{cases} T_0 & \text{if } 0 < x < b, \\ 0 & \text{otherwise.} \end{cases}$$

2.

$$f(x) = \begin{cases} \sin x & \text{if } 0 < x < \pi, \\ 0 & \text{otherwise.} \end{cases}$$

3. $f(x) = \frac{x}{1+x^2}$.

4. $f(x) = xe^{-x^2/2}$.

5. **A Neumann type condition.** Let $u(x, t)$ denote the solution of (1) and (2) with the Neumann type condition: $\frac{\partial u}{\partial x}(0, t) = 0$. Use the Fourier cosine transform to derive the solution

$$u(x, t) = \sqrt{\frac{2}{\pi}} \int_0^\infty \hat{f}_c(\omega) e^{-c^2 \omega^2 t} \cos \omega x \, d\omega \,.$$

6. Do Exercise 5 for the specific case when $c = 1$ and $f(x)$ is as in Exercise 1.

7. Show that the solution (4) of Example 1 can be put in the form

$$u(x, t) = \frac{1}{2c\sqrt{\pi t}} \int_0^\infty f(s) \left[\exp\left(-\frac{(x-s)^2}{4c^2 t}\right) - \exp\left(-\frac{(x+s)^2}{4c^2 t}\right) \right] ds \,.$$

[Hint: In (4), write $\hat{f}_s$ explicitly in terms of f.]

8. (a) Show that the solution of the heat problem of Example 1 with $f(x) = T_0$ if $0 < x < a$ and 0 otherwise can be expressed in the form

$$u(x, t) = T_0 \operatorname{erf}\left(\frac{x}{2c\sqrt{t}}\right) - \frac{T_0}{2} \left\{ \operatorname{erf}\left(\frac{x-a}{2c\sqrt{t}}\right) + \operatorname{erf}\left(\frac{x+a}{2c\sqrt{t}}\right) \right\} \,.$$

[Hint: Use Exercise 7.]

(b) Take $c = 1$, $T_0 = 5$, and plot the solution curves at $t = .2, .5, 1, 2, 3, 4, 5$. Repeat the graphs with $c = 2, 5$. What do you observe as the value of c increases?

9. Project Problem: A nonhomogeneous boundary condition. Consider the heat equation (1) with the following conditions $u(x,0) = 0$ for $x > 0$ and $u(0,t) = T_0$ for $t > 0$.
(a) Use the Fourier sine transform to show that

$$\frac{d}{dt}\widehat{u}_s(\omega,t) + c^2\omega^2\widehat{u}_s(\omega,t) = c^2\sqrt{\frac{2}{\pi}}\,\omega T_0; \quad \widehat{u}_s(\omega,0) = 0.$$

(b) Derive the solution by establishing that

$$\widehat{u}(\omega,t) = \sqrt{\frac{2}{\pi}}\frac{T_0}{\omega} - \sqrt{\frac{2}{\pi}}\frac{T_0}{\omega}e^{-c^2\omega^2 t};$$

$$u(x,t) = \frac{2T_0}{\pi}\int_0^\infty \frac{\sin\omega x}{\omega}\,d\omega - \frac{2T_0}{\pi}\int_0^\infty \frac{e^{-c^2\omega^2 t}}{\omega}\sin\omega x\,d\omega;$$

$$u(x,t) = T_0 - \frac{2T_0}{\pi}\int_0^\infty \frac{e^{-c^2\omega^2 t}}{\omega}\sin\omega x\,d\omega\,.$$

(c) Take $T_0 = 100$ and plot the solution at various values of t.   What do you observe as $t \to \infty$?

10. Time-dependent boundary condition. Consider the boundary value problem

$$\frac{\partial^2 u}{\partial t^2} = c^2\frac{\partial^2 u}{\partial x^2}, \qquad x > 0,\ t > 0,$$

$$u(x,0) = f(x), \qquad \frac{\partial u}{\partial t}(x,0) = 0,$$

$$u(0,t) = s(t),$$

where $s(t)$ is a given function of t.
(a) Give a possible physical interpretation of the problem.
(b) Let f^* denote the odd extension of f to the entire line. By direct verification, show that the solution is

$$u(x,t) = \begin{cases} \frac{1}{2}[f^*(x+ct) + f^*(x-ct)] & \text{if } x > ct, \\ \frac{1}{2}[f^*(x+ct) + f^*(x-ct)] + s(t - \frac{x}{c}) & \text{if } x < ct. \end{cases}$$

(c) Take $c = 1$, $f(x) = 0$, $s(t) = \sin t$, and plot the solution at $t = 0$, $\frac{\pi}{4}$, $\frac{\pi}{2}$, π, 2π, 4π.

In Exercises 11–16, solve the boundary value problem described by the picture.

11.

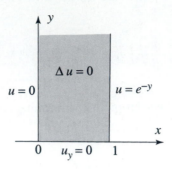

12.

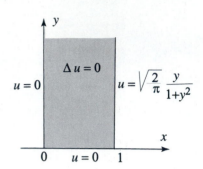

13.

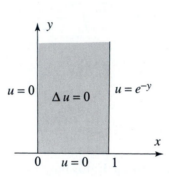

14.

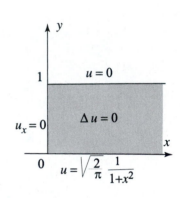

15.

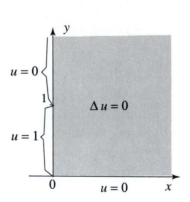

16.

12

THE LAPLACE AND HANKEL TRANSFORMS WITH APPLICATIONS

Should I refuse a good dinner simply because I do not understand the process of digestion?

-Oliver Heaviside

[Criticized for using formal mathematical manipulations, without understanding how they worked.]

Topics to Review

Sections 12.1 and 12.2 contain the basic properties of the Laplace transform and are self-contained. The applications of the Laplace transform to partial differential equations are presented in Section 12.3 and assume a familiarity with related boundary value problems for the heat and wave equations that have been treated earlier in the book. Section 12.4 develops the Hankel transform and requires some knowledge of the basic properties of Bessel functions from Sections 9.6 and 9.7. The applications in this section are also related to boundary value problems treated previously.

Looking Ahead

This chapter adds to the diversity of problems and applications that we have treated thus far. With it we complete the treatment of the standard boundary value problems that arise in the classical areas of heat conduction and wave motion. The reader who has encountered the Laplace transform in a course in ordinary differential equations will see it here fulfilling, next to the Fourier transform, a major role in the solution of boundary value problems for partial differential equations. And those of you who have enjoyed Bessel functions will see them again here at the heart of the definition of yet another important transform, the Hankel transform.

In the previous chapter we introduced the Fourier transform and the Fourier sine and cosine transforms and showed their utility in solving various boundary value problems for partial differential equations on unbounded domains. The problems to which these transforms applied were typically treated in Cartesian coordinates. Another transform that can frequently be applied with success is the Laplace transform, our first topic in this chapter. If one of the variables occurring in a problem ranges over a half-line $[0, \infty)$, we can often make progress by performing a Laplace transform with respect to this variable, in much the same way that we did with the sine and cosine transforms. Because of the importance of this transform in other settings, we present a self-contained treatment, including the solution of initial value problems for ordinary differential equations. For problems with other than Cartesian geometry, there are yet other transforms that are more natural and therefore more useful. For example, in unbounded problems with radial symmetry in either the plane or the space, so that the appropriate coordinates are polar, cylindrical, or spherical, the natural transform for the radial variable (r or ρ) involves Bessel functions. This transform, which depends on the order ν of the Bessel function involved, is known as the Hankel transform of order ν.

12.1   The Laplace Transform

In this section we present the definition and basic properties of the Laplace transform. As a warm-up for the applications with partial differential equations, we will use it to solve some simple ordinary differential equations.

Suppose that $f(t)$ is defined for all $t \geq 0$. The **Laplace transform** of f is the function

As a convention, functions $f, g, \ldots$ are defined for $t \geq 0$ and their transforms $F, G, \ldots$ are defined on the s-axis.

(1)
$$\mathcal{L}(f)(s) = \int_0^\infty f(t)e^{-st}\, dt.$$

Another commonly used notation for $\mathcal{L}(f)(s)$ is $F(s)$. For the integral to exist f cannot grow faster than an exponential. This motivates the following definition. We say that f is of **exponential order** if there exist positive numbers a and M such that

(2)
$$|f(t)| \leq Me^{at} \quad \text{for all } t \geq 0.$$

For example, the functions 1, $4\cos 2t$, $5t\sin 2t$, e^{3t} are all of exponential order. We can now give a sufficient condition for the existence of the Laplace transform.

**THEOREM 1
EXISTENCE OF THE
LAPLACE
TRANSFORM**

Suppose that f is piecewise continuous on the interval $[0, \infty)$ and of exponential order with $|f(t)| \leq Me^{at}$ for all $t \geq 0$. Then $\mathcal{L}(f)(s)$ exists for all $s > a$.

Proof We have to show that for $s > a$

$$\mathcal{L}(f)(s) = \int_0^\infty f(t)e^{-st}\, dt < \infty.$$

With M and a as before, we have

$$\left| \int_0^\infty f(t)e^{-st}\, dt \right| \leq \int_0^\infty |f(t)|e^{-st}\, dt \leq M\int_0^\infty e^{at}e^{-st}\, dt$$

$$= M\int_0^\infty e^{-(s-a)t}\, dt = \frac{M}{s-a} < \infty. \qquad \blacksquare$$

Note that the function $\frac{1}{\sqrt{t}}$ is not of exponential order, because of its behavior at $t = 0$. However, we will show in Example 2 that its Laplace transform $\mathcal{L}\left(\frac{1}{\sqrt{t}}\right)(s)$ exists for all $s > 0$. Thus Theorem 1 provides sufficient but not necessary conditions for the existence of the Laplace transform.

EXAMPLE 1 $\mathcal{L}(1)$, $\mathcal{L}(t)$, and $\mathcal{L}(e^{\alpha t})$
We compute these transforms using (1). We have

$$\mathcal{L}(1)(s) = \int_0^\infty e^{-st}\, dt = -\frac{1}{s} e^{-st}\Big|_0^\infty = \frac{1}{s}, \quad s > 0;$$

$$\mathcal{L}(t)(s) = \int_0^\infty t e^{-st}\, dt = \left(-\frac{t}{s} e^{-st} - \frac{1}{s^2} e^{-st}\right)\Big|_0^\infty = \frac{1}{s^2}, \quad s > 0;$$

and finally, for $s > \alpha$,

$$\mathcal{L}(e^{\alpha t})(s) = \int_0^\infty e^{-(s-\alpha)t}\, dt = -\frac{1}{s-\alpha} e^{-(s-\alpha)t}\Big|_0^\infty = \frac{1}{s-\alpha}.$$

Note that $\mathcal{L}(e^{\alpha t})(s)$ is not defined for $s \le \alpha$.   ■

In computing $\mathcal{L}(t)$ we had to integrate by parts once.   Similarly, we could compute $\mathcal{L}(t^n)$ (n a positive integer) by integrating by parts n times. Rather than doing this, we shall take advantage of an interesting connection between the Laplace transform and the gamma function.

EXAMPLE 2 $\mathcal{L}(t^a)$ **via the gamma function**
(a)   Evaluate $\mathcal{L}(t^a)(s)$ when $a > -1$ and $s > 0$.
(b)   Derive from (a) the transforms $\mathcal{L}(t)$, $\mathcal{L}(t^2)$, and, more generally $\mathcal{L}(t^n)$, where n is a positive integer.
(c)   What is $\mathcal{L}\left(\frac{1}{\sqrt{t}}\right)$?

Solution   (a)   From (1) we have

$$\mathcal{L}(t^a)(s) = \int_0^\infty t^a e^{-st}\, dt.$$

To compare with the definition of the gamma function (Exercise 24, Section 4.3), we make the change of variables $st = T$, $dt = \frac{1}{s}\, dT$. Then

$$\mathcal{L}(t^a)(s) = \int_0^\infty \left(\frac{T}{s}\right)^a e^{-T} \frac{dT}{s} = \frac{1}{s^{a+1}} \underbrace{\int_0^\infty T^a e^{-T}\, dT}_{=\Gamma(a+1)}$$

$$= \frac{\Gamma(a+1)}{s^{a+1}}.$$

(b)   Using (a),

$$\mathcal{L}(t) = \frac{\Gamma(2)}{s^2} = \frac{1}{s^2};$$

$$\mathcal{L}(t^2) = \frac{\Gamma(3)}{s^3} = \frac{2}{s^3};$$

and, more generally,

$$\mathcal{L}(t^n) = \frac{\Gamma(n+1)}{s^{n+1}} = \frac{n!}{s^{n+1}}.$$

(c) Using (a), and Exercise 25(a), Section 4.3,

$$\mathcal{L}\left(\frac{1}{\sqrt{t}}\right) = \mathcal{L}\left(t^{-1/2}\right) = \frac{\Gamma(1/2)}{s^{1/2}} = \sqrt{\frac{\pi}{s}}\,.$$

■

Operational Properties

We will derive in the rest of this section properties of the Laplace transform that will assist us in solving differential equations. We are particularly interested in those formulas involving a function, its transform, and the transform of its derivatives. These formulas are similar to the operational properties of the Fourier transform. Because the Laplace transform is defined by an integral over the interval $[0, \infty)$, some of the formulas will involve the values of the function and its derivatives at 0.

THEOREM 2
LINEARITY

If f and g are functions and α and β are numbers, then

$$\mathcal{L}(\alpha f + \beta g) = \alpha\mathcal{L}(f) + \beta\mathcal{L}(g)\,.$$

The proof is left as an exercise. You should also think about the domain of definition of $\mathcal{L}(\alpha f + \beta g)$ in terms of the domains of definition of $\mathcal{L}(f)$ and $\mathcal{L}(g)$.

EXAMPLE 3 $\mathcal{L}(\cos kt)$ **and** $\mathcal{L}(\sin kt)$

These transforms can be evaluated directly by using (1). Our derivation will be based on Euler's identity $e^{ikt} = \cos kt + i\sin kt$ and the linearity of the Laplace transform. We have

$$
\begin{aligned}
\mathcal{L}(\cos kt) + i\mathcal{L}(\sin kt) &= \int_0^\infty (\cos kt + i\sin kt)e^{-st}\,dt \\
&= \int_0^\infty e^{-t(s-ik)}\,dt = -\left.\frac{e^{-t(s-ik)}}{s - ik}\right|_0^\infty = \frac{1}{s - ik} \\
&= \frac{s + ik}{s^2 + k^2} = \frac{s}{s^2 + k^2} + i\,\frac{k}{s^2 + k^2}\,.
\end{aligned}
$$

Equating real and imaginary parts, we get

$$\mathcal{L}(\cos kt) = \frac{s}{s^2 + k^2} \quad \text{and} \quad \mathcal{L}(\sin kt) = \frac{k}{s^2 + k^2}\,.$$

■

The next result is very useful. It states that the Laplace transform takes derivatives into powers of s.

THEOREM 3
LAPLACE
TRANSFORMS OF
DERIVATIVES

(i) Suppose that f is continuous on $[0, \infty)$ and of exponential order as in (2). Suppose further that f' is piecewise continuous on $[0, \infty)$ and of exponential order. Then

$$(3) \qquad \mathcal{L}(f') = s\,\mathcal{L}(f) - f(0)\,.$$

(ii) More generally, if $f, f', \ldots, f^{(n-1)}$ are continuous on $[0, \infty)$ and of exponential order as in (2), and $f^{(n)}$ is piecewise continuous on $[0, \infty)$ and of exponential order, then

$$(4) \qquad \mathcal{L}(f^{(n)}) = s^n \mathcal{L}(f) - s^{n-1} f(0) - s^{n-2} f'(0) - \cdots - f^{(n-1)}(0)\,.$$

Proof Since f is of exponential order, then (2) holds for some positive constants a and M. The transform $\mathcal{L}(f')(s)$ is to be computed for $s > a$. Before we start the computation, note that for $s > a$

$$\lim_{t \to \infty} |f(t)|\, e^{-st} = \lim_{t \to \infty} \underbrace{|f(t)|}_{\leq Me^{at}} e^{-at} e^{-(s-a)t} \leq M \lim_{t \to \infty} e^{-(s-a)t} = 0,$$

because $s - a > 0$. We now compute, using (1) and integrating by parts,

$$\begin{aligned}
\mathcal{L}(f')(s) &= \int_0^\infty f'(t) e^{-st}\, dt \quad (s > a) \\
&= f(t) e^{-st}\Big|_0^\infty - (-s) \underbrace{\int_0^\infty f(t) e^{-st}\, dt}_{\mathcal{L}(f)(s)} \\
&= -f(0) + s\,\mathcal{L}(f),
\end{aligned}$$

which proves (i). Part (ii) follows by repeated applications of (i). ∎

When $n = 2$, (4) gives

$$(5) \qquad \boxed{\mathcal{L}(f'') = s^2 \mathcal{L}(f) - s f(0) - f'(0)\,.}$$

The following is a counterpart of Theorem 3 showing that the Laplace transform takes powers of t into derivatives.

THEOREM 4
DERIVATIVES OF
TRANSFORMS

(i) Suppose $f(t)$ is piecewise continuous and of exponential order. Then

$$(6) \qquad \mathcal{L}(t f(t))(s) = -\frac{d}{ds}\, \mathcal{L}(f)(s)\,.$$

(ii) In general, if $f(t)$ is piecewise continuous and of exponential order, then

$$(7) \qquad \mathcal{L}(t^n f(t)) = (-1)^n \frac{d^n}{ds^n}\, \mathcal{L}(f)(s)\,.$$

Proof Differentiation under the integral sign gives

$$[\mathcal{L}(f)]'(s) = \frac{d}{ds}\int_0^\infty f(t)e^{-st}\,dt = \int_0^\infty f(t)\frac{d}{ds}e^{-st}\,dt$$

$$= -\int_0^\infty tf(t)e^{-st}\,dt = -\mathcal{L}(tf(t))(s),$$

and (i) follows upon multiplying both sides by -1. Part (ii) is obtained by repeated applications of (i). ∎

EXAMPLE 4   Derivatives of transforms

(a) Evaluate $\mathcal{L}(t\sin 2t)$.   (b) Evaluate $\mathcal{L}(t^2\sin t)$.

Solution   (a) Using (6) and Example 3, we find

$$\mathcal{L}(t\sin 2t) = -\frac{d}{ds}\frac{2}{s^2+4} = \frac{4s}{(s^2+4)^2}.$$

(b) Similarly, using (7) and Example 3, we find

$$\mathcal{L}(t^2\sin t) = \frac{d^2}{ds^2}\frac{1}{s^2+1} = \frac{2(-1+3s^2)}{(s^2+1)^3}.$$

The following theorem states that multiplication of a function by $e^{\alpha t}$ causes the transform to be shifted by α units on the s-axis. This very important property has a counterpart which involves a shift on the t-axis (see Theorem 1, Section 12.2).

THEOREM 5
SHIFTING ON THE
s-AXIS

Suppose that f is of exponential order. Let α be a real number and a be as in (2). For $s > a + \alpha$, we have

$$\mathcal{L}(e^{\alpha t}f(t))(s) = F(s-\alpha),$$

where $F(s) = \mathcal{L}(f(t))(s)$.

Proof Note that $e^{\alpha t}f(t)$ is also of exponential order and (2) holds with a replaced by $a + \alpha$. Thus Theorem 1 guarantees the existence of $\mathcal{L}(e^{\alpha t}f(t))$ for $s > a + \alpha$. We have

$$\mathcal{L}(e^{\alpha t}f(t))(s) = \int_0^\infty f(t)e^{\alpha t}e^{-st}\,dt = \int_0^\infty f(t)e^{-(s-\alpha)t}\,dt = F(s-\alpha).$$ ∎

Taking $f = 1$ in Theorem 5, we obtain the third transform in Example 1, $\mathcal{L}(e^{\alpha t}) = \frac{1}{s-\alpha}$, since $\mathcal{L}(1) = \frac{1}{s}$.

The Inverse Laplace Transform

Given a function $F(s)$, if we can find a function $f(t)$ such that $\mathcal{L}(f) = F$, we will call $f(t)$ the **inverse Laplace transform** of $F(s)$ and denote it by

$$f(t) = \mathcal{L}^{-1}(F(s)) \quad \text{or simply} \quad f = \mathcal{L}^{-1}(F).$$

We will use the Laplace transform as a tool to solve differential equations, just like we used the Fourier transform. The method will consist of applying the Laplace transform to a problem, solving the transformed problem, and then taking the inverse of the Laplace transform to find the solution of the original problem. This process assumes that an inverse exists and that it is unique. Indeed, just like the Fourier transform has an inverse transform, the Laplace transform has an inverse. The formula involves integration in the complex plane and requires a certain amount of complex analysis. We will present it at the end of the section; for now we will take the uniqueness of the inverse transform for granted and compute the inverse transform by using known Laplace transforms, as illustrated by the following examples. We note that the inverse of any linear transform is itself linear. In particular, we have

$$\mathcal{L}^{-1}(\alpha F + \beta G) = \alpha \mathcal{L}^{-1}(F) + \beta \mathcal{L}^{-1}(G).$$

EXAMPLE 5   Inverse Laplace transforms

(a) Evaluate $\mathcal{L}^{-1}(\frac{2}{4+(s-1)^2})$.   (b) Evaluate $\mathcal{L}^{-1}(\frac{1}{s^2+2s+3})$.

Solution   (a) From the table of Laplace transforms, Appendix B.4 (or by using Example 3 and Theorem 5), we find that

$$\mathcal{L}(e^{at}\sin kt) = \frac{k}{(s-a)^2 + k^2}.$$

Taking $a = 1$ and $k = 2$, we get

$$\mathcal{L}(e^t \sin 2t) = \frac{2}{(s-1)^2 + 4}.$$

Hence

$$\mathcal{L}^{-1}\left(\frac{2}{4+(s-1)^2}\right) = e^t \sin 2t.$$

(b) Motivated by part (a), we first write

$$\frac{1}{s^2 + 2s + 3} = \frac{1}{(s+1)^2 + (\sqrt{2})^2} = \frac{1}{\sqrt{2}}\frac{\sqrt{2}}{(s+1)^2 + (\sqrt{2})^2}.$$

Now using the transform in (a) with $a = -1$ and $k = \sqrt{2}$, we get

$$\mathcal{L}^{-1}\left(\frac{1}{s^2 + 2s + 3}\right) = \frac{1}{\sqrt{2}}\mathcal{L}^{-1}\left(\frac{\sqrt{2}}{(s+1)^2 + (\sqrt{2})^2}\right) = \frac{1}{\sqrt{2}}e^{-t}\sin\sqrt{2}t. \qquad \blacksquare$$

EXAMPLE 6   Partial fractions

Evaluate $\mathcal{L}^{-1}\left(\frac{1}{s^2+2s-3}\right)$.

Solution   First Method   We can compute as we did in Example 5(b). First, write

$$\frac{1}{s^2 + 2s - 3} = \frac{1}{(s+1)^2 - 2^2} = \frac{1}{2}\frac{2}{(s+1)^2 - 2^2}.$$

From the table of Laplace transforms, Appendix B.4, we have

$$\mathcal{L}(e^{at}\sinh kt) = \frac{k}{(s-a)^2 - k^2}.$$

Taking $a = -1$ and $k = 2$, it follows that

$$\mathcal{L}^{-1}\left(\frac{1}{s^2 + 2s - 3}\right) = \frac{1}{2}\mathcal{L}^{-1}\left(\frac{2}{(s+1)^2 - 2^2}\right) = \frac{1}{2}e^{-t}\sinh 2t.$$

Second Method   Partial fractions are useful in evaluating inverse Laplace transforms of rational functions. To apply this method, we first factor the denominator as $s^2 + 2s - 3 = (s+3)(s-1)$. Now write

$$\frac{1}{(s+3)(s-1)} = \frac{A}{(s+3)} + \frac{B}{(s-1)},$$

so that

$$1 = A(s-1) + B(s+3).$$

Setting $s = 1$ and then $s = -3$ yields $B = \frac{1}{4}$ and $A = -\frac{1}{4}$, respectively. Thus

$$\mathcal{L}^{-1}\left(\frac{1}{s^2 + 2s - 3}\right) = -\frac{1}{4}\mathcal{L}^{-1}\left(\frac{1}{s+3}\right) + \frac{1}{4}\mathcal{L}^{-1}\left(\frac{1}{s-1}\right) = -\frac{1}{4}e^{-3t} + \frac{1}{4}e^t.$$

It is easy to see that this transform is also equal to $\frac{1}{2}e^{-t}\sinh 2t$, matching our earlier finding.   ∎

Laplace Transform and Differential Equations

The key to solving differential equations via the Laplace transform method is to use the operational properties, particularly those related to differentiation. We begin with a simple initial value problem. In what follows, we will denote the Laplace transform of $y(t)$ by $Y(s)$.

EXAMPLE 7   Solving a differential equation with the Laplace transform: Solve $y'' + y = 2, \quad y(0) = 0, \ y'(0) = 1$.

Solution   Taking the Laplace transform of both sides of the equation and using Theorem 3, we find

$$s^2 Y - sy(0) - y'(0) + Y = \mathcal{L}(2) = \frac{2}{s}.$$

Using the initial conditions, we obtain

$$(s^2 + 1)Y - 1 = \tfrac{2}{s};$$
$$Y = \tfrac{1}{s^2+1} + \tfrac{2}{s(s^2+1)}.$$

Using partial fractions on the second term, we find

$$Y = \frac{1}{s^2 + 1} + \frac{2}{s} - \frac{2s}{s^2 + 1}.$$

Finally, taking the inverse Laplace transform, we get

$$y = \sin t + 2 - 2\cos t.$$ ■

This example is a typical illustration of the Laplace transform method. Starting from a linear ordinary differential equation with constant coefficients in y, the Laplace transform produces an algebraic equation that can be solved for Y. The solution y is then found by taking the inverse Laplace transform of Y. The Laplace transform method is most compatible with initial value problems where the initial data is given at $t = 0$, due to the way the transform acts on derivatives. If the initial data is given at some other value t_0, the Laplace transform still applies: We simply make the change of variables $\tau = t - t_0$. The next example illustrates this process.

EXAMPLE 8   Shifting the time variable

Solve $y'' + 2y' + y = t$, $y(1) = 0$, $y'(1) = 0$.

Solution Making the change of variables $\tau = t - 1$, we arrive at the initial value problem

$$y'' + 2y' + y = \tau + 1,   y(0) = 0, \ y'(0) = 0,$$

where now a prime denotes differentiation with respect to τ. From this point, we proceed as in Example 7. Transforming yields

$$s^2 Y + 2sY + Y = \frac{1}{s^2} + \frac{1}{s};$$

$$Y = \frac{1}{s^2(s + 1)}.$$

Using partial fractions, we get

$$Y = \frac{1}{s + 1} - \frac{1}{s} + \frac{1}{s^2}.$$

Taking the inverse Laplace transform, we arrive at

$$y = e^{-\tau} - 1 + \tau.$$

Hence the solution is $y(t) = e^{1-t} + t - 2$.   ■

Recall from Section 11.2 an interesting way to compute the Fourier transform of e^{-ax^2} was to take an indirect approach and consider a differential equation that is satisfied by e^{-ax^2}. These ideas work also with the Laplace transform, as we now illustrate by giving a simple derivation of the Laplace transform of $\cos kx$.

EXAMPLE 9   Using differential equations to compute transforms

The function $y = \cos kx$ is the unique solution of the initial value problem

$$y'' + k^2 y = 0, \ y(0) = 1, \ y'(0) = 0.$$

Transforming this problem with the Laplace transform, we find that

$$s^2 Y - s + k^2 Y = 0 \quad \Rightarrow \quad Y = \frac{s}{s^2 + k^2},$$

which gives the Laplace transform of $\cos kx$. Knowing the transform of $\cos kx$, we can get the transform of $\sin kx$ quite easily. Note that $y = \sin kx$ is the unique solution of the initial value problem $y' = k \cos kx$ and $y(0) = 0$. Transforming this first order initial value problem with the Laplace transform, we find that

$$sY = \mathcal{L}\left(k \cos kx\right) \quad \Rightarrow \quad sY = \frac{ks}{s^2 + k^2} \quad \Rightarrow \quad Y = \frac{k}{s^2 + k^2},$$

which gives the Laplace transform of $\sin kx$. ∎

We end this section by deriving a formula for the inverse Laplace transform.

A Formula for the Inverse Laplace Transform

To describe the inverse of the Laplace transform, we will first extend the transform to complex numbers as follows. Suppose that f is piecewise smooth and of exponential order with $|f(t)| \leq M e^{at}$ for all $t \geq 0$ $(a > 0)$. For a complex number z in the right half-plane, $\operatorname{Re} z > a$, define

$$(8) \qquad \mathcal{L}(f)(z) = \int_0^\infty f(t) e^{-zt} \, dt.$$

When z is real, this definition reduces to (1). Also, for $\operatorname{Re} z > a$ (say $\operatorname{Re} z = a + \epsilon$, $\epsilon > 0$) and all $t \geq 0$, we have

$$\left| f(t) e^{-zt} \right| \leq M e^{at} \left| e^{-\operatorname{Re}(z)t} \right| \leq M e^{at} e^{-(a+\epsilon)t} \leq M e^{-\epsilon t},$$

and so the integral in (8) exists. We next argue that $\mathcal{L}(f)(z)$ is analytic in the half-plane $\operatorname{Re} z > a$. To simplify the proof, we will further suppose that f is continuous on $[0, \infty)$. Write

$$(9) \qquad \mathcal{L}(f)(z) = \int_0^\infty f(t) e^{-zt} \, dt = \sum_{n=0}^\infty \int_n^{n+1} f(t) e^{-zt} \, dt.$$

Each term in the sum is analytic, by Theorem 4, Section 3.5 (differentiation under the integral sign). Moreover, if K is a closed and bounded subset of the half-plane $\operatorname{Re} z > a$, then there is a $\delta > 0$ such that $|f(t) e^{-zt}| \leq M e^{-\delta t}$ for all z in K and all t (see Figure 1). Hence, for all z in K and all t,

$$\left| \int_n^{n+1} f(t) e^{-zt} \, dt \right| \leq M \int_n^{n+1} e^{-\delta t} \, dt = \frac{M}{\delta} \left(e^{-n\delta} - e^{-(n+1)\delta} \right) = M_n.$$

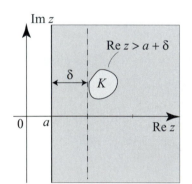

Figure 1 Since K is closed and bounded and disjoint from the line $\operatorname{Re} z = a$, its distance to this line is $\delta > 0$. For all z in K, $\operatorname{Re} z > a + \delta$; so

$$|f(z) e^{-zt}| \ \leq \ M e^{at} e^{-t \operatorname{Re} z}$$
$$\leq \ M e^{-\delta t}.$$

Since $\sum_{n=0}^{\infty} M_n < \infty$, it follows from the Weierstrass M-test that the series in (9) converges uniformly for all z in K. From Corollary 2, Section 4.2, we conclude that $\mathcal{L}(f)(z)$ is analytic in the half-plane $\operatorname{Re} z > a$.

We now proceed to invert the Laplace transform by appealing to the inverse Fourier transform. Pick $b > a$, and define $g(t) = e^{-bt} f(t)$ if $t \geq 0$ and $g(t) = 0$ if $t < 0$. Then because f is piecewise smooth and of exponential order with $a < b$, it follows that g is integrable and piecewise smooth on the real line, and so we may apply the inverse Fourier transform (Theorem 1, Section 11.1). Using $g(t) = 0$ for $t < 0$, we have the Fourier transform

$$(10) \qquad \widehat{g}(\omega) = \frac{1}{\sqrt{2\pi}} \int_0^\infty e^{-bt} f(t) e^{-i\omega t}\, dt = \frac{1}{\sqrt{2\pi}} \mathcal{L}(f)(b + i\omega),$$

and the inverse Fourier transform

$$(11) \qquad g(t) = \lim_{R \to \infty} \frac{1}{\sqrt{2\pi}} \int_{-R}^R \widehat{g}(\omega) e^{i\omega t}\, d\omega = \lim_{R \to \infty} \frac{1}{2\pi} \int_{-R}^R \mathcal{L}(f)(b + i\omega) e^{i\omega t}\, d\omega,$$

where the left side in (11) should be interpreted as the average of g at the points of discontinuity. Making the change of variables $z = b + i\omega$, $dz = i\,d\omega$, and recalling that $g(t) = e^{-bt} f(t)$, we get

$$(12) \qquad e^{-bt} f(t) = \lim_{R \to \infty} \frac{e^{-bt}}{2\pi i} \int_{b-iR}^{b+iR} \mathcal{L}(f)(z) e^{-zt}\, dz,$$

where the integral is over the vertical line segment from $b - iR$ to $b + iR$. Simplifying, and replacing f by its average to account for the possible discontinuities, we obtain the inversion formula for the Laplace transform:

$$(13) \qquad \frac{f(t+) + f(t-)}{2} = \lim_{R \to \infty} \frac{1}{2\pi i} \int_{b-iR}^{b+iR} \mathcal{L}(f)(z) e^{-zt}\, dz.$$

Note that the integral is independent of the choice of b as long as $b > a$. To see this, given any two vertical line segments, we can close the contour as in Figure 2. The integral over the closed contour is 0 by Cauchy's theorem. A straightforward estimate on the integrand shows that the integrals over the horizontal sides tend to zero as $R \to \infty$. This implies that

$$\lim_{R \to \infty} \frac{1}{2\pi i} \int_{b-iR}^{b+iR} \mathcal{L}(f)(z) e^{-zt}\, dz = \lim_{R \to \infty} \frac{1}{2\pi i} \int_{b'-iR}^{b'+iR} \mathcal{L}(f)(z) e^{-zt}\, dz,$$

as claimed. One important consequence of (13) is the **uniqueness** of the inverse Laplace transform. Since this inverse is given by an integral, any two inverses must be equal to the integral and thus must be the same.

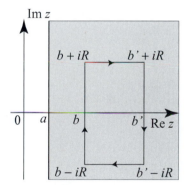

Figure 2

Exercises 12.1

In Exercises 1–6, show that the given function is of exponential order by establishing (2) with an appropriate choice of the numbers a and M.

1. $f(t) = 11 \cos 3t$.

2. $f(t) = \sin 2t + 3 \cos t$.

3. $f(t) = 5 e^{3t}$.

4. $f(t) = t^n$.

5. $f(t) = \sinh 3t$.

6. $f(t) = e^{5t} \sinh t$.

In Exercises 7–24, evaluate the Laplace transform of the given function using appropriate theorems and examples from this section.

7. $f(t) = 2t + 3$.

8. $f(t) = t^2 + 3t^4$.

9. $f(t) = \sqrt{t} + \frac{1}{\sqrt{t}}$.

10. $f(t) = t^2 + 3t + t^{3/2}$.

11. $f(t) = t^2 e^{3t}$.

12. $f(t) = t^4 e^{-3t} + e^t$.

13. $f(t) = t \sin 4t$.

14. $f(t) = t^2 \cos 2t$.

15. $f(t) = \sin^2 t$.

16. $f(t) = \cos t \sin t$.

17. $f(t) = e^{2t} \sin 3t$.

18. $f(t) = t \sinh 3t$.

19. $f(t) = te^{-t} \sin t$.

20. $f(t) = e^{t+1} \cos t$.

21. $f(t) = (t + 2)^2 \cos t$.

22. $f(t) = e^{\alpha t} t^{3/2}$.

23. $f(t) = e^{\alpha t} \sin \beta t$.

24. $f(t) = e^{\alpha t} \cos \beta t$.

In Exercises 25–38, evaluate the inverse Laplace transform of the given function.

25. $F(s) = \frac{1}{s^2}$.

26. $F(s) = \frac{1}{s^2 - 2}$.

27. $F(s) = \frac{4}{3s^2 + 1}$.

28. $F(s) = \frac{1}{(s-1)^2 - 2}$.

29. $F(s) = \frac{1}{(s-3)^5} + \frac{s-3}{1 + (s-3)^2}$.

30. $F(s) = \frac{1}{(s-1)(s+1)}$.

31. $F(s) = \frac{s}{s^2 + 2s + 1}$.

32. $F(s) = \frac{3s+1}{s^2 - 2s}$.

33. $F(s) = \frac{2s-1}{s^2 - s - 2}$.

34. $F(s) = \frac{s-1}{(s+1)(s^2+1)}$.

35. $F(s) = \frac{5s^2 + 2s - 4}{2s(s^2 + s - 2)}$.

36. $F(s) = \frac{s^2 + s + 3}{(s+2)(s^2+1)}$.

37. $F(s) = \frac{1}{s^2 + 3s + 2}$.

38. $F(s) = \frac{s}{s^2 + 3s + 2}$.

In Exercises 39–46, solve the given initial value problem with the Laplace transform.

39. $y' + y = \cos 2t$, $\quad y(0) = -2$.

40. $y' + 2y = 6e^{\alpha t}$, $\quad y(0) = 1$, α is a constant.

41. $y'' + y = \cos t$, $\quad y(\pi) = 0$, $y'(\pi) = 0$.

42. $y'' - y = 1 + t^2$, $\quad y(1) = 0$, $y'(1) = 1$.

43. $y'' + 2y' + y = te^{-2t}$, $\quad y(0) = 1$, $y'(0) = 1$.

44. $y'' + y' + 4y = 0$, $\quad y(0) = 1$, $y'(0) = 1$.

45. $y'' - y' - 6y = e^t \cos t$, $\quad y(0) = 0$, $y'(0) = 1$.

46. $y'' + 4y' + 5y = t^2$, $\quad y(0) = 0$, $y'(0) = 1$.

12.2   Further Properties of the Laplace Transform

We continue our study of the Laplace transform and start by computing the transform of special functions that arise naturally in studying operational properties of the transform. Recall the **Heaviside unit step function**

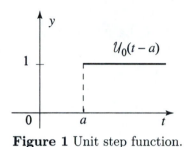

Figure 1 Unit step function.

$$(1) \qquad \mathcal{U}_a(t) = \mathcal{U}_0(t-a) = \begin{cases} 0 & \text{if } t < a, \\ 1 & \text{if } t \geq a \end{cases}$$

(Figure 1). Given a function $f(t)$, consider the product $\mathcal{U}_0(t-a)f(t-a)$. Written explicitly, we have

$$\mathcal{U}_0(t-a)f(t-a) = \begin{cases} 0 & \text{if } t < a, \\ f(t-a) & \text{if } t \geq a. \end{cases}$$

Thus, if $f(t)$ represents, say a message, then $\mathcal{U}_0(t-a)f(t-a)$ represents the same message, but delayed by a units of time.

THEOREM 1
SHIFTING ON THE
t-AXIS

If a is a positive real number, then

$$\mathcal{L}(\mathcal{U}_0(t-a)f(t-a))(s) = e^{-as}F(s),$$

where $F(s) = \mathcal{L}(f(t))(s)$.

Proof We have

$$\begin{aligned}
\mathcal{L}(\mathcal{U}_0(t-a)f(t-a))(s) &= \int_a^\infty f(t-a)e^{-st}\, dt \\
&= \int_0^\infty f(T)e^{-s(T+a)}\, dT \quad \text{(where } t - a = T,\ dt = dT) \\
&= e^{-as}\int_0^\infty f(T)e^{-sT}\, dT = e^{-as}F(s). \qquad \blacksquare
\end{aligned}$$

EXAMPLE 1   Transforms involving unit step functions
(a) Evaluate $\mathcal{L}(\mathcal{U}_0(t-a))$.
(b) Evaluate $\mathcal{L}(f(t))$, where

$$f(t) = \begin{cases} 2 & \text{if } 1 \leq t < 4, \\ 0 & \text{otherwise} \end{cases}$$

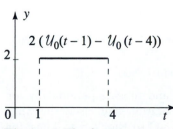

Figure 2 The function in Example 1.

(see Figure 2).

Solution (a) Using Theorem 1 with $f(t) = 1$, and Example 1 of the previous section, we find

$$\mathcal{L}(\mathcal{U}_0(t-a)) = \frac{e^{-as}}{s}, \quad s > 0.$$

(b) Write $f(t) = 2(\mathcal{U}_0(t-1) - \mathcal{U}_0(t-4))$ (check it!). Then,

$$\mathcal{L}(f(t)) = 2\left(\mathcal{L}(\mathcal{U}_0(t-1)) - \mathcal{L}(\mathcal{U}_0(t-4))\right) = \frac{2}{s}\left(e^{-s} - e^{-4s}\right), \quad s > 0. \qquad \blacksquare$$

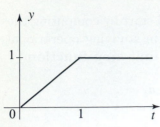

Figure 3 A ramp function.

EXAMPLE 2   A ramp function

Evaluate the Laplace transform of the ramp function shown in Figure 3.

Solution   For $t > 1$, we have $f(t) = \mathcal{U}_0(t - 1)$, and for $0 \leq t \leq 1$, we have $f(t) = t\left(\mathcal{U}_0(t) - \mathcal{U}_0(t - 1)\right)$. We can combine these two formulas and simply write

$$f(t) = t\left(\mathcal{U}_0(t) - \mathcal{U}_0(t - 1)\right) + \mathcal{U}_0(t - 1).$$

(Check it!) We are not quite ready to apply Theorem 1. To be able to do so, we rewrite $f(t)$ as follows

$$f(t) = -(t - 1)\mathcal{U}_0(t - 1) + t\mathcal{U}_0(t) = -(t - 1)\mathcal{U}_0(t - 1) + t.$$

Now recall that $\mathcal{L}(t) = \frac{1}{s^2}$. Also, by Theorem 1, $\mathcal{L}((t - 1)\mathcal{U}_0(t - 1)) = \frac{e^{-s}}{s^2}$. Hence

$$\mathcal{L}(f(t)) = -\frac{e^{-s}}{s^2} + \frac{1}{s^2}.   \blacksquare$$

In the next example we solve a nonhomogeneous differential equation involving a ramp function.

EXAMPLE 3   A nonhomogeneous differential equation

Solve $y'' + y = f(t)$, $y(0) = 0$, $y'(0) = 0$, where $f(t)$ is as in Example 2.

Solution   Taking the Laplace transform of both sides of the equation and using the result of Example 2, we find

$$s^2 Y + Y = \frac{1}{s^2} - \frac{e^{-s}}{s^2};$$

$$Y = \frac{1 - e^{-s}}{s^2(s^2 + 1)}.$$

Using partial fractions, or simply noticing that

$$\frac{1}{s^2(s^2 + 1)} = \frac{1}{s^2} - \frac{1}{(s^2 + 1)},$$

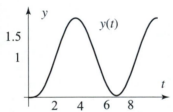

Figure 4 Solution in Example 3.

we obtain

$$\begin{aligned} Y &= (1 - e^{-s})\left(\frac{1}{s^2} - \frac{1}{s^2 + 1}\right) \\ &= \frac{1}{s^2} - \frac{1}{s^2 + 1} - e^{-s}\left(\frac{1}{s^2} - \frac{1}{s^2 + 1}\right). \end{aligned}$$

Taking the inverse Laplace transform, and using Theorem 1 for the term involving e^{-s}, we get

$$y(t) = t - \sin t - \mathcal{U}_0(t - 1)[(t - 1) - \sin(t - 1)].$$

Figure 4 shows that the solution is bounded for all t and repeats periodically for $t > 1$. The term $t - \mathcal{U}_0(t-1)(t-1)$ that appears in the solution is precisely the ramp function (see Example 2). In particular, this term is equal to 1 for $t > 1$. Thus, for $t > 1$, the solution is a sum of two 2π-periodic sine waves and the function that is identically 1. This explains the boundedness and the periodicity of the solution for $t > 1$. $\blacksquare$

Our next example involves a differential equation with a discontinuous term.

EXAMPLE 4   Discontinuous forcing term

Solve $y'' + 4y = f(t)$, $y(0) = 1$, $y'(0) = 0$, where the forcing term $f(t)$ is as in Figure 5.

Solution The forcing term can be expressed as

$$f(t) = 1 - \mathcal{U}_0(t - 1).$$

Taking the Laplace transform of both sides of the equation, we get

$$s^2 Y - s + 4Y = \frac{1}{s} - \frac{e^{-s}}{s};$$

$$Y = \frac{s}{s^2 + 4} + \frac{1 - e^{-s}}{s(s^2 + 4)},$$

and hence, using partial fractions,

$$Y = \frac{s}{s^2 + 4} + \frac{1}{4}(1 - e^{-s})\left(\frac{1}{s} - \frac{s}{s^2 + 4}\right).$$

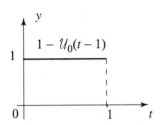

Figure 5 $f(t)$ in Example 4.

Taking the inverse Laplace transform, and using Theorem 1 for the term involving e^{-s}, we get

$$
\begin{aligned}
y &= \cos 2t + \frac{1}{4}(1 - \cos 2t) - \frac{1}{4}\mathcal{U}_0(t - 1)[1 - \cos 2(t - 1)] \\
&= \frac{1}{4} + \frac{3}{4}\cos 2t - \frac{1}{4}\mathcal{U}_0(t - 1)[1 - \cos 2(t - 1)]
\end{aligned}
$$

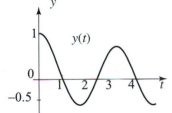

Figure 6 Solution in Example 4.

(Figure 6). Note that even though the forcing term is discontinuous at $t = 1$, the solution is continuous everywhere. A discontinuity at $t = 1$ will appear in the graph of the second derivative of the solution, since $y'' = f(t) - y$. ■

Convolutions and Laplace Transforms

Given two functions f and g, defined for all $t \geq 0$, we define their **convolution** $f * g(t)$ by

$$(2) \qquad \boxed{f * g(t) = \int_0^t f(t - \tau)g(\tau)d\tau \quad \text{for all } t \geq 0.}$$

Clearly this definition is related to the definition of convolutions that we presented in Section 11.2. In fact, if we think of f and g as being defined for all t with values 0 for $t < 0$, then (2) becomes $\int_{-\infty}^{\infty} f(t - \tau)g(\tau)\,d\tau$, which differs by a constant multiple from the convolution that we introduced in Section 11.2. It should be clear from the context which convolution we are talking about, and so there is no risk of confusion in using the same notation for the two different operations. In particular, all our discussion in this section concerns the convolution (2) and not the one in Section 11.2.

Note that since the convolution is defined by an integral over a finite interval, there is no problem in computing $f * g(t)$ if, say, f and g are piecewise continuous functions, no matter how fast they grow at infinity.

**THEOREM 2
TRANSFORMS OF
CONVOLUTIONS**

Suppose that f and g are piecewise continuous and of exponential order; then

$$\mathcal{L}(f * g) = \mathcal{L}(f)\mathcal{L}(g).$$

Proof If we extend the functions f and g to be zero for $t < 0$, then the integral in (2) is the same as

$$(3) \qquad \int_0^\infty f(t - \tau)g(\tau)\,d\tau .$$

Thus, throughout this proof, we assume that f and g are extended to the whole line with $f(t) = 0$ and $g(t) = 0$, for all $t < 0$. We have

$$
\begin{aligned}
\mathcal{L}(f * g)(s) &= \int_0^\infty \int_0^\infty f(t - \tau)\,g(\tau)\,d\tau e^{-st}\,dt \\
&= \int_0^\infty \int_0^\infty f(t - \tau)e^{-st}\,dt\,g(\tau)\,d\tau \quad \text{(Interchange order of integration.)} \\
&= \int_0^\infty \int_0^\infty f(u)e^{-s(u+\tau)}\,du\,g(\tau)\,d\tau \quad (u = t - \tau,\ du = dt) \\
&= \int_0^\infty \int_0^\infty f(u)e^{-su}\,du\,g(\tau)e^{-s\tau}\,d\tau = \mathcal{L}(f)(s)\mathcal{L}(g)(s). \quad \blacksquare
\end{aligned}
$$

EXAMPLE 5   Transforms involving convolutions
(a) Evaluate $\mathcal{L}(\int_0^t (t - \tau)\sin(\tau)\,d\tau)$.
(b) Evaluate $\mathcal{L}^{-1}(\frac{1}{s^2(s^2+4s+5)})$ using convolutions.

Solution (a) From Theorem 2, we have

$$\mathcal{L}\!\left(\int_0^t (t - \tau)\sin\tau\,d\tau\right) = \mathcal{L}(t)\mathcal{L}(\sin t) = \frac{1}{s^2}\frac{1}{s^2 + 1}.$$

(b) Treat the expression $\dfrac{1}{s^2(s^2 + 4s + 5)}$ as a product of the two Laplace transforms $\dfrac{1}{s^2}$ and $\dfrac{1}{s^2 + 4s + 5}$. Since

$$\mathcal{L}^{-1}\left(\frac{1}{s^2}\right) = t,$$

and

$$\mathcal{L}^{-1}\left(\frac{1}{s^2 + 4s + 5}\right) = \mathcal{L}^{-1}\left(\frac{1}{(s + 2)^2 + 1}\right) = e^{-2t}\sin t,$$

it follows that

$$\mathcal{L}^{-1}\left(\frac{1}{s^2(s^2+4s+5)}\right) = t * e^{-2t}\sin t = \int_0^t (t-\tau)e^{-2\tau}\sin\tau\,d\tau\,.$$

The integral in τ can be computed explicitly, using integration by parts. As a result, we get

$$\mathcal{L}^{-1}\left(\frac{1}{s^2(s^2+4s+5)}\right) = \frac{1}{25}(5t-4) + \frac{4}{25}e^{-2t}\cos t + \frac{3}{25}e^{-2t}\sin t. \qquad \blacksquare$$

EXAMPLE 6   Solving differential equations with convolutions

Express the solution of the initial value problem

$$y'' - 2y' + 5y = f(t), \quad y(0) = 0, \ y'(0) = 0,$$

as a convolution.

Solution Transforming the equation and then solving for Y, we find

$$Y = \frac{F(s)}{s^2 - 2s + 5},$$

where $F(s)$ is the Laplace transform of $f(t)$. We have

$$\frac{1}{s^2 - 2s + 5} = \frac{1}{(s-1)^2 + 2^2} = \frac{1}{2}\frac{2}{(s-1)^2 + 2^2}.$$

Hence

$$\mathcal{L}^{-1}\left(\frac{1}{2}\frac{2}{(s-1)^2 + 2^2}\right) = \frac{1}{2}e^t\sin 2t\,.$$

Taking the inverse Laplace transform of Y, and using Theorem 2, we get

$$y = \frac{1}{2}e^t\sin 2t * f(t) = \frac{1}{2}\int_0^t e^{t-\tau}\sin 2(t-\tau)f(\tau)\,d\tau. \qquad \blacksquare$$

Example 6 shows that in solving the differential equation with zero initial data all nonhomogeneous terms are treated equally: we simply integrate $f(t)$ against $\frac{1}{2}e^{t-\tau}\sin 2(t-\tau)$ on the interval 0 to t. Note that the latter function is a translate of the inverse Laplace transform of $1/(s^2 - 2s + 5)$, where the differential equation determines the denominator as follows: y'' corresponds to s^2, $-2y'$ corresponds to $-2s$, and $5y$ corresponds to 5. In other words, the response of the system to the input function (nonhomogeneous term) $f(t)$ is always related to that function via convolution with a "response function" that is determined solely by the associated homogeneous differential equation. It is clear that this remark applies for any linear nonhomogeneous differential equation, and it is therefore a general principle governing all such equations.

We close the section with examples that involve the Dirac delta function and its translates. Recall from Section 11.2 the effect of integrating against a Dirac delta is given by

$$(4) \qquad \int_a^b f(t)\delta_0(t - t_0)\, dt = \begin{cases} f(t_0) & \text{if } t_0 \text{ is in } [a, b], \\ 0 & \text{if } t_0 \text{ is not in } [a, b]. \end{cases}$$

With this formula in hand, we can compute Laplace transforms involving the Dirac delta function. We illustrate with a basic example.

EXAMPLE 7   Laplace transform of the Dirac delta function

Let $a \geq 0$. From the definition of the Laplace transform, we have

$$\mathcal{L}(\delta_0(t - a))(s) = \int_0^\infty e^{-st}\delta_0(t - a)\, dt.$$

Since a belongs to the interval $[0, \infty)$, it follows from (4) that the integral is equal to e^{-as}. Thus

$$\mathcal{L}(\delta_0(t - a)) = e^{-as}. \qquad \blacksquare$$

The Dirac delta function is a unit for the operation of convolution in the sense that

$$f * \delta_0(t) = \int_0^t f(t - \tau)\delta_0(\tau)\, d\tau = f(t).$$

Indeed, since 0 is in the interval $[0, t]$, we use (4) to infer that the integral is equal to the value of $f(t - \tau)$ at $\tau = 0$, or $f(t)$.

Our final example is a differential equation involving the delta function.

EXAMPLE 8   Effect of impulse functions

Solve $y'' + 4y = \delta_0(t - 2)$, $y(0) = 0$, $y'(0) = 0$. (This represents an oscillator, initially at rest, which receives a unit impulse at time $t = 2$.)

Solution Taking the Laplace transform of both sides of the equation, we get

$$s^2 Y + 4Y = e^{-2s};$$
$$Y = \frac{e^{-2s}}{s^2 + 4}.$$

Taking the inverse transform and using Theorem 1, we obtain the solution

$$y(t) = \frac{1}{2}\, \mathcal{U}_0(t - 2) \sin 2(t - 2).$$

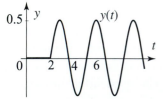

Figure 7 Solution in Example 8.

The graph in Figure 7 shows that the solution is identically 0 up to time $t = 2$, which corresponds to the fact that the oscillator started at rest and no forces acted upon it until that time. For $t > 2$, the solution oscillates periodically. $\blacksquare$

Exercises 12.2

In Exercises 1–6, evaluate the Laplace transform of the given function.

1. $f(t) = \mathcal{U}_0(t-1) - t + 1$.

2. $f(t) = (t-1)\mathcal{U}_0(t-1)$.

3. $f(t) = e^{2t}\mathcal{U}_0(t-2)$.

4. $f(t) = t\,\mathcal{U}_0(t-\pi)$.

5. $f(t) = \mathcal{U}_0(t-\pi)\sin t$.

6. $f(t) = \mathcal{U}_0(t-\pi)\cos t \sin t$.

In Exercises 7–14, (a) plot the given function. (b) Express it using unit step functions. (c) Evaluate its Laplace transform.

7. $f(x) = \begin{cases} 1 & \text{if } 0 < t < 2, \\ 0 & \text{if } t > 2. \end{cases}$

8. $f(x) = \begin{cases} t & \text{if } 0 \le t \le 1, \\ 2-t & \text{if } 1 \le t \le 2, \\ 0 & \text{if } t > 2. \end{cases}$

9. $f(x) = \begin{cases} 2 & \text{if } 2 \le t \le 3, \\ 0 & \text{otherwise.} \end{cases}$

10. $f(x) = \begin{cases} t & \text{if } 0 \le t \le 1, \\ 0 & \text{otherwise.} \end{cases}$

11. $f(x) = \begin{cases} t-1 & \text{if } 1 \le t \le 2, \\ 0 & \text{otherwise.} \end{cases}$

12. $f(x) = \begin{cases} -\sin t & \text{if } \pi \le t \le 2\pi, \\ 0 & \text{otherwise.} \end{cases}$

13. $f(x) = \begin{cases} 1 & \text{if } 1 \le t \le 4, \\ t-5 & \text{if } 4 \le t \le 5, \\ 0 & \text{otherwise.} \end{cases}$

14. $f(x) = \begin{cases} t & \text{if } 0 \le t \le 1, \\ 1 & \text{if } 1 \le t \le 3, \\ 4-t & \text{if } 3 < t \le 4, \\ 0 & \text{otherwise.} \end{cases}$

In Exercises 15–22, evaluate the inverse Laplace transform of the given function.

15. $F(s) = \dfrac{e^{-s}}{s^2}$.

16. $F(s) = \dfrac{e^{-\pi s}}{s^2 - 2}$.

17. $F(s) = \dfrac{e^{-s}}{s^2 + 1}$.

18. $F(s) = \dfrac{e^{-s}}{(s-1)^2 - 1}$.

19. $F(s) = \dfrac{s-3}{1 + (s-3)^2}$.

20. $F(s) = \dfrac{e^{-3s}}{(s-1)(s+1)}$.

21. $F(s) = \dfrac{e^{-s}}{s^{3/2}}$.

22. $F(s) = \dfrac{e^{-s}}{(s-2)^{3/2}}$.

In Exercises 23–28, evaluate the given convolution.

23. $1 * t$.

24. $e^t * e^{-t}$.

25. $t * t$.

26. $t * \sin t$.

27. $\sin t * \sin t$.

28. $e^t * \delta_0(t)$.

In Exercises 29–32, express the inverse Laplace transform of the given function as a convolution. Evaluate the integral in your answer.

29. $F(s) = \dfrac{1}{s(s^2 + 1)}$.

30. $F(s) = \dfrac{1}{s^2(s^2 + 1)}$.

31. $F(s) = \dfrac{1}{(s^2 + 1)(s^2 + 1)}$.

32. $F(s) = \dfrac{s}{(s^2 - 1)(s^2 - 1)}$.

In Exercises 33–38, solve the given initial value problem.

33. $y'' + y = \delta_0(t-1)$, $y(0) = 0$, $y'(0) = 0$.

34. $y'' - y = (t - 2)\,\mathcal{U}_0(t - 2), \quad y(0) = 0,\ y'(0) = 0.$

35. $y'' + 2y' + y = 3\delta_0(t - 2), \quad y(0) = 1,\ y'(0) = 0.$

36. $y'' - y = t * \cos t, \quad y(0) = -1,\ y'(0) = 0.$

37. $y'' + 4y = \mathcal{U}_0(t - 1)e^{t-1}, \quad y(0) = 0,\ y'(0) = 0.$

38. $y'' + y' + y = \delta_0(t - \pi), \quad y(0) = 0,\ y'(0) = 0.$

In Exercises 39–42, express the solution of the given initial value problem as a convolution.

39. $y'' + y = f(t), \quad y(0) = 0,\ y'(0) = 0.$

40. $y'' + y' + y = f(t), \quad y(0) = 0,\ y'(0) = 0.$

41. $y'' + 4y = \cos t, \quad y(0) = 0,\ y'(0) = 0.$

42. $4\,y'' + 4y' + 17y = t, \quad y(0) = 0,\ y'(0) = 0.$

Project Problem: Periodic functions. As you know, in computing Fourier series all we need is knowledge of the function on an interval of length equal to one period. In Exercise 43 you are asked to derive a formula for the Laplace transform of a periodic function, which enables you to compute the transform from just knowing the function over one period. As a project, do Exercise 43 and any one of Exercises 44–47.

43. Suppose that $f(t + T) = f(t)$ for all $t > 0$; that is, f is T-periodic.

(a) Show that

$$\mathcal{L}(f(t)) = \sum_{k=0}^{\infty} \int_{kT}^{(k+1)T} e^{-st} f(t)\, dt\,.$$

(b) Make a change of variables $\tau = t - kT$ and conclude that

$$\mathcal{L}(f(t)) = \int_0^T e^{-st} f(t)\, dt \sum_{k=0}^{\infty} e^{-ksT}\,.$$

(c) Sum the (geometric) series and conclude that

$$\mathcal{L}(f(t)) = \frac{\int_0^T e^{-st} f(t)\, dt}{1 - e^{-sT}}\,.$$

In Exercises 44–47, compute the Laplace transform of the given function using the result of Exercise 43.

44.   **45.**

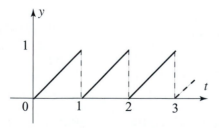

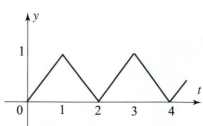

46.

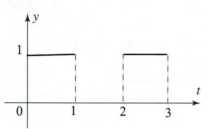

47.

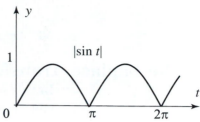

48. A function is given by its power series expansion, $f(t) = \sum_{k=0}^{\infty} a_k t^k$ for all t. Show that

$$\mathcal{L}(f(t)) = \sum_{k=0}^{\infty} a_k \frac{k!}{s^{k+1}}\,.$$

49.   Use the result of Exercise 48 to show that $\mathcal{L}(\frac{\sin t}{t}) = \tan^{-1}(\frac{1}{s})$. (For an alternative derivation, see Exercise 56.)

50.   **Laplace transform of** J_0. Recall from Section 4.3 that

$$J_0(t) = \sum_{k=0}^{\infty} \frac{(-1)^k t^{2k}}{2^{2k}(k!)^2}\,.$$

(a) Use the binomial theorem to show that

$$\frac{1}{\sqrt{1+s^2}} = \frac{1}{s}\left(1 + \frac{1}{s^2}\right)^{-1/2} = \sum_{k=0}^{\infty} \frac{(-1)^k (2k)!}{2^{2k}(k!)^2 s^{2k+1}}, \quad s > 1\,.$$

(b) Use the result of Exercise 48 and (a) to show that

$$\mathcal{L}(J_0(t)) = \frac{1}{\sqrt{1+s^2}}\,.$$

(For an alternative derivation using residues, see Exercise 37, Section 5.1.)

51.   Proceed as in Exercise 50 to show that $\mathcal{L}(J_0(\sqrt{t})) = \frac{e^{-1/4s}}{s}$.

52. The error function.   Use the Definition of the Laplace transform to show that

$$\mathcal{L}(\mathrm{erf}(t)) = \frac{2}{s\sqrt{\pi}} e^{\frac{s^2}{4}} \int_{s/2}^{\infty} e^{-u^2}\, du = \frac{1}{s} e^{\frac{s^2}{4}} \mathrm{erfc}\left(\frac{s}{2}\right)\,.$$

For the definition of the functions erf and erfc, see Section 11.4, (5) and Exercise 16. [Hint: After setting up the integrals, interchange the order of integration.]

53. A Gaussian function.   Use the definitions of the Laplace transform and the complementary error function to show that

$$\mathcal{L}\left(\frac{1}{a\sqrt{\pi}} e^{-t^2/4a^2}\right) = e^{a^2 s^2} \mathrm{erfc}(as)\,.$$

54.   Derive entry 37 in the table of Laplace transforms, Appendix B.4.

55. Let $F(s)$ denote the Laplace transform of $f(t)$. Establish the identity

$$\mathcal{L}\left(\frac{f(t)}{t}\right) = \int_s^\infty F(u)\, du\,.$$

56. Derive the Laplace transform in Exercise 49 using the result of Exercise 55.

12.3   The Laplace Transform Method

In this section we will use the Laplace transform to solve partial differential equations in the same way we used the Fourier transform. Before turning to examples, we set the notation for this section. The Laplace transform of $u(x, t)$ *with respect to the variable t* is

$$\mathcal{L}(u(x,t))(s) = U(x, s) = \int_0^\infty u(x,t)e^{-st}\, dt\,.$$

Using Theorem 3, Section 12.1, we find that

(1) $\mathcal{L}\left(\dfrac{\partial u}{\partial t}\right) = sU(x, s) - u(x, 0)$

and

(2) $\mathcal{L}\left(\dfrac{\partial^2 u}{\partial t^2}\right) = s^2 U(x, s) - su(x, 0) - \dfrac{\partial u}{\partial t}(x, 0)\,.$

Keeping in mind that we are taking the Laplace transform with respect to t, we also have

(3) $\mathcal{L}\left(\dfrac{\partial u}{\partial x}\right) = \dfrac{dU}{dx}(x, s),$

and

(4) $\mathcal{L}\left(\dfrac{\partial^2 u}{\partial x^2}\right) = \dfrac{d^2 U}{dx^2}(x, s)\,.$

These formulas are obtained by differentiating under the integral sign as we did in deriving the corresponding formulas with the Fourier transform in Section 11.2. Also, to emphasize the fact that s will be treated as a parameter in the transformed differential equations, we have used the notation for ordinary derivatives in (3) and (4).

We are now ready for the applications. We treat typical examples dealing with the heat and wave equations. The first problem models heat transfer in an infinite insulated rod whose initial temperature is $0°$, given that heat from a reservoir is introduced through one end of the rod.

EXAMPLE 1   Heat equation for a semi-infinite rod
Solve the heat problem

$$(5) \qquad \frac{\partial u}{\partial t} = c^2 \frac{\partial^2 u}{\partial x^2}, \qquad 0 < x < \infty, \ t > 0,$$

$$(6) \qquad u(0, t) = f(t), \qquad t > 0,$$

$$(7) \qquad u(x, 0) = 0, \qquad 0 < x < \infty.$$

The problem is illustrated in Figure 1.

Solution We transform (5) with the Laplace transform with respect to t. Using (1) and (4), we get

$$sU(x, s) - u(x, 0) = c^2 \frac{d^2 U}{dx^2}(x, s);$$

$$c^2 \frac{d^2 U}{dx^2}(x, s) - sU(x, s) = 0, \qquad \text{by (7).}$$

The general solution of this differential equation is

$$U(x, s) = A(s)e^{\sqrt{s}x/c} + B(s)e^{-\sqrt{s}x/c},$$

where $A(s)$ and $B(s)$ are constants that depend on s (see Appendix A.2). Expecting the transform to be bounded as $s \to \infty$, we set $A(s) = 0$. To determine $B(s)$, we transform (6) and obtain $U(0, s) = B(s) = F(s)$, where $F(s)$ denotes the Laplace transform of f. Hence, $B(s) = F(s)$, and so

$$U(x, s) = F(s)e^{-\sqrt{s}x/c}.$$

It is now clear from Theorem 2, Section 12.2, that $u(x, t)$ is the convolution of $f(t)$ with the function whose Laplace transform is $e^{-\sqrt{s}x/c}$. To find this transform, we use entry 41 of the table of Laplace transforms in Appendix B:

$$\mathcal{L}\left(\frac{a}{2\sqrt{\pi}t^{3/2}} e^{-\frac{a^2}{4t}} \right) = e^{-a\sqrt{s}}.$$

Setting $a = \frac{x}{c}$, it follows that

$$\mathcal{L}^{-1}\left(e^{-\sqrt{s}x/c} \right) = \frac{x}{2c\sqrt{\pi}t^{3/2}} e^{-\frac{x^2}{4c^2 t}}.$$

Thus

$$u(x, t) = f(t) * \frac{xe^{-x^2/4c^2 t}}{2c\sqrt{\pi}t^{3/2}}.$$

More explicitly,

$$(8) \qquad u(x, t) = \frac{x}{2c\sqrt{\pi}} \int_0^t \frac{f(\tau)}{(t - \tau)^{3/2}} e^{-\frac{x^2}{4c^2(t-\tau)}} d\tau.$$

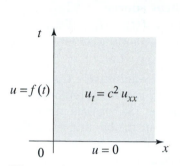

t

$u = f(t)$ $\qquad u_t = c^2 u_{xx}$

0 $\qquad u = 0$ $\qquad x$

Figure 1

Note that the solution depends on the value of $f(\tau)$ only for $0 < \tau < t$. We would expect this, because the temperature $f(\tau)$ of the heat reservoir in the future $(\tau > t)$ cannot possibly affect the temperature of the rod *now*.   ■

The convolution integral in (8) is hard to compute in general. In the following example we carry out the computations in the case of a constant heat source.

EXAMPLE 2   Heat problem with a constant heat source

Solve the problem of Example 1 in the special case when $f(t) = T_0$. Take $T_0 = 100$, $c = 1$, and illustrate the solution graphically by plotting $u(x, t)$ for various values of t.

Solution Before we embark on the solution, let us examine intuitively the consequences of the initial and boundary conditions. Given that the initial temperature is $0°$, and given that we are introducing heat at the origin at the constant temperature T_0, we expect the temperature to propagate throughout the bar and reach the temperature T_0 at all points. However, at any given time t, if we go far enough from the source of heat, the temperature is expected to be near the initial temperature $0°$. Thus at any time, the temperature distribution in the bar should look like the graph in Figure 2.

Now let us see if all this can be derived analytically, using the result of Example 1. Substituting $f(t) = T_0$ in (8) and making the change of variables $z = x/(2c\sqrt{t - \tau})$, $dz = \frac{x}{4c(t-\tau)^{3/2}} d\tau$, we get

$$u(x, t) = T_0 \frac{2}{\sqrt{\pi}} \int_{x/(2c\sqrt{t})}^{\infty} e^{-z^2} dz = T_0 \operatorname{erfc}\left(\frac{x}{2c\sqrt{t}}\right),$$

where erfc is the complementary error function introduced in Section 11.4, Exercise 16. Figure 3 illustrates the solution when $T_0 = 100$, and $c = 1$ and $u(x, t) = 100 \operatorname{erfc}\left(\frac{x}{2\sqrt{t}}\right)$.   ■

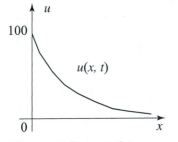

Figure 2 Expected temperature distribution.

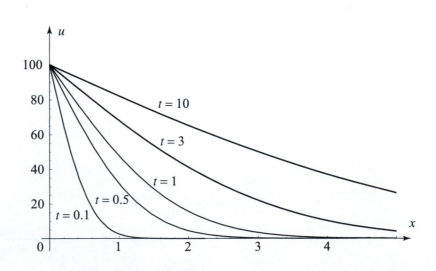

Figure 3 Temperature distribution in Example 2.

The next example illustrates the use of the Laplace transform in solving wave equations.

EXAMPLE 3   Forced vibrations of a semi-infinite string

A semi-infinite string is initially at rest on the x-axis with one end fastened at the origin. The string is set in motion by releasing it from rest in the presence of an external force. The motion of the string is modeled by the wave equation

$$\frac{\partial^2 u}{\partial t^2} = c^2 \frac{\partial^2 u}{\partial x^2} + f(t), \qquad x > 0,\ t > 0,$$

where $f(t)$ denotes the amount of force per unit length. Solve this differential equation subject to the boundary condition

$$u(0,t) = 0, \quad t > 0,$$

and the initial conditions

$$u(x,0) = 0, \quad \frac{\partial u}{\partial t}(x,0) = 0, \quad x > 0.$$

Solution Transforming both sides of the differential equation with respect to t and taking into account the initial conditions, we get

$$s^2 U(x,s) - s\,u(x,0) - \frac{\partial u}{\partial t}(x,0) = c^2 \frac{d^2 U}{dx^2}(x,s) + F(s);$$

$$(9) \qquad -c^2 \frac{d^2 U}{dx^2}(x,s) + s^2 U(x,s) = F(s).$$

This is a second order linear ordinary differential equation with constant coefficients in the variable x. Unlike the differential equation that we encountered in Example 1, this one is nonhomogeneous. Its general solution is the sum of the general solution of the associated homogeneous equation and any particular solution (see Appendix A.1, Theorem 5). The general solution of the associated homogeneous equation is

$$A(s)e^{-\frac{s}{c}x} + B(s)e^{\frac{s}{c}x}.$$

Recalling that $F(s)$ is constant as a function of x, it is easy to see that a particular solution of (9) is $\frac{F(s)}{s^2}$. Thus the general solution of (9) is

$$U(x,s) = A(s)e^{-\frac{s}{c}x} + B(s)e^{\frac{s}{c}x} + \frac{F(s)}{s^2}.$$

Expecting the transform to be bounded for $s > 0$ and $x > 0$, we take $B(s) = 0$. The boundary condition implies that $A(s) = -\frac{F(s)}{s^2}$. So

$$U(x,s) = F(s)\frac{1 - e^{-\frac{s}{c}x}}{s^2}.$$

Note that
$\mathcal{L}^{-1}(\frac{1}{s^2}) = t$
and
$\mathcal{L}^{-1}(\frac{e^{-\frac{s}{c}x}}{s^2}) = (t - \frac{x}{c})\,\mathcal{U}_0(t - \frac{x}{c}).$

To compute the inverse Laplace transform, we use Theorems 1 and 2 of the previous section, and we find

$$u(x,t) = f(t) * \mathcal{L}^{-1}\left(\frac{1 - e^{-\frac{s}{c}x}}{s^2}\right) = f(t) * \left[t - \left(t - \frac{x}{c}\right)\mathcal{U}_0\left(t - \frac{x}{c}\right)\right],$$

where $\mathcal{U}_0$ is the Heaviside step function. We can write the solution more explicitly as

(10) $\qquad u(x,t) = \int_0^t f(t-\tau)\left[\tau - \left(\tau - \frac{x}{c}\right)\mathcal{U}_0\left(\tau - \frac{x}{c}\right)\right] d\tau.$ ∎

The next example is a particularly interesting case of Example 3.

EXAMPLE 4   Vibrations of a string with gravitational acceleration

If the only external force in Example 3 is due to the gravitational acceleration g, then the differential equation becomes

$$\frac{\partial^2 u}{\partial t^2} = c^2 \frac{\partial^2 u}{\partial x^2} - g.$$

Under the same initial and boundary conditions as in the previous example, the solution (11) becomes

$$
\begin{aligned}
u(x,t) &= -g\int_0^t \left[\tau - \left(\tau - \frac{x}{c}\right)\mathcal{U}_0\left(\tau - \frac{x}{c}\right)\right] d\tau \\
&= -g\left[\frac{1}{2}t^2 - \int_0^t \left(\tau - \frac{x}{c}\right)\mathcal{U}_0\left(\tau - \frac{x}{c}\right) d\tau\right].
\end{aligned}
$$

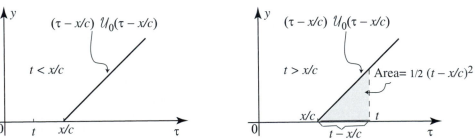

Figure 4

In evaluating the integral $\int_0^t (\tau - \frac{x}{c})\mathcal{U}_0(\tau - \frac{x}{c})\, d\tau$, we use some simple geometric considerations. For a fixed value of $x > 0$, the graph of the function $(\tau - \frac{x}{c})\mathcal{U}_0(\tau - \frac{x}{c})$ (as a function of τ) is the translate of the graph of $y = \tau$ by $\frac{x}{c}$ units to the right (see Figure 4). Thus, the integral of $(\tau - \frac{x}{c})\mathcal{U}_0(\tau - \frac{x}{c})$ from 0 to t is 0 if $0 < t < \frac{x}{c}$. If $t > \frac{x}{c} > 0$, this integral is equal to the triangular area shown in Figure 4. You can check that this area is $\frac{1}{2}(t - \frac{x}{c})^2$. We thus have

$$u(x,t) = \begin{cases} -\frac{g}{2}\left(t^2 - (t - \frac{x}{c})^2\right) & \text{if } 0 < x < ct, \\ -\frac{gt^2}{2} & \text{if } x > ct. \end{cases}$$

Figure 5 illustrates the solution at various values of t. This example models a semi-infinite string falling from rest under the influence of gravity with one end held fixed. Recalling that the position of a body that falls from rest is given by $-gt^2/2$, we see that for x larger than ct the string falls as if it were freely falling. Portions of the string at smaller values of x, however, fall less rapidly due to the restraining effect of the fixed end. Note that this effect propagates outward from the fixed end exactly at velocity c. ■

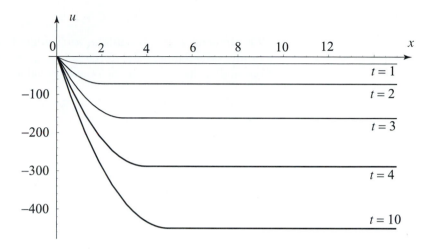

Figure 5 String falling under the influence of gravity.

Exercises 12.3

In Exercises 1–10 use the Laplace transform to solve the given boundary value problem. Give your answer in the form of an integral and simplify as much as possible. Whenever possible, use the examples from this section without repeating the derivations.

1.
$$\frac{\partial u}{\partial t} = \frac{\partial^2 u}{\partial x^2}, \ 0 < x < \infty, \ t > 0,$$
$$u(0, t) = 70, \ t > 0,$$
$$u(x, 0) = 0, \ 0 < x < \infty.$$

2.
$$\frac{\partial u}{\partial t} = \frac{\partial^2 u}{\partial x^2}, \ 0 < x < \infty, \ t > 0,$$
$$u(0, t) = 100 \left(1 - \mathcal{U}_0(t - 1)\right), \ t > 0,$$
$$u(x, 0) = 0, \ 0 < x < \infty.$$

3.
$$\frac{\partial u}{\partial t} = \frac{\partial^2 u}{\partial x^2}, \ 0 < x < \infty, \ t > 0,$$
$$u(0, t) = 100 \, \mathcal{U}_0(t - 2)), \ t > 0,$$
$$u(x, 0) = 0, \ 0 < x < \infty.$$

4.
$$\frac{\partial u}{\partial t} = \frac{\partial^2 u}{\partial x^2}, \ 0 < x < \infty, \ t > 0,$$
$$u(0, t) = 100 \left(\mathcal{U}_0(t - 1) - \mathcal{U}_0(t - 3)\right),$$
$$u(x, 0) = 0, \ 0 < x < \infty.$$

5.
$$\frac{\partial^2 u}{\partial t^2} = \frac{\partial^2 u}{\partial x^2} + t, \ x > 0, \ t > 0,$$
$$u(0, t) = 0, \ t > 0,$$
$$u(x, 0) = 0, \ \tfrac{\partial u}{\partial t}(x, 0) = 0, \ x > 0.$$

6.
$$\frac{\partial^2 u}{\partial t^2} = \frac{\partial^2 u}{\partial x^2} + e^{-t}, \ x > 0, \ t > 0,$$
$$u(0, t) = 0, \ t > 0,$$
$$u(x, 0) = 0, \ \tfrac{\partial u}{\partial t}(x, 0) = 0, \ x > 0.$$

7.
$$\frac{\partial^2 u}{\partial t^2} = \frac{\partial^2 u}{\partial x^2} - g, \ x > 0, t > 0,$$
$$u(0, t) = 0, \ t > 0,$$
$$u(x, 0) = 0, \ \tfrac{\partial u}{\partial t}(x, 0) = 1, \ x > 0.$$

8.
$$\frac{\partial^2 u}{\partial t^2} = \frac{\partial^2 u}{\partial x^2} + t^2, \ x > 0, \ t > 0,$$
$$u(0, t) = 0, \ t > 0,$$
$$u(x, 0) = 0, \ \tfrac{\partial u}{\partial t}(x, 0) = 0, \ x > 0.$$

9.
$$\frac{\partial^2 u}{\partial t^2} = \frac{\partial^2 u}{\partial x^2}, \quad x > 0, \ t > 0,$$
$$u(0, t) = \sin t, \ t > 0,$$
$$u(x, 0) = 0, \ \frac{\partial u}{\partial t}(x, 0) = 1, \ x > 0.$$

10.
$$\frac{\partial^2 u}{\partial t^2} = \frac{\partial^2 u}{\partial x^2}, \quad x > 0, \ t > 0,$$
$$u(0, t) = 0, \ t > 0,$$
$$u(x, 0) = 0, \ \frac{\partial u}{\partial t}(x, 0) = 1, \ x > 0.$$

11. Imagine a long (semi-infinite) insulated bar with initial temperature $0°$. To determine the temperature of the points after a brief application of a welding torch to one end of the bar, solve the boundary value problem in Example 1 with $f(t) = \delta(t)$.

12. Refer to Example 2 with the given numerical data: $c = 1, T_0 = 100$. Approximate how long it will take to raise the temperature of the point $x = 5$ to 10. How long does it take to raise the temperature to 20, 40, 80? What do you conclude from your answers? How high can the temperature reach?

13. The initial temperature in a semi-infinite bar is $70°$. At time $t = 0$, one end of the bar is given the temperature $100°$. To determine the temperature at any point in the bar, solve equation (1), with $c = 1$, subject to the following conditions

$$u(0, t) = 100, \ t > 0, \quad u(x, 0) = 70, \ x > 0.$$

What will eventually happen to the temperature throughout the bar? Illustrate your answer graphically. [Hint: Proceed as in Examples 1 and 2.]

14. Use the Laplace transform to solve the boundary value problem

$$\frac{\partial^2 u}{\partial t^2} = \frac{\partial^2 u}{\partial x^2} + \sin \pi x, \quad 0 < x < 1, \ t > 0,$$
$$u(0, t) = 0, \quad u(1, t) = 0, \ t > 0,$$
$$u(x, 0) = 0, \quad \frac{\partial u}{\partial t}(x, 0) = 0, \ x > 0.$$

15. Show that the solution of the boundary value problem

$$\frac{\partial u}{\partial t} = c^2 \frac{\partial^2 u}{\partial x^2}, \quad x > 0, \ t > 0,$$
$$u(0, t) = T_0, \quad t > 0,$$
$$u(x, 0) = T_1, \quad x > 0,$$

is given by

$$u(x, t) = (T_0 - T_1) \operatorname{erfc}\left(\frac{x}{2c\sqrt{t}}\right) + T_1 = (T_1 - T_0) \operatorname{erf}\left(\frac{x}{2c\sqrt{t}}\right) + T_0.$$

16. Illustrate graphically the solution in Exercise 15 when $c = 1$, $T_0 = 100$, and $T_1 = 70$.

12.4   The Hankel Transform with Applications

As we saw in Chapter 9, Bessel functions and Bessel series are instrumental in solving problems in polar coordinates. It should not surprise you, then, to see Bessel functions arising in the treatment of similar problems over unbounded regions. Instead of Bessel series, here we will need a transform, defined in terms of Bessel functions as follows.

Suppose that $f(x)$ is defined for all $x \geq 0$. The **Hankel transform of order** $\nu \geq 0$ of $f(x)$ is given by

$$(1) \qquad \boxed{\mathcal{H}_\nu(f)(s) = \int_0^\infty f(x) J_\nu(sx) x \, dx, \quad s \geq 0,}$$

where J_ν is Bessel's function of order ν (see Section 9.6 for background on Bessel functions). Under certain conditions on f (which we are going to assume hold) one can prove the **inversion formula**

$$(2) \qquad \boxed{\mathcal{H}_\nu \mathcal{H}_\nu f = f \, .}$$

Thus the Hankel transform is its *own inverse*. This property is also shared by the cosine and sine transforms (see Exercise 21, Section 11.6). Indeed, there is a close connection between these transforms and the Hankel transform of order $\nu = \frac{1}{2}$ (see Exercise 16).

The following operational properties are needed in the applications of this section:

OPERATIONAL PROPERTIES

$$(3) \qquad -s\mathcal{H}_0(f)(s) = \mathcal{H}_1(f')(s),$$

$$(4) \qquad \mathcal{H}_0\left(f' + \frac{1}{x}f\right)(s) = s\mathcal{H}_1(f)(s),$$

$$(5) \qquad \mathcal{H}_0\left(f'' + \frac{1}{x}f'\right)(s) = -s^2\mathcal{H}_0(f)(s) \, .$$

For the proofs, we will need the following properties of Bessel functions from Section 9.7:

$$\frac{d}{dx}(J_0(sx)) = -sJ_1(sx) \quad \text{and} \quad \int J_0(sx)x \, dx = \frac{1}{s}x\, J_1(sx) + C \, .$$

We also assume the following conditions on f: For any s,

$$\lim_{x \to 0} f(x)J_0(sx)x = 0 \quad \text{and} \quad \lim_{x \to \infty} f(x)J_0(sx)\, x = 0.$$

Since J_0 is bounded on the entire real line, these conditions are met if, for example, f is bounded near zero and decays to zero faster than $1/x$ as $x \to \infty$.

Now, to prove (3), we use (1) with $\nu = 0$ and integrate by parts:

$$
\mathcal{H}_0(f)(s) = \int_0^\infty f(x) J_0(sx) x \, dx
$$

$$
\begin{aligned}
& (u = f(x), \ du = f'(x) \, dx, \\
& \quad dv = J_0(sx) x \, dx, \ v = \tfrac{1}{s} x \, J_1(sx))
\end{aligned}
$$

$$
= \left. \frac{1}{s} f(x) J_0(sx) x \right|_0^\infty - \frac{1}{s} \underbrace{\int_0^\infty f'(x) J_1(sx) \, x \, dx}_{\mathcal{H}_1(f')} .
$$

By our assumption on f the first term is zero, and (3) follows. The proof of (4) is similar. First let us note that

$$
f' + \frac{1}{x} f = \frac{1}{x} \frac{d}{dx} [x f(x)] .
$$

Hence

$$
\mathcal{H}_0 \left(\frac{1}{x} \frac{d}{dx} [x f(x)] \right) (s) = \int_0^\infty \frac{1}{x} \frac{d}{dx} [x f(x)] J_0(sx) \, x \, dx
$$

$$
\begin{aligned}
& (u = J_0(sx), \ du = -s J_1(sx) dx \\
& \quad dv = \tfrac{d}{dx} [x f(x)] \, dx, \ v = x f(x))
\end{aligned}
$$

$$
= \left. f(x) J_0(sx) x \right|_0^\infty + s \int_0^\infty f(x) J_1(sx) x \, dx
$$

$$
= s \mathcal{H}_1(f)(s) .
$$

Finally, to prove (5), we apply (4), then (3), as follows:

$$
\mathcal{H}_0 \left(f'' + \frac{1}{x} f' \right) (s) = s \mathcal{H}_1(f')(s) = -s^2 \mathcal{H}_0(f)(s) .
$$

One very useful property of the Hankel transform is its **linearity**:

$$
\mathcal{H}_\nu(af + bg) = a \mathcal{H}_\nu(f) + b \mathcal{H}_\nu(g) .
$$

The proof is simple and is left as an exercise. We are now in a position to treat some applications. We start with a hanging chain problem.

EXAMPLE 1   Oscillations of an infinitely long hanging chain
The oscillations of a very long heavy chain can be modeled by assuming the chain is semi-infinite, with the fastened end sent to infinity. We will further assume that the transverse vibrations take place in one vertical plane. We place the chain on

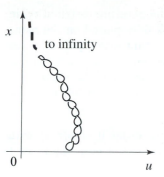

Figure 1 An infinite hanging chain, end fastened at ∞.

the x-axis, which we assume directed vertically and pointing upward (Figure 1). The boundary value problem governing the free motion of the chain becomes

(6)
$$\frac{\partial^2 u}{\partial t^2} = g[x\frac{\partial^2 u}{\partial x^2} + \frac{\partial u}{\partial x}], \qquad 0 < x < \infty, \quad t > 0,$$
$$u(x,0) = f(x), \qquad u_t(x,0) = v(x), \ x > 0.$$

Here g is the gravitational acceleration, $f(x)$ is the initial displacement of the chain, and $v(x)$ is its initial velocity. To proceed with the solution, we make the change of variables $z^2 = x$, $2z\,dz = dx$, and transform the equations into

$$\frac{\partial^2 u}{\partial t^2} = \frac{g}{4}[\frac{\partial^2 u}{\partial z^2} + \frac{1}{z}\frac{\partial u}{\partial z}],$$
$$u(z^2,0) = f(z^2), \quad u_t(z^2,0) = v(z^2).$$

(See Exercise 1 for details.) Let $U(s,t)$ denote the Hankel transform of order 0 of $u(z^2,t)$ with respect to the variable z. Transforming the new set of equations and using (5), we get

$$U_{tt}(s,t) = -s^2\frac{g}{4}U(s,t),$$
$$U(s,0) = \mathcal{H}_0(f(z^2))(s), \quad U_t(s,0) = \mathcal{H}_0(v(z^2))(s).$$

Solving the differential equation in t and using the transformed initial conditions to determine the arbitrary constants, we find

$$U(s,t) = A(s)\cos\left(\frac{\sqrt{g}}{2}st\right) + B(s)\sin\left(\frac{\sqrt{g}}{2}st\right),$$

with

$$A(s) = \mathcal{H}_0(f(z^2))(s), \quad \text{and} \quad B(s) = \frac{2}{\sqrt{g}s}\mathcal{H}_0(v(z^2))(s).$$

The solution is now obtained by taking the (inverse) Hankel transform of U:

$$u(z^2,t) = \int_0^\infty \left[A(s)\cos\left(\frac{\sqrt{g}}{2}st\right) + B(s)\sin\left(\frac{\sqrt{g}}{2}st\right)\right] J_0(zs)s\,ds.$$

Hence, in terms of x, we have

$$u(x,t) = \int_0^\infty \left[A(s)\cos\left(\frac{\sqrt{g}}{2}st\right) + B(s)\sin\left(\frac{\sqrt{g}}{2}st\right)\right] J_0(\sqrt{x}s)s\,ds. \qquad \blacksquare$$

Numerical illustrations for the hanging chain problem are presented in the exercises. Our next example is a Dirichlet problem in the *upper half-space*.

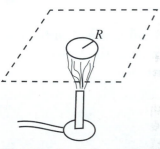

Figure 2 A heated large body.

EXAMPLE 2   Steady-state temperature in a large body
Solve the boundary value problem in cylindrical coordinates

$$(\Delta u =) \ u_{rr} + \frac{1}{r}u_r + u_{zz} = 0, \quad r > 0, z > 0;$$
$$-c^2 u_z(r,0) = \begin{cases} k & \text{if } 0 < r < R, \\ 0 & \text{if } r > R. \end{cases}$$

This boundary value problem models the steady-state temperature distribution in a very large body in the upper half-space, with one of its sides placed on the plane $z = 0$ (Figure 2). Heat is introduced in the body at a constant rate k per unit area per unit time through a circular area of radius R, centered at the origin, in the plane $z = 0$, and the rest of this plane is unheated. Because of circular symmetry, the steady-state temperature at a point in the upper half-space is a function of r and z only.

Solution We write $U(s, z)$ for the Hankel transform of order 0 of $u(r, z)$ with respect to r. Transforming the equations, using (5) and the first entry of Table 1 below, we get

$$-s^2 U + \tfrac{d^2 U}{dz^2} = 0,$$
$$-c^2 U_z(s, 0) = \mathcal{H}_0(k\,\mathcal{U}_0(R - r))(s) = k\,\tfrac{R}{s} J_1(Rs),$$

where $\mathcal{U}_0$ denotes the Heaviside step function. The solution of the first differential equation in z is

$$U(s, z) = A(s)e^{-sz} + B(s)e^{sz}.$$

Expecting the transform to remain bounded, we take $U(s, z) = A(s)e^{-sz}$. To determine $A(s)$, we use the transformed boundary condition and get

$$-c^2 U_z(s, 0) \;=\; sc^2 A(s) = k\frac{R}{s}J_1(Rs);$$
$$A(s) \;=\; k\frac{R}{c^2 s^2}J_1(Rs);$$
$$U(s, z) \;=\; k\frac{R}{c^2 s^2}J_1(Rs)e^{-sz}.$$

The solution is the (inverse) Hankel transform of order 0,

$$u(r, z) = \frac{kR}{c^2} \int_0^\infty \frac{e^{-sz}}{s} J_1(Rs) J_0(rs)\, ds. \qquad \blacksquare$$

The following table of Hankel transforms of order 0 is useful in the exercises. Because the Hankel transform is its own inverse, the table can be read both ways. For example, from the first entry we have

$$\mathcal{H}_0\left(\frac{a}{x} J_1(ax)\right) = \mathcal{U}_0(a - s).$$

Note the interesting fact that the function $e^{-x^2/2}$ is its own Hankel transform. (Compare with Example 5, Section 11.1.)

$f(x), \quad x \geq 0$	$\mathcal{H}_0(f)(s) = \int_0^\infty f(x) J_0(sx) x\, dx, \quad s \geq 0$
1. $\mathcal{U}_0(a - x), \ a > 0$	$\frac{a}{s} J_1(as)$
2. $x^{2N}\mathcal{U}_0(a - x), \ a > 0$	$\sum_{n=0}^N (-1)^n 2^n a^{2N-n+1} \frac{N!}{(N-n)!} \frac{J_{n+1}(as)}{s^{n+1}}$
3. $\frac{1}{x}$	$\frac{1}{s}$
4. $e^{-ax}, \ a > 0$	$\frac{a}{\sqrt{(s^2+a^2)^3}}$
5. $\frac{1}{x}e^{-ax}, \ a > 0$	$\frac{1}{\sqrt{s^2+a^2}}$
6. $e^{-a^2x^2}, \ a > 0$	$\frac{1}{2a^2}e^{-s^2/4a^2}$
7. $\frac{1}{x}\sin ax, \ a > 0$	$\frac{\mathcal{U}_0(a-s)}{\sqrt{a^2-s^2}}$

Table 1 Hankel Transforms of order 0.

Exercises 12.4

1. Let $z^2 = x$, and let $\tilde{u}\,(z, t) = u(z^2, t) = u(x, t)$.

(a) Use the chain rule to obtain

$$u_x = \frac{1}{2z}\,\tilde{u}_z; \quad u_{xx} = \frac{1}{4z^2}\,\tilde{u}_{zz} - \frac{1}{4z^3}\,\tilde{u}_z .$$

(b) Refer to Example 1 and justify the derivation of the new differential equation from (6).

In Exercises 2–3, use (1) and various properties of Bessel functions found in Sections 9.6 and 9.7 to establish the following identities. Take $\nu \geq 0$ and $a > 0$.

2. $\mathcal{H}_\nu(f(ax))(s) = \frac{1}{a^2}\mathcal{H}_\nu(f(x))(\frac{s}{a})$.

3. $\mathcal{H}_\nu[\frac{1}{x}\,f](s) = \frac{s}{2\nu}[\mathcal{H}_{\nu+1}(f)(s) + \mathcal{H}_{\nu-1}(f)(s)]$

4. (a) Use properties of Bessel functions to prove that, for $a > 0$,

$$\mathcal{H}_\nu[x^\nu \mathcal{U}_0(a - x)] = \frac{a^{\nu+1}}{s} J_{\nu+1}(as).$$

(b) Deduce entry 1 of Table 1.

5. Use Exercise 9 of Section 9.3 to derive entry 2 of Table 1.

6. (a) Show that $\mathcal{H}_0(\frac{1}{x}e^{-ax}) = \mathcal{L}\,(J_0(ax))\,(s)$.
(b) Obtain entry 5 of Table 1 from the table of Laplace transforms.

7. Justify entry 3 of Table 1 by letting $a \to 0$ in entry 5.

8. In this exercise we outline a proof of entry 6 of Table 1.
    (a) Use the series expansion of J_0 and obtain that

$$\mathcal{H}_0(e^{-a^2x^2})(s) = \frac{1}{2a^2} \sum_{n=0}^{\infty} \left[\frac{(-1)^n}{(n!)^2} \left(\frac{s^2}{4a^2} \right)^n \int_0^{\infty} t^n e^{-t}\, dt \right].$$

(b) Use the gamma function to show that the integral in the sum is $n!$.
(c) Complete the proof by comparing the series with the Taylor series expansion of $\frac{1}{2a^2} e^{-s^2/4a^2}$.

9. (a) Refer to Example 1 and describe the physical phenomenon corresponding to the data $g = 9.8$, $f(x) = 0$, $v(x) = \frac{1}{\sqrt{x}}$.
    (b) Write down the solution in the form of an integral.
    (c) Take $x = 1$ and approximate the solution when $t = .1, .2, .3$.

10. Repeat Exercise 9 for the data $g = 9.8$, $f(x) = \frac{1}{x}$, $v(x) = 0$.

11. Approximate the steady state temperature $u(1,1)$ in Example 2, given the numerical data: $R = 2$, $k = .5$, $c = .8$.

12. **Transverse vibrations of a large sheet of metal.** A very large sheet of metal vibrates under the action of radially symmetric pressure exerted at its surface. The boundary value problem in polar coordinates is described as follows

$$\begin{aligned} u_{tt} &= c^2 \left(u_{rr} + \frac{1}{r} u_r \right) + p(r,t), \quad r > 0, \ t > 0, \\ u(r,0) &= f(r), \quad u_t(r,0) = g(r), \qquad r > 0, \end{aligned}$$

where $p(r,t)$ corresponds to the outside pressure divided by the surface mass density of the sheet, f is the radially symmetric initial shape of the sheet, and g its radially symmetric initial velocity.

Show that the Hankel transform of order 0 (with respect to r) of the solution, $U(s,t)$, satisfies

$$\begin{aligned} U_{tt} &= -c^2 s^2 U + P(s,t), \\ U(s,0) &= F(s), \quad U_t(s,0) = G(s). \end{aligned}$$

13. Refer to Exercise 12. Show that if the pressure is 0, then the Hankel transform of the solution is

$$U(s,t) = F(s) \cos(cst) + \frac{1}{cs} G(s) \sin(cst).$$

14. Do Exercise 12 with $c = 1$, $p \equiv 0$, $f(r) = (1 - r^2)\mathcal{U}_0(1 - r)$, $g(r) = 0$.

15. (a) Do Exercise 12 with $c = 1$, $p \equiv 0$, $f(r) = \sqrt{1 - r^2}\,\mathcal{U}_0(1 - r)$, $g(r) = 0$.
    (b) Plot the initial shape of the membrane at various values of t.
    (c) Discuss the motion of the point at the origin.

16. Recalling from Section 9.6 that $J_{1/2}(x) = \sqrt{\frac{2}{\pi x}} \sin x$, show that $\mathcal{H}_{1/2}(f) = \frac{1}{\sqrt{s}} \mathcal{F}_s(f(x)\sqrt{x})$, where $\mathcal{F}_s$ is the Fourier sine transform. Argue from the fact that $\mathcal{F}_s$ is its own inverse transform that $\mathcal{H}_{1/2}$ is its own inverse transform.

APPENDIX A

ORDINARY DIFFERENTIAL EQUATIONS: REVIEW OF CONCEPTS AND METHODS

Topics to Review
This appendix is self-contained and requires basic knowledge of differential and integral calculus.

Looking Ahead
This appendix presents standard techniques for solving linear ordinary differential equations. They are needed throughout this book. Our primary use of the power series and Frobenius methods of Sections A.4 and A.5 will be to establish the basic properties of Bessel functions and Legendre polynomials.

Among all the mathematical disciplines the theory of differential equations is the most important. It furnishes the explanation of all those elementary manifestations of nature which involve time.

-Sophus Lie

A basic knowledge of ordinary differential equations is essential for the development of the main topics of this book. For example, solving a partial differential equation by standard methods leads naturally to ordinary differential equations. In this appendix we provide a review of the basic techniques that will be useful in this book. Our brief presentation covers the essentials of linear ordinary differential equations as taught in a first course on differential equations. In the first section we present the fundamental properties of existence, uniqueness, and linear independence of solutions. The rest of the appendix is devoted to the treatment of the basic tools for solving linear ordinary differential equations, including series methods. This appendix is not intended as a comprehensive treatment, but rather as a convenient reference for topics that are needed in the book.

A.1   Linear Ordinary Differential Equations

An ordinary differential equation is an equation that involves an unknown function y and its derivatives $y^{(j)}$ of order j. A **linear ordinary differential equation in standard form** is an equation

$$(1) \qquad y^{(n)} + p_{n-1}(x)y^{(n-1)} + \cdots + p_1(x)y' + p_0(x)y = g(x)$$

where n is called the **order** of the equation and $p_j(x)$ and $g(x)$ are given functions of x. The expression "standard form" refers to the fact that the leading coefficient is 1. A **solution** of (1) on an interval I is any n times differentiable function $y = y(x)$ satisfying (1) on I. The equation (1) is **homogeneous** if $g(x) \equiv 0$, otherwise it is **nonhomogeneous**.

The first and second order equations are of particular interest to us. We can describe completely the solution of the first one, while discussion of the second will occupy the rest of this appendix.

THEOREM 1
SOLUTION OF THE FIRST ORDER LINEAR DIFFERENTIAL EQUATION

Suppose that $p(x)$ and $g(x)$ are continuous on the interval I. Then any solution of the first order linear differential equation

$$y' + p(x)y = g(x), \quad x \text{ in } I,$$

is of the form

$$(2) \qquad y = e^{-\int p(x)\,dx}\left[C + \int g(x)e^{\int p(x)\,dx}\,dx\right],$$

where $\int p(x)\,dx$ represents the same antiderivative of $p(x)$ in both occurences.

The expression $\int p(x)\,dx$ represents an antiderivative of $p(x)$, and so it is defined up to a constant C. In applying (2), you should take the same value of C in both occurences of $\int p(x)\,dx$. Usually, the most convenient choice is $C = 0$. (See Example 1.)

Proof Suppose y is a solution. Multiply the equation through by the **integrating factor** $\mu(x) = e^{\int p(x)\,dx}$ and get

$$\mu(x)\left[y' + p(x)y\right] = g(x)\mu(x).$$

Since $\mu'(x) = p(x)\mu(x)$, by the product rule, the equation can be rewritten as $[\mu(x)y]' = g(x)\mu(x)$. Integrating both sides and dividing by $\mu(x)$ yields the desired solution. (Note that, since $\mu(x)$ is an exponential function, it is nonzero for all x, and so we may divide by it.) ∎

EXAMPLE 1   A first order differential equation

Using Theorem 1, we find all solutions of the differential equation

$$y' - y = 2.$$

The integrating factor is $\mu(x) = e^{-\int dx} = e^{-x}$ and hence the general solution is $y = e^x[C - 2e^{-x}] = Ce^x - 2$, where C is an arbitrary constant. ∎

In treating higher order equations we are not so fortunate, in that, in general, we do not have a closed form for the solutions. However, there are fundamental results concerning the general form of the solutions which we discuss in the remainder of this section. Important cases where solutions are given explicitly are presented in Sections A.2 and A.3. Our discussion focuses first on the homogeneous equation

$$(3) \qquad y^{(n)} + p_{n-1}(x)y^{(n-1)} + \cdots + p_1(x)y' + p_0(x)y = 0$$

where all the coefficient functions are continuous on some fixed interval I.

We begin with a definition that is central to our treatment.

THE WRONSKIAN

> Let $y_1, y_2, \ldots, y_n$ be any n solutions to the homogeneous linear differential equation (3). The **Wronskian** $W(y_1, y_2, \ldots, y_n)$ of these solutions is given by the following $n \times n$ determinant:
>
> $$W(y_1, y_2, \ldots, y_n) = \begin{vmatrix} y_1 & y_2 & \cdots & y_n \\ y_1' & y_2' & \cdots & y_n' \\ \vdots & \vdots & \vdots & \vdots \\ y_1^{(n-1)} & y_2^{(n-1)} & \cdots & y_n^{(n-1)} \end{vmatrix}$$

We will sometimes use the alternative notation $W(y_1, y_2, \ldots, y_n)(x)$, or simply $W(x)$, to emphasize that the Wronskian is a function of x.

EXAMPLE 2   Computing Wronskians

(a) The equation $y'' - y = 0$ has solutions $y_1 = e^x$ and $y_2 = e^{-x}$. Their Wronskian is

$$W(e^x, e^{-x}) = \begin{vmatrix} e^x & e^{-x} \\ e^x & -e^{-x} \end{vmatrix} = e^x(-e^{-x}) - e^x(e^{-x}) = -2.$$

(b)   The equation $y'' + y = 0$ has solutions $y_1 = \cos x$ and $y_2 = \sin x$. Their Wronskian is

$$W(\cos x, \sin x) = \begin{vmatrix} \cos x & \sin x \\ -\sin x & \cos x \end{vmatrix} = \cos^2 x + \sin^2 x = 1.$$

∎

The next theorem gives an explicit form of the Wronskian in terms of the coefficient functions in (3).

**THEOREM 2
ABEL'S FORMULA
FOR THE
WRONSKIAN**

Let $y_1, y_2, \ldots, y_n$ be any n solutions to the homogeneous linear differential equation (3). Then the Wronskian $W(x)$ satisfies the first order differential equation

$$W'(x) + p_{n-1}(x)W(x) = 0, \quad \text{for } x \text{ in } I,$$

and hence

(4)
$$W(x) = Ce^{-\int p_{n-1}(x)\,dx}.$$

Consequently, either $W(x) \neq 0$ for all x in I, or $W(x) \equiv 0$ on I.

The point of this theorem is that determining that $W(x) \neq 0$ for some x in I allows us to conclude that $W(x) \neq 0$ for all x in I.

Proof For clarity's sake, we give a proof only for $n = 2$. (The case $n = 3$ is treated in Exercise 31.) The same approach generalizes to higher dimensions but requires a greater knowledge of linear algebra. For $n = 2$, the equation is $y'' + p_1(x)y' + p_0(x)y = 0$, and $W(x) = y_1y_2' - y_1'y_2$. Since y_1 and y_2 are solutions, we have

Since the key issue is whether W vanishes or not, we cannot divide by W to separate variables. An appeal to Theorem 1 is necessary.

$$
\begin{aligned}
W'(x) &= y_1y_2'' - y_1''y_2 \\
&= y_1[-p_1(x)y_2' - p_0(x)y_2] - [-p_1(x)y_1' - p_0(x)y_1]y_2 \\
&= -p_1(x)W(x).
\end{aligned}
$$

This is a first order differential equation for $W(x)$, and (4) is an immediate consequence of (2) in Theorem 1. ∎

It is crucial in applying Theorem 2 to put the equation in standard form and verify the continuity of the coefficients. See Exercise 23.

We can now state a fundamental result in the theory of ordinary differential equations.

**THEOREM 3
EXISTENCE OF
FUNDAMENTAL
SETS OF SOLUTIONS**

The homogeneous equation (3)

$$y^{(n)} + p_{n-1}(x)y^{(n-1)} + \cdots + p_1(x)y' + p_0(x)y = 0,$$

where the coefficient functions $p_j(x)$ are all continuous on an interval I, has n solutions $y_1, y_2, \ldots, y_n$ with nonvanishing Wronskian on I. Furthermore, given any such set $y_1, y_2, \ldots, y_n$ and any solution y, then $y = c_1y_1 + c_2y_2 + \cdots + c_ny_n$ for a unique choice of constants $c_1, c_2, \ldots, c_n$. The set of solutions $y_1, y_2, \ldots, y_n$ with nonvanishing Wronskian is called a **fundamental set of solutions**.

Theorem 3 asserts that any linear homogeneous differential equation of order n has n solutions that *span* the set of all solutions of the equation. Note that the theorem does not assert the uniqueness of a fundamental set. If $y_1, y_2, \ldots, y_n$ is a fundamental set of solutions, then the **linear combination** $y = c_1y_1 + c_2y_2 + \cdots + c_ny_n$ is referred to as the **general solution**.

We have been using without verification that a linear combination of solutions of a homogeneous linear equation is again a solution. This simple fact, called the superposition principle, can be checked directly by appealing to (3). We state it here for ease of reference.

THEOREM 4
SUPERPOSITION
PRINCIPLE

Suppose that $u(x)$ and $v(x)$ are solutions of the linear homogeneous equation (3) and let c and d be any two numbers. Then the linear combination $c\,u(x) + d\,v(x)$ is also a solution of (3).

Nonhomogeneous Differential Equations

Let us recall from (1) the general nonhomogeneous equation

$$(5) \qquad y^{(n)} + p_{n-1}(x)y^{(n-1)} + \cdots + p_1(x)y' + p_0(x)y = g(x),$$

where the coefficient functions and $g(x)$ are all assumed to be continuous on some interval I. If $g(x)$ is replaced by 0, we call the resulting equation the **associated homogeneous equation**. We know from Theorem 3 that the associated homogeneous equation has a fundamental set of solutions y_1, $y_2, \ldots, y_n$. We will show in Section A.2, using the method of variation of parameters, that (5) has at least one solution that we will denote by y_p and call a **particular solution**. Now suppose that y is any other solution of (5). It is a simple exercise to check that $y - y_p$ is a solution of the associated homogeneous equation. By Theorem 3 we must have $y - y_p = c_1y_1 + c_2y_2 + \cdots + c_ny_n$. Thus, if y_h denotes the general solution of the associated homogeneous equation, it follows that $y = y_h + y_p$ (known as the **general solution** of the nonhomogeneous equation). We have proved the following important result.

THEOREM 5
GENERAL
SOLUTION OF NON-
HOMOGENEOUS
EQUATIONS

Let y_h and y_p denote, respectively, the general solution of the homogeneous equation associated with (5) and a particular solution of (5). Then any solution y of (5) has the form

$$y = y_h + y_p.$$

Often in applications we have to solve differential equations subject to certain conditions; for example,

$$y'' + y = e^x, \qquad y(0) = 0, \ y'(0) = 1.$$

Such a problem is called an **initial value problem**, and the conditions are called **initial conditions**. Typically we impose enough conditions to specify a unique solution of the problem. Since the general solution of an nth order linear differential equation has n arbitrary constants, we expect n conditions to suffice.

THEOREM 6
EXISTENCE AND
UNIQUENESS FOR
INITIAL VALUE
PROBLEMS

Consider the initial value problem consisting of the linear differential equation (5) and the initial conditions

$$y(x_0) = y_0, \; y'(x_0) = y_0', \; \ldots, \; y^{(n-1)}(x_0) = y_0^{(n-1)},$$

where x_0 is in I and y_0, y_0', $\ldots$, $y_0^{(n-1)}$ are prescribed values. Then this problem has a unique solution y on the interval I.

Proof Let y_p be a particular solution to (5) and let $y_1, y_2, \ldots, y_n$ be a fundamental set of solutions to the associated homogeneous equation. For clarity's sake we take $n = 2$, although the proof generalizes for arbitrary n. We need to solve the 2×2 system

$$\begin{cases} c_1 y_1(x_0) + c_2 y_2(x_0) + y_p(x_0) = y_0 \\ c_1 y_1'(x_0) + c_2 y_2'(x_0) + y_p'(x_0) = y_0' \end{cases} \Leftrightarrow \begin{cases} c_1 y_1(x_0) + c_2 y_2(x_0) = y_0 - y_p(x_0) \\ c_1 y_1'(x_0) + c_2 y_2'(x_0) = y_0' - y_p'(x_0) \end{cases}$$

for the unknowns c_1 and c_2. Since the determinant of the matrix of this system is precisely the Wronskian and is nonzero by definition (see Theorem 3), it follows that the system has a unique solution pair c_1, c_2. This proves existence of the solution of the initial value problem. To establish uniqueness, suppose that $u(x)$ and $v(x)$ are two solutions to the initial value problem. Let $w = u - v$. Then w satisfies the associated homogeneous equation with 0 initial values. We know from Theorem 3 that $w = ay_1 + by_2$ for some choice of a and b. Now using the initial values of w and the fact that $W(y_1, y_2)(x_0) \neq 0$, we infer that $a = b = 0$ and hence $w \equiv 0$ implying that $u \equiv v$. ∎

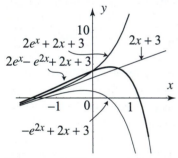

Figure 1 Various solutions from Example 3.

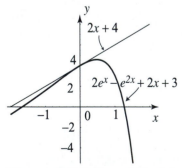

Figure 2 Solution of the initial value problem in Example 3. Notice the slope of the tangent line is equal to $y'(0) = 2$.

EXAMPLE 3   **An initial value problem**

You can (and should) check that the equation $y'' - 3y' + 2y = 4x$ has $y_p = 2x + 3$ as a particular solution and that the associated homogeneous equation $y'' - 3y' + 2y = 0$ has general solution $y_h = c_1 e^x + c_2 e^{2x}$. (General techniques for deriving these solutions will be developed in the next two sections.) Hence the general solution of the nonhomogeneous equation is

$$y = y_h + y_p = c_1 e^x + c_2 e^{2x} + 2x + 3.$$

Now suppose that we want to solve the initial value problem

$$y'' - 3y' + 2y = 4x, \quad y(0) = 4, \; y'(0) = 2.$$

Using the initial conditions, we get

$$y(0) = 4 \Rightarrow c_1 + c_2 = 1,$$
$$y'(0) = 2 \Rightarrow c_1 + 2c_2 = 0.$$

This system has a unique solution $c_1 = 2$ and $c_2 = -1$. Hence the (unique) solution of the initial value problem is $y = 2e^x - e^{2x} + 2x + 3$. ∎

We finally present a brief discussion of linear independence of solutions and its connection to fundamental sets.

The functions f_1, f_2, $\ldots$, f_n defined on an interval $[a, b]$ are said to be **linearly independent** on $[a, b]$ if the only choice of the constants c_1, c_2, $\ldots$, c_n for which the linear combination $c_1 f_1 + c_2 f_2 + \cdots + c_n f_n$ vanishes identically on $[a, b]$ is $c_1 = c_2 = \cdots = c_n = 0$. Otherwise, the functions are said to be **linearly dependent**. When $n = 2$, we note that linear dependence is equivalent to one function being a constant multiple of the other. The following result shows an intimate connection between this notion and fundamental sets.

THEOREM 7
WRONSKIAN
CRITERION FOR
LINEAR
INDEPENDENCE

Let $y_1, y_2, \ldots, y_n$ be any n solutions to the homogeneous linear differential equation (3) with coefficients continuous on an interval I. The following are equivalent:

(i) $y_1, y_2, \ldots, y_n$ are linearly independent on I;

(ii) $y_1, y_2, \ldots, y_n$ form a fundamental set of solutions of (3);

(iii) $W(y_1, y_2, \ldots, y_n)(x_0) \neq 0$ for some x_0 in I;

(iv) $W(y_1, y_2, \ldots, y_n)(x) \neq 0$ for all x in I.

Proof Clearly (ii) $\Leftrightarrow$ (iii)$\Leftrightarrow$ (iv) follow from Theorems 2 and 3 and the definition of a fundamental set.   We conclude by proving (i) $\Rightarrow$ (iii) and (iii) $\Rightarrow$ (i). We start with (iii) $\Rightarrow$ (i). Suppose that $y_1, y_2, \ldots, y_n$ are any n solutions such that $c_1 y_1 + c_2 y_2 + \cdots + c_n y_n = 0$. By differentiating this equation $n-1$ times in succession and then setting $x = x_0$, we get a system of n linear homogeneous equations in the n unknowns $c_1, c_2, \ldots, c_n$. The determinant of the matrix of this system is precisely $W(x_0)$, which is nonzero by assumption.   The nonvanishing of this determinant implies that $c_1 = c_2 = \cdots = c_n = 0$, proving (iii) $\Rightarrow$ (i). To complete the proof it is enough to show (i) $\Rightarrow$ (iii). The proof is by contradiction. We assume that $y_1, y_2, \ldots, y_n$ are linearly independent on I and that $W(y_1, y_2, \ldots, y_n)(x_0) = 0$. The vanishing of the Wronskian implies that there are constants $c_1, c_2, \ldots, c_n$, not all zero, such that all the initial values of $y = c_1 y_1 + c_2 y_2 + \cdots + c_n y_n$ are 0 at x_0 (for $n = 2$, this fact is clear). However, the solution $w \equiv 0$ has the same initial values, and so by the uniqueness part of Theorem 6, $w \equiv y \equiv 0$, implying that $c_1 = c_2 = \cdots = c_n = 0$ by the definition of linear independence, which is a contradiction. ∎

EXAMPLE 4   **Fundamental sets of solutions, linear independence**

We saw in Example 2 that e^x and e^{-x} are solutions of $y'' - y = 0$ and that $W(e^x, e^{-x}) = -2 \neq 0$.   Thus, by Theorem 7, the general solution of the differential equation is of the form $y = c_1 e^x + c_2 e^{-x}$; and so any solution is a linear combination of the functions e^x and e^{-x}, which form a fundamental set of solutions.

It is worthwhile to note that the fundamental set of solutions is not unique. For the equation at hand, you can easily check that $\cosh x$ and $\sinh x$ are also solutions.

Their Wronskian is

$$W(\cosh x, \sinh x) = \begin{vmatrix} \cosh x & \sinh x \\ \sinh x & \cosh x \end{vmatrix} = \cosh^2 x - \sinh^2 x = 1 .$$

Thus, by Theorem 7, the general solution of the differential equation is also of the form $y = c_1 \cosh x + c_2 \sinh x$.   ∎

Thus far we have developed an understanding of the general form of the solutions of linear differential equations. However, aside from the first order case, we have not developed techniques for finding explicit solutions. This will occupy us throughout the remainder of this appendix, where we will develop methods for solving certain important classes of higher order linear differential equations.

Exercises A.1

In Exercises 1–10, solve the given first order differential equation.

1. $y' + y = 1$.   **2.** $y' + 2xy = x$.   **3.** $y' = -.5\,y$.

4. $y' = 2y + x$.   **5.** $y' - y = \sin x$.   **6.** $y' - 2x\,y = x^3$.

7. $xy' + y = \cos x$.   **8.** $y' - \frac{2}{x}y = x^2$.   **9.** $y' + \tan x\,y = \cos x$.

10. $y' + \tan x\,y = \sec^2 x$.

In Exercises 11–20, solve the given initial value problem.

11. $y' = y$, $y(0) = 1$.   **12.** $y' + 2y = 1$, $y(0) = 2$.

13. $y' + xy = x$, $y(0) = 0$.   **14.** $y' + \frac{y}{x} = \frac{\sin x}{x}$, $y(\pi) = 1$.

15. $xy' + 2y = 1$, $y(-1) = -2$.   **16.** $xy' - 2y = \frac{1}{x}$, $y(1) = 0$.

17. $y' + y\tan x = \tan x$, $y(0) = 1$. **18.** $y' + y\tan x = \tan x$, $y(0) = 2$.

19. $y' + e^x y = e^x$, $y(0) = 2$.   **20.** $y' + y = e^x$, $y(3) = 0$.

21. (a) Check that the functions

$$e^x, \ e^{-x}, \cosh x, \ \sinh x$$

are solutions of

$$y'' - y = 0 .$$

(b) We know from Example 4 that $\{e^x, \ e^{-x}\}$ and $\{\cosh x, \sinh x\}$ are fundamental sets of solutions. Express e^x as a linear combination of the functions $\cosh x$, $\sinh x$.
(c) Can you think of other fundamental sets of solutions?

22. Check that the functions $e^x, \ e^{-x}, \cosh x$ are solutions of

$$y''' - y' = 0 .$$

(b) Show directly that $W(e^x, e^{-x}, \cosh x) \equiv 0$ and conclude that these functions do not form a fundamental set.
(c) Find a fundamental set of solutions by inspection.   Justify your answer by computing the Wronskian.

23. (a) Check that the functions x, x^2 are solutions of

$$x^2 y'' - 2xy' + 2y = 0.$$

(b) Compute the Wronskian of the solutions.
(c) Note that the Wronskian vanishes at $x = 0$. Does this contradict Theorem 2?

24. (a) Check that the functions e^x, $1 + x$ are solutions of

$$xy'' - (1 + x)y' + y = 0.$$

(b) Check that both of these functions satisfy $y(0) = 1$ and $y'(0) = 1$. Does this contradict the uniqueness part in Theorem 6?
(c) Compute the Wronskian of the solutions and conclude that they are linearly independent on $(0, \infty)$.

In Exercises 25–30, solve the initial value problem consisting of the differential equation in Example 3.

25. $y(0) = 0$, $y'(0) = 0$.   **26.** $y(0) = 1$, $y'(0) = -1$.

27. $y(1) = 0$, $y'(1) = 2$.   **28.** $y(1) = 1$, $y'(1) = -1$.
[Hint: In Exercises 27 and 28, it is easier to work with $y_h = c_1 e^{(x-1)} + c_2 e^{2(x-1)}$. Why is this possible?]

29. $y(2) = 0$, $y'(2) = 1$.   **30.** $y(3) = 3$, $y'(3) = 3$.
[Hint: Use a y_h that simplifies the computations as in the previous exercises.]

31. Project Problem: Abel's formula for $n = 3$.
(a) Let y_1, y_2, y_3 be any three solutions of the third order equation

$$y''' + p_2(x)y'' + p_1(x)y' + p_0(x)y = 0.$$

Derive the formulas for the Wronskian

$$W(y_1, y_2, y_3) = (y_2 y_3' - y_2' y_3)y_1'' - (y_1 y_3' - y_1' y_3)y_2'' + (y_1 y_2' - y_1' y_2)y_3''$$

and

$$W'(y_1, y_2, y_3) = (y_2 y_3' - y_2' y_3)y_1''' - (y_1 y_3' - y_1' y_3)y_2''' + (y_1 y_2' - y_1' y_2)y_3'''.$$

(b) Follow the proof of Theorem 2 to show that

$$W' = -p_2(x)W.$$

(c) Derive Abel's formula for $n = 3$.

32. (a) Show that $W(x^3, |x^3|) = 0$ for all x. (The function $|x^3|$ is differentiable for all x. Can you see why?)
(b) Show, however, that $x^3, |x^3|$ are linearly independent on the real line.
(c) Does this contradict Theorem 7? Justify your answer.

A.2   Linear Ordinary Differential Equations with Constant Coefficients

The general form of the **nth order homogeneous linear differential equation with constant coefficients** is

(1) $a_n y^{(n)} + a_{n-1} y^{(n-1)} + \cdots + a_1 y' + a_0 y = 0,$

where each a_j is a constant and $a_n \neq 0$. The equation can be put in standard form by dividing through by the leading coefficient a_n. This important class of equations is needed throughout this book. Our presentation will emphasize the second order case, which is by far the most useful.

The simple first order case of (1) suggests that we try

$$y = e^{\lambda x}$$

in solving the general case. Since

$$y' = \lambda e^{\lambda x}, \ y'' = \lambda^2 e^{\lambda x}, \ \ldots, \ y^{(n-1)} = \lambda^{n-1} e^{\lambda x}, \ y^{(n)} = \lambda^n e^{\lambda x},$$

substituting these into (1), we get

$$(a_n \lambda^n + a_{n-1} \lambda^{n-1} + \cdots + a_1 \lambda + a_0) e^{\lambda x} = 0.$$

Thus $y = e^{\lambda x}$ is a solution to (1) if λ is any root of the **characteristic equation**

$$\boxed{a_n \lambda^n + a_{n-1} \lambda^{n-1} + \cdots + a_1 \lambda + a_0 = 0.}$$

The roots of this equation are called the **characteristic roots**. For future use, we define the **characteristic polynomial**

$$p(\lambda) = a_n \lambda^n + a_{n-1} \lambda^{n-1} + \cdots + a_1 \lambda + a_0.$$

In the case when this polynomial has n distinct characteristic roots λ_1, λ_2, $\ldots$, λ_n, the general solution of (1) is

$$\boxed{y = c_1 e^{\lambda_1 x} + c_2 e^{\lambda_2 x} + \cdots + c_n e^{\lambda_n x}.}$$

Equivalently, a fundamental set of solutions is given by

(2) $$\boxed{e^{\lambda_1 x}, \ e^{\lambda_2 x}, \ \ldots, \ e^{\lambda_n x}.}$$

This follows from the nonvanishing of the Wronskian $W(e^{\lambda_1 x}, e^{\lambda_2 x}, \ldots, e^{\lambda_n x})$. See Exercise 77.

EXAMPLE 1   Characteristic equation with distinct real roots

The third order equation $y''' - y'' - 6\,y' = 0$ has characteristic equation

$$\lambda^3 - \lambda^2 - 6\,\lambda = 0\,.$$

Since this equation factors as $\lambda(\lambda - 3)(\lambda + 2) = 0$, we get the characteristic roots $\lambda_1 = 0, \lambda_2 = 3, \lambda_3 = -2$. Thus the general solution of the differential equation is $y = c_1 + c_2 e^{3x} + c_3 e^{-2x}$. Observe that $\{1, e^{3x},\ e^{-2x}\}$ is a fundamental set of solutions. You should check that $W(1, e^{3x},\ e^{-2x}) \neq 0$.   ∎

In the case of n distinct roots, the functions in (2) can still be used as a fundamental set even when some or all of the characteristic roots are complex numbers. However, for practical purposes, it is preferable to have real-valued functions. This can be done by using the fact that for any nonreal root $\mu = \alpha + i\beta$ ($\beta \neq 0$) its complex conjugate $\overline{\mu} = \alpha - i\beta$ is also a root (Figure 1).

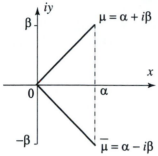

Figure 1

EXAMPLE 2   Complex characteristic roots

The characteristic roots of the equation $y'' - 4\,y' + 5\,y = 0$ are $\lambda_1 = 2 + i$ and $\lambda_2 = 2 - i$. From (2), the functions $e^{(2+i)x}$ and $e^{(2-i)x}$ form a fundamental set of solutions. We next show how to obtain real-valued solutions. Using Euler's identity, $e^{i\theta} = \cos\theta + i\sin\theta$, we have

$$e^{(2+i)x} = e^{2x}e^{ix} = e^{2x}(\cos x + i\sin x)\,.$$

Similarly, $e^{(2-i)x} = e^{2x}(\cos x - i\sin x)$. By taking sums and differences of these solutions, we arrive at the solutions $2\,e^{2x}\cos x$ and $2i\,e^{2x}\sin x$. Thus $e^{2x}\cos x$ and $e^{2x}\sin x$ are two real-valued solutions. We have

$$
\begin{aligned}
W(e^{2x}\cos x, e^{2x}\sin x) &= \begin{vmatrix} e^{2x}\cos x & e^{2x}\sin x \\ e^{2x}(2\,\cos x - \sin x) & e^{2x}(2\,\sin x + \cos x) \end{vmatrix} \\
&= e^{4x}\left((2\,\sin x + \cos x)\cos x - (2\,\cos x - \sin x)\sin x\right) \\
&= e^{4x}(\cos^2 x + \sin^2 x) = e^{4x} \neq 0\,.
\end{aligned}
$$

We conclude that $e^{2x}\cos x$ and $e^{2x}\sin x$ form a fundamental set of solutions. Note that in the case of two solutions, we could have verified that the set is fundamental by simply observing that one is not a constant multiple of the other.   ∎

In general, a real fundamental set of solutions is obtained by replacing the pair of functions $e^{\mu x}$, $e^{\overline{\mu} x}$ in (2) by

(3)
$$\boxed{e^{\alpha x}\cos\beta x, e^{\alpha x}\sin\beta x}$$

for each nonreal characteristic root $\mu = \alpha + i\beta$.

Thus far we have not dealt with the case of a repeated root μ of the characteristic polynomial of multiplicity $m > 1$. If we try to use (2), we get the function $e^{\mu x}$ repeated m times, and hence we do not obtain a fundamental set. The following example illustrates how to resolve this problem.

EXAMPLE 3   Repeated roots of the characteristic polynomial

(a) The second order equation $y'' + 2y' + y = 0$ has characteristic polynomial $(\lambda + 1)^2$, which has -1 as a root of multiplicity two. One solution of the equation is e^{-x}. A simple verification shows that the function xe^{-x} is also a solution of the differential equation, which is also independent of e^{-x}. We conclude that e^{-x} and xe^{-x} form a fundamental set of solutions (Figure 2). (We note that the method of reduction of order, presented in the next section, provides a straightforward way to obtain the second solution xe^{-x}.)

(b) The third order equation $y''' - 6y'' + 12y' - 8y = 0$ has characteristic polynomial $(\lambda - 2)^3$, which has 2 as a root of multiplicity three. One solution of the equation is e^{2x}. Taking a hint from the previous example, we try $y = xe^{2x}$ for a second solution. We have

$$y' = e^{2x}(1 + 2x), \quad y'' = 4e^{2x}(1 + x), \quad y''' = 4e^{2x}(3 + 2x).$$

Plugging into the equation, we obtain

$$4e^{2x}(3 + 2x) - 24e^{2x}(1 + x) + 12e^{2x}(1 + 2x) - 8xe^{2x} = 0.$$

Thus $y = xe^{2x}$ is a solution. For a third solution of the differential equation, we modify the first solution by multiplying by x^2, and try $y = x^2 e^{2x}$. You can check that

$$y' = 2xe^{2x}(1 + x), \quad y'' = 2e^{2x}(1 + 4x + 2x^2), \quad y''' = 4e^{2x}(3 + 6x + 2x^2),$$

and that $x^2 e^{2x}$ is a solution. It is easy to check that the solutions $e^{2x}, xe^{2x}, x^2 e^{2x}$ are linearly independent, and hence they form a fundamental set of solutions of the differential equation. ∎

This example motivates the following prescription for obtaining a fundamental set of solutions of (1) in the case when multiple roots occur. Let μ be a repeated root having multiplicity m. In (2), we replace the m occurrences of $e^{\mu x}$ by the following m functions:

Note that there are m linearly independent solutions corresponding to repeated roots of order m.

(4)
$$\boxed{e^{\mu x}, \ xe^{\mu x}, \ \ldots, \ x^{m-1} e^{\mu x}.}$$

To obtain a fundamental set of solutions of (1), make this replacement for each repeated characteristic root. (Keep in mind that the multiplicity m may vary from one root to another.)

Finally, if $\mu = \alpha + i\beta$ is a nonreal repeated characteristic root of multiplicity m, to get a fundamental set of real-valued solutions we combine the

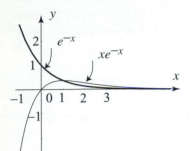

Figure 2 Solutions from Example 3(a).

previous methods and find that the $2m$ functions associated with the roots μ and $\overline{\mu}$ are

$$(5) \qquad
\begin{aligned}
&e^{\alpha x}\cos\beta x,\ x\,e^{\alpha x}\cos\beta x,\ \ldots,\ x^{m-1}e^{\alpha x}\cos\beta x;\\
&e^{\alpha x}\sin\beta x,\ xe^{\alpha x}\sin\beta x,\ \ldots,\ x^{m-1}e^{\alpha x}\sin\beta x\,.
\end{aligned}$$

For ease of reference, we restate in the following box the results of our discussion in terms of general solutions of (1). For convenience we view a nonrepeated characteristic root as a root of multiplicity $m = 1$.

GENERAL SOLUTION OF THE nth ORDER LINEAR HOMOGENEOUS EQUATION WITH CONSTANT COEFFICIENTS

Consider the equation

$$a_n y^{(n)} + a_{n-1}y^{(n-1)} + \cdots + a_1 y' + a_0 y = 0$$

with characteristic equation

$$a_n \lambda^n + a_{n-1}\lambda^{n-1} + \cdots + a_1 \lambda + a_0 = 0\,.$$

The general solution of the differential equation is constructed according to the nature of the roots of the characteristic equation as follows.

Case I   For each real root μ of multiplicity $m \geq 1$, we include a linear combination of the form

$$c_1 e^{\mu x} + c_2 x e^{\mu x} + \cdots + c_m x^{m-1}e^{\mu x}\,.$$

Case II   For each nonreal root $\mu = \alpha + i\beta$ and its complex conjugate $\overline{\mu} = \alpha - i\beta$ of multiplicity $m \geq 1$, we include a linear combination of the form

$$\begin{aligned}
&c_1 e^{\alpha x}\cos\beta x + c_2 x e^{\alpha x}\cos\beta x + \cdots + c_m x^{m-1}e^{\alpha x}\cos\beta x\\
&+ d_1 e^{\alpha x}\sin\beta x + d_2 x e^{\alpha x}\sin\beta x + \cdots + d_m x^{m-1}e^{\alpha x}\sin\beta x\,.
\end{aligned}$$

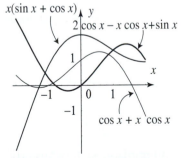

Figure 3 Various solutions from Example 4(b).

EXAMPLE 4   Fourth order equations

Find the general solution of the following equations:
(a) $y^{(4)} - 16\,y = 0$;
(b) $y^{(4)} + 2y'' + y = 0$.

Solution   (a) The characteristic equation is $\lambda^4 - 16 = 0$, with characteristic roots ± 2 and $\pm 2i$. Thus, the general solution is $y = c_1 e^{2x} + c_2 e^{-2x} + d_1 \cos 2x + d_2 \sin 2x$.

(b) The characteristic equation is $(\lambda^2 + 1)^2 = 0$. Thus the characteristic roots are i and $-i$ with multiplicity 2 each. Accordingly, the general solution is $y = c_1 \cos x + c_2 x \cos x + d_1 \sin x + d_2 x \sin x$. Figure 3 shows some specific solutions that are obtained by assigning different values to the constants c_1, c_2, d_1, d_2.   ■

In the second order case, we have a simpler form of the general solution which we now describe. You will be asked to verify these solutions in the exercises.

GENERAL SOLUTION OF THE SECOND ORDER LINEAR HOMOGENEOUS EQUATION WITH CONSTANT COEFFICIENTS

Consider the equation

$$(6) \qquad\qquad ay'' + by' + cy = 0$$

with characteristic equation

$$a\lambda^2 + b\lambda + c = 0.$$

Let λ_1 and λ_2 denote the characteristic roots. The general solution y of this differential equation is given by one of the following cases.

Case I If λ_1 and λ_2 are distinct real roots, then

$$y = c_1 e^{\lambda_1 x} + c_2 e^{\lambda_2 x}.$$

Equivalently, write $\lambda_1 = \frac{-b}{2a} + \frac{\sqrt{b^2 - 4ac}}{2a} = \alpha + \beta$ and $\lambda_2 = \alpha - \beta$, then

$$y = e^{\alpha x}(c_1 \cosh \beta x + c_2 \sinh \beta x).$$

Case II If $\lambda_1 = \lambda_2$, then

$$y = c_1 e^{\lambda_1 x} + c_2 x e^{\lambda_1 x}.$$

Case III If λ_1 and λ_2 are complex conjugate roots with $\lambda_1 = \alpha + i\beta$, then

$$y = c_1 e^{\alpha x} \cos \beta x + c_2 e^{\alpha x} \sin \beta x.$$

We now turn our attention to nonhomogeneous second order equations with constant coefficients.

The Method of Undetermined Coefficients

The study of various physical systems often leads to the equation

$$(7) \qquad\qquad ay'' + by' + cy = g(x).$$

The general solution of the associated homogeneous equation y_h has already been given explicitly. Thus, to find the general solution of (7), it is enough by Theorem 5 of the previous section to find a particular solution y_p. Finding y_p depends on the nonhomogeneous term g. In many interesting cases, the form of g allows us to guess the form of y_p up to a set of unknown coefficients. This method is called the method of **undetermined coefficients**. We illustrate with examples.

EXAMPLE 5   A simple undetermined coefficients problem

Find the general solution of the given nonhomogeneous equation.
(a) $y'' - 4y' + 5y = e^x$.   (b) $y'' + 2y' + y = e^{-x}$.

Solution (a) The associated homogeneous equation is solved in Example 2. We have $y_h = c_1 e^{2x} \cos x + c_2 e^{2x} \sin x$. To determine the general solution, it remains to find y_p. Since the right side of the equation is e^x it makes sense to try $y_p = A e^x$, where A is an unknown constant yet to be determined. A computation shows that $y_p'' - 4y_p' + 5y_p = 2A e^x$. For y_p to be a solution we must set $A = \frac{1}{2}$. Thus the general solution is $y = c_1 e^{2x} \cos x + c_2 e^{2x} \sin x + \frac{1}{2} e^x$.

(b) The associated homogeneous equation is solved in Example 3.   We have $y_h = c_1 e^{-x} + c_2 x e^{-x}$. To find y_p it is pointless to try $A e^{-x}$ or $A x e^{-x}$ since both of these solve the homogeneous equation. We modify our guess to $y_p = A x^2 e^{-x}$. Differentiating y_p, we find

$$y_p' = 2A x e^{-x} - A x^2 e^{-x}, \qquad y_p'' = 2A e^{-x} - 4A x e^{-x} + A x^2 e^{-x}.$$

Hence $y_p'' + 2y_p' + y_p = 2A e^{-x}$. For y_p to be a solution we must choose $A = \frac{1}{2}$. Thus the general solution is $y = c_1 e^{-x} + c_2 x e^{-x} + \frac{1}{2} x^2 e^{-x}$. ∎

The procedure we have used above is covered by the following general rules.

THE METHOD OF UNDETERMINED COEFFICIENTS

To find a particular solution of (7) when

$$(8) \qquad g(x) = (a_n x^n + a_{n-1} x^{n-1} + \cdots + a_0) e^{\alpha x} \begin{cases} \cos \beta x \\ \sin \beta x \end{cases}$$

we use

$$(9) \qquad y_p = (A_n x^n + A_{n-1} x^{n-1} + \cdots + A_0) e^{\alpha x} \cos \beta x$$
$$+ (B_n x^n + B_{n-1} x^{n-1} + \cdots + B_0) e^{\alpha x} \sin \beta x$$

provided that no term in the expression of y_p is a solution of the associated homogeneous equation. If not, we modify this expression by multiplying by x or x^2. We use x if the characteristic polynomial of the associated homogeneous equation has distinct roots, and x^2 if it has a double root.
Superposition rule If the right side of (7) is a sum of several different functions of the form (8), we use a corresponding sum of terms of the form (9).

EXAMPLE 6   The method of undetermined coefficients

Use the method of undetermined coefficients to find the general solution of

$$y'' - 3y' + 2y = x^3.$$

Solution It is straightforward to see that $y_h = c_1e^x + c_2e^{2x}$. From (9), we see that $y_p = Ax^3 + Bx^2 + Cx + D$. Note that none of these terms appears in the expression of y_h. We have

$$
\begin{aligned}
y_p'' - 3y_p' + 2y_p &= 6Ax + 2B - 3(3Ax^2 + 2Bx + C) + 2(Ax^3 + Bx^2 + Cx + D) \\
&= 2Ax^3 + (-9A + 2B)x^2 + (6A - 6B + 2C)x + 2B - 3C + 2D.
\end{aligned}
$$

For y_p to be a solution, A, B, C, and D must satisfy

$$
\begin{array}{rrrrl}
2A & & & & = 1 \\
-9A & +2B & & & = 0 \\
6A & -6B & +2C & & = 0 \\
& 2B & -3C & +2D & = 0
\end{array}
$$

The solution of this linear system is $A = \frac{1}{2}$, $B = \frac{9}{4}$, $C = \frac{21}{4}$, $D = \frac{45}{8}$. Thus

$$y_p = \frac{1}{2}x^3 + \frac{9}{4}x^2 + \frac{21}{4}x + \frac{45}{8},$$

and the general solution is obtained by adding on y_h.   ∎

EXAMPLE 7   The method of undetermined coefficients

Use the method of undetermined coefficients to find the general solution of

$$y'' - 3y' + 2y = e^x \cos x.$$

Solution According to (9), we take $y_p = Ae^x \cos x + Be^x \sin x$. Since neither of these terms appears in the expression of y_h (see Example 6), there is no need for modification. Now $y_p' = (A+B)e^x \cos x + (-A+B)e^x \sin x$ and $y_p'' = 2Be^x \cos x - 2Ae^x \sin x$, and hence

$$y_p'' - 3y_p' + 2y_p = -(A+B)e^x \cos x + (A-B)e^x \sin x.$$

For y_p to be a solution, we must solve $A + B = -1$ and $A - B = 0$. From this we get $y_p = -\frac{1}{2}e^x \cos x - \frac{1}{2}e^x \sin x$ and hence the general solution is

$$y = c_1e^x + c_2e^{2x} - \frac{1}{2}e^x(\cos x + \sin x).$$ ∎

EXAMPLE 8   Using the superposition rule

Consider the equation $y'' + 4y = e^{-2x} + 3\sin 2x$. It is easy to see that

$$y_h = c_1 \cos 2x + c_2 \sin 2x.$$

The term e^{-2x} requires a term Ae^{-2x} in y_p. The term $3\sin 2x$ suggests the expression $B \cos 2x + C \sin 2x$, but because these terms appear in y_h we must introduce an extra factor of x (the characteristic polynomial has distinct roots). Hence we take

$$y_p = A\,e^{-2x} + B\,x \cos 2x + C\,x \sin 2x\,.$$

Solving for A, B, and C as before, we find $y_p = \frac{1}{8}\,e^{-2x} - \frac{3}{4}\,x \cos 2x\,.$ ∎

Note that even though the nonhomogeneous term $g(x)$ in the equation of Example 8 has no cosine term in it, the particular solution does have one. This emphasizes the necessity of including both cosine and sine terms in y_p even when only one of these appears in $g(x)$.

Exercises A.2

In Exercises 1–24, find the general solution of the given equation.

1. $y'' - 4\,y' + 3\,y = 0.$ **2.** $y'' - y' - 6\,y = 0.$ **3.** $y'' - 5\,y' + 6\,y = 0.$

4. $2\,y'' - 3\,y' + y = 0.$ **5** $y'' + 2\,y' + y = 0.$ **6.** $4\,y'' - 13\,y' + 9\,y = 0.$

7. $4\,y'' - 4\,y' + y = 0.$ **8.** $4\,y'' - 12\,y' + 9\,y = 0.$ **9.** $y'' + y = 0.$

10. $9\,y'' + 4\,y = 0.$ **11.** $y'' - 4\,y = 0.$ **12.** $y'' + 3\,y' + 3\,y = 0.$

13. $y'' + 4\,y' + 5\,y = 0.$ **14.** $y'' - 2\,y' + 5\,y = 0.$ **15.** $y'' + 6\,y' + 13\,y = 0.$

16. $2\,y'' - 6\,y' + 5\,y = 0.$ **17.** $y''' - 2\,y'' + y' = 0.$ **18.** $y''' - 3\,y'' + 2\,y = 0.$

19. $y^{(4)} - 2\,y'' + y = 0.$ **20.** $y^{(4)} - y = 0.$

21. $y''' - 3\,y'' + 3\,y' - y = 0.$ **22.** $y''' - y = 0.$

23. $y^{(4)} - 6\,y'' + 8\,y' - 3y = 0.$ **24.** $y^{(4)} + 4\,y''' + 6\,y'' + 4y' + y = 0.$

In Exercises 25–44, find the general solution of the given equation using the method of undetermined coefficients.

25. $y'' - 4\,y' + 3\,y = e^{2x}.$ **26.** $y'' - y' - 6\,y = e^x.$

27. $y'' - 5\,y' + 6\,y = e^x + x.$ **28.** $2\,y'' - 3\,y' + y = e^{2x} + \sin x.$

29. $y'' - 4\,y' + 3\,y = x\,e^{-x}.$ **30.** $y'' - 4\,y = \cosh x.$

31. $y'' + 4\,y = \sin^2 x.$ **32.** $y'' + 4\,y = x \sin 2x.$

33. $y'' + y = \cos^2 x.$ **34.** $y'' + 2y' + 2\,y = e^{-x} \cos x.$

35. $y'' + 2y' + y = e^{-x}.$ **36.** $y'' + 2y' + y = x\,e^{-x} + 6.$

37. $y'' - y' - 2\,y = x^2 - 4.$ **38.** $y'' + y' - 2\,y = 2\,x^2 + x\,e^x.$

39. $y' + 2\,y = 2\,x + \sin x.$ **40.** $y' + 2\,y = \sin x.$

41. $2\,y' - y = e^{2x}.$ **42.** $y' - 7y = e^x + \cos x.$

43. $y'' + 9\,y = \sum_{n=1}^{6} \frac{\sin nx}{n}.$ **44.** $y'' + y = \sum_{n=1}^{6} \frac{\sin 2nx}{n}.$

In Exercises 45–54, find the solution of the associated homogeneous equation and state the form of a particular solution. Do not solve for the coefficients. Be sure to modify by x or x^2 as appropriate.

45. $y'' - 4\,y' + 3\,y = e^{2x} \sinh x.$ **46.** $y'' - 4\,y' + 3\,y = e^x \sinh 2x.$

47. $y'' + 2y' + 2\,y = \cos x + 6x^2 - e^{-x} \sin x.$

48. $y^{(4)} - y = \cosh x + \cosh 2x$.

49. $y'' - 3y' + 2y = 3x^4 e^x + x e^{-2x} \cos 3x$.

50. $y'' - 3y' + 2y = x^4 e^{-x} + 7x^2 e^{-2x}$.

51. $y'' + 4y = e^{2x}(\sin 2x + 2\cos 2x)$.

52. $y'' + 4y = x \sin 2x + 2 e^{2x} \cos 2x$.

53. $y'' - 2y'' + y = 6x - e^x$.   **54.** $y'' - 3y' + 2y = e^x + 3e^{-2x}$.

In Exercises 55–60, find the general solution of the given equation. Since an arbitrary parameter appears in the equation, you have to distinguish separate cases.

55. $y'' - 4y' + 3y = e^{\alpha x}$.   **56.** $y'' - 4y' + 4y = e^{\alpha x}$.

57. $y'' + 4y = \cos \omega x$.   **58.** $y'' + 9y = \sin \omega x$.

59. $y'' + \omega^2 y = \sin 2x$.   **60.** $y'' + by' + y = \sin x$.

In Exercises 61–70, solve the given initial value problem.

61. $y'' - 4y = 0$, $y(0) = 0$, $y'(0) = 3$.

62. $y'' + 2y' + y = 0$, $y(0) = 2$, $y'(0) = -1$.

63. $4y'' - 4y' + y = 0$, $y(0) = -1$, $y'(0) = 1$.

64. $y'' + y = 0$, $y(\pi) = 1$, $y'(\pi) = 0$.

65. $y'' - 5y' + 6y = e^x$, $y(0) = 0$, $y'(0) = 0$.

66. $2y'' - 3y' + y = \sin x$, $y(0) = 0$, $y'(0) = 0$.

67. $y'' - 4y' + 3y = x e^{-x}$, $y(0) = 0$, $y'(0) = 1$.

68. $y'' - 4y = \cosh x$, $y(0) = 1$, $y'(0) = 0$.

69. $y'' + 4y = \cos 2x$, $y(\pi/2) = 1$, $y'(\pi/2) = 0$.

70. $y'' + 9y = \sum_{n=1}^{6} \frac{\sin nx}{n}$, $y(0) = 0$, $y'(0) = 2$.

Integrals via undetermined coefficients *In Exercises 71–74, compute $\int g(x)\, dx$ by solving $y' = g(x)$ using the method of undetermined coefficients.*

71. $g(x) = e^x \sin x$.   **72.** $e^x \cos 2x$.

73. $g(x) = e^{ax} \cos bx$.   **74.** $e^{ax} \sin bx$.

Given n numbers $\lambda_1, \lambda_2, \ldots, \lambda_n$, we define the **Vandermonde determinant** $V(\lambda_1, \lambda_2, \ldots, \lambda_n)$ to be

$$
\begin{vmatrix}
1 & 1 & \cdots & 1 \\
\lambda_1 & \lambda_2 & \cdots & \lambda_n \\
\vdots & \vdots & \vdots & \vdots \\
\lambda_1^{n-1} & \lambda_2^{n-1} & \cdots & \lambda_n^{n-1}
\end{vmatrix}.
$$

75. Compute $V(\lambda_1, \lambda_2)$ and show that it is nonzero if and only if $\lambda_1 \neq \lambda_2$.

76. Show that $V(\lambda_1, \lambda_2, \lambda_3) = (\lambda_3 - \lambda_2)(\lambda_3 - \lambda_1)(\lambda_2 - \lambda_1)$ and conclude that $V(\lambda_1, \lambda_2, \lambda_3) \neq 0$ if and only if all the λ's are distinct.

77. It is a fact that $V(\lambda_1, \lambda_2, \ldots, \lambda_n) \neq 0$ if and only if all the λ's are distinct (see Exercise 78). Use this fact to prove that the Wronskian $W(e^{\lambda_1 x}, e^{\lambda_2 x}, \ldots, e^{\lambda_n x}) \neq 0$ if all the λ's are distinct.

78. (a) Based on Exercise 76, guess the formula for $V(\lambda_1, \lambda_2, \ldots, \lambda_n)$. Verify your guess for $n = 2, 3, 4$.
(b) Using (a), show that $V(\lambda_1, \lambda_2, \ldots, \lambda_n) \neq 0$ if and only if all the λ's are distinct.

A.3   Methods for Solving Ordinary Differential Equations

In the previous two sections we saw that the general solution of the second order linear differential equation

$$(1) \qquad y'' + p(x)y' + q(x)y = g(x)$$

is of the form $y = y_h + y_p$, where y_h is the general solution of the associated homogeneous equation and y_p is any particular solution of (1). We obtained a complete description of the solution when the coefficient functions are constants and g is of a special form. In this section we present two general methods for handling the cases not covered by the previous section. The first one is usually applied when solving for y_h, and the second one can be used to find y_p. We end the section by applying these methods to solve the important class of Euler equations.

Reduction of Order

For second order linear differential equations, we know that

$$y_h = c_1 y_1 + c_2 y_2,$$

where y_1 and y_2 are linearly independent solutions of the homogeneous equation

Note that the equation is in standard form. That is, the leading coefficient is 1.

$$(2) \qquad y'' + p(x)y' + q(x)y = 0.$$

Suppose that we know one nontrivial solution to (2), say y_1. The method of reduction of order allows us to find a second solution y_2 such that y_1 and y_2 are independent.

REDUCTION OF ORDER FORMULA

A second linearly independent solution of (2) given y_1 is

$$(3) \qquad y_2 = y_1 \int \frac{e^{-\int p(x)\,dx}}{y_1^2}\,dx.$$

Before you apply this formula, be sure that your equation is in standard form.

Since we are seeking only one independent solution, we can assign any fixed values to the constants of integration appearing in this formula. We will often neglect these constants.

Proof We know that cy_1 is a solution of (2). But of course this solution is linearly dependent with y_1. The idea is to find a nonconstant function $v(x)$ such that $v(x)y_1$ is a solution. Substituting $y = v(x)y_1$ into (2), we get

$$
\begin{aligned}
y'' + p(x)y' + q(x)y &= (y_1 v'' + 2y_1' v' + y_1'' v) + p(x)(y_1 v' + y_1' v) + q(x)y_1 v \\
&= y_1 v'' + (2y_1' + p(x)y_1)v' + (y_1'' + p(x)y_1' + q(x)y_1)v \\
&= y_1 v'' + (2y_1' + p(x)y_1)v'
\end{aligned}
$$

because $y_1'' + p(x)y_1' + q(x)y_1 = 0$. Hence, for $y = vy_1$ to be a solution to (2), v must satisfy $y_1 v'' + (2y_1' + p(x)y_1)v' = 0$. This is a first order equation in v' (hence the name reduction of order). Equivalently,

$$
v'' + \left(2\frac{y_1'}{y_1} + p(x) \right) v' = 0,
$$

and thus by Theorem 1, Section A.1, we find

$$
v'(x) = \frac{e^{-\int p(x)\,dx}}{y_1^2}.
$$

(We have taken the constant in this theorem to be 1, since we are only interested in one solution.) Hence (3) follows by integrating and multiplying by y_1. The linear independence of y_1 and y_2 can be checked by computing the Wronskian. See Exercise 41. ∎

EXAMPLE 1   Reduction of order in the presence of a double root

Given that $y_1 = e^x$ is a solution to $y'' - 2y' + y = 0$, a second linearly independent solution is obtained from (3):

$$
y_2 = e^x \int \frac{e^{2x}}{(e^x)^2}\,dx = xe^x.
$$
∎

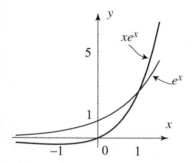

Figure 1 Solutions in Example 1.

The characteristic equation in Example 1 has 1 as a double root. We could have used the methods of the previous section to write down the second solution. However, the point of Example 1 is to show how reduction of order can be used to derive such solutions.

EXAMPLE 2   Applying the reduction of order formula

Given that e^x is a solution of $xy'' - (1+x)y' + y = 0$, a second linearly independent solution is obtained by appealing to (3). We have

$$
y_2 = e^x \int \frac{e^{\int (\frac{1}{x}+1)\,dx}}{(e^x)^2}\,dx = e^x \int \frac{xe^x}{(e^x)^2}\,dx = e^x \int xe^{-x}\,dx.
$$

$$
\int xe^{-x}\,dx
$$

$$
= -xe^{-x} - e^{-x} + C.
$$

Observe that before evaluating (3), we must put the equation in standard form. Integrating by parts, we find

$$
y_2 = e^x(-xe^{-x} - e^{-x}) = -x - 1.
$$

At this point, we may use $y_2 = x + 1$. ∎

Variation of Parameters

In this part, we suppose that we know the solution of the associated homogeneous equation (2) and describe a general method for generating a particular solution of (1). This method is called **variation of parameters**.

VARIATION OF PARAMETERS FORMULA

A particular solution of (1) is given by

$$(4) \qquad y_p = y_1 \int \frac{-y_2 g(x)}{W(y_1, y_2)} \, dx + y_2 \int \frac{y_1 g(x)}{W(y_1, y_2)} \, dx,$$

where y_1 and y_2 are linearly independent solutions of (2), and $W(y_1, y_2) = y_1 y_2' - y_1' y_2$ is the Wronskian of y_1 and y_2.

As in the reduction of order method, we can neglect the constants of integration.

Proof Since $c_1 y_1 + c_2 y_2$ is a solution of the homogeneous equation, our hope is that by allowing functions instead of constants we can generate a solution of the nonhomogeneous equation. We thus try

$$(5) \qquad y = u_1(x) y_1 + u_2(x) y_2$$

and solve for u_1 and u_2. Since we have two unknown functions and only one equation to satisfy, we are free to impose one additional relation between u_1 and u_2. As you will see, the following condition simplifies the computation:

$$(6) \qquad u_1'(x) y_1 + u_2'(x) y_2 = 0 \,.$$

We now compute using (5) and (6):

$$y' = u_1(x) y_1' + u_2(x) y_2'; \quad y'' = u_1(x) y_1'' + u_2(x) y_2'' + u_1'(x) y_1' + u_2'(x) y_2' \,.$$

Substituting these and (5) in the left side of (1) we get

$$
\begin{aligned}
y'' + p(x) y' + q(x) y &= (u_1(x) y_1'' + u_2(x) y_2'' + u_1'(x) y_1' + u_2'(x) y_2') \\
&\quad + p(x)(u_1(x) y_1' + u_2(x) y_2') + q(x)(u_1(x) y_1 + u_2(x) y_2) \\
&= u_1(x) \underbrace{(y_1'' + p(x) y_1' + q(x) y_1)}_{=\,0} \\
&\quad + u_2(x) \underbrace{(y_2'' + p(x) y_2' + q(x) y_2)}_{=\,0} + u_1'(x) y_1' + u_2'(x) y_2' \,.
\end{aligned}
$$

Therefore, recalling (6), we will have a solution if u_1 and u_2 satisfy

$$
\begin{cases}
y_1 u_1'(x) + y_2 u_2'(x) = 0, \\
y_1' u_1'(x) + y_2' u_2'(x) = g(x).
\end{cases}
$$

The determinant of this system is precisely $W(y_1, y_2)$, which is nonzero since y_1 and y_2 are linearly independent. Solving this system, we get the unique solutions

$$u_1' = \frac{-y_2 g(x)}{W(y_1, y_2)} \quad \text{and} \quad u_2' = \frac{y_1 g(x)}{W(y_1, y_2)} \,.$$

Integrating and substituting in (5) yields (4).   ■

EXAMPLE 3   Variation of parameters
Given that $y_1 = x$ and $y_2 = x^4$ are solutions of $x^2 y'' - 4xy' + 4y = 0$, find a particular solution of
$$x^2 y'' - 4xy' + 4y = 3x^2 .$$

Solution We have
$$W(y_1, y_2) = W(x, x^4) = 3x^4 .$$

We put the equation in standard form, apply (4), and get
$$y_p = x \int \frac{-3x^4}{3x^4} \, dx + x^4 \int \frac{3x}{3x^4} \, dx = -x^2 - \frac{1}{2} x^2 = -\frac{3}{2} x^2 .$$ ■

We end this section with a discussion of a class of differential equations with important applications.

Euler Equations

The differential equation

(7) $$x^2 y'' + \alpha\, xy' + \beta y = 0,$$

where α and β are constants, is known as **Euler's equation**. It is the simplest example of a second order linear differential equation with nonconstant coefficients for which we have an explicit solution.   Motivated by the first order version of this equation, $xy' + \alpha y = 0$, which has solution $y = x^{-\alpha}$, we try

(8) $$y = x^r$$

as a solution of (7). Plugging x^r into the equation, it follows that r must be a root of

(9) $$r(r - 1) + \alpha r + \beta = 0 .$$

This quadratic equation, known as the **indicial equation**, is the key to solving (7), in the same way that the characteristic equation is the key to solving an equation with constant coefficients.   As expected, the solutions will depend on the nature of the roots, referred to as the **indicial roots**.

If we put the general Euler's equation in standard form, the coefficients of y' and y are not defined at $x = 0$.   Because of this problem at 0, the cases $x > 0$ and $x < 0$ are to be treated separately.   In fact, in most applications we are only interested in the case $x > 0$. For completeness, in the following box we present the solution of Euler's equation in both cases using the absolute value.

GENERAL SOLUTION OF EULER'S EQUATION

Consider Euler's equation (7) with indicial equation (9), which we write as

$$(10) \qquad\qquad r^2 + (\alpha - 1)r + \beta = 0 \,.$$

Let r_1 and r_2 denote the indicial roots. The general solution y of this equation is given by the following cases.

Case I If r_1 and r_2 are distinct real roots, then

$$y = c_1 |x|^{r_1} + c_2 |x|^{r_2} \,.$$

Case II If $r_1 = r_2$, then

$$y = (c_1 + c_2 \ln |x|)|x|^{r_1} \,.$$

Case III If r_1 and r_2 are complex conjugate roots with $r_1 = a + ib$, then

$$y = |x|^a [c_1 \cos(b \ln |x|) + c_2 \sin(b \ln |x|)] \,.$$

Clearly when $x > 0$ we may drop the absolute values.

Proof For clarity's sake, we take $x > 0$. Case I follows immediately from (8) and the fact that $W(x^{r_1}, x^{r_2}) = (r_2 - r_1)x^{r_1 + r_2 - 1} \neq 0$ for $x > 0$. Case II is derived via reduction of order. Using $y_1 = x^{r_1}$ in (3), we get

$$y_2 = x^{r_1} \int \frac{e^{-\int \frac{\alpha}{x} \, dx}}{x^{2r_1}} \, dx = x^{r_1} \int x^{-\alpha - 2r_1} \, dx \,.$$

But because r_1 is a double root of the indicial equation (10), we have $2r_1 = r_1 + r_2 = -(\alpha - 1)$. So $x^{-\alpha - 2r_1} = x^{-1}$, and the integral evaluates to $\ln x$, implying $y = x^{r_1} \ln x$.

In Case III, two linearly independent complex-valued solutions are formally given by x^{a+ib} and x^{a-ib}. We interpret these as $e^{\ln x(a+ib)}$ and $e^{\ln x(a-ib)}$ and proceed to derive real-valued solutions. From Euler's identity, we have

$$e^{\ln x(a+ib)} = e^{a \ln x} e^{ib \ln x} = x^a [\cos(b \ln x) + i \sin(b \ln x)],$$

and, similarly,

$$e^{\ln x(a-ib)} = x^a [\cos(b \ln x) - i \sin(b \ln x)] \,.$$

Taking linear combinations, we arrive at the desired real solutions as we did previously when dealing with constant coefficient equations having complex characteristic roots. Linear independence follows by computing the Wronskian and is left as Exercise 42. ∎

There is a close similarity between the solution to Euler's equation and the solution to the constant coefficient equation. Indeed, the change of variables $t = \ln x$ in Euler's equation transforms it to an equation with constant coefficients.   This provides an alternative derivation of the general solution of Euler's equation.   See Exercise 43.

EXAMPLE 4   Euler's equation

Solve $2\,x^2 y'' + 5\,xy' - 2y = 0$ for $x > 0$.

Solution   We rewrite the equation as $x^2 y'' + \frac{5}{2}\,xy' - y = 0$.   From (10), the indicial equation is $r^2 + \frac{3}{2}r - 1 = 0$, which factors as $(r - \frac{1}{2})(r + 2) = 0$.   We get the indicial roots $r_1 = \frac{1}{2}$, $r_2 = -2$.   Thus by Case I, the general solution is $y = c_1\sqrt{x} + \frac{c_2}{x^2}$. The general solution is illustrated in Figure 2.   ∎

Note that when $x < 0$, the solution in Example 4 becomes

$$y = c_1\sqrt{-x} + \frac{c_2}{x^2}.$$

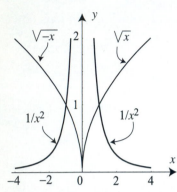

Figure 2 Solutions in Example 4. Notice that $\frac{1}{x^2}$ is not defined at $x = 0$ and $\sqrt{x}$ is not differentiable at $x = 0$. These solutions are valid for $x > 0$, where the differential equation is defined.

Exercises A.3

In Exercises 1–20, verify that the given function is a solution of the given equation, and then find the general solution using the reduction of order formula.

1. $y'' + 2\,y' - 3\,y = 0$, $y_1 = e^x$.

2. $y'' - 5\,y' + 6\,y = 0$, $y_1 = e^{3x}$.

3. $x\,y'' - (3 + x)y' + 3y = 0$, $y_1 = e^x$.

4. $xy'' - (2 - x)y' - 2\,y = 0$, $y_1 = e^{-x}$.

5. $y'' + 4\,y = 0$, $y_1 = \cos 2x$.

6. $y'' + 9\,y = 0$, $y_1 = \sin 3x$.

7. $y'' - y = 0$, $y_1 = \cosh x$.

8. $y'' + 2\,y' + y = 0$, $y_1 = e^{-x}$.

9. $(1 - x^2)y'' - 2xy' + 2y = 0$, $y_1 = x$.

10. $(1 - x^2)y'' - 2xy' = 0$, $y_1 = 1$.

11. $x^2 y'' + xy' - y = 0$, $y_1 = x$.

12. $x^2 y'' - xy' + y = 0$, $y_1 = x$.

13. $x^2 y'' + xy' + y = 0$, $y_1 = \cos(\ln x)$.

14. $x^2 y'' + 2\,xy' + \frac{1}{4}\,y = 0$, $y_1 = 1/\sqrt{x}$.

15. $x\,y'' + 2\,y' + 4x\,y = 0$, $y_1 = \frac{\sin 2x}{x}$.

16. $x\,y'' + 2\,y' - x\,y = 0$, $y_1 = \frac{e^x}{x}$.

17. $x\,y'' + 2\,(1 - x)y' + (x - 2)\,y = 0$, $y_1 = e^x$.

18. $(x - 1)^2\,y'' - 3(x - 1)\,y' + 4y = 0$, $y_1 = (x - 1)^2$.

19. $x^2 y'' - 2x\,y' + 2\,y = 0$, $y_1 = x^2$.

20. $(x^2 - 2x)y'' - (x^2 - 2)\,y' + 2(x - 1)y = 0$, $y_1 = e^x$.

In Exercises 21–30, find the general solution of the given equation using the method of variation of parameters.   Take $x > 0$.

21. $y'' - 4y' + 3y = e^{-x}$.

22. $y'' - 15y' + 56y = e^{7x} + 12x$.

23. $3y'' + 13y' + 10y = \sin x$.

24. $y'' + 3y = x$.

25. $y'' + y = \sec x$.

26. $y'' + y = \sin x + \cos x$.

27. $xy'' - (1+x)y' + y = x^3$.
[Hint: $y_1 = 1 + x, y_2 = e^x$.]

28. $xy'' - (1+x)y' + y = x^4 e^x$.
[Hint: Exercise 27.]

29. $x^2 y'' + 3xy' + y = \sqrt{x}$.

30. $x^2 y'' + xy' + y = x$.

In Exercises 31–40, solve the given Euler equation.   Take $x > 0$, unless otherwise stated.

31. $x^2 y'' + 4xy' + 2y = 0$.

32. $x^2 y'' + xy' - 4y = 0$.

33. $x^2 y'' + 3xy' + y = 0$.

34. $4x^2 y'' + 8xy' + y = 0$.

35. $x^2 y'' + xy' + 4y = 0$.

36. $4x^2 y'' + 4xy' + y = 0$.

37. $x^2 y'' + 7xy' + 13y = 0$.

38. $x^2 y'' - xy' + 5y = 0$.

39. $(x-2)^2 y'' + 3(x-2)y' + y = 0$
$(x > 2)$.
[Hint: Let $t = x - 2$.]

40. $(x+1)^2 y'' + (x+1)y' + y = 0$
$(x > -1)$.
[Hint: Let $t = x + 1$.]

41. Compute $W(y_1, y_2)$ with y_2 given by (3) and conclude that the reduction of order formula yields a second linearly independent solution.

42. Let y_1 and y_2 be the solutions of Euler's equation in Case III.   Show that $W(y_1, y_2) = x^{2a-1}$ and conclude that y_1 and y_2 are linearly independent for $x > 0$.

43 Show that the change of variables $t = \ln x$ $(x > 0)$ transforms Euler's equation (7) to the equation

$$\frac{d^2 y}{dt^2} + (\alpha - 1)\frac{dy}{dt} + \beta y = 0$$

with constant coefficients.

44. An alternative solution of Euler's equation. Using Exercise 43 and the solution of (6), Section A.2, derive the three cases of the general solution of Euler's equation.

45.   Reduction of order formula from Abel's formula.
(a) Use Abel's formula (Theorem 2, Section A.1) to conclude that

$$y_1 y_2' - y_1' y_2 = Ce^{-\int p(x)\, dx},$$

where y_1 and y_2 are any two solution of (2).
(b) Given y_1, set $C = 1$ in (a) and solve the resulting first order differential equation in y_2, thereby deriving (3).

46. Reduction of order for nonhomogeneous equations. In this exercise we demonstrate that the method of reduction of order also applies to nonhomogeneous equations given a solution y_1 to the associated homogeneous equation. Thus, given y_1, we may solve (1) directly without recourse to the method of variation of parameters.

(a) Show that if we want to solve (1) and carry out the proof of (3) we arrive at the equation

$$v'' + \left(\frac{y_1'}{y_1} + p(x) \right) v' = \frac{g(x)}{y_1} \, .$$

(b) Solve the equation using Theorem 1, Section A.1, and conclude that the general solution of (1) is

$$y = c_1 y_1 + c_2 \, y_1 \int \frac{e^{-\int p(x)\, dx}}{y_1^2} \, dx$$

$$+ y_1 \int \frac{e^{-\int p(x)\, dx}}{y_1^2} \, dx \left(\int y_1 e^{\int p(x)\, dx} g(x)\, dx \right) \, dx$$

where the last two occurrences of $\int p(x)\, dx$ represent the same antiderivative of $p(x)$.

In Exercises 47–50, find the general solution of the given equation by using the method of Exercise 46. To get a good feel for Exercise 46, we suggest that you repeat its proof with at least one of the following exercises.

47. $y'' - 4y' + 3y = e^x, \quad y_1 = e^x$.

48. $x^2 y'' + 3xy' + y = \sqrt{x}, \quad y_1 = \frac{1}{x}$.

49. $3\, y'' + 13\, y' + 10\, y = \sin x, \quad y_1 = e^{-x}$.

50. $x\, y'' - (1+x)y' + y = x^3, \quad y_1 = e^x$.

A.4   The Method of Power Series

In this and the next section we use power series to solve ordinary differential equations. These methods are very useful and can be applied to solve many important differential equations with nonconstant coefficients, such as the Legendre and Bessel equations. As you will soon find out, these methods are quite tedious to carry out by hand. But they can be implemented with ease on most computer algebra systems, which are capable of carrying out the laborious computations.

The idea behind solving a differential equation using power series is simple. First, assume that the solution can be written as a power series. Second, write each term in the differential equation as a power series. Third, equate coefficients of the resulting series on both sides of the equation. Finally, solve for the unknown coefficients in the series representation of the assumed solution.

If we assume that the differential equation has a power series solution

$$y = \sum_{m=0}^{\infty} a_m x^m = a_0 + a_1 x + a_2 x^2 + a_3 x^3 + \cdots,$$

then the derivatives of y are obtained by term by term differentiation:

$$y' = \sum_{m=1}^{\infty} m\, a_m\, x^{m-1} = a_1 + 2a_2 x + 3a_3 x^2 + \cdots,$$

$$y'' = \sum_{m=2}^{\infty} m\,(m-1)\, a_m\, x^{m-2} = 2a_2 + 2 \cdot 3\, a_3\, x + 3 \cdot 4\, a_4 x^2 + \cdots$$

and so on. We illustrate with a simple first order differential equation.

EXAMPLE 1   A first order equation

Solve the differential equation $y' - y = 0$.

Solution We assume that y has a power series representation, substitute the power series of y and y' into the equation, and obtain

To combine the series together, we need to have the same powers of x appearing in both. For this purpose, we reindex the first series by changing m to $m + 1$. The index of summation has to be lowered by 1, and so the first series will have to start at 0.

$$\sum_{m=1}^{\infty} m a_m x^{m-1} - \sum_{m=0}^{\infty} a_m x^m = 0;$$

$$\sum_{m=0}^{\infty} (m+1)a_{m+1} x^m - \sum_{m=0}^{\infty} a_m x^m = 0;$$

$$\sum_{m=0}^{\infty} [(m+1)a_{m+1} - a_m] x^m = 0.$$

Thus $(m+1)a_{m+1} - a_m = 0$ for all m, and so

(1)
$$a_{m+1} = \frac{a_m}{m+1}.$$

We have

$$a_1 = a_0, \quad a_2 = \frac{a_1}{2} = \frac{a_0}{1 \cdot 2}, \quad a_3 = \frac{a_2}{3} = \frac{a_0}{1 \cdot 2 \cdot 3}, \quad a_4 = \frac{a_3}{4} = \frac{a_0}{1 \cdot 2 \cdot 3 \cdot 4}, \ldots,$$

and in general $a_m = a_0/m!$. Hence the solution is

$$y = a_0 \sum_{m=0}^{\infty} \frac{x^m}{m!} = a_0 \left(1 + \frac{x}{1!} + \frac{x^2}{2!} + \frac{x^3}{3!} + \cdots \right) = a_0 e^x. \qquad \blacksquare$$

Reindexing a series works like a change of variables in an integral. If k is a positive integer and you change m to $m + k$ inside the series, then the starting point of the series must be lowered by k. If you change m to $m - k$, then the starting point of the series must be raised by k.

An expression like (1) that gives each coefficient in terms of preceding ones is called a **recurrence relation**. In Example 1 the recurrence relation determines the coefficients in steps of one. That is, from a_0 we get a_1, from a_1 we get a_2, and so on. In the next example, the recurrence relation determines the coefficients in steps of two.

EXAMPLE 2   A familiar second order equation

Solve the differential equation $y'' + y = 0$.

Solution Assuming that y has a power series representation and substituting into the equation gives

Change m to $m+2$ in the first series, and start the new series at 0.

$$\sum_{m=2}^{\infty} m(m-1)a_m x^{m-2} + \sum_{m=0}^{\infty} a_m x^m = 0,$$

$$\sum_{m=0}^{\infty} (m+2)(m+1)a_{m+2}\, x^m + \sum_{m=0}^{\infty} a_m x^m = 0,$$

$$\sum_{m=0}^{\infty} [(m+2)(m+1)a_{m+2} + a_m]x^m = 0.$$

Thus

$$a_{m+2} = -\frac{1}{(m+2)(m+1)}\, a_m.$$

It is clear from this relation that a_0 determines a_2, which in turn determines a_4 and so on. Similarly, a_1 determines a_3, which determines a_5, and so on. Thus the even coefficients are determined from a_0 and the odd coefficients from a_1:

$$a_2 = -\frac{1}{2!}a_0, \qquad\qquad a_3 = -\tfrac{1}{3!}a_1,$$

$$a_4 = -\tfrac{1}{4\cdot 3}a_2 = \tfrac{1}{4!}a_0, \qquad a_5 = -\tfrac{a_3}{5\cdot 4} = \tfrac{1}{5!}a_1,$$

$$a_6 = -\tfrac{1}{6\cdot 5}a_4 = -\tfrac{1}{6!}a_0, \quad a_7 = -\tfrac{a_5}{7\cdot 6} = -\tfrac{1}{7!}a_1.$$

The pattern we see gives

$$a_{2k} = (-1)^k \tfrac{a_0}{(2k)!}, \quad a_{2k+1} = (-1)^k \tfrac{a_1}{(2k+1)!}.$$

Writing the even-indexed and the odd-indexed terms separately, we get the solution

$$y = a_0 \sum_{k=0}^{\infty} (-1)^k \frac{x^{2k}}{(2k)!} + a_1 \sum_{k=0}^{\infty} (-1)^k \frac{x^{2k+1}}{(2k+1)!}.$$

We recognize these series as the Taylor series for the sine and cosine; thus, as you may have already guessed, $y = a_0 \cos x + a_1 \sin x$. ∎

Note that a_0 and a_1 account for the two arbitrary constants that we expect in the general solution to a second order differential equation.

In Examples 1 and 2 the solutions of the differential equations were familiar functions. The usefulness of the power series method, however, is more appreciated when we use it to find solutions that are not combinations of elementary functions.

EXAMPLE 3   A second order initial value problem

Solve the initial value problem

$$y'' + xy' + y = 0, \quad y(0) = 0, \quad y'(0) = 1.$$

Solution As before, we assume a solution of the form

$$y = \sum_{m=0}^{\infty} a_m x^m = a_0 + a_1 x + a_2 x^2 + \cdots .$$

At the outset, we may use the initial conditions to find a_0 and a_1. We have

$$y(0) = a_0 + a_1 0 + a_2 0^2 + \cdots = a_0$$

$$y'(0) = a_1 + 2\, a_2 0 + 3\, a_3 0^2 + \cdots = a_1$$

and so the initial conditions imply that $a_0 = 0$ and $a_1 = 1$. Substituting the series for y and its derivatives into the differential equation yields

$$\sum_{m=2}^{\infty} m(m-1)\, a_m\, x^{m-2} + \sum_{m=1}^{\infty} m a_m x^m + \sum_{m=0}^{\infty} a_m x^m = 0 .$$

Shifting the index of summation in the first series yields

Note how the index of summation in the first sum dropped by 2.

$$\sum_{m=0}^{\infty} (m+2)(m+1)a_{m+2}x^m + \sum_{m=1}^{\infty} m a_m x^m + \sum_{m=0}^{\infty} a_m x^m = 0;$$

$$(2a_2 + a_0) + \sum_{m=1}^{\infty} [(m+2)(m+1)\, a_{m+2} + m a_m + a_m]x^m = 0$$

Equating the constant term to zero gives $a_2 = -a_0/2$ and equating the coefficients of the series to zero yields the recurrence relation

$$a_{m+2} = \frac{-1}{m+2}\, a_m .$$

As in the preceding example, a_0 determines the coefficients with even indices and a_1 those with odd indices. Since $a_0 = 0$, we get $a_2 = 0$, $a_4 = 0$, and so on. For the coefficients with odd indices, the recurrence relation gives

$$a_1 = 1,$$

$$a_3 = -\tfrac{1}{3}a_1 = -\tfrac{1}{3},$$

$$a_5 = -\tfrac{1}{5}a_3 = \tfrac{1}{5 \cdot 3},$$

$$a_7 = -\tfrac{1}{7}\, a_5 = -\tfrac{1}{7 \cdot 5 \cdot 3} .$$

The pattern that emerges is

$$a_{2k+1} = \frac{(-1)^k}{1 \cdot 3 \cdot 5 \cdots (2k+1)} .$$

Substituting these values for the coefficients into y, we get the solution to the given initial value problem

$$y = x - \frac{1}{3}x^3 + \frac{1}{15}x^5 - \frac{1}{105}x^7 + \cdots + \frac{(-1)^k}{1 \cdot 3 \cdot 5 \cdots (2k+1)}x^{2k+1} + \cdots.$$

This solution can be expressed in a more compact form (see Exercise 17).   ■

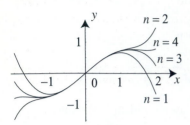

Figure 1 Polynomial (degree $2n + 1$) approximation of the solution in Example 3.

Polynomial approximations of the solution of Example 3 are graphed in Figure 1. The more terms we include, the better the approximation is on a larger interval. Note that all the polynomials satisfy the initial conditions.

It is useful to remark that when solving initial value problems like Example 3, using the power series method, we always have

$$\boxed{y(0) = a_0 \quad \text{and} \quad y'(0) = a_1.}$$

We now state a theorem that justifies using the power series method.

THEOREM 1
POWER SERIES
SOLUTIONS

Suppose that the functions $p(x)$, $q(x)$, and $g(x)$ have power series expansions at $x = a$; then any solution of

$$(2) \qquad\qquad y'' + p(x)y' + q(x)y = g(x)$$

has a power series expansion at $x = a$. Thus any solution y can be expressed as a power series centered at a,

$$y = \sum_{m=0}^{\infty} a_m(x - a)^m.$$

Moreover, the radius of convergence of this power series is at least as large as the minimum of the radii of convergence of p, q, and g.

When $p(x)$, $q(x)$, and $g(x)$ have power series expansions with positive radius of convergence at $x = a$, the point a is called an **ordinary point** for the equation (2).

Since the existence of the solutions of (2) is guaranteed by Theorem 3, Appendix A.1, the real purpose of Theorem 1 is to address the existence of power series expansions of the solutions. For a proof of Theorem 1, see Chapter 3 of *Ordinary Differential Equations*, by Garrett Birkhoff and Gian-Carlo Rota, 2nd ed., Wiley, 1969. To appreciate the power of Theorem 1, note that it asserts that a complicated equation such as $y'' + e^x y' + \cos xy = 1/(1+x^2)$ has a power series solution about $x = 0$ with radius of convergence at least 1.

As should be expected, it is not always possible to find simple patterns for the coefficients in the solution. In such cases we find as many terms of the series as we need. Finding these terms is routine but can be quite tedious. A computer program capable of symbolic manipulations is invaluable here. We give an example.

EXAMPLE 4   Initial value problems and the power series method

Solve the initial value problem

$$y'' + e^x y = \cos x, \quad y(0) = 1, \quad y'(0) = 0.$$

Solution   Here $p(x) = 0$, $q(x) = e^x$, and $g(x) = \cos x$, and each of these functions has a power series expansion at 0, valid for all x. By Theorem 1 there is a power series solution

$$y = \sum_{m=0}^{\infty} a_m x^m$$

about $x = 0$ with an infinite radius of convergence. As in Example 3, the initial conditions imply

$$a_0 = 1 \quad \text{and} \quad a_1 = 0.$$

Now, writing each term in the equation as a series gives

$$(2a_2 + 6a_3\, x + 12\, a_4 x^2 + \cdots) + (1 + x + \tfrac{1}{2!}x^2 + \cdots)(a_0 + a_1 x + a_2 x^2 + \cdots)$$
$$= 1 - \tfrac{1}{2!}x^2 + \tfrac{1}{4!}\, x^4 - \cdots$$

Multiplying the series in the second term and combining like terms gives

$$(2a_2 + a_0) + (a_0 + a_1 + 6\, a_3)x + (\tfrac{1}{2}\, a_0 + a_1 + a_2 + 12a_4)x^2 + \cdots$$
$$= 1 - \tfrac{1}{2!}x^2 + \tfrac{1}{4!}\, x^4 - \cdots$$

Equating coefficients, we find

$$2\, a_2 + a_0 = 1$$
$$a_0 + a_1 + 6\, a_3 = 0$$
$$\tfrac{1}{2}\, a_0 + a_1 + a_2 + 12\, a_4 = -\tfrac{1}{2}$$
$$\vdots$$

Using the values of a_0 and a_1, we solve this system of equations to get $a_2 = 0$, $a_3 = -1/6$, and $a_4 = -1/12$. Thus the solution is

$$y = 1 - \frac{1}{6}\, x^3 - \frac{1}{12}\, x^4 + \cdots .$$

With the help of a computer, we can continue this procedure and find as many terms as we please. Here are the first ten nonzero terms of the series:

$$y = 1 - \frac{1}{6}\, x^3 - \frac{1}{12}\, x^4 + \frac{1}{120}\, x^6 + \frac{19}{5,040}\, x^7 + \frac{1}{960}\, x^8 + \frac{43}{362,880}\, x^9 - \frac{83}{1,814,400}\, x^{10} - \cdots .$$

On the interval $[-1, 1]$, the graph of the approximating polynomial of degree 4 is indistinguishable from the graph of the polynomial of degree 7 (Figure 2). This

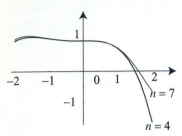

Figure 2 Polynomial (degree n) approximation of the solution.

suggests that on this interval we have a good approximation with a polynomial of degree 4. However, as we move away from the center of the series, more terms are needed to get a good approximation. ∎

In second order homogeneous problems, it is instructive to identify linearly independent solutions y_1 and y_2. As we saw in Examples 2 and 3, the recurrence relations allowed us to solve for the coefficients a_m, $m \geq 2$, in terms of a_0 and a_1. We can define y_1 by taking $a_0 = 1$, $a_1 = 0$ and y_2 by taking $a_0 = 0$, $a_1 = 1$. Then the general solution can be written as

$$y = c_1 y_1 + c_2 y_2 \,.$$

In terms of initial value problems, it is easily seen that y_1 satisfies the initial values $y_1(a) = 1$, $y_1'(a) = 0$. Similarly, y_2 satisfies $y_2(a) = 0$, $y_2'(a) = 1$.

Exercises A.4

In Exercises 1–12, use the power series method to solve the given equation. In each case, write down the recurrence relation for the coefficients, and compute at least three nonzero terms from each power series solution.

1. $y' + 2xy = 0$.

2. $y' + y = 0$.

3. $y' + y = x$.

4. $y' + (\cos x)y = 0$.

5. $y'' - y = 0$.

6. $y'' - y = x$.

7. $y'' - x\,y' + y = 0$.

8. $y'' + 2\,y' + 2\,y = 0$.

9. $y'' + 2\,x\,y' + y = 0$.

10. $y'' + x\,y' + y = 0$.

11. $y'' - 2\,y' + y = 0$, $y(0) = 0$, $y'(0) = 1$.

12. $y'' - 2y' + y = x$, $y(0) = 2, y'(0) = 1$.

In Exercises 13–16, (a) use the power series method to solve the given differential equation. (b) Plot several partial sums and decide from the graphs which one provides a good approximation of the solution on the interval $[-1/2, 1/2]$.

13. $y'' - y' + 2y = e^x$, $y(0) = 0$, $y'(0) = 1$.

14. $y'' + y = x + 1$, $y(0) = 0, y'(0) = 1$.

15. $y'' + e^x y = 0$, $y(0) = 1, y'(0) = 0$.

16. $y'' + x\,y = \sum_{n=1}^{\infty} \frac{x^n}{n(n+1)}$, $y(0) = 0, y'(0) = 1$.

17.   Show that the solution in Example 3 is

$$y = \sum_{k=0}^{\infty} \frac{(-1)^k 2^k k!}{(2k+1)!} x^{2k+1} \,.$$

18. Airy's differential equation. In this exercise, we find a series solution of Airy's equation

$$y'' - xy = 0 \,.$$

(a) State why the solutions of the differential equation have power series solutions with infinite radius of convergence.

(b) Assume that $y = \sum_{m=0}^{\infty} a_m x^m$. Show that $a_2 = 0$ and derive the *three-step recurrence relation*

$$(m+2)(m+1)a_{m+2} = a_{m-1}, \qquad m = 1, 2, 3, \dots .$$

(c) Show that the general solution is

$$y = a_0 \left[1 + \sum_{n=1}^{\infty} \frac{x^{3n}}{(3n)(3n-1)(3n-3)\dots 3\cdot 2} \right]$$

$$+ a_1 \left[x + \sum_{n=1}^{\infty} \frac{x^{3n+1}}{(3n+1)(3n)(3n-2)\dots 4\cdot 3} \right].$$

(d) Write $y = a_0 y_1 + a_1 y_2$ and plot several polynomial approximations of y_1 and y_2. Note that for negative values of x the graphs look like those of $\sin x$ and $\cos x$, and for positive values of x the graphs look like those of e^x. Examine the differential equation and give a reason for these observations.

19. Solve the following initial value problem for Airy's equation

$$y'' - xy = 0, \quad y(0) = 1, \ y'(0) = 0.$$

20. The sine and cosine functions. In this exercise, we show how a differential equation can be used to derive properties of its solutions. We refer to the differential equation of Example 2 with the two linearly independent solutions

$$C(x) = \sum_{k=0}^{\infty} (-1)^k \frac{1}{(2k)!} x^{2k} \quad \text{and} \quad S(x) = \sum_{k=0}^{\infty} (-1)^k \frac{1}{(2k+1)!} x^{2k+1}.$$

We pretend for now that we do not recognize these functions as the sine and cosine functions, and we derive their basic properties from the differential equation.
(a) Show that $C'(x) = -S(x)$ and $S'(x) = C(x)$.
(b) Using Abel's form of the Wronskian, show that $C^2(x) + S^2(x) = 1$.
(c) Using the fact that any solution of the differential equation $y'' + y = 0$ is a linear combination of $C(x)$ and $S(x)$, show that $S(x+a) = S(x)C(a) + C(x)S(a)$. [Hint: Show that $S(x + a)$ is a solution and conclude that $S(x + a) = c_1 S(x) + c_2 C(x)$. Determine the constants by evaluating at $x = 0$.]
(d) Establish the identity

$$C(x + a) = C(x)C(a) - S(x)S(a).$$

(e) Which trigonometric identities are implied by (c) and (d)?

A.5   The Method of Frobenius

While the method of power series is useful in many situations, and perhaps the most widely applicable method that we have developed thus far, there are still important differential equations that cannot be fully understood by this method. For example, consider Bessel's equation

$$x^2 y'' + xy' + (x^2 - p^2)y = 0 .$$

Putting it in standard form, we see that $x = 0$ is not an ordinary point. Hence, Theorem 1 of Appendix A.4 does not apply. For applications, it is of particular importance to understand the behavior of the solutions at $x = 0$. To this end, we will develop a generalization of the power series method, known as the **method of Frobenius**.

Consider the homogeneous differential equation

$$(1) \qquad y'' + p(x)y' + q(x)y = 0 .$$

Recall that a is an ordinary point of the differential equation if p and q have power series expansions at a. Otherwise, a is called a **singular point**. If a is an ordinary point, then we know from our work in the preceding section that the solutions of this equation have power series expansions at the point $x = a$. If a is a singular point, then the method of power series fails, as our next example shows.

EXAMPLE 1   An equation for which the power series method fails
Consider the differential equation $xy'' + 2y' + xy = 0$.
(a) Determine whether $x = 0$ is an ordinary or singular point.
(b) Apply the method of power series at $x = 0$ to the differential equation.
Solution (a) We have $p(x) = 2/x$, and $q(x) = 1$. Since $p(x)$ does not have a power series expansion at $x = 0$, the point $x = 0$ is a singular point of the equation.
(b) We assume that y has a power series representation centered at $x = 0$, $y = \sum_{m=0}^{\infty} a_m x^m$. Substituting into the differential equation and reindexing as needed gives

$$\sum_{m=2}^{\infty} m(m-1)a_m x^{m-1} + \sum_{m=1}^{\infty} 2m a_m x^{m-1} + \sum_{m=0}^{\infty} a_m x^{m+1} = 0;$$

$$\sum_{m=1}^{\infty} (m+1)m a_{m+1} x^m + \sum_{m=0}^{\infty} 2(m+1)a_{m+1} x^m + \sum_{m=1}^{\infty} a_{m-1} x^m = 0;$$

$$2a_1 + \sum_{m=1}^{\infty} [m(m+1)a_{m+1} + 2(m+1)a_{m+1} + a_{m-1}]x^m = 0 .$$

Equating coefficients, we find that $a_1 = 0$ and

$$m(m+1)a_{m+1} + 2(m+1)a_{m+1} + a_{m-1} = 0, \quad \text{for all } m,$$

from which we get the two-step recurrence relation

$$a_{m+1} = -\frac{1}{(m+2)(m+1)}a_{m-1}\,.$$

Since $a_1 = 0$, it follows that $a_3 = a_5 = a_7 = \cdots = 0$. For the even indexed coefficients we have

$$a_2 = -\tfrac{1}{3\cdot 2}a_0 = -\tfrac{1}{3!}a_0$$

$$a_4 = -\tfrac{1}{5\cdot 4}a_2 = \tfrac{1}{5!}a_0$$

$$a_6 = -\tfrac{1}{7\cdot 6}a_4 = -\tfrac{1}{7!}a_0\,.$$

It is clear how this sequence continues. Thus one solution is

$$y = a_0\left(1 - \frac{x^2}{3!} + \frac{x^4}{5!} - \frac{x^6}{7!} + \cdots\right) = \frac{a_0}{x}\left(x - \frac{x^3}{3!} + \frac{x^5}{5!} - \frac{x^7}{7!} + \cdots\right) = a_0\frac{\sin x}{x}$$

(Figure 1).

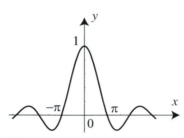

Figure 1

$$\frac{\sin x}{x} = \sum_{m=0}^{\infty}(-1)^m\frac{x^{2m}}{(2m+1)!}.$$

(Note that the function $\frac{\sin x}{x}$ has a power series expansion at $x = 0$ which converges for all x.) Since the differential equation is of second order, we need two linearly independent solutions. The method of power series yielded only one solution and hence failed to give the general solution. To get a second linearly independent solution, we can use the method of reduction of order, which gives (you should check this)

$$y = \frac{\cos x}{x}\,.$$

(See also Exercise 29.) ∎

Put simply, the method of power series failed to generate both linearly independent solutions in Example 1 because one of those solutions is not a power series. To see why, we write

$$\frac{\cos x}{x} = x^{-1}\left(1 - \frac{x^2}{2!} + \frac{x^4}{4!} - \frac{x^6}{6!} + \cdots\right)\,.$$

This is not a power series, since it contains a negative power of x. It turns out that if the singular point of a differential equation is not "very bad" then the solutions follow a pattern similar to that of Example 1. That is, at least one solution is a power series multiplied by x^r. We devote the rest of this section to making this idea precise. For simplicity, we consider only series expansions about $x = 0$, although analogous methods apply for series expanded about an arbitrary $x = a$.

Suppose that $x = 0$ is a singular point of the equation

(2) $$y'' + p(x)y' + q(x)y = 0\,.$$

We say that $x = 0$ is a **regular singular point** of the equation if both of the functions $xp(x)$ and $x^2q(x)$ have power series expansions at $x = 0$. (This is what we mean by a singularity that is not "very bad.") The Frobenius method that we now describe applies to equations for which $x = 0$ is a regular singular point. For clarity's sake, we restrict our attention to the case $x > 0$. The case $x < 0$ is handled similarly (or can be reduced to the case $x > 0$ by the change of variables $t = -x$). Motivated by Example 1, we try a series solution of the form

$$(3) \qquad y = x^r \left(a_0 + a_1 x + a_2 x^2 + \cdots \right) = \sum_{m=0}^{\infty} a_m x^{r+m}$$

with $a_0 \neq 0$. Such a series is called a **Frobenius series**. We have

$$y' = \sum_{m=0}^{\infty} a_m (r+m) x^{r+m-1}$$

and

$$y'' = \sum_{m=0}^{\infty} a_m (r+m)(r+m-1) x^{r+m-2} .$$

(Observe that the index of summation begins at 0 in both these series. Why?) Thus (2) becomes

$$\sum_{m=0}^{\infty} a_m (r+m)(r+m-1) x^{r+m-2}$$

$$+ p(x) \sum_{m=0}^{\infty} a_m (r+m) x^{r+m-1} + q(x) \sum_{m=0}^{\infty} a_m x^{r+m} = 0 .$$

We factor x from the second series and x^2 from the third to make all exponents the same and get

$$(4) \quad \sum_{m=0}^{\infty} a_m (r+m)(r+m-1) x^{r+m-2}$$

$$+ x p(x) \sum_{m=0}^{\infty} a_m (r+m) x^{r+m-2} + x^2 q(x) \sum_{m=0}^{\infty} a_m x^{r+m-2} = 0 .$$

Since by assumption $x = 0$ is a regular singular point, the functions $xp(x)$ and $x^2 q(x)$ have power series expansions about 0, say

$$xp(x) = p_0 + p_1 x + p_2 x^2 + \cdots$$

and

$$x^2 q(x) = q_0 + q_1 x + q_2 x^2 + \cdots .$$

Substituting these into (4) gives

$$\sum_{m=0}^{\infty} a_m (r+m)(r+m-1)x^{r+m-2}$$

$$+ (p_0 + p_1 x + p_2 x^2 + \cdots) \sum_{m=0}^{\infty} a_m (r+m) x^{r+m-2}$$

$$+ (q_0 + q_1 x + q_2 x^2 + \cdots) \sum_{m=0}^{\infty} a_m x^{r+m-2} = 0.$$

The total coefficient of each power of x on the left side of this equation must be 0, since the right side is 0. The lowest power of x that appears in the equation is x^{r-2}. Its coefficient is $a_0 r(r-1) + p_0 a_0 r + q_0 a_0 = a_0 [r(r-1) + p_0 r + q_0] = 0$. Since $a_0 \neq 0$, r must be a root of the **indicial equation**

(5)
$$\boxed{r(r-1) + p_0 r + q_0 = 0.}$$

The roots of this equation are called the **indicial roots** and are denoted by r_1 and r_2 with the convention that $r_1 \geq r_2$ whenever they are real. Note that p_0 and q_0 are easily determined, since they are the values of $xp(x)$ and $x^2 q(x)$ at $x = 0$. Once we have determined r_1 and r_2, we substitute r_1 in (4) and solve for the unknown coefficients a_n as we would do with the power series method. This will determine a first solution of (2). Summing up, we have the following important result.

**THEOREM 1
FROBENIUS
METHOD FIRST
SOLUTION**

If $x = 0$ is a regular singular point of the equation

$$y'' + p(x)y' + q(x)y = 0,$$

then one solution is of the form

$$y_1 = |x|^{r_1} \left(a_0 + a_1 x + a_2 x^2 + \cdots \right), \qquad a_0 \neq 0,$$

where r_1 is a root of the indicial equation (5), with the convention that r_1 is the larger of the two roots when both roots are real.

For a proof of this theorem, see Chapter 9 of *Ordinary Differential Equations* by Garrett Birkhoff and Gian-Carlo Rota, 2nd ed., Wiley, 1969. It can be shown that if $xp(x)$ and $x^2 q(x)$ have power series expansions that converge for $|x| < R$, then the series solution will converge for $0 < |x| < R$.

Note that Theorem 1 says nothing about a second solution of (2). This solution may have one of three forms, as described in the following box. The proofs are presented in an appendix at the end of this section.

THEOREM 2
THE FROBENIUS
METHOD

Suppose that $x = 0$ is a regular singular point of the differential equation

$$y'' + p(x)y' + q(x)y = 0,$$

and let r_1 and r_2 denote the indicial roots. The differential equation has two linearly independent solutions y_1 and y_2, as we now describe.

Case 1. If $r_1 - r_2$ is not an integer, then

$$y_1 = |x|^{r_1} \sum_{m=0}^{\infty} a_m x^m, \quad y_2 = |x|^{r_2} \sum_{m=0}^{\infty} b_m x^m, \quad a_0 \neq 0 \text{ and } b_0 \neq 0.$$

Case 2. If $r = r_1 = r_2$, then

$$y_1 = |x|^{r} \sum_{m=0}^{\infty} a_m x^m, \quad y_2 = y_1 \ln|x| + |x|^{r} \sum_{m=1}^{\infty} b_m x^m, \quad a_0 \neq 0.$$

Case 3. If $r_1 - r_2$ is a positive integer, with $r_1 \geq r_2$, then

$$y_1 = |x|^{r_1} \sum_{m=0}^{\infty} a_m x^m, \quad y_2 = ky_1 \ln|x| + |x|^{r_2} \sum_{m=0}^{\infty} b_m x^m,$$

where $a_0 \neq 0$, $b_0 \neq 0$ (k may or may not be 0).

Thus we find two linearly independent Frobenius series solutions in Case 1 and also in Case 3 when $k = 0$. Otherwise, we can find only one Frobenius series solution, and any linearly independent solution must involve a logarithmic term. In computing y_1 and y_2 it is convenient to take $a_0 = 1$ and $b_0 = 1$; then the two arbitrary constants will appear when writing the general solution in the form of an arbitrary linear combination of y_1 and y_2.

EXAMPLE 2   Frobenius method: $r_1 - r_2$ is not an integer
Solve the differential equation

$$y'' + \frac{1}{2x}y' - \frac{1}{4x}y = 0, \quad x > 0.$$

Solution Theorem 1 applies, since both $xp(x) = \frac{1}{2}$ and $x^2 q(x) = -\frac{1}{4}x$ have power series expansions at $x = 0$. Moreover, $p_0 = \frac{1}{2}$ and $q_0 = 0$, so the indicial equation and indicial roots are

$$r(r - 1) + \tfrac{1}{2}r = 0,$$
$$r^2 - \tfrac{1}{2}r = 0,$$
$$r_1 = \tfrac{1}{2} \quad \text{and} \quad r_2 = 0.$$

(Following our convention, we choose $r_1 \geq r_2$.) Since $r_1 - r_2$ is not an integer, the solutions are given by Case I of the Frobenius method. The next step is to determine the unknown coefficients in the series solutions. It is more convenient to work with the differential equation if we multiply both sides by $4x$:

$$4xy'' + 2y' - y = 0.$$

Substituting $y = \sum_{m=0}^{\infty} a_m x^{m+r}$ into the equation gives

$$\sum_{m=0}^{\infty} 4a_m(m+r)(m+r-1)x^{m+r-1}$$

$$+ \sum_{m=0}^{\infty} 2a_m(m+r)x^{m+r-1} - \sum_{m=1}^{\infty} a_{m-1}x^{m+r-1} = 0.$$

(We have shifted the index in the third series to match all the exponents of x.) Adding the terms corresponding to $m = 0$ and setting the coefficient of x^{0+r-1} equal to 0 gives us the indicial equation

$$4a_0 r(r-1) + 2a_0 r = 0,$$

$$r^2 - \frac{1}{2}r = 0 \quad (\text{since } a_0 \neq 0).$$

Setting the sum of the coefficients of x^{m+r-1} equal to 0 gives us

(6) $$\qquad 4a_m(m+r)(m+r-1) + 2a_m(m+r) - a_{m-1} = 0.$$

At this point, we treat separately the case $r_1 = \frac{1}{2}$ and $r_2 = 0$. When $r = \frac{1}{2}$, the recurrence relation (6) becomes

$$4a_m\left(m + \frac{1}{2}\right)\left(m + \frac{1}{2} - 1\right) + 2a_m\left(m + \frac{1}{2}\right) - a_{m-1} = 0$$

$$a_m = \frac{1}{2m(2m+1)}a_{m-1}.$$

From this recurrence relation we get the pattern

$$a_1 = \frac{1}{2 \cdot 3}a_0 = \frac{1}{3!}a_0, \qquad a_4 = \frac{1}{8 \cdot 9}a_3 = \frac{1}{9!}a_0,$$

$$a_2 = \frac{1}{4 \cdot 5}a_1 = \frac{1}{5!}a_0, \qquad a_5 = \frac{1}{10 \cdot 11}a_4 = \frac{1}{11!}a_0,$$

$$a_3 = \frac{1}{6 \cdot 7}a_2 = \frac{1}{7!}a_0, \qquad \vdots$$

Thus $a_m = \frac{1}{(2m+1)!}a_0$, and hence the solution corresponding to $r = \frac{1}{2}$ and $a_0 = 1$ is

$$y_1 = \sum_{m=0}^{\infty} \frac{1}{(2m+1)!}x^{m+\frac{1}{2}}.$$

For $r = 0$ we get the recurrence relation (6):

$$4a_m m(m-1) + 2a_m m - a_{m-1} = 0$$

$$a_m = \frac{1}{2m(2m-1)}a_{m-1}.$$

This gives the following pattern for the coefficients

$$a_1 = \frac{1}{2 \cdot 1} a_0 = \frac{1}{2!} a_0, \qquad a_4 = \frac{1}{8 \cdot 7} a_3 = \frac{1}{8!} a_0,$$

$$a_2 = \frac{1}{4 \cdot 3} a_1 = \frac{1}{4!} a_0, \qquad a_5 = \frac{1}{10 \cdot 9} a_4 = \frac{1}{10!} a_0,$$

$$a_3 = \frac{1}{6 \cdot 5} a_2 = \frac{1}{6!} a_0, \qquad \vdots$$

It is clear that $a_m = \frac{1}{(2m)!} a_0$, and hence the solution corresponding to $r = 0$ and $a_0 = 1$ is

$$y_2 = \sum_{m=0}^{\infty} \frac{1}{(2m)!} x^m .$$

So we obtain the general solution $y = c_1 y_1 + c_2 y_2$, where c_1 and c_2 are arbitrary constants.   ∎

The solutions in Example 2 can be simplified as

$$y_1 = \sinh \sqrt{x} \quad \text{and} \quad y_2 = \cosh \sqrt{x}, \quad x > 0 .$$

Do you see why? [Hint: Write x^m as $(\sqrt{x})^{2m}$.] We can verify the validity of these solutions by plugging them into the differential equation. In Figure 2 we compared the graphs $\cosh \sqrt{x}$ and the partial sums of y_2. Notice how the partial sums of y_2 approximate the real solution. With the aid of a computer system, you can do the same for y_1.

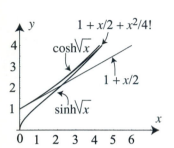

Figure 2 Solutions and partial sum approximation from Example 2.

EXAMPLE 3   Case of a double root $r = r_1 = r_2$

Solve the differential equation

(7) $$x^2 y'' + 3xy' + (1 - x)y = 0, \quad x > 0 .$$

Solution We easily verify that $p_0 = 3$, $q_0 = 1$, and so the indicial equation is $r^2 + 2r + 1 = 0$. We have a double root $r = -1$, and so we are in Case 2. Let $y_1 = \sum_{m=0}^{\infty} a_m x^{m-1}$ with $a_0 \neq 0$. Substituting in (7), after some computations as in Example 2, we arrive at the recurrence relation

$$a_{m+1} = \frac{a_m}{(m+1)^2} .$$

Taking $a_0 = 1$, we get the first solution

$$y_1 = \sum_{m=0}^{\infty} \frac{1}{(m!)^2} x^{m-1} = \frac{1}{x} \left(1 + x + \frac{x^2}{4} + \frac{x^3}{36} + \frac{x^4}{576} + \cdots \right) .$$

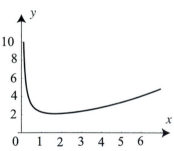

Figure 3 Approximation of y_1 near 0. Notice the blow up due to the term $\frac{1}{x}$.

(An approximation of y_1 is shown in Figure 3, using a partial sum of the infinite series. The details of the derivation of y_1 are left to Exercise 30.) According to Case 2, we try for a second solution

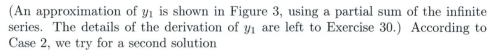

$$y = y_1 \ln x + \sum_{m=1}^{\infty} b_m x^{m-1} .$$

Substituting in (7) and simplifying, we get

(8)
$$\overbrace{[x^2 y_1'' + 3xy_1' + (1-x)y_1]}^{=\,0} \ln x + (2xy_1' + 2y_1)$$
$$+ \sum_{m=1}^{\infty} b_m(m-1)(m-2)x^{m-1} + \sum_{m=1}^{\infty} 3b_m(m-1)x^{m-1}$$
$$+ \sum_{m=1}^{\infty} b_m x^{m-1} - \sum_{m=2}^{\infty} b_{m-1} x^{m-1} = 0 \,.$$

The coefficient of $\ln x$ in (8) is 0, since y_1 is a solution of (7).   From the series representation of y_1 we see that

$$
\begin{aligned}
2xy_1' + 2y_1 &= 2\left[\sum_{m=0}^{\infty} \frac{m-1}{(m!)^2} x^{m-1} + \sum_{m=0}^{\infty} \frac{1}{(m!)^2} x^{m-1} \right] \\
&= \sum_{m=0}^{\infty} \frac{2m}{(m!)^2} x^{m-1} = \sum_{m=1}^{\infty} \frac{2}{(m-1)!m!} x^{m-1} \,.
\end{aligned}
$$

After substituting this into (8) and combining terms with like powers, we arrive at the equation

$$(2 + b_1) + \sum_{m=2}^{\infty} \left(\frac{2}{(m-1)!m!} + m^2 b_m - b_{m-1} \right) x^{m-1} = 0 \,.$$

Thus $b_1 = -2$, and the recurrence relation for the b_m's is

$$b_m = \frac{1}{m^2} \left(b_{m-1} - \frac{2}{(m-1)!m!} \right) \,.$$

From this relation we can calculate as many coefficients as we wish, although it would be difficult to find a closed formula for b_m in terms of m.   Using the first five coefficients, we have

$$y_2 = \ln x \sum_{m=0}^{\infty} \frac{1}{(m!)^2} x^{m-1} - \left(2 + \frac{3}{4} x + \frac{11}{108} x^2 + \frac{25}{3,456} x^3 + \frac{137}{432,000} x^4 + \cdots \right) \,.$$

Hence the general solution of (7) is

$$
\begin{aligned}
y &= c_1 y_1 + c_2 y_2 \\
&= (c_1 + c_2 \ln x) \sum_{m=0}^{\infty} \frac{1}{(m!)^2} x^{m-1} \\
&\quad - c_2 \left(2 + \frac{3}{4} x + \frac{11}{108} x^2 + \frac{25}{3,456} x^3 + \frac{137}{432,000} x^4 + \cdots \right) \,,
\end{aligned}
$$

where c_1 and c_2 are arbitrary constants.   ∎

EXAMPLE 4   Case $r_1 - r_2$ is an integer

Solve the differential equation

$$(9) \qquad x^2 y'' - xy' + (x - 3)y = 0, \quad x > 0.$$

Solution We have $p_0 = -1$, $q_0 = -3$, and so the indicial equation is $r^2 - 2r - 3 = 0$. The indicial roots are $r_1 = 3$ and $r_2 = -1$. We are in Case 3. For a first solution we try $y_1 = \sum_{m=0}^{\infty} a_m x^{m+3}$. Substituting in (9) and solving for the coefficients a_m, we arrive at the recurrence relation

$$a_m = \frac{-1}{m^2 + 4m} a_{m-1},$$

from which we get

$$(10) \qquad y_1 = a_0 x^3 \left(1 - \frac{1}{5} x + \frac{1}{60} x^2 - \frac{1}{1,260} x^3 + \frac{1}{40,320} x^4 - \cdots \right).$$

The details are left as an exercise (Exercise 31). The second solution is of the form

$$y = k y_1 \ln x + \sum_{m=0}^{\infty} b_m x^{m-1},$$

where k may or may not be 0. We substitute this in (9), simplify, use the fact that y_1 is a solution, and get

$$\overbrace{[x^2 y_1'' - x y_1' + (x - 3)y_1]}^{= 0} \ln x + 2k(x y_1' - y_1)$$

$$+ \sum_{m=0}^{\infty} b_m (m - 1)(m - 2)x^{m-1} - \sum_{m=0}^{\infty} b_m (m - 1)x^{m-1}$$

$$- \sum_{m=0}^{\infty} 3 b_m x^{m-1} + \sum_{m=1}^{\infty} b_{m-1} x^{m-1} = 0.$$

The lowest power of x appearing in the term $2k(x y_1' - y_1)$ is x^3, so we obtain the following equation upon simplification:

$$(11) \qquad (b_0 - 3b_1) + (b_1 - 4b_2)x + (b_2 - 3b_3)x^2 + 2k(x y_1' - y_1)$$

$$+ \sum_{m=4}^{\infty} [b_m (m^2 - 4m) + b_{m-1}]x^{m-1} = 0.$$

Taking $b_0 = 1$, setting the first three coefficients equal 0, and solving, we obtain

$$b_1 = \frac{1}{3}, \quad b_2 = \frac{1}{12}, \quad b_3 = \frac{1}{36}.$$

Now from (10) it follows that

$$2k(x y_1' - y_1) = 2k \left(2x^3 - \frac{3}{5} x^4 + \frac{1}{15} x^5 - \frac{1}{252} x^6 + \cdots \right),$$

and so, setting the coefficient of x^3 in (11) equal to 0, we get

$$4k + b_3 = 0.$$

Thus $k = -\frac{1}{4}b_3 = -\frac{1}{144}$. We can now continue to use (11) to find as many of the coefficients b_4, b_5, ... as we wish, by setting the coefficients of x^4, x^5, ... in turn equal to zero.   Omitting the details, we find that b_4 is arbitrary (so we choose $b_4 = 1$), and so $b_5 = -\frac{121}{600}$.   Hence

$$y_2 = -\frac{1}{144}\, y_1 \ln x + x^{-1}\left(1 + \frac{1}{3}\,x + \frac{1}{12}\,x^2 + \frac{1}{36}\,x^3 + x^4 - \frac{121}{600}\,x^5 + \cdots\right).$$

So the general solution of (9) is

$$
\begin{aligned}
y \;=\;& c_1 y_1 + c_2 y_2 \\
\;=\;& x^3\left(c_1 - c_2\,\frac{\ln x}{144}\right)\left(1 - \frac{1}{5}\,x + \frac{1}{60}\,x^2 - \frac{1}{1,260}\,x^3 + \frac{1}{4,032}\,x^4 - \cdots\right) \\
& + c_2 x^{-1}\left(1 + \frac{1}{3}\,x + \frac{1}{12}\,x^2 + \frac{1}{36}\,x^3 + x^4 - \frac{121}{600}\,x^5 + \cdots\right),
\end{aligned}
$$

where c_1 and c_2 are arbitrary constants.   ∎

In the preceding example the second solution y_2 contained a logarithmic term, since the constant k turned out to be nonzero.   On the other hand, Example 1 deals with a Case 3 equation in which this does not happen. Indeed, the indicial equation is $r(r - 1) + 2r = 0$ with roots $r_1 = 0$ and $r_2 = -1$ that differ by a positive integer, and the solutions of the equation, $y_1 = \frac{\sin x}{x}$ and $y_2 = \frac{\cos x}{x}$, do not contain a logarithmic term.

Use of the Reduction of Order Formula

In all three cases of the Frobenius method, once we have found y_1, a second solution can be derived by using the reduction of order formula

$$(12) \qquad\qquad y_2(x) = y_1(x) \int \frac{e^{-\int p(x)\,dx}}{y_1^2(x)}\,dx.$$

Thus Theorem 1 and the reduction of order formula are all that is needed to solve any Frobenius-type problem.   To get the series for y_2, we expand the integrand and integrate term by term. This process is illustrated by our next example and is justified in the appendix of this section. The task of expanding is greatly simplified with the help of a computer system.

EXAMPLE 5   The reduction of order formula
Find the general solution of the differential equation

$$x^2 y'' + x y' + (x^2 - 1)y = 0, \quad x > 0.$$

Solution The indicial equation is $r^2 - 1 = 0$ with indicial roots $r_1 = 1$ and $r_2 = -1$. (Note that the indicial roots differ by an integer, so we cannot assume at the outset that the solutions are both of the form (3).) By Theorem 1, one solution is of the form $y_1 = x^{r_1} \sum_{m=0}^{\infty} a_m x^m$ with $r_1 = 1$. Plugging in the differential equation and solving for the coefficients as in the previous examples, we find

$$y_1 = x \left(1 - \frac{x^2}{8} + \frac{x^4}{192} - \frac{x^6}{9,216} + \frac{x^8}{737,280} - \frac{x^{10}}{88,473,600} + \cdots \right)$$

(Exercise 32). We find a second linearly independent solution with the help of the reduction of order formula. Putting y_1 in (12) we get

$$
\begin{aligned}
y_2(x) &= y_1 \int \frac{e^{-\int 1/x\,dx}}{y_1^2}\,dx \qquad \left(e^{-\int 1/x\,dx} = \frac{1}{x} \right) \\[2mm]
&= y_1 \int \frac{1}{x^3 \left(1 - \frac{x^2}{8} + \frac{x^4}{192} - \frac{x^6}{9,216} + \frac{x^8}{737,280} - \frac{x^{10}}{88,473,600} + \cdots \right)^2}\,dx \\[2mm]
&= y_1 \int \frac{1}{x^3 \left(1 - \frac{x^2}{4} + \frac{5\,x^4}{192} - \frac{7\,x^6}{4,608} + \frac{7\,x^8}{122,880} - \frac{11\,x^{10}}{7,372,800} + \cdots \right)}\,dx
\end{aligned}
$$

(Squaring)

$$= y_1 \int \left(\frac{1}{x^3} + \frac{1}{4x} + \frac{7\,x}{192} + \frac{19\,x^3}{4,608} + \frac{149\,x^5}{368,640} + \frac{803\,x^7}{22,118,400} + \cdots \right) dx$$

(Series expansion or long division)

$$= y_1 \left(\frac{-1}{2\,x^2} + \frac{1}{4}\ln x + \frac{7x^2}{384} + \frac{19\,x^4}{18,432} + \frac{149\,x^6}{2,211,840} + \frac{803\,x^8}{176,947,200} + \cdots \right)$$

(Integrate term by term)

$$= y_1 \frac{\ln x}{4} + y_1 \left(\frac{-1}{2x^2} + \frac{7x^2}{384} + \frac{19x^4}{18,432} + \frac{149x^6}{2,211,840} + \frac{803x^8}{176,947,200} + \cdots \right).$$

This determines a second linearly independent solution. ∎

Example 5 shows that the reduction of order formula is a handy tool for checking for the presence of a logarithmic term in the solution.

Appendix: Further Proofs Related to the Frobenius Method

In this appendix we use the reduction of order formula to derive the second solution in the Frobenius problems, given the first one by Theorem 1. We restrict our attention to the case $x > 0$.

Recall that $xp(x)$ has a power series at $x = 0$. Thus we have

$$
\begin{aligned}
xp(x) &= p_0 + p_1 x + p_2 x^2 + \cdots, \\
p(x) &= \frac{p_0}{x} + p_1 + p_2 x + \cdots.
\end{aligned}
$$

We denote the function $p_1 + p_2 x + \cdots$ by $a(x)$ (note that $a(x)$ has a power series expansion at $x = 0$), and so

$$p(x) = \frac{p_0}{x} + a(x).$$

Integrating and taking the exponential, we get

$$(13) \qquad e^{-\int p(x)\,dx} = x^{-p_0} G(x),$$

where G is the function $e^{-\int a(x)\,dx}$, and, in particular, $G(0) \neq 0$. (Note that G has a power series expansion at $x = 0$.) Now, using Theorem 1 we can write

$$(14) \qquad \frac{1}{y_1^2} = \frac{1}{x^{2r_1}} H(x)$$

where $H(x) = 1 \big/ \left(\sum_{m=0}^{\infty} a_m x^m \right)^2$. Since $a_0 \neq 0$, the function $H(x)$ has a power series expansion at 0 and $H(0) \neq 0$. Combining (12), (13), and (14), we find that

$$(15) \qquad y_2 = y_1 \int x^{-p_0 - 2r_1} K(x)\,dx,$$

where $K(x)$ has a power series expansion about 0, and $K(0) \neq 0$. We write

$$K(x) = k_0 + k_1 x + k_2 x^2 + \cdots, \qquad k_0 \neq 0.$$

At this point, we establish a connection between r_1, r_2, p_0. Since r_1 and r_2 are the roots of equation (5), we have $r_1 + r_2 = 1 - p_0$. Thus (15) becomes

$$(16) \qquad y_2 = y_1 \int x^{-(r_1 - r_2 + 1)} (k_0 + k_1 x + k_2 x^2 + \cdots)\,dx, \qquad k_0 \neq 0.$$

Note that a logarithmic term in y_2 will appear if and only if a term in x^{-1} occurs in the integrand (after multiplying through by $x^{-(r_1 - r_2 + 1)}$). This is why $r_1 - r_2$ determines the nature of the second solution.

Case 1: If $r_1 - r_2$ is not an integer, clearly there can be no term in x^{-1}, and hence no logarithmic term in y_2.

Case 2: If $r_1 = r_2$, then the very first term yields the logarithmic term in y_2, since $k_0 \neq 0$.

Case 3: If $r_1 - r_2$ is a positive integer, say $r_1 - r_2 = n$, we may or may not have a logarithmic term, depending on whether k_n is 0.

Exercises A.5

In Exercises 1–6, decide whether $x = 0$ is an ordinary point or a singular point. In case it is a singular point, determine if it is a regular singular point.

1. $y'' + (1 - x^2)y' + xy = 0$.

2. $xy'' + \sin x\, y' + \frac{1}{x} y = 0$.

3. $x^3 y'' + x^2 y' + y = 0$.

4. $\sin x\, y'' + y' + \frac{1}{x} y = 0$.

5. $x^2 y'' + (1 - e^x)y' + xy = 0$.

6. $3\,xy'' + 2\,y' - \frac{1}{3x} y = 0$.

In Exercises 7–22, (a) check that $x = 0$ is a regular singular point. (b) Determine which case of the Frobenius method applies. (c) Determine at least three nonzero terms in each of two linearly independent series solutions. Take $x > 0$.

7. $4\,xy'' + 6\,y' + y = 0$.

8. $4\,xy'' + 6\,y' - y = 0$.

9. $4\,x^2\,y'' - 14\,x\,y' + (20 - x)\,y = 0$.

10. $4x\,y'' + 2\,y' + y = 0$.

11. $2\,x\,y'' + (1 + x)\,y' + y = 0$.

12. $y'' - \frac{1}{2x}\,y' + \frac{1}{x}\,y = 0$.

13. $x\,y'' + (1-x)\,y' + y = 0.$

14. $x\,y'' + 2\,y' - x\,y = 0.$

15. $x\,y'' + 2\,(1+x)\,y' + (x+2)\,y = 0.$

16. $x^2\,y'' + x\,y' + (x^2 - \frac{1}{4})y = 0.$

17. $x^2 y'' + 4\,x\,y' + (2-x^2)\,y = 0.$

18. $x\,y'' + (x+2)\,y' + (x+1)\,y = 0.$

19. $x\,y'' + 3\,y' + \frac{1+x}{x}\,y = 0.$

20. $x^2\,y'' + 4\,x\,y' + (2+x^2)\,y = 0.$

21. $x\,y'' + y' - \frac{1+x}{x}y = 0.$

22. $4\,x\,y'' + 2\,y' + x\,y = 0.$

In Exercises 23–28, determine which case of the Frobenius method applies, and then solve the equation using Theorem 1 and reduction of order as we did in Example 5. Does your second solution contain a logarithmic term?

23. $4\,x^2\,y'' - 14\,x\,y' + (20+x)\,y = 0, \quad x > 0.$

24. $y'' - \frac{1+x^2}{2x}\,y' + \frac{1}{2x^2}\,y = 0, \quad x > 0.$

25. $y'' + \frac{1}{x}\,y' + \frac{x-1}{x^2}\,y = 0, \quad x > 0.$

26. $y'' - \frac{1+x}{2\,x}\,y' + \frac{1+2x}{2x^2}\,y = 0, \; x > 0.$

27. $x(1-x)\,y'' + (1-3x)\,y' - y = 0, \quad 0 < x < 1.$

28. $x(1-x)\,y'' - \frac{1}{4}\,y = 0, \quad 0 < x < 1.$

29. (a) Make the change of variables $u = xy$ in the differential equation of Example 1 and show that it becomes $u'' + u = 0.$
(b) Use (a) to derive the solutions $y_1 = \frac{\sin x}{x}$ and $y_2 = \frac{\cos x}{x}$.

30. Derive the first solution in Example 3.

31. Supply the details leading to the first solution in Example 4.

32. Supply the details leading to the first solution in Example 5.

Euler's equation can be used to illustrate the solutions in all three cases of the Frobenius method. In Exercises 33–36, (a) show that $x = 0$ is a regular singular point, and determine which case of the Frobenius method applies. (b) Find two linearly independent solutions, using the methods of Section A.3.

33. $y'' + \frac{1}{x}y' - \frac{2}{x^2}\,y = 0.$

34. $y'' + \frac{2}{x}\,y' - \frac{3}{x^2}\,y = 0.$

35. $y'' + \frac{1}{x}\,y' + \frac{1}{9\,x^2}\,y = 0.$

36. $y'' + \frac{1}{2x}\,y' + \frac{1}{2x^2}\,y = 0.$

37. (a) Show that $x = 0$ is a regular singular point of the differential equation

$$(\sin x)^2 y'' + \tan x\,y' + (\cos x)^2 y = 0.$$

(b) Find the indicial roots and determine which case of the Frobenius method applies.
(c) Verify that $y_1 = \sin(\ln(\sin x))$ and $y_2 = \cos(\ln(\sin x))$ are solutions of the differential equation for $0 < x < R$, where R is a positive number. How large can R be?
(d) Make a change of variables $u = \sin x$ and transform the given equation to an Euler equation. Then derive the solutions given in (c).
(e) Plot the graphs of the solutions. Explain the behavior of the graphs near 0 and π.

38. Consider the equation

$$x\,y'' - (2+x)\,y' + 2\,y = 0, \quad x > 0.$$

(a) Show that $x = 0$ is a regular singular point. Find the indicial equation and the indicial roots, and conclude that we are in Case 3 of the Frobenius method.

(b) Use the method of Frobenius to find a Frobenius series solution corresponding to the larger root r_1.

(c) Identify the solution in (b) as $3!(e^x - 1 - x - \frac{x^2}{2})$.

(d) Even though the roots differ by an integer, there exists a second Frobenius series solution. Using the method of Frobenius, show that e^x or $1 + x + x^2/2$ can be taken as a second solution. [Hint: Argue that the coefficient b_3 is arbitrary.]

39. Consider the equation

$$xy'' - (n + x)\, y' + n\, y = 0, \quad x > 0,$$

where n is a nonnegative integer.

(a) Show that $x = 0$ is a regular singular point. Find the indicial equation and the indicial roots, and conclude that we are in Case 3 of the Frobenius method.

(b) Use the method of Frobenius to find a Frobenius series solution corresponding to the larger root r_1.

(c) Identify the solution in (b) as

$$(n + 1)!(e^x - 1 - x - \frac{x^2}{2} - \cdots - \frac{x^n}{n!}).$$

(d) Using the method of Frobenius, show that e^x or $1 + x + x^2/2 + \cdots + \frac{x^n}{n!}$ can be taken as a second solution. [Hint: Argue that the coefficient b_{n+1} is arbitrary.]

Table of Fourier Transforms

$$f(x) = \frac{1}{\sqrt{2\pi}} \int_{-\infty}^{\infty} \widehat{f}(\omega)e^{ix\omega}\, d\omega, \quad -\infty < x < \infty \qquad \widehat{f}(\omega) = \mathcal{F}(f)(\omega) = \frac{1}{\sqrt{2\pi}} \int_{-\infty}^{\infty} f(x)e^{-i\omega x}\, dx,$$
$$-\infty < \omega < \infty$$

1. $\begin{cases} 1 & \text{if } |x| < a \\ 0 & \text{if } |x| > a \end{cases}$ $\qquad\qquad \sqrt{\dfrac{2}{\pi}}\dfrac{\sin a\omega}{\omega}$

2. $\begin{cases} 1 & \text{if } a < x < b \\ 0 & \text{otherwise} \end{cases}$ $\qquad\qquad \dfrac{i\left(e^{-ib\omega} - e^{ia\omega}\right)}{\sqrt{2\pi}\,\omega}$

3. $\begin{cases} 1 - \frac{|x|}{a} & \text{if } |x| < a \\ 0 & \text{if } |x| > a \end{cases} \quad a > 0$ $\qquad 2\sqrt{\dfrac{2}{\pi}}\dfrac{\sin^2(\frac{a\omega}{2})}{a\omega^2}$

4. $\begin{cases} x & \text{if } |x| < a \\ 0 & \text{if } |x| > a \end{cases} \quad a > 0$ $\qquad i\sqrt{\dfrac{2}{\pi}}\dfrac{a\omega\cos(a\omega) - \sin(a\omega)}{\omega^2}$

5. $\begin{cases} \sin x & \text{if } |x| < \pi \\ 0 & \text{if } |x| > \pi \end{cases}$ $\qquad\qquad i\sqrt{\dfrac{2}{\pi}}\dfrac{\sin(\pi\omega)}{\omega^2 - 1}$

6. $\begin{cases} \sin(ax) & \text{if } |x| < b \\ 0 & \text{if } |x| > b \end{cases} \quad a,b > 0$ $\qquad i\sqrt{\dfrac{2}{\pi}}\dfrac{\omega\cos(b\omega)\sin(ab) - a\cos(ab)\sin(b\omega)}{\omega^2 - a^2}$

7. $\dfrac{1}{a^2 + x^2}, \ a > 0$ $\qquad\qquad\qquad \sqrt{\dfrac{\pi}{2}}\dfrac{e^{-a|\omega|}}{a}$

8. $\dfrac{x}{a^2 + x^2}, \ a > 0$ $\qquad\qquad\qquad -i\sqrt{\dfrac{\pi}{2}}\,\mathrm{sgn}\,\omega\, e^{-a|\omega|}$

9. $\sqrt{\dfrac{2}{\pi}}\dfrac{a}{1 + a^2x^2}, \ a > 0$ $\qquad\qquad e^{-\frac{|\omega|}{a}}$

10. $\dfrac{\sin ax}{x}, \ a > 0$ $\qquad\qquad\qquad \begin{cases} \sqrt{\frac{\pi}{2}} & \text{if } |\omega| < a \\ \frac{1}{2}\sqrt{\frac{\pi}{2}} & \text{if } |\omega| = a \\ 0 & \text{if } |\omega| > a \end{cases}$

11. $\dfrac{4}{\sqrt{2\pi}}\dfrac{\sin^2(\frac{1}{2}ax)}{ax^2}, \ a > 0$ $\qquad\qquad \begin{cases} 1 - \frac{|\omega|}{a} & \text{if } |\omega| < a \\ 0 & \text{if } |\omega| > a \end{cases}$

12. $\dfrac{4}{\sqrt{2\pi}}\dfrac{\sin^2(ax) - \sin^2(\frac{1}{2}ax)}{ax^2}, \ a > 0$ $\qquad \begin{cases} 1 & \text{if } |x| < a \\ (-x + 2a)/a & \text{if } a < x < 2a \\ (x + 2a)/a & \text{if } a < x < 2a \\ 0 & \text{if } |x| > 2a \end{cases}$

13. $e^{-a|x|}, \ a > 0$ $\qquad\qquad\qquad\qquad \sqrt{\dfrac{2}{\pi}}\dfrac{a}{a^2 + \omega^2}$

14. $\begin{cases} e^{-ax} & \text{if } x > 0 \\ 0 & \text{if } x < 0 \end{cases}, \ a > 0$ $\qquad \dfrac{1}{\sqrt{2\pi}}\dfrac{1}{a + i\omega}$

15. $\begin{cases} 0 & \text{if } x > 0 \\ e^{ax} & \text{if } x < 0 \end{cases}, \ a > 0$ $\qquad \dfrac{1}{\sqrt{2\pi}}\dfrac{1}{a - i\omega}$

16. $|x|^n e^{-a|x|}, \ a > 0, \ n > 0$ $\qquad \dfrac{\Gamma(n+1)}{\sqrt{2\pi}}\left(\dfrac{1}{(a - i\omega)^{1+n}} + \dfrac{1}{(a + i\omega)^{1+n}}\right)$

Table of Fourier Transforms (continued)

$f(x) = \frac{1}{\sqrt{2\pi}} \int_{-\infty}^{\infty} \hat{f}(\omega) e^{ix\omega}\, d\omega, \ -\infty < x < \infty$	$\hat{f}(\omega) = \mathcal{F}(f)(\omega) = \frac{1}{\sqrt{2\pi}} \int_{-\infty}^{\infty} f(x) e^{-i\omega x}\, dx,$ $-\infty < \omega < \infty$
17. $e^{-\frac{a}{2}x^2}, \ a > 0$	$\dfrac{1}{\sqrt{a}} e^{-\frac{\omega^2}{2a}}$
18. $e^{-ax^2}, \ a > 0$	$\dfrac{1}{\sqrt{2a}} e^{-\frac{\omega^2}{4a}}$
19. $xe^{-\frac{a}{2}x^2}, \ a > 0$	$\dfrac{-i\omega}{a^{3/2}} e^{-\frac{\omega^2}{2a}}$
20. $x^2 e^{-\frac{a}{2}x^2}, \ a > 0$	$\dfrac{a-\omega^2}{a^{5/2}} e^{-\frac{\omega^2}{2a}}$
21. $x^3 e^{-\frac{a}{2}x^2}, \ a > 0$	$\dfrac{-i\omega(3a-\omega^2)}{a^{7/2}} e^{-\frac{\omega^2}{2a}}$
22. $e^{-\frac{x^2}{2}} H_n(x),$ $H_n, \ n$th Hermite polynomial	$(-1)^n i^n e^{\frac{-\omega^2}{2}} H_n(\omega)$
23. $J_0(x)$, Bessel function of order 0	$\begin{cases} \sqrt{\frac{2}{\pi}} \frac{1}{\sqrt{1-\omega^2}} & \text{if } \lvert\omega\rvert < 1 \\ 0 & \text{if } \lvert\omega\rvert > 1 \end{cases}$
24. $J_n(x)$, Bessel function of order $n \geq 0$	$\begin{cases} \sqrt{\frac{2}{\pi}} \frac{(-i)^n}{\sqrt{1-\omega^2}} T_n(\omega) & \text{if } \lvert\omega\rvert < 1 \\ 0 & \text{if } \lvert\omega\rvert > 1 \end{cases}$ T_n, Chebyshev polynomial of degree n.

Special Transforms

25. $\mathcal{F}(\delta_0(x))(\omega) = \dfrac{1}{\sqrt{2\pi}}$

26 $\mathcal{F}(\delta_0(x-a))(\omega) = \dfrac{1}{\sqrt{2\pi}} e^{-ia\omega}$

27. $\mathcal{F}(\sqrt{\frac{2}{\pi}} \frac{1}{x})(\omega) = -i\,\mathrm{sgn}\,\omega$

28. $\mathcal{F}(e^{iax})(\omega) = \sqrt{2\pi}\,\delta_0(\omega - a)$

Operational Properties

29. $\mathcal{F}(af+bg)(\omega) = a\mathcal{F}(f) + b\mathcal{F}(g)$

30. $\mathcal{F}(f')(\omega) = i\omega\mathcal{F}(f)(\omega)$

31. $\mathcal{F}(f'')(\omega) = -\omega^2\mathcal{F}(f)(\omega)$

32. $\mathcal{F}(f^{(n)})(\omega) = (i\omega)^n\mathcal{F}(f)(\omega)$

33. $\mathcal{F}(xf(x))(\omega) = i\frac{d}{d\omega}\mathcal{F}(f)(\omega)$

34. $\mathcal{F}(x^n f(x))(\omega) = i^n\frac{d^n}{d\omega^n}\mathcal{F}(f)(\omega)$

35. $\mathcal{F}(f*g)(\omega) = \mathcal{F}(f)(\omega)\mathcal{F}(g)(\omega)$

36. $\mathcal{F}(f\,g)(\omega) = \mathcal{F}(f) * \mathcal{F}(g)(\omega)$

37. $\mathcal{F}(f(x-a))(\omega) = e^{-ia\omega}\mathcal{F}(f)(\omega)$

38. $\mathcal{F}(e^{iax}f(x))(\omega) = \mathcal{F}(f)(\omega - a)$

39. $\mathcal{F}(\cos(ax)f(x))(\omega) = \frac{\mathcal{F}(f)(\omega-a)+\mathcal{F}(f)(\omega+a)}{2}$

40. $\mathcal{F}(\sin(ax)f(x))(\omega) = \frac{\mathcal{F}(f)(\omega-a)-\mathcal{F}(f)(\omega+a)}{2i}$

41. $\mathcal{F}(f(ax))(\omega) = \frac{1}{\lvert a\rvert}\mathcal{F}(f)(\frac{\omega}{a}), \ a \neq 0$

42. $f(x) = \mathcal{F}(\mathcal{F}(\hat{f}))(-x), \ \mathcal{F}(\mathcal{F}(f)) = f(-x)$

Table of Fourier Cosine Transforms

$$f(x) = \sqrt{\frac{2}{\pi}} \int_0^\infty \mathcal{F}_c(f)(\omega) \cos \omega x \, d\omega, \qquad \mathcal{F}_c(f)(\omega) = \widehat{f}_c(\omega) = \sqrt{\frac{2}{\pi}} \int_0^\infty f(x) \cos \omega x \, dx,$$

$$0 < x < \infty \qquad\qquad\qquad 0 < \omega < \infty$$

1.	$\begin{cases} 1 & \text{if } 0 < x < a \\ 0 & \text{otherwise} \end{cases}$	$\sqrt{\frac{2}{\pi}} \dfrac{\sin a\omega}{\omega}$
2.	$e^{-ax}, \quad a > 0$	$\sqrt{\frac{2}{\pi}} \dfrac{a}{a^2 + \omega^2}$
3.	$x e^{-ax}, \quad a > 0$	$\sqrt{\frac{2}{\pi}} \dfrac{a^2 - \omega^2}{(a^2 + \omega^2)^2}$
4.	$e^{-a x^2/2}, \quad a > 0$	$\dfrac{1}{\sqrt{a}} e^{-\omega^2/2a}$
5.	$\cos ax \, e^{-ax}, \quad a > 0$	$\sqrt{\frac{2}{\pi}} \dfrac{a\omega^2 + 2a^3}{4a^4 + \omega^4}$
6.	$\sin ax \, e^{-ax}, \quad a > 0$	$\sqrt{\frac{2}{\pi}} \dfrac{2a^3 - a\omega^2}{4a^4 + \omega^4}$
7.	$\dfrac{a}{a^2 + x^2}, \quad a > 0$	$\sqrt{\frac{\pi}{2}} e^{-a\omega}$
8.	$x^p, \quad 0 < p < 1$	$\sqrt{\frac{2}{\pi}} \dfrac{\Gamma(p) \cos(p\omega/2)}{\omega^p}$
9.	$\begin{cases} \cos x & \text{if } 0 < x < a \\ 0 & \text{otherwise} \end{cases}$	$\dfrac{1}{\sqrt{2\pi}} \left[\dfrac{\sin a(1-\omega)}{1-\omega} + \dfrac{\sin a(1+\omega)}{1+\omega} \right]$

Operational Properties

10.	$\alpha f(x) + \beta g(x)$	$\alpha \mathcal{F}_c(f)(\omega) + \beta \mathcal{F}_c(g)(\omega)$
11.	$f(ax), \quad a > 0$	$\dfrac{1}{a} \widehat{f}_c\!\left(\dfrac{\omega}{a}\right)$
12.	$f'(x)$	$\omega \widehat{f}_s(\omega) - \sqrt{\frac{2}{\pi}} f(0)$
13.	$f''(x)$	$-\omega^2 \widehat{f}_c(\omega) - \sqrt{\frac{2}{\pi}} f'(0)$
14.	$x f(x)$	$\left[\widehat{f}_s\right]'(\omega)$
15.	$\mathcal{F}_c(\mathcal{F}_c f)$	f

Table of Fourier Sine Transforms

$$f(x) = \sqrt{\frac{2}{\pi}} \int_0^\infty \mathcal{F}_s(f)(\omega) \sin \omega x \, d\omega, \qquad \mathcal{F}_s(f)(\omega) = \widehat{f}_s(\omega) = \sqrt{\frac{2}{\pi}} \int_0^\infty f(x) \sin \omega x \, dx,$$

$$0 < x < \infty \qquad\qquad\qquad\qquad 0 < \omega < \infty$$

1.	$\begin{cases} 1 & \text{if } 0 < x < a \\ 0 & \text{otherwise} \end{cases}$	$\sqrt{\frac{2}{\pi}} \dfrac{1 - \cos a\omega}{\omega}$
2.	$e^{-ax}, \quad a > 0$	$\sqrt{\frac{2}{\pi}} \dfrac{\omega}{a^2 + \omega^2}$
3.	$x\, e^{-ax}, \quad a > 0$	$\sqrt{\frac{2}{\pi}} \dfrac{2a\omega}{(a^2 + \omega^2)^2}$
4.	$\dfrac{e^{-ax}}{x}, \quad a > 0$	$\sqrt{\frac{2}{\pi}} \tan^{-1} \dfrac{\omega}{a}$
5.	$\frac{1}{2} x\, e^{-a x^2}, \quad a > 0$	$\dfrac{\omega}{a^{3/2}} e^{-\omega^2/2a}$
6.	$\cos ax\, e^{-ax}, \quad a > 0$	$\sqrt{\frac{2}{\pi}} \dfrac{\omega^3}{4a^4 + \omega^4}$
7.	$\sin ax\, e^{-ax}, \quad a > 0$	$\sqrt{\frac{2}{\pi}} \dfrac{2a^2 \omega}{4a^4 + \omega^4}$
8.	$\dfrac{x}{a^2 + x^2}, \quad a > 0$	$\sqrt{\frac{\pi}{2}} e^{-a\omega}$
9.	$x^{p-1}, \quad 0 < p < 1$	$\sqrt{\frac{2}{\pi}} \dfrac{\Gamma(p) \cos(\pi p/2)}{\omega^p}$
10.	$\begin{cases} \sin x & \text{if } 0 < x < a \\ 0 & \text{otherwise} \end{cases}$	$\dfrac{1}{\sqrt{2\pi}} \left[\dfrac{\sin a(1-\omega)}{1-\omega} - \dfrac{\sin a(1+\omega)}{1+\omega} \right]$

Operational Properties

11.	$\alpha f(x) + \beta g(x)$	$\alpha\, \mathcal{F}_s(f)(\omega) + \beta\, \mathcal{F}_s(g)(\omega)$
12.	$f(ax), \quad a > 0$	$\frac{1}{a} \widehat{f}_s\!\left(\frac{\omega}{a}\right)$
13.	$f'(x)$	$-\omega \widehat{f}_c(\omega)$
14.	$f''(x)$	$-\omega^2 \widehat{f}_s(\omega) + \sqrt{\frac{2}{\pi}}\, \omega f(0)$
15.	$x f(x)$	$-\left[\widehat{f}_c\right]'(\omega)$
16.	$\mathcal{F}_s(\mathcal{F}_s f)$	f

Table of Laplace Transforms

$f(t), \quad t \geq 0$	$F(s) = \mathcal{L}(f)(s) = \int_0^\infty f(t)e^{-st}\,dt,$
1. 1	$\frac{1}{s}, \quad s > 0$
2. $t^n, \quad n = 1, 2, \ldots$	$\frac{n!}{s^{n+1}}, \quad s > 0$
3. $t^a \quad (a > -1)$	$\frac{\Gamma(a+1)}{s^{a+1}}, \quad s > 0$
4. e^{at}	$\frac{1}{s-a}, \quad s > a$
5. $t^n e^{at}$	$\frac{n!}{(s-a)^{n+1}}, \quad s > a$
6. $\frac{e^{at} - e^{bt}}{a - b}$	$\frac{1}{(s-a)(s-b)}, \quad s > \max(a, b)$
7. $\frac{ae^{at} - be^{bt}}{a - b}$	$\frac{s}{(s-a)(s-b)}, \quad s > \max(a, b)$
8. $\sin kt$	$\frac{k}{s^2 + k^2}, \quad s > 0$
9. $\cos kt$	$\frac{s}{s^2 + k^2}, \quad s > 0$
10. $e^{at} \sin kt$	$\frac{k}{(s-a)^2 + k^2}, \quad s > a$
11. $e^{at} \cos kt$	$\frac{s-a}{(s-a)^2 + k^2}, \quad s > a$
12. $t \sin kt$	$\frac{2ks}{(s^2 + k^2)^2}, \quad s > 0$
13. $t \cos kt$	$\frac{s^2 - k^2}{(s^2 + k^2)^2}, \quad s > 0$
14. $\frac{1}{2a^3}(\sin at - at \cos at)$	$\frac{1}{(s^2 + a^2)^2}, \quad s > 0$
15. $\sinh kt$	$\frac{k}{s^2 - k^2}, \quad s > |k|$
16. $\cosh kt$	$\frac{s}{s^2 - k^2}, \quad s > |k|$
17. $e^{at} \sinh kt$	$\frac{k}{(s-a)^2 - k^2} \quad s > a + |k|$
18. $e^{at} \cosh kt$	$\frac{s-a}{(s-a)^2 - k^2} \quad s > a + |k|$
19. $t \sinh kt$	$\frac{2ks}{(s^2 - k^2)^2}, \quad s > |k|$
20. $t \cosh kt$	$\frac{s^2 + k^2}{(s^2 - k^2)^2}, \quad s > |k|$
21. $\frac{1}{2k^3}(kt \cosh kt - \sinh kt)$	$\frac{1}{(s^2 - k^2)^2}, \quad s > |k|$

Table of Laplace Transforms (continued)

$f(t), \quad t \geq 0$	$F(s) = \mathcal{L}(f)(s) = \int_0^\infty f(t)e^{-st}\,dt,$
22. $\delta(t - t_0), \quad t_0 \geq 0$	$e^{-t_0 s}, \quad s > 0$
23. $\mathcal{U}(t - a) = \begin{cases} 0 & \text{if } t < a \\ 1 & \text{if } t \geq a \end{cases} \quad (a > 0)$	$\frac{e^{-as}}{s}, \quad s > 0$
24. $f(t + T) = f(t) \quad (T > 0)$	$\frac{1}{1 - e^{-Ts}} \int_0^T e^{-st} f(t)\,dt$
25. $f(t + T) = -f(t) \quad (T > 0)$	$\frac{1}{1 + e^{-Ts}} \int_0^T e^{-st} f(t)\,dt$
26. Triangular Wave	$\frac{1}{as^2}\left[\frac{1 - e^{-as}}{1 + e^{-as}}\right] = \frac{1}{as^2}\tanh\left(\frac{as}{2}\right), \quad s > 0$
27. Square Wave	$\frac{1}{s}\left[\frac{1 - e^{-as}}{1 + e^{-as}}\right] = \frac{1}{s}\tanh\left(\frac{as}{2}\right), \quad s > 0$
28. Sawtooth	$\frac{1}{as^2} - \frac{e^{-as}}{s(1 - e^{-as})}, \quad s > 0$
29. $\frac{\sin at}{t}$	$\tan^{-1}\left(\frac{a}{s}\right), \quad s > 0$
30. $J_0(at)$	$\frac{1}{\sqrt{s^2 + a^2}}, \quad s > 0$
31. $J_n(t) \quad (n \text{ integer} \geq 0)$	$\frac{1}{\sqrt{s^2 + 1}}\left(\sqrt{s^2 + 1} - s\right)^n, \quad s > 0$
32. $J_0(a\sqrt{t})$	$\frac{e^{-a^2/4s}}{s}, \quad s > 0$
33. $t^p J_p(at) \quad (p > -\frac{1}{2})$	$\frac{2^p a^p \Gamma(p + \frac{1}{2})}{\sqrt{\pi}(s^2 + a^2)^{p + \frac{1}{2}}}, \quad s > 0$
34. $\frac{\sqrt{\pi}}{\Gamma(k)}\left(\frac{t}{2a}\right)^{k - \frac{1}{2}} J_{k - \frac{1}{2}}(at) \quad (k > 0)$	$\frac{1}{(s^2 + a^2)^k}, \quad s > 0$
35. $\frac{\sqrt{\pi}}{\Gamma(k)} a \left(\frac{t}{2a}\right)^{k - \frac{1}{2}} J_{k - \frac{3}{2}}(at) \quad (k > \frac{1}{2})$	$\frac{s}{(s^2 + a^2)^k}, \quad s > 0$
36. $2 \sum_{m=1}^{n} \binom{2n - m - 1}{n - 1} \frac{t^{m-1}\cos(at - \frac{m\pi}{2})}{(2a)^{2n-m}(m - 1)!}$ $(n \text{ an integer} \geq 1)$	$\frac{1}{(s^2 + a^2)^n}, \quad s > 0$

Table of Laplace Transforms (continued)

$f(t), \quad t \geq 0$	$F(s) = \mathcal{L}(f)(s) = \int_0^\infty f(t)e^{-st}\,dt,$
37. $\operatorname{erf}(at) \quad (a > 0)$	$\frac{1}{s}e^{s^2/4a^2}\operatorname{erfc}\left(\frac{s}{2a}\right), \quad s > 0$
38. $\operatorname{erf}(a\sqrt{t})$	$\frac{a}{s\sqrt{s+a^2}}, \quad s > 0$
39. $e^{-a^2t^2} \quad (a > 0)$	$\frac{\sqrt{\pi}}{2a}e^{s^2/4a^2}\operatorname{erfc}(\frac{s}{2a}), \quad s > 0$
40. $\frac{1}{\sqrt{\pi t}}e^{-a^2/4t} \quad (a \geq 0)$	$\frac{e^{-a\sqrt{s}}}{\sqrt{s}}, \quad s > 0$
41. $\frac{a}{2\sqrt{\pi}t^{3/2}}e^{-a^2/4t} \quad (a > 0)$	$e^{-a\sqrt{s}}, \quad s > 0$
42. $\operatorname{erfc}\left(\frac{a}{2\sqrt{t}}\right) \quad (a \geq 0)$	$\frac{1}{s}e^{-a\sqrt{s}}, \quad s > 0$

Operational Properties

43. $\alpha f(t) + \beta g(t)$	$\alpha F(s) + \beta G(s)$
44. $f'(t)$	$sF(s) - f(0)$
45. $f''(t)$	$s^2 F(s) - sf(0) - f'(0)$
46. $f^{(n)}(t)$	$s^n F(s) - s^{n-1}f(0) - \ldots - f^{(n-1)}(0)$
47. $-tf(t)$	$F'(s)$
48. $t^n f(t)$	$(-1)^n F^{(n)}(s)$
49. $\int_0^t f(\tau)\,d\tau$	$\frac{1}{s}F(s), \quad s > 0$
50. $\int_0^t \int_0^\tau f(\rho)\,d\rho d\tau$	$\frac{1}{s^2}F(s), \quad s > 0$
51. $\frac{f(t)}{t}$	$\displaystyle\int_s^\infty F(u)\,du$
52. $\frac{f(t)}{t^2}$	$\displaystyle\int_s^\infty \int_\sigma^\infty F(u)\,du d\sigma$
53. $\mathcal{U}(t-a)f(t-a) \quad (a > 0)$	$e^{-as}F(s)$
54. $e^{at}f(t)$	$F(s-a)$
55. $f(ct) \quad (c > 0)$	$\frac{1}{c}F(\frac{s}{c})$
56. $f * g(t) = \displaystyle\int_0^t f(\tau)g(t-\tau)\,d\tau$	$F(s)G(s)$

Bibliography

[1] L. Ahlfors, *Complex Analysis* (3rd ed.), McGraw-Hill, New York, 1979.

[2] N. Asmar, *Partial Differential equations and Boundary Value Problems*, Prentice Hall, New Jersey, 1999.

[3] G. Birkhoff and G. C. Rota, *Ordinary Differential Equations* (4th ed.), John Wiley, New York, 1989.

[4] B. Belhoste, *Augustin-Louis Cauchy*, Springer-Verlag, New York, 1991.

[5] J. Brown and R. Churchill, *Complex Variables and Applications* (6th ed.), McGraw-Hill, New York, 1996.

[6] J. Brown and R. Churchill, *Fourier Series and Boundary Value Problems* (5th ed.), McGraw-Hill, New York, 1993.

[7] D. M. Burton, *The History of Mathematics* (3rd ed.), McGraw-Hill, New York, 1997.

[8] G. Carrier and C. Pearson, *Partial Differential Equations* (2nd ed.), Academic Press, New York, 1988.

[9] H. S. Carslaw, *An Introduction to the Theory of Fourier's Series and Integrals* (3rd ed.), Dover, New York, 1950.

[10] J. B. Conway, *Functions of One Complex Variable* (2nd ed.), Springer-Verlag, New York, 1978.

[11] R. Courant and D. Hilbert, *Methods of Mathematical Physics*, Vols. 1 and 2, Wiley-Interscience, New York, 1953 and 1962.

[12] G. B. Folland, *Introduction to Partial Differential Equations*, Princeton University Press, Princeton, N.J., 1976.

[13] G. B. Folland, *Fourier Analysis and its Applications*, Wadsworth and Brooks/Cole, 1992.

[14] L. Grafakos, *Modern and Classical Fourier Analysis*, Prentice Hall, to appear.

[15] E. Hewitt and R. Hewitt, "The Gibbs-Wilbraham phenomenon: an episode in Fourier analysis," *Archive for the History of the Exact Sciences* **21** (1979), 129-160.

[16] H. Jeffreys and B. S. Jeffreys, *Method of Mathematical Physics* (3rd ed.), Cambridge University Press, London, New York, 1956.

[17] J. Kevorkian, *Partial Differential Equations* (2nd ed.), Springer-Verlag, New York, 2000.

[18] T.W. Körner, *Fourier Analysis*, Cambridge University Press, Cambridge, U.K., 1988.

[19] N.N. Lebedev, *Special Functions and their Applications*, Dover, New York, 1972.

[20] Z. Nehari, *Conformal Mapping*, Dover Publications, New York, 1975.

[21] A. Papoulis, *The Fourier Integral and its Applications*, McGraw-Hill, New York, 1962.

[22] M. Pinsky, *Partial Differential Equations and Boundary Value Problems* (2nd ed.), McGraw-Hill, New York, 1991.

[23] G. Sansone, *Orthogonal Polynomials*, Interscience Publishers, New York, 1959.

[24] F. Smithies, *Cauchy and the Creation of Complex Numbers*, Cambridge University Press, Cambridge, U. K., 1997.

[25] W. A. Strauss, *Partial Differential Equations*, Wiley, New York, 1992.

[26] K. Stromberg, *Introduction to Classical Real Analysis*, Wadsworth, Belmont, Calif., 1981.

[27] D. J. Struik, *A Concise History of Mathematics*, Dover Publications, New York, 1967.

[28] G. Szegö, *Orthogonal Polynomials* (4th ed.), American Mathematical Society, Providence, R.I., 1975.

[29] G. P. Tolstov, *Fourier Series*, Dover, New York, 1976.

[30] G. N. Watson, *A Treatise on the Theory of Bessel Functions* (2nd ed.), Cambridge University Press, Cambridge, U.K., 1944.

[31] E. T. Whittaker and G. N. Watson, *A Course of Modern Analysis* (4th ed.), Cambridge University Press, London, New York, 1950.

[32] N. Wiener, *The Fourier Transform and Certain of its Applications*, Dover, New York, 1958.

[33] A. Zygmund, *Trigonometric Series* (2nd ed.), Cambridge University Press, Cambridge, U.K., 1977.

Answers to Selected Exercises

CHAPTER 1
SECTION 1.1

1. $0 + i(-1)$. **3.** $7 + i(-1)$. **5.** $3 + 4i$. **7.** $x^2 - y^2 + 2xyi$.
9. $\frac{25}{28} - \frac{2}{7}i$. **11.** $0 + 32i$. **12.** $1 + 8i$. **13.** $3 + 8i$.
15. $\frac{10101}{10001} - \frac{i}{10001}$. **17.** $x = \pm i\sqrt{6}$. **19.** $x = -\frac{1}{2} \pm i\frac{\sqrt{3}}{2}$.
21. $x = i, i, -i, -i$. **23.** $x = 5 \pm i\sqrt{15}$. **30.** (a) $z = -1, i, -i$. (b) $z = -6, -2 + i, -2 - i$. (c) $z = 1 + i, 1 - i, -1 + i, -1 - i$. (d) $z = 1 + i, 1 - i, 2 + i, 2 - i$.

32. (a) $\pm \frac{1+i}{\sqrt{2}}$. (b) $\pm \left(\sqrt{-1 + \frac{\sqrt{5}}{2}} + i \frac{1}{2\sqrt{-1 + \frac{\sqrt{5}}{2}}} \right)$.

(c) $\pm \left(\frac{1}{2} - i\frac{\sqrt{3}}{2} \right)$. **34.** $x = -2, 1 + i\sqrt{2}, 1 - i\sqrt{2}$.

35. $x = -4, i, -i$. **36.** $x = 3, -\frac{1}{2} + i\frac{\sqrt{15}}{2}, -\frac{1}{2} - i\frac{\sqrt{15}}{2}$.
37. $x = -1, \frac{3}{2} + i\frac{\sqrt{3}}{2}, \frac{3}{2} - i\frac{\sqrt{3}}{2}$.

SECTION 1.2

1. $-z = -1 + i$, $\overline{z} = 1 + i$, $|z| = \sqrt{2}$. **3.** $-z = -5 - 3i$, $\overline{z} = 5 - 3i$, $|z| = \sqrt{34}$. **5.** $-z = -1 - i$, $\overline{z} = 1 - i$, $|z| = \sqrt{2}$.
7. (b) $|z_1| = |z_2| = |z_3| = 1$, $|z_4| = \frac{\sqrt{5}}{2}$. z_1, z_2, z_3 tie as being closest to the origin. (d) $|z_1 - z_2|^2 = 2 - \sqrt{2} \approx .6$, $|z_1 - z_4|^2 = .5$. z_4 is closer to z_1. **9.** $2\sqrt{10}$. **11.** 1.
13. $\frac{1}{\sqrt{5}}$. **15.** Circle of radius 3 centered at 4. **17.** Empty set. **19.** Ellipse with foci 0, i, and semi-major axis 1.
21. $(x - \frac{1}{4})^2 + \frac{4y^2}{15} = 1$. **25.** (a) Directrix $x = -1$, focus $1 + i$. Parabola with vertex at i and opening toward the focus $1 + i$. (b) Directrix $x = a$, focus z_0. Parabola with vertex at $(\frac{\text{Re } z_0 + a}{2}, \text{Im } z_0)$ and opening toward the focus z_0. **31.** (a) $i^2 = -1 \neq |i|^2 = 1$. (b) $z^2 = |z|^2 \Leftrightarrow xy = 0$ and $x^2 - y^2 = x^2 + y^2 \Leftrightarrow -y^2 = y^2 \Leftrightarrow y = 0$. **33** (b) $|\cos\theta + i\sin\theta| = \sqrt{(\cos\theta)^2 + (\sin\theta)^2} = 1$. **35.** $|z - 4| \leq |z - 3i| + |3i - 4| \leq 6$. **36.** $|z - 4| \geq ||z - 3i| - |3i - 4|| \geq 4$.
37. $|z - 4| \geq ||z - 1| - 3| \geq 2$. So, $\frac{1}{|z-4|} \leq \frac{1}{2}$. **38.** $|1 - z| \geq ||-z + i| - |1 - i|| \geq \sqrt{2} - \frac{1}{2}$. So, $\frac{1}{|1-z|} \leq \frac{1}{\sqrt{2} - \frac{1}{2}} = \frac{2}{2\sqrt{2} - 1}$.
39. $|z^2 + z + 1| \geq 1 - |z| - |z|^2 \geq \frac{1}{4}$. So $\frac{1}{|z^2+z+1|} \leq 4$.
40. $|z^2 + z + 1| \leq \frac{1}{4} + \frac{1}{2} + 1 = \frac{7}{4}$. So $\frac{1}{|z^2+z+1|} \geq \frac{4}{7}$.

SECTION 1.3

5. $3\sqrt{2}(\cos\frac{5\pi}{4} + i\sin\frac{5\pi}{4})$. **7.** $2(\cos(-2\frac{\pi}{3}) + i\sin(-2\frac{\pi}{3}))$.
9. $\frac{1}{2}(\cos(-\frac{\pi}{2}) + i\sin(-\frac{\pi}{2}))$. **11.** $1 \cdot (\cos\frac{\pi}{2} + i\sin\frac{\pi}{2})$.
13. $\text{Arg } z = \tan^{-1}\frac{2}{13} \approx .15$, $\arg z = \tan^{-1}\frac{2}{13} + 2k\pi$.
15. $\text{Arg } z = \pi - \tan^{-1}\frac{1}{2} \approx 2.68$, $\arg z = \pi - \tan^{-1}\frac{1}{2} + 2k\pi$. **17.** $8(\cos\frac{\pi}{2} + i\sin\frac{\pi}{2})$. **19.** $\cos\pi + i\sin\pi$.
21. $\text{Re}(1 + i)^{30} = 0$, $\text{Im}(1 + i)^{30} = -2^{15}$.

23. $\text{Re}\left(\frac{1-i}{1+i}\right)^4 = 1$, $\text{Im}\left(\frac{1-i}{1+i}\right)^4 = 0$. **33.** Principal root $z_1 = \cos\frac{\pi}{8} + i\sin\frac{\pi}{8}$, $z_2 = \cos\frac{5\pi}{8} + i\sin\frac{5\pi}{8}$, $z_3 = \cos\frac{9\pi}{8} + i\sin\frac{9\pi}{8}$, $z_4 = \cos\frac{13\pi}{8} + i\sin\frac{13\pi}{8}$. **35.** Principal root $z_1 = 2^{7/6}$, $z_2 = 2^{7/6}(\cos\frac{\pi}{3} + i\sin\frac{\pi}{3})$, $z_3 = 2^{7/6}(\cos\frac{2\pi}{3} + i\sin\frac{2\pi}{3})$, $z_4 = -2^{7/6}$, $z_5 = 2^{7/6}(\cos\frac{4\pi}{3} + i\sin\frac{4\pi}{3})$, $z_6 = 2^{7/6}(\cos\frac{5\pi}{3} + i\sin\frac{5\pi}{3})$. **37.** Principal root $z_1 = \cos\frac{\pi}{7} + i\sin\frac{\pi}{7}$, $z_2 = \cos\frac{3\pi}{7} + i\sin\frac{3\pi}{7}$, $z_3 = \cos\frac{5\pi}{7} + i\sin\frac{5\pi}{7}$, $z_4 = -1$, $z_5 = \cos\frac{9\pi}{7} + i\sin\frac{9\pi}{7}$, $z_6 = \cos\frac{11\pi}{7} + i\sin\frac{11\pi}{7}$, $z_7 = \cos\frac{13\pi}{7} + i\sin\frac{13\pi}{7}$. **39.** $z_1 = -2 + 3^{1/3}(\cos\frac{\pi}{6} + i\sin\frac{\pi}{6})$, $z_2 = -2 + 3^{1/3}(\cos\frac{5\pi}{6} + i\sin\frac{5\pi}{6})$, $z_3 = -2 + 3^{1/3}(\cos\frac{9\pi}{6} + i\sin\frac{9\pi}{6})$. **41.** $z_1 = 5 + 5(\cos\frac{\pi}{3} + i\sin\frac{\pi}{3})$, $z_2 = 0$, $z_3 = 5 + 5(\cos\frac{5\pi}{3} + i\sin\frac{5\pi}{3})$. **43.** $z = 1 + i \pm (\frac{1}{\sqrt{2}} + \frac{i}{\sqrt{2}})$.
45. $z = -\frac{1}{2} + \pm 2^{-3/4}(\cos\frac{\pi}{8} - i\sin\frac{\pi}{8})$.

SECTION 1.4

1. $3 + 5i$. **3.** $-4 - 2i$. **5.** $2^{3/4}(\cos\frac{\pi}{8} + i\sin\frac{\pi}{8})$.
7. $2 - y + i(x - 1)$. **9.** $x^3 - 3xy^2 + i(3x^2y - y^3)$.
11. $\frac{x+1}{(x+1)^2+y^2} + i\frac{-y}{(x+1)^2+y^2}$. **13.** $\mathbb{C} \setminus \{0\}$. **15.** $\mathbb{C} \setminus \{-i, i\}$.
17. $\mathbb{C}$. **19.** $f(z) = (2 + i)z + 1$. **21.** $f[S] = \{z : |z| < 4\}$.
23. $f[S] = \{z : \text{Re } z < 0 \text{ and } \text{Im } z < 2\}$. **26.** $f(z) = (\cos\phi + i\sin\phi)(z - z_0) + z_0$. **29.** $f[S] = \{0, \pi\}$.
31. $f[S] = \{z : |z| \geq 1\}$. **33.** $f[S] = \{z : |z| \geq \frac{1}{3} \text{ and } -\frac{2\pi}{3} \leq \text{Arg } z \leq -\frac{\pi}{3}\}$. **39.** $f(z) = \frac{2}{z+i}$.
41. ± 1. **43.** $\pm i\sqrt{2}$.

SECTION 1.5

1. $-1 + 0 \cdot i$. **3.** $1 + 0 \cdot i$. **5.** $-\frac{\sqrt{2}}{2} + i\frac{\sqrt{2}}{2}$. **7.** $\frac{\sqrt{3}}{2e} - \frac{i}{2e}$.
9. $0 + 3e^3i$. **11.** $3\sqrt{2}e^{-\frac{3\pi}{4}i}$. **13.** $2e^{-\frac{2\pi}{3}i}$. **15.** $\frac{1}{2}e^{-\frac{\pi}{2}i}$. **17.**

$e^{\frac{\pi}{2}i}$. **19.** $\cos(1) + i\sin(1)$. **21.** 1. **23.** $e^4(\cos(3) - i\sin(3))$.
25. (a) $u(x, y) = e^{3x}\cos(3y)$, $v(x, y) = e^{3x}\sin(3y)$. (b)
$u(x, y) = e^{x^2 - y^2}\cos(2xy)$, $v(x, y) = e^{x^2 - y^2}\sin(2xy)$. (c)
$u(x, y) = e^x\cos y$, $v(x, y) = -e^x\sin y$. (d) $u(x, y) =$
$e^{-y}\cos x$, $v(x, y) = e^{-y}\sin x$. **33.** $\sin^4\theta = \frac{1}{24}\left(e^{i\theta} - e^{-i\theta}\right)^4$
$\frac{1}{16}(e^{4i\theta} + e^{-4i\theta}) - \frac{1}{4}(e^{2i\theta} + e^{-2i\theta}) + \frac{3}{8} = \frac{1}{8}\cos 4\theta - \frac{1}{2}\cos 2\theta +$
$\frac{3}{8}$. $\int\sin^4\theta\,d\theta = \frac{1}{32}\sin 4\theta - \frac{1}{4}\sin 2\theta + \frac{3}{8}\theta + C$. **35.** $\cos^6\theta =$
$\frac{1}{64}(e^{6i\theta} + e^{-6i\theta}) + \frac{3}{32}(e^{4i\theta} + e^{-4i\theta}) + \frac{15}{64}(e^{2i\theta} + e^{-2i\theta}) +$
$\frac{5}{16} = \frac{1}{32}\cos 6\theta + \frac{3}{16}\cos 4\theta + \frac{15}{32}\cos 2\theta + \frac{5}{16}$. $\int\cos^6\theta\,d\theta =$
$\frac{1}{192}\sin 6\theta + \frac{3}{64}\sin 4\theta + \frac{15}{64}\sin 2\theta + \frac{5}{16}\theta + C$. **39.** $|e^z| = e^x =$
$e^{\mathrm{Re}\,z} \leq 1 \Leftrightarrow \mathrm{Re}\,z \leq 0$. $|e^z| = 1 \Leftrightarrow \mathrm{Re}\,z = 0$.

SECTION 1.6

1. (a) $\cos i = \cosh 1$, $\sin i = i\sinh 1$. **3.** (a) $\cos(\frac{\pi}{2} + 2i) = -i\sinh 2$, $\sin(\frac{\pi}{2} + 2i) = \cosh 2$. **5.** (a) $\cos(1 + i) = \cos 1\cosh 1 - i\sin 1\sinh 1$, $\sin(1 + i) = \cosh 1\sin 1 + i\cos 1\sinh 1$, $\tan(1 + i) = \frac{\sin 2}{\cos 2 + \cosh 2} + i\frac{\sinh 2}{\cos 2 + \cosh 2}$,
(b) $|\cos(1 + i)| = \sqrt{(\cos 1\cosh 1)^2 + (\sin 1\sinh 1)^2}$,
$|\sin(1 + i)| = \sqrt{(\cosh 1\sin 1)^2 + (\cos 1\sinh 1)^2}$.
7. (a) $\cos(\frac{3\pi}{2} + i) = i\sinh 1$, $\sin(\frac{3\pi}{2} + i) = -\cosh 1$, $\tan(\frac{3\pi}{2} + i) = i\coth 1$,
(b) $|\cos(\frac{3\pi}{2} + i)| = \sinh 1$, $|\sin(\frac{3\pi}{2} + i)| = \cosh 1$.
9. $\sin(2z) = \cosh(2y)\sin(2x) + i\cos(2x)\sinh(2y)$.
11. $\sin(z) + 2z = 2x + \cosh y\sin x + i(2y + \cos x\sinh y)$.
13. $\tan z = \frac{\sin 2x}{\cos(2x) + \cosh(2y)} + i\frac{\sinh(2y)}{\cos(2x) + \cosh(2y)}$.
14. $\sec z = \frac{2\cos x\cosh y}{\cos(2x) + \cosh(2y)} + 2i\frac{\sin x\sinh y}{\cos(2x) + \cosh(2y)}$.

SECTION 1.7

1. $\log(2i) = \ln 2 + i\frac{\pi}{2} + 2k\pi i$. **3.** $\log(5e^{i\frac{\pi}{7}}) = \ln 5 + i\frac{\pi}{7} + 2k\pi i$. **5.** $\frac{\ln 12}{2} + i\frac{\pi}{6}$. **7.** $-\frac{6\pi}{7}i$. **8.** $\ln\alpha + i\pi$. **9.** $2\pi i$.
11. $1 + i\frac{7\pi}{2}$. **13.** $z = \ln 3 + 2k\pi i$, k integer. **15.** $z = -3 + i(\frac{\pi}{2} + 2k\pi)$, k integer. **17.** $z = \frac{1}{2}(\ln 5 + (2k + 1)\pi i)$, k integer. **19.** (a) $\mathrm{Log}\,e^{i\pi} = i\pi$, $\mathrm{Log}\,e^{3i\pi} = i\pi$, $\mathrm{Log}\,e^{5i\pi} = i\pi$.
23. $\cos(\ln 5) + i\sin(\ln 5)$. **25.** $5e^\pi(-\cos(\ln 5) + i\sin(\ln 5))$.
27. 3^4. **29.** $e^{-(\frac{\pi}{2} + 2k\pi)}$, k integer. **31.** $z = \frac{\pi}{4} + k\pi$.

CHAPTER 2
SECTION 2.1

1. Interior points $\{z : |z| < 1\}$; boundary points $\{z : |z| = 1\}$. **3.** The set of interior points is empty; boundary points $[0, 1]$. **5.** Open, connected, region. **7.** Open, connected, region. **9.** Neither open nor closed, connected, not a region. **11.** Open not connected, not a region. **13.** $B_1(0) \cup B_1(5)$ is not connected. **15.** $A = \{z : |z| < 1\}$, $B = \{z : |z| < 4\}$. Then, $A \subset B$ but boundary of $A = \{z : |z| = 1\}$ is not contained in boundary of

$B = \{z : |z| = 4\}$. **17.** If S is open and $\neq \emptyset$, let $T = \mathbb{C}\setminus S$. Suppose that S is open. To show that T is closed, we must show that every boundary point z_0 of T is in T. Since every neighborhood of z_0 contains points of T and points of S, z_0 is not an interior point of S. But S is open, every point in S is an interior point. So z_0 is not in S hence it is in T and hence T is closed. To prove the converse, suppose that T is closed. Let z_0 be in S. Show that z_0 is not a boundary point of T. Conclude that z_0 must e an interior point of S. Hence S is open. **18.** $(\Rightarrow)$ S is open and $\neq \emptyset$, $T = \mathbb{C}\setminus S$. If z_0 is a boundary point of S, then S cannot contain z_0 because all points in S are interior points. To prove the converse, show that for any nonempty set A any point in A is either an interior point or a boundary point. Conclude that if S contains none of its boundary points then all the points in S are interior points and hence S is open. **19.** If $z \in \bigcup_{n=1}^{\infty} A_n$, then $z \in A_{n_0}$ for some n_0. Since A_{n_0} is open, we can find $r > 0$ such that $z \in B_r(z) \subset A_{n_0} \subset \bigcup_{n=1}^{\infty} A_n$. Hence $\bigcup_{n=1}^{\infty} A_n$ is open. **20** (a) If $z \in A_n$ for all $n = 1, 2, \ldots, N$ and each A_n is open, let $B_{r_n}(z)$ be such that $z \in B_{r_n}(z) \subset A_n$, and let $r = \min(r_1, r_2, \ldots, r_N)$. Then $z \in B_r(z) \subset A_n$ for all n. Hence $z \in B_r(z) \subset \bigcap_{n=1}^{N} A_n$ and hence the finite intersection is open. **21.** $A \cup B$ is open because A is open and B is open. To show that $A \cup B$ is connected, let z_0 be a point in the (nonempty) intersection of A and B. Given any two points in $A \cup B$. Say $z_1 \in A$ and $z_2 \in B$. Since A is connected, we can join z_1 to z_0 by a polygonal line segment in A. Since B is connected, we can join z_2 to z_0 by a polygonal line segment in B. Hence we can connect z_1 to z_2 by a polygonal line segment in $A \cup B$. **23.** (a) Circle in the plane centered at the origin. (b) All the points on the sphere above a parallel of latitude, except the north pole. These together with N form a neighborhood of N on the sphere, which corresponds to a neighborhood of infinity in the plane. (c) Line through the origin. (d) Line not containing the origin. (e) Use parallels of latitude and check the images of points above and below a parallel of latitude.

SECTION 2.2

1. $-4 + 2i$. **3.** 0. **5.** ∞. **7.** 0. **9.** π^2. **11.** 0. **13.** $-\frac{i}{3}$. **15.** 1. **17.** ∞. **19.** $\lim_{z\to-3}\mathrm{Arg}\,z = \pi$ if z approaches -3 from the upper half-plane, and $\lim_{z\to-3}\mathrm{Arg}\,z = -\pi$ if z approaches -3 from the lower half-plane. **21.** $\lim_{z\to\infty}e^{-z} = \infty$ if $z = x < 0$, and $\lim_{z\to\infty}e^{-z} = 0$ if $z = x > 0$. **23.** $\lim_{z\to 0}e^{\frac{1}{z}} = \infty$ if $z = x > 0$, and $\lim_{z\to 0}e^{\frac{1}{z}} = 0$ if $z = x < 0$. **25.** $\lim_{z\to 0}\frac{z}{|z|} = 1$ if $z = x > 0$, and

$\lim_{z \to 0} \frac{z}{|z|} = -1$ if $z = x < 0$. **33.** Continuous for all $z \neq -1 - 3i$. **35.** Continuous for all z. **37.** Continuous for all z. **38.** Continuous for all z. **39.** Continuous for all z. **40.** Continuous for all $z \neq 0$.

SECTION 2.3

1. Analytic for all z, $f'(z) = 6z - 4$. **3.** Not analytic anywhere. **5.** Analytic for all $z \neq -1$, $e^{i\frac{\pi}{3}}$, $e^{i\frac{5\pi}{3}}$. **7.** Analytic for all $z \neq -1 - i$, $2 - i$. **8.** Analytic for all $z \neq -1 - 3i$, i. **9.** Analytic for all z except $z = 0$ and z negative. **11.** Analytic for all z except the half-line $\text{Im } z = -1$ and $\text{Re } z \leq 3$. **13.** $f(z) = z^{100}$, $f'(1) = 100$. **15.** $f(z) = \frac{1}{\sqrt{1+z}}$, $f'(0) = -\frac{1}{2}$. **17.** Analytic for $|z| \neq 1$, $f'(z) = 1$ if $|z| < 1$ and $f'(z) = 2z$ if $|z| > 1$.

SECTION 2.4

1. $u(x, y) = x$, $v(x, y) = y$, $u_x = 1$, $u_y = 0$, $v_x = 0$, $v_y = 1$; $u_x = v_y$ and $u_y = -v_x$ for all z; $f' = u_x + iv_x = 1$. **3.** $u(x, y) = e^{x^2-y^2}\cos(2xy)$, $v(x, y) = e^{x^2-y^2}\sin(2xy)$, $u_x = 2e^{x^2-y^2}\big(x\cos(2xy) - y\sin(2xy)\big)$, $u_y = 2e^{x^2-y^2}\big(-y\cos(2xy) - x\sin(2xy)\big)$, $v_x = 2e^{x^2-y^2}\big(y\cos(2xy) + x\sin(2xy)\big)$, $v_y = 2e^{x^2-y^2}\big(x\cos(2xy) - y\sin(2xy)\big)$; $u_x = v_y$ and $u_y = -v_x$ for all z; $f' = u_x + iv_x = 2ze^{z^2}$. **5.** $u(x, y) = e^x\cos y$, $v(x, y) = -e^x\sin y$, $u_x = e^x\cos y$, $u_y = -e^x\sin y$, $v_x = -e^x\sin y$, $v_y = -e^x\cos y$; the equations $u_x = v_y$ and $u_y = -v_x$ do not hold simultaneously for any z; the function is not analytic for any z. **7.** $u(x, y) = \frac{(x+1)^2-y^2}{D^2}$, $v(x, y) = \frac{-2(x+1)y}{D^2}$, where $D = (x + 1)^2 + y^2$; $u_x = \frac{-2(1+x)(1+2x+x^2-3y^2)}{D^3}$, $u_y = \frac{2y(-3-6x-3x^2+y^2)}{D^3}$, $v_x = \frac{2y(3+6x+3x^2-y^2)}{D^3}$, $v_y = \frac{-2(1+x)(1+2x+x^2-3y^2)}{D^3}$; analytic at all $z \neq -1$; $f' = \frac{-2}{(1+z)^3}$. **9.** $u(x, y) = xe^x\cos y - ye^x\sin y$, $v(x, y) = xe^x\sin y + ye^x\cos y$, $u_x = e^x\big((1+x)\cos y - y\sin y\big)$, $u_y = -e^x\big(y\cos y + (1+x)\sin y\big)$, $v_x = e^x\big(y\cos y + (1+x)\sin y\big)$, $v_y = e^x\big((1+x)\cos y - y\sin y\big)$; $u_x = v_y$ and $u_y = -v_x$ for all z; $f' = u_x + iv_x = e^z(z+1)$. **11.** $u(x, y) = \frac{\sin 2x}{\cos(2x)+\cosh(2y)}$, $v(x, y) = \frac{\sinh(2y)}{\cos(2x)+\cosh(2y)}$; $u_x = \frac{2(1+\cos(2x)\cosh(2y))}{D^2}$, $u_y = -\frac{2\sin(2x)\sin(2y)}{D^2}$, where $D = \cos(2x) + \cosh(2y)$, $v_x = -u_y$, $v_y = u_x$; $u_x = v_y$ and $u_y = -v_x$ for all z where $D \neq 0$; equivalently, for all $z \neq \frac{\pi}{2} + k\pi$, k an integer. $f' = u_x + iv_x = \frac{1}{(\cos x \cosh y - i \sin x \sinh y)^2} = \frac{1}{\cos^2 z}$. **13.** $u = x^2 + y^2$, $v = 0$; $u_x = v_y$, $u_y = -v_x$ only at $z = 0$; not analytic at any z. **15.** Analytic for all z, $f' = e^{z^2}(1+2z^2)$.

17. Analytic for all z, $f' = \cos(2z)$. **19.** Analytic for all $z \neq \pm i$ or $z = x$ real and $x \leq \frac{1}{3}$; $f' = \frac{1}{1+z^2}\big(\frac{3}{3z-1} - \frac{2z\,\text{Log}\,(3z-1)}{1+z^2}\big)$. **21.** Analytic for all z; $f' = 2iz\sin(3 - iz^2)$. **23.** Analytic for all $z \neq x$ where x is real and $x \leq 0$; $f' = iz^{i-1}$. **25.** Analytic for all $z \neq x + i$ where x is real and $x \leq 0$; $f' = -\frac{1}{2}(z - i)^{-3/2}$. **27.** 1. **29.** 1. **31.** $\frac{1}{1+z^2}$.

SECTION 2.5

1. Harmonic at all (x, y); real part of $z^2 + (2+i)z$. **3.** Harmonic at all (x, y); real part of e^z. **5.** Not harmonic for any (x, y). **7.** Harmonic for all (x, y); real part of $\sin z$. **9.** Harmonic for all (x, y); real part of e^{z^2}. **11.** Harmonic for all $(x, y) \neq (1, 0)$. **13.** $y - 2x$. **15.** $-e^y\sin x$. **17.** (a) $r^n\cos n\theta$ is real part of z^n and $r^n\sin n\theta$ is imaginary part of z^n. (b) (a) Harmonic conjugate of $r^n\cos n\theta$ is $r^n\sin n\theta$; and harmonic conjugate of $r^n\sin n\theta$ is $-r^n\cos n\theta$. **21.** $u + iv$ is analytic, so $(u + iv)^2$ is analytic, and hence $u^2 - v^2$ and $2uv$ are harmonic, being the real and imaginary parts of an analytic function. **25.** (b) $u(x, y) = 15x - 35$. (c) For $10 < T < 40$, the isotherm is the vertical line $x = \frac{T+35}{15}$. **27.** (a) $u(z) = \frac{60}{\pi}\text{Arg}\,z + 100$. (b) For $50 < T < 150$, the isotherm is the ray $\text{Arg}\,z = \frac{\pi}{60}(T - 100)$. **31.** (a) $u(r) = \frac{\ln r}{\ln 100}$. (b) For $0 < T < 100$, the isotherm is the circle $r = 100^{\frac{T}{100}}$. (c) Lines of heat flow are the rays $\text{Arg}\,z = C$, where C is a constant. **36.** (a) $u(x, y) = \frac{100}{\pi}\big[\text{Arg}\,(z - 1) - \text{Arg}\,(z + 1)\big]$. (b) All the points in the upper half-plane on the arc of the circle through -1 and 1 with center on the y-axis subtend the same angle over the interval $[-1, 1]$. Hence $\alpha(z)$ is constant. (c) Harmonic conjugate $v(x, y) = -\frac{100}{\pi}\big[\ln|z - 1| - \ln|z + 1|\big]$. Curves of heat flow are the circles $c^2 = \left|\frac{z+1}{z-1}\right|^2$.

37. $u(x, y) = T_2 + \frac{(T_1-T_2)}{\pi}\big(\text{Arg}\,(z - b) - \text{Arg}\,(z - a)\big)$. **39.** $\frac{20}{\pi}\big(\text{Arg}\,z - \text{Arg}\,(z + 1)\big) + \frac{100}{\pi}\big(\text{Arg}\,(z - 1) - \text{Arg}\,z\big)$. **43.** (a) $u(z) = -\frac{50}{\pi}\cot^{-1}(\tan x \coth y) + 100$. (b) Isotherms $\text{Arg}\,(\sin z) = \frac{\pi}{50}(100 - T)$. (c) $v(z) = \ln(\sin z)$. **45.** (a) $u(z) = \frac{100}{\pi}\big(\text{Arg}\,(\sin z - 1) - \text{Arg}\,(\sin z + 1)\big)$. (b) $v(z) = -\frac{100}{\pi}\big(\ln|\sin z - 1| - \ln|\sin z + 1|\big)$.

SECTION 2.6

1. $\lim_{(x,y)\to(0,0)} u(x, y) = \frac{1}{2}$ if $x = y$ and $= -\frac{1}{2}$ if $y = -x$. So u is not continuous at $(0, 0)$. $u_x(0, 0) = 0 = u_y(0, 0)$. **3.** $u(z) = u(z_0) + A(x - x_0) + B(y - y_0) + \epsilon(z)$, where $\epsilon(z) = 0$. **7.** $u(r, \theta) = \cos\theta\sin\theta = \frac{1}{2}\sin 2\theta$ if $r > 0$ and $u(0, \theta) = 0$. u is independent of r and takes values between $-\frac{1}{2}$ and $\frac{1}{2}$ in any neighborhood of 0. So it cannot converge to $0 = u(0, 0)$ as $r \to 0$.

CHAPTER 3
SECTION 3.1

1. $\gamma(t) = (1-t)(1+i) + t(-1-2i)$, $0 \le t \le 1$. **3.**
$\gamma(t) = 3i + e^{it}$, $0 \le t \le 2\pi$. **5.** $\gamma(t) = e^{it}$, $-\frac{\pi}{4} \le t \le \frac{\pi}{4}$.
7. $\gamma(t) = 3it$, for $0 \le t \le \frac{1}{3}$; $\gamma(t) = (2-3t)i - (3t-1)$,
for $\frac{1}{3} \le t \le \frac{2}{3}$; $\gamma(t) = (3t-3)$, for $\frac{2}{3} \le t \le 1$. **9.** $\gamma(t) =$
$-3+2i+5e^{it}$, $-\frac{\pi}{2} \le t \le 0$. **11.** $\gamma(t) = t+it^2$, $-1 \le t \le 1$.
13. $\gamma_r(t) = e^{-it}$, $0 \le t \le \frac{3\pi}{2}$. **19.** $f'(t) = e^{-it}(1-it)$. **21.**
$f'(t) = (3-6i)\sin(3it)$. **23.** $f'(t) = -4i\frac{t+i}{(t-i)^3}$.

25. $y_p = \sin t$. **27.** Use $e^{(1+i)t}$ on the right side. $y_p =$
$-\frac{1}{5}e^t \cos t + \frac{2}{5}e^t \sin t$. **29.** For $y'' - 2y' - 3y = \cos 4t$,
$y_p = -\frac{19}{425}\cos 4t - \frac{8}{425}\sin 4t$. For $y'' - 2y' - 3y = \sin 4t$,
$y_p = \frac{8}{425}\cos 4t - \frac{19}{425}\sin 4t$.
31. $\gamma(t) = (a-b)e^{it} + be^{-i\frac{a-b}{b}t}$. **33.** $\gamma(t) = ae^{it} - be^{i\frac{at}{2}}$.

SECTION 3.2

1. 0. **3.** $i(-1 + \cosh 1)$. **5.** $2 - \pi$. **7.** $-\frac{3}{2} - \frac{2}{3}i$. **9.** $F(x) =$
$\frac{3+2i}{2}x^2$ if $-1 \le x \le 0$, $F(x) = \frac{i}{3}x^3$ if $0 \le x \le 1$. **17.** 1.
19. -2π. **21.** 0. **23.** 0. **25.** 0. **27.** $\frac{2}{3}\left(-1 - \frac{\sqrt{2}}{2} + i\frac{\sqrt{2}}{2}\right)$.
29. $-\frac{8}{3} - 2i$. **31.** $\frac{2+2\sqrt{2}}{15}$. **33.** e. **35.** $l(C_1(0)) = 2\pi$; and
on $C_1(0)$, $\text{Log}\, z = i\,\text{Arg}\, z$, so $|\text{Log}\, z| \le \pi$. **37.** $l(\gamma) =$
$2\sqrt{5} + 2$; and on γ, $|z| \ge \frac{1}{\sqrt{5}}$, so $\frac{1}{|z|^5} \le \left(\sqrt{5}\right)^5$. **39.**
$l(C_1(0)) = 2\pi$; and on $C_1(0)$, $|e^{z^2+1}| = e \cdot e^{\cos(2x)} \le e^2$.

SECTION 3.3

1. $\frac{1}{3}z^3 + \frac{1}{2}z^2 - z$, $\Omega = \mathbb{C}$. **3.** $\frac{1}{2}(\text{Log}\, z)^2$, $\Omega = \mathbb{C} \setminus$
$(-\infty, 0]$. **5.** $\frac{1}{2}\text{Log}\frac{z-1}{z+1}$, $\Omega = \mathbb{C}\setminus(-\infty, 1]$. **7.** $\frac{1}{3}\sin(3z+$
$2)$, $\Omega = \mathbb{C}$. **9.** $\frac{1}{2}\cosh(z^2)$, $\Omega = \mathbb{C}$. **11.** $\frac{1}{2}z^2\text{Log}\, z -$
$\frac{1}{4}z^2$, $\Omega = \mathbb{C}\setminus(-\infty, 0]$. **13.** $z\log_0 z + z\log_{\frac{\pi}{2}} z - 2z + \text{Log}\, z$,
$\Omega = \mathbb{C}$ minus the real line and the ray at angle $\frac{\pi}{2}$. **15.** $-i$.
17. $-\frac{73+14\sqrt{2}}{3} - i(9+\frac{13\sqrt{2}}{3})$. **19.** $2e^{-1}+e^{\pi}-\pi e^{\pi}+i\pi e^{-1}$.
21. $\cos 2 - \cosh 2$. **23.** $i\frac{3\pi}{4}$. **25.** 0.

SECTION 3.4

1. Mutually continuously deformable. **2.** Not mutually
continuously deformable. **3.** Mutually continuously deformable. **4.** Mutually continuously deformable. **5.** Mutually continuously deformable. **6.** Mutually continuously
deformable. **7.** $H(t, s) = (1-s)\gamma_0(t) + s\gamma_1(t)$, where
$\gamma_0(t) = 3(1-2t) + 4it$ if $0 \le t \le \frac{1}{2}$, $\gamma_0(t) = (1-2(t-\frac{1}{2}))(2i) + 2(t-\frac{1}{2})(-3)$ if $\frac{1}{2} \le t \le 1$; and $\gamma_1(t) = 3e^{i\pi t}$,
$0 \le t \le 1$. **9.** $H(t, s) = (1-s)\gamma_0(t) + 3is$, where $\gamma_0(t) =$
$3\cos(2\pi t)+i(3+\sin(2\pi t))$, $0 \le t \le 1$, $0 \le s \le 1$. **10.** Only
the region in Figure 36 is convex. **11.** The regions in Figures 39, 41, 43, and 44 are simply connected. The regions
in Figures 40 and 42 are not simply connected. **12.** 0, by

Corollary 1. **13.** 0, by Corollary 1. **14.** $\frac{13}{3}$, independence
of path, Section 3.3. **15.** 0, by Corollary 1 or Theorem 5.
16. 0, by Corollary 1. The integrand is analytic on and inside $C_{\frac{3}{2}}(0)$, which contains the path. **17.** 0, by Corollary 1
or Theorem 5. **18.** $2\pi i$, by Example 4. **19.** 0, by Theorem 2, Section 3.3. **20.** -2π. **21.** 0, by Theorem 5. **22.**
π, use $\frac{1}{(z-i)(z+i)} = \frac{i}{2}\left(\frac{1}{z+i} - \frac{1}{z-i}\right)$ and Example 4. **23.**
0, use $\frac{1}{(z-i)(z+i)} = \frac{i}{2}\left(\frac{1}{z+i} - \frac{1}{z-i}\right)$ and Example 4. **24.**
0, use $\frac{1}{(z-1)(z-i)(z+i)} = \frac{1}{2}\frac{1}{z-1} + \frac{-1+i}{4(z-i)} - \frac{1+i}{4(z+i)}$. **25.** $2\pi i$,
use $\frac{z}{(z-1)(z+1)} = \frac{1}{2}\frac{1}{z-1} + \frac{1}{2}\frac{1}{z+1}$. **26.** 0, use $\frac{1}{(z-i)(z+i)} =$
$\frac{i}{2}\left(\frac{1}{z+i} - \frac{1}{z-i}\right)$. **27.** $-\frac{4}{3}\pi i$, use $\frac{z^2+1}{(z-2)(z+1)} = 1 + \frac{5}{3(z-2)} -$
$\frac{2}{3(z+1)}$. **28.** $2\pi i$, use $\frac{z}{(z-i)(z+i)} = \frac{1}{2(z-i)} + \frac{1}{2(z+i)}$. **29.** $\frac{\pi i}{2}$,
use $\frac{1}{(z+1)^2(z-i)(z+i)} = -\frac{1}{4(z-i)} - \frac{1}{4(z+i)} + \frac{1}{2(z+1)^2} + \frac{1}{2(z+1)}$.
30. $-2\pi i$. **31.** 0. **32.** 0 if both α and β are inside or
outside C; $\frac{2\pi i}{\alpha-\beta}$ if α is inside C and β is outside C; $\frac{2\pi i}{\beta-\alpha}$
if β is inside C and α is outside C.

SECTION 3.5

1. $K = [0, 1]$, $S = [-1, 0)$. **2.** K is the real line, S is the
graph of $y = \frac{1}{x}$ or any other nonzero function that tends
to 0 as $x \to \infty$. **3.** Any point z_0 in the region will work.

SECTION 3.6

1. $2\pi i$. **2.** $2\pi i e^{-1}\cosh 1$. **3.** $-\frac{1}{3}$. **4.** 0. **5.** π^2. **6.** $\frac{62}{13} - \frac{93}{13}i$.
7. $38\pi i$. **8.** $2\pi i$. **9.** 0. **10.** $-\frac{i}{\pi}$. **11.** $-2i$. **12.** $\frac{16}{9}\pi i$. **13.**
$-\pi - 2\pi i + \pi i \cosh \pi$. **14.** $-2\pi i$. **15.** $2\pi i$. **16.** 0. **17.** $-\frac{2\pi}{9}i$.
18. $-\frac{1}{3}$. **19.** $\frac{\pi i}{2}$. **20.** 0.

SECTION 3.7

1. $M = 1$, attained at all points on the circle $C_1(0)$; $m = 0$
attained at $z = 0$.
2. $|f|^2 = (2x + 3)^2 + (2y)^2$, $M = \sqrt{29}$, attained at $1 \pm i$;
$m = 1$ attained at $z = -1$. **3.** $|f| = \frac{e^{y^2}}{e^{x^2}}$, $M = e^4$, attained
at $\pm 2i$; $m = e^{-4}$ attained at $z = \pm 2$.
4. $M = \sinh 1 \approx 1.18$, attained at $z = i$; $m = 0$ attained at $z = \pi$. **5.** $M = 1$, attained at $z = 2e^{i\frac{\pi}{2}}$;
$m = \frac{3}{11}$ attained at $z = \pm 3$. **7.** $m = 0$ attained at
$z = \frac{1}{2}$. $M = 1$ is attained at all points on the bound-
ary $|z| = 1$. To see this multiply $|f(z)|$ by $|\bar{z}| = 1$ and
use $z\bar{z} = 1$. Then $|f(z)| = \left|\frac{2-\bar{z}}{2-z}\right| = 1$. **9.** $m = 0$ at-
tained at $z = 1$; $M = \left((\ln 2)^2 + (\frac{\pi}{4})^2\right)^{1/2}$, attained at
$z = 2e^{i\frac{\pi}{4}}$. **22.** $|e^z|$ does not tend to infinity as $z \to \infty$.
23. (a) If f does not vanish in Ω then the maximum
and minimum of $|f|$ occur on the boundary. But $|f|$

is constant on the boundary, so $|f|$ is constant in Ω, which implies that f is constant in Ω. (b) The condition $|z| < |f(z)|$ implies that $f(z) \neq 0$ for all z, and hence f is constant by (a). The constant is necessarily unimodular. **25.** $|g(z)| = 1$ for $|z| = 1$, because f is real-valued. By Exercise 22, $g(z) = e^{\operatorname{Re}(f(z)) + i \operatorname{Im}(f(z))} = e^{i\theta}$ is constant for all z. Hence $\operatorname{Re}(f(z))$ is constant for all $|z| \leq 1$, implying that $f(z)$ is constant. **26.** Consider $h(z) = \frac{f(z)}{g(z)}$. **27.** By Schwarz's Lemma, $|f(z)| \leq |z|$, and since $|f(\frac{i}{2})| = \frac{1}{2} = |\frac{i}{2}|$, $f(z) = Az$, where $|A| = 1$. **28.** No. Apply Schwarz's Lemma to $\frac{f(z)}{3}$. Then $\left|\frac{f'(z)}{3}\right| \leq 1$.

SECTION 3.8

1. (a) $u = \frac{\operatorname{Im} z^2}{2}$, harmonic by Theorem 1, Section 2.5. (b) $\phi = -iz$. **2.** (a) $u = \operatorname{Re}(e^{-iz})$, harmonic by Theorem 1, Section 2.5. (b) $\phi = -ie^{-iz}$. **3.** (a) $u = -\operatorname{Im}(\frac{1}{z})$, harmonic by Theorem 1, Section 2.5. (b) $\phi = -\frac{i}{z^2}$. **4.** (a) $u = 2\operatorname{Re}(\operatorname{Log} z)$, harmonic by Theorem 1, Section 2.5. (b) $\phi = \frac{2}{z}$. **5.** $\phi = u_x - iu_y = u_x + iv_x = f'(z)$ by the Cauchy-Riemann equations. **6.** Let v be the harmonic conjugate of u. The entire function $f = u + iv$ omits the strip $\{z: a < \operatorname{Re} z < b\}$. So f is constant by Exercise 17, Section 3.7. **8.** $u = \frac{y}{x^2 + y^2}$, $v = \frac{x}{x^2 + y^2}$.
11. $M = e$, when $z = 1$; $m = e^{-1}$, when $z = -1$.
12. No maximum or minimum in Ω. This does not contradict the corollary because Ω is not bounded.
13. $u(r, \theta) = 1 - r\cos\theta + r^2 \sin 2\theta$.
14. $u(r, \theta) = r\cos\theta - \frac{1}{2}r^2 \sin 2\theta$.
15. $u(r, \theta) = 50 + \frac{25}{2}r^2 \cos(2\theta)$.
16. $u(r, \theta) = \sum_{n=1}^{10} r^n \frac{\sin n\theta}{n}$.
17. Parabolas $x = y^2 + 1 - T$.

CHAPTER 4
SECTION 4.1

1. Converges to 0. **2.** Converges to i. **3.** Converges to 0. **4.** Converges to 0. **5.** Converges to 0. **6.** Converges to $\frac{2-i}{3}$. **7.** Let $\epsilon > 0$ be given. Use $a_n \to A$ and $a_n \to B$ and the triangle inequality to show $|A - B| < 2\epsilon$. Conclude $A = B$. **8.** Use $\sin(n+1)\theta = \sin n\theta \cos\theta + \cos n\theta \sin\theta$ and $\cos(n+1)\theta = \cos n\theta \cos\theta - \sin n\theta \sin\theta$ to show $\cos n\theta$ converges if and only if $\sin n\theta$ does, for θ not an integer multiple of π. **9.** (a) Use the definition of a limit. (b) If $a_n \to a$ then $a^2 + 2a - 3 = 0$, so $a = 1$ or $a = -3$. Use induction to show $\operatorname{Re} a_n \geq 0$; thus $\operatorname{Re} a \geq 0$ so $a = 1$.
10. (a) $a_n > 0$, $n = (1 + a_n)^n = \sum_{j=0}^{n} \binom{n}{j} a_n^j >$

$\binom{n}{2} a_n^2 = \frac{n(n-1)}{2} a_n^2 > 0$. (b) Thus $0 < a_n^2 < \frac{2}{n-1}$; by the squeeze theorem $a_n \to 0$. **11.** Geometric, converges to $\frac{3}{3-i}$. **12.** Geometric, converges to $\frac{2}{1-i}$. **13.** Geometric, converges to $-\frac{3}{2} + \frac{i}{2}$. **14.** Converges to $\operatorname{Re}\sum_{n=0}^{\infty} \frac{e^{in\theta}}{3^n} = \frac{1 - \frac{\cos\theta}{3}}{\frac{10}{9} - \frac{2\cos\theta}{3}}$. **15.** Converges to $\sum_{n=0}^{\infty} \frac{3}{10^n} + \operatorname{Im}\sum_{n=0}^{\infty} \frac{e^{in\theta}}{10^n} = \frac{10}{3} + \frac{\sin\theta}{10(\frac{101}{100} - \frac{\cos\theta}{5})}$. **16.** Converges to $\frac{1 - \frac{\cos\theta}{3}}{\frac{10}{9} - \frac{2\cos\theta}{3}} + \frac{1}{1 - \frac{2i}{3}}$. **17.** $\frac{1}{(n+i)(n-1+i)} = \frac{1}{n-1+i} - \frac{1}{n+i}$. Series telescopes and converges to $\frac{1}{1+i}$. **18.** Diverges by nth term test. **19.** $\frac{\sin in}{n^2} = \frac{i\sinh n}{n^2}$. Series diverges by nth term test. **20.** Diverges by nth term test. **21.** Converges by root test. **22.** Diverges by nth term test. **23.** Converges by comparison test. **24.** Converges by root test. **25.** Diverges by root test. **26.** $|\tan in| \leq 1$. Converges by comparison test. **27.** Converges by comparison and root tests. **28.** Converges by comparison test. **29.** Converges by root test. **30.** Diverges by nth term test. **31.** Converges by ratio test. **32.** Diverges by nth term test. **33.** $|z| < 1$; $\frac{2}{2-z}$. **34.** $|z+1| < 1$; $-1 - \frac{1}{z}$. **35.** $|z| < \sqrt{\frac{17}{10}}$; $\frac{4-i}{4-i-(3+i)z}$. **36.** $|z| > \sqrt{5}$; $\frac{z}{z-2-i}$. **37.** $|z - \frac{1}{5}| > \frac{1}{10}$; $\frac{1}{1-10z}$. **38.** $|z - 2 - i| > 2$; $\frac{2(2+i-z)}{1+i-z}$. **39.** $2 < |z| < 3$; $\frac{z}{z-2} + \frac{3}{3-z}$. **40.** $|z - 1| > 1$ and $|z| < 1$; $1 - \frac{1}{z} - \frac{1}{1-z}$. **41.** $s_n \to 0$ thus the series converges to 0. **42.** For any N, $|\sum_{n=0}^{N} a_n| \leq \sum_{n=0}^{N} |a_n| \leq \sum_{n=0}^{\infty} |a_n|$. Take the limit as $N \to \infty$. **47.** Converges by ratio test. **48.** Converges by ratio test. **49.** Establish and use $\cos z_1 \cos z_2 = \sum_{k=0}^{\infty} \sum_{j=0}^{k} (-1)^k \frac{z_1^{2j}}{(2j)!} \frac{z_2^{2k-2j}}{(2k-2j)!}$ and

$$\sin z_1 \sin z_2 = -\sum_{k=0}^{\infty} \sum_{j=0}^{k-1} (-1)^k \frac{z_1^{2j+1}}{(2j+1)!} \frac{z_2^{2k-2j-1}}{(2k-2j-1)!}.$$

SECTION 4.2

1. f_n converges uniformly to 0 on $[0, \pi]$. **3.** f_n converges uniformly to 0 on $[0, 1]$. **5.** f_n converges pointwise but not uniformly to 0 on $[0, 1]$. $f_n(\frac{1}{n}) = \frac{1}{2 - \frac{1}{n}} > .5$. Uniform convergence on any subinterval of the form $[a, 1]$, where $0 < a < 1$. **7.** f_n converges pointwise but not uniformly to 0 on $[0, 1]$. $f_n(\frac{1}{\sqrt{2}n}) = \frac{\sqrt{2}}{5} > .2$. Uniform convergence on any subinterval of the form $[a, 1]$, where $0 < a < 1$. **9.** f_n converges pointwise to 0. Uniformly convergent on $|z| \leq 1$. **11.** f_n converges pointwise to 0 if $\operatorname{Im} z = 0$ and diverges otherwise. Not uniformly convergent for $|z| \leq 1$. **13.** $M_n = \frac{1}{n^2}$, $\sum M_n < \infty$.

15. $M_n = \left(\frac{3.3}{4}\right)^n = r^n$, where $0 < r < 1$, $\sum M_n < \infty$.

17. $M_n = \left(\frac{4}{5}\right)^n = r^n$, where $0 < r < 1$, $\sum M_n < \infty$.

19. $M_n = \left(\frac{1.5}{4}\right)^n = r^n$, where $0 < r < 1$, $\sum M_n < \infty$.

21. $M_n = \left(\frac{2.9}{3}\right)^n + \left(\frac{2}{2.01}\right)^n = r^n + s^n$, where $0 < r, s < 1$, $\sum M_n < \infty$. **23.** (a) Yes. (b) $z + \sum_{k=2}^{\infty}\left(\frac{z^k}{k} - \frac{z^{k-1}}{k-1}\right)$.

25. (a) Yes, because $|z - \frac{1}{2}| < \frac{1}{6} \Rightarrow |z| < \frac{1}{6} + \frac{1}{2} = 1$. (b) No, but it converges uniformly on any closed subset of $|z - \frac{1}{2}| < \frac{1}{6}$. **29.** (a) $\left|\frac{1}{n^z}\right| \leq \frac{1}{n^\delta} = M_n$ if $\operatorname{Re} z \geq \delta > 1$, $\sum M_n < \infty$, so series converges uniformly by the Weierstrass M-test. (b) Apply Corollary 2. (c) $\zeta'(z) = -\sum_{n=2}^{\infty}\frac{\ln n}{n^z}$.

SECTION 4.3

1. $R = 1$, $|z| < 1$, $|z| = 1$. **3.** $R = \infty$. **5.** $R = \frac{1}{2}$, $|z + \frac{i}{2}| < \frac{1}{2}$, $|z + \frac{i}{2}| = \frac{1}{2}$. **7.** $R = 1$, $|z + 6| < 1$, $|z + 6| = 1$. **9.** $R = \frac{1}{2}$, $|z| < \frac{1}{2}$, $|z| = \frac{1}{2}$. **11.** $R = \frac{1}{2}$, $|z| < \frac{1}{2}$, $|z| = \frac{1}{2}$. **13.** $\frac{2}{(1-z)^2}$, $|z| < 1$. **15.** $\frac{1}{z^2}(z - (1+z)\operatorname{Log}(1+z))$, $|z| < 1$. **17.** $\frac{3}{-3z+3+i}$, $|z - \frac{i}{3}| < 1$.

SECTION 4.4

1. $R = \infty$. **3.** $R = 3$. **5.** $R = 2\sqrt{2}$. **7.** $R = 1$. **9.** $R = \frac{\pi}{2}$. **11.** $R = \sqrt{2 - \sqrt{3}}$. **13.** $\sum_{n=0}^{\infty} z^{n+1}$, $|z| < 1$. **15.** $-i\sum_{n=1}^{\infty} n(n+1)(iz)^n$, $|z| < 1$. **17.** $e + e\sum_{n=1}^{\infty}\left(\frac{1}{(n-1)!} + \frac{1}{n!}\right)(z-1)^n$, all z. **19.** $\sum_{n=0}^{\infty}\frac{(-1)^n}{(2n)!}\frac{z^{2n+1}}{2^{2n}}$, all z. **21.** (a) $\frac{1}{1-z} - \frac{1}{2-z} = \sum_{n=0}^{\infty} z^n\left(1 - \frac{1}{2^{n+1}}\right)$, $|z| < 1$. (b) and (c) $\sum_{n=0}^{\infty} z^n \sum_{n=0}^{\infty}\frac{1}{2^{n+1}} z^n = \sum c_n z^n$, where $c_n = \sum_{k=0}^{n}\frac{1}{2^{k+1}} = 1 - \frac{1}{2^{n+1}}$. **23.** $f(z) = \frac{1}{2i\sqrt{3}}\sum_{n=0}^{\infty}\left[\left(\frac{-1+i\sqrt{3}}{2}\right)^{n+1} - \left(\frac{-1-i\sqrt{3}}{2}\right)^{n+1}\right] z^n$, $|z| < 1$. **25.** $-\frac{1}{4}\sum_{n=0}^{\infty}\frac{(n+1)(n+2)}{2}\frac{z^n}{(2i)^n}$, $|z| < 2$. **27.** $\sum_{n=1}^{\infty}(-1)^n\frac{z^{2n-1}}{(2n)!}$, all z; f is entire. **29.** $\sum_{n=1}^{\infty}(-1)^n\frac{z^{2n-2}}{n!}$, all z; f is entire.

SECTION 4.5

1. $-\sum_{n=1}^{\infty}\frac{(-1)^n}{z^{2n}}$. **3.** $-1 - \sum_{n=1}^{\infty}\frac{1}{z^n}$. **5.** $\sum_{n=1}^{\infty}\frac{(-1)^{n-1}}{n z^n}$. **7.** $\sum_{n=0}^{\infty}\frac{1}{n!(1-z)^{n+1}}$. **9.** $\sum_{n=0}^{\infty}\frac{(-1)^n 2^{2n}}{(2n+1)! z^{2n+2}}$. **11.** $\sum_{n=0}^{\infty}\frac{2^{2n}B_{2n}}{(2n)!} z^{2n-1}$. **13.** $\frac{z}{(z+2)(z+3)} = \frac{3}{3+z} - \frac{2}{2+z}$, $\frac{z}{(z+2)(z+3)} = \sum_{n=0}^{\infty}(-1)^n\left(\frac{z}{3}\right)^n + \sum_{n=1}^{\infty}(-1)^n\left(\frac{2}{z}\right)^n$, for $2 < |z| < 3$. **15.** $\frac{1}{(3z-1)(2z+1)} = -\frac{2}{5(1+2z)} + \frac{3}{5(-1+3z)}$, $\frac{1}{(3z-1)(2z+1)} = -\frac{2}{5}\sum_{n=0}^{\infty}(-1)^n(2z)^n + \frac{3}{5}\sum_{n=1}^{\infty}\frac{1}{(3z)^n}$, for $\frac{1}{3} < |z| < \frac{1}{2}$.

17. $\frac{z^2+(1-i)z+2}{(z-i)(z+2)} = 1 + \frac{1}{z-i} - \frac{2}{2+z}$, $\frac{z^2+(1-i)z+2}{(z-i)(z+2)} = 1 - \sum_{n=0}^{\infty}(-1)^n\left(\frac{z}{2}\right)^n - i\sum_{n=1}^{\infty}\left(\frac{i}{z}\right)^n$, for $1 < |z| < 2$. **19.** $\frac{2z^3-4z^2-5z+11}{z-1} = 2z^2 - 2z - 7 + \frac{4}{z-1}$, $\frac{2z^3-4z^2-5z+11}{z-1} = 2(z-2)^2 + 2(z-2) - 11 - 4\sum_{n=1}^{\infty}\frac{(-1)^n}{(z-2)^n}$, for $1 < |z-2|$. **21.** $f(z) = \frac{a}{z-1} - \frac{a}{z+i} = -\frac{a}{2}\frac{1}{1-\left(\frac{z+1}{2}\right)} - \frac{a}{-1+i}\frac{1}{1+\left(\frac{z+1}{-1+i}\right)}$, where $a = \frac{1}{2} - \frac{i}{2}$. We have three Laurent series. In the disk $|z+1| < \sqrt{2}$, $f(z) = -\frac{a}{2}\sum_{n=0}^{\infty}\frac{(z+1)^n}{2^n} - \frac{a}{-1+i}\sum_{n=0}^{\infty}(-1)^n\left(\frac{z+1}{-1+i}\right)^n$. In the annulus, $\sqrt{2} < |z+1| < 2$, $f(z) = -\frac{a}{2}\sum_{n=0}^{\infty}\frac{(z+1)^n}{2^n} + \frac{a}{-1+i}\sum_{n=1}^{\infty}(-1)^n\left(\frac{-1+i}{z+1}\right)^n$. In the annulus $2 < |z+1|$, $f(z) = \frac{a}{2}\sum_{n=1}^{\infty}\left(\frac{2}{z+1}\right)^n - \frac{a}{-1+i}\sum_{n=1}^{\infty}(-1)^n\left(\frac{-1+i}{z+1}\right)^n$. **23.** (a) Use (5) with $w = -z$. (b) $\sum_{n=1}^{\infty}(-1)^{n-1}\frac{n}{z^{n+1}}$. (c) $\sum_{n=1}^{\infty}(-1)^{n-1}\frac{n}{z^n}$. (d) $\sum_{n=1}^{\infty}(-1)^{n-1}\frac{n(n+1)}{2z^n}$. **25.** $2\pi i$. **27.** $2\pi i\sum_{n=0}^{\infty}\frac{1}{(2n+1)!(2n)!} \approx 6.8i$. **29.** $2\pi i$.

SECTION 4.6

1. $z = \pm 1$ (simple), $z = k\pi$ (k integer) (simple). **2.** $z = 0$ (order 4), $z = 2k\pi i$ ($k \neq 0$ integer) (simple). **3.** $z = 0$ (simple), $z = 1$ (order 2). **4.** $z = \frac{1}{k\pi}$ ($k \neq 0$ integer) (simple). **5.** $z = 0$ (order 3), $z = k\pi$ ($k \neq 0$ integer) (order 7). **6.** $z = 1$ (order 3), $z = k\pi i$ (k integer) (order 2). **7.** $z = -1 \pm \sqrt{2}$ (order 3). **8.** $z = k\pi i$ (k integer) (order 1). **9.** order 4. **10.** order 2. **11.** order 3. **12.** order 1. **13.** $z = -1$ (order 1 pole), $z = k\pi$ (k integer) (order 1 pole). **14.** $z = i$ (order 1 pole), $z = 1$ (order 1 pole). **15.** $z = 0$ (order 1 pole), $z = 1$ (removable, set function to $= 0$), $z = k\pi$ (k integer) (order 1 pole), $z = k$ ($k \neq 0$, $k \neq 1$ integer) (order 1 pole). **16.** $z = 1$ (essential singularity). **17.** $z = 0$ (essential singularity), $z = \frac{1}{\frac{\pi}{2}+k\pi}$ (k integer) (order 1 pole). **18.** $z = 0$ (removable, set function to $= 1$), $z = 2k\pi i$ ($k \neq 0$ integer) (order 1 pole). **19.** $z = 0$ (order 3 pole), $z = e^{\frac{ik\pi}{2}}$ ($k = 0, 1, 2, 3$) (order 1 pole). **20.** $z = 0$ (essential singularity). **21.** $z = 0$ (essential singularity). **22.** $z = 2k\pi i$ (k integer) (order 2 pole). **23.** $z = k\pi$ (k integer) (order 1 pole), $z = \frac{\pi}{2}$ (removable, set function to $= 0$). **25.** Removable, zero. **26.** Removable. **27.** Removable, zero. **28.** Essential singularity. **29.** Removable, zero. **30.** Singularity at ∞ is not isolated. **31.** (c) Fix z, then find a region Ω that includes z and part of the real axis, such that each side of the proposed identity is analytic in Ω. **35.** (a) Use power series. (b) Use Laurent series. (c) Use the definition of a removable singularity.

(d) Let $h(z) = (z - z_0)^m g(z)$ for suitable m and remove h's singularity at z_0 so that $h(z)$ is analytic and nonvanishing in a neighborhood of z_0. The function $f(z)h(z)$ has an essential singularity if and only if $f(z)g(z)$ does. Now use the following: If $\lim_{z \to z_0} f(z)h(z)$ exists and is finite then $\lim_{z \to z_0} f(z)$ exists and is finite. Also, if $\lim_{z \to z_0} |f(z)h(z)| = \infty$ then $\lim_{z \to z_0} |f(z)| = \infty$. For the case where g has an essential singularity, the assertion is not true; look at $e^{\frac{1}{z}} e^{-\frac{1}{z}} = 1$, $z \neq 0$.

CHAPTER 5

SECTION 5.1

1. $\operatorname{Res}(f, 0) = 1$. **2.** $\operatorname{Res}(f, -1 - i) = \frac{1}{2}$; $\operatorname{Res}(f, -1 + i) = \frac{1}{2}$. **3.** $\operatorname{Res}(f, 0) = 3$. **4.** $\operatorname{Res}(f, 0) = 0$; $\operatorname{Res}(f, -i) = \frac{i}{2} \sin 1$; $\operatorname{Res}(f, i) = -\frac{i}{2} \sin 1$. **5.** $\operatorname{Res}(f, -3i) = -3 - 9i$. **6.** $\operatorname{Res}(f, 0) = \frac{1}{2}$. **7.** $\operatorname{Res}(f, 0) = 0$. **8.** $\operatorname{Res}(f, -1) = 0$. **9.** If $g(z) = \frac{z+1}{z-1}$, then for m an integer,
$\operatorname{Res}(g(z)\csc(\pi z), m) = \frac{(-1)^m}{\pi} g(m)$ $(m \neq 1)$ and
$\operatorname{Res}(g(z)\csc(\pi z), 1) = \frac{-1}{\pi}$. **10.** $\operatorname{Res}(f, 0) = 0$.
11. $\operatorname{Res}(f, 0) = \sum_{k=0}^{\infty} \frac{1}{k!(k+1)!} = -iJ_1(2i) \approx 1.59$, where J_1 is the Bessel function of order 1. **12.** $\operatorname{Res}(f, 0) = 1$.
13. $\frac{2\pi i}{3}$. **14.** $\frac{2\pi i}{5}$. **15.** -4π. **17.** $-\frac{18\pi i}{10!}$. **18.** $4\pi i$. **19.** 0.
20. $-2i$. **21.** 0. **22.** $2\pi i$. **23.** $\frac{2\pi i}{5!}$.
24. $\frac{2}{\pi} \sum_{k=1}^{15} k^2$. **25.** $\frac{2\pi i}{5!}$. **26.** $\frac{-1}{720}$.

SECTION 5.2

1. $\frac{2\pi}{\sqrt{3}}$. **2.** π. **3.** $\frac{\pi}{3}$. **4.** $\sqrt{2}\pi$. **5.** $\frac{\pi}{6}$. **6.** $-\frac{7\pi}{24}$. **7.** $\frac{\pi}{15}$.
8. $\frac{\pi}{2}(-7 + 4\sqrt{3})$. **9.** $\frac{\pi}{3}$. **10.** $\frac{\pi}{2\sqrt{5}}$.

SECTION 5.7

1. 0. **3.** 2. **5.** 2. **7.** 1. **9.** 0. **11.** 2. **13.** 1. **17.** By Rouché's theorem, $z^5 + 3z + 5$ has no zeros in $|z| < 1$, so the integral is zero by Cauchy's theorem. **19.** $2\pi i \cosh 1$.
29. $p_n(z) = p(z) + \frac{1}{n}$, then $p_n \to p$ uniformly on closed subsets of $|z| < 1$ (in fact on $\mathbb{C}$). By Rouché's theorem, $p_n(z)$ has no zeros in $|z| < 1$, so $p(z)$ has no zeros in $|z| < 1$, by Hurwitz's theorem.

CHAPTER 6

SECTION 6.1

1. $f'(z) = -e^{-z}(z - 1)^2$, f conformal at all $z \neq 1$.
3. $f'(z) = e^{-z}(\cos z - \sin z)$, f conformal at all $z \neq \frac{\pi}{4} + k\pi$, k an integer. **5.** $f'(z) = \frac{z^2 - 1}{z^2}$, f conformal at all $z \neq 0, \pm 1$. **7.** $f'(1) = 1 \cdot e^{i\pi}$, $|f'(1)| = 1$, $\arg f'(1) = \pi$; $f'(i) = 1$, $|f'(i)| = 1$, $\arg f'(i) = 0$; $f'(1 + i) =$

$\frac{e^{i\frac{\pi}{2}}}{2}$, $|f'(1 + i)| = \frac{1}{2}$, $\arg f'(1 + i) = \frac{\pi}{2}$. **9.** $f'(0) = 1$, $|f'(0)| = 1$, $\arg f'(0) = 0$; $f'(\pi + ia) = e^{i\pi} \cosh a$, $|f'(\pi + ia)| = \cosh a$, $\arg f'(\pi + ia) = \pi$; $f'(i\pi) = \cosh \pi$, $|f'(i\pi)| = \cosh \pi$, $\arg f'(i\pi) = 0$. **11.** $f'(\pi) = e^{-i\frac{\pi}{2}}$, $|f'(\pi)| = 1$, $\arg f'(\pi) = -\frac{\pi}{2}$; $f'(i\pi) = e^{-\pi} e^{i\frac{\pi}{2}}$, $|f'(i\pi)| = e^{-\pi}$, $\arg f'(i\pi) = \frac{\pi}{2}$; $f'(\frac{\pi}{2}) = e^{i\pi}$, $|f'(\frac{\pi}{2})| = 1$, $\arg f'(\frac{\pi}{2}) = \pi$. **13.** Images are orthogonal where $f'(a + ib) \neq 0$, $f'(z) = e^z \neq 0$ for all $z = a + ib$. **15.** Images are orthogonal where $f'(a + ib) \neq 0$, $f'(z) = 1 + i \neq 0$ for all $z = a + ib$. **19.** (a) $f'(z) = \frac{1+z}{(1-z)^3}$ for all $z \neq 1$, $f'(z) \neq 0$ for $z \neq -1$. (b) $\frac{z_1}{(1-z_1)^2} = \frac{z_2}{(1-z_2)^2} \Leftrightarrow (z_1 - z_2)(1 + z_1 z_2) = 0$. If $|z_1| < 1$ and $|z_2| < 1$, then the only solutions are $z_1 = z_2$, and so f is one-to-one in $|z| < 1$. If $|z_1| \geq 1$ or $|z_2| \geq 1$, say $|z_1| \geq 1$ and $z - 1 \neq i$, then $z_2 = -\frac{1}{z_1} \neq z_1$, $(|z_2| \leq 1)$, is a solution and so f is not one-to-one in $|z| \leq \rho$ for any $\rho \geq 1$.

SECTION 6.2

1. (a) $w_1 = 0$, $w_2 = i$, $w_3 = 1$. (b) $\phi[L_1]$ is the imaginary axis, $\phi[L_2]$ is the unit circle. **3.** (a) $w_1 = -i$, $w_2 = \infty$, $w_3 = 1 - 2i$. (b) $\phi[L_1]$ is the imaginary axis, $\phi[L_2]$ is the horizontal line through $-2i$. **5.** $\psi(w) = \frac{i-w}{i+w}$, $\psi(0) = 1$, $\psi(i) = 0$, $\psi(1) = i$. **7.** (a) $\frac{1}{z}$. (b) The unit circle; the region outside the unit disk; the unit disk. **9.** If $c \neq 0$, then $z = \frac{az+b}{cz+d} \Leftrightarrow cz^2 + (d - a)z - b = 0$ has at most two solutions. If $c = 0$ and $d \neq 0$, then $z = \frac{a}{d}z + \frac{b}{d}$ has one solution or no solution or infinitely many solutions depending on whether $a \neq d$, $a = d$ and $b \neq 0$, $a = d$ and $b = 0$. If $c = 0$ and $d = 0$, then $z = az + b$ has one solution or no solution or infinitely many solutions depending on whether $a \neq 1$, $a = 1$ and $b \neq 0$, $a = 1$ and $b = 0$. **12.** $w_1 = f_1(z) = z - i$, $w_2 = f_2(w_1) = \frac{e^{i\frac{\pi}{4}}}{\sqrt{2}} w_1$, $w_3 = f_3(w_2) = \frac{i-w_2}{i+w_2}$. **13.** $w_1 = f_1(z) = z - 1 - i$, $w_2 = f_2(w_1) = \frac{1}{w_1}$, $w_3 = f_3(w_2) = i\frac{1-w_2}{1+w_2}$.
14. $w_1 = f_1(z) = \frac{2+z}{2-z}$, $w_2 = f_2(w_1) = w_1^6$, $w_3 = f_3(w_2) = \frac{i-w_2}{i+w_2}$. **15.** $w_1 = f_1(z) = i\frac{1-z}{1+z}$, $w_2 = f_2(w_1) = \pi w_1$, $w_3 = f_3(w_2) = e_2^w$.
16. $w_1 = f_1(z) = \frac{1}{2}(z + \frac{1}{z})$, $w_2 = f_2(w_1) = \frac{i-w_1}{i+w_1}$.
17. $w_1 = f_1(z) = \sin z$, $w_2 = f_2(w_1) = \frac{i-w_1}{i+w_1}$.
18. $w_1 = f_1(z) = \frac{1}{z}$, $w_2 = f_2(w_1) = w_1 + \frac{1+i}{2}$, $w_3 = f_3(w_2) = (-i w_2)^{2/3}$.
19. $w_1 = f_1(z) = -\frac{i}{2}(z - 1)$, $w_2 = f_2(w_1) = \frac{i-w_1}{i+w_1}$.
20. $w_1 = f_1(z) = \frac{z^2}{4}$, $w_2 = f_2(w_1) = i\frac{1-w_1}{1+w_1}$.
21. $w_1 = f_1(z) = i\frac{1-z}{1+z}$, $w_2 = f_2(w_1) = \operatorname{Log} w_1$.
22. $w_1 = f_1(z) = i\frac{1-z}{1+z}$, $w_2 = f_2(w_1) = w_1^2$.

23. $w_1 = f_1(z) = e^z$, $w_2 = f_2(w_1) = iw_1$.

24. $w_1 = f_1(z) = i\frac{1-z}{1+z}$, $w_2 = f_2(w_1) = 1 + w_1^2$,

$w_3 = f_3(w_2) = (w_2)^{1/2}$, $\log_0$ branch.

26. $\alpha = \frac{1}{2}$, $\phi_\alpha(z) = \frac{2z-1}{2-z}$, $\phi_\alpha(b) = \frac{2}{3}$.

27. $\alpha = \frac{1}{4}$, $\phi_\alpha(z) = \frac{4z-1}{4-z}$, $\phi_\alpha(b) = \frac{1}{4}$.

28. $w_1 = f_1(z) = \frac{1}{z}$, $w_2 = f_2(w_1) = e^{i\frac{\pi}{4}}w_1$,

$w_3 = f_3(w_2) = \phi_\alpha(w_2)$ where $\alpha = \frac{1}{2}$, $\phi_\alpha(w_2) = \frac{2w_2-1}{2-w_2}$,

$f(z) = \frac{2e^{i\frac{\pi}{4}}-z}{2z-e^{i\frac{\pi}{4}}}$. **29.** $w_1 = f_1(z) = z - \frac{3}{2}i$, $w_2 = f_2(w_1) = \frac{1}{w_1}$, $w_3 = f_3(w_2) = \phi_\alpha(-\frac{i}{2}w_2)$ where $\alpha = 3 - 2\sqrt{2}$,

$f(z) = \frac{-i-2\alpha(z-\frac{3}{2}i)}{2(z-\frac{3}{2}i)+i\alpha}$.

SECTION 6.3

1. Use $w = \phi(z) = i\frac{1-z}{1+z}$. In the w-plane, $U(w) = -\frac{20}{\pi}\operatorname{Arg} w + 70$, and so

$u(z) = U(\phi(z)) = -\frac{20}{\pi}\operatorname{Arg}\left(i\frac{1-z}{1+z}\right) + 70$.

2. Use $w = \phi(z) = i\frac{1-z}{1+z}$. In the w-plane, $U(w) = \frac{100}{\pi}(\operatorname{Arg}(w-1) - \operatorname{Arg} w)$, and so $u(z) = U(\phi(z)) = \frac{100}{\pi}(\operatorname{Arg}(\phi(z)-1) - \operatorname{Arg}(\phi(z)))$.

3. Use $w = \phi(z) = i\frac{1-z}{1+z}$. Then $u(z) = U(\phi(z))$ where $U(w) = \frac{100}{\pi}(\operatorname{Arg}(w-1) - \operatorname{Arg} w + \operatorname{Arg}(w+1))$.

4. Use $w = \phi(z) = i\frac{1-z}{1+z}$. Then $u(z) = U(\phi(z))$ where $U(w) = 100 - \frac{180}{\pi}\operatorname{Arg} w$. **5.** Use $w_1 = z^4$, $w_2 = i\frac{1-z}{1+z}$. $w = \phi(z) = i\frac{1-z^4}{1+z^4}$ Then $u(z) = U(\phi(z))$ where $U(w) = 100 - \frac{200}{\pi}\operatorname{Arg} w$. **6.** Use $w = \phi(z) = \frac{2+i}{2-i}\frac{2+z}{2-z}$ (from Example 3, Section 6.2). Then $u(z) = U(\phi(z))$ where $U(w) = 80 - \frac{120}{\pi}\operatorname{Arg} w$. **7.** Use $w = \phi(z) = z^2 - 1$. Then $u(z) = U(\phi(z))$ where $U(w) = \frac{200}{\pi}\operatorname{Arg} w$. **8.** Use $w = \phi(z) = \sin z$. Then $u(z) = U(\phi(z))$ where $U(w) = 50 + \frac{50}{\pi}\operatorname{Arg}(w-1) - \frac{100}{\pi}\operatorname{Arg} w$.

9. Use $w = \phi(z) = \frac{1}{2}(z+\frac{1}{z})$. Then $u(z) = U(\phi(z))$ where $U(w) = 100 - \frac{100}{\pi}(\operatorname{Arg}(w-1) - \operatorname{Arg}(w+1))$.

10. Use $w = \phi(z) = \frac{2+i}{2-i}\frac{2+z}{2-z}$ (from Example 3, Section 6.2). Then $u(z) = U(\phi(z))$ where $U(w) = \frac{300}{\pi}\operatorname{Arg} w$. **11.** Use $w = \phi(z) = (i-1)\frac{z+2}{z-2}$. Then $u(z) = U(\phi(z))$ where $U(w) = \frac{400}{\pi}\operatorname{Arg} w$. **12.** Use $w_1 = \phi(z) = i\frac{z-1}{z+1}$, $w_2 = w_1^2$, $w = \phi(z) = -\left(\frac{z-1}{z+1}\right)^2$. Then $u(z) = U(\phi(z))$ where $U(w) = 100 - \frac{40}{\pi}\operatorname{Arg} w - \frac{60}{\pi}\operatorname{Arg}(w+1)$.

13. Use $w = \phi_\alpha(z) = \frac{4z-1}{4-z}$, $\alpha = \frac{1}{4}$, $\phi_\alpha(\frac{8}{17}) = \frac{1}{4}$ (see Exercise 27, Section 6.2). Then $u(z) = U(\phi(z))$ where $U(w) = -100\frac{\ln|w|}{\ln 4}$. **14.** Use $w = \phi_\alpha(z) = \frac{2z-1}{2-z}$, $\alpha = \frac{1}{2}$, $\phi_\alpha(\frac{7}{8}) = \frac{2}{3}$ (see Exercise 26, Section 6.2). Then $u(z) = U(\phi(z))$ where $U(w) = 100\frac{\ln|w|}{\ln\frac{2}{3}}$. **15.** Use $w = f(z) = $

$\frac{-i-2\alpha(z-\frac{3}{2}i)}{2(z-\frac{3}{2}i)+i\alpha}$, $\alpha = 3 - 2\sqrt{2}$, (see Exercise 29, Section 6.2). The transformed region is an annulus with inner radius $\phi_\alpha(\frac{1}{3}) = 3 - 2\sqrt{2}$. Moreover, $T_1 = 100$ and $T_2 = 0$. So $u(z) = U(f(z))$ where $U(w) = 100\frac{\ln|w|}{\ln(3-2\sqrt{2})}$.

17 (a) $u(x+iy) = \frac{1}{\pi}\tan^{-1}\left(\frac{x+1}{y}\right) + \frac{1}{\pi}\tan^{-1}\left(\frac{x-1}{y}\right) + \frac{y}{2\pi}\ln\left(\frac{(1-x)^2+y^2}{(1+x)^2+y^2}\right) + \frac{x}{\pi}\left(\tan^{-1}\left(\frac{1-x}{y}\right) + \tan^{-1}\left(\frac{1+x}{y}\right)\right)$.

(b) In your answer in (a), replace x by $\operatorname{Re}\sin z = \sin x\cosh y$ and y by $\operatorname{Im}\sin z = \cos x\sinh y$.

24. $\frac{-1+i\sqrt{3}}{2} = e^{i\frac{2\pi}{3}}$, $u(x+iy) = $

$2y\left[\frac{e^{i\frac{2\pi}{3}}}{\sqrt{3}\left((x-e^{i\frac{2\pi}{3}})^2+y^2\right)} + \frac{x+iy}{2y((x+iy)^2+(x+iy)+1)}\right]$.

25. Let $h(z) = \frac{1}{z^4+1}\frac{1}{(z-x)^2+y^2}$, $R_1 = \operatorname{Res}(h(z), x+iy) = \frac{1}{(x+iy)^4+1}\frac{1}{2iy}$;

$R_2 = \operatorname{Res}(h(z), e^{i\frac{\pi}{4}}) = \frac{1}{(e^{i\frac{\pi}{4}}-x)^2+y^2}\frac{1}{4e^{i\frac{3\pi}{4}}}$;

$R_3 = \operatorname{Res}(h(z), e^{i\frac{3\pi}{4}}) = \frac{1}{(e^{i\frac{3\pi}{4}}-x)^2+y^2}\frac{1}{4e^{i\frac{\pi}{4}}}$;

$u(x+iy) = 2yi(R_1 + R_2 + R_3)$.

26. $u(x+iy) = \frac{y}{\pi}\operatorname{Im}\left(\text{P.V.}\int_{-\infty}^{\infty}\frac{e^{is}}{s(s-x)^2+y^2}\,ds\right)$

$= \frac{(x\sin x - y\cos x)e^{-y}}{x^2+y^2} + \frac{y}{x^2+y^2}$. **27.** Use Exercise 28 with $a = 1$. **28.** $u(x+iy) = \cos ax\,e^{-|a|y}$. **29.** $u(x+iy) = \sin ax\,e^{-|a|y}$. **30.** Use $w = \phi(z) = (z^2+1)^{1/2}$, $\log_0$-branch to transform to a Dirichlet problem in the upper half-plane with boundary data $f(w) = 100$ if $|w| < 1$ and $f(w) = 0$ otherwise. In the w-plane, $U(w) = \frac{100}{\pi}(\operatorname{Arg}(w-1) - \operatorname{Arg}(w+1))$ and in the z-plane, $u(z) = U(\phi(z))$.

31. Use $w = \phi(z) = (z^2+1)^{1/2}$, $\log_0$-branch to transform to a Dirichlet problem in the upper half-plane with boundary data $f(w) = 100$ if $w > 0$ and $f(w) = 0$ if $w < 0$. In the w-plane, $U(w) = 100 - \frac{100}{\pi}\operatorname{Arg} w$ and in the z-plane, $u(z) = U(\phi(z))$. **32.** Use $w_1 = e^z$ to transform to a region as in Exercise 30. Then use $w_2 = \phi(w_1)$ where ϕ is as in Exercise 30. So $w = (e^{2z}+1)^{1/2}$ ($\log_0$-branch) will transform to a Dirichlet problem in the upper half-plane with boundary data $f(w) = 100$ if $|w| < 1$ and $f(w) = 0$ otherwise. In the w-plane, $U(w) = \frac{100}{\pi}(\operatorname{Arg}(w-1) - \operatorname{Arg}(w+1))$ and in the z-plane, $u(z) = U(\phi(z))$. **33.** Use $w = \phi(z) = z^2$ to transform to a Dirichlet problem in the upper half-plane with boundary data $f(w) = w$ if $0 < w < 1$ and $f(w) = 0$ otherwise. In the w-plane, $U(w) = \frac{\operatorname{Im} w}{2\pi}\ln\left(\frac{(1-\operatorname{Re} w)^2+(\operatorname{Im} w)^2}{(\operatorname{Re} w)^2+(\operatorname{Im} w)^2}\right) + \frac{\operatorname{Re} w}{\operatorname{Im} w}\left(\tan^{-1}\left(\frac{1-\operatorname{Re} w}{\operatorname{Im} w}\right) + \tan^{-1}\left(\frac{\operatorname{Re} w}{\operatorname{Im} w}\right)\right)$, and in the z-plane, $u(z) = U(\phi(z))$. **34.** Use same ϕ as in Exercise 33 to transform to a Dirichlet problem in the w-plane with boundary data $f(w) = |w|$ if $|w| < 1$ and $f(w) = 0$ otherwise. Let $U_1(w)$ denote the solution in Exercise 33. The solution in the w-plane is $U(w) = U_1(w) + U_1(-\operatorname{Re} w + i \operatorname{Im} w)$. In the z-plane $u(z) = U(\phi(z))$. **35.** Use $\phi(z) = \cos z$ to transform to a Dirichlet problem in the w-plane with boundary data $f(w) = T_n(w)$ if $|w| < 1$ and $f(w) = 0$ otherwise, where T_n is the Chebychev polynomial. In the w-plane,

$$U(w) = \frac{\operatorname{Im} w}{\pi} \int_{-1}^{1} \frac{T_n(s)}{(s - \operatorname{Re} w)^2 + (\operatorname{Im} w)^2}\, ds.$$

In the z-plane $u(z) = U(\phi(z))$. For $n = 2$,

$$U(w) = \frac{4 \operatorname{Im} w}{\pi} + \frac{2 \operatorname{Re} w \operatorname{Im} w}{\pi} \ln\left(\frac{(1 - \operatorname{Re} w)^2 + (\operatorname{Im} w)^2}{(1 + \operatorname{Re} w)^2 + (\operatorname{Im} w)^2}\right) +$$
$$\frac{(\operatorname{Re} w)^2 - (\operatorname{Im} w)^2 - 1}{\pi}\left(\tan^{-1}\left(\frac{1 - \operatorname{Re} w}{\operatorname{Im} w}\right) + \tan^{-1}\left(\frac{1 + \operatorname{Re} w}{\operatorname{Im} w}\right)\right).$$

SECTION 6.4

1. $\theta_1 = \frac{\pi}{2}$, $\theta_2 = \pi$, $\theta_3 = -\frac{\pi}{2}$; $A = -i$, $B = 0$; $f(z) = (1 - z^2)^{\frac{1}{2}}$. **2.** $\theta_1 = -\pi$, $\theta_2 = \frac{\pi}{2}$; $A = \frac{3i}{8\sqrt{2}}$, $B = 1$; $f(z) = \frac{3i}{8\sqrt{2}}\left(\frac{2}{3}(z - 1)^{\frac{3}{2}} + 4(z - 1)^{\frac{1}{2}}\right) + 1$. **3.** $\theta_1 = -\frac{\pi}{2}$, $\theta_2 = \frac{\pi}{2}$; $A = \frac{1}{\pi}$, $B = -\frac{i}{2}$; $f(z) = \frac{1}{\pi}\left(i(1 - z^2)^{\frac{1}{2}} - i\sin^{-1} z\right) - \frac{i}{2}$. **4.** $\theta_1 = \frac{\pi}{2}$, $\theta_2 = -\frac{\pi}{2}$; $A = \frac{1}{\pi}$, $B = \frac{i}{2}$; $f(z) = \frac{1}{\pi}\left(i(1 - z^2)^{\frac{1}{2}} + i\sin^{-1} z\right) + \frac{i}{2}$. **5.** $\theta_1 = -\frac{\pi}{2}$, $\theta_2 = -\frac{\pi}{2}$; $A = \frac{4i}{\pi}$, $B = 0$; $f(z) = -\frac{2}{\pi}\left(z(1 - z^2)^{\frac{1}{2}} + \sin^{-1} z\right)$. **6.** $\theta_1 = \pi$, $\theta_2 = -\frac{\pi}{2}$; $A = -\frac{i}{\pi\sqrt{2}}$, $B = i$;

$$f(z) = i - \frac{i\sqrt{2}}{\pi}(z - 1)^{\frac{1}{2}} - \frac{1}{\pi}\operatorname{Log}\frac{(z - 1)^{\frac{1}{2}} + i\sqrt{2}}{(z - 1)^{\frac{1}{2}} - i\sqrt{2}}.$$

SECTION 6.5

1. To map Ω onto the unit disk, use $\phi(z) = \frac{1 - z}{1 + z}$. Then, for $z = x + iy$ and $\zeta = s + it$, $G(z, \zeta) = \frac{1}{2}\ln\frac{(s - x)^2 + (t - y)^2}{(s + x)^2 (t - y)^2}$. **2.** To map Ω onto the unit disk, use $\phi(z) = z - 2i$. Then, $G(z, \zeta) = \ln\left|\frac{\zeta - z}{1 - (\zeta - 2i)(z - 2i)}\right|$. **3.** To map Ω onto the unit disk, use $\phi(z) = \frac{i - z^2}{i + z^2}$. Then, $G(z, \zeta) = \ln\frac{|z - \zeta||z + \zeta|}{|\overline{z} - \zeta||\overline{z} + \zeta|}$. **4.** To map Ω onto the unit disk, use $\phi(z) = \frac{i - z^4}{i + z^4}$. Then, $G(z, \zeta) = \ln\left|\frac{z^4 - \zeta^4}{(\overline{z})^4 - \zeta^4}\right|$. **5.** To map Ω onto the unit disk, use $\phi(z) = \frac{i - e^z}{i + e^z}$. Then, $G(z, \zeta) = \ln\left|\frac{e^z - e^\zeta}{e^{\overline{z}} - e^\zeta}\right|$. **6.** To map Ω onto the unit disk, use $\phi(z) = \frac{i - e^{\frac{\pi}{b}z}}{i + e^{\frac{\pi}{b}z}}$. Then, $G(z, \zeta) = \ln\left|\frac{e^{\frac{\pi}{b}z} - e^{\frac{\pi}{b}\zeta}}{e^{\frac{\pi}{b}\overline{z}} - e^{\frac{\pi}{b}\zeta}}\right|$. **7.** To map Ω onto the unit disk, use $\phi(z) = -\frac{z^2 - 2iz + 1}{z^2 + 2iz + 1}$. Then, $G(z, \zeta) = \ln\left|\frac{(z - \zeta)(1 - \zeta z)}{(\overline{z} - \zeta)(1 - \overline{z}\zeta)}\right|$. **8.** To map Ω onto the unit disk, use $\phi(z) = \frac{(1 - z)^2 + i(1 + z)^2}{-(1 - z)^2 + i(1 + z)^2}$. Then, $G(z, \zeta)$ same as in Exercise 7.

SECTION 6.6

1. To map Ω onto the unit disk, use $\phi(z) = i(z - 1)$. Then, $N(z, \zeta) = \ln|(z - \zeta)(\overline{z} + \zeta - 2)|$. **2.** To map Ω onto the unit disk, use $\phi(z) = \frac{i + z^2}{i - z^2}$. Then, $N(z, \zeta) = \ln\left|4\frac{(z^2 - \zeta^2)(\zeta^2 + (\overline{z})^2)}{(z^2 - i)(\zeta^2 - i)^2(i + (\overline{z})^2)}\right|$.

3. To map Ω onto the unit disk, use $\phi(z) = \frac{1 + z^2}{1 - z^2}$. Then, $N(z, \zeta) = \ln\left|4\frac{(z^2 - \zeta^2)(\zeta^2 - (\overline{z})^2)}{(z^2 - 1)(\zeta^2 - 1)^2((\overline{z})^2 - 1)}\right|$.

4. To map Ω onto the unit disk, use $\phi(z) = \frac{i - z^4}{i + z^4}$. Then, $N(z, \zeta) = \ln\left|4\frac{(z^4 - \zeta^4)(\zeta^4 + (\overline{z})^4)}{(i + z^4)(i + \zeta^4)^2((\overline{z})^4 - i)}\right|$.

5. To map Ω onto the unit disk, use $\phi(z) = \frac{i - \sin z}{i + \sin z}$. Then, $N(z, \zeta) = \ln\left|4\frac{(\sin z - \sin \zeta)(\sin \zeta + \overline{\sin z})}{(1 + \sin z)(i + \sin \zeta)^2(\overline{\sin z} - i)}\right|$.

6. To map Ω onto the unit disk, use $\phi(z) = \frac{i - e^z}{i + e^z}$. Then, $N(z, \zeta) = \ln\left|4\frac{(e^z - e^\zeta)(e^\zeta + e^{\overline{z}})}{(1 + e^z)(i + e^\zeta)^2(e^{\overline{z}} - i)}\right|$.

CHAPTER 7

SECTION 7.1

1. (a) $T = 2\pi$, (b) $T = 2$, (c) $T = 3\pi$, (d) $T = 2\pi$. **3.**

$$f(x) = \begin{cases} 1 & \text{if } 0 \le x < 1/2, \\ 0 & \text{if } 1/2 \le x < 1, \\ f(x + 1) & \text{otherwise.} \end{cases}$$

11. $f(x) = |\sin x|$ for all x, $\int_{-\pi/2}^{\pi/2} |\sin x|\, dx = 2$. **13.** $\pi/2$. **17.** (a)

$$F(x) = \begin{cases} x - \frac{x^2}{2} & \text{for } 0 \le x \le 2, \\ F(x + 2) & \text{otherwise.} \end{cases}$$

SECTION 7.2

1. (b) The graph of the Fourier series differs from the graph of the function at the points of discontinuity $x = k\pi$. At these points, the Fourier series takes the value $1/2$. **3.** (b) At the points $x = k\pi$, the Fourier series takes on the value 0. At all other points, the Fourier series agrees with the function. **19.** $\frac{1}{3 + 2\cos\theta} = \frac{1}{\sqrt{5}} + \frac{2}{\sqrt{5}}\sum_{n=1}^{\infty}\frac{1}{2^n}(-3 + \sqrt{5})^n \cos n\theta$. **21.** $\frac{1}{3 + \cos\theta} = \frac{1}{2\sqrt{2}} + \frac{1}{\sqrt{2}}\sum_{n=1}^{\infty}(-3 + 2\sqrt{2})^n \cos n\theta$. **23.** $\frac{1}{3 + \cos\theta + \sin\theta} =$
$$\frac{1}{\sqrt{7}}\left[1 + 2\sum_{n=1}^{\infty}\left(\frac{\sqrt{7} - 3}{\sqrt{2}}\right)^n \cos n(\theta - \frac{\pi}{4})\right].$$

SECTION 7.3

1. (a) Odd. **3.** (a) Even. **5.** (a) Even.
7. (a) Odd. **9.** (a) Even. **11.** (a)

$$1 - \frac{1}{2}\cos x - 2\sum_{n=2}^{\infty} \frac{(-1)^n}{n^2 - 1}\cos nx.$$

13. (a)

$$\frac{1}{2} + \frac{2}{\pi}\sum_{k=0}^{\infty} \frac{1}{2k+1}\sin(2k+1)\pi x.$$

20. (b) First show that the only function that is both even and odd is the zero function. Now suppose that $f = f_1 + f_2$ and $f = f_e + f_o$ where f_1 is even and f_2 is odd. Subtract to get $f_1 - f_e = f_2 - f_o$. The left side is even and the right side is odd. So, $f_1 - f_e = 0$ and $f_2 - f_o = 0$, which is what we want.
21. (a) $f_e(x) = |x|$ for $-1 \le x \le 1$,

$$f_o(x) = \begin{cases} -1 & \text{if } -1 < x < 0, \\ 1 & \text{if } 0 < x < 1. \end{cases}$$

(b) From Example 1 with $p = 1$,

$$f_e(x) = \frac{1}{2} - \frac{4}{\pi^2}\sum_{k=0}^{\infty} \frac{1}{(2k+1)^2}\cos(2k+1)\pi x.$$

From Exercise 1 with $p = 1$,

$$f_o(x) = \frac{4}{\pi}\sum_{k=0}^{\infty} \frac{1}{2k+1}\sin(2k+1)\pi x,$$

$$f(x) = \frac{1}{2} - \frac{4}{\pi}\sum_{k=0}^{\infty}\left\{\frac{1}{\pi(2k+1)^2}\cos(2k+1)\pi x - \frac{1}{2k+1}\sin(2k+1)\pi x\right\}.$$

23. (a) $f_e(x) = |x|$ for $-1 \le x \le 1$,

$$f_o(x) = \begin{cases} -x - 1 & \text{if } -1 < x < 0, \\ 1 - x & \text{if } 0 < x < 1. \end{cases}$$

(b) From Example 1 with $p = 1$,

$$f_e(x) = \frac{1}{2} - \frac{4}{\pi^2}\sum_{k=0}^{\infty} \frac{1}{(2k+1)^2}\cos(2k+1)\pi x.$$

From Exercise 7 with $p = 2$,

$$f_o(x) = \frac{4}{\pi}\sum_{k=1}^{\infty} \frac{1}{2k}\sin k\pi x,$$

$$f(x) = \frac{1}{2} - \frac{4}{\pi^2}\sum_{k=0}^{\infty} \frac{1}{(2k+1)^2}\cos(2k+1)\pi x + \frac{4}{\pi}\sum_{k=1}^{\infty} \frac{1}{2k}\sin k\pi x.$$

29. $\dfrac{2}{\pi}\sum_{n=1}^{\infty} \dfrac{1}{n}\left(1 - \cos\dfrac{dn\pi}{p}\right)\sin\dfrac{n\pi}{p}x.$

SECTION 7.4

1. (a) Cosine series expansion: 1.

Sine series expansion: $\dfrac{4}{\pi}\sum_{k=0}^{\infty} \dfrac{1}{(2k+1)}\sin(2k+1)\pi x.$

3. (a) Cosine series expansion:

$$\frac{1}{3} + \frac{4}{\pi^2}\sum_{n=1}^{\infty} \frac{(-1)^n}{n^2}\cos n\pi x.$$

Sine series expansion:

$$\frac{2}{\pi^3}\sum_{n=1}^{\infty} \frac{1}{n^3}\left[2((-1)^n - 1) - (-1)^n(n\pi)^2\right]\sin n\pi x.$$

5. (a) Cosine series expansion:

$$\frac{1}{p}(b - a) + \frac{2}{\pi}\sum_{n=1}^{\infty} \frac{1}{n}\left(\sin\frac{n\pi}{p}b - \sin\frac{n\pi}{p}a\right)\cos\frac{n\pi}{p}x.$$

Sine series expansion:

$$\frac{2}{\pi}\sum_{n=1}^{\infty} \frac{1}{n}\left(\cos\frac{n\pi}{p}a - \cos\frac{n\pi}{p}b\right)\sin\frac{n\pi}{p}x.$$

7. (a) Cosine series expansion:

$$\frac{2}{\pi} - \frac{4}{\pi}\sum_{n=1}^{\infty} \frac{(-1)^n}{4n^2 - 1}\cos 2nx.$$

Sine series expansion: $\dfrac{8}{\pi}\sum_{n=1}^{\infty} \dfrac{n}{4n^2 - 1}\sin 2nx.$

9.

$$\frac{8}{\pi^3}\sum_{k=0}^{\infty} \frac{1}{(2k+1)^3}\sin(2k+1)\pi x.$$

11. $\sin \pi x.$ **13.** $\frac{1}{2}\sin 2\pi x.$
15. $\sum_{n=1}^{\infty} \frac{2n\pi}{1+n^2\pi^2}\left((1 - (-1)^n e)\right)\sin n\pi x.$

SECTION 7.5

1. $\dfrac{\sinh \pi a}{\pi}\sum_{n=-\infty}^{\infty} \dfrac{(-1)^n a}{a^2 + n^2}e^{inx}, \quad -\pi < x < \pi.$

3. $\cos ax = \dfrac{\sin \pi a}{\pi}\sum_{n=-\infty}^{\infty} \dfrac{(-1)^n a}{a^2 - n^2}e^{inx}, \quad -\pi < x < \pi.$

5. $e^{-3ix} + \frac{1}{2}e^{-2ix} + \frac{1}{2}e^{2ix} + e^{3ix}.$
7. For $-\pi < x < \pi$,

$$\cos ax = \frac{\sin \pi a}{\pi a} + \sum_{n=1}^{\infty} \frac{2(-1)^n a \sin \pi a}{\pi(a^2 - n^2)}\cos nx.$$

15. $\dfrac{e^{i\theta}}{2+e^{2i\theta}}=\displaystyle\sum_{n=0}^{\infty}(-1)^n\dfrac{e^{(2n+1)i\theta}}{2^{n+1}}.$

17. $e^{e^{i\theta}}=\displaystyle\sum_{n=0}^{\infty}\dfrac{e^{in\theta}}{n!}.$

19. From the real part:

$\dfrac{3\cos\theta}{5+4\cos2\theta}=\displaystyle\sum_{n=0}^{\infty}(-1)^n\dfrac{\cos[(2n+1)\theta]}{2^{n+1}}.$

From the imaginary part:

$\dfrac{\sin\theta}{5+4\cos2\theta}=\displaystyle\sum_{n=0}^{\infty}(-1)^n\dfrac{\sin[(2n+1)\theta]}{2^{n+1}}.$

23. $\widehat{g}(1)e^{ix}+\widehat{g}(-1)e^{-ix}.$

CHAPTER 8
SECTION 8.1

1. Linear homogeneous equation, linear homogeneous boundary condition, order 2. **3.** Linear nonhomogeneous equation, linear nonhomogeneous boundary condition, order 2. **5.** Nonlinear equation, linear homogeneous boundary condition, order 2. **9.** (b) Any function of the form $u(x,y)=e^{ax}e^{by}$ with $a+b=-1$.
For example, $u(x,y)=e^{-y}$, or $u(x,y)=e^{x}e^{-2y}$.
11. (a) Hyperbolic, (c) parabolic,
(e) elliptic, (f) elliptic, (g) hyperbolic.
13. $u(x,t)=e^{\frac{x}{t+1}}.$

14. $u(x,t)=\left(\dfrac{-1+\sqrt{1-4t(t-x)}}{2t}\right)^2.$

15. $u(x,t)=\left(\dfrac{-1+\sqrt{1-4t(2t-x)}}{2t}\right)^2.$

16. $u(x,t)=\dfrac{-1+\sqrt{1+4tx}}{2t}.$ **17.** $u(x,t)=\sqrt{\dfrac{x}{t+1}}.$

SECTION 8.2

1. $u(x,t)=.05\sin\pi x\cos t.$ **3.** $u(x,t)=\sin\pi x\cos\pi t$ $+3\sin2\pi x\cos2\pi t-\sin5\pi x\cos5\pi t.$

5. $u(x,t)=\displaystyle\sum_{k=0}^{\infty}\dfrac{8(-1)^k}{\pi^2(2k+1)^2}\sin(2k+1)\pi x\cos4(2k+1)\pi t.$

7. $u(x,t)=\sum_{n=1}^{\infty}\dfrac{8(\sin(n\pi/4)+\sin(3n\pi/4))}{\pi^2n^2}\sin n\pi x\cos4n\pi t$ $+\sum_{k=0}^{\infty}\dfrac{\sin(2k+1)\pi x}{\pi^2(2k+1)^2}\sin4(2k+1)\pi t.$

9. $u(x,t)=\sum_{k=0}^{\infty}\dfrac{8\sin(2k+1)\pi x}{\pi^3(2k+1)^3}\cos(2k+1)\pi t$ $+\frac{1}{\pi}\sin\pi x\sin\pi t.$

13. $u(x,t)=e^{-\frac{t}{2}}\left(\cos\dfrac{\sqrt3}{2}t+\dfrac{1}{\sqrt3}\sin\dfrac{\sqrt3}{2}t\right)\sin x.$

14. $u(x,t)=\frac{\pi}{2}e^{-.5t}\sin x\left\{\cos\frac{\sqrt3}{2}t+\frac{1}{\sqrt3}\sin\frac{\sqrt3}{2}t\right\}$
$-\frac{16}{\pi}e^{-.5t}\sum_{k=1}^{\infty}\frac{k\sin2kx}{(4k^2-1)^2}$
$\times\left\{\cos\left(\sqrt{4k^2-\frac{1}{4}}t\right)+\frac{.5}{\sqrt{4k^2-\frac{1}{4}}}\sin\left(\sqrt{4k^2-\frac{1}{4}}t\right)\right\}.$

15. (a) $u(x,t)=\frac{16\sqrt5}{\pi}e^{-1.5t}\sin x\sinh\frac{\sqrt5}{2}t+\frac{40}{\pi}e^{-1.5t}$
$\times\sum_{k=1}^{\infty}\frac{\sin(2k+1)x\sin\left(\sqrt{(2k+1)^2-9/4}t\right)}{(2k+1)\sqrt{(2k+1)^2-9/4}}.$

SECTION 8.3

1. $u(x,t)=\dfrac{312}{\pi}\displaystyle\sum_{k=0}^{\infty}\dfrac{e^{-(2k+1)^2t}\sin(2k+1)x}{2k+1}.$

3. $u(x,t)=\dfrac{132}{\pi}\displaystyle\sum_{k=0}^{\infty}\dfrac{(-1)^k e^{-(2k+1)^2t}\sin(2k+1)x}{(2k+1)^2}.$

5. $u(x,t)=\dfrac{2}{\pi}\displaystyle\sum_{n=1}^{\infty}\dfrac{(-1)^{n+1}e^{-n^2\pi^2t}\sin n\pi x}{n}.$

9. (a) $u(x)=100x$ (b) $u(x)=100.$
11. $u(x,t)=100(1-x)+30e^{-\pi^2t}\sin\pi x$
$-\frac{200}{\pi}\sum_{n=1}^{\infty}\frac{e^{-n^2\pi^2t}\sin n\pi x}{n}.$
13. $u(x,t)=100-\frac{50x}{\pi}$
$+\frac{132}{\pi}\sum_{k=0}^{\infty}\frac{(-1)^k e^{-(2k+1)^2t}\sin(2k+1)x}{(2k+1)^2}$
$-\frac{100}{\pi}\sum_{n=1}^{\infty}\frac{2-(-1)^n}{n}e^{-n^2t}\sin nx.$

SECTION 8.4

1. $u(x,t)=100.$

3. $u(x,t)=25\pi-\dfrac{200}{\pi}\displaystyle\sum_{m=0}^{\infty}\dfrac{e^{-4(2m+1)^2t}\cos2(2m+1)x}{(2m+1)^2}.$

5. $u(x,t)=e^{-\pi^2t}\cos\pi x.$ **9.** In Exercise 1, average temperature $=100$. In Exercise 5, average temperature $=0$.

11. $u(x,t)=4\displaystyle\sum_{n=1}^{\infty}\dfrac{\sin\mu_n}{\mu_n^2(1+\cos^2\mu_n)}e^{-\mu_n^2t}\sin\mu_n x,$ where
μ_n is the nth positive root of $\tan\mu=-\mu.$

13. $u(x,t)=200\displaystyle\sum_{n=1}^{\infty}\dfrac{1-\cos\frac{\mu_n}{2}}{\mu_n(1+\cos^2\mu_n)}e^{-\mu_n^2t}\sin\mu_n x,$ where
μ_n is the nth positive root of $\tan\mu=-\mu.$

15. $u(x,t)=\displaystyle\sum_{n=1}^{\infty}d_n e^{-\mu_n^2c^2t}\cos\mu_n x,$ where μ_n is the nth
positive root of $\cot\mu L=\frac{\mu}{\kappa}$ and $d_n=\dfrac{\int_0^L f(x)\cos\mu_n x\,dx}{\frac12\left(L+\frac1\kappa\sin^2\mu_n L\right)}.$

SECTION 8.5

1. $u(x,y,t)=\sin3\pi x\sin\pi y\cos\sqrt{10}t.$
3. $u(x,y,t)$ is the sum of the answer in Example 1 plus $\frac{2}{\sqrt5}\sin\pi x\sin2\pi y\sin\sqrt5t.$ **5.** $u(x,y,t)=$
$\frac{16}{\pi^2}\sum_{n\text{ odd}}\sum_{m\text{ odd}}\frac{\sin m\pi x\sin n\pi y}{mn\sqrt{m^2+n^2}}\sin\sqrt{m^2+n^2}t.$

9. Nodal lines: $x = \frac{1}{4}$, $x = \frac{1}{2}$, $x = \frac{3}{4}$. **11.** $u(x, y, t) = \frac{1600}{\pi^2} \sum_{n \text{ odd}} \sum_{m \text{ odd}} \frac{\sin m\pi x \sin n\pi y}{mn} e^{-(m^2 + n^2)\pi^2 t}$.

13. $u(x, y, t) = \sin \pi x \sin \pi y e^{-2\pi^2 t}$.

14. $u(x, y, t) = \sum_{n=1}^{\infty} \sum_{m=1}^{\infty} A_{mn} \sin m\pi x \sin n\pi y \, e^{-\lambda_{mn}^2 t}$,

where $\lambda_{mn}^2 = \pi^2 (m^2 + n^2)$,

$A_{mn} = \frac{4((-1)^m (-1+(-1)^n) m^2 + (-1+(-1)^m) n^2)}{m (m-n) n (m+n) \pi^2}$ if $m \neq n$

and $A_{mm} = \frac{2(-1+(-1)^m)^2}{m^2 \pi^2}$.

SECTION 8.6

1. $u(x, y) = \frac{2}{\pi} \sum_{n=1}^{\infty} \frac{(-1)^{n-1} \sin n\pi x \sinh n\pi y}{n \sinh 2n\pi}$.

3. $u(x, y) = \frac{400}{\pi} \sum_{m \text{ odd}} \frac{\sin \frac{m\pi x}{2} \sinh \frac{m\pi(1-y)}{2}}{m \sinh \frac{m\pi}{2}}$

$+ \frac{200}{\pi} \sum_{n=1}^{\infty} \frac{\sinh n\pi x \sin n\pi y}{n \sinh 2n\pi}$.

5. $u(x, y) = \frac{\sin 7\pi x \sinh(7\pi(1-y))}{\sinh 7\pi} + \frac{\sin \pi x \sinh \pi y}{\sinh \pi}$

$+ \frac{\sinh 3\pi(1-x) \sin 3\pi y}{\sinh 3\pi} + \frac{\sinh 6\pi x \sin 6\pi y}{\sinh 6\pi}$.

SECTION 8.7

1. $u(x, y) = \frac{8}{\pi^4} \sum_{k=0}^{\infty} \sum_{m=1}^{\infty} \frac{(-1)^m \sin m\pi x \sin(2k+1)\pi y}{[m^2 + (2k+1)^2] m(2k+1)}$.

3. $u(x, y) = u_1(x, y) + u_2(x, y)$, where

$u_1(x, y) = \frac{-4 \sin \pi x}{\pi^3} \sum_{k=0}^{\infty} \frac{\sin(2k+1)\pi y}{(1+(2k+1)^2)(2k+1)}$,

$u_2(x, y) = \frac{2}{\pi} \sum_{n=1}^{\infty} \frac{(-1)^{n+1} \sin n\pi x \sinh n\pi y}{n \sinh n\pi}$.

4. $u(x, y) = u_1(x, y) + u_2(x, y)$, where u_2 is as in Exercise 3, and

$u_1(x, y) = \frac{4}{\pi^4} \sum_{n=1}^{\infty} \sum_{m=1}^{\infty} \frac{(-1)^{m+n+1} \sin m\pi x \sin n\pi y}{mn(m^2+n^2)}$.

5. $u(x, y) = \frac{16}{\pi^2} \sum_{l=0}^{\infty} \sum_{k=0}^{\infty} \frac{\sin((2k+1)\pi x) \sin((2l+1)\pi y)}{(2l+1)(2k+1)(3+\pi^2((2l+1)^2+(2k+1)^2))}$.

7. $u(x, y) = \left(c_1 e^{2\pi y} + c_2 e^{-2\pi y} - \frac{1}{4\pi^2}\right) \sin 2\pi x$, where

$$c_1 = \frac{e^{-2\pi} - e^{-4\pi}}{4\pi^2(1 - e^{-4\pi})}, \quad c_2 = \frac{1 - e^{-2\pi}}{4\pi^2(1 - e^{-4\pi})}.$$

SECTION 8.8

1. $u(x, y) = \sum_{m=1}^{\infty} A_m \sin mx \, (\cosh my - \tanh m \sinh my)$

where $A_m = \frac{400}{\pi(\cosh m - \tanh m \sinh m)}$ if m is odd and 0 if m is even.

3. $u(x, y) = \sum_{m=1}^{\infty} A_m \sin mx \, (\cosh my - \tanh m \sinh my)$

where $A_m = \frac{2}{\pi(1 - m \tanh m)} \int_0^{\pi} \sin mx g(x) \, dx$.

5. $u(x, y) = A_0 + \sum_{m=1}^{\infty} A_m \cos \frac{m\pi x}{a} \cosh \frac{m\pi(b-y)}{a}$ where

$A_0 = \frac{1}{a} \int_0^a g(x) \, dx$, $A_m = \frac{2}{a \cosh \frac{m\pi b}{a}} \int_0^a \cos \frac{m\pi x}{a} g(x) \, dx$.

7. $u(x, y) = \frac{1}{2 \cosh 2\pi} \sin 2x \sinh 2y$.

CHAPTER 9

SECTION 9.1

1. $u(r, \theta) = 100 + r \cos \theta = 100 + x$.

3. $u(r, \theta) = \sum_{n=1}^{\infty} \frac{r^n}{n} \sin n\theta$. **5.** $u(r, \theta) = \frac{25}{2} + \frac{100}{\pi} \sum_{n=1}^{\infty} \frac{r^n}{n} \left[\sin \frac{n\pi}{4} \cos n\theta + \left(1 - \cos \frac{n\pi}{4}\right) \sin n\theta \right]$.

14. (a) $u(r, \theta) = $
$50 + \frac{100}{\pi} \sum_{n=1}^{\infty} \frac{r^{-n}}{n} \sin n\theta (1 - (-1)^n)$
$= 50 + \frac{100}{\pi} \left[\tan^{-1} \left(\frac{\sin \theta}{r - \cos \theta} \right) + \tan^{-1} \left(\frac{\sin \theta}{r + \cos \theta} \right) \right]$.

(b) Isotherms: $x^2 + (y + \tan \frac{\pi T}{100})^2 = \sec^2 \frac{\pi T}{100}$.

15. (a) $u(r, \theta) = \frac{25}{2} + \frac{100}{\pi} \sum_{n=1}^{\infty} \frac{r^{-n}}{n} \left[\sin \frac{n\pi}{4} \cos n\theta + \left(1 - \cos \frac{n\pi}{4}\right) \sin n\theta \right]$
$= \frac{25}{2} + \frac{100}{\pi} \left[\tan^{-1} \left(\frac{\sin \theta}{r - \cos \theta} \right) - \tan^{-1} \left(\frac{\sin(\theta - \pi/4)}{r - \cos(\theta - \pi/4)} \right) \right]$.

16. (b) $a_0^* \frac{\ln r - \ln R_1}{\ln R^2 - \ln R_1}$
$+ \sum_{n=1}^{\infty} [a_n^* \cos n\theta + b_n^* \sin n\theta] \left(\frac{R_2}{r} \right)^n \left(\frac{r^{2n} - R_1^{2n}}{R_2^{2n} - R_1^{2n}} \right)$,
where a_0^*, a_n^*, and b_n^* are the Fourier coefficients of $f_2(\theta)$.
(c) Add the solutions in (a) and (b).

SECTION 9.2

1. $u(r, t) = 4 \sum_{n=1}^{\infty} \frac{J_0\left(\frac{\alpha_n r}{2}\right)}{\alpha_n^2 J_1(\alpha_n)} \sin\left(\frac{\alpha_n}{2} t\right)$.

3. $u(r, t) = \sum_{n=1}^{\infty} \frac{J_1\left(\frac{\alpha_n}{2}\right)}{\alpha_n^2 J_1(\alpha_n)^2} J_0(\alpha_n r) \sin(\alpha_n t)$.

5. $u(r, t) = J_0(\alpha_1 r) \cos(\alpha_1 t)$.

7. $u(r, t) = J_0(\alpha_3 r) \cos(\alpha_3 t) + 8 \sum_{n=1}^{\infty} \frac{J_0(\alpha_n r)}{\alpha_n^4 J_1(\alpha_n)} \sin(\alpha_n t)$.

11. (a) $u(r, t) = 200 \sum_{n=1}^{\infty} \frac{J_0(\frac{\alpha_n r}{a})}{\alpha_n J_1(\alpha_n)} e^{-c^2 \alpha_n^2 t/a^2}$.

This represents the time evolution of the temperature distribution of a disk with uniform initial temperature distribution of 100° and with its boundary kept at 0°.
(b) The maximum temperature should occur at the center of the disk, since this point is affected the least by the heat loss which is occuring along the boundary.

SECTION 9.3

1. $u(r,\theta,t) = 24 \sum_{n=1}^{\infty} \dfrac{J_2(\alpha_{2,n}r)}{\alpha_{2,n}^3 J_3(\alpha_{2,n})} \sin 2\theta \cos(\alpha_{2,n}t).$

3. $u(r,\theta,t) = 128 \sum_{n=1}^{\infty} \dfrac{J_1(\frac{\alpha_{1,n}r}{2})}{\alpha_{1,n}^3 J_2(\alpha_{1,n})} \sin\theta \cos(\dfrac{\alpha_{1,n}t}{2}) +$

$4 \sum_{n=1}^{\infty} \dfrac{J_0(\frac{\alpha_{0,n}r}{2})}{\alpha_{0,n}^2 J_1(\alpha_{0,n})} \sin(\dfrac{\alpha_{0,n}t}{2}).$

5. $u(r,\theta,t) = 24 \sum_{n=1}^{\infty} \dfrac{J_2(\alpha_{2,n}r)}{\alpha_{2,n}^4 J_3(\alpha_{2,n})} \sin 2\theta \sin \alpha_{2,n}t.$

12. (a) $u(r,\theta,t) = 16 \sum_{n=1}^{\infty} \dfrac{J_1(\alpha_{1,n}r)\sin\theta}{\alpha_{1,n}^3 J_2(\alpha_{1,n})} e^{-\alpha_{1,n}^2 t}.$

(b) Hottest point at $r = \frac{1}{\sqrt{3}}$ and $\theta = \frac{\pi}{2}$. Hottest value $f(\frac{1}{\sqrt{3}}, \frac{\pi}{2}) = \frac{2}{3\sqrt{3}}$.

(c) Hottest points are located on the ray $\theta = \frac{\pi}{2}$.

13. $u(r,\theta,t) = r^3 \sin 3\theta +$

$\sin 3\theta \sum_{n=1}^{\infty} \dfrac{-2}{\alpha_{3,n} J_4(\alpha_{3,n})} J_3(\alpha_{3,n}r) e^{-\alpha_{3,n}^2 t}.$

SECTION 9.4

1. $u(\rho,z) = 200 \sum_{n=1}^{\infty} \dfrac{J_0(\alpha_{0,n}\rho)}{\alpha_{0,n} J_1(\alpha_{0,n})} \dfrac{\sinh(\alpha_{0,n}z)}{\sinh(2\alpha_{0,n})}.$

2. $u(\rho,z) = 2 \sum_{n=1}^{\infty} \dfrac{99\alpha_{0,n}^2 + 4}{\alpha_{0,n}^3 J_1(\alpha_{0,n})} J_0(\alpha_{0,n}\rho) \dfrac{\sinh(\alpha_{0,n}z)}{\sinh(2\alpha_{0,n})}.$

3. $u(\rho,z) = 100 \sum_{n=1}^{\infty} \dfrac{J_1(\alpha_{0,n}/2)J_0(\alpha_{0,n}\rho)}{\alpha_{0,n} J_1(\alpha_{0,n})^2} \dfrac{\sinh(\alpha_{0,n}z)}{\sinh(2\alpha_{0,n})}.$

4. $u(\rho,z) = 140 J_0(1) \sum_{n=1}^{\infty} \dfrac{\alpha_{0,n} J_0(\alpha_{0,n}\rho)}{(\alpha_{0,n}^2 - 1)J_1(\alpha_{0,n})} \dfrac{\sinh(\alpha_{0,n}z)}{\sinh(2\alpha_{0,n})}.$

5. (a) $u(\rho,z) = \sum_{n=1}^{\infty} c_n J_0(\dfrac{\alpha_{0,n}\rho}{a}) \dfrac{\sinh(\frac{\alpha_{0,n}}{a}(h-z))}{\sinh(\frac{\alpha_{0,n}}{a}h)},$

where c_n are the Bessel coefficients of order 0 of the function $f(\rho)$, $0 < \rho < a$.

(b) $u(\rho,z) = \sum_{n=1}^{\infty} a_n J_0(\dfrac{\alpha_{0,n}\rho}{a}) \dfrac{\sinh(\frac{\alpha_{0,n}}{a}z)}{\sinh(\frac{\alpha_{0,n}}{a}h)} +$

$\sum_{n=1}^{\infty} c_n J_0(\dfrac{\alpha_{0,n}\rho}{a}) \dfrac{\sinh(\frac{\alpha_{0,n}}{a}(h-z))}{\sinh(\frac{\alpha_{0,n}}{a}h)},$

where a_n is the Bessel coefficient of order 0 of the $f_2(\rho)$, $0 < \rho < a$; and c_n is the Bessel coefficient of order 0 of the $f_1(\rho)$, $0 < \rho < a$.

6. $u(\rho,z) =$

$200 \sum_{n=1}^{\infty} \dfrac{J_0(\frac{\alpha_{0,n}\rho}{a})}{\alpha_{0,n} J_1(\alpha_{0,n}) \sinh(\frac{\alpha_{0,n}h}{a})} [\sinh(\dfrac{\alpha_{0,n}}{a}z)$

$+ \sinh(\dfrac{\alpha_{0,n}}{a}(h-z))].$

9. $u(\rho,z) = \dfrac{40}{\pi} \sum_{n=1}^{\infty} \dfrac{(-1)^{n+1}}{n} \dfrac{I_0(\frac{n\pi\rho}{2})}{I_0(\frac{n\pi}{2})} \sin \dfrac{n\pi z}{2}.$

11. $u(\rho,z) = 156 \sum_{n=1}^{\infty} \dfrac{J_0(\frac{\alpha_{0,n}\rho}{10})}{\alpha_{0,n} J_1(\alpha_{0,n})} \dfrac{\sinh(\frac{\alpha_{0,n}}{10}z)}{\sinh(\frac{3}{5}\alpha_{0,n})} +$

$112 \sum_{n=1}^{\infty} \dfrac{J_0(\frac{\alpha_{0,n}\rho}{10})}{\alpha_{0,n} J_1(\alpha_{0,n})} \dfrac{\sinh(\frac{\alpha_{0,n}}{10}(6-z))}{\sinh(\frac{3}{5}\alpha_{0,n})} +$

$\dfrac{2}{\pi} \sum_{n=1}^{\infty} [\dfrac{56 + 22\cos\frac{2n\pi}{3} - 78(-1)^n}{n}] \dfrac{I_0(\frac{n\pi\rho}{6})}{I_0(\frac{5n\pi}{3})} \sin \dfrac{n\pi z}{6}.$

SECTION 9.5

3. $\int_0^{2\pi} \int_0^a \phi_{mn}^2(r,\theta) r\, dr\, d\theta = \pi a^2 J_1^2(\alpha_{0n})$ for $m = 0$, and $\frac{\pi}{2}a^2 J_{m+1}^2(\alpha_{mn})$ for $m = 1,2,\ldots$.

5. $u(r,\theta) = 2 \sum_{n=1}^{\infty} \dfrac{J_0(\alpha_{0n}r)}{\alpha_{0n}(1-\alpha_{0n}^2) J_1(\alpha_{0n})}.$

7. $u(r,\theta) = -2 \sum_{n=1}^{\infty} \left\{ \dfrac{2J_0(\alpha_{0n}r)}{\alpha_{0n}^3 J_1(\alpha_{0n})} + \dfrac{J_3(\alpha_{3n}r)}{\alpha_{3n}^3 J_4(\alpha_{3n})} \cos 3\theta \right\}.$

9. $u(r,\theta) = -\sin\theta \sum_{n=1}^{\infty} \dfrac{J_2(\frac{\alpha_{1n}}{2})}{2\alpha_{1n}^3 J_2^2(\alpha_{1n})} J_1(\alpha_{1n}r).$

11. $u(r,\theta) = r^2 \sin 2\theta - 2 \sum_{n=1}^{\infty} \dfrac{J_0(\alpha_{0n}r)}{\alpha_{0n}^3 J_1(\alpha_{0n})}.$

14. a_{mn}, b_{mn} given by (12)–(14), Section 9.3. $c_{mn}(t)$, $d_{mn}(t)$ given by (12)–(14), Section 4.3, with $f(r,\theta)$ replaced by $q(r,\theta,t)$, with t fixed.

15. $u(r,\theta,t) = \sum_{n=1}^{\infty} J_0(\alpha_{0n}r) A_{0n}(t),$

where $A_{0n}(t) = \dfrac{2e^{-\alpha_{0n}^2 t}}{\alpha_{0n} J_1(\alpha_{0n})} [1 + \frac{1}{\alpha_{0n}^2 - 1}(e^{(\alpha_{0n}^2 - 1)t} - 1)].$

SECTION 9.6

1. $p = 3$, $J_3(x) = \frac{1}{3!}\left(\frac{x}{2}\right)^3 - \frac{1}{4!}\left(\frac{x}{2}\right)^5 + \frac{1}{2\cdot 5!}\left(\frac{x}{2}\right)^7 - \cdots$.

3. $p = \frac{1}{2}$, $J_{1/2}(x) = \sqrt{\frac{2}{\pi x}}\left(x - \frac{x^3}{3!} + \frac{x^5}{5!} - \cdots\right)$.

5. $y(x) = c_1 J_{3/2}(x) + c_2 J_{-3/2}(x)$

$= c_1 \sqrt{\frac{2}{\pi x}}(\frac{x^2}{3} - \frac{x^4}{30} + \cdots) + c_2 \sqrt{\frac{2}{\pi x}}(-\frac{1}{x} - \frac{x}{2} + \cdots).$

7.

$y(x) = c_1 J_{1/4}(x) + c_2 J_{-1/4}(x)$

$= c_1 \left(\frac{x}{2}\right)^{1/4} \left(\dfrac{1}{\Gamma(5/4)} - \dfrac{x^2}{4\Gamma(9/4)} + \cdots\right)$

$+ c_2 \left(\frac{2}{x}\right)^{1/4} \left(\dfrac{1}{\Gamma(3/4)} - \dfrac{x^2}{4\Gamma(7/4)} + \cdots\right).$

9. $-\frac{384}{x^3} - \frac{48}{x} - 6x + \cdots + (x^3 - \frac{x^5}{16} + \cdots) \ln x.$

11. $y = c_1 x J_1(x) + c_2 x Y_1(x).$

13. $y = c_1 x^{3/2} J_{3/2}(x) + c_2 x^{3/2} Y_{3/2}(x).$

17. $y = c_1 x^{-p} J_p(x) + c_2 x^{-p} Y_p(x).$

19. $y = c_1 J_0(2e^{-\frac{x}{2}}) + c_2 Y_0(2e^{-\frac{x}{2}}).$

27. $y = 0.123 J_0(e^{-t}) + 0.069 Y_0(e^{-t}).$

SECTION 9.7

3. $xJ_1(x) + C$. **4.** $x^4 J_4(x) + C$. **5.** $-J_0(x) + C$.

6. $-x^{-2}J_2(x) + C$. **7.** $x^3 J_3(x) + C$. **8.** $2x^2 J_0(x) +$
$(x^3 - 4x)J_1(x) + C$. **9.** $J_0(x) - \frac{4}{x}J_1(x) + C$.

10. $1/(n+1)$. **13.** $J_5(x) = -\frac{12}{x}(\frac{16}{x^2} - 1)J_0(x) + (\frac{384}{x^4} - \frac{72}{x^2} + 1)J_1(x)$. **16.** (a) $j - 1$ roots in $(0, a)$: $x_k = \frac{\alpha_{pk}}{\alpha_{pj}}a$,
$k = 1, 2, \ldots, j - 1$. **21.** (a) $\alpha_{mj} = j\pi, j = 1, 2, \ldots$.

(b) $x^{1/2} = \frac{2}{\pi}\sum_{j=1}^{\infty}\frac{(-1)^{j-1}}{j}\frac{\sin(j\pi x)}{\sqrt{x}}$.

(c) Multiplying both sides in (b) by $x^{1/2}$,
$x = \frac{2}{\pi}\sum_{j=1}^{\infty}\frac{(-1)^{j-1}}{j}\sin(j\pi x)$, for $0 < x < 1$.

23. $x^2 = 2\sum_{j=1}^{\infty}\frac{1}{\alpha_{2,j}J_3(\alpha_{2,j})}J_2(\alpha_{2,j}x)$.

25. $f(x) = -2\sum_{j=1}^{\infty}\frac{J_0(\alpha_{1j}) - J_0(\frac{\alpha_{1j}}{2})}{\alpha_{1j}J_0(\alpha_{1j})^2}J_1(\alpha_{1j}x)$.

27. $f(x) = -2\sum_{j=1}^{\infty}\frac{J_1(\alpha_{2,j}) - 2J_1(\frac{\alpha_{2,j}}{2})}{\alpha_{2,j}J_1(\alpha_{2,j})^2}J_2(\alpha_{2j}x)$.

28. $f(x) = -2\sum_{j=1}^{\infty}\frac{J_2(\alpha_{3,j}) - 900J_2(\frac{\alpha_{3,j}}{30})}{\alpha_{3,j}J_2(\alpha_{3,j})^2}J_3(\alpha_{2j}x)$.

31. For $j = 1, 2, \ldots$, $\lambda = \lambda_j = \alpha_{1j}$, and
$y = y_j = c_j J_1(\alpha_{1j}x)$.
33. For $j = 1, 2, \ldots$, $\lambda = \lambda_j = j$, and
$y = y_j = c_j\sqrt{\frac{2}{\pi x}}\sin jx$.

CHAPTER 10
SECTION 10.1

3. $P_0(\cos\theta) = 1$, $P_1(\cos\theta) = \cos\theta$,
$P_2(\cos\theta) = \frac{1}{2}(3\cos^2\theta - 1)$.
5. $u(r,\theta) = r\cos\theta, u_r = \cos\theta, u_{rr} = 0$,
$u_\theta = -r\sin\theta, u_{\theta\theta} = -r\cos\theta$.

SECTION 10.2

1. $u(r,\theta) = 20 + 20r\cos\theta$.
2. $u(r,\theta) = \frac{7}{3} + \frac{2}{3}r^2 P_2(\cos\theta)$.
3. $u(r,\theta) = 60 +$
$20\sum_{n=0}^{\infty}(-1)^n\frac{(2n)!(4n+3)}{2^{2n}(n!)^2(n+1)}r^{2n+1}P_{2n+1}(\cos\theta)$.
4. $u(r,\theta) = 25 +$
$50\sum_{n=1}^{\infty}(P_{n-1}(1/2) - P_{n+1}(1/2))r^n P_n(\cos\theta)$.
5. $u(r,\theta) = \frac{1}{4} + \frac{1}{2}r\cos\theta +$
$\sum_{n=1}^{\infty}(-1)^{n+1}\frac{n(2n-2)!}{2^{2n+1}(n!)^2}\left(\frac{4n+1}{n+1}\right)r^{2n}P_{2n}(\cos\theta)$.
6. $u(r,\theta) = \frac{1}{2} +$

$\sum_{n=1}^{\infty}(-1)^{n+1}\frac{n(2n-2)!}{2^{2n}(n!)^2}\left(\frac{4n+1}{n+1}\right)r^{2n}P_{2n}(\cos\theta)$.
9. $u(r,\theta) = 65 +$
$\frac{15}{2}\sum_{n=1}^{\infty}(P_{n-1}(-1/3) - P_{n+1}(-1/3))(\frac{r}{3})^n P_n(\cos\theta)$.
11. $u(r,\theta) = 60r^{-1} +$
$20\sum_{n=0}^{\infty}(-1)^n\frac{(2n)!(4n+3)}{2^{2n}(n!)^2(n+1)}r^{-(2n+2)}P_{2n+1}(\cos\theta)$.

SECTION 10.3

5. (a) For $m = 0$, $2\pi^2$; for $m \geq 1$, $\frac{2\pi i}{m}$.
(c) For $n = 1, 2, \ldots$, $A_{n,0,0} = 0$, because $\int_{-1}^{1}P_n(x)\,dx = 0$, by orthogonality of the Legendre polynomials.
9. $u(r,\theta,\phi) = \frac{1}{2\sqrt{\pi}}$.
10. $u(r,\theta,\phi) = rY_{1,0}(\theta,\phi) + 3rY_{1,1}(\theta,\phi)$.

11. $u(r,\theta,\phi) = \sum_{n=0}^{\infty}\sum_{m=-n}^{n}A_{nm}r^n Y_{n,m}(\theta,\phi)$, where

$A_{00} = \frac{100\sqrt{\pi}}{3}$, $A_{n0} = 0$, for $n > 0$; and for $m > 0$,

$A_{nm} = \frac{100}{m}\sqrt{\frac{2n+1}{4\pi}\frac{(n-m)!}{(n+m)!}}\sin\frac{m\pi}{3}I_{nm}$,

where I_{nm} is as in Exercise 6.

12. $u(r,\theta,\phi) = \sum_{n=0}^{\infty}\sum_{m=-n}^{n}A_{nm}r^n Y_{n,m}(\theta,\phi)$ where

$A_{00} = 50\sqrt{\pi}$, $A_{n0} = 0$, for $n > 0$; and for $m > 0$,

$A_{nm} = \frac{50 i\pi}{m}(e^{im\frac{\pi}{2}} - 1)\sqrt{\frac{2n+1}{4\pi}\frac{(n-m)!}{(n+m)!}}I_{nm}$,

where I_{nm} is as in Exercise 6.

SECTION 10.4

5. For $0 < r < 1$, $1 = 2\sum_{j=1}^{\infty}(-1)^{j+1}\frac{\sin(j\pi r)}{j\pi r}$.

6. For $0 < r < 1$,
$f(r) = \sum_{j=1}^{\infty}\frac{(-j\pi\cos(\frac{j\pi}{2}) + 2\sin(\frac{j\pi}{2}))\sin(j\pi r)}{j^2\pi^2 r}$.
7. For $0 < r < 1$,
$r^2 = 2\sum_{j=1}^{\infty}(-1)^{j+1}(-6 + j^2\pi^2)\frac{\sin(j\pi r)}{j^3\pi^3 r}$.

9. $u(r,\theta,\phi) = 2\sum_{j=1}^{\infty}(-1)^j\frac{\sin(j\pi r)}{j^2\pi^2 r}$.

15. (a) $u(r,\theta,\phi,t) = 60\sum_{j=1}^{\infty}(-1)^{j+1}\frac{\sin(j\pi r)}{j\pi r}e^{-j^2\pi^2 t}$.

(b) $u(0,\theta,\phi,t) = 30(1 - \theta_4(0, e^{-\pi^2 t}))$. (c) $t \approx 0.18$.

17. $u(r,\theta,\phi,t) = \frac{1}{2\sqrt{\pi}}\sum_{j=1}^{\infty}B_{j00}(t)j_0(\pi j r)$, where

$B_{j00}(t) = e^{-\pi^2 t}(f_{j00} - \frac{q_{j00}}{\pi^2 j^2}) + \frac{q_{j00}}{\pi^2 j^2}$, $f_{j00} = 400\sqrt{\pi}(-1)^{j+1}$,
$q_{j00} = \frac{80(-1)^j(-6 + j^2\pi^2)}{j^2\pi^{\frac{3}{2}}}$.

SECTION 10.5

5. 0. **6.** 0. **7** 0. **8.** $\frac{2}{5}$. **9.** Legendre's differential equation, $n = 5$, $y(x) = c_1 P_5(x) + c_2 Q_5(x)$

$= c_1\left(\frac{15\,x}{8} - \frac{35\,x^3}{4} + \cdots\right) + c_2\left(-\frac{8}{15} + 8\,x^2 + \cdots\right).$

10. Legendre's differential equation, $n = 6$,

$y(x) = c_1 P_6(x) + c_2 Q_6(x)$

$= c_1\left(-\frac{5}{16} + \frac{105\,x^2}{16} + \cdots\right) + c_2\left(\frac{-16\,x}{5} + \frac{64\,x^3}{3} + \cdots\right).$

11. Legendre's differential equation, $n = 0$,

$y(x) = c_1 P_0(x) + c_2 Q_0(x)$

$= c_1 + c_2\left(x + \frac{x^3}{3} + \cdots\right).$

12. Legendre's differential equation, $n = 2$,

$y(x) = c_1 P_2(x) + c_2 Q_2(x)$

$= c_1\left(-\frac{1}{2} + \frac{3\,x^2}{2} + \cdots\right) + c_2\left(\frac{-16\,x}{5} + \frac{64\,x^3}{3} + \cdots\right).$

13. Unbounded. **14.** Bounded. **15.** Unbounded.

16. Unbounded. **17.** $1 + x - \frac{3\,x^2}{8} + \frac{5\,x^3}{24} - \frac{21\,x^4}{128} + \cdots.$

18. $1 - x^2 - \frac{x^4}{3} + \cdots.$

19. $x - \frac{x^3}{6} + \cdots.$ **20.** $1 + x - x^2 - \frac{x^4}{3} + \cdots.$

SECTION 10.6

1. 0. **3.** $\frac{6}{35}$. **5.** 0. **13.** 0. **14.** 0. **15.** $\frac{2^{n+1}(n!)^2}{(2n+1)!}$. **16.** 0.

17. $-\frac{1}{3}$. **18.** $-\frac{1}{10}$. **19.** $-\frac{1}{2}$. **20.** $\frac{1}{2}$.

21. $I_0 = \ln 4 - 2$, $I_n = \frac{-2}{n(n+1)}, n > 0.$

22. $I_0 = \ln 4 - 2$, $I_n = \frac{(-1)^{n+1}2}{n(n+1)}, n > 0.$

23. $I_0 = -1$, $I_1 = \frac{2}{9}(\ln 8 - 4)$, $I_n = \frac{-2}{n^2 + n - 2}, n > 1.$

24. $I_0 = 1$, $I_1 = \frac{2}{9}(\ln 8 - 4)$, $I_n = \frac{(-1)^n 2}{n^2 + n - 2}, n > 1$

27. $A_0 = \frac{1}{2}$, $A_2 = \frac{5}{8}$ $A_4 = -\frac{3}{16}$, $A_1 = A_3 = 0.$

29. $\frac{1}{4} + \frac{1}{2}x + \sum_{n=1}^{\infty}(-1)^{n+1}\frac{n(2n-2)!}{2^{2n+1}(n!)^2}\left(\frac{4n+1}{n+1}\right)P_{2n}(x).$

31. $\ln 2 - 1 - \sum_{n=1}^{\infty}(-1)^n\frac{2n+1}{n(n+1)}P_n(x).$

SECTION 10.7

1. $-3\,x\sqrt{1-x^2}$, $\frac{x\sqrt{1-x^2}}{2}$. **3.** $105\,x\sqrt{1-x^2}\,(-1+x^2).$

5. $n = 1$, $m = 1$, $P_1^1(x) = -\sqrt{1-x^2}.$

7. $n = 2$, $m = 1$, $P_2^1(x) = -3\,x\sqrt{1-x^2}.$

CHAPTER 11

SECTION 11.1

1. $\sqrt{\frac{2}{\pi}}\frac{\sin\omega - \omega\cos\omega}{\omega^2}$. **3.** $\frac{i}{\sqrt{2\pi}}\frac{e^{i\omega}-1}{\omega}.$

5. $\sqrt{\frac{2}{\pi}}\frac{1-\cos\omega}{\omega^2}$ **7.** $\sqrt{\frac{2}{\pi}}\frac{e^2}{1+\omega^2}$.

9. $\sqrt{\frac{2}{\pi}}\frac{\cos\frac{\pi\omega}{2}}{1-\omega^2}.$

SECTION 11.2

1. $\widehat{f}(\omega) = e^{-\frac{\omega^2}{4}}\frac{2-\omega^2}{4\sqrt{2}}$. **3.** $\widehat{f}(\omega) = \frac{i}{2}\sqrt{\frac{\pi}{2}}e^{-|\omega|}|\omega|.$

5. $\widehat{f}(\omega) = \sqrt{\frac{2}{\pi}}\frac{1-12i\omega+\omega^2}{(1+\omega^2)^2}$. **7.** $\widehat{f}(\omega) = e^{-\frac{1}{2}(\omega+i)^2-3}.$

9. $\widehat{f}(\omega) = \frac{-1+(1+i\omega)\cos\omega + (\omega-i)\sin\omega}{\sqrt{2\pi}\omega^2}.$

11. $\widehat{f}(\omega) = \frac{-i}{|a|}\sqrt{\frac{\pi}{2}}\omega e^{-|a\omega|}$. **13.** $\widehat{f}(\omega) = e^{-\frac{\omega^2}{4}}\frac{2+\omega^2}{4\sqrt{2}}.$

15. $\widehat{f}(\omega) = e^{-\frac{\omega}{2}(\omega+2i)}(1-i\omega).$ **17.** $\frac{x}{4}e^{-\frac{x^2}{2}}.$

19. $\frac{1}{2}e^{-\frac{x^2}{2}}$. **21.** $f * g$ is a piecewise linear continuous tent function, such that $f * g(x) = 0$ if $|x| > a + b$, $f * g(x) = \sqrt{\frac{2}{\pi}}a$ if $a - b < x < b - a$, $f * g(x) = \frac{1}{\sqrt{2\pi}}(x + (a + b))$ if $-(a + b) < x < a - b$, $f * g(x) = -\frac{1}{\sqrt{2\pi}}(x - (a + b))$ if $b - a < x < a + b.$

27. $\frac{1}{2\sqrt{2}}\left(e^{-(1-\omega)^2/4} + e^{-(1+\omega)^2/4}\right).$

29. $\frac{\sqrt{\pi}}{2\sqrt{2}}\left(e^{-|\omega-1|} + e^{-|\omega+1|} + e^{-|\omega-2|} + e^{-|\omega+2|}\right).$

31. $\frac{1}{\sqrt{2\pi}}\left(\frac{\sin(\omega-1)}{\omega-1} + \frac{\sin(\omega+1)}{\omega+1}\right).$

33. $\mathcal{F}(f) = \frac{i}{\sqrt{2\pi}}\frac{e^{-4i\omega}-e^{-2i\omega}}{\omega}.$

35. $\mathcal{F}(f) = \frac{i}{\sqrt{2\pi}}\sum_{j=0}^{5}\frac{e^{-i(j+1)\omega}-e^{-ij\omega}}{\omega}.$

37. $\mathcal{F}(f) = \frac{1}{\sqrt{2\pi}}\frac{-1+2e^{-i\omega}-e^{-2i\omega}}{\omega^2}.$

SECTION 11.3

1. $\frac{1}{2}\int_{-\infty}^{\infty}e^{-|\omega|}\cos\omega t e^{i\omega x}\,d\omega.$

3. $\frac{1}{2\sqrt{\pi}}\int_{-\infty}^{\infty}e^{-\omega^2(1+t)/4}e^{i\omega x}\,d\omega.$

5. $\frac{1}{\sqrt{2\pi}}\int_{-1}^{1}\cos(ct\omega)e^{i\omega x}\,d\omega.$ **7.** $f\left(x - \frac{t}{3}\right).$

9. $3\cos\left(x + \frac{t^3}{3}\right).$ **11.** $f(x + t).$

13. $\frac{1}{\sqrt{2\pi}}\int_{-\infty}^{\infty}\widehat{f}(\omega)e^{-\frac{t^2}{2}\omega^2}e^{i\omega x}\,d\omega.$

15. $f(x)(e^{-t} + te^{-t}) + te^{-t}g(x).$

17. $\frac{100}{\pi}\int_{-\infty}^{\infty}\frac{\sin 2\omega}{\omega}e^{-\omega^2 t}e^{i\omega x}\,d\omega.$

19. $\frac{1}{\sqrt{2\pi}}\int_{-\infty}^{\infty}\left(\widehat{f}(\omega) + \frac{\widehat{g}(\omega)}{\omega^2}(1 - e^{-\omega^2 t})\right)e^{i\omega x}\,d\omega.$

23. $\frac{1}{\sqrt{2\pi}}\int_{-\infty}^{\infty}\widehat{f}(\omega)e^{-c^2\omega^2 t}e^{i\omega(x+kt)}\,d\omega.$

25. $u(x,t) = \dfrac{1}{\sqrt{2\pi}} \displaystyle\int_{-\infty}^{\infty} \widehat{u}(\omega,t) e^{i\omega x}\, d\omega$, where

$\widehat{u}(\omega,t) = \dfrac{1}{2}(\widehat{f}(\omega) - i\dfrac{\widehat{g}(\omega)}{c\omega^2}) e^{ic\omega^2 t} + \dfrac{1}{2}(\widehat{f}(\omega) + i\dfrac{\widehat{g}(\omega)}{c\omega^2}) e^{-ic\omega^2 t}$.

27. $\dfrac{1}{\sqrt{2\pi}} \displaystyle\int_{-\infty}^{\infty} \widehat{f}(\omega) e^{-i\omega^3 c^2 t} e^{i\omega x}\, d\omega$.

SECTION 11.4

1. $u(x,t) = 10\left(\operatorname{erf}[\dfrac{x+1}{\sqrt{t}}] - \operatorname{erf}[\dfrac{x-1}{\sqrt{t}}] \right)$.

3. $u(x,t) = \dfrac{70}{\sqrt{2t+1}} e^{-\frac{x^2}{2(2t+1)}}$.

5. $u(x,t) = \dfrac{50}{\sqrt{\pi t}} \displaystyle\int_{-\infty}^{\infty} \dfrac{1}{1+s^2} e^{-(x-s)^2/4t}\, ds$.

7. $u(x,t) = f * g_t(x)$ where $g_t(x) = \sqrt{\dfrac{3}{2}} \dfrac{e^{-\frac{3x^2}{4t^3}}}{t^{3/2}}$.

8. $u(x,t) = \dfrac{1}{2\sqrt{\pi t}} \displaystyle\int_{-\infty}^{\infty} f(s) e^{-(x-s)^2/4t}\, ds$.

9. $u(x,t) = f * g_t(x)$, where

$g_t(x) = \dfrac{1}{\sqrt{2(1-e^{-t})}} e^{-\frac{x^2}{4(1-e^{-t})}}$.

10. $u(x,t) = \dfrac{1}{c\sqrt{2t}} e^{-x^2/4c^2 t} * f(x+kt)$.

11. $u(x,t) = f * g_t(x)$, where

$g_t(x) = \dfrac{1}{\sqrt{2\int_0^t a(s)\, ds}} e^{\frac{-x^2}{4\int_0^t a(s)\, ds}}$.

25. $u(x,t) = \dfrac{1}{\sqrt{4c^2 t + 1}} e^{-(x-1)^2/(4c^2 t+1)}$.

27. $u(x,t) = \dfrac{1}{\sqrt{2c^2 t + 1}} e^{-(x-2)^2/2(2c^2 t+1)}$.

SECTION 11.5

1. $u(x,y) = \dfrac{50}{\pi}\left\{ \tan^{-1}\left(\dfrac{1+x}{y}\right) + \tan^{-1}\left(\dfrac{1-x}{y}\right) \right\}$.

3. $u(x,y) = \dfrac{y}{\pi} \displaystyle\int_{-\infty}^{\infty} \dfrac{\cos s}{(x-s)^2 + y^2}\, ds$.

7. $u(x,y) = \dfrac{1}{2}\sqrt{\dfrac{\pi}{2}} P_{2+y}(x) = \dfrac{1}{2} \dfrac{2+y}{x^2 + (2+y)^2}$.

SECTION 11.6

1. $f_c(\omega) = \sqrt{\dfrac{2}{\pi}} \dfrac{\sin\omega}{\omega}$.   **3.** $f_c(\omega) = \sqrt{\dfrac{2}{\pi}} \dfrac{6}{4+\omega^2}$.

5. $f_c(\omega) = \sqrt{\dfrac{2}{\pi}} \dfrac{\omega \sin(2\pi\omega)}{\omega^2 - 1}$.   **7.** $f_s(\omega) = \sqrt{\dfrac{2}{\pi}} \dfrac{1-\cos\omega}{\omega}$.

9. $f_s(\omega) = \sqrt{\dfrac{2}{\pi}} \dfrac{\omega}{4+\omega^2}$.   **11.** $f_s(\omega) = \sqrt{\dfrac{2}{\pi}} \dfrac{2\sin(\pi\omega)}{\omega^2 - 4}$.

13. $f_c(\omega) = \sqrt{\dfrac{\pi}{2}} e^{-\omega}$.   **15.** $f_s(\omega) = \sqrt{\dfrac{\pi}{2}} e^{-\omega}$.

17. $f_c(\omega) = \sqrt{\dfrac{\pi}{2}} \dfrac{1}{2}\left(e^{-|\omega-1|} + e^{-(\omega+1)} \right)$.

SECTION 11.7

1. $u(x,t) = \dfrac{2T_0}{\pi} \displaystyle\int_0^{\infty} \dfrac{1-\cos b\omega}{\omega} e^{-\omega^2 t} \sin\omega x\, d\omega$.

3. $u(x,t) = \displaystyle\int_0^{\infty} e^{-\omega^2 t - \omega} \sin\omega x\, d\omega$.

11. $u(x,y) = \dfrac{2}{\pi} \displaystyle\int_0^{\infty} \dfrac{1}{1+\omega^2} \dfrac{\sinh\omega x}{\sinh\omega} \cos\omega y\, d\omega$.

13. $u(x,y) = \dfrac{2}{\pi} \displaystyle\int_0^{\infty} \dfrac{\omega}{1+\omega^2} \dfrac{\sinh\omega x}{\sinh\omega} \sin\omega y\, d\omega$.

15. $u(x,y) = \dfrac{2}{\pi} \displaystyle\int_0^{\infty} \dfrac{1-\cos\omega}{\omega} e^{-\omega x} \sin\omega y\, d\omega$.

CHAPTER 12
SECTION 12.1

1. $a = 0, M = 11$.   **3.** $a = 3,\ M = 5$.   **5.** $a = 3, M = 1$.

7. $\dfrac{2}{s^2} + \dfrac{3}{s}$.   **9.** $\dfrac{\sqrt{\pi}}{2s^{3/2}} + \dfrac{\sqrt{\pi}}{s^{1/2}}$.   **11.** $\dfrac{2}{(s-3)^3}$.

13. $\dfrac{8s}{(16+s^2)^2}$.   **15.** $\dfrac{1}{2s} - \dfrac{s}{2(4+s^2)}$.

17. $\dfrac{3}{9+(s-2)^2}$.   **19.** $\dfrac{2(1+s)}{(1+(1+s^2))^2}$.

21. $\dfrac{2(-2 - s + 5s^3 + 2s^4 + 2s^5)}{(1+s^2)^3}$.

23. $\dfrac{\beta}{\beta^2 + (s-\alpha)^2}$.   **25.** t.   **27.** $\dfrac{4}{\sqrt{3}} \sin\left(t/\sqrt{3}\right)$.

29. $\dfrac{e^{3t} t^4}{24} + e^{3t}\cos t$.   **31.** $e^{-t}(1-t)$.

33. $e^{-t} + e^{2t}$.   **35.** $1 + e^{-2t} + \dfrac{e^t}{2}$.   **37.** $-e^{-2t} + e^{-t}$.

39. $y(t) = -\dfrac{11}{5}e^{-t} + \dfrac{1}{5}\cos 2t + \dfrac{2}{5}\sin 2t$.

41. $y(t) = \left(-\dfrac{\pi}{2} + \dfrac{t}{2}\right)\sin t$.

43. $y(t) = (2+t)e^{-2t} + (-1 + 3t)e^{-t}$.

45. $y(t) = -\dfrac{7}{50}e^{-2t} + \dfrac{7}{25}e^{3t} + \dfrac{e^t}{50}(-7\cos t + \sin t)$.

SECTION 12.2

1. $-\dfrac{1}{s^2} + \dfrac{1}{s}(1 + e^{-s})$.   **3.** $\dfrac{e^{2(2-s)}}{s-2}$.   **5.** $-\dfrac{e^{-\pi s}}{1+s^2}$.

7. $\dfrac{1}{s} - \dfrac{e^{-2s}}{s}$.   **9.** $\dfrac{2}{s}\left(e^{-2s} - e^{-3s}\right)$.

11. $\dfrac{1}{s^2}\left(e^{-s} - e^{-2s}\right) - \dfrac{e^{-2s}}{s}$.

13. $\dfrac{1}{s^2}\left(e^{-5s} - e^{-4s}\right) + \dfrac{e^{-s}}{s}$.

15. $(t-1)\mathcal{U}_0(t-1)$.   **17.** $\sin(t-1)\mathcal{U}_0(t-1)$.

19. $e^{3t}\cos t$.

21. $\dfrac{2}{\sqrt{\pi}}\sqrt{t-1}\mathcal{U}_0(t-1)$.

23. $\dfrac{t^2}{2}$.   **25.** $\dfrac{t^3}{6}$.   **27.** $-\dfrac{t\cos t}{2}+\dfrac{\sin t}{2}$.

29. $1*\sin t=\displaystyle\int_0^t\sin(t-\tau)d\tau=1-\cos t$.

31. $\sin t*\sin t=\displaystyle\int_0^t\sin\tau\sin(t-\tau)\,d\tau=-\dfrac{t\cos t}{2}+\dfrac{\sin t}{2}$.

33. $\sin(t-1)\mathcal{U}_0(t-1)$.

35. $e^{-t}+te^{-t}+\mathcal{U}_0(t-2)(3t-6)e^{2-t}$.

37. $\mathcal{U}_0(t-1)(\dfrac{1}{5}e^{-1+t}-\dfrac{1}{5}\cos(2(t-1))-\dfrac{1}{10}\sin 2(t-1))$.

39. $f*(1-\cos t)$.

41. $\dfrac{\sin 2t}{2}*\cos t=\dfrac{\cos t}{3}-\dfrac{\cos 2t}{3}$.   **45.** $F(s)=\dfrac{e^s-1}{s^2(e^s+1)}$.

47. $F(s)=\dfrac{1+e^{\pi s}}{(s^2+1)(e^{\pi s}-1)}$.

SECTION 12.3

1. $u(x,t)=70\,\mathrm{erfc}\left(\dfrac{x}{2\sqrt{t}}\right)$.

3. $u(x,t)=0$ if $0<t<2$, and

$$u(x,t)=100\,\mathrm{erfc}\left(\dfrac{x}{2\sqrt{t-2}}\right)\text{ if }t\geq 2.$$

5. $u(x,t)=\dfrac{1}{3!}t^3-\dfrac{1}{3!}\mathcal{U}_0(t-x)(t-x)^3$.

7. $u(x,t)=t-(t-x)\mathcal{U}_0(t-x)-\dfrac{g}{2}(t^2-(t-x)^2\mathcal{U}_0(t-x))$.

9. $u(x,t)=t+\sin(t-x)\mathcal{U}_0(t-x)-(t-x)\mathcal{U}_0(t-x)$.

11. $u(x,t)=\dfrac{x}{2c\sqrt{\pi}t^{3/2}}e^{-x^2/4c^2t}$.

13. $u(x,t)=30\,\mathrm{erfc}\left(\dfrac{x}{2\sqrt{t}}\right)+70$.

APPENDIX A

APPENDIX A.1

1. $y=1+ce^{-x}$.   **3.** $y=ce^{-x/2}$.   **5.** $y=-\dfrac{1}{2}(\cos x+\sin x)+ce^x$. **7.** $y=\dfrac{\sin x}{x}+\dfrac{c}{x}$.   **9.** $y=x\cos x+c\cos x$.

11. $y=e^x$.   **13.** $y=1-e^{-x^2/2}$.   **15.** $y=\dfrac{1}{2}-\dfrac{5}{2x^2}$.

17. $y=\cos x+\sin x$.   **19.** $y=1$. **21.** (b) $W(e^x,e^{-x})=-2\neq 0$, $W(\cosh x,\sinh x)=1\neq 0$. Hence by Theorem 7, $\{e^x,e^{-x}\}$ and $\{\cosh x,\sinh x\}$ are fundamental sets of solutions. (c) $e^x=\cosh x+\sinh x$. (d) $y_1=ae^x+be^{-x}$, $y_2=ce^x+de^{-x}$ with $ad-bc\neq 0$. **23.** (b) $W(x,x^2)=x^2$. (c) The coefficient functions in the equation in standard form are not continuous at 0. **25.** $y=e^{2x}-4e^x+2x+3$.

27. $y=5e^{2(x-1)}-10e^{x-1}+2x+3$.

29. $y=7e^{2(x-2)}-14e^{x-2}+2x+3$.

APPENDIX A.2

1. $y=c_1e^x+c_2e^{3x}$.   **3.** $y=c_1e^{2x}+c_2e^{3x}$.

5. $y=c_1e^{-x}+c_2xe^{-x}$.   **7.** $y=c_1e^{x/2}+c_2xe^{x/2}$.

9. $y=c_1\cos x+c_2\sin x$. **11.** $y=c_1e^{2x}+c_2e^{-2x}$.

13. $y=e^{-2x}(c_1\cos x+c_2\sin x)$.

15. $y=e^{-3x}(c_1\cos 2x+c_2\sin 2x)$.

17. $y=c_1+e^x(c_2+c_3x)$.

19. $y=c_1e^{-x}+c_2xe^{-x}+c_3e^x+c_4xe^x$.

21. $y=c_1e^x+c_2xe^x+c_3x^2e^x$.

23. $y=c_1e^{-3x}+c_2e^x+c_3xe^x+c_4x^2e^x$.

25. $y=-e^{2x}+c_1e^x+c_2e^{3x}$.

27. $y=\dfrac{5}{36}+\dfrac{1}{2}e^x+\dfrac{x}{6}+c_1e^{2x}+c_2e^{3x}$.

29. $y=\dfrac{3}{32}e^{-x}+\dfrac{x}{8}e^{-x}+c_1e^x+c_2e^{3x}$.

31. $y=\dfrac{1}{8}-\dfrac{x}{8}\sin 2x+c_1\cos 2x+c_2\sin 2x$.

33. $y=\dfrac{1}{2}-\dfrac{1}{6}\cos 2x+c_1\cos x+c_2\sin x$.

35. $y=\dfrac{x^2}{2}e^{-x}+c_1e^{-x}+c_2xe^{-x}$.

37. $y=\dfrac{5}{4}+\dfrac{x}{2}-\dfrac{x^2}{2}+c_1e^{-x}+c_2e^{2x}$.

39. $y=-\dfrac{1}{2}+x-\dfrac{1}{5}\cos x+\dfrac{2}{5}\sin x+c_1e^{-2x}$.

41. $y=\dfrac{1}{3}e^{2x}+c_1e^{x/2}$.

43. $y=c_1\cos 3x+c_2\sin 3x+\sum_{n\neq 3,\ n=1}^{6}\dfrac{\sin nx}{n(n^2-9)}-\dfrac{x}{8}\cos 3x$.

45. $y_h=c_1e^x+c_2e^{3x}$, $y_p=Axe^x+Bxe^{3x}$.

47. $y_h=e^{-x}(c_1\cos x+c_2\sin x)$, $y_p=xe^{-x}(A\cos x+B\sin x)+C\cos x+D\sin x+Ex^2+Fx+G$.

49. $y_h=c_1e^x+c_2e^{2x}$, $y_p=(A+Bx+Cx^2+Dx^3+Ex^4)e^x+(F+Gx)e^{-2x}\cos 3x+(H+Jx)e^{-3x}\sin 3x$.

51. $y_h=c_1\cos 2x+c_2\sin x)$, $y_p=e^{2x}(A\cos 2x+B\sin 2x)$.

53. $y_h=c_1e^x+c_2xe^x$, $y_p=Ax+B+Cx^2e^x$.

55. $y_h=c_1e^x+c_2e^{3x}$, $y_p=Ae^{\alpha x}$ if $\alpha\neq 1$ or 3; $y_p=Axe^x$ if $\alpha=1$; $y_p=Axe^{3x}$ if $\alpha=3$.

57. $y_h=c_1\cos 2x+c_2\sin 2x$, $y_p=A\cos\omega x+B\sin\omega x$ if $\omega\neq 2$ or 3; $y_p=x(A\cos\omega x+B\sin\omega x)$ if $\omega=2$ or 3.

59. $y_h=c_1\cos\omega x+c_2\sin\omega x$, $y_p=A\cos 2x+B\sin 2x$ if $\omega\neq 2$; $y_p=x(A\cos 2x+B\sin 2x)$ if $\omega=2$.

61. $y=\dfrac{3}{2}\sinh 2x$. **63.** $y=-e^{x/2}+\dfrac{3}{2}xe^{x/2}$.

65. $y=\dfrac{1}{2}e^x-e^{2x}+\dfrac{1}{2}e^{3x}$.

67. $y=\dfrac{3}{32}e^{-x}-\dfrac{5}{8}e^x+\dfrac{17}{32}e^{3x}+\dfrac{1}{8}xe^{-x}$.

69. $y=-\cos 2x+(-\dfrac{\pi}{8}+\dfrac{x}{4})\sin 2x$.

71. Try $e^x(A\cos x+B\sin x)$ and get $A=-\dfrac{1}{2}$, $B=\dfrac{1}{2}$.

73. Try $e^{ax}(A\cos bx+B\sin bx)$ and get $A=-\dfrac{b}{a^2+b^2}$, $B=\dfrac{a}{a^2+b^2}$.   **75.** $V(\lambda_1,\lambda_2)=\lambda_2-\lambda_1$, so $V(\lambda_1,\lambda_2)\neq 0$ if and only if $\lambda_1\neq\lambda_2$.   **77.** $W(e^{\lambda_1 x},e^{\lambda_2 x},\dots,e^{\lambda_n x})=e^{\lambda_1+\lambda_2+\dots+\lambda_n}\times V(\lambda_1,\lambda_2,\dots,\lambda_n)$. Since the exponential function is never zero, $W\neq 0$ if and only if $V\neq 0$ if and only if all the λ's are distinct.

APPENDIX A.3

1. $y = c_1 e^x + c_2 e^{-3x}$.   **3.** $y = c_1 e^x + c_2(6 + 6x + 3x^2 + x^3)$.
5. $y = c_1 \cos 2x + c_2 \sin 2x$. **7.** $y = c_1 \cosh x + c_2 \sinh x$.
9. $y = c_1 x + c_2 x \ln\left(\dfrac{1-x}{1+x}\right)$.   **11.** $y = c_1 x + c_2 \frac{1}{x}$.

13. $y = c_1 \cos(\ln x) + c_2 \sin(\ln x)$.   **15.** $y = c_1 \frac{\cos 2x}{x} +$
$c_2 \frac{\sin 2x}{x}$.   **17.** $y = c_1 e^x + c_2 \frac{e^x}{x}$.   **19.** $y = c_1 x + c_2 x^2$.

21. $y = c_1 e^x + c_2 e^{3x} + \frac{1}{8} e^{-x}$.   **23.** $y = c_1 e^{-\frac{10}{3}x} + c_2 e^{-x} + \frac{1}{218}(-13 \cos x + 7 \sin x)$.   **25.** $y = c_1 \cos x + c_2 \sin x + \cos x \ln(\cos x) + x \sin x$.   **27.** $y = c_1(1+x) + c_2 e^x + 4x^2 + \frac{4}{3}x^3 + \frac{x^4}{3}$.   **29.** $y = \frac{c_1}{x} + c_2 \frac{\ln x}{x} + \frac{4}{9}\sqrt{x}$.

31. $y = c_1 x^{-2} + c_2 x^{-1}$.   **33.** $y = c_1 x^{-1} + c_2 \frac{\ln x}{x}$ **35.** $y = c_1 \cos(2 \ln x) + c_2 \sin(2 \ln x)$.   **37.** $y = x^{-3}(c_1 \cos(2 \ln x) + c_2 \sin(2 \ln x))$.   **39.** $y = \frac{c_1}{x-2} + c_2 \frac{\ln(x-2)}{x-2}$.

47. $y = c_1 e^x + c_2 e^{3x} - \frac{1}{2} x e^x$.   **49.** Same as Exercise 23.

APPENDIX A.4

1. $a_1 = 0, a_{m+2} = \dfrac{-2}{m+2} a_m, y = a_0(1 - x^2 + \frac{1}{2}x^4 - \cdots) = a_0 e^{-x^2}$.   **3.** $a_{m+1} = \dfrac{-1}{m+1} a_m, y = a_0 - a_0 x + \frac{1}{2}(1 + a_0)x^2 - \cdots) = -1 + x + (a_0 + 1)e^{-x}$.   **5.** $a_{m+2} = \dfrac{1}{(m+1)(m+2)} a_m, y_1 = a_0 \cosh x, y_2 = a_1 \sinh x$.

7. $a_{m+2} = \dfrac{m-1}{(m+1)(m+2)} a_m, y_1 = a_0(1 - \frac{1}{2}x^2 - \frac{1}{4!}x^4 - \cdots), y_2 = x$.   **9.** $a_{m+2} = \dfrac{-(1+2m)}{(m+1)(m+2)} a_m, y_1 = a_0(1 - \frac{1}{2}x^2 + \frac{5}{4!}x^4 - \cdots)$ $y_2 = a_1(x - \frac{1}{2}x^3 + \frac{7}{40}x^5 - \cdots)$.

11. $a_{m+2} = \dfrac{2(m+1)a_{m+1} - a_m}{(m+1)(m+2)}, a_0 = 0, a_1 = 1, y = x + x^2 + \frac{1}{2}x^3 + \cdots = x(1 + x + \frac{1}{2}x^2 + \cdots) = x e^x$.

13. (a) $y = \dfrac{e^x}{2} - \dfrac{e^{x/2}}{2}[\cos \dfrac{\sqrt{7}x}{2} + \dfrac{3}{\sqrt{7}} \sin \dfrac{\sqrt{7}x}{2}]$.

15. (a) $y = 1 - \dfrac{1}{2}x^2 - \dfrac{1}{6}x^3 + \dfrac{5}{40}x^5 + \cdots$.

19. $y = x + \dfrac{1}{12}x^4 + \dfrac{1}{504}x^7 + \cdots$.

APPENDIX A.5

1. Ordinary point.   **3.** Singular point, not regular.   **5.** Regular singular point.   **7.** $r_1 = 0, r_2 = -\frac{1}{2}, r_1 - r_2 = \frac{1}{2}$, Case I, $y_1 = 1 - \frac{1}{3!}x + \frac{1}{5!}x^2 - \frac{1}{7!}x^3 + \cdots = \frac{\sin\sqrt{x}}{\sqrt{x}}$, $y_2 = \frac{1}{\sqrt{x}} - \frac{1}{2}\sqrt{x} + \frac{1}{4!}x^{3/2} - \frac{1}{6!}x^{5/2} + \cdots = \frac{\cos\sqrt{x}}{\sqrt{x}}$.   **9.** $r_1 = \frac{5}{2}, r_2 = 2, r_1 - r_2 = \frac{1}{2}$, Case I, $y_1 = x^2 + \frac{1}{2}x^3 + \frac{1}{4!}x^4 + \cdots = x^2 \cosh\sqrt{x}, y_2 = x^{5/2} + \frac{1}{3!}x^{7/2} + \frac{1}{5!}x^{9/2} + \cdots = x^2 \sinh\sqrt{x}$.
11. $r_1 = \frac{1}{2}, r_2 = 0, r_1 - r_2 = \frac{1}{2}$, Case I, $y_1 = \sqrt{x} - \frac{1}{2}x^{3/2} +$

$\frac{1}{8}x^{5/2} - \cdots = \sqrt{x}e^{-x/2}$, $y_2 = \sum_{n=0}^{\infty}(-1)^n \frac{2^n}{(2n)!} n! x^n$.
13. $r_1 = r_2 = 0$, Case II, $y_1 = 1 - x, y_2 = (1-x)\ln x + 3x - \frac{1}{4}x^2 - \frac{1}{36}x^3$.   **15.** $r_1 = 0, r_2 = -1, r_1 - r_2 = 1$, Case III, $y_1 = 1 - x + \frac{1}{2}x^2 - \frac{1}{3!}x^3 + \cdots = e^{-x}, y_2 = \frac{1}{x} - 1 + \frac{1}{2}x - \frac{1}{3!}x^2 + \cdots = \frac{e^{-x}}{x}$.   **17.** $r_1 = -1, r_2 = -2, r_1 - r_2 = 1$, Case III, $y_1 = \frac{1}{x} + \frac{1}{3!}x + \frac{1}{5!}x^3 + \cdots = \frac{\sinh x}{x^2}$, $y_2 = \frac{1}{x^2} + \frac{1}{2} + \frac{1}{4!}x^2 + \frac{1}{6!}x^4 + \cdots = \frac{\cosh x}{x^2}$.
19. $r_1 = r_2 = -1$, Case II, $y_1 = \frac{1}{x} - 1 + \frac{1}{4}x - \frac{1}{36}x^2 + \cdots$, $y_2 = 2 - \frac{3}{4}x + \frac{11}{108}x^2 - \cdots + y_1 \ln x$.   **21.** $r_1 = 1, r_2 = -1, r_1 - r_2 = 2$, Case III, $y_1 = x + \frac{1}{3}x^2 + \frac{1}{4!}x^3 + \cdots, y_2 = -\frac{2}{9} + 2 - \frac{4}{9}x^2 - \cdots + y_1 \ln x$.   **23.** $r_1 = \frac{5}{2}, r_2 = 2, r_1 - r_2 = \frac{1}{2}$, Case I, $y_1 = x^{5/2} - \frac{1}{3!}x^{7/2} + \frac{1}{5!}x^{9/2} - \cdots = x^2 \sin\sqrt{x}$, $y_2 = x^2 - \frac{1}{2}x^3 + \frac{1}{4!}x^4 - \cdots = x^2 \cos\sqrt{x}$.   **25.** $r_1 = 1, r_2 = -1, r_1 - r_2 = 2$, Case III, $y_1 = x - \frac{1}{3}x^2 + \frac{1}{4!}x^3 - \cdots$, $y_2 = -\frac{1}{2x} - \frac{1}{2} + \frac{291}{144}x + \cdots + \frac{1}{4}y_1 \ln x$.   **27.** $r_1 = r_2 = 0$, Case II, $y_1 = \frac{1}{1-x}$, $y_2 = \frac{\ln x}{1-x}$.   **33.** $y = c_1 x^{-\sqrt{2}} + c_2 x^{\sqrt{2}}$.
35. $y = c_1 \cos\left(\frac{\ln x}{3}\right) + c_2 \sin\left(\frac{\ln x}{3}\right)$.

Index